S0-BNF-952

How To Control Plant Diseases in Home and Garden

ART BY ROGER D. ALBERTSON AND MALCOLM C. SHURTLEFF

How To Control

Plant Diseases

in Home and Garden

Malcolm C. Shurtleff

PROFESSOR OF PLANT PATHOLOGY
AND EXTENSION PLANT PATHOLOGIST
UNIVERSITY OF ILLINOIS

Iowa State University Press, Ames, Iowa

ABOUT THE AUTHOR

MALCOLM C. SHURTLEFF, professor of plant pathology and extension plant pathologist at the University of Illinois, holds the B.S. degree from the University of Rhode Island and the M.S. and Ph.D. degrees from the University of Minnesota. He is a member of the American Association for the Advancement of Science; the American Phytopathological Society; the Botanical Society of America; and the Illinois and Iowa Academies of Science. Besides this book, he has to his credit over two hundred publications, mostly popular. He has written numerous articles and a regular column, "Plant Disease Clinic," for *Flower and Garden* magazine. He edited the pest control section and wrote the plant disease material for the *New Better Homes and Gardens Garden Book,* and has had articles published in *Successful Farming, Hoard's Dairyman, Golf Course Reporter, Phytopathology,* and others.

Contents

See detailed Table of Contents at beginning of each section

Page

Section 1

INTRODUCTION 3

Section 2

"WHAT IS IT?" 18

Environmental Factors 19

General Diseases 48

A. Foliage Diseases 48

B. Stem Diseases 76

C. Flower and Fruit Diseases 86

D. Root and Bulb (Corm) Diseases 87

E. Parasitic Flowering Plants 93

Section 3

"WHAT CAN I DO ABOUT IT?" 95

Measures 97

Materials 100

Equipment 112

Section 4

HOME AND GARDEN PLANTS AND THEIR DISEASES . . 125

Other Useful Information

APPENDIX 515

GLOSSARY 553

BIBLIOGRAPHY 571

INDEX 583

A Note to the Reader

All living plants grown in and around the home, yard, and garden are subject to attack by disease-producing organisms and agents. Plant diseases—the despair of gardeners the world over—should be considered as much a part of nature as sun, wind, rain, weeds, and insects.

This book is intended to acquaint gardeners—amateur and professional alike, as well as the many people who advise them—with the numerous types of disease problems that flowers, vines, trees, shrubs, lawngrasses, vegetables, and fruit may contract. The cultural and chemical practices necessary to keep them in check are outlined. Naturally a book of this size cannot exhaust the subject of more than 80,000 diseases. But it does describe the diseases—common and uncommon—of more than 900 genera of home and garden plants grown in the United States and Canada. Ailments of the same plant which look much alike have often been lumped together. Closely related plants which are damaged by the same general group of diseases, like those in the cabbage, cucumber, and carnation families, are also placed together. Since all plants are listed by both common and scientific names, you as a reader and user should have little difficulty finding what you are looking for.

To cover as much material as possible in a brief space, the style is terse and pointed. The material has been organized into sections; each section has its own index. Each is a unit by itself and should answer such questions as "What is it?" "What can I do about it?" and "How serious is it?"

Do not expect this book to be a panacea for all your garden ills. Occasionally diseases are found which can be easily confused with other diseases, soil deficiencies, insects, or mechanical injury. In cases like this, check over the disease descriptions given under the specific plant in Section 4 and under the disease in Section 2. Think back over the past history of the plant and the area of the yard or garden where it is growing. Still no answer? Now it is time to call in a specialist for his evaluation. Do not be afraid to talk over your problem with a successful grower, your local nurseryman, florist, or the people at the garden supply center. Ask them for their advice. There is also the local county extension office or extension specialist at your land-grant college or university whom you can call on for assistance.

No attempt has been made to include insect injuries except in a few instances when the effects of such injuries lead to definite disease-like conditions. Where insects are important in transmitting disease-producing organisms and agents, control measures are suggested.

Intentionally, I have avoided use of scientific names for the causal pathogens and agents. Common names generally suffice and are much more meaningful to all (except a handful of biologists) than is *Colletotrichum lindemuthianum, Gymnosporangium juniperi-virginianae,* or *Belonolaimus longicaudatus*. Besides, common names usually lend stability to nomenclature The scientific names of many bacteria, viruses, and nematodes, particularly, are in a state of flux.

Use this book as a handy reference when you are in trouble. Or, better still, *before* you are in trouble. And remember that even the best gardeners occasionally have disease problems.

Before plunging in, read the "How to use this book" section. From then on we hope it will be clear sailing.

Preface to the Second Edition

IN THE FIVE YEARS since the first edition of *How To Control Plant Diseases in Home and Garden* appeared, over 7,300 scientific and popular publications on plant diseases and their control have been printed. Trying to keep abreast of the flood of new, disease-resistant varieties of home and garden plants (over 780), new pesticides (165), newly described diseases (633), and making specific chemical and cultural control recommendations for more than 5,200 diseases is a never-ending task.

Of the 520 pages of the original edition, very few remain unchanged in the current edition. Much new information has been added and the present edition contains about 20 per cent more material than the original. The "expansion" includes new material on fungi; viruses (including new methods of spread and lists of weed hosts); nematodes (and how to send in samples for nematode determination); air pollution; pesticide safety—and what is behind the package label; new spray programs for fruits; factors affecting disease development; list of new disease-resistant crabapples; new fungicides and fungicidal action; plus a more extensive list of pesticide manufacturers, distributors, and spraying and dusting equipment manufacturers.

The glossary has over 100 new terms, a bibliography has been added, and the list of new host plants includes 98 genera and almost 1,000 species. New plants include those grown in the home, greenhouse or conservatory, rock garden or wild garden, plus a wider range of subtropical plants, e.g., cacti, orchids, palms, bamboos, and mango.

The Appendix now includes expanded sections on seed and soil treatment methods and materials, a list of disease-resistant trees, a new spray compatibility chart, useful units of measure, percentage solution, tractor speed conversions, and numbers of plants needed to plant an acre when set at different distances apart.

The section on cultural practices and plant injuries has new material on soil and potting mixtures, planting, soil insecticides, nutrient deficiences and their control, fertilization and a plant-food utilization chart, soluble salts injury, light and temperature requirements for house plants, chemical and mechanical injuries to plants, pruning, and soil reaction (pH) requirements for plants.

As a result of numerous requests, this edition includes 314 illustrations of which 149 are new. The 149 include: 9 of causal or transmitting agents, 7 of culture or injuries, 14 of pesticide equipment, and 146 new diseases.

The general format has been largely unchanged because numerous users report that it provides easy, quick reference to the information desired. Some, of course, would prefer a full discussion of symptoms and control measures for each disease given under every specific host plant. Unfortunately, this would fill several volumes while providing little additional basic information.

In the preparation of this edition, the writer has profited greatly by the help and interest of his fellow plant pathologists, both in and out of the Cooperative Extension Service, who have been generous with publications, suggestions for improving the current edition, or who furnished illustrative materials. A few of these individuals include W. M. Bever, M. P. Britton, J. C. Carter, J. W. Courter, J. L. Forsberg, T. E. Freeman, E. B. Himelick, F. W. Holmes, C. W. Horne, E. L. Knake, M. B. Linn, H. Jedlinski, I. C. MacSwann, L. I. Miller, P. W. Miller, D. E. Munnecke, Dan Neely, P. E. Nelson, L. P. Nichols, J. D. Paxton, H. S. Potter, Dwight Powell, D. P. Taylor, and H. H. Thornberry. Credit for photographs is given in the legends to the figures.

Limitations of space make it impossible to acknowledge adequately my indebtedness to thousands of research and extension workers and writers for the enormous storehouse of information from which to draw the material presented.

I should also like to give credit to my former teachers and professors who guided and encouraged my early steps into the realm of plant pathology: Drs. Frank L. Howard, James G. Horsfall, E. C. Stakman, Helen Hart, Carl J. Eide, Clyde M. Christensen, and the late J. J. Christensen.

My most sincere thanks go to the many gardeners who report that the first edition of *How To Control Plant Diseases* has been helpful in solving their plant disease problems. If the present volume merits as cordial and continued a reception as accorded the original edition, the thousands of hours spent in its preparation will have been considered well spent.

MALCOLM C. SHURTLEFF

Urbana, Illinois
November, 1966

How To Control
Plant Diseases
in Home and Garden

SECTION ONE

Introduction

How to use this book 3
Where you can get additional help . . 5
Land-grant institutions and agricultural experiment stations in the United States 6
How to send in plant specimens . . . 7
Extent of plant diseases 8
What is a plant disease? 8
Classification of diseases 8
Causes of plant diseases 9
Unfavorable growing conditions . . . 9
Parasites 9
Bacteria 9
Fungi 11
Viruses 13
Nematodes 14

How To Use This Book

THIS BOOK is written primarily for the home gardener. It also should prove useful for the commercial vegetable grower, orchardist, berry grower, nurseryman, house plant grower, florist, turf specialist, student, extension specialist in plant sciences, county agent, vocational agriculture teacher, biology teacher, specialist for commercial concerns or state departments of agriculture, garden writer, and others who grow, know, and love plants.

An attempt has been made to write in easy-to-understand language, omitting as much technical terminology (e.g., most scientific names of disease-causing organisms, mycological or pathological terms) as practical. An extensive glossary (pages 553–569) explains the technical terms used.

Two questions invariably asked about a plant disease are, "What is it?" and "What should I do about it?" Sometimes such questions as "How serious is it?" or "Will it kill my plants?" follow. This book answers these four basic questions.

The answer to "What is it?" or disease diagnosis is based on plant responses expressed as symptoms. These result from some disease-inducing factor. Most general types of diseases, based on external and internal symptoms, are given in Section 2. Symptoms are confined to those that can be observed with the naked eye or with a hand lens (magnifying glass). The listings of *Plants Attacked,* and diseases of these plants, in this section come largely from USDA Handbook No. 165, *Index of Plant Diseases in the United States,* and the second edition of Cynthia Westcott's *Plant Disease Handbook* published by the D. Van Nostrand Company, Inc.

Section 3, "What Can I Do About It?" covers essential points in control that govern most plant diseases. Usually several types of control measures are needed to protect against, check, or eradicate an infection. The amount of disease control is dependent upon the timeliness, completeness, and type of control measures used.

Cross references are made to general

types of diseases pictured and described in Section 2, and to control measures outlined in Section 3, listed under individual plants, or given in the Appendix 515–552 pages).

Information found in Section 4 (pages 125–513) includes those trees, shrubs, vines, house plants, flowers, fruit, vegetables, and lawngrasses likely to be grown in and around the home, yard, and garden in any area of the United States. Certain native plants sometimes grown in wild gardens are listed as well as forest trees sometimes used as ornamentals. Strictly field crops have been omitted.

Plants are listed alphabetically from Aaronsbeard to Zygopetalum under both common and scientific names. Important diseases are listed under each plant. To reduce bulkiness, plants have been put together that have similar disease problems. Where practical, all members of a plant family are placed under one or several plants in that family.

For example, plants related to carnation, e.g., garden pinks, sweet-william, babysbreath, Maltese cross, catchfly, and others in the family Caryophyllaceae having economic value are put together under carnation (see example below), probably the best known member of the family. Plants listed with carnation are cross-indexed under both common (Maltese cross) and scientific *(Lychnis)* names and refer you to carnation.

Where several species or horticultural types of a genus are widely grown, e.g., cottage, grass, maiden, and rainbow garden pinks, these are listed alphabetically within brackets after *GARDEN PINKS* (see below). Members of a plant family are listed alphabetically by the scientific name of the genus after the first common name *(CARNATION)*. *Agrostemma* comes before *Arenaria* with the remaining genera *Gypsophila, Lychnis,* and *Silene* following.

Example:

CARNATION [FLORIST'S, HARDY], *GARDEN PINKS* [ALPINE, ALWOOD HYBRIDS, CHEDDAR, CHINA, CLUSTERHEAD, COTTAGE, GRASS, MAIDEN, RAINBOW, SUPERB], *SWEET-WILLIAM (Dianthus); CORNCOCKLE (Agrostemma); SANDWORT* [CORSICAN, MOUNTAIN] *(Arenaria); BABYSBREATH (Gypsophila); FLOWER-OF-JOVE, MALTESE CROSS* or *SCARLET LYCHNIS* or *JERUSALEM-CROSS, EVENING CAMPION, RED* and *ROSE CAMPION, RAGGED-ROBIN* or *CUCKOOFLOWER, MULLEIN-PINK, GERMAN CATCHFLY, ROSE-OF-HEAVEN (Lychnis); CUSHION-PINK, FIRE-PINK, WILD PINK, CAMPION* [MOSS, SEA, STARRY], *CATCHFLY* [ALPINE, ROYAL, SWEET-WILLIAM] *(Silene)*

Where a disease infects certain plants in a listing and not others, the plants attacked are listed in parentheses after the name of the disease. Following the carnation example above we find on page 203 that Fusarium Wilts infect carnation, pinks, and sweet-william; Bacterial and Phialophora (Verticllium) Wilts attack only carnation; while Alternaria Blight, Collar and Branch Rot infect campion, carnation, Maltese cross, pinks, and sweet-william.

The diseases listed in this book are those reported from the continental United States. Diseases peculiar to Hawaii, Alaska, Puerto Rico, the Virgin Islands, and the Canal Zone are omitted principally for lack of space. Many of the same diseases are found more or less generally throughout the world.

The geographic range, in nature or in cultivation, of the diseases should be taken only as a rough guide. Diseases listed as "General" are usually coextensive with the host plant; "Widespread" means the disease is reported from many scattered locations in the United States but is not prevalent; "Frequent" and "Occasional" denote both intensity and range of occurrence.

Certain regional designations (e.g., eastern, southern, or southeastern states; Pacific Coast, Midwest) are also used to denote specific geographical areas where certain diseases are prevalent or have been reported.

If appropriate, prevalence of disease, potential destructiveness of the ailment, and weather conditions that favor or check disease development are mentioned. This should answer "How serious is it?" and "Will it kill my plant?"

The information you seek should be easy to find, especially with the extensive index. The index, generally under the common name of each plant, gives an alphabetical listing of diseases recorded on that plant in the United States. Two numbers are usually given after each disease. The first refers to the page where the disease is discussed under General Diseases in Section 2; the second number to the page in Section 4 where various diseases of that plant are described and control measures outlined.

The Appendix (515–552 pages) contains average spray programs for common fruits and seed and soil treatments with methods, materials, and precautions. Also included are conversion tables for measuring dry and liquid chemicals, useful units of measure, methods for converting Fahrenheit to Centigrade and vice versa, measurements and rates of application equivalents, a pesticide compatibility chart, an operating chart for tractor boom sprayers, amount of spray needed for fruit trees of various sizes, listing of trees relatively free of diseases, glossary of terms, and a bibliography.

To illustrate, let us take a plant disease and find all we can. To be specific, suppose the apple tree in the back yard has alarming spots on its leaves. As the fruit matures, somewhat similar spots appear. A neighbor spoke of various apple diseases when you purchased the tree and you vaguely remember his mentioning a disease that sounded like skob or scab might produce the symptoms that are evident.

So you begin in Section 2 looking under disease (14) Scab. After reading the introductory material, you come to material on control and observe that a spray program is suggested. Section 4 presents plants in alphabetical order, so looking through the A's you come to apple. Here disease 2 is Scab and controls are mentioned.

It is also possible that you only know that it is a disease of apple since it is on your apple tree. Here you would immediately turn to information on apples in Section 4 and start comparing your diseased specimen with the explanations and illustrations presented. Once evidence indicates that the disease is scab, you will turn back to Section 2 and read material there on scab as well as reading that in Section 4 and also turn to Section 5, the Appendix, to check the apple-spray program. From here, it is only necessary to follow directions and observe suggestions given in various parts of the book to which you are directed.

Where You Can Get Additional Help

You can get help on plant problems by contacting your county agent (sometimes called farm advisor (or adviser), extension agent, or county extension director) or from your state land-grant university. A list of these institutions is below. Write to your extension plant pathologist concerning diseases, the extension entomologist for information about insects, or to the extension horticulturist regarding cultural problems.

All states publish free pest control recommendations. A wide variety of printed information on plant pest problems may be obtained from your county extension office or land-grant institution.

There are local authorities or plant experts in your community who would be only too happy to talk with you about your plant disease problems. These people include florists, nurserymen, garden supply dealers, commercial fruit and vegetable growers, turf specialists, and personnel of nearby arboretums or botanical gardens. In Canada, contact the Science Service of the Dept. of Agriculture, Ottawa.

Land-Grant Institutions and Agricultural Experiment Stations in the United States

ALL STATES have Extension Horticulturists to answer questions on cultural management of garden plants, an Extension Entomologist (insects, mites) and practically all also have an Extension Plant Pathologist (diseases). Write to the specialist in care of the College of Agriculture at your state land-grant institution.

For free bulletins, circulars, pamphlets, spray schedules, etc., write to the Bulletin Room, College of Agriculture, at your state university.

ALABAMA: Auburn University, Auburn 36830.

ALASKA: University of Alaska, College 99735 (or Experiment Station, Palmer 99645).

ARIZONA: University of Arizona, Tucson 85721.

ARKANSAS: University of Arkansas, Fayetteville 72701 (or Cooperative Extension Service, 1201 McAlmont Ave., P.O. Box 391, Little Rock 72203).

CALIFORNIA: University of California, Berkeley 94720; Riverside 92502; or Davis 95616.

COLORADO: Colorado State University, Fort Collins 80521.

CONNECTICUT: University of Connecticut, Storrs 06268 (or Connecticut Agricultural Experiment Station, New Haven 06504).

DELAWARE: University of Delaware, Newark 19711.

FLORIDA: University of Florida, Gainesville 32603.

GEORGIA: University of Georgia, Athens 30601 [or Agricultural Experiment Station (State), Experiment 30212; Coastal Plain Station, Tifton 31794].

HAWAII: University of Hawaii, Honolulu 96822.

IDAHO: University of Idaho, Extension Service, Boise 83702; Agricultural Experiment Station, Moscow 83843.

ILLINOIS: University of Illinois, Urbana (or Illinois Natural History Survey, Urbana) 61801.

INDIANA: Purdue University, W. Lafayette 47907.

IOWA: Iowa State University, Ames 50010.

KANSAS: Kansas State University, Manhattan 66502.

KENTUCKY: University of Kentucky, Lexington 40506.

LOUISIANA: Louisiana State University, University Station, Baton Rouge 70803.

MAINE: University of Maine, Orono 04473.

MARYLAND: University of Maryland, College Park 20742 (or Vegetable Research Farm, Quantico Road, Salisbury 21801).

MASSACHUSETTS: University of Massachusetts, Amherst 01003.

MICHIGAN: Michigan State University, East Lansing 48823.

MINNESOTA: Institute of Agriculture, University of Minnesota, St. Paul 55101.

MISSISSIPPI: Mississippi State University, State College 39762.

MISSOURI: University of Missouri, Columbia 65202.

MONTANA: Montana State University, Bozeman 59715.

NEBRASKA: College of Agriculture, University of Nebraska, Lincoln 68503 (or Scott's Bluff Experiment Station, Mitchell 69357).

NEVADA: University of Nevada, Reno 89507.

NEW HAMPSHIRE: University of New Hampshire, Durham 03824.

NEW JERSEY: State College of Agriculture, Rutgers, The State University, New Brunswick 08903.

NEW MEXICO: New Mexico State University, University Park 88070.

NEW YORK: New York State College of Agriculture, Cornell University, Ithaca 14850 (or Agricultural Experiment Station, Geneva 14456; Ornamentals Research Laboratory, Farmingdale 11735).

NORTH CAROLINA: North Carolina State University, State College Station, Raleigh 27607 (or A. & T. College of North Carolina, P.O. Box 1014, Greensboro 28053; Horticultural Crops Research Station, Rte. 1, Box 121A, Castle Hayne 28429).

NORTH DAKOTA: North Dakota State University, State University Station, Fargo 58103.

OHIO: The Ohio State University, Columbus 43210 (or Ohio Agricultural Research & Development Center, Wooster 44691).

OKLAHOMA: Oklahoma State University, Stillwater 74075.

OREGON: Oregon State University, Corvallis 97331.

PENNSYLVANIA: The Pennsylvania State University, University Park 16802.

PUERTO RICO: Agricultural Extension Service, University of Puerto Rico, Rio Piedras 00927.

RHODE ISLAND: University of Rhode Island, Kingston 02881.

SOUTH CAROLINA: Clemson University, Clemson 29631.

SOUTH DAKOTA: South Dakota State University, Brookings 57007.

TENNESSEE: University of Tennessee, P.O. Box 1071, Knoxville 37901.

TEXAS: Texas A & M University, College Station 77841 (or Box 476, Weslaco 78596; Tyler Experiment Station No. 2, R. 6, Tyler 75703; Texas Agricultural Experiment Station, Rte. 3, Lubbock 79414).

UTAH: Utah State University, Logan 84321.

VERMONT: University of Vermont, Burlington 05401.

VIRGIN ISLANDS: Virgin Islands Agric. Project, Kingshill, St. Croix 00801 (officer in charge).

VIRGINIA: Virginia Polytechnic Institute, Blacksburg 24061 [or Virginia Truck Experiment Station (truck crops), Norfolk 23501; Piedmont Fruit Research Laboratory, Charlottesville 22903; Winchester Fruit Research Laboratory, Winchester 22601].

WASHINGTON: Washington State University, Pullman 99163 (or Western Washington Experiment Station, Puyallup 98371; Irrigation Experiment Station, Prosser 99350).

WEST VIRGINIA: West Virginia University, Morgantown 26506 (or Fruit Experiment Station, Kearneysville 25430).

WISCONSIN: University of Wisconsin, Madison 53706 (or Peninsular Branch Experiment Station, Sturgeon Bay 54235).

WYOMING: University of Wyoming, Laramie 82071.

How To Send in Plant Specimens

TO HELP in diagnosing plant pests (diseases, nematodes, insects, mites, or weeds), wrap fresh plant specimens, which show a range of symptoms, in cellophane, plastic bags, wax paper, or aluminum foil. Seal the wrapper tightly and mail in a crush-proof carton or mailing tube. Do *not* add moisture. Do not send fleshy fruit in advanced stages of decay. Enclose or attach a letter giving as much history about the problem as possible. This should include the date collected, variety and kind of plant attacked, prevalence of pest, degree of severity, description of problem, part diseased or injured, extent of garden area involved, cropping history when known, weather and soil conditions, recent fertilization, watering, pest control measures, etc. Do not forget your name and return address! Remember that correct diagnosis is essential before control measures can be suggested. A diagnosis can be only as good as the specimen you send!

For nematode determination, collect soil from the root zone of growing plants (4 to 12 inches deep). Include feeder roots. Sample at least 10 places in the suspicious area, saving about one-half to one cup of soil from each spot sampled. Thoroughly mix the soil collected and save at least one quart for mailing. Place the one-quart soil sample in a plastic freezer bag, close the open end securely, and place in a *strong* container. Be sure to include with the package your name and address, location of field, crop sampled, symptoms observed, cropping history, and approximate size of area sampled. Mail the package as soon after collecting as possible. Do *not* let it sit overnight! Nematodes are living animals and must reach the laboratory *alive.*

Extent of Plant Diseases

ALL garden plants are attacked at one time or another by disease. There are over 80,000 different diseases. Each species of plant is subject to its own characteristic diseases.

The annual loss in the United States from plant diseases and nematodes is about $4 billion. This loss means an extra $600 per year in higher grocery bills for the average family, plus increased costs for clothing and shelter for all of us. For example, diseases cause more loss—20 billion board feet—to our nation's forests each year than does fire. This is enough lumber to make a wooden sidewalk a mile wide, with all its underpinnings, stretching from New York to San Francisco! Each year in the United States, tree diseases ruin 40 per cent as much timber as is cut.

Plant diseases are a *normal* part of nature and are one of the many environmental factors that help keep the many hundreds of thousands of living organisms in balance with each other in undisturbed nature. When man selects and cultivates plants he must recognize that diseases must be considered one of the many expected hazards. Cultivated plants are usually *more* liable to disease than their wild cousins—partly because large numbers of the same kind are grown closely together so that disease spreads rapidly, once established. Also, many of our most valuable crop and garden plants are basically very susceptible to disease and would have a hard job surviving the "live or die" competition in a manless nature. In addition, cultivation—be it roses, cabbages, or sycamores—constantly is disturbing nature.

Plant diseases are not new. They undoubtedly arose and developed as life arose and developed on earth. Many injurious pests including rusts, mildew, and blast are mentioned in the Bible. These diseases and others have plagued man and have caused famines since the dawn of recorded history. Fossils have been found that prove plants had disease enemies 50 to 100 million years or more before man appeared on earth.

This book is not designed to scare you about the many thousands of diseases you will never see in your garden. It is to enable you to know and recognize the occasional disease that may require prompt action. The photographs, drawings, and disease descriptions should help you know the common ones.

What Is a Plant Disease?

WHEN a plant is *continuously* affected by some factor that interferes with its normal structure or activities, it is said to be diseased. Injury, in contrast, results from a momentary damage. Broadly speaking, a plant is considered diseased when it does not develop or produce normally, considering the conditions of its growth. Often there is no sharp distinction between healthy and diseased plants. Disease in some of its aspects is merely an extreme case of poor growth.

Classification of Diseases

A SICK PLANT may not be as different from a sick human or animal as you think. For instance, increases in temperature and rate of respiration may occur when plants become infected.

Plant ailments, like those of humans or animals, are often classified by their effects or visible symptoms. Humans have fevers and plants have wilts. People suffer from colds, sore backs, an unbalanced diet, or measles, while plants are weakened with spots or blights, rots, mildews, cankers and rusts, or an infertile, compacted soil.

Many plant diseases, however, that appear alike by external symptoms may be caused by widely different microorganisms or agents and require completely different methods of control. This is where a careful laboratory diagnosis by a trained plant doctor comes in handy. It is obvious that the causal organism, agent or environmental factor be known positively before proper control measures can be taken. Such diagnosis is essential.

In Section 2, "What is it?," we have classified plant diseases by symptoms divided conveniently into those that affect foliage; stems, twigs, branches, or trunk; flowers and fruits; or roots and other underground parts.

Plant diseases may also be grouped according to host plant (Section 4) or by their causes.

Causes of Plant Diseases

THE CAUSES of plant diseases may be divided into two principal groups: those caused by unfavorable growing conditions (nonparasitic or physiogenic), and those caused by parasites (bacteria, fungi, viruses, nematodes, and parasitic flowering plants).

UNFAVORABLE GROWING CONDITIONS

Nonparasitic or noninfectious diseases are those caused by an excess, deficiency, nonavailability, or improper balance of light, ventilation, air humidity, water, or essential soil nutrients (e.g., nitrogen, phosphorus, potassium, calcium, iron, magnesium, manganese, boron, copper, molybdenum, zinc, sulfur, etc.), soil moisture-oxygen disturbances, extreme acidity or alkalinity of the growing medium, pesticide injury, extremely high or low temperatures, injurious impurities in air or soil, soil grade changes, girdling tree roots, mechanical and electrical agents, plus unfavorable preharvest and storage conditions for fruits, vegetables, bulbs, roots, etc. Plants in poor health because of unfavorable growing conditions outnumber plants attacked by disease-producing organisms. Noninfectious or physiogenic troubles do *not* spread from sick to healthy plants. They arise, sometimes very suddenly, at about the same time—often on a variety of plants growing in a given area or environment.

In this book we do not attempt to cover in detail the wide range of nonparasitic ailments that affect garden plants. We simply suggest you follow the best cultural practices recommended in state and federal garden bulletins, nursery and seed catalogs, books, and magazines (pages 571–581). These sources include information on varieties and plants to grow, planting depth, shade or sun, type of soil, fertilization programs, water requirements, winter or summer protection, pruning, insect control, weed control, and other practices. Some general information on plant culture in relation to plant diseases is given in Section 2. Additional suggestions may be obtained from experienced and reliable garden supply dealers, nurserymen, florists, arborists, or fellow gardeners in your community. Garden clubs and plant societies also offer a means of exchanging helpful gardening hints.

Several of the more common nonparasitic diseases are shown in Figure 1.1.

PARASITES

The diseases we shall consider primarily are those caused by microorganisms (bacteria, fungi, and nematodes) and viruses. These organisms attack plants and live at the expense of their hosts. Parasites cause infectious diseases that often spread easily from diseased to healthy plants.

Bacteria

Bacteria are minute, one-celled plants (although staining techniques show that the bacterial body may actually be composed of two to four cells) which lack chlorophyll and hence cannot make their own food.

Placed end to end it would take about 20,000 bacteria to make one inch. Twenty trillion bacteria may weigh only an ounce. Yet the top foot of soil in 1,000 square feet of your garden or field contains about 20 pounds of bacteria! One

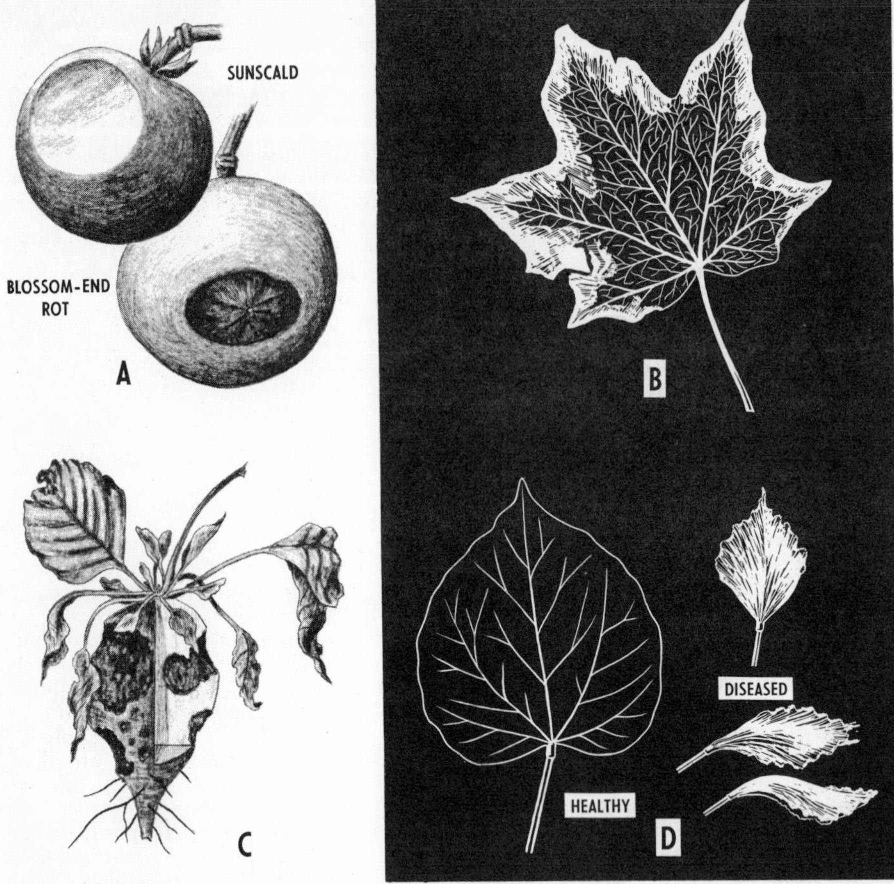

Fig. 1.1. Nonparasitic diseases. A. Sunscald and blossom-end rot of tomato. B. Leaf scorch of maple. C. Boron deficiency of beet. D. 2,4-D injury to redbud leaves.

pound of ordinary garden soil contains 4,500 to 2,270,000 million bacteria. A cubic foot of soil weighs an average of 85 to 90 pounds.

Most bacteria reproduce simply by dividing in half every 20 minutes to an hour or more when temperature, moisture, and food conditions are favorable. If a single bacterium divided in half, and all its descendants did likewise every 20 minutes for just 12 hours, there would be nearly 70 billion bacteria produced. Is it any wonder that bacteria cause iris, calla lily, vegetables, and fruits to sometimes rot so quickly?

Bacteria enter plants through wounds and small natural openings (primarily stomates, lenticels, hydathodes, and leaf scars) that occur over the surface of plants. Once inside, bacteria multiply rapidly, break down tissue, and often migrate throughout the plant. Many types move about in water or plant sap by means of whipping one or more threadlike "tails" called flagella (singular, flagellum), or by a rhythmic pulsation of the bacterial body. See Figure 1.2B.

Bacteria are spread by man through cultivating, pruning, and transporting diseased plant material such as seed, nursery stock, and transplants. Animals, insects, splashing rain, flowing water, and wind-blown dust are also common disseminating agents.

Fortunately, most bacteria are harmless or beneficial to man. Some cause human or animal disease and plant disease. The most common plant diseases caused by bacteria are slimy soft rots, leaf spots or blotches, leaf and stem blights, stem cankers, wilts, and galls. See Figures 2.18, 2.20, 2.33, 2.42, 2.45, and 2.46.

Bacteria, the simplest of plants, overwinter (or oversummer) on or inside perennial or winter annual plants, seeds, storage organs (e.g., potato tubers, flower bulbs, corms, rhizomes), plant refuse, garden tools, or in soil. A few may even live for several months or longer in the bodies of living insects.

Most disease-causing bacteria are quickly killed by high temperatures (10 minutes at 125° F. when exposed), dry conditions, and strong sunlight. Many bacteria in soil, capable of causing plant disease, are inhibited by antibiotic substances secreted by other soil-inhabiting organisms (chiefly bacteria, actinomycetes, and fungi).

Fungi

Fungi, like bacteria, also are simple plants that lack chlorophyll. They obtain their food from living plants and animals or from decaying, nonliving, organic material. Together with bacteria, fungi break down organic matter into nutrients that can be utilized by garden plants. If it were not for fungi and bacteria, our world would probably be piled many feet deep with dead plant and animal remains.

The top foot of soil in 1,000 square feet of garden contains about 30 pounds of fungi. One pound of soil contains 4½ to 225 million fungi and 450 to 227,000 million actinomycetes (filamentous bacteria).

A typical fungus usually begins life as a microscopic spore that can be compared in function to the seed of a higher plant. Under moist conditions the spore may germinate and produce one or more branched threads called hyphae. The hyphae grow and branch to form a fungus body called a mycelium. The mycelium may be an interlacing tangle of hyphae, a loose woolly mass, or even a compact solid body. A parasitic mycelium may grow on the surface of its plant host (powdery mildew or sooty mold), appearing as delicate, whitish cobweb-like threads or as sooty-brown to black filaments. Or the mycelium may be completely within the host plant (wilt-producing fungus) and not evident on the plant surface. Fungi that grow on, or in, a living plant and obtain nourishment from it are called parasites, and the living

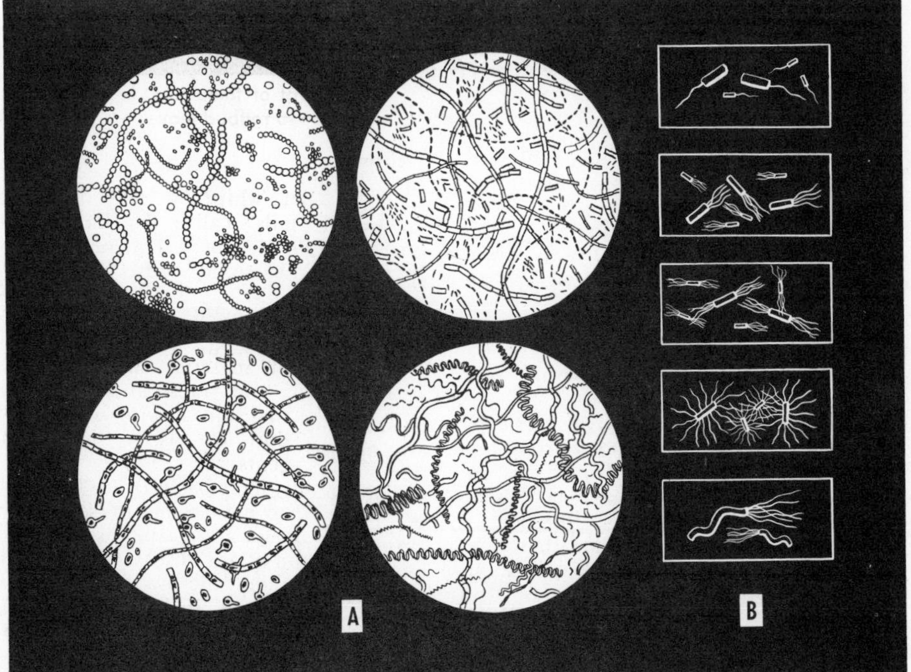

Fig. 1.2. Bacteria. A. Different forms as you might see them under a powerful laboratory microscope. B. Bacteria showing various types of flagella. No one knows how many species of bacteria there are — probably several thousands.

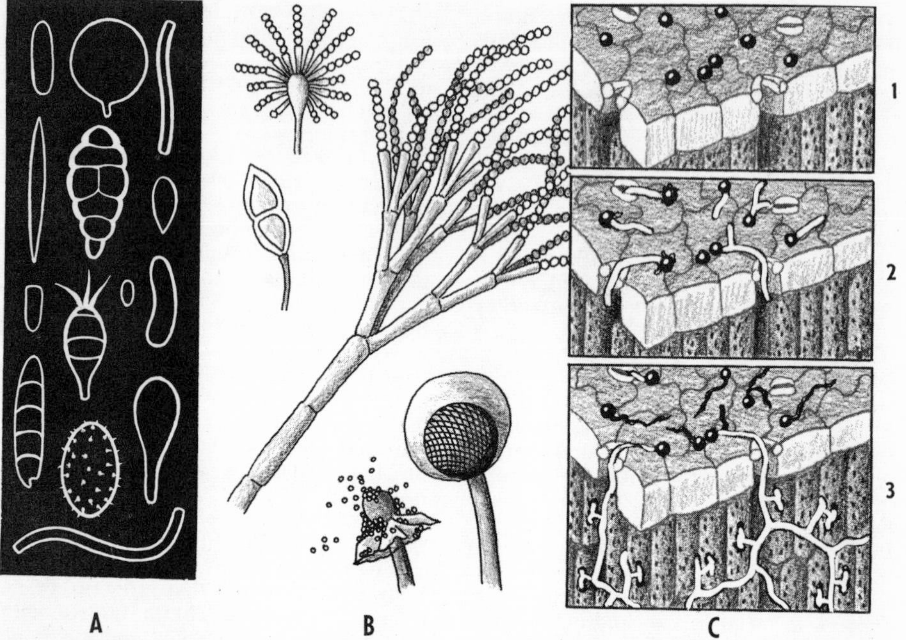

Fig. 1.3. Fungi. A. Various spore forms visible with a microscope. B. Different structures on which fungus spores are borne. C. Successive stages in germination of spores (1) followed by penetration into a leaf, (2) through stomates, and establishment of infection. (3) Fungus spores usually need a film of moisture on the surface of a leaf in order to germinate and penetrate plant tissue.

plant is called the host. Fungi that feed on dead organic matter are called saprophytes.

Certain fungi are able to attack living plants at certain times and live as saprophytes for the remainder. Fungus hyphae may penetrate a plant by growing into a wound, through a natural opening, or by forcing their way directly through a plant's protective skin or epidermis (see Figure 1.3C).

The fungus body usually gives rise to spores or spore-bearing bodies, completing the life cycle. The life cycles of certain fungi (e.g., rusts) are extremely complex and may involve a number of different spore stages and more than one plant host. See (8) Rust, under General Diseases in the next section.

Most fungi are rather inconspicuous, but certain molds, mildews, mushrooms and bracket fungi are known to almost everyone. Some have fruiting bodies which are several feet in diameter and weigh fifty pounds or more! Some large puffballs and bracket fungi contain many billions of spores. Certain large fruiting bodies are known to release as many as 100 billion spores in one day.

Spores are of many different shapes, sizes, colors, and borne in a variety of ways (see Figure 1.3). Perhaps "average" fungus spores, if laid end to end, would total 2,500 to an inch.

Spores play an important part in the multiplication, dissemination, and survival of fungi. Spores are easily carried by air currents, water, insects, man, animals, plant parts (e.g., seeds, bulbs, transplants, nursery stock, etc.) and equipment.

Certain fungus spores have been known to blow a thousand miles or more, sometimes at altitudes up to 80,000 feet or more, before descending (frequently in a rainstorm) and infecting plants.

Resting spores often allow the fungus to withstand unfavorable growing conditions such as extreme heat, cold, drying, and flooding. Spores of certain fungi may lie dormant for many years. When they persist in soil, they are extremely difficult to kill.

Certain fungi do not produce spores. They multiply by forming compact masses of hyphae called sclerotia or the fungus body divides into fragments that are broken off and spread by water, wind, man, and other agents.

Bacteria and fungi are more prevalent and damaging to plants in damp areas or seasons than in dry ones. Moisture is usually essential to their rapid reproduction, spread, penetration, and infection of plant parts.

Fungi cause the majority of infectious or parasitic plant diseases. They include all rusts, smuts, mildews, and scabs; many leaf spots, cankers, and blights; root, stem and fruit rots; wilts; leaf galls; and others.

Many parasitic fungi live alternately on dead and living plant tissues; others, like those causing rusts and mildews, exist only on living plants. Fungi, like bacteria, overwinter on and in plant refuse, soil, perennial plants, and seed, or occasionally in insects. Most fungus spores and hyphae are easily killed by adverse conditions. Knowledge of these habits guides the development of effective control measures (Section 3).

Species of fungi, like other types of parasites, are further subdivided into strains or physiologic races that may differ in (1) the species or varieties of plants they can attack, (2) the virulence of attack—ranging from nonparasitic to highly virulent, (3) the temperature at which disease attack occurs, (4) ability to develop in lower levels of soil, etc.

Viruses

Plant-infecting viruses are complex molecules (composed of nucleic acid with a protein "overcoat") that infect, multiply, mutate, and otherwise act like living organisms when in living plant cells. They are much smaller than bacteria (perhaps 200,000 or more to an inch) and cannot be seen with the aid of an ordinary laboratory (light) microscope. See Figure 1.4. Viruses are closely related in both structure and form to the chromosomes found in all living plant and animal cells. Both are nucleoproteins—composite molecules found exclusively in chromosomes and viruses.

A number of viruses have been crystallized in the laboratory. Yet they multiply only within living cells at the expense of normal plant proteins. Hence viruses usually lead to abnormal growth expressed in various ways. Virus-infected plants are often more susceptible to root rot and possibly other types of diseases.

The most common types of virus-caused diseases in North America are mosaics, yellows, curly-top, spotted wilt, ringspots, stunts, mottles, streaks, and phloem necrosis of elm. See Figures 2.34, 2.35, 2.36, and 2.37. Many crop plants and weeds may harbor viruses but show no external symptoms, especially in hot weather. Viruses are divided into strains that may differ greatly in virulence or produce widely variable symptoms on different varieties of the same or related plants. Two virus strains may be indistinguishable in one plant, but produce strikingly different diseases in another (orchid viruses). Diseases caused by altogether different viruses may resemble one another more closely than diseases caused by different strains of the same virus. Certain variegated plants (e.g., Abutilon, Rembrandt tulips) are inherently virus-infected.

Viruses produce symptoms that are often greatly variable even on different varieties of the same plant at different seasons (e.g., stone fruits). Virus diseases are consequently often grouped together generally by symptoms, regardless of true virus relationships. Symptoms of virus diseases pertain only to diseases with visible symptoms.

Some plant viruses are quite infectious, and spread easily from diseased to healthy plants by mere contact. Others are transmitted in nature only by the feeding and plant-to-plant movement of insects (primarily leafhoppers (Figure 4.26), aphids (Figure 4.78B), thrips, mealybugs, whiteflies, plant hoppers, and certain beetles). Practically all can be spread by propagating (e.g., grafting, budding, air layering, cuttings, root and rhizome divisions), by virus-infected planting stock (e.g., tubers, corms, bulbs), and a very few by infected seed, pollen, soil, mites, slugs and snails, fungi, nematodes, birds, or possibly other minute animal life in the soil.

Virus diseases are generally most serious in crops that are vegetatively propagated such as potato, iris, gladiolus, dahlia, chrysanthemum, carnation, gera-

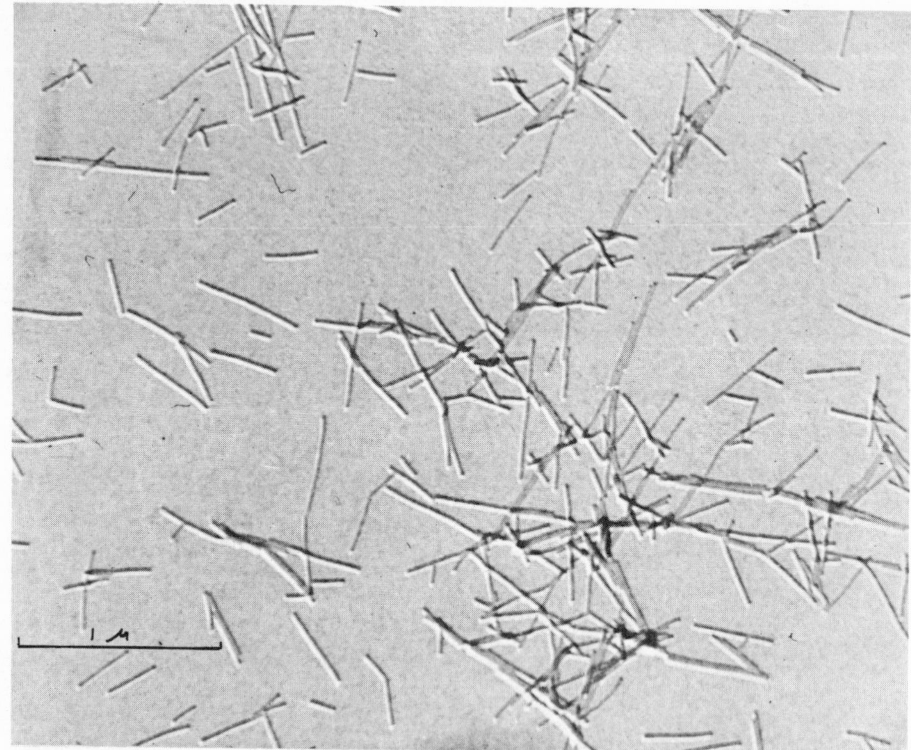

Fig. 1.4. Rods of tobacco mosaic virus from diseased tomato, shadowed with palladium. Picture taken with an electron microscope. The black line at the lower left equals one micron (1 μ) or 1/1,000 of a millimeter. (Courtesy Dr. H. H. Thornberry and Mary R. Thompson)

nium, lilies, other bulb crops, tree and cane or bush fruits, strawberry, etc.

Viruses often overwinter in biennial and perennial crops and weeds, in the bodies of insects, and in plant debris. Plants once infected normally remain so for life.

Virus-caused diseases are receiving more attention by research workers now than formerly. This is partly due to a better understanding of viruses and to an apparent increase in the number of new, virus-caused diseases. How new viruses originate is not fully known. They are probably products of evolution like higher organisms.

Nematodes

Nematodes that attack plants are active, slender, unsegmented, microscopic roundworms (often called nemas or eelworms). The majority cannot be seen with the naked eye because they are translucent and rarely exceed 1/20 of an inch in length (see Figure 1.5). Nematodes are common in water, decaying organic matter, all moist garden soils, and tissues of other living organisms. Most types are harmless because they feed primarily on decomposing organic material and other soil organisms. About 100 are even beneficial to man since they are parasitic on plant-feeding types, fungi, or other pests (Figure 1.6).

Parasitic nematodes usually obtain food by sucking juices from living plants. Feeding is accomplished by a hollow, needle-like mouth part called a spear or stylet used to puncture cells and extract the contents (Figures 1.5 and 1.6). Nematode feeding reduces the vigor and yield of plants and affords easy entrance for wilt- or rot-producing fungi and bacteria. Nematode-damaged plants also may be more susceptible to winter injury and drought.

Nematodes may live part of the time (sometimes in winter) free in soil around

Fig. 1.5. Typical female spiral nematode showing important anatomical details. (Courtesy Dr. D. P. Taylor)

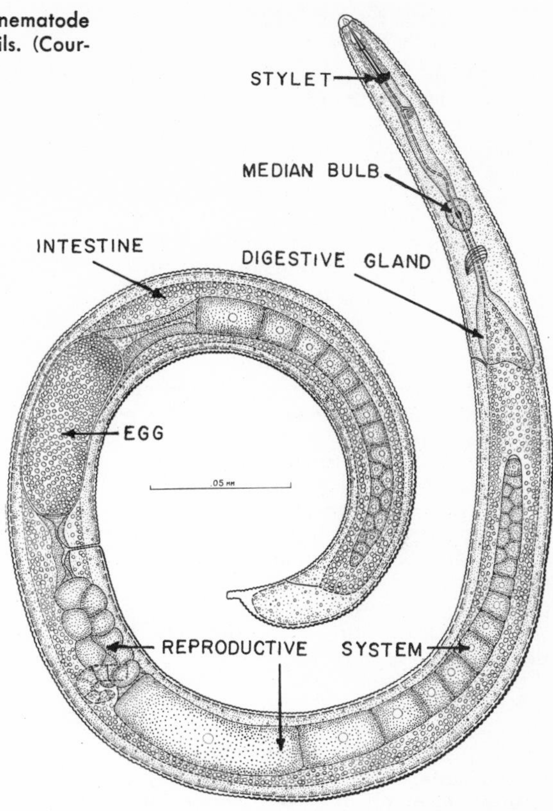

Fig. 1.6. A predator nematode (*Seinura*) feeding on another species (*Aphelenchus*). The stylet of the predator can be seen within the victim. Not all nematodes are bad! (Illinois Agricultural Experiment Station)

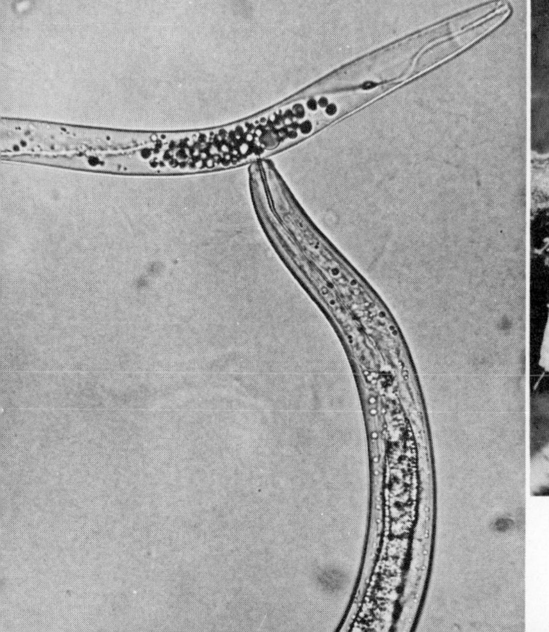

Fig. 1.7. Lance (*Hoplolaimus*) nematodes with heads partially embedded, feeding in mass on a root. (Illinois Agricultural Experiment Station)

roots or in fallow gardens and fields. Parasitic nemas tunnel inside plant tissues or feed externally from the plant surface, especially roots (Figure 1.7). Nemas may enter plants through wounds, natural openings, or by penetrating roots and pushing in between cells.

All plant-parasitic nematodes reproduce by laying eggs. These are deposited in plant tissues or in soil. Eggs hatch, sometimes after months or even years, releasing young, wormlike nematodes (larvae; better called juveniles) that usually are born ready to start feeding. Nematodes multiply much faster than higher animals, but much more slowly than bacteria and fungi. Reproduction without males is not uncommon.

Most parasitic species require about 3 weeks or longer to complete a generation from egg through the usual 4 larval stages to adult and back to egg again. Some nematodes have only one generation a year. But the offspring in this one generation may number many hundreds.

Soil populations and the developmental rate of plant-parasitic nematodes are affected by temperature; soil moisture, type, texture, and structure; populations of antagonistic or parasitic bacteria, viruses, nematode-trapping fungi, and cannibal nematodes (Figure 1.6); toxic chemicals in soil or secreted by plant roots; past cropping history (especially the last crop); current plants growing and their nutrition; and other factors.

Certain nematodes live strictly in light, sandy soils. Some build up high populations in muck soils; a few seem to thrive best in heavy soils. High populations of nematodes, as well as greater crop damage, are much more common in light sandy soils than in heavy clay soils.

Many plant-infecting nematodes become inactive at temperatures of 41° to 59° F. and over 86° to 104° F. The optimum temperature for most nematodes is 59° to 86° F., but varies greatly with the species, stage of development, activity, plant host, and other factors.

Many species of nemas are killed easily by air-drying soil after harvest or before planting. Other types remain alive but in a dormant state under the same conditions. When dormant (or as cysts) they are much more difficult to kill by chemicals (nematocides: see Table 22 in the Appendix) or heat than when they are moist and actively wiggling. Exposure to moist heat, such as steam or hot water at 120° F. for 30 minutes, is sufficient to kill most nematodes and nematode eggs. At higher temperatures, shorter periods provide control.

Watering during droughts, fertilizer applications to promote vigorous growth, clean cultivation, and the use of nematode-free planting material help greatly in reducing nematode losses. Crop rotation is often an important control measure. Because complete control is impossible or unlikely, periodic checks are most desirable whenever troublesome types have been found in large numbers.

A cover crop of African or American marigolds *(Tagetes)* or crotalaria is effective in reducing populations of certain plant-parasitic nematodes in soil. Scientists are also studying the effects of a number of chemosterilants in reducing soil populations.

After a plant-parasitic nema has been accidentally introduced into a garden or field, several years must pass before the nema population can build up sufficiently (several billion or more active nematodes per acre) to cause conspicuous disease symptoms in a large number of plants. This is because nematodes move very slowly through soil under their own power—rarely more than 30 inches a year.

Nematodes, however, are easily spread about by any agency involving moist infested soil, plant parts, or contaminated objects. These include all types of garden equipment and machinery, bags and other containers, running water, shoes, feet of animals, and movement of infested planting stock especially with soil around the roots.

Only a few garden-infesting species cause typical plant disease symptoms. These include root-knot and cyst nematodes; leaf, bud, stem, and bulb nematodes; and possibly a few others. See Figures 2.38, 2.52 and 2.53.

The best known nematode, the root-knot nematode, causes galls on roots of over 2,000 kinds of plants. Most root-feeding species, however, cause no specific symptoms. Infested plants are weakened and often appear as if suffering from drought, excessive soil moisture, malnutrition, or disease (e.g., wilt, dieback, crown rot, or root rot). Common signs of nematode injury include deficiency symptoms, stunting, loss of green color and yellowing, dieback, slow general decline, wilting on hot, bright days, and lack of normal response to water and fertilizer. Root systems are reduced, and may be "stubby" or excessively branched and often decayed.

Many times the first indication of nematode injury in a garden or field is the appearance of circular or irregular areas of stunted plants with yellow or bronzed foliage. The areas are small and gradually enlarge. Plants in the center may gradually decline in growth and die.

It has been conservatively estimated that nemas cost U.S. farmers in excess of a billion dollars each year in crop losses and higher production costs. Much of this loss could be avoided. All crop and ornamental plants are attacked by one or more species of nematodes.

The roots of unthrifty or abnormal plants, and also surrounding soil should be examined for known plant-parasitic types. This can be done only by a trained nematologist working in a well-equipped laboratory. If you suspect nematode injury, contact your local county extension office for further information regarding collection of samples for identification of parasitic types.

SECTION TWO

"What Is It?"

ENVIRONMENTAL FACTORS

1. Planting 19
2. Soil 21
3. Soil Mixes 22
4. Adding Organic Matter 23
5. Apply a Soil Insecticide 23
6. Loosening Hard Soil 23
7. Loose Soil Surface 24
8. Soil pH 24
9. Soil Test 25
10. Nutrient Deficiencies 25
11. Fertilizing Plants 29
12. Soluble Salt Injury (Salinity) 31
13. Pruning 32
    Shrubs 32
    Trees 32
    Flowers and Potted Plants . . 33
14. Tree Removal 33
15. Treatment of Wounds 34
16. Staking Trees and Shrubs . . . 36
17. Soil Drainage 36
18. Watering 37
19. Light 38
20. Oedema 39
21. Air Humidity 39
22. Temperature 40
    Light and Temperature Requirements for House Plants . . . 40
    Light and Temperature Requirements for Vegetables . . . 41
23. Leaf Scorch or Sunscorch 41
24. Winter Injury 41
25. Chemical Injuries 43
    Air Pollution 43
    Salt Injury 44
    Gas Injury 44
    Excess Care 45
    Pesticide Damage 45
26. Mechanical Injuries 45
27. Electrical Injuries 46
28. Check and Double Check . . . 48
29. Here We Go 48

GENERAL DISEASES

A. Foliage Diseases 48
  (1) Fungus Leaf Spot 48
  (2) Bacterial Leaf Spot or Blight, Bud Rot 50
  (3) Leaf Blight, Leaf Blotch, Anthracnose, Needle Blight, or Cast of Evergreens 51
  (4) Shot-hole 52
  (5) Botrytis Blight, Gray-mold Blight, Bud Rot, Blossom Blight, Twig Blight 52
  (6) Downy Mildew 55
  (7) Powdery Mildew 56
  (8) Rust-Leaf, Stem, Needle . . . 58
  (9) White-Rust, White Blister . . . 62
  (10) Leaf Curl or Gall, Leaf Blister, Witches'-broom, Plum Pockets . 62
  (11) Smut-Leaf, Stem, Anther, and Seed 62
  (12) Sooty Mold or Blotch, Black Mildew 64
  (13) White Smut, Leaf Smut . . . 65
  (14) Scab 65
  (15) Wilts 65
    A. Fusarium Wilt or Yellows . . 66
    B. Verticillium Wilt 68
    C. Bacterial Wilt, Brown Rot, or Blight 69
  (16) Mosaic, Mottle, Crinkle, Streak,

Calico, Virus Leaf Curl, Infectious Variegation, Flower Breaking . 70
(17) Spotted Wilt, Ringspot 72
(18) Yellows, Aster Yellows, Rosette, Dwarf, Stunt 73
(19) Curly-Top, Western Yellow Blight 75
(20) Leaf, Bud, Stem, and Leaf Gall Nematodes 75
B. Stem Diseases 76
(21) Crown, Foot, Stem, Stalk, Collar, or Rhizome Rot; Stem Blight, Southern Blight, Damping-off . 76
(22) Stem, Twig, Branch, or Trunk Canker; Dieback; Stem, Cane, or Limb Blight 77
(23) Wood, Butt, Wound, Heart, or Sapwood Rot 79
(24) Fire Blight, Bacterial Shoot Blight, Bacterial Canker, Gummosis . . 79
(25) Black Knot 81
(26) Rust 81
(27) Smut 81
(28) Leafy Gall, Fasciation, Witches'-broom 81
(29) Bacterial Soft Rot, Bacterial Stem Rot, Collar Rot 81
(30) Crown Gall, Cane Gall, Hairy Root, Bacterial Root Gall . . . 83
C. Flower and Fruit Diseases 86
(31) Flower or Blossom Blight, Ray or Inflorescence Blight 86
(32) Fruit Spot, Speck, Rot, or Blotch; Seed, Berry, or Tuber Rot; Storage Rot 86
(33) Smut 87
D. Root and Bulb (Corm) Diseases . . 87
(34) Root Rot, "Decline," Cutting Rot 87
(35) Clubroot 89
(36) Bulb (Corm) or Rhizome Rot . . 90
(37) Root-knot, Root Gall, Cyst Nematode 90
(38) Bulb Nematode, Ring Disease, Onion Bloat 92
E. Parasitic Flowering Plants 93
(39) Mistletoes 93
(40) Dodder, Strangleweed, Love Vine, Goldthread 94

Accurate disease diagnosis is essential to protecting and curing the plant. It is much like detective work. First you must find all the facts, then fit them together like pieces in a jigsaw puzzle to make your case and reach a logical decision.

Let us take the sad case of Brother Juniper's wilting aster plants. Is it because a wilt-producing fungus or bacterium has invaded the water-conducting tissues? Is it a root rot? Has a stem canker or rot shut off water to the foliage? Temperature too cold or hot? Can it be pesticide or fertilizer injury? How about borers or root-feeding insects? Is the soil too dry, too compact, and hard? Is soil drainage poor? Any one of these factors could cause the wilting. But which one? By careful observation of plants, close checking of each factor, and with experience, you have an excellent chance of reaching the right conclusion.

Before discussing general types of infectious diseases, let us briefly review some problems common to the culture of garden plants. These may cause nonparasitic ailments or lead to infection by microorganisms.

A complete discussion on the proper culture of plants would in itself constitute a volume this size. Thus, discussion will be limited to difficulties that result *from* improper culture and which *resemble* plant diseases. For a good discussion on the culture of garden plants read a book such as *The Care and Feeding of Garden Plants* published by the American Society for Horticultural Science and National Plant Food Institute, Washington, D.C.

ENVIRONMENTAL FACTORS

1. Planting

Plants may wither, die, or lack vigor (fail to grow, leaves change color, wilt, often drop early) because planting instructions were not followed. Death usually occurs during the first year.

You can get information on the average date of the last killing frost in your area by contacting your local county extension office. You can also get help on when and how specific vegetables, flowers, fruits, trees, and shrubs should be planted. How early you can safely plant

"Funny, you don't look a bit like your pictures!"

(Courtesy Publishers Syndicate)

depends on the hardiness of the plant and the climate in your area. Certain vegetables and flowers can withstand frost while others cannot. Mail-order nurseries will ship their stock to you at the time it should be planted. Follow the instructions which come with your order. There is no scientific evidence to support "planting by the moon."

Container-grown roses and other plants provide an extra planting period—most of the growing season.

Never allow the roots of plants to dry. On digging or receiving bare-rooted plants from a nursery, soak the roots in warm water for an hour or two and keep them damp until ready for planting. If you must hold plants for more than 2 or 3 days before planting, heel them in: place in a trench and cover roots with moist soil. Keep them out of wind and away from heat. Whenever possible, prepare the planting holes in advance so plants can be set out as soon as received.

For balled and burlapped evergreens, dig the hole at least a foot wider and deeper than the ball. Be sure to handle by the ball only, since handling plants may cause it to break or roots to pull loose. Set plants *at the same depth* they were in the nursery. If soil is poor, excavate and prepare a good soil mix composed of rich topsoil, sand, and peatmoss or compost. Fill the hole ⅔ full, water thoroughly, cut the twine, and peel back the burlap. Fill and pack firmly.

For bare-root plants, dig the hole large enough to prevent crowding and twisting of roots. Loosen the subsoil. Put a 4-inch mound of topsoil mixed with peatmoss in the bottom of the hole and spread roots in a natural position over it. Sever roots that are dead, broken, or bruised. Set plant *at the same depth* it grew in the nursery. Work a crumbly, rich topsoil mixture, composed of ⅓ peatmoss, manure or compost, among roots by hand as hole is filled with soil. Fill hole gradually. Soak hole thoroughly when about ½ full. Then settle plant firmly by shaking gently. This assures contact of roots with soil and prevents air pockets. Pack soil firmly. Leave a rim of soil at the edge to form a water-holding basin. Water deeply once a week (if less than ½ inch of rain has fallen per week during the growing season) for the first 2 years. Apply a loose surface mulch (e.g., peatmoss, salt hay, grass clippings, leaves, wood chips, ground corn cobs, etc.). See No. 7 Loose Soil Surface (below).

Shrubs next to the home should be planted at a distance of a little less than one-half the spread of a mature shrub. Do not crowd! Plant shrubs outside the eave line or overhang on a ranch-style home, especially on the north and east sides.

When soil is poor around the foundation, dig out the entire bed to a depth of 1½ to 2 feet and replace with good soil.

Garden vegetables, fruits, and most flowers do best in sunny, well-drained sites 50 feet or more removed from large trees that compete for nutrients, water, and soil space.

Always plant more than one variety of each type of fruit such as apples, pears, sweet cherries, some peaches, plums and prunes, some grapes, and most cultivated blueberries. This is necessary since many fruits are self-sterile and require pollination with another variety that blooms at the same time.

Planting shock—loss of too much water before roots are established—can be prevented on many plants by spraying the foliage with Wilt Pruf, Plantcote, Foligard, or Stop-Wilt. These materials are antidesiccants.

2. Soil

Soil is probably the most important factor in success or failure of growing adapted or hardy plants. Healthy plants

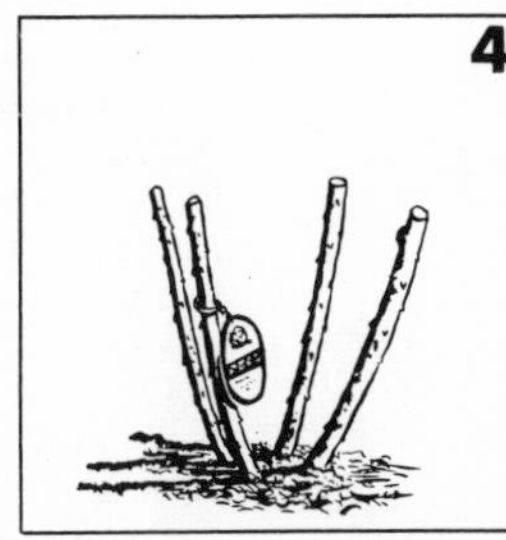

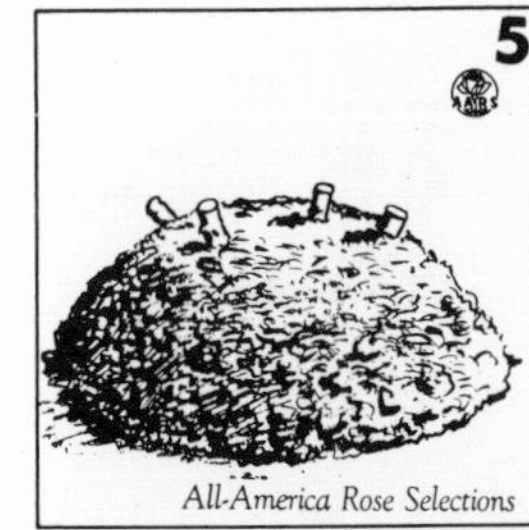

Fig. 2.1. How to plant a rose. 1. Dig a hole 18″ wide and about 15″ deep in well-drained, neutral soil. Replace several inches of loose, prepared soil (soil mixed well with a cup of balanced rose food — e.g., 5-10-5 — and several spadesful of peatmoss, leaf mold, or compost). Form a mound in the center. Place rosebush on mound and spread its roots naturally down slope. Position plant on mound so that bud union (knotlike, swollen area) is at about ground level. (In warm climates, like Florida and California, the bud union should be about 1″ above the final soil line.) 2. Cover roots with soil, working the soil in around roots to eliminate air pockets. Fill hole about ¾ full and tamp down firmly. 3. Pour several pailsful of water into hole and allow to drain. 4. After water has completely settled, fill in remainder of hole with soil. 5. Mound up around canes to a height of about 8 inches. Remove protective mound after a week or 10 days. Later, cover rooting area several inches deep with a mulching material to preserve moisture and keep roots cool.

need a vigorous root system, ample soil nutrients (more correctly called elements or raw materials), and a porous soil containing the proper mixture of clay, sand, silt, and humus or organic matter. Any soil can be improved!

A good soil mixture for most house plants is made up of equal volumes of good garden loam soil, organic matter (leafmold, shredded sphagnum moss, well-rotted manure or compost), and coarse sand. Ferns, gloxinia, gardenia, African-violet, many foliage plants, tuberous begonia, and azaleas require soils with a higher content of organic matter, while cacti and many succulents do best in a loose sandy soil low in organic matter. For *cacti* a good mixture is:

½ part garden loam,
2 parts sand,
½ part peatmoss or leafmold,
and a sprinkling of charcoal.

Other special soil mixtures may be desirable for certain plants (e.g., African-violet, geranium, orchids, bromeliads) but they are the exceptions. No matter what the potting soil mix, it should be pasteurized before planting (pages 538–548).

3. Soil Mixes

If good potting soil is scarce you may wish to try one of the following soil mixes.

Cornell University Peat-Lite Mix A. A light-weight, sterile, easily watered and handled mix with good physical characteristics. Quality is easily controlled.

A one-peck recipe includes:

1. Horticultural No. 2 or 4 Grade Vermiculite *or* Perlite . . . . 4 quarts
2. Shredded sphagnum peatmoss . . . . 4 quarts
3. 20% superphosphate (powdered) . . . . 1 tablespoon
4. Ground limestone . . . . . . 1½ tablespoons
5. 5-10-5 fertilizer . . . . 2 tablespoons

Mix ingredients on a clean floor or table area. Place mixture in pot, flat, or pack and thoroughly soak with water. Soak again after 30 minutes and set in transplants, seed, or cuttings. Water only as needed.

If using mix for plants that will be grown in containers for a long period, supplemental feedings should be made with any good soluble fertilizer at the recommended rate and frequency. This mix, minus lime and fertilizer, is sold ready-mixed as Sphag-lite.

UC Soil Mix C. A good multipurpose soil mix, developed at the University of California, for pots and beds. It is quite heavy but has excellent physical properties. Requires supplemental feeding within a short time. Good for rooted cuttings.

A one-cubic yard (= 86.84 pecks) recipe includes:

1. Fine sand (*washed* plaster grade) . . . . 10¾ bushels
2. Shredded sphagnum peatmoss . . . . 10¾ bushels
3. Potassium nitrate . . . . 4 ounces
4. Potassium sulfate . . . . 4 ounces
5. Superphosphate (20%) . . . . 2½ pounds
6. Dolomite limestone . . . . 7½ pounds
7. Calcium carbonate limestone . . . . 2½ pounds

The peatmoss—Canadian or German—should be wetted 1 or 2 days before use. The fine sand (particle size range 0.05 to 0.5 mm.), peatmoss, and mixed fertilizer ingredients are placed in convenient piles. Add proper amounts of sand and peat to a low, level pile and scatter the fertilizer components evenly over the surface. Turn with a shovel, progressively working through the mass from one side and form a second pile as the shovels of soil are turned. This second pile is turned back again and the process repeated until blending is complete. A concrete or ribbon mixer may be used and ingredients added by hand. Moist, but not excessively wet, sand and peatmoss are essential to uniform blending and reliability. The mix is then pasteurized using heat or chemicals (pages 538–548). Other advantages of the UC Mix: reliable and reproducible, no composting necessary, plant growth is uniform in size and time, no toxicity after steaming.

Other UC Mixes (A-E) are available for a variety of uses. For full details, read *The U.C. System for Producing Healthy Container-Grown Plants,* California Agricultural Experiment Station Manual 23.

John Innes Composts. A heavy, multipurpose mix widely used in Europe and the United States.

A one-bushel recipe includes:

I. Seed Compost
   1. Garden loam soil . . . . ½ bushel
   2. Coarse, well-washed river sand . . . . ¼ bushel

3. Shredded sphagnum peatmoss (coarse grade) . . . . ¼ bushel
4. 20% superphosphate (powdered) . . . . 1½ ounces
5. Builder's lime (chalk or calcium carbonate) . . . . . . ¾ ounce

II. Potting Compost
1. Garden loam soil . . . . 6/10 bushel
2. Shredded sphagnum peatmoss (coarse grade) . . ¼ bushel
3. Coarse, well-washed river sand . . . . 2/10 bushel
4. Hoof and horn meal . . . . . 1½ ounces
5. 20% superphosphate (powdered) . . . . 1½ ounces
6. Sulfate of potash (48% potash) . . . . ¾ ounce
7. Builder's lime (chalk or calcium carbonate) . . . . . . ¾ ounce

The garden loam should be humusy. Avoid sandy, heavy, or calcareous soil. Sift through a screen and steam at 180° F. for *10 minutes;* do not oversteam. Dump on a clean surface to cool. The sand should be fairly coarse grade (particle size range 0.25 to 1.0 mm.). Can be sterilized by solution of 3 tablespoons of household bleach (e.g., Clorox) to a gallon of water. Mix fertilizer components with the sand. Then combine pasteurized loam, peat, and sand mixture.

Sow seeds in the seed compost. When seedlings are showing their first true leaves, transplant into the potting compost. When old enough, transplant right into the garden.

No soil mix is "perfect." Certain ones may lack buffering capacity or quickly need supplemental feeding with a balanced fertilizer. If problems develop, check with your extension horticulturist or local florist.

4. Adding Organic Matter

The addition of more organic matter will aid most garden soils, especially heavy clay and sandy soils. Mix in well-decomposed compost, peatmoss, or well-rotted barnyard manure. A compost pile (composed of alternate layers of vegetable matter, topsoil, and commercial fertilizer) in a back corner of the yard is an excellent investment.

Compost can be made from leaves, weeds, grass clippings, or other disease-free waste vegetable matter. To make compost, pile these materials in layers 6 to 8 inches thick as they accumulate during the season. Add about 1 pound of a lime-fertilizer mixture to each 10 pounds of dry refuse (¼ pound to each 10 pounds of green material). The mixture can be made from 5 pounds of 10-10-10 or 12-12-12 fertilizer plus 2 pounds of fine limestone. This fertilizer treatment hastens decay and improves fertility of the compost. Spread soil over the material to hold it in place. Water the pile to keep damp, and occasionally turn and mix the soil and decaying material. The compost should be ready to spread over and then mix with garden soil in a year to 18 months.

If land is available, a green manure or cover crop (e.g., ryegrass, soybeans, cowpeas, crimson clover, a cereal, buckwheat, or Sudangrass) sowed in spring and plowed under in late summer makes excellent organic matter. In all but the most northern states you can sow ryegrass, rye, or winter wheat in late summer and turn it under in early spring. Fertilization before turning under hastens decomposition of the green manure. This is often desirable.

5. Apply a Soil Insecticide

A soil insecticide applied at or before planting along with fertilizer or organic matter prevents many later problems caused by soil insects. One application of most materials lasts 4 years or more. Use aldrin, dieldrin, chlordane, or Diazinon.

If vegetables or small fruits are to be grown in this treated soil during the next 5 years, apply only Diazinon or chlordane. Soil insecticides are sold in liquid, dust, or granular form and should be worked into the soil. Follow manufacturer's directions. The new granular soil insecticides are easy to use and can be applied along with fertilizer. Check with your extension entomologist regarding the latest soil insect control recommendations.

6. Loosening Hard Soil

If soil is too hard, grass refuses to grow while growth of trees and shrubs is slowed, foliage becomes sparse, and leaves may wilt. Hard, compact soil can be loosened by incorporating compost; sawdust; corncobs; bagasse; wood shavings or chips; weed-free hay or straw; peatmoss; buckwheat, cocoa-bean, peanut, or cottonseed hulls; grass clippings; leaves; chopped corn fodder; and fresh manure with straw bedding. When the added material is low in nitrogen (e.g., sawdust, corncobs, wood shavings, and straw), apply a nitrogen-containing fertilizer to prevent crop injury.

7. Loose Soil Surface

Plants do best with a loose soil surface. This can be supplied by cultivation or, better still, by a surface mulch of organic material (see above). Mulching also helps control weeds, conserve moisture, regulate soil temperature, keep fruit off damp soil, eliminate injury to crops by cultivation, and prevent erosion. Special types of paper and black plastic specifically for mulching are now on the market. These materials come in rolls of various widths and are unrolled over the prepared, well-fertilized seedbed before planting. Edges of the material are anchored in small furrows (about 2 inches deep) with soil on top. Seeds and transplants are then planted through holes cut in the mulch at the desired spacing.

8. Soil pH

Is your soil acid or alkaline? This can be easily tested by taking a composite soil sample and having it tested at your county extension office. The proper test will show soil to be alkaline, neutral, slightly acid, or moderately acid. Perhaps your test will be returned showing numbers on a pH scale. A pH of 7 is considered neutral. A pH of 6 to 6.5 is slightly acid, and a pH of 4.5 to 5.5 indicates the soil is moderately acid or sour. A pH above 7 means the soil is alkaline. Below pH 4 and above 9 most plants have a hard time growing. Soils in the United States range in pH from about 3.6 for certain acid peat soils to 9.5 for some "black alkali" soils. The most favorable pH range for most crop, garden, and house plants is from 6.0 to 7.5.

The great bulk of garden and crop plants are not particular to soil reaction (pH) and will grow under a wide range of conditions. Exceptions are alyssum, amaryllis, andromeda, arbutus, azalea, bayberry, bleedingheart, bluebeard-lily, blueberry, boxwood, bugbane, bunchberry, camellia, Carolina-jessamine, chokeberry, clarkia, clethra, cypress, dewberry, dogwood, fennel, certain ferns, fir, fringetree, galax, gardenia, gayfeather, goldthread, hardhack, heath, heather, hemlock, holly, huckleberry, hydrangea, inkberry, Jack-in-the-pulpit, Japanese iris, some junipers, Laborador-tea, leatherleaf, leucothoë, certain lilies, loblolly-bay, magnolia, mountain-laurel, New Jersey-tea, some oaks, oconee-bells, certain orchids, partridge-berry, peanut, pine, pittosporum, polygala, rhododendron, scotch broom, silverbell, sour gum, sourwood, spruce, styrax, tamarack, white-cedar, and a few others. The foliage of these plants may turn yellow (chlorotic or chlorosis) because of unavailable iron or other elements that may be due to excessive lime. These plants are best grown in an acid soil with a pH of about 4.2 to 5.5 or 6.0. Soil can be acidified by adding equal parts of powdered sulfur and iron sulfate, pine needles, or acid peatmoss (e.g., German peat or Michigan reed peat) to the soil at the same time that you apply fertilizer or compost. Addition of ¾ pound of sulfur per 100 square feet of garden increases the acidity of average soil about one pH point. Check with your extension horticulturist, county agent, or nurseryman regarding acidifying soil. Replacing soil about roots with an acid soil may be preferable to acidifying old soil.

Liming is necessary in some areas to make the soil less acid, increase availability of potassium, phosphorus, sulfur, calcium, and magnesium to avoid aluminum and manganese toxicity, and to speed up decomposition of plant residues. Avoid overliming. Too much lime may upset the balance of magnesium or potassium inside plants. A good rule to follow is "lime by test—not by guess." Lime, when needed, is spread on soil and then worked into the top 6 inches or so. Finely ground dolomite limestone containing calcium and magnesium is preferred in some areas needing these nutrients. Hydrated lime, finely ground oyster shells, and marl may also be used.

In alkaline soils of large areas of the Midwest and arid parts of the United States (with a soil pH above 6.7), many types of plants including pin oak, sweetgum, hackberry, pines, roses, gardenias, camellias, rhododendron, and azalea suffer from "iron chlorosis" unless soil is acidified or iron-containing salts are placed in the soil root zone, injected into plants, or sprayed on foliage. For full information on control read USDA H & G Bulletin 102, *Iron Deficiency in Plants: How to Control It in Yards and Gar-*

dens. Chlorosis may also be caused by a deficiency of calcium, zinc, magnesium, manganese or sulfur, viruses, flooding, winter injury, insect feeding, or other parasites. Excessive applications of lime may also induce iron, manganese, and zinc deficiencies.

9. Soil Test

If you suspect soil or fertilizer problems, contact your county extension office for advice. Frequently a soil or tissue test will be suggested to determine what, if any, fertilizers or other treatments are advisable. County agents and their assistants know soils in your area and what types of soil problems are most likely to occur. Only by a soil or plant-tissue test can you be reasonably sure of the amount and availability of nutrients in your soil. You can often save money by testing soil, since one or more nutrients may not be needed. Plants, like humans, differ in their individual nutrient needs. Plants in different stages of maturity also vary greatly in their tolerance to soil problems. Because these tests are not infallible, certain plants often respond to fertilization although the unfertilized soil has sufficient nutrients for "less particular" plants. Different varieties of the same plant may even react differently in the same soil!

Special forms and mailing tubes are available at your county extension office for you to have your soil tested, usually for a nominal fee. There are also a number of commercial soil testing laboratories that analyze soil for various major and minor elements.

10. Nutrient Deficiencies

Plants vary greatly in their ability to feed on the nutrients found in chemical combinations in soil. Plants growing in unfertile soil usually grow slowly, may appear stunted and weak, and fruiting is reduced. Abnormal foliage color or shape may be due to a deficiency of one or more soil nutrients. Nutrient shortages in soil may be intensified by drought, excess moisture, unseasonably low temperatures, insect damage, disease, or mechanical injury. Although more than 50 elements are absorbed and used by plants, only the more important nutrient deficiencies are discussed in this book. For additional information, contact your extension horticulturist or plant pathologist. Good, well-illustrated books on the subject include the 3rd edition of *Hunger Signs in Crops*, published by David McKay Company, New York; *The Care and Feeding of Garden Plants*, published by the American Society for Horticultural Science and the National Plant Food Institute; and the 3rd edition of *The Diagnosis of Mineral Deficiencies in Plants by Visual Symptoms*, published by Her Majesty's Stationary Office, London.

Stunted plants with small, pale green leaves fading to yellow and drying to a light brown—often indicate a *nitrogen* deficiency. Affected leaves often drop early, starting with older leaves. Nitrogen deficiency is probably the most common "hunger" sign in plants. Correct by applying a nitrogen-containing fertilizer, using legumes in the rotation, growing green-manure crops, or spraying with nitrogen materials. Check with your county agent or extension horticulturist. Nitrogen gives dark green color to plants, promotes leaf growth, produces *rapid* growth, increases protein content of food crops, and aids in decomposition of organic matter.

A superabundance of nitrogen may cause potassium or other deficiency, stunting, chlorosis, lack of or delay in flower and fruit development, and bud drop of rose, sweetpea, tomato, and other plants, and lead to winter injury.

Phosphorus-deficient plants usually have dark or bluish-green leaves, followed by bronzing, reddening, or purpling, especially along veins and margins. Later, leaves may develop purplish blotches. Lower leaves are sometimes yellow, drying to a greenish-brown or black. Plants and fruit are often stunted and spindly, mature late, and have shrunken seeds. Control by applying a complete commercial fertilizer or add separately as superphosphate or rock phosphate. Mycorrhizic fungi, associated with the roots of many higher plants, especially trees, may help overcome phosphorus deficiency. Phosphorus stimulates early root formation and growth, gives a rapid and vigorous start to plants, hastens maturity, stimulates blooming, aids in seed and kernel formation, and provides winter hardiness protection. Phosphorus is a component of many plant protein complexes.

Excessive phosphorus fertilization may produce iron and zinc deficiencies in corn, beans, tomatoes, citrus, and other plants.

Potassium deficiency often is evident as a curling, yellowing, scorching, browning, or bronzing of leaf margins and tips. Small dead areas may form at margins and between the leaf veins. Older leaves are usually affected first. Stems are weak; roots and tubers are undeveloped. Plants are often stunted and appear "rusty." Sandy soils are usually low in potash. Correct by applying a complete fertilizer containing about 5 to 10 per cent potash. Commercial growers may use muriate of potash or sulfate of potash. Potassium increases root growth and plumpness of seed; produces strong, stiff stalks; reduces water loss and wilting; imparts increased vigor and disease resistance; is essential to formation (photosynthesis) and transfer of starches and sugars; improves drought and winter hardiness; and increases the protein content of plants.

Nitrogen, phosphorus, and potassium are the 3 elements in every complete fertilizer. They are also the elements removed in largest amounts from soil by plants (Table 1). Crop plants differ greatly in the amounts of nutrients they need and also in the ratio of nutrients in crops.

Iron, manganese, and *zinc* deficiencies frequently cause mottling, yellowing, or scorching of tissues between veins in the *youngest* leaves (chlorosis). See Figures 4.37, 4.147 and 4.234. The entire leaf may turn yellow, then cream-colored to white, and finally brown and scorched. Leaves and plants often stunted and grow poorly. If severe, foliage and growing tips may die. Yields may be drastically reduced. *Zinc* deficiency also delays maturity; leaves become thick and brittle. Plants deficient in *manganese* are more susceptible to freeze damage.

Iron deficiency is usually associated with neutral or alkaline (high-lime) soils or acid soils high in manganese. Correct by acidifying the soil, by spraying plants with a weak solution of iron sulfate (0.25% by weight), or use a soluble organic iron complex (Irn-Gro), following manufacturer's directions.

Excess iron may cause a deficiency of phosphorus or manganese. Too much manganese, copper, or zinc may intensify iron chlorosis.

Manganese shortage is most common in humid regions in peat and muck soils low in total manganese and pH above 6.2. Manganese availability is low in neutral and alkaline soils and is reduced by liming acid soils. Manganese influences movement of iron within plants and synthesis of chlorophyll. Manganese deficiency can be controlled by adding manganese sulfate to soil (1 pound per 1,000 square feet) or by spraying with a weak solution of manganese sulfate (0.25% by weight) or apply manganese oxide (Nu-Manese). Follow manufacturer's directions.

TABLE 1

PLANT-FOOD UTILIZATION CHART*

Crop Plant	Nitrogen (N)	Phosphoric Acid (P_2O_5)	Potash (K_2O)
(per acre yields)	*(amounts in pounds in total plant with good acre yields)*		
Apples (600 bu.)	115	40	135
Peaches (600 bu.)	95	40	120
Grapes (10 tons)	60	30	80
Oranges (800 boxes)	120	40	175
Beets, Sugar (30 tons)	275	85	550
Cabbage (25 tons)	225	30	210
Celery (75 tons)	280	165	750
Corn (150 bushels)	200	75	195
Lettuce (20 tons)	90	30	185
Potatoes, Irish (400 cwt.)	200	55	310
Sweetpotatoes (500 bu.)	115	35	175
Tomatoes (30 tons)	250	80	480
Peanuts (3,000 lbs. nuts)	220	45	120
Lawngrass (2.5 tons dry)	225	60	140

* Adapted from one prepared by J. D. Romaine, American Potash Institute.

Excessive manganese produces puckered, mottled, and partially yellowed leaves with dead areas forming along and between the veins. Affected leaves may become brittle and ragged along the edges.

Zinc deficiency occurs under a very wide variety of soil conditions, being most common in strongly weathered, coarse-textured, and alkaline soils in regions of limited rainfall. Addition of lime and phosphorus may reduce zinc availability and cause zinc deficiency. Control by applying zinc sulfate to soil at the rate of about ½ pound per 1,000 square feet or by spraying affected plants with 0.5% zinc sulfate. NuZ contains 52% zinc in the form of zinc oxysulfate. Follow manufacturer's instructions.

Iron, manganese, and *zinc* deficiencies can also be controlled by acidifying neutral or alkaline soil (see under Soil pH above), by adding lime if soil is quite acid (below about 5.5), improving soil drainage, and use of chelates as sprays, ground applications, or trunk injections into woody plants. See under plant involved. Iron chelates are sold as Danitra, Nu-Iron, Ree-Green, Re-Nu, Sequestrene 138 Fe, 330 Fe and NaFe Iron Chelates, and Triangle Iron Chelate Granules. Manganese chelates are sold as Danitra and Sequestrene Na_2Mn Manganese Chelate. Zinc chelate is sold as Danitra and Sequestrene Na_2Zn Zinc Chelate. Follow manufacturer's directions to the letter. Remember that certain chelates are formulated for use in acid soils; others are strictly for alkaline soils.

Chelated Nutramin contains a completely soluble blend of iron, manganese, and zinc as well as boron, molybdenum, and copper.

There are a number of products available that contain various mixtures of several minor elements. These include Raplex-Fe, Raplex-Mn and Raplex-Zn (mixtures of iron, manganese, and zinc in powdered or granular form), Es-Min-El (iron, manganese, zinc, boron, magnesium, and copper), Fertminal (various trace elements), Flag's Citrus Nutritional Sprays (4 combinations of various trace elements), FTE (various combinations of boron, iron, manganese, copper, zinc, and molybdenum in slowly soluble fritted form), I-F-N Mixture (insecticide-fungicide and nutritional mixture containing variable amounts of zinc, copper, manganese, boron, sulfur, and molybdenum), Kilgore's Citrus Nutritional Sprays (8 combinations of various trace elements), S-P-M (Sul-Po-Mag) containing sulfur, potash and magnesium, K-Mag (sulfate of potash magnesia), Nutramin 6 (stabilized water-soluble blend of manganese, iron, copper, boron, molybdenum, and zinc), Nutra-Phos 3–25 (zinc manganese nutritional foliar spray containing 25% zinc, 25% manganese, 25% P_2O_5), Nutra-Spray (combination of 17.5% zinc, 4% manganese, 4% copper), Nutri-Sperse (complete nutritional spray), "PERK" Nutritional (combination of copper, zinc, manganese, and sulfur for use on spraying citrus, ornamentals, vegetables), and XXX Mineral Mix (combination of 9% zinc, 3.5% manganese, 4.5% iron). Before purchasing any trace element product check with your county extension office or extension horticulturist to see if your plants would likely benefit from ground or spray applications. In most agricultural areas, little or no response will likely be noted from their use if ordinary fertilization and good cultural management practices have been followed.

A lack of *boron* often causes plants to be stunted, "bushy," and brittle with a scorching of tips and margins of younger leaves. Older leaves are malformed and distorted while edible shoots, roots, and fruits are corky and discolored. Flower buds die. Twig tips may die back. Symptoms are most severe in dry areas or seasons and in sandy, overlimed, or alkaline soils. Only a very small amount of boron (a few parts per million) is needed for normal plant growth. Boron is highly important in reproductive processes including cell division, carbohydrate and nitrogen metabolism, and water relations. Boron influences the conversion of nitrogen and carbohydrates into more complex substances such as proteins. Boron also affects the transfer of carbohydrates within the plant. Control boron deficiency by applying borax (sodium borate) to soil or boron-containing sprays. See under Apple. Boron is sold as Agribor (Stauffer, Potash Co. of America), Polybor (U.S. Borax), Boro-Spray (American Potash), Tronabor (U.S. Borax), and Fertilizer Borate 46 and 65 (U.S. Borax). Apply 2 to 50 pounds of boron per acre as borax or boric acid, or directly to plants as a spray.

In some areas boron is excessive and injury is not uncommon. Excess boron may cause plants to be stunted and yellowed; leaves are scorched and plants may die prematurely. Germination may be delayed or prevented. Beans, peas, and corn are very sensitive to boron. For additional information, read USDA IF 211, *Boron Injury to Plants.*

Magnesium deficiency symptoms usually appear on older and lower leaves as a gradual fading or mottling of normal green color at margins, tips, and between the veins. Later these areas turn yellow (pink, red, orange, or purple on some plants) and finally brown. When severe, lower leaves die and may drop prematurely. Discoloration of leaves progresses *upward* until only tip leaves appear normal. In corn, or other plants with parallel veins, the yellowing appears as interveinal stripes. Magnesium deficiency is most prevalent in deep, sandy, acid soils of the Atlantic and Gulf Coastal Plains in wet seasons, or where sodium and potassium are excessive. A magnesium deficiency aggravates winter injury. Magnesium is essential to chlorophyll production, regulates uptake of other foods, acts as a carrier of phosphorus in plants, influences earliness and uniformity of maturity, promotes formation of oils and proteins in seeds, size of roots, fruits, and other marketable portions of crops, and stimulates growth of soil bacteria. Control magnesium shortage by using dolomitic limestone, magnesium-containing fertilizers containing magnesium sulfate (Emjeo), or Epsom salts.

Molybdenum deficiency causes Whiptail of cauliflower and broccoli (see under Cabbage). Leaves on other thin-leaved plants may be stunted, crinkled, curled upward, pale green or yellow, and malformed. Molybdenum is required in the nitrogen metabolism of all plants. Healthy plants may need only 10 parts per billion of their green weight as molybdenum. Correct by applying sodium or ammonium molybdate at rates of 0.1 to 1 pound per acre.

Plants also require *copper* in minute amounts. Lack of copper (principally in neutral to alkaline (pH 7 to 8) muck or peat soils of Florida and California) causes leaves to be darker green or have a grayish, olive, or bluish tint. Leaf edges curl upward and the green gradually fades until it borders just the principal veins. Beginning with the lowest leaves, pale areas in affected leaves gradually turn brown and die. Young leaves may wilt permanently without spotting or marked chlorosis. Shoot development is slow. Twigs may die back. Plants are stunted; flowering and fruit set is either delayed or checked altogether. Carrots are poorly colored and bitter. A deficiency of copper is easily checked by applying copper sprays to control disease or by applying copper sulfate to the soil (20 to 200 pounds to the acre). Follow local recommendations. Six ounces of copper may actually be sufficient per acre. Copper toxicity can be more troublesome than copper deficiency. Remember that the addition of copper to soils usually produces strong residual effects. The newest treatments for copper deficiency are chelates (Danitra and Sequestrene Na_2Cu Copper Chelate). Copper functions in plant growth as an enzyme activator, or as a part of certain enzymes. It is important in protein utilization.

A *calcium* deficiency causes the flower stem of gladiolus and tulip to topple. Roots of many plants are short and stubby. Stalk and twig tips may die back. Young terminal leaves are distorted and curled, finally dying back at tips and margins. Flowers are often abnormal. Calcium promotes early root formation and growth, improves general vigor, increases disease resistance and stiffness to stem, influences uptake of other plant foods, acts as a neutralizing agent on toxic substances formed within the plant, and increases calcium content of food crops. Correct by spraying growing plants with calcium nitrate or add calcium carbonate (ground limestone), dolomite, or calcium sulfate (gypsum) to the soil.

If calcium is present in excessive amounts, a deficiency of iron, zinc, manganese, magnesium, potassium, or boron may result. Similarly, an apparent excess of potassium, magnesium, or boron may really be a deficiency of calcium.

Sulfur deficiency symptoms are most evident during dry periods. *Young* leaves turn pale green to yellow (except for veins on some plants), while older leaves usually remain green. (With nitrogen, yellowing starts with *older* bottom leaves and proceeds upward.) Sulfur-deficient plants are characteristically stunted or dwarfed and spindly. Sulfur is a major plant nutrient useful to plants in the sulfate form, e.g., calcium sulfate. In industrial areas, rain and snow return 10 to 60 pounds or more of sulfur per acre each year—largely picked up by atmospheric moisture from smoke produced by burning coal, oil, and other fuels. Irrigation water and decomposition of plant material also supplies some sulfur as do many commercial fertilizers, e.g., ammonium sulfate and superphosphate.

Sulfur is associated with the formation of chlorophyll and thus affects carbohydrate metabolism. Growth-regulating hormones in plants also contain sulfur.

Deficiency symptoms will not appear if soil contains an ample, balanced supply of available plant food, and the soil pH is favorable for growing plants.

Starving plants, like starving animals or people, make poor growth and may become more susceptible to attack by certain disease organisms. Causes of poor growth are often complex and frequently cannot be traced back to a lack or excess of some plant nutrient. Freezing temperatures, hot dry winds, drought, mechanical injury, nematodes, insects or diseases, presence of nearby tree roots, lack of organic matter, damage from excess fertilizer or spray, too much sun or shade, waterlogging and poor soil drainage, etc., sometimes produce effects comparable to nutrient deficiencies.

11. Fertilizing Plants

Use fertilizer carefully and as directed. Careful use can reduce maintenance and pruning requirements of shrubs. Do not overfeed. Excessive fertilizer applications may cause serious injury (e.g., curling, yellowing, browning, stunting, and dropping of leaves, slow growth, wilting, and dieback). Late summer applications of fertilizer encourage tender growth that usually leads to severe winter injury.

Growth is usually stimulated by an application of fertilizer, provided there is ample moisture in the soil. Well-maintained and vigorous plants are also more resistant to drought, disease, and insect damage. If injured by pests, they also recover more rapidly and completely. The wounds of well-fertilized trees and shrubs heal more quickly.

Commercial fertilizers should always have a label that gives an analysis like 10-10-10, 6-10-4, or 5-10-5. The first figure denotes the percentage of nitrogen (N), the second, phosphorus (P)—actually "phosphoric acid" or phosphorus pentoxide (P_2O_5), and the third, potassium (K)—actually potash or potassium oxide (K_2O).

The rate and methods of application of commercial fertilizers vary considerably with individual needs of plants involved, purpose of application, and natural fertility of the soil.

The time to fertilize garden plants depends on their characteristics.

When preparing the lawn, vegetable, or flower seedbed, broadcast 2 to 4 pounds of a complete commercial fertilizer, e.g., 5-10-5, 7-7-7, or 10-6-4 per 100 square feet. More may be needed on light, easily leached soils. Work fertilizer into the top 4 to 6 inches or more of soil before planting.

For established lawns, spread fertilizer *evenly* on the dry grass surface. Use a carefully calibrated lawn spreader and water it in immediately. Do not overlap or miss strips!

For vegetables, flowers, small fruits, and other crops, fertilizer is frequently placed in a band or bands below or near seeds or plants at time of planting. The fertilizer is usually placed about 2 to 3 inches to one side of the seed in continuous bands 3 or 4 inches deep.

Fertilizer is also often applied as a side dressing in a narrow furrow along one side of the row or around these plants while they are growing. Use something like a 10-6-4 fertilizer, 10 pounds per 1,000 square feet. Check with your extension horticulturist or county extension office on what and how much to use for various plants. Avoid getting dry fertilizer on plant leaves as it will injure them.

Leafy vegetables such as broccoli, cabbage, lettuce, kale, and spinach do well when fertilizers fairly high in nitrogen are used. Root crops (beet, carrot, parsnip, turnip, radish, sweetpotato, white potato) respond to large quantities of both nitrogen and potassium. Beans, melons, onions, celery, lettuce, and tomatoes require considerable quantities of nitrogen, phosphorus, and potassium (Table 1). Many other elements are needed as explained above.

A starter solution of commercial fertilizer—high in phosphorus—dissolved in water is used around many garden plants when they are transplanted. The use of ½ pint of starter solution for a vegetable or flower plant speeds up growth, firms the soil, prevents air pockets that dry out the roots, and often speeds maturity and yield. Set such plants as cabbage and tomato in place and fill the hole partly with water. Starter solution is then poured in and allowed to soak into the soil. Finally, fill the hole with dry soil.

A multipurpose starter solution can be prepared by adding 2 or 3 ounces (4 tablespoons) of a 5-10-5, 6-10-4, or 4-12-4 fertilizer to a gallon of water. Stir until most of the fertilizer is dissolved or in suspension. While using, stir to keep the plant food from settling. If using a quick-dissolving fertilizer that is concentrated (e.g., 10-52-17, 10-50-10, or similar analysis), use about 2 tablespoons per gallon or follow manufacturer's directions.

Foliar application of fertilizers in dilute concentration is particularly valuable for applying trace elements (e.g., zinc, iron, manganese, magnesium, boron, calcium, copper, etc.). Many times these materials can be added to regular pest sprays. This is the latest and easiest way to fertilize plants. Some of the complete chemical fertilizers on the market for use in foliage feeding include Ra-Pid-Gro, Hygro, and Miracle-Gro. Check with your extension horticulturist regarding their use.

Avoid use of "miracle" fertilizers that claim to contain mysterious plant foods and are sold at outrageous prices.

Applications of fertilizer to trees and shrubs should be made in early spring or *late* fall after they go dormant. For most purposes, use fertilizers with a 2-1-1, 1-1-1, or 1-2-1 ratio. A 2-1-1 ratio would be approximated by a 16-8-8, 12-6-6, or 10-6-4 fertilizer. Examples of a 1-1-1 ratio are 10-10-10, 12-12-12, and 7-7-7. Examples of the 1-2-1 ratio include 10-20-10, 6-10-4, 12-24-12, and 5-10-5.

Fertilizer rates for shade trees are often based on trunk diameter at chest height. The amount of tree food needed is based on the amount of nitrogen required to maintain uniform growth. For young trees with trunk diameters up to 6 inches (or large shrubs), use 1 to 2 pounds of fertilizer for each inch of diameter. One pound of commercial fertilizer is enough if it contains 10 to 12 per cent nitrogen; use 2 pounds per inch of trunk diameter if the fertilizer contains 5 or 6 per cent nitrogen.

Heavier rates are normally used for older broadleaf trees with trunk diameters over 6 inches at shoulder height. With fertilizers containing 10 to 12 per cent nitrogen, use 1 to 3 pounds per inch. Apply 3 or 4 pounds per inch of trunk diameter if the fertilizer has 5 or 6 per cent nitrogen.

Mix fertilizer with 2 or 3 times its volume in topsoil, sand, or peatmoss and pour into a series of holes 6 to 12 inches deep for young trees and 12 to 18 or even 24 inches for larger trees. Bore holes (1½ to 2 inches in diameter) in soil with a punch-bar, crowbar, or soil auger.

The hand or electric auger drill is preferable since it does not compact the soil. Arborists commonly use a feeding needle, or lance, and a liquid tree food (e.g., 23-21-17, 21-18-17, 20-20-20, 19-28-14, or 15-15-15). The holes are made about 2 feet apart in the ring and each ring is about 2 feet from the next (Figure 2.2). Proportion the total application of fertilizer-topsoil or other mixture between the holes as given above. Then keep area thoroughly wet for several days using a lawn sprinkler or fine spray from the garden hose. Trees are best fed in spring or late fall when soil contains ample water.

Less fertilizer is generally applied for evergreens than for deciduous trees. Fruit and nut trees should be fertilized regularly. Follow local recommendations.

Fertilize shrubs, bush fruits, and most garden flowers by spreading fertilizer in an area under spread of the branches (Figure 2.3). Work the fertilizer into the top several inches of soil and water the area thoroughly. In most cases lawn fertilization is adequate for shrubs growing in the lawn. In borders and foundation plantings, apply fertilizer at the same rate as the lawn (about 2 to 4 pounds of 10-6-4 or 8-5-3 per 100 square feet of bed area).

House plants should be fed *only* when making active growth. From November through February they need very little, if any, additional feeding. Generally a complete fertilizer is recommended at 3- or 4-month intervals. A level teaspoon to a quart of soil (or 6-inch pot) before planting is usually enough. Put a level tablespoon of a complete fertilizer (e.g., 5-10-5, 10-6-4, or 7-7-7) in a quart of water. Let it stand overnight before applying. Water regularly to carry nutrients to all roots in the pot. Avoid overfertilizing! Do

Fig. 2.2. Fertilizing a tree by boring holes in the soil. Make the holes with a crowbar or soil auger. Punch holes 6 to 12 inches deep for young trees and 12 to 18 inches or deeper for large trees. The holes should be 2 feet apart and each ring about 2 feet from the next one. Place about 1 to 3 tablespoons of fertilizer in each hole — or enough to give the total application needed. Keep the rings at least 3 feet away from the trunk of young trees and at least 6 feet on larger trees. The outermost ring should be beyond the drip line of the tree.

not use fertilizer high in nitrogen when plants normally bloom—or no flowers may be produced.

12. Soluble Salt Injury (Salinity)

Begonias, African-violets, ferns, and other house plants are very susceptible to an excess of water-soluble salts, while carnation and stock are highly resistant. An excess of salts (salinity) may produce a scorching of leaf tips or margins, stunting, killing back of roots, wilting and yellowing of foliage, or poor seed germination. Salts may build up from those introduced by watering (and left behind after evaporation), excessive application of fertilizers (especially those that leave substantial unused residues in the soil), and use of leaf mold or manure from places where large amounts of salt have accumulated. Soluble salts are easily tested with a simple instrument called the Solubridge. This service is commonly available at the horticulture or floriculture department at your land-grant institution (pages 6–7).

Fig. 2.3. Fertilizing a shrub by spreading fertilizer in ring under tips of branches and beyond. Fertilizer is then worked into top several inches of soil and watered in.

To avoid accumulation of excessive soluble salts in the root zone, periodically flush house or greenhouse-grown plants with an amount of soft water equivalent to 5 or 6 times the soil volume. Check with your local florist or extension horticulturist for more information regarding fertilizing container-grown plants and control of soluble salts.

For additional information regarding fertilizing, watering, soil mixes, lighting, propagating, summer care, and culture of house plants, read USDA H & G Bulletin 82, *Selecting and Growing House Plants.*

13. Pruning

When bare-rooted trees and shrubs are transplanted, it is advisable to remove about ⅓ of the branch system. Even container-grown plants should be pruned a little. This reduces top growth and compensates for roots lost in moving. Broken, weak, rubbing, diseased, dead, and overcrowded branches should be removed whenever found. Do not cut back vigorous plants that have been thoroughly thinned at the nursery. Follow instructions outlined by your nurseryman.

Prune just above a strong bud (Figure 2.4). Each cut should be smooth. A smooth wound heals more rapidly. Prune to maintain and improve the natural shape and general appearance of each plant. Correct pruning restores vigor to older plants.

Peaches and nectarines must be pruned each year. Cherries, especially the sour, require little special pruning. Apples and pears may not need much pruning for 5 to 8 years after trees come into bearing. Later, of course, such trees should be pruned each year. Prune fruit trees to open up the center, space fruiting wood, control height, and make spraying easier. Most flowering trees, once they attain blooming stage, require little pruning.

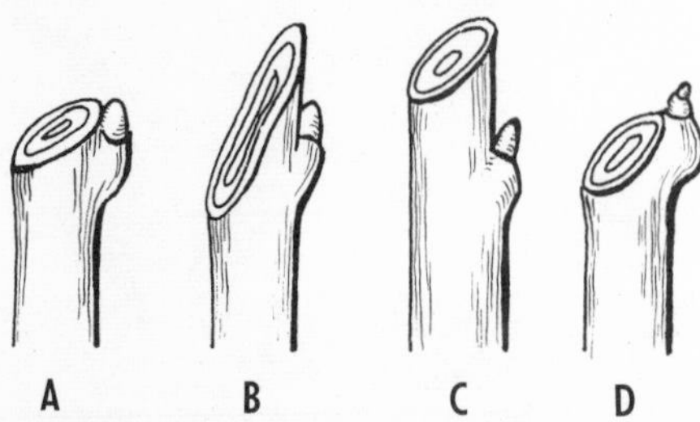

Fig. 2.4. Pruning in relation to buds. A. Correct surface. B. Too much surface. C. Too long a stub. D. Too close to bud.

Judicious pruning results in more vigorous plants, larger blooms, and more fruit as well as control of plant size and shape. Prune in dry weather to avoid possible spread of disease.

Shrubs. Prune sucker-type shrubs (e.g., spirea) by cutting about ⅓ of the older mature stems to the ground (Figure 2.5). Select a few of the better distributed stems (canes) to remain. Leave young vigorous growth. Cut suckers below ground and close to the parent stem.

Prune shrubs that come from a single main stem (e.g., honeysuckle) by removing some old branches at the base and others part way up. Leave young wood. But keep shrubs from becoming overgrown and straggly.

Spring-flowering shrubs and trees (e.g., forsythia, lilac, honeysuckle, spirea, beautybush, weigela, viburnums, mockorange, magnolias, flowering crabs and cherries, dogwood, cotoneaster, deutzia, and goldenrain-tree) should be pruned right *after* flowering. Prune summer- and fall-flowering types (e.g., rose-of-Sharon, hibiscus, hydrangea, snowberry, and butterflybush) in the fall, winter, or early spring *before* flower buds form. Leave young wood but keep shrubs within bounds. Remove some old wood annually plus about ⅓ of the length of new growth.

If plants are grown partly for their attractive fruits (e.g., cotoneaster, pyracantha, and viburnum), delay pruning until fruit fall.

Prune clipped hedges frequently when young to insure heavy, compact growth. The top should be kept narrower than the base. This helps keep the base compact as the bottom gets more light. Do not allow hedges to get taller and wider each year after they reach desired size. Renovate old hedges by cutting stems back to the ground.

Trees. Most deciduous trees can be pruned almost anytime "when the knife is sharp"—except excessive bleeders, i.e., maples, elms, dogwood, birch, beech, and yellowwood that is best trimmed in summer. To promote quicker and better healing of wounds prune in the dormant season and in midsummer.

Prune grapes in early winter to prevent excessive loss of sap. Old raspberry canes

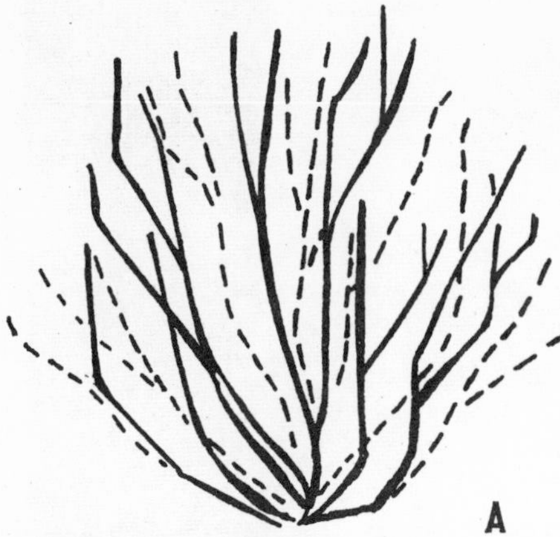

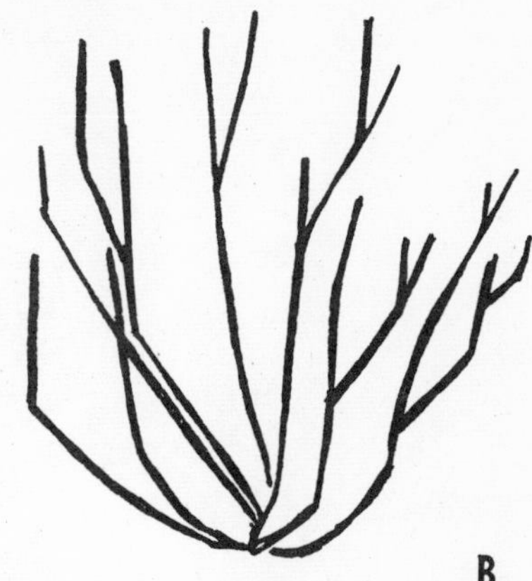

Fig. 2.5. Pruning shrubs. A. Before. B. After. Prune to maintain and improve the natural beautiful shape of each plant. Remove old, dead, diseased, damaged, and interfering branches.

should be removed right after harvest; new ones thinned out in early spring.

Branches that are diseased, dead, dying, broken, crossing, or rubbing should be removed as soon as possible and burned to improve appearance and prevent spread of disease-producing organisms and insects. Remove weaker, crowded, and rubbing branches while young. For fruit trees, remove all side branches except those desired to make permanent limbs.

Follow the procedure outlined in Figure 2.6 when removing limbs or branches too large to hold with one hand. Cuts 1 and 2 prevent stripping the bark. Make each cut clean and flush with stem or trunk. Do not leave short, useless stubs! These often lead to eventual wood rot and premature death of the plant.

Trees with narrow, V-shaped crotches (less than 45° angle) such as certain maples, willows, and honeylocust are subject to wind-splitting and later wood decay. One branch in the weak union is often removed while the tree is young. The narrow crotch angles of certain trees (e.g., elms) cannot be corrected in this manner. Here pruning back to lower branches to reduce their weight, plus cabling and bracing is often necessary.

Lower branches of shade trees growing in lawns or along streets and sidewalks should be removed while still small enough to remove with hand shears. Continue each year until the lowest branch is the height you want when the tree is mature (usually 8 to 12 feet).

Narrow-leaved evergreens (e.g., arborvitae, juniper, and yew) may need an occasional light shearing to thicken the plants and keep them within bounds. Shear in late spring or early summer.

Prune pines frequently to keep compact. Prune in late spring when new "candle" growth is full-grown but still soft. Firs and spruces require little pruning. Do it from late summer to winter. A little pruning of evergreens every year or two prevents a drastic operation at any one time.

For additional help with your pruning problems contact your county extension office or extension horticulturist. Many states have excellent pamphlets on this subject. The services of an experienced, responsible tree specialist should be employed for major pruning and for all hazardous aerial work. Select an arborist or tree surgeon who is insured against personal injury and property damage.

Flowers and Potted Plants. Some plants need "pinching" to make them short and bushy, produce smaller but more numerous blooms, and extend the flowering season. They include browallia, chrysanthemum, fuchsia, geranium, ivies, marigold, pètunia, phlox, snapdragon, vinca, and other trailing vines.

14. Tree Removal

All dead, hollow, seriously diseased and structurally weakened shade trees are potential hazards to life and property.

Fig. 2.6. Pruning trees — the right way and the wrong way. A. Tree topping "butchery." This is *not* a recommended way to prune trees (Courtesy Ottumwa, Iowa, *Courier*). B-D. The correct way to remove a large branch. B. Cut no. 1. C. Cut no. 2 severs the main part of the branch. These preliminary cuts prevent stripping of the bark. D. Final cut, made flush with the trunk, removes the stub. The wound should be promptly painted with a tree wound dressing.

Remove such trees. As this job is usually hazardous, secure the services of a competent arborist.

15. Treatment of Wounds

Wounds are treated to (1) prevent drying of tissues, (2) avoid infection by rot-producing organisms and insects, and (3) promote faster healing.

Bark wounds and pruning scars should be promptly treated. Pruning cuts less than about an inch in diameter are not normally treated. Vigorous, well-maintained trees and shrubs heal faster than sickly, undernourished ones.

To heal quickly and properly, large wounds should be shaped. All injured, splintered, or diseased wood and bark should be removed cleanly with a sharp-edged knife or chisel (Figure 2.7). Avoid leaving pockets where water may collect. If the job looks too big, call in a trained arborist. Margins of large wounds should be painted with shellac or wound dressing to prevent drying. After protecting margins, and excavating the cavity, all exposed wood should be sterilized by swabbing with household bleach (Clorox, Purex, or Saniclor) diluted 1 to 5 with water, 70 per cent alcohol, or a 1:1,000 solution of mercuric chloride. See page 101 for details on how to prepare this solution properly and what precautions should be followed.

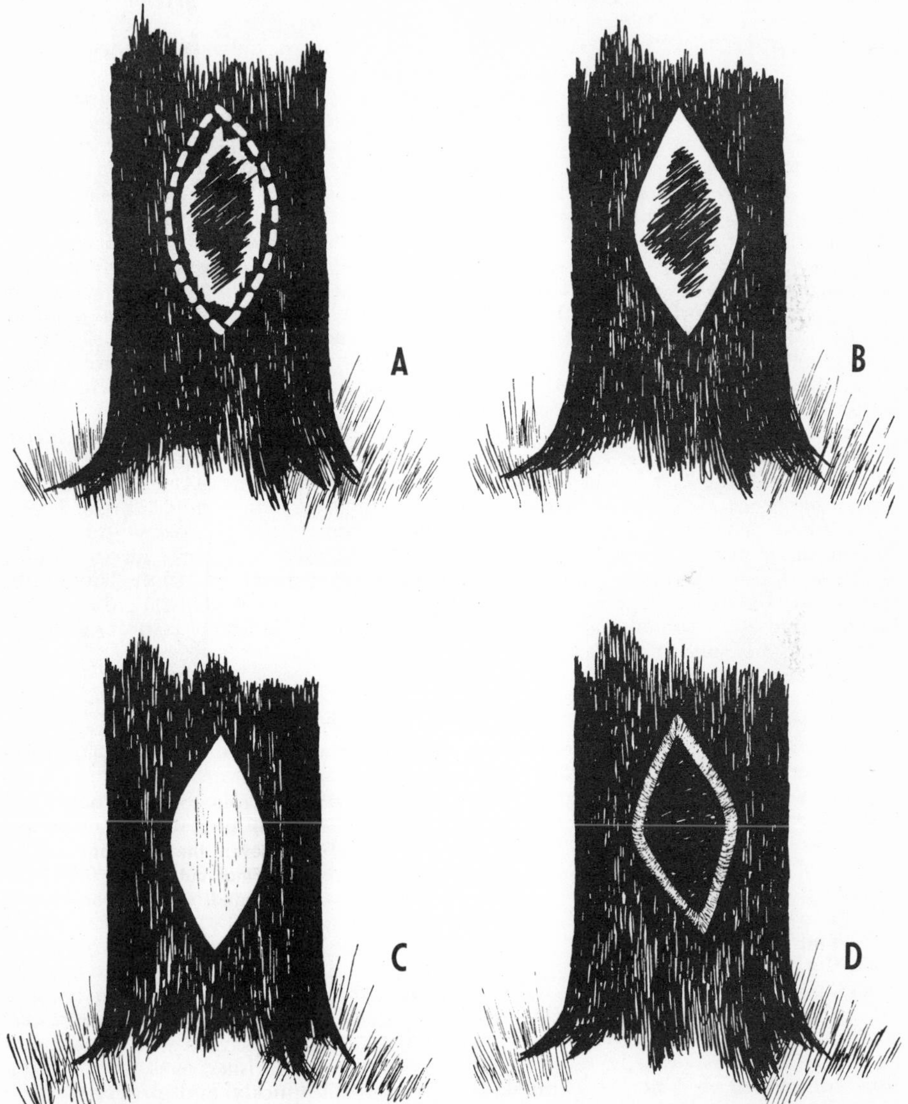

Fig. 2.7. Treating a tree wound. A. The dotted line indicates proper shaping for an irregular trunk wound. All margins should be cut back to live, healthy bark tissues. B. The bark has been removed cleanly together with some of the discolored or rotted wood beneath. The margins of the cut should be painted now with orange shellac. C. The wound has been completely cleaned and is ready to be painted with a tree wound dressing. D. Some time later. The old wound has been painted and rolls of callus growth are closing over the cavity.

Finally, all wounds should be painted with a permanent-type tree wound dressing.

(1) *Asphalt-varnish tree paints* are widely used and are available from many horticultural supply houses, garden supply stores, and nurseries. Those containing turpentine or coal tar will injure growing tissues. Buy type containing a disinfectant (e.g., 0.25 per cent phenyl mercuric nitrate or 6 per cent phenols).

(2) *Outside-type house paints* are fairly satisfactory if they do *not* contain creosote and are properly mixed with raw linseed oil. Use *only* after first coating wound with orange shellac.

(3) *Bordeaux paint* makes a good, semipermanent, and inexpensive tree dressing. Prepare by slowly stirring 1½ pints of raw linseed oil into 1 pound of fresh, dry, commercial bordeaux powder until a thick, creamlike paint is produced. If you object to the blue-green color, add lampblack suspended in oil. Disadvantages of bordeaux paint are that it hinders rapid healing during the first few years, and it has rather poor weathering qualities.

(4) *Cerano* is a new wound dressing containing mercury, developed by University of California researchers. It contains 8 parts glycerol, 2 parts anhydrous lanolin, and 0.3% phenyl mercuric nitrate. Cerano is sold by the John Taylor Fertilizer Co., Sacramento, California.

(5) Probably the best, all-round tree wound dressings contain a mixture of *lanolin, rosin,* and *gum.* The better ones also have a disinfectant added (see above). You can prepare a good tree paint by melting and stirring together 10 parts by weight of lanolin, 2 parts of rosin, and 2 parts of crude gum.

Tree wound dressings are also available in convenient aerosol-type cans.

Wound dressings should be checked periodically. Recoat when the surface cracks, peels, or blisters.

Needle-leaf evergreens usually seal small wounds with natural gums or resins. If resin forms over the pruning or other wound, it need not be painted. If no resin forms, treat as described above.

For additional information on treating wounds and cavities, as well as other types of tree injuries such as a split trunk or crotch or uprooted trees, read USDA H & G Bulletin 83, *Pruning Shade Trees and Repairing Their Injuries.*

The U.S. Department of the Interior, National Park Service, Tree Preservation Bulletin 3, *Tree Bracing,* gives an excellent account of the proper ways to cable and brace trees.

16. Staking Trees and Shrubs

Just after planting in windy locations, treelike shrubs and small evergreen trees should be guyed (Figure 2.8) or supported by means of a single, stout stake driven into the ground on the windward side about a foot from the trunk. Care should be exercised to prevent root damage. Attach tree to stake with guy wire threaded through scrap garden hose or larger tubing. Burlap or canvas with sash cord or light rope works nearly as well. Stake street trees, up to 3 inches in diameter, with two stakes on opposite sides about 18 inches from trunk. Street trees of 4- to 6-inch diameter should have 4 stakes in a box formation 18 inches from the tree.

Guy shade trees up to about 5 inches in diameter from three directions as shown in Figure 2.8. Guy wires are usually removed the second or third year after transplanting, except in windy areas. Permanent guys (one or double strands of 12-gauge wire) should be attached by eyebolts or lag hooks inserted into the trunk or branches, *never* wrapped around them.

17. Soil Drainage

Soil underlaid with an impervious layer of clay, hardpan, or rock that drains poorly may kill sensitive plants (e.g., roses, most evergreens). When water collects above this impervious layer, the soil becomes saturated (waterlogged). Home owners usually assume that sloping land provides sufficient soil drainage. This is often a false assumption because soil below the surface may be tight, resulting in poor drainage. Water remaining in the root area forces air out, causing roots to die from suffocation. This condition may kill plants outright or cause stunted and yellowish or reddish growth with reduced yields of flowers and fruit. Unless tolerant plants are grown, such soils should be drained by deep tillage (using a subsoiler or lister). Installing gently sloping lines of agricultural drain tile every 20 to 25 feet at a depth of 2 or 3 feet usually provides a permanent cure for waterlogged soil. Check with your county agent

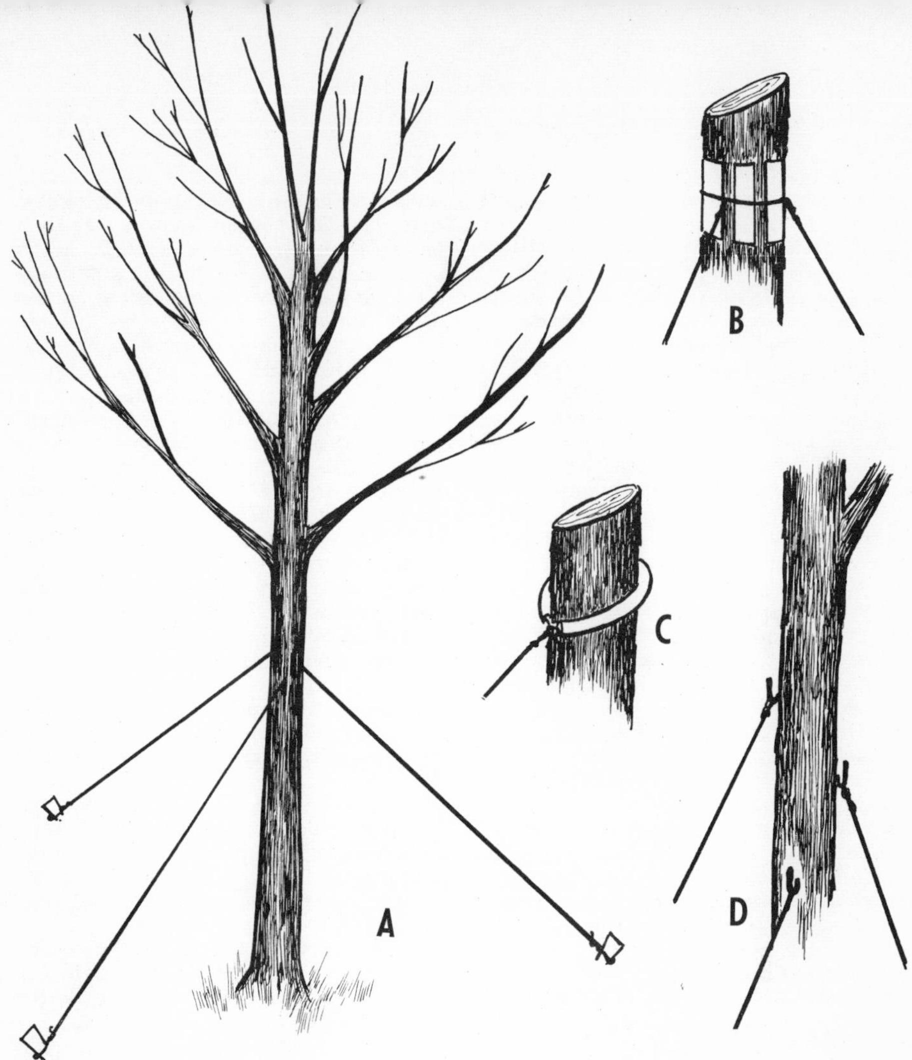

Fig. 2.8. Guying trees. A. All three guy wires should have the same tension. D. The best method of fastening wires to trees using lag hooks, eyebolts, or screw eyes. B and C. Less satisfactory methods.

if you have soil drainage problems.

If soil is permanently waterlogged or swampy (oxygen deficiency or asphyxiation) and cannot be drained, it may be necessary to grow plants adapted to these conditions (e.g., bog plants). Oxygen-deficient soil usually has a sour, disagreeable odor.

Lack of vigor, frequent wilting, and off-colored foliage are signs of poor soil drainage. Your soil should have good natural drainage to a depth of at least 2 feet, preferably deeper for trees. Willow, poplar, birch, alder, American ash, and bald cypress survive in poorly aerated soil. Elm, beech, oak, elder, and most conifers are susceptible to flooding.

18. Watering

Excessive moisture, droughts, or unfavorable temperatures cannot always be prevented, but their injurious effects can often be reduced. During hot, dry periods shade trees, evergreens, and deciduous shrubs need extra water, especially when growing in a lawn. Plants lose water rapidly under such conditions and unless watered, severe wilting, withering, and death often result. Almost the same symptoms may follow excessive rainfall and overwatering. It is not true that plants are injured by watering on a hot, sunny day. A tree may require 500 pounds of water to grow one pound of dry weight.

Shrubs and trees planted in spring usually die during hot summer weather if soil is not kept sufficiently moist. During droughts, plants in unfavorable locations (light sandy soils, slopes, winds, or full sun) are the first to suffer. Leaves may wilt, become scorched, and drop prematurely. If many feeder roots are killed, the entire plant may die. Injured trees and shrubs usually die slowly over a period of 1 or 2 or more years. Timely detection and treatment could save many valuable plants.

Drying out of soil is often disastrous for potted plants, evergreens, newly planted trees, shrubs, transplants, and new lawns. Grass and garden plants growing beneath trees, or areas filled with tree and shrub roots are difficult to keep watered.

Maximum growth from most plants in roughly the eastern half of the United States comes when they have an equivalent of at least an inch of water (from rainfall or irrigation) a week during the growing season. This is equivalent to 900 gallons on a 30- by 50-foot lawn or garden area. Much more water than this is needed in arid areas of the western states where moisture is greatly deficient. During droughts, water enough so that it soaks soil to a depth of 6 to 12 inches, preferably more for deeper-rooted plants such as trees. After the sprinkler or perforated watering hose has been on several hours in the same spot, use a trowel to check depth of water penetration. Or put several coffee or juice cans in the sprinkler path so you will know exactly how much water is being applied. Try for 1 to 2 inches to get good penetration. Water early enough in the day so that plants will dry off before late afternoon. This helps prevent foliage diseases.

House plants should be planted in light, well-drained soil, and not overwatered. Overwatering is probably the commonest cause of trouble in growing house plants; it encourages root and stem rots and cuts off essential air from roots. Nearly all indoor plants require a steady supply of water with the exception of cacti. A general rule is to apply moisture when the topsoil feels dry. Use water of room temperature and apply enough so that the soil of each plant is wet thoroughly from top to bottom. Do not water again until the topsoil feels dry. A drainage hole should be present in the bottom of the growing container. Only experience will tell you how much water your house plants need. Lack of water causes wilting, leaf scorching, early leaf and flower drop, root injury, and stunting, among other troubles. Too much water often results in sudden leaf drop, wilting, yellowing or spotting of leaves, and a soft dark rotting of the stems and roots. This is a common problem with begonias, African-violet, gloxinia, and geraniums. The amount of water your house plants need will vary with the weather, kind of plant and soil, type and size of container, plus the amount of exposure to sun and wind.

Does your home have wide, overhanging eaves? Do you always remember to give the soil underneath extra water? If you live in a windy location or in a dry region, it is even more necessary that perennial plants, especially evergreens, have adequate moisture during late fall and winter.

19. Light

Plants vary greatly in their requirements for light. Some thrive best in full sunlight; others need shade or partial shade to do their best.

Extremely high light intensity, especially during hot weather, may cause sunscalding of fruits and vegetables, or result in fading of deep flower colors. Valuable plants may be at least partially protected by spraying the foliage with an antidesiccant material like Wilt Pruf, Plantcote, Good-rite latex VL-600, D-Wax, Stop-Wilt, or Foli-gard before scorching occurs.

Duration of light also affects flowering and fruiting. Short-day plants such as asters, chrysanthemum, cosmos, dahlia, Christmas begonia, kalanchoë, and poinsettia bloom when the day length is 12 hours or shorter. Long-day plants such as corn, violets, pansies, peas, and lettuce are stimulated to flower when the day length is 14 hours or longer. African-violet, gloxinia, geraniums, and wax begonia are day-neutral or "indifferent," and do not respond to short or long days.

When planning your foundation and

garden plantings, check on light requirements for the various plants you plan to grow. This will save both money and later grief. Shrubs for the north side of a home must be selected for shade tolerance. Remember, too, that light-colored pavements and walls mirror heat and light rays onto sensitive foliage causing it to turn brown. Summer and winter are the most critical periods.

Most flowers, fruits, vegetables, lawngrasses, and flowering shrubs (e.g., lilac, viburnum, spirea) require full sun for at least part of the day. Such plants as laurel, rhododendron, and azalea will blossom in partial shade.

Give most house plants as much light as possible, especially during winter months. Such plants should be placed by a window, except plants that do not thrive in strong sun (e.g., ferns, begonias, cyclamen, African-violet, caladium, and foliage plants). To produce flowers, most house plants need 3 to 5 hours or more of sunlight a day. Lack of sufficient light often leads to pale green leaves, spindly growth, leaf drop, and loss of flowers. A sudden change in lighting intensity may cause general defoliation. Too much sun often causes yellowish-brown or silvery spots on leaves.

Artificial light may be supplied by fluorescent tubes to provide 300 to 600 foot candles at the top of plants. See Table 2. Give most plants 12 to 18 hours per day of this supplementary light, if needed, depending on the type of plants grown. Flowering plants need more light than foliage types (Table 3).

Most house plants can be moved outdoors for the summer. The best location depends on the amount of sunlight—direct, diffused, or subdued—they can tolerate. Grouping plants and sinking pots in the ground up to their rims reduces watering and care. Lift or twist pots monthly to prevent roots from growing through the drainage hole.

TABLE 2

ILLUMINATION OF HOUSE PLANTS IN FOOT-CANDLES (f-c) AT VARIOUS DISTANCES FROM TWO OR FOUR 40-WATT STANDARD, COOL, WHITE FLUORESCENT LAMPS MOUNTED ABOUT TWO INCHES BELOW A WHITE-PAINTED REFLECTING SURFACE

Distance From Lamps (inches)	Illumination		
	Two Lamps[1]	Four Lamps[1]	
	Used[2] (f-c)	Used[2] (f-c)	New (f-c)
1	1100	1600	1800
2	860	1400	1600
3	680	1300	1400
4	570	1100	1300
5	500	940	1150
6	420	820	1000
7	360	720	900
8	330	660	830
9	300	600	780
10	280	560	720
11	260	510	660
12	240	480	600
18	130	320	420
24	100	190	260

[1] Center-to-center distance between lamps is 2 inches.

[2] These lamps were used for approximately 200 hours.

20. Oedema

A common problem with indoor plants. When moisture conditions are excessive, small masses of leaf or stem tissue may expand and break out, causing watery swellings or galls. Later the exposed surface may become rusty and corky in texture. See Figure 4.103. Control by reducing air humidity, increasing light and air circulation, plus avoiding overwatering, especially during overcast periods. Oedema is a nonparasitic disease.

21. Air Humidity

The low humidity (often 15 to 30 per cent) in an apartment or modern home is often responsible for leaves and flower buds of begonias, African-violets, ferns, rubber plants *(Ficus)*, and others becoming spotted or scorched and falling prematurely (Figure 2.9). Plants taken from a cool, moist greenhouse or florist's shop to the hot (75° to 85° F.) dry air of a home or apartment are commonly affected. Many times flowers suddenly wither and drop off. Control by increasing humidity around susceptible plants and reducing temperature. Buy a ready-made humidifier, install a humidistat, or build a "humidifier" yourself. Set plants on an inverted pot or brick, over a large pan of water, in a glass or plastic case, or in a planter filled with sphagnum moss, sand, pebbles or small stones, coarse vermiculite, Perlite, or other filler that is thoroughly wet down occasionally. Evap-

TABLE 3

LIGHT AND TEMPERATURE REQUIREMENTS OF HOUSE PLANTS*

Direct Sun***	(South and west windows giving about 6 hours' sunlight)		
Agathea	**Citrus	**Jerusalem-cherry	**Primrose
Air-plant	**Coleus	**Kafir-lily	Sedum
Aloe	Crown-of-thorns	Kalanchoë	Silk-oak
Aluminum plant	**Cyclamen (50° night)	**Lantana	**Spring bulbs
Amaryllis	Echeveria	Moses-in-the-bulrushes	**Star-of-Bethlehem
Artillery-plant	English ivy	Octopus-plant	**Strawberry-begonia
Bloodleaf	Flowering maple	Orchid (Cattleya)	**Tabasco red pepper
Browallia	Gardenia	**Oxalis	Earth-star
Cacti	**Geraniums	Palm, dwarf	Ti plant
**Calla lily	Gold-dust Dracaena	**Paris daisy	Tradescantia
Centuryplant	**Hydrangea	Poinsettia	Velvet plant
Chrysanthemum	Jade plant	Prayer plant	Wax begonia
Partial Shade	(Just beyond sun's rays or north and east windows)		
Achimenes	Boston fern	German ivy	**Poormans-orchid
African-violet	Caladium	Gloxinia	Prayer plant
Amorphophallis	**Camellia (in flower)	Grape ivy	Rubber-plant
Anthurium	Ceriman	Holly fern	Schefflera
Aphelandra	Cineraria	**Hydrangea (in flower)	Scindapsus
Ardisia	*Clivia	Hydrosome	Shrimp-plant
**Azalea (in flower)	Croton	Orchids (most)	**Snapweed
**Begonias (most)	Fiddleleaf fig	Peperomia	Wandering-Jew
Bromeliads	Flame violets	Philodendrons (most)	Wax-plant
	**Fuchsia	Piggyback plant	
Shade	(Reflected light only, 8 or more feet from window or in north windows)		
Arrowhead-plant	Cast-iron plant	Kangaroo vine	Pteris fern
Birdsnest fern	Chinese evergreen	Moth orchid	Sansevieria
Birdsnest-hemp	Dieffenbachia	Norfolk Island pine	Screw-pine
Bowstring hemp	Dracaena	Pittosporum	Swiss-cheese-plant
	Gold-dust Aucuba	Pothos	

* From Pennsylvania Ext. Circ. 491 and Cornell Exp. Bul. 1073.

** Require cool night temperatures, 50–60° F. All other plants do best with 60–65° F. night temperatures. Most plants listed under Shade and Partial Shade (above) will do better with some sun or artificial light, provided temperatures stay moderate.

*** Approximate *minimum* light needed to maintain house plants attractively for at least a year. *Direct Sun* = 50 to 100 foot-candles or more; *Partial Shade* = 25 to 50 foot-candles; *Shade* = 15 to 25 foot-candles. See also Table 2.

oration of a gallon of water a day for the average room should moisten air enough for plants. Your florist can also provide tips on increasing air humidity during dry periods such as winter. A range of 40 to 60 per cent relative humidity satisfies most house plants. The most nearly perfect combination for plants *and* people is a temperature near 72° F. with a humidity about 50 per cent.

Potted plants should be brought indoors in early fall at least several weeks before heat is turned on. This enables plants to better adapt to indoor living. Before bringing indoors, spray or dip plants with an insecticide to rid them of insect or mite pests. The outsides of all pots should be scrubbed clean with a brush and water.

The average home is too dry for fungus or bacterial diseases to develop on the foliage of house plants.

22. Temperature

Light and Temperature Requirements for House Plants. Most house plants do best if the day temperature is about 70° to 75° F. and at night 60° to 65° F. Most modern homes and apartments are too warm for the best growth of many house plants. Since it is often not practical to keep the temperature within limits outlined, there are two alternatives: (1) select plants that "get along" at higher temperatures, or (2) replace flowering house plants after several weeks when blossoming is complete. A sudden change in temperature causes leaves and flower

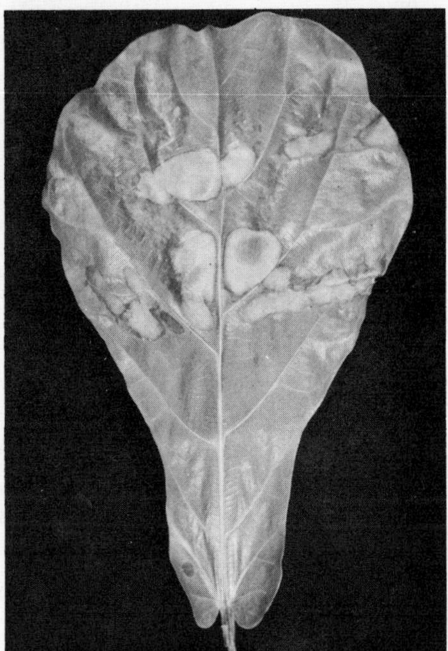

Fig. 2.9. Sun or heat scald in a *Ficus* leaf. This is a common indoor problem.

buds of certain house plants to turn yellow and drop suddenly. Avoid drafts; they may cause brown leaf tips.

For top results, place your flowering plants in a cool spot at night, away from radiators and heat registers. Avoid putting such sensitive plants as African-violet and gardenia close to windows during cold, winter weather.

Light and Temperature Requirements for Vegetables. Vegetables, too, are sensitive to temperature. Cauliflower, spinach, kale, parsley, mustard, and head lettuce grow best in cool weather. Celery, cabbage, and certain other vegetables may shoot up seed stalks if the temperature is 50° F. or below for a period of time when plants are young. Tomatoes, beans, and peppers often drop their blossoms during temperature extremes when night temperatures are 55° F. or below and day temperatures are about 95° F. or above. High temperatures are also responsible for premature flowering of spinach, broccoli, radishes, and lettuce and low yield of garden peas and beans.

23. Leaf Scorch or Sunscorch

During hot (80° F. or above), dry, windy weather the tips, margins, and tissues between the veins of leaves on many plants, especially trees and shrubs, turn yellow to brown and wither (Figures 1.1B and 4.148). Plants very susceptible to leaf scorch include maples, dogwood, horsechestnut, beech, linden, oaks, ash, hawthorn, elm, crabapples, and privet. Injury occurs when water is lost by leaves faster than it is replaced by roots. Scorching of leaf edges and between the veins may progress inward until entire leaves turn brown and wither. If severe, shoots and branches may die back.

Such symptoms may also be caused by other conditions (e.g., altered water table or soil grade, injured root system, compacted or shallow soils, paved surface over roots, girdling roots, salt drift near the ocean, low temperatures, or by a parasite). Sunscorch can often be checked by watering plants during summer droughts, fertilizing plants in low vigor, pruning to open up trees and shrubs, mulching, shallow cultivating, and spraying with materials such as Wilt Pruf, Foli-gard, Stop-Wilt or Plantcote; or planting in a more protected location.

24. Winter Injury

Trees and shrubs growing in exposed, windy locations or in poorly drained soils may be injured by low temperatures, repeated freezing and thawing, or drying winter winds. Plants overfed with high-nitrogen fertilizer or still actively growing in late fall are most commonly injured.

Frost cracks, a vertical separation of bark and wood, are common on the south or southwest sides of the trunk. Trees most subject to frost crack are apple, elm, London plane, linden, horsechestnut, maple, poplar, and willow. The cracks may reopen each winter, providing entry for wood decay fungi. Similar cracks may be caused by prolonged drought.

Blackened sapwood, death of twigs and branches, discolored cankers on exposed limbs or at the trunk base (sunscald), injured or dead taproot and side roots, plus death or injury to leaf and flower buds are other common ways trees and shrubs express winter injury.

Damage to needle and broadleaf evergreens (e.g., arborvitae, hollies, junipers, pines, rhododendrons) occurs following extreme and rapid fluctuations in temperature or by early fall or late spring freezes. See Figure 4.112. Isolated plants

Fig. 2.10. Typical "burning" and winterkill of juniper following dry winter winds, direct sun, and low temperatures. Evergreens growing in exposed windy locations in poor or dry soil are most susceptible to this type of winter injury. (Ross Daniels, Inc. photo)

growing in dry, sunny, windy locations are often most subject to injury (Figure 2.10).

Plants prone to winter injury and sunscald may be grown where shaded from midday or late afternoon sun—or shade may be provided by a cheesecloth or lath screen. Protect trunks of young, thin-barked trees from winter sunscald and frost cracks by wrapping with burlap strips, sisalkraft paper, or aluminum foil; by applying a coat of whitewash; or by tying a 6-inch board upright on the south side (Figure 2.11).

Protect exposed evergreens from leaf scorching, caused by drying winter winds and sun, by erecting canvas, plastic, burlap, or slat screens placed 2 feet away on the south and southwest sides. Better still, plant in a more protected location! Try covering foliage in fall with a special "no-wilt" latex or plastic antidesiccant (e.g., Wilt Pruf, Stop-Wilt, Foli-gard, or Plantcote). Follow manufacturer's directions. Repeat the spray during a midwinter thaw when the temperature is above 40° F. Evergreens and deciduous trees and shrubs planted within the past 2 years should be watered thoroughly during late fall and early winter. Then apply a 2- to 4-inch mulch of sawdust, wood shavings, salt hay, peatmoss, ground corncobs, or leaves. Mulching helps conserve moisture and prevent deep freezing or alternate freezing and thawing after ground is frozen, allowing more water absorption by roots.

Boxwood, junipers, yews and other multiple-stemmed evergreens that tend to spread apart and break under a load of ice or snow can be protected by tying the branches together with strong cord. See Figure 2.12.

A sharp variation in soil and air temperatures may cause abnormal growth. Low winter temperatures and late spring or early fall freezes may cause injury similar to burning, especially if plants have made late, tender growth.

Apply winter protection against mice and rabbits to apple, crabapple, blueberries, brambles, pines and many other susceptible plants. A good rabbit-proof fence is a cylinder of chicken wire or hardware cloth 3 feet or more high and a foot away from the trunk. Deeper snows require use of a rabbit repellent (e.g., Arasan 42-S) sprayed or brushed on. Protect against mice chewing the bark off the crowns and roots of trees during the winter by keeping trunk bases free of grass and debris, putting poison baits in holes and runways, and banking soil against the trunk. Or put ¼-inch hardware cloth guards around the trunk base.

Check with your extension horticulturist or county extension office on what, how, and when to apply winter and frost protection to your garden plants. He can help you select plants that are adapted

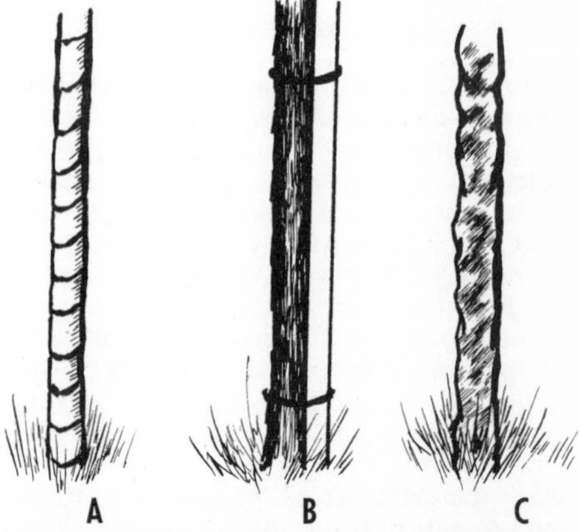

Fig. 2.11. Three methods of protecting trees against winter injury. A. Wrapping with sisalkraft paper or burlap. B. Use of a 6-inch board on the south side. C. Wrapping young trees with aluminum foil.

Fig. 2.12. Trees, such as soft maple, Chinese elm, linden, tree-of-Heaven, and boxelder, with soft brittle wood may be seriously damaged by coatings of ice. (Illinois Natural History Survey photo)

and will do well in your particular location.

Additional help can be obtained by studying the USDA Misc. Publ. 814 (revised), *Plant Hardiness Zone Map*. Numerous indicator plants are listed for each of the 10 hardiness zones in the United States.

25. Chemical Injuries

Air Pollution in and near large cities from pollutants emitted by smelters; pulp mills; coal- and petroleum-burning furnaces; incinerators; cement, aluminum, chemical, power or fertilizer plants; motor vehicles; etc. commonly results in plant damage, e.g., foliage discoloration, delayed maturity, defoliation, lack of vigor, abortion of blossoms, stunting, dieback of twigs and branches, or death. The *visible* damage to vegetation in California, due to air pollutants, is estimated at more than $13 million annually (only $1/2 million in 1953), affecting over 12,000 square miles. An estimated $18 million in visible damage occurs annually along the Atlantic Seaboard from Boston to Norfolk, Virginia. Estimated national losses to crops from air pollution due to growth suppression, delayed maturity, and reduction in yield and quality amount to $325 million annually.

The extent of injury and areas in which air pollution occurs are governed by the kind and concentration of the pollutant (type of industry, fuel use, number of motor vehicles), distance from the source, length of time exposed, city size and location, meteorological factors such as wind direction and speed, temperature, humidity, barometric pressure (conditions that lead to air inversions), land contour and air drainage, soil fertility and moisture, age of leaf tissues, time of year, photoperiod, plant species and varieties grown, etc. Air pollution damage is usually the worst in clear, still, humid weather when barometric pressure is high.

Air pollution injury is easily confused with disease, insect damage, nutritional imbalance, care or abuse by man, and adverse effects of temperature or wind.

The most important plant-affecting air pollutants are sulfur dioxide, fluorides, ozone, "smog," and ethylene.

Sulfur Dioxide injury to broad-leaved plants results in dry, papery-white to straw-colored, irregular, marginal or interveinal blotches in leaves (Figure 2.13). Yellowing (chlorosis) and a gradual bleaching of surrounding tissues is fairly common. Grass blades develop light tan to white streaks, while conifer needles turn reddish-brown starting at the tips. Very sensitive plants include cabbage, spinach, rhubarb, apple, crabapple, hawthorn, sumac, blackberry, raspberry, tulip, violet, larch, alfalfa, smartweed, ragweed, and curly dock. Somewhat resistant plants: maple, privet, pine, onion, corn, and potato.

Fluorides typically cause a killing (necrosis) of the tips and margins of the leaves of broad-leaved plants and a "tipburn" of grasses and confier needles. A sharp reddish-brown or yellowish line

Fig. 2.13. Sulfur dioxide injury to rhubarb.

may occur between living and injured tissues. Citrus and corn leaves develop a yellow mottling or spotting prior to the typical "burning." Injured areas in stone fruit leaves may drop out leaving shot-holes. Young succulent growth is most easily injured. Very sensitive plants: gladiolus, iris, corn, grape, crabgrass, Chinese apricot, Italian prune, white pine, and citrus. Resistant plants: rose, squash, tobacco, alfalfa, and pigweed. A tremendous variation exists in susceptibility to fluorides among varieties of the same plant (e.g., gladiolus and white pine).

Ozone causes collapse, killing of tissue, and markings on the *upper* leaf surface known as stipple (red-brown to black), flecking (white to straw-colored), and chlorosis or bleaching. Growth is stunted and affected leaves drop early. Ozone damage is often found in combination with oxidant damage or "smog" in the Los Angeles area. Very sensitive plants: tomato, beans, spinach, potato, grape, white pine, grasses, tobacco, and alfalfa. Resistant plants: gladiolus, geranium, pepper, citrus, and mint. Great differences in susceptibility to ozone are often expressed by varieties of the same plant (tobacco, bean). An active breeding program to develop ozone-resistant varieties is in progress.

Oxidant Damage or "Smog." Peroxyacetyl nitrate (PAN), formed by nitrogen oxides reacting with hydrocarbons, is a serious air pollutant in areas where organic fuels are combusted in large quantities (vehicular traffic is heavy). Injury appears as a collapse of leaf tissue on the *lower* surface. Typical affected leaves show a distinctive silvering, glazing, or bronzing. In grasses, the collapsed tissue is bleached, with tan to yellow streaks. Conifer needles turn yellow. PAN exposure also results in stunting, early maturity, and leaf drop. Very sensitive plants: beans, endive, escarole, spinach, Swiss chard, Romaine lettuce, pepper, tomato, petunia, certain grasses, and alfalfa. Resistant plants: cabbage, corn, and pansy.

Ethylene injury to broad-leaved plants occurs as drooping of leaves and shoots, early leaf and petal fall, chlorosis, and stunting of growth. Specific diseases due to ethylene are "sleepiness" in carnation, dry sepal in Cattleya and other orchids, and blasting of rose buds. Ethylene was once a problem in greenhouses where manufactured gas was used (see below). The solution was to shift to another type of fuel for heating. Very sensitive plants: tomato, sweetpea, narcissus, snapdragon, orchids, and carnation. Resistant plants: lettuce and grasses. Orchids are injured at ethylene concentrations as low as 5 parts per billion.

The solution to air pollution is not easy and involves enforced use of "blow-by" devices on automobiles, stopping emission at the source (e.g., smoke stack or combustion chamber), controlling legislation, plant breeding, shift to less susceptible plants, and spraying of high value crops. Ozoban (Charles Pfizer & Co.) or other trade products containing ascorbic acid, as well as fungicides containing zineb, maneb, ferbam, thiram, and dichlone, often reduce damage from oxidized hydrocarbons. Calcium oxide sprays prevent fluoride injury. For additional details, check with your nurseryman, extension horticulturist, or plant pathologist.

Salt Injury, whether it occurs along a seacoast where salt in gale winds is blown inland up to 50 miles, or along sidewalks and driveways where salt is applied for weed control, may cause scorching and killing of leaves similar to that caused by certain diseases, drought, extremes in temperature, or industrial fumes. Wide differences in susceptibility occur between different trees and shrubs. In general, evergreens are more resistant than deciduous trees. Salt applied on streets, roads, or drives to keep down dust or speed melting of snow and ice, may damage tree roots when washed or swept down into soil. Maples are injured more severely than most roadside trees.

Gas Injury. Trees, shrubs, and other plants often show scorched leaves and wilt or die along streets or where *manufactured illuminating gas* lines are buried. When lines develop leaks, gas penetrates and poisons the soil, killing nearby plant roots. Sudden or gradual severe wilting, dieback, an irregular yellow or brown discoloration and dropping of leaves, or other peculiar symptoms frequently occur depending on size of leak and period of time gas has been escaping. If you suspect poisoning, call the gas company. Natural gas, which has almost entirely replaced manufactured or illminating gas, does not contain the plant-toxic materials (ethylene and other unsaturated hydrocarbons, hydrogen cyanide, and carbon monoxide) found in manufactured gas. It may, however, injure trees by displacing oxygen from soil in which roots are growing, or by the drying of soil by the gas. For additional information read, *Effect of Natural Gas on House Plants and Vegetation,* distributed by American Gas Association, 420 Lexington Ave., New York 17, N.Y. Affected trees and shrubs can sometimes be saved by forced soil aeration followed by heavy watering.

Excess Care. Well-established plants sometimes are damaged by *too much care* (excess water or fertilizer or both) by overanxious home owners and careless workers. Fertilizer injury is usually indicated by browning and yellowing of leaves, especially along margins and between veins. Leaves are often stunted.

Pesticide Damage. Careless use of *weed-killers* may result in severe distortion (Figures 1.1D, 4.97, 4.109), dieback, or even death of trees, shrubs, other perennials, and annual plants. Weed-killing chemicals (herbicides) containing 2, 4-D, 2, 4, 5-T, Silvex, M.C.P.A., Ammate, Amitrol, or arsenicals should be used with extreme care. Spray injury is greatly affected by the type of plant grown, condition of plants, and weather at time of application. Follow manufacturer's directions to the letter! For additional information read USDA Farmers' Bulletin 2183, *Using Phenoxy Herbicides Effectively.*

The pesticides suggested in this book will not normally cause damage to plants if applied according to manufacturer's instructions on the *plants* specified, at *rates* specified, and at *times* specified. Injury is most apt to occur in hot weather or slow drying conditions, on tender growth, or when plants are wilted. Bordeaux mixture and other copper-containing fungicides may cause scorching and spotting of leaves on certain plants during cool, wet weather. Copper-injured plants may be stunted, with blossoming and fruit-setting delayed. Sulfur, Karathane, Acti-dione, dodine, dichlone, and dinitro materials may cause scorching in hot (above 85° F.), dry weather. Insecticides, including dormant oils, arsenicals, malathion, parathion, DDT, sevin, chlorobenzilate, and nicotine sulfate may also russet fruit and scorch leaves of certain types of plants. Precautions are usually listed on the package label. See the compatibility chart for fungicides, insecticides, and miticides on page 552. Emulsifiable pesticides are more likely to cause injury than wettable powders or dusts, particularly when mixtures of powders and emulsions are applied.

26. Mechanical Injuries

Mechanical damage caused by lawnmowers, automobiles, hailstones, wind, ice and snow, birds, browsing animals, insects, girdling wires, ground fires, axes, and boys' knives often results in poor growth, weakened plants, and wood rot. Such injuries should be treated promptly (Figure 2.7).

Strangling tree roots that wind tightly around the trunk and other roots may weaken and kill trees such as maples, elms, oaks, and pines (Figure 2.14). Such girdling roots, which may be above or below ground, should be cut off with a chisel and mallet and the exposed surface painted. You can help decrease the possibility of girdling roots by spreading the roots out naturally when planting (see 1 above and Figure 2.1). Treating of girdling roots may also be a job for a good arborist.

Digging a basement, a foundation, or utility (gas, sewer, water) trenches near large trees results in cutting many valuable feeding roots. If wounds are not

Fig. 2.14. Girdling roots have killed some bark at the base of this maple, as shown by the brown discoloration of inner bark where a narrow strip of outer bark has been removed. White living bark is visible at top of area where the outer bark has been removed. Girdling roots may be so injurious that affected trees may gradually decline and finally die. (Illinois Natural History Survey photo)

promptly treated, wilt and root-rotting fungi may enter. The water table for remaining roots may be changed. The effect may cause death of trees.

Construction Damage. Construction damage to shade trees is common by workmen building a home. Tree roots are broken, cut, or exposed—or trunks are scraped. Tractors and bulldozers compact soil, making conditions unfavorable for root development. Injury to tree roots also occurs when trenches for utility lines are dug or lawn grading occurs.

Cuts and bruises permit easy entrance for rot-producing fungi that commonly attack weakened trees and shrubs. Be sure wooden or metal fences are put up to protect valuable shade trees while construction is going on. For additional information read USDA H & G Bulletin 104, *Protecting Shade Trees During Home Construction.*

Changing the Soil Grade. Another common construction damage problem for new homes is a change in soil grade around trees (Figures 2.15 and 2.16). Roots are destroyed or exposed when soil is removed, while a heavy, compacted, clay earthfill smothers the root system. Even several inches of this type of fill can kill old, shallow-rooted trees. Most conifers, birches, hickories, oaks, elms, maples, tulip poplar, and beeches are very susceptible. Trees suffering from too much fill soil have smaller and yellower leaves than normal, and a dying back of outer and upper twigs and branches occurs. Sucker growth on the trunk is common. Fill-injured trees may take up to 10 years to die. A deep earthfill tends to raise the water table or increase moisture and reduce the oxygen content of soil until many roots are suffocated.

Nothing can be done to save trees buried under fill for a long time, and have dead or dying tops. Recent fills, or trees that are apparently not suffering seriously from older fills can be treated. Start corrective treatments at once. Attempt to recreate, as much as possible, the prefill conditions. Where fill is absolutely necessary, use a rock or brick well around the tree capped with screen (Figure 2.16). Install a wagon wheel design of tile drainage over root area before making deeper fills. No single method of constructing fills over tree roots fits all circumstances. An excellent USDA Information Bulletin 285, *Protecting Trees Against Damage From Construction Work,* covers this subject in detail. Before making grade changes around large trees, call in an experienced landscape architect or arborist for advice.

27. Electrical Injuries

Lightning damage is common, particularly on tall, isolated trees such as elm, maple, oak, pine, pecan, poplar, and tuliptree or yellow-poplar. (Beech, horsechestnut, and birch are rarely struck by lightning.) Trees may suffer no permanent damage, show streaks of split or peeled bark and wood that extend to the soil line, or be completely shattered. The leaves turn yellow and later drop. If you suspect lightning damage, call a competent arborist. Exposed wood on injured trees should be painted promptly. All shattered parts and dangerous hanging limbs should be removed. Lightning-protection equipment (fine-mesh copper ca-

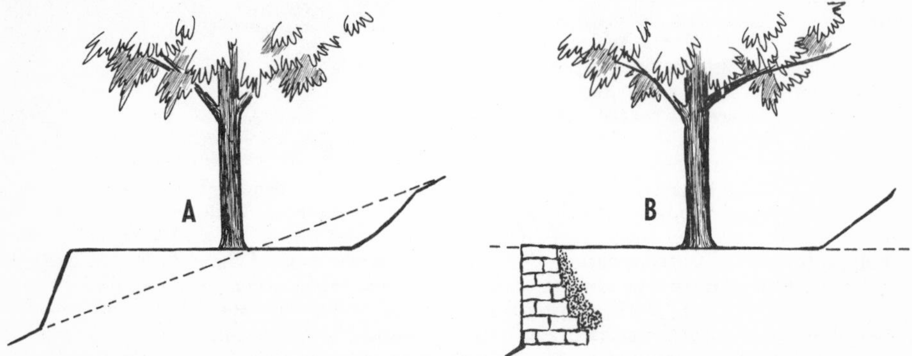

Fig. 2.15. Preserving a maximum of roots when lowering a soil grade. A. By terracing. B. By erecting a retaining wall. The original soil grades are shown by the dotted lines.

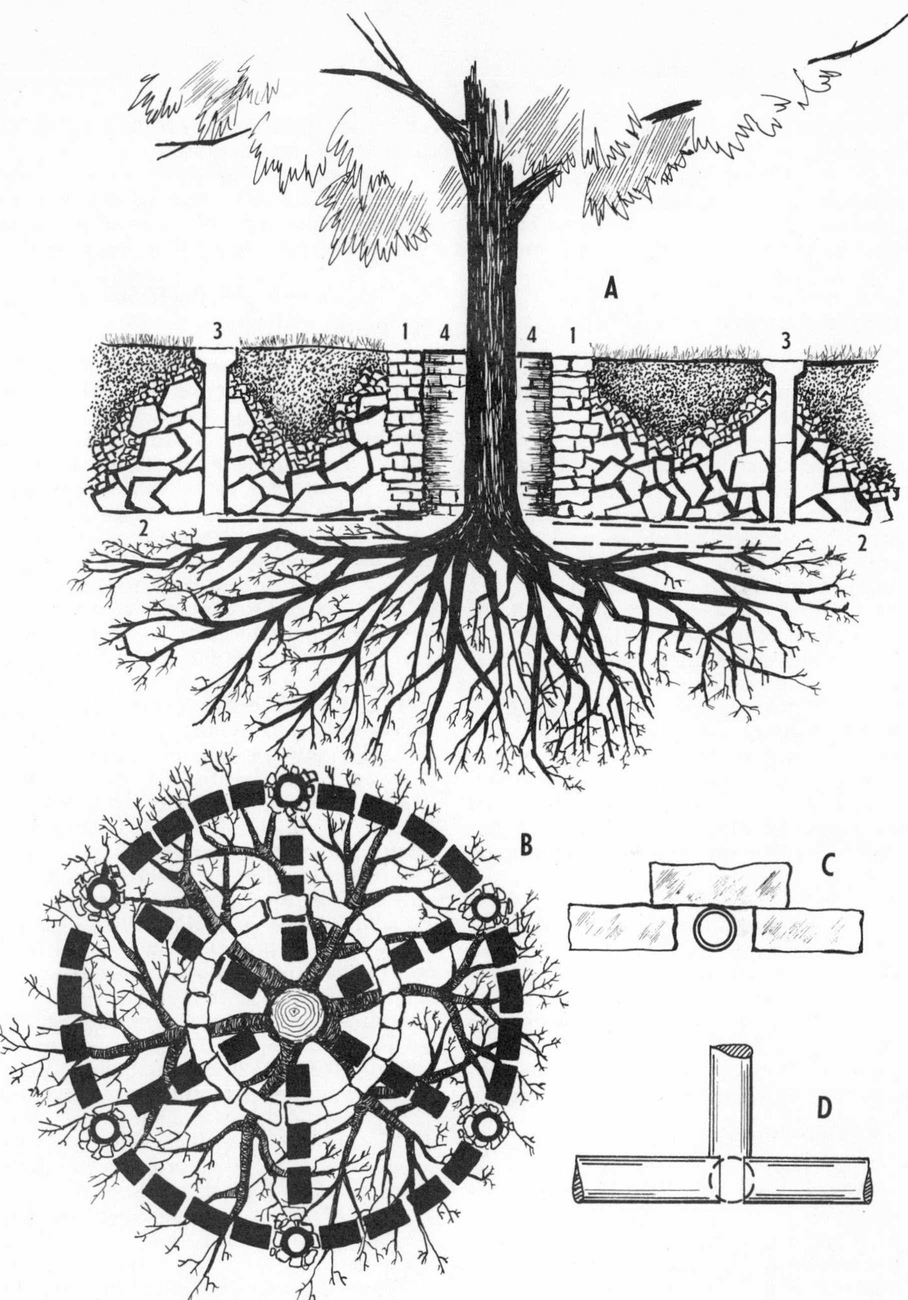

Fig. 2.16. Preventing injury when constructing a deep soil fill. A. Side view showing (1) the dry well; (2) ground tile, sloped to drain away from the trunk and off the roots; (3) vertical bell tile, connected with the drain; (4) metal grating to prevent falling into dry well. The tiles (2) and (3) are covered with rock and coarse gravel except for a foot of topsoil. B. Top view of A showing the trunk at the center surrounded by a dry well with 6 lines of tile radiating out to the ends of the branches. The ground tiles are connected by a row of tiles around the outside. Upright bell tiles are at all intersections. C. Protecting arch of stone is placed over drain tiles to prevent breakage. D. Vertical bell tile rest on the ends of horizontal ground tile that are spaced to allow for air circulation.

ble) is available for trees. Installation is *no* task for the man of the house.

Wires carrying electric current through trees, or near trunks and branches, should be covered with nonconductors at critical points. Be sure that lights used on outdoor Christmas trees are properly placed and equipment is not worn. Do not damage that prized evergreen!

28. Check and Double Check

Before blaming an infectious disease for an ailing plant, check soil, water supply, light requirements, winter protection, and other factors mentioned above. Did you carry out recommended cultural practices in planting, pruning, and fertilizing? Is an insect, mite, or rodent pest involved? How about spray or fume injury? If you eliminate these and the other possibilities outlined above—then one or more infectious diseases may be the answer. Find out what is wrong and start control measures earlier next year.

When investigating a plant trouble, examine symptoms carefully. If you suspect an infectious disease, read disease descriptions listed under the plant. Think back and consider what else might have gone wrong. But do not rush out and buy a new spray or dust "because the old one didn't work." Remember, by the time the blight, rot, or leaf spot is serious enough to be noticed, it is probably too late for spraying or dusting to do much good this season anyway.

Dr. Cynthia Westcott (the Plant Doctor) once said, "The chief hazard any garden plant has to endure is its owner or gardener." She also says do not jump to conclusions when a plant becomes sick.

29. Here We Go

Most plant diseases are named in accordance with their most conspicuous symptom or symptoms, just as many human and animal diseases are described.

Diseases that look alike have often been lumped even though they may be caused by different organisms. This is possible where control measures for each are the same. Where control measures are different, these are pointed out under the plant involved.

Different general types of diseases are divided into those that principally attack foliage, stems (including trunks, branches, and twigs), roots, flowers, and fruit. The same organism often attacks more than one plant part and hence may be listed in more than one category.

Although the diseases listed are based on records in the continental United States, a majority of these diseases is found wherever these plants are grown.

GENERAL DISEASES

A. Foliage Diseases

(1) Fungus Leaf Spot. Usually a rather definite spot—of varying size, shape, and color depending on cause. Often with a distinctive margin or sprinkled with black dots (fungus fruiting bodies). Spots sometimes marked with conspicuous concentric zones ("frogeyes"). If numerous, or if spots enlarge, diseased areas may join to form large irregular *blotches* or a *blight* (Figure 2.17). Infected leaves may wither and die prematurely. See (3) Leaf Blight. Certain leaf spots have special names such as *black spot, tar spot, spot anthracnose,* or *anthracnose.* The centers of some spots may fall out leaving holes. See (4) Shot-hole.

Leaf spots are the most common of diseases, being favored by wet seasons, high humidity, and water splashed on foliage.

Plants Attacked: Practically all.

Control: Most leaf spots do not warrant special measures. Where practical (or possible), collect and burn infected plant parts when first evident and at the season's end. Rotate garden plants avoiding members of the same family in the same soil in successive years. Sow disease-free seed or treat as directed under plant in question (Table 20 in Appendix). Control insects and mites (that may carry the fungus around). Use malathion plus DDT, sevin, or methoxychlor. Follow suggested spray or dust program for plant involved. Use captan, zineb, maneb, ziram, or copper-containing fungicide. See Section 3 for information about these chemicals.

Protection for trees and shrubs is usually needed during wet periods when leaves are expanding. Some plants such as roses, tomatoes, and apples may need regular applications through the growing season. Keep plants vigorous by fertilizing and watering. Grow resistant varieties when available. Indoors, or in arid regions, keep water off foliage, and humidity between 40 and 85 per cent. Too low humidity probably causes more damage

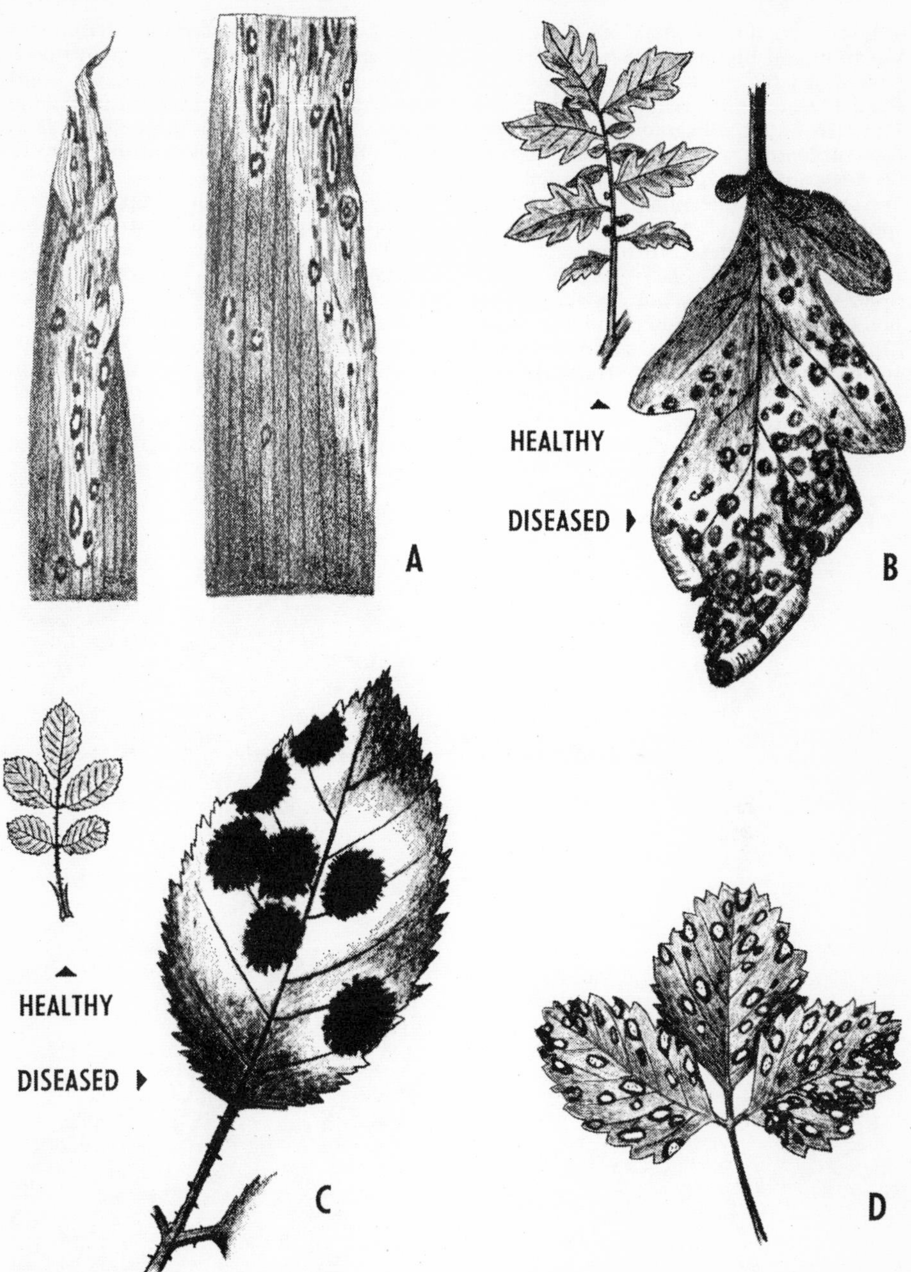

Fig. 2.17. Four leaf spots caused by fungi. A. Iris leaf spot. B. Septoria leaf spot of tomato. C. Black spot of rose. D. Strawberry leaf spot.

to house plants than organisms producing leaf spots and blights. Increase air circulation by spacing plants, especially those in shaded areas, plus removing the lower 4 to 6 inches of foliage on certain plants (e.g., phlox, chrysanthemum, and roses).

(2) Bacterial Leaf Spot or Blight, Bud Rot. Symptoms variable; dark, water-soaked spots or streaks often develop on leaves and stems that later turn gray, brown, reddish-brown, or black (Figure 2.18). Spots may even drop out leaving ragged holes. Leaves may wither and die early. On *crucifers* (cabbage, cauliflower, and related crops) V-shaped, yellow, brown, or dark green areas may develop in leaves with blackened veins. See also (15C) Bacterial Wilt, (24) Fire Blight, and (29) Bacterial Soft Rot.

Plants Attacked: Aconitum, African-violet, almond, apple, apricot, arrowwood, artemisia, asparagus-bean, avocado, bamboos, barberry, bean, beet, begonia, belamcanda, black-eyed pea, blueberry, boxelder, broccoli, Brussels sprouts, bryonopsis, butternut, cabbage, cacti, calabash, California-laurel, California-poppy, canna, cantaloup, carnation, carrot, casaba, castorbean, catnip, cauliflower, celery, cherry, cherry-laurel, chicory, Chinese cabbage, Chinese evergreen, Chinese hibiscus, Chinese lanternplant, chrysanthemum, citrus, collard, corn, cotton-rose, cucumber, cranesbill, currant, daffodil, dandelion, delphinium, dieffenbachia, eggplant, endive, English ivy, escarole, European cranberry-bush, ferns, filbert, flowering almond, flowering cherry, for-

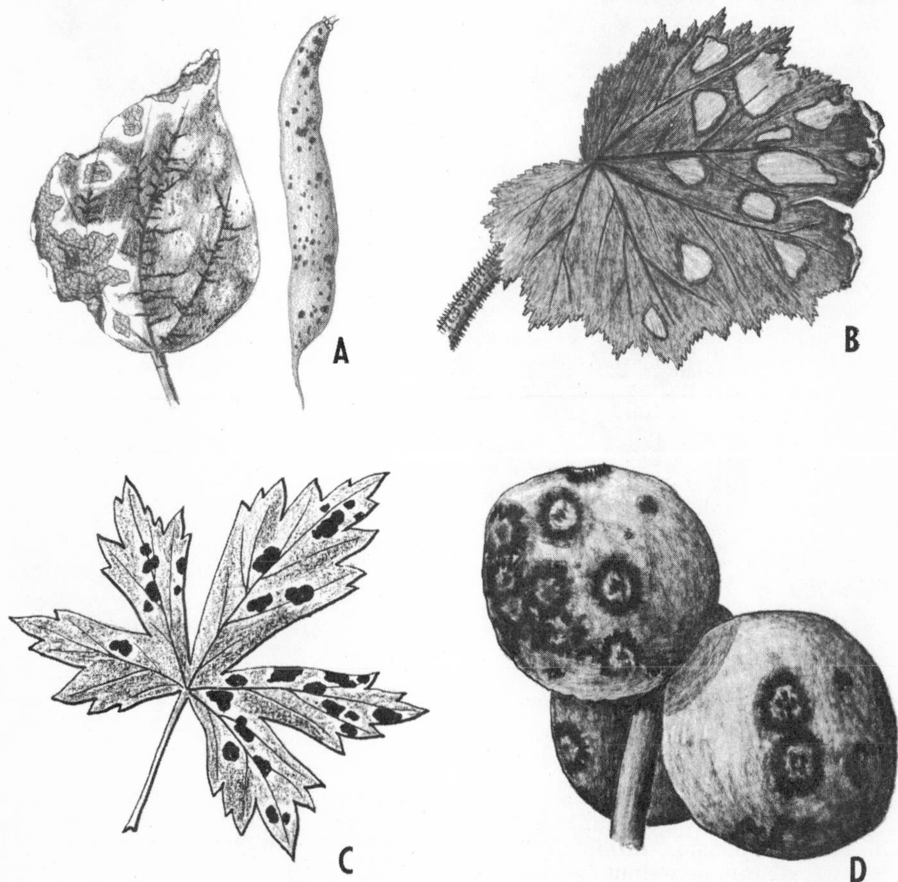

Fig. 2.18. Bacterial leaf spots and blights. A. Bacterial blight of bean. B. Bacterial leaf spot of begonia. C. Delphinium black blotch. D. Bacterial blight of Persian walnut.

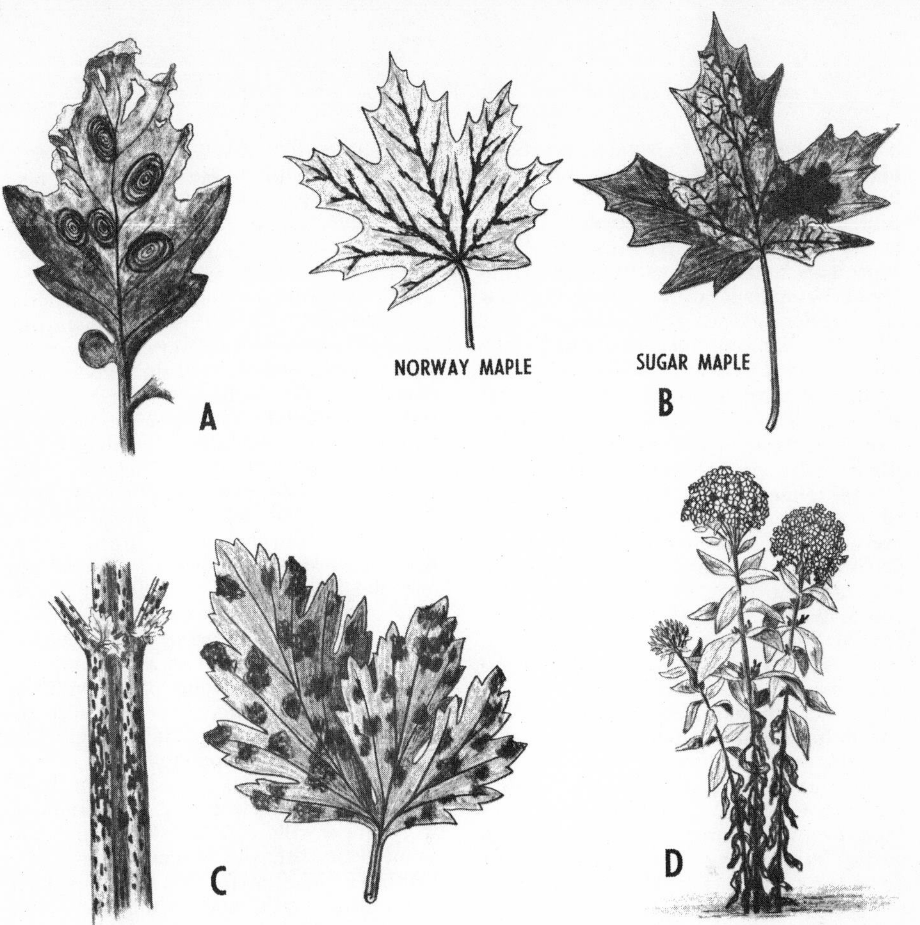

Fig. 2.19. Leaf blights. A. Early blight of tomato. B. Anthracnose of Norway and sugar maples. C. Late blight of celery. D. Leaf blight of phlox.

sythia, gardenia, geranium, gladiolus, gloxinia, golden currant, gourds, grapefruit, groundcherry, hazelnut, heronsbill, hibiscus, honeydew melon, horseradish, hyacinth, hyacinth-bean, iris, Jerusalem-artichoke, kale, kohlrabi, larkspur, lettuce, lilac, magnolia, maple, monkshood, mulberry, muskmelon, mustard, narcissus, nasturtium, nectarine, nephthytis, New Zealand spinach, onion, orange, orchids, palms, pea, peach, pear, pepper, philodendron, plum, poinsettia, poppy, potato, primrose, privet, proboscisflower, pumpkin, pyracantha, radish, rape, rhubarb, rose, rose-of-Sharon, rutabaga, safflower, scarlet runner bean, seakale, snowball, squash, stock, strawberry, sunflower, sweetpea, syngonium, tigerflower, tomato, tree-tomato, turnip, vegetable sponge, viburnum, walnut, watermelon, and West India gherkin.

Control: Same as for (1) Fungus Leaf Spot except use sprays or dusts containing copper or streptomycin or combination of both. See under plant involved.

(3) Leaf Blight, Leaf Blotch, Anthracnose (a term originally associated with diseases caused by certain types of fungi), **Needle Blight, or Cast of Evergreens.** Leaves often suddenly and conspicuously spotted. Spots often enlarge, may merge, and usually become angular to irregular in shape (Figure 2.19). Affected leaves and stems often wilt, wither, die, and may fall prematurely. Fruit may sunscald. Stems and twigs may die. Tops of vegetables and flowers may be killed. See (21) Crown Rot and (24) Fire Blight.

Plants Attacked: Abutilon, African daisy, allium, almond, alternanthera, amaranth, amaryllis, Amazon-lily, amelanchier, apple, apricot, arborvitae, arbutus, arctotis, artemisia, ash, asparagus, asparagus-fern, aspen, aspidistra, aster, aucuba, avocado, azalea, balloonflower, balsam, balsam-apple, balsam-pear, barberry, bean, beet, begonia, bentgrass, Bermudagrass, birch, blackberry, black-eyed pea, blueberry, blue-eyed grass, bluegrass, Boston ivy, buffalograss, boxelder, boxwood, boysenberry, broom, buckeye, butter-and-eggs, buttercup, butterfly-flower, butternut, buttonbush, cabbage, cacti, caesalpinia, calabash, calendula, California-laurel, calla lily, camellia, camphor-tree, campion, cantaloup, cape-cowslip, cape-honeysuckle, carnation, carpetgrass, carrot, casaba, cassabanana, castorbean, catalpa, cauliflower, celery, celeriac, centipedegrass, century-plant, ceriman, chamaecyparis, chayote, cherry, cherry-laurel, chestnut, chicory, China-aster, Chinese cabbage, Chinese evergreen, Chinese waxgourd, chives, Christmas-rose, chrysanthemum, cinnamon-tree, citron, clarkia, clematis, cockscomb, collard, coralberry, coriander, corn, cosmos, cotoneaster, crabapple, crapemyrtle, crassula, crinum, croton, cryptomeria, cucumber, currant, curuba, cyclamen, cypress, daffodil, dahlia, daisy, daphne, daylily, delphinium, dewberry, dieffenbachia, dill, dogwood, Douglas-fir, dracaena, eggplant, elm, endive, English ivy, erythronium, euonymus, European cranberry-bush, evening-primrose, fall daffodil, false-garlic, feijoa, ferns, fescue grass, fig, fir, flowering currant, flowering quince, flowering tobacco, forsythia, foxglove, furcraea, garden cress, garlic, gaultheria, gentian, geranium, gherkin, giant sequoia, ginkgo, gladiolus, gloxinia, goldenrod, gooseberry, gourds, grape, grapefruit, guava, Guernsey-lily, hackberry, hardy orange, hawthorn, heliopsis, hemlock, hen-and-chickens, hibiscus, hickory, holly, hollyhock, holodiscus, honeysuckle, horsechestnut, horseradish, hosta, houseleek, huckleberry, incense-cedar, India rubber tree, iris, Jack-in-the-pulpit, Japanese plum-yew, Japanese quince, Jerusalem-cherry, jetbead, juniper, kerria, Labrador-tea, larch, larkspur, laurel, lavatera, leek, lemon, lemon-verbena, lentil, lettuce, leucothoë, lilac, lily, lily-of-the-valley, lima bean, lime, linden, lippia, locust, loganberry, London plane, loosestrife, loquat, lupine, lycoris, lyonia, madrone, magnolia, mahonia, mallow, mango, maple, mayapple, medlar, mignonette, mint, mitella, mock-cucumber, monstera, montbretia, mountain-ash, mountain-laurel, muskmelon, mustard, nandina, narcissus, nectarine, nephthytis, nightshade, oak, okra, oleander, olive, onion, orange, orchids, Oregon-grape, Osage-orange, pachysandra, painted-tongue, palms, pansy, parsley, parsnip, passionflower, pawpaw, pea, peach, pear, pea-tree, pecan, peony, peperomia, pepper, peppermint, persimmon, petunia, philodendron, phlox, photinia, pine, pinks, pistachio, planetree, plum, poinciana, poinsettia, polygala, pomegranate, poplar, potato, potentilla, primrose, privet, pumpkin, pyracantha, pyrethrum, quince, radish, rape, raspberry, redcedar, redtop, redwood, retinospora, rhododendron, rhubarb, rollinia, rose, roselle, rosemallow, rutabaga, ryegrass, safflower, St. Augustinegrass, salal, salpiglossis, salsify, scarborough-lily, scarlet runner bean, sedum, sequoia, shallot, Shasta daisy, sicana, silver threads, snapdragon, snowball, snowberry, snowdrop, snowflake, soapberry, sorreltree, spider-lily, spinach, spruce, squash, stock, strawberry, sugarberry, sweetbay, sweetpea, sweetpotato, sweet-william, sycamore, tanbark-oak, toadflax, tomato, treemallow, tritonia, trumpetvine, tuberose, tulip, tuliptree, tupelo, turnip, udo, vegetable-marrow, vegetable sponge, vetch, viburnum, Virginia-creeper, vinca, violet, wallflower, walnut, waterlily, watermelon, wheatgrass, willow, yam, yew, yucca, zephyranthes, zinnia, and zoysiagrass.

Control: Same as for (1) Fungus Leaf Spot. Prune trees and shrubs for better air circulation.

(4) Shot-hole. Small spots on leaves of stone fruits and related plants that later drop out leaving typical shot-holes. Infected leaves often change color and drop prematurely. Young fruit may be spotted, deformed, and fall early. Sunken, reddish cankers may develop on twigs (Figure 2.20). Fruit buds and fruiting wood may die during winter. Shot-hole may be caused by bacteria, fungi, viruses, or spray injury. Often a secondary symptom of fungus or bacterial leaf spot on many kinds of plants. See (1) Fungus Leaf Spot and (2) Bacterial Leaf Spot above.

Plants Attacked: Almond, apricot, cherry, cherry-laurel, flowering almond, flowering cherry, nectarine, peach, and plum.

Control: Same as for (1) Fungus Leaf Spot. Follow spray program for fruits in question (Table 16 in Appendix). Collect and burn fallen leaves, where possible. Plant virus-free stock from reliable nursery.

(5) Botrytis Blight, Gray-mold Blight, Bud Rot, Blossom Blight, Twig Blight. Generally

Fig. 2.20. Shot-hole of peach and cherry. These same leaf symptoms may be caused by various fungi, bacteria, viruses, or pesticide injury.

distributed on dead and dying plant parts. Soft, tan to brown spots or blotches on leaves, stems, flowers, ripening fruit, tubers, or roots during or following cool, damp, cloudy periods. Affected parts often covered with coarse, tannish-gray mold in damp weather. Seedlings or young shoots may wilt and collapse. Buds may rot. Flowers often distorted with irregular flecks or spots (Figure 2.21). Older flowers rot quickly. (May be confused with thrips injury.) Common in greenhouses and cool, humid areas. See (21) Crown Rot, (31) Flower Blight, and (32) Fruit Spot. Fungus enters through wounds, dying leaves, or old flower petals.

Plants Attacked: African-violet, alder, almond, amaryllis, Amazon-lily, anemone, apple, apricot, arborvitae, aristolochia, artemisia, artichoke, ash, asparagus, asparagus-bean, aster, aucuba, azalea, babysbreath, barberry, bean, beet, begonia, bellflower, bignonia, blackberry, black-eyed pea, bleedingheart, bloodroot, blueberry, blue cohosh, boysenberry, broccoli, Brussels sprouts, buttercup, cabbage, cacti, caladium, calceolaria, calendula, California-poppy, calla lily, camass, camellia, candytuft, cape-marigold, carnation, carrot, castorbean, catalpa, cauliflower, celery, centuryplant, chamaecyparis, cherry, chicory, China-aster, Chinese hibiscus. chives, Christmas-rose, chrysanthemum, cigarflower, cineraria, clarkia, colchicum, coleus, columbine, coralberry, cornflower aster, cotton-rose, cranesbill, crotalaria, crown imperial, cucumber, currant, cyclamen, cypress, dahlia, dandelion, daphne, daylily, delphinium, dewberry, dogwood, dogstooth-violet, Douglas-fir, dracaena, dusty-miller, Dutchmans-pipe, eggplant,

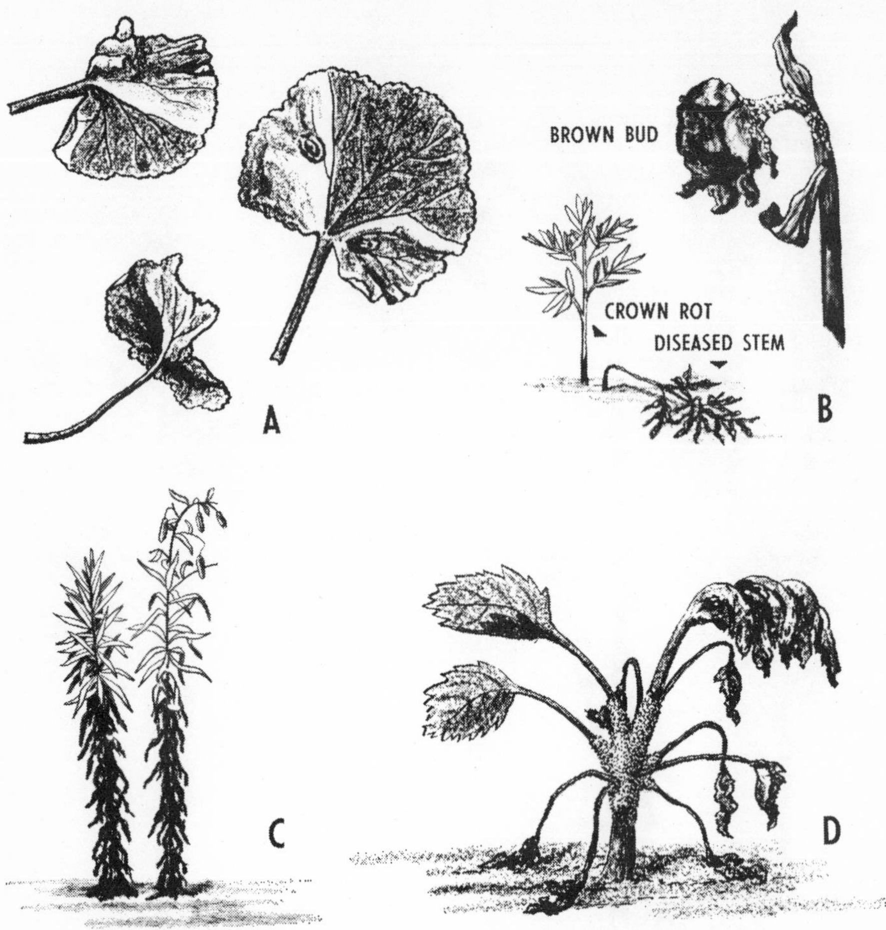

Fig. 2.21. Botrytis blight or gray-mold blight. A. Geranium. B. Peony. C. Lily. D. Aster. Note gray mold on peony bud and aster stem.

endive, English daisy, erigeron, escarole, eucalyptus, eupatorium, European cranberry-bush, exacum, feijoa, fennel, ferns, fig, fir, flax, flowering almond, flowering currant, forget-me-not, foxglove, freesia, fritillary, fuchsia, gardenia, garlic, gentian, geranium, gladiolus, globe-amaranth, globe artichoke, gloxinia, godetia, goldenchain, gooseberry, grape, gum, gypsophila, hawthorn, heath, heliotrope, hemlock, holly, honeysuckle, hyacinth, hydrangea, iris, ixia, Jack-in-the-pulpit, jasmine, juniper, kale, kohlrabi, laburnum, larch, leek, lentil, lettuce, lilac, lily, lily-of-the-valley, lobelia, lupine, magnolia, Maltese cross, marigold, mayapple, mertensia, mistflower, mockorange, narcissus, nasturtium, okra, onion, orchids, pansy, parsnip, passionflower, pea, peach, peanut, pear, peony, pepper, persimmon, petunia, phlox, pine, pinks, plantainlily, plum, poinsettia, pomegranate, poppy, potato, primrose, pumpkin, pyrethrum, quince, rape, raspberry, redwood, rhododendron, rhubarb, rockcress, rose, roselle, rose-of-Sharon, rutabaga, safflower, sea-lavender, senecio, sequoia, shallot, skullcap, snapdragon, snowberry, snowball, snowdrop, snow-on-the-mountain, spruce, squash, statice, stock, stokesia, strawberry, sumac, sunflower, sunrose, sweetpea, sweetpotato, thimbleberry, tomato, tradescantia, Transvaal daisy, tree-of-Heaven, tuberose, tulip, turnip, vegetable-marrow, verbena, viburnum, vinca, violet, Virginia bluebell, wallflower, and zinnia.

Control: Cut and burn infected plants and plant parts, where practical. Carefully remove and burn fading flowers before petals fall. Avoid overcrowding, overfeeding (especially with nitrogen), overhead irrigation, wet mulches, and shady or low spots with poor air circulation. Keep down weeds. Cure bulbs, corms, tubers, etc. rapidly at high temperatures before storing at the recommended temperature. Avoid high humidities indoors. Increase air circulation and temperature. Spray at 5-day intervals during cool, wet weather using captan, zineb, ferbam, thiram, Daconil 2787, Polyram, dichloran (Botran), maneb, folpet, Dyrene, dichlone, zineb-thiram, or copper-containing fungicide. Follow manufacturer's directions. Take cuttings from healthy plants and propagate in sterilized medium. Spray flowers just before storage using zineb or captan (1 tablespoon per gallon). Avoid wounding fruits and vegetables. Keep temperature as close to 32° F. as practical.

(6) Downy Mildew. Pale green or yellowish areas usually appear on upper leaf surface with corresponding light gray, downy, or purplish patches of mildew below. Affected areas enlarge and turn yellow or brown. Leaves often wilt, wither, and die early. Stems, flowers, and fruits sometimes infected (Figure 2.22). Seedlings may wilt and collapse. Attacks are most severe in cool, humid, or wet weather (warm days and cool nights).

Plants Attacked: Acalypha, agrimony, alyssum, ampelopsis, anemone, arrowwood, artemisia, artichoke, aster, avens, avocado, bachelors-button, balsam, balsam-apple, balsam-pear, bean, bedstraw, beet, Bermudagrass, blackberry, black-eyed pea, Boston ivy, boysenberry, broccoli, Brussels sprouts, bryonopsis, butter-

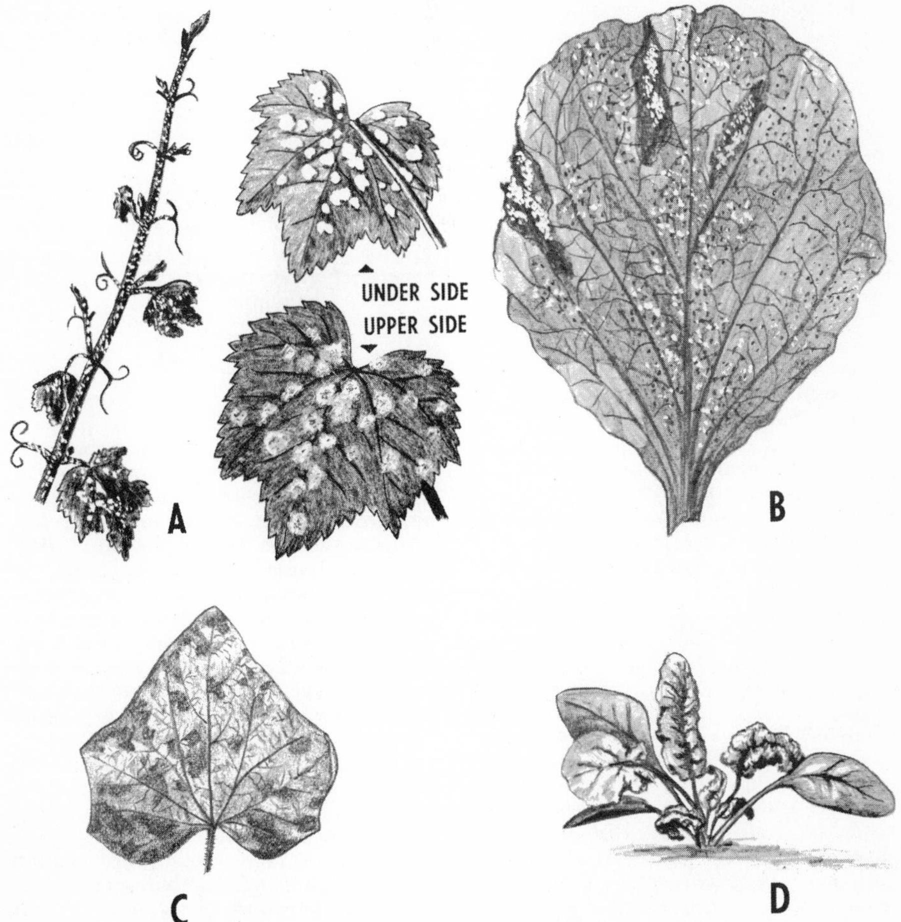

Fig. 2.22. Downy mildew. A. Grape. B. Lettuce. C. Cucumber. D. Spinach.

and-eggs, buttercup, butternut, cabbage, calabash, caladium, candytuft, cantaloup, cape-marigold, carnation, carrot, casaba, cauliflower, celery, celtuce, centaurea, chervil, chicory, China-aster, chinaberry, Chinese cabbage, Chinese waxgourd, chives, Christmas-rose, cimicifuga, cineraria, cinquefoil, citron, clarkia, collard, cornflower, cornflower aster, corydalis, cranesbill, crownbeard, cucumber, currant, cushion-pink, damesrocket, dewberry, dragonhead, Dutchmans-breeches, eggplant, endive, erysimum, escarole, eupatorium, European cranberry-bush, evening-primrose, everlasting, false-dragonhead, fennel, fleabane, flowering tobacco, forget-me-not, four-o'clock, foxglove, gaillardia, garden cress, garlic, gaura, germander, gilia, godetia, goldenglow, goldenrod, gooseberry, gourds, grape, hackberry, heronsbill, horseradish, houndstongue, houstonia, Jack-in-the-pulpit, Jerusalem-artichoke, Joe-pye-weed, kale, kohlrabi, leek, lettuce, liverleaf, lupine, mangold, marigold, meadowrue, meconopsis, melothria, mertensia, mignonette, mock-cucumber, mock-strawberry, monkshood-vine, mullein, muskmelon, mustard, nicotiana, onion, pansy, parsley, parsnip, pea, pecan, pepper, peppergrass, phlox, physostegia, poppy, potentilla, prairie-coneflower, prickly-poppy, primrose, privet, pumpkin, radish, rape, raspberry, redbud, rhubarb, rockcress, rockjasmine, rose, rudbeckia, rutabaga, salvia, sand-verbena, senecio, shallot, silene, silphium, snapdragon, snowball, speedwell, spiderflower, spinach, squash, squirrel-corn, stock, strawberry, strawflower, sugarberry, sugar beet, sunflower, sweet alyssum, sweetpea, Swiss chard, teasel, toadflax, tomato, toothwort, trailing four-o'clock, turnip, umbrellawort, vegetable-marrow, vegetable sponge, verbena, vetch, viburnum, violet, Virginia-creeper, wallflower, walnut, watercress, watermelon, wayfaring-tree, West India gherkin, whitlowgrass, and wild sweet-william.

Control: Practice 2- to 3-year crop rotation. Collect and burn infected plant parts when first evident. Burn tops after harvest. Maintain balanced soil fertility, based on a soil test. Use disease-free seed from healthy stock plants. Resistant varieties offer hope for some types. Avoid overcrowding, overhead sprinkling, excessive nitrogen fertilization, and high humidities indoors and in seedbed. Dust or spray at 5-day intervals in cool, wet weather. Use zineb, maneb, or fixed-copper fungicide. Or follow spray schedule outlined under plant in question.

(7) Powdery Mildew. Superficial, white to light grayish, powdery to mealy coating or felt on leaves, buds, flowers, and young shoots (Figure 2.23). Affected parts may be dwarfed and curled. Leaves may yellow, wither, and die prematurely. Mildew spots often enlarge until they eventually cover whole leaf. Tiny black fruiting bodies of the mildew fungus are often embedded in the mildew patches, especially late in the season. Common when cool nights follow warm days, in crowded low areas where air circulation is poor, or in damp, shaded locations. Powdery mildew is most common on many plants from midsummer on. Often more unsightly than harmful, especially on trees.

Plants Attacked: Abelia, acacia, acalypha, acanthopanax, achillea, aconitum, African-violet, ageratum, agrimony, alder, almond, amelanchier, amorpha, ampelopsis, anchusa, anemone, anoda, apple, apricot, aralia, arenaria, arnica, arrowwood, artemisia, artichoke, artillery-plant, ash, asparagus-bean, aspen, aster, astilbe, aucuba, avens, avocado, azalea, babytears vine, bachelors-button, balsam-apple, balsam-pear, balsamroot, barberry, basket-flower, bauhinia, bean, bearberry, bedstraw, beech, beet, begonia, bellflower, Bermudagrass, betony, birch, bishopscap, bittersweet, blackberry, black-eyed pea, black locust, bladder-senna, blazing-star, blueberry, blue daisy, bluegrass, boltonia, Boston ivy, boxelder, boysenberry, broccoli, broom, buckeye, buckthorn, buffaloberry, bundleflower, bur-marigold, burnet, bush-honeysuckle, buttercup, butterfly-flower, butterflyweed, buttonbush, cabbage, calabash, calendula, California-poppy, calycanthus, campanula, camphor-tree, candytuft, cantaloup, Canterbury-bells, cardoon, carnation, cassia, catalpa, cauliflower, centaurea, checkerberry, cherry, cherry-laurel, chestnut, chicory, China-aster, chinaberry, Chinese cabbage, chinquapin, chrysanthemum, cigarflower, cineraria, cinquefoil, citron, clematis, collard, collomia, columbine, coralbells, coralberry, cordia, coreopsis, cornflower, cornsalad, cosmos, cotoneaster, cottonwood, crabapple, cranesbill, crapemyrtle, crotalaria, crownbeard, cucumber, culversroot, currant, dahlia, dandelion, delphinium, dewberry, dogwood, dusty-miller,

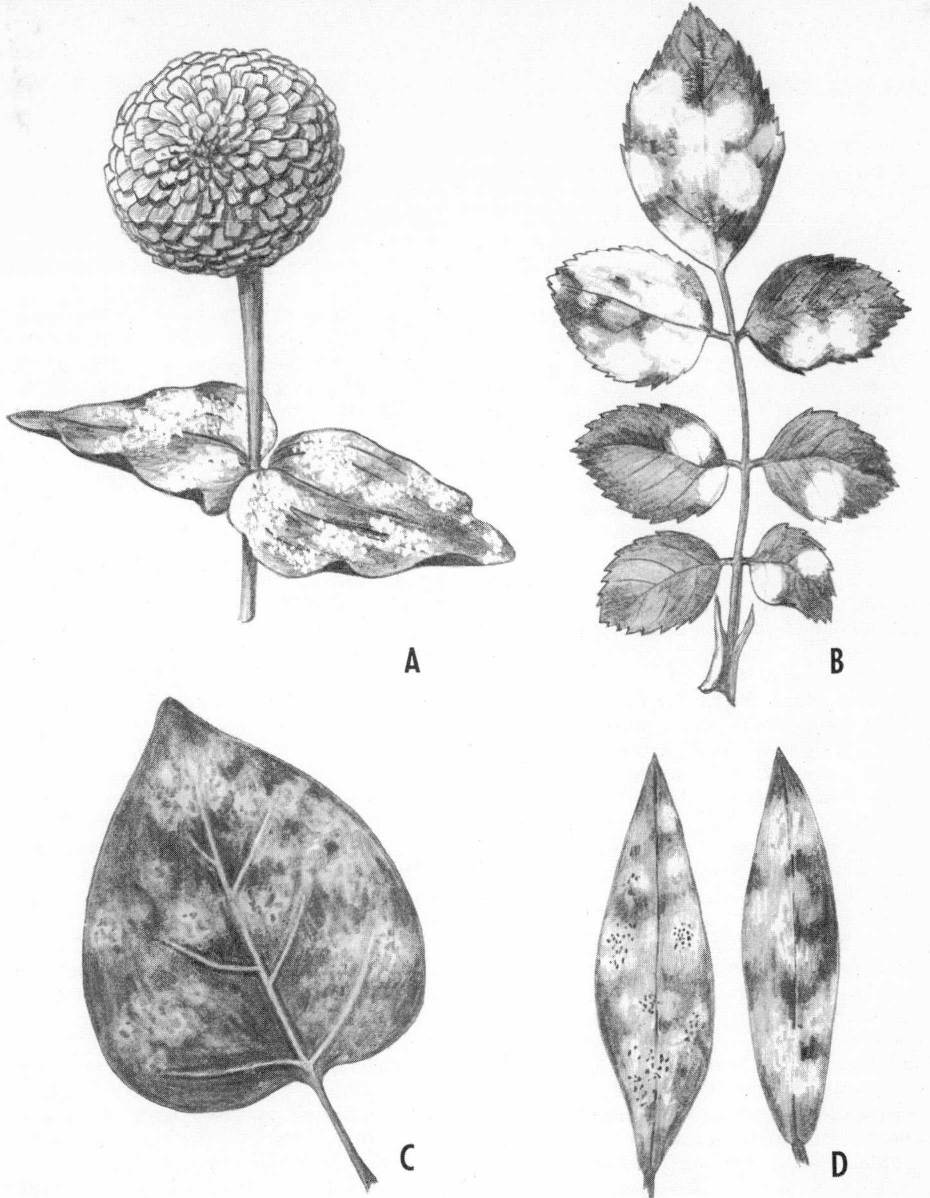

Fig. 2.23. Powdery mildew. A. Zinnia. B. Rose. C. Lilac. D. Phlox. The black specks in the left phlox leaf are the sexual fruiting bodies (cleistothecia) of the powdery mildew fungus. These are often produced on plants late in the growing season.

dwarf cornel, eggplant, elder, elecampane, elm, endive, erysimum, euonymus, eupatorium, English ivy, European cranberry-bush, evening-primrose, false-indigo, fescue grass, feverfew, filbert, filipendula, fleabane, flowering almond, flowering cherry, flowering raspberry, flowering tobacco, foamflower, forestiera, forget-me-not, fringetree, fuchsia, gaillardia, gardenia, gaultheria, gaura, genista, germander, gilia, globe artichoke, globemallow, glowing gold, gloxinia, golden-aster, golden cinquefoil, goldenglow, goldenrod, gooseberry, gourds, grape, groundsel, hackberry, hardhack, hawksbeard, hawthorn, hazelnut, heath, heliopsis, heuchera, hickory, highbush cranberry, holly, hollyhock, holodiscus, honeylocust, honeysuckle, hophornbeam, hoptree, hornbeam, horsechestnut, horseradish, houndstongue, huckleberry, hyacinth-bean, hydrangea, hypericum, indigo, indigobush, inula, Japanese pagodatree, Jerusalem-artichoke, Joe-pye-weed, kalanchoë, kale, kohlrabi, Labrador-tea, leopardsbane, lettuce, liatris, lilac, linden, lithospermum, lobelia, locust, London plane, lupine, lyonia, magnolia, mahonia, mallow, mango, maple, marguerite, matricaria, matrimony-vine, mayflower, meadowrue, meadowsweet, meconopsis, melothria, menziesia, mertensis, Michaelmas daisy, mint, mistflower, mitella, mock-cucumber, mockorange, monkeyflower, monkshoodvine, moonseed, mountain-ash, mountain-holly, mountain-laurel, mountain-mint, mulberry, mullein, muskmelon, mustard, myrtle, nectarine, nemophila, New Jersey-tea, nicotiana, ninebark, oak, okra, orange, osoberry, Oswego-tea, oxalis, painted-cup, painted-tongue, pansy, parsnip, pea, peach, pear, pecan, penstemon, peony, pepper, peppermint, persimmon, petunia, phacelia, philibertia, phlox, photinia, piggy-back plant, piqueria, planetree, plum, plumed thistle, polemonium, poplar, poppy, potato, potentilla, prairie-coneflower, prickly-ash, primrose, privet, prunella, pumpkin, purple-flowered groundcherry, pyrethrum, queen-of-the-prairie, quince, radish, rape, raspberry, ratibida, redtop, rhododendron, rose, roselle, rose-mallow, rudbeckia, rue-anemone, Russian-olive, rutabaga, safflower, St.-Johns-wort, salal, salpiglossis, salsify, salvia, sandwort, sarsaparilla, sassafras, saxifrage, scabiosa, scarlet runner bean, sea-lavender, sedum, senna, serviceberry, silphium, silverberry, silver king, skullcap, snapdragon, sneezeweed, snowball, snowberry, snow-on-the-mountain, soapberry, sophora, spearmint, speedwell, spiderflower, spirea, spurge, squash, stachys, statice, Stokes-aster, strawberry, sugar beet, sugarberry, sumac, sunflower, sunrose, swede, sweet alyssum, sweetpea, sycamore, tamarisk, tansy, teasel, tepary bean, thermopsis, thistle, tickseed, tidytips, toadflax, tomato, trailing-arbutus, Transvaal daisy, tree-tomato, trumpetvine, tuliptree, turnip, turtlehead, valerian, vegetable-marrow, vegetable sponge, verbena, vetch, viburnum, violet, Virginia-creeper, wallflower, walnut, watermelon, weigela, West India gherkin, wheatgrass, wild sweet-william, willow, winterberry, wisteria, witch-hazel, wolfberry, woodwaxen, yellow ironweed, yellowwood, and zinnia.

Control: Resistant varieties of certain flowers and vegetables are available. Avoid overcrowding and damp, shady locations. Avoid sprinkling foliage in late afternoon or evening. Indoors, increase air circulation and night temperature. Where needed, dust or spray several times, 7 to 10 days apart. Use Karathane, Morocide, Daconil 2787, folpet, sulfur, or Acti-dione. Avoid applications when temperatures are 85° F. or above. Collect and burn affected plant parts in fall.

(8) Rust—Leaf, Stem, Needle. Bright yellow, orange, orange-red, reddish-brown, chocolate-brown, or black, powdery, raised pustules. Some rusts may have as many as 5 different spore forms in a single season's life cycle that vary in size, shape, and color. Pustules are most common on the lower leaf surface and on stems (Figure 2.24). Leaves often wither and die early; plants may be stunted. When severe, they may wilt, wither, and die (e.g., snapdragon).

True rusts are of two general types:

(a) Fungus completes life cycle on the *same type of plant* (autoecious or monoecious rusts). *Examples:* Asparagus, bean, beet, blackberry, carnation, chrysanthemum, hollyhock, mint, pea, plum, raspberry, rose, snapdragon, sunflower, Swiss chard, and violet.

(b) Fungus requires two different kinds of plants (or an alternate host) to complete the life cycle. Such rusts are heteroecious. *Examples:*

1. *Juniper or redcedar* and apple, chokeberry, crabapple, Japanese quince, quince, flowering quince, mountain-ash, hawthorn, pear, photinia, amelanchier, squaw-apple, fendlera, mockorange, etc.
2. *Spruce* and bog-rosemary, creeping snowberry, Labrador-tea, pyrola, woodnymph, crowberry, rhododendron, etc.
3. *Pine* and bellflower, goldenrod, sunflower, moonflower, Jerusalem-artichoke, morning-glory, quamoclit, gaillardia, currant, gooseberry, blazing-star, marigold, senecio, asters, viburnum, Indian paintbrush, sweetfern, sweetgale, oak, buckleya, amsonia, sweetpotato, etc.
4. *Hemlock* and poplar, blueberry, azalea, rhododendron, menziesia, hydrangea, lyonia, etc.
5. *Douglas-fir* and poplar.

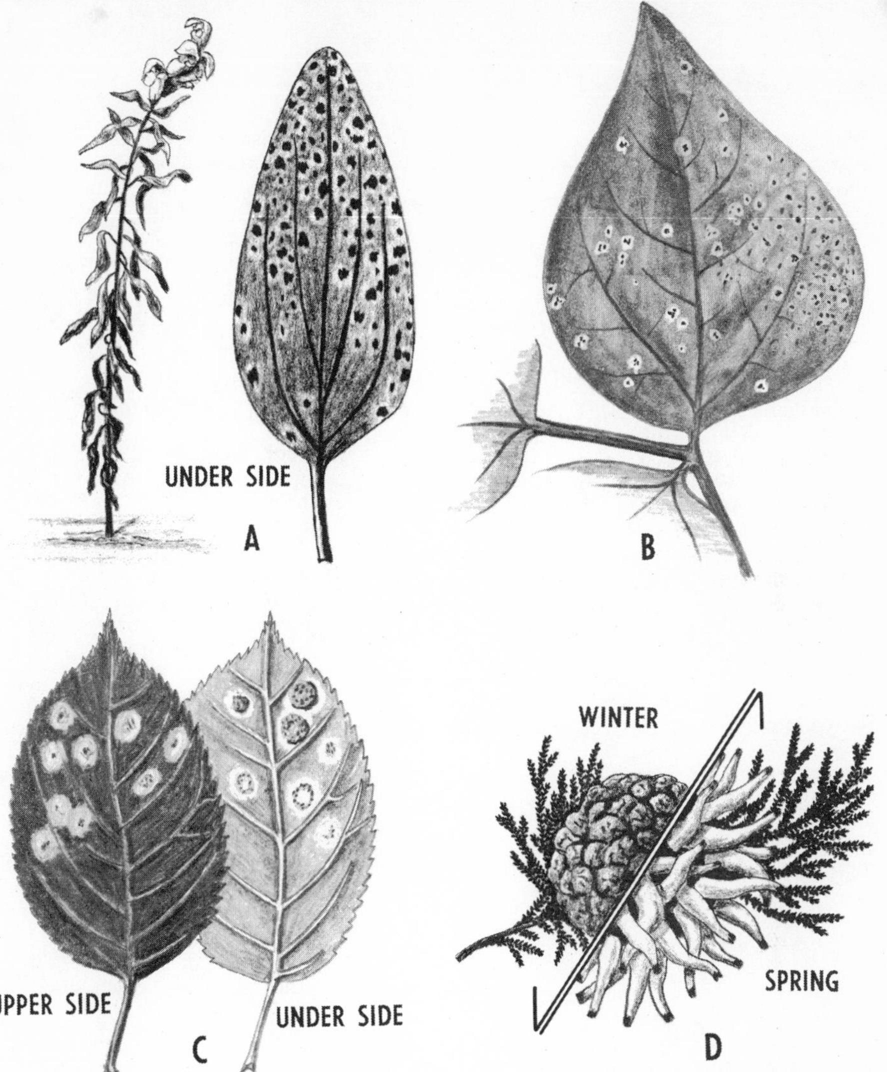

Fig. 2.24. Rust. A. Snapdragon. B. Bean. C. Apple. D. Juniper. The rust on juniper and apple (cedar-apple rust) are different disease symptoms caused by the same rust fungus.

6. *Larch* and willow, poplar, or birch.
7. *Lupine* and spinach, garden cress, and radish.
8. *Currant* and *gooseberry* and carex, willow, etc.
9. *Wild grasses* and aconite, anemone, barberry, beet, buckthorn, California-bluebell, campanula, clematis, columbine, coralberry, dalea, delphinium, fleabane, four-o'clock, garden cress, heliotrope, Indian paintbrush, meadowrue, mertensia, mint, primrose, queens-delight, radish, sand verbena, snowberry, spiderflower, spinach, turtlehead, wolfberry, etc.
10. *Corn* and oxalis.
11. *Fir* and ferns, willow, huckleberry, blueberry, chickweed, birch, etc.

Rust on juniper (redcedar) appears as reddish-brown, bean-shaped galls from which orange, gelatinous tendrils protrude during spring rains (Figure 2.24). *Plants Attacked:* Abutilon, acacia, acanthopanax, achillea, aconitum, ageratum, agrimony, alder, allium, almond, amelanchier, American spikenard, amorpha, ampelopsis, amsonia, anchusa, anemone, angelica, anise, anise-root, anoda, apple, apricot, aralia, arbutus, arenaria, armeria, arnica, arrowroot, arrowwood, artemisia, ash, asparagus, aspen, aster, atamasco-lily, avens, azalea, babysbreath, babytears vine, bachelors-button, balsam, balsamroot, bamboos, baneberry, barberry, basilweed, bayberry, bean, bearberry, bedstraw, beebalm, beet, bellflower, bellwort, bentgrass, Bermudagrass, birch, bishopscap, blackberry, black-eyed pea, black gum, bladder-senna, blazing-star, blueberry, blue-eyed grass, bluegrass, boisduvalia, boltonia, bouvardia, boysenberry, brodiaea, broom, buckeye, buckthorn, buffaloberry, buffalograss, bundleflower, bur-marigold, burnet, buttercup, butterflyweed, buttonbush, caesalpinia, calendula, California-bluebell, California-rose, calochortus, campion, canna, Canterbury-bells, cape-marigold, cardinal climber, carnation, carpetgrass, carrot, centaurea, chamaecyparis, chamaedaphne, checkermallow, cherry, chestnut, chicory, China-aster, Chinese lanternplant, chives, chokeberry, chrysanthemum, chuperosa, cimicifuga, cinquefoil, cissus, clarkia, clematis, climbing hempweed, coffeeberry, collinsia, collomia, columbine, columbo, coralbells, coralberry, cordia, coreopsis, corn, cornel, cornflower, corydalis, cosmos, cottonwood, cowania, crabapple, cranesbill, croton, crownbeard, culversroot, cunila, currant, cypress, cypress-vine, dalea, dandelion, datura, dayflower, delphinium, desertplume, dewberry, dittany, Douglas-fir, dogwood, dogstooth-violet, Dutchmans-breeches, dwarf cornel, dyschoriste, echeveria, eggplant, elder, elecampane, emilia, encelia, endive, erysimum, erythronium, eupatorium, evening-primrose, false-boneset, false-dragonhead, false-garlic, false-indigo, false-mallow, false-mesquite, fendlera, ferns, fescue grass, fig, fir, flax, fleabane, filipendula, Florida yellowtrumpet, flowering currant, flowering quince, flowering raspberry, foamflower, forestiera, forget-me-not, four-o'clock, frangipani, fritillaria, fuchsia, gaillardia, garden cress, garlic, gaura, genista, gentian, germander, giant night white bloomer, gilia, gladiolus, globemallow, godetia, golden-aster, goldenglow, goldenrod, gooseberry, grape, groundcherry, groundpink, groundsel, harebell, hawksbeard, hawthorn, hearts and honey vine, heath, heliopsis, heliotrope, hemlock, hen-and-chickens, Hercules-club, heuchera, hibiscus, holly, hollyhock, honeylocust, honeysuckle, hophornbeam, hoptree, horsechestnut, houseleek, houstonia, huckleberry, hyacinth, hydrangea, hypericum, incense-cedar, Indian paintbrush. indigo, indigobush, iris, Jack-in-the-pulpit, jacquemontia, jasmine, Jerusalem-artichoke, Joe-pyeweed, juniper, kochia, Labrador-tea, lantana, larch, lavatera, leadtree, leatherwood, lebbek, leek, leonotis, lettuce, liatris, lily, lithospermum, liverleaf, lobelia, locust, loganberry, loosestrife, lupine, lyonia, madrone, mahonia, malacothrix, mallow, Maltese cross, malvastrum, manfreda, mangel, mangold, manzanita, maranta, marbleseed, marigold, matricaria, matrimony-vine, mayapple, meadowrue, meadowsweet, medlar, mentzelia, menziesia, mertensia, mint, mistflower, mitella, mockorange, mock-strawberry, monarda, monardella, monkeyflower, monkshood, moonflower, morning-glory, mountain-ash, mountain-mint, mulberry, mustard, nasturtium, nectarine, New Jersey-tea, oak, okra, onion, orchids, Oregon-grape, Osage-orange, osier, oxalis, painted-cup, pansy, pea, peach, peanut, pear, pearl-everlasting, penstemon, peppergrass, peppermint, petunia, phacelia, phalaris, philibertia, phlox, photinia, physostegia, pine, pinks, plum, plumed thistle, poinciana, poinsettia, polemonium, polygala, poplar, poppy-mallow, potentilla, prairie-coneflower, prickly-ash, prickly-poppy, primrose, prune, purple-flowered groundcherry, quamoclit, queen-of-the-prairie, queens-delight, quince, radish, rainlily, raspberry, ratibida, redcedar, redtop, rhododendron, rhubarb, rockcress, rockjasmine, romanzoffia, rose, rosemallow, rose-of-Sharon, rougeplant, rudbeckia, rue-anemone, ruellia, Russian-olive, ryegrass, safflower, St.-Andrews-cross, St. Augustine-grass, St.-Johns-wort, salsify, salvia, sand-verbena, sarsaparilla, satin-flower, saxifrage, scarlet runner bean, scilla, sea-lavender, sedum, senecio, sensitive plant, serviceberry, shallot, shootingstar, sida, sidalacea, silene, silphium, silverberry, silver king, silver lacevine, smelowskia, smoketree, snapdragon, sneezeweed, snowberry, snow-on-the-mountain, sophora, sour gum, southern leatherwood, spearmint, speedwell, spiderflower, spinach, spruce, spurge, squirrelcorn, stanleya, stachys, statice, stenanthium, stenolobium, sumac, sunflower, sweetfern, sweetgale, sweet-jarvil, sweetpea, sweetpotato, sweet-william, Swiss chard, synthyris, taenidia, tamarack, tanbark-oak, tansy, tepary bean, thimbleberry, thistle, tickseed, toadflax, toothwort, tradescantia, trailing four-o'clock, trillium, trumpettree, tupelo, turtlehead, valerian, venus-

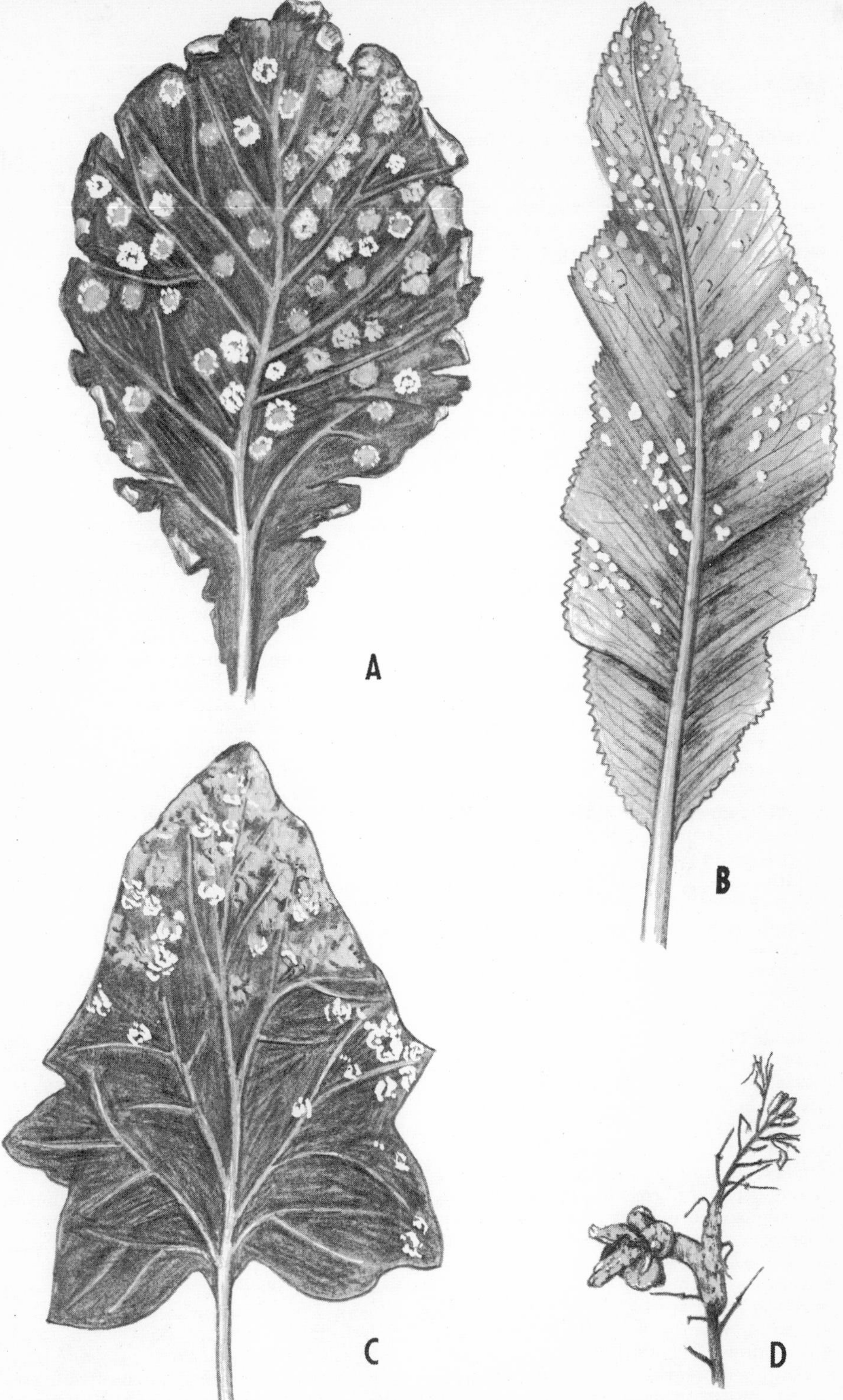

Fig. 2.25. White-rust. A. Cabbage. B. Horseradish. C. Spinach. D. Radish with aborted flower parts.

lookingglass verbena, vetch, viburnum, vinca, violet, watercress, waxmyrtle, western wallflower, wheatgrass, white-cedar, whitlowgrass, wild sweet-william, willow, wirelettuce, wolfberry, woodwaxen, wyethia, yarrow, yellow-cedar, yellow ironweed, yerba-buena, yucca, zauschneria, and zephyranthes.

Control: If possible, destroy nearby alternate host plants, especially weeds that show rust. A distance of several hundred yards or more between alternate hosts is usually necessary for some measure of control. Collect and burn infected plant parts when first seen. Follow spray program for fruits (Table 16 in Appendix) and certain flowers. Dust or spray with ferbam, Dithane M-45, zineb, maneb, Polyram, folpet, thiram, or sulfur. Start several weeks *before* rust normally is seen. Prune and burn infected tree branches showing rust cankers, galls, spindle-shaped swellings, or witches'-brooms. Use resistant varieties or species when available. Plant only healthy stock. Where rust is perennial in roots (e.g., orange rust of brambles), destroy plants before pustules appear in spring. Indoors, raise temperature, reduce air humidity, and keep water off foliage.

(9) White-Rust, White Blister. Pale yellow areas develop on upper leaf surface with whitish pustules that may yellow with age forming on corresponding underside of leaves (Figure 2.25). Yellowish areas later turn brown. Infections may spread to stems and flowers causing malformation and abortion. Leaves may eventually die; plants are dwarfed.

White rusts are not related to the true rusts above.

Plants Attacked: Alternanthera, alyssum, amaranth, antennaria, artemisia, aubretia, bachelors-button, beet, black-salsify, boltonia, broccoli, Brussels sprouts, cabbage, California-rose, candytuft, cardinal climber, cauliflower, centaurea, Chinese cabbage, collard, cornflower, cypressvine, damesrocket, erysimum, eupatorium, everlasting, feverfew, four-o'clock, froelichia, garden cress, giant night white bloomer, globe-amaranth, groundsel, hearts and honey vine, honesty, horseradish, jacquemontia, kale, lettuce, matricaria, moonflower, morning-glory, mustard, parsnip, peppergrass, plumed thistle, quamoclit, radish, rape, rockcress, rose-moss, rutabaga, salsify, scurvyweed, senecio, spinach, stock, sunflower, sweet alyssum, sweetpotato, thistle, toothwort, trailing four-o'clock, turnip, umbrellawort, wallflower, watercress, and whitlowgrass.

Control: Collect and burn infected plant parts. Destroy tops after harvest. Destroy nearby cruciferous weeds (e.g., mustards, shepherds-purse, charlock, lambsquarters, and pigweed). Spray plants several times, 10 to 14 days apart. Use thiram, maneb, zineb, or copper-containing fungicide.

(10) Leaf Curl or Gall, Leaf Blister, Witches'-broom, Plum Pockets. Conspicuous white, yellow, red, brown, or gray "blisters" on leaves. Leaves may become puffy, puckered, thickened, and curled. Tend to drop early. Often fail to set fruit. Young *peach* fruit may be distorted and cracked. *Plum* fruits turn into enlarged swollen "bladders." Witches'-brooms (brushlike development of many weak shoots arising at or close to the same point) occur on alder, amelanchier, birch, cherry, cherry-laurel, California buckeye, and rhododendron stems (Figure 2.26).

Plants Attacked: Alder, almond, amelanchier, apricot, azalea, bearberry, birch, blueberry, box sandmyrtle, buckeye, camellia, cassiope, catalpa, chamaedaphne, cherry, cherry-laurel, chinquapin, cottonwood, elm, ferns, filbert, flowering cherry, hazelnut, hophornbeam, hornbeam, horsechestnut, huckleberry, Labrador-tea, leucothoë, lyonia, madrone, manzanita, maple, mountain-laurel, nectarine, oak, peach, pear, plum, poplar, rhododendron, sumac, and willow.

Control: Prune to increase air circulation. Remove witches'-brooms. Apply single *dormant* spray to stone fruit trees before buds "break open" in spring. Then follow regular spray program (Table 16 in Appendix); see plant involved.

(11) Smut—Leaf, Stem, Anther, and Seed. Dark brown to black, sooty, spore masses formed inside swollen, whitish blisters or galls. Appear on leaves, stems, bulbs, flower parts, and seed. Affected parts may wither and die (Figure 2.27). Plants may be stunted. Many smut-producing fungi enter plants in the seedling stage and develop systemically within the plant as it grows. Smut may not be evident externally until near maturity. Other smuts (e.g., corn and violet) are localized and any actively growing tissue (shoots, leaves, tassels, young ears) may become infected. See (13) White Smut.

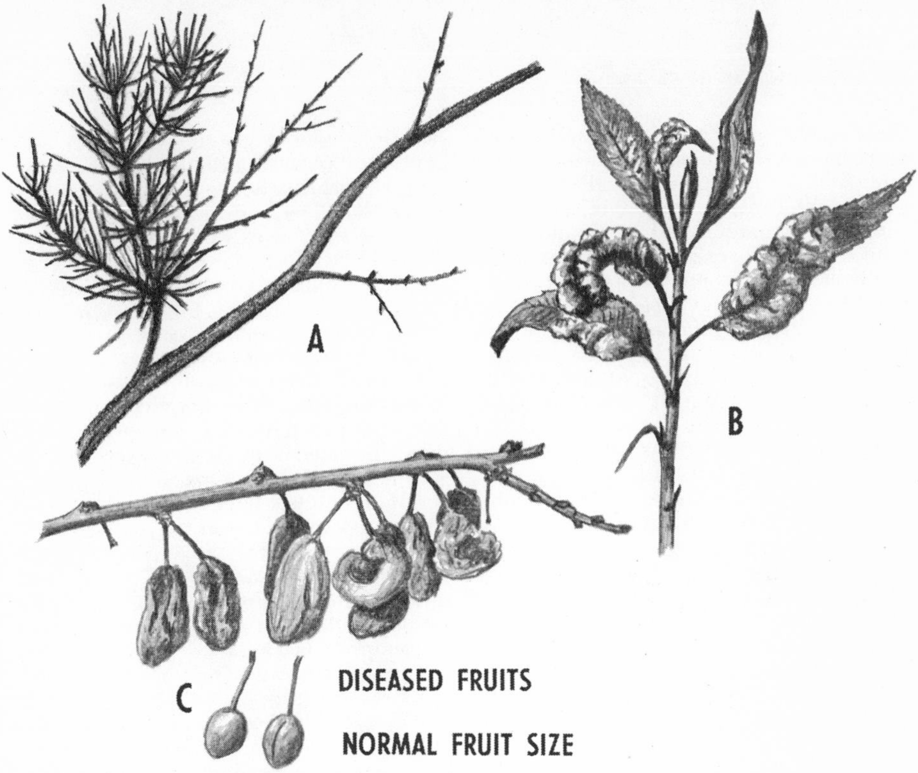

A
B
DISEASED FRUITS
C
NORMAL FRUIT SIZE

Fig. 2.26. A. Witches'-broom of cherry. B. Peach leaf curl. C. Plum pockets. These diseases are all caused by very closely related fungi.

EAR
STALK
A
YOUNG
DISEASED
HEALTHY
B

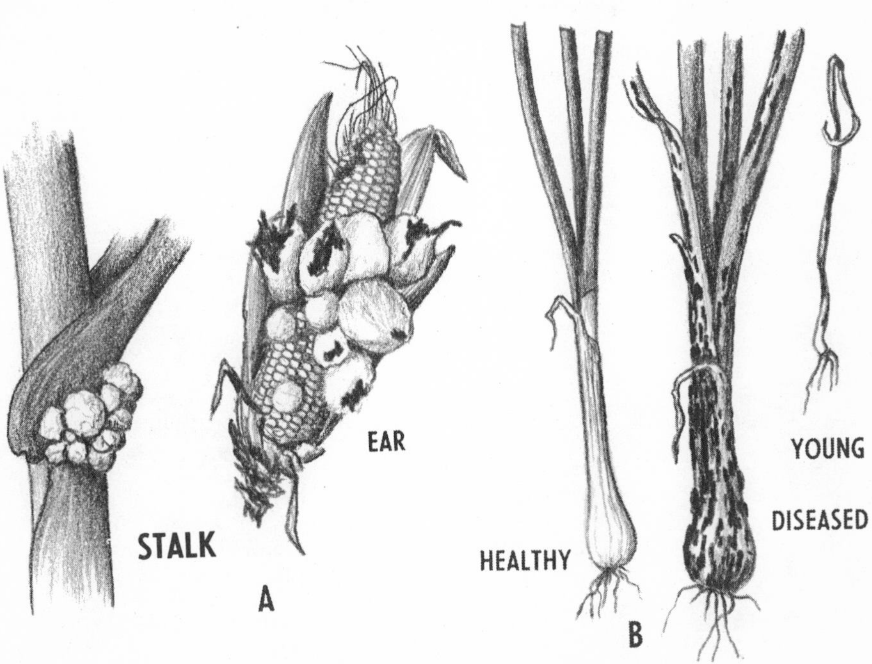

Fig. 2.27. Smut. A. Corn. B. Onion.

Plants Attacked: Aconitum, anemone, arenaria, avens, bamboos, baneberry, bentgrass, Bermudagrass, bloodleaf, bluegrass, buffalograss, buttercup, camass, campion, carnation, chives, cimicifuga, cissus, clematis, colchicum, columbine, coralbells, corn, cushion-pink, delphinium, dianthus, dogstooth-violet, ferns, fescue grass, fig, garlic, gilia, gladiolus, globeflower, grape-hyacinth, hepatica, heuchera, leek, liverleaf, lupine, Maltese cross, meadowrue, monkshood, onion, oxalis, pansy, plumed thistle, red campion, redtop, rue-anemone, ryegrass, shallot, silver lacevine, speedwell, spurge, squill, sweetwilliam, thistle, trillium, tulip, venuslookingglass, violet, and wheatgrass.

Control: Pick off and burn infected parts before blisters open. Use disease-free transplants. Treat seed or grow resistant varieties, where available. See plant in question and Table 20 in the Appendix.

(12) Sooty Mold or Blotch, Black Mildew. Unsightly, superficial, dark brown or black blotches or coating on leaves, fruit, and stems (Figure 2.28). Can be removed easily by rubbing. Fungus usually grows on "honeydew" excretions made by insects (e.g., aphids, scales, whiteflies, mealybugs and others) or in flowing sap. Causes little if any damage to most plants. Subtropical plants, however, are sometimes permanently disfigured.

Plants Attacked: Alder, American bladdernut, anisetree, apple, arborvitae, ash, avocado, aster, azalea, bamboos, bearberry, beech, bignonia, blackberry, blueberry, boxelder, buckthorn, buckwheattree, butterflybush, butternut, cacti, calendula, California-laurel, callicarpa, camellia, camphor-tree, Carolina-jessamine, catalpa, checkerberry, chinaberry, cimicifuga, citrus, clematis, coffeeberry, columbo, confederate-jasmine, cottonwood, crabapple, crapemyrtle, dahoon, dewberry, dogwood, duranta, elm, English ivy, eugenia, ferns, fig, figmarigold, fir, flowering cherry, Franklin-tree, gardenia, gaultheria, grape, grapefruit, hawthorn, hazelnut, holly, huckleberry, hyacinthbean, iceplant, inkberry, ixora, juniper, Kentucky coffeetree, lantana, leatherwood, lemon, lemon-verbena, leucothoë, linden, linnaea, lippia, lyonia, loblollybay, magnolia, mango, manzanita, maple, meconopsis, mockorange, mountain-ash, oak, oleander, olive, onion, orange, orchids, osmanthus, palms, pansy, parkinsonia, partridge-berry, pawpaw, peach, pear, penstemon, persimmon, philodendron, phlox, pine, plum, poplar, pricklyash, privet, raspberry, redbay, redcedar, rhododendron, salal, sassafras, serviceberry, silktassel-bush, spicebush, swampbay, sweetpotato, sycamore, tasseltree, tree-of-Heaven, tuliptree, twinflower, viburnum, violet, walnut, waxmyrtle, willow, wintergreen, yaupon, and yellowjessamine.

Fig. 2.28. Sooty mold on A. Magnolia. B. Apple. C. Basswood or Linden.

Control: Control insects by applying dusts or sprays containing malathion plus lindane, DDT, sevin, or methoxychlor. Check with your county agent, extension entomologist, florist, or nurseryman for latest information on insecticides. Follow spray schedules for fruit (Table 16 in Appendix). Avoid wounds. Paint injured areas promptly with tree wound dressing (pages 34–36).

(13) White Smut, Leaf Smut. Pale, colorless, white, yellowish to yellowish-green leaf spots that later turn dark brown to black (Figure 2.29). Spots may drop out leaving ragged holes. Usually a minor problem. Disease is apparently favored in some cases by acid soils and late planting. See also (11) Smut.

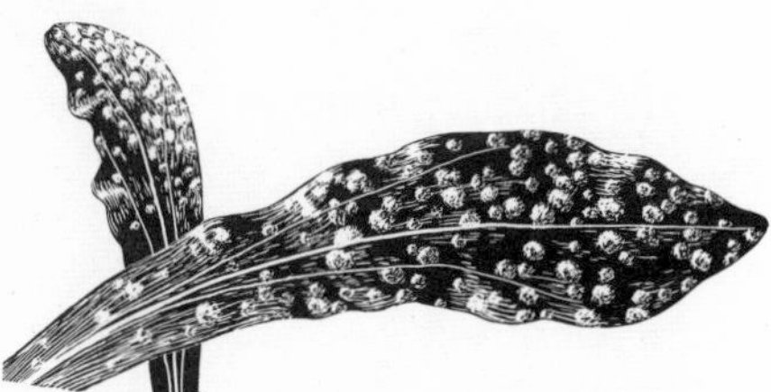

Fig. 2.29. White smut on calendula.

Plants Attacked: Anemone, arnica, aster, bentgrass, bluegrass, boltonia, browallia, buttercup, butter-and-eggs, calendula, California-poppy, chicory, Chinese lanternplant, chrysanthemum, collinsia, dahlia, delphinium, erigeron, eryngium, eupatorium, firewheel, fleabane, gaillardia, goldenglow, groundcherry, groundsel, lobelia, meadowrue, mertensia, moonseed, pondlily, poppy, prairie-coneflower, ratibida, redtop, rudbeckia, senecio, silphium, sneezeweed, speedwell, spinach, sunflower, treepoppy, waterlily, and wood anemone.

Control: Collect and burn diseased parts as they appear. Burn debris after harvest. Spray with captan or copper-containing fungicide at 7- to 10-day intervals. Follow suggested cultural practices.

(14) Scab. Symptoms variable depending on plant. Usually appears as roughened, crustlike (often raised or sunken) areas on surface of leaves, stem, fruit, root, tuber, or corm (Figure 2.30). Leaves may wither and drop early. Twigs often die back (willow). Caused by a few bacteria and wide range of fungi. Easily confused with certain leaf or fruit spots and blights. See (1) Fungus Leaf Spot, (3) Leaf Blight, and (32) Fruit Spot and Rot.

Plants Attacked: Almond, apple, apricot, avocado, beet, cabbage, cacti, camphortree, cantaloup, carrot, casaba, cherry, citron, citrus, coreopsis, cotoneaster, crabapple, crocus, cucumber, dahlia, eggplant, English ivy, euonymus, flowering cherry, freesia, gherkin, gladiolus, gooseberry, grapefruit, guava, hardy orange, hawthorn, Hercules-club, hickory, honeydew melon, iris, jasmine, lemon, lima bean, lime, loquat, mangel, mango, mangold, maple, mountain-ash, muskmelon, nectarine, oleander, onion, opuntia, orange, palms, pansy, parsnip, pea, peach, pear, pecan, persimmon, photinia, plum, poinsettia, poplar, potato, pumpkin, pyracantha, quince, radish, rape, rutabaga, salsify, snowberry, spinach, squash, sugar beet, sweetpotato, Swiss chard, tickseed, tigerflower, turnip, violet, walnut, watermelon, and willow.

Control:

A. For *root crops*—Use resistant varieties. Plant disease-free tubers (potatoes). Acidify soil (to about pH 4.5) where practical. Avoid liming, applications of wood ashes, other alkaline materials, or barnyard manure. Practice long rotation with nonsusceptible crops. Fertilize liberally. Keep down weeds. Turn under green manure crop (page 23) where practical.

B. For *other vegetables*—Long crop rotation. Treat seed (Table 20 in Appendix); plant resistant varieties; spray with zineb, maneb, ziram, or captan.

C. For *fruits and trees*—Use resistant varieties if available. Remove and destroy dead twigs and branches and rotted fruit in dormant season. Follow spray program (Appendix, Table 16). Use captan, sulfur, zineb, or other fungicide.

D. *Flowers*—Sanitation. Spray as for vegetables. Plant disease-free corms, tubers, bulbs, etc. (Table 20 in Appendix).

(15) Wilts. Wilting is due to temporary or permanent water deficiency in foliage and fruit. Wilt diseases (causing permanent wilt) are confused with root or crown rots, stem cankers, grub or borer injury, drought, baking or compacting of soil, etc. The overall result is the same—a wilting, withering, and dying of foliage *beyond* the point of injury. Only by close observation and experience can you de-

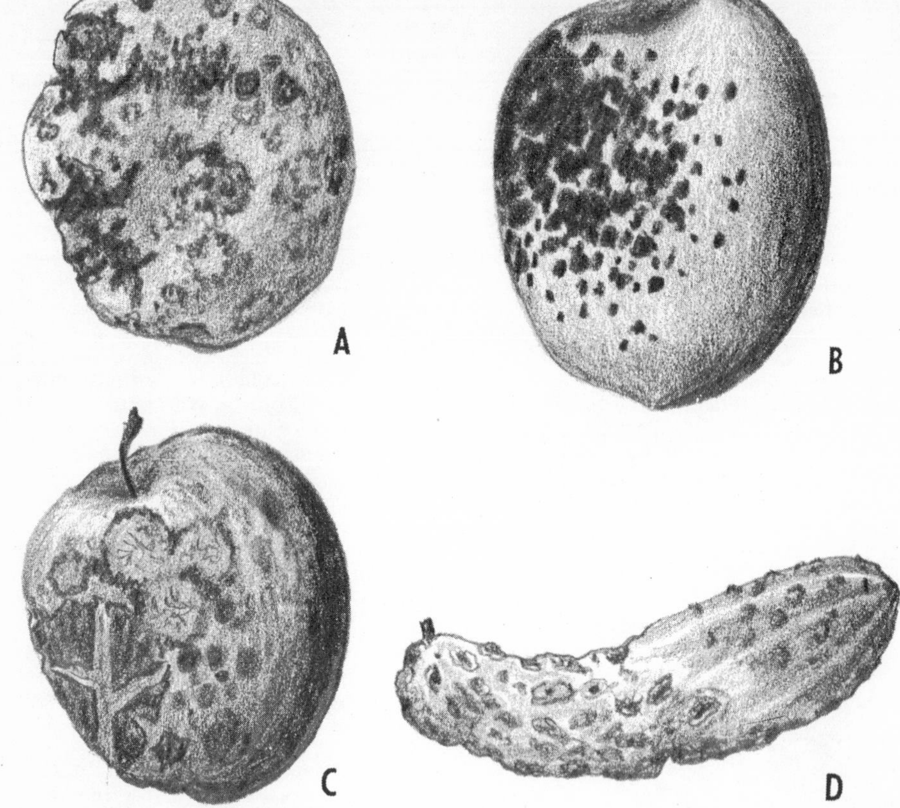

Fig. 2.30. Scab. A. Potato. B. Peach. C. Apple. D. Cucumber. The name scab is given to diseases caused by entirely different organisms. Control measures also vary greatly.

termine the true cause. See under Stem and Root Diseases. Sprays or dusts are ineffective in controlling wilts. Most wilt diseases are specific to one group of plants.

There are three common types of wilts that may look identical, but are caused by altogether different organisms. These are Fusarium Wilt, Verticillium Wilt, and Bacterial Wilt. Wilt-producing fungi and bacteria invade water and food-conducting vessels (vascular system) inside stems and roots. Vessels become plugged and killed or nonfunctional. Normal flow of liquid from roots to foliage is greatly reduced or stopped. Wilting of infected branches then follows.

Cutting into infected stems or branches commonly shows discolored streaks. Some plants die quickly from wilts; others withstand attacks for months or years. For wilts that attack only a few plants, like Oak Wilt and Dutch elm disease, see under plant involved.

A. Fusarium Wilt or Yellows. Plants usually stunted and yellow. Disease symptoms often start at base of stem and progress upwards causing leaves and flower heads to wilt and die (Figure 2.31). Lower parts of stem (inside always, outside sometimes) are dark and discolored. When stems are cut lengthwise, brown to black streaks are evident inside. Infected seedlings wilt and collapse. See (21) Crown Rot. Root injury may induce symptoms to appear more rapidly on certain plants. Nematodes (e.g., burrowing, lesion, root-knot, sheath, sting, stubby-root) often provide wounds by which the fusarium fungus may enter roots.

Plants Attacked: Alternanthera, asparagus, asparagus-fern, astilbe, bachelors-button, bean, beet, black-eyed pea, bleeding-heart, broccoli, browallia, Brussels sprouts, cabbage, cacti, cantaloup, cape-marigold, carnation, carrot, casaba, catnip, cauliflower, celery, centaurea, China-aster, chrysanthemum, cineraria, citron, clarkia, collard, coriander, cornflower,

cosmos, crotalaria, cucumber, cyclamen, dahlia, daphne, delphinium, dill, eggplant, eupatorium, fig, foxglove, freesia, garlic, gladiolus, gourds, groundsel, hebe, ixia, kale, kohlrabi, lantana, leek, lentil, lettuce, lupine, marigold, "mimosa" tree, mock-cucumber, morning-glory, muskmelon, mustard, okra, orchids, painted-tongue, pansy, parsley, pea, peanut, pepper, petunia, pinks, poinsettia, polemonium, potato, pumpkin, radish, rutabaga, safflower, seakale, sedum, "smilax," snapdragon, speedwell, spinach, squash, stock, sumac, sunflower, sweetpea, sweetpotato, sweet-william, tithonia, tomato, tritonia, turnip, vegetable-marrow, vetch, watermelon, and zinnia.
Control: Use normally resistant varieties

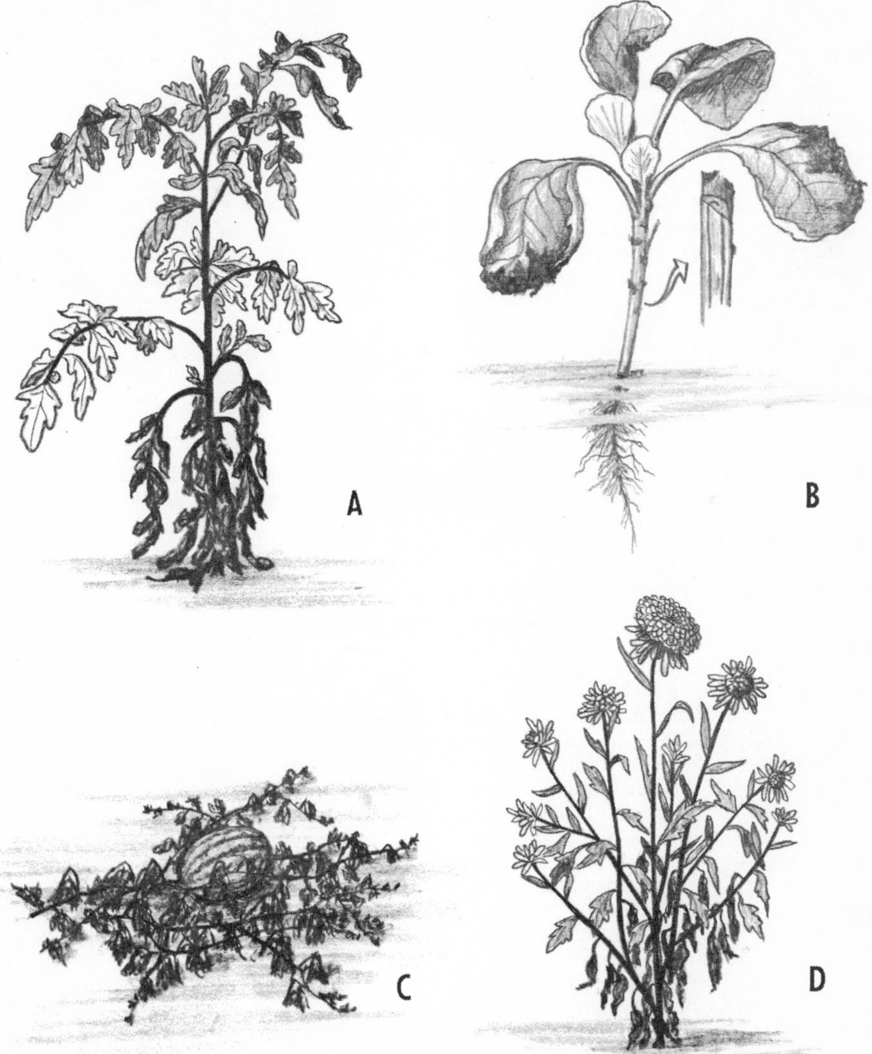

Fig. 2.31. Fusarium wilt. A. Tomato. B. Cabbage (often called cabbage yellows). C. Watermelon. D. Aster. The inside of diseased stems shows dark streaks when cut (see cabbage insert) where the fusarium fungus causes a partial to complete stoppage of liquids in the water-conducting tissue.

if available: African marigold, bean, carnation, China-aster, chrysanthemum, cabbage, gladiolus, "mimosa" tree, muskmelon, pea, pepper, sweetpea, tomato, and watermelon. Treat seed or plant in disinfested soil (Tables 20 and 22 in Appendix). Collect and burn infected plants. Practice long crop rotation excluding members of same family. Fertilize and water to encourage vigorous growth.

B. Verticillium Wilt. Leaves turn pale green to yellow and later brown—usually falling first from lower part of plant, then progressively upward. Infected trees and shrubs show one or more wilted branches with severe dropping of leaves from midsummer on. Annual plants and young trees are often stunted, wilted, and usually die. Perennials may die or recover. Twigs and branches often die back. Internal tissues of infected stems show dark streaks (greenish, gray, yellowish-brown, brown, bluish-brown, purplish, or black) when cut lengthwise (Figure 2.32). Verticillium Wilt is more prevalent in cooler climates and at lower soil temperatures than Fusarium Wilt. Infections often occur through wounds; e.g., those made by lesion or meadow nematodes, cultivating, transplanting, frost injury, etc.

Plants Attacked: Abutilon, acanthopanax, aconitum, almond, American spikenard, apricot, aralia, artichoke, ash, asparagus, asparagus-bean, aspen, aster, aucuba, avocado, azalea, bachelors-button, balsam, barberry, garden beans, beech, beet, begonia, bellflower, blackberry, black-eyed pea, black gum, black locust, blazing-star, boxelder, boxwood, boysenberry, broccoli, Brussels sprouts, butterfly-flower, cabbage, calceolaria, California-poppy, campanula, camphor-tree, cantaloup, cape-marigold, carnation, casaba, castorbean, catalpa, cauliflower, celeriac, celery, centaurea, chayote, cherry, cherry-laurel, chervil, chestnut, chicory, China-aster, chinaberry, Chinese lanternplant, chives, chrysanthemum, cineraria, citron, clarkia, cockscomb, coleus, collard, cornflower, cotoneaster, crimson daisy, crown-of-thorns, cucumber, currant, dahlia, damesrocket, daphne, delphinium, deutzia, dewberry, dogwood, eggplant, elder, elm, endive, erigeron, erythrina, fleabane, foxglove, fremontia, fuchsia, garlic, geranium, gooseberry, goldenrain-tree, grape, groundsel, heath, hebe, heliotrope, honeylocust, honeysuckle, horsechestnut, horseradish, Jerusalem-cherry, kale, Kentucky coffeetree, kohlrabi, larkspur, leek, lentil, lettuce, liatris, lilac, linden, locust, lupine, magnolia, maple, marguerite, marigold, Michaelmas daisy, mignonette, mint, mockorange, monarda, monkshood, muskmelon, mustard, New Zealand spinach, oak, okra, olive, onion, Osage-orange, osmanthus, painted-tongue, parsley, pea, peach, peanut, pecan, peony, pepper, peppermint, peppertree, persimmon, petunia, phlox, pistachio, pittosporum, plum, polemonium, poplar, poppy, poppy-mallow, potato, privet, prune, pumpkin, quince, radish, raspberry, redbud, rhubarb, rose, Russian-olive, rudbeckia, rutabaga, safflower, salsify, sassafras, scarlet eggplant, serviceberry, Shasta daisy, slipperwort, smoketree, snapdragon, sneezeweed, snowball, sophora, spearmint, spinach, spurge, squash, stock, strawberry, strawflower, sumac, sunflower, sweetpea, sweetpotato, sycamore, tickseed, tuliptree, tomato, Transvaal daisy, tree-of-Heaven, trumpetvine, tupelo, turnip, udo, vegetable-marrow, viburnum, walnut, watermelon, wayfaringtree, wildbergamot, yellowwood, and zinnia.

Control: Dig and burn infected flowers, small fruits, and vegetables. Propagate *only* from disease-free plants or use disease-free seed. Grow in well-drained, clean, or sterilized soil treated with methyl bromide, chloropicrin, Vorlex or Trizone (pages 538–49). Control soil insects using Diazinon, chlordane, etc. Practice long rotation with nonrelated crops. Keep down weeds, especially lambsquarters, purslanes, pigweed, horsenettle, nightshades, and groundcherries. Fertilize and water to encourage vigorous growth. Varieties of certain plants such as potato, blackberry, pepper, eggplant, strawberry, and tomato differ in resistance. For *trees and shrubs:* Remove wilted branches. Sterilize tools between cuts by dipping or swabbing in 70 per cent alcohol. Then paint wound surfaces with tree wound dressing (pages 34–36). Avoid wounding roots or trunk (stem) when planting or cultivating.

Plants resistant or immune to Verticillium wilt include: Cypress, fir, ginkgo, larch, juniper, pine, sequoia, spruce, gladiolus, lawngrasses, iris, lily, onion, orchids, palms, ferns, eucalyptus, fig, walnut, sweetgum, apple, pyracantha, pear, oaks, willows, citrus, California-laurel, manzanita, oleander, boxwood, mulberry,

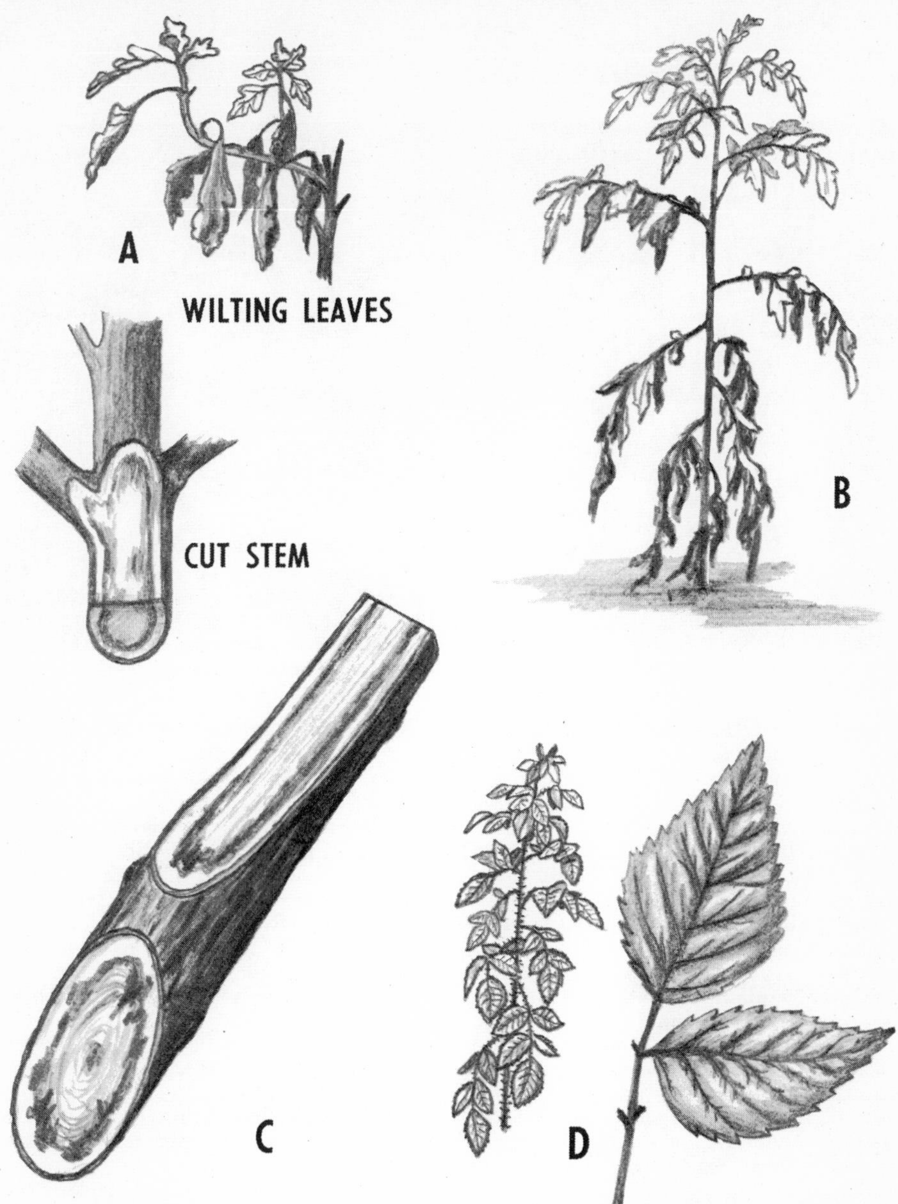

Fig. 2.32. Verticillium wilt. A. Eggplant. B. Tomato. C. Maple. D. Raspberry. Like Fusarium, the Verticillium wilt fungus plugs or otherwise disrupts the water-conducting tissue (see eggplant and maple). The result of this appears as dark streaks inside the stem. It is often difficult, without a microscopic examination, to tell which wilt-producing fungus is involved.

celery, asparagus, carrot, corn, sweetpotato, lettuce, beans, pea, ceanothus, soybean, cereals, and grasses.

C. Bacterial Wilt, Brown Rot, or Blight. Symptoms variable. See under plant involved. Some plants show dark green, water-soaked areas in leaves that expand rapidly. Leaves turn brown and dry. A shiny "crust" or "scale" may be evident on affected parts. Plants may be stunted, wilt suddenly or gradually, starting with younger leaves (Figure 2.33), or there may be a slight yellowing of older leaves. Stems often shrivel and dry out. Yellowish slime or water-soaked browning may be evident when stem base is cut through

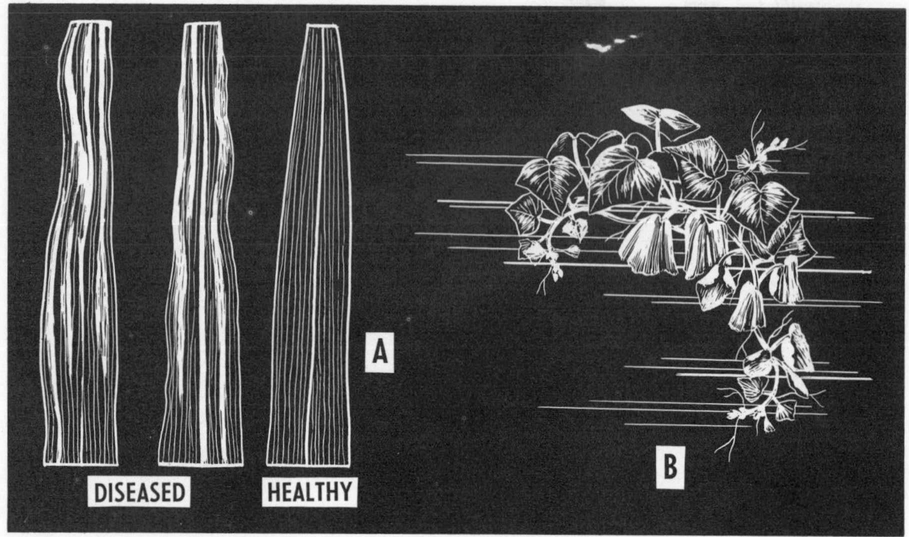

Fig. 2.33. Bacterial wilt. A. Corn. B. Cucumber. A sticky ooze is often evident when bacterial wilt-infected stems are cut and squeezed.

and squeezed. Bacterial Soft Rot often follows in wet weather. See (29) Bacterial Rots. Wilt-producing bacteria often enter through insect, nematode, and other mechanical wounds.

Plants Attacked: Aster, balsam, bean, beet, bird-of-paradise flower, canna, cantaloup, carnation, carrot, casaba, castor-bean, Chinese lanternplant, chrysanthemum, corn, cosmos, croton, cucumber, dahlia, datura, delphinium, dieffenbachia, eggplant, forsythia, geranium, gherkin, gourds, groundcherry, hibiscus, hollyhock, hydrangea, larkspur, lettuce, lilac, marigold, muskmelon, nasturtium, nicotiana, okra, palm, pea, peanut, pepper, petunia, potato, pumpkin, rhubarb, squash, sunflower, tomato, vegetable-marrow, verbena, wallflower, watermelon, and zinnia.

Control: Collect and burn infected plants. Control insects that may spread the bacteria. Use mixture of Dibrom (naled) or sevin plus malathion. Resistant varieties are available for a few plants including cucumber, corn, carnation, and potato. Sow disease-free seed or other propagative parts, or treat before planting. See plant in question and Table 20 in Appendix. Long crop rotation (6 years or more), excluding susceptible plants and weeds, e.g., horsenettles, cocklebur, ragweeds, horseweed, Spanish-needles, and physalis. Grow in well-drained, fertile soil that is clean or sterilized. Indoors, raise temperature until disease is under control. Take cuttings from healthy plants.

(16) Mosaic, Mottle, Crinkle, Streak, Calico, Virus Leaf Curl, Infectious Variegation, Flower Breaking. This virus disease or complex shows variable symptoms. Most commonly leaves have mild to severe yellowing or pattern of light and dark green areas forming a mosaic or mottle. Sometimes yellowish or white, ring or line patterns or both may be seen. Leaves commonly curled, distorted, puckered, crinkled, leathery, or even cupped sharply downward or upward. Leaf veins may be lighter than normal (cleared) or be banded with dark green or yellow areas.

Flowers may be blotched or streaked with white or yellow, distorted, or fail to open normally (Figure 2.34).

Plants may be stunted; fruit deformed, stunted, fewer in number, usually lack flavor. Infected plants often show no external symptoms, especially at high temperatures (85° F. or above).

Mosaiclike diseases are commonly spread by many species of aphids (Figure 4.78), some by contact—all by propagation (grafting, budding, slips) from infected plants.

Mosaics are often confused with plant nutrient deficiencies.

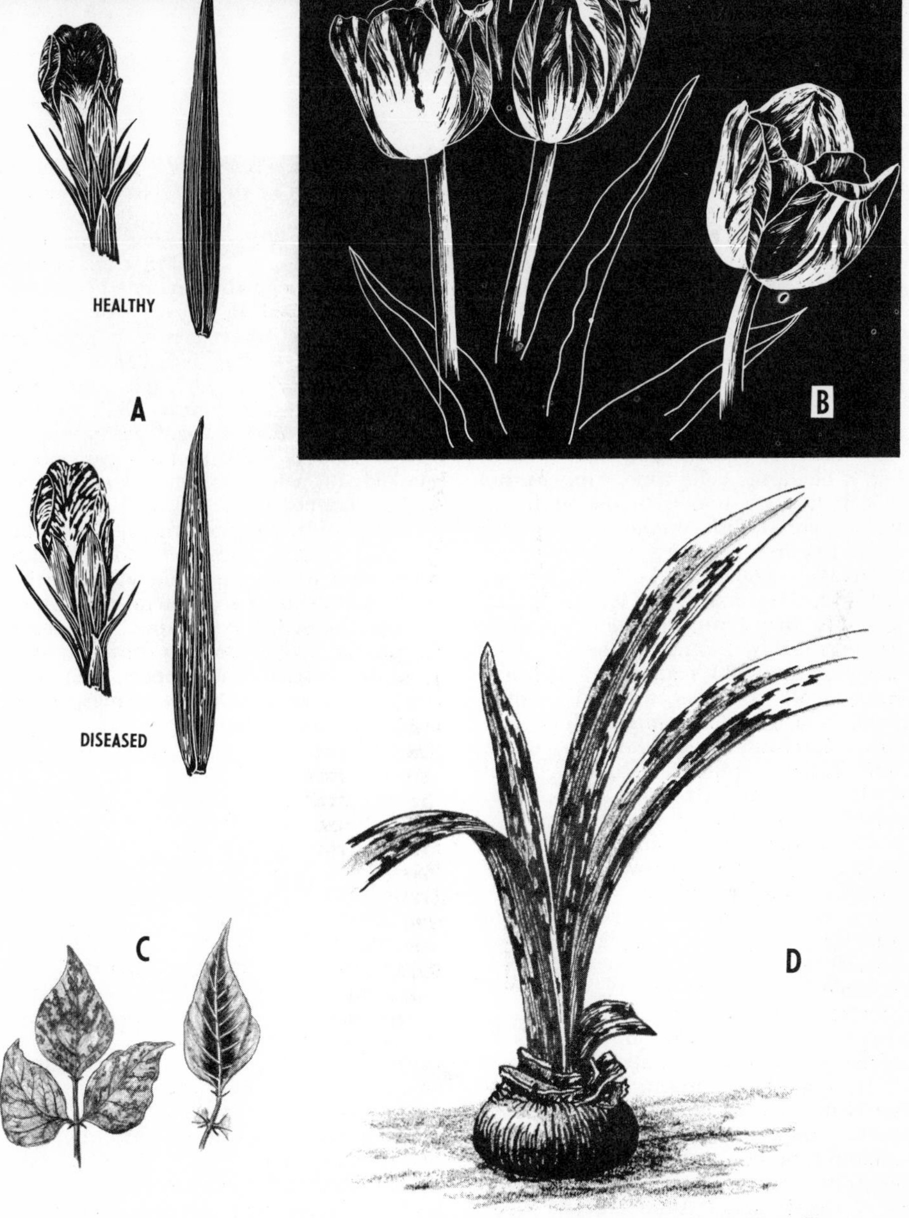

Fig. 2.34. Mosaic. A. Carnation. B. Tulip flower-breaking. C. Bean. D. Amaryllis.

Plants Attacked: Abutilon, aconitum, African-lily, African-violet, agrimony, allium, almond, amaranth, amaryllis, Amazon-lily, amsonia, anchusa, anemone, apple, apple-of-Peru, apricot, asparagus-bean, aster, babiana, balsam, balsam-apple, barberry, basil, bean, beet, begonia, belamcanda, bellflower, blackberry, black-eyed pea, blackberry-lily, blueberry, bluegrass, bougainvillea, boysenberry, broccoli, Brussels sprouts, buttercup, butterfly-bush, butterflyweed, cabbage, calabash, calceolaria, calendula, calla lily, candytuft, canna, cantaloup, cape-cowslip, cape-marigold, caraway, carnation, carrot, casaba, catnip, cauliflower, celeriac, celery, chayote, cherry, chervil, chicory, China-aster, Chinese cabbage, chincherinchee, Chinese hibiscus, Chinese lantern-plant, chrysanthemum, cineraria, citron, clematis, cockscomb, coleus, columbine, coriander, corn, cornflower aster, cosmos, cranesbill, crimson daisy, crinum, crocus, crotalaria, cucumber, currant, daffodil, dahlia, damesrocket, daphne, datura, delphinium, dewberry, dill, eggplant, elm, emilia, endive, erysimum, eucalyptus, euonymus, evening-primrose, false-garlic, fennel, fig, fleabane, flowering cherry, flowering raspberry, flowering tobacco, foxglove, freesia, fritillaria, gaillardia, garden cress, garlic, geranium, gherkin, gilia, gladiolus, goldenchain, goldenglow, goldenrod, gooseberry, gourds, grape, groundcherry, ground-ivy, gum, hackberry, heliopsis, heliotrope, hibiscus, hollyhock, honesty, honeydew melon, honeysuckle, horseradish, houndstongue, hyacinth, hyacinth-bean, iris, ixia, Jerusalem-cherry, kalanchoë, kale, kohlrabi, lantana, larkspur, lavatera, leek, lentil, lettuce, lilac, lily, lima bean, lithospermum, lobelia, loganberry, lupine, mallow, mangel, mangold, marigold, matrimony-vine, mertensia, mock-cucumber, monarda, monkshood, morning-glory, muskmelon, mustard, narcissus, nasturtium, nectarine, New Zealand spinach, nicotiana, nightshade, onion, orchids, oxeye daisy, pansy, parsley, parsnip, passionflower, pea, peach, peanut, pear, penstemon, peony, pepper, peppergrass, periwinkle, petunia, phacelia, phlox, pinks, pittosporum, plum, potato, primrose, privet, proboscisflower, pumpkin, purple-coneflower, radish, rape, raspberry, rhoea, rhubarb, rockcress, rose, rudbeckia, rutabaga, safflower, salvia, sassafras, scabiosa, scarlet pimpernel, seakale, senecio, shallot, sida, sidalacea, skyrocket, snapdragon, soapberry, spiderlily, spinach, squash, squill, star-of-Bethlehem, stock, stokesia, strawberry, streptanthera, summer-hyacinth, sunflower, sweet alyssum, sweet marjoram, sweetpea, sweetpotato, sweet-william, Swiss chard, teasel, thimbleberry, tigerflower, tobacco, tomato, tritonia, tulip, turnip, vegetable-marrow, verbena, vetch, vinca, violet, wallflower, wandering-Jew, wandflower, watercress, watermelon, watsonia, wheatgrass, wisteria, and zinnia.

Control: Use resistant varieties where adapted. Control insects, especially aphids, using malathion, lindane, Diazinon, or Dibrom (naled). Use Thiodan for resistant forms. Destroy infected garden plants when first found. They will not recover. Keep down weeds in and around garden area that may harbor viruses. Use virus-free seed and stock—certified or indexed if possible. Destroy crop debris after harvest. Burn or plow under cleanly. Do not propagate from infected plants.

(17) Spotted Wilt, Ringspot. Symptoms vary with virus-plant combination. Leaves often show yellowish or dead concentric rings, oakleaf, zigzag, or watermark patterns. Sometimes with green or yellow centers (Figure 2.35). Young leaves are usually puckered and malformed. Plants are often stunted; yield is sharply reduced. Yet some plants apparently recover. Symptoms are often masked in some species, especially at higher temperatures (85° F. or above), and on older plants. Foliage may be bronzed, with dead spots developing. Stems and petioles may show lengthwise dark streaks or rings. Stem tips often appear blighted, may collapse. Viruses are spread by thrips, possibly other insects (e.g., flea beetles, grasshoppers, etc.), and seed. Tobacco and tomato ringspot viruses are also transmitted by dagger (*Xiphinema*) nematodes; peach ringspot virus, by pollen and seed.

Plants Attacked: Almond, amaranth, amaryllis, Amazon-lily, anemone, apple-of-Peru, apricot, ash, aster, babysbreath, bean, beet, begonia, blueberry, broccoli, Brussels sprouts, browallia, buttercup, butterfly-flower, cabbage, calceolaria, calendula, California-bluebell, California-poppy, calla lily, campanula, candytuft, cantaloup, Canterbury-bells, carnation, carrot, casaba, cauliflower, celeriac, celery, centaurea, cherry, chicory, China-aster, Chinese lanternplant, chrysanthemum, cineraria, cockscomb, columbine, corn, cosmos, crassula, cucumber, currant, dahlia, dandelion, datura, delphinium, eggplant, elder, emilia, endive, erigeron, erysimum, fleabane, flowering tobacco, forget-me-not, foxglove, fuchsia, gaillardia, garden cress, geranium, gladiolus, globe-amaranth, gloxinia, godetia, gourds, groundcherry, honesty, hydrangea, iris, Jerusalem-cherry, kale, kohlrabi, lettuce, lilac, lily, lobelia, lupine, lychnis, mallow, Maltese cross, mayapple, mignonette, moonflower, morning-glory, muskmelon, mustard, narcissus, nasturtium, New Zealand spinach, nicotiana, nightshade, okra, orchids, painted-tongue, pansy, parsnip, pea, peach, penstemon, peony, peperomia, pepper, petunia, pinks, plum, poppy, potato, primrose, privet, pumpkin, radish, rape, raspberry, rhubarb, rockcress, rose, sage, salvia, satin-flower,

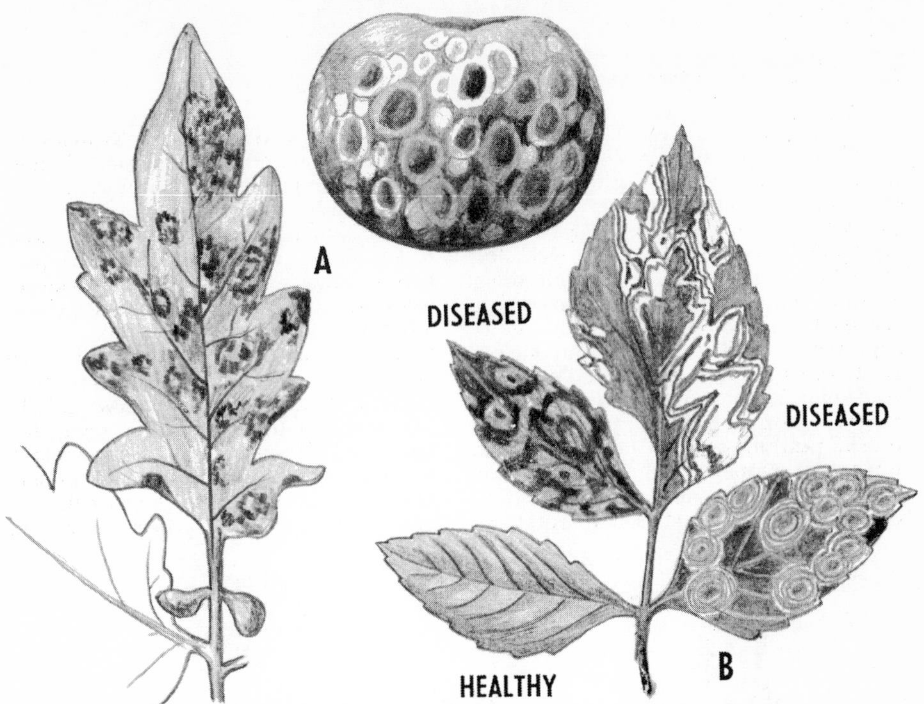

Fig. 2.35. Spotted wilt. A. Tomato. B. Dahlia. Note the wide range of symptoms produced. On dahlia the viruses are sometimes called ringspot, yellow ringspot, and oakleaf disease.

scabiosa, sea-lavender, senecio, slipperwort, snapdragon, spearmint, spinach, squash, stock, strawflower, sunflower, sweet alyssum, sweetpea, sweetpotato, sweet-william, Swiss chard, tidytips, tobacco, tomato, Transvaal daisy, tulip, turnip, vegetable-marrow, verbena, vetch, vinca, violet, wallflower, watermelon, yellow ironweed, and zinnia.

Control: Destroy infected plants when first seen. Keep down weeds in and around garden area, e.g., bindweed, sheep-sorrel, pigweed, lambsquarters, wild groundcherry, purslanes, wild lettuce, false madder, black nightshade, jimsonweed, chickweed, nettle, mallow, galinsoga, small ragweed, daisy fleabane, clovers, milkweeds, thistles, dandelion, etc. Control insects, especially thrips and aphids. Use Dibrom (naled), sevin, or methoxychlor plus malathion. Grow disease-free stock. Destroy crop debris after harvest.

(18) Yellows, Aster Yellows, Rosette, Dwarf, Stunt. Symptoms variable depending on virus-plant combination and weather factors. Entire plants, or certain parts, are often more or less uniformly yellow (sometimes red or purple), stunted, or dwarfed. They may wilt and die early. Leaves and shoots are often slender and stunted forming upright "rosettes." Fruit may ripen prematurely and usually lacks flavor or may be "warty." Flowers may be greenish-yellow, dwarfed, aborted, or absent (Figure 2.36). Viruses spread primarily by leafhoppers (Figure 4.26) —or aphids for a few viruses—and by propagating infected stock that may appear normal.

Plants Attacked: Allium, almond, alyssum, amaranthus, anagallis, anchusa, anemone, anise, apricot, avens, babysbreath, bachelors-button, basketflower, bayberry, bean, beet, begonia, bellflower, blackberry, black-salsify, blueberry, blue laceflower, broccoli, browallia, bur-marigold, buttercup, butterfly-flower, cabbage, calceolaria, calendula, California-poppy, camomile, campanula, canna, Canterbury-bells, cape-marigold, caraway, cardoon, carnation, carrot, cauliflower, celeriac, celery, centaurea, cherry, chicory, China-aster, Chinese cabbage, chrysanthemum, cineraria, clarkia, clockvine, corn, cornflower, corn-marigold, cosmos, crimson daisy, cucumber, dahlia, daisies, dandelion, datura, delphinium, dewberry, dianthus, dill, eggplant, endive, English daisy, erigeron, escarole, eupatorium, fennel, firewheel, flax, fleabane, forget-me-not, flowering tobacco, gaillardia, garlic, gaura, geranium, gilia, gladiolus, globe-amaranth, globe artichoke, gloxinia, godetia, goldenglow, grape, groundsel, gypsophila, heronsbill, kochia, leek, lettuce, lily, loganberry, love-lies-bleeding, mallow, mangel, mangold, marguerite, marigold, matricaria, mignonette, monkeyflower, mullein-pink, muskmelon, mustard, nasturtium, nectarine, New Zealand spinach, onion, oxeye daisy, painted-tongue, pansy, parsley, parsnip, pea, peach, peanut, pepper, petunia, phlox, pimpernel, piqueria, plum, potato, primrose, prune, pumpkin, pyrethrum, radish, rape, raspberry, rudbeckia, rutabaga, sage, salsify, salvia, sassafras, satinflower, scabiosa, sea-lavender, senecio, shallot, slipperwort, sneezeweed, speedwell, spinach, squash, statice, strawberry, strawflower, summer-cypress, sunflower, Swan River daisy, sweet alyssum, sweetpea, sweet-william, Swiss chard, tickseed, toadflax, tomato, vegetable-marrow, vinca, wallflower, watercress, and zinnia.

Control: Same as for Mosaic and Spotted Wilt (both above). Control insects, especially leafhoppers (and a few aphids) that transmit the viruses. Use sevin, lindane, or Dibrom (naled) plus malathion at least weekly, when insects are present or expected. Asters, certain other flowers, and lettuce are often grown under fine

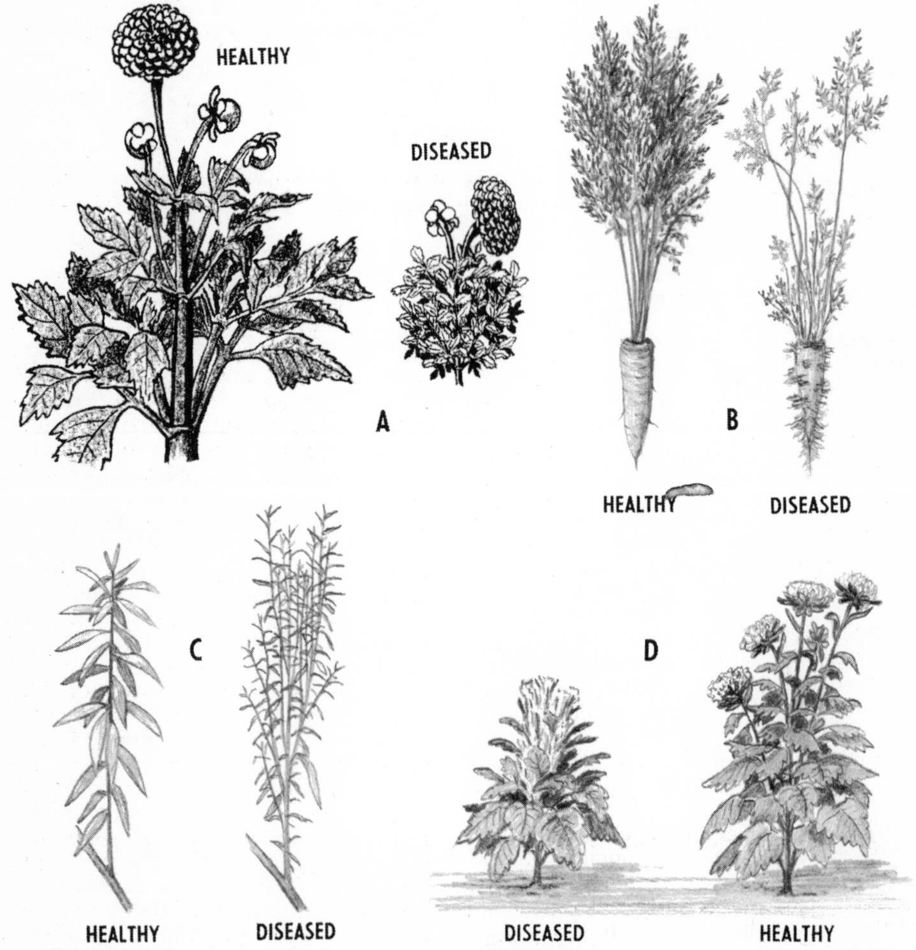

Fig. 2.36. Yellows or stunt. A. Dahlia stunt, dwarf, or mosaic. B. Aster yellows on carrot. C. Peach yellows. D. Aster yellows on aster.

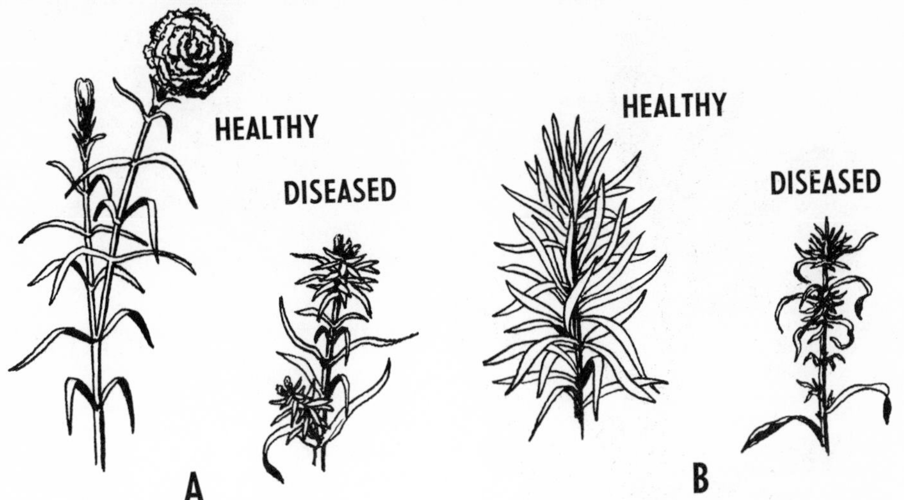

Fig. 2.37. Curly-top. A. Carnation. B. Strawflower.

cheesecloth (22 threads per inch) or wire screening (18 threads per inch) to keep out insects.

(19) Curly-Top, Western Yellow Blight. Common and destructive to many garden and weed plants in the western United States and other areas where light intensity and summer temperatures are high and relative humidity is low. Symptoms vary with virus-plant combination. Plants are usually stunted or dwarfed with leaves mottled, bunched, curled downward, rolled, thickened, and yellowed (Figure 2.37). Flowers and buds may drop early.

Plants Attacked: Alyssum, amaranthus, anchusa, bean, beet, black-eyed pea, broccoli, buttercup, cabbage, campanula, cantaloup, carnation, carrot, casaba, cauliflower, celeriac, celery, chervil, China-aster, Chinese cabbage, Chinese lantern-plant, citron, cockscomb, collard, columbine, coriander, cornsalad, cosmos, cress, cucumber, delphinium, dill, eggplant, fennel, flowering flax, flowering tobacco, four-o'clock, foxglove, geranium, gherkin, globe-amaranth, godetia, groundcherry, gourds, heliotrope, heronsbill, honeydew melon, horseradish, kochia, larkspur, lettuce, lima bean, lobelia, mallow, mangel, marguerite, mignonette, mock-cucumber, morning-glory, muskmelon, mustard, nasturtium, New Zealand spinach, nicotiana, okra, oxalis, pansy, parsley, parsnip, pepper, peppergrass, petunia, phacelia, pinks, poppy, potato, pumpkin, radish, rhubarb, rose-moss, rutabaga, safflower, salad chervil, salsify, scabiosa, Shasta daisy, spiderflower, spinach, squash, stock, strawberry, strawflower, sweetpotato, sweet-william, Swiss chard, tepary bean, tickseed, tomato, turnip, vegetable-marrow, veronica, vetch, vinca, violet, watermelon, and zinnia.

Control: Same as for Yellows. Plant as early as possible or at time recommended for your area. Avoid migrations of beet leafhoppers (Figure 4.26) that transmit the virus. Resistant varieties are available for certain crops.

(20) Leaf, Bud, Stem, and Leaf Gall Nematodes. Symptoms variable with plant host. Nematodes live over winter in soil or in infested leaves and stems. During the growing season, nematodes may swim up stem in film of water. Infestation only occurs when plants are wet from rain, sprinkling, heavy dew, and fog (Figure 2.38).

1. *Begonia, chrysanthemum, calendula, ferns, lantana, orchids*—Dark brown to black, angular, or wedge-shaped areas on leaves. Often delimited by leaf veins. Leaves may wither and die starting at base of plant.
2. *Strawberry*—Leaves often stunted, narrow, twisted, crinkled, and cupped (*spring and summer crimp or dwarf*).
3. *Bellflower, sweet-william*—Leaves very narrow, crinkled, wavy, often brittle. Stems may be swollen near tops, or curved sideways. Infested plants are stunted, fail to bloom, or may die prematurely.
4. *Lily*—Leaves bronzed, gnarled, blotched, and curled tightly downward *(bunchy top)*, later die back.

See (38) Bulb Nematode.

Plants Attacked: African-violet, anemone, aster, balsamroot, beet, begonia, bellflower, bentgrass, bouvardia, butter-and-

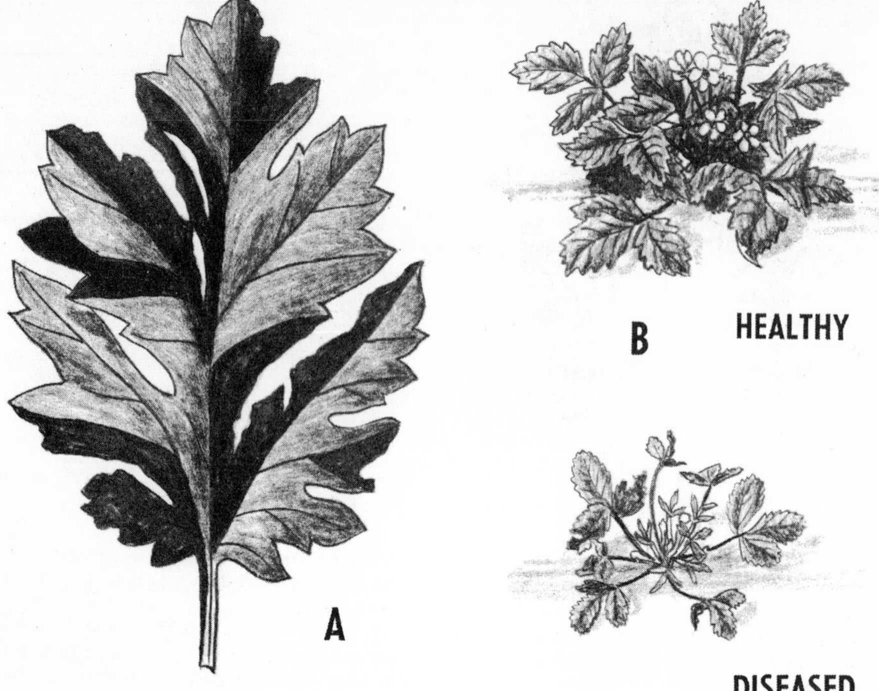

Fig. 2.38. Leaf, stem, and leaf gall nematodes. A. Chrysanthemum foliar or leaf nematode. B. Strawberry dwarf or crimp.

eggs, buttercup, butterfly-flower, calceolaria, calendula, campanula, carrot, carnation, celeriac, celery, Chinese lantern-plant, chrysanthemum, coleus, coralbells, crassula, cyclamen, daffodil, dahlia, daisy, dandelion, delphinium, evening-primrose, ferns, four-o'clock, foxglove, garden cress, garlic, geranium, gladiolus, glory-of-the-snow, gloxinia, gooseberry, grape-hyacinth, groundcherry, ground-pink, groundsel, heuchera, hyacinth, hydrangea, iris, lantana, leopardsbane, lily, loosestrife, lupine, monkeyflower, moonflower, narcissus, onion, orchids, oxalis, oxeye daisy, pansy, parsley, parsnip, penstemon, peony, peperomia, peppergrass, phlox, pinks, poppy, primrose, potato, privet, pyrethrum, radish, rape, rhubarb, salsify, salvia, schizanthus, senecio, slipperwort, snowdrop, strawberry, sunflower, sweetpotato, sweet-william, teasel, toadflax, tomato, tulip, verbena, violet, wyethia, and zinnia.

Control: Buy clean plants, certified and heat-treated if possible. Take *tip* cuttings from healthy plants. Collect and burn infested plants or plant parts as soon as noticed. Burn tops at end of season. Avoid overhead watering. Keep water off foliage. Apply a dry mulch. Practice 3- or 4-year rotation. Where practical, apply malathion or Diazinon twice weekly until nematodes are controlled. Certain potted plants, such as African-violet, begonia, and ferns may be dipped in hot water to rid them of nematodes. Chrysanthemum varieties differ in resistance.

B. Stem Diseases

(21) Crown, Foot, Stem, Stalk, Collar, or Rhizome Rot; Stem Blight, Southern Blight, Damping-off. Plants generally first unthrifty with leaves smaller and lighter green than normal. Leaves may later yellow or wilt, wither, and curl during hot, dry periods. Base of stem (or trunk) may be water-soaked, discolored, and decayed. Leaves then wilt, yellow, wither, and eventually die. Roots may decay (Figure 2.39). Seedlings usually wilt, collapse, and die (damping-off) before or after emergence. Stand may be poor. Seedlings are susceptible longer to damping-off in soil infested with parasitic nematodes. A cottony, or other type of mold, may grow over affected plant parts. Variously shaped, tan to black bodies (sclerotia) may form in cottony growth. Plants wilt and gradually or suddenly die when basal rot shuts off supply of water and nutrients to aboveground parts. See (2) Bacterial Spot, (6) Downy Mildew,

(29) Bacterial Soft Rot, (32) Fruit Spot, and (34) Root Rot.
Plants Attacked: Practically all.
Control: Plant in light, well-drained, well-prepared soil or sterile rooting medium (pages 538–548), such as sand, soil, vermiculite, Perlite, or sphagnum moss. Where possible, keep soil on dry side. Avoid overcrowding, overwatering, too deep planting, and overfertilizing (especially with nitrogen). Water seedlings at 5- to 7-day intervals with ferbam, ziram, captan, thiram, zineb, or folpet (1 to 1½ tablespoons per gallon of water). Use about ½ pint per square foot of bed surface. Where possible, sterilize soil with heat or chemicals before planting. Buy disease-free, crack-free seed of vegetables and flowers. Treat seed with captan, thiram, dichlone, or chloranil before planting. See plant involved and Table 20 in Appendix. Practice as long a crop rotation as practical. Carefully collect and burn infected plants and several inches of surrounding soil. Soak flower bed soil in infected areas with a 1:1,000 solution of mercuric chloride (see page 101 for precautions), or apply PCNB 75 per cent plus thiram, captan, or folpet (⅓ pound each per 100 square feet) to soil surface *before* disease starts and mix into top 5 to 6 inches of soil. Morton Soil Drench, Pano-drench, and Dexon are effective as seedling drenches when cuttings are stuck or transplanted. Follow manufacturer's directions carefully. Collect and burn all crop debris immediately after harvest, or plow under deeply and cleanly. Valuable plants may often be saved by taking tip cuttings or buds to start new plants.

(22) Stem, Twig, Branch, or Trunk Canker; Dieback; Stem, Cane, or Limb Blight. Both fungi and bacteria are responsible for producing cankers on stems, twigs, limbs, and trunks. Cankers are usually definitely marked, oval to irregular, often sunken or swollen, discolored, dead areas (especially in the bark) that slow normal healing of wounds (Figure 2.40). Many cankers crack open, exposing wood beneath. If canker enlarges and girdles

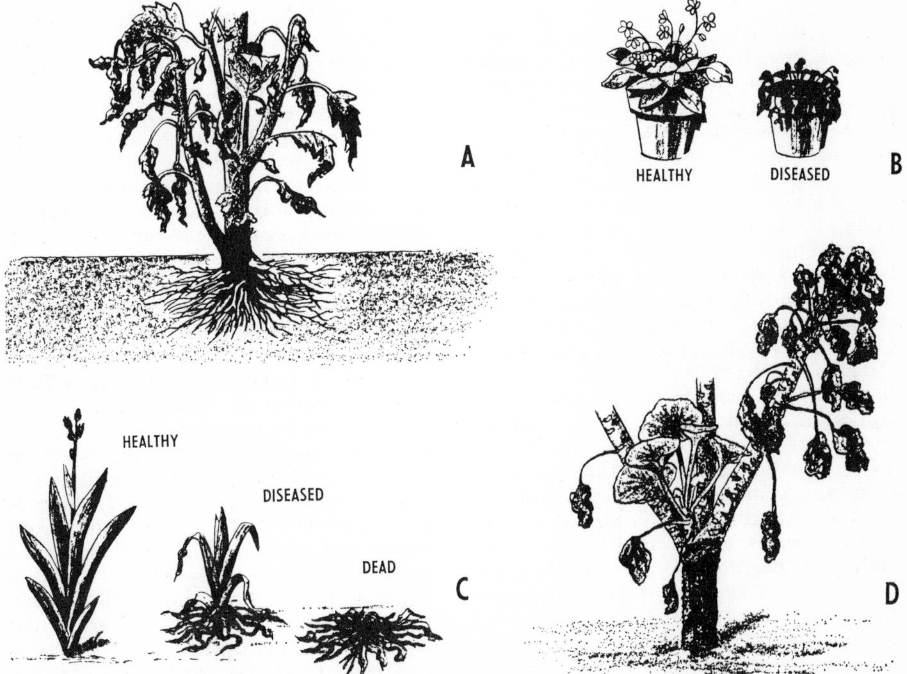

Fig. 2.39. Crown or stem rot. A. Chrysanthemum crown and root rot. B. African violet. C. Crown, rhizome, or bulb rot of iris. D. Blackleg of geranium.

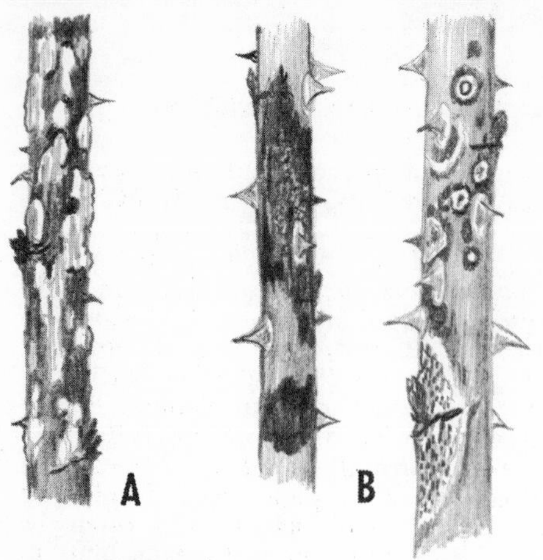

Fig. 2.40. Stem canker and anthracnose. A. Raspberry anthracnose. B. Various rose cankers.

stem, parts above diseased area usually lack good color; make little new growth; later wilt, wither, and die back from the tip. Infected branches often start growth later in spring. Cankers are most common on plants weakened by nematodes or insects, mechanical injuries, root rots, drought, nutritional deficiencies, severe winter conditions, etc. See also under (14) Scab, (15) Wilts, (23) Wood Rot, and (24) Fire Blight.

Plants Attacked: Abutilon, acacia, alder, almond, amelanchier, American bladdernut, amorpha, ampelopsis, andromeda, apple, apricot, araucaria, arborvitae, arbutus, arrowwood, ash, asparagus, asparagus-fern, aspen, aster, aucuba, avocado, azalea, baldcypress, barberry, bean, beech, beet, bignonia, birch, bittersweet, blackberry, black gum, black locust, bladdersenna, blueberry. Boston ivy, boxelder, boxwood, boysenberry, broccoli, broom, Brussels sprouts, butterflybush, butternut, cabbage, cacti, caesalpinia, calendula, California-laurel, callicarpa, calycanthus, camellia, camphor-tree, carissa, carnation, carrot, cassia, catalpa, cauliflower, ceanothus, cedar, chamaecyparis, cherry, cherry-laurel, chestnut, chicory, China-aster, chinaberry, Chinese cabbage, chinquapin, chokeberry, chrysanthemum, clarkia, clematis, columbine, coralberry, cornel, cosmos, cotoneaster, cottonwood, crabapple, crotalaria, cryptomeria, cucumber, currant, cypress, daphne, delphinium, dewberry, dogwood, Douglas-fir, eggplant, elder, elm, English ivy, euonymus, exacum, fennel, fig, filbert, fir, flowering almond, flowering cherry, flowering currant, flowering quince, flowering raspberry, forsythia, foxglove, fuchsia, gardenia, garlic, gentian, geranium, goatsbeard, goldenchain, golden chinquapin, goldenlarch, goldenrain-tree, goldenrod, gooseberry, grape, grapefruit, hackberry, hardhack, hardy orange, hawthorn, hazelnut, hemlock, Hercules-club, Hiba arborvitae, hibiscus, hickory, highbush cranberry, holly, hollyhock, holodiscus, honeydew melon, honeylocust, honeysuckle, hophornbeam, hornbeam, horsechestnut, India rubber tree, incense-cedar, indigobush, inkberry, Japanese pagodatree, Japanese plum-yew, jasmine, jetbead, juniper, kalanchoë, kale, kerria, larch, lemon, lilac, lily, lime, linden, locust, London plane, loquat, lupine, magnolia, mahonia, mallow, mango, maple, meconopsis, "mimosa" tree, mint, monkshoodvine, morning-glory, mountain-ash, mountain-laurel, mulberry, muskmelon, nectarine, oak, okra, oleander, onion, orange, Oregon-grape, osier, pachysandra, pagodatree, palms, paper-mulberry, parsnip, pawpaw, pea, peach, peanut, pear, pecan, pepper, persimmon, phlox, pine, pinks, planetree, plum, poinciana, poinsettia, poplar, potato, prairiegentian, prickly-ash, privet, pumpkin, pyracantha, quince, radish, rape, raspberry, redbay, redbud, redcedar, redwood, rhododendron, rhubarb, rollinia, rose, roselle, rosemallow, rubber plant, Russian-olive, rutabaga, sassafras, senna, sequoia, serviceberry, silk-oak, silverberry, silver threads, smoketree, snapdragon, snowberry, soapberry, sophora, sorreltree, spicebush, spinach, spirea, spruce, squash, sugarberry, sumac, stock, sweet alyssum, sweetgale, sweetgum, sweetpea, sweetpotato, sweetwilliam, sycamore, tamarisk, Texas silver leaf, thimbleberry, tomato, tree-of-Heaven, tuliptree, tupelo, turnip, umbrella-pine, viburnum, vinca, Virginia-creeper, walnut, watermelon, weigela, white-cedar, willow, wisteria, wolfberry, woodwaxen, yellow-cedar, yellowwood, yew, and zinnia.

Control: Carefully prune and burn infected or dead plant parts, cutting one to several inches behind canker. Include *all* discolored wood. On trees and shrubs, where practical, sterilize pruning shears between cuts by dipping or swabbing with 70 per cent alcohol. Follow spray schedules for fruits listed in Table 16 in Appendix. A multipurpose spray may be beneficial to certain flowers, shrubs, and vegetables. Treat seed of vegetables and flowers (Table 20 in Appendix). Plant in well-drained, sterilized soil. Avoid overwatering. Treat bark and wood injuries of trees and shrubs promptly. Cover with tree wound dressing (pages 34–36). Keep plants vigorous through proper fertilization, pruning, and watering during droughts. Avoid wounding plants. Grow resistant varieties where possible.

(23) Wood, Butt, Wound, Heart, or Sapwood Rot. Certain fungi cause water-soaked, spongy, stringy, crumbly, or hard rots in both living and dead woody plants. Damage usually occurs slowly, often over period of many years. Infection occurs almost entirely through unprotected wounds (e.g., pruning cuts, mowing bruises, sunscald, fire scars, lightning, frost and drought cracks, branch stubs, dwarf mistletoe cankers, insect wounds, sapsucker punctures, breaks due to ice and windstorms, etc.). Wood rots often indicated by external, corky, leathery, fleshy, punky, woody, hoof- or shelf-shaped fungus structures (conks or brackets). A single conk may shed *over 100 billion fungal spores* in one day! Clusters of small toadstools may form at trunk base or at wounds (Figures 2.41 and 2.49B). Affected wood may be discolored or stained for a distance of several feet or more above and below these conks. See also (34) Root Rot.

Plants Attacked: Practically all woody plants.

Control: Promptly treat bark and wood injuries with tree wound dressing (pages 34–36) to prevent wood-rotting fungi from becoming established. If rot is evident, remove cleanly all diseased wood and bark. Sterilize cavity by painting with household bleach (diluted 1 to 5 with water) or 1:1,000 solution of mercuric chloride (see page 101 for precautions). Then cover with tree wound dressing. Keep plants vigorous by fertilization, pruning, and watering. Control borers by spraying trunk and scaffold branches with DDT, dieldrin, or Thiodan. Check with your extension entomologist, county agent, or nurseryman for proper timing.

(24) Fire Blight, Bacterial Shoot Blight, Bacterial Canker, Gummosis. Blossoms, leaves, and young fruits suddenly turn brown or black and shrivel as if scorched by fire, but cling to twigs (Figure 2.42). Twigs are shrunken and brown to black. Twig tips often bent to form "shepherds' crooks." Rapidly growing shoots (e.g., watersprouts and suckers) are especially susceptible. Fire blight bacteria live over in buds and branch and trunk cankers. These cankers are slightly sunken, discolored areas with a sharp margin (or slight crack) separating healthy and diseased bark tissues. Fire blight is spread from buds and overwintering cankers to blossoms, leaves, and young twigs by insects,

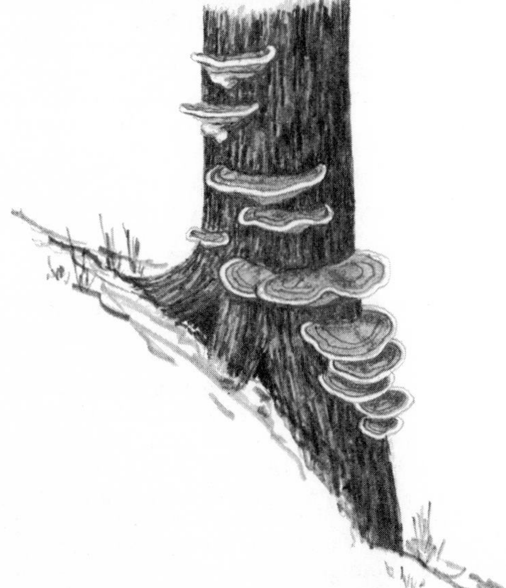

Fig. 2.41. Several types of fungus fruiting bodies (conks) which indicate wood rot is within.

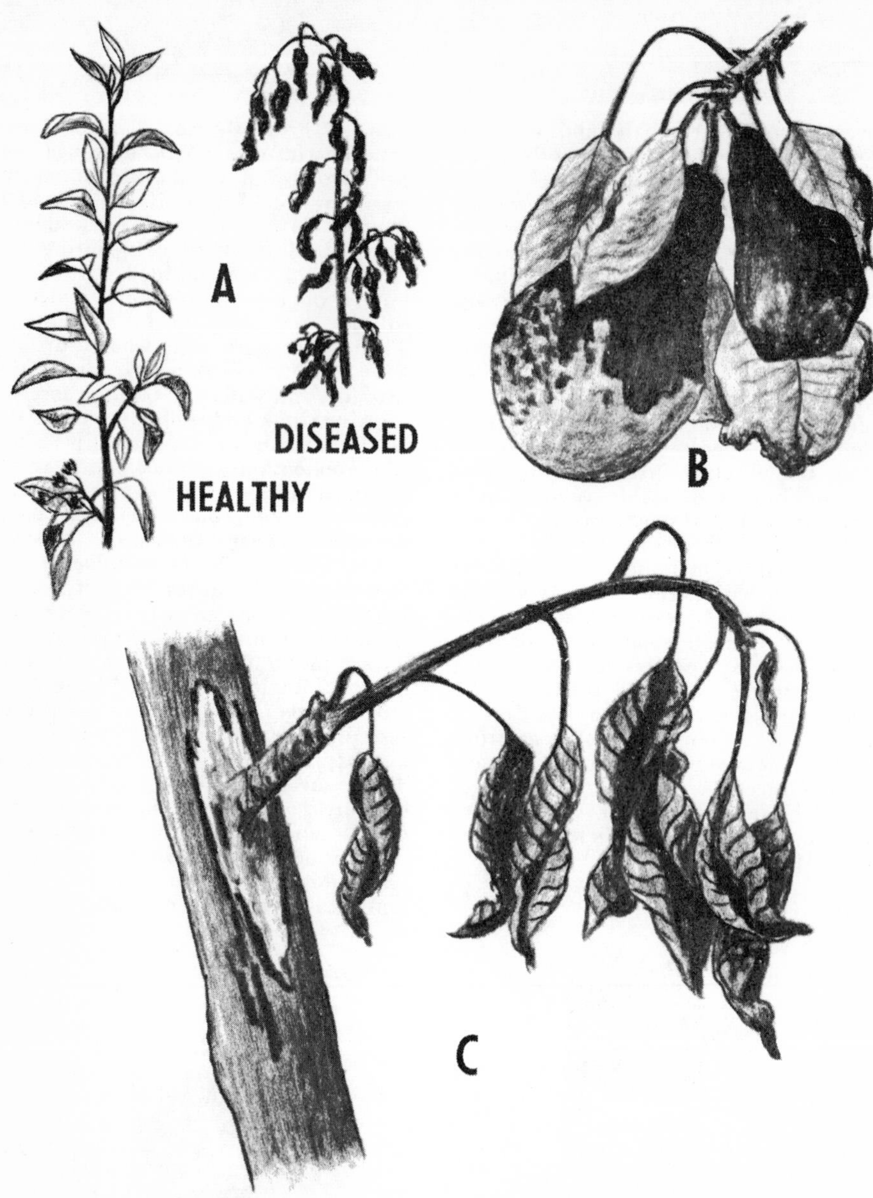

Fig. 2.42. Fire Blight. A and C. Apple. B. Pear. Note that in C infection has progressed down the shoot and is producing a canker in the branch.

splashing rain, wind, and pruning tools. See (2) Bacterial Leaf Spot and Blight. Blight is active from about blossoming time until rapid shoot growth ceases. *Plants Attacked:* Apricot, almond, amelanchier, apple, apricot, avens, blackberry, cherry, cherry-laurel, chokeberry, Christmasberry, cinquefoil, cotoneaster, crabapple, filbert, flowering almond, flowering cherry, flowering quince, goatsbeard, hawthorn, holly, holodiscus, jetbead, kerria, lilac, loquat, medlar, mountain-ash, nectarine, ninebark, peach, pear, photinia, plum, potentilla, pyracantha, quince, raspberry, rose, serviceberry, spirea, stranvaesia, strawberry, and walnut. *Control:* Use resistant varieties of apple, crabapple, and pear, or species of cotoneaster and pyracantha. See under Apple. *Avoid* heavy fertilization (especially using nitrogen), heavy pruning, and other practices that stimulate excessive growth.

Grow trees in sod in well-drained soil. Follow spray program outlined in Table 16 in Appendix to keep insects in check. Cut out infected twigs and small limbs during dormant season or late in summer. Prune at least 4 inches back from canker margin. Cut out cankers on large limbs or trunk and disinfect wound before coating with tree wound dressing to which Elgetol is added as a disinfectant (pages 34–36). Disinfect pruning tools between cuts by dipping in 70 per cent alcohol, 5 per cent Lysol, 4 per cent amphyl solution, household bleach, or a 1:1,000 solution of mercuric chloride (page 101). For additional information read USDA Leaflet 187, *Blight of Pears, Apples, and Quinces.*

If practical, spray with streptomycin at 3- to 7-day intervals beginning at very early bloom. Continue through blossoming period. Sprays are expensive and will not control the serious twig blight stage that follows. Follow manufacturer's directions. A *dormant* copper sulfate spray (5 pounds in 10 gallons of water), applied just before bud break in spring is beneficial.

(25) Black Knot. Rough, black, girdling swellings on twigs, branches, and even the trunk (Figure 2.43). Knots covered with olive-green, velvety surface in late spring. If undisturbed, black knot may stunt and kill trees.

Fig. 2.43. Black Knot. A. Cherry. B. Plum.

Plants Attacked: Apricot, cherries, mayday-tree, peach, prune, and plum.

Control: Cut and burn infected twigs and branches during dormant season. Make cuts 4 to 6 inches back of knot. Knots on large limbs should be removed by surgery (page 34). Cuts should go back into healthy wood. Cover pruning wounds with tree wound dressing (pages 34–36). Destroy nearby wild plums and cherries and any worthless fruit trees. Follow spray program as outlined (Table 16 in Appendix). Japanese *plums* are less susceptible than most American varieties.

(26) Rust. Powdery, orange-yellow galls or swellings on twigs, limbs, and trunk. Or bean-shaped galls on junipers that become orange and gelatinous in spring wet periods. See (8) Rust.

(27) Smut. Sooty, powdery masses on stems and branches. See (11) Smut.

(28) Leafy Gall, Fasciation, Witches'-broom. Symptoms variable. Dwarfed, thick, aborted shoots with distorted leaves (or cauliflower-like growth) form near soil line (Figure 2.44). Main stem is stunted. Few flowers are produced. See (30) Crown Gall.

Plants Attacked: Babysbreath, buddleia, butterfly-flower, carnation, chrysanthemum, coralbells, dahlia, delphinium, flowering tobacco, forsythia, geranium, gladiolus, heuchera, hollyhock, lily, nasturtium, nicotiana, pea, petunia, phlox, piqueria, primrose, pyrethrum, Shasta daisy, strawberry, sweetpea, viburnum, and wallflower.

Control: Where practical, dig up and burn infected plants. Sow disease-free seed or dip 1 minute in alcohol, then soak 20 minutes in a 1:1,000 solution of mercuric chloride (see page 101 for precautions). Some seeds cannot stand this treatment. Rinse and wash well in running water. Dry and dust with thiram or captan seed protectant. Sterilize soil in flower beds or use fresh soil (pages 538–548). Maintain good cultural practices. Plant disease-free stock. Practice 3-year rotation.

(29) Bacterial Soft Rot, Bacterial Stem Rot, Collar Rot. Rapid, mushy, slimy, or "cheesy" rot, usually with putrid odor. Roots, stems, fleshy tubers, rhizomes, bulbs, buds, leaves, and fruit become soft, watery, and pulpy. Foliage wilts, shrivels, and may collapse when lower stem or underground parts rot (Figure 2.45). Infections occur through wounds (e.g., insect or nematode injuries, other diseases, freezing or hail injury, harvest or cultivator wounds, etc.). Rot is most destructive in heavy, poorly drained soils in warm, moist weather. See also under

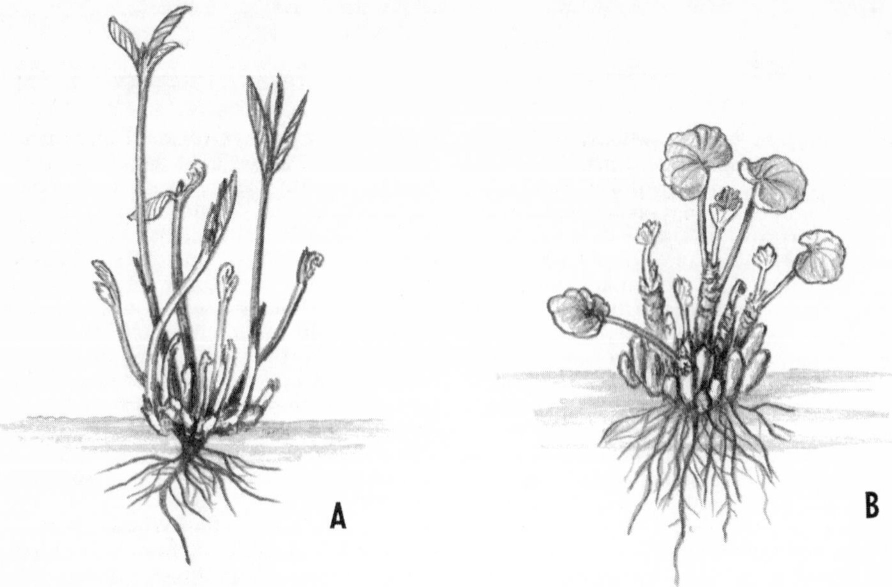

Fig. 2.44. Leafy gall or fasciation. A. Sweetpea. B. Geranium.

(21) Crown Rot, (32) Fruit Spot, and (36) Bulb Rot.

Plants Attacked: Apple, artichoke, asparagus, bean, beet, broccoli, Brussels sprouts, cabbage, cacti, caladium, calla, calla lily, canna, cantaloup, carrot, casaba, cauliflower, celeriac, celery, chicory, Chinese cabbage, Chinese evergreen, chrysanthemum, citron, collard, corn, cucumber, cyclamen, dahlia, dandelion, dasheen, delphinium, dieffenbachia, eggplant, elephants-ear, endive, escarole, fennel, ferns, finocchio, fittonia, garlic, geranium, gladiolus, gourds, grape-hyacinth, horseradish, hyacinth, iris, Jerusalem-cherry, kale, kohlrabi, larkspur, leek, lettuce, lily, mangel, mangold, muskmelon, mustard, nightshade, okra, onion, opuntia, orchids, pansy, parsley, parsnip, pea, pear, pepper, poinsettia, poppy, potato, pumpkin, radish, rape, rhubarb, rutabaga, salsify, sansevieria, shallot, spinach, squash, strawberry, sweetpotato, tomato, tuberose, tulip, turnip, vegetable-marrow, violet, watermelon, and xanthosoma.

Control: Avoid planting in poorly drained, unfertile soil. Treat soil before planting using Diazinon, chlordane, dieldrin, aldrin, etc., to control soil insects. Carefully dig up and burn infected plants. When rot starts in flower beds, disinfect soil by drenching with a 1:1,000 solution of mercuric chloride (see page 101 for precautions). Repeat treatment 10 days later. Spray to control foliage-feeding insects using Dibrom (naled), sevin, or DDT plus malathion. Control foliage blights, fruit rots, and other dis-

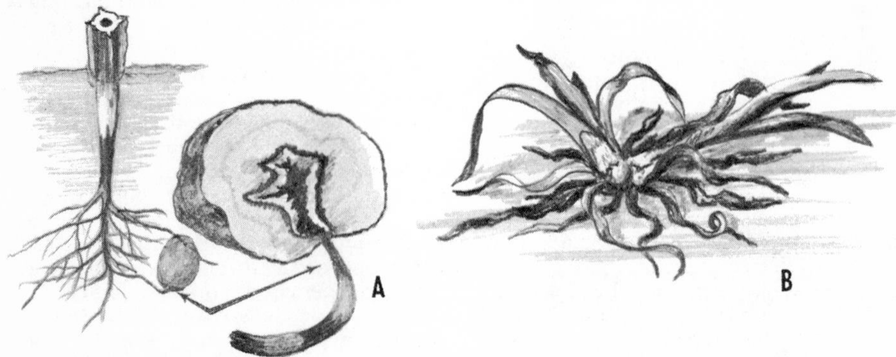

Fig. 2.45. Bacterial soft rot. A. Soft rot and blackleg of potato. B. Iris.

eases. Practice long crop rotation. Avoid wounding plants when cultivating, digging, handling at or after harvest, etc. Store *only* dry, sound, blemish-free vegetables and fruits in a dry, well-ventilated room at the recommended temperature and humidity. Storage area should first be swept clean. Then spray all surfaces from ceiling to floor with formaldehyde solution (1 pint of 40 per cent formalin in 10 gallons of water) or scrub down with copper sulfate solution (1 to 2 pounds in 10 gallons of water) before storing fruits, vegetables, roots, etc. Precool leafy vegetables to 45° F. or below. Then place in cold storage as soon after harvest as possible.

For *calla lily* and *iris:* Cut out rotted portion in bulb, corm, or rhizome. Then dry thoroughly for a day or two. Before planting, soak in a 1:1,000 solution of mercuric chloride.

(30) Crown Gall, Cane Gall, Hairy Root, Bacterial Root Gall. Rough-surfaced, hard or soft and spongy, swollen tumors or galls, up to several inches or more in diameter. May be flesh-colored, greenish or dark (Figure 2.46). Galls are usually found at or near soil line, graft or bud union, or on roots and stems. Small, fibrous roots are sometimes profuse, may resemble "witches'-brooms" (Hairy Root). These occur at base of trunk, crown, or on larger roots. As disease progresses, plants often become stunted, weak, and may eventually die. As the causal bacteria enter through wounds, galls are easily confused with callus overgrowths formed at wounds or graft unions that are perfectly normal. They may also be caused by fungi, viruses, or wounding. Corn, onion, asparagus, grasses, and cereals are immune.

Plants Attacked: Achillea, ageratum, almond, apple, apricot, araucaria, arbutus, artemisia, ash, asparagus, asparagus-fern, aster, avocado, azalea, babysbreath, bean, beet, begonia, bittersweet, blackberry, blueberry, boxwood, boysenberry, cabbage, cacti, caesalpinia, calendula, calycanthus, camellia, cape-jasmine, carnation, carrot, castorbean, catalpa, chamaecyparis, cherry, chrysanthemum, cinquefoil, clematis, clockvine, cotoneaster, cottonwood, crabapple, cucumber, cypress, dahlia, daisies, delphinium, dewberry, dogwood, Douglas-fir, eucalyptus, euonymus, fig, filbert, flowering quince, forsythia, foxglove, gardenia, geranium, gladiolus, gooseberry, grape, grapefruit, gum, hazelnut, hibiscus, hickory, holly, hollyhock, honeylocust, honeysuckle, horseradish, incense-cedar, India rubber tree, jasmine, Jerusalem-artichoke, Jerusalem-cherry, juniper, kalanchoë, lemon, lettuce, lilac, lippia, loganberry, loquat, lupine, mallow, mangel, mangold, maple, marguerite, mountain-ash, mulberry, muskmelon, nectarine, New Jersey-tea, nicotiana, nightshade, oak, orange, pansy, parsnip, peach, pear, pea-tree, pecan, peony, persimmon, phlox, plum, poinciana, poinsettia, poplar, potato, potentilla, privet, quince, radish, raspberry, redcedar, redwood, rhododendron, rhubarb, rose, rosemallow, Russian-olive, rutabaga, saguaro, Shasta daisy, snapdragon, snowball, snowberry, spirea, Sprenger asparagus, strawberry-tree, sugar beet, sumac, sunflower, sycamore, sweetpea, tomato, turnip, viburnum, walnut, weigela, white-cedar, willow, wisteria, witch-hazel, wormwood, yarrow, yew, and yucca.

Control: Carefully dig up and burn infected plants, especially woody ones. Remove as many infected roots as possible. Practice at least a 5-year rotation or avoid replanting in same location for that period or longer. Reject nursery and other plants showing suspicious bumps near crown, former soil line, or graft union. Budding is preferable to grafting. Where suspicious, dip grafting or pruning knives in 70 per cent alcohol between cuts. Avoid wounding plants when cultivating, transplanting, etc. Disinfestation of soil using steam or chemicals (e.g., Vapam or VPM Soil Fumigant, methyl bromide, Vorlex, D-D, EDB, chloropicrin) or formaldehyde (1 pint of 40 per cent formalin in 6 gallons of water) applied ½ gallon per square foot, is seldom complete unless soil is in confined containers (pages 538–548). Maintain soil as acid as practical (below pH 5.5) for vegetables. Control insects by sprays of malathion plus methoxychlor, sevin, or DDT. Some nurserymen dip woody planting stock immediately after digging in Terramycin solution (about 400 parts per million) for 15 to 60 minutes, followed by air drying. Others dip stock in a 1:1,000 solution of mercuric chloride for 1 to 10 minutes, or streptomycin (Agristrep or Agrimycin 100—following manufacturer's instructions). If galls are severe on larger trees, call in a reputable arborist. He will probably excavate, chisel off

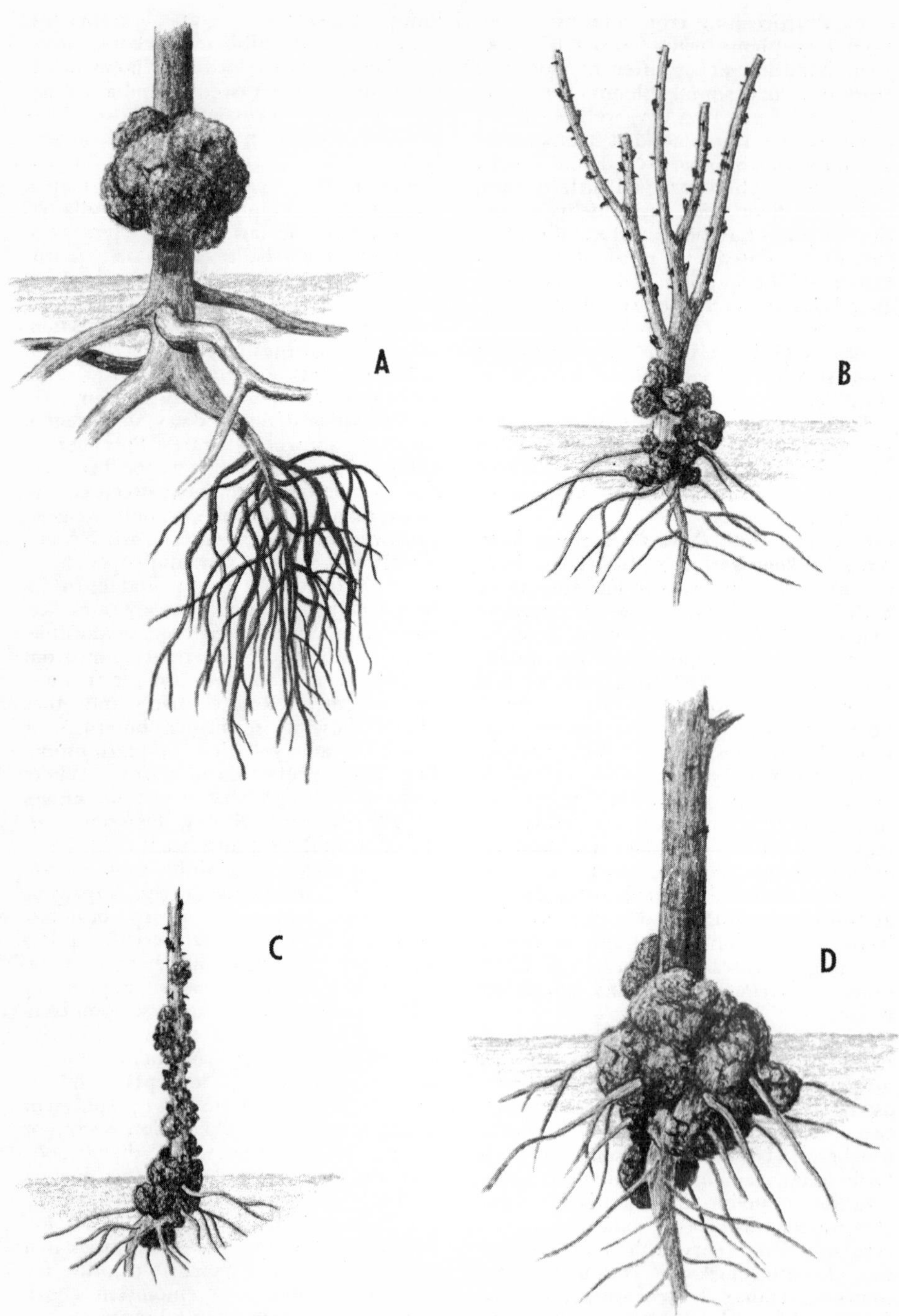

Fig. 2.46. Crown gall. A. Apple. B. Rose. C. Crown and cane gall of raspberry. D. Peach.

Fig. 2.47. Flower blight. A. Tulip fire. B. Gladiolus. C. Botrytis blight on geranium. D. Head blight of chrysanthemum.

the outer gall tissue, and paint the gall and its margins with a mixture of Elgetol (1 part) and methanol (4 or 5 parts). The damaged areas should then be painted with tree wound dressing (pages 34–36). The soil should be treated with a soil insecticide, e.g., Diazinon, chlordane, etc., to control chewing insects.

C. Flower and Fruit Diseases

(31) Flower or Blossom Blight, Ray or Inflorescence Blight. Flowers spotted, often wither or rot, causing fruit not to set, or young fruit may rot and drop early. Flowers and young fruit may be covered with dense mold during and following moist weather (Figure 2.47). See also under (5) Botrytis Blight, (6) Downy Mildew, (7) Powdery Mildew, (9) White Rust, (11) Smut, (16) Mosaic and Flower Breaking, (18) Yellows, (19) Curly-top, (24) Fire Blight, and (38) Bulb Nematode.

Plants Attacked: Achillea, African-violet, almond, amaranth, amelanchier, anemone, apple, apricot, aster, aucuba, avocado, azalea, begonia, blackberry, black-eyed pea, blueberry, calceolaria, calla lily, camellia, camomile, canna, carnation, castorbean, cherry, cherry-laurel, China-aster, Chinese hibiscus, chokeberry, Christmas-rose, chrysanthemum, cornflower aster, crabapple, crassula, cucumber, cyclamen, daffodil, dahlia, delphinium, dewberry, dogwood, English daisy, flowering almond, flowering cherry, flowering quince, forsythia, foxglove, gardenia, geranium, gerbera, gladiolus, gloxinia, gourds, hawthorn, hibiscus, holly, hyacinth, hydrangea, inula, iris, Japanese quince, jasmine, lilac, lily, loquat, magnolia, Maltese cross, mango, maple, marigold, matricaria, mistflower, mockorange, morning-glory, mountain-ash, mountain-laurel, narcissus, nectarine, okra, orchids, palms, pansy, pea, peach, pear, peony, pepper, petunia, plum, poinsettia, primrose, pumpkin, pyrethrum, quince, raspberry, rhododendron, rose, roselle, rose-of-Sharon, safflower, sea-lavender, snapdragon, snowberry, squash, statice, stock, stokesia, strawberry, sumac, sunflower, sweetpea, thimbleberry, tomato, Transvaal daisy, tuberose, tulip, vegetable-marrow, verbena, viburnum, violet, yucca, and zinnia.

Control: Treat flower and vegetable seed before planting (Table 20 in Appendix). Follow spray or dust schedule for vegetables and fruit. Spray others with captan, thiram, zineb, maneb, Daconil 2787 ferbam, Polyram, or copper-containing fungicide. Apply light, misty spray of zineb or maneb, captan, or dichloran (Botran) at 3- to 5-day intervals during bloom. Rotate garden plantings. If practical, carefully remove affected and fading flowers and young fruit when first noticed. Place in paper sack and burn. Space plants. Avoid overhead sprinkling. Grow in well-drained soil.

(32) Fruit Spot, Speck, Rot, or Blotch; Seed, Berry, or Tuber Rot; Storage Rot. Fruit, tuber, berry, etc. show one or more spots. Spots often enlarge and merge. Whole fruit may later rot and shrivel (Figure 2.48). Frequently starts at blossom or stem-end or underside of fruit when resting on damp soil. Bacterial Soft Rot may follow, causing mushy or watery, foul-smelling decay. See also under (6) Downy Mildew, (10) Leaf Curl, (11) Smut, (12) Sooty Mold, (14) Scab, (16) Mosaic, (21) Crown Rot, (29) Bacterial Soft Rot, (31) Flower Blight, and (36) Bulb Rot.

Plants Attacked: Almond, amelanchier, apple, apricot, artichoke, asparagus-bean, avocado, barberry, bean, beet, blackberry, black-eyed pea, blueberry, boysenberry, butternut, cabbage, canna, cantaloup, carrot, casaba, cauliflower, celery, chayote, checkerberry, cherry, chestnut, China aster, chrysanthemum, chokeberry, citron, coralberry, crabapple, cucumber, currant, dahlia, dasheen, datura, dewberry, eggplant, elephants-ear, endive, escarole, feijoa, fig, flowering almond, flowering currant, flowering quince, fragrant glad, gaultheria, gladiolus, gooseberry, gourds, grape, grapefruit, guava, hardy orange, hawthorn, hazelnut, hickory, honeydew melon, huckleberry, jackbean, kumquat, lemon, lentil, lettuce, lima bean, lime, loquat, mango, mock-cucumber, mountain-ash, mulberry, muskmelon, nectarine, okra, olive, onion, orange, palms, parsnip, pawpaw, pea, peach, peanut, pear, pea-tree, pecan, pepper, persimmon, plum, pomegranate, potato, pumpkin, pyracantha, quince, radish, rape, raspberry, rollinia, rutabaga, snowberry, squash, strawberry, sweetpotato, thimbleberry, tomato, turnip, vegetable-marrow, walnut, watermelon, yam, and yautia.

Control: Mulch vegetables to keep fruit

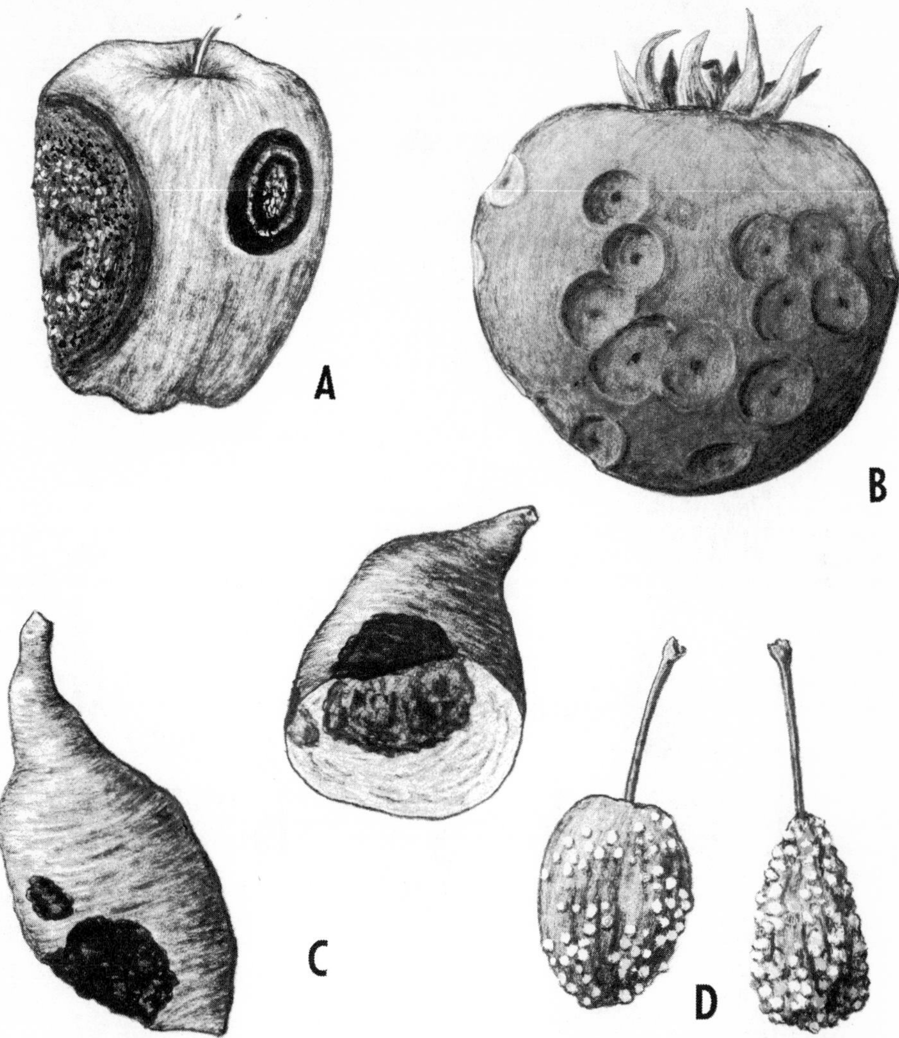

Fig. 2.48. Fruit rots. A. Bitter rot of apple. B. Tomato anthracnose. C. Black rot of sweetpotato. D. Brown rot of plum.

off soil. Follow spray or dust programs for vegetables and fruit. Control insects that transmit disease organisms and provide entrance wounds. Use malathion plus methoxychlor, sevin, or Dibrom (naled). Promptly collect and burn (or eat) spotted and rotting fruit. Guard against wounding fruit and vegetables from harvest through storage period. Store only sound, blemish-free fruit and vegetables at recommended storage temperature and humidity. Check with your extension horticulturist. Read USDA Farmers' Bulletin 1939 *Home Storage of Vegetables and Fruits.* Treat seed of vegetables and flowers as given under plant involved and Table 20 in Appendix. Sow only high-quality seed. Resistant varieties are available for some plants.

(33) Smut. Flower parts or seed may break open to release black, powdery mass. See (11) Smut.

D. Root and Bulb (Corm) Diseases

(34) Root Rot, "Decline," Cutting Rot. Symptoms are variable. Plants may gradually or suddenly lose vigor. Become pale or yellowed and stunted. Plants tend to wilt or die back; do not respond normally to water and fertilizer (Figure 2.49). Young plants wilt and collapse. Affected plants are more subject to drought and wind damage—may blow over or lodge. Root rot is often difficult to diagnose

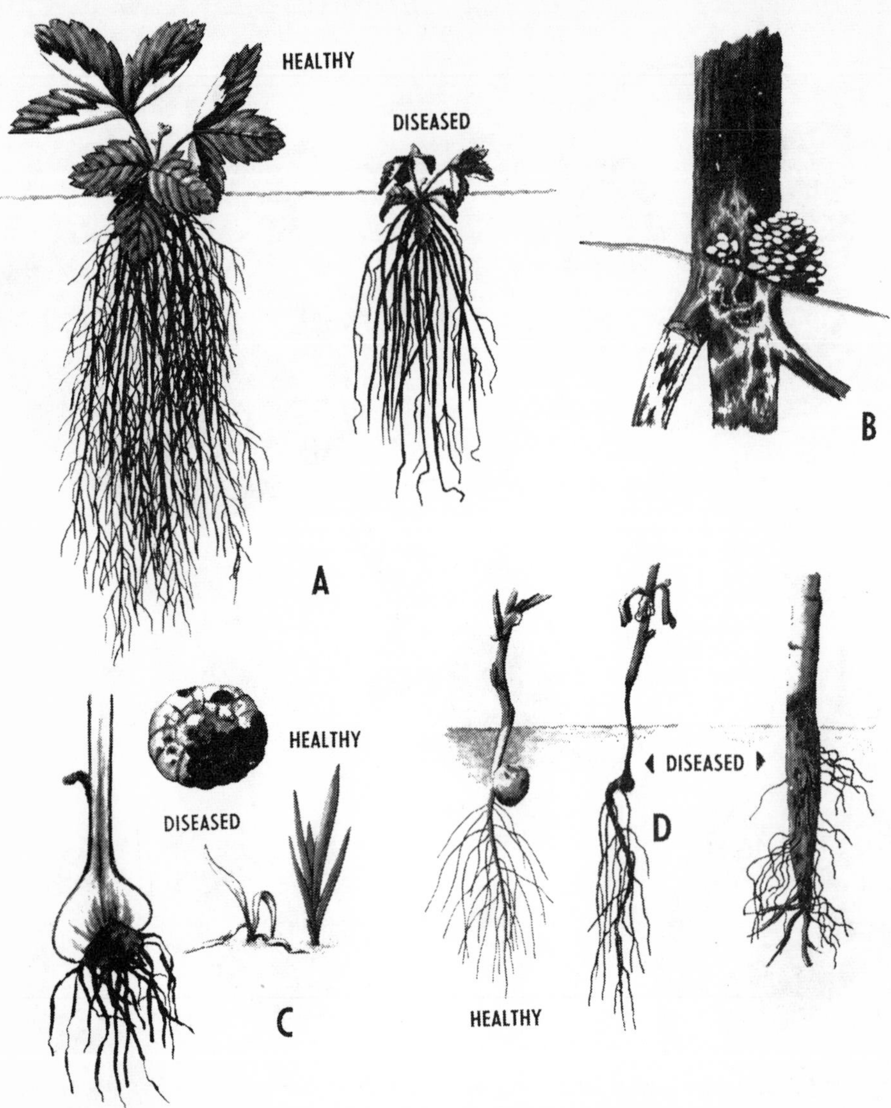

Fig. 2.49. Root rots. A. Black root rot of strawberry. B. Armillaria root and wood rot, a common disease of woody plants. C. Root and corm rot of gladiolus. D. Root and stem rot of pea.

because trouble is hidden. Roots decay. May be covered with mold (brown, black, white, gray) or flat, brown-to-black, stringlike strands or rhizomorphs (*Armillaria*). Decay may be water-soaked, mushy, spongy, or firm. Easily confused with wilts, bacterial soft rot, crown rots, crown gall, root-feeding insect or rodent damage, or nematodes. Nematodes often provide wounds by which root-rotting fungi and bacteria enter. Usually most serious on annual plants in cold, wet, poorly drained soils. See under (15) Wilts, (21) Crown Rot, (23) Wood Rot, (30) Crown Gall, (35) Clubroot, (36) Bulb Rot, and (37) Root-knot.

In the southwestern states (Oklahoma, Arkansas, and Texas to southeastern Utah, Nevada, and California) a widespread soil fungus (*Phymatotrichum*) causes a disease known as Cotton, or Texas, Root Rot (Figures 4.60 and 4.62). Over 1,700 kinds of plants including fruit, nut, and shade trees; shrubs; flowers; small fruits; vegetables; even weeds and native vegetation are attacked from midsummer on. Phymatotrichum often kills plants in more or less circular

patches, up to an acre or more in diameter. A firm brown rot of lower stem and roots occurs. Affected roots may be covered with a dirty-yellow, brown, or gray network (weft) of mold strands and brown or black bodies (sclerotia). A snow-white to buff, powdery spore mat, 2 to 12 inches in diameter, often appears on moist soil. The fungus is only serious in alkaline soils (pH 7.3 and above) low in humus, and where winters are mild. If you live in the Cotton Root Rot area, contact your state or extension plant pathologist for a listing of resistant plants (members of the grass, sedge, palm, cycad, lily, amaryllis, iris, onion, and orchid families are usually immune) and other control measures, including *deep* plowing and the turning under of large amounts of dry or green organic matter.
Plants Attacked: Practically all.
Control: Plant disease-free stock in well-drained, well-prepared soil high in organic matter (page 23). Where feasible, practice systematic crop rotation. Burn tops of annual plants after harvest. Dig and burn affected garden plants. Sterilize soil for seedbeds and house plants (pages 538–548). Start seed of very susceptible plants in sterile or uninfested medium (e.g., vermiculite, sifted sphagnum moss, Perlite, or soil). Treat seed before planting. See Table 20 in Appendix. Avoid overcrowding and overwatering in seedbed. Keep perennials growing vigorously through proper fertilization, watering, cultivating, and pruning. Avoid deep and close cultivations. Check with your extension plant pathologist for plants and rootstocks resistant to Armillaria Root Rot (e.g., Marianna 2624 plum, persimmon, fig, Northern California black walnut, domestic French pear). Avoid injuries to roots and lower stem (trunk). Practice balanced soil fertility. Keep down weeds. If possible, do not replant trees in same location where previous woody plants have died from Root Rot. Applying DBCP (Fumazone, Nemagon), VC-13 Nemacide, or Zinophos around certain living plants may be beneficial if root-feeding nematodes are present in large numbers. Applications of PCNB and Dexon to soil in nursery seedbeds is often beneficial for flowers, shrubs, and trees. Follow manufacturer's directions.

(35) Clubroot. Attacks crucifers. Plants may not form a heart, but remain stunted and yellowish. Often wilt on hot, dry days and partially recover at night. Outer leaves may turn yellow and drop. Roots form mass of small to large, distorted, club-shaped swellings; later rot from secondary organisms (Figure 2.50).

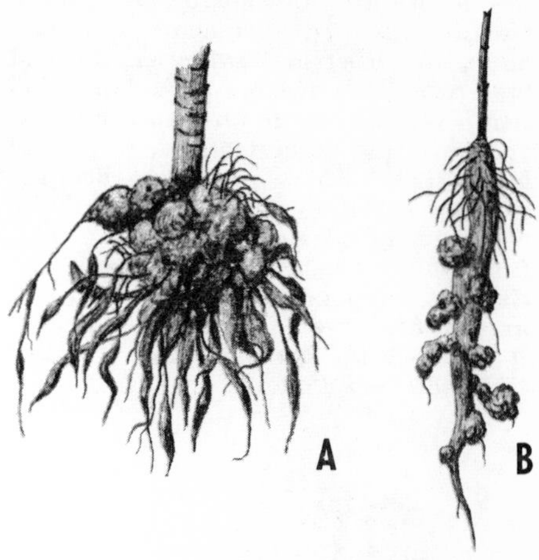

Fig. 2.50. Clubroot. A. Cabbage. B. Mustard.

Plants Attacked: Alyssum, broccoli, Brussels sprouts, cabbage, candytuft, Chinese cabbage, cauliflower, collard, damesrocket, erysimum, garden cress, honesty, horseradish, kale, kohlrabi, mustard, peppergrass, radish, rape, rockcress, rutabaga, seakale, stock, sweet alyssum, turnip, wallflower, and watercress.
Control: If practical, apply enough hydrated lime to soil at least 6 weeks before planting to reach pH of 7.2 or higher. Liming is *not* effective in light, sandy, or peat soils. Use a *minimum* of 30 pounds per 1,000 square feet. Where possible, avoid or drain wet soils. Plant disease-free seed and seedlings in sterilized soil (pages 538–548), water seedbed with a 1:2,000 solution (1 ounce in 15 gallons of water) of mercuric chloride (see page 101 for precautions), work PCNB into top 4 to 6 inches of soil about 2 weeks before planting or apply broadcast or as a band. Follow manufacturer's directions. Or add PCNB or Daconil 2787 to transplant water (see under Cabbage). Varieties differ in resistance. Some turnip, rutabaga, peppergrass, horseradish, kale, damesrocket, wallflower, stock, and garden cress varieties are usually very resistant. Practice long rotation. Keep out all cruciferous weeds (e.g., charlock, wild radishes and mustards, pennycress, shep-

herds-purse, yellow-rocket, or wintercress). Collect and burn crop debris after harvest.

(36) Bulb (Corm) or Rhizome Rot. Shoots fail to emerge or are weak with yellow leaves that die back progressively from tips. Roots often discolored and decayed. Usually associated with nematodes, bulb mites, and insects. Decay often starts at bulb base (root or basal plate, bottom of stem) progressing upward and outward. Spots may also develop on side or neck of bulb (Figure 2.51). There may be little external evidence of rot although bulbs often are lightweight and soft or punky. Rot usually progresses in storage, especially if temperature and moisture are uncontrolled. See (29) Bacterial Soft Rot, (32) Fruit Rot, (34) Root Rot, and (38) Bulb Nematode.

Plants Attacked: Acidanthera, allium, amaryllis, chives, colchicum, crocus, daffodil, freesia, fritillary, garlic, gladiolus, glory-of-the-snow, grape-hyacinth, hyacinth, iris, ixia, leek, lily, lily-of-the-valley, lycoris, narcissus, onion, shallot, snowdrop, snowflake, squill, tigerflower, tritonia, tuberose, tulip, and zephyranthes.

Control: Plant or store disease-free bulbs (and corms) free of rot. Cure bulbs thoroughly and rapidly after digging. Sort carefully at digging, before storage, and again just before planting. Treat bulbs or corms before planting. See under plant involved and Table 20 in Appendix. Rotate or grow in sterilized, well-drained soil (pages 538–548).

(37) Root-knot, Root Gall, Cyst Nematode. Nematode-infested plants usually lack

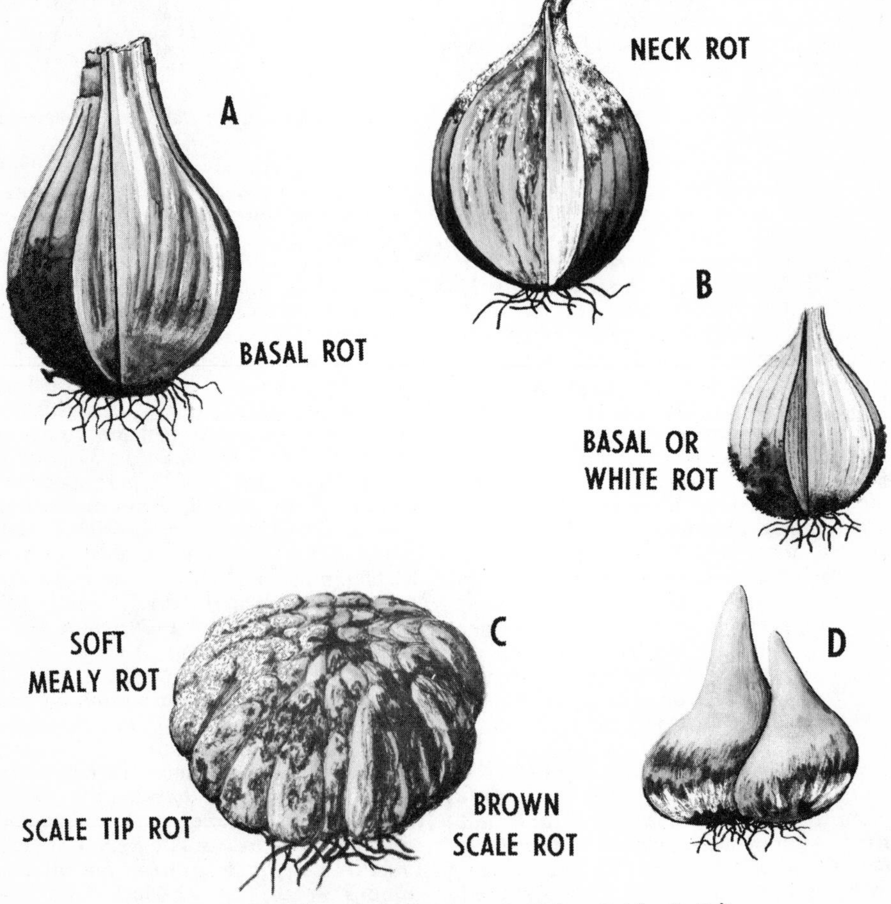

Fig. 2.51. Bulb rots. A. Narcissus. B. Onion. C. Lily. D. Tulip.

vigor. Often stunted and yellowish. Symptoms may resemble those of a soil deficiency, crown rot, root rot, etc. Severely infected plants may wilt in dry weather but recover at night for a time, then wither and die. Small to large swellings, galls, or knots, develop on roots. These are more or less round, or long and irregular, and cannot be broken off like nodules on legume roots. Roots may become "beaded," swollen, or greatly distorted (Figure 2.52). Disease is widespread and damaging in many southern soils, especially light, well-aerated ones. Both yield and quality are reduced. Primarily an indoor problem in most northern states, although there is one species of root-knot (*Meloidogyne hapla*) that is found mostly in northern states. Root-knot is distributed in infested nursery stock, in transplants, root crops, and soil transported on shoes, sacks, crates, tools, or equipment. Root-knot and cyst nematodes may increase severity of crown gall, certain wilt diseases, and root rots (e.g., *Pythium, Verticillium,* and *Fusarium*).

Plants Attacked: Practically all except certain grasses, small grains, hairy vetch, crotalarias, and weeds. The susceptibility of garden plants varies greatly depending on root-knot species and biotypes infesting the soil.

Control: Where practical, sterilize (fumigate) soil or rooting medium. Use heat or chemicals (e.g., methyl bromide, chloropicrin, EDB, Telone, Vorlex, DBCP, or Mylone; see pages 538–548 in Appendix). Soil temperature at treating time should generally be 60° to 65° F. or above. Soil is commonly fumigated in fall after harvest

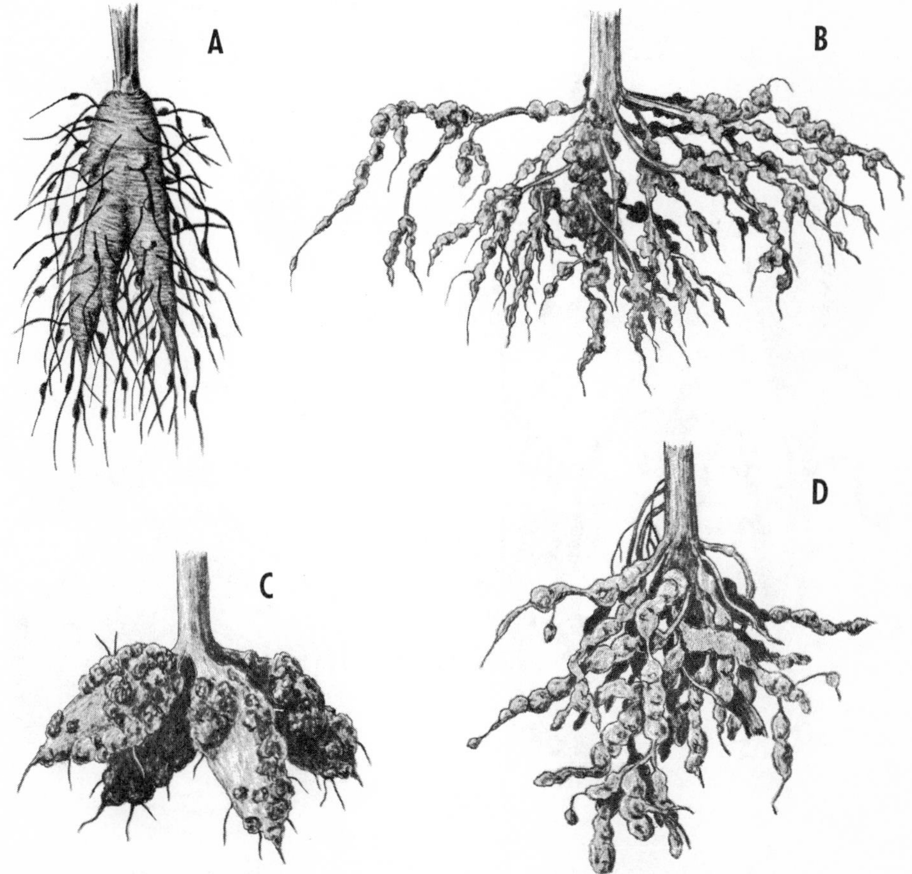

Fig. 2.52. Root-knot. A. Carrot. B. Watermelon. C. Dahlia. D. Tomato.

while it is still warm. DBCP (Nemagon, Fumazone) and VC-13 Nemacide may be used around many types of infested living plants. Follow manufacturer's directions. Plant disease-free nursery stock, leaf and stem cuttings, and certified transplants of cabbage, pepper, tomato, and certain other plants. Keep down weeds. Varieties resistant to certain root-knot species and subspecies are available for a number of plants (e.g., pepper, tomato, sweetpotato, beans, peach). See (34) Root Rot. Incorporate large quantities of organic matter into soil wherever possible. Practice rotation. In the home garden, remove roots from soil and burn when vegetables or annual flowers are harvested. Do *not* use roots for compost. Plow soil and leave fallow for a few days. Then plant a winter cover crop such as rye, wheat, tall fescue, or oats (South only). Plant early and encourage early growth by maintaining high soil fertility.

(38) Bulb Nematode, Ring Disease, Onion Bloat. Symptoms variable. Plants usually stunted with foliage twisted, crinkled, and yellow (Figure 2.53). Petioles often swollen and cracked. Main roots may be discolored; develop furrows and cracks.

1. *Hyacinth*—Yellow to brown flecks or blotches on leaves that become twisted, short, and split. Flower stalks stunted and flowers malformed. Bulb scales become thickened and turn brown. Cut bulbs are same as for Narcissus.
2. *Narcissus*—Small, yellowish, blisterlike swellings on leaves. Badly infested bulbs produce only few twisted and

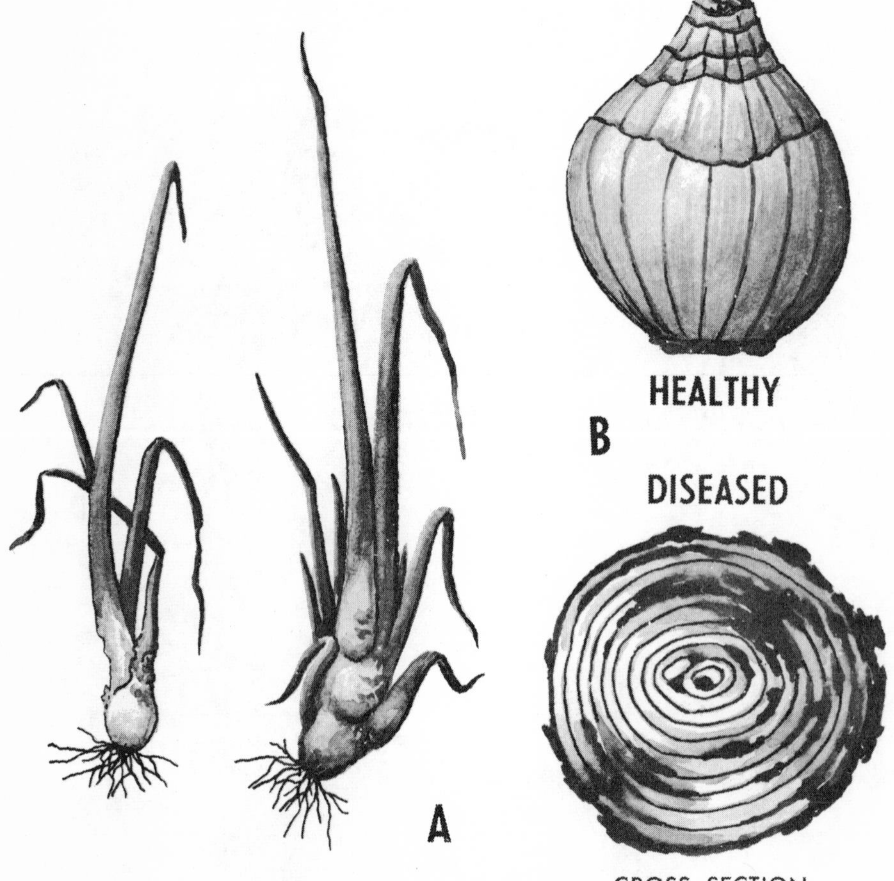

Fig. 2.53. Bulb nematodes. A. Onion bloat. B. Ring Disease of hyacinth.

bent leaves or none. Infested bulb scales are brown. When such bulbs are cut through, one or more dark rings are evident.

3. *Onion*—Seedlings dwarfed and twisted. Young bulbs swollen and misshapen (bloated); become soft.

Nematodes are spread by tools, running water, animals, infested soil, and planting infested bulbs. The nematode has many races specific for certain plants.
Plants Attacked: Chives, collomia, daffodil, dahlia, galtonia, garlic, gladiolus, glory-of-the-snow, grape-hyacinth, hyacinth, iris, lycoris, narcissus, onion, parsley, pea, potato, shallot, snowdrop, squill, strawberry, summer-hyacinth, sweetpotato, tigerflower, tulip, and wallflower.
Control: Destroy infested plants. Grow disease-free bulbs or seeds in sterilized or clean soil (pages 538–548 in Appendix). Avoid heavy, poorly drained soil. Soak *fully dormant* narcissus and hyacinth bulbs in hot water and formaldehyde. See Daffodil and Tulip. Collect and burn tops, infested bulbs, and other crop debris in the fall. Practice 3-year or longer rotation with noninfected crops. Keep down weeds.

E. Parasitic Flowering Plants

(39) Mistletoes. These are seed plants with green leaves that grow on trees and obtain nourishment by sending rootlike structures (haustoria) into food-conducting tissue of the inner bark. There are two general types: American or true mistletoes and dwarf mistletoes (Figure 2.54). Once established, a mistletoe may live as long as the woody plant it feeds upon.
American or True. Usually leafy, evergreen tufts of shoots with yellow to dark green, leathery, or scaly leaves. White, yellow, or pinkish to red berries are produced. The sticky seeds are easily spread by birds and animals. Infections commonly occur after seeds pass through birds. Leafy masses of these mistletoes may be up to 3 feet in diameter. They are most conspicuous after leaves drop in the fall.

Found as far north as southern New

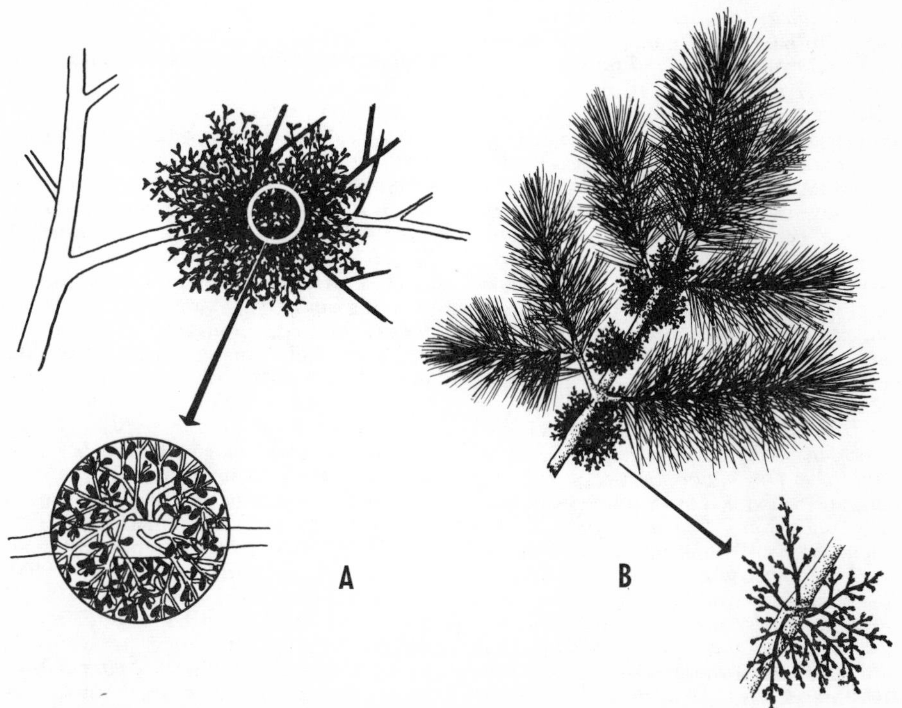

Fig. 2.54. Mistletoe. A. American or true mistletoe on apple. B. Dwarf mistletoe on pine.

Jersey in the east to Oregon in the west, where winters are not severe. Only relatively young branches are attacked. Tree branches beyond the mistletoe may be stunted, even die. This mistletoe is used for Christmas greens.

Dwarf. Perennial, scaly-leaved, simple or branched shoots usually scattered along conifer limbs and small branches 3 years old or older. Mistletoe shoots vary from yellow to brown to olive-green; berries olive-green to dark blue. The sticky seeds are explosively shot from the fruit for horizontal distances of 15 to more than 60 feet. Birds also distribute the seeds by wiping their beaks on branches. This mistletoe type is often less than an inch in diameter, but may be up to 12 inches. Dwarf mistletoes may seriously stunt, deform, or kill evergreens, especially seedlings and young trees. Conspicuous witches'-brooms form in the crown or spindle-shaped swellings (later cankers) in trunk. Branches or entire trees may die. Most common in western states in forested areas. Canker and wood-rotting fungi may enter through mistletoe wounds.

Plants Attacked: Acacia, alder, apple, ash, beech, birch, black gum, black locust, buckeye, camphor-tree, chestnut, chinaberry, cherry, cherry-laurel, citrus, cottonwood, cypress, dalea, dogwood, Douglas-fir, elaeagnus, elm, fir, forestiera, frangipani, hackberry, hawthorn, hemlock, Hercules-club, hickory, honeylocust, horsechestnut, incense-cedar, juniper, larch, linden, locust, manzanita, maple, oak, Osage-orange, osmanthus, paper-mulberry, parkinsonia, pear, pecan, persimmon, pine, planetree, plum, poplar, prickly-ash, redcedar, sassafras, smoketree, soapberry, sophora, spicebush, spruce, sugarberry, sweetgum, sycamore, tamarack, trumpetvine, tupelo, walnut, and willow.

Control: Cut off young infected branches a foot or more beyond any evidence of mistletoe. For older branches cut out bark and wood a foot or more away from each infection. Prune *before* berries ripen. Apply disinfectant to wound surface and paint with tree wound dressing (page 36). If a young evergreen trunk is infected with dwarf mistletoe, the tree should be cut down whenever practical.

(40) Dodder, Strangleweed, Love Vine, Goldthread. Orange to yellow, "leafless," slender, twining vines parasitic on a wide range of garden plants (Figure 2.55).

Fig. 2.55. Dodder on aster.

Occurs in tangled, yellowish-orange patches as dodder chokes out vigor of garden plantings.

Control: Plant clean vegetable and flower seed free of rough, irregularly round, flat-sided, gray to reddish-brown dodder seed. When dodder is found on garden plants, carefully remove and burn all infested plants *before* dodder forms seeds. Locate patches early, before plant spreads. When breaking new soil suspected of being dodder-infested, fumigate before planting. Use a soil fumigant (see next section and pages 538–548 in Appendix). Areas known to be heavily infested with dodder seed should be planted to resistant plants (e.g., grass, corn, or small grains) for at least two years in succession. Where this cannot be done, use of a soil fumigant is advisable. Mow dodder-infested borders early. Carefully remove and burn all infested plants before dodder forms seed. A glove, moistened with dilute 2,4-D spray and dried, may be used to stroke dodder plants strangling a valuable plant. If any dodder survives after 3 days, repeat treatment. Check with your county agent, extension weed specialist, or horticulturist for details.

SECTION THREE

"What Can I Do About It?"

Factors Affecting Disease Development 96

MEASURES

Plant adapted varieties and types of plants 97
Buy disease-free planting stock . . . 97
Treat the seed 97
Control damping-off 97
Plant in a well-prepared and well-drained seedbed 98
Follow a recommended crop rotation . 98
Pasteurization of infested soil and compost 98
Control weeds 98
Control insects, mites, and rodents . . 98
Avoid deep and close cultivations . . 98
Sanitary measures are important . . 98
Changing a cropping practice . . . 99
Observe quarantine regulations . . . 99
Destroy or remove alternate hosts of rust diseases 99
Store only sound, dry fruits and vegetables 99
Apply protective fungicides 100

MATERIALS

What Is a Fungicide? 100
- Protective fungicides 100
- Eradicant fungicides 100
- Protective and eradicant fungicides 100
- Chemotherapeutants 100

Modern Fungicides 100
Other Useful Fungicides 101
- Dinocap 101
- Mercuric chloride 101
- PCNB 101
- Daconil 2787 104
- Dichloran (Botran) 104
- Metiram (Polyram) 104
- Lanstan 104
- Difolatan (Folcid) 104
- Dexon 104
- Fixed copper fungicides 104
- Bordeaux mixture 104
- Burgundy mixture 105
- Sulfur products 105
- Antibiotics 105
  - A. Streptomycin 105
  - B. Cycloheximide 106
- Phenyl (organic) mercury materials . 106
- Lawn fungicides 106

Chemical Soil Treatments 106
- Soil fumigants 106
- Soil fungicides 106
- Miscellaneous fungicides 106

Safety Precautions When Handling Pesticides 107
- What's behind the pesticide label? . 107

Measuring Apparatus 109
When Spraying or Dusting 109
Multipurpose Sprays and Dusts . . . 110
- For vegetables, fruit, flowers, trees, and shrubs 111

"Shot-gun" Soil Drench 111
To Spray or To Dust? 111

EQUIPMENT

Sprayers and Dusters 112
- A. Sprayers 112
  - Household sprayers 112
  - Compressed air sprayers . . . 112
  - Knapsack sprayers 113
  - Slide pump or trombone sprayers 113
  - Wheelbarrow, cart, and barrel sprayers 113
  - Garden hose sprayers 114

Small power sprayers 114
Maintenance of Sprayers 114
B. Dusters 115
Plunger type dusters 115
Small bellows, crank, or rotary-fan dusters 116
Knapsack dusters 116
Small power dusters 117
Maintenance of Dusters 117
Spreaders, Stickers, and Wetting Agents 118
Fungicide Manufacturers and Distributors Plus Leading Spraying and Dusting Equipment Manufacturers . 120

ONCE a plant problem has been diagnosed as a disease (see also Sections 2 and 4), the next question is, "What can I do about it?" When a disease has progressed sufficiently to be easily recognized, it is often too late to start a spray or dust control program. Prevention and protection are usually the answer. Start *before* disease appears. Tree surgery, however, is useful as a curative measure in control of many cankers, blights, and decays of woody plants.

Successful disease control is based on an understanding of the cause, the pathogen's life cycle, time of infection, parts of plant involved, how the disease agent is distributed, and cost. Control should start with the purchase of the best seed or planting stock available and should continue in the seedbed, through the season in the garden or field, and even after harvest until the product is completely disposed of. There is much more to disease control than just dusting and spraying.

FACTORS AFFECTING DISEASE DEVELOPMENT

Diseases vary in severity from year to year and from one locality or geographical area to another, depending on the environment (principally amount and frequency of rains or heavy dews, relative humidity, air and soil temperature, and plant nutrition), relative resistance or susceptibility of the host plant, and the causal or disease-producing organism. All factors must be present and "in balance" for disease to develop. This can be put as a simple equation:

Susceptible Plant	+	Disease-Producing Organism	
+		+	= *DISEASE*
Proper Environment	+	Method of Distribution	

No infectious disease develops if *any one* of the above ingredients is lacking.

For example, if the environment is favorable for a disease and the causal organism is present, but the plant is highly resistant, little or no disease will develop. Similarly, if the disease organism is present and the plant is susceptible, but the environment is unfavorable, disease probably will not appear.

Environment—Many diseases develop best when there is abundant moisture during the growing season. Rain, irrigation, or heavy dew is necessary for spores of disease-producing fungi to germinate and attack a plant. Some diseases, such as seed rot and seedling blights, are favored by low soil temperatures after planting. Warm, dry weather favors build-up of insects, such as aphids, thrips, and leafhoppers, so virus diseases spread by these insects are usually most severe during warm, dry seasons.

Soil fertility is another environmental factor that may affect the severity of some infectious diseases. Soils maintained at highly productive levels by the wise use of proper fertilizers tend to produce vigorous plants. However, some diseases are seemingly unaffected by soil fertility, and a general statement cannot be made that fertile soil means healthy plants.

Host Resistance—Great differences exist among plants, varieties, and cultivars in their ability to resist various diseases. Inherent resistance or susceptibility often determines whether an epidemic (or epiphytotic) of a given disease will occur. Resistance to most diseases is determined by several genes. It is the genetic resistance of the host that can be manipulated by the plant breeder. He can modify hybrids to produce high levels of disease resistance combined with other desirable horticultural characters.

Causal Organism—No disease epidemic will occur if the specific disease-causing organism—a fungus, bacterium, virus, or nematode—is absent, even though the environment is favorable and the plant susceptible. The causal organisms are as variable in their virulence as the plants they attack are in their relative resistance. There are, for example, a number of strains or physiologic races of many fungi causing rust and smut diseases. These vary in their virulence on different varieties.

Effective disease control measures are aimed at "breaking" the above equation in one of three basic ways: (1) the susceptible plant is made more resistant or immune; (2) the environment is made less favorable for the causal organism and more favorable for the host plant; and (3) the disease organism is killed or prevented from reaching the plant, penetrating it, and producing disease.

These three basic methods of control can be divided into a number of different practices to help keep diseases in check. These include:

Plant adapted varieties and types of plants recommended by your state agricultural experiment station, cooperative extension service, and reputable seedsmen or nurserymen as being adapted to your area. Such varieties should include those having resistance to one or more common and serious diseases, plus possessing other desirable qualities. Unfortunately, resistant types are often less valuable because of some undesirable quality (e.g., poor foliage, inferior flowers or fruit, low yield). There are different degrees of resistance varying from tolerance or partial resistance to complete immunity. New resistant varieties of vegetables, fruits, flowers, lawngrasses, and other garden plants are released each year.

Progress with control by resistant varieties is hindered because parasitic organisms mutate, hybridize, or otherwise produce new races that attack varieties formerly resistant or immune. Certain fungi, bacteria, nematodes, and viruses are known to have few to many hundreds of physiologic races or strains.

Buy disease-free planting stock from a reputable nursery whose stock is carefully checked one or more times each season by experienced nursery inspectors. This insures comparative freedom from damaging insects, mites, and diseases. If available, buy clean, certified, disease-free seed, propagating material, or plants. Disease-free seed is often grown in arid parts of the western United States, under irrigation, where many diseases are unknown (e.g., black rot and blackleg of cabbage, bean anthracnose, and bacterial blights of peas and beans). Use seed, bulbs, corms, tubers, cuttings, or other plant parts, *only* from healthy plants. "Cultured," disease-free cuttings are available, for example, for chrysanthemums, carnations, geraniums, and possibly other plants.

Treat the seed of most flowers and vegetables against seed rot and damping-off by dusting seed lightly with thiram, captan, or chloranil (see Table 20 and Figure A.1 in Appendix) before planting. Bean, corn, crucifer, cucurbit, pea, and onion seed should also be treated with an insecticide to destroy seed-feeding insects (page 531). Follow manufacturer's directions. Seed can be treated in small packets or in larger quantities using Mason jars. See page 527 in Appendix. Seed treatment is good insurance for increased stands and bigger yields, especially if soil is cold and wet after planting.

Seeds, bulbs, corms, roots, and rhizomes may be disinfected by hot water or mercury which kills organisms (bacteria, fungi, nematodes) on the surface as well as within. See Table 20 in Appendix. Care should be taken with small seeds that are more readily injured by disinfectants than larger seeds.

Control damping-off—Occasionally captan, ferbam, zineb, folpet, Dexon, thiram, oxyquinoline sulfate (1 teaspoon to 3 gallons of water), Pano-drench, or Semesan, 1 to 2 tablespoons in a gallon of water, applied at ½ to 1 pint per square foot, checks advance of damping-off, root and crown rots. PCNB is also useful for certain soil-borne fungi, especially those that produce sclerotia. Before soaking infested area with fungicide, the diseased plants and about 6 inches of surrounding soil should be removed. Manufacturer's directions should be carefully followed. Do *not* use mercury-containing fungicides (e.g., Semesan, Pano-drench) in confined areas or on such susceptible plants as pansy, petunia, rose, snapdragon, and violet.

For a good multipurpose "shot-gun" soil drench, see page 111.

Plant in a well-prepared, fertile, well-drained seedbed, at times most suitable for your area. If possible, avoid heavy clays and very sandy soils. In most parts of the United States, water plants once a week during dry periods when less than an inch of rain falls during a week. Water early in the day so that plant foliage and stems will dry off before dew appears. Soak the soil thoroughly to a depth of 6 inches or more using a sprinkler, perforated hose, furrow irrigation, etc. Follow planting instructions and cultural practices published by your state agricultural experiment station and cooperative extension service and those given in nursery and seed catalogs. These instructions should include information on seedbed preparation, depth and rate of planting, spacing for efficiency and care of seedlings.

Follow a recommended crop rotation that excludes the same or closely related crops in the same garden area for 3 or 4 years or more. Most disease-causing organisms persist in soil, or in decaying crop debris, from one year to the next. Proper rotation "starves out" many of these organisms. The fungi causing clubroot of cabbage, verticillium and fusarium wilts, however, live almost indefinitely in soil without their favorite host plants being present. Some disease-causing organisms can remain alive through a compost pile or passage in the digestive tract of farm animals.

Pasteurization (usually called sterilization) *of infested soil and compost* is often an important control practice. See pages 538–548 in Appendix regarding when and how to disinfest soil using heat or chemicals (fumigants). All plant-parasitic organisms are killed by heating soil to 180° F. for 30 minutes.

Control weeds (by cultural or chemical means), especially perennials and winter annuals. Weeds are important reservoirs of insects, viruses (e.g., mosaics, curly-top, and aster yellows), and other disease-producing agents. Viruses can easily be carried to nearby garden plants by insects (e.g., leafhoppers, aphids, thrips, whiteflies, mealybugs) and mites. Weeds also decrease space, light, air circulation and slow normal drying of foliage following wet periods. This leads to more severe injury from leaf spots, blights, and mildews. The serious drain of plant nutrients and water from soil by weeds is well known to all gardeners. Check with your county agent or extension weed specialist regarding the latest weed control recommendations.

Control insects, mites, and rodents—Many diseases are spread largely or entirely by insects and mites. Control most foliage-feeding insects and mites by a multipurpose spray or dust containing malathion plus methoxychlor, DDT, lindane, sevin, Dibrom (naled), rotenone, or related materials. Excellent insecticides for controlling underground-feeding insects contain Diazinon, chlordane, aldrin, or dieldrin (page 23). If insect, mite, and rodent feeding could be prevented, losses from fruit rots, wood decays, root and crown rots, certain wilts and foliage diseases could be materially decreased. Rodents are especially fond of tulip and lily bulbs and crocus corms. Check with your county agent or extension entomologist regarding the latest insect and rodent control recommendations.

Avoid deep and close cultivations—Cultivator wounds weaken plants and provide easy entry for fungi and bacteria-producing root and crown rots, certain wilts, root rots, and crown gall. Mulch if possible or practical.

Sanitary measures are as important in keeping plant diseases in check as they are for animal and human diseases. Collect and burn infected plants or plant parts as they become infected. A disease may start in one or several plants and then spread throughout a garden when conditions (moisture, right temperature, susceptible plants) are favorable. Destroy (rogue) the first infected plants or plant parts as soon as found. This is very important with certain virus diseases where *entire* plants should be destroyed. Burn tops after harvest or plow under such debris cleanly. Sanitary measures are equally important in controlling insect, mite, and rodent pests. Eliminate trash piles, weedy fence rows, and bushy hedges. Avoid transferring bacteria, fungi, mites, certain viruses, and insect eggs on hands, clothing, or tools. Tools may be disinfected by dipping in 70 per cent alcohol (methanol, ethanol, or rubbing), 5 per cent formaldehyde, household bleach (e.g., Clorox, Purex, Saniclor), or a 1:1,000 solution of mercuric chloride (page 101). Hands should be scrubbed with soap and running hot water before handling healthy plants. Work among plants only when they are dry.

Sanitary measures are often the *only* ones needed in a home garden. But it helps if you can convince your neighbors to do likewise! Pruning of infected twigs, branches, and shoots helps to control many diseases. But prune with discretion. Excessive pruning may stimulate excess shoot growth that is often more susceptible to reinfection. Prune at least several inches below any sign of infection.

Changing a cropping practice is often a good method of disease control. Staking or mulching tomato vines, instead of letting them lie on the ground, reduces losses from fruit rots. Pruning to open up trees and shrubs lets in sun, increases air circulation, and hence reduces the chance of foliage diseases getting a foothold. Shallow planting often means a better stand of seedlings under cold, wet conditions. Mulches of leaves, corncobs, straw, earth, or snow protect plants against winter injury. Aeration of soil, application of a suitable fertilizer, planting to escape migrations of leafhoppers, aphids, or other pests, and watering during drought periods are other means of escaping injury from certain diseases.

Observe quarantine regulations—The majority of our most damaging plant diseases and other pests are alien. Some were brought to America by early settlers. Others came in with later emigrants. If allowed unrestricted entry, new invasions of plant enemies from foreign countries and offshore islands could cost U.S. citizens hundreds of millions of dollars in lost and damaged crops, reduced production and efficiency of our farms, ranches, gardens, nurseries, and forests—plus increased expenditures for pest control.

Since its beginning in 1912, the plant quarantine system has saved farmers and consumers billions of dollars by reducing sharply the rate at which destructive plant pests have gained entry into the United States from other countries or from offshore islands. The rapidly growing volume of travel, speedup in transportation, and multiplication of international ports, interior airports, and border-crossing points to Mexico and Canada have greatly increased the danger that harmful pests may be inadvertently brought into this country unless additional precautions are taken.

An ever-increasing volume of plant pests is being intercepted by the technically-trained staff of over 600 USDA plant quarantine inspectors each year. Pests capable of causing many millions of dollars worth of damage are being intercepted and stopped at the rate of about one entry every 12 minutes around the clock throughout the year. Over 470,000 lots of prohibited plant material is intercepted each year!

Federal plant quarantine laws and regulations prohibit or restrict entry into the United States of foreign plant pests and of plants, plant products, soil, or other materials or conveyances carrying or likely to carry plant pests. Special provisions permit entry—under strict quarantine conditions—of otherwise prohibited plants, pests, soil, or other materials for research purposes.

For information regarding quarantine regulations, and answers to specific questions about them, write to the Plant Quarantine Division, Agricultural Research Service, Washington 25, D.C. If you wish to import plants by mail, first obtain a permit by writing to: U.S. Department of Agriculture, Plant Quarantine Division, Plant Importations Branch, 209 River St., Hoboken, N.J.

Destroy or remove alternate hosts of rust diseases—See (8) Rust under General Diseases. Which type of plant is destroyed depends on such factors as the number and value of plants involved. Many such plants can now be protected against rusts by the proper and timely application of fungicides.

Store only sound, well-matured, dry fruits, and vegetables free of cuts, bruises, rot, insect or rodent injuries. Requisites of good storage are the proper temperature, aeration, humidity (and moisture content, where applicable) plus cleanliness (page 208). If in doubt concerning recommended storage conditions for your garden produce, check under the crop involved and with your county agent or extension horticulturists. The USDA Farmers' Bulletin 1939, *Home Storage of Vegetables and Fruits,* covers this subject in detail. For additional information on the most favorable temperature and approximate length of the storage period for cut flowers, rhizomes, tubers, roots, bulbs, corms, and nursery stock, see a book such as USDA Handbook 66, *Commercial Storage of Fruits, Vegetables, Florist, and Nursery Stocks.*

Apply protective fungicides—This means applying the *right* chemicals in the *right* way at the *right* concentration and at the *right* time. If this is done, there will be no danger of damaging tender plants or poisonous pesticides remaining on ripening fruits and vegetables.

It usually pays to follow a regular preventive schedule. Applications should generally be made every 3 to 7 days in wet weather and at 10- to 14-day intervals if weather is dry. If rains of an inch or more wash the fungicide away, reapply. Insecticides should be applied on a planned, protective schedule every 7 to 10 days, before large numbers of insects appear and damage occurs. For certain insects closer spacing of sprays is needed. Complete coverage of foliage and fruit is essential.

WHAT IS A FUNGICIDE?

A fungicide is any chemical that kills or inhibits fungi that land by chance on the plant, as spores or mycelium. No single compound is suitable for all purposes or is effective against all fungi.

Fungicides can be conveniently divided into four groups, according to their action.

Protective fungicides are applied as foliage and fruit sprays or dusts to keep disease-causing fungi from entering plants. These materials provide protection, but they *do not* (1) kill fungi established within a growing plant or seed (Exceptions: powdery mildew and sooty mold fungi, that are superficial and largely on the surface of plants, can be killed by surface dusts or sprays, after infection has occurred, without injuring the host plant); (2) protect against disease-causing organisms entering through roots, e.g., root rots, wilts, and clubroot; (3) control bacterial diseases—since most fungicides are poor bactericides; (4) protect against viruses, which are frequently injected into plants by insects; and (5) control nematodes.

Most fungicides in use today possess protective qualities. Those that are *only* protective include glyodin, zineb, sulfur, thiram, Polyram, ferbam, ziram, and possibly the inorganic copper materials. These chemicals *must* be applied before an infection starts. This means frequent applications at about 7- to 14-day intervals, depending on weather conditions.

All dust formulations function as protective fungicides and should be used accordingly. Dusts should be applied when wind is low and foliage is covered with moisture. Early morning or evening is an ideal time.

Eradicant fungicides are applied as foliage sprays, seed treatments, or soil drenches to kill or inhibit fungi *after* they have penetrated plants and become established. Examples are phenyl (organic) mercury materials used by commercial apple growers to "burn out" apple scab infections, and the mercury-containing chemicals used on certain types of seed, bulbs, tubers, and rhizomes to kill organisms under the seed coat or within propagative plant parts. These fungicides have limited uses and are often dangerous to use on green foliage and fruit. Dichlone is used as an eradicant, but it mainly inhibits growth of the organism without killing it. When temperatures are cool, dichlone becomes a fairly effective protectant.

Protective and eradicant fungicides are used to control foliage and fruit diseases and may also perform well as seed treatments. These materials may not be quite so residually effective as protective fungicides, but in addition to offering protection they are sufficiently toxic to fungus spores and mycelium to eradicate or "burn out" established infections. Captan, folpet, and dodine possess both fungicidal qualities. Other fungicides that have good protective characteristics and that may also partly eradicate established infections are maneb, Niacide M, and nabam.

Chemotherapeutants are chemicals that are absorbed and distributed within the plant to control certain diseases. Very few chemicals (examples are Acti-dione and oxyquinoline materials) now available work in this way, but chemotherapy is a promising field of research.

MODERN FUNGICIDES

Since World War II, a great many new fungicides have been introduced into American agriculture. These chemicals have largely replaced such old standbys as bordeaux mixture, fixed or neutral coppers, lime-sulfur, and wettable or paste sulfurs. These older materials are messy to handle and corrosive to spray equipment, cause injury to plants, and often reduce the quality and quantity of crops they were designed to protect. Unfortunately, many retail pesticide outlets still stock these outmoded fungicides and exclude the generally safer and more effective modern chemicals, e.g., captan, zineb, maneb, ferbam, thiram, dodine, folpet, ziram, etc. (Table 4).

If you cannot get some chemicals mentioned in this book or those recommended by your state cooperative extension service and agricultural experiment station, check with your county extension office or the list of names and addresses of leading fungicide manufacturers and distributors at the end of this section. If you cannot get a pesticide locally, write the chemical manufacturer or distributor.

Fungicides are marketed under a bewildering assortment of trade names. To relieve confusion, a set of common or "coined" names has been officially adopted and is now used on package labels with the more complicated chemical names (called active ingredients). Table 4 summarizes the common names, active ingredients, trade names, major distributors, and principal uses of the more common modern fungicides. This listing is necessarily incomplete as there are over 61,000 pesticides now registered by the Pure Food and Drug Administration in Washington, D.C., involving some 900 pesticidally active ingredients. About 35,000 of these pesticides are registered for agricultural use or for uses closely related to agriculture.

Where specific trade names or manufacturers are mentioned it is to be understood they are *not* listed to the exclusion of similar and competitive products or firms. They are simply representative of the most generally available products in the United States.

The most useful modern fungicides for the home gardener are captan, zineb, folpet, maneb, dinocap, PCNB, and mercuric chloride.

OTHER USEFUL FUNGICIDES

Dinocap (sold as Karathane-WD, Karathane Liquid Concentrate, and Karathane Dust). Contains 2-(1-methylheptyl)-4-6-dinitrophenyl crotonate and isomers as the active ingredients. This fungicide is *specific* for control of powdery mildews. Karathane has replaced sulfur in many multipurpose sprays and dusts. Compatible with practically all fungicides, insecticides, and miticides in these combination mixtures (page 111). Do *not* use in hot weather (above 85° F). Apply when foliage will dry rapidly. Has good eradicant action, but little residual. Used normally at rate of ½ to ⅔ teaspoonful per gallon of water.

Mercuric chloride is also known as corrosive sublimate and bichloride of mercury. Used as a general disinfectant, soil drench, dip treatment, and tree wound dressing. Controls rots of gladiolus, calla lily, canna, and iris besides killing disease-causing organisms carried inside pepper, cucumber, melons, cabbage, and other seed. The dipping time in mercuric chloride varies from 5 minutes to 2 hours or longer, depending on the plant part treated.

Mercuric chloride is a *caustic, deadly poison.* Handle with caution. Usually sold by pharmacists as a white powder or as coffin-shaped tablets (e.g., Diamond Poison Antiseptic, Mallinckrodt Corrosive Sublimate, Centerchem Mercuric Chloride). Commonly prepared to make a 1 in 1,000 solution by dissolving 1 ounce in 7½ gallons of water, or one 7.3-grain tablet in a pint of water. Handle *only* in nonmetallic containers such as wood, glass, enamel, or earthenware. Dissolve chemical in a small amount of hot water and add to rest of water. Seeds, bulbs, and corms should be washed in clear, running water for 10 to 15 minutes after treating. Dry, then plant. See "Seed Treatment Methods and Materials" in Appendix.

Mercuric chloride and mercurous chloride are combined in several useful lawn fungicides—Calo-clor, Calocure (Mallinckrodt), Fungchex, and Woodridge Mixture "21."

PCNB contains pentachloronitrobenzene as its active ingredient. A long-lasting soil fungicide applied as a dust, wettable powder, emulsifiable concentrate, or as granules. Sold as Terraclor, PCNB, Terrafume-2, Turf-Tec T, Terra Tox 10, Brassicol, and Fungiclor. Controls various soil-borne root, stem, and crown rots of vegetables, lawngrasses, and ornamentals,

TABLE 4–MODERN FUNGICIDES

Common Name and Active Ingredient	Trade Names and Distributors	Principal Uses and Remarks
CAPTAN N-(trichloromethylthio)-4-cyclohexene-1,2-dicarboximide *or* N-(trichloromethylthio) tetrahydrophthalimide	Captan 50-W, Captan 75 Seed Protectant, Captan-Dieldrin 60-15 Seed Protectant, Captan Garden Spray, Captan 80-WP Spray-Dip (Stauffer), Orthocide 50 or 80 Wettable, Orthocide Fruit and Vegetable Wash, Orthocide 75 Seed Protectant, Orthocide Garden Fungicide (Chevron), Chipman Captan Dust (Chipman), Agway Captan 5D and 7.5D (Agway), F & B Captan 7.5 Dust (Faesy & Besthoff), Miller Captan Dust (Miller), Captan 7.5 Dust (Chevron), etc.	Excellent, safe fungicide to control leaf spots, blights, fruit rots, etc., on fruits ornamentals, vegetables and turf. Seed protectant (often mixed with insecticide) for vegetables, flowers, and grasses. Postharvest dip for fruits and vegetables. Does *not* control powdery mildews and rusts. Soil treatment on plant beds to control crown rot and seedling blights. Widely used in multipurpose sprays and dusts. Both a protectant and an eradicant.
CHLORANIL Tetrachloro-*p*-benzoquinone	Spergon, Spergon Wettable, Spergon Seed Protectant (U.S. Rubber), Spergon Spray Powder (Niagara, General Chem.), Geigy SP 50 (Geigy), Spergon Dust (General), etc.	Seed and bulb treatment for flowers, vegetables, and grasses. Soil drench for crown rot of flowers. Corm and bulb dip for flowers. Sprays and dusts for certain foliage diseases.
DICHLONE 2,3-Dichloro-1,4-naphthoquinone	Phygon, Phygon-XL, Phygon Seed Protectant (U.S. Rubber), Corona Phygon-XL (Pittsburgh Plate Glass), Niagara Phygon (Niagara), Green Cross Phygon-XL Fungicide (Green Cross), Stauffer Phygon-XL (Stauffer), etc.	Seed treatment for certain vegetables and flowers. Spray for certain blights and fruit rots of vegetables and fruits. Soil drench to control damping-off. Treat as directed. Injurious at 85° F. or above. Mostly eradicative.
DODINE N-dodecylguanidine acetate	Cyprex Dodine 65-W, Cyprex Dodine Dust (Cyanamid), Miller's Cyprex Dusts (Miller), Pennsalt Cyprex 4% Dust (Pennsalt), etc.	Controls certain foliage diseases of apple, cherry, strawberry, pecan, and roses. Gives long-lasting protection. Good eradicant.
FERBAM Ferric dimethyldithiocarbamate	Fermate Ferbam Fungicide (DuPont), Karbam Black (Sherwin-Williams), Carbamate (Niagara), Ortho Ferbam 76 (Chevron), Orchard Brand Ferbam (General), Coromate Ferbam Fungicide (Pittsburgh Plate Glass), Chipman Ferbam W-76 (Chipman), etc.	General, safe fungicide to control many foliage diseases of flower, trees, shrubs, and fruits. Soil drench to control damping-off and seedling blights. Used in some multipurpose sprays. May leave objectionable black spray deposit on flowers, woodwork, etc. Mostly protective.

FOLPET (Phaltan) N-trichloromethylthiophthalimide	Corona Folpet 50 Wettable (Pittsburgh Plate Glass), Ortho Phaltan Rose and Garden Fungicide, Ortho Phaltan 50 Wettable (Chevron), Niagara Phaltan 50 Wettable (Niagara), Folpet 50-WP and 75-WP (Stauffer)	A close relative of captan and used for many of the same purposes on fruits, flowers, turf, vegetables, trees, and shrubs. Controls many powdery mildews. Excellent for roses. Both a protectant and eradicant. In multipurpose mixes.
MANEB Manganese ethylenebis (dithiocarbamate) also special formulation containing zinc	Manzate Maneb Fungicide, Manzate D Maneb (DuPont), Dithane M-22 and M-22 Special (Rohm & Haas), Kilgore's Maneb 80 Wettable (Kilgore), Dithane M-45 (Rohm & Haas), Twin Light Maneb Dust (Seacoast), 2% Maneb Dust (Carolina), Maneb 4.5D (Agway), etc.	Excellent general fungicide to control foliage and fruit diseases of vegetables, trees, turf, flowers, and some fruits. Very useful for tomato, potato, and vine crops. In multipurpose mixes. Controls rusts but not powdery mildews. Mostly protective.
THIRAM (TMTD) Bis (dimethylthiocarbamoyl) *or* Tetramethylthiuram disulfide	Tersan 75, Thylate Thiram Fungicide, Delsan A-D, Arasan 50-Red and 75 (DuPont), Panoram 75 and D-31 (Morton), Thiram 50 Dust (U.S. Rubber and DuPont), Thiram 4.8D (Agway), Thiram-65 and 75, Thiram S-42 (Pennsalt)	Seed and bulb treatment for vegetables, flowers, and grasses. Controls certain lawn, fruit, and vegetable diseases. Controls rusts. Soil drench for crown rot and damping-off. Only protective.
ZINEB Zinc ethylenebis (dithiocarbamate)	Dithane Z-78 (Rohm & Haas), Parzate Zineb Fungicide, Parzate C (DuPont), Ortho Zineb 75 Wettable, Ortho Dust (Chevron), Niagara Zineb (Niagara), Stauffer Zineb 75-W (Stauffer), Chipman Zineb (Chipman), Corona Zineb (Pittsburgh Plate Glass), Flight Brand 10% Dithane Dust (Carolina), Zineb 19.5D TF (Agway), 6.5% Zineb Dust (Flag Sulphur)	Excellent, safe fungicide for vegetables, fruits, flowers, trees, and shrubs to control leaf spots, blights, fruit rots, etc. Also useful on lawns as a soil drench to control crown and root rots. Controls rusts but not powdery mildews. In many multipurpose mixes for vegetables and flowers. Only protective.
ZIRAM Zinc dimethyldithiocarbamate	Zerlate Ziram Fungicide (DuPont), Karbam White (Sherwin-Williams), Z-C Spray or Dust (Niagara), Orchard Brand Ziram (Gen.), Ziram (Chevron, Chipman, Stauffer), etc.	General, safe fungicide. Useful for vegetables and ornamentals, especially tender seedlings. In many vegetable and flower multipurpose mixtures. Only protective.

clubroot of crucifers, potato scab and scurf, pink rot of celery, and damping-off of many plants. Often mixed with captan, maneb, dichlone, ferbam, thiram, Dexon, or folpet. We will see more of this and other soil fungicides in the future. Captan-PCNB mixtures are sold as Terracap, PCNB-Captan 10-10 Dust, PCNB-Captan 25-25 Wettable Powder, and Orthocide Soil Treater "X." *Daconil 2787* is a new, safe, broad-spectrum fungicide containing tetrachloroisophthalonitrile. Controls leaf spots and blights, blossom blights, fruit spots and rots, gray-mold or botrytis blights, and certain powdery mildews on lawngrasses, flowers, vegetables—especially potato, tomato and vine crops—and certain fruits. A special formulation is sold for use on turf and ornamentals.

Dichloran (DCNA), sold as Botran, contains 2,6-dichloro-4-nitroaniline. Various dust and wettable powder formulations are available. Useful as a foliar and soil fungicide for crown and root (bulb) rots, and gray-mold or botrytis blights of certain vegetables and flowers. One of the few fungicides effective against such soil-borne fungi as *Sclerotium, Sclerotinia,* and *Stromatinia.* Used as a post-harvest dip for sweetpotato roots and stone fruits to control *Botrytis* and *Rhizopus.* Follow manufacturer's directions.

Metiram (Polyram) is a new general foliar and seed protectant fungicide containing zinc polyethylene thiuram disulfide complex. Sold as an 80 per cent wettable powder. Useful for controlling rusts, downy mildews, leaf spots, and blights of vegetables and ornamentals. Similar to maneb in range of effectiveness. May be useful as a soil fungicide to control damping-off *(Pythium, Rhizoctonia).*

Lanstan is a new soil fungicide containing 1-chloro-2-nitropropane. Useful for vegetables and ornamentals to control damping-off, crown and root rots, clubroot of crucifers, scab, and certain wilts. Apply in planting furrow or incorporate into top several inches of soil. Follow manufacturer's directions.

Difolatan (Folcid) is chemically related to captan and folpet. Available as an 80 per cent wettable powder or 7.5 per cent dust. Contains N-(1,1,2,2-tetrachloroethylsulfenyl) -cis-Δ-4-cyclohexene-1,2-dicarboximide. A broad-spectrum fungicide that controls downy mildews, anthracnose, leaf spots and blights, and gray-mold on a number of vegetables, fruits, and ornamentals. May be used as a seed or soil treatment to control seed decay and damping-off. Follow manufacturer's directions.

Dexon is a new seed and soil fungicide that contains p-(dimethylamino) benzenediazo sodium sulfonate. Available as a 35 or 70 per cent wettable powder and as granules. Controls damping-off and root rots of many ornamentals, vegetables, and fruits, caused by water molds *(Pythium, Aphanomyces, Phytophthora).* Also sold in combination with PCNB and Dyrene.

Fixed or neutral copper fungicides are multipurpose fungicides represented by a large group of trade names (e.g., Corona Micronized Tri-Basic Copper Sulfate, Corona "26" Copper Fungicide, Microgel, Orchard Brand 530, Duo Copper, Cuprocide, Basic Copper Fungicide, Farmrite M-53 Fixed Copper, Basi-Cop, Tri-Basic Copper Sulphate, Coposil, Spraycop, Niagara C-O-C-S, Tricop, Ortho-K, Tennessee "26" Copper Fungicide, Ortho Copper Fungicide "53," Micro Nu-Cop, and many others). Similar in usage to bordeaux mixture (sprays, dusts, soil drenches), except fixed coppers often do not require mixing with lime, thus do less damage to young tender plants or others in cool, moist weather. Fixed coppers are also easier to handle and use. Two new, completely water-soluble copper fungicides are Sol-Kop 10 and TC-90. These products are less injurious to copper-susceptible plants (e.g., vine crops) and are more effective against a wider range of diseases including those caused by bacteria—angular leaf spot of cucurbits and bacterial spot of pepper and tomato. These and other coppers are still recommended to control various blights and leaf spots of vegetables, flowers, trees, and shrubs. Materials used as sprays usually contain 53 to 55 per cent metallic copper; dusts 5 to 11 per cent.

Bordeaux mixture has largely been replaced by new fungicides that do not "burn" leaves or "russet" fruit. Bordeaux is a mixture in water of copper sulfate (bluestone or blue vitriol) and hydrated spray lime. The formula is written in figures (e.g., 4-4-50). The first figure is copper sulfate in pounds, the second spray lime in pounds, and the third number water in gallons. Bordeaux can be made stronger than this or can be reduced in strength if it should "scorch" plants.

For preparing small amounts of 4-4-50 bordeaux mixture, dissolve 2 ounces (or 2 level tablespoons) of copper sulfate crystals ("snow") in a gallon of water. Then dissolve 2 ounces (2 heaping tablespoons) of fresh, hydrated spray lime in 2 gallons of water. Add the copper sulfate solution to the lime water. Strain mixture into the sprayer through several layers of cheesecloth. This makes 3 gallons of 4-4-50 bordeaux mixture. Use immediately.

Various dry products, ready to mix with water, are available (e.g., Acme Bordeaux Mixture, Bor-dox, Copper Hydro Bordo, and Ortho Bordo Mixture) but are generally inferior to homemade bordeaux. Bordeaux is still used to control certain fungus leaf spots, blights, anthracnose, and as a general disinfectant for storage cellars, work surfaces, and other areas. Bordeaux paint is used as a tree wound dressing (page 36).

Burgundy mixture solution is sometimes used instead of bordeaux, especially in Europe, to avoid spotting foliage. It is composed of copper sulfate, sodium carbonate (sal or washing soda) and water. Five gallons of this spray can be made by using 2 ounces of copper sulfate, 3 ounces of sodium carbonate, and 5 gallons of water. The two chemicals should be dissolved separately, then mixed before using. Spray while fresh for same uses as bordeaux. Very adhesive, but often more toxic to plants than bordeaux.

Sulfur products are available as wettable powders, pastes, liquids, and dusts under a wide variety of trade names. Twenty-five to 97 per cent wettable or colloidal sulfur makes a fine suspension in water. For best disease control use wettable sulfurs with an average particle size of not more than 5 to 7 microns. Sulfur dusts should be fine enough to pass through a 300- or 325-mesh screen. Sulfur is used primarily to control powdery mildews on many plants, brown rot of stone fruits, certain rusts, leaf blights, and fruit rots. May cause injury in hot, dry weather especially to sulfur-sensitive plants (e.g., viburnum, melons, tomato, gooseberry, raspberry, grape, blueberry). Sulfur is being rapidly replaced by newer fungicides—but it was the *first* one used. The "pest-averting sulphur" was mentioned in Homer's *Iliad* and *Odyssey,* written in about the ninth century B.C. Sulfur was probably used to control rusts, powdery mildews, and probably insects on cereals.

A green-colored sulfur that leaves no unsightly stain on the foliage is available.

Lime-sulfur is a reddish-brown liquid effective against both fungi and certain insect and mite pests. It is made by boiling lime and sulfur together. Most useful as a dormant or delayed dormant spray for trees, shrubs, and fruits to control blight or anthracnose, powdery mildew, apple scab, mites, and scale insects. Use one pint of concentrated lime-sulfur in 9 or 16 pints of water. Lime-sulfur is too toxic to foliage and fruit to use for summer sprays. Keep it away from white paint unless you want the paint blackened! Do *not* expose lime-sulfur to freezing temperatures. It is incompatible with many modern pesticides (page 552).

Antibiotics—Recently antibiotics have been widely used to control plant diseases. Antibiotics may be absorbed through plant surfaces to check or eradicate an infection, plus protecting against other diseases becoming established. The future of these materials in therapeutic control of certain, hard-to-control diseases is promising. We can look forward to even more useful antibiotics in the future. Two of the most widely available are:

A. *Streptomycin*—An antibacterial antibiotic sold commercially for plant use as Agrimycin 17 (Charles Pfizer), Phytomycin (Squibb, Olin Mathieson), Ortho Streptomycin (Chevron), Streptomycin Antibiotic Spray Powder (Miller), and Agri-strep (Niagara, Stauffer, Merck, Chevron). Agrimycin 500 (Charles Pfizer) is a mixture of 10 per cent streptomycin sulfate, basic copper sulfate and 1.0 per cent terramycin (oxytetracycline). Agrimycin 100 (Charles Pfizer, Chipman) contains 15.0 per cent streptomycin sulfate and 1.5 per cent oxytetracycline in combination.

Streptomycin formulations are used to control the blossom blight stage of fire blight [see (24) Fire Blight under General Diseases], bacterial spot of pepper and tomato, bacterial wilts, blights, and rots of various trees and ornamentals, and blackleg of potato. May cause severe chlorosis of some plants, e.g., geranium (***Pelargonium***).

B. *Cycloheximide*—An antifungal antibiotic, sold as Acti-dione by the Tuco Products Division, the Upjohn Company. It is absorbed through plant surfaces and distributed locally within a plant to check or eradicate infections. Acti-dione is also protective. It is effective against powdery mildews, cherry leaf spot, certain rusts, and several lawn diseases. Various formulations (e.g., Acti-dione PM, RZ, S, Ferrated, Acti-dione-Captan, and Acti-dione-Thiram) are sold for different purposes. Actispray (Upjohn and Niagara) comes as a tablet that dissolves in water. Acti-dione is used at concentrations as low as one part in one million parts of water! Do *not* overdose with this material.

Other new antifungal and antibacterial antibiotics will undoubtedly be widely available in the future.

Phenyl (organic) mercury materials are useful in controlling a number of lawn diseases, certain leaf blights and spots of trees and shrubs, bulb and corm rots, seedling blights, and a few fruit diseases (e.g., apple scab, strawberry foliage diseases). These materials act primarily as eradicant fungicides. Organic mercuries are sold as *liquids:* PMAS, Puratized Agricultural Spray, Puratized Apple Spray, Ortho LM Apple Spray, Tag Fungicide, Coromerc Liquid, Panogen Turf Spray, Pano-drench, Morton Soil Drench, Phenyl Mercury Lactate, and 10% Phenyl Mercury Acetate; and as *powders:* Phix and Coromerc. Do *not* apply phenyl mercury materials (1) to crop plants after fruit forms; (2) during bloom—they kill pollinating insects. Ceresan, Panogen, Chipcote, Semesan, Elcide 73, and Memmi are useful for treating certain types of seed, tubers, bulbs, and corms.

Lawn fungicides—The trend is toward use of a multipurpose fungicide mixture that controls a number of lawn and turf diseases. The more widely available products are Ortho Lawn and Turf Fungicide, Tersan OM, Thimer, Panogen Turf Spray, Acti-dione-Thiram, Dyrene, Fore, Daconil 2787, and Kromad. Zineb also controls a number of turf diseases. More specific lawn and turf fungicides often contain cadmium (e.g., Cadminate, Caddy, Cadtrete) or mercury chlorides (Calo-clor, Calocure, Woodridge Mixture "21," Fungchex).

CHEMICAL SOIL TREATMENTS

Soil fumigants are applied to the soil several weeks before planting. Apply when the soil moisture is relatively high and the soil temperature is at least 60° F., 4 inches deep. These chemicals generally break down in soil to release a toxic gas that kills not only fungi but also bacteria, nematodes, weed seeds, insects, and other animal life in the soil. Certain fumigants move through the soil slowly and require only a water "seal" after application. Other fast-acting ones must be confined with a plastic tarp or other covering to retain the fumes. The most useful fumigants to control *fungi* causing wilts, damping-off, root and crown rots, and other diseases include chloropicrin or tear gas (Larvacide, Picfume, etc.), methyl bromide (Bromex, Brozone, Pano-Fume, Weedfume, Edco MBX, Pano-Brome, Trizone, Trifume, Dowfume MC-2, Bed Fume, Pestmaster, Kolker Methyl Bromide, Mumfume, etc.), SMDC (Vapam, VPM Soil Fumigant, Chem-Vape), DMTT (Mylone, Barber "Pre-Plant" 50-D, Soil Fumigant M, Miller Mico-Fume), Bedrench, and Vorlex. All of these materials should be used *strictly* according to manufacturer's recommendations. Observe all safety precautions listed on the package label. For a full discussion on Soil Treatment Methods and Materials see pages 538–548 in Appendix.

Soil fungicides are usually applied as dusts, drenches, or granules to control damping-off, seedling blights, crown and root rots, wilts, and other diseases. A number of these fungicides have been mentioned above, e.g., captan, chloranil, dichlone, ferbam, folpet, Lanstan, PCNB, thiram, zineb, and ziram. Other chemicals applied to the soil are Pano-drench, Morton Soil Drench, Semesan, Dexon, Fulex A-D-O, Gerox, Bioquin, Natriphene, Wilson's Anti-Damp, and Sunox. Use these chemicals according to manufacturer's instructions.

Miscellaneous fungicides—A large number of fungicides are available that at present have limited uses. Some of these are CM-19, Dithane A-40 and S-31, glyodin (Crag Liquid Glyodin, Glyoxide Dry), Mico-Ban 531, Mildew King, Miller 658 Fungicide, Morsodren, Polytrap, Omazene, Amobam, etc.

SAFETY PRECAUTIONS WHEN HANDLING PESTICIDES

There is no pesticide (any chemical that kills pests) mentioned in this book that cannot be used with perfect safety if you follow necessary precautions. The chemicals were carefully chosen not only because of their general effectiveness but also for their relative safety to humans, animals, and plants.

Where food crops or soils to be planted with food crops are to be treated with a pesticide, check label information for crops approved and required interval beween treatment and harvest. Check with your county agent, extension entomologist, plant pathologist, or nematologist for the latest FDA regulations.

For excellent discussions on how pesticides are safety-tested, and how agricultural chemicals are used to protect our food, property, and health, read booklets such as *Open Door to Plenty, The Search for Abundance,* and *Pesticides and Public Policy* published by the National Agricultural Chemicals Association, 1155 15th St. N.W., Washington, D.C. 20005.

What's Behind the Pesticide Label?

The Federal Insecticide, Fungicide, and Rodenticide Act, administered and enforced by the Pesticides Regulations Division, Agricultural Research Service, of the U.S. Department of Agriculture, regulates all commercial pesticides sold in interstate commerce. State laws regulate the same products at the local level and also regulate pesticide products that are manufactured and sold within the same state.

1. *Every pesticide label must show:*
   - A. Name of product, brand, or trademark under which the product is sold.
   - B. Name and address of manufacturer, registrant, or person for whom manufactured.
   - C. The net content statement.
   - D. The registration number assigned to the product.
   - E. An ingredient statement—name and percentage, by weight, of each active ingredient, and total per cent of inert ingredients, or name of each active and each inert ingredient in descending order and relative abundance in each category and the total percentage of inert ingredients.
   - F. *Warning or Caution Statement* (required in a prominent place on the front panel).
     - (1) An economic poison *must* show warnings pertaining to ingestion, skin absorption, inhalation, and flammability or explosion. Appropriate warnings are also provided for protection of wildlife and fish.
     - (2) Economic poisons that are highly toxic to man must show "POISON" in red on a contrasting background; the statement "KEEP OUT OF REACH OF CHILDREN" or its equivalent; a "signal" word —such as "DANGER," "CAUTION," or "WARNING"—that draws attention to the need to handle the material carefully; skull and crossbones; and a statement of antidote (including directions to call a physician immediately) in the immediate vicinity of the skull and crossbones and "POISON."
   - G. Directions for use—optional on label, may appear on accompanying printed or graphic matter. Registered labels that bear directions for use give the names of crops or sites to which the material is intended to be applied, the diseases or pests to be controlled, dosage rates, a schedule of applications that provides effective control of the diseases or pests claimed, and any appropriate limitations in dosage or time of application consistent with residue requirements.
2. *Other label information that may be required:*
   - A. Data to support any or all claims on the labeling.
   - B. A complete statement of the composition of the product, including the percentage, by weight, of each of the active and inert ingredients, if such information does not appear on the label.
   - C. Any pertinent information about inert ingredients.
   - D. Any other information pertaining to physical or biological properties

of the product, e.g., caution against use on certain plant species or varieties known to give a toxic response to the chemical and against known incompatible combinations with other pesticides.

3. *It is unlawful to:*
   A. Represent a fungicide or other pesticide for use different from what the registration specifies.
   B. Sell products in containers other than the manufacturer's original one.
   C. Destroy, alter, deface, or detach the label from a container.
   D. Dilute or change the fungicide in the manufacturer's original container.

The label on each pesticide container is an important legal document. It represents four to 10 years or more of painstaking research, many thousands of hours of labor, involves about 20 pounds of technical data, typed double space on paper, and costs up to $3 million. The user has every reason to expect all claims on the label to be fair and reasonable. The law imposes many safeguards for the public but the most important act depends upon the user. All the research and precautions in the world are of no value if the consumer refuses to "READ THE LABEL." It is your responsibility to read it, follow the directions listed, and heed the precautions.

Approximately 80 per cent of all accidents from misuse of pesticides occur to children under 9 years of age. About 90 per cent of these happen to children under 5. The great majority of these accidents could be prevented if the individuals involved, or older members of their families, would follow these simple precautions:

Read and understand the entire package label before purchase. The information is printed for your protection. Reread the instructions, precautions, and warnings again before opening. Fungicides, like other pesticides, should be used strictly according to package directions, on *crops* specified, in *amounts* specified, and at *times* specified. Observe other precautions listed, especially safe handling, frequency of application, and interval between last application and harvest. Handle all pesticides with proper care and common sense.

Store chemicals in a locked, orderly-kept cabinet, outside the home, closed to irresponsible adults, children, and pets. Never leave pesticides open to children, pets, or birds. Do not store them near foods of any kind. Running water should be handy to flush away any spilled chemicals. Promptly destroy (burn or bury) empty or old pesticide containers so that they are not a hazard to man or animals. Wash out glass and metal containers before putting in the trash can.

Pour leftover spray into a gravel drive, down a drain or hole in the soil. *Never* leave puddles on an impervious surface to attract birds and pets.

Never use or store unlabeled chemicals, or those not in their original containers. Keep pesticide container tightly closed except when preparing the mix.

Never breath dusts, mists, or vapors of pesticides. Avoid spilling on shoes or other clothing. Immediately flush with large amounts of warm soapy water any body area contacted. Remove contaminated clothing and shoes.

Wear full protective clothing, where called for, when applying pesticides. Keep your sleeves and trouser legs rolled down, collar buttoned, and wear a washable cap. If respirator, goggles, rubber gloves, or other special preventive clothing is called for, leave this pesticide alone unless you are fully experienced and have notified your doctor what you are using and the antidote listed on the label.

Wash hands and face thoroughly before eating or smoking. Bathe promptly after spraying and change to fresh clothing. Launder clothing before reusing.

If you suspect pesticide poisoning CONTACT YOUR PHYSICIAN. He will probably direct you to the nearest poison control center, usually located in a hospital.

A Directory of Poison Control Centers has been published by the U.S. Department of Health, Education and Welfare, Public Health Service, Division of Accident Prevention, Washington, D.C. 20201.

Cover bird baths, pet dishes, and fish pools before spraying or dusting.

Avoid spray drift to fish ponds, streams, or other water supplies. Do not discard spray materials so they may contaminate water supplies.

Do not plant herbs, fruits, vegetables, and other edibles near ornamentals that may be dusted or sprayed with highly toxic, long-lasting pesticides.

Use wettable powder formulations or prepared dusts when combining insecticides and fungicides. Do *not* mix emulsion concentrates with wettable powders. Check the pesticide compatibility chart (page 552) before mixing chemicals together.

Do not apply any spray when the *temperature is 85° F. or above.*

Do not apply *oil sprays* if the temperature is below 45° F.

Do not contaminate your sprayer with weedkillers, especially those of the "hormone" type such as 2,4-D, 2,4,5-T, or MCPA. It is better to get a second sprayer and paint "WEED KILLERS ONLY!" in red on the sides.

Keep the sprayer or duster in good repair by following a regular maintenance program (pages 114, 117). Use oil-resistant hose and gaskets if applying dormant or summer oils.

MEASURING APPARATUS

For help in measuring out small to large amounts of liquids or powders (or conversion from one to the other) see pages 516–517, 520–521.

To avoid guesswork and possible plant injury in mixing pesticides, the following equipment, where applicable, is suggested:

1. For measuring wettable powders or liquids get a set of standard household measuring spoons and a measuring cup.
2. A letter balance is useful for weighing small amounts up to about 4 ounces.
3. A spring balance for weighing up to 4 pounds is something that many home gardeners will need. A larger scale weighing up to 25 pounds may be needed if large areas need spraying.
4. If using liquids, assemble a set of standard containers (¼ pint, ½ pint, 1 pint, 1 quart, 1 gallon, and 5 gallons) and a medicine dropper.

Do *not* use this equipment in the kitchen or for other household purposes. Keep it locked up with your pesticides.

WHEN SPRAYING OR DUSTING

For good disease and insect control it is important that *all* parts of the plant be uniformly coated. When spraying, be sure to spray from the top of the plant down, from the bottom up, and from the inside out (Figure 3.1). Remember that many fungi and bacteria penetrate only the underleaf surface. Apply a fine, misty spray that wets foliage evenly. Keep spray stream moving. Wet all surfaces until drops just start to fall (run-off). Do not drench the soil. Whenever possible, choose a mild, dry, windless day for spraying. Avoid a windy day or one when the temperature is below 45° F. or likely to drop to freezing before the spray dries.

One quart of spray mix should cover a 50-foot row of most flowers and vegetables when plants are young, and 20 to 25 feet when full grown.

Most sprays will be more effective if a wetting agent or spreader-sticker (see page 118) is added to the spray solution. Most household detergents and soaps are satisfactory. Add a sufficient amount (usually ¼ to 1 teaspoonful per gallon of spray) so that the spray solution will spread out on the leaf surfaces and not run off as large drops. Good coverage is essential and a spreader will help. A spreader-sticker is essential on glossy, hard-to-wet leaves.

The adjustable hollow cone-type nozzle is generally preferred when spraying garden plants. The flat fan-type nozzle is often used for applying insecticides on building surfaces and to protect outdoor living areas against mosquitoes and flies. The flat-fan spray nozzle is preferred for applying weed sprays (Figure 3.1). Drift is less than that of cone nozzles.

Use the correct spray pattern. Appropriate nozzles or nozzle adjustments should always be used for a specific job. When treating a flat, exposed surface such as a lawn area from a distance of 20 inches or less, use a fan nozzle to "broom" the spray evenly on the surface. For surfaces such as foliage from a distance of 20 inches or less, a cone nozzle will provide the needed penetration. When the spray must travel more than 20 inches, an adjustable cone nozzle appropriately set for the distance should be used. See Figure 3.1.

If dusting, apply a thin dust film to all aboveground plant surfaces. Adjust the nozzles to insure puffing the dust up and through the plant. Apply dusts when plants are damp and air is calm. Early morning or evening is often best.

Fig. 3.1. Good disease and insect protection requires proper timing of applications, full coverage of foliage (top), and the correct spray pattern (below). The adjustable cone-type nozzle is generally preferred for garden plants. (Courtesy National Sprayer and Duster Association)

One ounce of dust usually covers a 50-foot row early in the season, 2 to 3 ounces are required later.

Always accurately measure the amounts of fungicide or insecticide spray materials you plan to use. The amounts recommended have been carefully calculated to provide the correct amount. That extra spoonful you may add is only wasted—and may cause injury! Always measure, never guess.

The measures given throughout this book are in LEVEL spoonfuls and cupfuls.

For additional information on spraying fruits, see pages 523–525 in the Appendix.

MULTIPURPOSE SPRAYS AND DUSTS

Multipurpose sprays or dusts control a wide range of pests. They are easy and generally safe to use and have encouraged more home gardeners to keep their plants as pest-free as possible. Multipurpose sprays and dusts have largely eliminated the necessity of stocking a large assortment of garden medicines. Many chemical companies and nurseries now have these mixes designed especially for use on tomatoes, potatoes, vegetables, fruits, roses, gladioli, and flowers. Most mixes can be used on a wide variety of plants.

Found below are several safe and effective mixes, now recommended by most states. Be sure to read the fine printing on the label before you buy. DDT, methoxychlor, malathion, lindane, sevin, rotenone, and Kelthane are added to kill insects and mites.

Notes on Multipurpose Mixes

1. Where a fungicide (e.g., captan, folpet, zineb, maneb, or Karathane) is mentioned in the text, a multipurpose spray or dust containing the chemical may be substituted for the fungicide alone.
2. Captan is the preferred fungicide for use on fruits and zineb or maneb on vegetables. Both are good for flowers, trees, shrubs, and lawns. Maneb is often substituted for zineb on tomatoes, potatoes, celery, and certain other plants.
3. Methoxychlor or sevin is preferred to DDT for use on food crops, especially within a month of harvest. DDT is generally more effective and longer-lasting.
4. Add Karathane or sulfur to the vegetable and fruit mixes if powdery mildew becomes a problem.

"SHOT-GUN" SOIL DRENCH

A "shot-gun" treatment for treating cold frames, hot beds, cutting benches, or flower beds for control of organisms causing damping-off, cutting rot, crown and root rots, and stem cankers consists of captan plus ferbam, thiram, or Dexon and PCNB 75 (1 tablespoon of each per gallon of water). Apply as soil drench using 1 pint to 1 quart per square foot. Some plants may be injured by this treatment. Check label directions before using.

TO SPRAY OR TO DUST?

There really is not *one* answer to the question, "Should I *spray* or *dust* my plants?" This depends on what you want to do, how many and how large are the plants you wish to protect, your budget, and how much time and interest you have.

Many gardeners compromise and do both. Spraying is done on a regular schedule. When extra applications are needed, as a result of unexpected troubles or frequent rains, dusting saves time.

Sprays are generally preferred to dusts by commercial growers and experienced gardeners because they are directed easily and specifically even on rather windy days. No worrying, either, about getting dust in your eyes or on plants that do not need it. Spray films are more effective and last longer than dusts, and materials cost less. A sprayer is also more versatile and can be used for more pest control jobs than can a duster.

Dusts are often the choice of the average home gardener. Dusts are quickly and easily applied without messy mixing or measuring. Dusts are less likely to "burn" and they do not discolor foliage as do some sprays. Dusting can be a nuisance if air is not calm. A thin, light film of dew makes dust adhere better to foliage. Dusters cost less than comparable sprayers and are easier to carry, use, and maintain. Dusting is generally impractical on large shrubs or trees. Many gardeners prefer to dust flowers in spring and vegetables and fruits throughout the season. Spraying is considered preferable for flowers in or near bloom.

Vegetables	Fruits	Flowers, Trees, and Shrubs
zineb or maneb methoxychlor or sevin rotenone or malathion	captan, ferbam, or thiram methoxychlor or sevin malathion	zineb, captan, ferbam, thiram, or folpet sulfur or Karathane DDT, lindane, or sevin malathion

Strive for even, thorough coverage whether dusting or spraying.

Modern pesticides, including fungicides, may often be used as sprays or dusts, although dust formulations generally have much less active ingredient than those formulated for spraying. For spraying, insecticides come as emulsions or wettable powders; fungicides are generally wettable powders. Aerosol "bombs" are available for treating a few indoor plants. Granular insecticides and fungicides are used for controlling soil pests or other special uses. There are even small mist blowers for home gardens. Spray-dusters and "fog" machines are somewhat in between spraying and dusting.

SPRAYERS AND DUSTERS

There is a type and size of sprayer or duster for everyone's needs—from the person with only a few house plants to the owner of a large estate with acres of gardens or orchards that need protection.

Some devices are designed to serve several purposes, others are for specific jobs. The choice for you depends on the job, type of application (spray or dust) desired, and type of pesticide you use. Select equipment within your budget and that suits you best. Check these points when talking with your garden supply dealer: Does it handle and operate easily? Is it simple to fill and clean? Is it big enough to avoid frequent refilling? Is the manufacturer reputable? Is it well made with noncorroding parts? How long will it last if given good care? Remember, price should not be your main consideration.

Sprayers

Most sprayers have a number of accessories and special fittings available to meet most situations. These include adjustable nozzles, curved extension rods, extra hose, spray booms, pressure tanks, and special rubber-tired carts. Rust-free models—stainless steel, brass, or copper—last longer than cheaper tinplate or galvanized models. For covering fruit and shade trees you will need high pressures (200 to 600 pounds or more per square inch). If sprayer does not have an agitator, shake the solution as you spray.

Experienced gardeners have a separate sprayer, properly labeled, just for chemically controlling weeds. Many weedkillers, especially types such as 2, 4-D and 2, 4, 5-T, are extremely difficult to get completely out of sprayers. Check with your county extension office on how to decontaminate sprayers in which 2, 4-D or related materials have been used. USDA Farmers' Bulletin 2183, *Using Phenoxy Herbicides Effectively,* covers this subject in detail.

Clogging of spray nozzles can often be prevented by making up solutions in a thin, smooth batter of water and spray powder, then wash through fine-mesh cheesecloth or a silk stocking into the spray tank. Do not lay sprayer parts (e.g., plunger cylinder and nozzle) where they can pick up dirt or grass clippings that may clog the nozzle.

Household Sprayers (capacity 4 ounces to 1 gallon). Cheap, versatile, easy to operate, but limited to small jobs. Buy type that shoots continuous mist or fine spray and has an adjustable or changeable nozzle. Spray carries only a short distance.

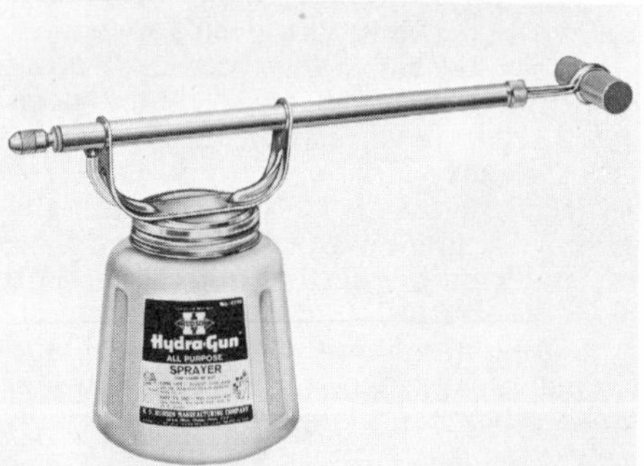

Fig. 3.2. A high pressure hand sprayer for use inside or out. Gives high pressure with easy strokes. (Courtesy H. D. Hudson Manufacturing Company)

Coverage difficult on underleaf surface. Frequent shaking is necessary to keep heavy suspensions from settling.

Compressed Air Sprayers (1 to 6 gallons). Popular, low-priced, easy to operate. Useful for most jobs around home, yard, and garden. Air is compressed into tank above the spray liquid by hand-operated air pump. Choose one with open or funnel-type mouth, pressure-relief top, chemical-resistant hose, curved extension rod, and adjustable nozzle. Uncomfortable to carry over shoulder, but some models have cart

Fig. 3.3A. A three-gallon, highly maneuverable, single-wheel, compressed air sprayer that turns on a dime. Legs of stand serve as guides for width of spray pattern. (Courtesy Universal Metal Products)

with rubber wheels. Normal operating pressure ranges from 30 to 80 pounds and is maintained by occasional pumping. Use caution in opening sprayer with *any* air pressure in tank. Each sprayer is usually equipped with assortment of nozzle discs to provide different spray patterns—solid cone, hollow cone, flat fan in fine or coarse spray, and solid stream.

Some models have carbon dioxide (CO_2) cylinders to provide operating pressure. These cylinders discharge up to 15 gallons of spray at uniform pressure. May be refilled at moderate cost.

Knapsack Sprayers (2 to 6 gallons). Very useful for large gardens and estates. Easy-pumping. Straps on back. Delivers fine, continuous mist further than compressed-air type but is more expensive. Uniform spray pressures range from 80 to 180 pounds. Buy type with agitation system to prevent settling out. Heavy for a lady to carry. Pump handle on some models may be attached at either side for right-hand or left-hand pumping. Some models have metal shield to prevent direct body contact with the tank surface.

Slide Pump or Trombone Sprayers. Relatively inexpensive, smooth, telescoping pumps that spray continuously. Delivers pressures up to about 180 pounds. Spray carries good distance (tops of 20-foot trees). Comes attached to ½ gallon jar or hose that dips into a bucket. Buy type with rust-resistant metal and an extension. Slide may become sticky. Tiring to use for long time and capacity is too small for more than a limited number of plants.

Wheelbarrow, Cart, and Barrel Sprayers (7 to 50 gallons). Similar to slide pump or trombone sprayer but generally have larger and more powerful pumps. Made in variety of designs mounted on various types of portable frames (e.g., wheelbarrow, horizontal or vertical cart, bucket, or barrel). Cart or wheelbarrow models

Fig. 3.3B. A three-gallon compressed air sprayer with adjustable nozzle and trigger-type shutoff. (Courtesy R. E. Chapin Manufacturing Works, Inc.)

have 2, 3, or 4 wheels for easy transport. A moving, large, full sprayer is difficult to maneuver on soft, wet, or unlevel soil. Continuous high pressures, up to about 250 pounds, may be developed for spraying medium-sized trees. Spraying easier with two people—one spraying, the other on the pump handle.

Some cart sprayers have a special pressure tank precharged with a tire pump or at a filling station. One "charge" gives about 10 minutes of continuous spraying at constant, desired pressure up to 200 pounds.

Garden Hose Sprayers. Attached to the garden hose so that water supplies the pressure. Add concentrated pesticides to glass or polyethylene jar screwed to the hose. The spray gun attached to the lid meters out concentrate from the jar by suction and mixes it with water from the hose. A quart jar of concentrate spray makes a number of gallons of dilute spray. Most models use liquids or wettable powders. Materials may not be applied at a constant rate. Hose pressure may not be sufficient to break spray into a fine mist. Adding detergent to spray will help. Inconveniences: Use is limited to garden area reached with the hose; dragging and injury by hose and stooping to cover underleaf surfaces of low-growing plants.

Small Power Sprayers (5 to 50 gallons). Wheelbarrow or estate types available that vary greatly in design and special features. A gasoline engine with powerful 1- or 2-cylinder hydraulic pump generally delivers 1 to 5 gallons of spray per minute at a pressure of 20 to 400 pounds. Electric-powered sprayers also available. Get large-wheeled, easily maneuverable model with extra hose, agitator, and adjustable, trigger-control spray gun. Excellent for small orchards, large gardens, and estates, but is relatively expensive. Not economical of spray material. Some types may have such features as a trailer or tractor hitch for hauling, spray boom for row-crop spraying, and skid mount.

MAINTENANCE OF SPRAYERS

Carefully follow manufacturer's instructions that accompany the sprayer for lubrication, operation, and maintenance. Before starting a new *power sprayer,* check all designated points for proper lubrication. Operate sprayer at slow speed, using water while checking delivery system, operation of control valves, and pressure regulator.

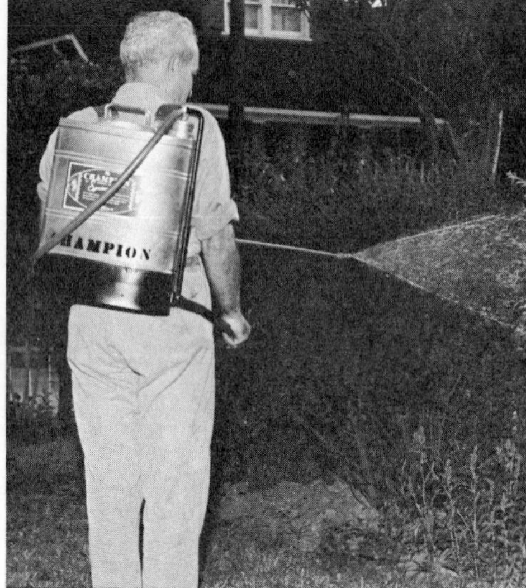

Fig. 3.4. Knapsack sprayer with five-gallon capacity. Pump handle gives high spray pressure with easy pumping. (Courtesy Champion Sprayer Company)

All types of sprayers should be thoroughly rinsed when changing to a different type of spray solution. Clean immediately after *each* use. Pump at least two changes of clean water through system. Store *funnel-top sprayers* upside down with hand pump removed.

Several times during the season, clean *hand sprayers* more thoroughly. Fill spray tank with hot water and let stand for a few minutes. Where possible, take apart and clean pump, extension tubes, nozzle parts, and shut-off valve. Wash strainers and nozzle parts with kerosene. Then rinse with hot soapy water. Use an old toothbrush or jet of compressed air. Replace worn washers and nozzle discs with enlarged holes. Reassemble and pump clean water through open nozzle head to flush out discharge line. Finally, pump to full pressure and check hose and gasket for leaks. Be sure shut-off valve is working properly. If pump fails to develop full pressure, remove plunger to the cylinder and reshape leather so that it seals tightly.

Household sprayers require little care or maintenance. If pump should lose compression, pull pump handle all the way

Fig. 3.5A. A slide pump brass sprayer. A built-in, 9-inch telescoping spray extension and rotable spray nozzle make it easy to put spray right where needed. (Courtesy H. D. Hudson Manufacturing Company)

Fig. 3.5B. A brass slide-type sprayer that shoots a fine mist to a 25- to 30-foot spray stream, up to 180 pounds pressure. Comes with plastic hose. Note the extension and adjustable nozzle to eliminate "bend and strain." (Courtesy H. D. Hudson Manufacturing Company)

out and add a few drops of oil in air hole at end of pump cylinder. This lubricates the plunger cup or leather.

Leave *hand sprayers* partially unassembled for winter. All metal parts should be lightly oiled and wrapped in newspaper. Tank must be clean and dry.

For *power sprayers,* follow manufacturer's directions regarding proper lubrication and special care required. Clean sprayer after each use. Drain and flush tank with clean water. Take apart nozzles and strainers and wash with kerosene followed by hot, soapy water. Before reassembling, pump clean water through discharge system until it comes out clean.

After spraying is completed in fall, soak nozzles and strainers (screens) in kerosene. Run wire through spray rods. Rinse hoses clean and put where they will not crack or freeze.

After cleaning, and before putting a power sprayer away for winter, pour a pint of oil in the tank. Fill with water and start pump. As water is discharged, a thin coating of oil covers inside of tank, pump, valves, and circulating system. Finally, drain sprayer completely and store where dry.

Dusters

Dusters are of simpler construction than comparable sprayers, with fewer parts to go wrong. For maximum protection and safety, dust only when air is calm (this may mean early morning or evening) and plants are damp. Do not dust if plants are wet—you will get unsightly deposits! Select duster large enough so refilling does not become a nuisance. Small dusters work best if they are shaken occasionally to "fluff" up the dust. Larger models are equipped with agitators to assure a steady flow.

Plunger Type Dusters (capacity ½ to 5 pounds of dust). Easy to operate. Handy for treating small areas with little waste. Choose one with dust chamber made of metal or glass. Larger models with extension tube and adjustable nozzle or deflector cap are much handier. Avoid "salt shaker," flick, plastic squeeze, and tele-

Fig. 3.6. Wheelbarrow sprayer (17½ gallons). This powerful, high pressure pump delivers up to 250 pounds pressure. Note the pressure tank (optional), pressure gauge, and tank cover to prevent spillage. (Courtesy H. D. Hudson Manufacturing Company)

scoping cardboard carton types. They do not provide uniform coverage.

Small Bellows, Crank, or Rotary-Fan Dusters (½ to 25 pounds). Easy to use and give better coverage than plunger type. Ideal for continuous dusting of small or large gardens. Cheap, light-weight, faster than spraying. May be difficult to direct dust exactly where you want it. Small models require frequent refilling. Get type with adjustable feed control, agitating device, extension tube, and flange nozzle (s). May be carried in front by shoulder straps.

Knapsack Dusters (5 to 25 pounds). Suitable for large gardens and estates. Throws up steady, fine cloud of dust. Covers large areas rapidly without refilling. Lightweight, simple to use. General types: *Bellows* (strapped on back) and *Rotary-fan* (carried in front by shoulder straps and operated with hand crank). Delivery tubes and nozzles are adjustable for both height and direction. Buy type with long extension tube, flaring "fishtail" nozzles, and agitating device. More expensive than hand dusters, but may be used for either intermittent or continuous dust applications.

Small Power Dusters (60 to over 200 pounds). Practical for large gardens, small orchards, or estates. Small gasoline engine models (weighing 40 to 67 pounds) are carried on the back; larger ones are powered by tractor. Duster may be mounted on tractor, trailer, truck, or other conveyance. The metering device (or feed regulator) may be adjusted to apply from 5 to 50 or 100 pounds of dust per acre. Covers large areas rapidly without refilling. Wasteful of material, as dust often blows onto other plants. Buy type with an agitator. Power dusters vary greatly in size of dust hopper, type and capacity of fan, type of distribution system, horsepower, and mounting.

MAINTENANCE OF DUSTERS

Follow manufacturer's directions regarding operating instructions and lubrication of moving parts. Use graphite for lubricating the steel rod and plunger in hand plunger dusters.

Empty and clean duster after using to prevent caking, clogging, and eventual corrosion. All slip joints should be given a protective coating in fall before winter storage in a dry place. Operate fans on power dusters at recommended speeds.

Fig. 3.7. Garden hose-end sprayer of ten-gallon capacity which is made of shatterproof polyethylene plastic and is lightweight, quick, low cost, and handy. Get one like this with finger-tip siphon control and capacity markings for accurate proportioning. (Courtesy H. D. Hudson Manufacturing Company)

For an excellent discussion of sprayers, dusters, their operation and uses, obtain copies of *Sprayer and Duster Manual* and *Outdoor Housekeeping,* published by the National Sprayer and Duster Association, 850 Wrigley Building North, 410 North Michigan, Chicago 60611 and USDA H & G Bul. 63, *Hand Sprayers and Dusters.*

SPREADERS, STICKERS, AND WETTING AGENTS

These materials help suspend pesticide in the spray solution, improve cohesiveness of spray, increase wetting of foliage by spray, or all three. All commercial pesticides have one or more of these materials already in the spray mix.

Adding a commercial or homemade spreader-sticker or wetting agent is recommended before spraying glossy, hard-to-wet foliage like roses, gladiolus, iris, ivies, carnation, tulip, onion, cabbage, pea, rhododendron, pine, spruce and mountain-laurel. These materials also aid in controlling powdery mildews and insects with a waxy coating, like woolly aphids and mealybugs.

Commercial *spreaders* (wetting agents) that ensure wetting of hairy or glossy foliage include Santomerse 80, Tween-20, Fluxit, Vatsol, and Sur-Ten Wetting

Fig. 3.8A. Horizontal, 10-gallon power sprayer that is mobile, easily filled and cleaned, and produces high, constant pressure. Ideal for most yard and garden jobs. (Courtesy H. D. Hudson Manufacturing Company)

Fig. 3.8B. Erect, 12½-gallon sprayer with large, easy-rolling wheel, adjustable pressure regulator, and constant, mechanical agitator. (Courtesy H. D. Hudson Manufacturing Company)

Agents, and household soaps or detergents (e.g., Dreft, Swerl, Tide, Vel, and Liquid Lux). Spreaders are used at ¼ to 1 teaspoon per gallon.

Common *stickers* that allow pesticides to adhere tenaciously to plant surfaces include powdered skim milk, wheat or soybean flour, fish oil, or casein. Use 1 tablespoon of wheat flour per gallon. Goodrite PEPS and De-Pester Sticker are good commercial stickers.

Commercial *spreader-stickers* are available under many trade names—among them DuPont Spreader-Sticker, Orchard Brand Filmfast Spreader-Sticker, Miller Nu-Film, Ortho (Spray-Sticker, X-77 Spreader, Spreader Sticker), Spreader-Activator (Chipman and De-Pester), Triton B-1956, Plyac Spreader-Sticker, Sherwin-Williams Spred-Rite, Multifilm X-77, etc. Most commercial spreader-stickers are used at ¼ to ½ teaspoon per gallon.

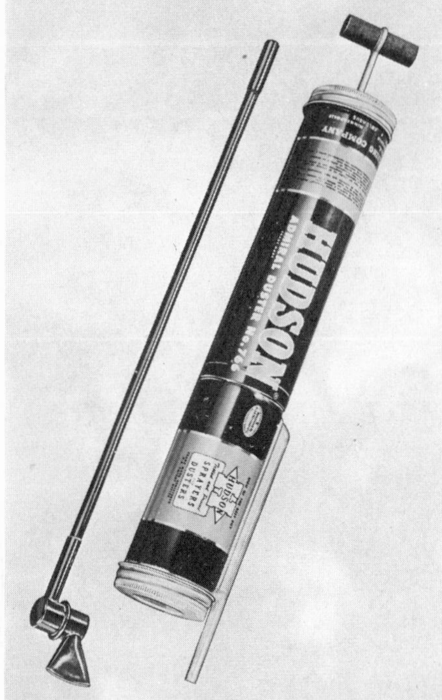

Fig. 3.9. Plunger type hand duster with long 21-inch extension and all-angle swivel deflector to make dusting simple. Discharges uniform cloud. (Courtesy H. D. Hudson Manufacturing Company)

PARTIAL LIST OF FUNGICIDE MANUFACTURERS AND DISTRIBUTORS PLUS LEADING SPRAYING AND DUSTING EQUIPMENT MANUFACTURERS

Acme Quality Paints, Inc., 8250 St. Aubin Ave., Detroit, Mich. 48211.

Agway, Inc., Terrace Hill, Ithaca, N.Y. 14850.

Allied Chemical Corp., General Chemical Div., 40 Rector St., New York, N.Y. 10006.

Amchem Products, Inc., Ambler, Pa. 19002.

American Cyanamid Co., Agric. Div., P.O. Box 400, Princeton, N.J. 08540; *or* 30 Rockefeller Plaza, New York, N.Y. 10020.

American Potash & Chemical Corp., 3000 W. 6th St., Los Angeles, Calif. 90054.

American Potash Institute, 1102 16th St., N. W., Washington, D.C. 20036.

Androc Chemical Co., 7301 W. Lake St., St. Louis Park, Minn. 55426.

Antrol Garden Products, Boyle-Midway, 22 East 40th St., New York, N.Y. 10016.

Arwell, Inc., 119 Glen Rock Ave., Waukegan, Ill. 60089.

Barco Mfg. Co., Inc., 119 Dewey St., Worcester, Mass. 01610.

Bartlett, N. M., Mfg. Co., Inc., Beamsville, Ontario, Canada.

Bean, John, Div., FMC Corp., P.O. Box 9490, Lansing, Mich. 48909.

Black Leaf Co., 6147 N. Broadway, Chicago, Ill. 60626.

Boyle-Midway, Inc., S. Ave. & Hale St., Cranford, N.J. 07016.

Bradson Company, 2165 Kurtz St., San Diego, Calif. 92110.

Brayton Chemicals, Inc., P.O. Box 56, Burlington, Iowa 52602.

Buffalo Turbine Agric. Equipment Co., Inc., 70 Industrial St., Gowanda, N.Y. 14070.

Canada Rex Spray Co., Ltd., Brighton, Ontario, Canada.

Carolina Chemicals, Inc., P.O. Box 128, West Columbia, S.C. 29169.

Castle Chemical Co., Castle Rock, Minn. 55010.

Chamberlain Corp., Dobbins Division, P.O. Box 610, Monroe, Ga. 30655.

Champion Sprayer Co., 6509 Heintz Ave., Detroit, Mich. 48211.

Chapin, R. E., Mfg. Works, Inc., Batavia, N.Y. 14021.

Chapman Chemical Co., P.O. Box 9158, Memphis, Tenn. 38101.

Chemagro Corp., P.O. Box 4913, Hawthorn Rd., Kansas City, Mo. 64120.

Chemical Formulators, Inc., P.O. Box 26, Nitro, W. Va. 25143.

Chemley Products Co., 5744 N. Western Ave., Chicago, Ill. 60645.

Chevron Chemical Co., Ortho Div., 200 Bush St., San Francisco, Calif. 94120; *or* Lucas St. and Ortho Way, Richmond, Calif. 94801; *or* 41 Kingshighway East, Haddonfield, N.J. 08033.

Chipman Chemical Co., Inc., P.O. Box 309, Bound Brook, N.J. 08805; *or* 1801 Murchison Dr., Burlingame, Calif. 94011.

Cleary, W. A., Corp., P.O. Box 749, New Brunswick, N.J. 08903.

Darworth, Inc., Chemical Products Div., Box 308, Simsbury, Conn. 06070.

Davison Chemical Div., W. R. Grace & Co., 101 N. Charles St., Baltimore, Md. 21200.

Fig. 3.10A. This easy-turning crank duster holds about ¾ pound of dust, is light and compact, emits a fine, uniform cloud of dust. Nozzle is adjustable to any angle. (Courtesy H. D. Hudson Manufacturing Company)

Fig. 3.10B. An easy-turning crank duster that puts out a steady cloud of dust. Note the fan-shaped adjustable nozzle. (Courtesy Champion Sprayer Company)

Fig. 3.11. Knapsack duster. Holds about 17 pounds of dust. It has a built-in filler scoop in cover. It is easily carried on the back and is ideal for spot dusting or complete coverage. (Courtesy H. D. Hudson Manufacturing Company)

Diamond Alkali Co., Agric. Chem. Dept., Union Commerce Bldg., Cleveland, Ohio 44114.

Doggett-Pfeil Co., 191 Mountain Ave., Springfield, N.J. 07081.

Douglas Chemical Co., 620 E. 16th St., North Kansas City, Mo. 64116.

Dow Chemical Co., Agric. Chemical Div., P.O. Box 512, Midland, Mich. 48640.

E. I. du Pont de Nemours & Co., Grasselli Chem. Div.; *or* Industrial & Biochemicals Dept., 1007 Market St., Wilmington, Del. 19898.

Edco Corp., Childs Road, Elkton, Md. 21921.

Elanco Products Co., Div. of Eli Lilly & Co., 640 South Alabama, Indianapolis, Ind. 46206; *or* P.O. Box 204, Greenfield, Ind. 46140.

E-Z-Flo Chemical Co., P.O. Box 808, Lansing, Mich. 48903.

Faesy & Besthoff, Inc., 25 E. 26th St., New York, N.Y. 10010.

Fairfield-Niagara, FMC Corp., 441 Lexington Ave., New York, N.Y. 10017.

Flag Sulphur & Chemical Co., 72nd St. & ACL RR., Tampa, Fla. 33601.

Floridin Co., P.O. Box 989, Tallahassee, Fla. 32302.

Florist Products, 1843 Oakton, Des Plaines, Ill. 60018.

Gallowhur Chemicals, Canada Ltd., 545 19th Ave., Lachine, Quebec, Canada.

Geigy Agric. Chem., Div. of Geigy Chemical Corp., P.O. Box 430, Yonkers, N.Y. 10707; *or* Saw Mill Road, Ardsley, N.Y. 10502.

General Aniline & Film Corp., 435 Hudson St., New York, N.Y. 10014.

Gillette Inhibitor Co., Inc., 1706 S. Canal St., Chicago, Ill. 60616.

Green Cross Products Div., Sherwin-Williams Co. of Canada Ltd., 2875 Centre St., P.O. Box 489, Montreal, Quebec, Canada.

Hardie Mfg. Co., 4200 Wissahickon Ave., Philadelphia, Pa. 19129; *or* P.O. Box 570, Wilkes-Barre, Pa. 18703.

Hayes Spray Gun Co., 98 N. San Gabriel Blvd., Pasadena, Calif. 91107.

Hercules Powder Co., 910 Market St., Wilmington, Del. 19899.

Hess and Clark, 7th & Orange Sts., Ashland, Ohio. 44805.

Howard, A. H., Chemical Co., Ltd., P.O. Box 740, Orangeville, Ontario, Canada.

Hubbard-Hall Chemical Co., 26 Benedict St., Waterbury, Conn. 06706.

Hudson, H. D., Mfg. Co., 589 East Illinois St., Chicago, Ill. 60611.

Imperial Chemical Co., Shenandoah, Iowa 51601.

Industrial Fumigant Co., 923 State Line Ave., Kansas City, Mo. 64101.

Lawrence, B. M., & Co., 24 California St., San Francisco, Calif. 94111.

Leffingwell Chemical Co., P.O. Box 1187, Perry Annex, Whittier, Calif. 90603.

Lorenz Chemical Co., 1024 N. 17th St., Omaha, Nebr. 68102.

Los Angeles Chemical Co., 4545 Ardine St., South Gate, Calif. 90280.

Mallinckrodt Chemical Works, 2nd and Mallinckrodt Sts., St. Louis, Mo. 63107.

McLaughlin Gormley King Co., 1715 S. E. 5th St., Minneapolis, Minn. 55414.

Merck & Co., Inc., Merck Chemical Div., P.O. Box M, Rahway, N.J. 07065.

Metalsalts Corp., 200 Wagraw Rd., Hawthorne, N.J. 07507.

Midvale Chemical Co., Lodi, N.J. 07644.

Miller Chemical & Fertilizer Corp., 3006 W. Cold Spring Lane, Baltimore, Md. 21215; *or* Trenton Ave. and William St., Philadelphia, Pa. 19134.

Miller Chemical Co., 525 N. 15th St., Omaha, Nebr. 68102.

Miller Products Co., 7737 N. E. Killingsworth, Portland, Oreg. 97218.

Monsanto Chemical Co., Agric. Chem. Div., 800 N. Lindbergh Blvd., St. Louis, Mo. 63166.

Mooney Chemicals, Inc., 2301 Scranton

Fig. 3.12. Power duster. Holds about 25 pounds of dust, yet is easily carried on a man's back. It dusts up to 25 feet high and can also be used for mist spraying. (Courtesy H. D. Hudson Manufacturing Company)

Rd., Cleveland, Ohio 44113.

Morton Chemical Co., Div. of Morton International, Inc., 110 N. Wacker Dr., Chicago, Ill. 60606; *or* 11710 Lake Ave., Woodstock, Ill. 60098.

Myers, F. E., & Bro. Co., Subsidiary of McNeil Corp., 400 Orange St., Ashland, Ohio 44805.

Nationwide Chemical Co., Ft. Myers, Fla. 33901.

Natriphene Co., P.O. Box C-26, Midland, Mich. 48641.

Niagara Chemical Div., FMC Corp., Middleport, N.Y. 14105.

Norwich Pharmaceutical Co., Norwich, N.Y. 13815.

Nott Mfg. Co., P.O. Box 832, Poughkeepsie, N.Y. 12602.

Oakes Mfg. Co., 516 Dearborn St., Tipton, Ind. 46072.

Olin-Mathieson Chemical Corp., Agric. Div., Mathieson Bldg., Baltimore, Md. 21203; *or* 745 Fifth Ave., New York, N.Y. 10022; *or* Squibb Inst. for Medical Research, 5 Georges Road, New Brunswick, N.J. 08902; *or* P.O. Box 991, Little Rock, Ark. 72203.

Pacific Supply Cooperative, Chemicals Div., P.O. Box 4380, Portland, Oreg. 97208.

Pearson-Ferguson Chemical Co., Inc., 1400 Union Ave., Kansas City, Mo. 64101.

Penick, S. B., & Co., 100 Church St., New York, N.Y. 10007.

Pennsalt Chemicals Corp., 2901 Taylor Way, Tacoma, Wash. 98421.

Pfizer, Charles A., & Co., Inc., 235 E. 42nd St., New York, N.Y. 10017; *or* 11 Bartlett St., Brooklyn, N.Y. 11206.

Phelps Dodge Refining Corp., 300 Park Ave., New York, N.Y. 10022.

Pittsburgh Plate Glass Co., Corona Chem. Div., Moorestown, N.J. 08057; *or* 420 Duquesne Blvd., Pittsburgh, Pa. 15222.

Ra-Pid-Gro Corp., P.O. Box 13, Dansville, N.Y. 14437.

Roberts Chemical Co., P.O. Box 446, Nitro, W. Va. 25143.

Rohm & Haas Co., Agricultural & Sanitary Chemicals Dept., Independence Mall West, Philadelphia, Pa. 19105.

Root-Lowell Corp., Div. of Root-Lowell Mfg. Co., 445 North Lake Shore Drive, Chicago, Ill. 60611; *or* 320 W. Main St., Lowell, Mich. 49331.

Seacoast Labs., Inc., 156–158 Perry St., New York, N.Y. 10014.

Shell Chemical Co., Agr. Chemicals Div., 110 W. 51st St., New York, N.Y. 10020; *or* Agric. Research Div., P.O. Box 3011, Modesto, Calif. 95353.

Sherwin-Williams Co., 101 Prospect Ave. N.W., Cleveland, Ohio 44101; *or* P.O. Box 489, Montreal, Quebec, Canada.

Smith, D. B., & Co., Inc., Smith Building, Main St., Utica, N.Y. 13502.

Spencer Chemical Co., 9009 W. 67th St., Merriam, Kans. 66221; *or* 610 Dwight Bldg., Kansas City, Mo. 64141.

Stauffer Chemical Co., Agric. Chemicals Div., 380 Madison Ave., New York, N.Y. 10017; *or* 636 California St., San Francisco, Calif. 94108.

Swift & Co., Agric. Chemical Div., 115 West Jackson St., Chicago, Ill. 60604; *or* Winter Haven, Fla. 33880.

Tennessee Corp., 44 Broad, Atlanta, Ga. 30303; *or* College Park, Ga. 30022.

Thompson-Hayward Chemical Co., P.O. Box 768, Kansas City, Mo. 64141.

Tuco Products Co., Division of The Upjohn Co., 7171 Portage Rd., Kalamazoo, Mich. 49001.

Union Carbide Chemicals Corp., Div. of Union Carbide Corp., 270 Park Ave., New York, N.Y. 10017.

Universal Metal Products Div., Leigh Products, Inc., Saranac, Mich. 48881.

United Co-Operatives, Inc., P.O. Box 836, 111 Glamorgan St., Alliance, Ohio 44601.

United States Rubber Co., Naugatuck Chemical Div., Naugatuck, Conn. 06771.

U.S. Borax Chemical Corp., 3075 Wilshire Blvd., Los Angeles, Calif. 90005.

Valley Chemical Co., P.O. Box 1317, Greenville, Miss. 38702.

Vanderbilt, R. T., Co., Inc., 230 Park Ave., New York, N.Y. 10017.

Vandermolen Export Co., 378 Mountain Ave., N. Caldwell, N.J. 07006.

Vaughan's Seed Co., 5300 Katrine Drive, Downers Grove, Ill. 60515.

Velsicol Chemical Corp., 341 East Ohio St., Chicago, Ill. 60611.

Virginia-Carolina Chemical Corp., 401 E. Main St., Richmond, Va. 23208.

Virginia Smelting Co., West Norfolk, Va. 23510.

Westbrook Mfg. Co., St. Joseph, Mich. 49085.

Woodbury Chemical Co., P.O. Box 788, St. Joseph, Mo. 64502.

Wood Ridge Chemical Corp., Park Place East, Wood Ridge, N.J. 07075.

Woolfolk Chemical Works, Inc., P.O. Box 922, East Main St., Fort Valley, Ga. 31030.

Xterminator Products Corp., 219 Monticello Ave., Jersey City, N.J. 07304.

SECTION FOUR

Home and Garden Plants and Their Diseases

African-violet 127
Anemone 131
Apple 133
Ash 144
Avocado 149
Barberry 153
Bean 155
Beet 162
Begonia 167
Bellflower 168
Birch 170
Bittersweet 173
Blueberry 175
Boxwood 181
Cabbage 186
Cacti 194
Calla lily 195
Camellia 198
Canna 202
Carnation 203
Carrot 206
Celery 213
Chrysanthemum 219
Citrus 227
Cockscomb 229
Corn 231
Cucumber 237
Currant 245
Cyclamen 248
Daffodil 249
Delphinium 254
Dogwood 258
Elm 263
Ferns 269
Fig 271
Forsythia 273
Fuchsia 275
Gardenia 275
Gentian 277
Geranium 278
Gladiolus 281
Grape 286
Heath 294
Holly 295
Hollyhock 297
Honeylocust 300
Horsechestnut 302
Hydrangea 304
Iris 307
Ivy 310
Juniper 312
Lantana 317
Larch 318
Lawngrass 319
Lettuce 328
Lily 334
Magnolia 340
Maple 343
Mertensia 348
Morning-glory 351
Oak 355
Oleander 360
Onion 361
Orchids 366
Pansy 374
Pea 377
Peach 382
Peanut 393

Phlox	399	Spirea	461
Pine	401	Strawberry	462
Poplar	411	Sumac	468
Poppy	413	Sweetpotato	470
Potato	415	Sycamore	476
Primrose	423	Tomato	479
Raspberry	426	Tulip	494
Rhododendron	433	Valerian	499
Rose	439	Viburnum	499
St.-Johns-wort	446	Vinca	500
Salvia	446	Walnut	501
Sedum	451	Waxmyrtle	507
Snapdragon	454	Willow	507
Snowberry	457	Yew	511

Garden plants are arranged alphabetically below, together with their more important diseases. Prevalent or serious diseases are listed first. For local or minor diseases, check with your extension plant pathologist or county agent. For additional information on different general diseases or control measures, see Sections 2 and 3 plus the Appendix. The diseases attacking each plant are also listed alphabetically in the Index (pages 583–649) under the common name of the plant.

AARONSBEARD — See St.-Johns-wort

AARONS-ROD — See Pea

ABELIA — See Snowberry

ABIES — See Pine

ABRONIA — See Four-o'clock

ABRUS — See Pea

ABUTILON — See Hollyhock

ABYSSINIAN WILDFLOWER — See Gladiolus

ACACIA — See Honeylocust

ACALYPHA, CHENILLE PLANT, COPPERLEAF [PAINTED, VIRGINIA] *(Acalypha)*

1. *Leaf Spots*—Leaves variously spotted. May wither and drop early. *Control:* Usually unnecessary. Pick and burn infected leaves. Space plants. Increase air circulation.
2. *Red Leaf Gall*—Reddish galls develop on leaves. *Control:* Same as for Leaf Spots.
3. *Downy Mildew*—See under Calla lily, and (6) Downy Mildew under General Diseases.
4. *Powdery Mildew*—See (7) Powdery Mildew under General Diseases.
5. *Oedema*—Indoor problem. Small, rust-colored spots or overgrowths on leaves. *Control:* Avoid overwatering and high humidity. Increase air circulation.
6. *Root Rots*—See under Geranium, and (34) Root Rot under General Diseases. May be associated with nematodes (e.g., lesion, root-knot).
7. *Root-knot*—See (37) Root-knot under General Diseases.

ACANTHOPANAX, FIVE-LEAF or FIVE-FINGERED ARALIA, CASTOR ARALIA *(Acanthopanax, Kalopanax)*; AMERICAN SPIKENARD, HERCULES-CLUB, SARSAPARILLA [BRISTLY, WILD], UDO, CHINESE ANGELICA *(Aralia)*

1. *Leaf Spots or Blight, Spot Anthracnose, Scab*—Spots of various colors, sizes, and shapes on leaves. See (1) Fungus Leaf Spot under General Diseases.
2. *Root Rots, Stem Rot, Watery Soft Rot*—See (21) Crown Rot, and (34) Root Rot under General Diseases.
3. *Verticillium Wilt* (acanthopanax, American spikenard, udo) —See (15B) Verticillium Wilt under General Diseases.
4. *Rust* (acanthopanax, American spikenard, sarsaparilla) —Dark, powdery pustules on leaves. *Control:* Pick and burn infected leaves.
5. *Powdery Mildew* (sarsaparilla) —Powdery, white mold on foliage. See (7) Powdery Mildew under General Diseases.
6. *Twig and Branch Cankers, Dieback* (Hercules-club) —Twigs and branches die back from discolored, girdling cankers. *Control:* Pick and burn affected parts. Make cuts several inches below any sign of infection.
7. *Wood Rot* (Hercules-club) —See under Birch, and (23) Wood Rot under General Diseases.

ACHILLEA —See Chrysanthemum

ACHIMENES — See African-violet

ACHLYS — See Barberry

ACIDANTHERA — See Gladiolus

ACONITE, ACONITUM — See Delphinium

ACROCOMIA — See Palms

ACTAEA — See Anemone

ACTINOMERIS — See Chrysanthemum

ADAM-AND-EVE — See Erythronium

ADAMS-NEEDLE — See Yucca

ADDERSTONGUE — See Erythronium

ADIANTUM — See Ferns

AERIDES — See Orchids

AESCULUS — See Horsechestnut

AETHIONEMA — See Cabbage

AFRICAN DAISY — See Chrysanthemum

AFRICAN FORGET-ME-NOT — See Mertensia

AFRICAN-LILY — See Tulip

AFRICAN-VIOLET *(Saintpaulia)*; CUPID'S-BOWER *(Achimenes)*; FLAME VIOLETS *(Episcia)*; STARFIRE *(Gesneria)*; GLOXINIA *(Sinningia)*

1. *Crown and Stem (Corm) Rots, Root Rots*—Widespread. Soft, mushy or dry, brown to black rot of crown, tuber, petioles, and roots. Plants sickly. Flowers fail to develop properly. Gradually or suddenly wilt, wither, and die. Plants easily pulled up. See Figure 2.39B under General Diseases. Often associated with water-logged soil, nematodes (e.g., lance, lesion, pin, root-knot, spiral, stubby-root, stunt), and high soluble salts content in soil. *Control:* Plant healthy stock in sterilized soil in sterilized containers (see pages 538–548). Take cuttings only from healthy plants. Root in sterile medium. Avoid overwatering, overfertilizing, deep planting, and temperature fluctuations. Destroy badly infected plants. Keep water off foliage. Increase air circulation. Soil should be loose and well-drained. Drenching soil with ferbam (2½ tablespoons per gallon), Dexon (100 ppm), or oxyquinoline sulfate (½ teaspoon per gallon) is often beneficial. Apply drenches at 2- or 4-week intervals.
2. *Botrytis Blight, Gray-mold Blight, Leaf Rot, Bud Rot*—Cosmopolitan. Soft, water-soaked to tan rot of crown, buds, flowers, petioles, or leaves. Coarse gray mold may

grow on diseased tissue. *Control:* Keep humidity down, keep water off foliage, and increase air circulation. Avoid overwatering and overcrowding. Promptly destroy fading blooms or other diseased or dead plant parts. Spray at 10-day intervals with zineb during humid weather. Control mites with Kelthane dips or sprays. Do *not* use malathion on African-violets.

3. *Ringspot*—General. Whitish to bright yellow or brown rings, arc, oakleaf patterns, and irregular spots or streaks on upper leaf surface (Figure 4.1). *Control:* Keep cold or hot water off foliage. Avoid sudden temperature changes. Keep plants out of direct sunlight for an hour before and after watering.

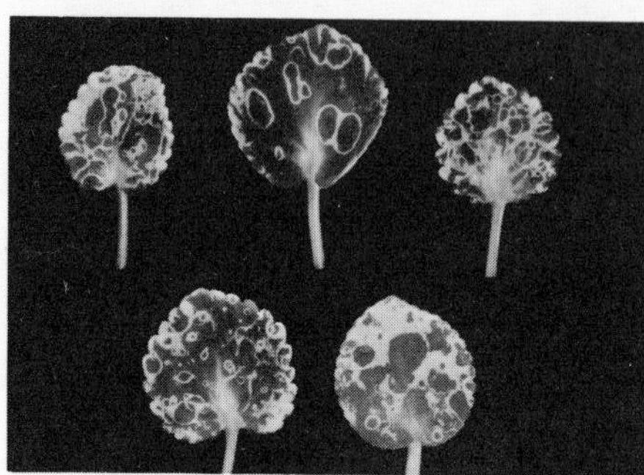

Fig. 4.1. Ringspots resulting from cold water coming in contact with African-violet leaves during watering. (Courtesy Illinois Natural History Survey)

4. *Root-knot*—General. Plants pale green, make poor growth. Flowers, if produced, may be small and distorted. Small, irregular, knotlike galls on roots, crowns, stems, and even leaves. Leaves often flaccid, thickened, and blistered. *Control:* Plant only disease-free cuttings in sterilized soil in sterilized containers. Remove and burn badly infested plants. Try DBCP as a soil application to plants in pots. Follow manufacturer's directions. Propagate using erect leaves. Cut petioles short.
5. *Leaf Scorch, Chlorosis*—Leaves pale to yellowish-green. May appear scorched. *Control:* Keep plants out of bright sun, and away from heating effects of large incandescent light bulbs.
6. *Petiole Rot*—Leaves wilt and wither from orange-brown to rust-colored rotting of petioles where they touch salt-encrusted pot rim. *Control:* Flush out salts with rainwater or distilled water. Avoid overfertilizing. Cover pots with aluminum foil, melted paraffin, waxed paper, or other material. Or mound up with sphagnum moss to keep petioles clear of pot.
7. *Powdery Mildew*—Whitish-gray patches on leaves, petioles, flower stems, and flowers. Flowers may be deformed and discolored (Figure 4.2). *Control:* Destroy old infected leaves (close to soil) and flowers. Space plants. Dip or spray with Karathane (½ teaspoon per gallon) plus spreader-sticker, dust with sulfur, or spray with wettable sulfur (1 tablespoon per gallon). Two applications, 10 days apart, should be sufficient. Otherwise same as for Botrytis Blight (above).

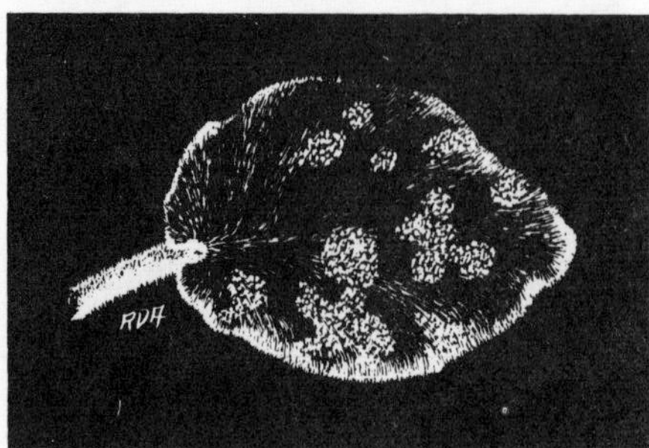

Fig. 4.2. Powdery mildew of African-violet.

8. *Mosaic*—Leaves crinkled and thickened. May show irregular light and dark green blotches. Plants may be stunted. Flowers reduced in size and number. *Control:* Destroy infected plants and start with new, virus-free ones.
9. *Bud Drop*—Flower buds shrivel, turn brown, and drop before opening. *Control:* Avoid low temperatures and air humidity, overwatering, gas injury, and extremes in soil temperature, moisture, and light. Control mites and other bud pests. Check with your local florist or extension entomologist.
10. *Leaf or Foliar Nematode*—Gradually enlarging, sunken, brown blotches between leaf veins. Spots mostly on underleaf surface. Few or no flowers are produced. Plants stunted, may die. *Control:* Same as for Root-knot (above). Remove and burn infested leaves. Soak potted plants in hot water (110° F.) for 30 minutes. Space plants. Keep water off foliage. Hot water soak controls mites, too.
11. *Spotted Wilt, Ringspot* (gloxinia)—Large, brown-ringed patterns with green centers on leaves. Leaves may die. *Control:* Destroy infected plants. Spray with DDT to control thrips that transmit virus.
12. *Aster Yellows* (gloxinia)—Plants slightly yellowish. Produce numerous secondary shoots and no flowers. *Control:* Same as for Spotted Wilt (above). Use DDT to control leafhoppers that transmit virus.
13. *Sclerotinia Blight* (gloxinia)—California. Soft, rapid rot of flowers that causes them to collapse. When infected flowers drop on leaves, rot moves into leaves and then petioles. Growing point may die, stunting plant. *Control:* Pick and burn rotting flowers when first found. Apply PCNB to soil. Follow manufacturer's directions. Same as for Crown Rot (above).
14. *Stunt Disease* (African-violet)—Leaves small, brittle, may be rolled or deformed. Symptoms may disappear then reappear. Growth is slow; flowers few. Patches on leaf surface may be relatively hairless. Sometimes leaf surface is mottled with lighter green, especially along midrib. Varieties differ in resistance and ease of becoming infected. *Control:* Destroy affected plants. Select and isolate healthy plants. Propagate only from known healthy plants. When propagating, snap off leaves with fingers. After handling each plant wash hands in hot, soapy, running water.
15. *Bacterial Leaf Blight* (African-violet, gloxinia)—Oregon. Water-soaked areas on leaves that may develop into black dead areas. Affected tissues may become mushy and watery. *Control:* Same as for Crown Rot (above).
16. *Boron Deficiency* (gloxinia)—Small, irregular, brownish-black areas near base of leaf that spread toward leaf tip and down into crown. Leaves wilt. Growing point is killed. Flowers and peduncles are dwarfed; appear wilted. Aboveground plant parts later turn brown. Plant dies. *Control:* Make several applications, 2 weeks apart, of a 6 per cent boric acid solution.

AFRICAN YELLOW-WOOD — See Yew
AGAPANTHUS — See Tulip
AGARITA — See Barberry
AGERATUM — See Chrysanthemum
AGLAONEMA — See Calla Lily
AGRIMONY *(Agrimonia)* — See Rose
AGROPYRON — See Lawngrass
AGROSTEMMA — See Carnation
AGROSTIS — See Lawngrass
AILANTHUS — See Tree-of-Heaven
AIRPLANT — See Sedum
AIR POTATO — See Yam
AJUGA, BUGLEWEED *(Ajuga)*

1. *Crown Rot, Southern Blight*—Serious in shady, wet, poorly drained areas. Plants suddenly turn yellow, wilt, and die in patches during warm, humid weather. Bases of stems rot, turn brown or black. Frequently covered with cottony mold in which small, reddish-tan, seedlike bodies develop. *Control:* Dig up and burn affected plants including 3 to 5 inches of surrounding soil. Apply PCNB 75 per cent (1 tablespoon per gallon) as soil drench, or use a 1:1,000 solution of mercuric chloride. Use 1 pint per square foot of bed surface.
2. *Root-knot*—Plants lack vigor, often stunted with knotlike galls on roots. See (37) Root-knot under General Diseases.

ALBIZZIA — See Honeylocust

ALDER — See Birch

ALEXANDERS — See Angelica

ALFILARIA — See Cranesbill

ALKANET — See Mertensia

ALLIONIA — See Four-o'clock

ALLIUM — See Onion

ALLSPICE — See Calycanthus

ALMOND — See Peach

ALOE [BARBADOS, HEDGEHOG, VARIEGATED], TORCH PLANT, PARTRIDGE-BREAST *(Aloe)*; HAWORTHIA, WART PLANT, CUSHION ALOE *(Haworthia)*

1. *Root Rot*—Serious nursery disease. See under Geranium. *Control:* Plant in light, fertile, well-drained soil. Avoid overwatering. Clean infected *aloe* and *haworthia* plants and soak in hot water (115° F.) for 20 to 40 minutes. Grow treated plants in clean or sterilized soil (pages 538–548).

ALPINE CURRANT — See Currant

ALTERNANTHERA — See Cockscomb

ALTHAEA — See Hollyhock

ALUMINUM PLANT — See Pilea

ALUMROOT — See Delphinium

ALYSSUM — See Cabbage

AMARANTH, AMARANTHUS — See Cockscomb

AMARYLLIS, AMAZON-LILY — See Daffodil

AMELANCHIER — See Apple

AMERICAN BLADDERNUT *(Staphylea)*

1. *Leaf Spots*—Spots of various sizes, shapes, and colors. *Control:* Collect and burn fallen leaves. Keep plants well pruned. If practical, spray during spring and summer rainy periods. Use zineb, maneb, or captan.
2. *Twig Blights, Dieback*—Twigs blighted. May die back. Small shrubs may be killed. *Control:* Prune and burn affected parts. Otherwise same as for Leaf Spots.
3. *Sooty Blotch*—See (12) Sooty Mold under General Diseases.

AMERICAN COWSLIP — See Primrose

AMERICAN SPIKENARD — See Acanthopanax

AMORPHA — See False-indigo

AMPELOPSIS — See Grape

AMSONIA — See Vinca

ANACAHUITA — See Cordia

ANAGALLIS — See Primrose

ANAPHALIS — See Chrysanthemum
ANCHUSA — See Mertensia
ANCISTROCACTUS — See Cacti
ANDROMEDA [FORMOSA, JAPANESE, MOUNTAIN] *(Pieris)*

1. *Leaf Spots, Tar Spot*—Small spots on leaves in which black dots may later be sprinkled. *Control:* Apply zineb, maneb, or ferbam at 2-week intervals. Start when leaves are half grown.
2. *Root Rot, Dieback*—Roots decay. Plants gradually decline, wither, die back, and finally die. Often associated with nematodes (e.g., bloat, lance, lesion, pin, reniform, ring, spiral, stubby-root, stunt). *Control:* See under Rhododendron, and (34) Root Rot under General Diseases. Keep soil quite acid (pH 4.5 to 5.5). Soil should be well-drained. Avoid overwatering.
3. *Twig Blight*—Twigs die back. *Control:* See (22) Stem Canker under General Diseases.

ANDROSACE — See Primrose
ANEMONE [CAROLINA, FLAME, GARDEN, JAPANESE, LANCE-LEAVED, MEADOW, POPPY, WOOD], PASQUEFLOWER, WINDFLOWER, SNOWDROP WINDFLOWER *(Anemone)*; BANEBERRY [RED, WHITE] *(Actaea)*; RUE-ANEMONE *(Anemonella)*; BUGBANE, SNAKEROOT, BLACK-SNAKEROOT or BLACK COHOSH *(Cimicifuga)*; HEPATICA [COMMON, ROUND-LOBED, SHARP-LOBED], LIVERLEAF *(Hepatica)*; GLOBEFLOWER [AMERICAN, ASIATIC, EUROPEAN], GOLDEN QUEEN *(Trollius)*

1. *Leaf Spots*—Spots of various sizes, shapes, and colors. Leaves may discolor and drop early; usually starting at base of plant. *Control:* Space plants. If serious, apply maneb, zineb, thiram, fixed copper, ferbam, or captan at about 10-day intervals during wet weather. Destroy infected plant parts when found. Burn tops in fall.
2. *Rusts*—Powdery, brown to black pustules or "orange cluster cups" on lower leaf surface. Affected plants rarely flower. Leaves may be stunted, thickened, crowded; turn pale and fleshy. Alternate hosts include plums, cherries, and various grasses. *Control:* Dig and burn infected *anemone* plants. They will not recover. Spray or dust remainder as for Leaf Spots (above). Propagate from disease-free plants.
3. *Downy Mildew*—Widespread on anemones. Large brown or black blotches on leaves. Corresponding undersides of leaves covered with delicate, white mildew patches. Leaves become lead-green, shrivel, erect, tend to roll upwards. Plants distorted. Infection may be systemic. *Control:* Same as for Leaf Spots (above). If severe, dig and burn affected plants. Four-year rotation.
4. *Leaf and Stem Smuts*—Irregular, dark brown to black, powdery blisters and streaks on swollen regions of leaves and leaf stalks (Figure 4.3). Common on wood anemone. *Control:* Samc as for Rusts (above).

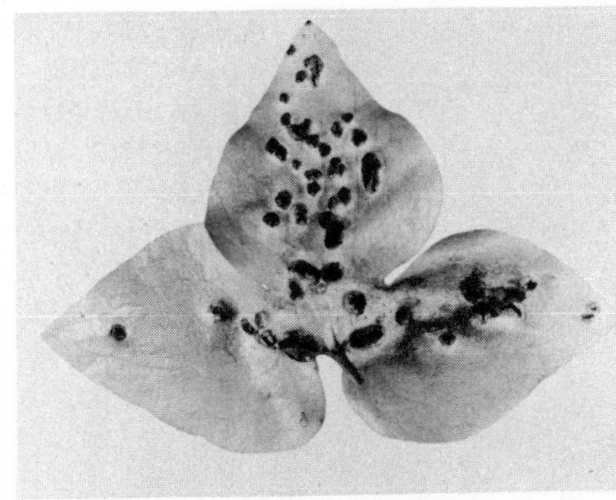

Fig. 4.3. Anemone smut.

5. *Leaf Gall, Spot Disease* (anemone)—Flowers, stems, and leaves spotted with small red warts. Flowers may become dwarfed, distorted, and fall early. *Control:* Same as for Leaf Spots (above).
6. *Leaf and Stem Nematode*—Dark brown or black blotches on leaves. Leaves may die. *Control:* See (20) Leaf and Stem Nematode under General Diseases.
7. *Powdery Mildew* (anemone, rue-anemone)—See (7) Powdery Mildew under General Diseases.
8. *Crown Rot, Rhizome Rot, Southern Blight* (anemone)—Underground parts or stem rots at soil line. Plants wilt, wither, and collapse. *Control:* Completely dig out and destroy infected plants together with 6 inches of surrounding soil. Sterilize remaining soil with heat or chemicals (538–548).
9. *Botrytis Blight, Collar Rot* (anemone)—Crowns rotted and destroyed near soil line. Flowers and flower buds may rot. Commonly occurs after injury from wet or cold weather or downy mildew. *Control:* Avoid crowding plants. Plant in light, well-drained soil. Spray as for Leaf Spots (above).
10. *Mosaic, Flower Breaking* (anemone)—Leaves mottled light and dark green. May be yellowish. Flowers may show light or off-color streaks and blotches. *Control:* Destroy infected plants. Keep down weeds. Control aphids that transmit virus. Use lindane or malathion.
11. *Aster Yellows*—See (18) Yellows under General Diseases.
12. *Spotted Wilt* (anemone)—See (17) Spotted Wilt under General Diseases. *Control:* Same as for Mosaic (above). Spray with DDT plus malathion to control thrips that transmit virus.
13. *Root-knot*—See (37) Root-knot under General Diseases. Anemone is very susceptible.
14. *Root Rot*—Uncommon. See (34) Root Rot under General Diseases. Dig corms, dry, discard rotted ones and replant in clean soil (pages 538–548).
15. *White Smut* (anemone)—See (13) White Smut under General Diseases.

ANEMONELLA — See Anemone

Anethum — See Celery

ANGELICA [PURPLE-STEMMED or ALEXANDERS, FILMY] *(Angelica)*; TAENIDIA

1. *Leaf Spots*—General. Spots of various sizes, shapes, and colors. If severe, leaves may wither. *Control:* Collect and burn tops in fall. Space plants. If severe, apply zineb, maneb, or ferbam at about 10-day intervals during rainy weather.
2. *Rusts*—Yellow, orange, reddish-brown or black, powdery pustules on leaves. *Control:* Same as for Leaf Spots (above).
3. *Root Rot*—See (34) Root Rot under General Diseases.

ANGELS-TRUMPET — See Tomato

ANGRAECUM — See Orchids

ANISE, ANISE-ROOT — See Celery

ANISETREE — See Magnolia

ANNUAL BLANKET-FLOWER — See Chrysanthemum

ANODA — See Hollyhock

ANTENNARIA, ANTHEMIS — See Chrysanthemum

ANTHONY WATERER — See Spirea

ANTHRISCUS — See Celery

ANTHURIUM — See Calla Lily

ANTIRRHINUM — See Snapdragon

APIUM — See Celery

APPLE, CRABAPPLE [AMERICAN, ARNOLD, BECHTEL, CARMINE, CHINESE FLOWERING, CUTLEAF, ELEY'S, GARLAND or WILD SWEET, HALLIS, HOPA, HYBRID, JAPANESE FLOWERING or SHOWY, KAIDO, LEMOINE, MANDSHURIAN, OREGON, PARKMAN, PEACH-LEAF, PEAR-LEAF, PRAIRIE, PURPLE, RIVERS', SARGENT, SCHEIDECKER, SIBERIAN, SOULARD, SOUTHERN, TEA, TIBETAN, TORINGO] *(Malus)*; SERVICEBERRY [ALLEGHENY or SMOOTH, APPLE, CLUSTER, CUSICK, DOWNY, RUNNING, WESTERN or SASKATOON, SUCCESS], JAPANESE JUNEBERRY, DWARF JUNEBERRY, JUNEBERRY, SHADBLOW, SHADBUSH *(Amelanchier)*; CHOKEBERRY [BLACK, PURPLE, RED] *(Aronia)*; FLOWERING QUINCE [DWARF JAPANESE, JAPANESE, CHINESE QUINCE] *(Chaenomeles)*; HAWTHORN or THORN [ARNOLD, BLACK, CHINESE, COCKSPUR, DOTTED, ENGLISH (many horticultural varieties), FLESHY, FROSTED, LAVALLE, PAUL'S DOUBLE SCARLET, RED HAW or SCARLET, SINGLE-SEED, TOBA, WASHINGTON] *(Crataegus)*; COTONEASTER [BEARBERRY, BOXLEAF, COIN-LEAF, CRANBERRY, CREEPING, EUROPEAN, HEDGE, HUPEH, MANYFLOWER, MULTIFLORA, NECKLACE, PEKING, ROCK or QUINCEBERRY, ROCKSPRAY, SMALL-LEAF, SPREADING, WILLOWLEAF] *(Cotoneaster)*; QUINCE *(Cydonia)*; LOQUAT *(Eriobotrya)*; MEDLAR *(Mespilis)*; PHOTINIA [CHINESE, JAPANESE, ORIENTAL], CHRISTMASBERRY or TOYON *(Photinia)*; FIRETHORN [ENGLISH, FORMOSA, SCARLET, YUNNAN] *(Pyracantha)*; PEAR [BIRCH-LEAF, CALLERY, COMMON, SAND, SNOW] *(Pyrus)*; MOUNTAIN-ASH [AMERICAN, CHINESE, COLUMNAR, EUROPEAN or ROWANTREE, KOREAN, PACIFIC, PYRAMIDAL, SHOWY, SHRUBBY CHINESE, SITKA, WEEPING, WESTERN], WHITE BEAMTREE, SERVICETREE *(Sorbus)*; CHINESE STRANVAESIA *(Stranvaesia)*

1. *Fire Blight, Bacterial Blast*—General and serious. Blossoms and fruit spurs are blasted. New shoots suddenly appear as if scorched by fire. Brown or blackened leaves cling to twigs. Slightly sunken, girdling, discolored cankers on twigs, branches, and trunk. Often followed by Black Rot and Wood Rots (both below). See Figure 2.42 under General Diseases and Figure 4.4. Infected fruit are first water-

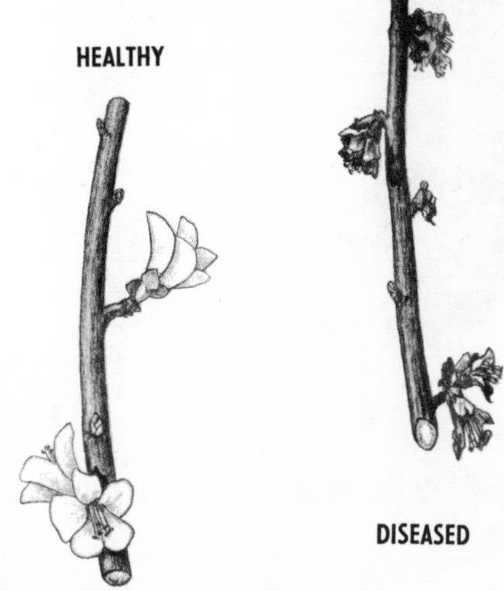

Fig. 4.4. Fire blight (blossom blight) of Japanese quince.

soaked, later turn brown, shrivel, and finally black. *Control:* Avoid overstimulation of trees. Plant in fertile, well-drained soil. Follow spray program in Appendix (Table 16). Prune out cankers 3 to 12 inches back into healthy wood. Prune during dormant season or in late summer if hot and dry. Apply disinfectant then tree wound dressing (pages 34–36) to cut surfaces. Swab or dip pruning tools with 70 per cent alcohol or household bleach (e.g., Clorox) between cuts. A 1:1,000 solution of mercuric chloride may also be used, as well as 5 per cent Lysol, or 4 per cent Amphyl. Fairly resistant *apples:* Anoka, Arkansas Black, Baldwin, Ben Davis, Blaze, Chestnut, Crimson Beauty, Crimson Winesap, Delicious, Early Red Bird, Early Winesap, Farmingdale, Fenton, Franklin, Giant York, Haralson, Jonadel, Julyred, King David, McIntosh, Minjon, Monroe, Northern Spy, Paducah, Red Flesh Crab, Redwell, Shalred, Sharon, Starking, Starkrimson, Stayman Red, Turley, Virginia, Wagener, Wellington, etc. Fairly resistant *crabapples:* See list on page 136. Others include, Alney, *M. arnoldiana, M. atrosanguinea, M. baccata gracilis, M. b. mandshurica, M. coronaria charlotte,* Crimson Brilliant, Dolgo, Double White Chinese Flowering, Katherine, Makamik, Marshall, Mary Potter, Nippon C., Nova, Red Jade, Red Silver, Sissipuk, Snowdrift, Whitney, etc. Fairly resistant *pears:* Ayres, Bantam, Campas, Carrick, Dabney, Ewart, Funks Colorado, Garber, Golden Spice, Hood, Kieffer, LeConte, Longworth, Magness, Maxine, Moe, Moonglow, Morgan, Nectar, Old Home, Oregon 18, Pai Li, Pine Apple, Pontotoc, Richard Peters, Seckel, Surecrop, Tyson, Vistica, Waite, Winter Nelis, Wurtenburg, etc. Numerous resistant *cotoneaster* and *pyracantha* species and varieties are also available. Check with your local nurseryman or extension horticulturist regarding adaptability of these varieties to your area. Spray 2 to 4 times, at 3- or 4-day intervals, during the bloom period when the temperature is above 65° F. Use streptomycin at 60 to 100 parts per million (ppm). Be sure the streptomycin was manufactured during the current season. Follow manufacturer's directions. Do *not* use streptomycin on *Cotoneaster racemiflora* or *Crataegus mollis.* See also (24) Fire Blight under General Diseases. Apply copper sulfate, ½ pound in 10 gallons of water, *before* buds break open in spring. Avoid overfertilizing with nitrogen. Use balanced fertility program based on a soil test.

2. *Scab*—Widespread and serious. Dull, smoky spots that change to velvety, olive-

Fig. 4.5. A. Photinia leaf spot. B. Photinia scab.

Fig. 4.6. Hawthorn rust (early infection).

green. Finally turn brown to black and often scaly. Spots occur on leaves, fruit, and flower parts. Fruit may be deformed, cracked, and russeted. Many young fruit and leaves drop early. See Figure 2.30C under General Diseases, and Figure 4.5B. *Control:* Collect and burn fallen leaves in autumn. Follow regular spray program. Use multipurpose fruit spray containing captan, zineb, folpet, difolatan, or thiram. Dodine, glyodin, phenyl mercury, and dichlone are widely used by commercial apple growers. See spray schedule in Appendix (Table 16). *Apple* and *pear* varieties differ in susceptibility. Check with your local nurseryman. Resistant *crabapples:* See list on page 136. Others include, Ames White, *M. baccata mandshurica, M. hupehensis, M. sieboldii arborescens, M. zumi calocarpa,* Bob White, Cathay, Douphin, Gloriosa, Gorgeous, Hall C., Helen, Hillier C., Hoopes, Kansu C., Kit Trio, Lee Trio, Liset, Makamik, Mary Potter, Morden Rosybloom, Mount Arbor, Orange, Oregon C., Peachblow, Profusion, Queen Choice, Redflesh, Red Tip, Rock C., Rondo, Rose Tea, Rudolph, Sargent, Scugog, Showy, Sikkim, Silvia, Sissipuk, Snowbank, Snowdrift, Sugar C., Sundog, Tea C., Toringo C., Wakpala, E. H. Wilson, Wisley C., Young America, Yunnan, and many more. Resistant *firethorns:* Shawnee and Yunnan.

3. *Rusts*—Widespread. Pale yellow, yellow-orange to orange-red or brownish-orange spots (with black specks) on upper leaf surface with mat of creamy-white, light orange to brown "cluster cups" forming on corresponding underleaf surface. Heavily spotted leaves turn yellow and drop early. Yellowish to reddish or greenish areas on fruit, usually near calyx end. Fruit may be distorted and drop early. Twigs and small branches may die back. See under Juniper Rusts. See Figure 2.24C under General Diseases and Figure 4.6. *Control:* Follow regular spray program as for Scab. Add 1 tablespoon of ferbam, zineb, thiram (Thylate), Polyram, or Niacide M to each gallon of spray from prebloom through second or third cover (see Table 16 in Appendix). Varieties differ in resistance to the several rusts. Check with your nurseryman, county agent, extension horticulturist, or plant pathologist. Resistant *apples:* Arkansas Black, Baldwin, Cortland, Charlamoff, Dutchess, Erickson, Fireside, Franklin, Golden Delicious, Grimes Golden, Haralson, Hibernal, King David, McIntosh, Macoun, Melrose, Northwestern Greening, Sharon, Stayman, Transparent, Turley, Wedge, Winesap, Wolf River, and York. Resistant *crabapples:* See list on page 136. Others include, Arnold C., *M. baccata gracilis, M. b. mandshurica,* Bob White, Carmine C., Columnaris C., Cutleaf C., Dawson C., Dolgo, Dorothea, Double White Chinese Flowering, Flame, Henry F. DuPont, *M. hupensis,* Japanese Flowering, Katherine, Lemoine C., Liset, Makamik, Manchurian C., Mary Potter, Marshall Oyama, Nippon C., Parkman C., Radiant, Red Jade, Red Silver, Rivers C., Sargent, Siberian C., *M. sieboldii arborescens,* Snowdrift, Tea C., Tree Toringo C., Van Eseltine, William Sim, and many others. Destroy nearby, worthless, erect junipers, redcedars, cypress, incense-cedar, or white-cedar which show rust galls.
4. *Powdery Mildews*—General. Whitish-gray, powdery mold or felty patches on young leaves, buds, blossoms, and twigs. Leaves may be crinkled, curled, dwarfed, narrowed, and erect. Shoots stunted with rosette-type growth (Figure 4.7). May die back. Fruits on certain susceptible varieties are russeted and may crack. Blossoms are brown and fruit set is reduced. *Control:* Where common, follow apple spray program using sulfur, Morestan, Morocide, or Karathane (plus wetting agent) in prebloom sprays through first cover. Complete coverage is essential. Grow resistant, adapted varieties. Check with your county agent, extension horticulturist, or plant pathologist. Somewhat resistant *apples:* Delicious, Monroe, and Winesap. *Very susceptible apples:* Cortland, Idared, Jonathan, and Rome Beauty. Resistant *crabapples:* See list on page 136. Others include, Arnold C., *M. baccata gracilis,* Bob White, Carmine C.. Cherry C., Columnaris C., Cutleaf C., Dawson C., Dolgo,

Fig. 4.7A (left). Powdery mildew of apple showing cupping, twisting, and rosetting of leaves. (Department of Botany & Plant Pathology and Cooperative Extension Service, Michigan State University photo)

Fig. 4.7B (top). Apple powdery mildew. These fruit show net russetting, a symptom of mildew. (Department of Botany & Plant Pathology, Michigan State University photo)

Dorothea, Double White Chinese Flowering, Henry F. DuPont, *M. hupensis,* Improved Bechtel, Japanese Flowering, Katherine, Marshall Oyama, Nippon C., Parkman C., Prince George, Radiant, Red Bud C., Red Jade, Rivers C., Sargent C., *M. scheideckeri,* Siberian C., Snowdrift, Tanner, Tea C., Van Eseltine, William Sim, *M. zumi calocarpa,* and many more.

Excellent modern *crabapples* resistant to fire blight, scab, rusts, and powdery mildew are:

Adams	David	Ormiston Roy
Ames White	Evelyn	Peachleaf C.
baccata Jackii	Ferrill's Crimson	Pixie
Baskatong	Golden Hornet	Professor Sprenger
Beauty	Goldfinch	Red Splendor
Blanche Ames	Gwendolyn	x *robusta* Persicifolia
Case Seedling	Jack C.	Seafoam
Centennial	Lady Northcliffe	Snowbank
Chestnut		Winter Gold

5. *Black Rots, Frog-eye Leaf Spot, Dieback* (primarily apple, crabapple, pear, quince, flowering quince, hawthorn, cotoneaster, mountain-ash) —Expanding, brown, often zoned spots on fruit that start at insect wounds or mechanical injuries. Fruits turn into shriveled, black mummies that cling to twigs. Leaf spots are round to irregular, purple, then gray to brown with a purple margin, in which black specks can later be found. Rough-barked cankers, with lobed margins, on twigs and large limbs, that often die back. Commonly follows Fire Blight or winter sunscald. *Control:* Prune out and burn all dead wood and destroy rotted fruit mummies. Follow regular spray program using captan, folpet, ferbam, difolatan, or zineb plus methoxychlor and malathion to control insects. Avoid wounding trees or fruit. Plant disease-free nursery stock. Protect trees against sunscald (see below). Keep trees growing vigorously. Refrigerate fruit promptly after picking.
6. *Crown Gall*—General. Rough, dark, corky gall, up to 4 inches in diameter, on trunk near soil line, at graft union, and on roots. Trees gradually lose vigor, may later die. See Figure 2.46A under General Diseases. *Control:* Plant disease-free stock with smooth graft union and free of root overgrowths or enlargements. Avoid wounding young trees, especially near soil line. Destroy badly infected trees.

Control root-feeding insects. See (30) Crown Gall under General Diseases.

7. *Infectious Hairy Root, Woolly Knot*—Widespread. Most common in nursery. Primary roots are swollen, often extend some distance from a basal gall before secondary fibrous roots are produced. Excessive growth (tufts) of these long, fine, fibrous roots gives woolly appearance. Bacteria enter only through wounds. Small fibrous roots occur singly or in clusters at base of trunk, crown, or roots. Often associated with Crown Gall, with mass of fleshy roots arising from a gall. May be confused with burr knot (a noninfectious proliferation of roots) characteristic of certain varieties. *Control:* See Crown Gall (above).
8. *Sooty Molds or Blotch, Fly Speck*—General in damp, shady areas. Round to irregular, sootlike spots, or clusters of 6 to 50 shiny "fly specks" on surface of fruit or twigs. Follows attacks by aphids, scales, and other insects. See Figure 2.28B under General Diseases. *Control:* Follow regular spray program (Table 16 in Appendix). Add zineb, ferbam, thiram, or folpet in midsummer and late summer sprays. Or alternate folpet, zineb, or thiram with captan from second cover on. Prune to open up trees.
9. *Blotch* (apple, crabapple)—Irregular, dark brown to black, shiny, somewhat sunken, irregularly lobed spots on fruit. Small, rough, light-colored cankers may girdle twigs and fruit spurs. Small, round, whitish spots occur on leaves in which a black speck develops. Or elongated, sunken, light-colored spots with several black dots may form on leaves. Severely infected leaves drop early. See Figure 4.8. *Control:* Same as for Black Rots (above). Resistant *apples:* Grimes Golden, Grove, Stayman Winesap, and Winesap.
10. *Root Rots*—Foliage and twig growth is thin and weak. Leaves often yellow, wither, and drop prematurely. Growth is slow. Twigs and branches die back. Fruit crop often is heavy just before tree dies. Coarse white to black mold fans or strands often grow just under bark of lower trunk and on root surfaces. Clumps of honey-colored mushrooms may appear near trunk base. See Figure 2.49B under General Diseases. May be associated with root-feeding nematodes, girdling by mice, winter injury, waterlogging of soil, or collar rot. *Control:* Plant in fertile, well-drained soil where root rot of woody plants has not occurred before. Remove and burn dying trees, including all roots. Do not replant in same location for number of years. Control rodents and borers. Check with your extension entomologist or county agent. Fertilize, water, and follow recommended spray program to maintain tree vigor. Soil fumigation in the fall with carbon disulfide has been used to control Armillaria Root Rot (Oak-root Fungus Disease) in light, well-prepared soils. This

Fig. 4.8. Apple blotch on leaf and twigs. (Courtesy Dr. V. H. Young)

is a job for an experienced arborist. Carbon disulfide is inflammable, explosive, and poisonous. The addition of PCNB (50 pounds per acre) plus Dexon (30 pounds per acre) or DBCP to the soil around newly planted apple trees has increased shoot growth 40 per cent in 2 years. Domestic French (from Bartlett or Winter Nelis seeds) *pear* is resistant to Armillaria Root Rot.

11. *Fruit Spots and Rots, Storage Rots*—General. Small to large, round to irregular, watersoaked to corky, light tan to black spots on and in fruit. Decay may later be covered with white, blue, green, pink, black, dark brown, or gray mold. Rot-producing organisms often enter insect and mechanical injuries, or follow scab and other diseases. Develop rapidly in warm moist storage. See Figure 2.48A under General Diseases. *Control:* Follow regular spray program, especially late cover sprays. Adding zineb or folpet to summer sprays often helps. Control insects using methoxychlor or sevin plus malathion. Apply captan or folpet alone just before harvest. Pick early before many fruit drop. Handle fruit carefully. Store only sound, blemish-free fruit just above freezing (33° F.) with high humidity (80 to 85 per cent). Storage area should be clean. Collect and destroy rotted fruits on ground or in tree. Keep trees well pruned and growing vigorously. *Apples* resistant to *Bitter Rot:* Arkansas Black, Delicious, Rome Beauty, Stayman Winesap, Winesap, Yates, and York Imperial. (*Apples very susceptible to Bitter Rot:* Golden Delicious, Grimes Golden, Jonathan, Northwestern Greening, and Yellow Newton.) *Apples* resistant to *Alternaria Cork Rot* in the orchard: Delicious, Rome Beauty, Stayman Winesap, and Winesap. *Apples* apparently resistant to *Black Pox:* Grano, Transparent, and York Imperial. For additional information read USDA Leaflet 406, *Apple Bitter Rot.*

12. *Twig, Branch, and Trunk Cankers, Dieback, Apple Anthracnose, Limb Blight*—Most common on weakened or injured trees low in vigor. Affected bark often sunken or swollen, cracked, roughened, and discolored with wood underneath dead and discolored. Round to irregular cankers often girdle twig, limb, or trunk killing parts beyond. Entire tree may wilt and die. Cankers often show zoned ridges (Figure 4.9). Certain organisms may also produce fruit spotting and rotting. Varieties differ in susceptibility. Check with your local nurseryman, extension horticulturist, or plant pathologist. *Control:* Prune and burn cankered twigs and branches in dry weather. Cut out large cankers and sterilize cuts with household

Fig. 4.9. Apple twig canker. (Iowa State University photo)

bleach or 1:1,000 solution of mercuric chloride (see page 101 for precautions). Then paint with tree wound dressing containing mercury (page 36). Follow regular spray program to control insects and other diseases. Avoid wounding twigs, branches, and trunk. Maintain vigor by fertilizing in spring. Water during dry periods. Wrap young trees to prevent sunscald. Where serious, a spray of 4-4-50 bordeaux is recommended after leaves drop in fall. Grow recommended varieties and species. See also under Wood Rots (below). *Apples* resistant to *Collar Rot (Phytophthora):* Antonovka, Delicious, McIntosh, Wealthy, and Winesap. (*Apples* very susceptible to *Collar Rot:* Dutchess, Grimes Golden, and Lodi.) *Quince* and *Oriental pear rootstocks* are more resistant than French roots. When Crown Rot or Canker is found, remove soil around tree base to depth of 8 or 10 inches. Remove all infected bark away to healthy wood. Sterilize surface with a 1:1,000 solution of mercuric chloride and then paint with tree wound dressing.

13. *Wood Rots, Butt Rots, Collar Rots, Heart Rots, Silver Leaf*—See under Birch, and (23) Wood Rot under General Diseases. Collar rot causes killing of bark and wood underneath at or near soil line. Foliage may be lighter colored on branches over cankered area. Trees may be girdled and killed. *Control:* Follow regular apple spray program. Control borers by spraying trunk and scaffold limbs monthly with lead arsenate, Thiodan, or DDT. Check with your county agent or extension entomologist regarding rates to use and dates of application for your area. Keep trees vigorous by pruning, fertilization, and regular spray program. Plant in well-drained soil. For resistant varieties and rootstocks see Cankers (above).
14. *Winter Injury, Collar Rot*—Roots, shoots, twigs, or buds may be killed. Splitting of bark is common, especially on trunk (Figure 4.10). Generally occurs on south or southwest side. Sometimes results from failure of wood to mature in the fall, excessively low temperatures, and other factors. *Control:* Same as for Wood Rots (above) and Sunscald (below).
15. *Sunscald*—Most common on young, exposed trees. Freezing injury to trunk and larger branches on south or southwest side and where fruit are exposed to direct sun during hot (90° F. or above), dry weather. Round, pale yellow to brown or

Fig. 4.10. Winter injury to a young apple tree. (Iowa State University photo)

black areas on fruit. Dark patches occur on *pear* and *apple* leaves in midsummer. *Control:* Wrap young trees and exposed larger branches (Figure 2.11) with burlap, or sisalkraft paper. Follow spray program (Table 16 in Appendix). Maintain trees in vigorous condition. Water during summer droughts. Do not prune off lower limbs for several years. Leave lowest limb on south side.

16. *Fruit Breakdowns in Storage* (Scald, Bitterpit or Stippen, Baldwin or Jonathan Spot, Stayman Spot, Black End, Brown Core or Heart, Freezing Injury, Internal Breakdown, Soggy Breakdown or Soft Scald, Water Core) —May be caused by an irregular water supply during the season, general unthrifty growth, unbalance or lack of essential soil nutrients (e.g., low calcium and high nitrogen, lack of boron or phosphorus), toxic vapors given off by fruit in storage, freezing injury, poor storage conditions, and other factors. *Control:* Check with a local grower or your extension horticulturist. Have soil tested. Prune, water, spray, cultivate, and fertilize trees to keep vigorous. Pick fruit when first mature. Place in cold (32–38° F.), well-ventilated storage as soon as possible. Avoid drastic pruning, over- or under-watering and overfertilizing, especially with nitrogen. Commercial apple and pear growers control scald by (1) oiled, diphenylamine (DPA) impregnated wraps, (2) dipping fruit in ethoxyquin (sold as Stop-Scald and Santoquin) or DPA (sold as No Scald DPA). DPA and ethoxyquin may also be applied as tree sprays, roller sprays, and by flooding or drip applicator. Follow manufacturer's directions carefully. Wayne *apple* is not subject to Bitterpit.

17. *Leaf Spots, Leaf Blight, Leaf Scorch, Anthracnose*—Leaves variously spotted. May turn color and drop early (Figures 4.5A and 4.11). Fruit and twigs may be spotted. *Control:* Cut and burn blighted twigs. Collect and burn fallen leaves. Follow spray program as for Scab (above) or use zineb. Maneb, thiram, ferbam, dodine, zineb, folpet, difolatan, Acti-dione, or phenyl mercury may be used on *hawthorn*. Apply at about 2-week intervals or follow manufacturer's directions. Start when new leaves emerge from buds. Partially resistant *pear* varieties to *Mycosphaerella Leaf Spot:* Campas, Conference, Garber, Lincoln, Mooers, Orient, and Pineapple. *Apple* varieties resistant to *Helminthosporium Leaf Spot* and *Bark Canker:* Cortland, Detroit Red, Lodi, and Yates. English *hawthorn,* especially the variety Paul's Scarlet, is very susceptible to *Fabraea (Entomosporium) Leaf Blight.* Cockspur and Washington hawthorns are resistant.

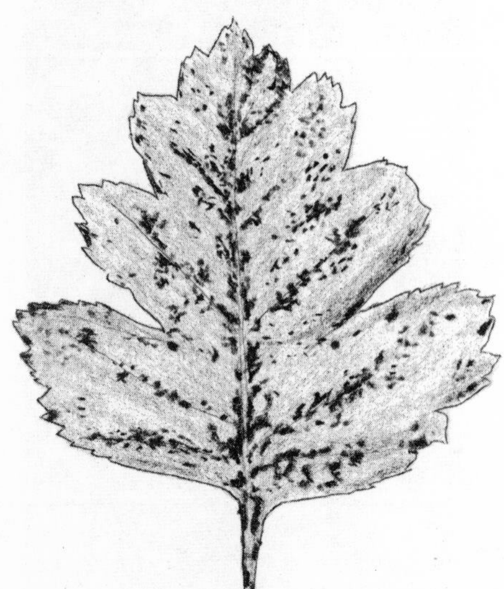

Fig. 4.11. Hawthorn leaf blight.

18. *Stony Pit of Pear*—Mostly Pacific Coast states and New York. Widespread in the Beurre Bosc variety, occasional in d'Anjou, Comice, Bartlett, Flemish Beauty, Kieffer, Seckel, and Worden. Symptoms variable. Small, dark green areas on young fruit that later develop into deep "pits." Fruit may become deformed, gnarled, stunted, and woody or "stony" at maturity. Often drop early. Foliage is reduced. *Control:* Destroy infected trees as they will not recover. Plant resistant varieties (e.g., Bartlett—usually a symptomless carrier) propagated from virus-free stock. Varieties showing mild symptoms include Clairgeau, Old Home, Packham's Triumph, and Waite.
19. *Witches'-broom, Black Mildew of Amelanchier (Serviceberry)*—Widespread. Mass of sturdy, dwarfed, new shoots form perennial witches'-brooms or slender, stiff branches are twisted into a loose tangle. Undersurfaces of leaves usually coated with black mold. Trees and shrubs gradually decline in vigor. *Control:* Cut off and burn witches'-brooms.
20. *Mistletoe* (apple, hawthorn, pear) —See (39) Mistletoe under General Diseases.
21. *Root Nematodes* (e.g., burrowing, dagger, lance, lesion, pin, ring, sheath, spear, spiral, stunt, stubby-root, tylenchus) —Tree vigor is reduced. Leaves stunted. Terminal growth may die back. May be associated with root rot and "little leaf" or rosette. Small, dead, dark spots on white rootlets. Affected roots may be stunted, distorted, and die back. *Control:* Keep trees vigorous by feeding, pruning, and watering. Fumigate soil (pages 538–548) before replanting. DBCP may be used around certain established plants. Follow label directions.
22. *Root-knot*—Common in southern states, especially in sandy soils and where cover crops are used. Knots and beadlike swellings on roots may be confused with those made by woolly apple aphids, crown gall, and hairy root. Plants may be stunted. Foliage yellowish with scorching of leaf margins. Leaves may wilt temporarily in hot, dry weather. *Control:* Same as for Root Nematodes (above).
23. *Zinc Deficiency, Little Leaf, Rosette*—Common in alkaline soils on apple and pear. Whorls of small, stiff, narrow, yellowish, sometimes mottled leaves (called rosettes) at tips of current season's growth. Twigs usually spindly and stunted. May die back after the first year. *Control:* Check with your county agent or extension horticulturist. They may recommend a dormant spray of zinc sulfate (about 2 pounds of 23% zinc sulfate in 10 gallons of water).
24. *Internal Cork, "Drought Spot" or Dieback, Boron Deficiency* (apple, pear, quince) —Widespread, may be serious in local areas, especially during droughts. Fruits often small and deformed. Large, more or less superficial dead areas, that become russeted and cracked (pitted in pear and quince), usually appear before fruit is half grown. Brown, spongy to corky spots and streaks form in fruit flesh around the core. Twigs die back in late summer. Leaves on current season twigs may be yellowish with red veins. Somewhat cupped and distorted. Dead areas develop at tips and margins. Leaves may curl, die, and drop off. Normal buds fail to develop or make poor growth. Twigs may form witches'-brooms (rosettes). *Pear* blossoms may be blasted. *Control:* Add borax to soil around trees in early spring (use about 2 oz. per 30 square yards). Or spray with boric acid (0.1 to 1 per cent), or sodium pentaborate (sold as Polybor, Solubor, Boro-Spray) following local recommendation of your extension horticulturist or a local grower. Usually 1 or 2 sprays are applied to apple, 1 and 2 weeks after petal-fall. Boron-containing materials (page 27) may also be added to pest sprays. Follow manufacturer's directions.
25. *Leaf Scorch, Potassium, Calcium, Magnesium or Copper Deficiency* (primarily apple and pear) —Leaves on many varieties develop marginal curling, yellowing, scorching (browning or blackening), and shriveling around middle of the growing season. Yellow or red discoloration is characteristic of certain varieties. Twigs slender and stunted. May die back. Fruit poor, pale in color, and "woody." Similar scorching

may be caused by drought; strong, hot, dry winds; mites; spray injury; and water-logged or "shallow" soil. *Control:* Have soil tested and apply a potash, magnesium, or calcium-containing fertilizer as recommended. Varieties differ greatly in susceptibility.

26. *Pear Decline*—Widespread and serious in Pacific Coast states. Trees, usually 10 to 40 years old are stunted, make little or no new growth. Trees gradually—over several years—or quickly decline, wilt, collapse, and die. Leaves on affected trees are small and sparse. Often turn dull to bright red prematurely. Pear varieties on oriental and imported French rootstocks are most susceptible. Failure or poor union between scion and rootstock. The phloem tissue turns brown. Roots die back. Pear decline virus is transmitted by pear psylla insect and grafting. *Control:* Replant using resistant, own-rooted rootstocks (e.g., virus-free Old Home, Bartlett, Winter Nelis). Check with your extension or state plant pathologist regarding latest information. Control pear psylla. Destroy wild or abandoned pear trees.
27. *Apple Mosaics*—Leaves variously mottled (light and dark green mosaic) with many small, irregular, white to creamy yellow flecks, spots, and patches. Leaf veins may be banded by white or creamy yellow stripes. Symptoms extremely variable on a single tree or single branch. May be masked at temperatures above 80° F. A few mottled leaves may appear on an otherwise healthy-appearing branch. Large numbers of leaves may turn brown and drop early. Tree vigor and yield may be reduced. Viruses spread by root grafts. Also by budding and grafting. Some varieties may be symptomless carriers. *Control:* Destroy wild apples near orchard. Plant certified, virus-free stock. Follow regular spray program (Table 16 in Appendix). Destroy severely affected trees.
28. *Flat Limb* (apple, pyracantha)—Symptoms may be latent in certain varieties. Slight depressions, grooves, pockets, or furrows on branches or trunk that become more pronounced with age. Bark over depression is smooth. Affected branches may become twisted and deformed; are often brittle. May break under a fruit load. *Control:* Same as for Apple Mosaics (above). Gravenstein is commonly affected.
29. *Rubbery Wood* (apple, pear)—Twigs and smaller branches bend over or "weep" from their own weight. Wood is soft and "cheesy" in texture. Affected trees often stunted, less vigorous than normal. Symptoms often restricted to certain branches of a tree. Fruit appear normal, but yield is reduced. Many varieties and rootstocks are symptomless carriers of the virus. *Control:* Same as for Apple Mosaics (above). *Apple* twig tips held at 99° F. for 4 weeks are free of the virus.
30. *Stem- or Wood-pitting of Apple, Pear and Quince*—Most apple varieties act as latent virus carriers. Symptoms variable. Wood under bark of trunk is longitudinally deeply pitted and grooved. Trunks of older trees are ridged and twisted. On susceptible *apple* varieties, trees may be dwarfed and less vigorous. Trees gradually decline; set heavy fruit crops. Twigs and branches die back. Are more spreading than normal with numerous suckers from the rootstock below the susceptible portion of the tree. Root system is poor. Fruit yield may be reduced 50 to 80 per cent. Severe symptoms are most common on crabapples (e.g., Beauty, Columbia, Florence, Hopa, Hyslop, Jay Darling, Purple Lemoine, Purple Wave, Red River, Robin, Spy 227, Sugar, and Virginia). *Control:* Same as for Apple Mosaics (above).
31. *Apple Russet Ring*—Fruit symptoms variable depending on variety, season, and temperature. Some varieties are symptomless carriers. Fruit covered with elaborate light brown, russet patterns, often ringlike. In storage, russeting becomes dark brown. Other apple varieties develop purple to brown blotches on the fruit skin. Early formed leaves are often dwarfed and puckered. Leaves often show yellowish-green flecking, mottling, and distortion. Leaves formed in hot weather appear normal. Twig dieback may occur. *Control:* Same as for Apple Mosaics (above).

32. *Dapple Apple*—Round to irregular, enlarging, patches of skin on fruit remain green while normal pigment develops on remainder of fruit surface. Blotchy areas are unsightly and somewhat flattened. All fruit on a diseased tree are affected. The virus has very little effect on tree growth or yield. Fruit quality is seriously affected. *Control:* Same as for Apple Mosaics (above).
33. *Apple Green Crinkle or False Sting*—Irregularly lobed depressions or creases, often elongate, develop in surface of fruit. Depressed areas remain green, later become superficially corky, russeted, and cracked. Affected fruit often severely stunted and "bumpy." Symptoms may be confined to one or a few branches on a tree for a number of years. Wood growth is greatly retarded; tree dies within several years. *Control:* Same as for Apple Mosaics (above).
34. *Apple Scar Skin*—Small, light green, water-soaked areas form on calyx end of young fruit. The "scar" tissue grows with the fruit and may affect up to 50 per cent of the surface. Diseased fruit are stunted. Reported on Jonathan, Red Delicious, and Turley. Golden Delicious is apparently tolerant. *Control:* Same as for Apple Mosaics (above).
35. *Thread Blight*—Southeastern states. See under Walnut.
36. *Verticillium Wilt* (cotoneaster, quince, serviceberry)—See (15B) Verticillium Wilt under General Diseases.
37. *2,4-D Injury*—Leaves may be thickened, curled, rolled, cupped downward, heart-shaped, or curled into "ram's-horns." *Control:* See under Grape.
38. *Chlorosis, Iron or Manganese Deficiency*—Leaves on shoot tips are yellowish except for the leaf veins. If severe, shoot tips may die back (iron deficiency). Young leaves on strongly growing shoots may be only little affected (manganese deficiency). *Control:* Have soil tested and follow the recommendations. Add ferrous or manganese sulfate (or chelate) to one or more pest sprays.
39. *Wetwood* (apple)—See under Elm.
40. *Leaf Blister* (pear)—Pacific Northwest. See (10) Leaf Curl under General Diseases. Green blisters form on upper leaf surface. Turn black and are later covered with whitish bloom. *Control:* See Peach, Leaf Curl.
41. *Seedling Blight* (hawthorn)—See under Pine.
42. *Leaf Blister, Witches'-broom* (amelanchier)—California. See (10) Leaf Curl under General Diseases.

APRICOT — See Peach
AQUILEGIA — See Delphinium
ARABIAN-TEA — See Bittersweet
ARABIS — See Cabbage
ARACHIS — See Peanut
ARALIA — See Acanthopanax
ARAUCARIA, NORFOLK ISLAND PINE, MONKEYPUZZLE TREE, BUNYA-BUNYA, MORETON BAY or HOOP PINE *(Araucaria)*

1. *Branch Blight, Dieback*—Tips of lower branches die back. Later, entire branches are killed. Tip ends bend and break off. Disease gradually spreads upward killing the tree. *Control:* Prune and burn infected branches when first noticed. Fixed copper sprays during spring and fall rainy periods may be beneficial. Avoid bark wounds.
2. *Crown Gall*—Smooth, roughly circular galls, up to an inch in diameter, occur at or near soil line. *Control:* Avoid wounding tree base when mowing, cultivating, etc.
3. *Root Rot*—See under Apple, and (34) Root Rot under General Diseases.
4. *Leaf Spots*—Leaves spotted. May shed early. Unimportant.
5. *Seedling Blight, Damping-off*—See under Pine.

ARBORVITAE — See Juniper
ARBUTUS, ARCTOSTAPHYLOS — See Blueberry
ARCHONTOPHOENIX — See Palms
ARCTOTIS — See Chrysanthemum
ARDISIA

1. *Algal Leaf Spot*—Florida. See under Magnolia.
2. *Black Mildew*—See (12) Sooty Mold under General Diseases.

ARECASTRUM — See Palms
ARENARIA — See Carnation
ARENGA — See Palms
ARGEMONE — See Poppy
ARGYREIA — See Morning-glory
ARIOCARPUS — See Cacti
ARISAEMA — See Calla Lily
ARISTOLOCHIA, BIRTHWORT, DUTCHMANS-PIPE, CALICO-FLOWER, PELICAN-FLOWER, VIRGINIA SNAKEROOT *(Aristolochia)*

1. *Leaf Spots*—Round to angular spots. *Control:* Apply ferbam, zineb, or maneb several times, 10 to 14 days apart. Start when leaves are $\frac{1}{4}$ inch out.
2. *Gray-mold Blight*—See (5) Botrytis Blight under General Diseases. *Control:* Spray as for Leaf Spots.
3. *Root Rot*—See under Geranium. Plant in light, well-drained soil. Avoid overwatering.

ARMERIA — See Sea-lavender
ARNICA — See Chrysanthemum
ARONIA — See Apple
ARROWROOT — See Maranta
ARROWWOOD — See Viburnum
ARTEMISIA — See Chrysanthemum
ARTICHOKE — See Lettuce and Chrysanthemum
ARTILLERY-PLANT — See Pilea
ARUM-LILY — See Calla Lily
ARUNCUS — See Rose
ASCLEPIAS — See Butterflyweed
ASCYRUM — See St.-Johns-wort
ASH [BLACK or HOOP, BLUE, EUROPEAN, FLOWERING, GREEN, MANCHURIAN, MODESTO, OREGON, RED, VELVET, WEEPING EUROPEAN, WHITE] *(Fraxinus)*; FRINGETREE [AMERICAN or WHITE, CHINESE or ORIENTAL] *(Chionanthus)*; FORESTIERA [COMMON or SWAMP-PRIVET, NEW MEXICAN] *(Forestiera)*

1. *Rusts* (ash, forestiera) —General. Ash twigs and petioles may be swollen; leaves distorted. Bright yellowish-orange, powdery pustules appear on affected parts (Figure 4.12). Infected leaves may wither and drop early. Another rust on forestiera, causes reddish-brown to black, powdery pustules on leaves. *Control:* Avoid growing ash and forestiera where alternate hosts, cord and marsh grasses (*Spartina* spp.), are abundant. Or destroy the grasses. If serious, spray three times at 2-week intervals. Start about when apples are in bloom. Use ferbam, zineb, ziram, dichlone, bordeaux, or sulfur. Fertilize injured trees (pages 30–32).

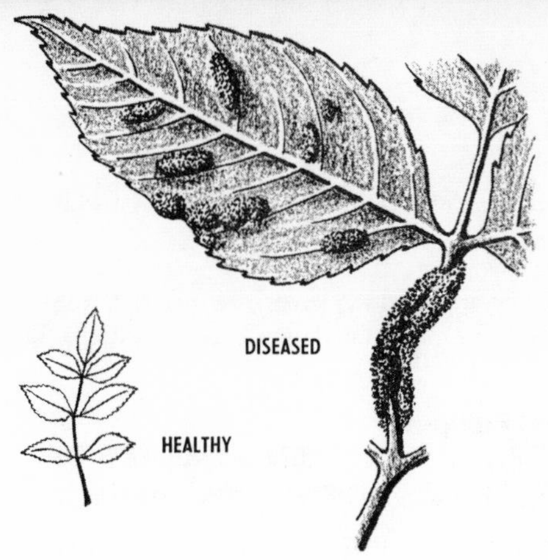

Fig. 4.12. Ash rust.

2. *Leaf Spots, Anthracnose*—Widespread during wet springs. Small to large, round to irregular spots and blotches of various colors. Develop particularly between leaf veins and along edges. Leaves may be distorted, wither, and drop early. A second crop of leaves is produced which may later be injured by anthracnose. *Control:* Collect and burn fallen leaves in autumn. If practical, spray several times, 10 days apart, starting as buds swell. Use zineb, ferbam, phenyl mercury, ziram, captan, or fixed copper.
3. *Leaf Scorch* (primarily ash)—Margins turn brown in hot, dry, windy weather; leaves may drop. Dead areas advance between leaf veins. Most prevalent through top of tree or on windward side during July and August. *Control:* Prune to open up tree. Fertilize to increase vigor. Water during dry periods.
4. *Twig Blights, Dieback, Branch and Trunk Cankers*—General. Twigs and branches or entire sections of affected trees die back. Foliage may be dwarfed, yellowish, and tufted before twigs and branches die back. Cankered area is swollen, flattened, or misshapen. Dark streaks are often evident in sapwood around canker. *Control:* Prune and burn infected twigs and branches. Keep trees vigorous. Fertilize and water during summer droughts. Control other disease and insect pests. Avoid wounding bark. Paint wounds promptly with tree paint.
5. *Wood Rots*—Widespread. See under Birch, and (23) Wood Rot under General Diseases. Control borers by monthly sprays, May to September, applied to trunk. Use DDT or Thiodan. Keep trees growing vigorously.
6. *Crown Gall, Hairy Root* (ash)—See under Apple, and (30) Crown Gall under General Diseases.
7. *Powdery Mildews*—Widespread from late summer on. Grayish-white, powdery patches on leaves and young twigs. Leaves may yellow and drop early. *Control:* If severe, and where practical, apply sulfur or Karathane one or two times, 10 days apart. Start when mildew is first seen. Avoid applications when temperature is above 85° F. Collect and burn fallen leaves.
8. *Root Rot*—See under Apple.
9. *Verticillium Wilt* (ash)—See (15B) Verticillium Wilt under General Diseases. Streaking of the sapwood may be absent or appear as flecks.
10. *Sooty Mold, Black Mildew*—Black mold patches on leaves. *Control:* Spray to control insects. Use DDT and malathion plus a fungicide.
11. *White Ash Flower Gall*—Clusters of very irregular, bunchy (¼–¾ inch in diameter), brown galls on male (staminate) flowers. Galls, caused by minute mites, are conspicuous during winter months. *Control:* Apply malathion, lime-sulfur, or a miscible oil after buds swell and before new growth emerges in spring. Prune out and destroy galls.
12. *Root-knot*—See under Peach.
13. *Mistletoe* (ash, forestiera)—See (39) Mistletoe under General Diseases.

14. *Felt Fungus*—Southeastern states. Purple-black, feltlike growth on bark. Associated with scale insects. See under Hackberry.
15. *Seedling Blight*—See under Pine.
16. *White Ash Ringspot*—New York. Leaves develop yellow, green, or reddish spots, rings, and line patterns along veins. May be associated with severe dieback. *Control:* Unknown.

ASIMINA — See Pawpaw

ASPARAGUS, GARDEN; ASPARAGUS-FERN or LACE-FERN, "SMILAX" OF FLORISTS, SPRENGER ASPARAGUS, SICKLE THORN *(Asparagus)*

1. *Rust* (asparagus) —General. Yellow to orange-red, then reddish-brown to black, powdery pustules on stems and leaves. Tops may turn yellow and die early, reducing next year's crop. *Control:* Grow somewhat rust-resistant Mary, Waltham, and Martha Washington or California 500. Destroy volunteer plants or wild asparagus near producing beds. Cut and burn tops in fall. Asparagus species vary greatly in resistance. If practical, apply zineb, maneb, Polyram, Dithane M-45 or S-31 just *after* harvest. Continue at 10-day intervals to midsummer. Start dusting or spraying *young, uncut beds* in midspring. Acti-dione may also be used. Follow manufacturer's directions. Elgetol or Krenite may be used in fall or spring on *dormant* plants. Follow manufacturer's directions.
2. *Fusarium Wilt, Yellows, Foot and Root Rot*—Plants stunted, slender, turn yellow, wilt, and gradually die (decline). Spears gradually decrease in size and number. Roots and shoot bases usually show red or reddish-brown streaks and flecks when cut through. Affected tips may be overgrown with pink mold. Poor stand. *Control:* Plant disease-free, treated seed (Table 20) or healthy crowns (soaked in captan or maneb solution, 4 tablespoons per gallon, for 5 minutes) in clean, well-drained, fertile soil as far from old beds as possible. Plow under unproductive beds. Keep plants vigorous. Water and fertilize regularly. Sprenger asparagus is apparently immune. Resistant lines of garden asparagus (var. Mary Washington) are available.
3. *Branchlet Blight, Stem Cankers, Dieback* (asparagus, asparagus-fern) —Branchlets wither and fall. Plants may die back to crown. Pale cankers, often sprinkled with black dots, may occur on stems. *Control:* Collect and burn tops in fall. If feasible, apply maneb or zineb sprays before wet periods, 7 to 10 days apart. Avoid overwatering. Plant in sterilized (fumigated) soil, where practical.
4. *Crown Gall*—Irregular, thick, pale green, fleshy gall at base of stem. *Control:* Plant healthy stock in soil free of the disease for 3 years or more. Dig up and destroy infected plants.
5. *Leaf (Branchlet and Stem) Spots, Anthracnose, Leaf Blight*—Small, generally oval to elliptic, spots on stems, branches, and needles. Spots tan to gray with reddish-brown border. Needles and youngest branches may die and drop early. *Control:* Same as for Rust (above). Keep plants vigorous by watering and fertilizing. Apply zineb, maneb, or captan weekly during rainy periods.
6. *Stem and Spear Rots, Crown Rot*—Watery, soft spots or rot of shoots and spears (Figure 4.14). Often near soil line. Tissues may be covered with gray, pink, cottony, or blue-green mold. Spots often enlarge rapidly. Tops wilt, may collapse. See (5) Botrytis Blight, (21) Crown Rot, and (29) Bacterial Soft Rot under General Diseases. *Control:* Keep soil surface loose and dry. Carefully dig and destroy infected plants and several inches of surrounding soil. Plant in fertile, well-drained soil. Avoid overcrowding and wounding stems. Applying PCNB to soil surface before rots start may help. See under Bean. Protect seedlings for winter by slight hilling in autumn. Avoid injury when harvesting. Precool spears to 40° F. Keep refrigerated.

Fig. 4.13. Conks of the white mottled rot fungus (*Fomes applanatus*) on white ash. (Illinois Natural History Survey photo)

Fig. 4.14. Asparagus "spear" rots. A. Phytophthora. B. Bacterial soft rot. C. Fusarium.

7. *Root-knot*—Enlarged areas or galls on roots. Plants may be stunted, pale, and weak. *Control:* See (37) Root-knot under General Diseases. For *asparagus-fern* may use DBCP or VC-13 as a post-planting treatment. Make new leaf cuttings in peat or vermiculite and plant in sterilized soil. Discard infested plants.
8. *Chlorosis*—Normal green color fades to yellow. *Control:* Avoid overwatering. Have soil tested. It should be near neutral (pH 7.0).
9. *Verticillium Wilt*—See (15B) Verticillium Wilt under General Diseases.
10. *Root Rots*—Plants spindly, weak, and yellowish. Roots decayed. See (34) Root Rot under General Diseases. May be associated with nematodes (e.g., lance, needle, pin, root-knot, spiral, stem, stunt).
11. *Damping-off*—Seedlings may die before or after emergence. *Control:* Treat seed with thiram or Calogreen (4 oz. per pound of seed) or Ceresan M (1 tablespoon per pound of seed). Sow in fertile, well-prepared, well-drained seedbed.
12. *Leaf Drop* (asparagus-fern)—Indoor problem. Leaves turn brown, wither, and drop. *Control:* Lower room temperature to 70–72° F. and increase air humidity.

ASPARAGUS-BEAN — See Bean
ASPARAGUS-FERN — See Asparagus
ASPASIA — See Orchids
ASPEN — See Poplar
ASPIDISTRA, CAST-IRON PLANT *(Aspidistra)*

1. *Chlorosis and Root Rot*—Leaves turn yellow and die. *Control:* Avoid overwatering and too much light. Plant in well-drained soil.
2. *Leaf Spots, Leaf Blight, Anthracnose*—Spots of various sizes, shapes, and colors. If severe, leaves may wither and die. *Control:* Pick and burn infected leaves. If practical, spray as for Leaf Spots of Chrysanthemum.

ASPLENIUM — See Ferns
ASTER (CHINA and HARDY) — See Chrysanthemum
ASTILBE

1. *Powdery Mildew*—See (7) Powdery Mildew under General Diseases.
2. *Fusarium Wilt*—Plants yellow, wilt, and die. May be associated with nematodes (e.g., lesion). *Control:* Set healthy plants in clean or sterilized soil (pages 538–548). Avoid injuring roots and crown.

ASTROPHYTUM — See Cacti
ATAMASCO-LILY — See Daffodil
ATHEL TREE — See Tamarisk
ATHYRIUM — See Ferns
AUBRIETIA — See Cabbage
AUCUBA [JAPANESE or GREENLEAF, SULPHUR-LEAF], GOLDDUST-TREE *(Aucuba)*

1. *Leaf Spots*—Brown or black spots, often zoned. Mostly near margins of leaves. Infected leaves may wither and drop early. *Control:* Spray several times, 10 days apart. Use zineb, maneb, or fixed copper. Combine with an insecticide (e.g., sevin, DDT, or malathion) if scale insects are present.
2. *Gray-mold Blight*—Problem where moist. Twigs blighted and die back. Affected areas often covered with dense gray mold in damp weather. *Control:* Space plants. Keep humidity down. Increase air circulation. Keep water off foliage.
3. *Anthracnose, Wither Tip*—Spots develop on leaves and flowers. Stem cankers cause

branch tips to wilt and die back. *Control:* Pick or cut off affected plant parts. Apply zineb, captan, or maneb before wet periods.

4. *Verticillium Wilt*—See (15B) Verticillium Wilt under General Diseases.
5. *Frost or Winter Injury*—Easily mistaken for disease. Young leaves nipped by spring frosts. Older leaves turn brown in winter in northern states unless protected. Check with your local nurseryman or extension horticulturist regarding winter protection.

AURICULA — See Primrose
AUSTRALIAN BRUSH-CHERRY — See Eucalyptus
AUSTRALIAN PEA — See Pea
AUSTRALIAN-PINE — See Casuarina
AUSTRALIAN UMBRELLA-TREE — See Schefflera
AUTUMN-CROCUS — See Colchicum
AVENS — See Rose
AVOCADO, REDBAY, SWAMPBAY *(Persea)*; CAMPHOR-TREE, CINNAMON-TREE *(Cinnamomum)*; SPICEBUSH *(Lindera or Benzoin)*; CALIFORNIA-LAUREL *(Umbellularia)*; POND-SPICE *(Litsea)*; SASSAFRAS

1. *Root Rots*—Cosmopolitan on avocado and camphor-tree. Trees of any age gradually wilt or suddenly decline in vigor and productivity. Leaves pale or yellowish-green, tend to wilt, and drop early. Twigs, branches, and roots die back. Trees may die. Often serious in wet, poorly drained soils. See under Apple. *Control:* Treat suspected *avocado* seed before planting. Soak 30 minutes in hot water (120° to 122° F.). Then wash in clean, cold water and dry. Or plant certified, disease-free avocado nursery stock. In California, Texas, and Florida, grow resistant strains of *avocado* (e.g., Duke or Scott rootstocks) developed for planting in Phytophthora-infested soils. Grow in deep, *well-drained* soil (fumigated, if possible) where root rot has not been present before. Indoors plant in sterilized soil (pages 538–548). Prevent movement of water, soil, and plants from infested to noninfested areas. Avoid root injuries. Monthly soil drenches of Dexon have proved beneficial. Follow manufacturer's directions. For more information read *Avocado Root Rot,* University of California Circular 511.
2. *Wood, Heart, Trunk and Collar Rots, Trunk Cankers*—Widespread. Trees may gradually decline in vigor or die suddenly due to rotting and girdling of trunk (collar rot), or encircling cankers. *Control:* See under Birch and Dogwood. Avoid covering bud union with soil.
3. *Anthracnose or Black Spot, Leaf Spots or Blotch, Leaf Blight, Leaf and Fruit Scab*—General, may be serious. Small to large, round to irregular spots and blotches of various colors on the leaves. Leaves or branch tips may wilt and wither. Spots may also occur on *avocado* flowers, fruit and flower stems, and fruit. Fruit may appear "scabby" and young fruit may wither and drop early (Figure 4.15). *Control:* Collect and burn fallen leaves and spotted or scabby fruit. Otherwise same as for Twig Canker (below). *Avocado* varieties resistant to Scab: Booth 1, Collins, Collinson, Fuchsia, Gottfried, Itzamna, Linda, Pollock, Ruehle, and Waldin. Winter Mexican *avocado* is resistant to Scab *(Sphaceloma)* while Kilgore Special *avocado* has some resistance to Anthracnose *(Glomerella)*. Ruehle *avocado* is fairly resistant to both Anthracnose and Cercospora Spot. Prevent bruises and skin breaks to the fruit.
4. *Twig and Branch Cankers, Dieback*—Widespread. Foliage on twigs and branches may wilt and wither. Twigs and branches die back from small to large, girdling, discolored cankers. Cankered areas may be slightly sunken or swollen and rough. *Control:* Remove and burn affected plant parts. Avoid wounds. Paint promptly

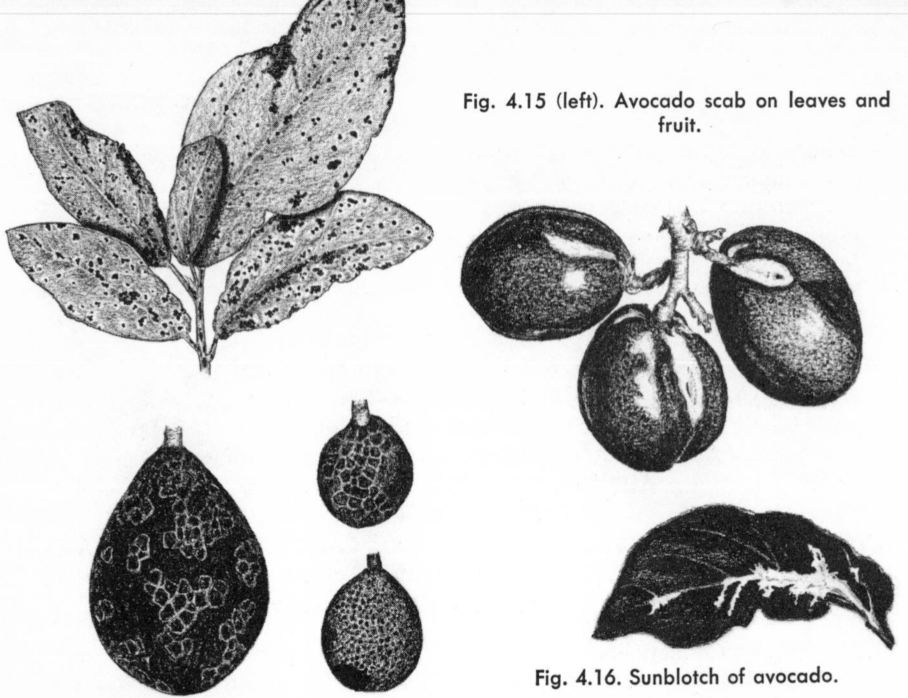

Fig. 4.15 (left). Avocado scab on leaves and fruit.

Fig. 4.16. Sunblotch of avocado.

with tree wound dressing. Spray weekly during rainy periods using ferbam, zineb, maneb, captan, or fixed copper. Start about time blossom buds appear and continue until fruits are ⅔ grown or more. Where avocados are grown commercially, follow recommended spray schedule for your area. Check with your county agent or extension plant pathologist.

5. *Oedema* (avocado)—Indoor problem. Small "scabby" or wartlike growths on upper leaf surface. Later crack and turn reddish-brown. *Control:* Avoid overwatering during moist, overcast weather. Increase air circulation and temperature.
6. *Verticillium Wilt* (avocado, camphor-tree, sassafras)—Leaves on one to several branches or entire plant, wilt, turn brown, wither, and remain on tree. Brown discoloration may be seen in wood just under the bark on branches and trunk. Trees may die or recover completely. *Control:* See under Maple.
7. *Sunblotch* (avocado)—Long, narrow, shallow streaks (creases) develop near stem end of fruit. Streaks are whitish or yellow in green fruits and red or purplish-red on purple or black varieties. White to yellow streaks and blotches occur on leaves, green stems, and branches (Figure 4.16). Older stems are crooked, uneven, and rough. Shoots prostrate, willowy, twisted, and lack vigor. Some infected trees may be symptomless; others may be dwarfed. *Control:* Grow seed, scion wood, and rootstocks from known virus-free trees.
8. *Little Leaf, Rosette, Zinc Deficiency* (avocado)—Leaves severely mottled at edges. Leaf tips may appear scorched without mottling. Avocado fruit may be deformed. If prolonged, branches may die back. *Control:* Sprays of zinc sulfate (8 ounces in 10 gallons) and hydrated lime (4 ounces in 10 gallons), applied soon after growth appears, are often used. A single application of zinc chelate (page 27) may give protection for 3 years or more. Check with your county agent or extension horticulturist.
9. *Powdery Mildew*—Powdery white mold on underside of leaves. Mostly on trees growing in shaded, damp locations. Shoot tips may die back. *Control:* If serious, apply two Karathane sprays, 10 days apart.
10. *Fruit Spots and Rots*—Small to large, enlarging spots of various colors on *avocado*

fruit. Fruit on tree or after harvest may appear russeted, scabby, or cracked. Fruit often covered with black, blue, gray, dark green, or white mold. *Control:* Follow recommended spray schedule for your area (see Twig Canker above). Keep trees well pruned. Pick fruit early, handle carefully, and store at recommended temperature and humidity. Check with your extension horticulturist. Store hard or firm fruit at 45° to 55° F.; ripe fruit as low as 32° F.

11. *Black Mildews, Sooty Blotch* (avocado, California-laurel, camphor-tree, redbay, sassafras, spicebush, swampbay) —Primarily in Gulf states and California. Black mold patches on foliage and branches. *Control:* Spray as for Twig Canker (above). Control insects with malathion sprays.
12. *Root-knot and Other Root-Feeding Nematodes* (burrowing, dagger, lesion, pin, ring, sheath, sphaeronema, stubby-root) —See under Peach. Avocado is usually resistant to Root-knot. Associated with decline and unthrifty foliage.
13. *Mistletoe* (camphor-tree, spicebush, sassafras) —See (39) Mistletoe under General Diseases.
14. *Yellows* (sassafras) —Leaves dwarfed and rolled. Branch tips bunched and fasciated. *Control:* None suggested.
15. *Chlorosis, Mottle Leaf*—Mostly in alkaline soils. Leaves turn pale green to yellowish-white except for veins. See under Maple. Avocado varieties differ in resistance.
16. *Seed Rot and Seedling Blights*—Seeds decay; stand is poor. Seedlings wilt and die. Roots decayed. *Control:* See under Pine.
17. *Bacterial Blast, Leaf Spot* (avocado, California-laurel) —Small, black, irregular, scabby areas and cracks on avocado fruit and angular black spots on California-laurel leaves. *Control:* Check with local authorities, i.e., county agent or extension plant pathologist.
18. *Downy Mildew* (avocado) —See (6) Downy Mildew under General Diseases.
19. *Crown Gall* (avocado) —Rare. Small to large, overgrowths; may be gnarled and irregular. Occurs at soil line. *Control:* Avoid wounding trunk base. See also Peach, Crown Gall.

AXONOPUS — See Lawngrass

AZALEA — See Rhododendron

AZARA

1. *Stem Rot*—See (21) Crown Rot under General Diseases. *Control:* Dig up and destroy infected plants. Set new plants in clean or sterilized soil (pages 538–548).

AZTEC LILY — See Gladiolus

BABIANA — See Iris

BABY-BLUE-EYES — See Phacelia

BABYSBREATH — See Carnation

BABYTEARS VINE *(Helxine)*

1. *Leaf Spot*—See (1) Fungus Leaf Spot under General Diseases.
2. *Rust*—See (8) Rust under General Diseases.
3. *Powdery Mildew*—See (7) Powdery Mildew under General Diseases.

BACHELORS-BUTTON — See Chrysanthemum

BAG-FLOWER — See Lantana

BAHIAGRASS — See Lawngrass

BALDCYPRESS — See Pine

BALLOONFLOWER — See Bellflower

BALM — See Salvia

BALM-OF-GILEAD — See Poplar

BALSAM [GARDEN, SULTAN]; PATIENCE PLANT *(Impatiens)*

1. *Leaf Spots, Anthracnose*—Small to large, circular to irregular spots or blotches. Leaves and young shoots may blight. *Control:* Apply zineb, maneb, thiram, or captan. Burn tops in fall.
2. *Wilts (Bacterial and Verticillium)*—See (15B and C) under General Diseases. *Control:* Destroy infected plants. Do not replant in same soil without first treating with heat or chemicals (pages 538–548).
3. *Stem Rots*—Watery, soft rot at soil line. Later covered with cottony mold. Seedlings wilt and collapse. Older plants wilt, yellow, wither, and die. *Control:* Same as for Wilts.
4. *Mosaic*—Plants often stunted. Leaves mottled light and dark green. May be deformed. Flower production is inhibited. May be striped and small. *Control:* Dig up and burn affected plants. Keep down weeds. Spray with malathion or lindane to control aphids that transmit virus.
5. *Root Rot*—Plants stunted, yellowed, with reduced foliage and late flowering. *Control:* See under Chrysanthemum. May be associated with nematodes (e.g., dagger, lesion, pin, ring, root-knot, reniform, stubby-root).
6. *Root-knot*—Primarily southern states. Plants stunted, unthrifty, with small to large galls on roots. *Control:* Dig up and burn infested plants. Grow disease-free plants in sterilized soil (pages 538–548).
7. *Rust*—Generally unimportant in gardens. See (8) Rust under General Diseases. Alternate hosts include *Adoxa* and wild grasses *(Agrostis* and *Elymus).*
8. *Damping-off*—Seedlings wilt and collapse from rot at ground level. *Control:* Sow seed in clean, well-drained soil or a sterile medium. Avoid overwatering.
9. *Downy Mildew*—See (6) Downy Mildew under General Diseases.

BALSAM-APPLE, BALSAM-PEAR — See Cucumber

BALSAMROOT *(Balsamorhiza)* — See Chrysanthemum

BALTIC IVY — See Ivy

BAMBOOS [BLACKJOINT, COMMON, DWARF, FEATHER, GOLDEN, HEDGE, JAPANESE TIMBER] *(Bambusa, Phyllostachys)*

1. *Bacterial Leaf Spot*—See (2) Bacterial Leaf Spot under General Diseases. Seriously affected leaves may drop early. *Control:* Apply copper-containing fungicide at 10-day intervals in rainy weather.
2. *Fungus Leaf and Culm Spots*—Spots of various sizes, shapes, and colors on leaves and culms. *Control:* Same as for Bacterial Leaf Spot.
3. *Black Mildew*—Black blotches on leaves. May follow insect attack. See (12) Sooty Mold under General Diseases.
4. *Rusts*—See (8) Rust under General Diseases.
5. *Stem Smut*—See (11) Smut under General Diseases.

BANEBERRY — See Anemone

BANGALAY — See Eucalyptus

BAPTISIA — See False-indigo

BARBADOS CHERRY *(Malpighia)*

1. *Leaf Spot*—See (1) Fungus Leaf Spot under General Diseases.
2. *Green Scurf, Algal Leaf Spot*—See under Magnolia.
3. *Root-knot*—See (37) Root-knot under General Diseases. Keep plants well mulched.
4. *Chlorosis*—May be due to deficiency of iron, zinc, manganese, magnesium, or copper. See pages 26–28.

BARBERRY [BLACK, BOX, GREEN-LEAF, DARWIN, DWARF MAGELLAN, JAPANESE or THUNBERGS, KOREAN, MENTOR, MINIATURE, NEUBERT, RED-LEAF JAPANESE, ROSEMARY, THREESPINE, UPRIGHT JAPANESE, WARTY, WILSON, WINTERGREEN] *(Berberis);* VANILLALEAF *(Achlys);* BLUE COHOSH *(Caulophyllum);* MAHONIA [CASCADES, CLUSTER, CALIFORNIA BARBERRY, CREEPING LAREDO, LAREDO, LEATHERLEAF or CHINESE, OREGON-GRAPE or HOLLYGRAPE], AGARITA, CHINESE HOLLYGRAPE *(Mahonia)*

1. *Verticillium Wilt* (barberry)—Leaves wilt, turn yellow, brown or red, wither, and fall on one or more branches. Inside of stems show green to brown streaks when cut. Entire plants may die. *Control:* Dig up and burn severely infected plants. Do not replant in same location for several years. Plant in well-drained soil. Avoid wounding roots and stem when transplanting and cultivating. Fertilize and water to stimulate vigor.
2. *Bacterial Leaf Spot and Twig Blight* (barberry)—General. Small, round to irregular, dark green, water-soaked spots that turn dark purplish to reddish-brown or black. Severely infected leaves drop early. Twigs may die back from girdling spots that are at first water-soaked. Buds are killed (Figure 4.17). *Control:* Prune out and burn infected twigs. Apply several fixed copper, bordeaux (4-4-50), or streptomycin sprays (100 parts per million) 10 days apart. Start when new leaves open.
3. *Anthracnose, Fungus Leaf Spots, Leaf Blotch*—Tan, brown, or purple spots on leaves. Often with distinct margin. *Control:* If serious, apply several sprays 7 to 10 days apart. Use copper, zineb, maneb, or ferbam. Collect and burn spotted leaves before new growth appears in spring.
4. *Powdery Mildew* (barberry, mahonia)—Whitish powdery mold on leaves. *Control:* If serious, apply several sprays 7 to 10 days apart. Use sulfur or Karathane.
5. *Root-knot*—Plants lack vigor and winter hardiness. Small swellings or knots on roots. See (37) Root-knot under General Diseases.
6. *Rusts*—Small, bright orange to blood-red spots on upper leaf surface in spring. Underside of infected leaves shows golden-yellow, cuplike growths with fringed margins. Or minute, brownish, powdery pustules (Figure 4.18). Cultivated

Fig. 4.17 (above). Bacterial leaf spot and twig blight of Japanese barberry.

Fig. 4.18. Stem rust on barberry. (USDA photo)

barberries (Japanese and Mentor types) and most evergreen varieties are highly resistant or immune to the serious stem rust of cereals and grasses of which common (European), Allegheny or Canada and Colorado barberries, Oregon-grape, and other Mahonias are alternate hosts. In many states it is unlawful to grow rust-susceptible varieties and species of barberry or Mahonia. Check with your nurseryman or extension plant pathologist before planting (Figure 4.19). For additional information read USDA Leaflet 416, *Barberry Eradication in Stem Rust Control.* Other alternate hosts of Mahonia rusts include *Koeleria* and *Oxalis.*

7. *Root Rot*—See under Apple, and (34) Root Rot under General Diseases. Avoid heavy, poorly drained, compact soil, low in organic matter. Space plants for good aeration and light.
8. *Mosaic* (barberry)—Irregular pattern of reddish blotches on leaves. *Control:* Destroy affected plants.
9. *Cankers, Dieback, Twig and Stem Blight, Heart Rot*—Twigs and branches die back. See under Maple. *Control:* Cut out and burn infected parts. Spray as for Anthracnose (above).
10. *Gray-mold Blight* (barberry, blue cohosh)—Blossoms and leaves blighted. May be covered with gray mold in damp weather. *Control:* Same as for Anthracnose (above).
11. *Leaf Scorch, Scald* (mahonia)—Large brown blotches in leaves scorched by winter winds and sun in northern states. *Control:* Plant in protected locations, where adapted. Erect barriers to ward off winter winds.
12. *Root-feeding Nematodes* (e.g., dagger, lesion, pin, ring, spiral, stem, stunt)—Associated with pale, weak, stunted plants in state of decline. *Control:* Same as for Root-knot (above).

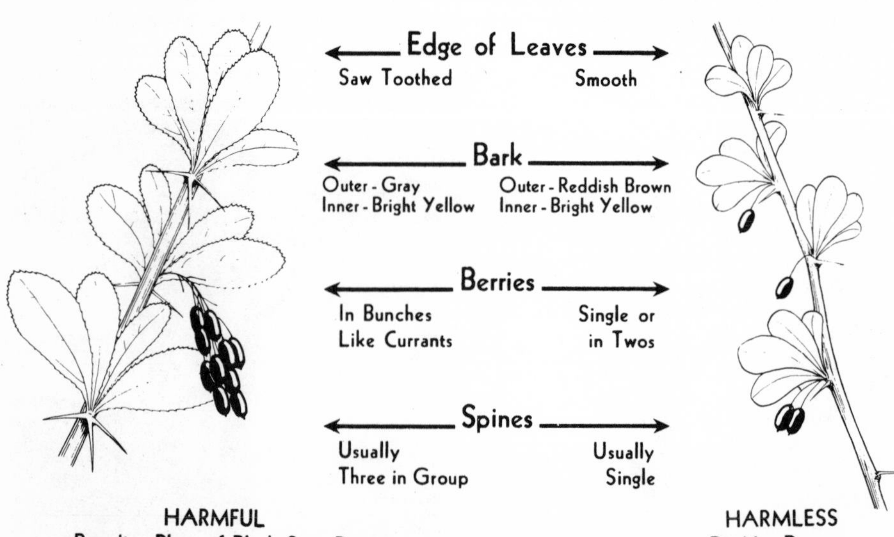

Fig. 4.19. How to identify the rust-spreading common barberry from the harmless Japanese barberry. (USDA photo)

13. *Damping-off* (barberry) —See under Pine.

BASIL, BASILWEED — See Salvia
BASKETFLOWER *(Centaurea)* — See Chrysanthemum; BASKETFLOWER *(Hymenocallis)* — See Daffodil
BASSWOOD — See Linden
BAUHINIA — See Honeylocust
BAYBERRY — See Waxmyrtle
BEACH ASTER — See Chrysanthemum
BEACH-PEA — See Pea
BEAKED CORNSALAD — See Valerian
BEAMTREE — See Apple

BEAN [ADZUKI, GARDEN (vine or pole; bush or dwarf, kidney, snap or string), LIMA, MUNG or URD, SIEVA or CIVET, SCARLET RUNNER, TEPARY, TEXAS] *(Phaseolus)*; JACKBEAN, SWORDBEAN *(Canavalia)*; ASPARAGUS-BEAN or YARDLONG BEAN, BLACK-EYED PEA or SOUTHERN PEA *(Vigna)*

1. *Bacterial Blights and Brown Spot*—General. Small, water-soaked spots and blotches on leaves, stems, and pods that usually enlarge, become irregular, and brown or reddish-brown with yellowish or red margins. (Brown Spot has no yellow border.) Leaves become dry, brittle, and ragged. Stems may be girdled, causing dwarfing, killing, or breaking over of plants. Diseased seed may be shriveled, discolored, and show varnish-like coating or appear healthy. Most severe after hail, frost, overhead irrigation, and splashing, wind-blown rains. See Figures 2.18A and 4.21C. *Control:* Plant only certified, disease-free, western-grown seed where soil fertility is balanced. Do *not* work among wet plants. Keep down weeds. Space plants. Grow in well-drained soil. Maintain 3-year rotation. Collect and burn or bury tops after harvest. Resistant *bean* varieties to one or more blights: Blue Lake strains, Columbia Pinto, Corneli 14, Fullgreen, Furore, Great Northern strains, Kentucky Wonder strains, Michelite, Perfect, Pinto strains, Red Mexican strains, Richmond Wonder, Robust, Seminole, SRS-PT, Starland Wax, US-1140, Tendergreen, and Tenderlong 15. *Scarlet runner* and *Tepary beans* are normally resistant to bacterial blights. Plant lima beans as far from lilac, pear, wild cherry, etc., as possible. Avoid overhead irrigation whenever possible. Copper-containing sprays are effective during wet weather. Refrigerate harvested beans promptly to 45° F.
2. *Mosaics, Mottle, Yellow Dot and Stipple, Chlorotic Mottle, Streak, Pod Mottle*—General. Symptoms variable depending on variety, age at time of infection, virus, virus strain, environmental conditions, and other factors. Leaves usually puckered, dwarfed, crinkled, and curled downward. Show irregular, yellow or light and dark green mottling. Plants often stunted, bunchy, and sickly yellow. Produce few pods. Pods often deformed, curled, mottled, rough, shiny (or greasy), and stunted. Mosaic (common and NY 15) bean viruses are transmitted in infected pollen. See Figure 2.34C under General Diseases. *Control:* Plant certified, virus-free seed. Resistant *bean* varieties to two or more mosaic-type viruses: Astro, Blue Lake strains, Choctaw, Columbia Pinto, Comet, Contender, Corneli 14, Dade, D-11, Earligreen, Earliwax, Early Harvest, Executive, Extender, Florigreen, Gallatin 50, Garden Green, Gratiot, Great Northern strains, Harter, Harvester, Highlander, Idaho Bountiful, Idagreen, Ideal Market, Improved Brittle Wax, Improved Higrade, Improved New Stringless, Improved Tendergreen, Kentucky 191, Kentucky Wonder strains, McCaslan, Michelite-62, Monroe, Mountaineer, Pearl Green, Pompano, Processor, Puregold Wax, Red Mexican strains, Red Shellout,

Resistant Asgrow Valentine, Resistant Cherokee, Resistant Kinghorn Wax, Rialto, Richgreen, Saginaw, Seaway, Seaway 65, Seminole, Sensation Refugee strains, Shipper, Slender White, Slimgreen, Small White 51 and 52, Spartan Arrow, Sprite, Tenderbest, Tendercrop, Tenderlong 15, Tenderwhite, Topcrop, Topmost, Wade, Wadex, White Seeded Contender, White Seeded Tendercrop, and Wisconsin Refugee. *Lima beans* resistant to mosaic: Burpee Best, Burpee Improved, Carpenteria, Challenger, Detroit Mammoth, Dreer Bush, Dwarf Large White, Early Jersey, Fordhook, Improved No. 243, King of the Garden, Leviathan, McCrea, New Wonder, and Seibert. *Scarlet runner beans* are resistant to a wide range of bean viruses. Keep down weeds. Control aphids and leafhoppers that transmit viruses. The bean leaf beetle can transmit the bean pod mottle, Arkansas cowpea mosaic, and Southern bean mosaic viruses. Apply malathion plus sevin at weekly intervals. Grow beans as far from alfalfa, clovers, gladiolus, lupine, and possibly corn, as practical. Destroy first infected plants, where practical.

3. *Root and Stem Rots, Stem Cankers*—General. Plants weak and stunted with few pods. May wilt and die. Leaves turn yellow and drop early. Roots and stem near soil line are discolored—gray, brown, black or brick-red—and decayed (Figure 4.20). May be associated with nematodes. *Control:* Sow certified, disease-free seed, treated with thiram, chloranil, or captan (Table 20 in Appendix). Plant crack-free seed, shallow, in warm (65° F. or above), well-drained, fertile soil where beans or related crops have not grown for 6 years or more. Plowing under a green manure crop the season before planting is often beneficial. Cultivate shallowly. Keep plants growing vigorously. Burn crop debris after harvest. *Bean* varieties differ in resistance to Fusarium Root Rot. *Scarlet runner beans* are resistant. Application of Lanstan or PCNB (5 pounds per acre active) dust or spray in planting furrow is beneficial in controlling Rhizoctonia Root Rot and Stem Canker. Difolatan is effective in controlling Fusarium Root Rot. Follow manufacturer's directions.
4. *White Mold, Watery Soft Rot, Sclerotiniose, Crown Rot, Sclerotinia Wilt, Southern*

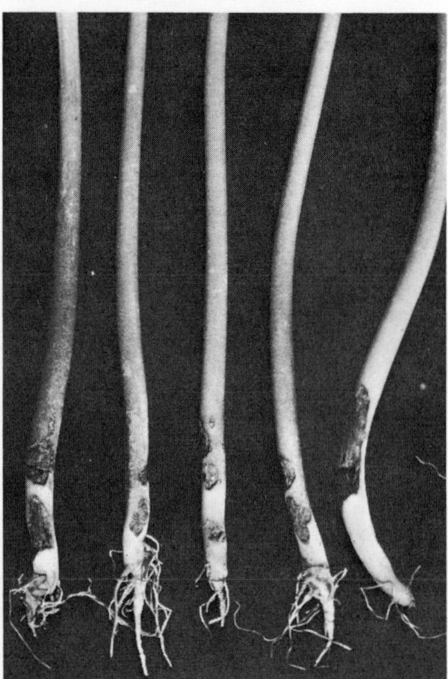

Fig. 4.20. Bean — root rot and stem cankers caused by *Rhizoctonia*. (Pennsylvania State University Extension Service photo)

Blight—General in humid, moist areas. Soft, dark, water-soaked spots on leaves, stems, and pods which soon enlarge and become covered with cottony mold in which irregular black bodies (sclerotia) may be embedded. Stem just below soil line is often water-soaked and darkened. Affected parts become mushy, wilt, and die. Most common when foliage is dense. Common on pods after harvest where crop is not refrigerated. *Control:* Avoid overcrowding and overhead sprinkling. Keep down weeds. Plant in wide rows in well-drained soil. Apply PCNB, Lanstan, or Botran as dust or spray, following manufacturer's directions. Rotate with nonsusceptible crops, e.g., onion, corn, small grains, and hay. Carefully dig up and burn affected plants before vines shade ground. Partially resistant or tolerant *bean* varieties: FM 1, Golden Wax, Gratiot, Sanilac, Seaway, and Stringless Green Pod.

5. *Anthracnoses*—General except in dry areas in western states. Sunken, reddish-brown to nearly black spots or blotches on pods (Figure 4.21A) and seed. Veins on underside of leaves develop dark purplish or reddish portions. Oval to long, dark red or brown, sunken, girdling cankers develop on stems. Young plants may die. *Control:* Same as for Bacterial Blights (above). Plant in well-drained soil and keep down weeds. No good resistant *bean* varieties are yet available to all fungus races. *Bean* varieties resistant to two or more races: Black Seeded Blue Lake, Charlevoix, Gratiot, Great Northern strains, Low's Champion, Michelite, Michelite-62, Monroe, Montcalm, Perry Marrow, Red Mexican strains, Saginaw, Sanilac, Seaway, Seaway 65, Tennessee Green Pod, and White Kidney. More highly resistant varieties should be available soon. Where practical, spray as for Downy Mildew (below), especially in seedbed, or use ferbam, captan, thiram, or dichlone. Control insects with multipurpose spray containing malathion plus sevin, methoxychlor, or rotenone. Avoid overhead irrigation if at all possible.

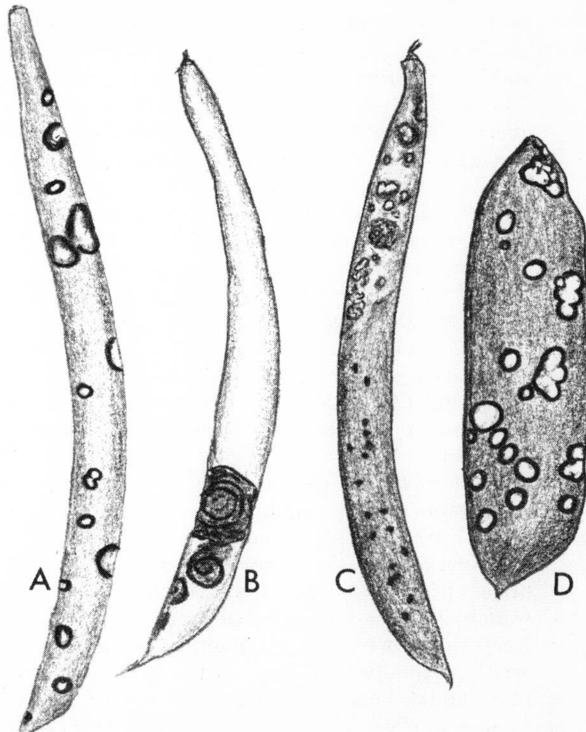

Fig. 4.21. A. Bean anthracnose (*Colletotrichum*). B. Bean soil rot (*Rhizoctonia*). C. Bacterial blight (common at top and "halo" blight on lower half). D. Lima bean scab (*Elsinoë*) — not found yet in the United States.

6. *Rust*—General in warm (70° to 80° F.), moist weather. Infrequent on lima bean and scarlet runner bean. White, reddish-brown, then chocolate-brown to black, pinhead-sized pustules on aboveground plant parts. Mostly on underleaf surface. Leaves may yellow, wither, and fall early. Yield is reduced. See Figure 2.24B under General Diseases. *Control:* Same as for Bacterial Blights (above): *Bean* varieties resistant or tolerant to many rust races: Borinquen, Criolla, Dade, Extender, Florigreen, FM 1, Golden Gate Wax, Golden "No Wilt," Great Northern 1140, Green Savage, Harvester, Hawaiian Wonder, Kentucky 191, Kentucky Wonder strains, Landreth Stringless Green Pod, Lualualei, Pencil Pod Black Wax, Pinto strains, Potomac, Rialto, Round Pod Kidney Wax, Rust-proof Golden Wax, Seminole, Stringless Black Valentine, Stringless Blue Lake strains, Tendergreen, Tennessee Green pod, and Wade. Where necessary, make several weekly applications of maneb, zineb, ziram, thiram, dichlone, Daconil 2787, Polyram, or sulfur. Start about 2 weeks before rust is expected. If dusting, making applications twice weekly. Dip or spray *used poles* with formaldehyde (1 gallon to 50 gallons of water).
7. *Powdery Mildews*—Widespread, especially in southern states and along Pacific Coast. Whitish, powdery patches on leaves, stems, and young pods. Young leaves may curl, turn yellow, shrivel, and fall early. Pods may be dwarfed and turn purplish-black. Yields are reduced. *Control:* Rotate. Burn or plow under plant debris after harvest. Keep down weeds. Apply Karathane or sulfur one to four times at about weekly intervals. Start when mildew is first seen. Normally resistant *bean* varieties: Columbia Pinto, Contender, Dixie Belle, Extender, Flight, Fullgreen, Idaho Refugee, Ideal Market, Kidney Wax, Lady Washington, Pinto strains, Ranger, Round Pod, Seminole, Stringless Green Refugee, Striped Hope, Tenderlong 15, Topcrop, U.S. No. 5 Refugee, Valentine, and Wade.
8. *Lima Bean Downy Mildew*—Most serious along Atlantic seaboard where days are moderately warm (below 85° F.), nights cool, dews are heavy and long, and humidity is high. Irregular, cottony mold patches on pods, young shoots, flowers, and leaves. Affected parts turn black and shrivel. Shoots and flowers are distorted. *Control:* Plant certified, western-grown seed. Maintain 4-year rotation. Burn crop debris after harvest. Plant in well-drained soil with good air drainage. Avoid overhead irrigation, whenever possible. The downy mildew fungus is transmitted by slugs. Normally resistant *lima bean* varieties: New Fordhook lines, Piloy, Thaxter, and resistant U.S. numbers. Spray at least weekly with maneb, zineb, folpet, Dyrene, captan, or fixed copper. Start when mildew is *first* seen or when blossoms show first color. Copper sprays may cause some pod spotting. Pick pods when dry.
9. *Seed Rots, Damping-off*—Seeds decay. Stand is thin and weak. Seedlings wilt and collapse from rot at or below soil line. Most serious in cool, wet weather. *Control:* Sow crack-free seed treated with thiram, chloranil, captan, or Dexon plus dieldrin or lindane (Table 20 in Appendix). Plant in warm (65° F. or above), dry, mellow, well-drained soil. Rotate. Plant shallow. After planting, treat as for Cabbage Wirestem.
10. *Ashy Stem Blight, Charcoal Rot*—Widespread. Mostly in southern states during warm rainy seasons. Seedlings blacken and collapse. Slightly sunken, reddish-brown to black areas on stem at soil line. Centers later turn ash-gray and become sprinkled with black dots. Disease spreads up stem and down into roots. Plants usually die before producing seed. Roots decayed and blackened. *Control:* Sow certified, western-grown seed in well-drained soil. Treat seed as for Seed Rots (above). Keep down weeds. Keep plants vigorous. Fertilize properly and water in dry weather. Rotate, excluding beans, gourds, watermelon, sweetpotato, pepper, potato, dahlia, chrysanthemum, and related plants. Collect and burn crop debris.
11. *Lima Bean Pod Spot or Blight*—Atlantic seaboard. Small to large, irregular, brown

patches on leaves. Pale brown to black watery spots may develop on older pods. Affected areas may be sprinkled with minute, greenish-gray to black "pimples." *Control:* Sow certified, western-grown seed. Maintain 4-year rotation. Spray as for Downy Mildew (above). Avoid working among wet plants. Collect and burn or plow down crop debris after harvest.

12. *Curly-top*—Western states. Leaves become puckered, cup downward. Become brittle, stunted, and either a darker green or yellowish. Plants dwarfed and bunchy (Figure 4.22). Older plants ripen early, while younger plants turn yellow, die quickly or gradually. Pods reduced in size and number. *Control:* Plant at proper time for your area (check with your county agent or extension entomologist) to escape spring migration of beet leafhopper (Figure 4.26) that transmits the Curly-top virus. Resistant or tolerant *bean* varieties: Burtner's Blightless, California Pink, California Red, Columbia Pinto, D-11, Earligreen, Earliwax, Golden Gem, Great Northern strains, Idaho Bountiful, Idelight, Improved Golden Wax, Improved Topnotch, Jenkins, Kentucky Wonder Wax, McCaslan, Mountaineer, Pink and Red Mexican strains, Pinto strains, Pioneer, Puregold, Slendergreen, Slimgreen, Stringless Green Pod, Tendergreen, and Yakima. Most *lima bean* varieties are fairly resistant. Remove and destroy diseased plants when first noticed.

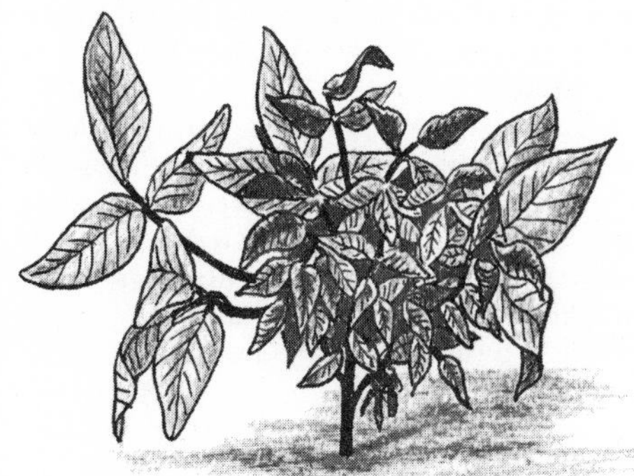

Fig. 4.22. Curly-top of bean.

13. *Bacterial Wilts*—May closely resemble Bacterial Blights (above). Plants stunted. Leaves hang limply during heat of day. Later turn brown, wither, and drop off. Badly infected seed are yellow and wrinkled, often shrunken and "varnished." Seedlings or older plants wilt and die. Rusty-brown areas may form at nodes. *Control:* Sow certified, western-grown seed. Rotate. Avoid wounding roots and stems. Grow resistant *bean* varieties: Black Wax, Crystal White Wax, Golden Wax, Great Northern strains, Kentucky Wonder strains, Michelite, Monroe, Refugee Wax, Tendergreen, and Valentine. *Lima beans* are generally resistant.
14. *Root-knot, Cyst Nematodes*—Mostly southern states in light sandy soils. Plants usually stunted, yellowish, and weak. Irregular, swollen galls which are *enlargements* of roots themselves. Cannot be rubbed off. *Control:* Rotate. Maintain vigor. Fertilize, water, keep down weeds. Where severe, fumigate soil in fall with DBCP, D-D, Telone, Vidden-D, EDB, etc. Follow manufacturer's directions. See (37) Root-knot under General Diseases. Resistant *bean* varieties: Alabama Pole No. 1 and 2, Blackeye No. 5, Spartan, and State. Resistant *lima beans:* Bixby, Hopi 155, 200, and 5989, Nemagreen, Nemagreen Bush, Westan, and certain Rico numbers. *Jackbean* and *swordbean* are also resistant.
15. *Other Fungus Leaf Spots, Pod and Stem Spots or Blotches, Gray-mold Blight, Stem Anthracnose, Lima Bean Scab*—Spots of various sizes, shapes, and colors on leaves, stems, and pods (Figure 4.21). Leaf spots may drop out leaving shot-holes. Leaves may wither and fall early. Plants may be yellowish, stunted, and die early. Seeds

shriveled. *Control:* Sow certified, disease-free, western-grown seed in well-drained soil. Avoid overcrowding and keep down weeds. Keep plants growing vigorously. Spray weekly with maneb, starting at about flowering time. Two-year rotation or longer. Collect and burn crop debris after harvest. Resistant varieties (e.g., *lima bean* to Stem Anthracnose) may be available shortly. *Black-eyed peas* highly resistant to Cladosporium Spot: Florida 133–0110 and Louisiana Purchase. Refrigerate promptly after harvest to 45° F.

16. *Baldhead, Snakehead*—Seedlings are stumps without growing tips. Plants die or remain stunted. *Control:* Sow certified, high-quality, disease-free, crack-free seed in well-prepared seedbed. Treat seed as for Seed Rots (above). Avoid deep planting.
17. *Sunscald, Leaf Scorch*—Pods exposed to hot sun, following cool overcast weather, show reddish or pale brown spots and streaks. Irregular, brown or bronze areas form on leaves. *Control:* Plant in well-drained soil. Control other diseases and insects.
18. *Web Blight*—Southeastern states. Definite, round to irregular, tan to dark brown spots with distinct, darker border on leaves and pods. Spots variable in size. When moist, spots expand and leaves are scalded and light green. Later, entire leaf turns gray to brown and dies. Other leaves and stems soon die. Stems girdled by enlarging, tan to brown cankers. Foliage may be webbed by delicate, whitish fungus hyphae. Seedlings may collapse and die before or after emergence. *Control:* Same as for Root Rots (above). Spray as for Downy Mildew (above).
19. *Fusarium Yellows, Wilts*—Western states and South Carolina. Lower leaves, especially on one side of plant, gradually turn yellow. Disease progresses upwards with older leaves turning yellow and dropping off. Plants may die. Stems show brown streaks when cut. *Control:* Sow disease-free seed in uninfested soil. *Lima beans* are highly resistant or immune. Otherwise same as for Root Rots (above).
20. *Verticillium Wilt*—See (15B) Verticillium Wilt under General Diseases.
21. *Spotted Wilt, Ringspot*—Plants may die outright or leaves die one by one. See (17) Spotted Wilt under General Diseases.
22. *2,4-Injury*—See under Grape. Beans are very susceptible. Leaves may be fan-shaped with prominent veins tending to parallel each other.
23. *Other Root-feeding Nematodes* (awl. dagger, lance, lesion, pin reniform, ring, sheath, spear, spiral, sting, stubby-root, stunt)—Mostly southern states. Associated with weak, pale, stunted plants. *Control:* See under Root-knot (above).
24. *Chlorosis*—Plants often stunted and yellowish. Leaf margins may be scorched. May die prematurely. In alkaline or very acid soils where there is a "deficiency" of zinc, iron, copper, potassium, magnesium, or manganese. *Control:* Have soil tested. Follow recommendations. *Bean* varieties differ in susceptibility to zinc deficiency.
25. *Crown Gall*—See (30) Crown Gall under General Diseases.
26. *Bacterial Soft Rot*—See (29) Bacterial Soft Rot under General Diseases.
27. *Blossom and Pod Drop*—Blossoms and pods drop during excessively hot, dry, windy weather. *Control:* If feasible, apply blossom-set hormone spray. Follow manufacturer's directions. Cultivate shallowly.
28. *Pseudo-curly-top* (beans)—Florida. Foliage forms dense rosettes with many axillary shoots. Plants stunted, develop leaf mottle. Virus is spread by treehoppers. *Control:* Keep down weeds. Spray to control treehoppers.

For additional information on bean diseases, read USDA Agr. Handbook 225, *Bean Diseases—How To Control Them.*

BEANTREE — See Goldenchain
BEARBERRY — See Blueberry
BEARGRASS — See Colchicum and Yucca
BEARD-TONGUE — See Snapdragon

Fig. 4.23. Hooflike conk of wood rot fungus (*Fomes igniarius*) at a seam covering an old branch stub on American beech. (USDA Forest Service photo)

BEAUTYBERRY — See Lantana
BEAUTYBUSH — See Viburnum
BEDSTRAW — See Buttonbush
BEEBALM — See Salvia
BEECH [AMERICAN or SILVER, EUROPEAN (many horticultural varieties), FERNLEAF, JAPANESE, ORIENTAL, PURPLE, PURPLE WEEPING, PYRAMIDAL, SIEBOLD'S, WEEPING] *(Fagus)*

1. *Wood, Heart, and Butt Rots*—Cosmopolitan (Figure 4.23). See under Birch, and (23) Wood Rot under General Diseases.
2. *Twig and Branch Cankers, Dieback*—Widespread. May be serious. Foliage wilts. Individual twigs and branches girdled and die back. Crowns open and thin. Entire tree may die. Leaves often yellow and cling to twigs. Bark may be roughened or cracked with coral-red or black "pimples" embedded in the bark. *Control:* Spray to control scale, woolly aphid, and weevil insects using DDT plus malathion. Or apply *dormant* lime-sulfur spray (1 to 20 dilution). Check with your extension entomologist regarding timing and latest spray information. Keep trees vigorous.
3. *Bleeding or Phytophthora Canker*—Northeastern states. Light brown to reddish-brown sap ("blood") oozes from bark usually near base. *Control:* Cut and burn severely infected trees. Avoid wounding bark and overfeeding. Keep trees vigorous. Fertilize and water during droughts. Injection of trees with "Carosel," containing mixture of helione orange and malachite green dyes, has helped check disease. Remove cankered bark and wood and shape the wound (Figure 2.7). Disinfect the wound, treat with shellac, and paint with tree wound dressing (pages 34–36).
4. *Powdery Mildew*—Powdery, grayish-white mold on leaves and young shoots. Leaves may yellow and wither. *Control:* If severe, spray twice, 10 days apart. Use sulfur or Karathane.
5. *Root Rots*—See under Apple, and (34) Root Rot under General Diseases. May be associated with nematodes (e.g., root-knot, spiral).
6. *Leaf Scorch*—Leaf margins turn brown. Scorching may progress until leaves wither

and fall. Often occurs when hot, dry, windy weather follows overcast periods. May follow excessive use of fertilizers, root injury, water-logged soil, etc. *Control:* Fertilize trees. Water in hot, dry weather. Grow resistant clones. Check with your nurseryman or extension horticulturist.

7. *Leaf Spots, Anthracnose*—Minor problem. *Control:* See under Maple.
8. *Mottle Leaf*—Northeastern states. Cause unknown, possibly a virus. Unfolding leaves sprinkled with small, semitransparent spots with yellowish-green or white halo. Spots enlarge, turn brown, and dry. Most common between veins near midrib and along leaf margins. Entire leaf may be scorched. Leaves drop early. A new set of normal leaves may form in midsummer. Bark on trunk and large branches may scald and die in patches. *Control:* Fertilize to maintain vigor. Protect bark from summer sun. Wrap with burlap or sisalkraft paper.
9. *Mistletoe*—See (39) Mistletoe under General Diseases.
10. *Sooty Mold*—Black moldy patches on foliage. See under Elm.
11. *Felt Fungus*—Southern states. See under Hackberry.
12. *Verticillium Wilt*—See under Maple.
13. *Seedling Blight, Damping-off*—Poor stand. Seedlings wither and collapse from rot at soil level and below. Most common in wet spring weather. *Control:* See under Pine.
14. *Chlorosis*—See under Oak. Occurs in alkaline soils.

BEEFWOOD — See Casuarina

BEET [GARDEN, SUGAR], SWISS CHARD, LEAF or SPINACH BEET, MANGEL, MANGOLD *(Beta)*; SPINACH *(Spinacea)*; BURNING-BUSH, SUMMER-CYPRESS *(Kochia)*; NEW ZEALAND SPINACH *(Tetragonia)*

1. *Cercospora Leaf Spot*—General in warm moist weather. Small, roundish spots with gray to light brown centers and dark brown to reddish-purple margins. Develop on leaves, leaf stems, flower parts, seed pods, and even seeds. Spots may drop out leaving ragged leaves that often scorch, wither, and die (Figure 4.24). *Garden* or *table beets* are somewhat resistant. *Spinach* and *mangold* are seldom seriously injured. *Sugar beet* varieties differ in resistance. Often most severe when plants also infected with Virus Yellows (below). *Control:* Three-year rotation. Burn tops or turn under cleanly after harvest. Sow disease-free seed or treat as for Seed Rot (below). Keep plants vigorous. Fertilize and water during dry periods. Space plants. Apply maneb, zineb, Dithane M-45, Dyrene, Polyram, Daconil 2787, fixed or soluble copper fungicides at 2-week intervals in warm, rainy weather. Keep down weeds (e.g., lambsquarters, pigweed, thistles).
2. *Seed Rot, Black Root Rot, Damping-off*—General. Seeds rot. Poor stand. Seedlings and older plants wilt and collapse (Figure 4.25) or produce sickly, yellowish, stunted plants. Roots decay. *Control:* Rotate. Plant in deep, fertile, well-drained soil. Avoid overwatering. Treat seed with thiram, captan, dichlone, or Dexon plus lindane (Table 20 in Appendix). If damping-off starts, water with captan, folpet, or maneb (1 tablespoon per gallon; use 1 pint per square foot). Or apply captan, Lanstan, or PCNB-Dexon mixture in furrow at planting time. Use 5 to 7 pounds of captan 50 in 20 to 30 gallons of water or 10 per cent dust, 25 to 30 pounds to 1 acre (40,800 feet of row). Follow manufacturer's directions.
3. *Curly-top*—Common and serious in western half of United States. Younger leaves curl upward. May later turn yellow. Plants become stunted or dwarfed with thickened, rolled, dull green, crimped, or brittle leaves. Roots "hairy" or "woolly." Brown to black rings appear when beet is cut across. Young plants may turn yellow and die quickly. *Control:* Plant early, or at time recommended for your area to escape spring migration of beet leafhopper (Figure 4.26) that transmits the curly-top virus. Keep down weeds. Destroy the first infected plants. Resistant *garden*

Fig. 4.24. Cercospora leaf spot of beet.

Fig. 4.25 (below). Spinach seedlings — damping-off; 1 healthy and 2 diseased. (Illinois Agricultural Experiment Station photo)

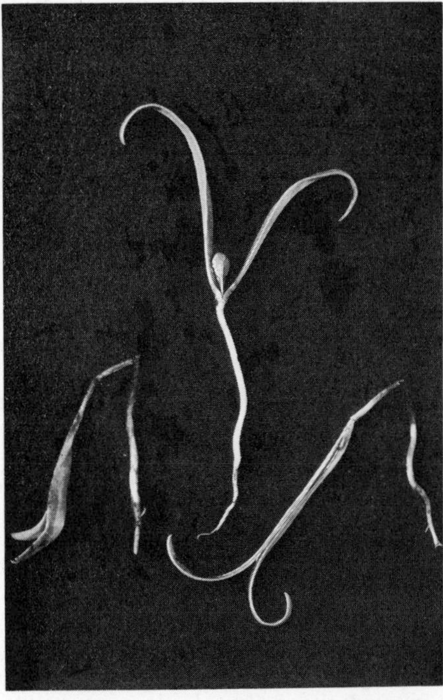

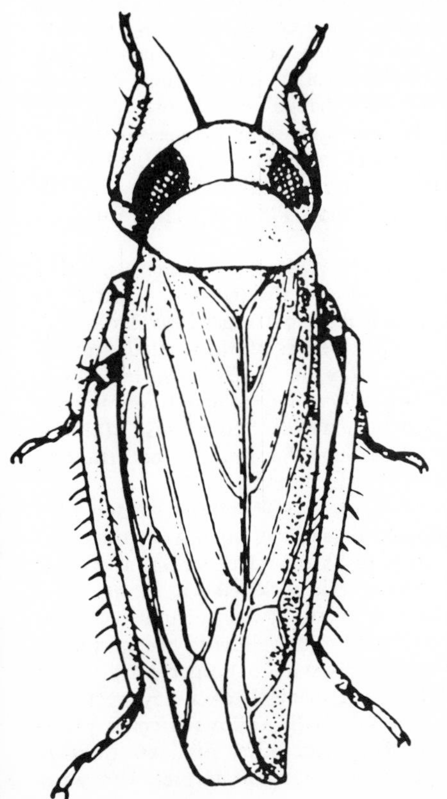

Fig. 4.26 (left). Beet leafhopper, carrier of the curly-top virus. (USDA photo)

beet: Gardeners Model; resistant *sugar beets* are widely available where curly-top is serious.

4. *Fusarium Yellows, Wilt* (spinach, sugar beet) —General. Young plants stunted, or may wilt and die. Leaves on older plants pale, turn yellow, wilt, and die slowly, starting with oldest leaves. Brown streaks inside root. Roots blacken and decay. *Control:* Long rotation. Harvest early. Destroy plant debris after harvest. If practical, plant in clean or sterilized soil (pages 538–548). Resistant varieties of *spinach* offer best hope (e.g., strains of Virginia Savoy and Domino).
5. *Blackleg, Crown or Heart Rot, Dry Rot*—Seedlings may darken and collapse. Brown, zoned rings may form on older leaves. Dark brown to black cankers at soil line or on taproot. Lower leaves turn yellow, then brown, and fall early. *Control:* Treat seed as for Seed Rot (above). Or sow disease-free seed in warm, well-drained soil in deep, well-prepared seedbed. Avoid overcrowding, excessive soil acidity, and deep planting. Adequately supply soil with boron, phosphorus, and lime. Keep plants growing vigorously.
6. *Phosphorus Deficiency, Blackheart* (primarily beet, sugar beet, mangold) —General. Locally severe where phosphorus supply in soil is low. Plants grow slowly, often slightly stunted. Leaves small and dark green with dark brown to black or bronze areas between veins. Leaf edges yellowed then brown to black. May be shriveled. *Spinach* leaves may be small and curled. Older leaves crinkled; shrivel. Taproot often small and internally discolored. *Control:* Have soil tested. Apply recommended amount of superphosphate-containing fertilizer.
7. *Boron Deficiency, Heart Rot, Cracked Stem*—Symptoms variable. Locally severe where boron is lacking in soil, and where nitrogen and potassium are high. Middle and older leaves abnormally crinkled. May wilt and become scorched. If severe, young leaves stunted, twisted, narrow, and darker green, redder, or blacker (pale green or yellow and one-sided in *spinach*) than normal. "Heart" may be dead. Leaves may wilt, turn brown, and die in midsummer. Irregular, hard or corky, gray, brown to black, internal or external dry rot, surface cankers and cracks of taproot. See Figure 1.1C. Phoma rot is associated with a lack of boron. *Control:* Have soil tested. Apply borax as recommended. Usually 1 ounce over 16 to 20 square yards. Resistant *table beets:* Detroit Dark Red and Long Blood. Plant in well-drained soil that is slightly acid (pH 6.0).
8. *Mosaics, Savoy, Yellow Net, Yellow Vein, Ringspot*—Widespread. Leaves mottled; light and dark green or yellow. Leaf surface often crinkled and curled (or savoyed). May show faint zigzag lines, rings, and other patterns. Leaves die back from tips. Growth often bunchy and stunted. Center leaves often mottled, stunted, and distorted. *Beet* roots smaller than normal. May be "hairy." *Control:* Plant virus-free seed from healthy mother plants. Plant early or late to avoid migration of virus-carrying aphids. Keep down weeds. Destroy first infected plants. Maintain balanced soil fertility. Control aphids and plant bugs (lacebugs), that transmit viruses. Apply malathion, Diazinon, or naled (Dibrom) weekly.
9. *Virus or Beet Yellows, Spinach Blight (Cucumber mosaic)*—Primarily garden beet, mangel, mangold, New Zealand spinach, spinach, sugar beet, summer-cypress, and Swiss chard. Symptoms variable depending on plant, virus, virus strain, and other factors. Sometimes confused with nitrogen or magnesium deficiency. Plants often dwarfed and yellowed. Veins in young leaves are usually conspicuously yellow (cleared). Outer and inner leaves gradually become bright yellow to orange-yellow, thickened, and brittle (Figure 4.27). Yellowing usually starts at upper margins and tips. *Table beet* leaves usually deep red with little yellowing. Leaves may later turn brown and die early, starting at tips. Root may be small. Yellows-infected plants may be more susceptible to Leaf Spot, Root Rots, and other diseases. *Control:* Plant late or early to avoid migrations of aphids that transmit viruses. Use virus-

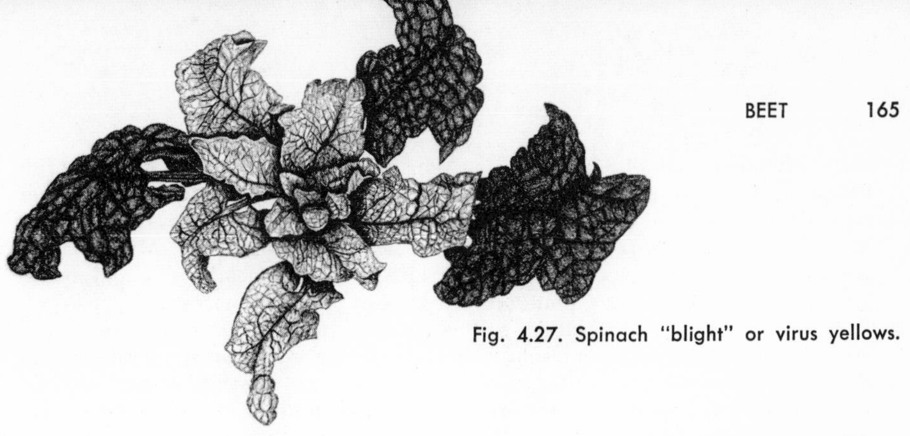

Fig. 4.27. Spinach "blight" or virus yellows.

free seed. Keep down weeds, e.g., shepherds-purse, catnip, milkweed, horsebean, plantains, pokeweed, motherwort, hollyhock, wild lettuce and Brassicas, etc. as well as wild beets and mangolds in and around garden. Many weeds are symptomless carriers. Spray with malathion, Diazinon, or naled (Dibrom) in spring to control aphids. Plant normally resistant *spinach* varieties: Badger Savoy, Basra, Blight Resistant Virginia Savoy, Chesapeake, Dixie Market, Dixie Savoy, Domino, Hybrids 7, 56, and 612, Old Diminion, Salma, and Virginia Blight Resistant. Check adaptability of spinach varieties with your county agent or extension horticulturist. *Beet* varieties also differ greatly in resistance.

10. *Downy Mildew, Blue Mold*—General during cool, humid seasons. Young leaves covered with violet- or yellowish-gray mold. Older leaves have small to large, irregular, pale green or yellow areas on upper leaf surface with mold on underside. Leaves may become thickened and curled, dry up, darken, or rot. Similar spots occur on seed stalks that may be stunted or killed. Crown infections may permit other root decay organisms to invade plant. Serious in cool, wet weather. Damaging to spinach and Swiss chard in the seedbed and field. See Figure 2.22D under General Diseases. *Control:* Three-year rotation. Avoid overcrowding. Keep down weeds. Plant disease-free seed in well-drained, fertile soil. Destroy old crop debris. Spray a week or two before downy mildew is expected. Use zineb, maneb, or copper-containing fungicide. Repeat at 5- to 10-day intervals. Soak *spinach* seed in hot water (122° F.) for 25 minutes. Dry, then dust with thiram, chloranil, dichlone, captan, or Semesan. Resistant *spinach* varieties to 1 or more races: Badger Savoy, Califlay, Chesapeake, Dixie Market, Dixie Savoy, Hybrids 7, 11, 424, 425, 530, and 612, Marathon, Resistoflay, Salma, Savoy Supreme, Viking, and Wisconsin Bloomsdale. Resistant *table beets:* Detroit Dark Red-F.M. and Short Top strains. *Sugar beets* also differ in resistance.
11. *Scab*—Widespread. Raised, rough, scablike areas or warts on root. See under Potato. Worst in lime-rich soils. *Control:* Do not grow in soil that has produced scabby potatoes. Keep soil acid. Varieties differ in resistance.
12. *Root Gall, Root-knot, Cyst Nematodes*—Plants stunted, wilt on hot days. If severe, plants may wilt and die. Small, beadlike galls on roots, and an excessive number of secondary roots (hairy root symptom). See (37) Root-knot under General Diseases. *Control:* Long rotation where possible (4 or 5 years or more) excluding beets, crucifers, rhubarb, tomato, and spinach. Plant early. Control weeds, especially mustards, lambsquarters, sowbane, black nightshade, saltbush, knotweed, dock, pigweed, purslane, shepherds-purse, New Zealand spinach, chickweed, and wild radish that are hosts for the sugar beet nematode. If serious, fumigate soil with D-D or Telone before planting (pages 538–548). For additional information read USDA Leaflet 486, *The Sugar Beet Nematode and Its Control.* There are *sugar beet* lines resistant to the cyst nematode.
13. *Watery Soft Rot, Drop, Stem Rot, Wilt, Storage Rots*—See under Carrot.

14. *Rusts*—Western half of United States. Small, yellow to bright orange, reddish-brown, or dark brown pustules. Mostly on underside of leaves. Leaves may wither. One rust is serious only near alternate hosts—salt and needle grasses *(Distichlis* and *Aristida). Control:* Spray as for Cercospora Leaf Spot and Downy Mildew (both above), or use dichlone. Keep potash level up. Destroy nearby salt and needle grasses. Burn in fall or plow under deeply. Destroy tops after harvest.
15. *Crown Gall*—Occasional. Swollen, gall-like growths near soil line. *Control:* Avoid cultivating injuries. Long rotation. Carefully dig up and destroy affected plants.
16. *Root Rots, Crown Rots, Southern Blight*—General. Leaves wilt, wither, and die. Roots and crowns on such plants are decayed; may be covered with mold growth. See under Bean, and (34) Root Rot under General Diseases. *Beet* varieties in future may have resistance. Apply chlordane or Diazinon over row at 4- to 6-leaf stage to control crown maggots.
17. *White-rust*—Mostly southern states. Small, white, powdery pustules mostly on underleaf surface (Figure 2.25C). If severe, leaves or entire plants may wither and die. *Control:* Three-year rotation. Sow disease-free seed. Spraying as for Downy Mildew and Cercospora Leaf Spot (both above) may be beneficial. Destroy infected plants and debris after harvest. Resistant *spinach* varieties as well as tolerant *sugar beet* lines should be available soon.
18. *Minor Leaf Spots, Anthracnose*—General. Small, round to irregular, variously colored spots; often enlarge and merge. Leaves may wither and die. Mold may develop on affected parts in damp weather. *Control:* Same as for Cercospora Leaf Spot (above). *Spinach* varieties resistant to Anthracnose should be available soon.
19. *Web Blight*—Southeastern states. See under Bean.
20. *Gray-mold Blight*—See (5) Botrytis Blight under General Diseases.
21. *Black Streak, Bacterial Spot* (beet, New Zealand spinach)—Western states. Irregular, dark brown to black, spots on leaves and streaks on petioles, midrib, and leaf veins. Leaves may bend sharply, wilt, and later wither. *Control:* Sow disease-free seed in soil free of beets or related crops for 3 years. Collect and burn or bury plant debris after harvest. Avoid overfertilizing with nitrogen.
22. *Bacterial Soft Rot*—General. See (29) Bacterial Soft Rot under General Diseases. *Control:* Same as for Black Streak (above).
23. *Bacterial Pocket, Beet Gall* (beets)—Deeply indented, tumor-like growths on crown of roots. Galls have irregular brown areas inside. Galls soon disintegrate. *Control:* Same as for Black Streak (above). Avoid cultivating and insect injuries.
24. *Bacterial Wilt*—See (15C) Bacterial Wilt under General Diseases.
25. *White or Leaf Smut* (spinach)—Indistinct white spots on lower leaf surface. When severe, leaves may whiten. *Control:* Not usually necessary. Same as for Cercospora Leaf Spot and Downy Mildew (both above).
26. *Spinach Yellow Dwarf*—Young leaves may be mottled light and dark green, curled, and puckered. Older leaves develop yellow blotches. Heart leaves are stunted. May yellow and die. *Control:* Avoid spring spinach near winter spinach. Destroy wild spinach. Control aphids that transmit the virus. Use malathion.
27. *Spotted Wilt, Ringspot*—Zigzag lines or other irregular markings on leaves. See (17) Spotted Wilt under General Diseases. The beet ringspot virus is transmitted by the needle *(Longidorus)* nematode.
28. *Chlorosis*—Leaves mottled, turn yellowish-green to golden-yellow (may be deep red to purple in *table beets*) at margins, later between the veins. Leaves may later wither and die. Usually due to a soil deficiency of iron, calcium, magnesium, manganese, or zinc in very acid, neutral, or alkaline soils. *Control:* Acidify or lime soil to get pH between 6.0 and 6.5. Or spray plants with iron sulfate, manganese sulfate, or zinc sulfate (or all 3). Use 1 to 2 ounces in 2½ gallons of water. Plants should recover in 1 to 2 weeks. Repeat as necessary. *Garden beets* differ in resistance.

29. *Other Root-feeding Nematodes* (e.g., lance, lesion, naccobus, pin, reniform, rot, sheath, spear, spiral, sting, stubby-root, stunt)—Associated with stunted, sickly plants. Roots short, stubby, discolored, and die back. Nematodes make wounds for root-rotting fungi. *Control:* Same as for Root Gall (above).
30. *Verticillium Wilt* (beet, New Zealand spinach, spinach)—See (15B) Verticillium Wilt under General Diseases. Sow disease-free seed.
31. *Powdery Mildew* (beet)—See (7) Powdery Mildew under General Diseases. Sugar *beet* varieties differ in resistance. *Control:* Space plants. Fertilize, based on a soil test. Spray with Karathane.

BEGONIA [BEEFSTEAK, CAMELLIA-FLOWERED, CHRISTMAS, CRESTED, DAFFODIL-FLOWERED, FIBROUS-ROOTED, HANGING BASKET, HARDY, HOLLYHOCK-FLOWERED, HYBRID, REX or FOLIAGE, RHIZOMATOUS, SEMITUBEROUS, TUBEROUS-ROOTED, WAX or EVERBLOOMING] *(Begonia)*

1. *Gray-mold Blight, Botrytis Blight, Blotch*—Cosmopolitan and serious. Enlarging, blotchy, soft, tan-colored rotted areas on leaves, stems, seedlings, cuttings, and flowers. Coarse, gray mold grows on affected parts in wet weather. Leaves and flowers turn brownish-black and die. Common in greenhouses, especially plants weakened by root rot, powdery mildew, sunburn, cold temperatures, etc. *Control:* Carefully destroy severely affected plants or plant parts when first found. Space plants. Improve air circulation. Keep water off foliage. Control weeds. Avoid wet mulch, overcrowding, and overwatering. Propagate cuttings from healthy plants and grow in well-drained sterilized soil (pages 538–548). Keep plants growing at *steady* rate. Indoors use sufficient light. Apply captan, maneb, thiram, ferbam, ziram, dichloran (Botran), or zineb to plants and soil.
2. *Crown, Stem, and Root Rots, Cutting Rot, Damping-off*—Cosmopolitan. Plants pale. Causes poor growth. Stems and crowns often water-soaked and darkened near soil line. May wilt and collapse. Lower leaves often water-soaked and flabby. Roots discolored and decayed. Seedlings and cuttings rot and collapse. Tubers may also rot. Often associated with root-feeding nematodes (e.g. aphelenchus, lesion, pin, spiral). *Control:* Same as for Gray-mold Blight (above). Use seed and leaves (cuttings) from disease-free plants. Soil drenches of Dexon or oxyquinoline sulfate are effective. Follow manufacturer's directions. Do *not* drench very young seedlings.
3. *Powdery Mildews*—Common. May be serious on tuberous- and fibrous-rooted begonias. White, powdery blotches on leaves (Figure 4.28). Leaves often deformed. *Control:* Apply Karathane, folpet, or sulfur, plus spreader-sticker, when mildew is

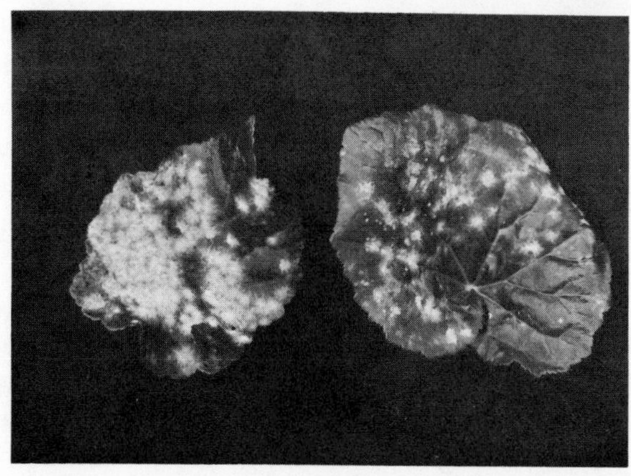

Fig. 4.28. Powdery mildew on begonia leaves. The white, powdery growth is more abundant on the upper leaf surface. (Illinois Natural History Survey photo)

first seen. Repeat as necessary. Acti-dione has also proven effective. Increase air circulation. Avoid sprinkling foliage.

4. *Bacterial Leaf Spot, Bacteriosis*—Widespread on tuberous- and hybrid begonias. Small, roundish, water-soaked (translucent) spots that turn yellow or brown. Spots enlarge and merge. Leaves may wither and fall early. Main stem may rot, killing plant. See Figure 2.18B under General Diseases. *Control:* Same as for Gray-mold Blight (above) except use streptomycin or fixed copper when necessary. Propagate only with cuttings from disease-free plants. Varieties differ in resistance. Indoors reduce temperature and humidity. Keep water off foliage. Space plants.
5. *Crown Gall*—More or less round to irregular galls develop on stem near soil line. Galls may later rot, killing plant. *Control:* Take tip cuttings and destroy mother plant. Pot in sterilized soil (pages 538–548). Avoid wounding stems. Space plants for good air movement. Begonias vary in resistance.
6. *Root-knot*—Swellings form on roots and crowns. Plants stunted and lack vigor. On tuberous begonias, root galls become quite large. *Control:* Make new leaf cuttings in peat or vermiculite and plant in sterilized soil. Discard infested plants. Soak *dormant* tuberous-begonia tubers in hot water (120° F.) for 30 minutes. Cool and plant.
7. *Leaf Nematode Blight*—Widespread and serious. Small to large, irregular, water-soaked then dark brown patches on leaves between the veins. Leaves curl up and drop off. Plants often stunted and unsightly. Produce few if any flowers. See (20) Leaf Nematode under General Diseases. *Control:* Keep water off foliage. Space plants. Propagate only from disease-free plants. Grow in sterilized soil. Where necessary, dip or spray plants weekly with malathion. Remove and burn all infested plant parts. Begonias in small pots may be disinfested by dipping in hot water (1 minute at 120° to 121° F.; 3 minutes at 117° to 119° F.; or 5 minutes at 115° F.).
8. *Spotted Wilt*—Zoned, yellowish to brown, or ringlike spots on leaves. Plants stunted and bronzed with poor flowers. *Control:* Destroy infected plants. Keep down weeds. Use DDT or lindane to control thrips that transmit the virus.
9. *Aster Yellows*—Plants stunted, bushy, and yellow. *Control:* Same as for Spotted Wilt. Virus is spread by leafhoppers.
10. *Corky Scab, Oedema*—Indoor plants show small, water-soaked spots that later become light brown, corky growths. Found on underside of leaves and along stems. *Control:* Avoid overwatering or sprinkling plants. Keep down humidity in cool, cloudy weather.
11. *Leaf Spots, Anthracnose*—Spots of various sizes, shapes, and colors. May merge to form irregular leaf blotches. *Control:* Same as for Gray-mold Blight (above).
12. *Mosaic*—Uncommon. Yellowish areas and sometimes brown spots, between leaf veins. *Control:* Destroy affected plants. Spray or fumigate to control aphids and other insects. Use DDT or methoxychlor plus malathion.
13. *Verticillium Wilt*—Lower leaves become pale green, gradually turn yellow and wilt. Later, upper leaves becomes affected. Plants may die. *Control:* Avoid using soil where wilt has occurred on potato, tomato, eggplant, pepper, brambles, strawberry, etc. Repot into sterilized soil in sterilized containers.

BEGONIA-LEAF — See Calla Lily

BELAMCANDA — See Iris

BELLADONNA LILY — See Chrysanthemum

BELLFLOWER [ADRIA, AMERICAN, CHIMNEY, DALMATIAN, LIGURIAN, SIBERIAN, TUSSOCK, WILLOW or PEACH BELLS], BLUEBELLS-of-SCOTLAND, CANTERBURY-BELLS, HAREBELL, COVENTRY-BELLS *(Campanula)*; BALLOONFLOWER, CHINESE BELLFLOWER *(Platycodon)*; VENUS-LOOKINGGLASS *(Specularia)*

1. *Leaf Spots*—Small to large, round to irregular, yellowish, brown, gray, or blackish spots. Similar spots may occur on flower petals. Infected leaves may wither and dry up. *Control:* Destroy tops in fall. Keep down weeds. Apply zineb, captan, thiram, or copper-containing fungicide, several times, 10 days apart.
2. *Stem Rot or Blight, Crown Rot, Southern Blight, Damping-off, Root Rots*—Grayish-white, cottony, or brown areas on stem that later enlarge. Stem base and roots decay, causing seedlings or older plants to wilt and collapse. *Control:* Dig up and burn infected plants and surrounding soil. Four- to 6-year rotation. Grow in clean or sterilized soil (pages 538–548) that is well-drained. A soil drench of mercuric chloride (1:2,000 dilution) or Morton Soil Drench may be beneficial if applied in time.
3. *Rusts* (bellflower, Canterbury-bells, harebell, venus-lookingglass) —Widespread. Yellow-orange, orange-red, reddish-brown, or black powdery pustules on underside of leaves. Plants stunted. Usually serious if certain infected pines (alternate host of one rust) are nearby. *Control:* Avoid growing near pines *(Pinus rigida* and *P. resinosa),* or apply zineb, thiram, ferbam, or maneb, three times, 10 days apart. Start when rust is *first* seen.
4. *Root-knot*—See (37) Root-knot under General Diseases.
5. *Powdery Mildew* (bellflower, Canterbury-bells) —Whitish, powdery spots on leaves and stems. *Control:* Apply sulfur or Karathane, several times, 10 days apart.
6. *Spotted Wilt* (Canterbury-bells) —Plants stunted. Growth is poor. Leaves show pale ringspots or wavy line markings. *Control:* Destroy infected plants. Control thrips that transmit the virus. Use DDT or malathion.
7. *Aster Yellows* (Canterbury-bells) —See (18) Yellows under General Diseases.
8. *Curly-top* (campanula) —See (19) Curly-top under General Diseases.
9. *Gray-mold Blight, Crown Rot*—See under Begonia, and (5) Botrytis Blight under General Diseases.
10. *Verticillium Wilt*—Lower leaves yellow, wilt, and die. Wilt gradually progresses up plant. *Control:* Destroy infected plants. Rotate. Grow in well-drained soil.
11. *Mosaic*—Irregular, yellowish-green and dark green blotches on leaves. Plants may be stunted. See (16) Mosaic under General Diseases.
12. *Seed Smut* (venus-lookingglass) —Seeds filled with dark, powdery masses. *Control:* Sow only disease-free seed.
13. *Leaf and Stem Nematode* (bellflower) —See (20) Leaf Nematode under General Diseases.

BELLIS — See Chrysanthemum

BELLS OF IRELAND — See Salvia

BELLWORT — See Lily

BELOPERONE — See Clockvine

BENINCASA — See Cucumber

BENT, BENTGRASS — See Lawngrass

BENZOIN —See Avocado

BERBERIS — See Barberry

BERMUDA BUTTERCUP — See Oxalis

BERMUDAGRASS — See Lawngrass

BETONY — See Salvia

BETULA — See Birch

BIDENS—See Chrysanthemum

BIGNONIA, TRUMPETFLOWER, CROSSVINE, CHINESE TRUMPETCREEPER *(Bignonia)*; FLAME-VINE *(Pyrostegia)*

1. *Leaf Spots, Spot Anthracnose*—Leaves variously spotted. *Control:* See under Chrysanthemum.
2. *Sooty Mold, Black Mildews*—Southern states. Black, moldy patches on foliage. Follows attack by aphids, mealybugs, scales, and whiteflies. See (12) Sooty Mold under General Diseases.
3. *Root-knot*—See (37) Root-knot under General Diseases.
4. *Dieback, Canker*—See under Apple.

BIRCH [CHINESE PAPER, CUTLEAF, DWARF, ELM-LEAVED, EUROPEAN WHITE (many horticultural varieties), GRAY or POPLAR, MONARCH, PAPER or CANOE, PYRAMIDAL, RIVER or RED, SWAMP or HAIRY DWARF, SWEET or CHERRY, WATER or MOUNTAIN, WEEPING, WESTERN PAPER, WHITE, YELLOW] *(Betula)*; ALDER [AMERICAN, BLACK, CUTLEAF BLACK, EUROPEAN GREEN, GREEN, HAZEL or SMOOTH, ITALIAN, SPECKLED or HOARY-LEAVED, JAPANESE, MANCHURIAN, NEW MEXICAN, OREGON, RED, SEASIDE, SIERRA, SITKA, SPECKLED or GRAY, THINLEAF, WEEPING] *(Alnus)*; HORNBEAM [AMERICAN, EUROPEAN (many horticultural varieties), HEARTLEAF, YEDDO] *(Carpinus)*; HAZELNUT [AMERICAN, BEAKED or CUCKOLD, CALIFORNIA, CHINESE, EUROPEAN, JAPANESE, TIBETIAN, TURKISH], FILBERT, PURPLE-LEAVED FILBERT *(Corylus)*; IRONWOOD, HOPHORNBEAM [AMERICAN, EASTERN, EUROPEAN, JAPANESE] *(Ostrya)*

1. *Wood Rots, Heart Rots, Butt Rots*—Cosmopolitan. Wood may be lightweight, soft, crumbly, and punky or powdery. Hoof- or shelf-shaped fruiting bodies (conks) develop along dead or dying trunk and branches from old wounds where a stub has been left or a limb broken off (Figures 2.41 and 4.29). Diseased wood may be

Fig. 4.29. Conks of a wood rot fungus (*Poria obliqua*) on yellow birch. (USDA photo)

discolored or stained. *Control:* Maintain good tree vigor. Fertilize and water during dry periods. Make flush pruning cuts (page 33). Control other diseases and insects, especially borers. Avoid wounding bark. Remove and burn all dead and dying branches. Paint all wounds promptly with good tree wound dressing (page 36). Wrap trunks of newly set trees (page 42) for at least the first two years. Use heavy paper, 3 to 4 inches wide, in a spiral wrap from the lowest branch to the ground. Help control borers by monthly sprays, May to September. Apply to trunk and scaffold branches. Use DDT, dieldrin, or Thiodan.

2. *Dieback, Twig, Branch and Trunk Cankers*—Upper branches progressively die back. Irregular, sunken, swollen or "catface," discolored cankers may develop on twigs, branches, and trunk. Infected trunk and branches may be flattened and bent (Nectria or European canker). Plant parts die when cankers girdle. Often associated with borers, drought, poor soil drainage, high soil temperatures, and other diseases. *Control:* Destroy trees with severe trunk cankers. Remove and burn young cankers and all dead wood. Make clean, flush cuts. Sterilize tools between cuts. Otherwise same as for Wood Rots. Plant adapted species in protected locations.
3. *Leaf Spots, Anthracnose*—Widespread. Round to irregular, brownish, yellowish or gray spots, following cool, wet, spring weather. Certain spots may drop out, leaving ragged holes. If numerous, leaves may drop early. *Control:* Seldom necessary. Collect and burn fallen leaves. If practical, spray as for Leaf Rust (below) or use captan.
4. *Leaf Rust*—Seldom serious. Numerous, small, round, bright reddish-yellow, orange or waxy yellow, later dark brown to nearly black, powdery pustules. Mostly on lower leaf surface. Leaves may drop early. *Control:* Do not plant near larch, the alternate host. If practical, apply zineb, maneb, or dichlone as buds burst open. Repeat 10 and 20 days later. Rake and burn fallen leaves.
5. *Leaf Blisters or Curls, Witches'-broom, Alder Catkin Gall*—Yellow or red blisters may cause leaves to be swollen, distorted, curl, and drop early. Mature female *alder* catkins become enlarged, elongated, and red. Later turn whitish and finally dark brown. Crowded clusters of weak twigs (witches'-brooms) may arise near same spots on twigs. Smaller twigs are killed. (Witches'-brooms may also be caused by mites.) *Control:* Prune and burn witches'-brooms and swollen catkins, where feasible. Collect and burn fallen leaves. If practical, apply ferbam or copper-containing fungicide, plus spreader-sticker, once *before* buds swell in early spring.
6. *Powdery Mildews*—Widespread. Powdery, grayish-white patches on leaves, buds, young twigs, and female catkins. Severely infected leaves may yellow, wither, and drop early. *Control:* Where practical, apply sulfur, Acti-dione, or Karathane one or two times 10 days apart. Start when mildew is first seen. Collect and burn fallen leaves.
7. *Bacterial Blight, Bacteriosis* (hazelnut, filbert)—Serious on young trees along Pacific Coast in wet seasons. Round to irregular, yellowish-green, water-soaked spots that later turn reddish-brown to black. On buds, leaves, and young shoots (up to 4 years old) in the spring (Figure 4.30). Girdling cankers develop on nut-bearing twigs, branches, and young tree trunks. Often kills young trees. Maturing nuts are discolored. *Control:* Keep trees vigorous. Water during droughts and protect against sunscald (see below). Resistant *filberts:* Bolwyller, Daviana, and Graham. Turkish *hazelnut* is also resistant. Apply bordeaux mixture (3-1½-50), or fixed copper and spreader-sticker, in August before fall rains. Repeat when ¾ of leaves have dropped. Sterilize pruning tools between cuts or trees with a 1:1,000 solution of mercuric chloride or 70 per cent alcohol. Grow disease-free nursery stock in well-drained soil.
8. *Root Rots*—Trees decline in vigor. Foliage thin and sickly. Leaves may yellow,

Fig. 4.30. Twigs and branches on Filbert tree killed by bacterial blight. (Courtesy Dr. P. W. Miller)

wither, and drop early. Twigs and branches die back, often starting on one side. *Control:* See under Apple, and (34) Root Rot under General Diseases.

9. *Twig Blights, Dieback, Bark Canker*—Widespread. Discolored cankers girdle stems. Foliage and stems beyond wither and die. Buds rot. *Control:* Prune and burn dead and cankered wood. Spraying as for Leaf Rust (above) may be beneficial. Apply spray of phenyl mercury in fall, just before leaves fall. Follow manufacturer's directions.
10. *Wetwood, Slime Flux*—See under Elm.
11. *Sooty Mold*—Dark sooty material on leaves. *Control:* Spray to control insects. If practical, wash leaves with strong stream of water.
12. *Crown Gall*—See under Apple, and (30) Crown Gall under General Diseases.
13. *Kernel Bitter Rot, Brown Spot or Stain* (filbert, hazelnut)—Problem on Pacific Coast. Nuts may rot and drop early. Check with local authorities (see above).
14. *Bleeding Canker*—Northeastern states. See under Beech and Maple.
15. *Felt Fungus* (hornbeam)—Southern states. Smooth, buff-colored growth on bark. See under Hackberry.
16. *Mistletoe* (alder, birch)—See (39) Mistletoe under General Diseases.
17. *2,4-D Injury*—Birches very susceptible. Severely injured leaves may become twisted, rolled, boat-shaped, or curled into ram's-horns. *Control:* See under Grape.
18. *Sunscald* (primarily filbert)—Young trees very susceptible to hot rays of sun when temperature is 95° F. and above. Damage occurs on south or southwest sides of trunk. Trees suffering from drought or those in low, poorly drained soil with sparse foliage are most susceptible. Leaves may be scorched. *Control:* Protect trunk from hot sun (page 42).

BIRD-OF-PARADISE-FLOWER *(Strelitzia)*

1. *Root and Seed Rots*—See (34) Root Rot under General Diseases. May be associated with root-feeding nematodes (e.g., burrowing, lesion, stubby-root).

Control: Plant disease-free seed or soak seed for a day in water (at room temperature) then in hot water (135° F.) for 30 minutes. Cool, dry, and plant in clean or sterilized soil that is fertile and well-drained.

2. *Bacterial Wilt*—Leaf margins are yellow to orange. Later entire leaves are discolored. Finally curl upward, turn brown, and wither. Petioles and leaf sheaths wither. Plants slowly wilt, starting with older leaves. Dark brown to black streaks inside rhizome and pseudostem. *Control:* Destroy affected plants. Otherwise, same as for Root and Seed Rot.

BIRD'S-EYES —See Phlox
BIRTHWORT — See Aristolochia
BISHOPSCAP — See Hydrangea
BITTERSWEET [AMERICAN or CLIMBING, ORIENTAL or CHINESE, LOESENER] *(Celastrus);* ARABIAN-TEA *(Catha);* EUONYMUS [BROADLEAF, BROOK or STRAWBERRY-BUSH, CLIMBING, EVERGREEN or EVERGREEN BURNING-BUSH, JAPANESE EVERGREEN, LACE, WINTERBERRY, YEDDO], SPINDLE-TREE [ALDENHAM, EUROPEAN, WINGED or WINGED BURNING-BUSH], BURNING-BUSH or WAHOO, and WINTERCREEPER [BIGLEAF, GLOSSY, SILVER-EDGE] *(Euonymus);* CLIFFGREEN or CANBY, MYRTLE BOXLEAF *(Pachistima)*

1. *Powdery Mildews*—Widespread. May be serious. White powdery mold or gray "felt" patches on leaves. Leaves may yellow, curl, and drop early. *Control:* Remove badly attacked shoots. Prune to thin out shrubs. Where practical, apply sulfur, Karathane, or Acti-dione several times, 10 days apart. Avoid applications above 85° F.
2. *Crown Gall*—Common. Euonymus may be severely infected. Large, rounded galls with an irregular rough surface, appear on roots or stems (Figure 4.31). Plants lack vigor. May die back. *Control:* Do not set out plants with galls. Avoid wounding stems or roots. If disease appears, prune out and burn affected plant parts. Dig up and burn severely affected plants.
3. *Leaf Spots, Anthracnoses, Leaf Scab*—Variously sized and colored spots on leaves. Common following wet weather. *Control:* Collect and burn fallen leaves. Prune to thin out shrubs. Apply zineb, maneb, or fixed copper several times, 7 to 10 days apart. Start as leaves unfold.
4. *Stem Cankers, Dieback*—See under Maple. Same as for Leaf Spots (above). Cut and burn dead and dying twigs.

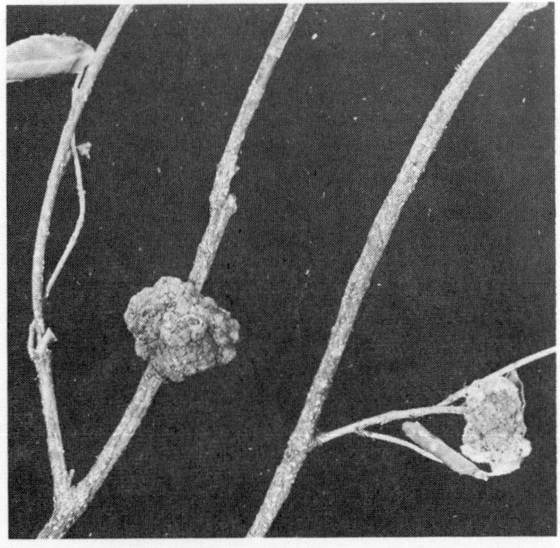

Fig. 4.31. Crown gall of euonymus. (Pennsylvania State University Extension Service photo)

5. *Root-knot*—See (37) Root-knot under General Diseases.
6. *Other Root-feeding Nematodes* (e.g., dagger, lesion, needle, pin, ring, spiral, stunt) —Associated with weak, stunted plants. *Control:* Same as for Root-knot (above).
7. *Euonymus Mosaic, Infectious Variegation*—Yellowing develops along leaf veins. *Control:* Avoid variegated plants for propagating. Destroy infected plants.
8. *Thread Blight* (euonymus) —Southeastern states. See under Walnut.
9. *Root Rot* (bittersweet, euonymus) —See (34) Root Rot under General Diseases.

BLACK-ALDER — See Holly
BLACKBERRY — See Raspberry
BLACKBERRY-LILY — See Iris
BLACK COHOSH — See Anemone
BLACK-EYED PEA — See Bean
BLACK-EYED-SUSAN — See Chrysanthemum and Clockvine
BLACK GUM — See Dogwood
BLACKHAW — See Viburnum
BLACK-SALSIFY — See Lettuce
BLACK SAMPSON — See Chrysanthemum
BLACK-SNAKEROOT — See Anemone
BLACKTHORN — See Peach
BLACK WALNUT — See Walnut
BLADDERNUT — See American Bladdernut
BLADDER-SENNA — See Honeylocust
BLAZING-STAR — See Chrysanthemum, Gladiolus, and Mentzelia
BLECHNUM — See Ferns

BLEEDINGHEART, DUTCHMANS-BREECHES, GOLDEN EARDROPS, SQUIRRELCORN, TURKEY CORN or WILD BLEEDINGHEART *(Dicentra)*; CORYDALIS [GOLDEN, PALE] *(Corydalis)*

1. *Rusts* (corydalis, Dutchmans-breeches, squirrelcorn) —Yellowish-orange spots on upper leaf surface. Clusters of slender, cuplike structures form on lower surface. Alternate hosts include wood-nettle and wild grasses. *Control:* Destroy nearby alternate hosts. If practical, apply zineb, maneb, or ferbam several times, 10 days apart. Start when rust is first seen.
2. *Crown Rot, Stem Rots, Wilt*—Plants wilt and die. Cobwebby white growth, in which black bodies (sclerotia) are later embedded, occurs on diseased tissue and over nearby soil surface. *Control:* Remove and destroy affected plants plus 3 inches of surrounding soil. For lightly affected plants, drench around crown with 1:1,000 solution of mercuric chloride (see page 101 for precautions). Grow plants in neutral, well-drained soil.
3. *Fusarium Wilt* (bleedingheart) —See (15A) Fusarium Wilt under General Diseases.
4. *Downy Mildew*—May be serious on plants growing in shade. See (6) Downy Mildew under General Diseases.
5. *Leaf Spot*—Small spots on leaves. *Control:* Spray or dust as for Rusts (above).
6. *Root-knot*—See (37) Root-knot under General Diseases.
7. *Gray-mold Blight*—See (5) Botrytis Blight under General Diseases.

BLESSEDTHISTLE — See Chrysanthemum
BLOODFLOWER — See Butterflyweed
BLOODLEAF — See Cockscomb
BLOODROOT — See Poppy

BLUEBEARD — See Lantana
BLUEBELLS — See Mertensia and Tulip
BLUEBELLS OF ENGLAND — See Tulip
BLUEBELLS-OF-SCOTLAND — See Bellflower
BLUEBERRY [BOX or EVERGREEN, CLUSTER, DRYLAND or LOW, GROUND, HIGHBUSH (BLACK, NORTHERN, SOUTHERN or SWAMP), LOWBUSH (RABBITEYE, SUGAR, UPLAND, MOUNTAIN CRANBERRY or LINGONBERRY)], SHORE COWBERRY, WORTLEBERRY *(Vaccinium)*; STRAWBERRY-TREE, MADRONE or MADRONA *(Arbutus)*; MANZANITA, BEARBERRY, WOOLLY MANZANITA *(Arctostaphylos)*; CASSANDRA or LEATHERLEAF *(Chamaedaphne)*; CASSIOPE; HUCKLEBERRY [BLACK, BOX, BUSH, GARDEN, DANGLEBERRY] *(Gaylussacia)*; BOG-LAUREL, MOUNTAIN-LAUREL, PALE LAUREL, SHEEP-LAUREL or LAMBKILL *(Kalmia)*; FETTERBUSH, MALEBERRY, STAGGERBUSH *(Lyonia)*; RUSTYLEAF, MINNIE-BUSH *(Menziesia)*

1. *Mummy Berry, Brown Rot, Blossom Blight, Twig Blight* (blueberry, mountain-laurel) —General eastern half of United States in wet, humid springs. Tips of new shoots wilt and turn brown. Blossoms blasted. Nearly full grown blueberry fruit turn gray or tan, then shrivel into hard mummies (Figure 4.32). *Control:* Avoid crowding and overfertilizing. Resistant *blueberry* varieties: Cabot, Stanley, and Weymouth. Apply captan, zineb, ziram, thiram, ferbam, or dichlone several times, 7 to 10 days apart. Start when buds swell. Practice *very clean* cultivation (disc, rake, sweep, hoe weekly or more often) in spring through bloom. Commercial blueberry growers work calcium cyanamide into the soil under plants *before* buds break, using 200 pounds of 57 per cent pulverized or granular material per acre (5 pounds per 1000 square feet).
2. *Botrytis Blight, Gray-mold Blight*—General. Blossoms blasted. Young blueberry fruit shrivel and turn dull bluish-purple. They soon fall. There are irregular brown

Fig. 4.32. Mummy berry of blueberry. The 5 dark berries to the upper left are "mummy berries." The rest of the fruit is healthy.

leaf blotches. Shoot tips die back. Dense gray mold may grow on affected parts. *Control:* Spray as for Mummy Berry (above). Avoid spring applications of fertilizer high in nitrogen. Prune shrubs annually.

3. *Red Leaf Gall, Swamp Cheese, "Rose Bloom," Shoot Hypertrophy*—General. Gall-like growths on twigs. Small, red blisters on leaves. Blossoms and small fruit swell abnormally. Affected parts turn pink to bright red before falling off. *Control:* Plant disease-free stock. Prune and burn infected parts when first seen. Apply a single dormant spray *before* buds swell. Use fixed copper, ferbam, or zineb.
4. *Powdery Mildew*—General on blueberry. Compact whitish mold on upper surface of leaves of certain susceptible varieties (Adams, Concord, Jersey, and Rubel). Etched, circular, water-soaked spots appear on lower surface of young leaves (Figure 4.33). Spots enlarge and leaves of susceptible varieties (e.g., Cabot, Pioneer, Wareham) gradually turn yellow and may drop early. Mildew usually occurs after harvest and causes little damage. *Control:* Grow resistant *blueberries:* Berkeley, Blue Crop, Coville, Dixi, Earliblue, Harding, Ivanhoe, Pemberton, Rancocas, Stanley, and Weymouth. More resistant varieties should be available in the future. Apply Karathane, plus spreader-sticker, twice, 10 days apart.

Fig. 4.33. Blueberry powdery mildew on lower leaf surface. Note the water-soaked (dark), etched, circular lesions. (Department of Botany & Plant Pathology and Cooperative Extension Service, Michigan State University photo)

5. *Crown Gall*—Sometimes called Cane Gall. Swollen, rough, irregular, light brown to black galls along stems and small twigs (Figure 4.34). Sometimes at base of stems. Galls may girdle and kill stems. *Control:* Set out disease-free plants. Destroy severely infected plants. When only slight infection is found, prune several inches below any sign of infection. Disinfect before each cut. Dip shears in 1:1,000 solution of mercuric chloride (see page 101 for precautions).

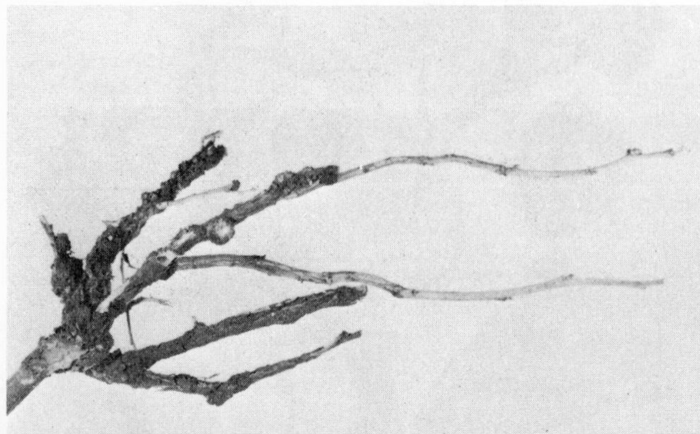

Fig. 4.34. Severe cane or crown gall on a young blueberry plant. (Department of Botany & Plant Pathology and Cooperative Extension Service, Michigan State University photo)

6. *Witches'-broom*—Short, swollen, crowded twigs giving bushy or broomlike appearance. No fruit produced. Found near firs, alternate host of the rust fungus. Witches'-broom of *mountain-laurel* is not caused by a rust. *Control:* Prune and burn infected branches. Spraying as for Mummy Berry and Gray-mold Blight (both above) may be beneficial. Destroy nearby, worthless true firs and infected blueberry plants since rust fungus is perennial and systemic.
7. *Leaf Rust*—Widespread. Small, irregular, reddish-brown, orange, reddish-yellow, brown or black spots and yellowish pustules on leaves. Leaves may wither and drop early. Usually severe only in certain years near hemlocks and spruces; alternate hosts of the rust fungus. *Control:* If practical, spray as for Mummy Berry and Gray-mold Blight (both above). Grow resistant *blueberry* varieties.
8. *Blueberry Stunt*—Symptoms vary with time of year, stage of growth, and variety. Tip leaves on young shoots in spring are first pale yellow at margins and tip. Leaves later turn completely yellow except for veins. These leaves become round, dwarfed, and cupped (Figure 4.35). Leaves on old canes turn prematurely red in early fall. The brilliant red color occurs in two lengthwise bands on the leaf, parallel to the midrib. Plants stunted and bushy with numerous, dwarfed, slender side twigs. Affected plants unproductive with small, hard, bitter fruit. Sharp-nosed leafhoppers transmit the virus. *Control:* Set out *only* certified, virus-free nursery stock. Promptly dig up and destroy infected bushes, including roots, when first found. Destroy wild blueberry plants in area. Follow spray program in Appendix. Rancocas is highly resistant.

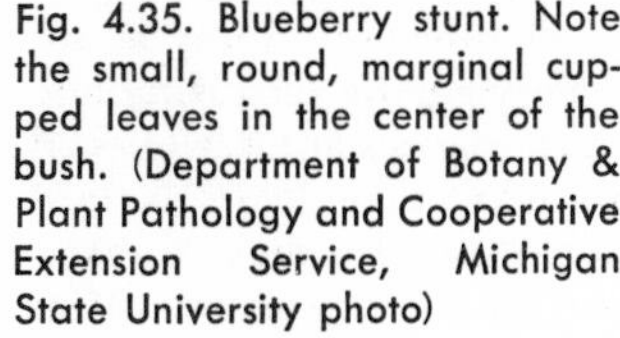

Fig. 4.35. Blueberry stunt. Note the small, round, marginal cupped leaves in the center of the bush. (Department of Botany & Plant Pathology and Cooperative Extension Service, Michigan State University photo)

9. *Blueberry Mosaic*—Leaves on one or more canes brilliantly mottled with yellow, yellow-green, and red areas. Lower leaves on a cane generally show more color. Fruit production gradually declines. Mosaic is sometimes confused with magnesium deficiency. Most common on Concord, Dixi, and Stanley. *Control:* Same as for Stunt (above).
10. *Blueberry Shoestring*—Symptoms variable, even on same plant. Lowbush blueberries may be symptomless. Brown to red bands develop along midrib of a leaf. Often extend partially into lateral veins. Such leaves are wavy, distorted, and crescent-shaped. Severely affected leaves are narrow, pointed, strap-shaped, and light green to dull red (Figure 4.36). On certain varieties (e.g., Burlington, Cabot, and Jersey) red streaks, bands, or oval patches occur along new canes and twigs. Cane growth is long and spindly. Such canes do not produce fruit. *Control:* Same as for Stunt (above).
11. *Blueberry Ringspots*—Widespread. Symptoms variable depending on virus, variety, and other factors. Leaves develop conspicuous, yellow to dead or red rings, spots,

Fig. 4.36. Blueberry shoestring affecting shoots of the Burlington variety. *Left* — severely affected showing narrow, pointed strap-shaped leaves; *center* — early symptoms of wavy, crescent-shaped leaves and recurving shoot growth; *right* — healthy shoot. (Department of Botany & Plant Pathology and Cooperative Extension Service, Michigan State University photo)

blotches, line, or jagged oakleaf patterns on one or more branches. Rings may drop out giving tattered appearance. Rings or blotches may be evident on stems throughout year. The viruses spread rather rapidly in the field. Plants may be stunted and unproductive. Twigs die back. Necrotic ringspot virus is transmitted by the dagger *(Xiphinema)* nematode. *Control:* Same as for Stunt (above).

12. *Twig Blight, Stem Blight, Dieback, Cane Cankers*—Often follows winter injury, sunscald, excessive soil moisture, insect or nematode attack, and other factors. Foliage on infected shoots wilts, withers, and dies due to discolored, girdling cankers on twigs and branches. *Control:* Plant in well-drained soil. Remove and burn weak, unthrifty plants. Prune and burn dead, blighted, or weakened twigs in winter. Make clean cuts 6 to 8 inches behind any sign of infection. Spray as for Mummy Berry and Botrytis Blight (both above). Increase vigor by fertilizing and watering during summer droughts. Plant disease-free stock of resistant varieties, if available. Check with your local nurseryman or extension plant pathologist. Named *rabbiteye blueberries,* except Locke, are resistant to Botryosphaeria Stem Canker. Normally canker-resistant *highbush blueberries:* Angola, Croatan, Morrow, Murphy, Scammell,* and Wolcott.
13. *Iron Chlorosis*—General in neutral and alkaline soils where these plants are not adapted. Leaves are first mottled, then turn yellow except for veins (Figure 4.37). Plants stunted, may die. *Control:* Keep soil acid (pH 4.2 to 5.2) and add iron as sulfate or chelate (pages 26–27).
14. *Leaf Spots, Leaf Blotch, Spot Anthracnose, Anthracnose, Tar Spot*—Widespread. Small to large, round to irregular spots of various colors. Black dots are sprinkled in centers of certain spots (Figure 4.38). Common in rainy seasons. If severe, leaves may wither and drop early. Tips of twigs may also be spotted and later blighted. Fruit may rot. *Control:* If practical, spray as for Mummy Berry (above), or use maneb or Dyrene. See also Table 16 in Appendix. For a few plants, pick or rake and burn infected leaves. Avoid crowding.
15. *Root Rots*—New leaves appear scorched or may be mottled green and yellow, with or without "burnt tips," accompanied by dieback of branches. Plants stunted. Affected leaves may drop early. Fine roots are decayed. Similar symptoms may be caused by excess fertilizer, root injury by insects or mice, soluble salt injury, etc. May be associated with root-feeding nematodes (below). *Control:* Treat soil before planting and grub-proof established plantings in sod every 5 years with aldrin, dieldrin, or chlordane. Use 2 ounces of *actual* aldrin or dieldrin, or 4 ounces of *actual* chlordane per 1,000 square feet. Follow manufacturer's instructions.
16. *Wood Rots, Trunk Canker*—General. Plants often stunted. Affected leaves may

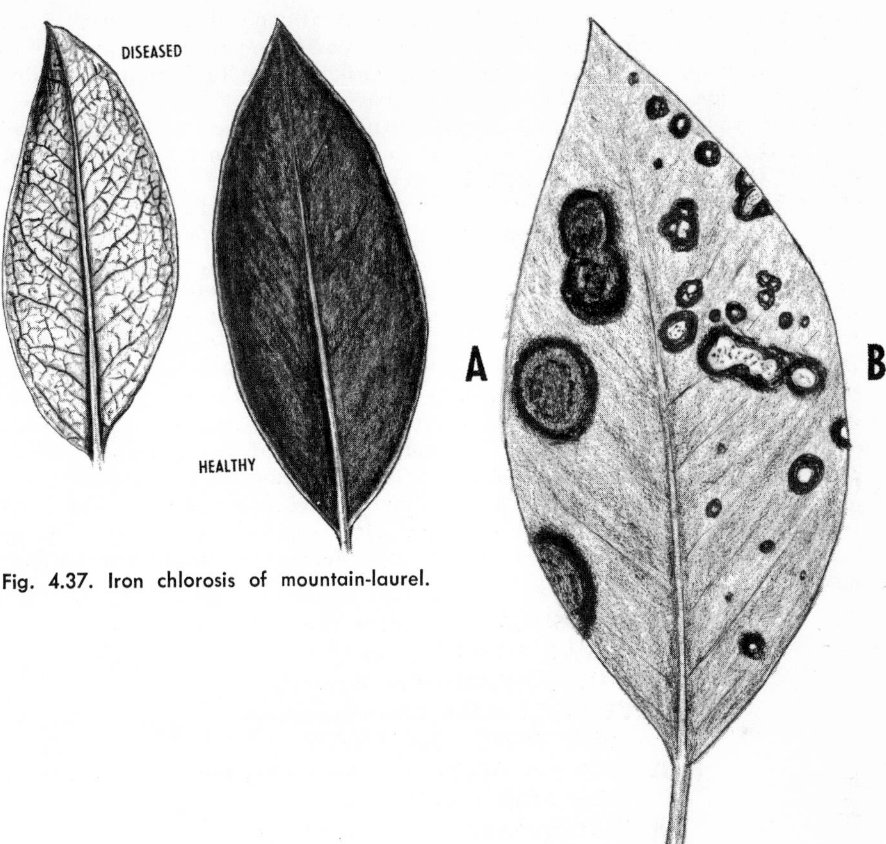

Fig. 4.37. Iron chlorosis of mountain-laurel.

Fig. 4.38. A. Leaf blight (*Phomopsis*). B. Leaf spot (*Phyllosticta*) of mountain-laurel.

drop early. Plants decline. See under Birch and Dogwood, and (23) Wood Rot under General Diseases.

17. *Mistletoe* (manzanita)—See (39) Mistletoe under General Diseases.

18. *Bacterial Stem Canker of Blueberry*—Serious on Pacific Coast. Irregular, water-soaked areas develop on last year's canes during winter. Infected areas soon turn into well-defined, reddish-brown to black cankers. All buds are killed. Stems may be girdled and killed. Young plants may die. *Control:* Prune and burn all diseased wood when found. Grow resistant varieties: Burlington, June, Pioneer, Rancocas, Rubel, and Weymouth. Apply bordeaux (4-4-50) or fixed copper twice in October and November.

19. *Drought and Winter Injury* (primarily mountain-laurel)—Leaves gradually turn brown, starting at tip or margins. Such leaves dry up and later drop off. *Control:* Plant in shady, protected locations where soil is acid, moist, and high in organic matter. Water plants during dry periods in summer and fall. Mulch plants in early winter after ground is frozen. Protect exposed plants against winter winds by burlap or canvas barriers.

20. *Black Mildew*—Primarily in Gulf states. See (12) Sooty Mold under General Diseases.

21. *Fruit Rots* (blueberry)—Fruits rot on plant or after harvest. May be covered with

gray, brown, or black mold. *Control:* Same as for Mummy Berry (above). Spray with multipurpose spray containing methoxychlor or sevin plus malathion to control blueberry maggot, fruitworms, and other insects. See Table 16 in Appendix. Use captan alone from midbloom through harvest.

22. *Blueberry Root Gall*—Galls white, later become dark brown, woody, and covered with bark. Stem cankers occur near soil line. Small galls may occur on twigs. *Control:* Grow resistant varieties: Dixi, Jersey, and Rubel.
23. *Blueberry Red Leaf Disease*—Northeastern states on lowbush blueberry. Red leaves appear on certain branches with white, feltlike layer on undersurface. Some shoots die back each year. Fruit set is reduced. The causal fungus is related to those causing Red Leaf Gall; however the red leaf disease fungus is systemic and perennial in the rhizome. *Control:* Dig out and destroy infected plants when first found.
24. *Bud-proliferating Gall*—Galls form at soil line. Buds abort to form clusters of weak shoots, 1 to 6 inches tall. *Control:* Same as for Crown Gall (above).
25. *Root-feeding Nematodes* (awl, dagger, lance, lesion, pin, ring, root-knot, sheath, spear, sphaeronema, spiral, stubby-root, stunt)—Associated with stunted, declining plants. *Control:* See (37) Root-knot under General Diseases.
26. *Magnesium Deficiency*—Widespread. Areas between veins are bright red. Leaf veins remain green; resembles a "Christmas tree." Dead spots and shades of yellow or brown may replace red. *Control:* Apply magnesium sulfate in regular fertilizer mixture, following a soil test.

BLUE-BLOSSOM — See New Jersey-tea

BLUE BONNET — See Pea

BLUE COHOSH — See Barberry

BLUE DAISY — See Chrysanthemum

BLUE DICKS — See Brodiaea

BLUE DAWN-FLOWER — See Morning-glory

BLUE-EYED GRASS — See Iris

BLUE-EYED-MARY — See Snapdragon

BLUE FLAG — See Iris

BLUE GILIA — See Phlox

BLUEGRASS — See Lawngrass

BLUE LACEFLOWER — See Celery

BLUELIPS — See Snapdragon

BLUE SPIREA — See Lantana

BLUETS — See Buttonbush

BOG-LAUREL — See Blueberry

BOISDUVALIA — See Fuchsia

BOLTONIA, BONESET — See Chrysanthemum

BORAGE *(Borago)* — See Mertensia

BOSTON IVY — See Grape

BOUGAINVILLEA

1. *Leaf Spots*—Round to irregular, light to dark spots. Sometimes zoned with distinct margin. Common during rainy seasons. *Control:* Collect and burn leaves in autumn. If needed, spray several times during rainy periods. Use zineb, maneb, or fixed copper. Grow in full sun and avoid overhead watering. Varieties with white, lavender, or light purple flower bracts are apparently highly resistant or immune to Cercospora Leaf Spot.
2. *Mosaic*—See (16) Mosaic under General Diseases.

3. *Chlorosis*—Deficiency of manganese, zinc, iron, or copper in acid or alkaline soils. Have soil tested and follow suggestions in report. See under Rose.

BOUSSINGAULTIA — See Lythrum

BOUVARDIA

1. *Rust*—Southern states. Yellow, yellowish-orange or dark, powdery pustules on leaves. *Control:* Pick and burn spotted leaves. If practical, spray several times, 10 days apart. Start 2 weeks before rust normally appears. Use ferbam, zineb, or maneb. Indoors keep water off foliage. Space plants.
2. *Leaf Nematode*—Leaves develop dark, unsightly blotches. Flower clusters deformed and stunted. See (20) Leaf Nematode under General Diseases.
3. *Root-knot*—Plants may be stunted and unthrifty with galls or knots on roots. See (37) Root-knot under General Diseases.

BOWSTRING HEMP — See Sansevieria

BOXELDER — See Maple

BOX SANDMYRTLE — See Labrador-tea

BOXWOOD [BOX, DWARF or ENGLISH, EDGING, HARLAND, JAPANESE, KOREAN, LITTLELEAF, MYRTLE-LEAF, ROSEMARY-LEAF, TREE or AMERICAN, VARIEGATED] *(Buxus)*

1. *Decline, Dieback, Root Rots, Twig Blight, Cankers*—General and destructive. Infected branches often start growth later in spring than normal ones. Leaves on such branches curl upward close to stem and turn light grayish-green or bronze and finally straw-colored. Leaves may wither and drop early leaving bare twigs. See Figure 4.39. Often follows winter or drought injury, nematodes, root rots, nutrient imbalance, and poor soil drainage. Twigs, branches, or main stems die back. Small, pinkish to black mounds often develop on affected parts. Roots may be decayed. The bark at base of larger branches may die and slough off. *Control:* Before growth starts in spring, remove and burn all leaves on ground and lodged in twig crotches. Break or prune out dead twigs and severely cankered branches as soon as noticed. Cut out cankers on larger branches, cover wound with shellac and then tree paint (page 36), then spray: (1) just after removing dead leaves and branches and before growth starts, (2) as new leaves break out of buds, (3) 2 and 4 weeks later.

Fig. 4.39. Boxwood decline or dieback. (Pennsylvania State University Extension Service photo)

Use bordeaux (3-3-50), lime-sulfur (5 level tablespoons per gallon of water), fixed copper, ziram, thiram, ferbam, or phenyl mercury plus spreader-sticker. Protect plants against Winter Injury (below). Maintain vigor. Fertilize in fall and water during droughts. Select disease-free plants grown in sterilized soils. Plant in clean, well-drained soil. *Phytophthora Root Rot* is controlled by several soil drenches of Dexon (1 teaspoonful in 4 gal. of water) applied 10 to 14 days apart.

2. *Winter Injury, Sunscald, Windburn*—Serious in northern states. Symptoms variable. Leaves may turn bronze-colored to rusty-brown or red with dead areas around margin. Leaves dry in late spring. Leaves, twigs, even entire plants may die back. Injured bark may split and peel. Stems girdled, with parts beyond later dying. *Control:* Grow in protected location or start with hardier plants. Erect burlap, canvas, or lattice windbreaks to ward off drying winter winds and sun. Try spraying with Wilt Pruf, Plantcote, Stop-Wilt, Foli-gard, or similar antidesiccant in late fall. Fertilize in late fall or very early spring. Water plants thoroughly during long droughts in spring and early summer and especially late in a dry fall just before soil freezes. Then mulch plants to prevent deep freezing. Check with your local nurseryman or extension horticulturist regarding mulching. In spring, prune back dead branches to healthy wood.
3. *Leaf Spots, Leaf or Tip Blights, Leaf Cast*—Leaves variously spotted; may become pale or straw-colored, sometimes dull tan or brown, starting at margins and tips. Conspicuous, raised black dots may be evident on upper leaf surface (Figure 4.40). Many leaves drop early. Young twigs may die. *Control:* Same as for Decline (above). Protect plants against Winter Injury (above). Control insects with malathion plus lindane or DDT.

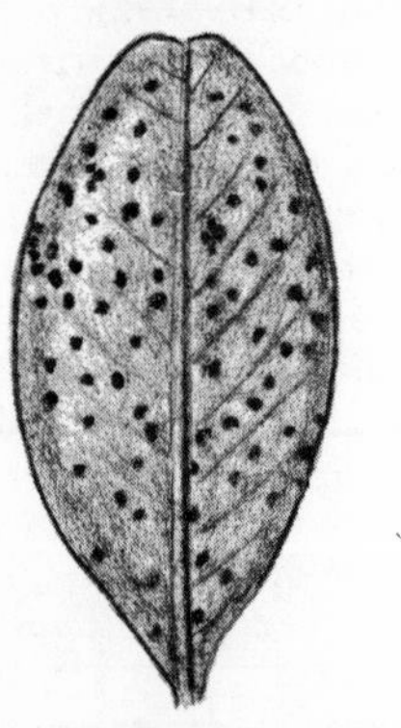
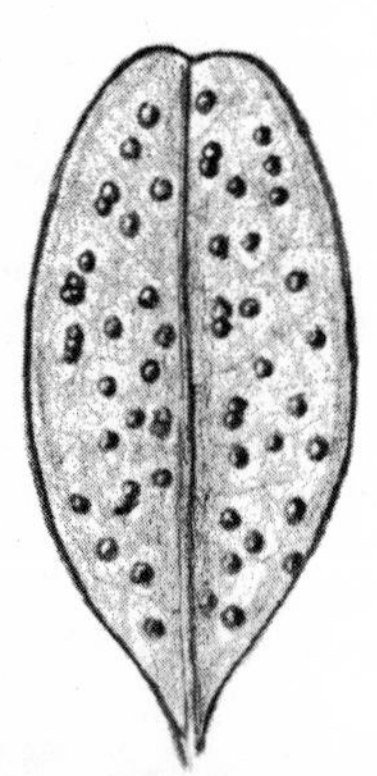

Fig. 4.40. Macrophoma leaf spot of boxwood.

4. *Nematodes* (burrowing, dagger, lance, lesion, needle, pin, ring, root-knot, sheath, spiral, stem, sting, stubby-root, stunt) —Plants weak, stunted, lack vigor. May wilt on hot, dry days. Leaves may be pale green to sickly bronze or orange. Plants gradually decline. Branches may die back. Roots stunted, often bushy and dark or galled. Root rot or winter injury may follow nematode damage. *Control:* Drench soil around plant roots with DBCP, Zinophos, or VC-13 following manufacturer's directions. Check also with your nurseryman, extension plant pathologist, or entomologist. Mulch, water, and fertilize to keep plants as vigorous as possible. Fumigate planting site before replanting boxwood in infested soil. Young bare-root plants may be disinfested of Root-knot (and possibly other nematodes) by dipping in hot water (118° F. for 30 minutes) or Zinophos at 1,000 parts per million for 30 minutes. Then plant in clean or fumigated soil.
5. *Heart Rots, Trunk Rot*—See Wood Rot under Birch, and (23) Wood Rot under General Diseases.

6. *Crown Gall*—See (30) Crown Gall under General Diseases.
7. *Thread Blight*—Southeastern states. See under Fig.
For additional information read USDA Farmers' Bulletin 1855, *Culture, Diseases, and Pests of the Box Tree.*

BOYSENBERRY — See Raspberry
BRACHYCOME — See Chrysanthemum
BRAKE — See Ferns
BRASSAVOLA, BRASSIA — See Orchids
BRASSICA — See Cabbage
BRIDALWREATH — See Spirea
BROCCOLI — See Cabbage
BRODIAEA, TRIPLET LILY, BLUE DICKS, PRETTY-FACE, SNAKE LILY, CALIFORNIA-HYACINTH, WILD HYACINTH, SPRING STARFLOWER *(Brodiaea)*

1. *Rusts*—Western states. Yellow, orange, reddish-brown or black, powdery pustules on leaves. Alternate host: none or wild grasses (*Agropyron* and *Elymus*). Control: Where serious, collect and burn rusted leaves after flowering. Apply zineb, maneb, or ferbam several times, 10 days apart. Start 2 weeks before rust normally appears.

BROOM [KEW, PORTUGUESE, PROVENCE, PURPLE, SCOTCH, SPIKE, SWEET, WARMINSTER], GENISTA OF FLORISTS, DYER'S GREENWEED or WOODWAXEN, DOUBLE-FLOWERED DYER'S GREENWEED, SPANISH BROOM or GORSE *(Cytisus, Genista)*; DALEA [BLACK, FEATHER, MESA, SMOKETREE] *(Dalea)*; BUNDLEFLOWER *(Desmanthus)*

1. *Leaf Spot, Blight, Diebacks*—Small, irregular black spots on leaves. Spots enlarge rapidly forming a blotch or blight. Leaves drop early. Brown spots may develop on stems. Shoots die back. Plants often killed in 2 weeks. *Control:* Destroy infected plant parts. Spray weekly. Start when disease first noticed. Use bordeaux mixture or fixed copper and spray lime.
2. *Powdery Mildew*—Powdery white patches on leaves. *Control:* Dust or spray twice, 10 days apart. Use sulfur or Karathane.
3. *Rust*—See (8) Rust under General Diseases.
4. *Root Rot*—See under Apple, and (34) Root Rot under General Diseases. May be associated with lesion nematodes.
5. *Mistletoe* (dalea) —See (39) Mistletoe under General Diseases.

BROUSSONETIA — See Fig
BROWALLIA — See Tomato
BROWN-EYED-SUSAN — See Chrysanthemum
BRUSH-CHERRY — See Eugenia
BRUSSELS SPROUTS — See Cabbage
BRYONOPSIS — See Cucumber
BRYOPHYLLUM — See Sedum
BUCHLOE — See Lawngrass
BUCKEYE — See Horsechestnut
BUCKTHORN [ALDER, CALIFORNIA or COFFEEBERRY, CAROLINA (YELLOW or INDIAN CHERRY), CASCARA, COMMON, DAHURIAN, GLOSSY or ENGLISH, HOLLYLEAF or RED-BERRIED] *(Rhamnus)*

1. *Leaf Spots*—Widespread, but not destructive. Small, round to irregular, gray, brown, or black spots. If severe, plants may be unsightly. *Control:* If needed, apply sprays at 10-day intervals during rainy periods. Use zineb, ferbam, maneb, or fixed copper.
2. *Rusts*—Widespread. Small, yellow to orange spots on leaves. Causes little damage. Rusts spread to nearby oats and several wild grasses where they cause the destructive Crown Rust disease. See Figure 4.41. Buckthorn is a noxious weed in Iowa. In California one rust produces black pustules on coffeeberry and hollyleaf buckthorn. *Control:* If practical, spray as for Leaf Spots (above). Start when leaves appear. Buckthorn should *not* be grown in commercial oat-growing areas.

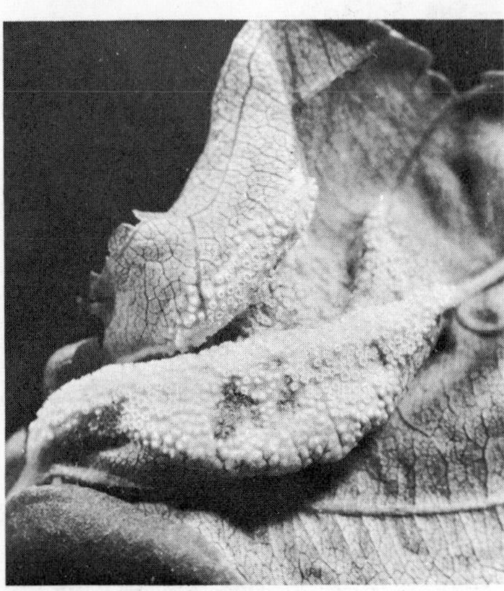

Fig. 4.41. Crown rust on common buckthorn (extreme closeup). Note aecial "cluster cups." (Courtesy Dr. W. H. Bragonier)

3. *Wood Rots*—See under Birch, and (23) Wood Rot under General Diseases.
4. *Powdery Mildew*—See (7) Powdery Mildew under General Diseases.
5. *Sooty Mold*—Black, powdery mold patches on leaves. Follows attacks of aphids or scales. *Control:* Apply malathion to control insects.
6. *Root Rot*—See (34) Root Rot under General Diseases.

BUCKWHEAT-TREE *(Cliftonia)*; SOUTHERN LEATHERWOOD *(Cyrilla)*

1. *Leaf Spots*—Spots of various sizes, shapes, and colors. *Control:* If serious, collect and burn fallen leaves. Spray during wet periods. Use zineb, maneb, or ferbam.
2. *Rust* (southern leatherwood)—Southeastern states. See (8) Rust under General Diseases. *Control:* Same as for Leaf Spots.
3. *Black Mildew*—See (12) Sooty Mold under General Diseases.
4. *Brown Felt Canker*—See under Hackberry.

BUDDLEIA — See Butterflybush
BUFFALOBERRY — See Russian-olive
BUFFALOGRASS — See Lawngrass
BUGBANE — See Anemone
BUGLEWEED — See Ajuga
BUNCHBERRY — See Dogwood
BUNDLEFLOWER — See Broom
BUNYA-BUNYA — See Araucaria

BUR-MARIGOLD — See Chrysanthemum
BURNET — See Rose
BURNING-BUSH — See Beet and Bittersweet
BURRO'S-TAIL — See Sedum
BUSHCHERRY — See Peach
BUSH-HONEYSUCKLE — See Snowberry
BUSH-MALLOW — See Hollyhock
BUSH MORNING-GLORY — See Morning-glory
BUSH-PEA — See Pea
BUTIA — See Palms
BUTTER-AND-EGGS — See Snapdragon
BUTTERCUP — See Delphinium
BUTTERFLYBUSH [FARQUHAR, FOUNTAIN, ILE DE FRANCE, JAPANESE, LINDLEY, ORANGE-EYE BUTTERFLYBUSH or SUMMER-LILAC, OXEYE] *(Buddleia)*; YELLOW-JESSAMINE, CAROLINA-JESSAMINE *(Gelsemium)*

1. *Mosaic* (Butterflybush) —Leaves mottled light and dark green, malformed, and tapered. See (16) Mosaic under General Diseases.
2. *Twig and Stem Canker*—See (22) Stem Canker under General Diseases.
3. *Root Rot*—See under Apple, and (34) Root Rot under General Diseases. Grow in full sun in rich, well-drained soil high in organic matter.
4. *Sooty Mold, Black Mildew, Leaf Spot*—See (1) Fungus Leaf Spot, and (12) Sooty Mold under General Diseases.
5. *Silky Thread Blight* (Carolina-jessamine) —Southeastern states. See under Walnut.
6. *Chlorosis*—Mineral deficiency in very acid or alkaline soils. See under Rose. *Control:* Have soil tested and follow suggestions in report.
7. *Root-knot*—See (37) Root-knot under General Diseases. Butterflybush is very susceptible.

BUTTERFLY-FLOWER — See Tomato
BUTTERFLY-PEA — See Pea
BUTTERFLYWEED, MILKWEED [FOUR-LEAVED, MEXICAN, PURPLE, SWAMP, WHORLED], BLOODFLOWER *(Asclepias)*; PHILIBERTIA

1. *Leaf Spots*—Spot of various sizes, shapes, and colors. *Control:* See under Chrysanthemum.
2. *Rusts*—Widespread. Yellow, orange-yellow, reddish-brown or black, powdery pustules on leaves. Alternate hosts: Grama (*Bouteloua* spp.), cord grasses *(Spartina)*, unknown, or none. *Control:* Pick and burn rusted leaves. If practical, spray with ferbam or zineb, several times, 10 days apart. Start about a week before rust normally appears.
3. *Mosaic* (butterflyweed) —Plants dwarfed with stunted, mottled, distorted leaves. Irregular, yellowish-green blotches on leaves. *Control:* Destroy infected plants when first found. Control aphids that transmit virus. Use malathion or lindane.
4. *Root Rot*—See under Geranium, and (34) Root Rot under General Diseases.
5. *Powdery Mildew*—Powdery, whitish mold on leaves. *Control:* If needed, apply Karathane or sulfur twice, 10 days apart. Start when mildew first evident.

BUTTERNUT — See Walnut
BUTTONBUSH *(Cephalanthus)*; CINCHONA; BEDSTRAW [NORTHERN, SWEET-SCENTED, WHITE, YELLOW] *(Galium)*; BLUETS, STAR VIOLET *(Houstonia)*; PARTRIDGEBERRY *(Mitchella)*; JUNGLEFLAME *(Ixora)*

1. *Leaf Spots, Leaf Blight*—Spots of various sizes, shapes, and colors. *Control:* Pick and burn affected leaves. If serious, apply zineb, maneb, ferbam, or fixed copper several times, 10 to 14 days apart. Start when leaves begin to open.
2. *Powdery Mildews* (bedstraw, buttonbush)—Widespread. White powdery coating on leaves, starting in midsummer. *Control:* Apply sulfur several times about 10 days apart.
3. *Rusts* (bedstraw, buttonbush, houstonia)—See under Bellflower, and (8) Rust under General Diseases. Alternate hosts: none or *Spartina, Distichlis, Aristida,* or *Sisyrinchium.*
4. *Downy Mildews* (bedstraw, houstonia)—Uncommon. See (6) Downy Mildew under General Diseases.
5. *Root-knot*—See (37) Root-knot under General Diseases.
6. *Black Mildew, Sooty Mold* (ixora, partridgeberry)—Unsightly black blotches on foliage. *Control:* Same as for Leaf Spots (above). Apply malathion to control insects.
7. *Root Rot*—See under Apple, and (34) Root Rot under General Diseases. Unthrifty, declining plants. May be infested with nematodes (e.g., burrowing).
8. *Thread Blight*—Southeastern States. See under Walnut.
9. *Stem Rot* (partridgeberry)—See (21) Crown Rot under General Diseases.
10. *Chlorosis*—Mineral deficiency in very acid or alkaline soil. See under Rose. *Control:* Apply mixture of iron, zinc, and manganese sulfate, 2 ounces each, in 3 gallons of water. Repeat as necessary.

BUTTON SNAKEROOT — See Chrysanthemum
BUTTONWOOD — See Sycamore
BUXUS — See Boxwood

CABBAGE, BROCCOLI, BRUSSELS SPROUTS, CAULIFLOWER, CHINESE CABBAGE, PAK-CHOI, COLLARD, KALE, FLOWERING KALE, KOHLRABI, MUSTARD [BLACK, LEAF, WHITE], RAPE, RUTABAGA, TURNIP *(Brassica)*; STONECRESS *(Aethionema)*; ALYSSUM, YELLOWTUFT, GOLDENTUFT or GOLDDUST *(Alyssum)*; ROCKCRESS, WALLCRESS *(Arabis)*; HORSERADISH *(Armoracia)*; PURPLE ROCKCRESS *(Aubrietia)*; WALLFLOWER *(Cheiranthus)*; SCURVYWEED *(Cochlearia)*; SEAKALE *(Crambe)*; TOOTHWORT, CRINKLEROOT, PEPPER-ROOT *(Dentaria)*; WHITLOWGRASS [ALLEN, SMOOTH] *(Draba)*; ERYSIMUM, WALL-FLOWER [ALPINE, SIBERIAN, WESTERN, PRAIRIE ROCKET] *(Erysimum)*; DAMESROCKET, ROCKET *(Hesperis)*; CANDYTUFT [EDGING or PERENNIAL, GIBRALTAR, GLOBE or ANNUAL, ROCKET] *(Iberis)*; GARDEN CRESS or PEPPERGRASS *(Lepidium)*; SWEET ALYSSUM *(Lobularia)*; HONESTY, PERENNIAL HONESTY *(Lunaria)*; STOCK [COMMON or TEN-WEEKS, EVENING- or NIGHT-SCENTED] *(Matthiola)*; WATERCRESS *(Nasturtium)*; RADISH *(Raphanus)*; SMELOWSKIA; DESERTPLUME *(Stanleya)*

1. *Yellows, Fusarium Wilt*—General. Most serious at high soil temperatures (over 70° F.). Leaves turn dull yellow, curl, die, and fall starting at base of plant. May show one-sided growth. Plants pale and stunted. Brown streaks inside stems. Seedlings yellow, wilt, and die. See Figure 2.31B under General Diseases. *Control:* Start disease-free seed or transplants in fertile, disease-free soil. Seed may be disinfected by treating in hot water (see under Blackleg below). Grow resistant varieties, where adapted: *Cabbage*—Allhead Select, Badger Ballhead, Badger Ballhead 14, Badger Market, Badger Shipper, Bugner, Charleston Wakefield,

Copenhagen Resistant, Early Jersey Wakefield, Empire Danish, Globe 62-M, Globe Y.R., Glory 61, Golden Acre Y.R., Improved Wisconsin All Seasons, Improved Wisconsin Ballhead, Jersey Queen, King Cole, Marion Market, Market Master, Market Topper, Polaris, Racine Market, Red Hollander, Red Yellows Resistant, Resistant Flat Dutch, Resistant Danish, Resistant Detroit, Resistant Glory, Resistant Golden Acre, Wisconsin Copenhagen, Wisconsin Golden Acre, Wisconsin Greenback Y.R., Wisconsin Hollander No. 8, Wisconsin Pride, and many more; *Cauliflower*—Early Snowball; *Broccoli*—Calabrese, Di Cicco, Early Green Sprouting, Grand Central, Midway, and Waltham 29; *Kale*—Siberian Kale; *Radish*—Red Prince. Plant radish in early spring or in the fall.

2. *Blackleg, Canker, Dry Rot*—General east of Rocky Mountains. Light brown or gray spots on stems, leaves, seed pods, and seed stalks in which black dots develop. Leaves may wilt, discolor, and die. Stem is girdled, blackens, and rots (Figure 4.42). Plants often stunted. May break over as top enlarges. Taproot often decays. *Control:* Collect and burn tops after harvest. Avoid overcrowding, overwatering, and overfertilizing—especially with nitrogen. Plant in well-drained soil. Four-year rotation or longer. Keep down cruciferous weeds. Plant disease-free, western-grown seed. If disease has been a problem in past, soak untreated *cabbage* and *Brussels sprouts* seed in hot water (exactly 122° F. for 25 minutes). For *broccoli, cauliflower, Chinese cabbage, collard, kale, kohlrabi, rape, rutabaga,* and *turnip* seed, soak at same temperature, but for 20 minutes. For *cress, mustard,* and *radish* seed soak only 15 minutes. Soak *stock* seed at 130° F. for 10 minutes. Then dry seed carefully at room temperature and dust with thiram, captan, chloranil, or Semesan before planting. See Table 20 in Appendix. Control cutworms and cabbage root maggots. Spray 10-inch strip of soil over row just after planting or transplanting. Repeat 10 days later. Use chlordane, Diazinon, or trichlorfon (Dylox). Control other insects using sevin or Dibrom (naled) plus malathion. Check with your county agent or extension entomologist for latest insect recommendations. Maintain balanced soil fertility. Treat seedbed as for Wirestem (below).

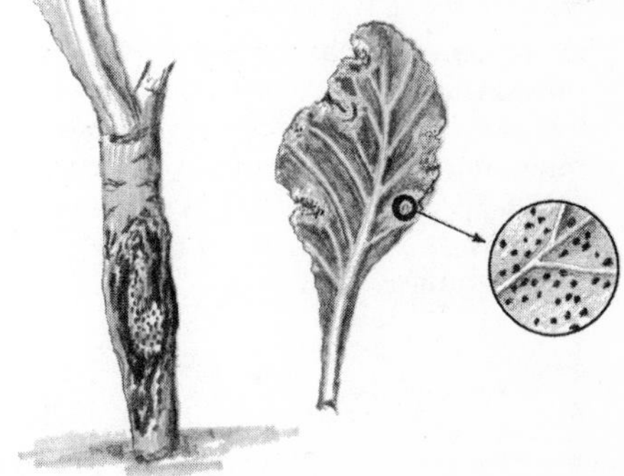

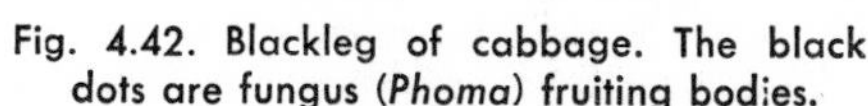

Fig. 4.42. Blackleg of cabbage. The black dots are fungus (*Phoma*) fruiting bodies.

3. *Black Rot, Bacterial Wilt, or Blight*—General in warm, moist seasons. Seedlings stunted, may turn yellow to brown, wilt, and collapse. V-shaped, yellow, brown, or dark green areas with blackened veins usually starting at leaf margin. Lower leaves of cabbage, cauliflower, and stock turn yellow or brown and drop early. One-sided growth is common. Plants and flowers are dwarfed. May rot quickly. Black or dark brown ring inside stem when cut across (Figure 4.43). *Control:* Same as for Blackleg (above). Avoid overhead irrigation whenever possible. *Cabbage* varieties differ in resistance. Resistant *mustard:* Florida Broadleaf. Resistant *kale:* Dwarf Siberian.

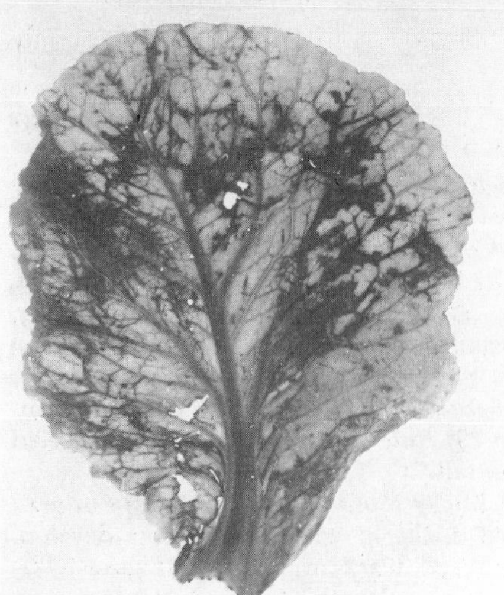

Fig. 4.43A. Cabbage black rot. Leaf showing darkened veins. (Pennsylvania State University Extension Service photo)

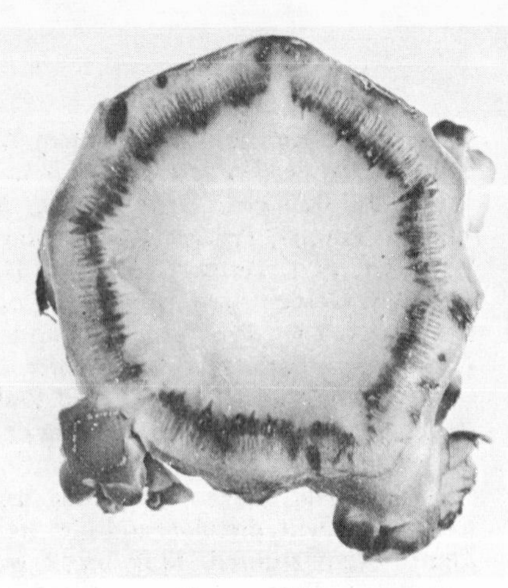

Fig. 4.43B. Cabbage black rot. Cut stem showing darkened water-conducting tissue. (Pennsylvania State University Extension Service photo)

4. *Wirestem, Seed Rot, Damping-off, Foot or Collar Rots, Rhizoctonia Disease*—General. Seeds rot. Seedlings wilt, curl, and collapse from rot at soil line. Older stems are girdled by brown or black cankers, shrivel, turn dark and woody (*wirestem*). Foliage above may wilt, wither, and die (Figure 4.44). Fine roots decay. Transplanted seedlings make slow growth or die. Most serious under cloudy, wet conditions. Dark, firm rot of base of cabbage head. Black scabby areas on *radish* root. Outer leaves wilt, darken at base. See (21) Crown Rot, and (22) Stem Blight under General Diseases. *Control:* Same as for Blackleg (above). If needed, dust seed of *radish, cress,* and *mustard* with thiram, captan, or Semesan. Avoid overcrowding, overfertilizing with nitrogen, and overwatering. Apply *one* of the following treatments to soil or around base of young plants: (a) soil drench of PCNB 75 and captan 50 (sold as Terracap, PCNB-Captan, and Orthocide Soil Treater "X"), ½ tablespoon of each per gallon, applied over 20 square feet just after planting; (b) spread ½ cup each PCNB 20 and captan 7½ to 10 per cent dust uniformly over 50 square feet and rake or rototill evenly into top 4 inches of soil before planting; (c) apply ziram or chloranil sprays in seedbed at 3- to 7-day intervals, wet both seedlings and soil; or (d) treat soil with chloropicrin, Bedrench, SMDC (Vapam or VPM), Vorlex, or Mylone 3 to 4 weeks before seeding to control diseases, insects, weeds, and nematodes. Globelle *cabbage* is normally free of Rhizoctonia Bottom Rot.

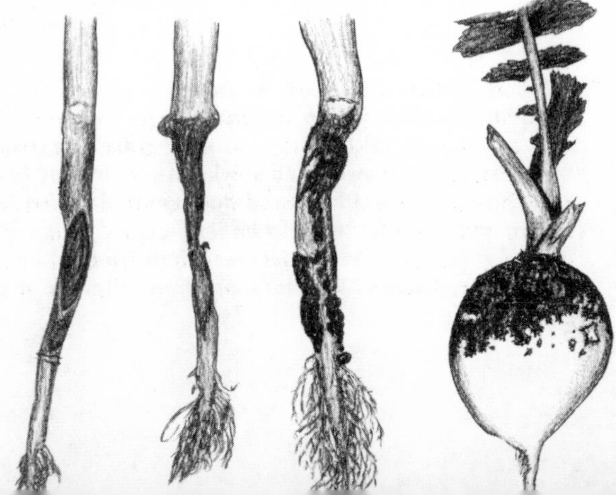

Fig. 4.44. Left, Wirestem of cabbage seedlings. Right, Rhizoctonia disease on radish root.

5. *Clubroot*—General. Yellowish leaves that wilt on hot days. Plants stunted. May die before maturing. Often fail to produce decent tops. Roots greatly enlarged and distorted with warty overgrowths or "clubs." See Figure 2.50 under General Diseases and Figure 4.45. *Control:* See (35) Clubroot under General Diseases. In addition, locate seedbed or flower bed in area where no infested soil can wash. Seedbed soil should be clean or pasteurized (pages 538–548). Apply PCNB 75 or Daconil 2787 in transplant water. Follow manufacturer's directions. Do *not* use PCNB emulsion! Apply ½ pint per plant. Drench flower bed areas. In-furrow drenches of SMDC (Vapam, VPM), 1 pint to 100 feet of row, 2 to 3 weeks before planting, give excellent Clubroot and weed control. Follow manufacturer's directions. Grow varieties resistant to one or more races, where adapted; for example, *cabbage:* Badger Shipper and two German varieties, Bindsachsener and Böhmerwald; stock *turnip:* Bruce, Dale's Hybrid, May, and Wallace; *rutabaga:* American Purple Top, Immuna II, Resistant Baugholm, and Wilhemsburger; some varieties of *garden cress* or *peppergrass, kale, radish, rape, stock,* and *wallflower* are also resistant to a number of races of the clubroot organism. Resistant *broccoli* and *cauliflower* varieties should be available soon. Keep down weeds in mustard family.

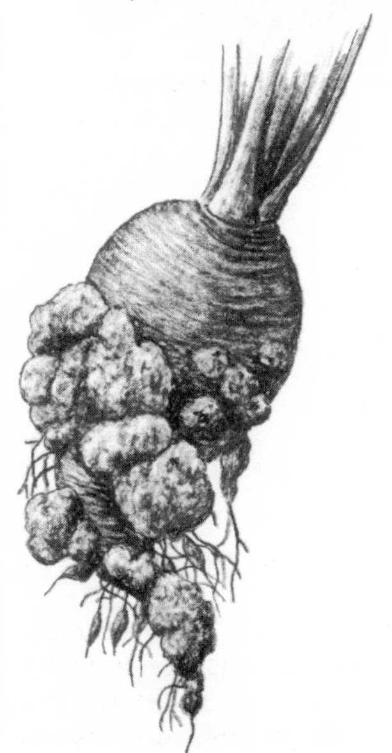

Fig. 4.45. Clubroot of turnip.

Fig. 4.46. Downy mildew of broccoli. (Courtesy Dr. H. H. Millsap)

6. *Downy Mildew*—General in cool, wet seasons. Seedling leaves appear moldy. Pale green to yellow spotting of upper leaf surface of older leaves, followed by purpling, browning, wilting, and dying of leaves. A white, bluish-white, or gray mold forms on corresponding undersurface of affected leaves. Spots also form on heads, stems, flower stalks, and flowers. *Broccoli* and *wallflower* heads may be distorted. Spots on heads or curds are brown to black (Figure 4.46). Bacterial Soft Rot may follow. Young plants may blacken and die. Irregular, brown, or bluish-black streaks may form on and in fleshy *turnip* or *radish* roots and curd stalks of *broccoli. Control:* Treat seed as for Blackleg (above). Maintain balanced soil fertility. Keep potash level up. Avoid overcrowding and sprinkling foliage. Three- to 4-year rotation. Plant in well-drained soil. Pick off and burn infected plant parts as they appear. In seedbed or field, during cool, rainy periods, apply maneb, fixed copper, Daconil 2787, or Dithane M-45, to which spreader-sticker is added, at about 5-day intervals.

Follow manufacturer's directions. Certain strains of *broccoli, cauliflower, Chinese cabbage, kale, mustard, radish, rutabaga,* and *turnip* are normally resistant. Resistant *cabbage* varieties may be available soon.

7. *Bacterial Soft Rot, Stump Rot, Wet Rot*—Cosmopolitan. Slimy, soft head, curd, stem, and root rot with foul odor. Top falls away easily leaving a slimy stump (Figure 4.47). Often follows other diseases, insects (worms and maggots), or freezing injury. *Control:* Store only dry, sound heads, curds, or roots just above freezing. Collect and burn, compost, or bury plant debris after harvest. Control insects and other diseases. See under Blackleg (above). Avoid injuries when cultivating or harvesting. *Chinese cabbage* varieties, and probably others, differ in resistance.

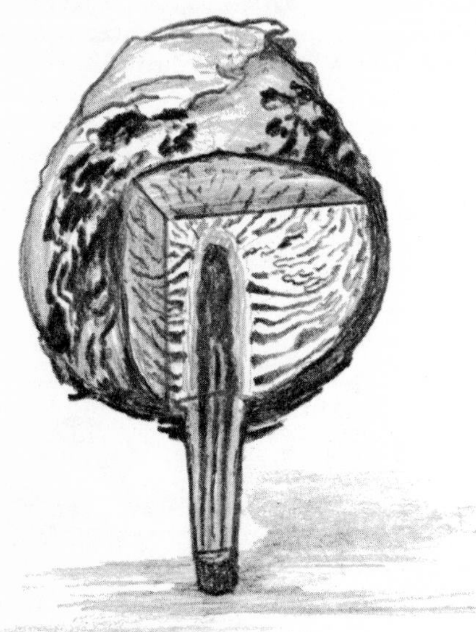

Fig. 4.47. Bacterial soft rot of cabbage.

8. *Head and Fleshy Root Rots*—Primarily storage problem. See under Carrot. Wax and store *rutabagas* and *turnips* at 32° to 40° F.
9. *Fungus Leaf Spots, Black (Alternaria) Leaf Spot, White Spot, Anthracnose*—General in wet weather. Pale yellow or white, tan, gray, brown, dark green, violet, black, or water-soaked spots on leaves, petioles, stalks, and seed pods. Certain leaf spots *(Alternaria, Mycosphaerella)* are concentrically zoned, or drop out leaving shot-holes. Leaves may wilt, shrivel, and die early. Seedlings may be killed. *Cauliflower* and *broccoli* heads are discolored. Common in seedbed, field, and in storage. *Control:* Same as for Blackleg (above). In addition, avoid injuring heads. Refrigerate heads promptly after harvest (30° to 34° F.) at low humidity. Spray in seedbed as for Downy Mildew (above). In field during rainy periods, apply zineb, thiram, difolatan, Polyram, Daconil 2787, or maneb, plus spreader-sticker, several times, 7 to 10 days apart. Southern Curled Giant *mustard* is highly resistant to Anthracnose.
10. *Bacterial Leaf Spots, Pepper Spot* (primarily broccoli, cabbage, cauliflower, Chinese cabbage, horseradish, leaf mustard, radish, rutabaga, turnip) —Widespread. Small, dark green, brown to purplish or black, water-soaked (or tan to white) spots on leaves between veins. Spots later enlarge, dry, become dark and angular. Dark spots may also occur on petioles, stems, and *cauliflower* curds. Leaves may wither and drop early. Small, sunken, brown to black spots develop in *radish* roots pack-

aged in plastic bags. *Control:* Same as for Blackleg (above). Apply fixed copper or streptomycin several times, 10 days apart. Start when spots are first evident. Follow manufacturer's directions. Add chlorine during washing and hydrocooling before packaging radish in plastic bags. Refrigerate promptly.

11. *White Mold or Blight, Drop, Cottony Rot, Watery Soft Rot, Stem or Crown Rot, Southern Blight*—General in cool, damp weather. Irregular, soft, water-soaked areas on stem and lower leaves. Leaves later wilt, often yellow, and drop. Plant collapses. Cottony mold on and in stem and head in moist weather. Head may become wet, slimy mass. Roots may decay. *Control:* Same as for Wirestem (above). Dig up and burn infected plants plus 6 inches of surrounding soil. In addition, store only dry, sound heads and roots. Handle carefully. If practical, pick off and burn fading flowers. Varieties differ in resistance.
12. *Root-knot, Root Gall, Cyst Nematodes*—Similar to Clubroot (above) but root galls smaller and usually more evenly distributed. Plants pale and stunted. *Control:* See under Bean, and (37) Root-knot under General Diseases. Adding DBCP to transplant water, ½ pint per plant, may be effective. Follow manufacturer's directions.
13. *Tipburn*—Primarily cabbage and cauliflower. Tips and margins of leaves turn pale, brown, or black and shrivel. If severe, outer leaves may die and tips of young leaves are scorched and "papery." Plants usually stunted. Head flabby and weak. *Control:* Maintain balanced soil fertility (especially ratio of phosphorus and potash) based on soil test. Early and kraut *cabbage* varieties are normally somewhat resistant (e.g., Wisconsin Copenhagen, Globe 62M, Globelle, Glory of Enkhuizen, Bonanza, Resistant Detroit, and Wisconsin Golden Acre). Danish types are commonly affected.
14. *White-rust, White Blister*—Widespread in cool, wet weather. Pale yellow to reddish-brown spots on upper leaf surface with white, powdery blisters on underside of leaves, smaller stems, seedpods, and flower parts. Affected parts may be swollen and distorted. Plants may be stunted. Normal root growth is prevented. See Figure 2.25 under General Diseases. *Control:* Destroy infected plant parts when seen and plant debris after harvest. Keep down weeds. Long rotation with plants outside cabbage or mustard family. Where serious, apply maneb, thiram, zineb, or fixed copper several times, 7 to 10 days apart. Start before White-rust normally appears. Plant disease-free *horseradish* "set" roots taken from *base* of primary root. Strains of Bohemian *horseradish* are resistant. Resistant *garden radishes:* China Rose Winter and Round Black Spanish.
15. *Boron Deficiency, Brown Heart or Rot*—Primarily broccoli, cauliflower, cabbage, radish, rutabaga, and turnip in alkaline soils. Leaves often stunted, mottled, or scorched at edges. May roll, become very brittle and deformed. Plants may be dwarfed with very narrow leaves. Stems (stalks) may be hollow and dry or show external cracks. Roots small, hollow, often "glassy," gray, brown, bronzed, or black inside. *Cauliflower* curd and *cabbage* head gradually turn brown. Affected heads and roots are bitter and tough. Affected plants are more susceptible to bacterial infection. *Control:* Have soil tested. Apply borax as recommended, about 1 ounce to 16 or 20 square yards. Avoid overliming. Fairly resistant *cabbage* varieties: Allhead Select, Wisconsin All Seasons, Wisconsin Ballhead, and Wisconsin Hollander No. 8.
16. *Gray-mold Blight, Botrytis Blight*—Water-soaked, grayish-green to brownish spots and rotting of outer leaves and stem. Young plants may wilt and die. Flowers may be brown-spotted and rotted. Affected areas may be covered with coarse gray mold. Serious storage problem. Botrytis infection of *stock* leaves commonly follows cold injury or tipburn caused by excess soluble salts in soil. *Control:* Same as for Bacterial Soft Rot (above). Space plants. Avoid overwatering and heavy nitrogen

fertilization. If practical, spray as for Downy Mildew (above). Indoors, keep humidity down and increase air circulation. Mist flowers lightly with captan, maneb, or zineb (1 tablespoon per gallon) at 3- or 4-day intervals.

17. *Mosaics, Flower Breaking*—Symptoms variable. May be masked in hot weather (above 80° F.). Leaves usually somewhat yellow, may drop prematurely, often distorted, curled, mottled, light and dark green or yellow, and crinkled. *Cabbage* leaves often show brown to black or purple flecks or spots (stippling). Plants may be stunted and bunchy. *Stock, damesrocket, sweet alyssum,* and *wallflower* flowers may show blotches or streaks. White and yellow *stock* varieties do not show flower breaking (Figure 4.48). At least one virus (Arabis mosaic) is transmitted by a dagger *(Xiphinema)* nematode. *Control:* Destroy affected plants when first found. Plow under cleanly after harvest. Keep down weeds (especially wild mustards, radish, charlock, shepherds-purse, yellow-rocket, and pennycress) in and around seedbed and garden area. Resistant or tolerant *cabbage* varieties: Badger Ballhead Y.R., Badger Market, Empire Danish, Globe, Improved Wisconsin Ballhead, Improved Wisconsin All Seasons, Racine Market, and Penn State Ballhead. *Stock* varieties also vary in resistance (e.g., Christmas Red is normally resistant to Turnip Mosaic virus). Control insects, especially aphids, cabbage worms, beetles, and grasshoppers that transmit viruses. Use malathion or Diazinon plus sevin or Dibrom (naled). Apply at about 5-day intervals. Follow manufacturer's directions. Or grow seedlings under fine screening. See under Blackleg (above).

Fig. 4.48. Mosaic or flower breaking of stock. A. Healthy. B. Diseased.

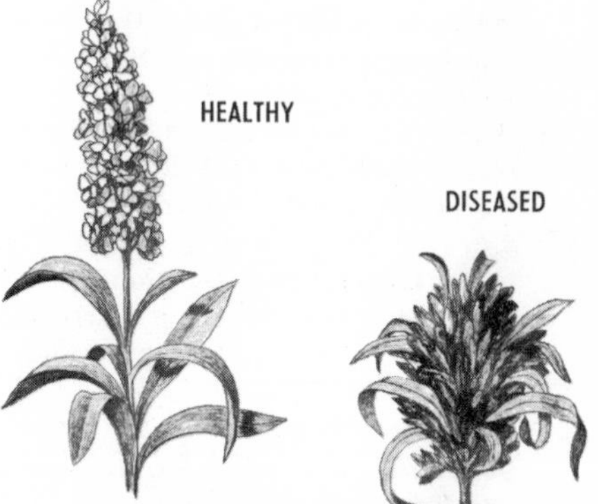

Fig. 4.49. Curly-top of stock.

18. *Black Ringspot, Ring Necrosis*—Symptoms variable. Small, yellow then black, concentric rings or spots on cabbage and older broccoli leaves. Leaves may curl, crinkle, and drop early. Small, light and dark green or yellowish, mottled areas on cauliflower, broccoli, stock, wallflower, arabis, and honesty leaves. Turnip and horseradish leaves yellowish, mottled, crinkled, and stunted. Irregular, dark green areas appear in yellowed leaves. Roots may show black flecks and streaks. *Control:* Same as for Mosaics (above). If practical, surround seedbed with screening to keep out insects.

19. *Curly-top, Brittleroot* (primarily horseradish and stock)—Western half of U.S. Outer and later inner leaves are narrow, curled, and puckered; roll inward and wilt; soon wither and die. *Stock* plants are stunted and bushy. May turn yellow, white or purple, wilt, and die in 2 or 3 weeks (Figure 4.49). When cut, *horseradish* roots are yellowish-tan with ring of black dots in center. Roots later become brown to black and brittle. *Control:* Destroy infected plants when first found. Plant virus-free horseradish roots. Plant early. Control leafhoppers that transmit the virus.

Spray weekly with sevin plus malathion. See under Blackleg (above).

20. *Aster Yellows*—See (18) Yellows under General Diseases.
21. *Scab* (primarily cabbage, radish, rape, rutabaga, and turnip)—Rough, raised, scabby areas on surface of root (Figure 4.50). *Control:* Work plenty of organic matter into soil. See under Beet and Potato.

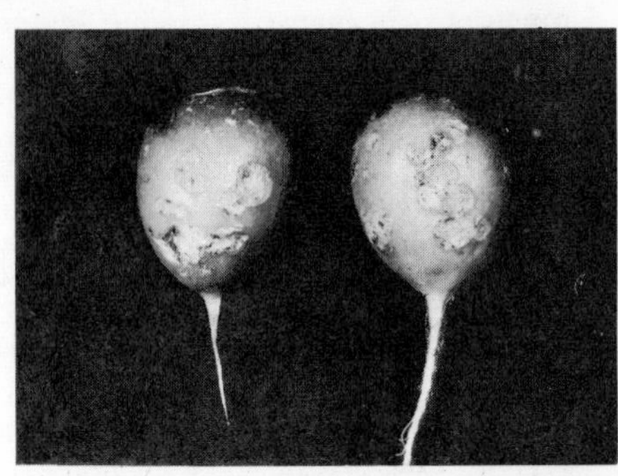

Fig. 4.50. Scab on radish roots. (Illinois Agricultural Experiment Station photo)

22. *Root Rots, Black Root*—General on radish and stock in warm soils. Seedlings or older leaves may discolor, wilt, and die. Plants may wilt gradually or suddenly collapse. Roots and crown often decay. Dark spots and blotches at base of side roots. Spots on *radish* enlarge to form metallic gray to black areas on fleshy root. Entire root system may die. Fleshy root may be distorted, constricted, and turn black. Most common on White Icicle-type radish. *Control:* Avoid heavy, wet, poorly drained soil and excessive irrigation. Treat soil with a soil fumigant (e.g., SMDC (Vapam, VPM), Mylone, or formaldehyde) before planting (pages 538–548). Grow resistant *radish* types (e.g., colored and late varieties). Avoid long or White Icicle-type radishes in infested soil. Rotate 3 or 4 years with plants outside cabbage family. Burn or plow under deeply all crop debris after harvest. Treat seed as for Blackleg (above).
23. *Whiptail, Molybdenum Deficiency*—Primarily cauliflower, broccoli, Brussels sprouts, and cabbage in very acid, heavily fertilized soils. Leaves long and narrow, ruffled, thickened, grayish-green, and very brittle. Plants stunted. If severe, head may be absent. *Control:* Have soil tested. Apply hydrated lime so soil reaction (pH) will be near neutral (pH 6 to 7). Addition of about 1 ounce of ammonium molybdate per 1,000 square feet has given good control. May apply with fertilizer, in transplant water (1 ounce to about 12 gallons of water), or by foliar sprays (page 28). Check with a local grower, your county agent, or extension horticulturist. Varieties differ considerably in resistance. Resistant *cauliflower* varieties: Holland Erfurt, Snowball X or Snowdrift, and Snowball Y.
24. *Powdery Mildew*—White, powdery mold patches on leaves and stems. If severe, leaves may be distorted, twisted, yellow to brown, and drop early. Small dark flecks may disfigure outer *cabbage* leaves. *Control:* If serious, apply sulfur or Karathane. Otherwise same as for Blackleg (above). *Cabbage* and *kale* varieties differ greatly in resistance.
25. *Crown Gall*—See under Asparagus, and (30) Crown Gall under General Diseases.
26. *Rust* (garden cress, mustards, radish, rockcress, smelowskia, stanleya, toothwort, watercress, western wallflower, and whitlowgrass)—Small, yellowish spots on leaves. Alternate host: wild grasses. *Control:* None usually necessary.
27. *Verticillium Wilt*—Plants may or may not be stunted. Lower leaves yellow, wilt, and die. Disease progresses up stem. Dark streaks occur inside stem. See (15B) Verticillium Wilt under General Diseases. *Control:* Plant in clean or sterilized soil (pages 538–548).

28. *Web Blight*—Southeastern states. See under Bean.
29. *Oedema, Intumescence*—Primarily indoor problem with broccoli, Brussels sprouts, cabbage, cauliflower, and kale. Small wartlike or ridgelike growths on underside of leaves. Corresponding upper side may be depressed. Growths become white, later turn yellow or brown and become corky. Permanent injury is rare if environmental conditions are changed. *Control:* Maintain even soil moisture supply. Increase air circulation. Avoid overwatering, forcing plants too rapidly, and copper sprays. Grow where sand-free winds will whip plants.
30. *Other Root-feeding Nematodes* (dagger, lance, lesion, naccobus, pin, reniform, rot, spear, spiral, sting, stubby-root, stunt) —Mostly in southern states. Associated with weak, stunted plants. Roots short, bushy, and die back. *Control:* Same as for Root-knot (above).
31. *Leaf and Stem Nematode* (radish) —See (20) Leaf Nematode under General Diseases.
32. *Chlorosis*—Plants stunted. Yellowish mottling or bronzing between veins of leaves. Later affected leaves may show tints of orange, red, purple, or are completely bleached. Leaf margins may be scorched (potassium or magnesium deficiency). May be due to deficiency of iron, magnesium, manganese, potassium, or zinc. *Control:* Have soil tested and follow recommendations in report.

For additional information, read USDA Agr. Handbook 144, *Diseases of Cabbage and Related Plants.*

CACTI: ANCISTROCACTUS (fishhook); ARIOCARPUS (living rock, seven-sisters); ASTROPHYTUM (star, biznaga, mitra); CARNEGIEA (saguaro or giant); CEPHALOCEREUS (old-man); CEREUS; CORYPHANTHA; ECHINOCACTUS or FEROCACTUS (barrel, star); ECHINOCEREUS; ECHINOMASTUS; ECHINOPSIS (sea-urchin); EPIPHYLLUM (crab, orchid); EPITHELANTHA; ESCOBARIA; HAMATOCACTUS (hook); HOMALOCEPHALA (manco caballo); LEMAIREOCEREUS; LOPHOPHORA; LYLOCEREUS (night-blooming); MAMMILLARIA (pincushion); MELOCACTUS (Turk's-head); NYCTOCEREUS (night-blooming); OPUNTIA (pricklypear or pad, cholla, Indian fig, tuna, nopal, tasajillo); PACHYCEREUS (hairbrush, organ-pipe); PEDIOCACTUS (snowball); PELECYPHORA (hatchet); ROSEOCACTUS; SCHLUMBERGERA (Easter, Christmas); SCLEROCACTUS; SELENICEREUS (night-blooming); STROMBOCACTUS; THELOCACTUS; TRICHOCEREUS (night-blooming); ZYGOCACTUS (crab or Christmas)

1. *Corky Scab*—Pale, yellowish-green spots on stems and shoots that often become irregular corky or rusty areas and may become sunken. Spots may remain smooth and grayish-white. Shoots may die. *Control:* Avoid overwatering, overcrowding, and applying too much fertilizer at one time. Avoid low potassium and high sodium content in soil; keep calcium level high. Plant in well-drained sandy soil (page 22). Increase light and air circulation. Decrease air humidity.
2. *Stem and Root Rots, Cutting Rots, Wilts, Anthracnose, Cladode Rot, Seedling Blight*—Cuttings, stems, and branches discolored (yellow, light to dark green, brown, or black) or spotted. Gradually or suddenly wilt and rot. May become slimy and collapse (Bacterial Soft Rot, Fusarium). Gray or black mold may grow on affected tissues. Roots and crowns decay. May be associated with nematodes (aphelenchoides, cyst, lance, lesion, pin, root-knot, spiral, stunt). *Control:* Cut out and destroy infected plant parts. Keep air humidity down. Sterilize seeds and disease-free cuttings by dipping in normal Semesan solution for 5 minutes before planting. Grow in sandy, sterilized soil (pages 538–548). For cacti potting mixture

see page 22. Avoid overwatering and wounding plants. Captan sprays at 10- to 14-day intervals may be beneficial (Helminthosporium Stem Rot). Keep water off aboveground parts.

3. *Bud Drop*—Buds fall early, especially on Christmas cacti. Plants may be stunted. *Control:* Fertilize adequately. Maintain uniform soil moisture. Avoid large temperature changes, cold drafts, and cold water on foliage.
4. *Glassiness*—Dark green, somewhat transparent spots that finally turn black. Shoots may die back. Otherwise same as for Corky Scab (above).
5. *Scorch, "Sunscald"*—Segments turn reddish-brown and die. Young spots are zoned with grayish-brown, cracked centers. *Control:* Cut out affected parts. Avoid high temperatures and too much sun. Otherwise same as for Bud Drop (above).
6. *Black Mildew*—Florida. See (12) Sooty Mold under General Diseases.
7. *Root-knot, Cyst Nematode*—See (37) Root-knot under General Diseases.
8. *Leaf Spot, Zonate Spot, Leaf Scorch*—See (1) Fungus Leaf Spot under General Diseases.
9. *Gray-mold Blight*—See (5) Botrytis Blight under General Diseases.
10. *Crown Gall* (mostly outdoor plants)—Large, spongy, cancer-like growths on stem or roots. *Control:* Cut out masses. Sterilize tools between cuts.
11. *Bacterial Blight*—See (2) Bacterial Leaf Spot under General Diseases.
12. *Scab*—See (14) Scab under General Diseases.
13. *Bacterial Necrosis* (saguaro, cholla, pricklypear, barrel, organ-pipe)—Most common on older plants. Dark, soft, enlarging spots that become purplish-black, split open, and "leak." Develop on trunk or branches. The soft areas may dry up and crack with diseased internal tissues being dark and dry. *Control:* Carefully remove all decayed tissue and ½ inch of healthy tissue beyond the rot pocket. Wash area with household bleach diluted 9 times with water. Let the pocket stand "open."

CAESALPINIA — See Honeylocust
CALABASH — See Cucumber
CALADIUM — See Calla Lily
CALAMONDIN — See Citrus
CALANTHE — See Orchids
CALATHEA — See Maranta
CALCEOLARIA — See Snapdragon
CALENDULA — See Chrysanthemum
CALICO-FLOWER — See Aristolochia
CALIFORNIA-BLUEBELL — See Phacelia
CALIFORNIA FREMONTIA — See Phoenix-tree
CALIFORNIA FUCHSIA — See Evening-primrose
CALIFORNIA-HYACINTH — See Brodiaea
CALIFORNIA-LAUREL — See Avocado
CALIFORNIA-PITCHERPLANT — See Pitcherplant
CALIFORNIA-POPPY — See Poppy
CALIFORNIA-ROSE — See Morning-glory
CALIFORNIA SWEETSHRUB — See Calycanthus

CALLA LILY [BABY WHITE, COMMON, DWARF PINK or ROSE, GOLDEN or YELLOW, RED or PINK, SPOTTED], ARUM-LILY *(Zantedeschia)*; CALLA; CHINESE EVERGREEN *(Aglaonema)*; ANTHURIUM, FLAMINGO-FLOWER, BEGONIA-LEAF *(Anthurium)*; DRAGONROOT, JACK-IN-THE-PULPIT *(Arisaema)*; CALADIUM; DASHEEN, ELEPHANTS-EAR *(Colocasia)*;

DIEFFENBACHIA, DUMBCANE *(Dieffenbachia)*; HOMALOMENA; CERIMAN *(Monstera)*; NEPHTHYTIS or SYNGONIUM; PHILODENDRON (many species and horticultural varieties); POTHOS or IVY-ARUM, DEVIL'S IVY, MARBLE QUEEN *(Scindapsus)*; YAUTIA, MALANGA *(Xanthosoma)*

1. *Bacterial Soft Rot, Leafstalk Rot* (caladium, calla lily, Chinese evergreen, dasheen, dieffenbachia, xanthosoma)—Common and destructive. Slimy, wet, often foul-smelling rot of stem, leaf stalks, flower stalks, and underground parts. Dark, water-soaked, and spreading spots on leaves. Leaves and flower stalks may suddenly wilt, turn yellow or brown, collapse, and die. *Control:* Destroy infected plants and rotted corms, rhizomes, or tubers. Rotate. Plant dormant, disease-free *calla lily* corms or rhizomes soaked 30 to 60 minutes in a 1:1,000 solution of mercuric chloride or formalin (1 to 50 dilution with water for 24 hours). Soak *dieffenbachia* canes in streptomycin (200 parts per million) for 15 minutes. Drain and plant immediately in sterilized soil in sterilized containers. Commercial growers dip 2-foot sections of hardened *dieffenbachia* canes (1 to 1½ inches in diameter) in hot water (120° F.) for 40 to 60 minutes, cool, and place in sterilized sphagnum moss until new growth starts. Then cut into pieces, each with a single bud, and plant in sterilized soil mix (page 538). Replant in soil when top is well started. Plant shallow in light, well-drained but moist, rich soil, sterilized if possible (pages 538–548), or where disease has not been present in past. Keep starting temperature above 70° F. and below 90° F. Avoid injuries to plants, heavy shade, poor air circulation, crowding plants, overwatering, and excessive nitrogen fertilization. Spraying every 4 to 5 days with streptomycin, starting when first symptoms appear, may be beneficial. Keep humidity down. Propagate only from disease-free plants, disinfect tools by dipping in household bleach or formalin solution (1 to 50 dilution with water). Or start *calla lily* from seed.
2. *Tuber, Corm, Stem (Cane), Root and Rhizome Rots, Cutting Rots, Southern Blight*—Plants often stunted; leaves turn yellowish, later wither and die. Plants may not blossom or flowers may be deformed and decayed. Underground plant parts, or stem at and near soil line, may be discolored and rot. Plants collapse or are easily pulled up (Figure 4.51). *Control:* Same as for Bacterial Soft Rot

Fig. 4.51. Root rot of calla.

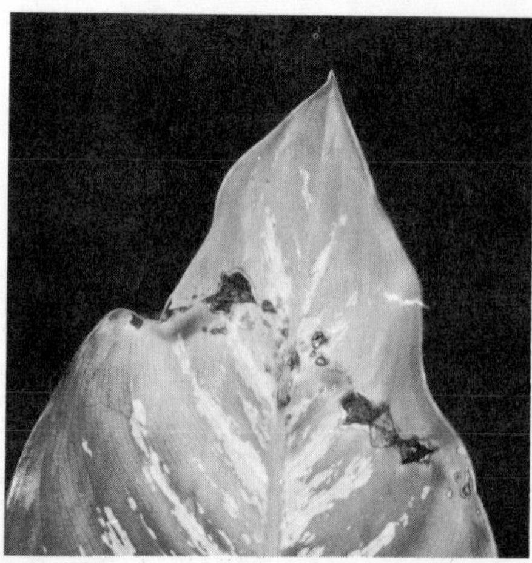

Fig. 4.52. Bacterial leaf spot of dieffenbachia. The young leaf was infected near the tip before it unrolled. (Illinois Natural History Survey photo)

(above). Soak dormant *caladium* tubers in hot water (122° F.) for 30 minutes, and dormant *calla lily* corms or rhizomes at same temperature but for 60 minutes. Cool, dry, and plant in clean soil. Treat *philodendron* canes as for Bacterial Leaf and Stem Rot (below). Dip hardened, 2-foot long cane pieces or bare-root *nephthytis (Syngonium)* and old *Chinese evergreen* cane pieces in hot water (120° F.) for 30 minutes. Cool and plant in sterilized soil. Or take tip cuttings from *Chinese evergreen*. Drench soil with mixture of PCNB 75 (1 tablespoon per gallon of water) and ferbam 76 per cent (1½ tablespoons per gallon), or Dexon. Follow manufacturer's directions. Use 1 pint per square foot. Store *dasheen* tubers at 50° F. with ventilation.

3. *Root-knot and Other Root-feeding Nematodes* (aphelenchoides, burrowing, lesion, spiral, stem, stubby-root)—Internal discolored spots may be evident in *caladium* tubers. Plants may be sickly and stunted, gradually decline in vigor due to stubby, discolored roots. *Control:* Nurserymen soak bare-root *philodendron* and *Chinese evergreen* roots in hot water (122° F.) for 10 minutes before planting. Or soak in Zinophos for 30 minutes, following manufacturer's directions. Soak dormant *caladium* tubers as for Tuber Rot (above). DBCP may be added to soil of potted plants. Follow manufacturer's directions.
4. *Spotted Wilt* (calla lily)—Numerous, whitish-yellow flecks, spots, streaks, and even zoned rings on leaves, flower stalks, and flower buds. Leaf spots may later turn brown. Leaves and flowers may be twisted, crinkled, and deformed. Pale greenish blotches and streaks form on white flowers and green buds. *Control:* Destroy infected plants when first found. Plant disease-free nursery stock. Control thrips with frequent DDT, sevin, or malathion sprays. Keep down weeds.
5. *Mosaics*—Leaves may be curled, show a yellowish mottle. Plants stunted. *Control:* Same as for Spotted Wilt (above). Control virus-carrying aphids with malathion sprays.
6. *Leaf Spots, Anthracnose, Leaf and Flower Blight*—Small to large, round to irregular, spots of various colors on leaves, leaf or flower stalks, and flowers. Spots may enlarge, and merge to form irregular blotches. Certain spots have border of a different color, or concentric ring patterns. Severely infected leaves may wilt, wither, and die prematurely. *Control:* Pick off and destroy severely spotted leaves. Keep down weeds. Indoors, avoid wetting foliage, overcrowding, heavy shade, overwatering, and chilling. Keep temperature and humidity as low as practical. Control insects with mixture of DDT or sevin and malathion. Apply ferbam, zineb, ziram, thiram, maneb, or captan sprays, plus spreader-sticker, during rainy periods.
7. *Bacterial Leaf Spot or Rot of Dieffenbachia and Nephthytis*—Small, round to irregular, yellow to yellowish-orange, tan, or reddish-brown spots on leaves (Figure 4.52). Centers of spots may be dull and watery-green. In wet weather the spots often enlarge and merge. Spots may become dry, brittle, and tear away. In dry weather the spots remain small, dry, reddish-brown specks. Leaves may turn yellow, wilt, and die. Irregular, brownish, water-soaked, sunken, soft dead areas may form on stem. *Control:* Pick and burn severely spotted leaves. Space plants. Keep water off foliage. Lower air temperature and avoid heavy shade. Spraying with streptomycin may be beneficial. Soak *dieffenbachia* canes as for Bacterial Soft Rot (above).
8. *Bacterial Leaf and Stem Rots of Philodendron*—Small, irregular, water-soaked spots on leaves and sometimes stem. Spots enlarge rapidly during warm, moist weather. Leaves and leaf stalks may rot, collapse, and become mushy. Or spots may become dry, yellow to tan, with yellow margins. *Control:* Remove and burn infected plant parts when first found. Destroy dilapidated plants. Scrub and thoroughly dry containers before reusing. Plant cuttings from disease-free plants in sterilized soil (pages 538–548). Where practical, spray at 4- to 10-day intervals, using streptomycin (200 parts per million) or use Agrimycin 100 following manufacturer's directions.

Commercial growers control Stem Rot (also *Rhizoctonia*) by soaking propagating canes for 30 minutes in hot water (120° F.). Canes are then cooled and rooted in sterile sphagnum moss. Avoid overcrowding and sprinkling foliage.

9. *Bacterial Leaf Spot of Chinese Evergreen*—Florida. Small, diffused, water-soaked spots with yellowish "halo." Infected plants appear blighted and gradually decline in vigor. Where severe, entire leaves and petioles may suddenly collapse with soft, slimy, foul-smelling rot (Bacterial Soft Rot). *Control:* Same as for Bacterial Soft Rot and Bacterial Leaf Spot of Dieffenbachia (both above). Certain species of *Aglaonema* are resistant.
10. *Philodendron Leaf Yellowing, Dieback*—Indoor problem. Leaves may be stunted, yellow, and drop early. Shoot tips may die back. *Control:* Increase light and humidity. Avoid overfertilizing, overwatering, and growing in heavy, poorly drained soil. Repot plants if needed.
11. *Physiological Leaf Spot of Philodendron*—Small, scattered, pale yellow spots form during hot periods of rapid plant growth. Droplets of sap often covered with sooty mold. *Control:* Prevent extreme temperature fluctuations during periods of rapid growth. If sooty molds are a problem, apply captan.
12. *Rust* (Jack-in-the-pulpit)—Small, lemon-yellow pustules on leaves and spathe. Foliage yellows and dies. Plants do not flower. *Control:* Pull up and burn infected plants. The rust fungus is perennial.
13. *Sooty Mold*—See (12) Sooty Mold under General Diseases.
14. *Downy Mildew, Gray-mold Blight* (caladium, calla lily, calla, Jack-in-the-pulpit)—See (5) Botrytis Blight, and (6) Downy Mildew under General Diseases.

CALLIANDRA, FALSE-MESQUITE, POWDER-PUFF TREE *(Calliandra)*

1. *Root Rot*—See (34) Root Rot under General Diseases.
2. *Rust* (false-mesquite)—Arizona. See (8) Rust under General Diseases.

CALLICARPA — See Lantana
CALLIRHOË — See Hollyhock
CALLISTEPHUS — See Chrysanthemum
CALLUNA — See Heath
CALOCHORTUS — See Mariposa Lily
CALONYCTION — See Morning-glory
CALYCANTHUS, CAROLINA ALLSPICE or STRAWBERRY-SHRUB, MOUNTAIN SPICEWOOD, CALIFORNIA SWEETSHRUB or SPICEBUSH *(Calycanthus)*

1. *Twig and Branch Canker*—See under Maple, Chestnut, and (22) Stem Blight under General Diseases. Prune and burn infected parts.
2. *Crown Gall*—See under Apple, and (30) Crown Gall under General Diseases.
3. *Powdery Mildew*—See (7) Powdery Mildew under General Diseases.

CAMASS *(Camassia)* — See Colchicum
CAMELLIA [COMMON or JAPONICA, RETICULATA, SASANQUA] *(Camellia)*

1. *Flower Blight*—Serious. Widespread in southern states and along Pacific Coast. Numerous, small, enlarging, tan to brown specks or spots on flowers. Veins are usually darker brown than surrounding petal tissue. Whole flower soon turns dull brown, withers, and drops (Figure 4.53). Large, hard, brown to black bodies (sclerotia) form in center of old flowers. *Control:* Best done on a community-wide basis. During *early* winter apply PCNB dust or spray to soil surface, *after* removal

Fig. 4.53. Camellia flower blight.

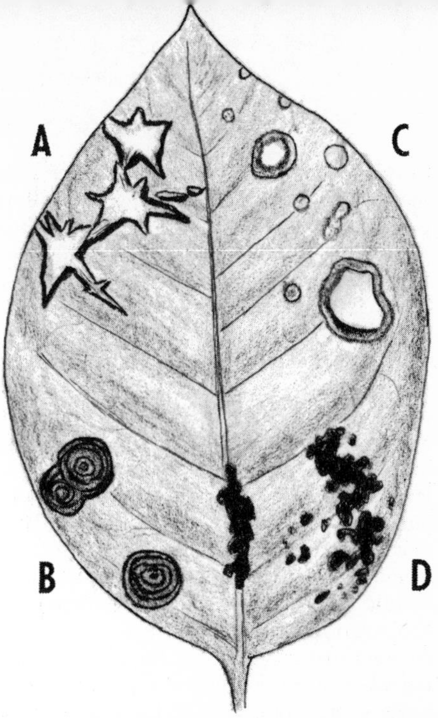

Fig. 4.54. Camellia leaf spots. A. Angular spot. B. Concentric spot. C. White spot. D. Black spot. All 4 types of spots would never be found on the same leaf.

and destruction of leaf litter, beneath plants and area 10 feet beyond. Follow manufacturer's recommendations (1 cup of PCNB 75 per cent wettable per 100 square feet, plus 1 teaspoonful of household detergent; or 2.5 pounds of PCNB 20 per cent dust per 100 square feet). Repeat one month later. Remove and burn *all* fading flowers as soon as disease is evident. During bloom apply zineb or Thylate (1 tablespoon per gallon), folpet, Acti-dione-Thiram, or captan at 3-day intervals if period is rainy. Also spray soil beneath plants. Buy only certified, disease-free, bare-rooted plants. Before planting remove and burn all flower buds showing color. Grow in soil where camellias have not previously been planted.

2. *Dieback, Cankers, Damping-off, Graft Blight*—Widespread and serious in southern states and California, especially of cuttings and young plants. Foliage wilts, turns dull green, and dies from slightly sunken, girdling, sometimes blackened and dead cankers on twigs, branches, or graft union. Affected parts turn brown and die back. Cuttings wilt, drop their leaves, and die from decay near soil line. *Control:* Cut out and burn diseased stems several inches below canker. Make flush cuts just below vigorous side branches. Cut out cankers on larger branches. If pruning cuts are over ½ inch in diameter, treat with a tree wound dressing (page 36). For propagation, use only healthy plants. Take cuttings high on the plant. Dip scions, cuttings, stocks, and grafting tools in ferbam or captan solution (3 tablespoons per gallon of water). Avoid wounding stems, overcrowding, overwatering, overfertilization, and too high a humidity. Keep plants vigorous. Apply adequate amounts of fertilizer, based on a soil test. Captan, zineb, thiram, maneb, or fixed copper sprays applied just before wet periods should prevent infections. Camellia varieties resistant to Dieback (*Glomerella*): Governor Moulton and Professor Sargent. Grow plants in sterilized soil (pages 538–548). Check with your nurseryman, extension horticulturist, or plant pathologist.

3. *Leaf Spots, Leaf Blight or Blotch, Spot Anthracnose or Scab*—Widespread outdoors

in rainy seasons. Most common on plants weakened by freezing, sunscald, insects, wounding, etc. Small to large, round to irregular, yellow, brown, black, gray, purplish or silvery spots on leaves. Often with distinct margin or sprinkled with black dots. Spots may enlarge and merge to form blotches (Figure 4.54). Infected leaves may drop early. Twigs may die back. *Control:* Collect and burn infected leaves. Prune and burn blighted twigs. Apply same fungicides as for Dieback (above). Maintain steady, even growth with good root system. Provide light shade or windbreak. Avoid overcrowding, overwatering, and too high a humidity. Keep soil acid (pH 5.0 to 5.5). Protect plants from weather (cold, sun, wind) or mechanical injury.

4. *Sunscald*—Primarily an outdoor problem. Silvery to yellowish, faded green, bronzed, or brown areas with irregular margins form in the centers of exposed leaves. Affected areas may be attacked by secondary fungi, change color, and drop out. *Control:* Provide light shade and protect against strong winds. Mulching may also help. Varieties differ in resistance.
5. *Bud Drop*—Widespread. Buds turn dark and drop. Due to unfavorable growing conditions (e.g., spring or fall frost, severe winter freezing, high or fluctuating temperature, cold drafts, heavy fertilizer application or malnutrition, bud mites, root rot, irregular water supply, and low air humidity). *Control:* Avoid overwatering when buds are forming. Keep soil moisture as uniform as possible; or plant in well-drained soil. Keep plants well supplied with nutrients. Avoid great fluctuations in temperature and soil moisture, cold drafts, low light, and excessive nitrogen fertilizer. Maintain air humidity over 50 per cent (page 39). Repot only when roots are somewhat pot-bound. Cover outdoor plants to protect against sharp frosts or freezes. Harden off plants by avoiding late watering and fertilizing. Mulch plants to maintain as uniform soil temperature and moisture as possible. Grow resistant varieties.
6. *Bud Rot, Bud Blight, Botrytis Flower Blight*—General. White or brown spots on flower petals in cool, humid weather. Buds and flowers rot. Often covered with dense gray mold in wet weather. Commonly follows frost injury. *Control:* Increase air circulation and decrease humidity. Provide slightly warmer temperature, water in morning, and space plants. Captan, folpet, Thylate, or Acti-dione-Thiram sprays are beneficial.
7. *Black Mold or Mildew, Sooty Mold*—Black, moldy patches on leaves and twigs. *Control:* Spray with malathion or lindane to control white flies, mealybugs, scales, aphids, and other insects. Do *not* use DDT on camellias—it is injurious to them.
8. *Leaf, Bud and Stem Galls, Leaf Curl*—Southeastern states. Buds and leaves enlarged, thickened, distorted, and discolored light green, white, or even nearly reddish. May be covered with whitish "bloom" on underside that cracks and peels. Stems of new shoots may be thickened (Figure 4.55). *Control:* Pick and burn affected parts *before* they turn white. Spray as for Dieback (above).
9. *Infectious Leaf and Flower Variegation, Yellow Mottle Leaf*—Symptoms differ greatly depending on virus strain and variety. Certain varieties develop round to *irregular* yellow specks, spots, and large blotches or yellow mottling on leaves. Certain colored varieties often show white specks, spots, or splotches on petals. White flowers show no symptoms. Plants off color. May die back. Plants gradually lose vigor. *Control:* Destroy suspected virus-infected plants or at least separate from healthy plants. Propagate only from virus-free, nonvariegated plants. Applications of iron chelate (page 27), ½ teaspoonful per plant in a 3-gallon container, every 6 to 8 weeks for 6 months reduces virus symptoms on both leaves and flowers. Greening effect may last 6 to 18 months. Iron sulfate may be substituted on acid soils. If plants are severely virus-infected, apply only small amounts of nitrogen.
10. *Chlorosis*—May be caused by soil deficiency. Areas between veins on leaves turn

Fig. 4.55. Camellia leaf and stem gall. (Courtesy Dr. V. H. Young)

yellow (iron, zinc, or magnesium deficiency) or dead spots occur (manganese deficiency). General yellowing or paling of leaves (nitrogen deficiency). Leaves curled. Plants grow slowly. *Control:* Fertilize adequately and regularly, based on soil test. See also Leaf and Flower Variegation (above). Soil should be kept acid (pH 5.0 to 5.5) by addition of sulfur or acid fertilizer. May also apply two tablespoons each of iron, magnesium, zinc, and manganese sulfate in 3 gallons of water plus 3 ounces of hydrated lime and a teaspoon of detergent. Or mix 4 parts of magnesium sulfate, 1 part manganese sulfate. 1 part iron sulfate, and 1 part sulfur. Use 1 ounce or 2 teaspoons of mixture per 3- to 4-foot plant or 1 to 1½ pounds for 100 square feet. You can also buy special Camellia mixes.

11. *Root Rots*—Roots decay causing foliage on one or more branches to wilt, turn yellow, wither, and die. Plants may be stunted and die back. Often associated with nematodes (e.g., burrowing, dagger, lance, lesion, needle, pin, reniform, ring, root-knot, sheath, sheathoid, spiral, stubby-root, stunt). Varieties differ in resistance. *Control:* Grow healthy stock in clean, disinfested soil that is acid, loose, fertile, and high in organic matter. *Phytophthora Root Rot* can be controlled by several soil drenches of Dexon (1 teaspoonful in 4 gal. of water), applied at 10- to 14-day intervals. Dig up and burn diseased plants. Avoid replanting in same area without disinfesting soil (pages 538–548).
12. *Root-knot*—See (37) Root-knot under General Diseases. *Control:* Same as for Root Rots (above). DBCP and VC-13 may be used as post-planting treatments. Follow manufacturer's directions.
13. *Crown Gall*—See (30) Crown Gall under General Diseases.
14. *Oedema, Corky Scab, Scurf*—Primarily an indoor problem. White or black, rough, corky swellings, mostly on undersides of leaves. *Control:* Maintain uniform soil moisture. Avoid overwatering during cloudy, humid weather. Avoid overfertilizing and excessive sunlight.
15. *Leaf Scorch*—Leaves appear scorched, especially at margins. May be caused by sunscald, cold weather, lack of water or fertilizer, high concentration of soluble salts, too much fertilizer, or deep planting. Any of these conditions may also cause plants not to bloom. *Control:* Plant in loose, acid soil. Avoid overfertilization. Leach excess salts with an occasional heavy irrigation.
16. *Algal Leaf Spot*—Southeastern states in damp, shady locations. Round, slightly raised, gray to orange spots, up to ½ inch or more in diameter, on mature foliage.

Control: Apply copper sprays at about 2-week intervals, as new growth is maturing, but before summer rainy season starts.

For additional information read USDA H & G Bulletin 86, *Growing Camellias.*

CAMOMILE — See Chrysanthemum

CAMPANULA — See Bellflower

CAMPHOR-TREE — See Avocado

CAMPION — See Carnation

CAMPSIS — See Trumpetvine

CAMPTOSORUS — See Ferns

CANARYBIRDFLOWER — See Nasturtium

CANARYGRASS — See Ribbongrass

CANAVALIA — See Bean

CANBY — See Bittersweet

CANDLEBERRY — See Waxmyrtle

CANDLES OF THE LORD — See Yucca

CANDYTUFT — See Cabbage

CANNA [EDIBLE and GARDEN], INDIAN SHOT *(Canna)*

1. *Bacterial Bud Rot*—Widespread on young plants early in season. Flower buds and stalks may blacken and rot. Irregular, yellowish to brown, water-soaked streaks or spots may appear on older leaves. Irregular, thin, expanding streaks develop along leaves. Areas white at first, then grayish-brown and finally black. Leaves appear ragged, spotted, or striped. May be distorted. Gummy sap may exude from blackened areas on stalks. Flowers often ruined. See Figure 4.56. *Control:* Discard badly diseased plants. Soak *dormant,* healthy-appearing rootstocks or tubers 2 hours in a 1:1,000 solution of mercuric chloride (page 101 for precautions). Avoid overcrowding, overwatering, and sprinkling foliage. Increase air circulation. Destroy infected buds early. Streptomycin bud and leaf sprays may be beneficial.

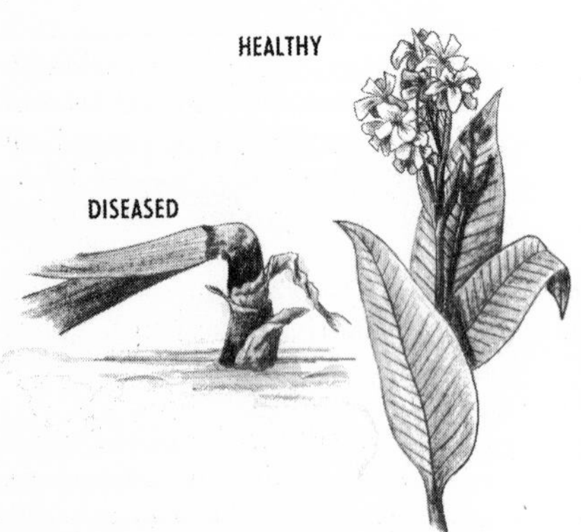

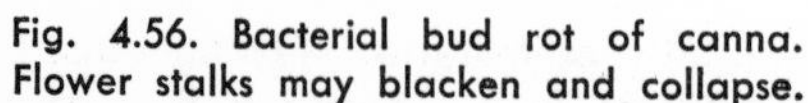

Fig. 4.56. Bacterial bud rot of canna. Flower stalks may blacken and collapse.

Fig. 4.57. Fusarium wilt of carnation in a greenhouse bench.

2. *Mosaic*—Irregular, light and dark green areas, or pale yellow stripes running outward from center of leaf to margins. Areas may later turn rusty-brown. Leaves somewhat wrinkled and curled. Stems and flower parts often show yellowish bands. Plants may be stunted and flower late. *Control:* Destroy infected plants as they will

not recover. Keep down weeds. Control aphids that transmit the virus, using malathion. The President variety is apparently immune.

3. *Yellows*—Plants dwarfed with young leaves developing an irregular, diffuse, dull yellowing that turns bronze-colored with age. Young leaves may wither and die back. *Control:* Same as for Mosaic (above). Spray to control leafhoppers that transmit the virus.
4. *Rust*—Yellowish to black powdery pustules on lower leaf surface. *Control:* Generally not necessary. If severe, apply zineb.
5. *Tuber or Rhizome Rot, Crown Rot, Southern Blight*—See under Calla Lily. May be associated with nematodes (e.g., burrowing).
6. *Bacterial Wilt*—See under Tomato, and (15C) Bacterial Wilt under General Diseases.
7. *Leaf Spot*—Small spots on leaves. *Control:* Same as for Rust (above).
8. *Petal Blight*—Irregular, light brown spots on petals that merge and kill large areas. *Control:* Same as for Rust. Destroy faded flowers promptly.

CANTALOUP — See Cucumber

CANTERBURY-BELLS — See Bellflower

CAPE COAST LILY — See Daffodil

CAPE-COWSLIP — See Tulip

CAPE-HONEYSUCKLE — See Trumpettree

CAPE-JASMINE — See Gardenia

CAPE-MARIGOLD — See Chrysanthemum

CAPSICUM — See Tomato

CARAGANA — See Honeylocust

CARAWAY — See Celery

CARDINAL CLIMBER — See Morning-glory

CARDINALFLOWER — See Lobelia

CARDOON — See Lettuce

CARISSA — See Oleander

CAROLINA-JESSAMINE — See Butterflybush

CARNATION [FLORIST'S, HARDY], GARDEN PINKS [ALPINE, ALWOOD HYBRIDS, CHEDDAR, CHINA, CLUSTERHEAD, COTTAGE, GRASS, MAIDEN, RAINBOW, SUPERB], SWEET-WILLIAM *(Dianthus)*; CORNCOCKLE *(Agrostemma)*; SANDWORT [CORSICAN, MOUNTAIN] *(Arenaria)*; BABYSBREATH *(Gypsophila)*; FLOWER-OF-JOVE, MALTESE CROSS or SCARLET LYCHNIS or JERUSALEM-CROSS, EVENING CAMPION, RED and ROSE CAMPION, RAGGED-ROBIN or CUCKOO-FLOWER, MULLEIN-PINK, GERMAN CATCHFLY, ROSE-OF-HEAVEN *(Lychnis)*; CUSHION-PINK, FIRE-PINK, WILD PINK, CAMPION [MOSS, SEA, STARRY], CATCHFLY [ALPINE, ROYAL, SWEET-WILLIAM] *(Silene)*

1. *Fusarium Wilts, Yellows* (carnation, pinks, sweet-william)—General and serious at soil temperature of 75° F. and above. Plants become grayish-green, wilt, turn greenish-gray, then yellow, and die gradually (Figure 4.57). Young shoots often yellowed and stunted. Plants often one-sided. Inside of lower stem may show dark brown to red streaks when split. Roots are generally healthy. Stems softened; may become dry and shredded. Root-feeding nematodes (e.g., lesion, root-knot) may increase severity of wilt. *Control:* Take tip cuttings only from known disease-free plants. (The Fusarium fungus may be carried in symptomless cuttings.) Cultured *carnation* cuttings are available free of Fusarium Wilt and other diseases. Plant in sterilized soil, using disease-free seed or cuttings. Dig and burn infected plants

when found. Indoors, keep temperature (50° to 60° F.) and soluble salts low. Avoid overwatering, deep planting, overfertilizing with nitrogen, and injuring plants. Varieties differ in resistance. Biweekly applications of oxyquinoline sulfate (1:4,000) for 3 weeks, starting when rooted cuttings are flatted or benched, is beneficial.

2. *Bacterial Wilts* (carnation) —Plants may be stunted, especially when young. In hot weather tops of infected plants suddenly or gradually wilt. Are grayish-green, then yellowish and finally straw-colored. Inside of cut stems and roots may show yellow to brown, sticky streaks and ooze. Occasionally, in cool weather, elongated, discolored stripes form on stems that split open. Roots rotted and sticky. Affected plants are easily pulled up. Root-feeding nematodes (e.g., root-knot, spiral) increase severity of wilt. *Control:* Same as for Fusarium Wilt (above). Keep potassium and calcium levels high and phosphorus level low. Keep foliage as dry as possible. Varieties differ in resistance.

3. *Phialophora or Verticillium Wilt* (carnation) —Uncommon. Occurs at lower temperatures (below 80° F.) than Fusarium Wilt. Symptoms much like Fusarium Wilt. Gradual wilting of plants; not one-sided as Fusarium Wilt may be. Infected tissues turn brown. Stems more or less solid. *Control:* Same as for Fusarium Wilt (above).

4. *Alternaria Blight, Collar and Branch Rot* (campion, carnation, Maltese cross, pinks, sweet-william) —General. Tiny purple spots on leaves, flower petals, and stems that enlarge to form ash-gray to grayish-brown, dead, sunken areas. Spots later become dark brown or black with yellow-green or purple margin. Base of leaves and branches may rot, killing parts beyond. Rotted stems firm and dark brown. Most common on lower leaves and branches. Infected cuttings die from water-soaked, dark green to black rot. The disease may appear first on one side of a plant. Shoots may be dwarfed, later turn yellow, wilt, and die. *Control:* Rotate. Destroy infected plant parts when seen and crop debris after harvest. Avoid fertilizers high in nitrogen and overcrowding. Take cuttings from upper half of disease-free plants. Indoors, keep water off foliage. Keep humidity below 85 per cent and increase air circulation. Apply zineb, maneb, captan, dichlone, ziram, thiram, or folpet, plus spreader-sticker, at weekly intervals during wet periods. Or use dodine, following manufacturer's directions. Start cuttings in sterilized rooting medium.

5. *Stem (Crown) and Root Rots, Southern Blight, Cutting (Foot) Rots, Damping-off* —General. Plants turn pale green or yellow, wilt, then brown and die. Stems rot, often break off near soil line without yellowing. Cuttings may develop soft, brown, mushy, or dry rot at base and collapse. Seedlings wilt and topple. Roots often discolored and rotted. May be associated with nematodes (e.g., lesion, root-knot). *Control:* Same as for Fusarium Wilt (above). Before sticking, dip cuttings in household bleach, captan, Pano-drench, ferbam, or zineb solution. Treat sterilized soil by raking in mixture of PCNB and captan 50 (page 188), Semesan, or Morton Soil Drench. Follow manufacturer's directions. Carnation varieties differ in resistance. Spray as for Alternaria Blight (above). Indoors, avoid overwatering, splashing water on foliage, and high humidity.

6. *Rusts* (babysbreath, carnation, evening campion, Maltese cross, pinks, red campion, silene, sweet-william) —General, especially where moist. Orange, reddish-brown, or chocolate-colored to dark gray, powdery pustules on leaves, stems, and buds (Figure 4.58). Plants often stunted and yellowish with curled-up leaves. Varieties differ greatly in resistance. *Control:* Same as for Alternaria Blight (above), except use only zineb, maneb, ferbam and sulfur, thiram, or dichlone. Grow resistant *carnation* varieties (with bluish foliage). Dip cuttings in ferbam (2 tablespoons per gallon of water plus spreader-sticker) before planting. Keep plants growing vigorously.

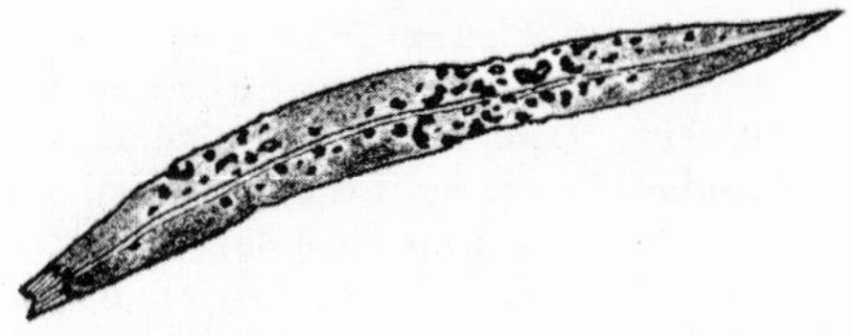

Fig. 4.58. Carnation rust.

7. *Mosaics, Mottles, Streak* (carnation, pinks, sweet-william) —Widespread. Symptoms variable. Leaves show light and dark green to yellowish-white, pink, red, purplish, or grayish-brown spots, irregular blotches, streaks, mottling, and flecking. Leaves may be curled and distorted. On seriously affected plants, lower leaves may yellow and die prematurely. Light streaks or blotches may develop in colored flower petals of some varieties. Plant vigor, size, and flowering often reduced. See Figure 2.34A under General Diseases. All commercial *carnations* are believed infected with Mottle virus, spread by handling. *Control:* Plant virus-free seed, plants, or cuttings. Destroy infected plants when first seen. By using malathion or lindane, control aphids that transmit the viruses. Wash hands and cutting knife with soap and plenty of hot running water before handling healthy plants. Cultured *carnation* cuttings are available. See also under Ringspot (below). (Remember that both Mottle and Ringspot viruses are spread by handling and cutting knife.)
8. *Ringspot*—Uncommon. Symptoms variable. *Carnation, babysbreath,* and *lychnis* leaves may show yellowish or gray rings, often zoned, or a pronounced mosaic mottling with some dead flecks. *Carnation* and *babysbreath* leaf margins are wavy. Older leaves may redden and curl. In *sweet-william,* zoned rings on leaves turn into a general mosaic with scattered, white spots. *Control:* Same as for Mosaics (above). Take short stem tips from selected plants grown at 100° F. for 60 days. Commercial growers can rid *carnation* of ringspot, streak, etched ring, mottle, and latent viruses by this treatment, or by culturing the root tips (0.5 millimeter long).
9. *Curly-top* (carnation, pinks, sweet-william) —See under Beet, and Figure 2.37A under General Diseases. *Control:* Same as for Mosaics (above). Use DDT plus malathion to control leafhoppers that transmit virus.
10. *Aster Yellows* (babysbreath, carnation, dianthus, mullein-pink, sweet-william) — See under Chrysanthemum, and (18) Yellows under General Diseases.
11. *Other Leaf Spots, Anthracnose, Leaf Rot, Greasy Blotch*—Small to large, round to irregular, pale, purplish, gray, brown, greasy, or grayish-green spots and blotches. Later often covered with mold or sprinkled with black dots. Sometimes on stems and flowers. More common on lower leaves that may wither and die prematurely. *Control:* Same as for Alternaria Blight (above).
12. *Gray-mold Blight, Botrytis Flower Blight*—Cosmopolitan in damp weather. Flower petals show soft, watery areas that turn brown. Some petals may mat together. Buds may rot and fail to open; tops of plants die back. Gray mold may grow on diseased areas in damp weather. Varieties differ in susceptibility. Common storage rot of flower petals and stems. *Control:* Pick and burn blighted flowers, buds, and stem tips. Apply captan, zineb, dichloran (Botran), or ferbam, 1 to 3 times just before bloom. Destroy tops in fall or after harvest. Indoors, keep water off foliage and humidity below 85 per cent. Increase air circulation and temperature. Space plants. Apply fine misty spray of zineb or captan (1 tablespoon per gallon), plus spreader-sticker, as flowers open.
13. *Fusarium Bud Rot* (primarily carnation) —Young buds fail to open. Interior of older buds is brown or pink, moist or dry, and decayed. Cottony mold and white mites may be present. White varieties are more susceptible than colored ones. *Control:* Pick and destroy infected buds when first found. Spray with malathion to control mites that spread the Fusarium fungus from plant to plant. Do *not* bring field-grown plants indoors.

14. *Anther or Flower Smut* (arenaria, campion, carnation, cushion-pink, dianthus, Maltese cross, red campion, silene, sweet-william) —Anthers in flowers filled with blackish powder. Pistillate (female) flowers aborted. Flower stalks stunted and produce flower buds that are thick, squat, often split. Infected plants grow slowly. Produce numerous secondary shoots. Appear "grassy" and bushy. *Control:* Destroy infected plants *before* buds open. Take cuttings *only* from healthy plants. Most modern *carnation* varieties are resistant.
15. *Witches'-broom, Fasciation, Leafy Gall* (babysbreath, carnation) —Fairly common but not serious. Leaves distorted; growth is stunted. Masses of short, spindly shoots develop at one place on lower stem. *Control:* Destroy affected plants. Propagate only from healthy plants. Stick cuttings in sterilized soil (pages 538–548).
16. *Powdery Mildew* (arenaria, carnation) —Mealy, whitish mold on leaves, stems, and flowers. Varieties differ in resistance. *Control:* See under African-violet, and (7) Powdery Mildew under General Diseases.
17. *Bacterial Leaf Spot* (carnation) —Widespread in humid weather. Small, elongated, dark water-soaked spots. Spots later enlarge, become brown and sunken. May be surrounded by purplish concentric rings. Spots may merge to kill a leaf. When severe, plants are killed. *Control:* Same as for Bacterial Wilt (above). Spray with phenyl mercury, plus spreader-sticker. Add malathion to control mites, aphids, and other insects. Pick off and destroy infected leaves.
18. *Crown Gall, Root and Stem Gall*—Soft, gall-like overgrowths at graft or soil line. Plants may be girdled, wilt, and die. *Control:* Dig up and burn affected plants. Do not propagate from infected plants. Dip newly grafted plants and grafting knives in household bleach solution (2 to 6 ounces per gallon of water) for 2 minutes between cuts. Carnation varieties differ in resistance.
19. *Root-knot, Cyst Nematode*—See (37) Root-knot under General Diseases.
20. *Leaf and Stem Nematode* (carnation, pinks, sweet-william) —Leaves crinkled and stems swollen. See (20) Leaf Nematode under General Diseases.
21. *Other Root-feeding Nematodes* (lance, lesion, pin, ring, spear, spiral, stunt) — Plants stunted, pale, weak, and unthrifty. Blooming and root growth are reduced. *Control:* Plant in sterilized soil (pages 538–548).
22. *Boron Deficiency*—Tips of shoots twisted and curled. May be deformed into witches'-brooms. Shoots get "stuck" trying to elongate. *Control:* Apply borax to soil (about 1 ounce to 100 square feet) and water in. Reapply as necessary (page 27). Check with your local florist or extension horticulturist.
23. *Downy Mildew* (carnation, cushion-pink, silene) —California. Leaves pale and curl downward. Whitish mold forms on underleaf surface in humid weather. Plants stunted. *Control:* Same as for Alternaria Blight (above).
24. *Web Blight*—Southeastern states. See under Bean.
25. *Chlorosis*—Mineral deficiency in alkaline or very acid soils. See under Rose.

CARNEGIEA — See Cacti
CAROLINA ALLSPICE — See Calycanthus
CAROLINA MOONSEED — See Moonseed
CAROLINA-JESSAMINE — See Butterflybush
CARPETGRASS — See Lawngrass
CARPINUS — See Birch
CARANDA — See Oleander
CARROT *(Daucus)*; PARSNIP *(Pastinaca)*

1. *Leaf Blights and Spots*—General in wet seasons. Outer leaves and petioles turn yellow, then brown and wither due to tan, gray, yellowish-green, reddish-brown, dark brown, or black spotting. Spots also may occur on petioles, stems, and flower

parts. Whole top may die when spots enlarge and merge. Yield is reduced. Bacterial blight may cause dark, scabby cankers on *carrot* roots. *Control:* Whenever possible, sow in well-drained soil in raised beds. Keep down weeds. Three- or 4-year rotation. Burn or bury tops after harvest. Plant disease-free seed or soak *carrot* seed in hot water (122° F.) for 15 to 20 minutes (or 126° F. for 10 min.). Dry seed, then dust with captan, thiram, dichlone, or chloranil. Apply maneb, zineb, difolatan, Dyrene, Daconil 2787, folpet, or ziram at 7- to 10-day intervals during rainy periods. If bacterial blight is a problem, add fixed copper or soluble copper (TC-90 or Sol-Kop 10) to regular fungicide. *Carrot* varieties differ in resistance to Alternaria Leaf Blight.

2. *Aster Yellows*—Widespread. Younger and inner leaves are yellow, stunted, and twisted. Plants stunted, appear bunchy. Outer and older leaves may have a bronzed, reddish, or purple color. The root is stunted, woody, and covered with many fine "hairy" roots. See Figure 2.36B under General Diseases. *Control:* Destroy first infected plants. Keep down weeds in and around garden area. Spray or dust at 5-day intervals with sevin or methoxychlor plus malathion to control leafhoppers that transmit virus. Start when seedlings are 2 inches tall. *Carrot* varieties differ in resistance.

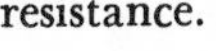

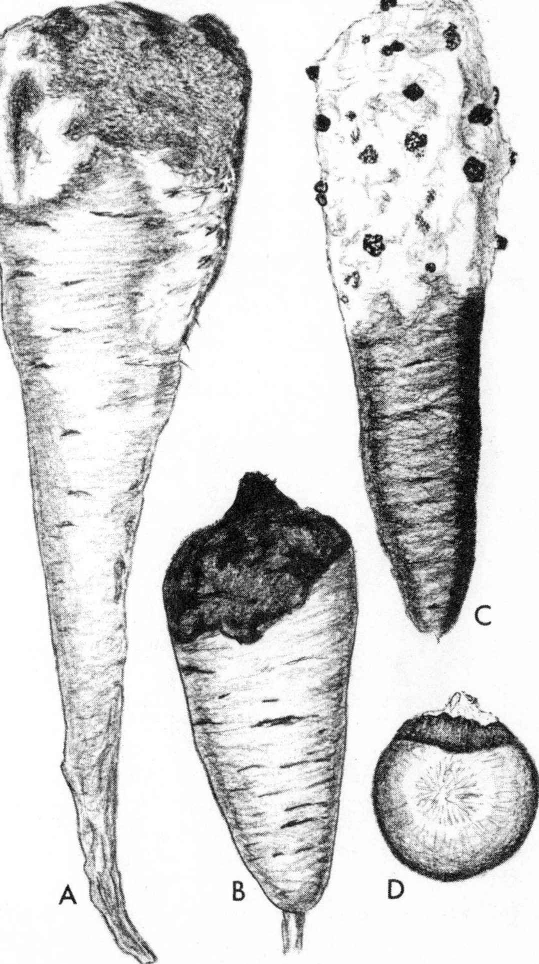

Fig. 4.59. Storage rots of carrot and parsnip. A. Parsnip gray mold rot (Botrytis). B. Carrot black rot (*Stemphylium*). C. Carrot watery soft rot (*Sclerotinia*). D. X-section of carrot with watery soft rot.

3. *Storage Rots, Crown and Root Cankers*—Cosmopolitan. Rots of various types—some dry; some watery, slimy, foul-smelling and wet (Bacterial Soft Rot) or "sour" (Sour Rot). May be covered with white, gray, tan, pink, blue, green, purplish, or black mold (Figure 4.59). Spread rapidly in damp, warm bins where roots are piled closely together. Rot may start at crown, root tip, or side wound. See (29) Bacterial Soft Rot under General Diseases. *Control:* Three-year rotation. Plant in well-drained soil and on raised beds, where possible. In field, keep nitrogen level on low side and potassium level high. Spray or dust in field as for Leaf Blights (above). Sweep storage areas clean, then spray with copper sulfate solution (1 pound in 10 gallons of water) or formaldehyde (1 pint to 10 gallons) before storing vegetables and fruit. Dust roots lightly with thiram or captan (or dip in Dowicide A) and then store. Store roots as close to 32° F. as possible without freezing, and with a high humidity (85 to 95 per cent). Avoid free moisture in storage. Store only dry, sound, blemish-free roots in layers with straw, dry leaves, sand, or other dry filler material in between. Avoid injuries. Do *not* store carrots and apples, pears, or other fruits together. For information on what to store and where, check with your extension horticulturist.
4. *Root Rots*—Plants may be stunted and sickly; roots discolored and decayed (Figure 4.60). May follow attacks by carrot weevil, carrot rust fly, growth cracks, nematodes (e.g., dagger, lance, lesion, naccobus, pin, reniform, root-knot, rot, sheath, stem, stubby-root, stunt), injuries, or other diseases. *Control:* Avoid overfertilizing. Plant late in loose, well-drained soil in raised beds. Long rotation. Control other diseases

Fig. 4.60. Phymatotrichum root rot of carrot. (Plant Science Department, Texas A & M University photo)

and weeds. Treat soil with Diazinon or chlordane. Follow manufacturer's directions. See under Seed Rot (below). Harvest early. Tolerant *carrot* varieties to Violet Root Rot: Chantenay and Chantenay Red Cored. Use disease-free seed.

5. *Seed Rot, Damping-off*—General. Poor stand. Seeds rot. Seedlings wilt and collapse from rot at soil level. *Control:* See under Leaf Blights (above). In addition, apply Diazinon or chlordane to soil surface before planting and immediately work into top 3 to 6 inches. This treatment controls carrot rust fly (maggot), wireworms, weevil, and other insects. Avoid overwatering, deep planting, and overcrowding.
6. *Root-knot, Cyst Nematode*—General in southern states. Occasional in rest of United States. Localized areas of stunted plants. Small, gall-like swellings on roots. Taproot may be forked, twisted, misshapen, and undersized. See Figure 2.52A under General Diseases. Yield is reduced. *Control:* Rotate with corn, members of the cabbage family, and grasses. If serious, fumigate in early fall using D-D, Vidden-D, EDB, DBCP, Dorlone, or Telone (pages 538–548). Follow manufacturer's directions.
7. *Watery Soft Rot, Cottony Rot, Sclerotinia Rot, Southern Blight*—Widespread. Cottony white mold on roots, crown and leaves (Figure 4.59C). See under Bean. Occurs in patches in field or garden.
8. *Scab*—Raised or sunken, greasy, areas on root, often near crown. Variable in color and shape. Young roots may be bleached yellowish-white. *Control:* See under Beet and Potato.
9. *Parsnip Root Canker, "Sore Head," Itersonilia Leaf Spot*—May be serious in rainy seasons. Irregular, brown to reddish, roughened surface on "shoulder" or crown of root that later becomes purplish-brown to black and sunken. Whole root may decay and become invaded by other rot organisms, especially the soft rot bacterium. Small, irregular, greenish-yellow leaf spots, often with brown centers, may develop. Entire leaves may die and fall early. *Control:* Plant in deep, neutral (pH 7.0), well-drained soil and on raised beds. Keep down weeds, especially wild carrot, wild parsnip, and related plants. The broad-shouldered varieties, Hollow Crown and All-America, are very susceptible while varieties with rounded crowns, e.g., Tender-and-True and Model, have some resistance. High ridging with soil to cover shoulders of root is helpful. Apply fixed copper, bordeaux (4-2-50), maneb, or zineb at 10-day intervals. Start when leaf spots are first seen. Control insects, such as carrot rust fly, and nematodes. See Seed Rot and Root Rots (both above). Rotate.
10. *Mosaics, Motley Dwarf, Ringspot*—Leaves mottled light to dark green and yellowed. Some rings or dead spots may appear in leaves. Plants may be stunted with twisted petioles. *Control:* Same as for Aster Yellows (above). Control aphids that transmit viruses. Use malathion. *Carrots* differ in resistance to Motley Dwarf.
11. *Curly-top*—Western states. See under Beet, and (19) Curly-top under General Diseases.
12. *Root Cracking, Boron Deficiency*—Young leaves yellowed and malformed. Plants wilt readily. Longitudinal cracks often occur in carrot (root) when heavy rains follow drought, or when boron is deficient in soil. Root may be stunted and woody. *Control:* Maintain as uniform soil moisture as possible. Water during dry periods. Have soil tested. If deficient in boron, apply borax as recommended by your county agent or extension horticulturist (page 27).
13. *Bacterial or Southern Wilt*—Southern states. Gradual or sudden wilting of plants. Leaves wither, turn black, and drop early. Some infected plants may die. Older plants dwarfed. Brown discoloration inside of stems. *Control:* See under Tomato, and (15C) Bacterial Wilt under General Diseases.
14. *Crown Gall*—Uncommon. Variously sized, rather hard galls occur on any part of root. Mostly in low wet areas. *Control:* Grow in clean soil in raised beds.

15. *Downy Mildew*—Pale, yellow spots on upper leaf surface that darken with age. Yellowish mildew forms on corresponding underside in damp weather. *Control:* Same as for Leaf Blights (above).
16. *Rust*—Small yellowish pustules on leaves. Alternate host is bulrush *(Scirpus). Control:* Same as for Leaf Blights (above).
17. *Web Blight*—Southeastern states. See under Bean.
18. *Powdery Mildew* (parsnip)—Minor problem. White, flourlike patches on leaves. *Control:* If serious, spray or dust with sulfur.
19. *White-rust* (parsnip)—Minor problem. See (9) White-rust under General Diseases.
20. *Irregular Roots*—Misshapen and twisted roots. Result from overseeding and crowded stand, rocks and other debris in soil interfering with normal growth. *Control:* Grow in deep, well-prepared soil relatively free of solid material (e.g., dense clods, stones, coarse trash). Avoid oversowing and thick stands.
21. *Potassium Deficiency*—Leaves curled. Plants stunted and "squat." Leaves somewhat chlorotic with scorched margins. *Control:* Have soil tested and apply potash as recommended.

CARTHAMUS — See Chrysanthemum
CARUM — See Celery
CARYA — See Walnut
CARYOPTERIS — See Lantana
CARYOTA — See Palms
CASHMERE BOUQUET — See Lantana
CASABA, CASSABANANA — See Cucumber
CASSANDRA — See Blueberry
CASSIA — See Honeylocust
CASSIOPE — See Blueberry
CASTANEA, CASTANOPSIS — See Chestnut
CASTILLEJA — See Snapdragon
CAST-IRON PLANT — See Aspidistra
CASTORBEAN *(Ricinis)*; CHINESE TALLOWTREE *(Sapium)*; QUEENS-DELIGHT *(Stillingia)*

1. *Gray-mold Blight, Botrytis Blight, Capsule Mold*—Eastern and southern states on castorbean. Occurs during humid, wet seasons. Small to large, pale brown to blackish spots on leaves, stems, flower stalks, fruit clusters, and capsules over which pale to olive-gray mold later grows in damp weather. Blight usually occurs at blooming time. *Control:* Sow seed from disease-free castorbean plants. Where practical, carefully pick off and destroy fading flowers and infected parts. Burn tops in fall. *Castorbean* varieties differ in resistance. Varieties with compact inflorescences are most susceptible to Botrytis Capsule Mold. Apply captan, zineb, thiram, or fixed copper one to three times, 5 to 7 days apart.
2. *Fungus Leaf Spots, Capsule Molds, Capsule Drop*—Serious in humid south. Round to irregular, white, gray, or brown spots and blotches on leaves and capsules. Leaves and seed may wither and fall early. *Control:* Same as for Gray-mold Blight (above). Tolerant or resistant *castorbean* varieties to Alternaria Leaf Spot: Baker 296, Cimarron, Dawn, Hale, Lynn, and MW-1.
3. *Bacterial Leaf Spot of Castorbean*—Southern states during rainy seasons. Brown to black, angular spots on leaves (Figure 4.61). *Control:* Sow disease-free seed. Somewhat resistant varieties: Anjou, Baker 195, Cimarron, Hale, Illinois 48–36, Lynn, and Western Oil Hybrid 9.
4. *Root Rots*—See under Bean, Apple, and (34) Root Rot under General Diseases

Fig. 4.61. Bacterial leaf spot of castorbean. (Courtesy Dr. R. D. Brigham)

Fig. 4.62. Cotton root rot (*Phymatotrichum*) of castorbean. (Courtesy Dr. R. D. Brigham)

(Figure 4.62). Often associated with root-feeding nematodes (e.g., burrowing, lesion). *Castorbean* varieties differ in resistance to Thielaviopsis Root Rot.

5. *Root-knot*—See (37) Root-knot under General Diseases.
6. *Bacterial Wilt or Brown Rot* (castorbean)—Southeastern states. Plants wilt but may recover temporarily. Leaves shrivel, turn black, and drop early. Stalks and branches are blackened. Plants infected when young may die; older plants may be dwarfed. Brown discoloration inside lower stem and roots. Roots decay. *Control:* Grow plants in clean soil. Use disease-free seed.
7. *Seed Rot, Seedling Blight*—Seeds rot. Seedlings become stunted, wilt, and collapse. Leaves blighted. *Control:* Collect and burn infected plants. Rotate. Grow in warm, well-drained, sterilized soil (pages 538–548). Spray during cool wet periods as for Gray-mold Blight (above). Treat seed with thiram, captan, or chloranil.
8. *Stem and Crown Rots, Southern Blight* (castorbean)—Stems discolored and rotted at or below soil line. Plants die early. *Control:* Fertilize and water to keep plants vigorous. Grow in well-drained soil. Avoid overwatering and overcrowding. Keep down weeds.
9. *Verticillium Wilt* (castorbean)—Yellowish, then dead, areas form in leaves between veins. Disease progresses up stem causing dying of branches and blighting of capsules. *Control:* Same as for Bacterial Wilt (above).

10. *Crown Gall*—More or less round to irregular galls on stem and crown. See (30) Crown Gall under General Diseases.
11. *Seedling Red Gall*—Texas. Small red galls develop on leaves, petioles, and stems. *Control:* Pick and burn affected parts.
12. *Rust* (queens-delight) —Small yellowish spots or pustules on foliage. Alternate host is a grass *(Panicum). Control:* Spray as for Gray-mold Blight (above) but use zineb, thiram, or maneb.

CASUARINA, AUSTRALIAN-PINE, BEEFWOOD or HORSETAIL-TREE *(Casuarina)*

1. *Root Rots*—See under Apple, and (34) Root Rot under General Diseases. *C. cunninghamiana* is reported more resistant to Clitocybe Root Rot than other species.
2. *Root-knot*—See under Peach, and (37) Root-knot under General Diseases.

CATALPA [CHINESE, COMMON or SOUTHERN, JAPANESE, MANCHURIAN, NORTHERN or WESTERN, UMBRELLA-] *(Catalpa);* DESERT-WILLOW *(Chilopus);* JACARANDA; PODRANEA

1. *Verticillium Wilt* (catalpa) —Widespread. Leaves on one or more branches wilt, turn brown, and hang downward or fall early (Figure 4.63). Purple to bluish-brown streaks are evident in sapwood under bark. Affected trees may die the first year or live for many years. *Control:* See under Maple.

Fig. 4.63. Verticillium wilt of catalpa. (Pennsylvania State University Extension Service photo)

2. *Leaf Spots, Anthracnose, Spot Anthracnose*—General in rainy seasons. Round to irregular, yellowish-brown to black spots. Spots may later fall out leaving holes. Leaves may wither and drop early. *Control:* Collect and burn fallen leaves. Where economical, apply three sprays, 2 weeks apart. Start as leaves begin to unfold. Use ferbam, thiram, captan, maneb, phenyl mercury, fixed copper, or bordeaux (4-4-50).
3. *Powdery Mildews* (catalpa) —Powdery white patches on leaves. If severe, leaves may wither and drop early. *Control:* If practical, spray twice, 10 to 14 days apart. Use sulfur or Karathane.
4. *Wood Rots*—Widespread. See under Birch, and (23) Wood Rot under General Diseases.
5. *Root Rots*—See under Apple, and (34) Root Rot under General Diseases. May be associated with root-feeding nematodes (e.g., burrowing, stunt).

6. *Chlorosis* (catalpa) —See under Maple. Occurs in alkaline soils.
7. *Leaf Scorch*—Primarily in Middle West. Follows hot, dry, windy periods. See under Maple.
8. *Sooty Mold*—See under Elm, and (12) Sooty Mold under General Diseases.
9. *Root-knot*—See under Peach, and (37) Root-knot under General Diseases.
10. *Crown Gall*—See (30) Crown Gall under General Diseases.
11. *Dieback, Canker*—See under Apple and Maple.
12. *Seedling Blight, Damping-off*—See under Pine.

CATASETUM — See Orchids

CATCHFLY — See Carnation

CATCLAW — See Honeylocust

CATHA — See Bittersweet

CATNIP — See Salvia

CAT'S-CLAW — See Trumpetvine

CATTLEYA — See Orchids

CAULIFLOWER — See Cabbage

CAULOPHYLLUM — See Barberry

CEANOTHUS — See New Jersey-tea

CEDAR *(Cedrus)* — See Pine

CEDAR [CREEPING, RED, WHITE-, YELLOW-] *(Juniperus, Chamaecyparis, Thuja)* — See Juniper

CEDRELA — See Chinaberry

CELANDINE — See Poppy

CELASTRUS — See Bittersweet

CELERY, CELERIAC *(Apium)*; DILL *(Anethum)*; SALAD CHERVIL *(Anthriscus)*, TURNIP-ROOTED CHERVIL *(Chaerophyllum)*; CARAWAY *(Carum)*; CORIANDER *(Coriandrum)*; ERYNGO, SEA HOLLY, RATTLESNAKE-MASTER *(Eryngium)*; FENNEL, FINOCCHIO *(Foeniculum)*; SWEET-JARVIL, ANISE-ROOT *(Osmorhiza)*; PARSLEY *(Petroselinum)*; ANISE *(Pimpinella)*; BLUE LACEFLOWER *(Trachymene or Didiscus)*

1. *Leaf Blights and Spots, Anthracnose*—General, may be serious. Round to irregular, brown, yellowish-brown, ash-gray, yellow, tan, reddish-brown, black, or water-soaked spots on leaves, petioles, leaf stalks, stems, and seed pods. If numerous, leaves may turn yellow to brownish-black, wilt, shrivel, and die. Quality and yield may be reduced. Infected seed may be discolored. See Figure 2.19C under General Diseases. *Control:* Grow disease-free transplants in deep, rich, well-drained soil. Space plants. Keep down weeds. Do *not* work among wet plants. *Celery* varieties differ in resistance: Florimart and Emerald are partially resistant to Early Blight *(Cercospora);* Emerson Pascal has moderate resistance to Early and Late Blights *(Septoria);* while Giant Pascal and White Plume have moderate resistance to Late Blight. Burn or bury tops after harvest. If practical, plant rows north and south. Sow certified, 2-year-old, disease-free celery seed, or soak *celery* seed (also *celeriac*) in hot water at exactly 118° F. for 30 minutes (soak *coriander* seed at 127–128° F. for 30 minutes). Dry seed and dust with thiram, captan, chloranil, or Semesan. Dust other seed with thiram or Semesan. Three-year rotation, especially of seedbed. Apply thiram, maneb, folpet, zineb, ziram, dichlone, Dyrene, Daconil 2787, Miller 658-Z, or fixed copper sprays at 5- to 7-day intervals in field during wet weather. Use ziram, ferbam, or thiram in seedbed. Thorough coverage is essential. If Bacterial Blight is a problem, use fixed copper or soluble copper (Sol-Kop 10, TC-90). Start in seedling stage. Avoid overfertilizing with nitrogen.

2. *Seed Rot, Damping-off*—Cosmopolitan. Seeds rot. Poor stand. Seedlings wilt, shrivel, and collapse. *Control:* Buy treated seed or treat as for Leaf Blights (above). Grow transplants in sterilized soil (fumigated before planting with chloropicrin, formaldehyde, Vorlex, SMDC (Vapam or VPM), Mylone, or methyl bromide) where feasible. Spray seedlings at 5- to 7-day intervals using ziram, ferbam, thiram, Semesan, or fixed copper. Do not overwater.
3. *Fusarium Yellows, Wilt* (celery, coriander, dill, parsley) —General, except in southern states. Symptoms vary with particular fungus strain. Plants stunted, turn pale green, straw-color to golden yellow. Growth often one-sided. Seedlings stunted, may wilt and die. Stalks brittle and have bitter taste. Leaves may be cupped downward. Yellowish to reddish-brown or nearly black strands inside stems. Roots may decay. *Control:* Plant healthy seedlings in clean soil. Grow resistant *celery* varieties in infested soil: Cornell No. 19, Easy Blanching, Emerson Pascal, Florida Golden, Forbes Golden Plume, Giant Pascal, Golden 99, Golden Pascal, Kilgore's Pride, Masterpiece, Michigan Improved Golden, Michigan State Green Gold, Pascal 284, Slow Bolting Green No. 12, Supreme Golden, Tall Golden Plume, Utah 15, and Woodruff's Beauty where adapted. Green *celeries* are generally resistant. Check with your extension horticulturist.
4. *Aster Yellows*—Widespread. Inner, heart leaves dwarfed, yellowed, curved, intertwined and twisted. Whole plant may be bushy, gradually turns yellow. Petioles upright, brittle, and commonly crack. *Control:* See under Carrot, and (18) Yellows under General Diseases.
5. *Root-knot*—Widespread. Dill and celery are very susceptible. Seedlings may be severely stunted and yellowed with small, beadlike galls or enlargements on roots. *Control:* Rotate. Plant disease-free seed and transplants grown in clean or pasteurized soil (pages 538–548). Use D-D, Telone, methyl bromide, etc. *Celery* transplants can be largely freed of root-knot for 100 days or more by dipping in a solution of Zinophos (1,000 to 2,500 parts per million) for 10 to 60 minutes. Follow manufacturer's directions.
6. *Stem or Stalk (Foot) Rots, Pink Rot, Storage Rots*—Cosmopolitan. Cut celery and parsley butts and stalks may become discolored or soft, slimy, and putrid (Bacterial Soft Rot). Diseased areas may be covered with cottony, pink, gray, bluish-green, or black mold. In the field, stems and leaf stalks may suddenly soften, wilt, and collapse. Cottony mold may cover rotted area at soil line. See (29) Bacterial Soft Rot, and (21) Crown Rot under General Diseases. *Control:* See Storage Rots under Carrot. Spray as for Leaf Blights (above). Apply PCNB, Botran, or ferbam dust or spray in field. Follow manufacturer's directions. Harvest quickly and refrigerate (32° to 34° F.) promptly. Avoid injuries when cultivating, spraying, etc.
7. *Mosaics, Calico, Yellow Spot, Motley Dwarf, Streak*—Symptoms variable. Plants often bunchy and stunted, with curled and twisted stalks. Leaves often brightly mottled, spotted, or striped with yellow and green, crinkled, twisted, narrower than normal, cupped downward, and sickly greenish-yellow or grayish. Green and yellow zigzag bands and green "islands" may develop in yellow areas of leaves *(Calico)*. Leaf stalks may develop sunken brown streaks, later turn brown and shrivel. *Control:* Plant virus-free seedlings. Destroy first infected plants. Keep down all weeds—especially wandering-Jew (*Commelina* spp.). This is particularly important in seedbed area. Control aphids that transmit the viruses. Use malathion or Diazinon.
8. *Blackheart, Calcium Deficiency, Heart Rot* (celery, fennel, parsley) —General in hot weather with rapid plant growth on soils high in fertilizer salts. Inner, heart leaves die at tips and turn brown or black. Later all center leaves, heart, and petioles are affected. Outer stalks are never involved. Bacterial Soft Rot may follow, resulting in a slimy decay. Common in low spots in the field. *Control:* Plant rows north and

south in well-drained soil. Water during dry periods to maintain as uniform soil moisture as practical. Harvest promptly at maturity. Avoid overfertilization with nitrogen (especially sodium nitrate) and potash. If practical, apply sprays of calcium chloride (3/4 to 1 1/2 ounces per gallon of water) or calcium nitrate (1 1/2 to 3 ounces per gallon), into heart leaves at weekly intervals. Start 5 to 7 weeks before harvest. Resistant *celery* varieties: Cornell 19, Emerald, Emerson, Florida Golden, Golden Pascal, Golden Phenomenal, Salt Lake, and Winter Queen.

9. *Boron Deficiency, Stem-cracking, Brown Checking*—Occasional in all commercial celery districts where soil is heavy and alkaline. Younger and inner leaves brownish and mottled along margins. Long, brown streaks may appear on stems. Ragged, crosswise cracking of leaf stems. Stems (petioles) stiff, brittle, and sometimes bitter. Roots turn brown. *Celeriac* crowns often hollow. *Control:* Have soil tested and apply borax or boric acid as recommended (page 27). Resistant *celery* varieties: Columbia No. 4130, Delmar, Dwarf Golden Self Blanching, Easy Blanching No. 5178, Giant Pascal, Golden Self Blanching, Utah 52–70, and 52–75. Avoid excessive use of fertilizer high in nitrogen or potassium in boron-deficient soil.
10. *Spotted Wilt*—Numerous small, yellow to orange spots on older leaves. Spots later turn brown. Leaves may wither and die. Plants often dwarfed. *Control:* See under Tomato, and (17) Spotted Wilt under General Diseases.
11. *Curly-top* (celery, celeriac, chervil, coriander, dill, fennel, and parsley)—See under Beet, and (19) Curly-top under General Diseases.
12. *Ringspot* (celeriac, celery)—Yellowish rings or spots, line or zigzag patterns on crinkled leaves. *Control:* Same as for Mosaics (above).
13. *Root Rots, Basal or Crown Rot*—Foliage often stunted and sickly. May wilt and collapse. Roots, leaf stalk bases, and crown may decay, turn red, brown, black, or bluish-green. Root-feeding nematodes (see below) may provide wounds for rot-producing organisms. *Control:* Same as for Leaf Blights and Root-knot (both above). Apply chlordane to soil before planting and work into top 4 inches of soil to control wireworms, cutworms, and carrot rust fly.
14. *Stem Nematode* (celeriac, celery, parsley)—See (20) Leaf and Stem Nematode under General Diseases.
15. *Verticillium Wilt* (celeriac, celery, chervil, parsley)—See (15B) Verticillium Wilt under General Diseases.
16. *Rust* (anise, anise-root, sweet-jarvil)—General. See (8) Rust under General Diseases.
17. *Leaf Smut, White Smut* (eryngium)—See (13) White Smut under General Diseases.
18. *Downy Mildew* (celery, chervil, fennel, parsley)—See under Carrot. *Control:* Same as for Leaf Blights (above).
19. *Bacterial Petiole Spot* (celery)—Water-soaked, sunken spots develop on inner and outer petioles. Spots later enlarge up to an inch or so in diameter. Turn yellow to deep brown. Cornell 19 and Utah 52–70 are very susceptible. *Control:* Use copper sprays throughout season to control foliage diseases. Otherwise same as for Leaf Blights (above).
20. *Root-feeding Nematodes* (awl, lance, lesion, pin, sheath, stem and bulb, sting, stubby-root, stunt)—Associated with stunted, unthrifty plants. Roots often short and stubby. Often found together with Root-knot. *Control:* Same as for Root-knot (above).
21. *Magnesium Deficiency*—Tips and leaf margins are mottled, then yellowed and scorched. *Control:* Apply sprays of magnesium sulfate (10 pounds per acre) at 10- to 14-day intervals, starting 2–3 weeks after transplanting. Or use magnesium chelates (page 28) following manufacturer's instructions. *Celery* varieties differ in

resistance: Utah 15 and 52–75 are resistant; Emerson Pascal and Utah 52–70 are intermediate; and Utah 10–73 is very susceptible.

CELOSIA — See Cockscomb

CELTIS — See Hackberry

CELTUCE — See Lettuce

CENTAUREA — See Chrysanthemum

CENTIPEDEGRASS — See Lawngrass

CENTRANTHUS — See Valerian

CENTROSEMA — See Pea

CENTURYPLANT, FALSE-ALOE *(Agave)*; FURCRAEA; WILD TUBEROSE, SPICELILY *(Manfreda)*

1. *Anthracnose, Black Rot*—Round, sunken, dark leaf spots with raised border. Spots may enlarge and merge causing whole leaf to die. *Control:* Pick and burn infected leaves. Indoors keep water off foliage. If practical, apply copper, zineb, or maneb several times, 10 days apart, during rainy periods.
2. *Leaf Spots, Leaf Blights, or Scorch*—Spots and blotches of various colors and shapes. Spots may enlarge and cover entire leaf. *Control:* Same as for Anthracnose. Indoors, raise temperature and maintain uniform soil moisture.
3. *Gray-mold Blight*—See under Begonia. Follows overwatering or chilling. *Control:* Avoid overwatering or chilling. Space plants to increase air circulation.
4. *Root-knot*—See (37) Root-knot under General Diseases.
5. *Rust* (manfreda)—Small yellowish pustules on leaves. *Control:* Same as for Anthracnose (above).

CEPHALANTHUS — See Buttonbush

CEPHALOCEREUS — See Cacti

CEPHALOTAXUS — See Japanese Plum-yew

CERCIS — See Honeylocust

CEREUS — See Cacti

CERIMAN — See Calla Lily

CHAENOMELES — See Apple

CHAEROPHYLLUM — See Celery

CHAMAECYPARIS — See Juniper

CHAMAEDAPHNE — See Blueberry

CHAMAEROPS — See Palms

CHARD — See Beet

CHASTE-TREE — See Lantana

CHAYOTE — See Cucumber

CHECKERBERRY — See Heath

CHECKERMALLOW — See Hollyhock

CHEIRANTHUS — See Cabbage

CHELIDONIUM — See Poppy

CHELONE — See Snapdragon

CHENILLE PLANT — See Acalypha

CHERRY, CHERRY-LAUREL — See Peach

CHERVIL — See Celery

CHESTNUT [AMERICAN, CHINESE, JAPANESE, SEGUIN, SPANISH or EUROPEAN], CHINQUAPIN *(Castanea)*; GOLDEN or GIANT CHINQUAPIN *(Castanopsis)*

1. *Chestnut Blight or Endothia Canker*—General and serious, practically eliminating

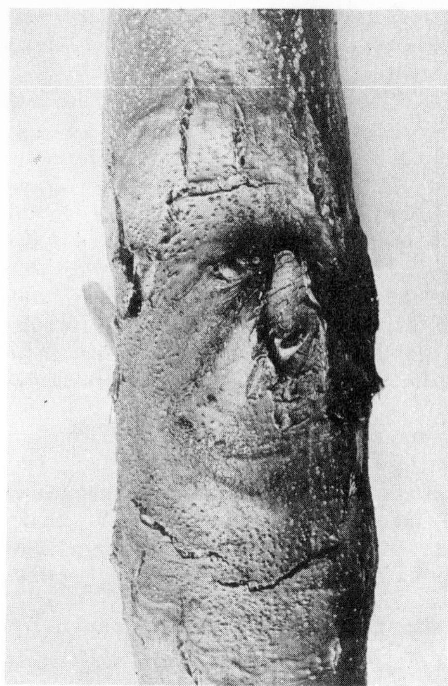

Fig. 4.64A (above). Blight or Endothia canker of chestnut. (Pennsylvania State University Extension Service photo)

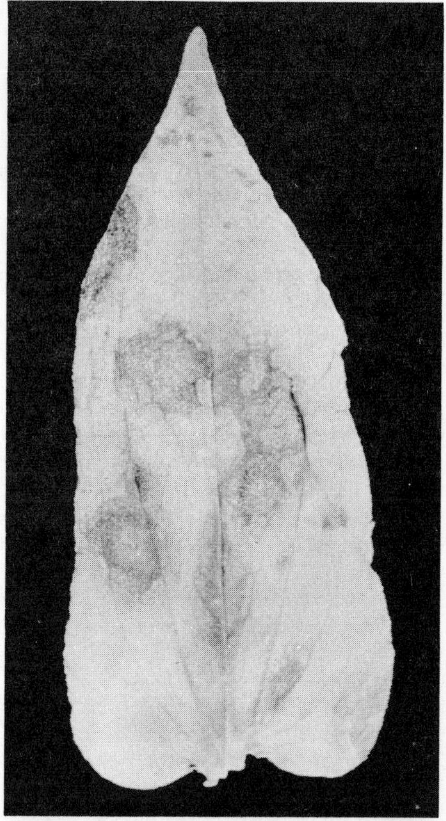

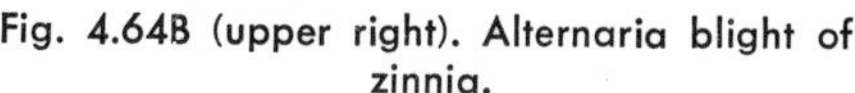

Fig. 4.64B (upper right). Alternaria blight of zinnia.

Fig. 4.64C. Septoria leaf spot and blight of chrysanthemum.

the American and Spanish or European chestnuts from their natural range in the United States. Also seriously infects chinquapins and post oak. The disease is largely absent in Pacific Coast states. Reddish-orange, yellowish-brown or brown, slightly sunken or swollen, cracked, girdling cankers on twigs, branches, and trunk (Figure 4.64A). Blight quickly spreads down into trunk. Leaves on affected parts suddenly turn yellow or brown, wilt, die, and hang downward. Spread by insects, splashing rain and wind, infected nursery stock, seed, lumber, and birds. *Control:* Plant adapted, resistant Chinese and Japanese chestnuts. Resistant varieties: Alaling, Alamoore, Blackbeauty, Ching Chow, Crane, and Orrin. Keep trees growing vigorously. Water and fertilize regularly. For additional information read USDA Farmers' Bulletin 2068, *Chestnut Blight and Resistant Chestnuts.*

2. *Leaf Spots, Anthracnose*—Spots of various sizes and colors. Not serious. Spots may drop out or enlarge and merge causing leaves to blight. *Control:* Collect and burn leaves in fall. If practical, spray as for Oak Anthracnose.
3. *Twig Blights, Cankers, Dieback*—Widespread. Discolored, sunken to swollen, oval to elongated cankers—later sprinkled with small, black pustules—on twigs, branches, and trunk. May girdle and kill parts beyond. Leaves on girdled branches wilt, turn brown, and die. Growth stunted. Trees may be deformed. Most serious on young trees. *Control:* Maintain vigor by fertilizing and watering. Grow in well-drained soil where air circulation is good. Prune and burn dead or cankered wood. Make cuts at least 6 inches below any sign of infection. Avoid leaving stubs. For large cankers on trunk contact a competent tree surgeon. Paint over wounds promptly with a good tree wound dressing. All varieties and seedlings are resistant to common twig cankers [*Cryptodiaporthe* (or *Fusicococcum*) and *Botryosphaeria*] when grown on proper sites. Remove and burn severely diseased trees. Spray as for Leaf Spots (above) or use captan.
4. *Wood Rots*—See (23) Wood Rot under General Diseases.
5. *Root Rots*—Trees decline in vigor. Foliage is thin. Leaves yellow, wither, and drop early. Roots decay. May appear "inky" or be covered with fan-shaped patches of mold and black rootlike strands. Branches commonly die back. *Control:* See under Oak, and (34) Root Rot under General Diseases. Most Asiatic varieties and seedlings are highly resistant to Phytophthora Root Rot. Grow in well-drained soil.
6. *Powdery Mildews*—Powdery, grayish-white mold on leaves and young shoots. If severe, leaves may yellow and wither. *Control:* If serious, spray twice, 10 days apart. Use Karathane or sulfur.
7. *Oak Wilt*—See under Oak.
8. *Leaf Blister* (chinquapin)—See under Oak.
9. *Blossom-end Rot of Nuts* (chestnut, Chinese chestnut)—Brown then black rot of blossom-end of fruit. Rot area often later covered with pale grayish mold. *Control:* Spray developing fruit with captan or zineb. Keep trees well pruned. Individual trees vary greatly in resistance.
10. *Rust* (chestnut)—See (8) Rust under General Diseases.
11. *Brown Felt Canker* (chinquapin)—See under Hackberry, Felt Fungus.
12. *Mistletoe* (chestnut)—See (39) Mistletoe under General Diseases.
13. *Verticillium Wilt*—See under Maple. Leaves on one or more branches turn yellow with brown margins. Leaves drop early and branches die back. Brown discoloration appears when branch is cut through on an angle.

CHICKEN GIZZARD PLANT — See Cockscomb

CHICORY — See Lettuce

CHILOPUS — See Catalpa

CHINA-ASTER — See Chrysanthemum

CHINABERRY or CHINA-TREE *(Melia)*; CEDRELA, CHINESE CEDRELA *(Cedrela)*

1. *Leaf Spots, Downy Mildew*—General. Rarely serious. Spots of various sizes, shapes, and colors on leaves. *Control:* None usually necessary. If needed, spray several times during wet periods. Use zineb, maneb, or fixed copper.
2. *Twig Blight, Cankers, Limb Blight*—Twigs and branches die back. Discolored cankers may form on twigs. Parts beyond wilt and die. *Control:* Remove dead and cankered twigs. If serious, spray as for Leaf Spots (above).
3. *Powdery Mildew*—Grayish-white mold patches on leaves. *Control:* If serious, spray once or twice, 10 days apart. Use Karathane or sulfur.
4. *Black Mildew, Sooty Mold*—Black, moldy patches on leaves following insect attacks.

Control: Apply malathion to control scales, whiteflies, and other insects.

5. *Wood Rots*—See (23) Wood Rot under General Diseases.
6. *Root-knot*—See under Peach, and (37) Root-knot under General Diseases.
7. *Mistletoe*—See (39) Mistletoe under General Diseases.
8. *Root Rots*—See under Apple, and (34) Root Rots under General Diseases.
9. *Thread Blight*—Southeastern states. See under Walnut. *Control:* Spray as for Leaf Spots (above).
10. *Verticillium Wilt* (chinaberry)—See (15B) Verticillium Wilt under General Diseases.

CHINCHERINCHEE — See Tulip

CHINESE ANGELICA — See Acanthopanax

CHINESE ARTICHOKE — See Salvia

CHINESE BEAUTY BUSH — See Viburnum

CHINESE BELLFLOWER — See Bellflower

CHINESE BITTERSWEET — See Bittersweet

CHINESE CABBAGE — See Cabbage

CHINESE CEDRELA — See Chinaberry

CHINESE EVERGREEN — See Calla Lily

CHINESE FAN — See Palms

CHINESE FORGET-ME-NOT — See Mertensia

CHINESE HIBISCUS — See Hollyhock

CHINESE HOUSES — See Snapdragon

CHINESE LANTERNPLANT — See Tomato

CHINESE PARASOLTREE — See Phoenix-tree

CHINESE PRIMROSE — See Primrose

CHINESE REDBUD, CHINESE SCHOLARTREE — See Honeylocust

CHINESE SACRED LILY — See Daffodil

CHINESE TALLOWTREE — See Castorbean

CHINESE TRUMPETCREEPER — See Bignonia

CHINESE WAXGOURD — See Cucumber

CHINESE WOLFBERRY — See Matrimony-vine

CHINQUAPIN — See Chestnut

CHIONANTHUS — See Ash

CHIONODOXA — See Tulip

CHIVES — See Onion

CHOKEBERRY, CHRISTMASBERRY — See Apple

CHRISTMAS CHERRY — See Tomato

CHRISTMAS-ROSE — See Delphinium

CHRYSALIDOCARPUS — See Palms

CHRYSANTHEMUM [ANEMONE-FLOWERED, ARCTICUM, CUSHION, FLORISTS', FOOTBALL MUMS, GARLAND, HARDY, ITALIAN, KOREAN, NIPPON, POMPON, TRICOLOR], ASTER or ARTIC DAISY, CORN-MARIGOLD, COSTMARY, FEVERFEW, GIANT DAISY, GOLDEN-STAR, OXEYE DAISY, SHASTA DAISY, PARIS DAISY or MARGUERITE, PYRETHRUM [COMMON or PAINTED DAISY, DALMATION] *(Chrysanthemum)*; SNEEZEWORT, YARROW [COMMON, SWEET] *(Achillea)*; YELLOW IRONWEED *(Actinomeris)*; AGERATUM; PEARLEVERLASTING *(Anaphalis)*; EVERLASTING, PUSSYTOES or LADIES'-TOBACCO *(Antennaria)*; CAMOMILE, GOLDEN MARGUERITE *(Anthemis)*;

AFRICAN DAISY *(Arctotis);* **ARNICA; WORMWOOD [BEACH, COMMON, MUGWORT, SOUTHERNWOOD or OLDMAN, ROMAN, RUSSIAN, SILVER KING, SWEET], TARRAGON, DUSTY-MILLER** *(Artemisia);* **ASTER [BLUE WOOD, FLAT-TOPPED, ITALIAN, NEW ENGLAND, NEW YORK, PERENNIAL (HARDY ASTER or MICHAELMAS DAISY), ROCK, SMOOTH, SEASIDE, SAVORY-LEAVED, SWAMP, TARTARIAN, WHITE HEATH, WHITE UPLAND]** *(Aster);* **BALSAMROOT** *(Balsamorhiza);* **ENGLISH or TRUE DAISY** *(Bellis);* **BUR-MARIGOLD** *(Bidens);* **BOLTONIA [VIOLET, WHITE], FALSE-CAMOMILE** *(Boltonia);* **SWAN RIVER DAISY** *(Brachycome);* **POT MARIGOLD or ENGLISH MARIGOLD** *(Calendula);* **CHINA-ASTER or ANNUAL ASTER** *(Callistephus);* **SAFFLOWER** *(Carthamus);* **CORNFLOWER or BACHELORS-BUTTON, BASKETFLOWER, DUSTY-MILLER, MOUNTAIN BLUET, SWEET SULTAN** *(Centaurea);* **GOLDEN-ASTER, ROSINWEED, GOLDEN-STAR, GROUND GOLD-FLOWER** *(Chrysopsis);* **THISTLE, PLUMED THISTLE** *(Cirsium);* **BLESSEDTHISTLE** *(Cnicus);* **COREOPSIS [COMMON, EARED, LANCE, ROSE, TALL, THREAD-LEAVED], TICKSEED, GOLDEN-WAVE** *(Coreopsis);* **COSMOS [BLACK, COMMON, YELLOW]** *(Cosmos);* **HAWKSBEARD** *(Crepis);* **DAHLIA [ANEMONE, BALL, CACTUS, COLLARETTE, COMMON, DECORATIVE, MIGNON, MINIATURE, PEONY, POMPON, TOPMIX or BABY]** *(Dahlia);* **CAPE-MARIGOLD** *(Dimorphotheca);* **LEOPARDSBANE** *(Doronicum);* **PURPLE-CONEFLOWER or BLACK SAMPSON, PURPLE DAISY** *(Echinacea);* **GLOBETHISTLE** *(Echinops);* **TASSELFLOWER, FLORAS-PAINTBRUSH** *(Emilia);* **ENCELIA; FLEABANE, ROBIN'S-PLANTAIN, DOUBLE ORANGE DAISY, BEACH ASTER** *(Erigeron);* **BONESET, WHITE SNAKEROOT, MISTFLOWER, JOE-PYE-WEED [BLUESTEM, COASTAL-PLAIN, PURPLESTEM, SPOTTED], THOROUGHWORT [HYSSOP, TALL]** *(Eupatorium);* **BLUE DAISY, KINGFISHER DAISY** *(Felicia);* **GAILLARDIA, FIREWHEEL or ANNUAL BLANKET-FLOWER** *(Gaillardia);* **AFRICAN DAISY** *(Gazania);* **TRANSVAAL DAISY** *(Gerbera);* **SNEEZEWEED, YELLOW STAR** *(Helenium);* **JERUSALEM-ARTICHOKE, SUNFLOWER [ASHY, COMMON, DARKEYE, GIANT, MAXIMILIAN, PRAIRIE, SAWTOOTH, STIFF, SWAMP, THINLEAF, WOODLAND, WILLOW-LEAVED]** *(Helianthus);* **STRAWFLOWER** *(Helichrysum);* **ORANGE SUNFLOWER or OXEYE, FALSE SUNFLOWER** *(Heliopsis);* **INULA, ELECAMPANE** *(Inula);* **FALSE-BONESET** *(Kuhnia);* **TIDYTIPS, WHITE DAISY** *(Layia);* **BLAZING-STAR [BUTTON, GRASS-LEAVED], GAYFEATHER, BUTTON SNAKEROOT** *(Liatris);* **MALACOTHRIX; FALSE-CAMOMILE or GERMAN CAMOMILE, SCENTLESS CAMOMILE, TURFING DAISY** *(Matricaria);* **CLIMBING HEMPWEED** *(Mikania);* **STEVIA** *(Piqueria);* **PRAIRIE-CONEFLOWER** *(Ratibida);* **BLACK-EYED-SUSAN, GLORIOSA DAISY, GOLDENGLOW, BROWN-EYED-SUSAN, THIMBLEFLOWER, CONEFLOWER [CUTLEAF, ORANGE, SWEET]** *(Rudbeckia);* **CINERARIA, DUSTY-MILLER, PARLOR- or GERMAN IVY, PEPPERMINT STICK, GROUNDSEL, MEXICAN FLAME-VINE, RAGWORT [GOLDEN, PURPLE, TANSY]** *(Senecio);* **SILPHIUM, CUP-PLANT or INDIAN-CUP, COMPASSPLANT** *(Silphium);* **GOLDENROD [BLUE-STEMMED, DWARF or GRAY, EUROPEAN, LANCE-LEAVED, OLD-FIELD, ROCK, SEASIDE or BEACH, STOUT, WHITE or SILVERROD, WREATH]** *(Solidago);* **WIRELETTUCE** *(Stephanomeria);* **STOKES-ASTER, CORNFLOWER ASTER** *(Stokesia);* **MARIGOLD [AZTEC (AFRICAN or AMERICAN),**

FRENCH, GUINEA GOLD] *(Tagetes)*; TANSY *(Tanacetum)*; DANDELION *(Taraxacum)*; TORCH FLOWER, MEXICAN SUNFLOWER *(Tithonia)*; COLTSFOOT *(Tussilago)*; CROWNBEARD *(Verbesina)*; WYETHIA; ZINNIA

1. *Leaf Spots, Leaf Blight or Blotch*—General in humid, wet weather. May be serious. Spots of various sizes, shapes, and colors. Spots may enlarge and merge to form large, irregular blotches. Leaves may discolor, gradually wither, and die (Figures 4.64B and 4.64C). Often starts at base of plant and progresses upward. Leaves may drop early or cling to stem. Certain spots also occur on stems, flower bracts, and flower petals. *Control:* Burn tops in fall. Rotate. Keep down weeds. If practical, pick and burn first infected leaves. Apply zineb, ferbam, maneb, thiram, Daconil 2787, captan, ziram, or folpet, plus spreader-sticker, at weekly intervals during wet weather. Use disease-free cuttings or dip in solution of ferbam, captan, zineb, ziram, or thiram (1½ tablespoons per gallon) before sticking in rooting medium. Soak *China-aster* seed for 30 minutes in a 1:1,000 solution of mercuric chloride or Semesan (2 teaspoons per quart), then wash (after mercuric chloride treatment) thoroughly in running water for 5 minutes, dry, and plant. Soak *zinnia* seed in hot water (125° F.) for 30 minutes, then dry and treat seed with captan, thiram, or chloranil. Treat other seed by dusting with captan, thiram, or chloranil. See "Seed Treatment Methods and Materials" in Appendix. Varieties of certain plants (e.g., *chrysanthemum* and *zinnia*) differ in susceptibility. Indoors, keep water off foliage and humidity below 85 per cent. Do not work among wet plants.
2. *Powdery Mildews*—General in damp, partly shady locations. Powdery, whitish-gray patches on leaves, stems, and flower buds especially in late summer. Mostly on lower half of infected plants. Leaves may wither and drop prematurely. Plants may be stunted, disfigured, and weakened (Figure 4.65). See also Figure 2.23A under General Diseases. *Control:* Avoid overcrowding. Collect and burn tops in fall or after harvest. Resistant varieties are available for some plants, e.g., *zinnia.* Apply Karathane, sulfur, or Acti-dione. Follow manufacturer's directions. Do not dust or spray when temperature is above 85° F. Indoors, keep water off foliage and humidity low.

Fig. 4.65. Chrysanthemum powdery mildew.

3. *Crown, Foot, Stem and Root Rots, Southern Blight, Cutting Rots, Damping-off*—General. May be serious in wet weather. Plants often stunted and sickly. Foliage wilts and discolors. Plants suddenly or gradually die. May collapse. Near soil line stem may be water-soaked or discolored, brown, bleached white, black, or covered with cottony mold. Roots (and tuber) often decayed or may rot in storage. Common on heavy, wet soils. Seedlings wilt and collapse (damp-off). See Figure 2.39A, (21) Crown Rot, and (34) Root Rot under General Diseases. Often associated with borers or root-feeding nematodes. *Control:* Dig and destroy infected plants. Take tip cuttings from vigorous, disease-free plants and grow in sterilized rooting medium (pages 538–548), or propagate from disease-free seed. Plant in warm, rich, well-drained soil in sunny location. Five-year rotation. Grow resistant varieties, if available. Spray as for Leaf Spots (above). Avoid overcrowding, overwatering, injuring stems or roots, overstimulation with fertilizer, wet mulches,

and planting in cold, wet, infested soil. Spot drench with PCNB 75 ($\frac{1}{2}$ pound per 100 square feet) and captan 50 (1 pound in 15 gallons per 100 square feet), ferbam 76 (2 pounds per 100 square feet), or apply Dexon or Morton Soil Drench following manufacturer's directions. PCNB-Dexon 35–35 mixture is very effective. Use $1\frac{1}{2}$ pounds per 400 square feet of bed. Water in well. Apply above treatments when crown rot is *first* evident. Plant disease-free seed treated with captan, thiram, or chloranil. See Table 20 in Appendix.

4. *Aster Yellows, Stunt*—General and serious. Symptoms variable. Leaves yellowish. Plants often stiff, stunted or dwarfed, and "bushy" with numerous, yellowish, upright, spindly shoots. Leaves often mottled, stunted, and distorted. Older leaves may be reddish or purplish. Flowers may be deformed and yellowish-green (leafy), dwarfed, or absent. Bloom often greatly reduced, especially in future years. Infected plants never recover. See Figures 2.36A and D under General Diseases and Figure 4.66. *Control:* Grow plants under fine cheesecloth (22 threads per inch), wire (18 wires per inch), or apply insecticide (e.g., DDT plus malathion) at least weekly. This controls leafhoppers which transmit the virus. Destroy infected plants. Keep down all weeds within 200 feet of flower beds, if possible. Plant certified, virus-free stock from a reputable nursery.

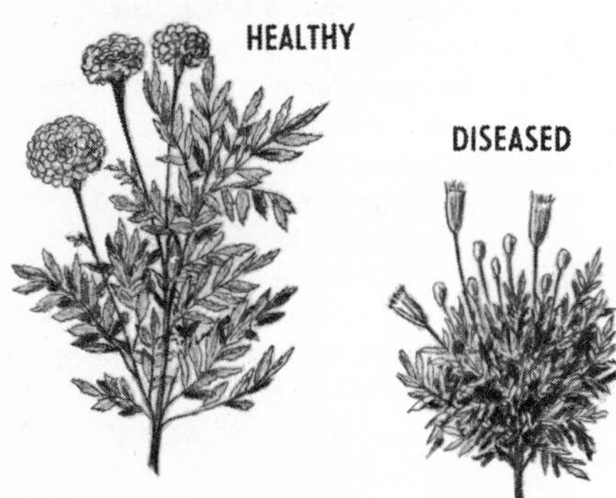

Fig. 4.66. Aster yellows of marigold.

Fig. 4.67. Chrysanthemum rust.

5. *Mosaics, Mottles, Dwarf, Stunt, Leaf Curl, Rosettes*—General. A virus complex. Plants often stunted or dwarfed and "bushy" with curled, crinkled, dwarfed, distorted, mottled, yellow, and light or dark green leaves. Bloom may be greatly reduced, especially in future years. Infected plants never recover; often winter-kill more easily than healthy plants. Flowers often stunted or open prematurely. May show streaks or develop distorted petals. Symptoms may be masked in hot weather or are never expressed in certain varieties. See (16) Mosaic, and (18) Yellows (Stunt) under General Diseases. *Control:* Remove and burn infected plants when first seen. Plant virus-free seed, tubers, cuttings, and transplants or propagate only from known disease-free plants. Cultured *chrysanthemum* cuttings, free from several viruses, are available. Keep down weeds. Control aphids and leafhoppers that transmit viruses. Use DDT, sevin, lindane or methoxychlor plus malathion. Wash hands and cutting knife thoroughly in hot running water after touching a stunted plant.

6. *Ringspot, Oakleaf Disease* (primarily aster, calendula, China-aster, chrysanthemum, dahlia, strawflower, sunflower, zinnia)—Symptoms variable. Irregular, pale green to bright yellow or yellowish-green spots, single or zoned rings, oakleaf or irregular zigzag markings, other line patterns or mottling on leaves. See Figure 2.35B under General Diseases. Leaves may tend to outgrow symptoms. *Control:* Same as for Mosaics (above).

7. *Spotted Wilt, Ringspot, Flower Distortion, Tomato Aspermy*—Symptoms variable. Pale yellow to dark green spots, ring or irregular line patterns develop in leaves of some varieties. *Chrysanthemum* plants frequently one-sided. Leaves may later be distorted or appear bronzed, often with slight mottling. Dead areas or streaks may appear in leaves and stem. Young plants commonly killed. Older plants may wilt, turn yellow, and die. Flowers often dwarfed, streaked, and distorted. See (17) Spotted Wilt under General Diseases. *Control:* Same as for Mosaics (above). Spotted wilt can be eliminated in *dahlias* by propagating with tip cuttings. Control thrips and aphids that transmit the viruses. Use mixture of DDT or sevin plus malathion. Persian Carpet *zinnia* is highly resistant to Tobacco Ringspot virus. Keep down weeds (e.g., knotweed, plantains, mallows, etc.).
8. *Streak* (cineraria)—Leaves roughened and curled upward. Reddish-brown, irregular areas develop in leaf veins, petioles, and stalks. These areas enlarge, killing large, triangular areas in leaves. Entire leaves or whole plant may die. Wilting, rolling, yellowing, and death of leaves may occur near blossoming. *Control:* Plant virus-free seed. Control thrips that transmit the virus. Destroy affected plants when first noticed.
9. *Curly-top*—Plants "bushy" with shoot tips turning yellow. Leaflets often curled and twisted with petioles turning down. Flower buds and flowers are dwarfed. See Figure 2.37B. *Control:* Same as for Aster Yellows (above).
10. *Fusarium Wilt, Stem Rot*—General and serious on China-aster and marigold at high soil temperatures. Symptoms variable. Plants may wilt, and gradually or suddenly wither and die at any age. Stem may be water-soaked or darkened near soil line. Plants often stunted, yellowish, show one-sided growth. Young plants suddenly wilt and die. Roots often decayed. Internal brown to black streaks when stem is cut. See Figure 2.31D under General Diseases. *Control:* Grow wilt-resistant (WR) varieties of *China-aster, chrysanthemum, safflower,* and *African marigold.* Plant disease-free seed, tubers, or plants in clean or sterilized soil (pages 538–548). Or soak *China-aster* seed 30 minutes in a 1:1,000 solution of mercuric chloride (page 101 for precautions). Then wash thoroughly for 5 minutes in running water, dry at room temperature, and dust with thiram, chloranil, or captan. Five- to 6-year rotation. Avoid wounding roots and crown. Grow in well-drained, disease-free soil where wilt has not been a problem.
11. *Phialophora or Verticillium Wilt*—Widespread. Symptoms variable. Whole plant may be stunted with pale green to yellow or brown leaves. Leaves wilt, wither, die, and cling to stem starting usually at base, resembling leaf nematode injury (see below). Often starts on one side of plant. *Safflower* leaves may be yellowed only on one side of midrib and distorted. Plants may ripen prematurely. Shoot tips may be blighted. Dark streaks occur inside lower stem. *Chrysanthemum* and *safflower* varieties differ in resistance. See (15B) Verticillium Wilt under General Diseases. *Control:* Dig up and burn infected plants. Otherwise same as for Fusarium Wilt (above). Cultured cuttings of *chrysanthemum* are available.
12. *Bacterial Wilt*—Mostly in southern states or in greenhouses. Plants suddenly wilt and collapse. "Recover" at night for a few days before plants dry up and die. There is usually a soft, wet rot inside stem near soil line. *Control:* Plant disease-free seed or cuttings in clean or sterilized soil. Dig up and burn infected plants. Indoors, keep temperature below 65° F. and humidity as low as practical. Cultured *chrysanthemum* cuttings are available.
13. *Rusts*—General. Small, bright orange, yellowish-orange, reddish-brown to chocolate-brown, dusty pustules mostly on underside of leaves (Figure 4.67). Leaves may turn yellow, wither, and die. If severe, plants may be stunted. Infected *safflower* seedlings swell, often bend and twist, collapse, usually die. Alternate hosts may include pines, sedges, rushes, and wild grasses. *Control:* If practical, pick and burn

infected leaves. Burn tops in fall or turn under cleanly. Rotate. Take cuttings from healthy plants. Destroy alternate hosts where feasible. Indoors avoid sprinkling foliage. Keep humidity below 85 per cent and increase air circulation. Plant in well-drained soil. Apply zineb, ferbam, thiram, maneb, or Dithane S-31 at 5- to 10-day intervals during wet periods. Start in early summer. Normally resistant varieties of *China-aster, safflower, sunflower,* and possibly other plants are available. Treat *safflower* seed with Panogen or Ceresan and store, if possible, for 1 to 2 months or longer before planting. Acti-dione (1 to 20 parts per million) as a 30-minute soak or dichlone dust are also recommended.

14. *Gray-mold Blight, Botrytis Petal Spot or Blight, Head or Blossom Blights, Ray or Speck Blight, Stem Blight, Bud Rot*—Cosmopolitan during humid, overcast weather. Small to large, water-soaked to tan, brown or black spots or streaks on flower petals, bracts, leaves, buds, and stems. Stems (cuttings) may rot causing foliage beyond to wilt, wither, and die. Coarse, tannish-gray or cottony mold may cover affected parts in damp weather. Seedlings wilt and collapse. Common on fading flower heads. Flowers may be deformed, one-sided, or blasted. Buds may rot. See Figures 2.21D and 2.47D under General Diseases and Figure 4.68. *Control:* Destroy fading flower heads promptly. Space plants and keep down weeds. Avoid overwatering, overfertilizing (especially with nitrogen), and forcing plants too rapidly. Indoors, same as for Rusts (above). Spray buds and blooms lightly (1 tablespoon per gallon) with maneb, zineb, or captan at 3- to 4-day intervals. Apply maneb, zineb, dichloran (Botran), Daconil 2787, ferbam, dichlone, or captan,

Fig. 4.68. *Stemphylium* petal specking and leaf blight of "Charm" chrysanthemum. The dying leaves produce large numbers of fungus spores. (Courtesy Dr. C. R. Jackson)

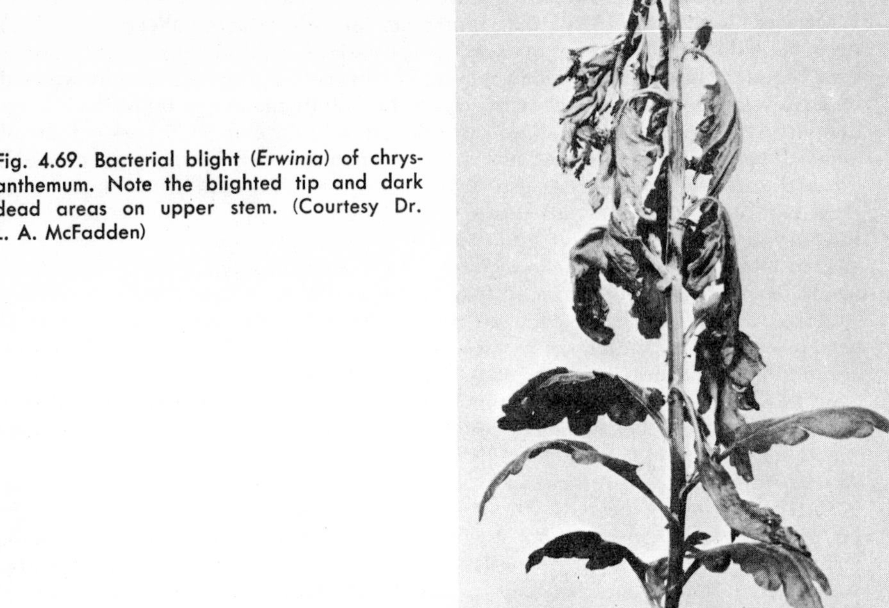

Fig. 4.69. Bacterial blight (*Erwinia*) of chrysanthemum. Note the blighted tip and dark dead areas on upper stem. (Courtesy Dr. L. A. McFadden)

plus wetting agent, twice weekly during wet weather. Spot drench soil with PCNB 75 just before planting. Destroy plant debris in fall after harvest. *Chrysanthemum* varieties differ in resistance to Itersoniiia Petal Blight and Botrytis.

15. *Bacterial Blight or Stem Rot* (chrysanthemum, safflower)—Symptoms variable. Growing tips often wilt, turn dark brown or black and collapse from one or more water-soaked to dark brown, black, or reddish-brown, rotted spots or streaks on stem that may extend to soil line. Stems may wilt and collapse (Figure 4.69). *Chrysanthemum* cuttings rot at base. Insides of stems are reddish-brown and soft at first; later are hollow. *Safflower* leaves develop reddish-brown spots with pale green margins. *Control:* Snap off cuttings from healthy plants. Grow in sterilized soil. Destroy infected plants and cuttings when first found. Cultured cuttings are available for *chrysanthemum.* Indoors, avoid high humidity and temperature, overwatering, and overfertilizing with nitrogen.
16. *Stem Cankers*—Discolored cankers form on stem near soil line or where branches arise. Stems may be girdled causing death to the top of the plant. Roots and crown are usually healthy. *Control:* Keep base of plants dry. Spray as for Leaf Spots and Rusts (both above). Burn tops in fall. Destroy infected plants when found.
17. *Root-knot*—Plants weak, pale, make poor growth. Knots or swellings on roots. *Dahlia, calendula* and *safflower* are highly susceptible; *marigolds* are highly resistant. See Figure 2.52 under General Diseases. *Control:* Rotate. Grow in sterilized soil (pages 538–548) where feasible. *Transvaal daisy* can be freed of root-knot by soaking bare-root plants in hot water (118° F.) for 30 minutes.
18. *Downy Mildews*—Widespread. Pale green or light yellow spots or mottling on upper leaf surface with delicate, whitish mold on corresponding underleaf surface. Leaves shrivel and die. Seedlings or older plants may wilt and die. *Control:* Same as for Gray-mold Blight (above). Sow disease-free seed.

19. *Leaf, Stem, or Leaf Gall Nematodes* (aster, balsamroot, calendula, chrysanthemum, dahlia, groundsel, leopardsbane, pyrethrum, senecio, sunflower, wyethia, zinnia) — Widespread and serious on chrysanthemum in wet seasons. Wedge-shaped to irregular, yellowish-brown to gray areas in leaves, bounded by larger veins. Blotches turn brown to black, enlarge, and merge. Leaves wither, die, and hang downward on stem starting at base. Plants lack vigor. May die prematurely. Buds and flowers develop improperly. Longitudinal, rust-brown cracks develop in tuberous roots of *dahlia*. Varieties differ in resistance. Plants stunted in early spring with dwarfed, distorted, and crinkled leaves. See Figure 2.38A under General Diseases. *Control:* Take cuttings from tips of tall, disease-free plants. Burn *all* tops and fallen leaves in autumn. Pick and burn infested leaves and two healthy-appearing leaves above, as they develop. Mulch plants to avoid splashing water on foliage. Rotate at least 3 years. Do not propagate from infested clumps. Space plants. Avoid overhead watering. Grow in disinfested soil (pages 538–548). Nurserymen soak *dormant chrysanthemum* stock plants ("stools") in hot water (118° F.) for 15 minutes or 112° F. for 30 minutes. The treatment is often injurious.
20. *Crown Gall*—Plants stunted with spindling shoots. Large, gall-like tumors at base of plant, on roots, or both places. *Control:* See (30) Crown Gall under General Diseases. Pull up and burn affected plants.
21. *Storage Rots* (dahlia, Jerusalem-artichoke) —Tubers or roots break down and decay in storage. Are often covered with white to pink, tan, or bluish mold. *Control:* Do not dig until tubers are mature, but before frost injury occurs. Store only sound, blemish-free, thoroughly dried tubers (roots) in a dry, cool (40° F.) location. Storage areas should be clean and washed with formalin (1 part in 50 of water) following heavy storage losses.
22. *White or Leaf Smuts* (arnica, aster, boltonia, calendula, chrysanthemum, dahlia, eupatorium, fleabane, gaillardia, prairie-coneflower, rudbeckia, senecio, silphium, sneezeweed, sunflower) —Round to irregular, yellowish-green leaf spots that turn brownish to black. See Figure 2.29. Spots may merge causing withering of leaves; or spots may drop out leaving shot-holes. Flowers may be few and worthless. Leaves may drop early. *Control:* See (13) White Smut under General Diseases. Indoors, keep water off foliage. If severe, apply captan or copper sprays. *Dahlia* varieties differ in susceptibility to Leaf Smut.
23. *White-rust*—Pale yellowish spots on leaves that later turn brown. Snow-white, powdery pustules develop on lower leaf surface. Leaves may die. Plants often dwarfed. *Control:* Collect and burn infected plant parts and destroy all plant debris after harvest. Keep down weeds. See also (9) White-rust under General Diseases.
24. *Fasciation, Leafy Gall* (primarily chrysanthemum, dahlia, piqueria, pyrethrum, Shasta daisy) —Mass of stems, shortened and thick with cauliflower-like growth at soil line. Leaves aborted and misshapen. Plants dwarfed with abundant buds and distorted shoots. Clumps may rot. *Control:* See (28) Leafy Gall under General Diseases.
25. *Black Mold*—Sootlike patches on foliage. See (12) Sooty Mold under General Diseases.
26. *Scab* (dahlia, tickseed) —See under Beet, and (14) Scab under General Diseases.
27. *Dodder*—See Figure 2.46 under General Diseases.
28. *2,4-D Injury*—Chrysanthemum and related plants are very susceptible. Leaves and stems are curled and distorted. See under Grape.
29. *Other Root-feeding Nematodes* (aphelenchoides, bulb and stem, dagger, lance, lesion, reniform, ring, spiral, sting, stubby-root, stunt, tylenchus) —Associated with stunted, sickly plants. Roots may be short, stubby, discolored, and decayed. *Dahlia* roots infested with the potato-rot nematode show abnormal transverse cracking and

flaking off of the outer surface. Internal rot is dark, dry, and granular. *Control:* See under Root-knot (above). *Marigolds* are highly resistant to root-feeding nematodes (e.g., dagger, lesion, pin, ring, root-knot, spiral, sting, stubby-root, stunt).

30. *Hopperburn* (primarily dahlia)—Leaf margins turn yellow, then scorched. Plants may be stunted and yellowish. *Control:* Spray with DDT plus malathion to control leafhoppers.
31. *Inflorescence Smut* (plumed thistle, thistle)—See under Carnation.
32. *Yellow Strapleaf* (chrysanthemum)—Cause unknown. Symptoms widely variable, depending on variety and age. Terminal leaves are strap-shaped and pale yellow or ivory. Growth is stunted. Several new leaves may be yellowed and cupped, followed by normal growth. Flowers on "recovered" plants may be distorted with green centers. *Control:* Unknown. Lifting and resetting young plants may be beneficial.
33. *Chlorosis*—Mineral deficiency in alkaline or very acid soil. See under Rose.

For additional information on chrysanthemums read USDA H & G Bulletin 65, *Growing Chrysanthemums in the Home Garden.*

CHRYSOBALANUS, COCOPLUM *(Chrysobalanus)*

1. *Leaf Spot*—See (1) Fungus Leaf Spot under General Diseases.
2. *Algal Spot*—See under Magnolia.

CHRYSOPSIS — See Chrysanthemum
CHUFA — See Umbrellaplant
CHUPEROSA — See Clockvine
CHYSIS — See Orchids
CIBOTIUM — See Ferns
CICHORIUM — See Lettuce
CIGARFLOWER, FIRECRACKER PLANT *(Cuphea)*

1. *Gray-mold Blight*—Occasional in greenhouses. See under Begonia, and (5) Botrytis Blight under General Diseases.
2. *Powdery Mildew*—See (7) Powdery Mildew under General Diseases.
3. *Root Rot*—See under Geranium, and (34) Root Rot under General Diseases.
4. *Root-knot*—See (37) Root-knot under General Diseases.
5. *Leaf Spot*—See (1) Fungus Leaf Spot under General Diseases.

CIMICIFUGA — See Anemone
CINCHONA — See Buttonbush
CINERARIA — See Chrysanthemum
CINNAMON-TREE *(Cinnamomum)* — See Avocado
CINNAMONVINE — See Yam
CINQUEFOIL — See Rose
CIRSIUM — See Chrysanthemum
CISSUS — See Grape
CITRON — See Cucumber
CITRUS: CALAMONDIN or MINIATURE ORANGE, LEMON [CHINESE, MEYER, PONDEROSA], LIME [ACID, SWEET], ORANGE [SOUR or SEVILLE, KING, OTAHEITE, SATSUMA, MANDARIN, COMMON or SWEET], GRAPEFRUIT, PUMMELO, TANGERINE, TANGELO *(Citrus)*; KUMQUAT [MEIWA, OVAL or NAGAMI, ROUND or MARUMI] *(Fortunella)*; HARDY or TRIFOLIATE ORANGE, CITRALDIN *(Poncirus)*

1. *Root Rots, Collar or Foot Rot, Tree Decline*—Foliage is stunted, sparse, often

yellowish. Leaves drop early. Growth is slow and reduced. Twigs and branches die back. Seedlings may wither and die. Roots dark and decayed; frequently slough off. Often associated with nematodes (e.g., burrowing, citrus-root, dagger, lance, lesion, needle, reniform, ring, root-knot, sheath, sheathoid, spiral, sting, stubby-root, stunt). *Control:* Plant certified nursery stock in sterilized or fumigated soil (pages 538–548) that is light and well-drained. Avoid overwatering and overfertilizing with an ammonia-containing fertilizer. Commercial growers soak bare-rooted nursery stock in hot water (122° F.) for 10 minutes before planting. *Citrus* rootstocks differ in resistance to Thielaviopsis and Phytophthora Root Rots as well as to citrus-root and burrowing nematodes.

2. *Chlorosis, Leaf Yellowing, Mottle-leaf*—Leaves turn yellow, yellowish-green or remain darker green along veins. May fall early. Plants stunted and lack vigor. Twigs may die back. See also under Gardenia. *Control:* Plant in fertile, slightly acid (pH 6.5), sterilized, coarse, well-drained soil. Avoid overwatering, overfertilizing, adding excess lime, and great temperature changes. Repot if roots are potbound. Control insects and mites with malathion sprays. Place in sunny location. *Trifoliate orange rootstock* is more susceptible to zinc and iron deficiencies than other rootstocks used for citrus.
3. *Leaf Spot, Anthracnoses, Wither Tip, Twig Blight*—General. Leaves spotted. Shoots and twigs may wither and die back. Gum may exude from wounds. Buds may not open and often drop. *Control:* Indoors, keep water off foliage. Keep trees vigorous. Fertilize and water, and control insects with malathion. Pick or prune off and burn blighted parts. Outdoors, spray during rainy periods. Use zineb, ferbam, maneb, or captan.
4. *Sooty Molds, Sooty Blotch*—Black mold patches on leaves, twigs, and fruits. Follows attacks by aphids, scales, mealybugs, whiteflies, and other insects. *Control:* Apply malathion to control insects.
5. *Crown Gall*—See under Apple, and (30) Crown Gall under General Diseases.
6. *Other Diseases*—If grown outdoors, in citrus-growing areas, many other diseases could occur. Check with your county agent or extension plant pathologist. Excellent bulletins, e.g., *Handbook of Citrus Diseases in Florida* and *Florida Guide to Citrus Insects, Diseases, and Nutritional Disorders,* can be obtained by writing to the Florida Agricultural Experiment Station. The University of California Press, Berkeley, has published *Color Handbook of Citrus Diseases,* another excellent guide.

CLADRASTIS — See Honeylocust

CLARKIA — See Fuchsia

CLEMATIS [ALPINE, BLUE, GOLDEN, HENRY, HILL, JACKMAN, JAPANESE or SWEET AUTUMN, MANCHURIAN, SCARLET], VIRGINSBOWER, LEATHER FLOWER, VINE BOWER *(Clematis)*; YELLOWROOT or SHRUB-YELLOWROOT *(Xanthorhiza)*

1. *Leaf Spots, Leaf Blight*—Widespread. Water-soaked, tan, gray, reddish-brown, or brown spots of various sizes and shapes. Leaves may wither and drop early. *Control:* If practical, remove and burn infected leaves. Space and support vines. Plant in new location. Apply zineb, ferbam, captan, maneb, thiram, sulfur, or fixed copper at 7- to 10-day intervals during wet weather. Start when new growth appears.
2. *Stem Rot, Wilt, Dieback* (clematis)—Foliage wilts, withers, and dies from girdling dark stem canker near soil line. Roots may decay. *Control:* Propagate from disease-free plants; or sow seed in new location. Otherwise, same as for Leaf Spots. Grow in rich, moist, well-drained, neutral soil. Remove and burn diseased stems, cutting well below diseased area. Avoid matting of vines.

3. *Powdery Mildews*—Widespread. Powdery, white mold on clematis foliage. Sometimes on flower petals. Large-flowered varieties appear more susceptible. *Control:* Spray several times, 10 days apart, using Karathane.
4. *Root-knot, Cyst Nematode*—Clematis is very susceptible. Plants stunted and pale with small galls formed on roots. Plants may wilt, wither, and die. *Control:* Remove and burn heavily infested vines. Plant disease-free plants in clean or sterilized soil (pages 538–548).
5. *Rusts* (clematis)—Small, yellowish spots on leaves. Alternate hosts include many native grasses. *Control:* Same as for Leaf Spots (above). *Clematis* varieties differ in resistance.
6. *Crown Gall* (clematis)—See (30) Crown Gall under General Diseases.
7. *Mosaic* (clematis)—See (16) Mosaic under General Diseases.
8. *Smut* (clematis)—See (11) Smut under General Diseases.
9. *Other Root-feeding Nematodes* (e.g., dagger, lesion, pin, spiral, stunt)—Often associated with weak, pale, stunted plants. *Control:* Same as for Root-knot (above).
10. *Black Mildew* (clematis)—See (12) Sooty Mold under General Diseases.

CLEOME — See Spiderflower

CLERODENDRON — See Lantana

CLETHRA — See Sweet-pepperbush

CLIFFGREEN — See Bittersweet

CLIFFROSE — See Rose

CLIFTONIA — See Buckwheat-tree

CLIMBING HEMPWEED — See Chrysanthemum

CLIMBING MIGNONETTE — See Lythrum

CLINOPODIUM — See Salvia

CLITORIA — See Pea

CLOCKVINE [BUSH, COMMON], BLACK-EYED SUSAN VINE, MOUNTAIN CREEPER, SKY-FLOWER *(Thunbergia)*; CHUPEROSA, SHRIMP-PLANT *(Beloperone)*; DYSCHORISTE; ERANTHEMUM; RUELLIA; SANCHEZIA

1. *Leaf Spots*—Spots of various sizes, shapes, and colors. *Control:* Pick and burn spotted leaves. Spray during rainy periods using ferbam or zineb.
2. *Rusts* (chuperosa, dyschoriste, ruellia)—Yellow, orange-yellow, reddish-brown or black, powdery pustules on leaves. *Control:* Same as for Leaf Spots.
3. *Crown Gall*—Rough, swollen galls form at or near soil line. See (30) Crown Gall under General Diseases.
4. *Oedema* (eranthemum)—Indoor problem. "Sandy," rust-colored, swollen spots. Mostly on upper leaf surface. *Control:* Avoid overwatering in overcast, damp weather. Increase air circulation.
5. *Root-knot*—Clockvine is very susceptible. See (37) Root-knot under General Diseases.
6. *Aster Yellows*—See (18) Aster Yellows under General Diseases.
7. *Root Rot*—See (34) Root Rot under General Diseases. May be associated with root-feeding nematodes (e.g., burrowing, lesion, root-knot).

CLOUDBERRY — See Raspberry

CLOVETREE — See Eucalyptus

CNICUS — See Chrysanthemum

COCCULUS — See Moonseed

COCHLEARIA — See Cabbage

COCKSCOMB *(Celosia)*; GLOBE-AMARANTH *(Gomphrena)*; AMARANTH,

LOVE-LIES-BLEEDING, JOSEPHSCOAT, PRINCESFEATHER, TAMPALA *(Amaranthus)*; GARDEN ALTERNANTHERA, MINIATURE JOSEPHSCOAT, INDOOR CLOVER *(Alternanthera)*; BLOODLEAF, CHICKEN GIZZARD PLANT *(Iresine)*; FROELICHIA

1. *Seed Rot, Damping-off, Blight*—Seeds rot. Seedlings wilt and collapse from water-soaked, blackish-brown spots or rot near soil line. Older plants show rusty-brown, zoned spots on leaves, petioles, and stems. Spots dry out and become cankers. *Control:* See under Beet. Pull and burn badly infected plants. Avoid overwatering. Rotate or grow in sterilized soil (pages 538–548). Apply Dexon as soil drench (½ teaspoon per gallon). Follow manufacturer's directions.
2. *Leaf Spots, Leaf Blight*—Spots of various sizes, shapes, and colors. Leaves may curl up and fall early. *Control:* See under Chrysanthemum.
3. *Curly-top, Yellows*—See under Beet, and (18) Yellows and (19) Curly-top under General Diseases. Plants stunted and yellowed.
4. *Root-knot*—See (37) Root-knot under General Diseases.
5. *White-rust* (alternanthera, amaranth, froelichia, globe-amaranth)—White, powdery spots on leaves that turn reddish-brown. Flowers and stems may be stunted and distorted. *Control:* See under Chrysanthemum.
6. *Leaf Roll* (amaranth, cockscomb)—See under Potato.
7. *Black Ringspot* (cockscomb)—See under Cabbage.
8. *Root and Crown Rots* (alternanthera, amaranth, bloodleaf, cockscomb, froelichia)—Cuttings and young plants are stunted, often wilt and die. Roots and crowns decayed. *Control:* Grow in sterilized soil, or treat soil with PCNB before planting.
9. *Fusarium Wilt* (alternanthera)—See under Chrysanthemum.
10. *Inflorescence Smut* (bloodleaf)—See (11) Smut under General Diseases.
11. *Blossom Blight* (amaranth)—See (31) Flower Blight under General Diseases.
12. *Other Root-feeding Nematodes* (e.g., lesion, pin, spiral)—Associated with pale, weak, stunted plants. *Control:* Same as for Root-knot (above).
13. *Mosaic* (amaranth, cockscomb)—See (16) Mosaic under General Diseases.
14. *Gray-mold Blight*—See (5) Botrytis Blight under General Diseases.
15. *Ringspot* (amaranth, globe-amaranth)—Yellowish or brownish spots and rings on leaves. See (17) Ringspot under General Diseases.
16. *Verticillium Wilt* (cockscomb)—See (15B) Verticillium Wilt under General Diseases.

COCONUT *(Cocos)* — See Palms

COCOPLUM — See Chrysobalanus

CODIAEUM — See Croton

COELOGYNE — See Orchids

COFFEEBERRY — See Buckthorn

COLCHICUM, AUTUMN-CROCUS, MEADOW SAFFRON *(Colchicum)*; CAMASS, WILD HYACINTH, BEARGRASS *(Camassia)*

1. *Smut*—Small, swollen "blisters" (spots or stripes) on leaves, leaf sheaths, corms, stems, and flowers. Blisters filled with black, powdery masses. Varieties differ in resistance. *Control:* Pick and burn affected parts before blisters open.
2. *Leaf Spot*—Small spots on leaves. *Control:* Same as for Botrytis Leaf Spot (below).
3. *Corm and Bulb Rots*—Corms and bulbs rot in field or storage. *Control:* Grow in well-drained soil. See also under Tulip.
4. *Botrytis Leaf Spot and Tip Blight*—See under Tulip, and (5) Botrytis Blight under General Diseases.

5. *Root Rot* (camass)—See (34) Root Rot under General Diseases. Grow in rich, moist, loamy soil. Plant in fall and do not disturb.

COLEUS — See Salvia
COLLARD — See Cabbage
COLLINSIA — See Snapdragon
COLLOMIA — See Phlox
COLOCASIA — See Calla Lily
COLTSFOOT — See Chrysanthemum
COLUMBINE — See Delphinium
COLUMBO — See Gentian
COLUTEA — See Honeylocust
COMMELINA — See Tradescantia
COMPASSPLANT — See Chrysanthemum
COMPTONIA — See Sweetfern
CONEFLOWER — See Chrysanthemum
CONFEDERATE-JASMINE — See Oleander
CONVALLARIA — See Lily
CONVOLVULUS — See Morning-glory
COOPERIA — See Daffodil
COPPERLEAF — See Acalypha
COPPER-TIP — See Gladiolus
COPTIS — See Delphinium
CORAL BEADS — See Sedum
CORALBEAN, CORAL-TREE — See Honeylocust
CORALBELLS — See Hydrangea
CORALBERRY — See Snowberry

CORDIA, ANACAHUITA, GEIGER-TREE *(Cordia)*

1. *Powdery Mildew*—See (7) Powdery Mildew under General Diseases.
2. *Root Rots*—See under Apple, and (34) Root Rot under General Diseases.
3. *Rust*—See (8) Rust under General Diseases.

CORDYLINE — See Dracaena
COREOPSIS — See Chrysanthemum
CORIANDER *(Coriandrum)* — See Celery
CORONILLA — See Pea

CORN [BROOM, ORNAMENTAL or INDIAN, POP, SWEET] *(Zea)*

1. *Bacterial Wilt, Stewart's Disease*—General. May be serious. Most severe in northern states following mild winters. Long, white, yellowish or pale green streaks or spots in leaves that later turn brown. Seedlings or older plants often stunted. May wilt rapidly and die. Plants often develop a premature white tassel. Leaves dry and appear frosted. Sweet corn and popcorn are quite susceptible. See Figure 2.33A under General Diseases and Figure 4.70. *Control:* Sow disease-free seed. Spray with sevin or DDT 1 to 6 times, 3 to 5 days apart. Start a day *before* corn emerges. This controls the small corn flea beetles that transmit the wilt-producing bacterium. Check with your county agent or extension entomologist regarding timing of sprays for your area. Somewhat resistant *sweet corn* varieties: Aristogold Bantam Evergreen, Atlas, Barbeque, Calumet, Carmelcross, Country Gentleman W-R, Duet, Early Gold Crest, Erie, Evergreen Hybrid, F-M Cross, Golden 22, 25, 50 and 60, Golden Bantam, Golden Beauty, Golden Cross W-R, Golden Fancy, Golden

Fig. 4.70. Bacterial wilt of corn. (Iowa State University photo)

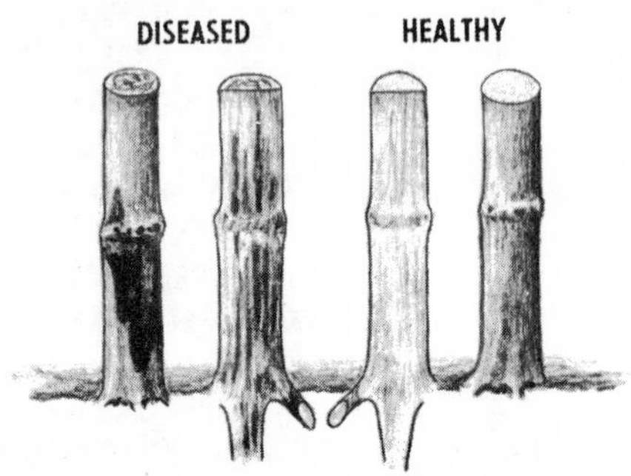

Fig. 4.71. Stalk rot of corn.

Harvest, Golden Jewell, Golden Pirate, Golden Security, Gold Rush, Honeycross, Hoosier Gold, Huron, Hybrid E 2125, Ioana, Iochief, Iogold, Marcross, Merit, Midway, N.K. 199, Northern Belle, Northern Cross, North Star, Seneca, Seneca Arrow, Seneca Chief, Seneca Dawn, Seneca Market, Spancross, Tendermost, Titan, Whipcross, White Cross Bantam, Wintergarden, Wintergreen, and many more. Somewhat resistant *popcorn* varieties: South American and Sunburst. Check with your county agent or extension horticulturist regarding adaptability of these varieties to your area.

2. *Smuts*—General, especially following hail, insect, or mechanical injury. Small to large, silvery-white galls or "boils" on tassel, ears, husks, leaves, stalk, or prop roots. Membrane breaks releasing dark brown to black, sooty masses. Stalks may be barren. Sweet corn is very susceptible. See Figure 2.27A under General Diseases. *Head smut,* found primarily in northwestern intermountain areas of the U.S., often causes plants to be dwarfed. Ears and tassel are aborted, barren, or converted to a smut gall. *Control:* Where practical, cut out and burn galls before they break. Rotate. Control insects (e.g., corn earworm, European corn borer) by sevin sprays. Check with your county agent for correct timing. Partially resistant *sweet corn* varieties: Aristogold Bantam Evergreen, Asgrow Golden 60, Burpee's Honeycross, Country Gentleman, Evertender, Giant Bantam Hybrid, Golden Cross Bantam, Golden Hybrid 2057, Ioana, Iochief, Mellow Gold, Pennlewis, Prospector, Seneca Brave, Tenderblonde, and Victory Golden. Differences also exist between *popcorn* hybrids. Avoid smutty plants in making compost.

3. *Root and Stalk Rots*—General. Plants often lean, break, or blow over. May die prematurely. Plants may be stunted, lack vigor. Stalks weak, rotted internally at base. Both fine and larger roots decay. Often associated with root-feeding nematodes (see below). When severe, plants may turn grayish-green, wither, and die prematurely, producing poorly filled or nubbin ears. Excessive nitrogen fertilizer, insect damage, hail, and severe leaf blight, increase stalk rotting and breakage. See Figure 4.71. *Control:* Treat seed as for Seed Rot (below). Grow in

warm, fertile, well-drained soil. Maintain balanced soil fertility based on a soil test. Water during droughts. Avoid excessively deep and close cultivations. Use locally adapted, resistant hybrids whenever available. Control soil insects (e.g., wireworms, grubs, rootworms, flea beetles, seed corn maggot, seed corn beetle, etc.) using Diazinon or chlordane. Apply to soil surface before planting (or with fertilizer) and work into soil. Follow manufacturer's recommendations and those of your county agent and extension entomologist. Rotate corn with other crops. Burn, bury, or compost stalks after harvest.

4. *Ear and Kernel Rots*—Cosmopolitan. Ears and husks may be partially or completely rotted, lightweight, and white, gray, blue-green, pink to reddish, or black. Plants may be stunted. Often associated with Root and Stalk Rots. Rotted shanks and ears may break over early. *Control:* Same as for Root and Stalk Rots (above). In addition, handle carefully at harvest. Cool immediately. Control earworms, Japanese beetles, and sapbeetles by sevin sprays into silks. Check with your county agent or extension entomologist regarding timing.

5. *Seed Rot, Seedling Blights*—General. Seeds rot. Stand is poor, especially in cold (50° F. or under), wet soil. Growth is uneven. Seedlings are yellow, stunted, may wilt and die. *Control:* Sow seed treated with captan, thiram, chloranil, or dichlone plus an insecticide (e.g., dieldrin, chlordane, lindane). See Table 20 in Appendix. Plant mature, high-quality, crack-free seed in warm, well-drained soil. Treat for soil insects as given under Root and Stalk Rots (above).

6. *Helminthosporium Leaf Blights*—General in humid areas. Small to large, grayish-green to tan or brown spots (often elliptical-shaped) on lower leaves. Upper leaves may later become infected. Plants often appear suffering from drought or frost. Ears may be immature and chaffy. See Figure 4.72. *Control:* Hybrids and varieties show differences in resistance. Often practical to spray sweet corn in very humid areas (e.g., Gulf Coast), using zineb, maneb, or Dithane M-45, plus spreader-sticker, at about weekly intervals. Start just before silking or when first spots are evident. Thorough coverage is essential. Plant early. Burn or bury deeply all plant debris after harvest. Avoid overhead sprinkling and poorly drained soil. Resistant or tolerant *sweet corn* varieties: Florigold 107, Gold Cup G, Seneca Wampum, Silver Queen, Staygold, Surecrop, Sweetex No. 2, Wampum, Wintergarden, and

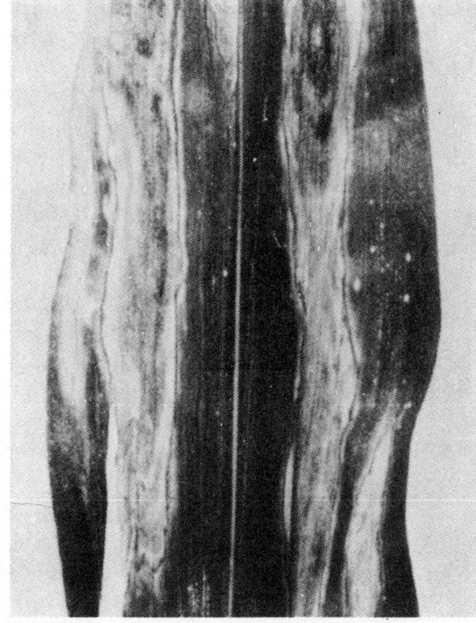

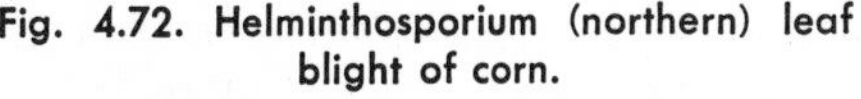

Fig. 4.72. Helminthosporium (northern) leaf blight of corn.

Wintergreen. Ladyfinger *popcorn* is highly resistant. Much more resistant varieties should be available shortly.

7. *Rusts*—General. Small, yellowish-orange to reddish-brown or cinnamon-brown, powdery pustules on both leaf surfaces that finally turn brownish-black. If numerous, leaves may die early. Alternate host of Common Corn Rust is wood sorrel *(Oxalis)*. *Control:* None usually needed. Spraying as for Helminthosporium Leaf Blights (above) is beneficial. Resistant *sweet corns:* Country Gentleman and Crosby. *Popcorn* is usually resistant.
8. *Bacterial Leaf Blights and Stalk Rots*—Minor problem. Small, yellow or tan to dark brown spots to long narrow stripes on leaves following hot, showery weather. Spots may enlarge and merge to form elongated or irregular blotches. Stalk may suddenly become water-soaked, slimy, and collapse. *Control:* None usually necessary. Seed treatment as for Seed Rot (above). Hybrids offer some hope. If irrigating, apply water at soil level only.
9. *Maize Dwarf Mosaic, Stunt*—Widespread. Most common near rivers. Plants stunted or dwarfed and bushy with yellowish-green spots, streaking, and mottling of youngest leaves. Older or upper leaves may show brilliant reddish-purple streaks, splotches, and bands. Tips and margins of leaves may be reddish-bronze. Leaves below ear are often dark green. Early-affected plants are barren, have a poor seed set, or may die early. Stunt virus is transmitted by leafhoppers; Maize Dwarf Mosaic virus by aphids. *Control:* Resistant varieties offer best hope. Destroy nearby Johnsongrass and gamagrass (*Tripsacum*), perennial hosts of the MDM virus.
10. *Mosaics, Leaf Fleck, Celery Stripe*—Southern states, but uncommon. Symptoms variable, may closely resemble Maize Dwarf Mosaic. Many small, broken, yellowish or bleached flecks, spots, and long streaks in leaves parallel with veins. Plants may be dwarfed and pale yellow. Others show new leaves almost free of stripes. Ears often absent, barren, or unmarketable. *Control:* None known. Resistant varieties may be available in the future. Aphids transmit the viruses.
11. *Minor Leaf Spots*—Mostly southern states. Small, round to elongate, light green, gray, tan to yellow, reddish-brown and water-soaked, or dark brown spots, blotches, or streaks on leaves and leaf sheaths. Spots may merge to form irregular blotches. Infected stalks may break easily. *Control:* Same as Helminthosporium Leaf Blights (above).
12. *Purple Sheath Spot*—Widespread, but minor problem. Irregular, purple-brown blotches on leaf sheaths. *Control:* Hybrids and varieties differ in resistance.
13. *Black Bundle*—Stalks and leaves are reddish-purple. Stalks often barren or produce small, poorly-filled ears. Black streaks inside stalk. Plants often tiller excessively. *Control:* None known.
14. *Witchweed (Striga)*—Eastern coastal sections of North and South Carolina. A total of some 262,500 acres and 11,380 farms in 34 counties is involved, but is being rapidly eradicated. Problem in light, sandy soils. Caused by a small, parasitic flowering plant *(Striga)*. Corn plants are severely stunted, yellowed. Wilt severely. Eventually leaves turn brown and plant dies. *Striga* feeds on roots of corn and related grassy plants (wheat, oats, barley, sorghum, sugarcane, and more than 60 species in the grass and sedge families). Witchweed plants rarely grow over 8 to 9 inches tall, with small, brick-red, yellowish-red, yellow, or white flowers. A single plant may produce ½ million seeds! Leaves are small, bright green, and slightly hairy. *Control:* If you suspect Witchweed, check with your county agent, extension weed specialist, or plant pathologist. Apply amine form of 2,4-D, where practical, *before* plants produce seeds. All crops must be kept free of weedy grasses, such as crabgrass. For further information read USDA Leaflet PA-331, *Watch out for Witchweed.*
15. *Root Nematodes* (awl, burrowing, dagger, lance, lesion, needle, pin, reniform, ring,

sheath, spiral, sting, stubby-root, stunt)—Most serious in southern states in light, sandy soils. Symptoms variable. Plants may be stunted and unthrifty. Do not respond normally to water and fertilizer. Often discolored; may be confused with one or more deficiency diseases. Roots often stunted (stubby), may be "bushy" and blackened, show dark spots, witches'-brooms, or galls. *Control:* Rotate. Plant early. Fertilize and water to keep plants vigorous.

16. *Crazy Top, Downy Mildew*—Widespread but uncommon. Tassel usually composed of a large, plumy mass of short leaves (Figure 4.73). Plants stunted, tiller excessively, with narrow straplike leaves. Stalks barren. *Control:* Avoid planting, where practical, in low, wet areas likely to be flooded after seeding. Or drain such areas. Corn varieties differ in resistance.

Fig. 4.73. Crazy top of corn.

17. *Chlorosis*—Deficiency of magnesium, manganese, zinc, iron, sulfur, or copper. See pages 26–28. Zinc, iron, magnesium, and manganese deficiencies appear as yellowing or white stripes between midrib and margin of affected leaves. These leaves commonly die early. Copper-deficient plants show yellowing of upper leaves. *Control:* Same as for Nitrogen Deficiency (below). See pages 29–30.
18. *Nitrogen Deficiency*—Leaves "fired" starting at bottom. Such leaves turn yellow starting at tip and proceeding down the midrib toward the stalk. Entire plant is light green. Nitrogen deficiency is accelerated by dry weather. *Control:* Practice balanced soil fertility based on a soil test.
19. *Potassium Deficiency*—Tips and leaf edges turn yellow, may later appear scorched. Midrib area remains green. Lower leaves are affected first. Ears are often small and pointed, with poorly developed tips. *Control:* See Nitrogen Deficiency (above).
20. *Phosphorus Deficiency*—Leaves develop strong purplish color at margins, starting at base of plant and progressing upward. Plants often unusually dark green and stunted. *Control:* Same as for Nitrogen Deficiency (above).

For additional information, read USDA Handbook 199, *Corn Diseases in the United States.*

CORNCOCKLE — See Carnation

CORNEL, CORNELIAN-CHERRY — See Dogwood

CORNFLAG — See Gladiolus

CORNFLOWER, CORNFLOWER ASTER, CORN-MARIGOLD — See Chrysanthemum

CORNSALAD — See Valerian

CORNUS — See Dogwood

CORYDALIS — See Bleedingheart
CORYLUS — See Birch
CORYPHANTHA — See Cacti
COSMOS, COSTMARY — See Chrysanthemum
COTINUS — See Sumac
COTONEASTER — See Apple
COTTON-ROSE — See Hollyhock
COTTONWOOD — See Poplar
COVENTRY-BELLS — See Bellflower
COWANIA — See Rose
COWBERRY — See Blueberry
COWSLIP — See Primrose
CRABAPPLE — See Apple
CRAMBE — See Cabbage
CRANBERRY, CRANBERRY-BUSH — See Viburnum

CRANESBILL, GERANIUM [BLOOD-RED, SPOTTED or WILD], HERB-ROBERT or RED-ROBIN *(Geranium)*; HERONSBILL, ALFILARIA or FILAREE *(Erodium)*

1. *Fungus Leaf Spots*—Spots of various sizes, shapes, and colors. *Control:* Pick and destroy spotted leaves. If serious, spray several times, 10 to 14 days apart, during wet periods. Use captan, zineb, or maneb.
2. *Bacterial Leaf Spots*—Small, round to angular, reddish-brown to black spots with colorless or water-soaked borders. Centers of spots later dry and resemble "frog-eyes." Young leaves may wither and drop. Petioles may also be spotted. Infected stems turn dull blackish-brown and shrivel with a semidry rot. *Control:* Pick and burn infected leaves. Where practical, keep water off leaves. Space plants. Keep down humidity. Propagate only from disease-free plants. Root in a sterile medium (pages 538–548).
3. *Mosaic* (geranium)—Plants stunted with mottled, distorted leaves. *Control:* Destroy infected plants when first seen. Keep down weeds. Use lindane or malathion to control aphids that transmit virus.
4. *Stem, Crown, Root and Rhizome Rots*—See Root Rot under Geranium. May be associated with root-feeding nematodes (e.g., lance, lesion, pin, root-knot, spiral, stem, stubby-root, stunt).
5. *Rusts* (geranium)—See under Chrysanthemum. Alternate host: *Polygonum* spp. or none.
6. *Downy Mildew*—Widespread. See (6) Downy Mildew under General Diseases.
7. *Botrytis Leaf Spot, Stem Rot*—See (5) Botrytis Blight under General Diseases.
8. *Aster Yellows* (heronsbill)—See (18) Yellows under General Diseases.
9. *Curly-top* (heronsbill)—See (19) Curly-top under General Diseases.
10. *Powdery Mildew* (geranium)—See (7) Powdery Mildew under General Diseases.
11. *Root-knot*—See (37) Root-knot under General Diseases.

CRAPE-JASMINE — See Oleander

CRAPEMYRTLE, QUEEN'S-FLOWER *(Lagerstroemia)*

1. *Powdery Mildews*—Widespread and serious during spring and fall. Leaves and young shoots may be heavily coated with powdery, white mold. Shoots, leaves, and flowers may later be distorted, thickened, and stunted. Flower buds may not open. Infected leaves and buds often drop early. *Control:* Apply sulfur or Karathane at 10-day intervals until mildew is checked. Acti-dione may be used. Follow

manufacturer's directions. Start when mildew is first seen, or apply lime-sulfur (diluted 1 to 80 with water) as buds open. Repeat 2 weeks later.

2. *Leaf Spots, Black Spot, Tip Blight*—Leaves spotted. If severe, leaves may wither and drop early. *Control:* If practical, apply zineb or maneb several times, 10 days apart. Start when first spots are evident.
3. *Sooty Mold*—Black, sootlike blotches on foliage. Follows attacks by aphids or scale insects. *Control:* Apply malathion to control insects.
4. *Root Rot*—See (34) Root Rot under General Diseases. Grow in full sun in well-drained soil high in organic matter.
5. *Chlorosis*—Mineral deficiency. Plants yellowish. *Control:* Fertilize adequately based on a soil test. Be sure soil is near neutral (pH 6.5 to 7.2).
6. *Thread Blight*—Southeastern states. See under Walnut. Plants may be defoliated early.
7. *Root-knot*—See (37) Root-knot under General Diseases.

CRASSULA — See Sedum

CRATAEGUS — See Apple

CREEPING CHARLIE — See Primrose

CREEPING MINT, CREEPING THYME — See Salvia

CREEPING SNOWBERRY — See Heath

CREPIS — See Chrysanthemum

CRESS — See Cabbage

CRIMSON DAISY — See Chrysanthemum

CRINKLEROOT — See Cabbage

CRINUM — See Daffodil

CROCANTHEMUM — See Sunrose

CROCOSMIA, CROCUS — See Gladiolus

CROSSVINE — See Bignonia

CROTALARIA — See Pea

CROTON *(Codiaeum)*

1. *Anthracnose, Leaf and Stem Spot*—Widespread. Yellowish-gray leaf spots that later turn whitish and dry. *Control:* Keep water off foliage. Apply zineb or captan before wet periods.
2. *Root Rot*—See (34) Root Rot under General Diseases. May be associated with nematodes (e.g., reniform). Grow in loose, well-drained soil high in organic matter. Avoid overwatering.
3. *Bacterial Wilt*—See (15C) Bacterial Wilt under General Diseases.

CROWFOOT — See Delphinium

CROWNBEARD — See Chrysanthemum

CROWN IMPERIAL — See Tulip

CROWN-OF-THORNS — See Poinsettia

CROWNVETCH — See Pea

CRYOPHYTUM — See Iceplant

CRYPTOGRAMMA — See Ferns

CRYPTOMERIA — See Juniper

CUBAN LILY — See Tulip

CUCKOO-FLOWER — See Carnation

CUCUMBER [COMMON and ENGLISH FORCING], MUSKMELON, CANTALOUP, HONEYDEW, CASABA, WINTER MELON, WEST INDIA GHERKIN, SNAKE MELON, MANGO MELON, POMEGRANATE MELON

(*Cucumis*); CHINESE WAXGOURD (*Benincasa*); BRYONOPSIS; WATERMELON, CITRON (*Citrullus*); SQUASH [SUMMER or BUSH, WINTER], PUMPKIN [WINTER CROOKNECK or CUSHAW, COMMON], VEGETABLE-MARROW (*Cucurbita*); GOURDS, CALABASH, VEGETABLE SPONGE, GUINEA BEAN, NEW GUINEA BEAN (*Cucumis, Cucurbita, Lagenaria, Luffa,* and *Trichosanthes*); MOCK- or WILD CUCUMBER (*Echinocystis*); MELOTHRIA; BALSAM-APPLE, BALSAM-PEAR (*Momordica*); CHAYOTE (*Sechium*); CURUBA or CASSABANANA (*Sicana*)

1. *Anthracnoses* (squashes and pumpkins usually almost immune) —General in warm (65° to 85° F.), humid weather. Often destructive. Round to angular, enlarging, reddish-brown to black spots on leaves. Spots may later dry and tear out. Older foliage may suddenly wither. Vines may be completely defoliated. Light brown to black, elongated streaks on stems and petioles. Some runners may be girdled and killed. Young fruit may blacken, shrivel, and drop off. Older fruits have small to large, round, sunken spots, water-soaked at first, later dark green to black, with flesh-colored, oozing centers (Figure 4.74). Bacterial Soft Rot often follows. Dark

Fig. 4.74. Anthracnose of muskmelon. (Purdue University photo)

sunken spots develop on seed leaves and stems of seedlings. Stems may be girdled and seedlings die. *Control:* Sow western-grown, certified, or disease-free seed; or soak seed for 5 to 10 minutes in a warm (60° to 80° F.), 1:1,000 solution of mercuric chloride. See page 101 for precautions. Wash seed 15 minutes in running water, dry, and dust with captan, thiram, chloranil, or Semesan plus dieldrin or lindane (Table 20 in Appendix). *Cucumber* seed can be soaked in hot water (122° F.) for 20 minutes as a substitute. Dry and dust as for mercuric chloride treatment. Avoid working among plants when they are wet, and overhead irrigation, whenever possible. Plow under or burn debris after harvest. Keep down weeds. Three- or 4-year rotation. Grow in well-drained soil. Apply captan, zineb, Dyrene, thiram, maneb, Daconil 2787, or Dithane M-45 at 7- to 10-day intervals. Start about when vines "begin to run." Use ziram or captan on young plants. Handle fruit carefully. *Watermelons* resistant to one or more races: Black Kleckley, Blackstone, Charleston Gray, Congo, Crimson Sweet, Dunbarton, Early Resistant Queen, Fairfax, Garrisonian, Graybelle, Hope Diamond, Jubilee, and Spaulding. *Cucumbers* tolerant or resistant to one or more races: Ashe, Fletcher, Palmetto, Pixie, Polaris, Santee, and Stono.
2. *Scab, Spot Rot, Pox* (primarily cucumber, gherkin, muskmelon, pumpkin, summer squash, watermelon) —General in moist areas with cool, damp nights. Small,

angular, water-soaked, or pale green spots on leaves that turn white to gray or brown. Spots tear out leaving ragged holes. Similar elongated spots occur on petioles and stems. Small, gray to brown, slightly sunken, oozing, gummy spots on fruit. Enlarge and become sunken, dark cavities lined with dark olive-green mold. Fruit may crack, often later destroyed by Bacterial Soft Rot. See Figure 2.30D under General Diseases. *Control:* Same as for Anthracnoses (above), except use maneb, Dithane M-45, folpet, difolatan, or Daconil 2787, plus spreader-sticker. Where serious, apply sprays at weekly intervals before bloom, then twice weekly through harvest. Thorough coverage is essential. Start spraying 3 weeks after seeding. Resistant *cucumbers:* Armour, Ashe, Crispy, CS-2M, Dark Green Slicer, Empress Hybrid, Fletcher, F.M. Hybrids 51 and 52, High Mark II, Highmoor, Hybrid Long Green Pickle, Hybrid 805 Pickle, Improved Highmoor, Improved Hycrop Pickling, Maine No. 2, Nappa 63, Princess, Robinson RX-SMR 58, Spartan 27, Spartan Champion, Spartan Dawn, Spartangreen, Windermoor, Wisconsin SMR-12, 15, 18, and 58. Resistant *muskmelon:* Edisto 47. Cucuzzi *gourd* is highly resistant.

3. *Leaf Spots, Leaf Blights* (often serious on muskmelon and cantaloup) —General. Small, round to irregular, water-soaked, yellow, tan, gray, brown, or black spots on leaves. Spots may enlarge, turn dark brown, and become zoned (target spots). See Figure 4.75. Center leaves may wither and fall early. Fruit may sunburn and ripen prematurely. Most serious in warm, moist weather. Spots may also occur on petioles, stems, and fruit. *Control:* Same as for Anthracnoses (above). Keep plants vigorous by fertilizing and watering. Normally resistant *muskmelons* to Alternaria Leaf Blight: Edisto, Edisto 47, Georgia 47, Hales Best 936, Harper's Hybrid, Healy's Pride, Netted Gem, Pride of Wisconsin, Purdue 44, Saticoy Hybrid, Smith's Perfect, and Sweetie.

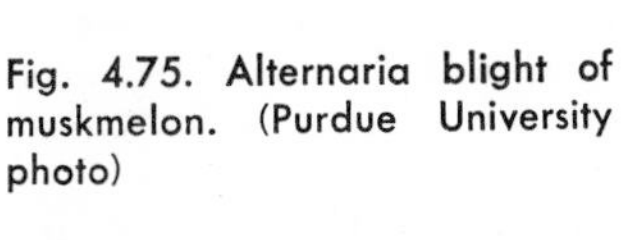
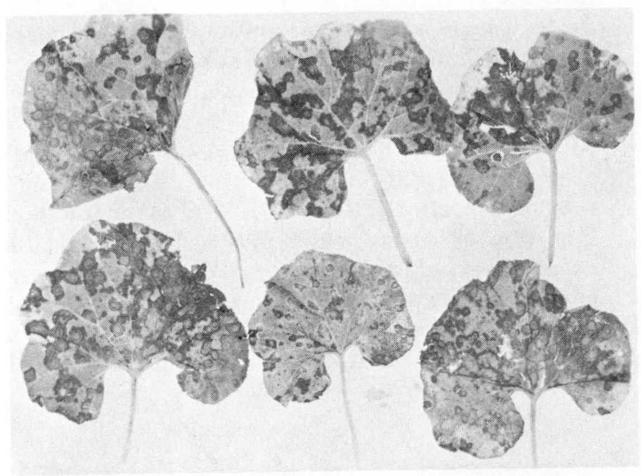

Fig. 4.75. Alternaria blight of muskmelon. (Purdue University photo)

4. *Angular Leaf Spot, Bacterial Spot or Blight* (primarily cucumber, muskmelon, gourds, squash, West India gherkin, bryonopsis) —General where summer rains are frequent. Small, angular to irregular, water-soaked spots on leaves that later dry, turn whitish-gray to brown, and drop out. Leaves ragged. Similar spots may occur on stems and petioles. Round, water-soaked spots on fruit (Figure 4.76). Fruits may be distorted into "crooks." Fruit drop often follows. A whitish crust may form on surface of leaf, stem, and fruit spots. *Control:* Sow certified, western-grown seed. Treat seed as for Anthracnoses (above). Three-year rotation. Apply mixture of zineb or maneb plus fixed copper or soluble copper (TC-90, Sol-Kop 10, Copoloid) at 5- to 7-day intervals during warm (75° F.), wet weather. Start when plants begin to vine. Avoid overhead irrigation. Destroy plant debris after harvest. Do not work

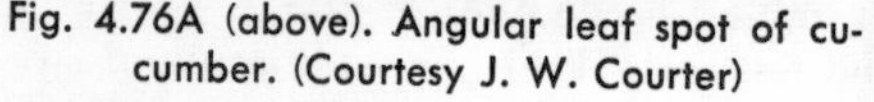

Fig. 4.76A (above). Angular leaf spot of cucumber. (Courtesy J. W. Courter)

Fig. 4.76B. Bacterial spot on "Bush" pumpkin. Note the whitish crust on the lesions. (Illinois Agricultural Experiment Station photo)

among wet plants. Control insects with rotenone, sevin, or methoxychlor plus malathion. *Watermelon, muskmelon,* and *squash* varieties differ in resistance. Santee *cucumber* has some resistance. If Anthracnose or Scab is present, alternate copper with maneb, zineb, or captan.

5. *Bacterial Wilt* (primarily cucumber, muskmelon, squashes, gherkin, pumpkins, gourds) —General and serious. Vines rapidly wilt, wither, and die starting with one or a few leaves on one vine. *Squash* vines are dwarfed. See Figure 2.33B under General Diseases. *Watermelon* is almost immune. *Control:* Rotate. Control cucumber beetles (Figure 4.77) and other insects that transmit the causal bacteria. Use sevin or methoxychlor plus malathion at 5- to 7-day intervals. *Start as plants crack soil. Apply to dry plants.* For a few garden plants, start under Hotkaps then enclose in cheesecloth tents to keep out insects. Pull up and destroy wilted plants. Resistant *cucumbers:* certain pickling strains (e.g., Chicago Pickling). Fairly resistant *squash:* Boston Marrow, Buttercup, Butternut, Delicious, Early Market,

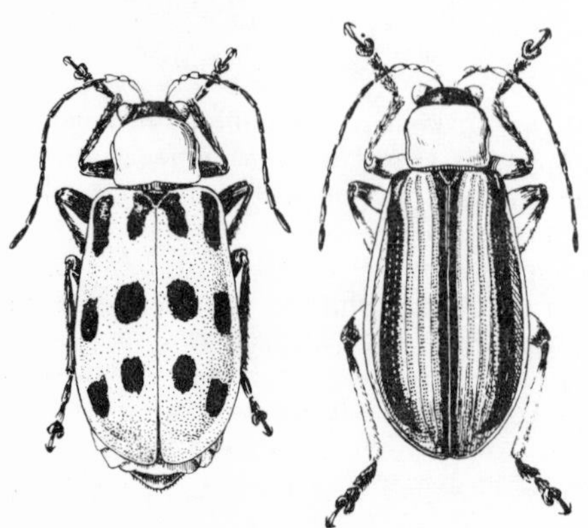

Fig. 4.77. Cucumber beetles. Left, spotted; right, striped. These insects transmit bacterial wilt. (USDA photo)

Mammoth Chili, Table Queen or Acorn, and Warren. Somewhat resistant *muskmelons:* Pride of Wisconsin or Queen of Colorado types.

6. *Fusarium Wilts, Fruit Rots* (primarily watermelon, citron, muskmelon or cantaloup, cucumber, mock-cucumber) —General in warm areas. Plants stunted and often yellow. Leaves suddenly wilt, wither, and runners gradually die. Yellow to dark brown streaks inside stems when cut. Cantaloup fruit show soft, sunken, irregular spots. May be covered with white or pink mold. Seeds may rot in the soil. Seedlings often wilt and collapse. Roots gradually decay. See Figure 2.31C under General Diseases. *Control:* Treat seed as for Anthracnoses (above). Sanitation. Normally resistant *watermelons,* where adapted: Baby Kleckley, Baby Klondike, Blacklee, Blue Ribbon, Blue Ribbon Striped Klondike, Bush Desert King, Calhoun Sweet, Charleston Gray, Charleston Gray 133, Congo W.R., Crimson Sweet, Crisscross (Chris-cross), Dixie Hybrid, Dixie Queen W.R., Early Resistant Queen, Fairfax, Garrisonian, Georgia W-R, Graybelle, Hawklee, Hawksbury (Purdue strain), Hope Diamond, Iowa King, Improved Kleckley Sweet No. 6, Improved Stone Mountain No. 5 and 19, Ironsides, Jubilee, Klondike R-7 and RS-57, Klondike Striped, Leesburg, Miles, Missouri Queen, Princeton 134, Purdue Hybrid No. 1, Queen Hybrid, Shipper, Spaulding, Summit, and White Hope. Resistant *muskmelons:* Delicious 51, Early Market, Golden Honey, Gold Star, Harper Market, Harvest Queen, Honey Rock Fusarium Resistant, Hybrid 26, Iroquois, Minnesota Honey, Minnesota Hybrid 26, Minnesota Midget, Samson, Saticoy Hybrid, Spartan Rock, and Supermarket. The Persian, Honeydew, Honeyball, and Casaba melons are also relatively resistant. Control nematodes (see Root-knot below). Keep plants vigorous. Fertilize and water. Handle fruit carefully and store at recommended temperature.

7. *Mosaics*—General and serious. Symptoms may be masked in hot weather. Yellow-green and dark green mottling or distortion of leaves. Leaves often stunted, wrinkled, and curled. Vines commonly stunted to dwarfed, may be yellowed and "bunchy." Fruits often show yellow to orange, yellow-green, or pale green spotting and mottling. May be stunted, warty or knobby, with bitter taste (Figure 4.78A). *Squash* and *watermelon* fruit may also show yellowish rings. If severe, all leaves except those at runner tips (rosettes) may die. Yield is reduced and often of poor quality. *Control:* Rigidly control weeds (e.g., bur and wild cucumber, catnip, horsenettle, pigweed, milkweed, matrimony-vine, motherwort, pokeweed, white cock
alfa'

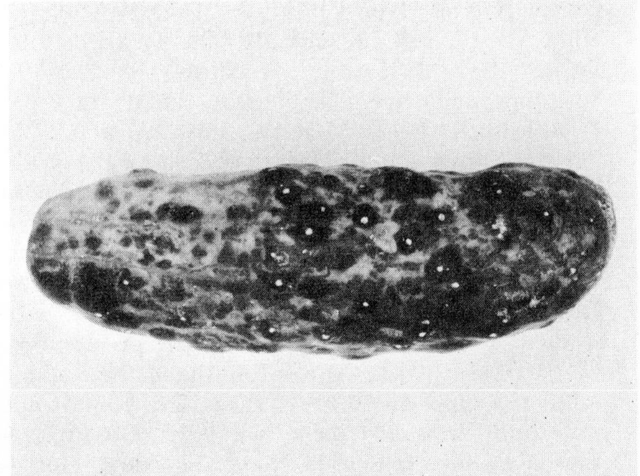

Fig. 4.78A. Cucumber mosaic. (Pennsylvania State University Extension Service photo)

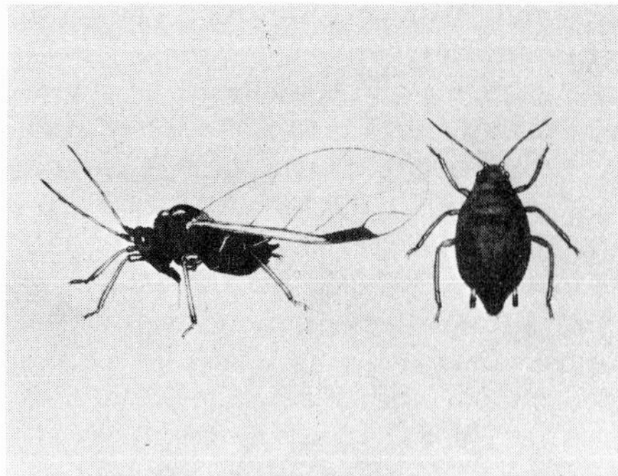

Fig. 4.78B. Melon aphid, carrier of a number of viruses that attack vine and other crops. (USDA photo)

Fig. 4.79. Powdery mildew of squash. (Iowa State University photo)

others). Destroy first infected plants *after* applying malathion. Control aphids (Figure 4.78B), cucumber beetles, and grasshoppers that transmit the viruses. Use malathion plus sevin, methoxychlor, or lindane at 5-day intervals. Plant virus-free *muskmelon, squash* and *vegetable-marrow* seed. Resistant *cucumbers:* Armour, Burpee Hybrid, Challenger, Crispy, CS-2M, Early Marketer, Early Surecrop, Empress Hybrid, High Mark II, Hybrid 51, Hybrid 805 Pickle, Hybrid Long Green, Hycrop Hybrid Pickling, M&M Hybrid, Jet, Nappa 61, N.K. 805 and 831, Ohio MR-17, MR-25, and MR-200, Princess, Puerto Rico 10 and 27, Resistant Burpee Hybrid, Saticoy, Sensation, Shamrock, Spartan Dawn, Spartangreen, Surecrop, Tablegreen, Tabletreat, Total Marketer, Triumph, Vaughan's Hybrid, Wisconsin SMR-12, SMR-15, SMR-18, and SMR-58, and Yorkstate Pickling. Resistant *muskmelons:* Harper and Saticoy Hybrids. Zucchini *squash* has tolerance. Virus-carrying aphids are repelled by an aluminum foil mulch.

8. *Powdery Mildew*—General. May be destructive near harvest, especially on melons, pumpkins, cucumber, and squash. White or brownish mealy growth on leaves and young stems (Figure 4.79). Occasionally on muskmelon and watermelon fruit. Older leaves and young stems may turn yellow, wither, and die. Plants weakened or stunted. Fruit may sunscald or ripen prematurely. *Control:* Rotate. Keep down weeds. Space plants. Apply Karathane, Morestan, Morocide, folpet, or fixed (or soluble) copper one to three times, 7 to 10 days apart, depending on severity. Start when mildew is *first* seen. Normally resistant *muskmelons:* Campo, Desert Sun, Edisto, Edisto 47, Florida No. 1, Floridew, Florigold, Florisun, Georgia 47, Gold Cup, Golden Gate 45, Homegarden, Honey Ball 306, Jacumba, Perlita, Powdery

Mildew Resistant (P.M.R.) 5, 6, 45, 45SJ, 88, and 450, Rio Gold, Rio Sweet, Seminole, Samson, Smith's Perfect, and Topmark 27–36SR. *Muskmelons tolerant to sulfur:* SR 59, SR 91, SR 91 Improved, Topmark 27–36 SR, and Wescan. Tolerant or resistant *cucumbers:* Ashe, Ashley, Fletcher, High Mark II, M&M Hybrid, Palmetto, Pixie, Polaris, Stono, Yates Conqueror, and Yates Invader. Resistant *pumpkins* and *squashes* and *Honeydew melon* types may be available soon.

9. *Downy Mildews*—General in warm, moist areas. Vague or sharply defined, irregular to angular, yellowish to brownish or black areas on upper side of older leaves, usually near center of hill. Underside of diseased leaves may show pale, grayish-purple to dark purple mold following damp weather. Mold may vary from white to nearly black. Spots enlarge rapidly causing leaves to wither and die. May resemble frost injury as entire vines are scorched and killed. Fruit often nubbins with poor flavor. See Figure 2.22C under General Diseases. *Control:* Rotate. Space plants. Apply maneb, zineb, Polyram, Dithane M-45, difolatan, Daconil 2787, Dyrene, or fixed copper thoroughly weekly in wet weather. Start about when vines "begin to run." Keep down weeds. Avoid overfertilizing. Control cucumber beetles as for Bacterial Wilt (above). Keep away from wet plants. *Muskmelons* resistant or tolerant to one or more strains: Early Market, Edisto, Edisto 47, Florida No. 1, Floridew, Florigold, Florisun, Georgia 47, Golden Model, Granite State, Homegarden, Jacumba, Perlita, Rio Gold, Rio Sweet, Seminole, Supermarket, Texas No. 1, Topmark 27–36 SR, and Weslaco F and H. Somewhat resistant or tolerant *cucumbers:* Ashe, Ashley, Barclay, Burpee Hybrid, Challenger, Dark Green Slicer, Early Marketer, Early Surecrop, Empress Hybrid, Fletcher, High Mark II, M&M Hybrid, P-51 DMR, Palmetto, Palomar, Pixie, Polaris, Saticoy, Sensation, Stono, Supermarket, Surecrop, Total Marketer, Triumph, and Wescan. *Watermelon* varieties also differ in resistance.

10. *Curly-top*—Western states. Infected seedlings may die. Older vines stunted. Young tip leaves often bunched together, bend upward, are often darker green than normal, and are dwarfed, while older leaves turn yellow. Leaves puckered and cupped downward. Fruit dwarfed and distorted with poor flavor. Yield is reduced. *Control:* Keep down weeds. Destroy infected plants when *first* found. Control leafhoppers that transmit the virus. Use sevin plus malathion as given under Bacterial Wilt (above). Resistant *pumpkins:* Big Tom, Calhoun, Chirimen, Cushaw, Japanese Pie, Kentucky Field, Sweet Cheese, and Tennessee Sweet Potato. Resistant or tolerant *squashes:* Long White Bush, Marblehead (Umatilla and Yakima strains), Table Queen, and Sweet Potato. Resistant *vegetable-marrows:* Boston Creek, Bush Green, Green of Milan, Long Bush Green, Long White Bush, and Zuchetta Nostrana Nana.

11. *Ringspot*—May closely resemble mosaic. Irregular, black, dead areas or small raised "pimples" on watermelon fruit. Leaves may be distorted, show small brown dots surrounded by light yellow borders. Concentric rings often form on fruit. Fruit rots may follow. See (17) Spotted Wilt under General Diseases. The Tobacco Ringspot virus may be transmitted by seed or dagger *(Xiphinema)* nematodes.

12. *Aster Yellows*—Plants stunted and yellowed. Numerous secondary shoots and green, aborted flowers may develop on squash and vegetable-marrow. *Control:* Keep down weeds. Apply insecticides to control leafhoppers that transmit the virus. See under Bacterial Wilt (above). Destroy infected plants when first seen.

13. *White Wilt, Cottony Rot, Southern Blight, Stem Rot, Watery Soft Rot*—Stem rots and dries at or near soil line. Cottony mold grows on rotted area in which dark brown or black bodies (sclerotia) are embedded. Leaves turn yellow and wilt. Infected fruits soft and watery. May be covered with cottony mold. *Control:* Rotate. Plow deeply. Carefully remove infected vines and fruit and destroy. Grow in well-drained soil. Avoid overhead sprinkling. Keep down weeds.

14. *Fruit Spots and Rots, Storage Rots*—Cosmopolitan. See under Carrot. Small to

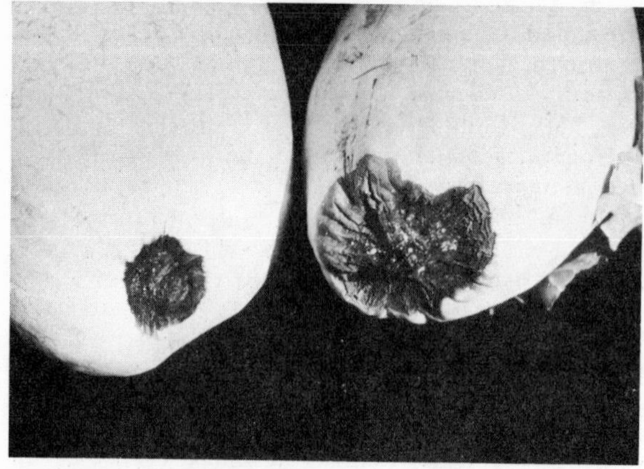

Fig. 4.80. Blossom-end rot of watermelon.

large spots in fruit, often sunken. Rot often starts at wounds or where fruit rest on damp soil. Rotted area may be covered with white, black, green, blue, or pink mold. Decay may be watery, slimy, and putrid (Bacterial Soft Rot), or dry and firm (fungus rots). *Control:* Same as for Anthracnoses (above) or apply folpet or difolatan as plants "start to run." *Cucumber* and *watermelon* varieties differ in resistance. Store mature, dry, sound, blemish-free *squash* and *pumpkin* fruit at 50° to 60° F. with low relative humidity (70 to 80 per cent) after curing at 75° to 85° F. for 10 days at 80 to 85 per cent humidity. Check with your extension horticulturist. Handle fruit with care. If possible, store on shelves. Do not pile fruit over three deep in storage. Where practical in the garden, rest fruit on a dry surface (e.g., a dry mulch). Refrigerate *cucumbers* promptly.

15. *Gummy Stem Blight, Stem-end Rot (Black Rot), Leaf Spot* (primarily cucumber, pumpkins, squashes, watermelon, gourds)—Widespread. Disease usually spreads outward from crown of plant. Small to large, water-soaked gray fruit spots that turn black and gummy or brown. Leaf spots are gray to reddish-brown and irregularly circular. Leaves may yellow, wither, and drop. Stem spots and streaks are oily green to dark brown and gummy. Ooze during wet or humid weather. Older vines (or seedlings) may wilt and die back. *Control:* Same as for Anthracnoses (above). Varieties differ in resistance. Store *cucumber* fruit at about 45° F.
16. *Seed Rots, Damping-off*—General during long periods of cool, wet weather. Seeds rot. Stand is poor. Seedlings wilt and collapse (Figure A.1 in Appendix). *Control:* Avoid planting too early before soil warms up. Select light, well-drained soil. Treat seed as for Anthracnoses (above). Apply seedbed spray of captan or ziram (2 tablespoons per gallon. Use 1 gallon per 125 square feet.) every 5 to 7 days when soil is below 75° F.
17. *Root Rots*—See under Bean, and (34) Root Rot under General Diseases. May be associated with nematodes (e.g., burrowing, dagger, lesion, naccobus, pin, ring, reniform, root-knot, spiral, stem, sting, stubby-root, stunt). *Control:* Same as for Anthracnoses (above). Grow in fertile, mellow, disease-free soil. Cultivate as shallow as possible. Fumigate as for Root-knot (below).
18. *Root-knot, Cyst Nematode*—General and serious in southern states. Plants unthrifty and stunted. May wilt on hot days; die before producing fruit. Swellings or small galls on roots. Yield may be greatly reduced. See Figure 2.52B under General Diseases. *Control:* Rotate. If necessary, fumigate soil after harvest using D-D, EDB, Vorlex, Telone, etc. Commercial growers often apply DBCP (Nemagon, Fumazone) or Zinophos at or just before planting with the fertilizer. Follow manufacturer's recommendations. Many varieties of *cucumber* are highly resistant.

Resistant *squashes:* Black Zucchini, Butternut, Caserta, and Early Prolific Straight Neck, also Small Gherkin. Resistance or susceptibility often depends on what Root-knot species or subspecies are present. Most *cantaloups* are very susceptible.

19. *Verticillium Wilt*—Plants suddenly wilt, wither, and die. Brown to black streaks inside stem. See (15B) Verticillium Wilt under General Diseases. *Melons* differ in susceptibility.
20. *Blossom Blight*—Blossoms blighted. May be covered with dense mold. Young fruit rot and drop off. Rot usually starts at blossom end. *Control:* Spray as for Anthracnoses and Downy Mildew (both above). Grow in well-drained soil. Rotate. Remove and destroy rotting flowers and fruits.
21. *Blossom-end Rot* (primarily squash and watermelon)—Dark brown, dry areas on blossom-end of fruit (Figure 4.80). *Control:* See under Tomato.
22. *Bacterial Spot* (primarily cucumber, gourds, pumpkin, squash)—Small, round to angular leaf spots between veins with bright yellow borders. Spots may enlarge and merge to blight whole leaf. Spots may crack, but do *not* drop out. *Control:* Same as for Angular Leaf Spot (above).
23. *Sooty Mold*—Black mold on foliage and fruit. Follows insect attack. *Control:* Spray to control insects. See under Bacterial Wilt (above).
24. *Stem Streak, Dieback*—Small pink to tan spots on stems and petioles that soon merge to form long streaks. Affected parts are girdled, with leaves on vine beyond collapsing and dying. Stems later turn brown. *Control:* Keep plants vigorous. Apply adequate amounts of balanced fertilizer based on a soil test. Otherwise same as for Anthracnoses (above). No resistant varieties are available.
25. *Potassium Deficiency*—Leaf margins and tips turn yellow; are later scorched. *Control:* Practice balanced fertility, based on a soil test report.
26. *Chlorosis*—Plants yellowish and stunted. Leaves may have dark green veins while tissue in between is yellowish-green to ivory. Deficiency of iron, zinc, magnesium, or manganese. See pages 26–28.
27. *Boron Deficiency* (primarily squash)—Leaves stunted, yellowed, cupped downward, and brittle. Petioles curled and thickened. *Control:* Have soil tested. Apply borax or other boron material as recommended (page 27).
28. *2,4-D Injury*—See under Grape. Melons and cucumber are very susceptible.
29. *Web Blight*—Southeastern states. See under Bean.
30. *Crown Gall*—See (30) Crown Gall under General Diseases.

For additional information read USDA Handbook 216, *Muskmelon Culture,* USDA Leaflet 509, *Muskmelons for the Garden,* USDA Farmers' Bulletin 2086, *Growing Pumpkins and Squashes,* USDA Leaflet 528, *Watermelons for the Garden,* and USDA Agr. Info. Bulletin 259, *Commercial Watermelon Growing.*

CUCUMBERTREE — See Magnolia

CULVERSROOT — See Speedwell

CUNILA — See Salvia

CUPFLOWER — See Tomato

CUPHEA — See Cigarflower

CUPID'S-BOWER — See African-violet

CUP-PLANT — See Chrysanthemum

CUPRESSUS — See Juniper

CURRANT [ALPINE or MOUNTAIN, CLOVE (BUFFALO or MISSOURI), COMMON or GARDEN RED, EUROPEAN BLACK or BLACK, FLOWERING, GOLDEN, NORTHERN RED, RED-FLOWERED or WINTER, SIBERIAN, WAX], WINTERBERRY, GOOSEBERRY [European or GARDEN, FUCHSIA-FLOWERED, HAIRYSTEM] *(Ribes)*

Fig. 4.81. Currant leaf spot and anthracnose.

1. *Leaf Spots, Anthracnose, Spot Anthracnose or Scab*—General in humid areas. Small, dark spots on older leaves that may develop gray centers (Figure 4.81). Spots may merge to form large brown blotches. Lower leaves affected first. Leaves often turn yellow or brown and drop early, starting on lower stems. Fruit may be spotted; reduced in size and number. Discolored sunken spots may occur on young canes and leaf stems. *Control:* Space plants. Prune to open up centers. Where practical, collect and burn fallen leaves. Apply fixed copper, zineb, maneb, captan, or folpet. Follow spray schedule in Appendix (Table 16). Currant and gooseberry varieties differ in resistance to Anthracnose *(Pseudopeziza)*. Welcome *gooseberry* is resistant, as are Cherry, Fay's Prolific, Perfection, Red Lake, White Imperial, and Wilder *currants*.
2. *Powdery Mildews*—General during cool, damp periods. Most common and serious on gooseberry. Leaves, young shoots, and berries covered with powdery, bluish-white, or light grayish patches that later turn rusty-brown. Leaves and shoot tips may be stunted and distorted. Leaves attacked by mildew drop early. Berries may be dwarfed, russeted, and cracked. *Control:* Apply Karathane plus spreader-sticker as buds start to swell, just before, and just after bloom. Avoid shade, weeds, and overcrowding. Prune to "open up" plants. Currant and gooseberry varieties differ in resistance. Practice *balanced* fertilization, based on a soil test. Avoid excessive use of nitrogen fertilizers.
3. *Fruit Rots*—General. Fruit rots near picking time. Mold may form over rot spots. Often serious in cool, humid weather. *Control:* Same as for Leaf Spots (above). Space plants—keep pruned. If preharvest period is wet, apply captan at 5-day intervals.
4. *Rusts* (primarily Blister and Cluster cup) —Widespread. Small, yellowish to orange, bright red or brown, dusty or "hairy" pustules on underside of leaves, twigs, and fruit. Bright orange to reddish-yellow or reddish-brown pustules appear on upper side. Leaves may be curled, thickened, and drop early. Most common on older leaves. Fruit may be distorted. *Control:* Do *not* plant susceptible currants and gooseberries (*Ribes*) within 1,000 feet of 5-needle pines. These plants are alternate hosts of the serious White Pine Blister Rust; see under Pine. Quarantine regulations prevent planting of *susceptible* currants and gooseberries in areas where white pine is an important lumber tree. Check with your state department of agriculture before planting currants and gooseberries. Destroy rust-susceptible wild *Ribes*. Destroy nearby sedges (*Carex* spp.), alternate hosts of Cluster Cup Rust. Apply sprays as for Leaf Spots (above). Several weekly applications of ferbam,

zineb, or maneb are effective between bud burst and flowering. Be sure to cover both leaf surfaces. Where feasible, remove and destroy infected leaves. Keep down weeds.

5. *Dieback, Twig Cankers, Black Pustule, Cane-knot Canker, Cane Blights*—Widespread. Tip leaves on young canes suddenly wilt and die. Canes blighted, dry up, and die back. Most conspicuous just before fruit ripens. Gray mold, or small, coral-pink to black "pimples" may be evident on blighted parts (Figure 4.82). May follow winter injury. Do not confuse with borers tunneling within the main shoots. *Control:* Prune out and burn blighted canes before leaves appear. Take cuttings from healthy plants. Grow in well-drained soil. Make flush pruning cuts. Avoid heavy applications of fertilizer high in nitrogen. Spray as for Leaf Spots (above).

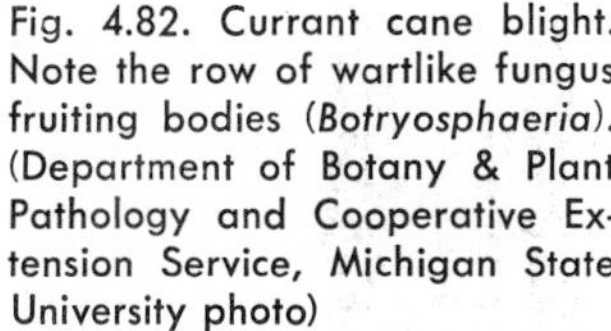

Fig. 4.82. Currant cane blight. Note the row of wartlike fungus fruiting bodies (*Botryosphaeria*). (Department of Botany & Plant Pathology and Cooperative Extension Service, Michigan State University photo)

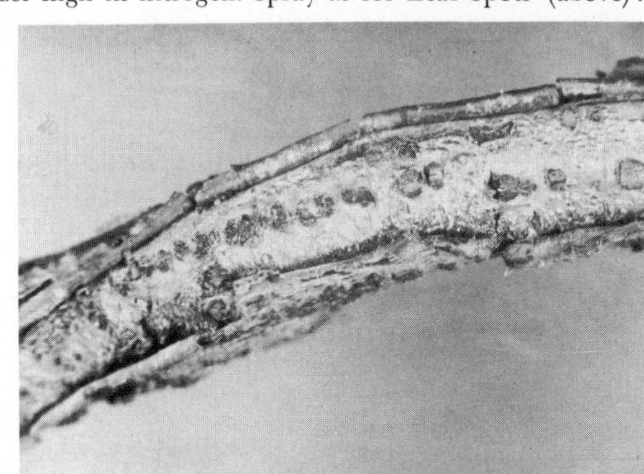

6. *Collar Rot*—Widespread. Plants die slowly over several years. Perennial, hooflike or shelflike structures, mostly at or near soil line on older bushes. *Control:* Remove and burn affected bushes. Avoid injuries when cultivating, etc.
7. *Root Rots*—Plants gradually decline in health. Produce little or no new growth. Roots decay. Groups of mushrooms may appear at base of infected plants. Often associated with nematodes (e.g., lesion, root-knot, stunt). *Control:* See under Apple, and (34) Root Rot under General Diseases.
8. *Mosaic* (primarily red currant) —Round, light green to yellowish spots on leaves. Spots enlarge and merge to form irregular bands along leaf veins. Leaves may become yellowish with scorched margins. Plants gradually decline in vigor, become stunted, produce less and less fruit. *Control:* Dig up and destroy affected plants when first seen. Control virus-spreading insects with malathion. Plant virus-free stock from a reputable nursery. Destroy nearby wild currants and gooseberries.
9. *Gooseberry Sunscald*—Fruit soft, pale, and drops early. Often covered with various types of secondary molds. *Control:* Spray as for Leaf Spots (above). Keep plants vigorous (fertilize, and water during dry periods). Como *gooseberry* is resistant.
10. *Downy Mildew, Dryberry*—See (6) Downy Mildew under General Diseases. Flowers may turn brown and drop. Berries may appear dried, leathery; usually drop early. *Control:* Same as for Leaf Spots (above).
11. *Verticillium Wilt*—Uncommon. See under Barberry, and (15B) Verticillium Wilt under General Diseases.
12. *Chlorosis*—Leaves yellowish, especially near tips of young shoots. *Control:* Have soil tested and follow recommendations. Apply iron or manganese sulfate in one or more pest sprays (pages 26–27).
13. *Thread Blight*—Southeastern states. See under Walnut. *Control:* Spray as for Leaf Spots (above).

14. *Bud Nematode*—California in humid areas. Young leaves are blighted and killed. See (20) Leaf Nematode under General Diseases. *Control:* Dip infested cuttings in hot water (110° F.) for 30 minutes. Destroy affected bushes.
15. *Bacterial Spot* (golden currant)—Round to irregular, dark brown spots, with water-soaked, reddish-brown margins on the leaves. Spotted leaves turn various shades of yellow and red. Severely affected leaves soon turn brown, wither, and fall early. Swollen spots occur on fruit which ripen prematurely. *Control:* Same as for Leaf Spots (above). Apply fixed copper sprays.
16. *Crown Gall*—See (30) Crown Gall under General Diseases.
17. *Potassium Deficiency, Leaf Scorch*—Growth is poor. Leaves purplish or yellowish between the veins. Later leaf margins and tips are scorched. Fruit may ripen unevenly. *Control:* Have soil tested and apply potash (1–2 ounces per square yard) around each bush.

CURUBA, CUSHAW — See Cucumber
CUSHION-PINK — See Carnation
CYATHEA — See Ferns
CYCLAMEN

1. *Gray-mold Blight, Botrytis Bud and Leaf Rot, Crown Rot, Petal Spot*—Cosmopolitan and serious. Leaves, stalks, and flower petals are spotted. Buds and flowers may rot. Dense gray mold may cover affected areas in damp weather. Rot may spread gradually into the center and tuber (corm), killing plant. *Control:* Avoid overcrowding and high rates of nitrogen fertilizer. Increase air circulation and lower humidity to 85 per cent or below. Pick and destroy affected plant parts. Spray plants at 10- to 14-day intervals and drench soil. Use captan, maneb, or zineb. Mist flowers lightly with zineb (1 tablespoon per gallon).
2. *Stunt, Wilt, Ramularia Leaf Disease*—Widespread. Plants and leaves stunted but not usually killed. Leaves may turn yellow and wilt. A frosty, mildew-like growth may develop on lower leaf surfaces. Irregular brown areas with indefinite margins form on leaves. Flowers characteristically open *below* the small leaves. Reddish-brown dead areas appear when corms are cut through. *Control:* Destroy infected plants. They will not recover. Plant disease-free tubers (corms) or start seed from healthy plants in clean or sterilized soil (pages 538–548). Spray plants as for Gray-mold Blight (above).
3. *Bacterial Soft Rot, Tuber Rot*—Common and destructive. Leaf and flower stems suddenly wilt, droop, may collapse. Soft, slimy, foul-smelling rot of petioles and tuber or corm. See Figure 4.83. *Control:* Avoid overshading, crowding plants, overstimulating with fertilizer, overwatering, and wounding underground parts.

Fig. 4.83. Bacterial soft rot of cyclamen.

Grow in moist, well-drained, sterilized soil. Destroy affected parts. Keep water off foliage.

4. *Root-knot*—Cosmopolitan. Cyclamen is high susceptible. See (37) Root-knot under General Diseases. DBCP may be applied to soil of potted plants. Follow manufacturer's directions.
5. *Root Rots*—Plants weak, pale; may wilt. Easily pulled up. Roots and tubers discolored and rotted. May be associated with grubs or nematodes (e.g., lesion, root-knot). See (34) Root-knot under General Diseases. *Control:* Same as for Bacterial Soft Rot (above). Some control, if early enough, by immersing root clump in 0.5 per cent thiram, captan, zineb, or maneb solution. Or apply monthly soil drenches of these chemicals.
6. *Leaf Spots, Leaf and Bud Blight, White Mold*—Various types of dead spots in leaves. Leaves may wither and fall early. Buds may be blasted. *Control:* Pick and destroy infected leaves. Increase air circulation. Keep water off foliage. Space plants. Spray or dip potted plants at 10- to 14-day intervals. Use captan, zineb, maneb, or ferbam. Sow disease-free seed in new or sterilized soil (pages 538–548).
7. *Seedling Blight, Damping-off*—Seedlings wilt and collapse from rot at soil line. Usually starts in a few plants and spreads outward. *Control:* Grow in well-drained, humusy, sterilized soil. Avoid overwatering. Keep humidity down. If disease starts, sprinkle affected areas with captan or zineb solution (1 tablespoon per gallon), or use PCNB-Dexon mixture following manufacturer's directions. Repeat 5 to 7 days later.
8. *Leaf Nematode*—Leaves and stems may collapse or buds may be aborted. *Control:* Plant nematode-free tubers in sterilized soil (pages 538–548).
9. *Fusarium Wilt*—See (15A) Fusarium Wilt under General Diseases. *Control:* Same as for Stunt (above).

CYCLOPHORUS — See Ferns
CYCNOCHES — See Orchids
CYDONIA — See Apple
CYMBIDIUM — See Orchids
CYNARA — See Lettuce
CYNODON — See Lawngrass
CYNOGLOSSUM — See Mertensia
CYPHOMANDRA — See Tomato
CYPERUS — See Umbrellaplant
CYPRESS — See Juniper
CYPRESSVINE — See Morning-glory
CYPRIPEDIUM — See Orchids
CYRILLA — See Buckwheat-tree
CYRTOMIUM, CYSTOPTERIS — See Ferns
CYTISUS — See Broom

DAFFODIL, JONQUILS, NARCISSUS [CYCLAMINEUS, DOUBLE, MINIATURE, POETAZ or CLUSTER FLOWERED, "PAPER WHITE," PETTICOAT, POET'S, POLYANTHUS or TAZETTA, SMALL and LARGE CUPPED, TRIANDRUS, TRUMPETS (BICOLOR, YELLOW, WHITE)], CHINESE SACRED LILY *(Narcissus)*; AMARYLLIS [HYBRID, MAGIC LILY], BELLADONNA-LILY *(Amaryllis, Hippeastrum)*; RAINLILY, EVENING STAR, FAIRY LILY *(Cooperia)*; CRINUM [COMMON, FLORIDA, JAMAICA], CAPE COAST LILY *(Crinum)*; AMAZON-LILY *(Eucharis)*; SNOWDROP [COMMON, GIANT] *(Galanthus)*;

SPIDERLILY, BASKETFLOWER, PERUVIAN DAFFODIL *(Hymenocallis)*; SNOWFLAKE [SPRING, SUMMER, AUTUMN] *(Leucojum)*; HARDY AMARYLLIS, GOLDEN SPIDERLILY *(Lycoris)*; TUBEROSE *(Polianthes)*; GUERNSEY-LILY *(Nerine)*; WINTER- or FALL-DAFFODIL *(Sternbergia)*; SCARBOROUGH-LILY *(Vallota)*; ATAMASCO-LILY, ZEPHYRLILY *(Zephyranthes)*

1. *Bulb Rots, Crown Rot, Root Rots*—General. Plants fail to emerge, or only pale, stunted shoots come up with weak yellowish or blighted leaves. Bulbs rot—dry and spongy, woody, moist, slimy or mushy—usually starting at base (root plate). Rot spreads through bulb to neck. Roots often rotted. Blue, green, black, gray, pink, or white mold (or small brown to black seedlike bodies, sclerotia) often evident on bulb or between bulb scales. Often associated with bulb mites, maggots (bulb flies), and nematodes (e.g., lance, lesion, spiral), high storage temperature, and poor air circulation. See Figure 2.51A under General Diseases. *Control:* Plant only high quality, well-cured, disease-free bulbs free of cuts, bruises, sunscald, or other injuries in fertile, well-drained soil treated with chlordane, dieldrin, or heptachlor before planting to control bulb flies. Discard infected bulbs. Four-year rotation. Space plants. Avoid overheating or wounding bulbs and overfertilization, especially with nitrogen or manure. Remove winter foliage mulch in early spring. Carefully remove and burn infected plants including all underground parts together with several inches of soil. Nurserymen treat mature, *dormant narcissus* and *snowdrop* bulbs by soaking in solution containing Memmi, Puratized, phenyl mercuric acetate (PMA), Mersolite-W, Dowicide B, or captan. Follow manufacturer's directions. Or soak *dormant* bulbs in hot water-formalin solution as for Stem and Bulb Nematode (below). Dry bulbs rapidly after treatment and plant immediately in light, well-drained soil that is clean or sterilized (pages 538–548). *Narcissus* varieties differ in resistance. Certain rots (Sclerotium and Black Rot) of *narcissus* are controlled by PCNB dust or spray applied in the furrow at planting time. Follow manufacturer's directions. Before planting, soak *amaryllis* bulbs for 2 hours in 1:1,000 solution of mercuric chloride or formaldehyde solution (1 pint to 12 gallons of water). Plant 1 to 2 days after treatment in sterilized soil, where possible.
2. *Mosaics, Flower Streak, Yellow Stripe, or Gray Disease*—General. Symptoms variable in narcissus, depending on species, variety, and virus (es) involved. Light green,

Fig. 4.84. Mosaic of narcissus.

grayish-green, dark green, or bright yellow streaks, an indefinite mottling, or blotching of leaves and flower stalk. Flowers often stunted, distorted, and mottled or streaked. Flower and bulb production decreases. Plants less vigorous and more stunted each year. A pronounced green and yellow mosaic, large ringspots, and chevron patterns may form in *amaryllis* leaves. *Narcissus* leaves may be spirally twisted and roughened. See Figures 2.34D and 4.84. *Control:* Collect and burn infected plants when first found, in bloom, and again late in the season. Keep down weeds. Control aphids that spread the viruses. Use malathion or lindane. Avoid growing near onions. Plant only largest, virus-free bulbs available.

3. *Narcissus Decline, White (Silver or Purple) Streak, Paper Tip*—A virus complex caused by two or more viruses in the same plant. General. White, gray, or yellowish-white streaks develop in leaves and flower stems after bloom. Tips dry up and become papery. Finally wilt and collapse. Bulbs small. Plants mature (decline) very rapidly. Chocolate-colored spots or purple streaks are produced by another virus frequently present in plants showing White Streak. *Control:* Same as for Mosaics (above) except destroy diseased plants after blooming. Replant using only largest, virus-free bulbs.
4. *Fire, Botrytis Blight, Gray-mold or Leaf Blight, Flower Spot*—Widespread. Often follows chilling. Watery, light brown spots on flowers. Large, ash-gray, watery spots form on flower stems. Later, bright yellow, elongated spots with chocolate-brown or reddish-brown centers develop on leaves. Flowers and leaves turn yellow or brown and rot rapidly in warm, damp weather. *Narcissus* varieties differ in resistance. *Control:* Same as for Bulb Rots (above). Keep down weeds. Collect and burn tops in fall. If practical, carefully pick off and burn affected parts *as* they occur. Apply zineb, maneb, ferbam, captan, dichlone, or fixed copper plus spreader-sticker several times, at about weekly intervals. Start when leaves are 4 to 8 inches tall. Spray just before wet periods when infections occur. Rotate.
5. *Stagonospora Leaf Scorch, Red Spot or Blotch, Red Fire Disease*—General. Leaf tips may appear blighted upon emerging, as if injured by frost. Small, red to reddish-brown or purplish spots, streaks, or cankers (often with a yellow border) develop on leaves and flower stalks. Flower stalks may be severely stunted and distorted; wither and die early. Spots may enlarge and merge to form large blotches. Leaves may wither and die prematurely. Flowers may be spotted dark red or brown. See Figure 4.85A. *Control:* Same as for Fire and Bulb Rots (both above).

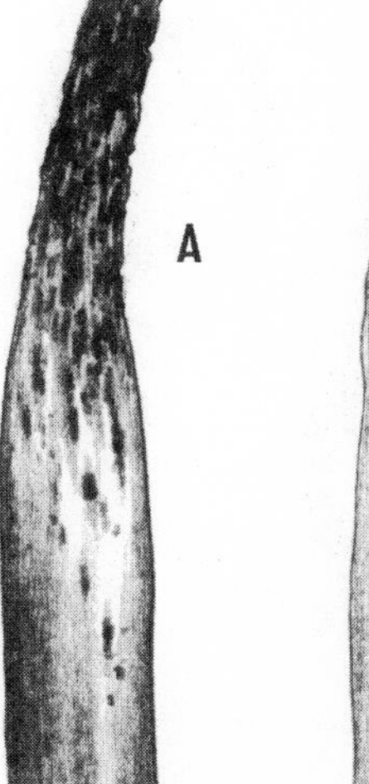

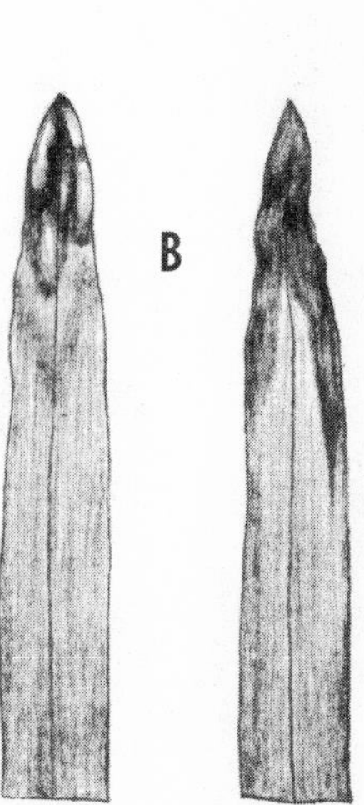

Fig. 4.85. A. Leaf scorch of narcissus. B. White mold of narcissus.

Indoors, keep water off foliage. Avoid overwatering, high humidity, high temperatures, and crowding or wounding plants. Increase light.

6. *Bud and Leaf Nematode, Stem and Bulb Nematode, Browning or "Ring" Disease* (lycoris, narcissus) —In all commercial bulb-growing areas. Leaves weak, stunted, thickened, and twisted with small, yellowish blisters or "spikkels." Bulbs when cut across show yellow to dark brown rings of infested scales. Bulbs may lose their firmness, rot, and disintegrate during storage. Badly infested bulbs fail to sprout. A large *narcissus* bulb may harbor 250,000 nematodes, and the population may increase 8,000 fold in 1 year! See Figures 2.53B and 4.86. *Control:* Dig up and burn infested bulbs and adjacent ones that appear healthy, plus 6 inches of surrounding soil. Three-year rotation with nonbulb plants. Keep down weeds. Plant only large, disease-free bulbs in light, well-drained soil fumigated with D-D or EDB, where practical. Commercial growers harvest early and treat properly cured, *dormant narcissus* bulbs by presoaking in water at room temperature (75° F.) for 2 hours followed by soaking 3 to 4 hours in a solution of hot water at exactly 110° to 111° F. and formalin in a proportion of 1:200 (1 pint of formalin in 25 gallons of water). Treatment also controls mites, bulb flies, and fungi. Treated bulbs should be planted as soon as possible.
7. *Narcissus "Smoulder," Neck Rot* (narcissus, snowdrop) —Widespread and serious in cold, wet weather. Plants stunted or missing. Young shoots may be brown and blighted with crumpled leaf tips. Quickly become slimy (Figure 4.87). Flower stems may rot and flowers develop brown spots. Yellowish-brown rot of bulb in storage. Small, flattened, black bodies (sclerotia) develop on "nose" of bulb, between husks, or as crust on affected leaves. *Control:* Same as for Bulb Rots (above). Destroy badly infected plants or plant parts. Spray as for Fire (above). Keep down weeds.
8. *Narcissus White Mold, Ramularia Blight*—Primarily Pacific Northwest in rainy, foggy areas. The emerging brown leaf tips may resemble frost injury. Sunken, yellow to gray spots or streaks appear on leaves, then enlarge and turn dark green to yellowish-brown with yellow margin. Leaves and flower stalks rot quickly. Affected areas covered with fluffy, white, powdery mold in wet weather. See Figure 4.85B. Varieties differ in susceptibility. *Control:* Same as for Fire and Bulb Rots (both above). In wet weather, spray with fixed copper or zineb plus spreader-sticker.

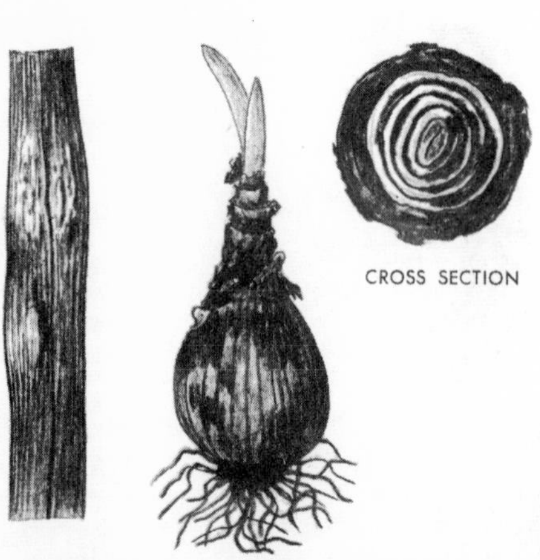

Fig. 4.86. Stem and bulb nematode damage to narcissus.

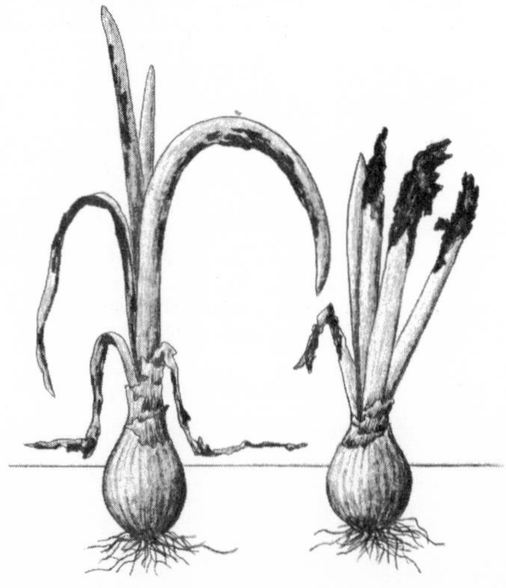

Fig. 4.87. Narcissus "smoulder."

9. *Amaryllis Spotted Wilt*—Numerous pale yellow or white spots on leaves. Reddish-brown spots or red lines may develop, especially along leaf edges. Leaves usually later turn yellow and die. *Control:* Same as for Mosaics (above). Control thrips that transmit virus. Use DDT plus malathion.
10. *Lesion and Other Root-feeding Nematodes* (e.g., lance, pin, sheath, spiral, stunt)—Plants, bulbs, and flowers often stunted or reduced in numbers. Foliage turns yellow and withers prematurely in certain areas. Eventually whole plant may decline, wilt, and die. Roots few, short, and stubby. Often show dark spots or rot. *Control:* Plant in clean or sterilized soil (pages 538–548). Use D-D, chloropicrin, or EDB. Follow manufacturer's directions.
11. *Root-knot*—See (37) Root-knot under General Diseases. *Control:* Same as for Lesion Nematodes (above). Soak *tuberose* tubers, offsets, or "seed" in hot water (120° F.) for 1 hour. Cool, dry, and plant.
12. *Yellow Dwarf*—Plants stunted and yellowish. See under Onion.
13. *Leaf Spot*—Small water-soaked spots appear after blooming is over. Spots enlarge, turn grayish-brown, and merge. Leaves often wither and die early. *Control:* Destroy spotted leaves. Spray as for Fire (above).
14. *Rust* (atamasco-lily, rainlily, zephyranthes)—Southern states. Yellow, orange, reddish-brown, or black powdery pustules on leaves. *Control:* Spray as for Fire (above). Destroy rusted leaves.
15. *Bacterial Blight, Stem Rot* (narcissus)—Flower stems rapidly collapse from a wet basal rot that is first grayish-pink. Later varies from pink to brown, often with a violet tint. Entire stem later turns yellow. *Control:* Same as for Bulb Rots (above).
16. *Bud Blast* (narcissus)—Flower buds turn brown and wither before opening. *Control:* If condition reappears year after year, dig and discard these bulbs. Plants should be well watered and fed. Bulbs should be well spaced outdoors. Indoors, keep bulbs cool with a good root system before being forced.

DAHLIA — See Chrysanthemum
DAHOON — See Holly
DAISIES — See Chrysanthemum
DALEA — See Broom
DAMESROCKET — See Cabbage
DANDELION — See Chrysanthemum
DANGLEBERRY — See Blueberry
DAPHNE [CAUCASIAN, LILAC, SOMERSET, WINTER], MEZEREON or FEBRUARY DAPHNE, GARLAND-FLOWER or ROSE DAPHNE, SPURGE LAUREL *(Daphne)*

1. *Leaf Spots, Anthracnose*—Widespread. Thick, brown, purplish, reddish, or irregular greenish spots. Leaves may yellow, wither, and fall early. *Control:* Remove and burn infected leaves and branches. Spray several times, 10 to 14 days apart. Use fixed copper, zineb, ferbam, thiram, maneb, or captan.
2. *Twig Blight, Canker, Dieback*—Twigs blighted. Twigs and branches die back from rough, sunken cankers. May be covered with small, coral-red "cushions" or gray mold. *Control:* Remove and burn infected parts. Spraying as for Leaf Spots may be beneficial.
3. *Crown Rots, Stem and Root Rots, Wilts*—Cuttings and established plants wilt and die. Dark brown to black, irregular decayed areas may develop on stems at or near soil line. Stems girdled and roots may be blackened and decayed. *Control:* Grow in well-drained, slightly alkaline, sterilized soil, if possible.
4. *Mosaic*—Leaves mottled yellowish-green. Plants may be stunted. Flowering is reduced. *Control:* Destroy infected plants. Control insects that spread viruses. Use

DDT plus malathion. See (16) Mosaic under General Diseases.

5. *Wilts* (*Verticillium and Fusarium*)—Rare. See (15A) and (15B) Fusarium and Verticillium Wilts under General Diseases.
6. *Winter Injury*—Foliage scorched following an ice crust or wind burn. May be coupled with crown or root rot, unbalanced nutrition, and growth late in the fall or Verticillium wilt infection.

DARLINGTONIA — See Pitcherplant
DASHEEN — See Calla Lily
DATE PLUM — See Persimmon
DATURA — See Tomato
DAVALLIA — See Ferns
DAYFLOWER — See Tradescantia
DAYLILY — See Hemerocallis
DECUMARIA — See Hydrangea
DEERGRASS, MEADOWBEAUTY *(Rhexia)*

1. *Leaf Spots*—Spots of various colors, sizes, and shapes. *Control:* Pick and burn spotted leaves. If serious, spray several times, 10 days apart, during rainy periods. Use zineb, maneb, or captan.

DEER'S-TONGUE — See Gentian
DELONIX — See Honeylocust
DELPHINIUM, LARKSPUR [BOUQUET, CANDLE, CAROLINA, CHINESE, FIELD, GARDEN, GARLAND, RED, ROCK, ROCKET or ANNUAL, SCARLET, TALL] *(Delphinium)*; MONKSHOOD, ACONITE, WOLFSBANE *(Aconitum)*; COLUMBINE *(Aquilegia)*; GOLDTHREAD *(Coptis)*; CHRISTMAS-ROSE, LENTEN ROSE *(Helleborus)*; PEONY [ANEMONE, CHINESE or COMMON, DOUBLE, HYBRID, JAPANESE, OFFICINALIS, SEMIDOUBLE, SINGLE, TREE], *(Paeonia)*; BUTTERCUP [COMMON, DOUBLE-FLOWERED, FLORISTS' or TURBAN], CROWFOOT *(Ranunculus)*; MEADOWRUE [ALPINE, EARLY, PURPLE, TALL] *(Thalictrum)*

1. *Stem Cankers, Southern Blight, Wilt, Stem, Crown, and Root Rots*—General and serious in poorly drained soils in wet seasons. Stems stunted, suddenly or gradually wilt. May turn yellow, darken, and die. Often collapse. Stems may be cracked, cankered, or rotted at or near soil line. May be slimy and have a foul odor (Bacterial Crown Rot). Roots often decayed. Cottony to bluish-gray fungus threads and small, tan to dark red or brown to black seedlike bodies (sclerotia) may form at crown and roots and inside affected stems (Southern Blight or Crown Rot). Infected plants easily pulled up. *Control:* Dig up and divide older clumps. Dig and burn severely diseased plants, including 3 to 5 inches of surrounding soil. Plant disease-free stock or seed from disease-free plants in ridges for good water drainage. Treat suspicious *delphinium* and *larkspur* seed by soaking in hot water (122° to 129° F. for 10 minutes). Cool, dry, and plant. Grow in clean or sterilized soil (pages 538–548) that is well-drained and in a sunny spot. Rotate. Avoid overwatering, close planting, wounding stems, and wet mulch around crowns. In fall after cutting and burning the tops, and again in early spring, drench crowns and surrounding soil several times, a week apart. Use Semesan or 1:2,000 solution of mercuric chloride. See page 101 for precautions. Certain fungus rots are controlled by working PCNB-Dexon mixture around plants a week or more before planting. Follow manufacturer's directions, or drench soil with this mixture. Apply 1 pint

per square foot. Spraying in spring as for Gray-mold Blights (below) is often beneficial.

2. *Gray-mold Blights, Botrytis Blight, Flower Blight, Bud Blast*—General in cool, moist weather. Young shoots may wilt, yellow, and collapse from soft brown to black rot near soil line. Buds turn brown or black and fail to open, or flowers are brown-spotted, watery, and matted. Large, irregular, dark brown blotches may occur on leaves. Coarse gray mold grows on affected areas in damp weather. *Peony* buds that turn brown and dry from pea to marble size may be due to feeding of thrips, frost injury (especially if soil potash supply is low), lack of water and soil nutrients, or other factors. See Figure 2.21B under General Diseases. *Control:* Avoid close, dense plantings and working with plants when foliage is damp. Cut and burn tops just below ground level in fall. Remove and destroy infected plant parts as they appear. Spray shoots and surrounding soil with captan, zineb, ferbam, or maneb (1½ to 2 tablespoons per gallon) plus spreader-sticker as shoots emerge. Repeat 10, 20, and 30 days later. Apply last spray as flowers start to open. Additional sprays at 2-week intervals after bloom may be needed if period is rainy or humid.
3. *Peony Bud Blast*—General. Buds reach size of small peas and turn brown or black; fail to develop further. Associated with low potassium in soil, late spring frost or dry periods, root-knot or other nematode infection, too deep planting in infertile soil, or excessive shade. *Control:* Avoid as many of these factors as possible.
4. *Mosaics, Ringspot*—General. Symptoms variable. Plants may be severely stunted with mottled, pale green and yellow leaves (Mosaics). Or with lemon-yellow to orange-amber spots, blotches, line patterns, bands, arcs, or striking, zoned rings with "green islands" (Ringspot or Spotted Wilt). Young leaves may be distorted. Flowering may be prevented (Figure 4.88). *Control:* Use virus-free planting stock. Do *not* propagate from infected plants. Destroy diseased plants when first seen. Keep down weeds. Control insects, especially aphids, with malathion.

Fig. 4.88. Peony ringspot.

5. *Delphinium and Aconitum Black Blotch, Bacterial Leaf Spot*—Widespread in cool, wet weather. Small, water-soaked spots that later are irregular, shiny, and tarlike with yellow borders. See Figure 2.18C under General Diseases. Mostly on leaves, but also on buds, stems, and blossoms. Lower leaves infected first. Black blotches may unite killing the leaf. *Control:* Plant disease-free stock in clean soil. Destroy infected leaves as they appear. Cut and burn tops in fall. If possible, keep water from splashing on foliage. Three- to 4-year rotation. Drench crowns and surrounding soil when plants 6 to 10 inches tall. Use zineb, maneb, fixed copper,

Semesan, or a 1:2,000 solution of mercuric chloride. Thereafter, if serious, spray weekly with fixed copper and spray lime.

6. *Powdery Mildews*—General. Powdery white mold on leaves. Leaves may be deformed, turn yellow, and drop early. Varieties differ greatly in resistance. *Control:* Spray two or three times, 10 days apart. Use Karathane, Acti-dione, or sulfur plus spreader-sticker. *Delphinium* varieties differ in resistance. Collect and burn fallen leaves. Space plants.
7. *Yellows, Stunt, Witches'-broom, "Greens"*—General. Plants "bunchy" and often stunted with numerous, slender, yellowish, upright shoots. Parts or all of certain flowers are green and leafy. *Control:* Same as for Mosaics (above). Control leafhoppers that transmit virus, using DDT plus malathion.
8. *Seed Rot, Damping-off*—Cosmopolitan. Seeds rot. Stand is poor. Seedlings wilt and collapse. *Control:* Sow disease-free seed in sifted sphagnum moss or other suitable starting medium. Soil drenches of Dexon may be effective.
9. *Wilts (Fusarium and Verticillium)*—Plants gradually wilt, wither, and die about blooming time, starting at base. Insides of stems and crowns show green to brown or black streaks. Fusarium also causes sunken, brown, water-soaked areas (cankers) on stems. *Control:* Dig up and destroy infected plants. Grow disease-free stock in clean or sterilized soil (pages 538–548) that has not grown wilted plants.
10. *Root-knot and Other Root-feeding Nematodes* (e.g., dagger, lance, lesion, reniform, ring, spiral, stem) —Plants lack vigor. May be stunted, spindly, pale, and do not bloom normally. Small galls, usually 1/8 to 1/4 inch in diameter form on finer roots (Root-knot). Roots may decay and die back or be stunted and bushy. *Control:* Plant nematode-free roots or planting stock in clean or sterilized soil (pages 538–548). Drench around established plants using DBCP or VC-13 Nemacide. Follow manufacturer's directions. Disinfest *dormant peony* roots by soaking in hot water (120° F.) for 30 minutes.
11. *Leaf Spots, Leaf Blight or Blotch, Anthracnose, Black Spot*—Common in damp weather. Small to large leaf spots or blotches of various colors. May merge to form irregular dead areas. Leaves and young stems may wither and die early. Similar spots, cankers, or streaks may occur on stems, petioles, flower petals, and seed pods. *Peony* buds may blast. See Figures 4.89A and 4.89B. *Control:* Cut and burn tops in fall. Spray foliage as for Gray-mold Blights (above). Propagate only from disease-free plants, where feasible. Rotate with nonrelated plants.
12. *Rusts* (aconitum, buttercup, columbine, delphinium, meadowrue) —Widespread. Small yellow spots on leaves. Alternate hosts include grasses, barley, *Prunus* spp., and alpine bistort *(Polygonum viviparum)*. *Control:* Spray as for Gray-mold Blights (above).

Fig. 4.89A (left). Cladosporium leaf blight or blotch (Measles) on stems and underside of leaves. (Illinois Natural History Survey photo)

Fig. 4.89B. Black spot of Christmas-rose.

13. *Crown Gall*—See under Begonia, and (30) Crown Gall under General Diseases.
14. *LeMoine Disease* (peony)—Common. Plants dwarfed with spindly shoots. Produce no flowers. Roots irregularly swollen and stubby with few feeding roots. *Control:* Dig and burn infected plants. Grow disease-free stock in new location.
15. *Peony Crown Elongation Disease, Witches'-broom*—Small leaves on dwarfed, slender shoots. Crowns elongated with small, weak buds at tips. Plants do not flower. *Control:* Same as for LeMoine Disease (above).
16. *Leaf and Stem Smuts, White Smut*—Blister-like swellings on leaves and leaf stalks later filled with black powder. *Control:* Same as for Stem Cankers (above). Pick and burn diseased parts before blisters open.
17. *Curly-top*—Western states. Plants stunted with younger leaves curled and bunched on main stem and side branches. See (19) Curly-top under General Diseases.
18. *Leaf and Stem Nematode*—Branches may be aborted and foliage distorted. See (20) Leaf Nematode under General Diseases.
19. *Spotted Wilt, Ringspot* (buttercup, columbine, dahlia, delphinium, peony)—See under Bellflower, and (17) Spotted Wilt under General Diseases.
20. *Flower Spot or Blight*—Flower petals spotted. Spots may enlarge, blighting complete flower and seed pod. See (31) Flower Blight under General Diseases.
21. *Downy Mildew* (aconite, buttercup, Christmas-rose, meadowrue)—Occasional. See (6) Downy Mildew under General Diseases. *Control:* Spray as for Gray-mold Blights (above).
22. *Chlorosis*—See under Rose. May also be caused by low temperatures and wet soil.
23. *Thread Blight*—Southeastern states. See under Fig.
24. *Delphinium "Blacks"*—Caused by minute cyclamen mites. Plants stunted, curled, and seriously deformed. Buds and flowers turn black; are deformed and distorted. Dark brown to black streaks and blotches occur on stems and petioles. *Control:* Check with your extension entomologist regarding suitable spray program.
25. *Leaf Curl* (peony)—Plants dwarfed with curled leaves. Flower stalks cracked. *Control:* Dig up and destroy infected plants.

DENDROBIUM — See Orchids

DENDROMECON — See Poppy

DENNSTAEDTIA — See Ferns

DENTARIA, DESERTPLUME — See Cabbage

DESERT-WILLOW — See Catalpa

DEUTZIA — See Hydrangea

DEVILSCLAW — See Proboscisflower

DEVIL'S IVY — See Calla Lily

DEVILWOOD — See Osmanthus

DEWBERRY — See Raspberry

DIANTHUS — See Carnation

DICENTRA — See Bleedingheart

DICKSONIA — See Ferns

DICTYOSPERMA — See Palms

DIDISCUS — See Celery

DIEFFENBACHIA — See Calla Lily

DIERVILLA — See Snowberry

DIGITALIS — See Snapdragon

DILL — See Celery

DIMORPHOTHECA — See Chrysanthemum

DIOSCOREA — See Yam

DIOSPYROS — See Persimmon
DIPSACUS — See Teasel
DIRCA — See Leatherwood
DITTANY — See Salvia
DODECATHEON — See Primrose
DOGSTOOTH-VIOLET — See Erythronium

DOGWOOD [BAILEY, BLOODTWIG, CHINESE or KOUSA, FLOWERING, GIANT, GRAY or PANICLED, JAPANESE CORNEL, KOREAN, PACIFIC, PAGODA or ALTERNATE-LEAVED, PINK, RED, RED-OSIER, RED-TWIGGED, SIBERIAN, TARTARIAN, VARIEGATED or CREEMEDGE, WHITE, WEEPING WHITE, YELLOW-TWIGGED], CORNEL [DWARF or BUNCHBERRY, ROUGHLEAF, SILKY], OSIER [GREEN, RED, WESTERN], CORNELIAN-CHERRY, JAPANESE CORNELIAN-CHERRY *(Cornus)*; TASSELTREE, SILKTASSEL-BUSH *(Garrya)*; TUPELO (SOUR-GUM, BLACK GUM), WEEPING TUPELO *(Nyssa)*

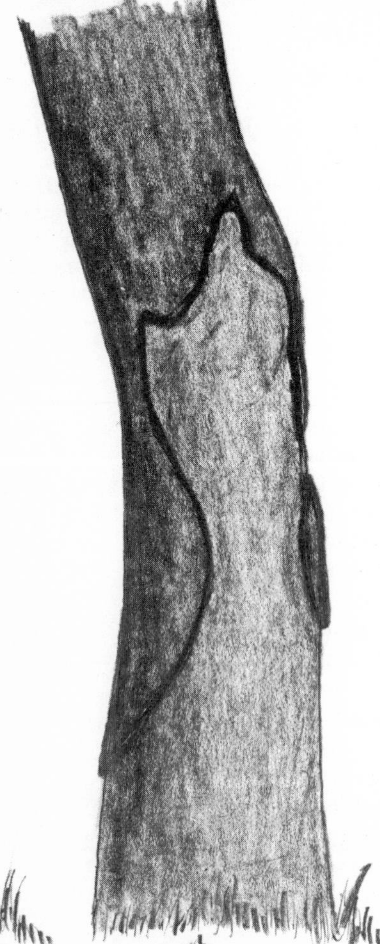

Fig. 4.90. Dogwood collar rot.

1. *Dogwood Collar Rot, Trunk or Bleeding Canker*—Widespread in eastern states and along Pacific Coast. Trees and shrubs lack vigor. Leaves dwarfed and pale green, turn yellow or prematurely red in late summer. Drop early. Twigs and branches stunted. May wilt and die back; frequently on one side of tree. Sunken canker on lower trunk, crown, or roots that enlarges slowly for several years. Girdles trunk or shrub base killing parts beyond (Figure 4.90). Diseased trees often bear large crops of flowers and fruits for one or more years prior to death. Cankers ooze sap or dark fluid in spring. Trees may decline and die in 1–5 years or more. *Control:* Dig up and burn trees showing large cankers (over halfway around). Do *not* replant in same soil for several years without drenching first with one part formalin in 50 parts of water. Or use zineb. Remove smaller cankers promptly. Cut out 1½ inches of surrounding "healthy" bark and discolored wood (see Figure 2.7). Paint wound edges with orange shellac. Swab remainder of wound with household bleach (diluted 1:5 with water), a 1:1,000 solution of mercuric chloride, or bordeaux paste (page 105). Finally paint with tree wound dressing. Keep trees growing vigorously. Avoid wounding trunk during transplanting, mowing, etc. Keep base of trunk dry. Grow in acid, well-drained soil high in organic matter. Control borers by painting or spraying trunk and main branches monthly with DDT or Thiodan, starting about in mid-May. Do *not* spray foliage with DDT. Check with your county agent or extension entomologist regarding timing of sprays for your area.
2. *Leaf Spots, Leaf Blotch, Spot Anthracnose*—Widespread in wet seasons. Spots of various sizes, shapes, and colors. Often with dark purple to brown borders (Figure 4.91). Spots on leaves may fall out leaving ragged holes. If severe, leaves may drop early. Certain spots also occur on young stems, flowers, and berries. Flowers may be stunted, malformed, or do not open. Shoots and berries are "scabby." *Control:* Apply same spray materials as for Gray-mold Blight (below) or use maneb, thiram, or phenyl mercury plus spreader-sticker. Spray when buds first open, just before, and after bloom. Then repeat monthly to August. Collect and burn fallen leaves in autumn. Keep trees pruned.
3. *Gray-mold Blight, Flower and Shoot Blight, Bud Blight*—Widespread in wet springs. Irregular brown areas on fading flowers and leaves where old flower bracts have dropped. May be covered with grayish mold in damp weather. Buds blasted. *Control:* Apply captan, zineb, or folpet plus spreader-sticker just before, during, and after bloom. If possible, spray just before wet periods. Prune to keep trees open.
4. *Powdery Mildews*—General. Leaves covered with white, powdery patches in late summer and fall. *Control:* If serious, add Karathane or sulfur to Leaf Spot sprays (above). Start when mildew *first* appears.
5. *Twig Blights, Branch and Trunk Cankers, Dieback*—Twigs and branches die back from girdling, discolored cankers. Cankers may be sunken or swollen, cracked, or target-like. *Control:* Prune and burn infected branches back to healthy wood. Avoid wounding bark. Paint wounds promptly as for Collar Rot (above). Keep trees vigorous by watering and fertilizing. Spraying as for Leaf Spots (above) should be beneficial.
6. *Root Rots*—See under Apple, and (34) Root Rot under General Diseases. Often associated with root-feeding nematodes (e.g., dagger, lance, pin, root-knot, spiral, sting, stubby-root). DBCP may be applied to soil under tree if nematodes are a problem. Follow manufacturer's directions. Grow in acid, well-drained soil high in organic matter.
7. *Wood Rots, Heart Rot*—See under Birch, and (23) Wood Rot under General Diseases. Control borers. Burn dead and dying branches and spray trunk at monthly intervals as for Collar Rot (above). Keep trees growing vigorously. Avoid wounding trunk.

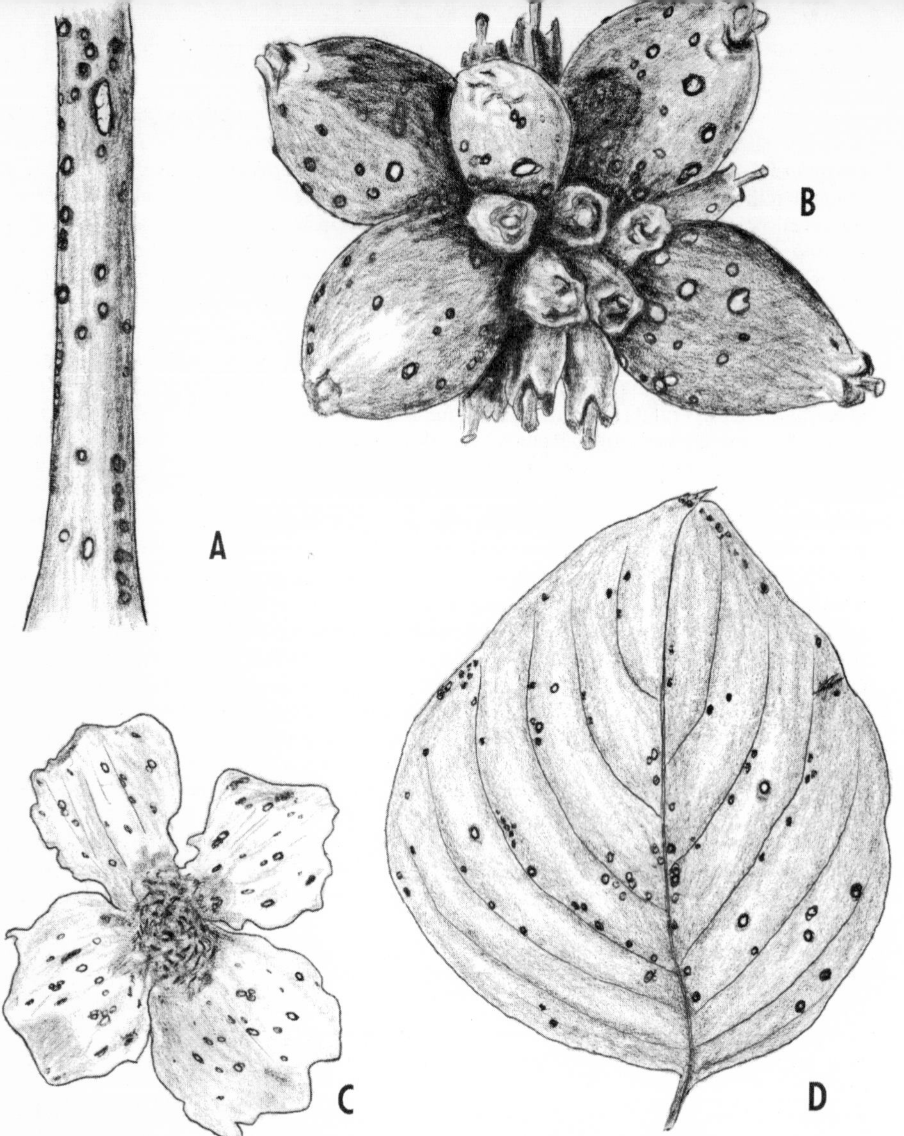

Fig. 4.91. Spot anthracnose of dogwood. Spots on: A. Stem. B. Berries. C. Flower. D. Leaf.

8. *Leaf Scorch*—Develops in July and August following hot, dry, windy weather. Margins of leaves turn light brown on trees growing in full sun and poor soil. If severe, entire branches may defoliate and die. *Control:* Have soil tested. Apply fertilizer as recommended. Prune trees to keep them vigorous. Water during summer dry periods. Transplant carefully into good soil early in spring (page 19). Apply mulch to keep soil cool and moist.
9. *Sunscald, Heat or Drought Injury*—Results in death of young trees during first few years following transplanting. Often due to sunscald, lack of soil moisture, and careless handling of trees too large for easy transplanting. Leaves often curl or cup and turn red or reddish-purple. Some leaves may drop. Twigs may die back. *Control:* Transplant small trees into partial shade. Wrap or shade south and

southwest sides of trunk. See under Apple, and Figure 2.11. Water and fertilize properly. Keep trees and shrubs well mulched.

10. *Sooty Mold, Black Mildew*—Mostly in southeastern states. See under Apple, and (12) Sooty Mold under General Diseases.
11. *Crown Gall*—See under Apple, and (30) Crown Gall under General Diseases.
12. *Verticillium Wilt*—See under Maple.
13. *Rust* (dwarf cornel, tupelo) —Widespread. Small, reddish-brown to black, powdery pustules on leaves. *Control:* If serious, spray as for Leaf Spots (above). Use zineb or maneb.
14. *Mistletoe* (black gum, dogwood, tupelo) —See (39) Mistletoe under General Diseases.
15. *Thread Blight*—Southeastern states. See under Walnut. *Control:* Spray as for Leaf Spots (above).
16. *Felt Fungus*—Southeastern states. Purple-black, feltlike growth. Associated with scale insects. See under Hackberry.
17. *Herbicide Injury (2,4-D, 2,4-5-T and Silvex)*—Dogwoods are very susceptible to weed-killing chemicals. See under Grape.

For additional information, read USDA H & G Bulletin 88, *Growing the Flowering Dogwood.*

DOLICHOS — See Pea
DOUGLAS-FIR — See Pine
DORONICUM — See Chrysanthemum
DOXANTHA — See Trumpetvine
DRABA — See Cabbage
DRACAENA, DRAGONTREE *(Dracaena)*; CORDYLINE, HAWAIIAN TI PLANT or RED DRACAENA *(Cordyline)*

1. *Leaf Spots, Tip Blight, Anthracnose*—General. Round to irregular spots of various colors. Centers of spots may be sprinkled with black dots. Lower and center leaves may die back from tips (Tip Blight). See Figure 4.92. If severe, all leaves may wither and die except a few at top of plant. *Control:* Keep water off foliage. Destroy infected leaves when first seen. If necessary, apply zineb, maneb, copper, or captan sprays, at 7- to 10-day intervals, during rainy weather.
2. *Chlorosis*—Plants yellowish and sickly. Most prevalent on poorly drained acid or alkaline soils. *Control:* Spray weekly for a month using iron sulfate or an iron chelate (page 27). Follow manufacturer's directions. Grow in well-drained soil that is near neutral (about pH 6.5).
3. *Gray-mold Blight*—May cause soft brown rotting in damp periods. Affected areas covered with coarse grayish mold. *Control:* Same as for Leaf Spots (above).
4. *Stem Rot*—Leaves turn yellow starting at base of stem or tip cutting. Plants wilt. Lower part of stem is rotted, black, and water-soaked. *Control:* Grow disease-free stock or tip cuttings (dipped in ferbam, 2 tablespoons per gallon), in clean or sterilized soil (pages 538–548). Watering with zineb, captan, thiram, or ferbam (1 tablespoon per gallon) during rooting may help.
5. *Root-feeding Nematodes* (lance, sheath) —Plants may appear unthrifty. Gradually decline. Often associated with Root Rot. *Control:* Same as for Root Rot and Root-knot (both below).
6. *Root Rot*—See (34) Root Rot under General Diseases.
7. *Root-knot*—See (37) Root-knot under General Diseases.

DRACOCEPHALUM, DRAGONHEAD — See Salvia
DRAGONROOT — See Calla Lily

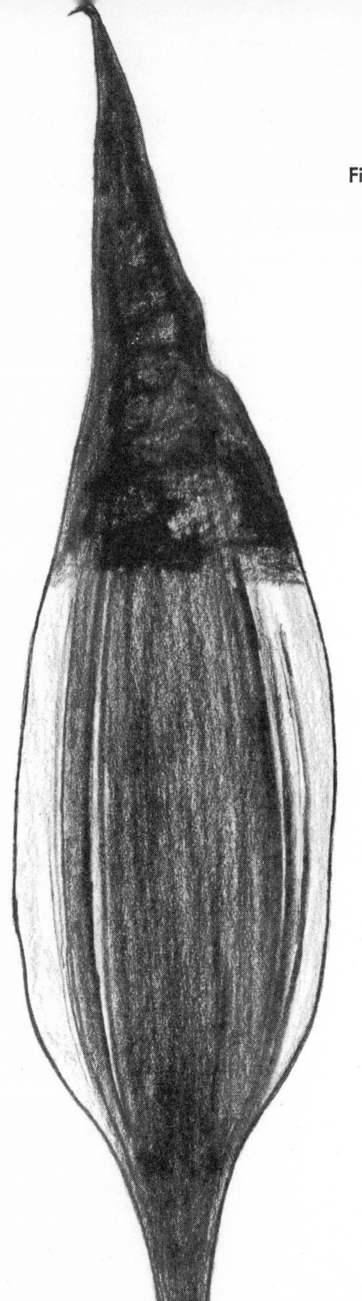

Fig. 4.92. Dracaena tip blight.

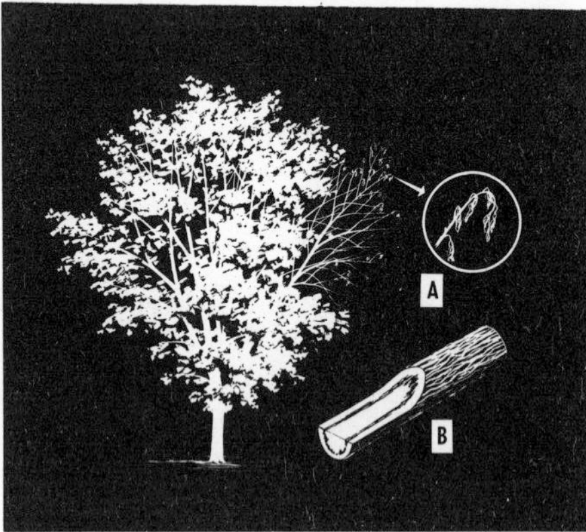

Fig. 4.93. Dutch elm disease. Note shepherd's crook and cut twig showing dark discoloration in the outer sapwood. The causal fungus makes the water-conducting tissue nonfunctional.

DRAGONTREE — See Dracaena
DROPWORT — See Rose
DRYOPTERIS — See Ferns
DUCHESNEA — See Rose
DUMBCANE — See Calla Lily
DURANTA — See Lantana
DUSTY-MILLER — See Chrysanthemum
DUTCHMANS-BREECHES — See Bleedingheart

DUTCHMANS-PIPE — See Aristolochia

DWARF INDIGO — See False-indigo

DWARF LACE PLANT — See Silver-lacevine

DYER'S GREENWEED — See Broom

DYSCHORISTE — See Clockvine

ECHEVERIA — See Sedum

ECHINACEA — See Chrysanthemum

ECHINOCACTUS, ECHINOCEREUS — See Cacti

ECHINOCYSTIS — See Cucumber

ECHINOMASTUS — See Cacti

ECHINOPS — See Chrysanthemum

ECHINOPSIS — See Cacti

ECHIUM — See Mertensia

EGGPLANT — See Tomato

EGYPTIAN PAPER PLANT — See Umbrellaplant

ELAEAGNUS — See Russian-olive

ELDER — See Snowberry

ELECAMPANE — See Chrysanthemum

ELEPHANTS-EAR — See Calla Lily

ELM [AMERICAN (many horticultural varieties), CAMPERDOWN, CEDAR, CHINESE, DUTCH, ENGLISH, HYBRID, JAPANESE, RED or SLIPPERY, ROCK or CORK, SCOTCH, SIBERIAN or DWARF, SMOOTH-LEAVED EUROPEAN, WEEPING, WINGED or WAHOO, WYCH] *(Ulmus)*; JAPANESE ZELKOVA *(Zelkova)*

1. *Dutch Elm Disease*—Roughly eastern half of United States and Canada (and Oregon) and spreading westward. Serious. Leaves wilt, turn dull green to yellow or brown, curl, and usually drop early on one or more branches. Twig tips may curve downward to form "crooks." Branches die back. Entire trees may die in 1 to 2 months; others survive 1 to 3 years or longer. A brown to almost black discoloration occurs in the white sapwood just under the bark in wilting branches. See Figure 4.93. Positive identification only possible through laboratory culturing. *Control:* A community-wide program is needed in threatened areas. Remove and burn *all* dead, weak, and dying elm trees when found. Clean up and burn *all* injured, weak, and dead elm wood (with tight bark) in trees or on ground *before* trees leaf out in early spring. Debark stumps and fireplace wood. Repair and paint tree wounds promptly. Keep trees vigorous by proper watering, fertilizing, and pruning. A single *dormant* spray of DDT or methoxychlor is recommended in many areas where disease is present. Check with your park department, city arborist or forester, local arborist or county agent. The spray kills elm bark beetles that transmit the Dutch elm disease fungus. Beetles are spread long distances by road and rail traffic. Applying SMDC (Vapam or VPM) in holes drilled in the soil between healthy and *early-infected* trees, growing within 50 feet of each other, stops spread through root grafts. Drill holes (3/4 inch in diameter, 3 feet deep and 6 to 9 inches apart) in a straight line, T- or L-shape (if obstructions are present), equidistant between trees. Use 1 part of SMDC to 9 parts of water. Fill each hole with diluted chemical to within 2 inches of soil surface. Resistant *elms:* Christine Buisman, Bea Schwarz, Chinese, Groeneveld or Green Field, Siberian, Hybrid, and European Field. These trees are *not* immune to the disease. *Infected trees cannot be saved or cured.* For more information read USDA Info. Bulletin 193, *The Dutch Elm Disease and its Control;* and USDA Leaflet 185, *Elm Bark Beetles.*

2. *Phloem Necrosis*—Serious in eastern and central United States below 40° North latitude, on American (and its varieties) or Winged elms. During June and July, leaves roll upward, turn yellow or brown, wither, and fall, starting throughout upper crown. Foliage is thin. Trees showing symptoms usually drop their leaves and die in a month or two. The inner bark, especially near the trunk base, is often yellow to butterscotch-colored; sometimes flecked with dark brown or black. When such freshly cut bark is warmed in a closed jar, an "oil of wintergreen" odor is often evident. Roots die first. Disease spread may occur through root grafts. *Control:* Highly resistant or immune *elms:* Chinese, Christine Buisman, and Hybrid. Spray susceptible elms with DDT or methoxychlor plus malathion to control leafhoppers that transmit virus. Two summer applications are usually made, 35 to 40 days apart. Check with your city forester, county agent, or extension entomologist regarding timing of sprays for your area. *Infected trees cannot be saved or cured.*
3. *Other Wilts, Diebacks (Verticillium, Dothiorella)*—Widespread. External and internal symptoms often closely resemble Dutch elm disease. Infected trees may live on for a number of years, slowly dying back and declining in health. Leaves may be dwarfed and yellowed. Sucker growth is common on trunk and larger branches. Laboratory culturing is needed for positive diagnosis. *Control:* Remove and burn severely infected trees. On others, remove infected, cankered, or dead branches flush with next larger limb or trunk. Disinfect pruning tools between cuts. Keep trees vigorous. Fertilize liberally and water during dry periods. Spray as for Black Leaf Spot (below). The Groeneveld *elm* is resistant to Verticillium Wilt. Spray to control defoliating insects, e.g., spring cankerworm and elm leaf beetle.
4. *Black Leaf Spot*—General, especially in wet seasons. Asiatic and American elms are very susceptible. Small, irregular, grayish spots on leaves. Spots later become shiny, tar-black, and raised (Figure 4.94). Infected leaves often turn yellow or brown and drop early in large numbers. Trees vary greatly in susceptibility. Twigs may die back. *Control:* Collect and burn fallen leaves. Prune out dead twigs. Where practical, spray as leaves unfold. Repeat 2 and 4 weeks later. Apply fixed copper, dichlone, zineb, captan, thiram, phenyl mercury, or ferbam, plus spreader-sticker. Thorough coverage is essential.
5. *Anthracnoses, Leaf Spots, Twig Blight*—General. Irregular, brown to reddish-brown spots or dead areas in leaves between veins and margins. Twigs may die back. *Control:* Same as for Black Leaf Spot (above). Spray as buds break open, 7 and 14 days later.
6. *Winter Injury, Sunscald, Frost Crack, Winter Drying*—Twigs and branches may die back from tips. Dwarfed, unthrifty leaves may unfold on affected branches and then die. Limbs fail to leaf out in spring. *Freezing Injury:* Inner bark may be water-soaked, then very dark. Several or all limbs may die. Long, up-and-down cracks (Frost Cracks) or large, dead, discolored areas (Frost Cankers) may form on exposed south or southwest sides of trunk and larger limbs. *Control:* Water trees during dry periods, especially in a dry fall. Keep tree crown pruned open. Fertilize in spring. Wrap young trees with burlap, sisalkraft paper or otherwise shade from winter sun. See Figure 2.11.
7. *Wetwood, Slime Flux*—Cosmopolitan on trees over 5 years old after heartwood has formed. Asiatic elms very susceptible. Fermenting, dark-colored sap (flux) flows from branch stub, split crotch, trunk crack, or other bark wound. Common in spring and summer or after wet weather. The fluxing sap dries to form a light gray to white stain on trunk and larger branches (Figure 4.95). On young trees, leaves on one or more branches may wilt, curl, turn color, and drop early. Branches on older trees die gradually. Foliage is sickly yellow or scorched. Trees gradually decline in health. Invaded wood becomes dark brown and water-soaked. Often

Fig. 4.94 (left). Black leaf spot of elm.

Fig. 4.95. Wetwood or slime flux of elm.

associated with wet soil, mechanical damage to roots, branches, trunk, and crotches, or frost cracks. *Control:* Fertilize and water to stimulate vigor. Remove dead and weak branches. Repair bark wounds promptly (page 34). Cover with tree wound dressing. Check with a good tree surgeon. Trees may need cabling, bracing, or drain pipes installed. For further information, read Illinois Natural History Survey Circular 50, *The Wetwood Disease of Elm.*

8. *Mosaics, Mottle-leaf*—General. Infected leaves abnormally large or small. Leaves stiff, often distorted. Dwarfed leaves usually mottled light green and yellow. Witches'-brooms may form on twigs. Tree gradually loses vigor over period of years. Foliage appears thinner and branches die back. Virus is seed- and pollen-transmitted. *Control:* Where practical, replace with virus-free trees. No control known.

Fig. 4.96. Dothiorella canker on elm branch. Note the black, raised pustules (fungus fruiting bodies) in the dead bark. Under moist conditions, spores of the fungus ooze out of the pustules and are deposited on the surface of the bark. (Illinois Natural History Survey photo)

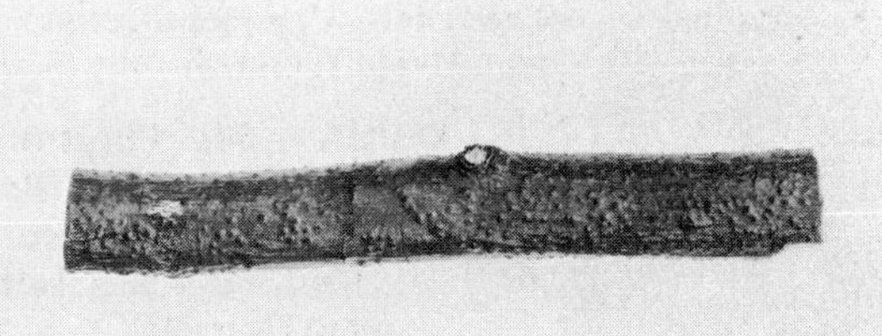

9. *Twig Blights, Diebacks, Cankers*—Widespread. Twigs and branches die back from cankers. Bark is often discolored, sunken, or shows small, coral-pink to black "pimples" on the surface (Figure 4.96). May follow winter, drought, or insect injury. Leaves on affected branches often dwarfed, turning yellow or wilting later in summer. *Control:* Prune and burn dead twigs and branches. Make cuts several inches beyond any sign of infection. Paint wounds promptly with good tree wound dressing. Fertilize and water to maintain vigor. The Groeneveld *elm* is somewhat resistant to Nectria Canker, while Siberian and Christine Buisman elms are very susceptible.
10. *Wood Rots, Heart Rots*—Cosmopolitan. See under Birch, and (23) Wood Rot under General Diseases. Control borers by pruning and burning dead and dying branches. Spray trunk at monthly intervals with DDT or Thiodan. Keep trees growing vigorously.
11. *Root Rot*—Trees gradually decline in vigor. Foliage is thin and sickly. Leaves may yellow, wither, and fall early. May be associated with nematodes. *Control:* See under Apple and Oak.
12. *Physiological Leaf Scorch*—Browning or scorching of leaves between veins and along margins. Follows hot, dry, windy weather. *Control:* Fertilize and prune trees to increase vigor. Water during summer dry periods.
13. *Bleeding Canker*—Northeastern states. See under Beech and Maple.
14. *Leaf Blister*—Small, grayish-white to brown blisters on thickened, puckered leaves. Leaves may yellow and drop early. *Control:* Same as for Anthracnoses (above).
15. *Powdery Mildews*—General. White, powdery patches on leaves. Leaves may later turn yellow and shrivel. *Control:* If serious, spray twice, 10 days apart. Use Karathane or sulfur.
16. *Root-knot*—Elm is highly susceptible. See under Peach, and (37) Root-knot under General Diseases. DBCP may be applied to the root zone following manufacturer's directions.
17. *Sooty Mold*—Black, sooty mold grows in "honeydew" produced by aphids and scale insects. *Control:* Not necessary. Control insects with DDT or lindane plus malathion.
18. *Seed Rot, Seedling Blight, Damping-off*—See under Pine.
19. *Virus Leaf Scorch of American Elm*—Southeastern states. Leaves appear scorched at margins and between veins. Growth stunted. Trees decline in vigor; crown gradually dies back. Tree later dies. Scorch-affected trees are more "susceptible" to Dutch elm disease. *Control:* Check with your city arborist or park department. Destroy infected trees. Seek virus-free stock from a reliable nursery.
20. *2,4-D Injury*—Elms very susceptible. Leaves may become thickened, cupped downward, or rolled. More severe injury causes leaves to be long and narrow with prominent veins. Leaves may be boat-shaped or curled into "ram's-horns"; later die (Figure 4.97). *Control:* See under Grape.
21. *Mistletoe*—See (39) Mistletoe under General Diseases.
22. *Thread Blight*—Southeastern states. See under Walnut.
23. *Other Root-feeding Nematodes* (dagger, lance, ring, spear, spiral, stem, stunt)—Often associated with weak, unthrifty, declining trees. See under Peach. DBCP may be applied to root zone following manufacturer's directions.
24. *Chlorosis*—See under Maple. Occurs mainly in alkaline soils.

EMILIA — See Chrysanthemum

EMPRESSTREE — See Paulownia

ENCELIA — See Chrysanthemum

ENDIVE — See Lettuce

ENDYMION — See Tulip

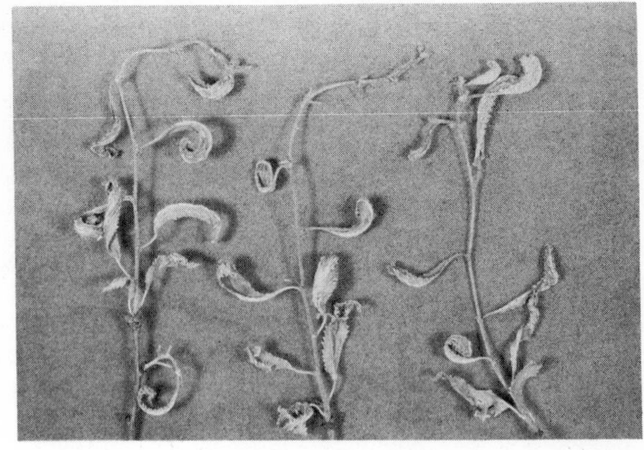

Fig. 4.97. 2,4-D injury to English elm leaves. Leaves of trees sensitive to 2,4-D injury grow long and narrow, and the veins become unusually prominent. When injury is severe the leaves become twisted and rolled. (Illinois Natural History Survey photo)

ENGELMANN IVY — See Grape
ENGLISH DAISY — See Chrysanthemum
ENGLISH IVY — See Ivy
ENGLISH MARIGOLD — See Chrysanthemum
EPIDENDRUM — See Orchids
EPIGAEA — See Heath
EPIPHYLLUM — See Cacti
EPISCIA — See African-violet
EPITHELANTHA — See Cacti
ERANTHEMUM — See Clockvine
EREMOCHLOA — See Lawngrass
ERICA — See Heath
ERIGERON — See Chrysanthemum
ERIOBOTRYA — See Apple
ERODIUM — See Cranesbill
ERVATAMIA — See Oleander
ERYNGIUM, ERYNGO — See Celery
ERYSIMUM — See Cabbage
ERYTHRINA — See Honeylocust

ERYTHRONIUM, DOGSTOOTH-VIOLET, YELLOW ADDERSTONGUE, TROUTLILY, FAWN-LILY, ADAM-AND-EVE or GLACIER-LILY *(Erythronium)*

1. *Leaf Spot, Leaf Blight, Black Spot*—Small specks or spots on leaves. Leaves may later yellow, wither, and collapse. *Control:* Pick and burn leaves when first found infected.
2. *Leaf Smuts*—Widespread. Large, blister-like swellings on leaves that break open to release brownish-black, powdery masses. Leaves may crack open and die. *Control:* Same as for Leaf Spot.
3. *Botrytis Blights*—See (5) Botrytis Blight under General Diseases.
4. *Rust*—Western states. See (8) Rust under General Diseases.

ESCAROLE — See Lettuce
ESCHSCHOLTZIA — See Poppy
ESCOBARIA — See Cacti

EUCALYPTUS, GUM [BLUE, DESERT, GRAY, LEMON, MAHOGANY,

MANNA, PEPPERMINT, SCARLET, SUGAR], BANGALAY, IRONBARK, YATE-TREE, TASMANIA STRINGY-BARK *(Eucalyptus)*; EUGENIA, AUSTRALIAN BRUSH-CHERRY, CLOVETREE, MALABAR-PLUM or ROSE-APPLE, PITANGA or SURINAM-CHERRY *(Eugenia)*

1. *Leaf Spots*—Spots of various sizes, shapes, and colors. If severe, leaves may wither and drop early. *Control:* Try applying zineb, maneb, or captan at 10-day intervals in rainy weather, where practical.
2. *Root Rots*—Eucalyptus is resistant to Armillaria Root Rot. See under Apple, and (34) Root Rot under General Diseases.
3. *Wood and Heart Rots*—See (23) Wood Rot under General Diseases.
4. *Crown Gall* (eucalyptus) —See (30) Crown Gall under General Diseases.
5. *Gray-mold Blight, Shoot Dieback*—Shoot tips may die back in wet weather following frost injury. Tips become covered with gray mold. *Control:* Same as for Leaf Spots (above).
6. *Seedling Blight*—See under Pine.
7. *Winter and Frost Injury*—Grow only adapted species in a protected location. Check with your extension horticulturist or nurseryman.
8. *Oedema*—Minor indoor problem; resembles mite injury or rust. Small, blister-like galls on leaves. May crack open and turn rust-colored. *Control:* Avoid overwatering, especially in cloudy, humid weather. Increase air circulation.
9. *Mosaic* (eucalyptus) —See (16) Mosaic under General Diseases.
10. *Black Mildew* (eugenia) —Sootlike spots on foliage. See (12) Sooty Mold under General Diseases.

EUCHARIS — See Daffodil
EUGENIA — See Eucalyptus
EUONYMUS — See Bittersweet
EUPATORIUM — See Chrysanthemum
EUPHORBIA — See Poinsettia
EUROPEAN CRANBERRY-BUSH — See Viburnum
EUSTOMA — See Gentian
EVENING-PRIMROSE [COMMON, WHITE], MISSOURI PRIMROSE, SUNDROPS, GOLDENEGGS *(Oenothera)*; CALIFORNIA FUCHSIA, FIRE-CHALICE *(Zauschneria)*

1. *Rusts*—General. Yellow-orange or reddish-brown, powdery pustules, mostly on underleaf surface. See (8) Rust under General Diseases. Alternate hosts include wild grasses and sedges (*Aristida, Distichlis,* and *Carex* spp.).
2. *Powdery Mildew*—General. See (7) Powdery Mildew under General Diseases. Usually causes little injury.
3. *Leaf Spots, Leaf Gall, Anthracnose*—Round to irregular, variously colored spots, often with dark margin. If severe, leaves may wither and drop early. *Control:* Collect and burn infected leaves. Apply zineb, several times, 10 days apart. Start when spotting is first evident.
4. *Downy Mildew*—Widespread. See (6) Downy Mildew under General Diseases.
5. *Root Rots*—See (34) Root Rot under General Diseases.
6. *Mosaic*—See (16) Mosaic under General Diseases.
7. *Stem Nematode*—See under Phlox.

EVENING STAR — See Daffodil
EVERLASTING — See Chrysanthemum
EXACUM — See Gentian

FAGUS — See Beech

FAIRY LILY, FALL-DAFFODIL — See Daffodil

FALLING STARS — See Gladiolus

FALSE-ACACIA — See Honeylocust

FALSE-ALOE — See Centuryplant

FALSE-BONESET, FALSE-CAMOMILE — See Chrysanthemum

FALSE-DRAGONHEAD — See Salvia

FALSE-GARLIC — See Onion

FALSE HOLLY — See Osmanthus

FALSE-INDIGO [BLUE, PLAINS, PRAIRIE, WHITE], YELLOW WILD INDIGO *(Baptisia)*; INDIGOBUSH, DWARF INDIGO, LEADPLANT *(Amorpha)*; INDIGO *(Indigofera)*

1. *Leaf Spots*—Widespread. Spots of various sizes, shapes, and colors. *Control:* Collect and burn tops in fall. Apply zineb, maneb, or fixed copper at 7- to 10-day intervals during rainy periods in spring and early summer.
2. *Powdery Mildews*—Common. White, powdery patches on leaves. *Control:* Apply sulfur or Karathane twice, 10 days apart.
3. *Rusts*—General. Small, yellowish to orange spots and pustules on leaves. Infected leaves may drop in large numbers. *Control:* Same as for Leaf Spots (above).
4. *Root Rots*—See (34) Root Rot under General Diseases. May be associated with root-feeding nematodes (e.g., burrowing).
5. *Twig Canker* (amorpha) —See under Maple.

FALSE-MALLOW — See Hollyhock

FALSE-MESQUITE — See Calliandra

FAREWELL-TO-SPRING — See Fuchsia

FAWN-LILY — See Erythronium

FEATHERBELLS — See Tulip

FEIJOA — See Myrtle

FELICIA — See Chrysanthemum

FENDLERA — See Hydrangea

FENNEL — See Celery

FERNS: MAIDENHAIR *(Adiantum)*; BIRDSNEST, SPLEENWORT [EBONY, GREEN, MAIDENHAIR, MOTHER, WALL-RUE] *(Asplenium)*; JAPANESE, LADY, SPLEENWORT [SILVERY, SWAMP] *(Athyrium)*; BLECHNUM [DEER, SAW] *(Blechnum)*; WALKING or WALKINGLEAF *(Camptosorus)*; CIBOTIUM; ROCKBRAKE or PARSLEY *(Cryptogramma)*; CYATHEA; TONGUE *(Cyclophorus)*; HOUSE HOLLY *(Cyrtomium)*; BERRY BLADDER, BRITTLE or FRAGILE *(Cystopteris)*; BALL *(Davallia)*; HAY-SCENTED or BOULDER *(Dennstaedtia)*; DICKSONIA; BEECH [BROAD, NARROW], BOOT'S, CLINTON, CRESTED, GOLDIE'S, LEATHER WOOD or MARGINAL SHIELD, MALE, MARSH, NEW YORK, OAK, SHIELD, TOOTHED WOOD, WOOD *(Dryopteris)*; BASKET, BOSTON, BURROW, PERSON, SCOTT, SWORD, TUBER, WHITMAN *(Nephrolepis)*; SENSITIVE *(Onoclea)*; CLAW *(Onychium)*; ADDERSTONGUE *(Ophioglossum)*; CINNAMON, INTERRUPTED, ROYAL *(Osmunda)*; CLIFFBRAKE *(Pellaea)*; GOLD [CALIFORNIA, JAMAICA] *(Pityrogramma)*; STAGHORN or ELK *(Platycerium)*; POLYPODY [COMMON or WALL, ROCK], BEAR'S-PAW, HARESFOOT, RESURRECTION *(Polypodium)*;

ANDERSON SHIELD, DAGGER or CHRISTMAS, HOLLY [BRAUN'S, GIANT, LEATHER, MOUNTAIN] *(Polystichum)*; OSTRICH [AMERICAN, EUROPEAN] *(Pteretis)*; BRACKEN or BRAKE *(Pteridium)*; BRAKE or SPIDER, TABLE *(Pteris)*; WOODSIA [BLUNTLOBED, RUSTY, ROCKY MOUNTAIN], ROCK *(Woodsia)*; CHAIN *(Woodwardia)*

1. *Anthracnose, Tip Blight*—Common on Boston fern. Growing tips of leaves (fronds) turn brown, shrivel, and die. Plants appear weak and blighted. *Control:* Remove and burn blighted fronds. Destroy badly infected plants. If possible, avoid sprinkling foliage. Indoors, regulate temperature, humidity, and ventilation. Avoid overwatering.
2. *Leaf Spots, Leaf Blight, Tar Spot, Botrytis Blight*—Spots of various colors and sizes, especially at or near margins. Spots may be zoned or merge to form large blotches. Leaves may roll and wither. *Control:* Same as for Anthracnose (above).
3. *Leaf Nematodes*—Symptoms variable. Water-soaked then dark reddish-brown to black bands, limited by leaf veins. Often extend from center to leaf margin. Or irregular blotches may occur (Figure 4.98). On *birdsnest* fern, base of frond turns dull brownish-black. Discoloration later moves upward. Plants may die. *Control:* Remove and burn infested leaves or severely diseased plants. Indoors, keep water off foliage. Disinfest *birdsnest* and similar ferns by immersing plants in hot water (110° F.) for 10 to 15 minutes.

Fig. 4.98. Leaf nematode of fern.

4. *Rusts*—Primarily an outdoor problem. Reddish-brown to black, powdery pustules, arranged *irregularly* on fronds. Most common near alternate hosts (alpine, balsam, grand, noble, Pacific silver, and lowland white firs). *Control:* None usually needed. Same as for Anthracnose (above).
5. *Leaf Blisters, Leaf Gall*—Well-marked yellow areas on both leaf surfaces. *Control:* None usually needed. Same as for Anthracnose (above).
6. *Leaf Scorch*—Tips and margins are scorched. Fronds tend to die back. *Control:* Avoid full sun and wind. Grow in shady spot in natural, moist "woods soil."
7. *Bacterial Leaf Spot or Blight* (birdsnest fern, asplenium)—Small, water-soaked spots on fronds. Enlarge rapidly in warm, humid weather. Entire fronds may be blighted. Plants quickly killed if crown becomes infected. *Control:* Same as for Anthracnose (above) and Damping-off (below).
8. *Sooty Mold, Black Mildew*—Black mold grows on leaves following scales, mealybugs, or other insects. *Control:* Destroy insects using malathion dips or sprays. Do *not* use Diazinon on ferns!
9. *Damping-off*—Seedlings (prothallia) become soft, dark, and collapse. *Control:* Sow fern spores from healthy plants in a sterile rooting medium (e.g., soil or sifted sphagnum moss). When transplanting prothallia, handle with forceps frequently dipped in alcohol and flamed. Transplant into sterilized soil.
10. *Inflorescence Smut* (osmunda)—See (11) Smut under General Diseases.

FERN, ASPARAGUS or LACE — See Asparagus
FEROCACTUS — See Cacti
FESCUE, FESCUEGRASS *(Festuca)* — See Lawngrass
FETTERBUSH — See Blueberry
FEVERFEW See Chrysanthemum
FIESTA-FLOWER — See Phacelia
FIG [BENJAMIN or WEEPING LAUREL, COMMON, CREEPING, FIDDLELEAF, FLORIDA STRANGLER, MISTLETOE, MORETON BAY], INDIAN or CUBAN LAUREL, INDIA RUBBER TREE or RUBBER-PLANT *(Ficus)*; PAPER-MULBERRY *(Broussonetia)*

1. *Anthracnose, Leaf Spots, Fruit Rot*—General. Tips and margins of leaves are yellowish, then tan, finally dark brown and "scorched." Minute pinkish "pimples," sometimes in zones, may be sprinkled in diseased areas. Dead tissue may later dry and fall out. Enlarging, sunken, discolored spots on fig fruit. Mature fruit may drop or immature fruit may "dry up." *Control:* Pick off and burn infected leaves. In commercial fig-growing area, check with your county agent or extension plant pathologist regarding a spray program. Indoors, avoid sprinkling foliage. If practical, apply zineb, maneb, or captan before wet periods. Collect and burn fallen leaves and fruit.
2. *Leaf Scorch, Leaf Fall*—Primarily an indoor problem. Tips and margins of leaves are scorched, or large blotches occur in leaves. Affected areas may curl and crack. See Figure 2.9. Leaves drop early. *Control:* Keep soil at uniform moisture level. Avoid high room temperatures and exposure to full sun. Do not keep air too dry. Repot plants using light, fast-draining potting mixture (page 98).
3. *Root-knot, Fig Cyst Nematode*—A limiting factor in fig production. Knotlike swellings and galls form on roots. Affected trees may be sickly, lack vigor. Stunted, with leaves and fruit falling prematurely. See under Peach, and (37) Root-knot under General Diseases. Fumigate before planting with D-D, Telone, EDB, SMDC (Vapam or VPM), DBCP, or VC-13. Or apply post-planting treatment of DBCP. Follow manufacturer's directions. *Fig* species are available that are resistant rootstocks.
4. *Root Rots*—Trees decline, often stunted and die back. See under Apple, and (34) Root Rot under General Diseases. May be associated with nematodes (e.g., burrowing, dagger, fig cyst, lesion, needle, pin, ring, root-knot, spiral, stubby-root, stunt). *Control:* Destroy badly infected indoor plants and replant in sterilized soil (pages 538–548). Grow outdoor plants in fertile, well-drained soil. Do not replant in same place. If nematodes are a problem, DBCP may be applied to the root zone of living plants. Follow manufacturer's directions. *Fig* is resistant to Armillaria Root Rot.
5. *Dieback, Cankers, Twig Blights, Limb Blight*—Rough, more or less oval, dead areas (cankers) develop around pruning cuts, branch stubs, or on limbs injured by frost or sunburn. *Limb Blight* (southeastern states) develops as bright salmon-pink, smooth, velvety incrustation on *fig* limbs. Later this becomes rough and broken into patches. Fades to a dirty-pink. Leaves suddenly wilt, yellow, and die. *Control:* Prune plants carefully late in the season and remove all dead branches, cankers, and dried-up fruit. Dip pruning shears in 70 per cent alcohol, 1:1,000 mercuric chloride, or phenyl mercury solution after *each* cut through a canker. Paint all cuts with tree wound dressing (page 36). Spray as for Rust (below). Protect trees against sunscald and winter injury. See under Apple and Elm.
6. *Fruit Rots, Souring* (fig)—Rot spots of various colors develop in fruit. Black, gray, or pink mold may grow on affected areas. *Control:* Pick fruit as soon as ripe. Promptly collect and destroy decaying or dropped fruit. *Fig* varieties differ in

resistance. Spraying as for Rust (below) or Anthracnose (above) should be beneficial. Control insects that carry fungi when entering young fruit.

7. *Rust* (fig) —Numerous small, reddish to brownish spots on underleaf surface. If severe, leaves may be distorted, turn yellowish-brown, and drop early. *Control:* Apply ferbam, maneb, fixed copper, or bordeaux mixture (4-4-50) at 2- to 3-week intervals. Start 1 to 2 weeks *before* rust normally appears, or when first leaves reach full size. Collect and burn fallen leaves, where practical.
8. *Mosaic* (fig) —Widespread. Symptoms variable. Leaves may be stunted, severely distorted, show irregular, yellowish-green blotches, spots, bands, or yellowish mottling. Fruit may be deformed and spotted. Both fruit and leaves may drop early. Transmitted by grafting and mites. *Control:* Plant virus-free stock. Destroy infected plants when found. Varieties differ in resistance.
9. *Sunscald, Winter Injury*—See under Apple and Elm. Disease organisms get into frost-weakened and sunburned trees. *Control:* Whitewash or wrap trees (page 42). Plant in location protected from winter winds or grow in containers and move indoors for winter.
10. *Oedema* (rubber-plant) —Indoor problem. Corky callus-growths on petioles and underleaf surface. *Control:* Increase light and temperature and decrease soil moisture, especially in humid weather.
11. *Web Blight, Thread Blight*—Southeastern states in warm, moist weather. White to brownish, threadlike, fungus hyphae grow over underleaf surface killing the leaves. Many remain hanging on tree matted together by spiderweb-like fungus threads. See also under Bean. *Control:* Apply 1 or 2 sprays of fixed copper or bordeaux. Pruning out of infected branches may be warranted.
12. *Crown Gall*—See under Begonia, Apple, and (30) Crown Gall under General Diseases.
13. *Other Leaf Spots, Leaf Blotch, Rusty Leaf*—Spots and streaks of various colors, sizes, and shapes. Spots may enlarge until entire leaf is blighted. *Control:* Same as for Anthracnose and Rust (both above).
14. *Sooty Mold*—See under Apple, and (12) Sooty Mold under General Diseases.
15. *Wood Rots* (fig) —See under Birch, and (23) Wood Rot under General Diseases.
16. *Fusarium Wilt* (fig) —See (15A) Fusarium Wilt under General Diseases.
17. *Mistletoe* (paper-mulberry) —See (39) Mistletoe under General Diseases.
18. *Southern Blight*—See (21) Crown Rot under General Diseases.
19. *Chlorosis, Zinc and Manganese Deficiency*—See under Walnut.
20. *Felt Fungus*—Gulf Coast states. See under Hackberry.

For additional information, read USDA Handbook 196, *Fig Growing in the South;* and H & G Bulletin 87, *Growing Figs in the South for Home Use.*

FIDDLENECK — See Phacelia

FIGMARIGOLD — See Iceplant

FILAREE — See Cranesbill

FILBERT — See Birch

FILIPENDULA — See Rose

FINOCCHIO — See Celery

FIR — See Pine

FIRE-CHALICE — See Evening-primrose

FIRECRACKER PLANT — See Cigarflower

FIRE-FERN — See Oxalis

FIRE-PINK — See Carnation

FIRETHORN — See Apple

FIREWHEEL — See Chrysanthemum

FIRMIANA — See Phoenix-tree

FITTONIA — See Silver Threads

FIVE-LEAF or FIVE-FINGERED ARALIA — See Acanthopanax

FIVE-SPOT — See Phacelia

FLAME-VINE — See Bignonia

FLAME VIOLETS — See African-violet

FLAMINGO-FLOWER — See Calla Lily

FLANNEL-BUSH — See Phoenix-tree

FLOWER-OF-JOVE — See Carnation

FLAX, FLOWERING [ANNUAL, BLUE, GOLDEN] *(Linum)*

1. *Stem and Root Rot, Damping-off*—Stand may be poor. Seedlings wilt and collapse. Stems rot off near soil line. May be covered with cottony mold. *Control:* Treat seed with captan, thiram, or chloranil. Avoid overwatering, overcrowding, and planting in poorly drained soil. Dig up and burn older, dying plants together with several inches of surrounding soil.
2. *Root-knot*—See under Bean, and (37) Root-knot under General Diseases.
3. *Curly-top, Yellows*—See (18) Yellows, and (19) Curly-top under General Diseases. Plants stunted and sickly. May be yellowed.

FLEABANE, FLORAS-PAINTBRUSH — See Chrysanthemum

FLORIDA YELLOWTRUMPET — See Trumpettree

FLOWERING ALMOND — See Peach

FLOWERING CRABAPPLE — See Apple

FLOWERING CURRANT — See Currant

FLOWERING MAPLE — See Hollyhock

FLOWERING QUINCE — See Apple

FLOWERING TOBACCO — See Tomato

FLOWER-OF-AN-HOUR — See Hollyhock

FOAMFLOWER — See Hydrangea

FOENICULUM — See Celery

FOGFRUIT — See Lantana

FORESTIERA — See Ash

FORGET-ME-NOT — See Mertensia

FORSYTHIA [BORDER, EARLY, GOLDENBELLS, KOREAN, SHOWY BORDER, SPRING GLORY, WEEPING (FORTUNE, SIEBOLD)] *(Forsythia)*

1. *Leaf Spots, Anthracnose*—Small to large, grayish, yellow, or brown spots. *Control:* Pick and burn infected leaves. If necessary, apply zineb or maneb several times, 10 days apart.
2. *Twig or Cane Blight, Dieback, Southern Blight, Blossom Blight*—Blossoms turn brown. Twigs wither and die back from girdling cankers or crown rot. Cottony mold may grow over plant near soil line. *Control:* Prune and burn infected twigs. Keep soil surface at crown loose and dry. PCNB dust or spray to crown may help. See under Bean, White Mold.
3. *Bacterial Blight*—Angular, brown to black spots in leaves. Shoots and entire branches may later blacken and die. Brown stain in wood. Often throughout an entire branch. *Control:* See under Lilac.
4. *Phomopsis Stem Gall*—Round to irregular, bunchy overgrowths along stems. Stems unsightly in winter. May die back. *Control:* Remove and burn infected branches.
5. *Crown Gall*—See under Apple, and (30) Crown Gall under General Diseases.

6. *Root-knot*—Forsythia is quite susceptible. See (37) Root-knot under General Diseases.
7. *Root Rot*—See (34) Root Rot under General Diseases. May be associated with root-feeding nematodes (e.g., dagger, lesion, ring, root-knot, stem, stunt).
8. *Chlorosis*—Mineral deficiency in alkaline soils. See under Maple.

FORTUNELLA — See Citrus

FOUR-O'CLOCK [COMMON, COLORADO] *(Mirabilis)*; TRAILING FOUR-O'CLOCK *(Allionia)*; SAND-VERBENA *(Abronia)*; UMBRELLAWORT *(Oxybaphus)*

1. *Rusts* (four-o'clock, sand-verbena, trailing four-o'clock)—Southwestern states. Small, yellow or yellowish-orange spots on leaves. Alternate hosts may include wild grasses *(Aristida* and *Distichlis)*. *Control:* If serious, apply zineb or maneb about 10 days before rust normally appears. Repeat at 10-day intervals.
2. *White-rust*—Pale yellow spots on upper leaf surface with white pustules on corresponding underside. *Control:* Collect and burn tops in fall or spotted leaves as they appear. Spray as for Rusts (above).
3. *Leaf Spots*—Small and indistinct or round, pale brown to tan spots, with dark borders. *Control:* Pick and burn spotted leaves. Spray as for Rusts (above).
4. *Root-knot*—See under Bean, and (37) Root-knot under General Diseases.
5. *Curly-top* (four-o'clock)—Western states. See under Beet, and (19) Curly-top under General Diseases.
6. *Root Rot*—See (34) Root Rot under General Diseases.
7. *Downy Mildew* (four-o'clock, sand-verbena, trailing four-o'clock, umbrellawort)—See (6) Downy Mildew under General Diseases.

FOXGLOVE — See Snapdragon
FRAGARIA — See Strawberry
FRAGRANT GLAD — See Gladiolus
FRAGRANT PINK — See Carnation
FRAGRANT STOCK — See Cabbage
FRAGRANT VIBURNUM — See Viburnum
FRANGIPANI — See Oleander
FRANKLIN-TREE, LOBLOLLY-BAY *(Franklinia, Gordonia)*

1. *Leaf Spot*—Small spots on leaves. *Control:* If serious, spray with zineb, maneb, or fixed copper in wet seasons.
2. *Black Mildew*—Southern states. Black, moldy spots or blotches on leaves. *Control:* Same as for Leaf Spot. Control insects with malathion sprays.
3. *Root Rot*—See (34) Root Rot under General Diseases. Grow in full sun in moist, well-drained, acid soil high in organic matter.

FRASERA — See Gentian
FRAXINUS — See Ash
FREESIA — See Gladiolus
FREMONTIA — See Phoenix-tree
FRENCH ENDIVE — See Lettuce
FRENCH-MULBERRY — See Lantana
FRINGEFLOWER — See Tomato
FRINGETREE — See Ash
FRITILLARIA, FRITILLARY — See Tulip
FROELICHIA — See Cockscomb

FROSTWEED, FROSTWORT — See Sun-rose

FUCHSIA; ROCKY MOUNTAIN GARLAND *(Clarkia)*; BOISDUVALIA, SPIKE-PRIMROSE *(Boisduvalia)*; GAURA; FAREWELL-TO-SPRING, SATIN-FLOWER *(Godetia)*

1. *Rusts*—Pale spots on upper leaf surface and yellowish-orange or brown to black, powdery pustules on lower leaf surface. Lower leaves may shrivel and die. *Control:* Pick and burn infected leaves. If serious, apply zineb, ferbam, maneb, or dichlone at weekly intervals. Destroy fireweed *(Epilobium)*, a host for *Fuchsia* rust.
2. *Gray-mold Blight, Stem Canker*—Leaves, flowers, and seed pods blighted in damp weather. Covered with dense gray mold. See (5) Botrytis Blight under General Diseases. *Control:* Pick and burn affected parts. Avoid overcrowding. Increase air circulation and reduce humidity.
3. *Root Rots*—Plants may wilt suddenly and die within a few days. See (34) Root Rot under General Diseases. Soil drenches of Dexon, applied at 2- or 4-week intervals, have proven beneficial.
4. *Verticillium Wilt*—See (15B) Verticillium Wilt under General Diseases.
5. *Root-knot*—See under African-violet, and (37) Root-knot under General Diseases.
6. *Stem (Foot) Rots*—Brown or black area on stem at or near soil line. Plants may be stunted, yellowed, wilt and die. *Control:* Sow disease-free seed or transplants in clean or sterilized soil (pages 538–548). Rotate. Collect and burn plant debris in fall. Avoid overwatering. Dexon soil drenches may be beneficial. See Root Rots (above).
7. *Damping-off*—Seedlings collapse from rot at soil line. *Control:* Same as for Stem Rots (above).
8. *Spotted Wilt*—See under Begonia, and (17) Spotted Wilt under General Diseases.
9. *Aster Yellows, Curly-top* (clarkia, gaura, godetia)—See under Chrysanthemum.
10. *Downy Mildew* (clarkia, gaura, godetia)—See under Chrysanthemum.
11. *Leaf Spots, Anthracnose* (clarkia, fuchsia, gaura)—See under Chrysanthemum.
12. *Powdery Mildew* (fuchsia, gaura)—See (7) Powdery Mildew under General Diseases.
13. *Fusarium Wilt* (clarkia)—See (15A) Fusarium Wilt under General Diseases.

FURCRAEA — See Centuryplant

GAILLARDIA — See Chrysanthemum

GALANTHUS — See Daffodil

GALAX; OCONEE-BELLS *(Shortia)*

1. *Leaf Spots*—Small to large, round to irregularly lobed spots and blotches. Leaves may wither and die early. *Control:* Pick and burn spotted leaves. Do not syringe plants. If needed, spray several times at weekly intervals. Use zineb, maneb, or dichlone.

GALIUM — See Buttonbush

GALTONIA — See Tulip

GARDEN BALSAM — See Balsam

GARDEN CRESS — See Cabbage

GARDEN-HELIOTROPE *(Valeriana)* — See Valerian

GARDEN HUCKLEBERRY — See Blueberry

GARDEN VERBENA — See Lantana

GARDENIA or CAPE-JASMINE, DWARF CAPE-JASMINE *(Gardenia)*

1. *Stem Canker or Gall*—Widespread. Leaves dwarfed, wilt, shrivel, turn pale green then yellow and fall early. Flower buds blasted and drop before opening. Oblong,

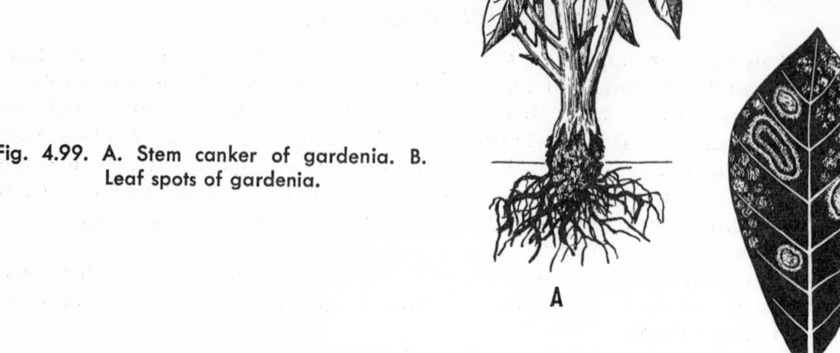

Fig. 4.99. A. Stem canker of gardenia. B. Leaf spots of gardenia.

sunken to swollen, brown cankers form on branches and at crown. May girdle affected parts causing slow stunting and death. Crown cankers appear as corky overgrowths (swollen, cracked ridges). See Figure 4.99A. *Control:* Avoid overwatering, poorly drained soil, and injuring plants during handling and potting. Remove cankers on branches by cutting stems 3 inches back of cankers. Swab cuts immediately with 70 per cent alcohol. Remove and burn severely infected plants. Keep water off foliage. Buy disease-free plants or take tip cuttings from healthy plants. Use a sharp knife. Commercial growers dip cuttings for propagation in potassium permanganate solution (1 ounce in 5 gallons of water) for 5 to 10 minutes before planting. Or dip cuttings in ferbam (2 tablespoons per gallon), Semesan, or phenyl mercury solution (1 ounce in 5 gallons) for 5 minutes before sticking. Use pasteurized rooting medium (pages 538–548). Spray stems and crowns weekly using zineb or ferbam. Place new plants in another location. Veitchii is more resistant to Phomopsis *(Diaporthe)* Canker than Belmont or Hadley.

2. *Chlorosis*—Widespread. May be caused by insufficient light, low temperatures, alkaline soil, too dry atmosphere, overwatering or poor drainage, iron deficiency, root rot combined with nematodes, and stem rot or canker. Leaves are stunted and pale green or yellowish between the veins. Young leaves may turn yellow, die, and fall early. Plants make poor growth. Tips may die. See Figure 4.37. *Control:* Grow in light, well-drained, slightly acid (pH 6) soil. Have soil tested if in doubt. Keep plants free of Stem Canker and Root-knot. Have soil above 60° to 62° F. Spray plants monthly with iron sulfate, 1 tablespoon per gallon of water. Or apply iron chelate (or iron sulfate) to soil in water solution. Follow manufacturer's directions.
3. *Fungus Leaf Spots*—Round to oval spots, sometimes zoned. Mostly on lower leaves following wet weather. Leaves may die. See Figure 4.99B. *Control:* Pick and burn infected leaves. Space plants. Avoid sprinkling water on foliage and wounding leaves. Spray foliage as for Stem Canker (above). Use disease-free plants for propagation. Varieties differ in resistance.
4. *Bacterial Leaf Spots*—Small to large, round to angular, brown or reddish-brown spots on leaves, surrounded by narrow, water-soaked to greasy or yellowish border. Leaves may yellow and drop early, starting at base of plant. *Control:* Same as for Fungus Leaf Spots (above). Plant only disease-free cuttings. Sterilize soil and containers before planting (pages 538–548).
5. *Tipburn*—Margins and tips of leaves become discolored and die. *Control:* Keep soil moisture level as uniform as possible. Otherwise, same as for Bud Rot (below) and

Chlorosis (above). Keep potash level up by adding potassium sulfate or potassium chloride.

6. *Bud Rot, Bud Drop*—Buds may turn pale green or yellow. Often soften, darken, and drop. Flower stalks may be discolored. *Control:* Same as for Fungus Leaf Spots (above). Indoors, avoid chilling, large temperature and soil moisture fluctuations, overwatering or fertilizing, poor drainage, and dry or extremely humid air. Add daytime lights during overcast periods. Night temperature should be about 62° to 65° F., and above 70° during the day. Pick off and burn affected buds. Increase humidity in home. Apply malathion *carefully* to control mealybugs, thrips, ants, and whiteflies. Varieties differ in resistance. Control Root Rot and Nematodes (both below).
7. *Root Rots*—Plants often stunted, yellowed, produce few flower buds. Plants gradually decline in vigor; may die. *Control:* See (34) Root Rot under General Diseases. Often associated with nematodes (e.g., burrowing, dagger, lance, lesion, needle, pin, reniform, root-knot, spiral, stubby-root, stunt). Grow in an acid, well-drained soil high in organic matter.
8. *Root-knot*—Widespread in southern states and northern greenhouses. Gardenia is highly susceptible. Leaves may wilt during the day; fall prematurely. Plants may be stunted with pale, mottled leaves. Swellings or galls on the roots. *Control:* Dig up and burn infested plants. Grow disease-free stock (or tip cuttings grown in peat or vermiculite) in sterilized soil. *Gardenia thunbergia* is a resistant rootstock. Dip bare-root stock in Zinophos (600 to 800 parts per million) for 30 minutes before planting.
9. *Gray-mold Blight, Botrytis Petal Blight*—Numerous, light brown spots on petals that enlarge and merge to form blotches. Gray mold may grow on infected tissues in damp weather. Buds may drop early. *Control:* Same as for Fungus Leaf Spots (above). Carefully pick and burn blighted flowers and buds. If practical, spray blooms at 2- or 3-day intervals in damp weather. Use fine mist of captan, Botran, or zineb (1 tablespoon per gallon).
10. *Sooty Mold*—Very common in Gulf states following attacks by scales, whiteflies, mealybugs, aphids, and other insects. Crusty black coating on leaves and stems. Leaves may be shaded or "smothered." *Control:* Wash off sticky coating. Apply malathion to control insects.
11. *Powdery Mildew*—See (7) Powdery Mildew under General Diseases.
12. *Dieback*—Indoor problem. Branches die back from black, girdling cankers. Leaves wither and fall off. Enlarging, black, dead areas form on twigs. *Control:* Avoid overwatering. Plant in acid, well-drained soil.
13. *Crown Gall*—See (30) Crown Gall under General Diseases.

For additional information, read USDA Leaflet 199, *Gardenia Culture.*

GARLAND-FLOWER — See Daphne

GARLIC — See Onion

GARRYA — See Dogwood

GAULTHERIA — See Heath

GAURA — See Fuchsia

GAYFEATHER — See Chrysanthemum

GAYLUSSACIA — See Blueberry

GAZANIA — See Chrysanthemum

GEIGER-TREE — See Cordia

GELSEMIUM — See Butterflybush

GENISTA — See Broom

GENTIAN [CLOSED, FRINGED, NARROW-LEAVED, YELLOW] *(Gentiana);*

PRAIRIEGENTIAN or TEXAS-BLUEBELL *(Eustoma)*; EXACUM; COLUMBO, DEER'S-TONGUE *(Frasera)*

1. *Leaf Spots, Leaf Blotch*—Spots of various colors, shapes, and sizes. If severe, leaves may wither. *Control:* Pick and burn spotted leaves. Spray at 10- to 14-day intervals during wet periods. Use zineb, maneb, or fixed copper.
2. *Botrytis Blight, Stem Canker* (gentian, exacum) —Light brown spots or blotches on leaves with darker margins. Cankers may form on stems. Gray mold often covers infected areas in damp weather. *Control:* Remove and destroy infected parts. Space plants. Spray as for Leaf Spots (above).
3. *Rusts* (columbo, gentian) —Yellow spots on lower leaves; may later turn into reddish-brown, dark brown, or black powdery pustules. Disease moves upward as season progresses. *Control:* Destroy infected plants. Spray remainder as for Leaf Spots (above).
4. *Root and Crown Rot, Damping-off*—Seedlings wilt and collapse. Older plants rot at base. Roots often decay. *Control:* Grow in clean, well-drained soil. Rotate. Avoid overwatering and wounding stems and roots.
5. *Stem Blights* (prairiegentian) —See (22) Stem Blight under General Diseases. *Control:* Spray as for Leaf Spots (above). Add malathion to control insects.
6. *Black Mildew* (columbo) —See (12) Sooty Mold under General Diseases.

GERANIUM [CACTUS, FISH, FLORISTS', IVY, LADY WASHINGTON or SHOW, NUTMEG, OAK-LEAVED, PEPPERMINT, ROSE, STORKSBILL] *(Pelargonium)* (See also Cranesbill)

1. *Pythium and Fusarium Blackleg, Stem and Cutting Rots*—Cosmopolitan. Leaves turn yellow or reddish and drop. Plants stunted. Die gradually. Base of cutting or stem is soft, brown, and water-soaked. Commonly turns coal-black and shiny then shrivels. May turn slimy. Rot works upward from soil line until plant wilts and dies. Plants easily pulled up. See Figure 2.39D under General Diseases and Figure 4.100. *Control:* Take *only* tip cuttings from disease-free plants sprayed 30, 20, and 10 days before taking cuttings. Use zineb or captan (1½ tablespoons per gallon of water). Dip cuttings for 10 minutes in a ferbam, zineb, or captan solution and plant in sterile medium (pages 538–548). Avoid overcrowding, overwatering, or sprinkling water on foliage. Sterilize cutting knife and other tools by dipping for 1 minute in a 1:1,000 solution of mercuric chloride or 70 per cent alcohol between cuts. Keep humidity down and increase air circulation. Separate healthy from diseased plants. If rot starts in cutting bed, remove infected plants or plant parts and apply zineb or Dexon as a soil drench. Follow manufacturer's directions. Destroy infected plants when first discovered. Cultured, disease-free cuttings of *geraniums* are available.
2. *Root Rots*—Widespread. Leaves, especially lower ones, turn yellow, wilt, die, and then fall off. New shoots pale green and stunted. Flowering is reduced, or blossoms fall soon after opening. Roots brown or black and rotted. May be associated with nematodes (e.g., lesion, pin, root-knot, spiral, stem, stubby-root, stunt). *Control:* Same as for Blackleg (above).
3. *Virus Complex* [*Mottle, Mosaics, Yellow Net Vein, Leaf Curl (Crinkle), Chlorotic Spot, Leaf Breaking, Spotted Wilt, Ringspots, Curly-top (Leaf Cupping)*]—General. Variable symptoms that often disappear in hot weather. Leaves may be stunted, ruffled, crinkled, cupped upward and inward (Leaf Cupping), and puckered. Small, round to irregular, pale yellow to white, red, brown or purple spots, rings, or arclike patterns may appear on leaves. See Figure 4.101. Leaf veins may turn yellow in a net pattern (Yellow Net Vein). Or leaves may be mottled with light and dark green areas (Mosaic). Normal purplish zones or horseshoe leaf

Fig. 4.100. Cutting rot of geranium. (Courtesy Dr. Donald E. Munnecke)

patterns tend to disappear (Leaf Breaking). Leaves may yellow and drop early. Symptoms most apparent on young leaves in cool weather. Plants usually stunted. May appear bushy. Flowering is reduced. *Control:* Take tip cuttings *only* from known virus-free plants. Discard plants that look suspicious. Handle plants as infrequently as possible. Varieties differ greatly in apparent resistance. Control insects and mites using malathion. Cultured, disease-free *geranium* cuttings are available. Grow in sterilized soil, pots, and benches.

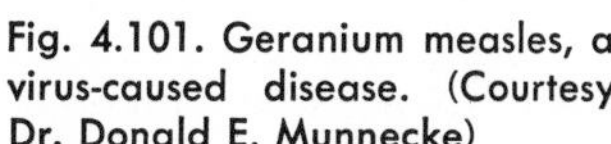

Fig. 4.101. Geranium measles, a virus-caused disease. (Courtesy Dr. Donald E. Munnecke)

4. *Fungus Leaf Spots, Blossom Blight, Gray-mold Blight*—Cosmopolitan in wet weather. Small to large, round to irregular spots that are water-soaked, light brown to tan, brick-red, reddish-brown or dark brown. Spots may enlarge and merge to kill the leaf. Flower petals may discolor, fade, and wilt. Flowers fall prematurely. Affected parts may be covered with olive-green, gray, dark brown, or black mold. See Figures 2.21A and 2.47C under General Diseases. *Control:* Same as for Blackleg (above). Pick and burn infected or dead leaves and blossoms when first seen. Apply captan, zineb, maneb, Daconil 2787, Botran (dichloran), or fixed copper during damp periods.
5. *Bacterial Stem Rot (Wilt) and Leaf Spots*—General in warm, humid weather. Often limiting factor in production. Small, dark green, water-soaked blisters on leaves of most varieties. Primarily on underleaf surface. Spots often enlarge and merge. Become angular with centers sunken and dark brown to black (Figure 4.102). May resemble "frogeyes." Leaves wilt, turn yellow, wither, and fall. Often starts on one side of plant. Infected stems turn dull blackish-brown and shriveled with semidry rot. Larger infected plants are upright and defoliated except for shoot tips. Cuttings wilt, fail to root, and rot progressively upwards from base. Roots are blackened. *Control:* Same as for Blackleg (above). Destroy infected leaves. *Geranium* varieties and *Pelargonium* species differ in resistance. Avoid forcing plants too rapidly, especially during warm, humid weather. Space plants. Maintain balanced fertility. Whiteflies may spread the causal bacteria from plant to plant. Use malathion to keep insects in check. Cultured, disease-free *geranium* cuttings are available.
6. *Verticillium Wilt*—Leaves may develop bright yellow, wedge-shaped areas or streaks along the principal veins. Leaves turn yellow at margins, wilt, wither, and drop early. Shoots die back. Wilt later progresses up stem. Plants may be greatly stunted, grow slowly. May closely resemble Bacterial Stem Rot (above). Stems blackened and shriveled from tip downward. *Control:* Destroy infected plants. Take only tip cuttings from known healthy plants. Then treat as for Blackleg (above). Keep plants growing vigorously.
7. *Root-knot*—Plants may be unthrifty and weak from small galls on roots. *Control:* Destroy infested plants and grow new ones in sterilized soil (pages 538–548). Start with tip cuttings.
8. *Crown Gall*—Common but not very damaging. Cauliflower-like galls or knots form on roots and crown. Growth is checked. *Control:* Destroy infected plants. Take only tip cuttings from healthy plants and grow in sterilized soil. Avoid wounding stems.
9. *Leafy Gall, Fasciation*—Plants often stunted. Clusters of greenish-white, short, fleshy, thick shoots and buds form near soil line. May closely resemble Crown Gall. See Figure 2.44B under General Diseases. *Control:* Same as for Crown Gall and Blackleg (both above). Treat cuttings regularly before planting. Use streptomycin plus oxyquinoline sulfate following manufacturer's directions. Miniature geraniums often serve as carriers.
10. *Oedema, Dropsy*—Common indoor problem in cloudy, cool weather. Small, water-soaked leaf "blisters" that later become reddish-brown, corky, and raised (Figure 4.103). Resemble an insect gall or blister. Mostly on larger leaf veins but also occur on stems and petioles as corky ridges. Severely affected leaves may turn yellow and drop early. *Control:* Avoid overwatering in cool, overcast, humid weather. Keep water off foliage. Lower humidity and increase heat, ventilation, and light. Space plants. Avoid low potassium and calcium levels in the soil. The Irene varieties of florist's geranium are very susceptible.
11. *Leaf Nematode*—See (20) Leaf Nematode under General Diseases.
12. *Potassium Deficiency*—Leaf margins yellowed; later may be scorched. *Control:*

Fig. 4.102. Bacterial leaf spot of geranium. (Herbarium, Department of Plant Pathology, Cornell University photo)

Fig. 4.103. Oedema of geranium. (Herbarium, Department of Plant Pathology, Cornell University photo)

Have soil test made and follow suggestions for adding fertilizer.

GERBERA, GERMAN CAMOMILE — See Chrysanthemum
GERMANDER — See Salvia
GERMAN IVY — See Chrysanthemum
GESNERIA — See African-violet
GEUM — See Rose
GHERKIN, WEST INDIA — See Cucumber
GIANT DAISY — See Chrysanthemum
GIANT WHITE NIGHT BLOOMER — See Morning-glory
GILIA — See Phlox
GINKGO, MAIDENHAIR-TREE *(Ginkgo)*

1. *Leaf Spots, Anthracnose*—Uncommon. Yellow to brownish spots. Leaves may turn yellow. *Control:* See Maple, Leaf Spots.
2. *Root-knot*—See under Peach, and (37) Root-knot under General Diseases.
3. *Wood Rots*—See under Birch, and (23) Wood Rot under General Diseases.
4. *Root Rot*—See (34) Root Rot under General Diseases. May be associated with root-feeding nematodes (e.g., lesion, pin, stunt).
5. *Leaf Scorch*—Margins of leaves turn brown. See Maple, Leaf Scorch.

GLACIER-LILY — See Erythronium
GLADIOLUS [COMMON, MINIATURE, NIGHT-BLOOMING], CORNFLAG *(Gladiolus)*; FRAGRANT GLAD, ABYSSINIAN WILDFLOWER *(Acidanthera)*; COPPER-TIP, FALLING STARS *(Crocosmia)*; CROCUS [CLOTH-OF-GOLD, COMMON, DUTCH, SAFFRON, SCOTCH] *(Crocus)*; FREESIA;

TIGERFLOWER, AZTEC LILY, SHELL-FLOWER *(Tigridia)*; MONTBRETIA, BLAZING-STAR *(Tritonia)*; WATSONIA

1. *Corm and Bulb Rots, Crown Rot, Southern Blight, Flower Blights, Wilt, Fusarium Yellows or Brown Rot*—General and serious. Leaves pale, turn yellow to brown, wither, die back, and may collapse from rotting of crown or other underground parts. Roots often decay. Corm (or bulb) often shows round to irregular, tan, yellowish-brown, reddish-brown, dark brown or black rotted areas that may be somewhat sunken. Whole corm may rot, becoming a dry, chalky-white to brownish-black "mummy." Rot often spreads into leaves that darken and rot at their bases. Corms continue to rot in storage. May show no symptoms at digging time. Bluish-green, gray, black, or cottony mold may develop on rotted areas during storage. Flowers may be spotted and blighted. See Figure 2.49C under General Diseases and Figure 4.104. *Control:* Plant only best quality, disease-free corms in well-drained, sterilized soil (pages 538–548) or where disease has not occurred before. Grow in sunny spot where air circulation is good. Avoid low, wet spots. Fertilize well with potassium and phosphorus but keep nitrogen on low side. Dig and destroy infected plants as they occur. Harvest corms (or bulbs) early, shake off loose soil, and cure rapidly at 80° to 95° F. with good ventilation for 1 to 2 weeks. Sort and discard those corms that are damaged or rotted. (Handle corms carefully at all times to prevent bruising.) Then remove tops and roots and dust corms thoroughly with thiram-dieldrin mixture (e.g., Delsan A-D or Panoram D-31). When treating, shake corms and dust together in a tight paper sack. For each quart of corms use an amount of dust at least equal in quantity to an aspirin tablet. Or soak *gladiolus* corms for 5 minutes in Dowicide B solution (¾ pound in 25 gallons). After treatment (dust or dip) store corms over winter in a well-ventilated, cool (35° to 40° F.) location with low humidity (about 75 per cent). In the spring, before planting, soak *gladiolus* corms in a 1:1,000 solution of mercuric chloride for 2 hours, or use Chipcote 25, Elcide 73, Ortho LM Seed Protectant, Ceresan L, or Panogen following manufacturer's directions. Plant immediately. Certain varieties may be injured by these mercury treatments. Use with caution. *Gladiolus* cormels may be freed of corm-rotting organisms and nematodes by soaking sound, hard, *fully dormant* cormels, kept in a warm dry room, in hot water (135° F.) for 30 minutes. Then cool rapidly, dry thoroughly, and place in cold storage after dusting with mixture of chloranil, thiram, or captan plus DDT or dieldrin. PCNB and dichloran (Botran) applied in the open furrow at planting time have given good control of Stromatinia *(Sclerotinia)* Rot. Follow manufacturer's directions. Keep down weeds. Burn all plant debris in the fall. In humid areas, spray as for Leaf Spots (below). *Gladiolus* varieties differ in resistance to Fusarium Yellows. Florida Pink *gladiolus* is resistant to Botrytis and Curvularia.
2. *Bacterial Scab, Neck Rot, Bacterial Leaf Spot*—General in warm wet weather. Pale yellow to brownish-black, varnish-like spots or streaks form on husks. Corm tissue underneath shows round, yellowish to tan, water-soaked spots that later become brownish-black, sunken, scabby, and gummy. Scabby corms are associated with bulb mites. Small, more or less round, reddish-brown, water-soaked leaf spots may enlarge and merge to form large blotches. Leaves may turn yellow at tips and die prematurely. Leaf bases (neck) may prematurely die (Figure 4.105). *Control.* Same as for Corm Rots (above). Rotate 3 years with nonrelated plants. Plant only scab-free corms. Apply soil insecticide, e.g., aldrin, dieldrin, or heptachlor (4 ounces of actual insecticide to 1,200 feet of row) in furrow before planting. Treatment controls wireworms, other soil insects, and bulb mites. Commercial growers rid *gladiolus cormels* of scab by soaking 48 hours in cool water, followed by 4 hours in a 1:200 solution of commercial formalin or 100 parts per million of Agrimycin 100.

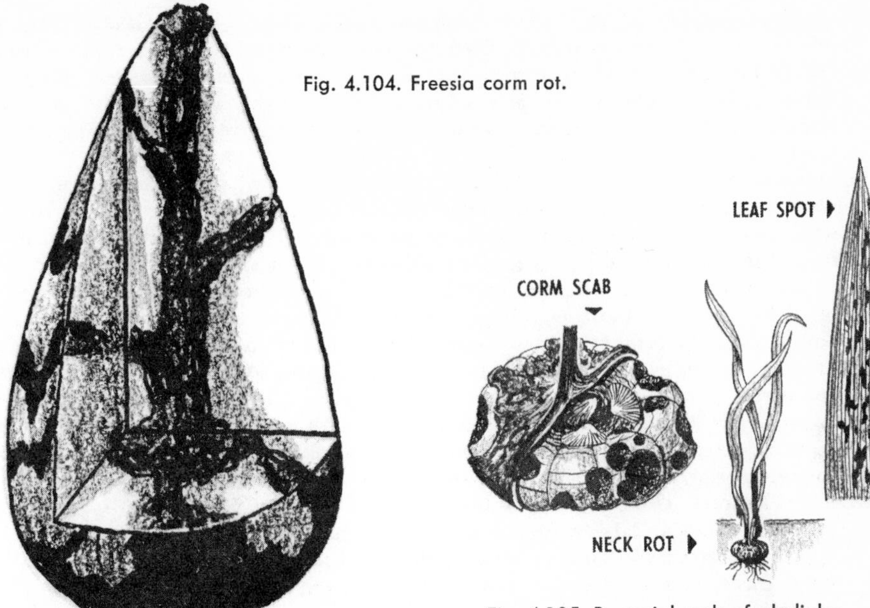

Fig. 4.104. Freesia corm rot.

Fig. 4.105. Bacterial scab of gladiolus.

This is then followed by a soak in hot water (135° F.) for 30 minutes. See Corm Rots (above).

3. *Bacterial Leaf Blight*—May be serious in wet seasons. Numerous, small, water-soaked, dark green leaf spots that later turn brown or purplish. Spots become gummy and square or rectangular. Leaves appear scorched. May die. Spots restricted to between leaf veins when young (Figure 4.105). Younger plants most severely attacked. *Control:* Soak corms in mercury-containing solution. See Corm Rots (above). Several weekly sprays of streptomycin (a formulation without copper), 200 parts per million, may be necessary in warm, wet weather. See page 105.
4. *Fungus Leaf Spots, Flower Spots or Blights*—General in rainy weather. Small, round to elongated or irregular spots of various colors—tan, yellow, red, pale brown, rusty-brown, purplish-brown or black—some with one or more dots or mold in center. Leaves may yellow, wither, and die back from tips. Plants may not bloom. Corm size and production is often decreased. Spots also may occur on corms, flower stems, bud sheaths, and flower petals. See Figure 2.47B under General Diseases and Figure 4.106. *Control:* Same as for Corm Rots (above). Spray one or more times

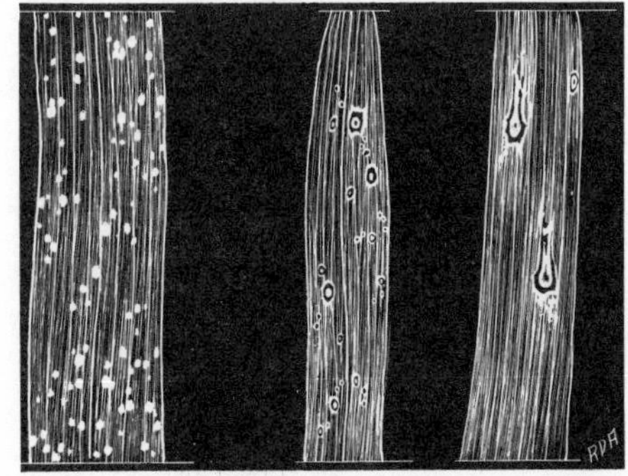

Fig. 4.106. Gladiolus leaf spots. Left to right: Botrytis, Septoria or Stemphylium, and Curvularia.

weekly during moist weather. Use maneb, zineb, Botran, Daconil 2787, captan, or folpet plus spreader-sticker. Start when leaves 6 to 10 inches tall or when disease first appears. Destroy tops after flowering. Commercial growers often dip flower spikes in phenyl mercury solution, plus wetting agent, before shipping. Varieties differ in resistance. Florida Pink *gladiolus* is resistant to Botrytis and Curvularia.

5. *Mosaics, White Break, Stunt*—Widespread. Symptoms variable depending on virus, virus strain, plant, variety, and environmental conditions. Plants may be stunted with mild to severe yellowish-green to white or pale gray, spotted, mottled, or striped leaves. Infected plants may bloom early. Flower spikes may be stunted with petals crinkled, deformed, and streaked, striped, or flecked with whitish, yellowish, or light greenish blotches (Figure 4.107). Diseased corms sometimes warty, "pitted," or deformed. Certain viruses (e.g., Tobacco Ringspot, Bean Yellow Mosaics, Cucumber Mosaic) may be spread by cutting shears in harvesting flowers and corms or bulbs. *Control:* Dig and burn infected plants when first found. Keep down weeds. Control aphids that transmit certain viruses. Use malathion. Avoid growing near beans, clovers, cucumber, melons, or tomatoes. Plant virus-free cormels, corms, or bulbs. Disinfect shears after cutting suspicious flowers or tops.
6. *Aster Yellows, Grassy-top*—Symptoms greatly variable. Young leaves may turn yellowish-green and be twisted. Flower spike may be spindly, green, twisted, or appear "corkscrew." Plants may turn yellow, mature, and die early. Often produce small corms and no color in flowers. Corms from infected plants may produce multiple shoots, commonly called "grassy-top." *Control:* Same as for Mosaics (above). Control leafhoppers that transmit the virus. Use DDT plus malathion at about weekly intervals.
7. *Chlorosis*—Plants unthrifty yellowish-green, yellow, or ivory color. Plants often stunted. *Control:* Have soil tested. It should be slightly acid (pH 5.5 to 6.5). Spray plants with iron sulfate (1 teaspoon per gallon) or iron chelate (page 27). Follow manufacturer's directions. May apply with regular pest sprays. Start when first noticed. Repeat as necessary.
8. *Root Rots, Neck or Collar Rots*—Plants may wilt, wither, collapse, and die. Easily pulled up. Leaf bases and roots are often shredded and rotted in wet soil. Plants often killed out in section of row. May be associated with nematodes (see below). *Control:* Same as for Corm Rots (above). A soil drench of PCNB or Botran plus DBCP (Nemagon, Fumazone) in infested areas may help. Follow manufacturer's directions. Start when disease first evident.

Fig. 4.107. White break of gladiolus. White or yellow blotching of flowers is the first symptom. (Illinois Natural History Survey photo)

9. *Root-feeding Nematodes* (e.g., lesion, pin, root-knot, sheath, spiral, summer crimp) —Leaves often die back from tips. Plants may be unthrifty and stunted with discolored, stubby roots. Roots may show small, knotlike galls (Root-knot). *Control:* Soak *gladiolus* corms 4 hours in hot water (110° F.) plus 0.5 per cent formalin. Or plant disease-free, high-quality corms in clean or sterilized soil (pages 538–548). Drenching over the row with DBCP is beneficial.
10. *Bulb Nematode* (gladiolus, tigridia)—See (38) Bulb Nematode under General Diseases.
11. *Blind Buds* (crocus)—Flower buds do not grow. Dry up or rot. *Control:* Water during dry periods. Avoid excessive heat after harvest and in storage.
12. *Topple, Calcium Deficiency* (gladiolus)—See under Tulip and page 28.
13. *Crown Gall* (gladiolus)—Cancer-like growth on corm. See (30) Crown Gall under General Diseases.
14. *Smut* (gladiolus)—Rare. Elongated, dark brown to black, powdery blisters or stripes in leaves, stems, and corms. Seedlings may shred and die early. *Control:* Same as for Corm Rots (above). Destroy infected plants when *first* found. Soak healthy corms as for Root-feeding Nematodes (above).
15. *Rust* (gladiolus) Very rare. See (8) Rust under General Diseases. Destroy affected plants when first found.
16. *Boron Deficiency*—Horizontal cracks form in leaves. Begin at margins and extend inward toward the midrib. Leaf tips may be blunt and curved. *Control:* Have soil tested and apply boron-containing fertilizer as recommended. Or add borax or Solubor to 3 or 4 regular pest sprays at 1- or 2-week intervals. See page 27.

GLEDITSIA — See Honeylocust

GLOBE-AMARANTH — See Cockscomb

GLOBE ARTICHOKE — See Lettuce

GLOBEFLOWER — See Anemone

GLOBEMALLOW — See Hollyhock

GLOBETHISTLE — See Chrysanthemum

GLOBE LILY, GLOBE-TULIP — See Mariposa Lily

GLORYBOWER — See Lantana

GLORY-OF-THE-SNOW — See Tulip

GLORYVINE — See Grape

GLOWING GOLD — See Pea

GLOXINIA — See African-violet

GOATSBEARD — See Rose

GODETIA — See Fuchsia

GOLDDUST — See Cabbage

GOLDDUST-TREE — See Aucuba

GOLDEN-ASTER — See Chrysanthemum

GOLDEN CHINQUAPIN — See Chestnut

GOLDEN-DEWDROP — See Lantana

GOLDENCHAIN or BEANTREE [COMMON, WATERER'S], SCOTCH LABURNUM *(Laburnum)*

1. *Leaf Spots*—Round to irregular, light gray, grayish-brown to dark brown spots and blotches. Black dots or mold may form in center of older spots. *Control:* If serious, apply zineb, maneb, or fixed copper at 10- to 14-day intervals during rainy weather.
2. *Twig Blights and Dieback*—In wet springs, discolored areas may develop on twigs and branches causing foliage beyond to blight. Twigs and branches may die back.

Control: Prune and burn infected twigs and branches. Spray as for Leaf Spots (above).

3. *Mosaic, Infectious Variegation*—Leaves mottled light and dark green. Often brightly variegated. Leaf veins may be yellow and prominent. Tree growth is apparently normal. *Control:* Use malathion or lindane to control aphids that probably transmit virus. Propagate only from mosaic-free plants.
4. *Root Rot*—See under Apple, and (34) Root Rot under General Diseases. May be associated with root-feeding nematodes (e.g., lesion, root-knot, stunt). Grow in moist, well-drained soil in sun or light shade.
5. *Root-knot*—See (37) Root-knot under General Diseases.
6. *Gray-mold Blight, Botrytis Blight*—See (5) Botrytis Blight under General Diseases.

GOLDEN EARDROPS — See Bleedingheart

GOLDENEGGS — See Evening-primrose

GOLDEN ELDER — See Snowberry

GOLDEN-EYED GRASS — See Iris

GOLDENGLOW — See Chrysanthemum

GOLDENLARCH — See Larch

GOLDEN MARGUERITE — See Chrysanthemum

GOLDEN-PEA — See Pea

GOLDEN QUEEN — See Anemone

GOLDENRAIN-TREE *(Koelreuteria)*

1. *Coral Spot, Twig Canker*—Coral-red cankers on twigs. Affected parts wither and die back. *Control:* Prune and burn blighted parts. Keep trees vigorous. Fertilize and water during dry periods.
2. *Verticillium Wilt*—See under Maple, and (15B) Verticillium Wilt.
3. *Leaf Spot*—Small, tan to gray spots. *Control:* None necessary.

GOLDENROD — See Chrysanthemum

GOLDEN ROSE OF CHINA — See Rose

GOLDEN-SHOWER — See Honeylocust

GOLDEN SPIDERLILY — See Daffodil

GOLDEN-STAR — See Chrysanthemum

GOLDENTUFT — See Cabbage

GOLDEN-WAVE — See Chrysanthemum

GOLDEN WREATH — See Honeylocust

GOLDFLOWER — See St.-Johns-wort

GOLDTHREAD — See Delphinium

GOMPHRENA — See Cockscomb

GOOSEBERRY — See Currant

GORDONIA — See Franklin-tree

GORSE — See Broom

GOURDS — See Cucumber

GRAMMATOPHYLLUM — See Orchids

GRANADILLA — See Passionflower

GRAND DUTCHESS — See Oxalis

GRAPE [BIRD, CALIFORNIA, CANYON, EUROPEAN WINE, FOX, FROST or RIVERBANK, MUSCADINE, POSSUM, SAND, SUMMER or PIGEON, SWEET WINTER, WINTER], GLORYVINE *(Vitus);* PEPPERVINE, MONKSHOOD-VINE, TURQUOISE or PORCELAIN BERRY *(Ampelopsis);*

Fig. 4.108A (left). Black rot of grape.

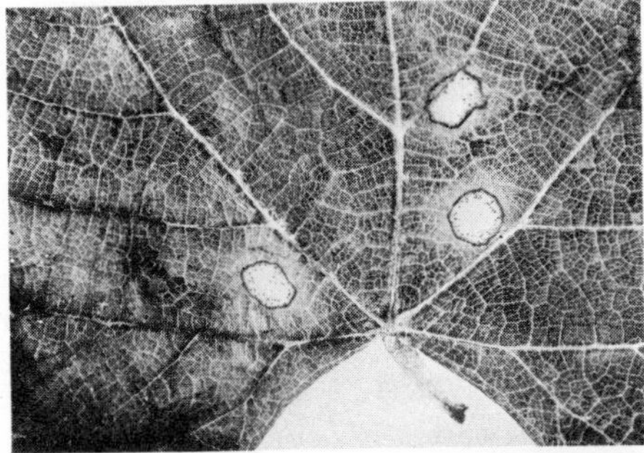

Fig. 4.108B. Close-up of black rot infection on a grape leaf. Note dark margin and the ring of black specks (fruiting bodies of the *Guignardia* fungus). (Department of Botany & Plant Pathology and Cooperative Extension Service, Michigan State University photo)

MARINE-IVY, GRAPE-IVY, KANGAROO VINE, REX-BEGONIA VINE *(Cissus)*; IVY [BOSTON, ENGELMANN, ST. PAUL], VIRGINIA-CREEPER or WOODBINE, PLUME HYACINTH *(Parthenocissus)*

1. *Black Rot, Leaf Spot*—Widespread and serious in wet seasons. Small, more or less circular to angular, reddish-brown to tannish-brown spots on leaves. Spots usually have dark brown margins and black specks arranged in a ring just inside the margin. Round, rapidly enlarging, tan spots on *grape* fruit that become sunken and surrounded with a brown ring, giving a "bird's-eye" effect. Rot later turns brownish-black and enlarges. Berry quickly rots and usually turns into a hard, shriveled, wrinkled, black mummy (Figures 4.108A and 4.108B). Such fruits drop early. Several or all berries in a cluster may become infected. Blossoms blasted. Elongated, depressed, purple to black spots occur on stems, stalks, and shoots. *Control:* Prune and retie grape vines annually. Burn prunings plus overwintering berries, leaves, and tendrils. Space plants. Keep down weeds. In humid areas, apply captan, zineb, fixed copper and spray lime, ferbam, or folpet. Follow grape spray schedule in Appendix (Table 16). For other plants spray just before bloom, just after bloom, and 10 days later. Use captan, zineb, or folpet. Check with your county agent or extension plant pathologist regarding timing of sprays for your area. Normally resistant *grapes:* America, Beaver, Beta, Carman, Champion, Cimarron, Clinton, Delaware, Dracut Amber, Eaton, Elvira, Everglades, Loretto, Lucile,

Lutie, Mills, Missouri Riesling, Moore Early, Muench, Norton, Ruby, Sheridan, Tarheel, and Topsail. Resistant *Muscadine grapes:* Albermarle, Chowan, Pamlico, Roanoke, and Tarheel. Check with your county agent or extension horticulturist regarding adaptability of these varieties to your area. Cultivate in early spring to cover old fruit mummies, leaves, and other debris.

2. *Downy Mildew*—Widespread and serious in warm, wet seasons. Greenish-yellow, irregular blotches on upper leaf surface, that later turn dark reddish-brown. Dense, white, downy mold grows on corresponding underleaf surface. Young fruits, shoots, and tendrils are also attacked. Affected leaves and fruit may drop early. Flower clusters and young berry clusters may be killed. Shoot growth is stunted and thickened; may turn brown and die. See Figure 2.22A under General Diseases. *Control:* Same cultural practices as for Black Rot (above). *Highly susceptible grapes:* Agawam, Campbell Early, Catawba, Champion, Delaware, Fredonia, Niagara, Urbana, Van Buren, and Worden. *Resistant grapes:* America, Carman, Clinton, Concord, Loretto, Lutie, Muench, Norton, Steuben, and many French Hybrids (e.g., Seyve-Villard (S-V) 5–276, 12–309, 12–375, 18–315, 23–18, Seibel 8357 and 4986, Joannes-Seyve 23–416, Bertille-Seyve 2667, Couderc 71–20, etc.). Where downy mildew is a problem, apply captan, zineb, fixed copper and spray lime, or folpet. Follow grape spray schedule in Appendix (Table 16). Cover underleaf surface thoroughly. Start when mildew is *first* seen. Repeat at about 2-week intervals. Collect and burn fallen leaves.

3. *2, 4-D Injury*—Cosmopolitan. Leaves, tendrils, and young shoots are misshapen. Leaves often dwarfed, have many sawtooth edges and close yellow veins. May appear narrow, fan-shaped, and stiff (Figure 4.109). Fruits ripen unevenly, if at all. Symptoms likely 1 to 3 weeks after exposure to fumes or spray drift (up to a mile or more). *Control:* Use only the *granular or amine* form of 2, 4-D, at low pressure, near grape, tomato, rose, melons, beans, and other garden plantings. Apply to

Fig. 4.109 (left). 2,4-D injury of grape. (Iowa State University photo)

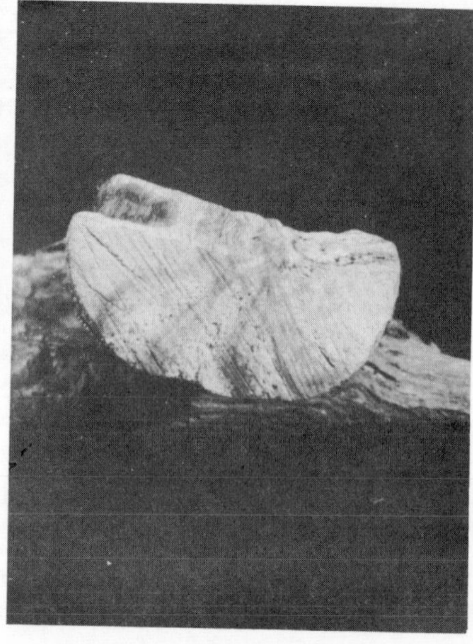

Fig. 4.110 (right). Grape dead arm. Cross section through a cane showing the typical, U-shaped, diseased areas. (Department of Botany & Plant Pathology and Cooperative Extension Service, Michigan State University photo)

lawns in fall. Apply only when air movement is sluggish and away from grapes and other susceptible plantings. Use separate sprayer for applying weed killers.

4. *Grape Fruit Rots, Fly Speck*—Widespread in rainy weather. Ripening fruits rot; later mummify. May be covered with gray, white, green, black, bluish-green, or pink mold. *Control:* Handle fruit carefully. Store fruit as cool (32° F.) and dry as practical. Prune and retie grape vines annually. Destroy rotting fruit. Spray as for Downy Mildew (above). Apply captan alone just before harvest. Control grape insects (e.g., Japanese beetles, grape berry moth, flea beetles, leafhoppers, leaf folder, etc.) with methoxychlor or sevin plus malathion. All California and Arizona table grapes are fumigated with sulfur dioxide before shipment to prevent decay.
5. *Powdery Mildew*—General, especially from midsummer on. Indistinct or thin, powdery, flourlike patches on leaves, young shoots (canes), tendrils, blossoms, and young fruit that may later turn brown or black. If severe, may cause stunting, yellowing, withering, and premature defoliation. Berries may be dwarfed, distorted, brownish, russeted, and cracked. Such fruit may fail to ripen or "shell off" vine before harvest. Mildew builds up following extended use of captan or zineb sprays. *Control:* Spray with fixed copper and spray lime as for Downy Mildew (above). If serious, add Karathane, micronized sulfur, Morocide or Morestan, to 2 consecutive sprays. Keep down weeds. Same cultural practices as for Black Rot (above). Resistant or tolerant *grapes:* Clinton, Delaware, Dutchess, Early Niabell, Elvira, Ives, Niabell, Royalty, and Scarlet.
6. *Grape Dead-arm, Fruit Rot*—Widespread. Locally severe in certain areas, especially on French Hybrids. Young shoots, trunks, and branches (or arms) may be weakened and killed by perennial, reddish-brown to purplish-black, elongated cankers. Cankers enlarge, may split open. Often become irregular, black, and crusty. Girdle trunk or "arm" (Figure 4.110). Portion beyond dies. Leaves on cankered canes often dwarfed, yellowish, cupped, tattered and "crimped," especially along edges. Such leaves usually drop early. Leaves that appear late in the season are stunted but otherwise normal. Suckers often form at base of killed spurs giving vines a bushy appearance. Fruit may rot. *Control:* Annually prune out and burn all dead and cankered (blighted) wood at least 6 inches below any sign of disease. Disinfect between cuts, in diseased vineyards, by dipping shears in 70 per cent alcohol. Follow spray program as for Black Rot and Downy Mildew (both above). Where common, apply a captan spray when buds swell. Repeat when new shoots are ½ to 1 inch long.
7. *Anthracnose, Bird's Eye Fruit Rot, Leaf Scab*—Widespread on grape. Small, sunken, grayish spots with dark margins. Found on leaves, young shoots, tendrils, fruit, and canes. Canes may be girdled, causing dwarfing and death of tips. Leaf spots may drop out leaving leaves ragged and distorted. *Control:* Same as for Black Rot and Downy Mildew (both above) or apply thiram or ziram: (1) as buds swell, (2) as buds break open, and (3) when new shoots are 7 to 9 inches long. Resistant *grapes:* Beacon, Concord, Delaware, Everglades, Fredonia, Herbemont, Lutie, Moore Early, Niagara, and President.
8. *Crown Gall*—Widespread. Rough, cream-colored, later reddish-brown to black galls form on canes and lower trunk and on roots. If severe, vines may be stunted with yellowish foliage. Vines may later die back and eventually die. *Control:* Avoid injuring roots and lower stem. Plant disease-free stock in soil where Crown Gall is unknown.
9. *Root Rots*—Plants decline in vigor, gradually die. Foliage is sparse and often yellowish. Fruit production declines sharply. Canes winter-kill. Roots die back. Often associated with root-feeding nematodes and insects. *Control:* See under Apple. Rootstocks vary in resistance.
10. *Wood Rots*—Vines gradually decline in vigor. Shoots may die back. Leaves may be

mottled, yellowish or bronzed, scorched, puckered, and distorted. Often drop early. Fruit may crack, be speckled, shrivel, and dry up. Trunks and branches have soft, spongy, internal decay. Older vines most severely infected. *Control:* Avoid wounding when cultivating or mowing, etc. Follow spray program as for Downy Mildew and Black Rot (both above). Prune and retie vines annually. Prune affected vines back to ground.

11. *Root-knot*—Southern states. Badly affected vines become stunted and lack vigor. Yields reduced. Small galls or swellings on roots. Roots often decay. *Control:* Root-knot resistant *grape* rootstocks are available (e.g., Dogridge, Salt Creek). Check with your local nurseryman or extension plant pathologist. Nurserymen disinfest *dormant,* strong, 1-year-old *grape* rootings by soaking in hot water (122° F. for 10 minutes; 125° F. for 5 minutes; *or* 127° F. for 3 minutes). Grow in clean or fumigated soil (pages 538–548). Use D-D, Telone, Vidden-D, or SMDC (Vapam or VPM). DBCP may be used around established plants. Follow manufacturer's directions. Keep vines vigorous. Prune, fertilize, water, control insects and other diseases.
12. *Boron Deficiency*—Symptoms variable. Terminal leaves develop enlarging, yellowish, then reddish, dead areas between leaf veins or along margins. Resembles magnesium deficiency and Pierce's Disease (both below), but the latter affects basal leaves first. Leaves often dwarfed, rolled, and distorted. Fall early. Growth may be stunted, bushy, and distorted. Or die in early summer. Fruit is tough and tasteless; may be small or shatter easily. *Control:* Have soil tested and follow recommendations. Apply borax in late winter and repeat once or twice more before bloom. *Excess* boron may produce bushy shoot growth, deformed leaves, or even death of vines (page 28).
13. *Zinc Deficiency or Little Leaf, Manganese Deficiency*—Leaves develop yellowish areas near margins and between veins (which remain green). Canes may be stunted; produce numerous lateral shoots. *Zinc deficiency* symptoms show first on the terminal leaves of main shoots and on leaves of lateral shoots that develop in summer. *Manganese deficiency* first shows on basal leaves. Grape yields are reduced. *Control:* Apply zinc sulfate or manganese sulfate with pest sprays (1 ounce of zinc or manganese sulfate and ½ ounce of spray lime in a gallon of water). Apply when grape shoots 12 to 15 inches long. Zinc sulfate (23 per cent metallic zinc) can be swabbed on fresh pruning wounds of spur-pruned vines within 3 hours after pruning during the dormant season. Use 1 to 2 pounds in a gallon of water. See also Iron Chlorosis (below).
14. *Potassium Deficiency, Leaf Scorch* (grape)—Leaf margins change from pale green to bronze. Margins and interveinal areas later become scorched. Sometimes confused with Magnesium Deficiency (below). Fruit in compact clusters of small berries. Maturity is delayed or uneven. *Control:* Have soil and petiole tissue tests made and follow the reports.
15. *Magnesium Deficiency* (grape)—Progressive yellowing of leaf margins and interveinal areas, starting with *basal* leaves. *Control:* Same as for Potassium Deficiency (above).
16. *Dieback, Canker, Wilt, Cane or Shoot Blight*—Stems or canes are cankered. Foliage wilts and dies back. *Control:* Prune and burn all infected branches. Spray as for Downy Mildew (above).
17. *Pierce's Disease of Grape, Grape Decline*—Primarily southern states. Symptoms very variable depending on variety, age, and locality. Vines decline in vigor. Die in several months to 4 to 5 years. Most varieties are slow to leaf out in the spring. Shoots remain dwarfed; tend to die back from the tips. Affected leaves usually drop early, starting at base of shoots. From midsummer into the fall, leaf margins and areas between leaf veins show more and more scalding. Fruit clusters often wilt and

dry up. *Control:* If suspicious, check with your county agent, a local grower, or extension plant pathologist. Grow only certified, virus-free plants. Destroying infected vines and spraying to control leafhoppers that transmit the virus have *not* been satisfactory. Keep down weeds, especially Bermudagrass. *Grape* species in the Gulf states are resistant. Resistant or tolerant *grapes:* Champanel, Herbemont, Lake Emerald, and Lenoir. Other more highly resistant varieties may be available shortly.

18. *Fanleaf, Infectious Degeneration* (grape) —Varieties are affected differently. (Fanleaf, Yellow Mosaic, and Vein-banding are all strains of the same virus.) Leaves and new growth severely stunted. Secondary shoots develop abnormally with 2 or 3 buds forming double or treble nodes. Leaves come out at an acute angle with the stem. Leaf margins more deeply cut than normal. May resemble half-closed fan. Tissues somewhat puckered, folded, and mottled. Symptoms tend to disappear during season. Fruit production decreases sharply. Vine gradually becomes dwarfed. Symptoms somewhat similar to 2,4-D injury (above). Transmitted by nematodes (aphelenchus, dagger, lesion, pin, ring, stem and bulb, stubby-root, tylenchus), grafting, and planting infected cuttings. *Control:* Destroy infected vines when first found. Replant with certified, virus-free stock in fumigated soil (pages 538–548). *All known grape viruses* can be inactivated in shoot tips propagated under mist from plants held 60 days or more at 100° F.
19. *Yellow Mosaic, Infectious Yellows* (grape) —Symptoms variable. Leaves variously mottled and yellowed. Shoots, leaves, stems, and even blossom clusters may be yellow. Blossoms drop early. See Fanleaf (above).
20. *Leafroll or Red-leaf of Grape (White Emperor Disease)*—Symptoms variable. A virus disease, primarily in California. Easily confused with potassium deficiency near harvest. Plants may be severely stunted. Leaves darker and thicker than normal. Turn bronze or reddish along veins and yellow between veins. Leaves turn color early with lower leaves scorched and rolled downward and inward. Fruit yield and quality are greatly reduced. Normally red fruit remains greenish-white, greenish-yellow, or pink. Virus interferes with movement of potassium within vine and accentuates potassium and manganese deficiency in grape. *Control:* Destroy infected plants. Replant with certified, virus-free stock. Keep potassium level up, based on a soil test.
21. *Other Leaf Spots, Blights or Blotches, Spot Anthracnose*—Widespread. Spots of various sizes, shapes, and colors. *Control:* Same as for Black Rot (above). Muscadine *grapes* resistant to Mycosphaerella Leaf Spot: Albermarle, Chowan, Pamlico, and Roanoke.
22. *Other Root-feeding Nematodes* (aphelenchus, citrus-root, dagger, lesion, pin, needle, ring, spiral, sting, stubby-root, stunt) —Plants may produce weak, stunted top growth. Vines gradually decline in vigor. Yields are reduced. Cuttings do not develop normally. Roots may be stunted, "stubby" or bushy, and dark in color. *Control:* Same as for Root-knot (above). Nematode-resistant rootstocks may be available shortly.
23. *Thread Blight* (peppervine, Virginia-creeper) —Southeastern states. See under Walnut.
24. *Rust* (ampelopsis, cissus, grape) —Southeastern states. See (8) Rust under General Diseases. *Control:* Same as for Black Rot and Downy Mildew (both above). Use zineb or ferbam.
25. *Sooty Mold* (grape) —See (12) Sooty Mold under General Diseases.
26. *Verticillium Wilt* (grape) —Sudden or gradual wilting of leaves and shoots followed by drying of leaves. Gray or brownish-black streaks under the bark. See (15B) Verticillium Wilt under General Diseases.
27. *Smut* (cissus) —Florida. See (11) Smut under General Diseases.

28. *Iron Chlorosis or Deficiency*—Occurs chiefly in alkaline soils. Youngest leaves are pale green, yellow, or ivory-colored except for the green veins. Easily confused with Zinc Deficiency (above). Shoot growth is weak. Fruit production is poor. If severe, vines may die. *Control:* Apply iron sulfate (as Copperas, etc.) in several pest sprays. Use about 1 pound for each 10 gallons of spray. Repeat as necessary. Use chlorosis-resistant rootstocks. Check with your extension horticulturist or extension plant pathologist.

For additional information read USDA Farmers' Bulletin 1893, *Control of Grape Diseases and Insects in the Eastern United States.*

GRAPEFRUIT — See Citrus

GRAPE-HYACINTH — See Tulip

GRAPE-IVY — See Grape

GREAT LAUREL — See Rhododendron

GREEK VALERIAN — See Phlox

GREVILLEA — See Silk-oak

GROMWELL — See Mertensia

GROUNDCHERRY — See Tomato

GROUND GOLD-FLOWER — See Chrysanthemum

GROUND HEMLOCK — See Yew

GROUND-IVY — See Salvia

GROUND-MYRTLE — See Vinca

GROUND-PINK — See Phlox

GROUND-RATTAN — See Palms

GROUNDSEL — See Chrysanthemum

GRU-GRU — See Palms

GUAVA — See Myrtle

GUERNSEY-LILY — See Daffodil

GUINEA BEAN — See Cucumber

GUINEA-HEN FLOWER — See Tulip

GUM TREE — See Eucalyptus

GUMBO — See Hollyhock

GYMNOCLADUS — See Honeylocust

GYPSOPHILA — See Carnation

HACKBERRY [CHINESE, COMMON, SOUTHERN or SUGARBERRY, WESTERN] *(Celtis)*

1. *Witches'-broom*—Widespread, especially in central states. Unsightly, tight clusters of dwarfed twigs evident in dormant season (Figure 4.111). Several hundred witches'-broom galls may be found on one tree. Twigs may die back. Trees vary greatly in susceptibility. *Control:* Grow somewhat resistant types, e.g., Chinese and Southern. Cut and burn "brooms" back to healthy wood. A dormant lime-sulfur spray may be beneficial. Fertilizing to stimulate vigorous growth may be desirable.
2. *Powdery Mildews*—Widespread. Powdery, white growth on both leaf surfaces. *Control:* If practical, apply dormant spray of lime-sulfur. When mildew appears apply one or two sprays of Karathane or sulfur.
3. *Wood Rots*—Cosmopolitan. See under Birch, and (23) Wood Rot under General Diseases.
4. *Root Rots*—See under Apple, and (34) Root Rot under General Diseases.
5. *Leaf Spots, Leaf Blight*—General. See under Elm.
6. *Winter Injury*—See under Elm.

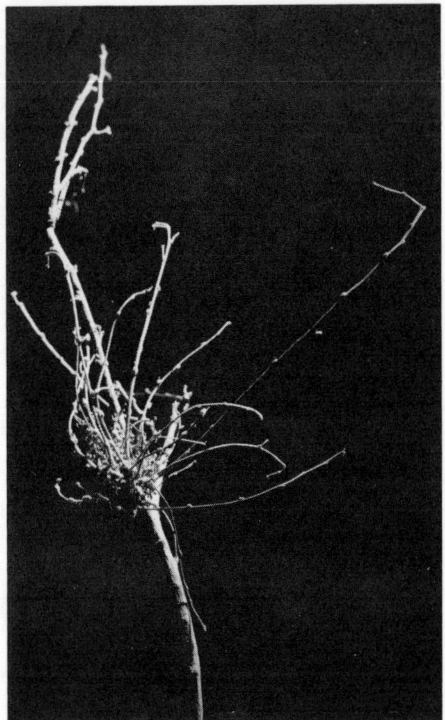

Fig. 4.111. Witches'-broom of hackberry. (Iowa State University photo)

7. *Downy Mildew*—See (6) Downy Mildew under General Diseases.
8. *Mosaic*—See under Elm.
9. *Felt Fungus*—Southern states on neglected trees infested with scale insects. Dark, sometimes ringed, crusty or spongy, felty to leathery growth over scales. *Control:* Spray with DDT plus malathion to control "crawler" stage of scales. Check with your extension entomologist or county agent regarding timing of sprays.
10. *Mistletoe*—See (39) Mistletoe under General Diseases.
11. *Thread Blight*—Southeastern states. See under Walnut.
12. *Seedling Blight*—See under Pine.
13. *Chlorosis*—Common in alkaline soils. Leaves become yellowish-green, especially between the veins. See under Maple, Chlorosis.
14. *2,4-D Injury*—See under Elm and Grape. Leaves may be thickened and leathery with tips and margins cupped downward or rolled.

HALESIA — See Silverbell

HAMAMELIS — See Witch-hazel

HAMATOCACTUS — See Cacti

HARDHACK — See Spirea

HARDY AMARYLLIS — See Daffodil

HARDY ORANGE — See Citrus

HAREBELL — See Bellflower

HAWAIIAN TI PLANT — See Dracaena

HAWKSBEARD — See Chrysanthemum

HAWORTHIA — See Aloe

HAWTHORN — See Apple

HAZELNUT — See Birch
HEAL-ALL — See Salvia
HEARTS AND HONEY VINE — See Morning-glory
HEATH [CORNISH, CORSICAN, CROSS-LEAVED, DARLEY, FRINGED or DORSET, GARLAND, MEDITERRANEAN, MOOR, SCOTCH or TWISTED, SPANISH, SPRING, TREE] *(Erica);* HEATHER *(Calluna);* TRAILING-ARBUTUS [AMERICAN or MAYFLOWER, JAPANESE] *(Epigaea);* WINTERGREEN (EASTERN or WESTERN), CHECKERBERRY or TEABERRY, CREEPING SNOWBERRY, SHALLON or SALAL *(Gaultheria)*

1. *Wilt, Root and Collar Rots, Stem Rot*—Plants wilt, wither, and die from rotting of stem, crown, and roots. Leaves are grayish before wilting. *Control:* Grow disease-free plants in clean or sterilized soil (pages 538–548) that is acid and well-drained. Avoid overwatering. Grow resistant species or varieties (e.g., *Erica persoluta*). May be associated with root-feeding nematodes (e.g., spiral).
2. *Powdery Mildew, Twist*—May be serious on salal. Youngest leaves redden, later turn yellow to brown and drop early. Infected leaves covered with whitish-gray mold patches especially on lower leaf surface. Plants dwarfed, bushy, and twisted. May die back. Blooming reduced. *Control:* Space plants. Dust or spray several times at weekly intervals. Use Karathane or sulfur.
3. *Chlorosis, "Yellows"*—Plants lack vigor. New growth stunted and pale yellow to whitish. Branches or entire plants die back. *Control:* Grow in light, well-drained soil that is quite acid. Spray plants monthly with iron sulfate. Avoid use of lime.
4. *Rust* (heath)—Orange, powdery pustules on leaves. Infected leaves turn yellow, wither, and drop early. *Control:* Spray or dust several times in early spring. Use zineb, maneb, or dichlone.
5. *Verticillium Wilt* (heath)—One or all branches may turn yellowish-green, wilt, and die. Disease usually progresses up plant. *Control:* Grow disease-free, resistant varieties in soil free of the fungus, or grow in sterilized soil.
6. *Damping-off, Cutting Rot*—Seedlings wilt and collapse. Cuttings rot at base. *Control:* Use clean or sterilized soil for starting seed or cuttings. Avoid overwatering and poorly drained soil. If disease strikes, apply drench of thiram or zineb (1 tablespoon per gallon). Repeat 5 to 7 days later.
7. *Leaf Spots, Fruit Spot, Spot Anthracnose*—General on gaultheria. Spots of various sizes, shapes, and colors on leaves and fruit. If severe, leaves may wither and fall early. *Control:* Same as for Rust (above).
8. *Sooty Mold, Black Mildew*—Black mold patches on foliage. Follows insect attacks. *Control:* Apply malathion to control insects.
9. *Gray-mold Blight*—See (5) Botrytis Blight under General Diseases.
10. *Red Leaf Gall* (gaultheria)—Northern states. Small, red galls on foliage. *Control:* Pick and burn affected parts.

HEATHER — See Heath
HEAVENLY BAMBOO — See Nandina
HEBE — See Speedwell
HEDERA — See Ivy
HEDGENETTLE — See Salvia
HEDGETHORN — See Oleander
HEDYSCEPE — See Palms
HELENIUM — See Chrysanthemum
HELIANTHEMUM — See Sunrose
HELIANTHUS, HELICHRYSUM, HELIOPSIS — See Chrysanthemum

HELIOTROPE *(Heliotropium)* — See Mertensia
HELLEBORUS — See Delphinium
HELXINE — See Babytears Vine
HEMEROCALLIS, DAYLILY *(Hemerocallis)*

1. *Leaf Spots, Blight*—Small, more or less round, tan to gray or black spots. One leaf spot has fairly broad, dark red border. *Control:* Collect and burn infected leaves as they appear. Cut and burn tops late in fall.
2. *Root-knot*—Gall-like growths on roots or discolored spots within fleshy roots. *Control:* Plant disease-free stock in clean or fumigated soil (pages 538–548).
3. *Root Rots*—See under Geranium, and (34) Root Rot under General Diseases. May be associated with nematodes (e.g., root-knot, spiral). Grow in fertile, well-drained soil. Winter-mulch evergreen kinds in north.
4. *Winter or Frost Injury*—Plants start slowly in spring with few yellowish, deformed leaves and little vigor. Most common with evergreen types. Crowns may decay. Shoots often stunted or spindling. Plants may or may not recover. Injury from late spring freezes appears as brown or tattered leaf tips followed by normal development and flowering. *Control:* Mulch will help delay growth until frost danger is past.
5. *Daylily Blight*—One or more flower stems suddenly wither and die. Foliage may turn yellowish-brown and die. Usually after flower stem has blighted. Roots decayed. Plants usually recover in same or following season. *Control:* Unknown.
6. *Russet Spot*—Greenish-yellow spots form on leaves. Spots enlarge and turn orange-brown. Tips or entire leaves may wither and die. Varieties differ greatly in susceptibility. *Control:* Grow plants in partial shade.
7. *Gray-mold Blight*—See (5) Botrytis Blight under General Diseases.

HEMLOCK — See Pine
HEMPTREE — See Lantana
HEN-AND-CHICKENS — See Sedum
HEPATICA — See Anemone
HERB-ROBERT — See Cranesbill
HERCULES-CLUB — See Acanthopanax and Hoptree
HERONSBILL — See Cranesbill
HESPERIS — See Cabbage
HEUCHERA — See Hydrangea
HIBA ARBORVITAE — See Juniper
HIBISCUS — See Hollyhock
HICKORY — See Walnut
HIGHBUSH CRANBERRY — See Viburnum
HINOKI CYPRESS — See Juniper
HIPPEASTRUM — See Daffodil
HOBBLEBUSH — See Viburnum
HOLLY [AMERICAN, BURFORD, CHINESE, ENGLISH or CHRISTMAS (HEDGEHOG, SILVER-EDGE), JAPANESE, LONG-STALK, LUSTER-LEAF, MOUNTAIN], DAHOON, INKBERRY, POSSUMHAW, WINTERBERRY (BLACK-ALDER) or NORTHERN HOLLY, JAPANESE WINTERBERRY, YAUPON, DWARF YAUPON *(Ilex)*; MOUNTAIN-HOLLY *(Nemopanthus)*

1. *Leaf Spots, Anthracnose, Tar Spots, Leaf and Twig Blight, Spot Anthracnose*—Round to irregular spots of various sizes on leaves. Spots are white, yellow, purple, gray, brown, reddish-brown, red-black, or black. Sometimes with a yellow, reddish,

brown, or purplish margin. Spots may drop out leaving shot-holes or merge to form blotches. Young twigs and berries may be spotted or girdled and die back. Often serious in nursery. May be associated with mechanical injury, insects (e.g., midges), copper spray injury, lack of boron, etc. *Control:* Pick and burn spotted leaves, where practical. Fertilize and water to maintain good tree vigor. Check with your nurseryman or extension horticulturist. If serious, spray with zineb, ferbam, or dichlone at 2-week intervals. Start as leaves unfold. Repeat during late summer or early fall wet periods. Space plants. Grow where air circulation is fair to good. Removal of lower branches plus burning fallen leaves is often beneficial.

2. *Spine Spot, Purple Spot*—Small, grayish-brown to dark red punctures and scratches, usually surrounded with purplish border, appear on both leaf surfaces in late winter or early spring. Spots caused by wounding of spines from nearby leaves during high winds. *Control:* Plant adapted tree species in sheltered location protected from severe winds, or erect barriers to prevent wind-whipping of leaves. Spraying with an antidesiccant (e.g., Wilt Pruf, Plantcote, Foli-gard or Stop-Wilt) is effective.
3. *Leaf Scorch, Scald*—Leaves, especially near margins, are scorched brown, white, gray, or purplish after transplanting into full sun; salt spray near ocean; in cold, wind-swept areas; or when sun shines brightly on ice-coated leaves (Figure 4.112). Occurs in late winter or early spring following low freezing temperatures. *Control:* Same as for Spine Spot (above). Pick affected leaves. Young trees should have a thick mulch over the root area, e.g., pine needles, sawdust, or oak leaves.

Fig. 4.112. Leaf scorch of holly. (Pennsylvania State University Extension Service photo)

4. *Powdery Mildews*—General. Grayish-white mold patches on foliage. *Control:* Where serious, spray with Karathane or sulfur several times, a week apart. Start when mildew is first noticed.
5. *Twig Blights, Cankers, Dieback*—Light brown to black, sunken cankers form on twigs and branches. Small "pimples" often develop in bark. Affected parts wilt, wither, and die back when girdled. Often follows wounding or winter injury. *Control:* Prune and burn dead or cankered twigs and branches. Spray as for Leaf Spots (above) during cool, rainy, spring and fall weather.
6. *Sooty Molds, Black Mildew*—Primarily Gulf states. Black mold patches on foliage. Follows insect attacks. *Control:* Use malathion to control scales, aphids, whiteflies, mealybugs, and other insects.
7. *Root Rots*—Plants decline in vigor; may die. Twigs die back. Foliage is sparse. See under Apple, and (34) Root Rot under General Diseases. May be associated with poorly drained soil and nematodes (e.g., burrowing, dagger, lance, lesion, pin, ring, root-knot, sheath, spiral, sting, stubby-root, stunt, tylenchus).
8. *Leaf Rot or Drop of Cuttings* (American and English holly)—Lower leaves

spotted, decay, and drop when cuttings are in rooting medium. Usually starts 2 to 3 weeks after cuttings are set. *Control:* Stick cuttings in clean, fresh sand or sterilize old sand with heat or chemicals (pages 538–548). Drench cutting bench with PCNB-Dexon mixture when disease starts. Follow manufacturer's directions. Take cuttings *only* from branches that do not touch the ground.

9. *Botrytis Flower Blight* (American holly)—Water-soaked tan spots on flower parts. Entire flower clusters may blight. Twigs die back in humid, wet weather. *Control:* Spray as for Leaf Spots (above). Many American and English *hollies* are apparently resistant. Space plants for good air circulation. Where practical, remove and burn affected parts.
10. *Green Algae*—Dull, yellowish-green, powdery layer on leaves, twigs, and trunk in wet or humid weather in moist, stagnant locations. *Control:* Prune to open up trees and increase air circulation. If serious, apply phenyl mercury, zineb, or maneb sprays, plus spreader-sticker, starting in early spring.
11. *Crown Gall* (holly)—See (30) Crown Gall under General Diseases.
12. *Rust*—American holly is attacked in southern states. See (8) Rust under General Diseases.
13. *Wood Rots*—General. See under Birch, and (23) Wood Rot under General Diseases.
14. *Chlorosis*—Occurs in alkaline soils. See under Maple.
15. *Thread Blights*—Southeastern states. See under Walnut. *Control:* Spray as for Leaf Spots (above).
16. *Felt Fungus*—Southern states. Smooth, shiny, chocolate-brown to almost black growth on bark. See under Hackberry.
17. *Boron Deficiency*—Irregular, reddish or reddish-purple spots on upper leaf surface that are water-soaked on lower surface. Spots later enlarge and form concentric rings with yellowish border. Leaves formed late in the season may be distorted. Affected leaves may drop early. *Control:* Apply borax (e.g., Polybor) to soil around trees (page 27). Follow manufacturer's directions.
18. *Bacterial Leaf and Twig Blight* (American holly)—Massachusetts. Leaves and shoot tips black and scorched in late spring or early summer. Leaves later drop leaving bare, withered shoots. Resembles fire blight of apple and pear (page 80). *Control:* Many American holly varieties, e.g., Cardinal, Clark, etc., appear to be immune. Blighted twigs should be removed and burned. Disinfect pruning shears by dipping in a 1:800 solution of phenyl mercury (7.5 per cent) fungicide or a 1:1,600 solution of a 10.2 per cent phenyl mercury. Avoid large applications of manure or fertilizers high in nitrogen.
19. *Winter Injury*—Foliage is brown and scorched. Young twigs may die back. Bark splitting may sometimes occur from low temperatures. *Control:* Same as for Spine Spot and Leaf Scorch (both above).

HOLLYGRAPE — See Barberry

HOLLYHOCK [ANTWERP, COMMON] *(Althaea)*; ABUTILON, FLOWERING MAPLE *(Abutilon)*; ANODA; POPPY-MALLOW *(Callirhoë)*; ROSELLE, ROSEMALLOWS, FLOWER-OF-AN-HOUR, ABELMOSK or MUSK MALLOW, CHINESE HIBISCUS or ROSE-OF-CHINA, COTTON-ROSE or CONFEDERATE-ROSE, ROSE-OF-SHARON or SHRUB-ALTHEA, FRINGED HIBISCUS, RED-LEAF HIBISCUS, OKRA or GUMBO *(Hibiscus)*; TREEMALLOW *(Lavatera)*; MALLOW [COMMON, CURLED, HIGH, MUSK- or MUSK ROSE, VERVAIN or EUROPEAN] *(Malva)*; RED FALSE-MALLOW, BUSH-MALLOW *(Malvastrum)*; SIDA; CHECKERMALLOW *(Sidalcea)*; GLOBEMALLOW *(Sphaeralcea)*

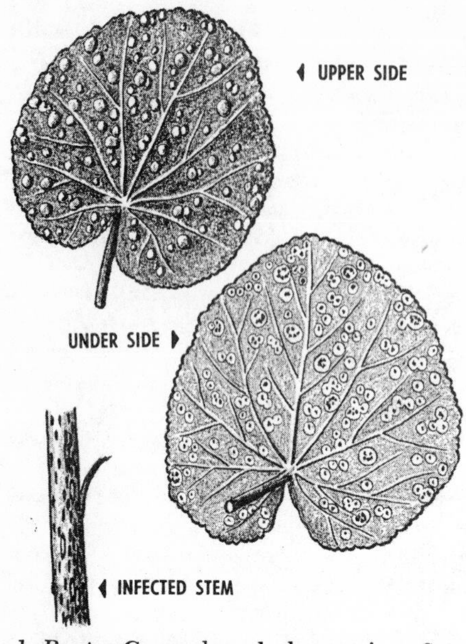

Fig. 4.113. Hollyhock rust.

1. *Rusts*—General and destructive. Small, orange-yellow, yellowish-brown, or rust-brown spots on leaves, bracts, and stems. Pustules may become powdery and reddish-brown to chocolate-brown. Mostly on underleaf surface (Figure 4.113). If severe, leaves may yellow, wither, and drop early. *Control:* Where practical, collect and burn all tops in fall. Pick and burn *first* infected leaves as they appear in early spring. Destroy nearby round-leaf mallow or cheeseweed *(Malva rotundifolia).* These plants may also be attacked, as are wild grasses, e.g., *Muhlenbergia, Sporobolus,* and *Stipa.* Keep down weeds. Plant only seed from healthy plants. *Hollyhock* varieties differ in resistance. Apply zineb, thiram, maneb, ferbam, dichlone, or sulfur several times at weekly intervals to keep new foliage covered. Start when growth begins. Repeat applications in late summer.
2. *Anthracnose, Dieback, Seedling Blight*—May be serious in hot, humid weather. Black blotches on stem, petioles, and even roots. Angular, grayish to black spots on leaves with a dark margin, that may later drop out, leaving "shot-holes." Shoot tips may die back. *Control:* Same as for Rusts (above).
3. *Leaf Spots, Stem Canker, Pod Spot, Blossom Blight, Gray-mold*—Spots of various colors, sizes, and shapes on leaves. Some spots may drop out. If severe, leaves may wilt, roll, wither, and drop early. Similar spots may occur on stems, pods, and flowers. May be covered with dense mold or sprinkled with black dots. *Control:* Same as for Rusts (above). Grow in well-drained soil. Rotate. Keep plants vigorous.
4. *Stem Canker, Stem or Crown Rots, Southern Blight*—Girdling cankers or internal decay of stems near soil line. Cankers often water-soaked, black, brown, or dark green. Later become covered with cottony mold or become white, tan, or yellowish-brown. *Control:* Dig up and burn infected plants and 6 inches of surrounding soil. If practical, sterilize infected soil (pages 538–548) or treat with PCNB before planting. Rotate. Grow in fertile, well-drained soil. Avoid overwatering, shading, and crowding plants.
5. *Powdery Mildews*—Powdery white growth on leaves. *Control:* If serious, spray or dust at 7- to 10-day intervals. Use sulfur or Karathane.
6. *Crown Gall and Hairy Root*—See (30) Crown Gall under General Diseases.

7. *Root Rots*—Plants often stunted. Wilt gradually or rapidly. Gradually decline and die. Roots discolored and decayed. See (34) Root Rot under General Diseases. May be associated with nematodes (e.g., burrowing, dagger, lance, lesion, pin, reniform, ring, root-knot, spiral, stem, sting, stubby-root, stunt). Grow in full sun, in fertile, well-drained soil enriched with organic matter.
8. *Root-knot*—Primarily southern states. Hollyhock and okra are very susceptible. Plants stunted and lack vigor with small to large galls on roots. *Control:* See (37) Root-knot under General Diseases. *Hibiscus* (Chinese) varieties differ in resistance.
9. *Bacterial Wilt*—Southern states. See under Tomato, and (15C) Bacterial Wilt under General Diseases.
10. *Verticillium Wilt* (abutilon, okra, poppy-mallow)—Widespread. Leaves first pale green, later wilt, yellow, then brown, and drop early. Dark brown streaks occur inside affected stems. See under Bellflower, and (15B) Verticillium Wilt under General Diseases.
11. *Fusarium Wilt* (okra)—Serious in commercial okra-growing areas. Plants progressively stunted and yellowed. Gradually die. Leaves wilt and roll. Dark streaks inside infected stems and roots. Often associated with nematodes (see Root Rots above). *Control:* If destructive, grow in wilt-free soil or practice 6-year rotation or longer.
12. *Mosaic, Infectious Variegation* (abutilon, Chinese hibiscus, hollyhock, lavatera, mallow, sida, sidalcea, treemallow)—Variegated forms of abutilon and treemallow are infected. Often increases ornamental value. Leaves mottled bright yellow and green. Leaf veins may be yellow. Severely affected leaves may be cupped. Propagated by grafting and occasionally by seed of certain species. The Abutilon Infectious Variegation virus is transmitted by a whitefly *(Bemisia)*. A branch or plant may apparently outgrow virus and appear normal.
13. *Seed Rot, Damping-off*—Seeds rot. Stand is poor. Seedlings may wilt and collapse. *Control:* Sow in warm, well-drained soil. Treat seed with thiram, captan, chloranil, or dichlone. Avoid overwatering.
14. *Curly-top* (mallow, okra)—See (19) Curly-top under General Diseases.
15. *Spotted Wilt and Yellows* (mallow, okra)—See (17) Spotted Wilt and (18) Yellows under General Diseases.
16. *Web Blight, Thread Blight*—Southeastern states. See under Bean and Walnut.
17. *Twig Blight* (hibiscus)—See under Maple.
18. *Gray-mold Blight*—See (5) Botrytis Blight under General Diseases.
19. *Chlorosis*—Common in alkaline or quite acid soils. Iron, manganese, zinc, or magnesium deficiency. See under Rose. Apply sprays of iron, zinc, magnesium or manganese sulfate or chelate (pages 26–28), using about 1 ounce in 2½ gallons of water. For chelates, follow manufacturer's directions. Adjust soil reaction to near neutral, pH 6.0 to 6.5, by adding lime or sulfur.
20. *Bacterial Leaf Spot* (Chinese hibiscus, cotton-rose, hibiscus, rose-of-Sharon)—See (2) Bacterial Leaf Spot under General Diseases. *Control:* Apply fixed copper or bordeaux mixture sprays during rainy periods. Prune to open up plants. Collect and burn fallen leaves.
21. *Flower Bud Drop* (primarily hibiscus)—May be caused by nutrient deficiency, damage by nematodes or insects, root rot, poor soil drainage or watering practices, drought, over or under fertilizing, mechanical injury, hot dry weather, and rapid drop in temperature. *Control:* Correct condition. *Hibiscus* varieties differ in resistance. Check with your nurseryman or extension horticulturist.
22. *Strapleaf, Molybdenum Deficiency* (hibiscus)—Florida. See under Cabbage.

HOLLY-OSMANTHUS — See Osmanthus

HOLODISCUS, OCEANSPRAY or ROCKSPIREA *(Holodiscus)*

1. *Leaf Spots, Leaf Blight*—Western states. Spots of various sizes, shapes, and colors. *Control:* If serious, apply several sprays at 10- to 14-day intervals, during wet periods. Use zineb, maneb, or fixed copper.
2. *Powdery Mildew*—Western states. Grayish-white, mold patches on foliage. *Control:* Apply 2 or 3 sprays of sulfur or Karathane, 10 days apart.
3. *Twig Canker, Dieback, Coral Spot*—Pacific Northwest. Twigs die back due to girdling cankers. Cankered areas may be covered with small, coral-pink fruiting bodies of the causal fungus. *Control:* Prune and burn cankered twigs. Keep trees growing vigorously. Spraying as for Leaf Spots (above) may be beneficial.
4. *Witches'-broom*—Pacific Northwest. New lateral branches are very slender and wirelike with small, crowded leaves. In later years, several to many laterals arise from same area on stem (witches'-broom) and are much branched. Nonshaded leaves turn bronzy-red in early summer. Affected plants do not bloom. *Control:* Remove and burn infected plants. Protect healthy plants by controlling aphids that transmit the virus.
5. *Fire Blight*—See under Apple.

HOMALOCEPHALA — See Cacti

HOMALOMENA — See Calla Lily

HONESTY — See Cabbage

HONEYDEW MELON — See Cucumber

HONEYLOCUST (many horticultural varieties), JAPANESE HONEYLOCUST *(Gleditsia)*; ACACIA [BLACK or LIGHTWOOD, CULTIVATED, KNIFE, STAR, SWEET or HUISACHE, SYDNEY, WINGED], KANGAROO THORN, GOLDEN WREATH, WEEPING MYALL, WATTLE [BLACK, COOTAMUNDRA, GREEN, GOLDEN, HAIRY, MOUNT MORGAN, SILVER], CATCLAW [LONG-FLOWERED, ROUND-FLOWERED, TEXAS] *(Acacia)*; SILKTREE or "MIMOSA," LEBBEK *(Albizzia)*; BAUHINIA, ORCHID-TREE or MOUNTAIN-EBONY *(Bauhinia)*; CAESALPINIA; PEASHRUB [DWARF, LITTLELEAF, PYGMY, RUSSIAN, SIBERIAN, SPRING], PEATREE *(Caragana)*; SENNA, GOLDEN-SHOWER *(Cassia)*; REDBUD [CHINESE, COMMON or EASTERN, TEXAS, WESTERN WHITE-FLOWERED], JUDAS-TREE *(Cercis)*; YELLOWWOOD, AMERICAN or KENTUCKY YELLOWWOOD *(Cladrastis)*; BLADDER-SENNA [CILICIAN, COMMON] *(Colutea)*; ROYAL POINCIANA *(Delonix)*; ERYTHRINA, RED-CARDINAL, CORAL-TREE, CORALBEAN *(Erythrina)*; KENTUCKY COFFEETREE *(Gymnocladus)*; LEADTREE *(Leucaena)*; MAACKIA; PARKINSONIA, JERUSALEM-THORN *(Parkinsonia)*; POINCIANA; LOCUST [BLACK or FALSE-ACACIA, BRISTLY or MOSSY (ROSE-ACACIA), CLAMMY, KELSEY, NEW MEXICO or THORNY] *(Robinia)*; SOPHORA [MESCALBEAN or FRIJOLITO, VETCHLEAF], JAPANESE PAGODATREE or CHINESE SCHOLARTREE *(Sophora)*; WISTERIA [AMERICAN, CHINESE, JAPANESE] *(Wisteria)*

1. *Twig Blights, Branch Cankers, Dieback, Wilt*—Discolored, slightly sunken, oval to elongated cankers girdle twigs and branches or trunk causing parts beyond to wilt, wither, and die. Small coral-pink to black "pimples" may be evident in affected bark. Diseased branches show sparse foliage. Trees may die back. Often follows freezing injury, borers, and bark wounds. *Control:* Severely cankered branches should be pruned and burned. Keep trees vigorous. Fertilize and water during summer dry periods. Grow varieties and species adapted and recommended for your area. Check with your nurseryman or extension horticulturist. Control borers with

DDT, Thiodan, or dieldrin sprays. Avoid bark injuries and pruning in wet weather. Sterilize pruning tools between cuts.

2. *Powdery Mildews*—Widespread. Powdery, grayish-white mold on leaves and young shoots. If severe, leaves may yellow, wither, and fall. *Control:* Where feasible, apply sulfur or Karathane twice, 10 days apart.
3. *Wood and Heart Rots*—General. See under Birch, and (23) Wood Rot under General Diseases. Wood rot fungi frequently follow borers and other insects. *Control:* Same as for Twig Blights (above). Paint over wounds promptly with tree wound dressing (page 36).
4. *Root and Crown Rots*—Trees decline in vigor. Foliage is thin and weak. Leaves may yellow, wither, and drop early. Often associated with nematodes (e.g., burrowing, lesion, pin, root-knot, spiral, stubby-root, stunt). See under Apple, and (34) Root Rot under General Diseases.
5. *Fusarium Wilt of "Mimosa" or Silktree*—Serious in eastern and southeastern states. Leaves rapidly turn yellow, wilt, shrivel, and hang downward on one or more branches. Leaves later fall and branch dies. Brown streaks in wood under bark. Trunks may "bleed" extensively in early stages of disease. Tree dies within a month to a year, branch by branch. Nematode injury (e.g., root-knot, lesion) speeds up wilting. *Control:* Remove and burn infected trees. Disinfect pruning cuts with a disinfectant (e.g., 70 per cent alcohol or 1:1,000 solution of mercuric chloride) and cover with tree wound dressing (page 36). Grow normally wilt-resistant strains: Charlotte, Tryon, U.S. No. 64. Better still, plant a different type of tree. Check with your extension horticulturist or nurseryman.
6. *Leaf Spots or Blights, Anthracnose, Tar Spot, Pod Blight*—General in wet seasons. May be serious in southern states. Small to large, round to irregular spots and blotches on leaves and pods. Leaves may be blighted and drop early. *Control:* Collect and burn fallen leaves. Prune out dead twigs and weak wood to increase air circulation. If practical, apply 3 sprays: as buds begin to unfold, 10 and 20 days later. Use thiram, ferbam, dichlone, captan, or phenyl mercury plus spreader-sticker. Fertilize and water to stimulate vigorous growth.
7. *Collar Rot, Trunk Canker*—See under Dogwood.
8. *Witches'-broom, Brooming Disease* (black locust, honeylocust, sophora)—Eastern half of United States. New leaves stunted with light-colored veins. Dense clusters or bunches (witches'-brooms) of short, spindling shoots with dwarfed, distorted leaves. Usually occurs in late summer. Brooms tend to die back during winter. Common on young sprouts following cuttings of twigs, branches, or trunk. Witches'-brooms also form on roots. *Control:* Where severe, remove and destroy infected trees. Trees have also appeared to recover naturally.
9. *Verticillium Wilt*—Foliage on affected branches wilts partially or completely. Branches or even entire trees may die. Greenish to brown or black streaks occur in sapwood of larger wilting or dead branches. *Control:* See under Maple.
10. *Sunscald, Winter Injury*—See under Apple.
11. *2,4-D Injury*—Redbud, honeylocust, black locust, and "mimosa" are very susceptible. Leaves may become thickened, cupped, twisted or rolled, appear boat-shaped or curled into ram's horns. Leaf veins are prominent. See Figure 1.1. *Control:* See under Grape.
12. *Crown Gall, Hairy Root*—See under Apple, and (30) Crown Gall under General Diseases.
13. *Root-knot*—See under Peach, and (37) Root-knot under General Disease. Nurserymen soak dormant *black locust* trees in hot water (118° F.) for 30 minutes.
14. *Rusts* (acacia, bladder-senna, caesalpinia, honeylocust, leadtree, lebbek, poinciana, sophora)—Reddish-brown or black, powdery pustules on leaves. *Control:* Same as for Leaf Spots (above).

15. *Damping-off, Seedling Blights*—Seedlings wilt and collapse. Stem at ground line decays. *Control:* Set out seedlings in clean or sterilized soil (pages 538–548) that is light and well-drained. See under Pine.
16. *Chlorosis*—Iron or zinc deficiency in alkaline soils. See under Maple and Walnut.
17. *Mosaic* (wisteria) —Yellowish blotches with scattered green "islands" on new leaves. Older leaves laterally twisted. *Control:* Do not propagate from infected plants.
18. *Mistletoe* (acacia, black locust, honeylocust, locust, parkinsonia, sophora) —See (39) Mistletoe under General Diseases.
19. *Thread Blights*—Southeastern states. See under Walnut.
20. *Sooty Mold* (Kentucky coffeetree, parkinsonia) —See (12) Sooty Mold under General Diseases.
21. *Felt Fungus*—Southern states. See under Hackberry.
22. *Wetwood, Slime Flux* (primarily locust and redbud) —See under Elm.
23. *Algal Leaf Spot* (acacia) —Far south and where damp. Velvety, reddish-brown patches on leaves and twigs. *Control:* See under Magnolia.
24. *Downy Mildew* (redbud) —See (6) Downy Mildew under General Diseases.

HONEYSUCKLE — See Snowberry

HOPHORNBEAM — See Birch

HOPTREE or WAFER ASH *(Ptelea)*; PRICKLY-ASH [COMMON, FLATSPINE, LIME], HERCULES-CLUB *(Zanthoxylum)*

1. *Leaf Spots, Tar Spot*—Small, round to irregular spots. *Control:* Pick and burn spotted leaves. If serious, spray at 10-day intervals during rainy periods. Use zineb or maneb. Collect and burn fallen leaves in autumn.
2. *Rusts*—Small, yellow or yellowish-orange spots on leaves. Alternate host is a wild grass *(Tridens* or *Andropogon)*. *Control:* Same as for Leaf Spots (above).
3. *Twig and Stem Canker*—Cankers girdle twigs causing them to die back. *Control:* Cut out cankered twigs. Spray as for Leaf Spots (above).
4. *Wood Rot* (Hercules-club, prickly-ash) —See under Birch, and (23) Wood Rot under General Diseases.
5. *Mistletoe* (Hercules-club, prickly-ash) —See (39) Mistletoe under General Diseases.
6. *Powdery Mildew* (hoptree, prickly-ash) —Widespread. Powdery, white mold on leaves. *Control:* See under Horsechestnut.
7. *Sooty Blotch*—See under Apple, and (12) Sooty Mold under General Diseases.
8. *Root Rot*—See (34) Root Rot under General Diseases.
9. *Root-knot*—See under Peach, and (37) Root-knot under General Diseases. Hercules-club is susceptible to one species *(Meloidogyne javanica)*.

HOREHOUND — See Salvia

HORNBEAM — See Birch

HORSEBALM — See Snapdragon

HORSECHESTNUT [BAUMANN'S, COMMON, DAMASK, JAPANESE, RED or RUBY, SCARLET], BUCKEYE [BOTTLEBRUSH, CALIFORNIA, OHIO, RED, TEXAS, YELLOW or SWEET] *(Aesculus)*

1. *Leaf Blotch*—Widespread and serious in wet seasons. Small to large, irregular, reddish-brown spots or blotches often with narrow, bright yellow borders. Small black spects later appear in centers of spots. Infected leaves usually curl, dry, and fall prematurely. Trees often appear scorched by midsummer. Spots also occur on nuts and leaf stalks. May closely resemble Leaf Scorch that results from drought and unfavorable growing conditions. See Figure 4.114. *Control:* Rake and burn fallen leaves. If practical, spray 2 to 4 times, 10 to 14 days apart (or just before rainy periods). Start when buds break open. Use dodine, ziram, zineb, thiram,

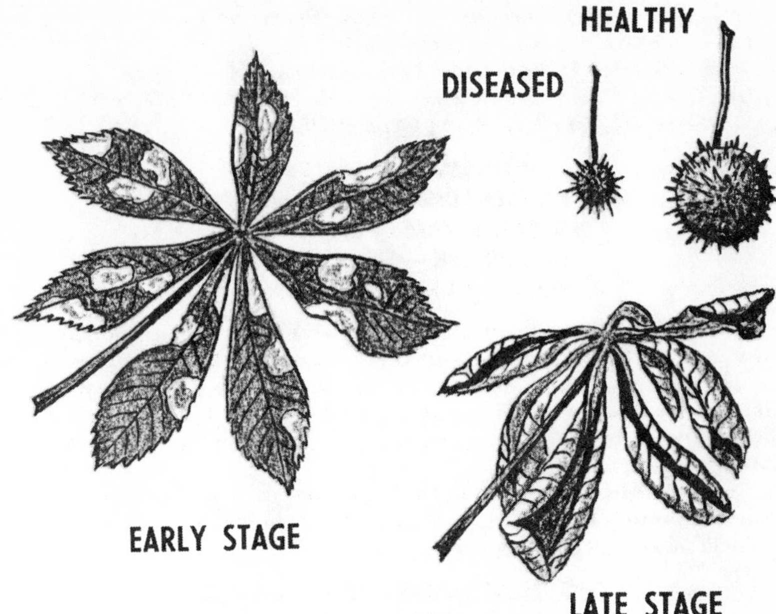

Fig. 4.114. Horsechestnut leaf blotch.

dichlone, or fixed copper plus spreader-sticker. Fertilize and water to stimulate vigorous growth. *Aesculus ambigua, A. arguta, A. glabra sargentii, A. glaucescens, A. mississippiensis, A. parviflora, and A. turbinata* are highly resistant or immune.

2. *Leaf Scorch*—Margins and tips of leaves become brown and curled in July or August. Scorch may spread over entire leaflet. Most evident in top and sides of tree exposed to wind and sun. *Control:* Prune susceptible trees to open them up. Water liberally during hot, dry periods. Fertilize to keep trees vigorous. The red or ruby horsechestnut is resistant.
3. *Leaf Spots, Anthracnose*—Tips of shoots may die back several inches. Spots on leaves are round to irregular, vary in color and size. *Control:* Same as for Leaf Blotch (above). Prune and burn infected twigs.
4. *Powdery Mildews*—General, especially in eastern and central states. Underside of leaves covered with white mold patches in late summer and fall. *Control:* If practical, spray two or three times, 7 to 10 days apart. Start when mildew is first seen. Use sulfur or Karathane.
5. *Wood Rots*—See under Birch, and (23) Wood Rot under General Diseases.
6. *Twig Blight, Branch and Trunk Cankers, Dieback*—General. Twigs and branches die back. Coral-pink to black "pimples" or gray mold form in affected bark. Leaves drop early. *Control:* See under Elm and Maple. Prune and burn affected parts, cutting several inches back of infected area.
7. *Verticillium Wilt*—Leaves on individual branches wilt, turn yellow or brown. Drop from late spring to September. New, stunted leaves may form later in the season. *Control:* See under Maple.
8. *Wetwood, Slime Flux*—See under Elm.
9. *Root Rot*—Cosmopolitan. See under Apple, and (34) Root Rot under General Diseases.

10. *Yellow Leaf Blister, Witches'-broom*—Yellow blisters on leaves that turn dull red. Witches'-brooms are formed. See under Birch.
11. *Rust*—Midwest. See (8) Rust under General Diseases.
12. *Bleeding Canker*—Northeastern states. See under Beech and Maple.
13. *Mistletoe*—See (39) Mistletoe under General Diseases.

HORSEMINT — See Salvia
HORSERADISH — See Cabbage
HORSETAIL-TREE — See Casuarina
HORTENSIA — See Hydrangea
HOSTA, PLANTAINLILY *(Hosta)*

1. *Crown Rot, Root Rot*—Plants wilt and collapse from rot at soil line or below. Crown may be covered with gray or cottony mold. *Control:* Carefully dig out and burn infected plants and 6 inches of surrounding soil. Set disease-free plants shallow in clean or sterilized soil (pages 538–548) that is rich, moist, but well-drained. A week before planting, work PCNB dust into top 4 to 6 inches of soil. Follow manufacturer's directions.
2. *Leaf Spots, Anthracnose*—Small to large spots on leaves and stems. Leaves may be disfigured. *Control:* Collect and burn plant debris in fall. Spray during wet periods. Use zineb, maneb, or fixed copper.

HOUNDSTONGUE — See Mertensia
HOUSELEEK — See Sedum
HOUSTONIA — See Buttonbush
HOWEA — See Palms
HUCKLEBERRY — See Blueberry
HUISACHE — See Honeylocust
HUNTSMANSCUP — See Pitcherplant
HUSK-TOMATO — See Tomato
HYACINTH *(Hyacinthus)* — See Tulip
HYACINTH-BEAN — See Pea
HYDRANGEA [BIGLEAF, CHINESE, CLIMBING, HILLS-OF-SNOW, HOUSE or HORTENSIA, OAKLEAF, PANICLE, PEEGEE (P.G.), PINK, SMOOTH or WILD, SNOWHILL] *(Hydrangea)*; DECUMARIA; DEUTZIA [FUZZY, LEMOINE, SLENDER] *(Deutzia)*; FENDLERA; CORALBELLS, ALUMROOT *(Heuchera)*; MITREWORT or BISHOPSCAP *(Mitella)*; MOCKORANGE [COMMON, GOLDEN, GORDON, LEMOINE, LEWIS, PEKIN, STAR, SWEET, VIRGINAL, ZEYHER] *(Philadelphus)*; STRAWBERRY-BEGONIA, SAXIFRAGE *(Saxifraga)*; FOAMFLOWER *(Tiarella)*

1. *Powdery Mildews* (coralbells, foamflower, heuchera, hydrangea, mitella, mock-orange, saxifrage) —General, especially in damp shady areas. White to grayish, powdery mold patches on leaves, stem tips, buds, and flowers (Figure 4.115). Leaves and flowers are stunted and distorted. May die early. *Control:* Apply Karathane, Acti-dione, or sulfur, plus spreader-sticker, 2 or 3 times, a week apart. Can combine with materials to control Leaf Spots or Gray-mold Blight (both below). *Avoid* using sulfur after color shows in buds. Indoors, avoid extreme temperature changes and reduce air humidity. *Hydrangea* varieties differ in resistance.
2. *Gray-mold Blight, Botrytis Blight, Bud Blight, Flower and Shoot Blight*—Cosmopolitan on hydrangea and mockorange, especially in humid areas. Flower clusters, buds, and shoots turn brown and rot. Often covered with grayish-brown mold in

Fig. 4.115. Powdery mildew on a hydrangea leaf. (Courtesy Dr. J. L. Forsberg)

damp weather or storage. May follow frost injury. Serious in wet seasons. *Control:* Indoors, keep water off foliage, space plants, and keep humidity down. Carefully collect and burn infected flower clusters. Spray flower buds just before opening. Use captan, maneb, Botran (dichloran), or zineb. Nurserymen usually spray or dip *hydrangea* plants with Botran, captan, or zineb before placing in storage.

3. *Leaf Spots, Flower Spot*—General. May be serious in rainy seasons. Small- to medium-sized spots, round to angular or irregular, of various colors on leaves and blossoms. Affected parts may be killed. *Control:* Same cultural practices as for Gray-mold Blight (above). Where practical, pick and burn spotted leaves and flowers. Apply captan, maneb, zineb, thiram, ferbam, or fixed copper at 7- to 10-day intervals. Start when new leaves unfold. Stop when flowers start to appear. Mist flowers lightly with zineb or captan (2 teaspoons per gallon) or Botran.
4. *Bacterial Wilt* (hydrangea)—Flowers, buds, and young leaves wilt and turn brown. Roots rot. Many plants may die. Most serious in hot, humid weather. *Control:* Destroy infected plant parts, or discard severely affected plants.
5. *Rusts* (coralbells, fendlera, foamflower, heuchera, hydrangea, mitella, mockorange, saxifrage)—Widespread. Small, yellowish or reddish-brown, powdery pustules mostly on underside of leaves. Severely infected leaves may wither and die early. Alternate hosts: hemlocks or junipers. *Control:* Same as for Leaf Spots (above). Apply zineb, ferbam, maneb, or dichlone.
6. *Root-knot*—Deutzia is commonly attacked. See (37) Root-knot under General Diseases.
7. *Chlorosis* (primarily hydrangea and mockorange)—Leaves yellowish, especially between the veins. Plants stunted; flower color is poor. Common in alkaline (lime-rich) soils, or soils high in nitrates and calcium. *Control:* Have soil tested. Grow plants in slightly acid soil. Add an acid fertilizer such as ammonium sulfate. Water plants with weak solution of ferrous (iron) or magnesium sulfate or mixture of the two. Or spray with an iron chelate (page 27). Follow manufacturer's directions.
8. *Dieback, Twig Canker* (hydrangea, mockorange)—Round to irregular, rough, discolored cankers on stems. Twigs and branches die back. *Control:* Cut and burn infected parts. Spray as for Leaf Spots (above).
9. *Stem or Crown Rot, Southern Blight* (coralbells, heuchera, hydrangea)—Primarily

an indoor problem except in southern states. Stem rots, wilts, and collapses from rot at soil line that may be covered with cottony mold. See under Cutting Rot (below).

10. *Cutting Rot, Damping-off*—Base of cuttings decay; fail to root. Seedlings wilt and collapse. *Control:* Grow in sterile rooting medium. Avoid overwatering and plunging cuttings too deeply. Destroy infected cuttings.
11. *Sunscald* (primarily hydrangea)—Exposed, tender leaves are "scorched" by hot sun, temperatures over 100° F., excessive wind, or toxic sprays. Leaf edges turn brown. *Control:* Keep plants out of direct sun and dry winds on hot days. Wrap trunks of young, recently transplanted trees. See under Apple and Elm. Varieties differ in resistance.
12. *Ringspots* (hydrangea)—Symptoms variable, depending on variety and virus or virus strain. Leaves often dull and yellowish with irregular, dark green to yellowish blotches; yellow, green, or brown rings; and oakleaf patterns. Plants may be dwarfed with stunted leaves. Leaves often irregular in shape, narrow, sometimes stiff and brittle. Flowers usually stunted. May open irregularly, with green and colored flowers in same cluster. *Control:* Destroy infected plants. Take cuttings *only* from healthy plants. Virus-free hydrangea tip cuttings can be obtained by heat-treating affected plants.
13. *Verticillium Wilt* (deutzia)—See (15B) Verticillium Wilt under General Diseases.
14. *Wood Rot*—See under Birch, and (23) Wood Rot under General Diseases.
15. *Sooty Mold or Blotch* (mockorange)—See (12) Sooty Mold under General Diseases.
16. *Leaf and Stem Nematodes* (coralbells, heuchera, hydrangea)—See (20) Leaf Nematode under General Diseases. Stems of greenhouse hydrangeas are swollen and split. Leaves drop off. *Control:* Keep water off foliage. Cutting potted plants back severely and repotting after removing infested parts, taking tip cuttings from healthy plants, and potting in sterilized soil usually cleans up an infestation.
17. *Leaf and Stem Smut* (coralbells, heuchera)—See (11) Smut under General Diseases.
18. *Root Rots*—See under Geranium, and (34) Root Rot under General Diseases. May be associated with root-feeding nematodes (e.g., aphelenchus, dagger, lesion, pin, ring, spiral, stem, stunt). Grow in full sun in well-drained soil high in organic matter.
19. *Leafy Gall* (coralbells, heuchera, mockorange)—See (28) Leafy Gall under General Diseases.

HYLOCEREUS — See Cacti
HYMENOCALLIS — See Daffodil
HYPERICUM — See St.-Johns-wort
HYSSOP *(Hyssopus)* — See Salvia
IBERIS — See Cabbage
ICEPLANT, FIGMARIGOLD *(Mesembryanthemum, Cryophytum)*

1. *Root-knot*—See (37) Root-knot under General Diseases.
2. *Sooty Mold*—Black moldy patches on foliage. *Control:* Apply malathion to control scales, mealybugs, and other insects.

ILEX — See Holly
ILLICIUM — See Magnolia
IMPATIENS — See Balsam
INCENSE-CEDAR — See Juniper
INDIAN CHERRY — See Buckthorn

INDIAN-CUP — See Chrysanthemum

INDIANCURRANT — See Snowberry

INDIAN FIG — See Cacti

INDIAN PAINTBRUSH — See Snapdragon

INDIAN SHOT — See Canna

INDIAN STRAWBERRY — See Rose

INDIAN-TOBACCO — See Lobelia

INDIA RUBBER-PLANT or TREE — See Fig

INDIGO *(Indigofera)*, INDIGOBUSH —See False-indigo

INDOOR CLOVER — See Cockscomb

INKBERRY — See Holly

INULA — See Chrysanthemum

IPOMOEA — See Morning-glory and Sweetpotato

IRESINE — See Cockscomb

IRIS [COPPER, CRESTED, CUBESEED, DANFORD, DANISH, DIXIE, DUTCH, DWARF or VERNAL, DWARF LAKE, ENGLISH, JAPANESE or KAEMPFERI, LAMANCE, LOUISIANA, NETTED, RETICULATED, ROOF, SIBERIAN, SPANISH, SPURIA, TALL BEARDED or GERMAN, VESPER, YELLOW, ZUA], BLUE FLAG *(Iris)*; BABIANA; BLACKBERRY-LILY *(Belamcanda)*; IXIA; BLUE-EYED GRASS, GOLDEN-EYED-GRASS, SPRING BELL *(Sisyrinchium)*; WANDFLOWER *(Sparaxis)*; STREPTANTHERA

1. *Crown, Rhizome, Bulb or Corm Rots, Root Rots*—General and serious, especially on iris. Young fans may fail to grow in spring. Leaves yellow, wither, and die. Leaves suddenly wilt and collapse or die back gradually from tips. Rhizome or bulb, crown, and leaf bases may be dark green to brown, slimy or mushy and foul-smelling *(Bacterial Soft Rot)*. Or covered with cottony, dark gray to purplish or bluish-green mold, shriveled, dried, and rotted (fungus rots). Roots may decay; be few or none. Newer *iris* hybrids are more susceptible than older varieties. See Figure 2.39C and Figure 2.45B under General Diseases, as well as Figure 4.116. *Control:* Plant firm, disease-free stock, shallow in clean or sterilized soil (pages 538–548). Just before planting, soak *dormant* rhizomes or bulbs for 10 minutes in a 1:1,000 solution of mercuric chloride (see page 101 for precautions), Semesan Bel S solution (2 tablespoons per gallon), a 1:200 solution of hot water and formalin (see under Bulb and Stem Nematode below), or a phenyl mercury solution for 30 minutes. Follow manufacturer's directions. Space plants and grow in very well-drained soil in a sunny spot where these plants have not grown for several years. Avoid wounding flower stalks and leaves, overwatering, and a wet mulch. Keep down weeds. Carefully collect and burn *all* flower stalks and leaves and other debris from flower beds in the fall. Carefully dig and divide clumps every 2 to 4 years. Dry rhizomes or bulbs thoroughly in the sun for several days after digging. If rot strikes, dig up and burn seriously infected plants. Cut out rotted areas of slightly infected rhizomes. Drench soil around infected plants with a 1:2,000 solution of mercuric chloride (1 pint per square foot) or use Semesan Bel S (1 ounce in 3 gallons of water). Phenyl mercury may be used following manufacturer's directions. Repeat drench treatment 10 days later. Control insects, especially iris borers. Spray or dust with DDT or sevin plus malathion, at least weekly, starting when new leaves are 2 to 4 inches tall. Continue through bloom period. May combine with Leaf Spot sprays (see below). Certain fungus rots (e.g., *Sclerotium* and *Sclerotinia*) are controlled by mixing—soil drench or dust—PCNB into top 2 to 4 inches of soil before planting. Follow manufacturer's directions.

2. *Leaf Spots, Blotch*—General and serious following wet weather. Small, grayish-brown to brown spots with water-soaked, yellowish, red, or dark brown margins. Spots often enlarge and merge causing leaves to yellow and die early from tip down. Spots may occur on stems and flower buds. Plants unsightly, gradually weakened. See Figure 2.17A under General Diseases. *Control:* Same cultural practices as for Crown Rots (above). Remove and burn heavily spotted leaves as they occur. Collect and burn all dead leaves in late fall and again in early spring. Spray four to six times, 7 to 10 days apart. Start when first growth is noted in spring. Use zineb, maneb, folpet, captan, Daconil 2787, or phenyl mercury plus spreader-sticker (page 118). Varieties and species differ greatly in susceptibility, especially iris. *Iris sibirica,* the parent species of many forms and hybrids, is remarkably resistant to Didymellina *(Heterosporium)* Leaf Spot. Avoid sprinkling foliage when watering.
3. *Mosaics, Stripe*—Most severe on bulbous iris and certain rhizomatous species. *Iris germanica* and its hybrids are only slightly stunted. General on dwarf and English iris and certain tall, bearded types. Also attacks other plants listed. Symptoms variable. Even masked in some varieties, especially in hot weather. Leaves and bud sheaths mottled light and dark green, or show light, yellowish-green streaks. Flowers often mottled and streaked or fail to open. Plants and flowers may be stunted. Wedgewood *iris* is very susceptible. See Figure 4.117. *Control:* Dig and burn infected plants when first found. Plant disease-free stock. Control aphids, that transmit viruses. Use malathion or lindane. May combine with Leaf Spot sprays (above).
4. *Scorch, Red Fire* (primarily tall and dwarf bearded iris)—Central leaves turn bright, reddish-brown at tips in early spring. The whole fan of leaves is soon "scorched" and withered. Roots are soft, dead, reddish, hollow, and later disintegrate. Rhizome becomes reddish but remains firm. Plants often die. Nematodes often found in roots of affected plants. *Control:* Dig and burn infected plants. Same as for Crown Rots (above). Grow clean stock in soil fumigated with D-D, EDB, or chloropicrin (pages 538–548).
5. *Bacterial Leaf Spot or Blight* (belamcanda, iris)—Irregular, dark green, water-soaked spots and elongated streaks on leaves and flower stem. Infected areas may enlarge, merge, turn yellowish-green, and finally brown. In wet weather, spots may be covered with sticky droplets that dry to a thin gluelike film. Leaves and flower stem may collapse in continued wet weather. See Figure 4.118A. *Control:* Same as for Leaf Spots (above), except that spraying is ineffective. Avoid damp, shaded locations and crowding plants.
6. *Rusts*—More common on wild iris and cultivated varieties of *Iris germanica* and related types. Small, orange-brown, reddish-brown, dark brown or black, powdery pustules on leaves and stems. Often surrounded by yellowish border. (Figure 4.118B). Leaves of certain varieties and species may wither and die early. *Control:* Collect and burn tops in fall. Most *iris* varieties are highly resistant or immune. Keep down weeds. If severe, spray as for Leaf Spots (above). Use zineb, dichlone, maneb, or ferbam.
7. *Blossom Blight, Gray-mold Blight* (iris)—Blossoms spotted or blighted in wet weather. *Control:* Spray 2 or 3 times, a week apart. Start as flowers begin to open. Use zineb or captan (1 tablespoon per gallon), Daconil 2787, or Botran. Pick and destroy spotted flowers when first found.
8. *Bulb and Stem Nematode*—Most serious on bulbous iris. Yellow spots or streaks on stem and sheath. Base of stem under outer coating may turn gray, brown, or lead-colored and streaked. Discolored areas appear as rings when bulbs are cut through. Infested plants may be stunted and dry up prematurely. Roots discolored, decayed, or lacking. Bulbs decay. See Figure 4.119. *Control:* Grow nematode-free stock in

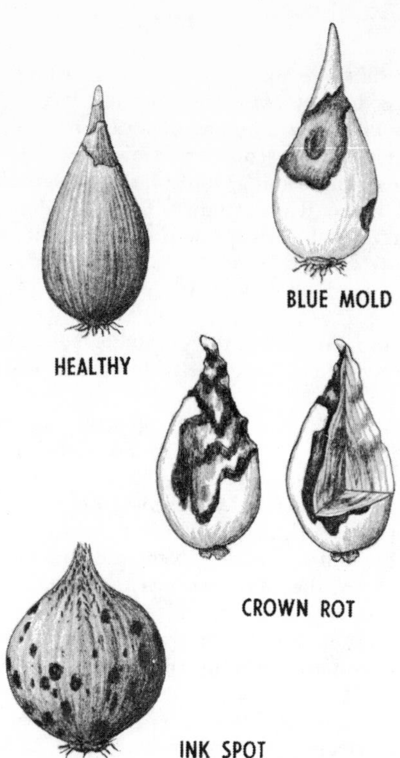

Fig. 4.117. Iris mosaic.

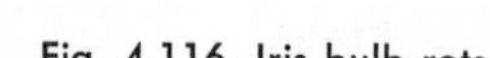

Fig. 4.116. Iris bulb rots.

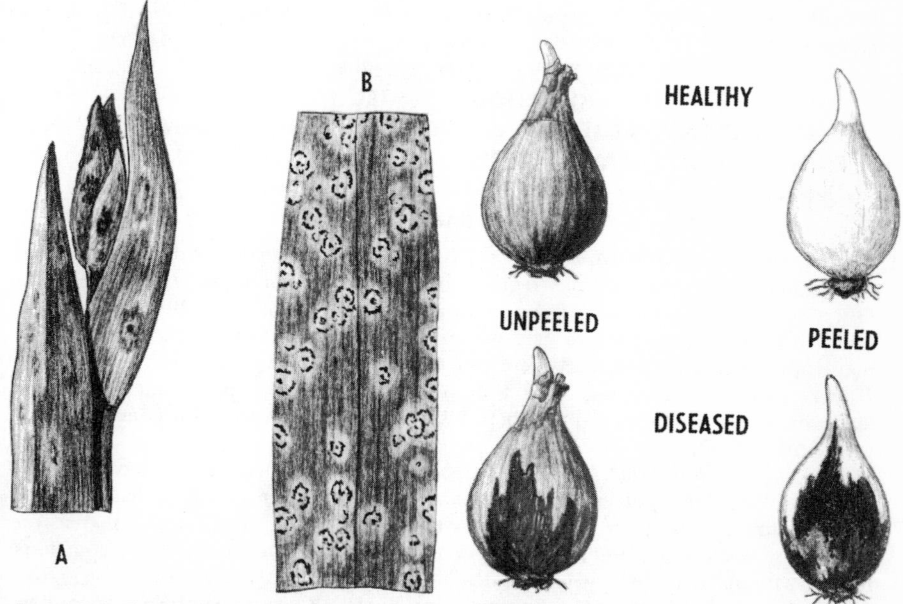

Fig. 4.118. A. Bacterial leaf spot of iris. B. Iris rust.

Fig. 4.119. Iris bulb and stem nematode.

clean soil. If suspicious, growers presoak *dormant iris bulbs* for 2 hours in water at room temperature. Then bulbs are soaked in hot formalin solution 1:200 (1 teaspoonful of 40 per cent commercial formaldehyde in 1 quart of water; or 1 pint in 25 gallons) at 110° F. for 3 hours. Dry and plant as soon as possible.

9. *Other Root-feeding Nematodes* (aphelenchus, lance, lesion, root plate, pin, spiral) —Plant dwarfed. May die from rotting of roots. Root system often matted and "tufted." Younger, newer roots have small, reddish-brown spots. Many roots rot off. Rot organisms and blue mold often later destroy infested bulb. *Control:* Grow nematode-free stock in soil pasteurized by heat or chemicals (pages 538–548) Destroy badly infested plants.
10. *Root-knot*—Plants may be dwarfed and weak. Leaves pale, die gradually, from tips down. Knobby, beadlike galls or knots on roots. *Control:* Same as for Other Root-feeding Nematodes (above).
11. *Ink Disease, Leaf Blight* (primarily bulbous iris: Dutch, English, Spanish and *Iris reticulata*) —Irregular, ink-black streaks and patches, often ringlike, on bulb scales. Blotches enlarge to cover whole scale. Bulbs may blacken, shrivel, and become hard. Irregular, black, sooty blotches on leaves that later turn reddish-brown. Leaves turn yellow, wither from tip down (Figure 4.116). *Control:* Dig and burn infected bulbs and leaves. Treat lightly infected bulbs as for Bulb and Stem Nematode (above). Spray as for Leaf Spots (above). Where serious, dig bulbs each year. Replant in new location.
12. *Blindness, Blasting* (forced iris) —Flowers do not develop or flower buds do not open after forming. May be caused by rotting of underground parts, late digging, low curing or precooling temperature, small bulb size, and other factors.
13. *Chlorosis*—Mineral deficiency in alkaline or very acid soils. See under Rose.
14. *Leaf Streak, Stunt* (iris) —Tomato ringspot virus. See (17) Spotted Wilt under General Diseases.
15. *Topple* (Dutch iris) —Flower stalk collapses from water-soaked rot. *Control:* See under Tulip.

For additional information, read USDA H & G Bulletin 66, *Growing Iris in the Home Garden.*

IRONBARK — See Eucalyptus

IRONWOOD — See Birch

IVY [ALBANY, ALGERIAN or CANARY, BALTIC, CLUSTERED, ENGLISH, HEART-LEAVED, IRISH, etc. many horticultural forms] (*Hedera*). IVY — See Grape for Boston, Engelmann, Grape-, and Marine-ivy. See Snapdragon for Kenilworth ivy.

1. *Fungus Leaf Spots, Stem Spots, Twig Blight, Spot Anthracnose or Scab, Anthracnose*—Small to large, round to irregular spots of various colors on leaves and stems. Often with conspicuous margins or concentric rings. Stems may be blighted, wither, and die back. Infected leaves may wither and fall early. See Figure 4.120. Plants may appear ragged. *Control:* Where practical, pick and burn infected plant parts. If severe, spray weekly using zineb, thiram, maneb, or fixed copper. Start when new leaf growth begins. Do not grow where high temperatures and moist conditions are prevalent.
2. *Bacterial Leaf Spot, Stem Canker*—Common indoor problem, where moist. Small, round to irregular, light green, water-soaked spots on lower or inside leaves. Spots enlarge, turn brown to brownish-black, with reddish-purple borders. Black areas on petioles or stems may cause lengthwise cracking, girdling, and withering of portions beyond. Plants may be dwarfed, with unthrifty, yellowish-green foliage (Figure 4.120). *Control:* Same as for Fungus Leaf Spots (above). Apply fixed copper, bordeaux, or streptomycin sprays. Indoors, keep water off foliage. Space plants.

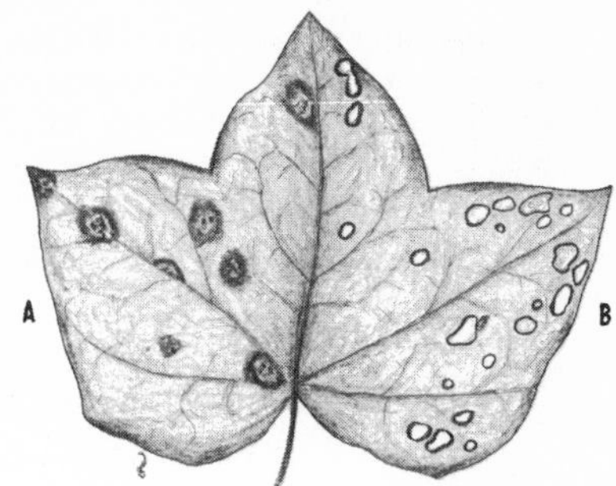

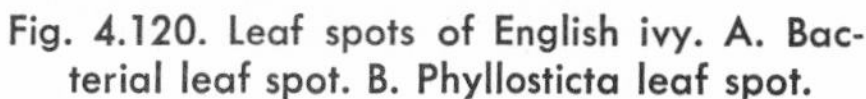
Fig. 4.120. Leaf spots of English ivy. A. Bacterial leaf spot. B. Phyllosticta leaf spot.

Avoid high humidity and high temperatures.

3. *Winter Injury, Sunscald*—Leaf tips and edges are scorched and browned in early spring. Whole stems may die back. *Control:* Follow best local cultural practices. Check with your extension horticulturist or nurseryman. Grow adapted varieties and species in protected location. Baltic *ivy* is less susceptible than English.
4. *Sooty Mold*—Common on ground cover under trees. Black mold grows on insect honeydew dropped from aphids feeding in trees above. See (12) Sooty Mold under General Diseases.
5. *Root Rots*—See (34) Root Rot under General Diseases. Often associated with root-feeding nematodes (e.g., lance, spiral, stem, stubby-root, stunt).
6. *Powdery Mildew*—See (7) Powdery Mildew under General Diseases.

IVY-ARUM — See Calla Lily
IXIA — See Iris
IXORA — See Buttonbush
JACARANDA — See Catalpa
JACKBEAN — See Bean
JACK-IN-THE-PULPIT — See Calla Lily
JACOBS-LADDER — See Phlox
JACQUEMONTIA — See Morning-glory
JADE PLANT — See Sedum
JAPANESE LAWNGRASS — See Lawngrass
JAPANESE PAGODATREE — See Honeylocust
JAPANESE PLUM-YEW *(Cephalotaxus)*

1. *Twig Blight*—See under Juniper.

JAPANESE SPURGE — See Pachysandra
JASMINE [ARABIAN, BEES, COMMON, DOWNY, ITALIAN, PRIMROSE, ROSY, SHOWY, SPANISH, WINTER] *(Jasminum)*

1. *Leaf Spots, Spot Anthracnose or Scab*—Leaves spotted; may fall early. *Control:* Pick and burn infected leaves. If serious, spray several times, 10 to 14 days apart. Use zineb or maneb.
2. *Root Rots*—See under Apple, and (34) Root Rot under General Diseases. May be associated with root-feeding nematodes (e.g., burrowing).
3. *Blossom Blight*—See (31) Blossom Blight under General Diseases.
4. *Crown Gall*—See (30) Crown Gall under General Diseases.
5. *Root-knot*—See under Peach, and (37) Root-knot under General Diseases.
6. *Crown Rot, Southern Blight*—See (21) Crown Rot under General Diseases.

7. *Rust*—Uncommon. Leaves, stem, flowers, and fruit deformed with raised, powdery pustules that later turn brown. *Control:* Same as for Leaf Spots (above).
8. *Variegation, Infectious Chlorosis*—See under Hollyhock.
9. *Stem Gall*—Small galls occur on stems. *Control:* Cut out and burn affected parts. Spray as for Leaf Spots (above).
10. *Botrytis Shoot Blight*—See (5) Botrytis Blight under General Diseases.
11. *Algal Leaf Spot*—Far south where damp. See under Magnolia.

JERUSALEM-ARTICHOKE — See Chrysanthemum
JERUSALEM-CHERRY — See Tomato
JERUSALEM-CROSS — See Carnation
JERUSALEM-THORN — See Honeylocust
JESSAMINE — See Butterflybush

JETBEAD or WHITE KERRIA *(Rhodotypos)*

1. *Leaf Spot, Anthracnose*—Leaves spotted in rainy seasons. *Control:* Pick and burn infected leaves. If serious, spray several times, 10 to 14 days apart. Use zineb, ferbam, or captan.
2. *Twig Blight, Coral Spot*—Twigs die back. Cankers on affected parts may be covered with bright, coral-colored pustules. *Control:* Cut and burn blighted twigs. Fertilize and water to maintain good vigor.

JEWELBERRY — See Lantana
JOE-PYE-WEED — See Chrysanthemum
JOSEPHSCOAT — See Cockscomb
JOSHUA-TREE — See Yucca
JUBAEA — See Palms
JUDAS-TREE — See Honeylocust
JUGLANS — See Walnut
JUNEBERRY — See Apple
JUNGLEFLAME — See Buttonbush

JUNIPER [ALLIGATOR, ANDORRA, CHINESE (COLUMNAR, GLOBE, HETZ BLUE, PYRAMIDAL), COMMON (DWARF SWEDISH, SWEDISH), CREEPING, FOUNTAIN, GREEK, HILL, IRISH, JAPANESE, JAPANESE SHORE, MOUNTAIN, MEYER'S, NEEDLE, ONESEED, PFITZER, PROSTRATE, ROCKY MOUNTAIN, SARGENT, SAVIN, SHORE, TAM, TAMARIX, UTAH, VONEHRON, WAUKEGAN, WEST INDIES, etc.], CREEPING CEDAR, RED CEDAR [CANAERT, CREEPING, EASTERN, FOUNTAIN, GLOBE, GOLDTIP, HILL'S PYRAMIDAL, KETELEER, KOSTER, SCHOTT, SILVER, SOUTHERN, WEEPING, WESTERN] *(Juniperus)*; WHITE-CEDAR [ATLANTIC, SOUTHERN], LAWSON CYPRESS or PORT ORFORD CEDAR (many horticultural forms), YELLOW-CEDAR [ALASKA or NOOTKA], HINOKI CYPRESS, SAWARA-CYPRESS, YELLOW CYPRESS *(Chamaecyparis or Retinospora)*; CRYPTOMERIA; CYPRESS [ARIZONA, COLUMNAR ITALIAN, ITALIAN, MACNAB, MONTEREY, MOURNING, PORTUGUESE] *(Cupressus)*; INCENSE-CEDAR *(Libocedrus)*; ARBORVITAE [AMERICAN (many horticultural forms), EASTERN or NORTHERN WHITE-CEDAR (numerous horticultural forms), BLACK AMERICAN, GIANT (DARK GREEN, COLUMNAR), GLOBE, GOLDEN, GOLDEN BIOTA, JAPANESE, KOREAN, ORIENTAL (numerous horticultural forms), PYRAMIDAL, SILVER] *(Thuja)*; HIBA ARBORVITAE, DWARF HIBA ARBORVITAE *(Thujopsis)*

1. *Rusts, Gall Witches'-broom* (primarily chamaecyparis, incense-cedar, juniper, redcedar) —General. Greenish-brown to reddish-brown, corky, round to irregular galls (up to 2 inches in diameter), elongated swellings, or witches'-brooms on leaves and small branches. Tips of branches may die back. Elongated, rough, dark-colored, swollen cankers, burls, or spindle-shaped swellings may develop on larger branches and trunk. Masses of bright orange to brown, jelly-like tendrils (spore horns) form in wet spring weather. Bright orange, fading to pale yellow, spore masses form on *Chamaecyparis* leaves. Seedlings may be severely stunted. See Figure 2.24D under General Diseases. A large number of alternate hosts, some of which are listed under Apple. Other hosts include bayberry, sweetfern, and waxmyrtle. *Control:* Destroy worthless, upright junipers or alternate hosts. Grow rust-resistant junipers. Horticultural forms and varieties of the following *Juniperus* species are resistant to cedar-apple and cedar-hawthorn rusts: *J. chinensis, J. communis, J. conferta, J. davurica, J. excelsa f. stricta, J. formosana, J. glauca, J. glaucescens, J. hetzii, J. horizontalis, J. japonica nana, J. macrocarpa, J. monosperma, J. procumbens, J. rigida, J. sabina, J. squamata, J. virginalis,* and *J. virginiana.* In general, upright junipers are susceptible while horizontal types are resistant. Check with your nurseryman or extension horticulturist. If practical, handpick galls or prune out witches'-brooms by late winter or spray monthly, May to September. Use Acti-dione, zineb, thiram, Niacide, or ferbam. Spray with Acti-dione, Elgetol, or bordeaux when tendrils 1/16 inch out. Follow manufacturer's directions.
2. *Twig Blights, Nursery or Juniper Blight (Phomopsis), Needle Blight, Dieback*—Widespread and serious. Needles, twigs, and smaller branches turn light brown to reddish-brown; gradually die back from tips (Figure 4.121). Old and new leaves and twigs may be involved. Serious on seedlings and young trees in wet seasons. Entire branches or even trees may die. Tiny, brown to black dots usually appear later on infected parts. Often confused with normal fall browning (see below) of inner leaves, winter injury, magnesium deficiency, spider mite damage, etc.

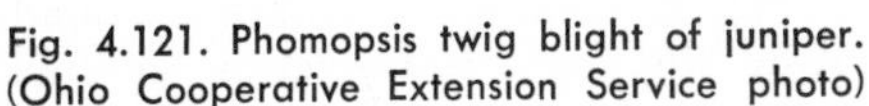

Fig. 4.121. Phomopsis twig blight of juniper. (Ohio Cooperative Extension Service photo)

Control: Keep plants growing steadily. Prune and burn blighted parts when plants are dry. Destroy infected plants in nursery. Avoid wounding when transplanting or cultivating. Apply phenyl mercury, fixed copper, Kromad, zineb, captan, maneb, Polyram, or Acti-dione, plus spreader-sticker (page 118) at about 1- to 2-week intervals during spring and fall wet seasons. Follow manufacturer's directions. Space plants for good air circulation. Avoid overhead sprinkling in nursery. *Junipers* somewhat resistant to Phomopsis Blight: Spiny Greek, Hill, and Keteleer redcedar. The Dwarf Greenspike *arborvitae* is fairly resistant to Phomopsis Blight. American and European species of *arborvitae* seem to be immune to Coryneum or Berckman's Blight.

3. *Leaf Spot or Blights, Seedling Blight, Nursery Blights, Needle Cast* (primarily American and giant arborvitae and western redcedar) —Widespread and damaging, especially on seedlings and young trees in wet years. Small, round to irregular, olive-brown to black cushions or gray mold *(Botrytis)* form on leaves in late spring. Infected leaves appear scorched. Later, leaves and smaller twigs drop. Branches are left bare. Most common on bottom ⅔ of young trees. Seedlings may be killed. *Control:* Same as for Twig Blights (above). Resistant *arborvitae* varieties may be available soon.
4. *Winter Injury, Winter Drying*—General. Last year's foliage or older shoot growth is scorched, turns brown and dies back from tips and margins. Injury evident in late winter and spring. If serious, branches or entire trees may be killed (Figure 2.10). Heavy coatings of ice and snow may seriously injure or kill if allowed to remain on trees and shrubs. *Control:* Water plants deeply late in fall, and during summer dry periods. Mulch for winter to conserve moisture, prevent deep freezing plus alternate thawing and freezing of soil. Grow in protected location from drying winter winds and sun or erect protective screens (burlap, canvas, slats, etc.). Control mites with malathion, Kelthane, or other sprays. Grow chamaecyparis only where adapted. Fertilize in spring to stimulate growth. Remove dead wood after growth starts in spring.
5. *Twig and Branch Cankers*—Serious on Monterey and Columnar Italian cypress in California. Discolored, slightly sunken, oval to elongated cankers on twigs and branches that gradually girdle and kill portions beyond. Branches or entire trees slowly drop their yellowed and browned leaves and finally die. *Control:* Prune off affected parts at least 6 inches below cankers. Protect pruning cuts with tree wound dressing (page 36). Disinfect pruning tools by dipping in formaldehyde solution (1:18 dilution with water). Keep trees growing vigorously. Spray as for Twig Blights (above). Remove and burn severely diseased trees.
6. *Natural Leaf-browning and Shedding*—Noticeable on arborvitae, certain junipers, and chamaecyparis. Older and inner leaves turn brown and fall in large numbers in early to late fall or spring. May occur quite suddenly in a week or two. *Control:* This is a normal plant reaction. May be accentuated by drought.
7. *Root Rots*—Foliage often wilts and fades to tan, yellow, or light brown. Branches or top dies back. Trees of all ages gradually decline and die from rot in trunk base and larger roots (Figure 4.122). Phytophthora is very serious on Port Orford cedar (Lawson cypress) and Hinoki cypress along Pacific Coast. *Control:* Set out healthy plants in soil known to be free of root-rotting fungi. Grow in well-drained soil free from surface drainage water. Nurserymen grow disease-free stock in new or sterilized soil (pages 538–548). Avoid large plantings of Port Orford cedar or Hinoki cypress as windbreaks or hedges. Remove and destroy affected plants, including roots. Grow junipers, arborvitae, and cypress varieties and species resistant to Cinnamon and Cypress root-rotting fungi (*Phytophthora*). Check with your state or extension plant pathologist.
8. *Wood and Heart Rots*—Cosmopolitan. See under Birch, and (23) Wood Rot under General Diseases.

Fig. 4.122. Phytophthora root rot of Lawson cypress. (Courtesy Dr. R. A. Young)

9. *Chlorosis*—Occurs in alkaline soils. Parts or all of the foliage turns yellow. May later turn brown and die. See under Maple.
10. *Sooty Mold, Black Mildews*—See (12) Sooty Mold under General Diseases.
11. *Crown Gall*—See under Apple, and (30) Crown Gall under General Diseases.
12. *Mistletoes* (cypress, incense-cedar, juniper, redcedar)—Widespread. See (39) Mistletoe under General Diseases. Serious on incense-cedar, causing spindle-shaped swellings in branches.
13. *Brown Felt Blight, Snow Blight* (arborvitae, incense-cedar, juniper)—See under Pine.
14. *Root-feeding Nematodes* (aphelenchoides, aphelenchus, dagger, lance, lesion, pin, ring, spiral, stubby-root, stunt, tylenchus)—Associated with stunted, weak trees in state of decline. See under Peach. To rid *redcedar* transplants of nematodes, nurserymen dip them in a hot water bath (104° F.) for 10 minutes, followed by a 2-minute dip in water at 125–126° F.
15. *Dog Injury*—Dead, oily foliage toward bottom of plants. Becomes dark brown in color. *Control:* Try one or more "dog-off" sprays.

JUPITERS-BEARD — See Valerian

KALANCHOE — See Sedum

KALE — See Cabbage

KALMIA — See Blueberry

KALOPANAX — See Acanthopanax

KANGAROO THORN — See Honeylocust

KANGAROO VINE — See Grape

KARANDA — See Oleander

KARO — See Pittosporum

KENILWORTH IVY — See Snapdragon

KENTIA — See Palms

KENTUCKY COFFEETREE — See Honeylocust

KERRIA

1. *Leaf and Twig Blight, Leaf Spots, Canker*—Widespread. Small, round to irregular, tan to reddish-brown spots with darker margin on leaves. If numerous, leaves turn yellow, wither, and drop early. Spots (cankers) on stems are round and reddish-brown to black. Spots may merge to form elongated cankers (Figure 4.123). Bark may split and twigs die back. Dwarf, variegated Kerria is very susceptible. *Control:* Prune and burn affected parts. Collect and burn fallen leaves. Spray or dust several times, 10 days apart. Start when leaves 1/4 inch long. Use zineb, ferbam, thiram, maneb, or fixed copper.
2. *Twig Blights, Canker*—Widespread. Oval to elongated or irregular, tan spots on stems. Centers of spots may be sprinkled with tiny black dots or bright, coral-red pustules. Twigs may die back. Leaves may become blighted. *Control:* Same as for Leaf and Twig Blight.
3. *Root Rot*—See (34) Root Rot under General Diseases.
4. *Fire Blight*—See under Apple.

KNIPHOFIA — See Redhot-pokerplant

KOCHIA — See Beet

KOELREUTERIA — See Goldenrain-tree

KOHLRABI — See Cabbage

KOLKWITZIA — See Viburnum

KUHNIA — See Chrysanthemum

KUMQUAT — See Citrus

LABRADOR-TEA, WILD ROSEMARY *(Ledum)*; LEUCOTHOË [COAST, DROOPING, KEISK'S, SWEETBELLS] *(Leucothoë)*; SANDMYRTLE or BOX SANDMYRTLE *(Leiophyllum)*

1. *Leaf Galls*—Gall-like growths and red spots on leaves. *Control:* Prune and burn infected parts when seen. Grow disease-free stock. If necessary, apply dormant spray *before* buds swell. Repeat during wet periods. Use ferbam, zineb, or maneb.
2. *Spot Anthracnose* (Labrador-tea, leucothoë)—Not serious. Grayish-white spots on leaves with reddish-brown borders, zoned with purplish ring. Infected areas appear on capsules, petioles, and branches. *Control:* Pick and burn infected parts. Zineb or maneb sprays should give good protection.
3. *Rusts* (Labrador-tea)—Causes little injury. Powdery pustules on leaves. Alternate hosts are spruces. *Control:* Not necessary. Spraying as for Spot Anthracnose and Leaf Galls (both above) should be beneficial.
4. *Leaf Spots, Tar Spot, Black Spot*—Spots of various sizes, shapes, and colors. Centers of spots may later be sprinkled with black dots (Figure 4.124). *Control:* Pick and burn infected leaves. Where serious, apply ferbam or zineb at 10- to 14-day intervals. Start when new leaves 1/4 inch out.
5. *Black Mildew* (leucothoë)—Gulf states. See (12) Sooty Mold under General Diseases.
6. *Felt Fungus* (leucothoë)—Southern states. See under Hackberry.
7. *Powdery Mildew* (Labrador-tea)—See (7) Powdery Mildew under General Diseases.

LABURNUM — See Goldenchain

LACHENALIA — See Tulip

LADIES'-TOBACCO — See Chrysanthemum

LADYS-SORREL — See Oxalis

LAELIA — See Orchids

LAGENARIA — See Cucumber

LAGERSTROEMIA — See Crapemyrtle

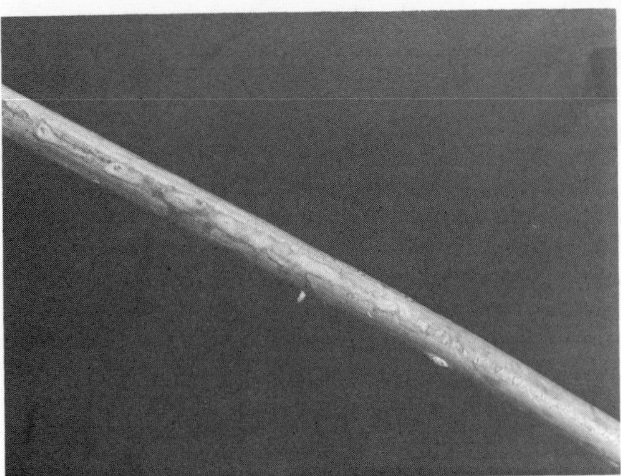

Fig. 4.123 (above). Phomopsis twig blight of Kerria. (Pennsylvania State University Extension Service photo)

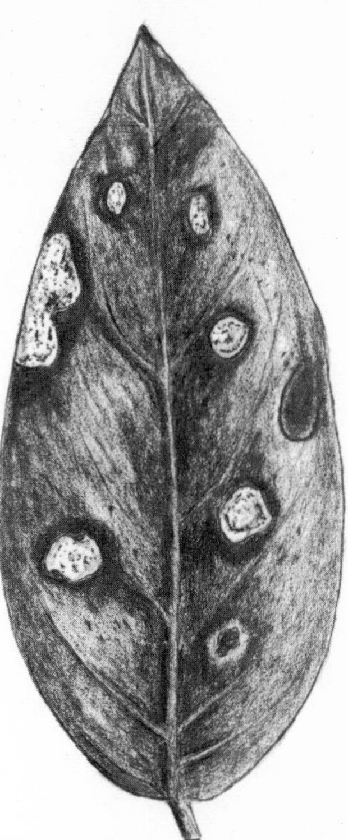

Fig. 4.124. Leucothoë leaf spots.

LAMBKILL — See Blueberry
LAMBS-EARS — See Salvia
LAMBSLETTUCE — See Valerian
LANTANA [COMMON, TRAILING or WEEPING] *(Lantana)*; BEAUTYBERRY [AMERICAN or FRENCH-MULBERRY, PURPLE], FRENCH-MULBERRY, JEWELBERRY [GIRALD'S or BODINIER BEAUTYBERRY, KOREAN, JAPANESE] *(Callicarpa)*; BLUEBEARD [COMMON, MONGOLIAN], BLUE SPIREA *(Caryopteris)*; BAG-FLOWER, CASHMERE BOUQUET, GLORYBOWER *(Clerodendron)*; DURANTA, GOLDEN-DEWDROP or TROPICAL LILAC *(Duranta)*; LEMON-VERBENA, FOGFRUIT, WHITEBRUSH *(Lippia)*; PHYLA; VERBENA [BLUE, CLUMP, GARDEN, ROSE, WHITE], VERVAIN *(Verbena)*; CHASTE-TREE [CUTLEAF, HARDY LILAC, LILAC or HEMPTREE] *(Vitex)*

1. *Leaf Spots, Anthracnose, Spot Anthracnose*—Small to large spots of various colors and shapes. *Control:* Pick and burn spotted leaves. Spray at 7- to 10-day intervals during wet periods. Use zineb, maneb, or captan.
2. *Root-knot*—Widespread. See (37) Root-knot under General Diseases. Severely infested plants are much more susceptible to winter-killing. *Control:* Apply DBCP (Nemagon, Fumazone) before planting and as a side-dressing during the growing season. Follow manufacturer's directions.
3. *Leaf Nematode* (lantana, verbena)—Brown blotches, first bounded by leaf veins. Leaves killed starting at the stem base. *Control:* See under Chrysanthemum, and (20) Leaf Nematode under General Diseases.

4. *Stem Rot, Southern Blight, Blight*—See under Geranium.
5. *Black Mold or Mildew*—See (12) Sooty Mold under General Diseases.
6. *Fusarium Wilt* (lantana)—See (15A) Fusarium Wilt under General Diseases.
7. *Powdery Mildew* (verbena)—General. Powdery, white mold on leaves and stems. *Control:* Space plants. Apply two sprays of sulfur or Karathane, 10 days apart.
8. *Rusts* (lantana, verbena)—Small, yellow to dark brown pustules on leaves. Alternate host may be wild grasses. *Control:* Same as for Leaf Spots (above).
9. *Bacterial Wilt* (verbena)—Leaves yellow, wilt, and die. Plants often die before blooming. Inside of stems is dark brown near soil line. *Control:* Set disease-free plants in clean or sterilized soil (pages 538–548).
10. *Flower Blight, Stem Blight* (verbena)—Flower petals and stem spotted, may rot. Affected areas covered with dense gray mold in damp weather. *Control:* Carefully pick and burn affected parts. Spray as for Leaf Spots (above).
11. *Spotted Wilt* (verbena)—See (17) Spotted Wilt under General Diseases.
12. *Downy Mildew* (verbena)—See (6) Downy Mildew under General Diseases.
13. *Gray Patch* (phyla)—California in hot weather. Ground cover plants dry and die in patches a foot or more across. *Control:* Try working PCNB dust into soil in affected areas or spot drench with PCNB 75. Follow manufacturer's directions.
14. *Mosaic* (lantana, verbena)—Light green and yellow mottling on leaves. Leaves may be somewhat crinkled and distorted. *Control:* Destroy infected plants.
15. *Dieback, Canker* (callicarpa)—See under Apple.
16. *Root Rots*—See (34) Root Rot under General Diseases. May be associated with nematodes (e.g., aphelenchoides, burrowing, root-knot).
17. *Crown Gall* (lippia)—See (30) Crown Gall under General Diseases.

LARCH [ALPINE, AMERICAN or TAMARACK, DAHURIAN, EASTERN, EUROPEAN, HYBRID, JAPANESE, WESTERN] *(Larix)*; GOLDENLARCH *(Pseudolarix)*

1. *Leaf Casts, Needle and Shoot Blights*—Primarily American, European, and western larches in wet seasons. Trees appear scorched at distance. Needles spotted, turn yellow, then reddish-brown. Needles and shoot tips die back. Needles usually drop early. Some may cling over winter. Infected needles often sprinkled with tiny white to black dots or gray mold *(Botrytis)*. *Control:* Gather and burn fallen needles in late autumn. If practical, spray ornamental larches several times, 2 weeks apart. Use zineb, captan, maneb, dichlone, fixed copper, bordeaux mixture, or lime-sulfur. Start when new growth commences. Keep trees growing vigorously.
2. *Needle Rusts*—Small, pale yellow to bright orange pustules on new needles. Needles yellow, may be distorted and fall early. Causes little damage. Alternate hosts include willows, poplars and aspens, birches, and alders. *Control:* Where practical, destroy nearby, worthless, alternate hosts. If serious, spray larches several times, 10 days apart. Use zineb or ferbam. Start just before apple trees bloom.
3. *Wood Rots*—Cosmopolitan. Mostly on older, neglected trees that lack vigor. See under Birch, and (23) Wood Rot under General Diseases. Maintain trees in good vigor. See under European Larch Canker (below).
4. *Twig Blight*—Tips of new growth develop dead, curled leaves during cold, wet weather. Gray mold may cover affected areas. *Control:* Prune and space plants for better air circulation.
5. *Root Rots*—Cosmopolitan. Trees decline in vigor. Foliage is thin and weak. Leaves may yellow, wither, and drop early. May be associated with nematodes (e.g., lesion, root-knot, stunt). *Control:* See under Apple, and (34) Root Rot under General Diseases.
6. *Frost Injury*—New, young needles suddenly turn brown. Shoot tips usually most seriously affected. Needles later shrivel; tend to *remain* on twigs for balance of

season. Twigs may die back. *Control:* Grow adapted species in a protected location. Avoid fertilizing or otherwise promoting growth in late autumn.

7. *European or Dasyscypha Larch Canker* (especially American and European larches, Golden-larch) —Northeastern states, but probably eradicated. Cankers on larger trees develop gradually from year to year. Appear as flattened, sunken, bowl-like, dead areas in bark on trunk or branches. Large amounts of resin may flow from cracked areas. Trees distorted, swollen, and weakened. Stems and branches of young or weakened trees are quickly girdled and killed. Japanese larch is relatively resistant. *Control:* Maintain trees in good vigor. Fertilize and water during droughts. Control insects. Grow in full sun, in moist, well-drained soil in areas relatively free of late frosts. Avoid dry locations (e.g., sandy hillsides) and bark injuries. Treat all wounds promptly with tree wound dressing (page 36). Cut down and burn all infected trees. If you suspect Dasyscypha Canker, contact your state or extension plant pathologist.
8. *Dwarf Mistletoes* (larch) —Northeastern and northwestern states. See (39) Mistletoe under General Diseases.
9. *Seedling Blight, Damping-off*—Seedlings discolored and stunted. Often wilt and collapse. Roots decay. *Control:* Treat seed with thiram, captan, or dichlone. Sow in clean, well-drained soil. Avoid overwatering. Keep down weeds. Where practical, sterilize soil before planting (pages 538–548).
10. *Chlorosis*—Common in alkaline soils. Foliage yellows. Twigs may die back. See under Pine.

LARKSPUR — See Delphinium

LATHYRUS — See Pea

LATUCA — See Lettuce

LAUREL, MOUNTAIN — See Blueberry

LAUREL, TRUE; SWEETBAY *(Laurus)*

1. *Leaf Spot, Anthracnose*—Small spots on leaves in which dark fungus fruiting bodies may develop. *Control:* Pick and burn spotted leaves.
2. *Thread Blight*—Southeastern states. See under Walnut.

LAUROCERASUS — See Peach

LAVATERA — See Hollyhock

LAVENDER *(Lavandula)* — See Salvia

LAVENDER QUEEN — See Snapdragon

LAWNGRASS: WHEATGRASS *(Agropyron)*; BENT, BENTGRASS, REDTOP *(Agrostis)*; CARPETGRASS *(Axonopus)*; BUFFALOGRASS *(Buchloë)*; BERMUDAGRASS *(Cynodon)*; CENTIPEDEGRASS *(Eremochloa)*; FESCUE, FESCUEGRASS *(Festuca)*; RYEGRASS *(Lolium)*; BAHIAGRASS *(Paspalum)*; BLUEGRASS *(Poa)*; ST. AUGUSTINEGRASS *(Stenotaphrum)*; ZOYSIA, ZOYSIAGRASS, JAPANESE LAWNGRASS, MANILAGRASS *(Zoysia)*

1. *Leaf Spots and Blights, Melting-out, Dying-out, Anthracnose*—General in wet weather from mid-spring to late fall. Purplish-black, dark reddish-brown, chocolate-brown, light gray, or tan spots on leaves, leaf sheaths, and stems. Spots may be round, oval, or oblong. With or without a prominent border. Spots may enlarge and develop light-colored centers. Older leaves or entire plants may turn yellow, then brown and die. Crowns, roots, and rhizomes turn reddish-brown to black and rot. See Figure 4.125. Infected areas may have a general brownish undercast. Turf is thin and weak or killed out in round to irregular spots that enlarge during summer. *Control:* Sow high quality, disease-free seed. Avoid heavy watering, frequent sprinkling, and overstimulation with fertilizer (especially with nitrogen)

Fig. 4.125. Leaf spot of Kentucky bluegrass. (Courtesy Upjohn Co.)

Fig. 4.126. Bluegrass powdery mildew. (Courtesy Upjohn Co.)

Fig. 4.127. Rust on Merion bluegrass. (Courtesy Upjohn Co.)

during summer months. If practical, collect clippings. Mow regularly at height recommended for your area. Grass varieties differ in resistance (e.g., Merion Kentucky *bluegrass* is resistant to leaf spots caused by *Helminthosporium vagans, H. dictyoides,* and *H. sativum*. Pennlawn *red fescue* is resistant to Helminthosporium Blight *(H. dictyoides)*. Newport and Park Kentucky *bluegrasses* are resistant to *H. sativum* and *H. dictyoides*. Suwannee *Bermudagrass* is highly resistant to Helminthosporium Leaf Spots. *Bluegrasses* also differ in resistance to Septoria Leaf Spot). Apply phenyl mercury (but not to Merion bluegrass), Panogen Turf Spray, Dyrene, difolatan, Thimer, Tersan OM, zineb, maneb, Dithane M-45 (or Fore), Daconil 2787, captan, folpet, Acti-dione-Thiram, or Ortho Lawn and Turf Fungicide at 1- or 2-week intervals during spring, wet weather. Continue through July. Follow manufacturer's directions. Mercury-containing fungicides frequently injure strictly southern grasses. Check with a lawn specialist. Soil drenches of zineb or captan (1 pound per 1,000 square feet), well watered in, are used by golf course superintendents to control Melting-out. Make several weekly applications starting in mid-spring. Apply chlordane ($2\frac{1}{2}$ pounds actual), as granules or spray per 1,000 square feet of lawn in early spring as grass starts to grow. *Water in thoroughly*. This treatment controls crabgrass, earthworms, grubs, cutworms, Cicada killers, webworms, ants, leafhoppers, chiggers, mites, and fleas. Remove excess mat and thatch in spring or fall. Use a "vertical mower" or "power rake." Provide good surface drainage when establishing a new turf area.

2. *Powdery Mildew*—General when nights are cool and days are warm. Mostly on bluegrasses, fescues, and wheatgrass. Flour-white, grayish-white to brown mold patches on leaves in shaded or poorly drained areas. Leaves may yellow and wither. Plants may be weakened and die. Most serious on new plantings. Merion bluegrass is very susceptible. See Figure 4.126. *Control:* If necessary, make several applications of sulfur, Karathane, or Acti-dione-Thiram mixture, a week apart. Water and fertilize to maintain vigor. Follow cultural practices as for Brown Patch (below).
3. *Rusts* (primarily bluegrasses, Bermudagrass, ryegrass, St. Augustinegrass and zoysia)—General. Yellow-orange, orange, reddish-brown, brownish-yellow or black, powdery pustules on leaves and leaf sheaths. If severe, leaves may yellow, wither, and die. Turf may be thinned, weakened, and more susceptible to drought, winter injury, and other diseases. See Figure 4.127. *Control:* Fertilize with nitrogen and water deeply so grass grows steadily during summer and early fall. Otherwise apply Acti-dione-Thiram mixture, zineb, maneb, Fore, Daconil 2787, dichlone, sulfur, or thiram several times 7 to 10 days apart. Remove clippings if possible. Grow rust-resistant grasses (e.g., Newport and Park Kentucky *bluegrasses* are more resistant to stem rust than the very susceptible Common, Merion, Delft, Delta, and Prato bluegrasses. Sunturf *Bermudagrass* is extremely susceptible to rust).
4. *Brown Patch, Rhizoctonia Disease*—Active in hot, humid weather when night temperatures above 70° F. Bentgrasses and St. Augustinegrass are more susceptible than coarser grasses. Roughly circular patches, up to 3 feet or more in diameter. Often with grayish-black margin (especially on bentgrass). Grass leaves first water-soaked, soon dry, wither, and turn light brown (white in St. Augustinegrass). Roots, rhizomes (stolons), and crowns may rot. Especially in southern states. Turf may thin out in large areas, especially on southern grasses. *Control:* Avoid overwatering and frequent sprinkling in late afternoon or evening. Prune dense trees and shrubs to increase air circulation and reduce shade. Avoid overfertilization during hot weather with quickly available, high-nitrogen fertilizer. Provide for good soil drainage when establishing new turf area. If possible, remove clippings after each mowing. Spray weekly in hot, humid weather. Use Calocure, Ortho Lawn and Turf Fungicide, Dyrene, Daconil 2787, PMA, Mercuram, Panogen Turf Spray, Dithane M-45 (Fore), difolatan, Tersan OM, Thimer, or Calocure-thiram mixture. Ortho Lawn Disease Control, PCNB, and Acti-dione RZ give good control on St. Augustinegrass. Follow manufacturer's directions.
5. *Sclerotinia Dollar Spot or Blight*—Active in warm (60° to 85° F.), moist weather. Round, brown, later bleached spots about size of a silver dollar, to somewhat larger. Spots may merge to form large, irregular, straw-colored, sunken areas (Figure 4.128). Small to large, straw-colored, girdling blotches with dark brown or red-brown borders girdle leaf blades. *Control:* Same cultural practices as for Brown Patch (above). Maintain adequate to high fertility. Follow recommended lawn-feeding program for your area. Collect clippings. Spray during spring, late summer, and fall using Tersan OM, Thimer, Cadtrete, Panogen Turf Spray, Daconil, PMA, Kromad, Caddy, Cadminate, Mercuram, Dyrene, Ortho Lawn and Turf Fungicide, Calocure, or Acti-dione-Thiram. Follow manufacturer's directions. Attacks practically *all* lawn grasses, but creeping bents, Sunturf Bermudagrass, and bahiagrass are most susceptible.

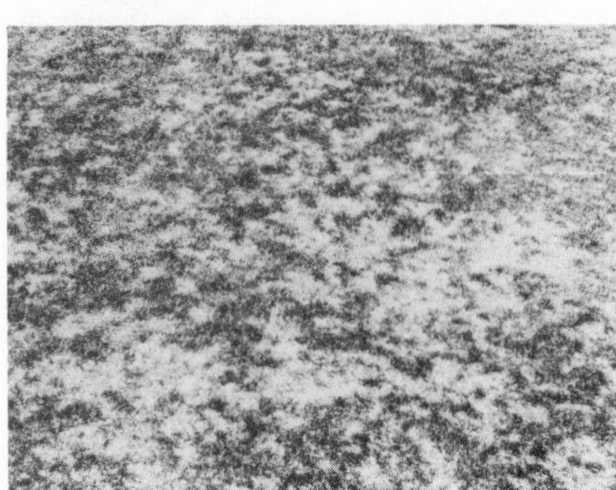

Fig. 4.128. Dollar spot. (Courtesy E. I. Du Pont de Nemours & Co., Inc.)

Fig. 4.129. Slime mold on Kentucky bluegrass. (Courtesy Upjohn Co.)

6. *Slime Molds*—General. Suddenly occur in warm weather following heavy watering or rains. Technically not a disease. Small, colorless, white, gray, or yellow slimy masses grow up and over grass surface in round to irregular patches, shading or discoloring otherwise healthy turf. Masses dry to form bluish-gray, gray, dirty-yellow, black, or white powdery growths that are easily rubbed off (Figure 4.129). *Control:* Mow, rake, or brush affected areas and wash down with water. Spray as for Leaf Spots or Rusts (both above) if desired.
7. *Fairy Rings, Mushrooms (or "Toadstools"), Puffballs*—Rings or arcs of dark green, fast-growing grass from several inches to 50 feet or more in diameter. Sometimes an inner ring of thin or dead grass occurs. Fairy rings expand outward at rate of a few inches to 2 feet or more per year. Toadstools, mushrooms, or puffballs may spring up around edge of ring following heavy watering or rains (Figures 4.130A and 4.130B). Dry patches in turf areas may actually be infested with the spawn of these fungi. Clumps of mushrooms (toadstools) may also spring up from buried organic matter such as pieces of construction lumber, logs, roots, or tree stumps. *Control:* Pump large amounts of water 10 to 24 inches deep using a root feeder. Pump water in the ring of dark green grass and the dead ring. Repeat if rings reappear. Or remove sod and sterilize soil beneath using a soil fumigant (pages 538–548) such as methyl bromide, chloropicrin, or formaldehyde. Keep lawn well fertilized so that rings do not show so distinctly. Before planting a new turf area, remove large roots, construction lumber, and other large pieces of organic matter.
8. *Typhula Blight, Snow Scald, Gray Snow Mold*—General in northern states. More or less circular, dead, bleached areas, up to about 2 feet in diameter. May be covered at first with fluffy, bluish-gray to almost black mold. Found in winter or early spring in shaded, wet areas where snow is slow to melt. Usually associated with melting snow. A crustlike mat may form where grass has been left long (Figure 4.131). *Control:* In northern states avoid fertilizing after about September 15. Before first heavy snow is forecast or before cold drizzly weather, apply Tersan OM, Dyrene, Thimer, Ortho Lawn and Turf Fungicide, Cadminate, Calocure, Calo-clor, Panogen Turf Spray, Polycide, thiram, or phenyl mercury. Follow manufacturer's directions. Repeat during midwinter thaw. Spray during growing season as for Leaf Spots, Brown Patch, and Dollar Spot (all above). Provide for good surface drainage when establishing a new turf area. Keep turf cut in fall.
9. *Fusarium Patch, Pink Snow Mold*—Small, round, bleached-brown patches, usually 1 to 4 inches in diameter. Spots sometimes enlarge to a foot or more across. May be

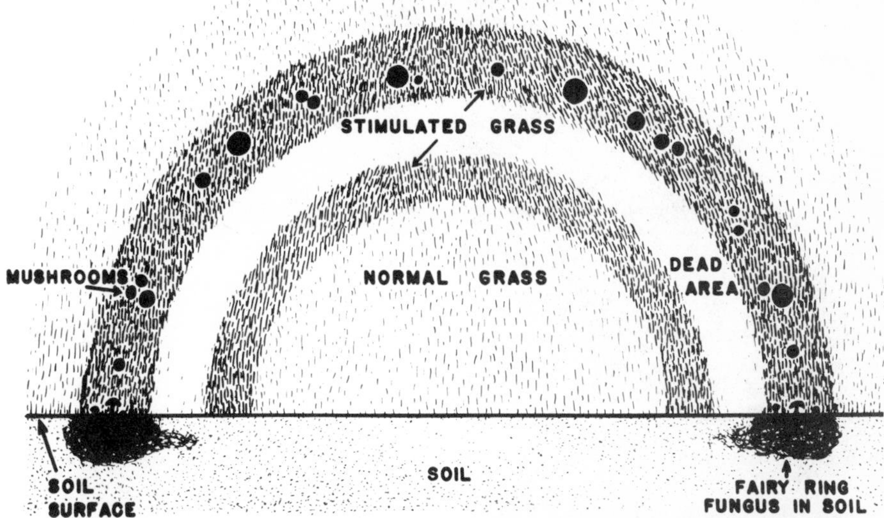

Fig. 4.130A (top). Mushrooms in a fairy ring. (Courtesy Dr. Ellery Knake)

Fig. 4.130B (center). Drawing of a fairy ring showing zones of stimulated and dead grass, mushrooms, and fairy ring fungus in soil.

Fig. 4.131. Snow mold. (Courtesy E. I. Du Pont de Nemours & Co., Inc.)

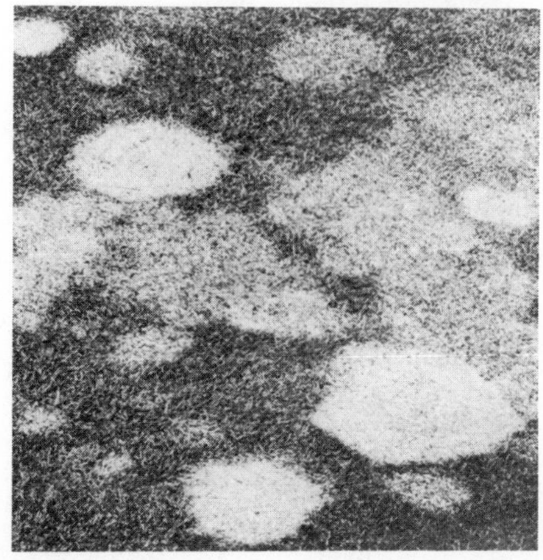

covered at first with white to pink mold. Usually associated with melting snow, but attacks may occur during prolonged, cool, drizzly weather (up to about 65° F.). *Control:* Same as for Typhula Blight or Snow Scald (above). Phenyl mercury and Calocure applied alternately at 2-week intervals during spring and autumn, and monthly during rest of the year in western Washington, give excellent control. Somewhat resistant *bentgrasses:* Cohansey, Penn-cross, and Pennlu. Liming increases severity of disease.

10. *Pythium Disease, Grease Spot, Spot Blight, Cottony Blight*—Widespread and destructive in hot, humid (75° to 110° F.) weather, on heavy, poorly drained soils. Round, reddish-brown spots, up to several inches in diameter, with blackened, greasy borders (Grease Spot or Spot Blight). Spots often merge to form streaks. Disease is spread by mowing and flowing water. Spots dry out and become bleached. Disease may spread rapidly, killing out turf. In southern states, small white spots that later become cottony are common (Figure 4.132), especially on overseeded areas during late fall, winter, and early spring (Cottony Blight). Roots often die back. *Control:* Follow cultural practices as for Brown Patch (above). Apply Dexon, Dexon-captan, or Dyrene-Dexon mixtures 1 to 3 times weekly during hot, humid periods. Follow manufacturer's directions. In southeastern states where Cottony Blight is severe, check with your extension or state plant pathologist for recommendations. Delay seeding until weather is cool and dry. Apply Dexon when planting ryegrass. Or use Dexon-captan mixture or Tersan OM. Repeat 1 week later.

Fig. 4.132. Close up of cottony blight (*Pythium*) on ryegrass. (Courtesy Dr. T. E. Freeman)

11. *Root-feeding Nematodes* (aphelenchus, awl, bud and leaf, burrowing, cyst, cystoid, dagger, false root-knot, lance, lesion, needle, pin, reniform, ring, root-knot, sheath, sheathoid, spiral, sting, stubby-root, stunt, tylenchus)—Turf declines; lacks vigor. Does not respond normally to water and fertilizer. Often appears stunted and yellowed with dead and dying areas. Injury easily confused with fertilizer burn, a soil deficiency, poor soil aeration, drought, compaction, insects, and other types of injury. Grass blades dying back from tips may be interspersed with apparently healthy leaves. Roots may be largely missing, swollen, stunted, bushy, "stubby," and dark in color. Turf may later thin out, wilt, and "melt out." Severity of symptoms varies with nematode population feeding on roots. Parasitic nematodes can be identified only by taking suspected turf plugs and subjecting them to analysis by a competent nematologist. *Control:* Keep grass vigorous. Water and fertilize as recommended for your area. If necessary, apply DBCP (Nemagon, Fumazone), Diazinon, Zinophos, Sarolex, or VC-13 Nemacide. Follow manufacturer's directions. Uganda *Bermudagrass* is resistant or immune to all root-knot species and subspecies. Tiflawn (Tifton 57) and Tifdwarf *Bermudagrasses* are resistant to several types of nematodes.

12. *Chlorosis, Yellowing*—Problem in moderately to highly acid or alkaline soils. Turf areas irregularly yellow-green or yellowed and stunted. If severe, plants may die. Caused by minor element deficiency, e.g., iron or magnesium. *Control:* Have soil tested for reaction (pH). Follow directions in report. Lime will be called for in acid soils; and sulfur, ammonium sulfate, or other acid-forming fertilizers in alkaline soil. Apply 2 to 3 ounces of ferrous (iron) sulfate, sold as Copperas or Sulfasoil, in 5 to 10 gallons of water per 1,000 square feet. Sprinkle in immediately. Repeat treatment as necessary to maintain good green turf color. Iron chelates (page 27) may be substituted. Follow manufacturer's directions. Some lawn fertilizers and fungicides (e.g., Kromad, Acti-dione Ferrated) have iron sulfate in the spray mixture. If magnesium is the problem, add magnesium sulfate (Epsom salts) at rate of ½ pound in 5 gallons of water per 1,000 square feet. Repeat as necessary to maintain color.
13. *Smuts*—General and increasing in importance. Long or short stripes in leaves that rupture and release dark brown or black powdery masses. Leaves may be shredded, wilted, and withered. Plants stunted and yellowed. Later die in patches 2 to 8 inches in diameter. *Control:* Where feasible, dig out and burn affected plants. Merion is more susceptible than most other bluegrasses. Clones of Kentucky *bluegrass* (e.g., Dwarf, Park, Pa. K5, etc.) are highly resistant or immune to flag smut *(Urocystis)*. Creeping bents apparently differ in reaction to stripe smut *(Ustilago)*. Pennlu, Penn-cross, Seaside, and Washington bents are susceptible. Chemically sterilizing heavily infested soil (pages 538–548) and reseeding, sprigging, or sodding to less susceptible varieties or species may be the most practical control.
14. *Gray Leaf Spot of St. Augustinegrass*—Destructive in warm to hot, humid, rainy periods. Small brown dots that later enlarge to form round to oblong, dirty-yellow, brown or ash-colored leaf spots and stem cankers with reddish-brown, purple, or water-soaked borders (Figure 4.133). Spots may be covered with grayish mold in humid weather. Turf is unsightly, "scorched," and more susceptible to drought and other damage. *Control:* Apply Ortho Lawn and Turf Fungicide, Thimer, Tersan

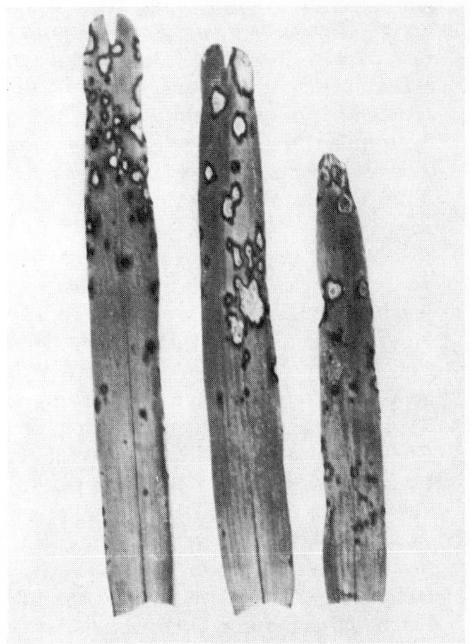

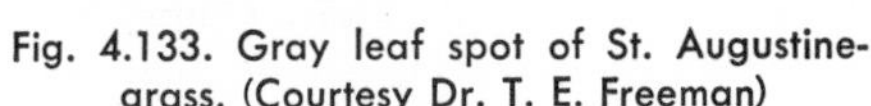

Fig. 4.133. Gray leaf spot of St. Augustinegrass. (Courtesy Dr. T. E. Freeman)

OM, captan, Daconil, Kromad, zineb, or thiram at 8- to 10-day intervals. Follow manufacturer's directions. Avoid heavy rates of nitrogen fertilizer. Yellow-green types of St. Augustinegrass have some resistance.

15. *Fusarium Blight*—Northeastern and central states, affecting bentgrasses, bluegrasses, and red fescue. Serious only when grass is weak and under stress, e.g., drought. Small, tan or straw-colored spots develop in lawns in early summer. Centers of green grass may remain within the diseased area giving characteristic "frogeye" or "doughnut" patterns. Spots later enlarge and merge. Large turf areas may turn brown. *Control:* Dithane M-45 (Fore), and Tersan OM, 4 ounces per 1,000 square feet in 5 gallons of water, have given good control. Other fungicides (see above) may work as well. Avoid high rates of nitrogen and low calcium. Keep turf well watered.
16. *Red Thread, Pink Patch*—Northern coastal regions, mostly on fescues, bents, bluegrasses, and ryegrass. Irregularly-shaped, pink patches of dead grass 2 to 6 inches or more in diameter develop during cool (65–75° F.), damp, weather in spring, winter, and fall. Usually only leaves are affected. If severe, patches turn brown and plants die. Characteristic bright coral-pink to red threads "bind" leaves together in moist weather. *Control:* Maintain adequate to high fertility. Same chemicals as for Sclerotinia Dollar Spot (above). Apply at about 2-week intervals. Collect clippings.
17. *Copper Spot* (bents, redtop)—Humid coastal regions in acid soils. Disease attacks occur during warm (65–85° F.), wet periods. More or less circular, coppery-red to orange spots, 1 to 3 inches in diameter. Spots may merge to form irregular, copper-colored areas. *Control:* Same as for Sclerotinia Dollar Spot (above).
18. *Algae, Green Scum*—Technically not a disease. Occurs in low, wet, shaded, or heavily tracked and compacted spots. Green to blackish scum forms on bare soil or thinned turf. Dries to form thin, black crust that later cracks and peels. *Control:* Same cultural practices as for Brown Patch (above). Provide for good soil drainage. Aerify. Maintain turf in vigorous condition. Where necessary, apply spray of copper sulfate, 2 ounces in 3 to 5 gallons of water to 1,000 square feet. Or use Dithane M-45 (Fore), 6 ounces per 1,000 square feet.
19. *Spring Dead Spot of Bermudagrass*—Fairly widespread in well-cared-for turf. Conspicuous, bleached, round dead spots present in spring. Spots vary in size from a few inches to 3 or more feet in diameter. Roots black and rotted. Sometimes center of spots may survive, resulting in "doughnuts." Spots usually remain dead for number of years. Such areas become invaded by weeds and other grasses. Disease is most severe on U-3 Bermudagrass although African, Common, Tiffine, and Tifgreen are also infected. May be confused with snow mold, winter, or insect injury. *Control:* Check with your extension plant pathologist. Spring-Bak (contains 85 per cent nabam) has proved effective.
20. *Seed Rot, Seedling Blight, Damping-off*—General. Seeds rot. Stand is thin in patches. Seedlings stunted, water-soaked and matted, then turn yellow to brown. May wilt and collapse. Surviving plants are weakened. Stand is slow to fill in. *Control:* Sow fresh, best quality seed in well-prepared, fertile seedbed. Provide for good soil drainage when establishing new lawn. Treat seed with captan or thiram. Avoid overwatering after planting. If possible, plant in late summer or early fall. After sowing, apply dilute spray of Kromad, captan, thiram, zineb, or folpet *plus* Dexon (4 ounces each in 3 to 5 gallons of water to cover 1,000 square feet). Avoid low spots in seedbed.
21. *Mosses*—Occur in lawn areas low in fertility with poor drainage, high soil acidity, too much shade, watered improperly, heavily compacted, or combination of these factors. *Control:* Remove by hand raking. Follow cultural practices as for Brown Patch (above) and Compacted Areas (below). Have soil test made and follow

instructions in report. Improve soil drainage and treat with ferrous ammonium sulfate (1 pound in 4 gallons of water per 100 square feet).

22. *Compacted Areas*—Thin turf or bare spots in heavily tracked areas. Waterlogged and heavy-textured soils become packed and bake hard if walked on constantly. *Control:* Aerify soil using hand aerifier. Aerating machines are sold or rented by garden supply stores, many nurserymen, and golf courses. If necessary, fertilize and reseed. Reduce foot traffic by putting area into a walk or patio or erect a fence.
23. *Dog Injury*—Urine injury may resemble brown patch, fairy ring, or snow mold. Affected areas usually roundish and up to a couple of feet or more in diameter. Often surrounded by a ring of lush green grass. Injured grass turns brown or straw-colored and usually dies. *Control:* Train dog.
24. *Insect Injury*—Numerous insects, including grubs, ants, sod webworms, mole crickets, wireworms, armyworms, billbugs, chinch bugs, leafhoppers, and others may damage turf. Injury may closely resemble one or more lawn diseases. *Control:* For information on lawn insects and their control, contact your county agent or extension entomologist. See under Leaf Spots (above). Read USDA H&G Bulletin 53, *Lawn Insects: How To Control Them.*
25. *Ophiobolus Patch*—Pacific Northwest or Northeast, near coast where cool and moist. Roundish, reddish-brown areas that turn brown to gray or straw-colored. From several inches to 2 feet or more in diameter. Spots may merge to kill fairly large, irregularly shaped areas. Centers often later invaded by weeds or turn green, leaving "frogeyes" or "doughnuts." Both shoots and roots are killed. *Control:* Same cultural practices as for Brown Patch (above). Maintain balanced fertility. Avoid overliming and too much nitrogen. Spray with Tersan OM or phenyl mercury at 1- or 2-week intervals during cool, wet weather in late fall or early spring. Apply 20 gallons of water per 1,000 square feet. Chlordane and ammonium sulfate also give good control. Resistant grasses: red fescue, Kentucky bluegrass, and *Poa trivialis.*
26. *Buried Debris*—Thin layer of soil over rocks, lumber, tree stumps, bricks, plaster, or concrete dries out rapidly in dry weather. May resemble disease. *Control:* Dig up suspicious areas and remove cause.
27. *Mosaic* (primarily bluegrass)—Yellowish or light and dark green mottling and striping of leaves. *Control:* None suggested.
28. *Downy Mildew* (Bermudagrass)—Southwestern states. Leaf tips have short, black, dead areas. Causes little damage.

LAWSON CYPRESS — See Juniper
LAYIA — See Chrysanthemum
LEADPLANT — See False-indigo
LEADTREE — See Honeylocust
LEATHERLEAF — See Blueberry
LEATHER FLOWER — See Clematis
LEATHERWOOD or WICOPY *(Dirca)*

1. *Rust*—Eastern half of United States. Small yellowish spots on leaves. Alternate host is *Carex. Control:* Destroy alternate host. If needed, apply ferbam or zineb several times, 10 days apart. Start 1 or 2 weeks before rust usually appears.
2. *Sooty Mold*—Black moldy blotches on leaves and twigs. *Control:* Apply malathion to control insects.

LEBANON SQUILL — See Tulip
LEBBEK — See Honeylocust
LEBOCEDRUS — See Juniper
LEDUM — See Labrador-tea

LEEK — See Onion

LEIOPHYLLUM — See Labrador-tea

LEMAIREOCEREUS — See Cacti

LEMON — See Citrus

LEMON MINT — See Salvia

LEMON-VERBENA — See Lantana

LENTEN ROSE — See Delphinium

LENTIL *(Lens)* — See Pea

LEONOTIS — See Salvia

LEOPARDSBANE — See Chrysanthemum

LEPIDIUM — See Cabbage

LETTUCE [BUTTERHEAD or LOOSE HEAD (BIBB), COS or ROMAINE, CRISPHEAD, LOOSELEAF, STEM or CELTUCE] *(Latuca)*; ENDIVE, ESCAROLE, CHICORY, WITLOOF CHICORY or FRENCH ENDIVE *(Cichorium)*; CARDOON, ARTICHOKE or GLOBE ARTICHOKE *(Cynara)*; BLACK-SALSIFY *(Scorzonera)*; SALSIFY or VEGETABLE OYSTER *(Tragopogon)*

1. *Bottom Rots, Drop or Watery Soft Rot, Head Rots, Southern Blight, Stem, Foot, or Crown Rots* (celtuce, chicory, endive, escarole, globe artichoke, lettuce) —General and destructive, especially in moist weather. Head wilts and rots. Often starts with outermost leaves. May become slimy and foul-smelling, or covered with dense, white, gray, or blue-green mold. Upright *lettuce* varieties (e.g., Cos or Romaine) often less seriously damaged. Heads continue to rot after harvest. *Control:* Avoid overwatering, overcrowding, and planting in wet, poorly drained soil. Grow in raised beds high in fertility and organic matter. Keep down weeds. Three- or 4-year rotation. Harvest and refrigerate promptly. Store only sound, disease-free heads. Destroy crop refuse after harvest. Plow or turn under cleanly and deeply. Polyethylene mulches in strips under plants largely prevent head rots. Or apply mixture of captan and PCNB (sold as Terracap, Soil Treater "X," PCNB-Captan) as dust or spray to *head lettuce* only, just before thinning (plants 2 to 3 inches tall). Repeat 10 days later. Follow manufacturer's directions. *Do not use* PCNB on looseleaf lettuce, endive, or escarole. Wet lower leaves and soil under plants. Control insects with sevin plus malathion. Grow upright lettuce varieties and types (e.g., Cos or Romaine, Iceberg). Treat seedbed soil as for Seed Rot (below). Pasteurize soil before planting, using heat or chemicals (pages 538–548). Indoors, increase air circulation and lower the humidity. Ferbam or thiram sprays at weekly intervals are effective against Sclerotiniose or Drop on *lettuce.*
2. *Tipburn*—General and destructive, especially on summer and indoor lettuce crops. Most severe on head lettuce types during cloudy, humid periods. Margins of tender leaves turn brown to black and dry (Figure 4.134). *Control:* Grow in mellow, well-drained soil. Cultivate deeply and frequently to avoid soil packing. Use fertilizers sparingly. Grow tolerant to resistant *lettuce* varieties: Alaska, Arizona Sunbright, Bibb strains, Blackpool, Butter King, Buttercrisp, Calmar, Climax, Cornell 456, Cosberg, Empire, Fulton, Golden State strains, Grand Rapids strains, Great Lakes strains, Imperial strains, Jade, Lakeland, Merit, Mesa 659, New York 515 and PW55, Oswego, Pennlake, Progress, Ruby, Salad Bowl, Slobolt, Summer Bibb, Sunblest No. 1, Sweetheart, and Vanguard. Keep harvested heads cool and market or eat promptly.
3. *Aster Yellows, White Heart, Curly Dwarf*—Widespread. Center leaves become dwarfed, twisted, narrow, and yellowed. Plants dwarfed, "bunchy," and yellowish. May die prematurely. *Lettuce* heads are loose and lightweight—or may not form.

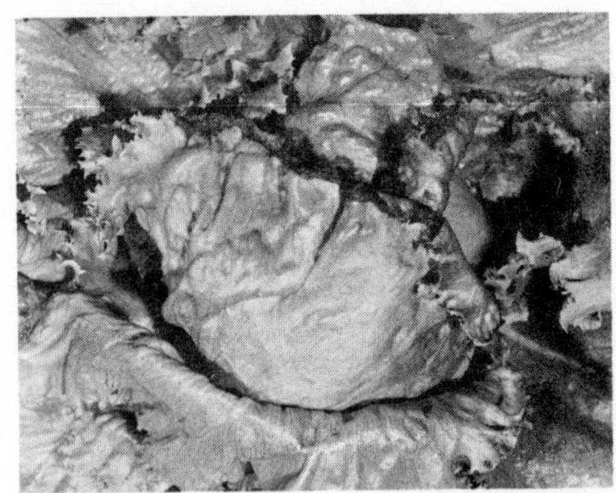

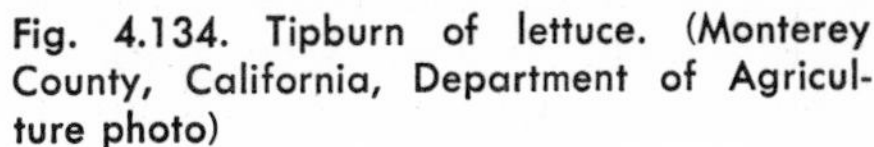

Fig. 4.134. Tipburn of lettuce. (Monterey County, California, Department of Agriculture photo)

Control: Keep down weeds in and around garden. Destroy first infected plants. Sow at time recommended for your area. Control leafhoppers that transmit virus. Apply sevin plus malathion at about 5-day intervals. Start when plants emerge. Four to 6 applications may be necessary. Turn under plant debris immediately after harvest.

4. *Mosaics* (primarily lettuce)—General. Leaves mottled yellow and light green, "blistered," ruffled, or bronzed; even distorted and dwarfed. Older plants stunted, dull green to yellow or brown. May die or not. May not form a head (heart). Symptoms often masked in hot weather and low light. Common Lettuce Mosaic virus is transmitted through the pollen and ovules. *Control:* Sow indexed, mosaic-free lettuce seed. Destroy young, infected plants when first seen. Keep down weeds (e.g., groundsel, prickly lettuce, shepherds-purse, lambsquarters, henbit, cheeseweed, chickweed, milk and sowthistles, etc.). Separate new and old plantings. Destroy plant refuse right after harvest. Rotate. Control aphids that transmit the viruses with malathion or Dibrom (naled). Apply at 3- to 5-day intervals. Tolerant *lettuce* varieties: Parris Island and Valmaine Cos. More resistant varieties will be available in future.
5. *Downy Mildew* (celtuce, chicory, endive, escarole, lettuce)—General and serious in cool (43° to 60° F.), humid weather. Pale green or yellowish areas develop on upper leaf surface. Fluffy, whitish to bluish-gray mold may form on corresponding underleaf surface during cool, moist weather. Later, the somewhat angular spots turn dark brown. When severe, plants may be killed or dwarfed, and yellow, with outer leaves turning brown and dying. Infections may be systemic on lettuce. See Figure 2.22B under General Diseases. *Control:* Sow disease-free seed grown in western states in steamed or fumigated (pages 538–548), seedbed soil. In seedbed, increase air circulation and light. Avoid overcrowding, overwatering, overfertilizing, and water splashing on leaves. Rotate. Collect and burn or turn under plant debris after harvest. Do not work among wet plants. Grow in well-drained soil where air circulation is good. Spray with maneb, thiram, Daconil 2787, Dithane M-45, Polyram, or zineb thoroughly at 5- to 7-day intervals in damp weather, especially in seedbed. Keep down air humidity. Normally resistant *lettuce* varieties, where adapted: Artic King, Bath Cos, Big Boston, Calmar, Grand Rapids strains, many Imperial strains, Salad Bowl, Valmaine Cos, and Valverde. More resistant varieties may be available soon. Keep down weeds and volunteer lettuce.
6. *Fusarium Yellows, Wilt* (lettuce)—Leaves yellow, wilt, and drop starting at base of plant. Dark brown streaks inside stems and larger veins. *Control:* Long rotation. Grow in well-drained soil.
7. *Gray-mold or Botrytis Blight*—General in lettuce hot beds and indoors in damp, overcast weather. Grayish-green, reddish-brown or dark brown, water-soaked areas on lower leaves and stem. Or flower head (buds) of globe artichoke. May be

covered with coarse gray mold in damp weather. Plants may collapse. Heads may rot in field or after cutting. Seedlings may wilt and collapse. *Control:* Treat in seedbed as for Downy Mildew (above) and Seed Rot (below). If necessary, dust or spray with mixture of PCNB and captan as for Bottom Rots (above). Or use dichloran (Botran). Follow manufacturer's directions. Indoors, keep water off foliage, add heat, and increase air circulation. Store heads as cool and dry as practical.

8. *Lettuce Big Vein*—Widespread. Most severe in cool seasons. Plants may be stunted with open, erect, crinkled, yellowed, mottled, and brittle leaves with frilled edges. Leaf veins are swollen and light yellow (Figure 4.135). Yellow spots may develop in young leaves with irregular, brown blotches in older leaves. Heads may not form or are reduced in size and weight, and delayed in maturity. The big-vein virus is transmitted by the fungus *Olphidium,* which also transmits the tobacco necrosis virus to lettuce. *Control:* Grow in clean or sterilized soil (pages 538–548) that is well-drained. Keep soil on dry side. Grow transplants at temperature of 75° F. or above, or plant outdoor lettuce as midsummer or fall crop (where feasible). Treat the soil with PCNB as for Seed Rot (below) and Bottom Rots (above), or thiram (0.2 per cent using 2 gallons per square yard) or Polyram-thiram combination (¼ ounce per square yard). Keep down weeds that may harbor the fungus and/or virus, e.g., *Veronica.* Moderately resistant or tolerant varieties: Calmar, Caravan, and Merit. More resistant varieties should be available soon.

Fig. 4.135. Big vein of lettuce. (Courtesy Dr. R. B. Marlatt and Mr. R. T. McKittrick)

9. *Bacterial Leaf Spots and Rots, Bacterial Wilt, Marginal Blight* (chicory, endive, escarole, lettuce)—Widespread in wet weather. Leaf margins may rot and turn brownish-black. Later become thin and brittle. Small, water-soaked, yellow to yellowish-red, brown, or black spots; or large, irregular blotches may develop in leaves. Spots may remain small and dry or enlarge and become soft and moist. Leaves and stems may wilt and rot. Plants may be yellowed and stunted. *Control:* Same as for Downy Mildew (above). Avoid splashing soil on plants when watering. Control insects by methoxychlor or sevin plus malathion sprays. Keep down related weeds, e.g., sowthistles, prickly lettuce, trumpet fireweed, etc.
10. *Seed Rot, Damping-off*—General in cold, wet soil or poorly ventilated cold frames. Seeds rot. Stand is poor. Seedlings wilt and collapse from rot at soil line. Roots brown and rotted. *Control:* Avoid overcrowding, overwatering, poor ventilation, and heavy, wet soil. In seedbed, mix combination of PCNB and captan (see Bottom Rots above) *before* planting. Or *after* planting apply drench of PCNB-captan. Follow manufacturer's directions. Otherwise treat as for Downy Mildew (above). If stand has been poor, dust *lettuce* seed with chloranil, thiram, captan, or Semesan. Dust *chicory, endive, escarole,* and *salsify* seed with thiram or Semesan.
11. *Rusts*—Small, bright yellow to yellow-orange spots on leaves. Pustules may later be reddish-brown and powdery. Leaves may wither. *Control:* Same as for Downy

Mildew (above). Alternate hosts: sedges (*Carex* spp.), or none. Where practical, destroy all sedges within 100 yards or more of garden area.

12. *Powdery Mildew*—General. A lettuce problem in California and Arizona. Small to large patches of whitish mildew on leaves. May cover foliage in overcast, damp weather. Leaves later curl and turn yellow, then brown and die. *Control:* In problem areas, grow resistant *lettuce* varieties: Artic King, Bath Cos, Big Boston, Imperial strains, and Salad Bowl. Several sulfur dusts or sprays, 10 days apart, should keep mildew in check. Apply as necessary.
13. *Root Rots, Stunt, Wilt*—Lower leaves a dull, dark green. Later wilt, wither, and die. Plants stunted with poor heads. Inside of roots is brown or black. Roots discolored and rooted. Seedlings wilt and collapse. Often associated with nematodes. *Control:* Avoid planting in heavy, cold, wet soil. Rotate. Fertilize and water plants to maintain vigor. Avoid fertilizers high in ammonia or nitrite. Resistant *lettuce* varieties and types: Big Boston, Cos, Grand Rapids, Simpsons Curled, and White Boston. Control insects by sevin or malathion sprays.
14. *Other Fungus Leaf Spots or Blights, Anthracnose*—Spots of various sizes and colors that may enlarge and merge to form irregular dead blotches. Spots may drop out leaving shot-holes. Leaves may wither and die starting usually with oldest ones. Plants may be stunted and lack vigor. *Control:* Same as for Downy Mildew (above). Spray weekly in field during rainy periods. Use zineb, maneb, captan, or fixed copper. Keep plants growing vigorously. Sow disease-free seed or treat as for Seed Rot (above). Seedborne *Septoria* and *Marssonina* (Ringspot) are controlled by soaking seed in hot water (118° F.) for 30 minutes. Collect and burn plant debris afer harvest. *Lettuce* varieties differ in resistance.
15. *Lettuce Brown Blight*—Pale yellow or brown, irregular spots and blotches develop on inner leaves. Spots later enlarge. Plants become stunted, flat, and rosetted. Leaves on such plants gradually turn brown starting at base. Many plants die before harvest. *Control:* Grow resistant varieties: Big Boston, Imperial, or Great Lakes strains.
16. *Spotted Wilt, Ringspot*—Widespread. Irregular, brown, dead pits and streaks in inner leaves of older plants. Outer leaves droop and are stunted. Leaves may twist causing head to lean one way. Leaves and entire plant may show general yellowing. "Heart" may turn black. Young plants often turn yellow, flatten out, and die early. *Control:* Same as for Mosaics (above). Control thrips that transmit the virus. Use malathion plus sevin at 5- to 7-day intervals. Mignonette *lettuce* may have some resistance. The Lettuce Ringspot virus is transmitted by the needle *(Longidorus)* nematode.
17. *White-rust* (primarily black-salsify and salsify)—General. Often destructive in warm, moist weather. Foliage is twisted and distorted. Numerous whitish pustules, that yellow with age, break out on all aboveground plant parts. *Control:* Rotate. Keep down goatsbeard and other related weeds. Pick and burn spotted leaves. Spraying as for Downy Mildew (above) should be beneficial.
18. *Rib Blight, Brown or Red Rib* (lettuce)—Occurs most commonly during periods of hot weather. Reddish-brown or black, elongated spots, streaks, or bands appear on midrib of outer leaves near harvest. A brownish discoloration may extend through the smaller veins of the leaf. *Control:* Varieties differ in resistance. Great Lakes strains are commonly affected. Store lettuce at 32° F. Cool rapidly after harvest.
19. *Brown Heart* (endive)—Immature heart leaves turn brown. At harvest, outermost 4 to 6 whorls may be affected. *Control:* Keep plants well supplied with available calcium. Try spraying with weak solution of calcium chloride (½ ounce per gallon).
20. *Boron Deficiency* (primarily chicory and lettuce)—Younger *lettuce* leaves stunted and malformed; may turn black. Leaf tips spotted and scorched. Growing point

may die. *Chicory* leaves turn red; tend to grow with twisted midribs. *Control:* Apply borax based on a soil test; see page 27.

21. *Root-knot*—General over much of United States. Lettuce is often stunted. See under Bean, and (37) Root-knot under General Diseases.
22. *Other Root-feeding Nematodes* (e.g., dagger, lance, lesion, naccobus, pin, reniform, ring, spiral, sting, stubby-root, stunt) —Associated with stunted, unthrifty plants. Roots often short, stubby, bushy, swollen, and discolored. *Control:* Same as for Root-knot (above).
23. *Curly-top* (lettuce) —See (19) Curly-top under General Diseases.
24. *Leaf and Stem Nematode* (salsify) —See (20) Leaf Nematode under General Diseases.
25. *Verticillium Wilt*—See (15B) Verticillium Wilt under General Diseases.
26. *Crown Gall* (lettuce) —See (30) Crown Gall under General Diseases.
27. *Scab* (salsify) —See (14) Scab under General Diseases.
28. *Slime Molds*—See under Lawngrass. Control measures are not needed.

For more information read USDA Handbook 221, *Lettuce and Its Production.*

LEUCAENA — See Honeylocust

LEUCOJUM — See Daffodil

LEUCOPHYLLUM — See Texas Silverleaf

LEUCOTHOË — See Labrador-tea

LIATRIS — See Chrysanthemum

LIBOCEDRUS — See Juniper

LIGHTWOOD — See Honeylocust

LIGUSTRUM — See Privet

LILAC [AMUR or MANCHURIAN, CHENGTU, CHINESE, COMMON or OLD-FASHIONED, CUTLEAF, DAPHNE, DWARF KOREAN, EVANGELINE, FRENCH HYBRID, HIMALAYAN, HUNGARIAN, JAPANESE, JAPANESE TREE, KOREAN EARLY, LITTLE-LEAF, MEYER'S, NODDING, PEKING, PERSIAN, PINK PEARL, PRESTON HYBRIDS, ROUEN] *(Syringa)*

1. *Powdery Mildew*—General but not serious. Grayish-white mold patches or coating on leaves from midsummer on. Most common on lower or shaded leaves. See Figure 2.23C under General Diseases. *Control:* Apply Karathane, sulfur, Morestan, fixed copper, or Acti-dione several times, 10 days apart, starting when mildew is first seen.
2. *Bacterial and Phytophthora Shoot Blights, Dieback, Stem and Twig Cankers*—Common and severe in wet spring weather on shaded or crowded plants. Brown to black, angular spots in leaves. Later entire leaves may turn blackish-brown and cling to the stem or drop early. Flower buds or blossoms wilt, and burn brown to black (Figure 4.136). Dark brown to black, elongated cankers on stem (Bacterial Blight). Shoots wilt, may die back to ground (Phytophthora Blight). Often associated with stem borers. *Control:* Prune shrubs annually for good air circulation. Avoid overcrowding. Keep down weeds. Prune and burn blighted portions when found. Cut well back into healthy tissue. Swab pruning tools with 70 per cent alcohol between cuts. Apply zineb, maneb, fixed copper, or bordeaux (4-4-50) before spring wet periods. Start as leaves begin to unfold. Spray after bloom as for Leaf Spots (below). Avoid overfertilizing with manure or high-nitrogenous fertilizer and planting close to rhododendron and privet. Keep soil reaction near neutral (pH 6.5–7.5). Spray with dieldrin, DDT, or Thiodan to control borers. Check with your county agent or extension entomologist regarding timing of sprays for your area. White-flowered lilac varieties are apparently more susceptible to Bacterial Shoot Blight than varieties with colored flowers.

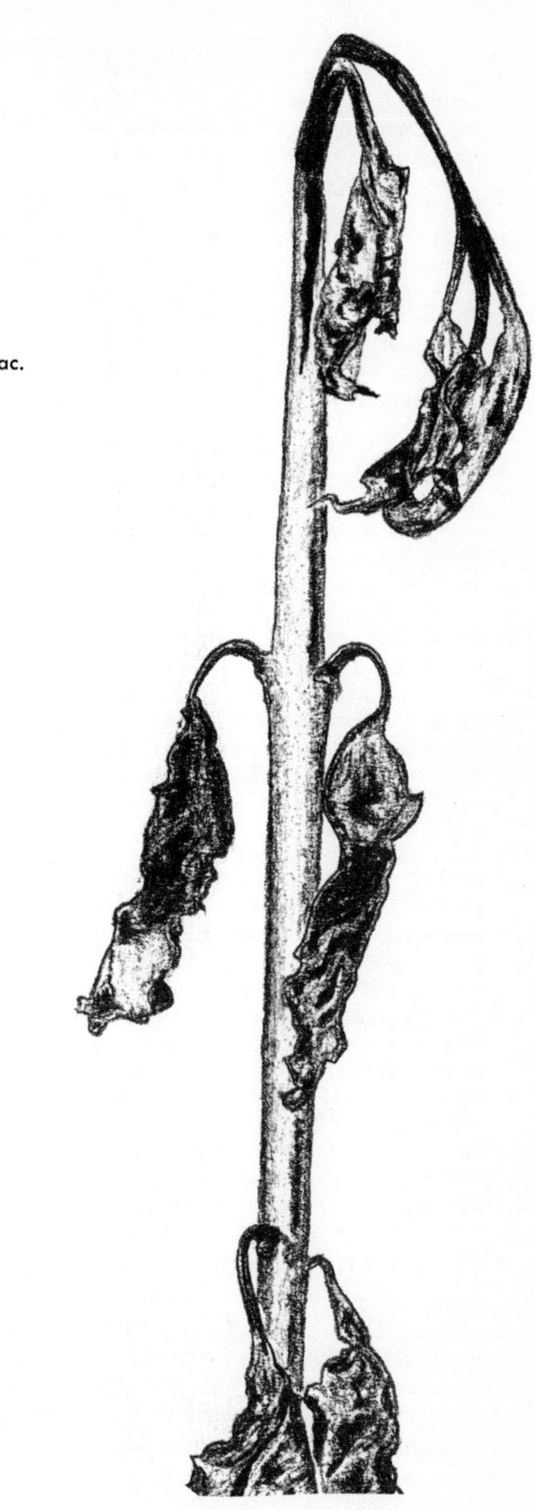

Fig. 4.136. Bacterial shoot blight of lilac.

3. *Leaf Spots, Leaf Blotch or Blights, Anthracnose*—Small to large, variously colored spots, often with dark margins. Spots may be zoned or later drop out leaving irregular ragged holes. *Control:* If serious, apply zineb, maneb, captan, or fixed copper several times, 7 to 10 days apart. Start just after lilacs bloom. In wet seasons, applications may be needed before summer and fall rainy periods.
4. *Gray-mold Blight, Botrytis Blight, Blossom Blight*—Common in humid, wet areas. May follow frost or other injury. Buds and blossoms may turn light brown, water-soaked, and rot. Become covered with coarse, tannish-gray mold. Twigs die back from tips. *Control:* Same as for Shoot Blights and Leaf Spots (both above).
5. *Root Rots*—Certain branches, or entire shrub may die. Underground parts are rotted. See under Apple. May be associated with root-feeding nematodes (e.g., citrus-root, dagger, lance, lesion, potato-rot, ring, spiral, stem, stunt). *Control:* Carefully dig up and burn affected plants. Include as many roots as possible. Do *not* replant in same area without drenching soil as for Verticillium Wilt (below).
6. *Verticillium Wilt*—Leaves on one or more stems are pale, wilt, and fall early starting at base and spreading upward. Clusters of leaves at stem tip may hang on for a long time. Stems show brownish-green streaks under bark. Branches die back. Slightly infected shoots are stunted and thickened with stunted flower clusters. *Control:* Dig up and burn infected plants. Roots and all. Avoid replanting in same location for 5 or 6 years or more without first drenching soil with SMDC (Vapam or VPM Soil Fumigant). Follow manufacturer's directions.
7. *Wood Rot*—See under Birch, and (23) Wood Rot under General Diseases. Commonly follows borers and other types of injuries. See under Shoot Blights (above). Keep shrubs growing vigorously.
8. *Mosaic*—Virus complex. Leaves dwarfed, puckered, folded, and yellow-mottled or spotted. May closely resemble Graft Blight (below) or soil deficiency. Symptoms tend to disappear during certain seasons. *Control:* Dig up and destroy infected plants when found to be mosaic.
9. *Ringspot*—Pale green to yellow rings, spots, lines, and broad bands on leaves. Leaves may be distorted and ragged with holes. *Control:* See Mosaic (above).
10. *Witches'-broom*—On *common lilac* leaf veins first yellow and clear. Several months to a year later, numerous, slender, lateral shoots with dwarfed leaves are formed producing a witches'-broom. On *Japanese lilac* several (2 to 6) slender shoots, usually near top of plant, branch freely and bear dwarfed leaves that are often twisted and rolled. *Control:* Same as for Mosaic (above).
11. *Graft Blight, Incompatibility Disease*—Occurs where lilac grafted on privet. Leaves dwarfed, irregular, rolled, curled or cupped, and brittle. Leaves usually yellowish, later brown, at margins and between veins (Figure 4.137). Resembles a nutrient deficiency or Mosaic. Plants stunted. *Control:* Propagate lilac on its own roots, use a piece root-graft, or graft on another lilac species or variety.
12. *Frost Injury*—Freezing temperatures in late spring may cause leaves to turn brown and become torn along veins in irregular pattern.
13. *Root-knot*—See under Peach, and (37) Root-knot under General Diseases.
14. *Crown Gall*—See (30) Crown Gall under General Diseases.

LILY [AURATUM HYBRIDS, AURELIAN HYBRIDS, BACKHOUSE HYBRIDS, BELLINGHAM HYBRIDS, BERMUDA, CAMEO HYBRIDS, CANDLESTICK, CAROLINA, CAUCASIAN, CHAPARRAL, CORAL, EASTER, FIESTA HYBRIDS, FORMOSA, GOLDENBANDED, GOLDEN CHALICE HYBRIDS, GRAY'S, HANSON, HARLEQUIN HYBRIDS, JAPANESE, KELLOGG, LEOPARD or PANTHER, MADONNA, MARTAGON, MEADOW, MICHIGAN, MID-CENTURY HYBRIDS, NANKEEN, OLYMPIC HYBRIDS,

Fig. 4.137. Graft incompatibility of lilac. (Iowa State University photo)

ORANGE, PAISLEY HYBRIDS, PRESTON HYBRIDS, REGAL or ROYAL, SARGENT, SCARLET TURKSCAP, SHOWY or JAPANESE, SIERRA, STAR, TIGER, TURKSCAP, WASHINGTON, WHITE-TRUMPET, WILD (ORANGE-RED or WOOD), WILD YELLOW or CANADA], also *Lilium columbianum, L. elegans, L. giganteum, L. henryi, L. hollandicum, L. humboldtii, L. japonicum, L. rubrum, etc. (Lilium);* LILY-OF-THE-VALLEY *(Convallaria);* BELLWORT, MERRYBELLS *(Uvularia)*

1. *Botrytis Blights, Gray-mold Blight, Stem Rot*—General and serious on most lilies and lily-of-the-valley, in cool (50° to 65° F.), damp weather. Symptoms variable. Soft, yellowish to reddish-brown or dark brown (sometimes light gray), oval to round spots on leaves, stems, flower stalks, and flowers. Spots may enlarge and merge to blight whole leaf. Leaves may blacken, wither, and hang limply. Often starts at base of stem. Tip of shoot or entire top may die and bend sharply downward. Bulb or rhizomes may rot. Seedlings damp-off. Plants often stunted. Buds rot or open to distorted, brown-flecked flowers. Coarse gray mold may grow on affected parts in damp weather. See Figure 2.21C under General Diseases. *Control:* Carefully collect and destroy infected parts as they occur. Burn all tops in fall. Space plants. Increase air circulation. Keep down weeds. Plant best quality, disease-free bulbs (or rhizomes) in a warm, sunny spot in well-drained soil. Avoid overcrowding, splashing water on foliage when watering, and overfertilizing with nitrogen. During cool, wet weather apply 4-2-50 bordeaux mixture (4 ounces of copper sulfate and 2 ounces hydrated spray lime in 3 gallons of water) or zineb, at 7- to 10-day intervals. Add detergent or spreader-sticker (page 118) to ensure wetting foliage. Cover all aboveground parts with each spray. Regal, Hanson, Martagon, Wild Yellow or Canada, Turkscap, *L. henryi,* and Goldenbanded *lilies* are usually more resistant than Easter, Formosa, Madonna, Nankeen, Scarlet Turkscap, Showy, and Tiger. Indoors, increase light and circulation. Keep humidity below 85 per cent, and water off foliage. Spray buds and early blooms lightly with zineb or Botran.
2. *Bulb Rots, Scale Rots, Root Rots*—Cosmopolitan and serious. Plants stunted and lack vigor with rotted bulb and roots. Lower leaves turn yellow or purple prematurely. Stems may dry up from base with leaves withering and falling. Buds or flowers may be blasted and fewer in number. Often associated with viruses, bacteria, nematodes, symphilids, and mites. Mold growth on stored bulbs. Bulbs may be wet, slimy, and foul-smelling (Bacterial Soft Rot), soft and mushy

(Rhizopus Bulb Rot), dry and punky (*Penicillium*), or "chalky." Marginal, brown areas (Tipburn or Leaf Scorch) may develop on certain leaves as they age. Lily bulb scales may show discolored "scabby" spots. Bulbs unsightly. See Figure 2.51C under General Diseases. *Control:* Plant only large, top quality bulbs or rhizomes in open, fertile, well-drained soil, sterilized if practical (using steam, methyl bromide, chloropicrin, Vorlex) with a pH of 6.5 to 7. Before planting, soak *lily* bulbs 30 minutes in mixture of PCNB 75 (1 ounce in 6 gallons of water) *plus* captan 50 or ferbam 76 (2 ounces in 6 gallons), or Dexon following manufacturer's directions. Dig up and destroy all infected bulbs, plants, and surrounding soil. Avoid wounding bulbs or growing plants. Five- to 6-year rotation. Avoid overwatering, high-nitrogen fertilizers, and manure. Practice balanced soil fertility based on a soil test. Mixing PCNB in the furrow before planting has proved beneficial, as has drenches of Dexon (30 parts per million) to potted plants, applied at 10-day intervals. Follow manufacturer's directions. Keep bulbs cool in storage. Nurserymen propagate from rot-free scales dusted with thiram or chloranil. *Lilies* resistant to Fusarium Bulb Rot (Basal Rot): Aurelian hybrids, Mid-Century hybrids, Showy or Japanese, Turkscap, Wild Yellow or Canada, and *L. henryi.*

3. *Mosaics, Mottle, Flower Breaking*—Serious and widespread on lilies. Symptoms variable, depending on virus or virus strain and lily species. Often masked at temperatures over 75° F. Leaves mottled, yellow, and light and dark green or with yellowish (or brown) streaks. Leaves may be stunted, curled, twisted, and narrowed. Flowers often blotched or "broken." Some species stunted or killed; die from root and bulb rots. Symptoms may fade as plants get older. Certain *lilies* (e.g., Candlestick, Madonna, Tiger) may carry viruses but show no symptoms. Often confused with various mineral deficiencies but symptoms usually more uniform. See Figure 4.138. *Control:* Dig out and destroy infected plants when first seen. They will not recover. Plant only largest, virus-free bulbs from a reputable nursery, or grow from seed. Control aphids that transmit the viruses. Use lindane or malathion. Keep down weeds. Resistant *lily* species: Coral, Hanson, Martagon, Tiger, and Leopard (*Lilium pardalinum*). Avoid growing Formosa lily and *L. rubrum* near other lilies or near "broken" or mosaic-infected tulips, Star-of-Bethlehem, and Fritillarias. Grow Goldenbanded and *L. rubrum* lilies separately. Grow *L. henryi* from seed.
4. *Rosette, Yellow Flat*—Symptoms vary with lily species and variety. Very destructive, especially to Easter lily. Plants and flowers may be dwarfed. Upper leaves pale green or yellowish and tightly curled downward forming a flat "rosette." Leaves sometimes twisted sideways and distorted. Bulbs grow smaller each year. Plants rarely flower and never recover. Often confused with frost injury, water-logged soil, aphid injury, or root and bulb rots. *Control:* Same as for Mosaics (above).
5. *Fleck* (lily)—A virus complex. Plants stunted, distorted, and short-lived. Blocklike, "translucent windows" or elongated, white, yellowish, or brown flecks in leaves. Leaves tend to curl and twist. Flowers smaller, fewer in number, and short-lived. May be twisted and streaked. *Control:* Same as for Mosaics (above).
6. *Ringspot* (Easter, regal, and tiger lilies)—Dark, ringlike patterns on leaves that soon become dead areas which spread throughout plant. No flowers produced. Plants twisted and stunted or killed outright. Hybrid lilies may show only faint mottling. *Control:* Same as for Mosaics (above).
7. *Noninfectious Chlorosis*—Plants stunted. Yellowish leaves with green veins. May closely resemble Mosaic complex (above) except yellowish patterns are more regular and more yellow. *Control:* Common in alkaline soils. Have soil tested. Add acid fertilizer and solution of ferrous sulfate to soil. For fast results, spray foliage with ferbam, iron (ferrous) sulfate, or an iron chelate (page 27). Repeat as necessary to maintain green color.

Fig. 4.138. Lily mosaic.

8. *Stem Rots or Canker, Foot Rots, Stump Rot, Root Rots*—Cosmopolitan. Plants often stunted and lack vigor. May wilt gradually or suddenly, wither, often collapse. Stem rots at ground line. Roots rotted. Tips of young plants may wither. *Control:* Same as for Bulb Rots (above). In addition, wash mud from crowns as plants emerge. Dipping bulbs just before planting in mixture of PCNB and ferbam or captan may be beneficial. Dexon drenches also aid control.
9. *Rusts*—Widespread, but uncommon. Yellow spots on upper leaf surface with yellow-orange, reddish-brown or black, powdery pustules on corresponding underleaf surface. Alternate hosts: *Sporobolus* (ribbon or reed grasses) or none. *Control:* Destroy infected leaves when found. Where serious, apply ferbam, zineb, maneb, or dichlone at 10-day intervals. Destroy weed hosts.
10. *Stalk Rot, Southern Blight*—May be serious in southern states in heavy, wet soils. Chalky white rot of bulb with white, fanlike mold patches. Plants wither and die in patches. Regal lily is very susceptible. *Control:* Same as for Bulb Rots (above). Resistant *lily* species: Easter and Showy.
11. *Leaf and Bud (Bulb) Nematode, "Bunchy Top," Dieback* (lily)—Primarily Pacific Northwest. Symptoms variable. Leaves may be blotched yellow or purplish and green. Leaves turn yellow, bronze to dark brown, and curl against stem. Affected leaves may be thick, pointed; produce bunchy top or "crooks." Lower leaf whorls usually are most seriously infested; may resemble frost damage. Forced, indoor plants often produce "blind" buds. Blooming is retarded or flowers are destroyed. *Control:* Three-year rotation. Avoid overhead sprinkling. Space plants. Dig and burn infested plants when first found. Plant disease-free, best quality, heat-treated

bulbs in clean, well-drained soil. Or soak *dormant* bulblets before planting in hot formaldehyde solution (1 tablespoon of 40 per cent commercial formalin in 2 quarts of water; 1 pint to 25 gallons) at *exactly* 111° F. for an hour. Then dip in chloranil solution (5 ounces of wettable Spergon in 3 gallons of water). Dry and plant immediately. For *Croft Easter lilies:* Cure bulbs at 95° F. and 95 per cent humidity for 1–2 weeks; presoak 2 days in cool water (about 75° F.). Then treat for 2 hours in hot formaldehyde solution (1 part 40 per cent formalin to 200 of water) at 115° F. Cool, then soak bulbs in Puratized Agricultural Spray (1 fluid ounce in 8 gallons of water) for 24 hours. Scale the bulbs and dust scales with ferbam; place on moist vermiculite at 75° F. and 95 per cent humidity. Remove bulbils and plant in treated soil (pages 538–548).

12. *Root-feeding Nematodes* (e.g., dagger, lance, lesion, root-knot, spiral, stubby-root, stunt)—Plants may be stunted. Foliage turns pale or yellow prematurely. Roots stunted and die back. May be bushy, "stubby," beady, or stunted with numerous dead spots. Nematodes often part of disease complex with root- and bulb-rotting fungi and mites. *Control:* Rotate. Same as for Leaf and Bud Nematode (above). Grow in clean or sterilized soil (pages 538–548). Root-prune infested roots before treating and planting.
13. *Leaf Spots, Blotch, Anthracnose*—Spots of various sizes and shapes on leaves. Spots may drop out, leaving ragged holes. Leaves may wither and die early. Spots may also occur on leaf and flower stalks. Young plants may be stunted. *Control:* Same as for Rusts (above). Rotate.
14. *Frost Injury*—Growing point is killed; plants are stunted by severe frosts. Leaves may be "puffy." Frost injury may stimulate Botrytis Blight. *Control:* Cover young shoots or mulch on cold nights when freezing temperatures are expected. Prune and burn injured parts.
15. *Damping-off*—Cosmopolitan. See under Beet.

LILY LEEK — See Onion
LILY-OF-THE-NILE — See Tulip
LILY-OF-THE-VALLEY — See Lily
LIME — See Citrus and Linden
LIMONIUM — See Sea-lavender
LINARIA — See Snapdragon

LINDEN or LIME [AMERICAN or BASSWOOD, COMMON, COMMON EUROPEAN, CRIMEAN, JAPANESE, LARGE-LEAVED, MANCHURIAN, MONGOLIAN, PYRAMIDAL, SILVER or WHITE, SMALL-LEAVED EUROPEAN, WEEPING or PENDENT SILVER, WEEPING WHITE] *(Tilia)*

1. *Leaf Blotch or Blight, Anthracnose*—General in wet seasons. Small to large, round to irregular, light brown spots with blackish-brown margins. Spots enlarge and form blotches along veins. Leaves may wither and drop early. Young twigs and branches may wilt and die back. *Control:* Prune and burn dead or cankered twigs. Collect and burn fallen leaves. Fertilize to increase vigor. Where serious, spray as buds start to swell. Repeat twice more at 10-day intervals. Use fixed copper, zineb, ferbam, thiram, captan, or bordeaux mixture (4-4-50).
2. *Twig Blight, Dieback, Trunk and Branch Cankers*—See under Elm and Maple.
3. *Powdery Mildews*—General. Powdery, grayish-white mold on leaves and young shoots. If severe, leaves may yellow and wither. *Control:* Where serious, spray twice, 10 days apart. Use sulfur or Karathane. Start when mildew is first seen.
4. *Leaf Scorch*—Margins turn brown in midsummer following hot, dry, windy weather. *Control:* Water during dry periods. Fertilize and prune to increase vigor.
5. *Leaf Spots, Spot Anthracnose*—Generally small spots of various colors and shapes;

may drop out leaving shot-holes. *Control:* Same as for Leaf Blotch (above).

6. *Wood Rots*—Cosmopolitan. See under Birch, and (23) Wood Rot under General Diseases.
7. *Sooty Mold*—Black, sooty mold patches on leaves. Mold grows on "honeydew" secreted by aphids, scales, whiteflies, and other insects. See Figure 2.28C under General Diseases. *Control:* Apply malathion to kill insects.
8. *Verticillium Wilt*—Leaves on certain branches wilt. Streaks in sapwood are dark gray to brown. General symptoms and control are the same as for Maple, Verticillium Wilt.
9. *Root Rot*—See under Apple, and (34) Root Rot under General Diseases.
10. *Sunscald, Winter Injury*—Common on newly planted trees with thin bark. See under Elm and Apple.
11. *Bleeding Canker*—Northeastern states. See under Beech and Maple.
12. *Wetwood, Slime Flux*—See under Elm.
13. *Mistletoe*—See (39) Mistletoe under General Diseases.
14. *Felt Fungus*—Southern states. See under Hackberry.
15. *Seed Rot, Damping-off*—See under Pine.

LINDERA — See Avocado

LINGONBERRY — See Blueberry

LINNAEA — See Twinflower

LINUM — See Flax

LIONS-EAR or LIONS-TAIL — See Salvia

LIPPIA — See Lantana

LIQUIDAMBAR — See Witch-hazel

LIRIODENDRON — See Magnolia

LITHOCARPUS — See Oak

LITHOSPERMUM — See Mertensia

LITSEA — See Avocado

LIVEFOREVER — See Sedum

LIVERLEAF — See Anemone

LIVISTONA — See Palms

LOBELIA [BLUE, EDGING], CARDINALFLOWER, INDIAN-TOBACCO *(Lobelia)*

1. *Root Rots, Stem and Crown Rot, Damping-off*—Lower leaves turn yellow. Crown and lower part of stem decay. Tops wilt and die, or may be stunted. Seedlings wilt and collapse. *Control:* Avoid overwatering, overcrowding, and planting in poorly drained soil. Take cuttings only from healthy plants. Root cuttings in sterile medium (pages 538–548). Drench crown and surrounding soil with 1:1,000 solution of mercuric chloride when plants are young. Or treat soil with PCNB before planting. Follow manufacturer's directions.
2. *Mosaic*—Leaves blotched and mottled pale and dark green (certain varieties). Young leaves distorted and twisted. Older ones somewhat malformed and brittle. *Control:* Plant healthy stock. Keep down weeds. Destroy first infected plants. Control aphids that transmit virus. Use malathion or lindane.
3. *Leaf Spots*—Round to irregular, pale tan to reddish-brown spots. Leaves may wither and drop early. *Control:* Pick and burn spotted leaves as they appear. Spray with zineb or maneb during wet periods.
4. *Gray-mold Blight*—See under Chrysanthemum, and (5) Botrytis Blight under General Diseases.
5. *Rust*—See under Chrysanthemum, and (8) Rust under General Diseases.

6. *Root-knot and Other Nematodes*—See (37) Root-knot under General Diseases. Lobelia is very susceptible.
7. *Curly-top*—Shoots have rosettes. Flowers reduced in size. See (19) Curly-top under General Diseases.
8. *Spotted Wilt*—See under Bellflower, and (17) Spotted Wilt under General Diseases.
9. *Powdery Mildew*—See under Bellflower, and (7) Powdery Mildew under General Diseases.
10. *Leaf Smut*—See (13) White Smut under General Diseases.

LOBLOLLY-BAY — See Franklin-tree
LOBULARIA — See Cabbage
LOCKHARTIA — See Orchids
LOCUST — See Honeylocust
LOGANBERRY — See Raspberry
LOLIUM — See Lawngrass
LONICERA — See Snowberry
LOOSESTRIFE — See Lythrum and Primrose
LOPHOPHORA — See Cacti
LOQUAT — See Apple
LOTUS — See Waterlily
LOVE-LIES-BLEEDING — See Cockscomb
LUFFA — See Cucumber
LUNARIA — See Cabbage
LUNGWORT — See Mertensia
LUPINE *(Lupinus)* — See Pea
LYCASTE — See Orchids
LYCHNIS — See Carnation
LYCIUM — See Matrimony-vine
LYCOPERSICON — See Tomato
LYCORIS — See Daffodil
LYLOCEREUS — See Cacti
LYONIA — See Blueberry
LYSIMACHIA — See Primrose

LYTHRUM, PURPLE and WINGED LOOSESTRIFE *(Lythrum)*; MADEIRA-VINE, CLIMBING MIGNONETTE *(Boussingaultia)*

1. *Leaf Spots*—See (1) Fungus Leaf Spot under General Diseases.
2. *Root Rot*—See under Chrysanthemum, and (34) Root Rot under General Diseases.
3. *Root-knot*—See (37) Root-knot under General Diseases.

MAACKIA — See Honeylocust
MACLURA — See Osage-orange
MADEIRA-VINE — See Lythrum
MADRONE — See Blueberry
MAGIC LILY — See Daffodil

MAGNOLIA [ANISE, BIG-LEAF, CAMPBELL, FRASER or MOUNTAIN, KOBUS, LENNEI, LILY-FLOWERED, OYAMA, PINK STAR, SAUCER, SOUTHERN or BULLBAY, STAR, SWEETBAY or LAUREL, UMBRELLA, WHITE-LEAF JAPANESE, WILSON, YULAN], CUCUMBERTREE

(Magnolia); TULIPTREE or YELLOW-POPLAR *(Liriodendron)*; ANISETREE [COMMON, JAPANESE, STAR or CHINESE] *(Illicium)*

1. *Leaf Spots, Tar Spot, Spot Anthracnose*—Widespread but rarely serious. Small spots to large, irregular blotches on leaves. Of various colors and shapes. Infected leaves may drop early. *Control:* Gather and burn fallen leaves. If practical, spray several times, 10 days apart. Start as buds break open. Use zineb, thiram, maneb, fixed copper, or phenyl mercury. Follow manufacturer's directions.
2. *Wood or Heart Rots*—Cosmopolitan. Foliage is thin. Upper branches die back. Leaves may appear to have nutrient deficiency. See under Birch, and (23) Wood Rot under General Diseases.
3. *Powdery Mildews*—Widespread. Powdery, grayish-white patches on leaves and young shoots. If severe, leaves may yellow and wither. *Control:* Apply two sulfur or Karathane sprays, 10 days apart.
4. *Twig Blights, Cankers, Dieback*—Top of tree dies back. Sunken, flattened, or discolored cankers form on twigs, larger limbs, and trunk. Bark over affected areas often discolored with longitudinal cracks. *Control:* Cut out and burn cankered wood on trunk and larger limbs. Paint with good tree wound dressing. Fertilize and water to keep trees vigorous. Spraying as for Leaf Spots (above) may be beneficial. *Magnolia* varieties differ in susceptibility.
5. *Verticillium Wilt*—Leaves on one or more branches commonly droop, roll inward, and turn yellow. Leaves on other branches or trees may wilt rapidly and turn brown or black. Such leaves usually drop early. Branches die back; may produce dwarfed, sickly leaves the second season. Outer sapwood in larger branches and trunk often shows a dark discoloration. *Control:* See under Maple.
6. *Sooty Mold, Black Mildew*—Cosmopolitan. Black moldy patches on leaves. See Figure 2.28A under General Diseases. *Control:* Spray with malathion to control aphids, scales, and other insects. Addition of a fungicide (see Leaf Spots above) is beneficial. Check with your county agent or extension entomologist regarding timing of sprays.
7. *Leaf Yellowing or Scorch, Sunscald*—Leaves yellow and drop during hot, dry periods. *Control:* Water trees during these periods. Fertilize to keep trees vigorous. Keep pruned.
8. *Root Rots*—See under Apple, and (34) Root Rot under General Diseases. May be associated with root-feeding nematodes (e.g., burrowing, dagger, lance, lesion, root-knot, spiral, sting, stubby-root, stunt, trophotylenchulus). If nematodes are primary problem, apply DBCP to root zone following manufacturer's directions. Grow in full sun in rich, humusy soil that holds moisture yet is well-drained.
9. *Root-knot*—See (37) Root-knot under General Diseases.
10. *Seedling Blight, Cutting Rot*—See under Pine.
11. *Thread Blight*—Southeastern states. See under Walnut.
12. *Felt Fungus*—Gulf states. Purple-black felt on bark. See under Hackberry.
13. *Algal Leaf Spot, Green Scurf, "Red Rust"*—Far south and where damp. Velvety, reddish-brown to orange, cushiony patches or greenish-brown spots on leaves. *Control:* Spray with fixed copper during rainy periods. Improve drainage.
14. *Bacterial Leaf Spot and Twig Blight* (magnolia)—See (2) Bacterial Leaf Spot under General Diseases.
15. *Chlorosis*—See under Maple.
16. *Petal Rot, Bud Rot* (magnolia)—See (31) Flower Blight and (5) Botrytis Blight under General Diseases.
17. *Wetwood, Slime Flux* (tuliptree)—See under Elm.

MAHONIA — See Barberry
MAIDENHAIR-TREE — See Ginkgo

MAJORANA — See Salvia
MALABAR-PLUM — See Eucalyptus
MALACOTHRIX — See Chrysanthemum
MALANGA — See Calla Lily
MALEBERRY — See Blueberry
MALLOW — See Hollyhock
MALPIGHIA — See Barbados Cherry
MALTESE CROSS — See Carnation
MALUS — See Apple
MALVA, MALVASTRUM — See Hollyhock
MAMMILLARIA — See Cacti
MANDRAKE — See Mayapple
MANFREDA — See Centuryplant
MANGEL, MANGOLD — See Beet

MANGO *(Mangifera)*

1. *Anthracnose, Blossom and Fruit Blight, Ripe Fruit Rot*—Cosmopolitan in humid, rainy weather from bloom to fruit half-grown. Minute, brown or black spots on flower panicles that gradually enlarge and often merge to kill flowers or entire panicle. Small, dark, circular to angular or irregular spots on leaves. Often merge to form large dead blotches that may crack and drop out leaving ragged holes. Black or brown dots form in centers. Become covered with pinkish spore masses in moist weather. Enlarging, roundish spots on fruits. Young infected fruit blacken and mummify. Fruit spots may merge and appear as a surface stain or russet on older fruits. Ripe fruit may decay on tree or after harvest. Twigs may blight and die back (Wither-tip). *Control:* Space trees. Prune regularly. Water and fertilize to keep trees vigorous. Spray with captan, zineb, or maneb, plus spreader-sticker (page 118), at 5- to 7-day intervals. Start at *very early bloom,* before individual flowers have opened. After fruit set apply fixed copper or bordeaux (3-3-50) at 2- to 4-week intervals until fruit is half-grown or more. Hot water soak of ripe fruit (5 minutes at 130° to 132.5° F.) reduces losses. Varieties differ in resistance to Macrophoma Blight.
2. *Spot Anthracnose or Scab*—Serious in nurseries. Small, round to angular, dark brown to grayish-brown or black spots and blotches on young leaves, blossom panicles, fruits, and twigs. If severe, leaves may be crinkled, distorted, shot-holed, and drop early. Fruit spots may crack and become corky. Velvety, grayish-brown to olive mold may form on affected parts in moist weather. *Control:* Same as for Anthracnose (above). Keep new flushes of growth protected with spray.
3. *Leaf Spots*—General. Spots of various sizes, shapes, and colors. *Control:* Same as for Anthracnose (above).
4. *Sooty Mold*—Cosmopolitan on unsprayed trees. Black sooty blotches on foliage and fruit. Follows attacks by scales. *Control:* Apply postharvest oil emulsion spray to control scale insects. Spray during season with malathion following local recommendations. Thorough coverage is essential.
5. *Powdery Mildew*—Minor problem in damp, shady locations. Leaves, flowers, flower stalks, and young fruits coated with white powdery mold. Flowers and young fruit may turn brown and fall. Poor fruit set. Older fruit and leaves develop irregular, purplish-brown blotches. *Control:* Apply dormant bordeaux spray followed by sulfur or Karathane during and after flowering and fruit set.
6. *"Red Rust," Green Scurf, Algal Leaf Spot*—Cosmopolitan on unsprayed trees. Roundish, greenish-gray leaf spots that turn rusty-red. Similar blotches form on cracked and thickened bark on stems. *Control:* Same as for Anthracnose (above).

7. *Zinc Deficiency, Little Leaf*—In alkaline soils. Small, recurved, thickened, stiff, yellowed leaves at branch tips. Twigs may die back. Trees unthrifty with poor fruit set. *Control:* Apply zinc sulfate (2 tablespoons per gallon), or mixed minor nutrients (page 27) with one or more pest sprays.
8. *Diplodia Stem-end Fruit Rot*—Light brown to almost black, postharvest fruit rot. *Control:* Harvest fruit when mature. Handle carefully. Avoid storing with poor ventilation and high temperature. Spray as for Anthracnose (above).
9. *Wood Rots*—See under Birch, and (23) Wood Rot under General Diseases.
10. *Root Rot*—See under Apple, and (34) Root Rot under General Diseases. Often associated with root-feeding nematodes (e.g., dagger, needle, spiral, stunt).
11. *Twig Blight, Dieback*—See under Maple. Spraying and cultural practices as for Anthracnose (above) should be beneficial.
12. *Tipburn*—Irregular, brown scorching of leaf tip and margins. Dead tissue may be invaded by secondary molds. Associated with salt in overhead irrigation water, salt spray near ocean, or high salts in soil. May also be caused by drought, hot drying winds, and damage to the root system by root rot and/or nematodes.
13. *Brown Felt Fungus*—Associated with scale insects. See under Hackberry.

MANILAGRASS — See Lawngrass

MANZANITA — See Blueberry

MAPLE [AMUR or GINNALA, BIGLEAF, BIGTOOTH, BLACK, COLUMNAR RED, COLUMNAR SUGAR, DEVIL, DOUGLAS, DRUMMOND, ENGLISH HEDGE, FERNLEAF, FLORIDA, GOLDEN MOON, HARLEQUIN, HAWTHORN, HEDGE or FIELD, JAPANESE, MANCHURIAN, MOUNTAIN, NIKKO, NORWAY (many horticultural varieties), PAPERBARK, PURPLEBLOW, OREGON, RED or SCARLET (many horticultural varieties), RED LEAF (CRIMSON KING, SCHWEDLER), REDUEIN, SIEBOLD, SILVER or SOFT, STRIPED or MOOSEWOOD, SUGAR or HARD, SYCAMORE (many horticultural varieties), TATARIAN, TRIDENT, VINE], BOXELDER [COMMON, CALIFORNIA] *(Acer)*

1. *Anthracnose, Leaf Blights*—General in wet springs on Japanese, Norway, sugar, and silver maples, and boxelder. Small to large, round to irregular, light to reddish-brown, purplish-brown, or black dead areas on leaves. Many along veins. Areas often enlarge and merge, killing the leaf. Leaves often appear scorched by frost or hot, dry weather. Many infected leaves drop prematurely. Twigs may die back. See Figure 2.19B under General Diseases. *Control:* Collect and burn fallen leaves. Prune out dead twigs and weak wood to increase air circulation and promote faster drying. If practical, apply 3 sprays: as buds unfold in spring and repeat 10 and 20 days later. Use phenyl mercury, zineb, dichlone, thiram, captan, or ferbam plus spreader-sticker (page 118). Try to time sprays just before wet periods when infections occur. Fertilize and water to stimulate vigorous growth.
2. *Verticillium Wilt, Maple Wilt*—Widespread and destructive, especially on bigleaf, hedge, Japanese, Norway, red, silver, sugar, and sycamore maples. Leaves suddenly discolor and wilt on one or more limbs, usually in midsummer. Often on one side of tree or in crown. Leaves die and fall or hang on dead branches. Later other limbs wilt and die. Olive-green to brown or bluish-black streaks may occur in sapwood, often at distance from wilting foliage. Infected trees may die within a few weeks or live on for years. See Figure 2.32C under General Diseases and Figure 4.139. *Control:* Cut and burn severely infected trees. Where only 1 to 6 limbs are dead or show wilt, prune affected parts. Swab tools with 70 per cent alcohol between cuts. Paint wounds promptly with tree wound dressing (page 36). Avoid wounding roots or lower trunk. Fertilize heavily in early spring and water trees

Fig. 4.139. Verticillium wilt of maple.

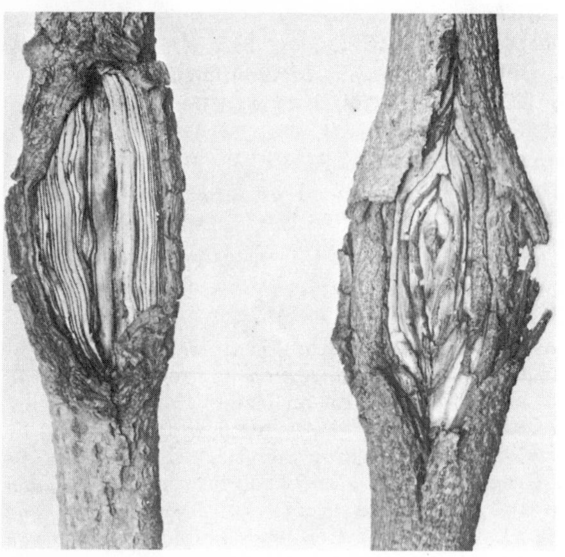

Fig. 4.140. Target or callus-roll type of canker on sugar maple, typical of older Nectria cankers on many trees. (USDA Forest Service photo)

during dry periods to stimulate vigor. *Maple* species and varieties differ in resistance.

3. *Leaf Scorch*—Light or dark brown areas between veins or along margins. Foliage appears bronzed, dried, and scorched, commonly from midsummer on (Figure 1.1B). Affected leaves may drop early. Causes: Late spring frost, salt run-off from nearby road or street, hot, dry summer winds, and drought. *Control:* Water trees during summer droughts. Grow in deep, fertile soil protected from drying winter and summer winds and bright, all-day sun (especially harlequin, Norway, red leaf, and sugar maples). Fertilize and prune to keep trees vigorous. Grow maples recommended for your area.
4. *Chlorosis*—May be caused by several factors, e.g., deficiency of one or more mineral elements (especially iron) or moisture, excessive soil moisture or plant nutrients,

high water table, winter injury, insect feeding, gas main leak, or change in soil level. See under "Environmental Factors" in Section 2.

In *alkaline soils,* iron or lime-induced chlorosis (page 24) is characterized by a light yellowish color of young leaves between the veins. Leaves may later become dwarfed and ivory-colored. Twigs stunted and die back. Affected plants quite susceptible to winter injury. *Control:* Determine cause of chlorosis and correct condition. Check iron chlorosis by adding iron (ferrous) citrate or sulfate (0.2 to 1 per cent) to one or more late spring or early summer pest sprays. Trees and shrubs can also be treated by applying a 1-to-1 mixture of iron sulfate (sold as Copperas) and sulfur when leaves are expanding. Or use an iron chelate (e.g., Danitra, Nu-Iron, Re-Nu, Sequestrene 138 Fe, 330 Fe and NaFe Iron Chelate, Ortho Greenol Liquid Iron, Perma Green 125 or 135, Iron Tetrine, Nullapon NaFe-13, Versenol Iron Chelate, and Versene-OL) following manufacturer's directions. The iron compound is placed in a series of holes in one or more rings in soil under outer ends of branches (see page 30 on how to fertilize trees). Three ounces worked into the soil is usually enough for an average-sized shrub. Trees take 1 to 3 pounds per inch of trunk diameter at shoulder height. Follow fertilizer manufacturer's directions. Such mixing can be done when plants are fertilized. The soil should be well watered after treatment. Check with your nurseryman, county agent, or extension horticulturist on what and how much to use. Gelatin capsules containing iron (ferrous or ferric) citrate, sulfate, phosphate, ammonium citrate, or tartrate are available in some areas. Insert into 3/8 inch diameter holes drilled in the trunk base. On larger trees, holes are drilled 1 to 2 inches deep and about 3 inches apart. The usual dosage is 5 grams of iron citrate per inch of trunk diameter. Call in a competent arborist since holes should be drilled correctly (on an oblique angle, slanting downward) and then sealed with cork stoppers. Finally, wounds are painted with tree paint or grafting wax to seal out moisture and wood rot organisms. Iron chelates may also be injected into trees. Treatment is usually undertaken a few weeks *before* growth starts in spring; but treatment may be made any time.

5. *Twig Blights, Cankers, Dieback*—Sunken, or swollen, flattened, discolored cankers appear on twigs, branches, and even trunk. Twigs and small branches die back. Entire tree may die. Affected bark is often sprinkled with small "pimples" that may erupt and show minute, black, reddish-brown or coral-red, cushion-shaped bodies. Thick callus, sometimes with concentric rings, may accompany canker (Figure 4.140). Wood beneath cankered bark is discolored. Decay may start at cankers on older trees. Trees gradually weakened. *Control:* Avoid wounding bark. Make flush pruning cuts (see Figure 2.6). Cover promptly with tree wound dressing (page 36). Water and fertilize during spring and early summer to stimulate vigor. Prune out cankers 6 inches or more below diseased areas and paint over cuts with tree wound dressing. Burn all infected wood. Extensive trunk cankers cannot be surgically removed. Grow winter-hardy types recommended for your area. *Maple* species differ in resistance to Nectria Canker and Dieback.
6. *Bleeding Canker or Phytophthora Canker*—Northeastern states. Infects Norway, red, sugar, and sycamore maples. Round to elongate, reddish-brown cankers form in inner bark of trunk and larger branches. Bark becomes sunken and furrowed over cankers. Light brown to reddish-brown to black sap ("blood") oozes through openings in outer bark. Foliage often sparse, dwarfed, and yellowish-green. Leaves may wilt and branches or tops die back. *Control:* See under Beech. Grow winter-hardy maples.
7. *Branch and Trunk Cankers*—Foliage thin, dwarfed, pale green or yellow. Branches or entire trees die back from girdling cankers that often are discolored and sunken, swollen, or flattened—may follow squirrel or other types of injury. Roots may

decay. *Control:* See under Apple and Beech. Grow in well-drained, rich soil. Avoid bark injuries.

8. *Wood Rots*—Cosmopolitan. See under Birch, and (23) Wood Rot under General Diseases. Maples are quite susceptible. Control borers by removing and burning dead and dying branches. Spray trunk at monthly intervals, May to September, using DDT, dieldrin, or Thiodan. Keep trees growing vigorously.
9. *Leaf Spots, Tar Spots, Leaf Scab*—General. Seldom serious. Small to large, round to irregular, light to reddish-brown, tan, gray, yellowish-gray, or shiny, black spots (Figure 4.141). Some leaves may wither and drop early. *Control:* If practical, same as for Anthracnose (above).

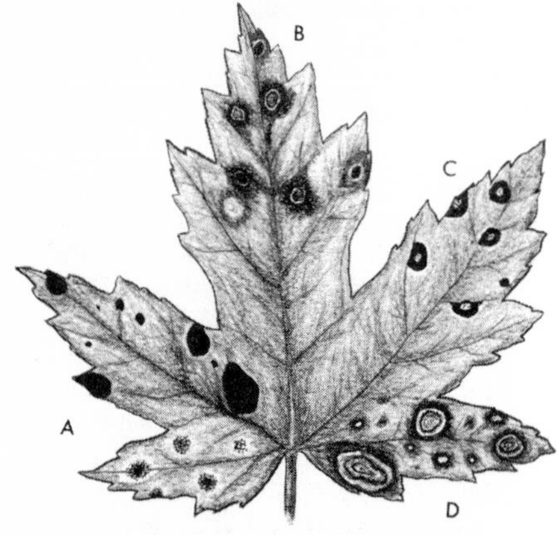

Fig. 4.141. Maple leaf spots. A. Tar spots. B. Septoria spot. C. Phyllosticta spot. D. Bulls-eye spot. All 4 types of spots would never be found on the same leaf.

10. *Root Rots*—General. Trees decline in vigor. Foliage thin and weak. Leaves may yellow, wither, and fall early. Trees tend to die back. See under Apple, and (34) Root Rot under General Diseases. May be associated with high water table or root-feeding nematodes (e.g., dagger, lance, lesion, pin, ring, root-knot, sheath, spear, spiral, stem, stubby-root, stunt). If nematodes are primary problem, apply DBCP to root zone area (page 546), following manufacturer's directions.
11. *Crown Gall*—Rough, irregular, swollen galls at trunk base or on roots. Trees lack vigor. Make poor growth. Leaves may turn yellow. *Control:* See under Apple, and (30) Crown Gall under General Diseases.
12. *Leaf Blisters* (primarily black, red, and sugar maples)—Minor. Round to irregular "blisters" that are ochre-buff to dark brown above and pinkish-buff underneath. Leaves may roll, twist, curl inward, and drop early. *Control:* If prevalent in past, apply single dormant spray. See under Peach, and (10) Leaf Curl under General Diseases.
13. *Wetwood, Slime Flux*—See under Elm. A toxic, fermenting sap flows or oozes from cracks or other bark injuries. A whitish-gray stain or brown streak is left on the bark after drying. Affected tree may show symptoms of decline; twigs and branches may die back and leaves are small. Affected trees often develop Verticillium Wilt (above). *Control:* See under Elm, Wetwood.
14. *Sooty Mold, Black Mildew*—Black mold on leaves and sometimes twigs and branches. *Control:* Spray with mixture of DDT or sevin plus malathion to control insects (e.g., aphids, scales) that secrete honeydew in which sooty mold fungi grow.
15. *Sunscald, Winter Injury*—Exposed trees are very susceptible. See under Elm and Apple. *Control:* Grow winter-hardy species in a protected location.
16. *Powdery Mildews*—Minor problem. Superficial, white patches on leaves from midsummer on. *Control:* Apply sulfur, Karathane, or Acti-dione 1 to 3 times, starting when mildew is first seen. *Maple* species differ in resistance.

17. *2,4-D Injury*—Maples and boxelders are very susceptible. Leaves twisted, curled, and cupped. Developing leaves are dwarfed and strap-shaped. *Control:* See under Grape.
18. *Thread Blight*—Southeastern states. See under Walnut.
19. *Felt Fungus*—Southern states. See under Hackberry.
20. *Seedling Blight*—See under Pine.
21. *Bacterial Leaf Spot*—Small water-soaked spots on leaves with yellowish margins. Later the spots become brown to black. *Control:* Space plants. Prune for good air circulation.
22. *Flower Blight*—Minor pest of red and silver maples. Flowers killed and seed formation is prevented. *Control:* Not necessary.
23. *Mistletoe*—See (39) Mistletoe under General Diseases.
24. *Maple Gall Mites*—Small wartlike galls on upper leaf surface; yellowish-green at first. Later turn blood-red to black. If numerous, leaves may be deformed. *Control:* Spray in spring during blossoming period or shortly after. Use a white mineral oil spray prepared for use as summer spray. Malathion applied to bark in the fall acts as a contact spray. Do *not* apply when there is danger of freeze. A *dormant* spray of malathion or lime-sulfur in spring, before budbreak, is also effective. Destroy infested leaves.

For additional information, read USDA Handbook 211, *Diseases of Shade and Ornamental Maples,* and USDA H&G Bulletin 81, *Maple Diseases and Their Control—A Guide for Homeowners.*

MARANTA, RABBIT TRACKS, PRAYER PLANT, ARROWROOT *(Maranta)*; CALATHEA, ZEBRA-PLANT *(Calathea)*

1. *Leaf Spots*—See (1) Fungus Leaf Spot under General Diseases.
2. *Rust*—See (8) Rust under General Diseases.
3. *Root Rot*—See (34) Root Rot under General Diseases. May be associated with root-knot or other root-feeding nematodes (below).
4. *Root-knot*—See (37) Root-knot under General Diseases.
5. *Other Root-feeding Nematodes* (burrowing, lesion, spiral)—Often associated with stunted, unthrifty plants. *Control:* Same as for Root-knot. Grow in sterilized soil (pages 538–548).

MARBLE QUEEN — See Calla Lily
MARBLESEED — See Mertensia
MARGUERITE, MARIGOLD — See Chrysanthemum
MARINE-IVY — See Grape
MARIPOSA LILY, GLOBE-TULIP, GLOBE LILY *(Calochortus)*

1. *Rust*—Western half of United States. Yellow to dark, powdery pustules on foliage. *Control:* Cut and burn infected foliage after plants bloom. Apply zineb, ferbam, or maneb at 10- to 14-day intervals.

MARJORAM, MARRUBIUM — See Salvia
MATRICARIA — See Chrysanthemum
MATRIMONY-VINE, CHINESE WOLFBERRY *(Lycium)*

1. *Powdery Mildews*—Powdery white patches on leaves in summer and fall. *Control:* Spray or dust 2 or 3 times. Start when mildew first appears. Use sulfur, Karathane, or Acti-done at 10-day intervals. Follow manufacturer's directions.
2. *Leaf Spots*—More or less round, tan, gray, or brown spots on leaves in rainy seasons. *Control:* Pick and burn infected leaves. If serious, spray several times, 10 days apart. Use zineb or maneb.

3. *Rusts*—Reddish-brown, then black, powdery pustules on leaves. *Control:* Not usually necessary. Same as for Leaf Spots (above).
4. *Mosaic*—See (16) Mosaic under General Diseases.

MATTHIOLA — See Cabbage
MAURANDYA — See Snapdragon
MAYAPPLE or MANDRAKE *(Podophyllum)*

1. *Rust*—General and destructive. Large areas of new leaves in spring turn yellowish in spots and later chocolate-brown. Leaves wither and die early. *Control:* Completely dig up and destroy infected plants.
2. *Leaf Spots, Leaf Blight*—Widespread in rainy seasons. Spots of various colors, sizes, and shapes. Common on leaves in shady spots. *Control:* Pick and burn spotted leaves. Where practical, apply several sprays, at about 10-day intervals. Use zineb, maneb, or fixed copper.
3. *Stem Rot*—Stem base may rot. Foliage wilts, withers, and dies. *Control:* Grow in well-drained soil. Avoid overwatering and heavy shade. PCNB applied to soil may be beneficial if applied early. Follow manufacturer's directions.
4. *Gray-mold Blight*—See (5) Botrytis Blight under General Diseases. *Control:* Same as for Leaf Spots (above).
5. *Ringspot*—See (17) Ringspot under General Diseases.

MAYDAY-TREE — See Peach
MAYFLOWER — See Heath
MAYPOP — See Passionflower
MEADOWBEAUTY — See Deergrass
MEADOWRUE — See Delphinium
MEADOW SAFFRON — See Colchicum
MEADOWSWEET — See Rose and Salvia
MECONOPSIS — See Poppy
MEDLAR — See Apple
MELIA — See Chinaberry
MELISSA — See Salvia
MELOCACTUS — See Cacti
MELON — See Cucumber
MENISPERMUM — See Moonseed
MENTHA — See Salvia
MENTZELIA, BLAZING-STAR, PRAIRIE LILY *(Mentzelia)*

1. *Leaf Spots*—Small spots usually with distinctive border. Centers of older spots sprinkled with black dots. *Control:* Pick and destroy infected leaves. If practical, spray during wet periods. Use zineb, maneb, or fixed copper.
2. *Root and Stem Rots*—Plants stunted and weak. May wilt and collapse from rot at soil line and below. *Control:* Avoid overwatering and planting in heavy, poorly drained soil. Drench or dust soil with PCNB before planting. Follow manufacturer's directions.
3. *Rusts*—Small yellow to orange spots on leaves. Alternate hosts: wild grasses or none. *Control:* Same as for Leaf Spots (above).

MENZIESIA — See Blueberry
MERRYBELLS — See Lily
MERTENSIA, BLUEBELLS or LUNGWORT, VIRGINIA COWSLIP *(Mertensia)*; ALKANET, BUGLOSS, AFRICAN FORGET-ME-NOT *(Anchusa)*; BORAGE

(Borago); HOUNDSTONGUE, CHINESE FORGET-ME-NOT (Cynoglossum); VIPER'S-BUGLOSS (Echium); HELIOTROPE (Heliotropium); PUCCOON, GROMWELL (Lithospermum); FORGET-ME-NOT (Myosotis); MARBLESEED (Onosmodium)

1. *Gray-mold Blight, Botrytis Blight* (forget-me-not, heliotrope, mertensia) —Cosmopolitan in cool, wet weather. Stem base turns brown, rots. Plant may collapse in damp weather. Flowers, leaves, buds, and shoot tips may be blighted. Affected parts often covered with coarse gray mold. *Control:* Carefully collect and burn infected parts. Space plants. Control weeds. Spray at 5- to 10-day intervals during cool, moist weather. Use captan, zineb, maneb, or fixed copper.
2. *Stem Rots, Crown Rot, Southern Blight, Wilt, Damping-off*—Plants wilt, later turn brown at soil line and rot. Affected area may be covered with cottony mold. Seedlings wilt and collapse. *Control:* Remove and destroy infected plants with 6 inches of surrounding soil. Avoid overwatering. Rotate. See also under Delphinium.
3. *Downy Mildew* (forget-me-not, houndstongue, mertensia) —Pale spots or yellowish blotches on upper leaf surface with delicate, grayish-white mold on corresponding underleaf surface in humid weather. *Control:* Spray at 5- to 10-day intervals during cool, rainy weather. Use zineb, maneb, or fixed copper.
4. *Mosaic* (anchusa, heliotrope, houndstongue, lithospermum, mertensia) —Widespread. Mottled, yellowish-green and dark green leaves. Plants may be stunted. *Control:* Dig up and burn first infected plants. Keep down weeds. Spray with malathion or lindane to control aphids that transmit virus.
5. *Aster Yellows, Curly-top* (anchusa, forget-me-not) —See (18) Yellows, and (19) Curly-top under General Diseases.
6. *Black Ringspot* (forget-me-not) —See under Cabbage.
7. *Powdery Mildew* (anchusa, forget-me-not, houndstongue, lithospermum, mertensia) —White, powdery blotches on stems and leaves. Leaves may yellow and wither. *Control:* Space plants. Increase air circulation. Dust or spray several times at 10-day intervals. Use sulfur or Karathane.
8. *Leaf Spots*—See under Chrysanthemum, and (1) Fungus Leaf Spot under General Diseases.
9. *Leaf Smut* (mertensia) —Round, pale spots or "blisters" on leaves. Later filled with black powdery masses. *Control:* See (13) White Smut under General Diseases.
10. *Root-knot*—See (37) Root-knot under General Diseases.
11. *Rusts* (anchusa, forget-me-not, heliotrope, lithospermum, marbleseed, mertensia) — See (8) Rust under General Diseases. Alternate hosts: wild grasses and rye.
12. *Root Rot*—See under Geranium, and (34) Root Rot under General Diseases. Grow in partial shade in moist, humusy, fertile soil. Avoid overwatering.
13. *Verticillium Wilt* (heliotrope) —Leaves and flowers turn black, wilt, and shrivel. Brown streaks appear inside stem when cut. *Control:* See (15B) Verticillium Wilt under General Diseases.
14. *Leaf Scorch* (heliotrope) —Leaf margins turn dark brown in hot, dry weather. *Control:* Grow in partial shade in cooler, more protected location.

MESCALBEAN — See Honeylocust

MESEMBRYANTHEMUM — See Iceplant

MESPILUS — See Apple

MEXICAN FIRE-PLANT — See Poinsettia

MEXICAN FLAME-VINE, MEXICAN SUNFLOWER — See Chrysanthemum

MEZEREON — See Daphne

MICHAELMAS DAISY — See Chrysanthemum

MICROMERIA — See Salvia

MIGNONETTE *(Reseda)*

1. *Leaf Spot, Blight*—Widespread in eastern half of United States. Small, pale tan to yellowish-brown spots with reddish-brown borders, mostly on lower leaves. Spots enlarge and quickly merge in wet weather. Leaves become blighted, reddish, and die. Stalks and seedpods may also be infected. *Control:* Sow disease-free seed. Spray weekly in wet weather. Use zineb, maneb, or fixed copper.
2. *Root Rot, Damping-off*—Seedlings wilt and collapse. Older plants stunted with yellowish leaves. Stem base and roots decay. *Control:* Destroy infected plants. Rotate. Avoid overwatering and growing in poorly drained soil. Where practical, replant in sterilized soil (pages 538–548).
3. *Downy Mildew*—Pale spots in upper leaf surface with delicate whitish mold on corresponding undersurface in moist weather. *Control:* Same as for Leaf Spot (above).
4. *Root-knot*—See (37) Root-knot under General Diseases.
5. *Verticillium Wilt*—See (15B) Verticillium Wilt under General Diseases.
6. *Black Ringspot*—See under Cabbage.
7. *Yellows*—See (18) Yellows under General Diseases.
8. *Curly-top*—See (19) Curly-top under General Diseases.

MIKANIA — See Chrysanthemum

MILK-BUSH — See Poinsettia

MILKWEED — See Butterflyweed

MILKWORT — See Phlox

MILTONIA — See Orchids

MIMOSA — See Pea

"MIMOSA" TREE — See Honeylocust

MIMULUS — See Snapdragon

MINIATURE JOSEPHSCOAT — See Cockscomb

MINNIE-BUSH — See Blueberry

MINT — See Salvia

MIRABILIS — See Four-o'clock

MISSOURI PRIMROSE — See Evening-primrose

MISTFLOWER — See Chrysanthemum

MITCHELLA — See Buttonbush

MITREWORT *(Mitella)* — See Hydrangea

MOCK-CUCUMBER — See Cucumber

MOCKORANGE — See Hydrangea

MOCK-STRAWBERRY — See Rose

MOLUCELLA — See Salvia

MOMORDICA — See Cucumber

MONARDA, MONARDELLA — See Salvia

MONEYWORT — See Primrose

MONKEY-COCONUT — See Palms

MONKEYFLOWER — See Snapdragon

MONKEYPUZZLE TREE — See Araucaria

MONKSHOOD — See Delphinium

MONKSHOOD-VINE — See Grape

MONSTERA — See Calla Lily

MONTBRETIA — See Gladiolus

MOONFLOWER — See Morning-glory

MOONSEED *(Menispermum)*; CAROLINA MOONSEED *(Cocculus)*

1. *Leaf Spots*—Widespread. Spots of various colors, sizes, and shapes on leaves and stems. *Control:* Pick and burn spotted leaves. If severe, spray several times during rainy periods, at 7- to 10-day intervals. Use zineb, maneb, or captan.
2. *Powdery Mildew* (menispermum) —Widespread. Grayish-white, powdery mold on foliage. *Control:* If serious, spray several times, 10 days apart, with sulfur or Karathane.
3. *Leaf Smut* (menispermum) —See (13) White Smut under General Diseases.
4. *Root Rot*—See (34) Root Rot under General Diseases.
5. *Burrowing Nematode*—Florida. Associated with weak, declining Carolina moonseed plants. *Control:* Set disease-free plants in sterilized soil (pages 538–548).

MORETON BAY or HOOP PINE — See Araucaria

MORNING-GLORY [BRAZIL, BUSH, CEYLON, COMMON, IVYLEAF, JAPANESE, TRICOLOR, WHITE-EDGE], BLUE DAWN-FLOWER, HEARTS AND HONEY VINE, GIANT NIGHT WHITE BLOOMER *(Ipomoea)*; ARGYREIA; MOONFLOWER *(Calonyction)*; CALIFORNIA-ROSE, BUSH or DWARF MORNING-GLORY *(Convolvulus)*; JACQUEMONTIA; CYPRESSVINE, STARGLORY, CARDINAL CLIMBER, SPANISH FLAG *(Quamoclit)*

1. *Leaf Spots*—Spots of various colors and sizes. Often with distinct margin. *Control:* Pick and burn spotted leaves. Cut and burn tops in fall. If serious, spray as for Rusts (below).
2. *Rusts*—Primarily in southern states. Unimportant in gardens. Yellowish spots on upper leaf surface and reddish-brown, powdery pustules on lower surface. Alternate hosts include pines. *Control:* Apply ferbam, fixed copper, or zineb several times, 10 days apart.
3. *Stem Rot, Stem Canker, Damping-off*—Sunken, brown, soft rot on stem or crown. May be covered with cottony mold. Seedlings wilt and collapse. *Control:* Destroy infected plants. Grow in clean, well-drained soil. Avoid overwatering.
4. *White-rust*—Widespread. Serious sometimes in southern states. See under Cabbage, and (9) White-rust under General Diseases.
5. *Root Rot*—See (34) Root Rot under General Diseases. May be associated with root-feeding nematodes (e.g., lesion, reniform, sting).
6. *Root-knot*—See (37) Root-knot under General Diseases.
7. *Leaf Nematode* (moonflower) —Brown blotches, bounded by leaf veins. Leaves may shrivel, die, and hang on plant. Variegated moonflower varieties considered more susceptible. *Control:* See under Chrysanthemum, and (20) Leaf Nematode under General Diseases.
8. *Mosaic*—Yellowish-white and greenish mottling of leaves. See (16) Mosaic under General Diseases.
9. *Curly-top*—Western states. See (19) Curly-top under General Diseases.
10. *Ringspot* (blue dawn-flower, moonflower, morning-glory) —Symptoms greatly variable depending on plant and virus strain. May be absent or produce mottling, chlorotic spotting, distortion, or small dead spots on cotyledonary or true leaves. See (17) Ringspot under General Diseases.
11. *Fusarium Wilt* (morning-glory) —See (15A) Fusarium Wilt under General Diseases.
12. *Blossom Blight*—See (31) Flower Blight under General Diseases.
13. *Thread Blight* (jacquemontia, morning-glory) —Southeastern states. See under Walnut. *Control:* Spray as for Leaf Spots (above).

MORUS — See Mulberry
MOSES-IN-A-BOAT — See Rhoea
MOSQUITO BILLS — See Primrose
MOSS-PINK — See Phlox
MOTHER-OF-THYME — See Salvia
MOUNDLILY — See Yucca
MOUNTAIN-ASH — See Apple
MOUNTAIN-BLUET — See Chrysanthemum
MOUNTAIN CRANBERRY — See Blueberry
MOUNTAIN CREEPER — See Clockvine
MOUNTAIN-EBONY — See Honeylocust
MOUNTAIN FLEECE — See Silver-lacevine
MOUNTAIN-HOLLY — See Holly
MOUNTAIN-LAUREL — See Blueberry
MOUNTAIN-MINT — See Salvia
MOUNTAIN-SPICEWOOD — See Calycanthus
MOUNTAIN SPURGE — See Pachysandra
MUGWORT — See Chrysanthemum

MULBERRY [RED or AMERICAN, RUSSIAN, WEEPING, WHITE or CHINESE] *(Morus)*

1. *Twig Blights, Cankers, Dieback*—Widespread. Small to long, sunken, girdling cankers on twigs and branches. Foliage beyond turns yellow, wilts, and dies. Branches die back. Trees may gradually die over period of years. Small, coral-pink to black "pimples" often evident in bark. *Control:* Prune and burn all twigs and branches showing cankers. Make cuts at least a foot below any sign of canker. Sterilize pruning tools between cuts (page 101). Keep trees vigorous. Fertilize and water during droughts. Varieties differ in resistance.
2. *Bacterial Spot or Blight, Dieback*—General. Small, angular, dark green, water-soaked spots on leaves and shoots that turn brown to black and become sunken. Young leaves may be distorted; turn yellow to brown. Elongated, black spots or stripes on shoots. Twigs may be blighted and die back. Trees often stunted and disfigured. *Control:* Prune and burn blighted twigs in late fall. If practical, apply copper fungicide before rainy periods. Avoid overhead sprinkling in nursery.
3. *Fungus Leaf Spots*—Common in rainy seasons. Cause little damage. Spots of various colors, sizes, and shapes. Some leaves may drop early. *Control:* Collect and burn fallen leaves. If practical, spray several times, 10 days apart. Use zineb, maneb, or fixed copper. Varieties and species differ in resistance to Mycosphaerella Leaf Spot.
4. *Powdery Mildews*—Powdery, white blotches on underleaf surface. *Control:* Keep plants pruned. Avoid overcrowding and excess shade. Spray two or three times, 10 days apart. Use sulfur or Karathane.
5. *False Mildew*—Southern states. Indefinite, whitish, cobwebby blotches on underleaf surface in midsummer. Yellowish areas later develop on upper side. Leaves may wither and fall early. Most serious in shady areas. *Control:* Same as for Fungus Leaf Spots (above).
6. *Wood and Heart Rots*—See (23) Wood Rot under General Diseases.
7. *Root-knot*—Mulberry is very susceptible. See under Peach, and (37) Root-knot under General Diseases.
8. *"Popcorn" (Berry-hardening) Disease*—Widespread but minor disease in southern states. Infected fruit carpels (drupelets) are small, hard, and remain green. Interferes with normal ripening. *Control:* None usually necessary. Spraying as for Fungus Leaf Spots (above) at flowering time may be beneficial.

9. *Root Rots*—See under Apple, and (34) Root Rot under General Diseases.
10. *Hairy Root*—See under Apple.
11. *Rust*—Southern states. Unimportant. Brownish pustules on lower leaf surface. *Control:* Same as for Fungus Leaf Spots (above).
12. *Wetwood, Slime Flux*—See under Elm.

MULLEIN-PINK — See Carnation
MUSCARI — See Tulip
MUSKMELON — See Cucumber
MUSK MALLOW — See Hollyhock
MUSTARD — See Cabbage
MYOSOTIS — See Mertensia
MYRICA — See Waxmyrtle

MYRTLE *(Myrtus)*; FEIJOA, PINEAPPLE GUAVA *(Feijoa)*; GUAVA [CATTLEY or STRAWBERRY, COMMON] *(Psidium)*

1. *Powdery Mildew* (myrtle)—Southern states. Grayish-white, powdery growth on leaves. Most common in crowded, shaded areas. *Control:* Spray two or three times, 10 days apart, with Karathane.
2. *Leaf Spots*—Small spots or blotches. *Control:* Pick and burn spotted leaves.
3. *Stem or Crown Rot, Seedling Blight*—Plants wilt and die from rot at soil line. *Control:* Avoid overwatering, overcrowding, and planting in poorly drained soil. See under Delphinium.
4. *Root Rots*—See under Apple, and (34) Root Rot under General Diseases. Declining, unthrifty plants may be associated with root-feeding nematodes (e.g., burrowing, dagger, root-knot, tetylenchus).
5. *Anthracnose, Spot Anthracnose or Scab, Leaf and Fruit Spots, Fruit Spots* (feijoa, guava)—Leaves and fruits variously spotted and rotted. Ripe fruits may rot and be covered with gray, brown, black, bluish-green, or pink mold. *Control:* Use malathion plus DDT or lindane to control insects. Spray during rainy periods using captan or zineb.
6. *Root-knot*—If severe, tops may be stunted and die back. Guava is very susceptible. See (37) Root-knot under General Diseases.
7. *Damping-off* (guava)—Seedlings wilt, wither, and die. *Control:* Sow seed in 50:50 mixture of vermiculite and peatmoss or sandy loam that has been pasteurized (pages 538–548). Treat seed with captan or thiram. Spray young seedlings with captan (1 ounce in 3 gallons of water).
8. *Wood Rots*—Attack guava through wounds. See (23) Wood Rot under General Diseases.
9. *Thread Blight* (feijoa, guava)—Southeastern states. Small irregular spots on upper leaves that merge and spread downward until whole plant may die. Spiderweb-like fungus strands may be seen on leaves in humid weather. *Control:* Keep foliage protected in wet weather. Use captan or zineb.
10. *Algal Leaf Spot* (guava)—See under Magnolia.
11. *Zinc Deficiency, Little Leaf* (guava)—Growth stunted, may die back. Leaves dwarfed, yellowish, crinkled and distorted. *Control:* Apply nutritional spray containing zinc sulfate or chelate (page 27).

MYRTLE BOXLEAF — See Bittersweet

NANDINA, HEAVENLY BAMBOO *(Nandina)*

1. *Leaf Spot, Anthracnose*—Southern states. Leaves spotted. Centers of red spots may later turn almost black. *Control:* Pick and burn spotted leaves. If serious, spray several times, 10 days apart. Use zineb, maneb, or fixed copper.

2. *Root-knot*—Plants may be weak and stunted with nodule-like galls on roots. See (37) Root-knot under General Diseases.
3. *Root Rot*—See (34) Root Rot under General Diseases. May be associated with nematodes (e.g., dagger, lance, lesion, root-knot, spiral).
4. *Chlorosis*—Foliage yellows in alkaline soils. Plants often stunted. See under Maple.

NANNYBERRY — See Viburnum
NARCISSUS — See Daffodil
NASTURTIUM (Watercress) — See Cabbage
NASTURTIUM [COMMON GARDEN, DWARF], CANARYBIRDFLOWER *(Tropaeolum)*

1. *Bacterial Wilt*—Mostly southern states. Plants may wilt, yellow, and die before blossoming. Stems near soil line often water-soaked. Black streaks appear when stem is cut through. Roots decay and turn black. *Control:* Remove and burn infected plants. Grow in clean soil that has not grown wilted potato, tomato, pepper, eggplant, tobacco, zinnia, dahlia, marigold, or other plants. Rotate plantings.
2. *Fungus Leaf Spots*—Spots of various sizes, shapes, and colors. Leaves may wither and drop early. Heterosporium spot may also cause reddish-brown rotting of stems. Seedlings may turn yellow, wilt, and collapse. It is severe in California along the coast. *Control:* If serious, apply zineb, maneb, or fixed copper at about weekly intervals during rainy weather. Commercial seedsmen control seed-borne *Heterosporium* by soaking seed in hot water, 30 minutes at 125° F. The seed is first presoaked in cool tap water for an hour. Rotate.
3. *Bacterial Leaf Spot*—Small, water-soaked, brownish spots on leaves. Leaves may later rot. *Control:* Pick and burn spotted leaves. Spray as for Fungus Leaf Spots, using fixed copper.
4. *Mosaics*—Symptoms highly variable. Yellow spotting and mottling of leaves. Young leaves may be ruffled and cupped. Large, yellow to brown, arrow-shaped blotches or white, dead ringspots may also develop on leaves. Flowers show distinctive color break. May be small and crinkled. *Control:* Destroy infected plants when first found. Keep down weeds. Spray or dust weekly with mixture of malathion plus DDT or sevin to control aphids that transmit viruses.
5. *Spotted Wilt*—Yellowish to brown spots and blotches form on leaves. Leaves distorted, cupped, and stunted. May show yellowish mottling. Plants usually stunted. *Control:* Same as for Mosaics (above). Virus is spread by thrips.
6. *Ringspot*—Leaves mottled light and dark green with yellowish-green and yellow ring and line patterns. Or yellowish spots border veins. Leaves may be crinkled, stunted, and partly dead. Plants may apparently recover and appear normal. *Control:* Same as for Mosaics (above).
7. *Curly-top*—Older leaves usually yellow. Numerous shoots produced with dwarfed, cupped leaves. Parts of immature flowers withered and dry. Flower buds usually dwarfed and yellow. May fail to open. *Control:* Same as for Mosaics (above).
8. *Aster Yellows*—Plants stunted, yellow, and bushy. *Control:* Same as for Mosaics (above). Virus is spread by leafhoppers.
9. *Root-knot, Root Gall*—Small knots or galls on roots. Plants may appear stunted and weak. *Control:* See (37) Root-knot under General Diseases.
10. *Fasciation*—Rare. See (28) Leafy Gall under General Diseases.
11. *Rust*—Rare. Small, yellowish pustules on leaves. Alternate hosts: wild grasses *(Aristida* and *Distichlis). Control:* Same as for Fungus Leaf Spots (above).
12. *Stem Rot*—See (21) Crown Rot under General Diseases.
13. *Gray-mold Blight*—See (5) Botrytis Blight under General Diseases.

NATAL-PLUM — See Oleander
NECKLACE PLANT or VINE — See Sedum
NECTARINE — See Peach
NELUMBO — See Waterlily
NEMOPANTHUS — See Holly
NEMOPHILA — See Phacelia
NEPETA — See Salvia
NEPHROLEPIS — See Ferns
NEPHTHYTIS — See Calla Lily
NERINE — See Daffodil
NERIUM — See Oleander
NEW GUINEA BEAN — See Cucumber
NEW JERSEY-TEA or JERSEY-TEA, BLUE-BLOSSOM, POINT REYES LILAC or CREEPER, RED-HEART, CEANOTHUS [DELISLE, FENDLER, HOLLYLEAF, INLAND, SANTA BARBARA] *(Ceanothus)*

1. *Leaf Spots*—Common. Small, more or less round spots. *Control:* Not usually necessary. If serious, spray during rainy periods. Use zineb or maneb.
2. *Powdery Mildew*—Widespread in late summer and fall. Powdery, white patches on leaves. *Control:* Apply sulfur or Karathane weekly. Start when mildew first appears.
3. *Rust*—Small, yellowish spots on leaves. Alternate hosts are wild grasses. *Control:* Same as for Leaf Spots (above).
4. *Crown Gall*—See under Apple, and (30) Crown Gall under General Diseases.
5. *Wood Rot*—See under Birch, and (23) Wood Rot under General Diseases.
6. *Root Rot*—See under Apple, and (34) Root Rot under General Diseases.
7. *Dieback, Canker*—See under Apple and Maple.

NEW ZEALAND SPINACH — See Beet
NICANDRA, NICOTIANA, NIEREMBERGIA, NIGHTSHADE — See Tomato
NINEBARK [COMMON, DWARF, GOLDLEAF, ILLINOIS] *(Physocarpus)*

1. *Leaf Spots*—Not serious. See under Maple.
2. *Powdery Mildew*—Not serious. See under Birch.
3. *Wood Rot*—See under Birch.
4. *Root Rot*—See (34) Root Rot under General Diseases.
5. *Fire Blight*—See under Apple.

NORFOLK ISLAND PINE — See Araucaria
NOTHOSCORDUM — See Onion
NUPHAR, NYMPHAEA — See Waterlily
NYCTOCEREUS — See Cacti
NYSSA — See Dogwood
OAK [BLACK, BLACKJACK, BUR, CALIFORNIA BLACK, CHERRYBARK, CHESTNUT (CHINESE, SWAMP or BASKET, CHINQUAPIN, and DWARF CHINQUAPIN), COLUMNAR ENGLISH, CORK, ENGLISH, EVERGREEN WHITE, HOLM or HOLLY, LAUREL, LIVE (CALIFORNIA, CANYON, INTERIOR, SOUTHERN), PIN and NORTHERN PIN, POST, OVERCUP, RED, SCARLET, SCRUB, SHINGLE, SHUMARD, TEXAS, TURKEY, WATER, WHITE (OREGON, ROCKY MOUNTAIN, SWAMP, VALLEY), WILLOW] *(Quercus)*; TANBARK-OAK *(Lithocarpus)*

1. *Oak Wilt*—Serious over much of the northern two-thirds of eastern half of United States. Spreading slowly both south and west. *All* oaks are susceptible. Chinese,

European, and American chestnut, tanbark-oak, and bush chinquapin may also be infected. Uncommon on planted specimen trees. Leaves on trees in red and black oak groups may turn pale or dull green or be water-soaked, curl, progressively become more yellowed or bronzed from tips and margins inward. Leaves on upper branches are usually affected and drop first. Red and black oaks die in a few weeks during late spring and summer. Individual white or bur oaks die back slowly (becoming stag-headed) over period of several years or longer. Leaves on wilting bur and white oaks usually turn light brown or straw-color, curl, wilt, and remain attached. Brown or black streaks often occur in wood of red and black oaks—occasionally in bur oak—just under bark of wilting branches. See Figures 4.142A, B and 4.142C. Causal fungus may spread to nearby trees by underground root grafts, sap-feeding insects, or squirrels. *Control:* Girdle or remove and burn all infected and dead oaks as soon as possible. If other oaks are nearby on property, dig a trench 3 to 4 feet deep, or treat equidistant between trees with SMDC (Vapam or VPM Soil Fumigant) diluted 1 to 4 with water as given under Elm, Dutch Elm Disease. Or poison all oaks with brush-killer (50:50 mixture of 2,4-D and 2,4,5-T in fuel oil) or Ammate within a 50-foot circle of the infected tree. Do not injure or prune trees from March through June. Paint wounds promptly with tree wound dressing (page 36). Check with your nurseryman, county agent, or extension plant pathologist. *Infected trees cannot be cured.*

2. *Anthracnose, Leaf and Twig Blight*—White, bur, and chestnut oaks are very susceptible. Most serious on bottom half of trees in shaded areas during moist spring weather. Leaves turn brown, brittle, and curl from margins giving a scorched

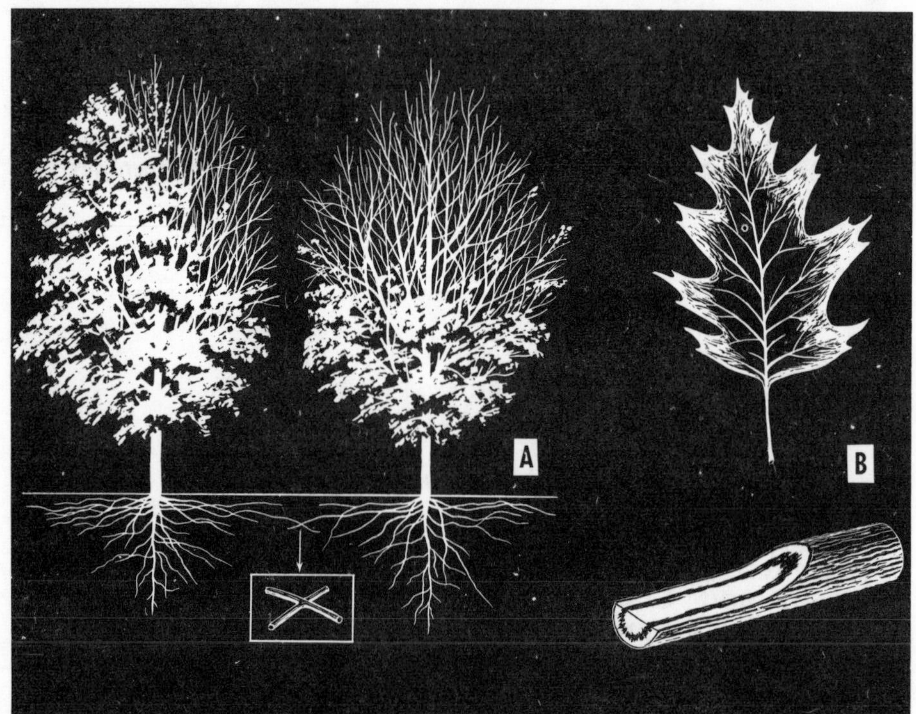

Figs. 4.142A and B. Oak wilt. A. Gross symptoms of red or black oaks (insert shows grafting of roots). B. Red oak leaf and twig symptoms.

Fig. 4.142C. "Stag-head" of white oak. (Iowa State University photo)

Fig. 4.143. Oak anthracnose.

appearance. Small to large, round to irregular, light brown to black areas occur on black, laurel, pin, red, Shumard, water, and willow oak leaves. Spots often enlarge along veins. Twigs may die back from cankers. Weakened trees may die if leaves drop early. A new set of leaves may appear later in the season. See Figure 4.143. *Control:* Collect and burn fallen leaves and diseased twigs. Thin out crown to increase air circulation. Fertilize severely injured trees to increase vigor. *Oak* species vary greatly in resistance. Spray when leaves half-grown and twice more at 10- to 14-day intervals. Use phenyl mercury, zineb, dichlone, or thiram. A phenyl mercury spray as buds swell in early spring has proven effective.

3. *Leaf Blister, Leaf Curl* (oaks) —General in cool, wet springs. Yellowish-green to gray, reddish, purple, yellow or brown, more or less round, raised blisters on upper leaf surface (Figure 4.144). Trees may appear scorched. Affected leaves may pucker

Fig. 4.144. Oak leaf blister or leaf curl. (Courtesy Dr. V. H. Young)

or curl and drop in large numbers, weakening affected trees. A new set of leaves may appear later in the season. *Control:* If practical, spray 1 to 2 weeks *before* buds swell. Use zineb, maneb, ferbam, dichlone, captan, or fixed copper. Collect and burn fallen leaves. *Resistant oaks:* Chinquapin, pin, post, and Shumard; *slightly susceptible:* black, blackjack, and southern; *highly susceptible:* Cherrybark, laurel, live, red, southern red, water, and willow oaks.

4. *Leaf Spots, Spot Anthracnose, Frosty Mildew*—Small to large, round to irregular, spots or patches of various colors develop on leaves in wet seasons. If severe, some leaves may wither and drop early. *Control:* Same as for Anthracnose (above).
5. *Wood Rots, Heart Rots, Butt Rot*—Cosmopolitan. See under Birch, and (23) Wood Rot under General Diseases. Prune and burn dead and dying limbs. Control borers by monthly trunk sprays, May to September. Use DDT, dieldrin, or Thiodan. Keep trees vigorous. Water during droughts and fertilize in spring or late fall, based on a soil test (page 25).
6. *Twig Blights, Dieback*—Cosmopolitan. Leaves scattered over tree suddenly turn pale green, later brown, curl, and hang downward. Twigs and small branches die back from discolored, sunken, girdling cankers in which black "pimples" are embedded (Figure 4.145). Commonly follows drought. *Control:* Prune infected twigs, where practical, back 6 inches into healthy wood. Keep trees vigorous by watering during droughts and fertilization. Prune and burn all dead wood. Spray as for Anthracnose (above).
7. *Branch and Trunk Cankers*—Widespread. Discolored, enlarging, flattened or sunken cankers on branches or trunk. Small trees may die. The causal fungi enter through dead branch stubs, insect injuries, or other wounds. See under Elm.
8. *Root Rots*—Cosmopolitan. Trees lack vigor, decline. Foliage thin and weak. Leaves may yellow, wither, and drop early. Stag-headed, dead branches are common (Figure 4.146). Similar symptoms are produced by drought or heavy, waterlogged

Fig. 4.145 (above). Dothiorella canker on oak branches. Fruiting bodies of the causal fungus develop as black, raised pustules in the dead bark. (Illinois Natural History Survey photo)

Fig. 4.146. Armillaria or shoestring root rot of white oak. (Illinois Natural History Survey photo)

soil. Trunk may break over during high winds. Roots and trunk base are partially or completely rotted. Coarse, fan-shaped strands of white or brownish-black mold, and black rootlike strands (*Armillaria*) growing just under the bark or over roots are common. Clumps of toadstools may appear near trunk base. See Figure 2.40B. *Control:* See under Apple.

9. *Crown Gall*—See under Apple, and (30) Crown Gall under General Diseases.
10. *Leaf Rusts*—Minor problem. Small, yellowish spots with brown, bristle-like tendrils on underleaf surface. Later, brown pustules develop. Alternate hosts: 2- and 3-needle pines. *Control:* None needed. Full-size leaves are resistant to infection.
11. *Powdery Mildews* (oaks)—General. May cause injury in southern and western states. White to brown, powdery mold on leaves, new shoots, and buds. Leaves may be stunted, distorted, yellowish; later wither, and drop early. Twigs may be stunted and bunchy. Produce witches'-brooms—short bushy shoots with dwarfed leaves—of live oak in California. Oaks vary greatly in susceptibility. *Control:* If practical, spray when mildew is first evident. Use sulfur or Karathane. Prune out witches'-brooms on live oak along Pacific Coast. Apply dormant spray of lime-sulfur (1 to 50 dilution). Or spray at bud break (February or early March) and again in April, with Acti-dione.
12. *Chlorosis, Iron Deficiency*—Pin and willow oaks are very susceptible in neutral and alkaline soils. Yellowish-green to ivory coloration of leaves, especially between the veins (Figure 4.147). Where severe, leaves curl and turn brown along the margins, or they develop brownish areas between the veins. Twig growth is stunted. May die back. *Control:* See under Maple.
13. *Sooty Mold, Black Mildew*—Primarily southeastern states. Purplish-black mold patches on foliage. Often follows attacks by scales and other insects. *Control:* See (12) Sooty Mold under General Diseases.
14. *Leaf Scorch*—Yellowing, browning, or scorching of leaves between veins or along margins. Follows hot, dry, windy weather in July and August (Figure 4.148). Leaves may curl upward. *Control:* Water during summer dry periods. Prune, water, and fertilize to increase vigor.
15. *Smooth Bark Patch of Post and White Oaks*—Small to large, round to irregular,

Fig. 4.147. Iron chlorosis of pin oak. Affected leaves are lighter green than normal; may be ivory- or cream-colored and stunted (right). Nearly normal leaves to left.

Fig. 4.148. Leaf scorch of oak. Scorch is a noninfectious disease that develops as browning or yellowing between the veins or along margins of leaves, or as complete browning and withering of leaves. (Illinois Natural History Survey photo)

smooth, light-gray, sunken patches where outer, rough, dead bark has sloughed away. Trees are not injured.

16. *Verticillium Wilt* (oaks) —See under Maple.
17. *Wetwood, Slime Flux*—See under Elm.
18. *Mistletoe*—See (39) Mistletoe under General Diseases.
19. *Bleeding Canker*—Northeastern states. See under Beech and Maple.
20. *Felt Fungus*—Deep south on neglected willow and water oaks. Smooth, shiny, chocolate-brown to almost black growth on bark. See under Hackberry.
21. *Root-feeding Nematodes* (dagger, lance, lesion, needle, pin, ring, root-knot, sheath, spear, spiral, stem, sting, stubby-root, stunt, trophotylenchulus) —May be associated with Root Rots (above) and weak, declining trees. *Control:* See under Peach. If nematodes are the primary problem, apply DBCP to root-zone area. Follow manufacturer's directions.
22. *Damping-off, Seed (Acorn) Rot*—Seedlings wither and die. Often occurs in patches. Acorn and roots decay. May be covered with mold. *Control:* See under Pine.
23. *2,4-D Injury*—Oaks are very susceptible. Leaves become thickened, cupped downward or rolled. More severe injury produces long, narrow leaves with prominent veins. *Control:* See under Grape.

OCEANSPRAY — See Holodiscus

OCIMUM — See Salvia

OCONEE-BELLS — See Galax

ODONTOGLOSSUM — See Orchids

OENOTHERA — See Evening-primrose

OKRA — See Hollyhock

OLEA — See Osmanthus

OLEANDER [COMMON, SWEET] *(Nerium[1]);* CARISSA, HEDGETHORN, CARANDA (KARANDA) or PERUNKILA, NATAL-PLUM *(Carissa);* FRANGIPANI, RED JASMINE *(Plumeria);* CRAPE-JASMINE, ERVATAMIA *(Tabernaemontana);* CONFEDERATE-JASMINE *(Trachelospermum)*

1. *Leaf Spots, Spot Anthracnose or Scab*—Small to large spots of various colors and shapes on leaves and seedpods. Infected leaves may wither and fall early. *Control:* Pick spotted leaves. Spraying at 10-day intervals during rainy periods should prove beneficial. Use zineb or maneb.
2. *Bacterial Gall or Knot* (oleander) —Wartlike galls form on branches, shoots, leaves, and even flowers. Young leaves and seedpods may be distorted and curled. Canker-like tumors on older branches are soft or spongy and rough. Such galls darken with age. Plants weakened and unsightly. *Control:* Prune infected parts 12 inches, if possible, below any sign of infection. Dip shears in 70 per cent alcohol, fresh household bleach diluted 1:9, 5 per cent amphyl, or 2 per cent Lysol between cuts. Select cuttings or propagate only from high branches of healthy plants. Control scales, aphids, and mealybugs by spraying regularly with malathion or lindane. Disinfect cuttings by dipping in streptomycin solution (300 parts per million). Keep water off foliage when sprinkling.
3. *Sooty Mold, Black Mildew*—Gulf states. Black, powdery patches on foliage. Follows insect attacks. *Control:* Apply malathion or lindane to control insects.
4. *Cankers, Dieback*—Shoots and twigs die back from discolored cankers. *Control:* Prune infected parts. Make cuts several inches below any sign of infection. Spraying as for Leaf Spots (above) should be beneficial.
5. *Root-knot*—See under Peach, and (37) Root-knot under General Diseases.

[1] *All parts* of the *oleander* are toxic. One leaf can kill an adult. Never gather up trimmed branches and leaves and burn them. Even the smoke is highly toxic!

6. *Root Rots*—See under Apple, and (34) Root Rot under General Diseases. May be associated with nematodes (e.g., burrowing, dagger, root-knot, stem, stubby-root). Grow in moist, well-drained soil well supplied with organic matter.
7. *Rust* (frangipani)—Reddish-brown, later black, powdery pustules on leaves. *Control:* Same as for Leaf Spots (above).
8. *Mistletoe* (frangipani)—See (39) Mistletoe under General Diseases.
9. *Chlorosis*—Yellowed foliage. Most common in alkaline or very acid soils. See under Maple.

OLEASTER — See Russian-olive
OLIVE — See Osmanthus
ONCIDIUM — See Orchids
ONION [COMMON, NAPLES, NODDING, POTATO or MULTIPLIER, WELSH or SPANISH, WILD], CHIVES, CHINESE CHIVES, GARLIC [COMMON, GIANT], LEEK, LILY LEEK, ORNAMENTAL ALLIUMS, SHALLOT, STARS OF PERSIA *(Allium)*; FALSE-GARLIC *(Nothoscordum)*

1. *Neck Rot, Gray-mold Blight, Leaf Blight*—Widespread; serious storage problem. Leaves may die in field. Soft, sunken, spongy areas on neck, in field or more commonly in storage, that spread into bulb. Gray mold may grow over and between bulb scales. Neck and bulb soften; appear somewhat brownish, water-soaked, and cooked. See Figures 2.51B, 4.149A, and 4.149B. Bulbs later mummify and may develop small to large, brown to black, crustlike masses (sclerotia). White *onions* usually much more susceptible than yellow or red types. Neck rot often follows Downy Mildew and Bacterial Soft Rot (both below). Bulbs may be covered with mold. *Control:* Sort bulbs well before storage. Handle carefully to avoid wounding. Cure mature bulbs at 90° to 115° F. for 2 to 3 days (or 60° to 80° F. for 2 weeks). Store *only* healthy, mature, well-dried onions at 32° to 36° F. with good ventilation in shallow boxes, trays, or slatted crates. Keep humidity as low as practical (65 to 75 per cent). Collect and burn plant debris after harvest. Spray in field as for Blast, Downy Mildew, and Purple Blotch (all below) during wet periods. Avoid late

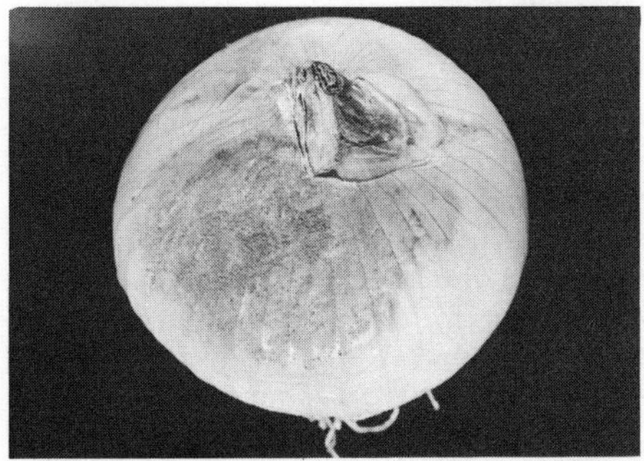

Fig. 4.149A (above). Onion neck rot. Gray mold (*Botrytis*) fruiting on the surface. (Pennsylvania Agricultural Extension Service photo)

Fig. 4.149B. Longitudinal section through an onion bulb affected with Botrytis neck rot. (USDA photo)

fertilizer applications, especially those containing nitrogen. Keep down weeds. Grow in well-drained soil.

2. *Bacterial Soft Rot*—Widespread and destructive. Bulb and leaves are water-soaked then mushy, slimy, and foul-smelling. Follows injury (sunscald, frost, hail, insects, cultivator or harvest wounds, other diseases). *Control:* Same as for Neck Rot (above).

3. *Bulb Rots and Molds, White Rot*—General in moist soils. Leaves often yellow or wilt, shrivel, and die back. May collapse. Roots and bulb progressively decay. Bulb may soften and be covered with white, pale pink, black, lemon-yellow, bluish-green, or gray mold. *Garlic* plants may be destroyed before emergence. Bacterial Soft Rot may follow. Often associated with nematodes, mites, maggots, and other insects. Loss occurs in both field and storage. See Figure 2.51B under General Diseases. *Control:* Grow blemish- and disease-free sets and transplants in clean, rich, well-drained soil. Dip *shallot* bulbs in Dowicide B (1 ounce in 3 gallons) for 15 minutes before planting. Long rotation. Keep down weeds. Collect and burn crop debris after harvest. Cure and sort bulbs well before storage. See Neck Rot (above). Avoid injuring bulbs or cloves in field, at harvest, or during storage. Keep plants vigorous throughout season. Control other diseases. *Onions* differ in resistance. Colored varieties are much less susceptible than white onions. Elba Globe, Grandee, Hickory, Iowa 44, and Premier *onions* have some resistance to Fusarium Basal Rot. Control onion maggots using seed pelleted with Diazinon or apply 4-inch band in planter furrow. Check with your county agent or extension entomologist. Follow manufacturer's directions. Apply PCNB to *garlic* or *shallot* sets (or cloves); or apply as dust or spray in planter furrow. Follow manufacturer's directions. May combine with Diazinon. Penicillium decay of *garlic* is controlled by granular Sorbistat K or Mycoban applied in planting furrow. Botran controls White Rot *(Sclerotium)*. Follow manufacturer's directions.

4. *Smut*—General, especially in northern states. Elongated, blister-like streaks occur in bulb scales or seedling leaves. Streaks filled with dark brown to black, powdery masses that later break out. Leaves curled, swollen, and distorted. Plants stunted; may fail to bulb properly. Most infected seedlings die early. See Figure 2.27B under General Diseases. *Control:* Set out disease-free sets or transplants or grow seedlings in clean soil. Treat seed using 1 ounce of 75 per cent thiram or captan for each 3 ounces of seed. Or use seed protectant containing captan or thiram plus dieldrin (page 531). Hexachlorobenzene (sold as Anti-carie 80) has given excellent control. Follow manufacturer's directions. Commercial growers used to drip formaldehyde solution into planter furrow (1 part of 40 percent formalin to about 100 parts of water). Use 1 pint to 35 feet of row when soil surface is rather dry. Nabam and urea-formaldehyde (UFC-85, N-dure, Uracide) are also effective liquid treatments at a 1 per cent concentration, and applied like formaldehyde. Smut-resistant *onions,* where adapted: Evergreen Bunching (Nebuka), Beltsville Bunching, White Welch, and Winterbeck. *Chives* is resistant; *garlic* is apparently immune. Other *Allium* species are highly resistant.

5. *Blast or Tip Blight, Botrytis Leaf Blight and Leaf Fleck, Tipburn*—Widespread. Small, pale, paper-like flecks or spots on leaves during or following cloudy, warm, humid weather. Leaf tips commonly wilt and die back. Leaves may turn light tan, then brown, collapse, and die. Bulbs undersized and immature. Accumulation of ozone at relatively high concentrations during rainstorms may lead to leaf flecking followed by tipburn. *Control:* Grow disease-free sets or transplants in well-drained soil where air circulation is good. Keep down all weeds, especially perennial and wild onions and garlic. Four-year rotation. Destroy plant debris after harvest. Avoid overcrowding and overfertilizing with nitrogen. Spray weekly with zineb, maneb, Dithane M-45, Dyrene, Daconil 2787, or captan, plus spreader-sticker. Start when seedlings are 3 to 4 inches tall. Three to 10 applications may be needed.

Control thrips and other insects with malathion or Diazinon. Control other diseases. *Onion* varieties resistant to Botrytis should be available soon.

6. *Downy Mildew*—General during cool, moist weather. Sunken, pale green to grayish areas develop in leaves that become covered with pale purplish to gray fuzzy mold. Leaves may yellow, die back, and break over (Figure 4.150). Other plants may be stunted with distorted, pale green leaves. Bulbs undersized, often soft and immature. Resistant onions (e.g., Calred and hybrids) appear promising. *Control:* Same as for Blast (above). Maneb, zineb, and Daconil 2787 give good control. Red *onions* have some resistance.
7. *Purple Blotch, Alternaria Leaf Blights*—Widespread in wet seasons. Small to large, gray to brown or purplish, sunken blotches on leaves, flower stalk, and bulb. Mold may grow on affected areas in moist weather (Figure 4.151). Leaves may yellow, die back, and collapse. Bulbs often rot starting at neck. Bulb tissue is deep yellow, gradually turning wine-red. May be serious in storage. Common following insect injury or other diseases. Sweet Spanish onion is very susceptible. *Control:* Same as for Neck Rot (above). Treat seed as for Smut (above). Four-year rotation. Apply maneb, zineb, Daconil 2787, difolatan, or Dyrene at 5- to 7-day intervals. Add detergent or spreader-sticker to insure wetting foliage. Control thrips with DDT, malathion, or Diazinon. *Onion* varieties differ in resistance. Those with waxy foliage are most resistant. Resistant *shallots* may be available soon.
8. *Pink Root, Fusarium Root Rot*—Widespread. Pink Root is most common on onions, garlic, and shallots. Seedlings wilt and die. Leaves often die back from tips. Plants stunted. Roots turn pink to red or yellowish-brown, shrivel, darken, and die (Figure 4.152). Bulbs may be dwarfed or undersized. Often follows cold, heat, drought, flooding, lack of fertilizer, or other unfavorable growing conditions. *Control:* Plant sets or transplants grown in disease-free soil. Grow in clean, mellow, well-drained, fertile soil. Turn under green manure crop several weeks before planting. Or fumigate with chloropicrin, Vorlex, or SMDC (Vapam or VPM Soil

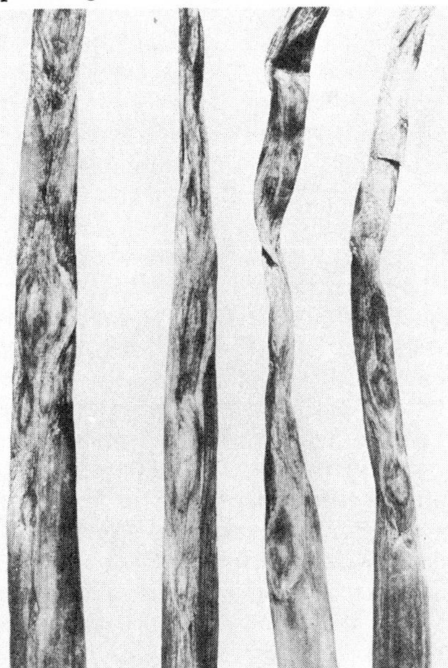

Fig. 4.150. Downy mildew on older onion leaves. (USDA photo)

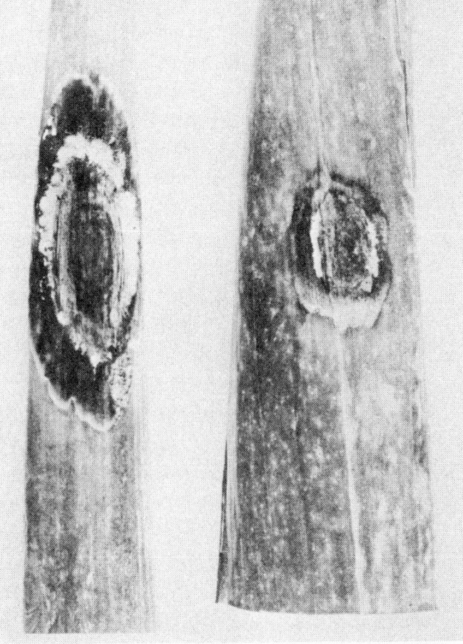

Fig. 4.151. Purple blotch stem lesions showing masses of fungus (*Alternaria*) spores and typical light and dark zonation. (USDA photo)

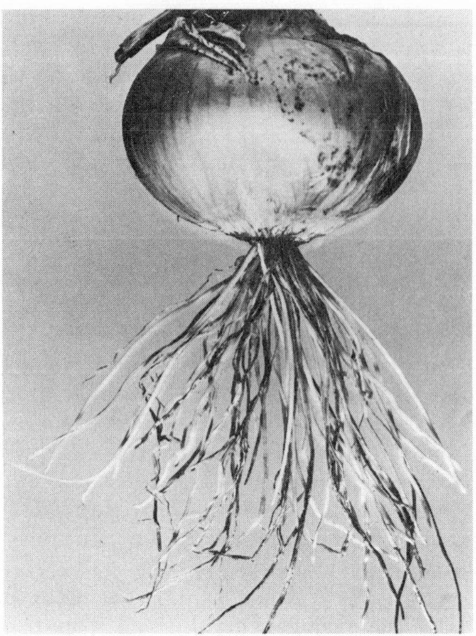

Fig. 4.152 (above). Onion affected with pink root. (USDA photo)

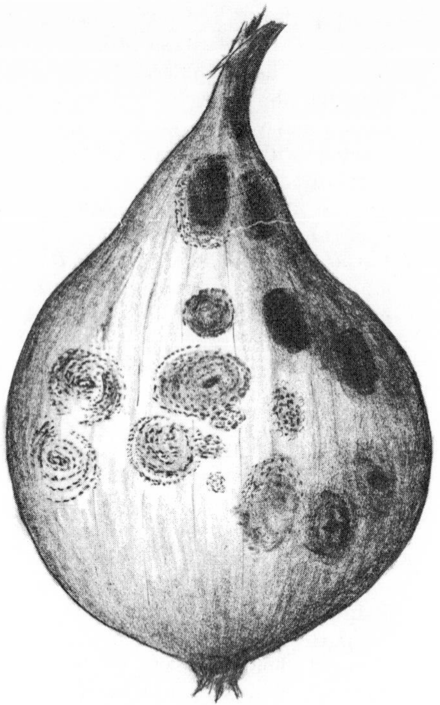

Fig. 4.153. Onion smudge.

Fumigant). See pages 538–548. Keep plants growing vigorously. Five- or 6-year rotation. *Onion* hybrids and varieties resistant or tolerant to Pink Root, where adapted: Asgrow Y2, Beltsville Bunching, Bermex 5, Brilliance, Colorado No. 6, Early Crystal 281, Eclipse, Evergreen Bunching (Nebuka), Excel, Granex, Granex 33, L-36, Texas Hybrid 28, White Granex, Yellow Bermuda, and strain of Yellow Sweet Spanish. Much more resistant hybrids should be available soon. Check with your state or extension plant pathologist. *Leek* and *chives* are highly resistant to Pink Root. Resistant *shallots:* Louisiana Pearl, Evergreen. Onions should *not* follow ladino clover, ryegrass, or corn.

9. *Bloat, Stem and Bulb Nematode* (primarily garlic, onion, and shallot)—Seedlings stunted, pale, twisted, wrinkled, and deformed. Infested sets develop much stunted, yellow, and wilted foliage. Outer bulb scales soft and mealy. Inner bulb scales swollen. Splitting or double bulbs are common. Plants die gradually. Root- and bulb-rotting fungi and bacteria often follow. Bulbs lightweight, punky, store poorly. Worse in wet seasons and on heavy soils. See Figure 2.53A under General Diseases. *Control:* Grow disease-free sets, transplants, or cloves in clean soil. Or plant in soil fumigated in fall before planting. Use D-D, Vorlex, or Telone. (Do *not* use EDB where onions will be grown within 3 years.) Disinfect tools and equipment. Dig up and destroy infested plants when first found. Thoroughly clean up and burn all plant debris after harvest. Rotate 3 or 4 years with resistant crops (e.g., potato, carrot, or lettuce). Do not save bulbs grown in infested soil! Soak suspicious *shallot* bulblets (cloves) in hot water (115° F.) for 1 hour. Soak *garlic* cloves in hot water (110° F.) and formalin (1:200 or 1 pint of 40 per cent formaldehyde in 25 gallons) for 3 hours just before planting. Or presoak *garlic* cloves at 100° F. and 1:200 formalin solution for 30 minutes followed by a 20-

minute soak in water-formalin (1:200) at 120° F. Soak *onion* sets in hot water (110° F.) for 2 hours. Cool, and plant.

10. *Root-knot*—Tops pale green and stunted. Small, thick-necked bulbs are produced. Small, round swellings form on roots. Crooked roots are common. *Control:* If serious, apply D-D, Vorlex, or Telone as for Bloat (above).
11. *Smudge, Anthracnose* (primarily white onions, false-garlic, leek, shallot)—General, especially in low areas. Dark green to black, often ringed, unsightly blotches (made up of small dots) on bulb or neck (Figure 4.153). Usually at side or top. Stored bulbs may shrivel slightly or sprout prematurely. Seedlings may damp-off. *Control:* Grow yellow or red *onions* or disease-free sets. Otherwise, same as for Neck Rot (above). Provide good ventilation in storage. Rotate.
12. *Yellow Dwarf, Mosaics, Yellows* (false-garlic, garlic, onion, ornamental alliums, shallot)—Widespread. Leaves yellowed, or with yellow and green stripes. Usually severely stunted, twisted, and crinkled. Flower stalks dwarfed, curled, and twisted. Bulbs undersized. *Shallot* leaves spindly and yellowed. Seed sets and edible parts may not form. Certain *onion* varieties may show no symptoms. *Control:* Grow virus-free sets or transplants. Destroy wild or volunteer onions. Isolate bulb- from seed-producing fields as far as possible. Keep down weeds. Control aphids that transmit the viruses. Use malathion or Diazinon. Tolerant *onions:* Beltsville Bunching, Burrell's Sweet Spanish, Colorado No. 6, Crystal Grano, Early Grano, Early Yellow Babosa, Evergreen Bunching (Nebuka), Lord Howe Island, Riverside Sweet Spanish, San Joaquin, Spanish Crystal Grano, Utah Sweet Spanish, White Babosa, White Sweet Spanish, and Yellow Sweet Spanish. Resistant *shallots* are also available.
13. *Aster Yellows* (garlic, leek, onion, shallot)—Plants gradually turn *bright yellow.* Often starts on one side of plant. All leaves may yellow at about same time. Leaves may be curved and twisted. If severe, bulbs may not form. *Control:* Destroy infected plants when first found. Turn under crop debris cleanly after harvest. Use virus-free sets.
14. *Rusts* (chives, false-garlic, garlic, leek, onion, ornamental alliums, shallot)—Occasional. Small, light yellow to orange or reddish, then lead-colored to black, powdery pustules on leaves and stalk. Leaves may yellow, wither, and die early. *Control:* Rotate. Collect and burn plant debris after harvest. Keep down weeds. Spraying as for Blast or Downy Mildew (both above) should keep Rusts in check. Use ferbam, zineb, or maneb. Avoid growing near asparagus, dense plantings, and excessive use of nitrogen fertilizers.
15. *Minor Leaf Spots and Blights, Leaf Blotch, Tip Dieback*—White, gray, pale green, yellow to pale brown or black spots and blotches on leaves. May be covered with dark mold in damp weather. Leaves often die back from tip or break over. *Control:* Same as for Blast, Downy Mildew, and Purple Blotch (all above).
16. *Sunscald*—Bleached, then soft, slippery areas, especially on immature bulbs of white onions. Bacterial Soft Rot often follows. *Control:* Protect bulbs from hot, bright sun during curing and harvesting. Cover with tops. Harvest later in day.
17. *Freezing Injury*—Tissues in cut bulbs are water-soaked and more or less transparent with scattered opaque areas. Bermuda and Sweet Spanish *onions* much more susceptible than Globe varieties. Bacterial Soft Rot commonly follows. *Control:* Avoid storage temperatures below 32° F.
18. *Damping-off*—Seedlings wilt and collapse from rot at soil line or below. Often occurs in more or less circular patches of various sizes. *Control:* Follow best cultural practices. Treat seed and soil as for Smut (above). Spray young seedlings at 5- to 7-day intervals using zineb (1½ tablespoons per gallon).
19. *Verticillium Wilt*—See (15B) Verticillium Wilt under General Diseases.
20. *Scab* (onion)—See (14) Scab under General Diseases.

21. *Bacterial Leaf Streak* (onion)—Colorado. Mottled brown streaks form on older leaves and leaf sheaths in warm weather. *Control:* Same as for Neck Rot (above).
22. *Other Root-feeding Nematodes* (lance, lesion, pin, spiral, sting, stubby-root, stunt) —Associated with stunted, declining plants. Roots stubby, short, and discolored. Bulbs may be smaller. Often provides wounds for root- and bulb-rotting organisms to enter. *Control:* Same as for Root-knot (above).
23. *Southern Blight*—Rotting of leaves at soil line. May be covered with white mold in which small "mustard" seed are embedded. *Control:* Same as for Neck Rot (above).
24. *Potassium Deficiency*—Tips of older leaves turn brown and die back. *Control:* Practice balanced fertility, based on a soil test.
25. *Chlorosis*—Leaves yellowed. May be stunted. May be due to zinc or some other minor element in alkaline soil. *Control:* See pages 26–28.

For additional information on onion diseases, read USDA Handbook 208, *Onion Diseases and Their Control.*

ONOCLEA — See Ferns

ONOSMODIUM — See Mertensia

ONYCHIUM, OPHIOGLOSSUM — See Ferns

OPUNTIA — See Cacti

ORANGE — See Citrus

ORANGE SUNFLOWER — See Chrysanthemum

ORCHIDS: AERIDES, ANGRAECUM, ASPASIA, BRASSAVOLA, BRASSIA, CALANTHE, CATASETUM, CATTLEYA, COELOGYNE, CYCNOCHES, CYMBIDIUM, CYPRIPEDIUM, DENDROBIUM, EPIDENDRUM, GRAMMATOPHYLLUM, LAELIA, LOCKHARTIA, LYCASTE, MILTONIA, ODONTOGLOSSUM, ONCIDIUM, PERISTERIA, PHAIUS, PHALAENOPSIS, RENANTHERA, SOPHRONITIS, STANHOPEA, VANDA, and ZYGOPETALUM

1. *Seed Rot, Mold, Seedling Blight*—Common in culture tubes, flasks, and bottles. Seeds rot. Seedlings weak and collapse. Small spots on leaves. May be covered with mold, often blue-green. *Control:* Sterilize seeds by soaking several hours in small bottles (½ inch by 2 inches) filled ½ inch deep with sterile distilled water. Then add ½ ounce of distilled water in which a chlorine tablet is dissolved. Shake 3 to 4 minutes and pour into sterile flasks containing agar and Knudson formula "B" or "C" (see in Glossary), to let seed germinate. Rotate culture flasks evenly to distribute seed. Then pour off solution. Culture flasks and tubes are available, already prepared, from orchid supply houses. Hydrogen peroxide may also be used as a sterilizing agent instead of household bleach.
2. *Leaf and Heart Blight, Pythium and Phytophthora Black Rot, Damping-off*—Common in community pots. Round to oval, translucent or water-soaked, dark brown to purplish-black spots, sometimes with zoned, yellowish margins, develop on leaves. Spots enlarge rapidly in moist weather until entire leaf may soften, blacken, wither, collapse, and fall early. Seedlings often wilt and topple over. Pseudobulbs, rhizomes, roots, and flower buds may also blacken and rot. Entire plants may die if pseudobulbs, rhizomes, and roots are attacked. *Control:* Destroy infected seedlings. Promptly cut out and burn infected parts on older plants. Carefully remove rotting leaves and pseudobulbs. When dividing plants, sterilize knives by flaming, plunging in hot water, dipping in formaldehyde solution (diluted 1:7 with water) or 70 per cent alcohol. Keep foliage as dry as possible. Avoid excess moisture. Keep humidity down and provide for better drainage. Space plants for good air circulation. Isolate diseased plants. Drench seedlings or dip infected plants, pot and all, for an hour (large plants for 2 or 3 hours) in a 1:2,000 solution of Bioquin 700 (8-hydroxyquinoline sulfate) or Natriphene (sodium salt

of o-hydroxyphenyl). A 1:2,000 solution is about ¾ teaspoonful per gallon of water. Wilson's Anti-Damp, Morton Soil Drench, or Dexon may be substituted for Bioquin or Natriphene, following manufacturer's directions. If rot continues, repeat treatment 3 to 7 days later. Grow in a sterile medium (e.g., shredded Douglas-fir or Pine bark or Osmunda fiber).

3. *Leaf Spots, Leaf Blotch, Anthracnoses, Black Spot*—Round to irregular, light to dark spots. Often sunken, sharply defined, or zoned. Spots may later be ringed, coalesce, or centers drop out leaving shot-holes. Disease may start at leaf tips and progress downwards. If severe, leaves may darken, die, and fall away. Spots may also occur on pseudobulbs and flowers. *Control:* Spray with Bioquin 700 (1:1,000), fixed copper, captan, thiram, ferbam, or zineb, plus spreader-sticker, several times at weekly intervals. Keep humidity down and foliage dry. Avoid overfertilizing with nitrogen. Increase both light and air circulation. Do not propagate from infected plants. Cut and burn infected leaves. Avoid injuring leaves (e.g., sunscald, cold, chemicals).
4. *Stem, Collar, and Root Rots*—Serious on many orchids. Leaves yellow, wilt, wither, and usually collapse from rot of leaf and stem base or root collar. Roots may decay; soft and watery or somewhat dry. White or brown mold may grow on rotting tissues. See also Root Nematodes (below). Destroy infected plants. Avoid overwatering or wounding plants, poor drainage, excessive humidity and temperature. Soak plant, pot and all, for an hour in a 1:2,000 solution of Bioquin 700 or Natriphene (about ¾ teaspoon per gallon), or PCNB 75 (1½ tablespoons per gallon) for 5 minutes. Follow manufacturer's directions. Repeat treatment a week later if necessary. Steam or chemically treat potting mixture (pages 538–548) before planting. Repot in clean pot with fresh potting medium with good drainage.
5. *Petal Blights, Brown Speck, Gray-mold Blight*—Small tan, brown, or black spots and streaks on flowers. May be bordered with delicate pink rings or water-soaked areas. Spots enlarge and merge in cool, damp weather to form brown blotches. Petals may be destroyed. Grayish-brown mold may develop on decaying tissue in moist weather. *Control:* Carefully cut off and burn infected flowers as soon as found. Spray newly opened flowers with 1:2,000 solution of Natriphene or Bioquin 700, plus spreader-sticker. Same cultural practices as for Leaf Spots (above). Avoid sprinkling flowers. Eliminate old flowers and plant debris both inside and outside growing area.
6. *Bacterial Soft Rot, Brown Spot or Rot*—Small, soft, water-soaked, dark green, amber-colored, brown or black spots and streaks on leaves. Leaves may soon collapse, turn yellow or brown, mushy, and foul-smelling. Rotting crowns shrivel. Leaves may drop early. Pseudobulbs and rhizomes may develop soft, water-soaked, brown to black, foul-smelling rot. Seedlings may die. *Control:* Avoid wounding plants. Remove rotted plant parts promptly. Disinfect knife between cuts by flaming, or dipping in boiling water, 1:7 dilution of formaldehyde, or mercury (e.g., Semesan or 1:1,000 solution of mercuric chloride). See page 101 for precautions. Separate diseased from healthy plants. Keep foliage as dry as possible. Avoid excess moisture. Observe strict sanitation. Dip plants as for Leaf and Heart Blight (above). Or swab small infected areas on *Cattleya* leaves with a 1:1,000 solution of mercuric chloride, or use 8-hydroxyquinoline sulfate or benzoate, 1:2,000 (sold as Chinosol, Bioquin 700, Fulex A-D-O, or Wilson's Anti-Damp). Follow manufacturer's directions. If necessary, repeat treatment a week later.
7. *Rusts*—Yellowish-green spots on upper leaf surface with bright, yellowish-orange or brown to black, powdery pustules on corresponding underside or both leaf surfaces. Blooms few, smaller in size, or may not develop. Plants become stunted, yellowish, and weak. *Control:* Cut off and burn rusted leaves when *first* found. Otherwise same as for Leaf Spots (above). Spray with zineb, captan, thiram, or ferbam.
8. *Fusarium Wilt*—Plants gradually decline in vigor. Leaves yellow, wilt, wither, and

drop off. Pseudobulbs, roots, and rhizomes rot and die. Internal, circular, purple band when rhizome is cut across. Flowers stunted, short-lived, and fewer in number. *Control:* Destroy infected plants. Grow in sterile rooting medium.

9. *Mosaics, Mottle, Leaf Necrosis* (most genera) —General. Symptoms greatly variable depending on virus (es) involved, orchids, and environmental conditions. The same virus may cause widely different symptoms on different orchids. In *Cattleyas* and related orchids yellowish blotches or streaks; later sunken, reddish-brown to purple-black patterns, brown or purple-black ringspots, irregular spots, mottling, or streaks form in leaves. Leaves may wither and die early. Fewer and smaller flowers are produced. New leaf growth is stunted, cupped, and often pock-marked. In *Cymbidiums* and *Phalaenopsis* symptoms are variable in pattern and severity. Spots often first small, elongated, and yellowish. Later enlarge and become more defined. Leaves mildly or severely mottled light and dark green. Dark streaks, spots, rings, and irregular patterns may develop in older leaves. Severely infected leaves may die and drop early. Growth may be stunted. In *Miltonia* leaves small, solid, elliptical, reddish to reddish-brown spots are formed. In *Aerides, Angraecum, Oncidium,* and *Vanda* conspicuous, irregular, yellowish blotches and streaks or light green mosaic mottle may develop in new growth. Elongated, purple-black streaks and blotches form in older leaves. Round to oval, dark brown to black, sunken, spots and streaks form on underside of *Epidendrum* leaves. Halo-like, brownish rings with dark centers form in very young leaves. In *Odontoglossum* an irregular, light green to yellowish mosaic pattern is formed, or streaks and small rings develop in leaves that become rust-colored or dead in some cases. Leaves may be stunted. Narrow, longitudinal, yellow-green to dark green streaks or mosaic mottle form in *Lycaste* and *Phaius* leaves. *Control:* Destroy infected plants; they will not recover. Separate healthy and virus-infected plants. Select virus-free propagation stock and seedlings. When dividing or harvesting flowers, disinfect pruning tools between cuts by flaming, dipping in household bleach or 70 per cent alcohol. Malathion sprays control aphids that transmit viruses. Keep down weeds around growing orchids. Hand-pull or use Telvar (80 per cent Monuron). Follow manufacturer's directions. Check with the plant pathology department at your land-grant university concerning special tests for determining infected plants. Use seedlings since viruses do *not* invade orchid seed.

10. *Flower Breaking* (cattleya, cymbidium, dendrobium, oncidium, vanda, etc.) —Leaf symptoms absent or show irregular, mild to severe mosaic mottling, streaking, and some malformation. Flower petals and sepals may be mottled, somewhat distorted, or show irregular and abnormal color streaks and blotches. Sepals and petals may be rolled and twisted (severe color-break). Virus symptomless in some plants. *Control:* Same as for Mosaics (above).

11. *Ringspots* (cymbidium, dendrobium, laelia, miltonia, odontoglossum, oncidium, phalaenopsis, stanhopea, vanda, etc.) —Symptoms variable. Small, dead (or light green to pale yellow or brown) flecks, spots, lines, or rings—partial or complete—develop on upper leaf surface of young and old leaves. Single or concentric rings may later be dead to purple-black enclosing a central light green or yellow "island." Rings may overlap or merge to form large compound patterns. Some leaves may yellow and drop early. *Cymbidium* leaves may be mottled. If severe, entire plant may be dwarfed and die. *Control:* Same as for Mosaics (above).

12. *Cattleya Blossom Brown Necrotic Streak*—A virus complex. Apparently widely distributed. Elongate brown spots and streaks appear a week or longer *after* blossoms open. Spots and streaks often increase in number and size until blooms dry up and drop 2 to 3 weeks after opening. Long, irregular, yellowish streaks may develop in leaves. Dark sunken areas or streaks develop on some infected plants. Virus (es) spread by cutting knife unless sterilized by flaming between cuts. *Control:* Same as for Mosaics (above).

13. *Leaf Nematodes, Yellow Bud Blight*—Angular, brown or blackish spots wedged between larger veins. Infested leaves later die. The stem and bulb nematode may cause some malformation of lower leaves that become brittle and break off easily. Flower buds may turn yellow, then brown, shrivel, and usually drop off without opening. Severely infested spikes blacken, shrivel, and do not set new flowers. *Control:* Propagate from healthy plants into sterile rooting medium. Keep foliage dry. Cut off and burn *all* infested leaves and flower spikes. Keep orchids away from ferns, phlox, and other susceptible plants. Disinfest *Vanda* cuttings by immersing in hot water (115° F.) for 10 minutes.
14. *Root Nematodes* (e.g., lesion) —Usually found associated with root and bulb rots and yellowish leaves. Older leaves die prematurely. *Control:* Same as for Stem Rots (above).
15. *Tipburn*—Tips of older leaves darken and die back. Various molds may grow on dead tissue in moist weather. *Control:* Grow in light, well-drained, sterile soil, Douglas-fir bark, or Osmunda fiber, low in soluble salts (page 31). Apply fertilizers frequently but in very dilute form. Use slowly available forms of nitrogen. Check with your florist, garden or orchid supply dealer, or extension horticulturist.
16. *Sooty Blotch, Sooty Mold, Flyspeck*—Superficial, dark brown to sootlike blotches or coating on both leaf surfaces and pseudobulbs (or cluster of 6 to 50 or more black specks—Flyspeck). Most common on plants in shade or under trees in which insects, such as aphids, are feeding. *Control:* Wipe affected parts gently with soft, damp cloth. Spray leaves as for Leaf Spots (above). Spray trees overhead with malathion.
17. *Snow Mold or Potting Fiber Mold*—White, powdery mold on and in tree fern and Douglas-fir bark potting media. Orchids decline in vigor; may die. *Control:* Repot plants and remove all mold and as much fiber as practical. Then immerse roots, rhizomes and lower part of pseudobulb in Shield 10 per cent (N-alkyl dimethyl benzyl ammonium chlorides, N-alkyl dimethyl ethyl benzyl ammonium chlorides, and N-alkyl methyl isoquinolinium chlorides). Use as drench or dip, 1 ounce in 4 gallons of water. Repeat 1 or more times, 2 weeks apart.
18. *Sunburn*—Oval to round spots, up to an inch or more in diameter, on leaves. Spots first grayish-green or yellowish-gray; soon turn yellow then grayish-black, sunken, and papery. *Control:* Grow plants in more subdued light, especially during hot summer months.

For additional information read, *Orchid Diseases,* State of Florida Department of Agriculture, Division of Plant Industry, Volume 1, No. 3, 1965, by Harry C. Burnett.

ORCHID-TREE — See Honeylocust
OREGON-GRAPE — See Barberry
ORNITHOGALUM — See Tulip
OSAGE-ORANGE *(Maclura)*

1. *Rust*—Southern states. Minor disease. Reddish-brown, powdery pustules on leaves. *Control:* Unnecessary.
2. *Leaf Spots, Leaf Blight*—Unimportant. Small to large, tan, gray, or "cottony" spots. *Control:* Collect and burn fallen leaves.
3. *Verticillium Wilt*—See (15B) Verticillium Wilt under General Diseases.
4. *Root Rot*—See (34) Root Rot under General Diseases.
5. *Damping-off*—See under Pine.
6. *Mistletoe*—See (39) Mistletoe under General Diseases.

OSIER — See Dogwood and Willow

OSMANTHUS [CHINESE, DELAVAY, HOLLY], SWEETOLIVE or TEA OLIVE, WILDOLIVE or DEVILWOOD, FALSE HOLLY *(Osmanthus)*; OLIVE *(Olea)*

1. *Leaf Spots, Black Leaf Spot, Anthracnose*—Small to large spots on leaves, often black in color. Spots may also occur on olive fruit. *Control:* Collect and burn fallen leaves. Keep trees pruned. If serious, spray during spring and fall rainy periods. Try using fixed copper, zineb, maneb, or captan.
2. *Sooty Molds, Black Mildew*—Gulf states. Black, moldy patches on leaves. Often follows insect attacks. *Control:* Apply malathion to keep insects in check.
3. *Root Rots*—See under Apple, and (34) Root Rot under General Diseases. May be associated with nematodes (e.g., citrus-root, lesion). Grow in well-drained, humusy soil in a protected location.
4. *Root-knot*—See under Peach, and (37) Root-knot under General Diseases.
5. *Bacterial Knot* (devilwood, olive) —California. Irregular to roundish, spongy to hard, knotty galls, up to several inches in diameter. May appear on leaf and fruit stems, twigs, branches, trunk, and roots. Most knots occur at wounds or leaf scars. Shoots may be stunted and die back. Entire trees may die. *Control:* Cut out galls carefully in midsummer. Disinfect tools between cuts. Dip in 70 per cent alcohol or household bleach (diluted 1:9), 5 per cent amphyl, 2 per cent Lysol, or 4 per cent formalin. Paint over larger wounds with bordeaux paste (page 105) or treat galls as outlined under Peach, Crown Gall. Grow healthy nursery stock. Follow spray program recommended for your area.
6. *Verticillium Wilt* (olive, osmanthus) —Leaves and twigs on affected branches roll, wither, and dry about flowering time. Later, entire trees may yellow, dry out, and die. Twigs may die back. Wilt symptoms may be aggravated by rootlet rots. *Control:* Rototill dry wood shavings (100 to 200 pounds) into soil under affected trees. Check with your state or extension plant pathologist.
7. *Olive Bitter Pit, Dry Rot of Fruit*—California. Dry spots may develop in fruit flesh, usually toward blossom-end, which becomes shriveled and shrunken in spots or one side. *Control:* Avoid overfertilizing trees.
8. *Olive Soft Nose or Blue Nose*—Mainly on Sevillavo variety. Ripe or nearly ripe fruit turns bluish at blossom end. Spot first appears like a bruise but soon turns dark and shrivels. Worse on young trees. *Control:* Keep trees growing vigorously. Water during drought periods.
9. *Mistletoe* (osmanthus) —See (39) Mistletoe under General Diseases.

OSMARONIA — See Rose

OSMORHIZA — See Celery

OSMUNDA — See Ferns

OSOBERRY — See Rose

OSTRYA — See Birch

OSWEGO-TEA — See Salvia

OXALIS, WOODSORREL [COMMON, VIOLET or PURPLE, YELLOW], LADYS-SORREL, SHAMROCK, BERMUDA BUTTERCUP *(Oxalis)*

1. *Leaf Spots, Tar Spot*—Spots of various shapes, colors, and sizes. *Control:* Pick and burn spotted leaves. If serious, spray at 10-day intervals. Use zineb, maneb, or fixed copper.
2. *Powdery Mildew*—White, powdery blotches on foliage. *Control:* Spray two or three times, 10 days apart, with Karathane.
3. *Rusts*—May be serious. Yellow-orange, reddish-brown, or black, powdery pustules on leaves. Alternate hosts: wild grasses, corn, and Oregon-grape. *Control:* If severe, apply ferbam or zineb at 10-day intervals.

4. *Root Rot*—See under Geranium, and (34) Root Rot under General Diseases.
5. *Curly-top*—See under Beet, and (19) Curly-top under General Diseases.
6. *Stem Nematode*—See (20) Leaf Nematode under General Diseases.
7. *Seed Smut*—See (11) Smut under General Diseases.

OXEYE, OXEYE DAISY — See Chrysanthemum
OXLIP — See Primrose
OXYBAPHUS — See Four-o'clock
OXYDENDRUM — See Sorreltree
OYSTER-PLANT — See Rhoea
PACHISTIMA — See Bittersweet
PACHYCEREUS — See Cacti
PACHYSANDRA, SPURGE [ALLEGHENY, JAPANESE, MOUNTAIN] *(Pachysandra)*

1. *Volutella Leaf Blight, Dieback, Stem Canker*—Large, brown blotches on leaves. Leaves later brown to black and blighted. Stems withered by dark cankers. Salmon-pink pustules cover affected parts in wet weather. Plants die out in patches. Common after injury (insects, low temperatures, etc.). *Control:* Apply malathion or Diazinon to control scales and other insects. Mulch plants with light material for winter that does not hold moisture. Avoid excess moisture. Thin out thick beds to increase light and air circulation. Remove and burn all infected plants. Spray remainder thoroughly with bordeaux or fixed copper. Repeat 7 to 10 days later. Or apply ferbam, zineb, or captan, plus spreader-sticker, three or four times, 7 to 10 days apart. Begin when new growth starts in spring.
2. *Leaf Spots*—Small spots on leaves. *Control:* Rarely necessary. Spray as for Leaf Blight (above). Thin out thick beds.
3. *Root Rot*—Plants wilt and die. Fine roots decay. Larger roots and stolons have dark brownish rot spots. Same as for Leaf Blight (above). Soil drenches of oxyquinoline sulfate (1:4,000) are effective if plants not too far gone.
4. *Root-knot*—Small galls form on roots. Plants lack vigor. See (37) Root-knot under General Diseases.
5. *Other Root-feeding Nematodes* (dagger, lance, lesion, pin, spiral, stunt)—Associated with stunted, unthrifty plants. *Control:* Same as for Root-knot (above).

PAEONIA — See Delphinium
PAGODATREE — See Honeylocust
PAINTED-CUP — See Snapdragon
PAINTED DAISY — See Chrysanthemum
PAINTED LADY — See Sedum
PAINTED-TONGUE — See Tomato
PAK-CHOI — See Cabbage
PALE LAUREL — See Blueberry
PALMS: GRU-GRU or TOTAI *(Acrocomia);* KING [ALEXANDRA, PICCABEEN] *(Archontophoenix);* QUEEN or PLUMY-COCONUT *(Arecastrum);* SUGAR *(Arenga);* BUTIA or PINDO *(Butia);* FISHTAIL *(Caryota);* DWARF FAN or EUROPEAN FAN *(Chamaerops);* ARECA or CANE *(Chrysalidocarpus);* COCONUT *(Cocos);* WHITE PRINCESS *(Dictyosperma);* UMBRELLA *(Hedyscepe);* *CURLY or BELMORE SENTRY,* FLAT or FORSTER SENTRY *(Howea, Kentia);* CHILEAN HONEY, COQUITA, WINE or MONKEY-COCONUT *(Jubaea);* AUSTRALIAN or CHINESE FAN

(Livistona); PHOENIX, DATE [CANARY ISLANDS, COMMON, DWARF, ROBELEN or PIGMY, SENEGAL, WILD] (Phoenix); BROAD-LEAF LADY or GROUND-RATTAN, SLENDER LADY (Rhapis); NEEDLE or PORCUPINE (Rhapidophyllum); PALMETTO [BERMUDA, COMMON or CABBAGE, DWARF or BUSH, HISPANIOLAN, PUERTO RICO HAT] (Sabal); SCRUB or SAW PALMETTO (Serenoa); HEMP or WINDMILL (Trachycarpus); WASHINGTON, CALIFORNIA FAN, MEXICAN (Washingtonia)

1. *False Smuts, Leaf Scab* (date, palmetto, queen, royal, sugar, Washington)—Widespread in humid areas. Numerous, small, yellow spots and small, dark brown to black, hard warts, pustules, or scabs on leaves. Severely infected leaves soon die. *Control:* Cut out and burn old infected leaves or parts of leaves. Indoors, keep water off foliage. When first evident, spray with copper fungicide. Repeat during cool, rainy weather. Thiram, zineb, or ziram may do just as well without causing injury.
2. *Leaf Spots or Blight, Anthracnose, Dieback*—Cosmopolitan. Small to large, round to irregular spots of various colors. Spots may merge to form large blotches or streaks. A brown blight may work downward from leaf tip. Infections may cause rotting of leaf bases. Leaves may die. See Figure 4.154. *Control:* Indoors, maintain sufficient light. Otherwise same as for False Smuts (above).
3. *Stem (Trunk) and Root Rots, Wilt, Decline*—Leaves may turn gray or yellow and wilt progressively up plant. Later wither and fall. Plants unthrifty, often stunted, lack vigor. Yellow to brown to almost black discoloration of the heartwood may develop. Gumming may be evident on trunk. Small, black, club-shaped fruiting bodies may grow out from decaying roots. May be associated with nematodes (e.g.,

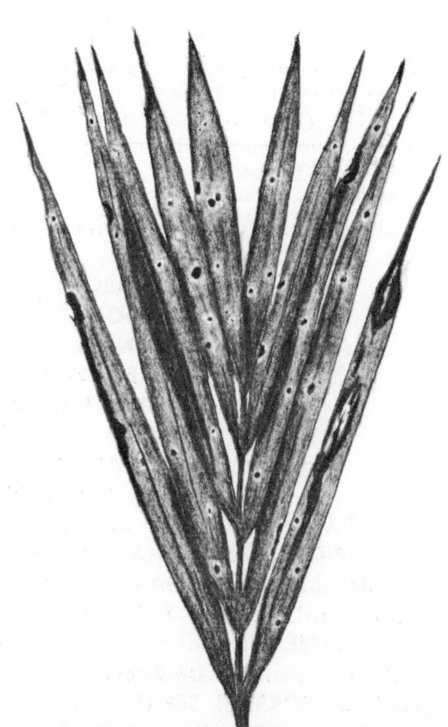

Fig. 4.154. Leaf spot of palm.

burrowing, dagger, lesion, spiral, stubby-root, tylenchus). *Control:* Grow potted palms in sterilized soil (pages 538–548). Keep plants vigorous through fertilizing and watering. Avoid wounding roots or stem (trunk) when cultivating or removing offshoots. Coat wounds promptly with tree wound dressing (page 36). *Date* varieties differ in resistance.

4. *Phytophthora Bud Rot, Phytophthora and Pythium Wilts, Trunk Rot, Leaf Drop* (primarily coconut, palmetto, queen, and Washington)—Soft, spongy rot of stem (trunk) base. Leaves droop, wilt, and die in several months. Terminal bud develops wet, putrid rot. Bud loosens and withers. *Control:* Cut out and burn crown, infected buds, and leaves from diseased trees. Avoid trunk injuries.
5. *Black Scorch, Heart Bud Rot* (coconut, date)—Dark brown to black hard spots and rot in leaves, leaf stalks, buds, and flowers. If severe, crown leaves may be distorted, dwarfed, and dry up ("black scorch"). *Control:* Cut out and bury or burn all infected plant parts. Disinfect pruning cuts. Copper sprays during rainy periods may be beneficial. Otherwise, same as for Stem (Trunk) and Root Rots (above). *Date* varieties differ in resistance.
6. *Bacterial Wilt* (coconut, Cuban royal)—Lower leaves turn gray and wilt. Trunk later exudes gum. Crown finally collapses. *Control:* Unknown. Grow disease-free plants in sterilized soil. Destroy infected plants when found.
7. *Penicillium Disease* [*Leaf Base Rot, Bud Rot, Trunk Canker, Gummosis*] (coconut, date, Mexican, queen, Washington, etc.)—Serious along southern Pacific Coast. Symptoms variable. *Date palms:* Linear streaks and leaf base rot occur. *Washington palms:* Stunted, deformed leaves, dwarfed terminal growth, wilted, drooping fronds, and bud rot. *Queen or Coconut:* Trunk canker. Leaf bases rot progressively upward on tree. Trunk weakened. Later breaks. *Control:* Treat trunk cankers early. Grow normally resistant Washington palms (e.g., *W. robusta*). Check with the Department of Plant Pathology, University of California, Riverside. Keep palms in high state of vigor. Avoid sprinkling foliage. Wettable sulfur sprays provide some protection.
8. *Root-knot and Other Nematodes* (e.g., burrowing, lance, lesion, needle, reniform, spiral, stem, stubby-root)—See under Peach, and (37) Root-knot under General Diseases. Nurserymen dip young bare-root stock in Zinophos or Bayer 25141 at 600 to 800 parts per million for 30 minutes.
9. *Black Mildews* (palmetto)—Gulf states. Black, powdery patches on leaves. *Control:* Spray with fungicide plus malathion or lindane to control scales, mealybugs, aphids, and other insects. Follow manufacturer's directions.
10. *Sunscald*—Indoor problem. Large, dry, leaf blotches with yellow centers and brown margins. *Control:* Provide more shade.
11. *Withered Leaf Tips*—Indoor problem. Tips or even entire leaves may wither and die after being moved to new location. *Control:* Avoid sudden changes in humidity (e.g., from damp greenhouse to dry living room). Buy plants with tough, dark green leaves. Feed lightly with balanced fertilizer. Avoid overwatering and strong drafts. Repot only when absolutely necessary. Raise air humidity or mist leaves occasionally with warm water (page 39).
12. *Bacterial Leaf Spot*—See (2) Bacterial Leaf Spot under General Diseases. *Control:* Apply copper sprays during rainy periods.
13. *Chlorosis*—Southeastern states in alkaline or very acid soils. See under Walnut.
14. *Thread Blight*—Southeastern states. See under Walnut.
15. *Fruit Spots and Rots* (date)—Fruit rot and may "sour" during ripening period or after packaging if weather is rainy or very humid. Dates may crack and become covered with mold growth. *Control:* Check with your state or extension plant pathologist.
16. *Felt Fungus*—Southeastern states. See under Hackberry.

17. *Damping-off, Seed Decay*—Seedlings may rot and die after emerging. *Control:* Avoid keeping soil overly wet after planting. Use potting soil composed of 1 part peat and 2 parts fine sand by volume. Add 1 pound of superphosphate per cubic yard.
18. *Manganese Deficiency* (canary date, Chinese fan, fishtail, queen, royal, etc.) — Leaves yellowish. Where advanced, dead areas appear between leaf veins. Leaves dwarfed; may appear "frizzled." *Control:* Apply manganese sulfate to soil under trees or as a 1 per cent spray to foliage (page 26).

PANDANUS — See Screwpine

PANDA PLANT — See Sedum

PANSY [HORNED, TUFTED or BEDDING PANSY, VIOLET], VIOLET [BIRDSFOOT, BLUE, CANADA, CONFEDERATE, GOLDEN, MEADOW, SWEET or FLORISTS', SWEET WHITE, WILD, YELLOW] *(Viola)*

1. *Anthracnose, Leaf Spots, Spot Anthracnose or Scab*—General and serious. Small to large leaf spots of various sizes, shapes, and colors. Often with distinct dark margin. Spots may enlarge and merge to form irregular blotches or concentric zoned areas. Some spots may drop out leaving ragged leaves. Similar spots may occur on stems, petioles, flower petals, flower stalks, and seed capsules (Figure 4.155). Infected petals distorted. Flowers may fail to produce seed. Entire plant may die. *Control:* Collect and burn infected plant parts. Space plants. Rotate. Destroy old tops in fall. Indoors, keep water off foliage and avoid dampness. Grow disease-free plants in warm, dry location in sterilized soil (pages 538–548). Spray at 5- to 7-day intervals during wet weather using zineb, captan, thiram, ferbam, maneb, or fixed copper. Control insects with malathion. Plant disease-free seed. If in doubt, treat seed as for Wilts (below). Varieties differ in resistance.

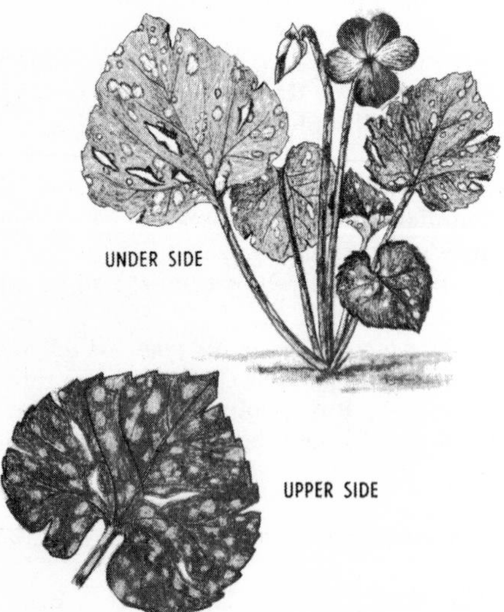

Fig. 4.155. Violet scab or spot anthracnose.

2. *Gray-mold Blight, Botrytis Blight*—Common in wet springs. Soft, grayish-brown, rotted spots on buds, leaves, stems, and flower clusters. Spots enlarge rapidly in damp weather. Affected parts may be covered with gray mold and become slimy. *Control:* Same as for Anthracnose (above).
3. *Powdery Mildew*—Whitish patches of mildew on leaves. If severe, leaves may turn

brown and wither. *Control:* If serious, apply sulfur or Karathane several times, 10 to 14 days apart. Do *not* use when temperature is 85° F. or above.

4. *Wilts, Root Rots, Stem or Crown Rots, Southern Blight, Damping-off*—General in wet weather. Leaves stunted, yellowish, finally die. Stem base and roots decay; may be dry, mushy or slimy, and foul-smelling (Bacterial Soft Rot). Plants yellow, wilt, and gradually or suddenly die. Seedlings wilt and collapse. May be associated with root-feeding nematodes. *Control:* Remove and destroy affected plants. Where practical, grow healthy plants in rich, moist, well-drained soil that is sterilized (pages 538–548). Mulch to keep soil cool and moist. Avoid overwatering. Rotate. Treat seed with Semesan, captan, thiram, dichlone, or chloranil.
5. *Mosaics, Calico, Flower Breaking*—Common. Flower petals develop abnormal streaks or blotches that vary with flower color. Flowers or entire plants often dwarfed and deformed (Figure 4.156). Leaves slightly curled and yellowed or mottled green and yellow. Young leaves may be dwarfed and malformed. *Control:* Buy virus-free plants or start from seed. Destroy infected plants when first found. Keep down weeds. Control aphids that transmit viruses. Use malathion.

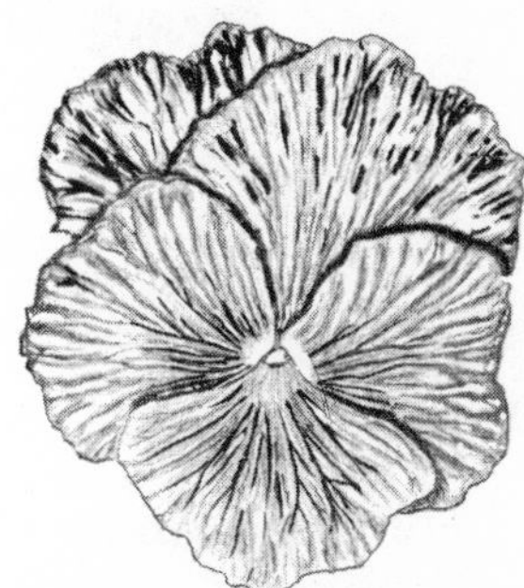

Fig. 4.156. Mosaic or flower breaking of Jumbo pansy.

6. *Smuts*—Leaves, stems, flowers, flower stalks, and seed may be infected. Large, elongated, dark purple blisters on foliage that later burst to release black, powdery masses. Petioles deformed. *Control:* Dig up and burn infected plants when first found. Treat seed as for Wilts (above) or plant only disease-free stock. Rotate.
7. *Downy Mildew*—Irregular, pale leaf spots with light grayish-purple mold on underleaf surface in damp weather. Leaves rot quickly in wet weather. Plants may droop and die without showing definite dead areas. *Control:* Same as for Anthracnose (above).
8. *Rusts*—General. Pale green spots on upper leaf surface and yellow "cluster cups" on corresponding underside. Rust also occurs on swollen parts of veins, petioles,. and stems. Pustules later turn light brown; finally dark brown or black. Flowering may be reduced. Alternate hosts: wild grasses *(Andropogon)* or none. *Control:* Same as for Anthracnose (above).
9. *Oedema, Corky Scab*—Indoor problem. Small, corky, or wartlike growths on leaves and flower stalks. Affected plant parts become dry and brittle. *Control:* Increase air circulation. Avoid overwatering. Lighten or cultivate soil. Control insects with malathion.
10. *Root Nematodes*—Plants may appear unthrifty and weak. Small wartlike galls on roots (Root-knot) or sunken brown areas (dagger, lance, lesion, spiral, stunt). Roots may be stunted, stubby, bushy, and decayed. *Control:* Set out disease-free plants in clean or sterilized soil (pages 538–548).
11. *Leaf Nematode*—New leaves dwarfed and distorted. Stalks stunted. Base of petiole may be swollen and irregular. Plants may not flower. *Control:* Move to a new bed. Start with new, disease-free plants. If you must use plants from an old, infested bed,

first soak plants in hot water (110° F.) for 30 minutes. Cool, then plant.

12. *Curly-top*—Western states. Shoots stunted. Form rosettes. Flowers dwarfed and produce few seed. Diseased plants stop flowering; gradually deteriorate. *Control:* Apply malathion plus DDT to control leafhoppers that transmit virus. Otherwise, same as for Mosaics (above).
13. *Ringspot*—Widespread. Numerous dead rings with green centers develop in leaves. *Control:* Same as for Mosaics (above).
14. *Aster Yellows*—Plants stunted and bright yellow. *Control:* Same as for Curly-top (above).
15. *Sooty Mold*—Black, moldy patches on leaves. Follows insect attacks. *Control:* Spray with malathion when insects first seen.

PAPAVER — See Poppy

PAPER-MULBERRY — See Fig

PARASOLTREE — See Phoenix-tree

PARKINSONIA — See Honeylocust

PARLOR-IVY, PARIS DAISY — See Chrysanthemum

PARSLEY — See Celery

PARSNIP — See Carrot

PARTHENOCISSUS — See Grape

PARTRIDGE-BERRY — See Buttonbush

PARTRIDGE-BREAST — See Aloe

PASPALUM — See Lawngrass

PASQUEFLOWER — See Anemone

PASSIONFLOWER [BLUE CROWN, RED], PASSION VINE, MAYPOP, GRANADILLA [GIANT, PURPLE, SWEET, YELLOW] *(Passiflora)*

1. *Leaf Spots*—Spots of various sizes, shapes, and colors. Spots may also occur on petioles, canes, and fruit, causing shriveling. *Control:* Apply zineb, maneb, or fixed copper just before wet periods. Prune annually to thin out plants. Keep plants growing vigorously.
2. *Stem or Collar Rot, Southern Blight*—Plants yellow and wilt from rot at soil line. Rot may be covered with white mold. The bark rots, exposing wood. *Control:* Remove and burn infected plants. Grow in well-drained, neutral (pH 7) soil.
3. *Mosaic, Woodiness Disease*—Northern California. Plants stunted with mottled, curled, distorted leaves. Rind of fruit may become "woody" and thick. Infected vines produce fewer fruit. *Control:* Destroy infected plants. Control aphids that transmit viruses; use malathion.
4. *Gray-mold Blight*—Plants may wilt and die. See under Begonia, and (5) Botrytis Blight under General Diseases.
5. *Root Rot*—See (34) Root Rot under General Diseases.
6. *Root-knot*—Southern states. Plants stunted and weak, especially on sandy soils. See (37) Root-knot under General Diseases.
7. *Seedling Wilt, Anthracnose, Stem and Leaf Spot*—Uncommon. Seedlings wilt and die. *Control:* Same as for Leaf Spots (above).

PASTINACA — See Carrot

PATIENCE PLANT — See Balsam

PAULOWNIA, EMPRESSTREE or PRINCESSTREE *(Paulownia)*

1. *Leaf Spots*—Round to irregular, yellowish-brown, gray or dark spots occur in rainy seasons. Collect and burn fallen leaves. Where practical, spray several times during rainy periods. Use zineb, maneb, or fixed copper.

2. *Wood Rot*—See under Birch, and (23) Wood Rot under General Diseases. *Control:* Avoid wounding bark on trunk. Paint wounds promptly with tree wound dressing (page 36). Fertilize and water to maintain vigor.
3. *Root Rot*—See (34) Root Rot under General Diseases. Grow in rich, loamy, well-drained soil in a sheltered spot.

PAWPAW *(Asimina)*; ROLLINIA

1. *Leaf Spots, Leaf Blotch*—Widespread. Small to large, round to irregular spots and blotches of various colors. *Control:* Collect and burn leaves in fall. Prune to keep trees thinned. If necessary, spray several times during rainy periods, 10 days apart. Try using maneb, zineb, or fixed copper.
2. *Twig and Branch Cankers, Diebacks*—Twigs and branches die back due to discolored cankers. *Control:* Prune and burn infected parts. Make cuts several inches beyond any sign of infection. See under Maple.
3. *Wood Rots*—See under Birch, and (23) Wood Rot under General Diseases.
4. *Sooty Mold*—Black, moldy patches on leaves and twigs. *Control:* Apply malathion to control insects.
5. *Fruit Rot*—Fruits spotted, later rot. *Control:* Apply zineb or captan during rainy periods as fruits are growing.

PEA, GARDEN *(Pisum)*; ROSARYPEA *(Abrus)*; BUTTERFLY-PEA *(Centrosema, Clitoria)*; CROWNVETCH *(Coronilla)*; CROTALARIA; HYACINTH-BEAN, AUSTRALIAN PEA, TWINFLOWER *(Dolichos)*; SWEETPEA, EVERLASTING or PERENNIAL PEA, PRIDE-OF-CALIFORNIA, BEACH-PEA *(Lathyrus)*; LENTIL *(Lens)*; LUPINE [ANNUAL, BLUE, PERENNIAL, RUSSELL, SUNDIAL or QUAKER BONNETS, TEXAS or BLUEBONNET, WASHINGTON, WHITE, YELLOW] *(Lupinus)*; SENSITIVE PLANT *(Mimosa)*; BUSH-PEA or GOLDEN-PEA, GLOWING GOLD, AARONS-ROD *(Thermopsis)*; VETCH [COW, WINTER] *(Vicia)*

1. *Fusarium Wilt, Near-Wilt, Root Rot*—Widespread in northern states, especially on garden pea. Caused by two races of the same or closely related fungi. Yellowish dwarfed plants that often wilt and die. Starts with lower leaves about blossoming time. Leaflets distorted, small, curved downward and inward. Plants may break over near soil line. Lemon-yellow to brownish-black (Pea Wilt) or brick-red (Near-Wilt) streaks inside lower part of stem. Wilt often occurs in roughly circular patches. Pods normally do not form. Roots may or may not decay. Seedlings damp-off. Often associated with nematodes (e.g., lesion, sheath). *Control:* Avoid overfertilization and overcrowding. Sow disease-free seed treated as for Seed Rot (below). Rotate 4 years or more. If practical, grow in sterilized soil (pages 538–548) or drench soil for *flowers* using a 1:2,000 solution of mercuric chloride (1 pint to 5 feet of row). Grow adapted, wilt-resistant *pea* varieties: Ace, Alaska strains, Alcross, Alderman, Alsweet No. 4683, Apex, Bridger, Bruce, Canner King, Cascade, Climax, Dark-Podded Thomas Laxton, Dark Skin Perfection, Delwiche Commando, Dwarf Alderman, Dwarf Gray Sugar, Early Freezer, Early Harvest, Early Perfection W.R., Early Sweet, Eureka, Extra Early, Famous, Freezer strains, Freezonian, Frosty, Giant Stride, Glacier, Green Giant, Hardy, Horal, Hyalite, Icer, Improved Gradus, Jade, King, Laurel, Laxton 7, Little Marvel W.R., Lolo, Mammoth Melting Sugar, Midfreezer, Midway, Morse Market, New Era, New Season, New Wales, Nome, Oracle, Pacemaker W.R., Pacific Market 40, Penin, Perfection strains, Perfected Freezer, Perfected Wales, Pixie, Pride, Profusion, Progress No. 9, Ranger, Ranier, Resistant Early Perfection 326, Resistant Surprise, Shasta, Shoshone, Signal, Signet, Small Late Canner, Small Sieve Freezer, Sprite,

Stratagem, Surpass, Teton, Thomas Laxton strains, Victory Freezer, Viking, Wando, Wasatch, Wisconsin Early Sweet, Wisconsin Merit, Wisconsin Perfection, Wyola, Yellow or Green Admiral, and Yukon. *Pea* varieties normally resistant to Near-Wilt: Delwiche Commando, Horal, Horsford, New Era, New Pride, New Season, New Wales, and Rogers 95. *Yellow lupines* also differ in resistance.

2. *Root Rots, Foot Rots, Southern Blight, Crown Rots, Stem Cankers, Sclerotinia White Mold*—General and serious in wet weather. Seeds rot. Stand is poor. Seedlings yellow, shrivel, and die. Older plants often stunted and yellow. May collapse. Plants often wilt and die at or near flowering. Crown and roots discolored and rotted. May be covered with mold. Occurs in patches that gradually enlarge during season. Plants easily pulled up. Often associated with Fusarium Wilt or Near-Wilt, Seed Rot, and Nematodes. See Figure 2.49D. *Control:* Treat seed as for Seed Rot (below). Plant *early* in fertile, well-prepared, well-drained soil. Four- to 6-year rotation. Avoid close and deep cultivating, overcrowding, and overfertilizing. Practice balanced soil fertility based on a soil test. Keep plants growing vigorously. If practical, dig up and burn infected plants. *Pea* varieties somewhat resistant to Fusarium or Aphanomyces Root Rots: Acquisition, Green Admiral, Freezonian, Horal, Pacific Freezer, Premier Gem, Rice No. 300, Resistant Thomas Laxton 251, Selkirk, Sutton's Ideal, Wando, and World's Record. For *flowers,* drench soil with mercuric chloride (see Fusarium Wilt, above), phenyl mercury (Panogen, PMAS, Morton Soil Drench), or grow in sterilized soil.
3. *Powdery Mildews*—General where nights are cool. White to gray powdery coating on leaves, petioles, pods, and stems of older plants. Leaves may be distorted, turn yellow, wither, and die. Plants may be dwarfed. *Pea* yield and quality may be reduced. See Figure 4.157B. *Control:* Sow plump, disease-free seed. Rotate. Increase air circulation and keep water off foliage. Space plants. *Pea* varieties and *Lupine* species differ in resistance. If severe, apply sulfur or Karathane several times, a week apart. Do *not* apply when temperature is over 80° F. or plants in flower. Burn or cleanly turn under all crop refuse after harvest.
4. *Virus Complex* (Mosaics, Streaks, Stunt, Mottle, Wilt)—General. Symptoms differ with plant, variety, virus (es) involved, age of plant at time of infection, weather conditions, and other factors. Leaves generally mottled (lightly to severely), netted, or spotted with yellow, light green, and dark green mosaic pattern. Plants may be dwarfed (Stunt) and yellowish with crinkled, curled, and distorted pods and leaves. Light brown, reddish-brown, or purplish streaks may develop on stems and

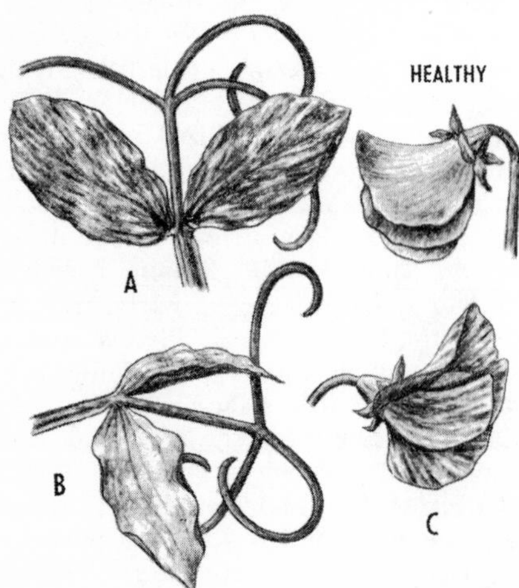

Fig. 4.157. A. Mosaic of sweetpea, leaves. B. Powdery mildew of sweetpea. C. Mosaic or flower breaking.

petioles (Streak). Young shoots may be stunted and "bushy." Plants may wilt, even die. Flowers often "broken" with white streaks and blotches (Figure 4.157C). Pods may be dwarfed, spotted, distorted, twisted, roughened, or few. Seed formation is often abortive. Virus-infected *pea* plants are more susceptible to Root Rots (see above). *Control:* Keep down weeds, especially clovers and alfalfa. Use malathion at least weekly to control aphids and other insects that transmit viruses. Sow seed from virus-free plants as early as possible. Destroy infected plants when first found. Burn plant refuse after harvest. *Pea* varieties resistant to one or more viruses: Bridger, Canner Prince, Foli, Green Giant Perfection, Horal, Hundredfold, Little Marvel, Midway, Morse Market, New Era, New Season, New Wales, Perfected Freezer 60, Pride, Resistant Early Perfection, Resistant Surprise, Small Sieve Freezer, Superior, Thomas Laxton 60, Wasatch, Wilt Resistant Perfection, Wisconsin Early Sweet, and Wisconsin Perfection. *Sweetpea* varieties differ in resistance to Common Pea Mosaic virus.

5. *Seed Rot, Damping-off*—General. Seeds rot. Seedlings weak, wilt, and collapse from rot at soil line and below. Stand is thinned. *Control:* Treat *sweetpea* seed by soaking 1 minute in alcohol followed by a 20-minute soak in a 1:1,000 solution of mercuric chloride. Wash treated seed in running tap water for 3 to 5 minutes. Dry and dust as for other seed. Treat other seed with thiram, captan, chloranil, dichlone, difolatan, Semesan, or Dexon to which dieldrin or lindane is added. Avoid deep planting and poorly drained soil. If damping-off starts, apply Shot-gun Soil drench (page 111). Or use Dexon or Lanstan following manufacturer's directions.
6. *Spotted Wilt, Ringspot*—Yellow, brown, or purple rings, often zoned. Or small, round, brown, or purple spots on leaves. Leaves may be somewhat mottled, distorted, and misshapen. Discolored, circular patterns may appear on flowers. Reddish-brown to dull purple streaks develop on stems and petioles of *sweetpea.* Leaves and shoots may yellow and die. *Control:* Use sevin plus malathion to control thrips that transmit the virus. Destroy infected plants when first found. Keep down weeds.
7. *Ascochyta and Mycosphaerella Blights, Stem Rot, Foot Rot, Leaf and Pod Spot*—General in wet seasons. Light brown, dark brown, gray, black, or purplish streaks on stem. Stems may be darkened and girdled, killing portions beyond. Leaf spots round to irregular with light or dark centers and darker margins. Spots may merge to form brownish-purple blotches. Sunken, tan or dark spots occur on pods. Black dots may be sprinkled in diseased areas. Young pods may be distorted and wither. See Figures 4.158 and 4.159A. *Control:* Sow large, plump, disease-free seed, grown in western states, in well-drained soil. Treat seed as for Seed Rot (above). Three- to 5-year rotation. Destroy plant debris after harvest. Burn or turn under deeply. Keep down weeds. *Pea* varieties Creamette and Perfection (Advancer) have some resistance. Avoid overwatering and overcrowding. Spray *sweetpeas* with ferbam, thiram, zineb, or captan.
8. *Downy Mildew*—General in cool, damp weather. Irregular, water-soaked, yellow to brown blotches on upper leaf surface and white, light violet, grayish-brown, or almost black mold on underleaf surface. Leaves curl down, gradually wither and die. Yellowish-green blotches on pea pods filled with mealy, white mold. Pods distorted with spots turning dark brown. Young plants very susceptible. Stems may be distorted, stunted, and killed. *Control:* Sow disease-free seed grown in western states. Four-year rotation. Keep down weeds. Grow in well-drained soil. Burn or cleanly turn under plant debris after harvest. Apply zineb, maneb, or fixed copper, plus spreader-sticker, one or more times weekly during cool, wet periods. Start at beginning of flowering. *Pea* varieties differ in resistance.
9. *Septoria Blight, Leaf Spot or Blotch*—Widespread in cool moist weather, but

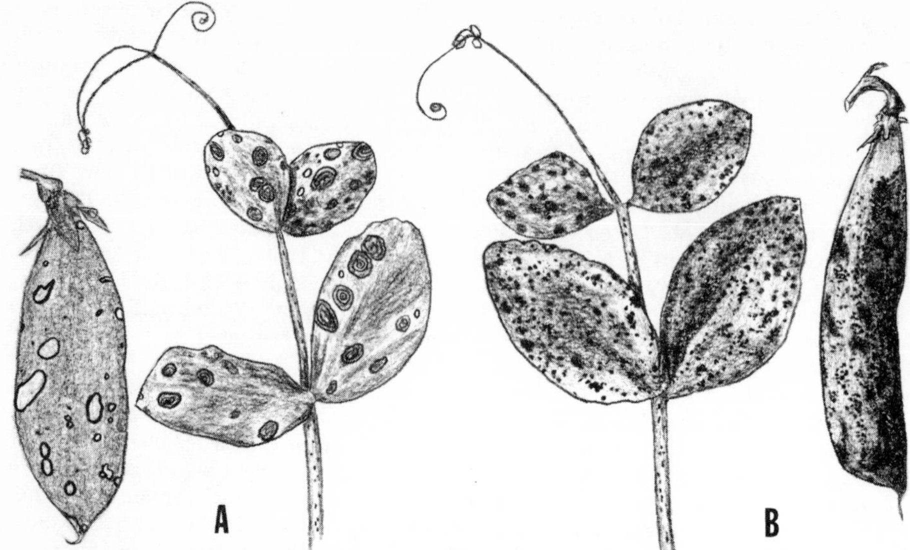

Fig. 4.158. Ascochyta and Mycosphaerella blights of garden pea. Left, Ascochyta; right, Mycosphaerella.

infrequent. Yellowish, indefinite spots at margins of leaves that later enlarge and turn brown. Centers may later be sprinkled with black dots. Leaves often wither and die. Young plants may die. *Control:* Same as for Downy Mildew (above). *Pea* varieties (e.g., Perfection strains) have some resistance.

10. *Bacterial Leaf Spots and Blights*—General in humid areas. Small to large, dark green, water-soaked spots (streaks on *sweetpea*) on pods, petioles, stems, leaves, flower stalks, and blossoms. Spots later dry, turn yellow to dark brown. Flowers are killed or young pods shrivel. Pods may appear spotted, scalded, and cracked (Figure 4.159B). Branches or entire plants may appear stunted, wilt, and die. Seedlings wilt and collapse. Often follows strong winds, late spring frosts, or hail. *Control:* Sow disease-free seed grown in western states. Treat seed as for Seed Rot (above). Rotate 3 years or more. Keep down weeds. Do not work among wet plants. Plant early in well-drained soil. Remove and destroy affected plants, where feasible. *Pea* varieties differ in resistance to Bacterial Blight *(Pseudomonas pisi).*
11. *Anthracnose, Leaf and Pod Spot*—Widespread and serious. White or gray to brown sunken spots with dark margins on leaves, pods, flowers, and stems. Spots may enlarge. Leaves wilt, wither, and fall early. Commonly follows Ascochyta Blights (above). Shoots and flower stalks may wilt and wither starting at tips. Drying pods blanch and shrivel. *Control:* Same as for Ascochyta Blights. Spray as for Downy Mildew (above) or use captan or ferbam. Fertilize and water to keep plants vigorous. Do not plant near apple, privet, or lilac. *Blue lupines* differ in resistance to Anthracnose *(Glomerella).*
12. *Other Leaf Spots and Blights, Ramularia White Blight, Pod Spots, Scab*—Spots of various sizes, shapes, and colors on leaves. Leaves may be stunted and distorted or wither and drop off. May start at base of plant and progress upwards. Pods may be spotted and distorted. May be covered with mold (Figure 4.159). *Control:* Pick and burn affected leaves. Sow disease-free seed in clean or sterilized soil (pages 538–548). Dust or spray as for Downy Mildew (above). Start about 3 weeks before leaf spot usually appears. Rotate. *Pea* varieties and *blue lupines* differ in resistance. Indoors, lower humidity and keep water off foliage. Space plants.

13. *Rusts*—Generally unimportant. Small, cinnamon-brown to dark brown, dusty pustules on leaves. Mostly on underleaf surface. Causes decreased vigor. *Pea* varieties and *lupine* species differ in resistance. Late-maturing varieties are more commonly infected. Alternate hosts: spurge *(Euphorbia)* or wheatgrass *(Andropogon)*. *Control:* If serious, apply fixed copper, sulfur, zineb, or maneb, 1 to 3 times, a week apart. Start about blooming time.
14. *Gray-mold Blight, Blossom and Shoot Blight, Pod Rot*—General on *sweetpea*. Flowers spotted and rot in damp weather. Brown blotches develop on lower leaves, stipules, and pods. Base of stem may rot causing plants to wilt and die. Seedlings wilt, collapse, and die. Gray mold may grow on affected parts in damp weather. *Control:* Same as for Downy Mildew (above).
15. *Fasciation, Leafy Gall, Crown Gall*—Primarily sweetpea. Many very short, thick, strap-shaped, fleshy stems (witches'-brooms) develop near soil line that form distorted and misshapen leaves. Plants stunted. Blossoming and fruit set is reduced. See Figure 2.44A under General Diseases. *Control:* Sow disease-free seed in clean or sterilized soil (pages 538–548). Treat soil for *sweetpea* as for Fusarium Wilt (above). Treat seed as for Seed Rot (above).
16. *Root-knot, Cyst Nematodes*—Mostly southern states in light, sandy soils and indoors in north. Oval or elongated, irregular, fleshy galls and cysts, that are

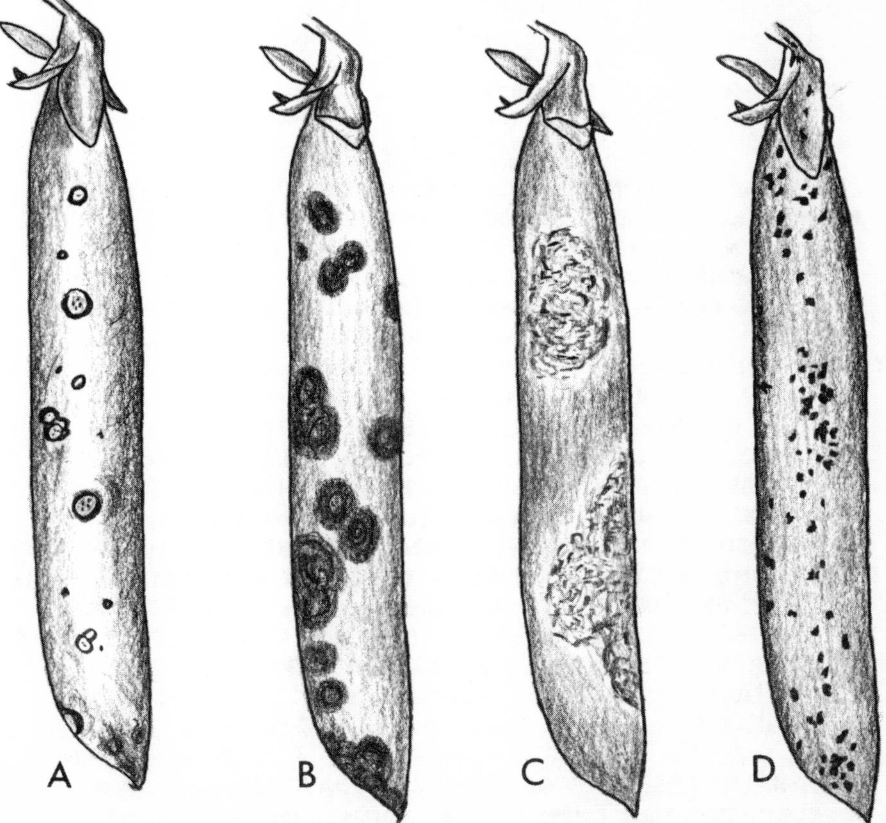

Fig. 4.159. Pod diseases of garden pea. A. Pod spot (*Mycosphaerella* or *Ascochyta*). B. Bacterial blight. C. Downy mildew. D. Scab (*Cladosporium*).

enlargements of roots themselves. Form along roots. Plants stunted, yellowish, grow slowly. Varieties and species differ in resistance. Check with your extension horticulturist or plant pathologist. See (37) Root-knot under General Diseases.

17. *Verticillium Wilt*—Plants yellow, wither, and die starting with the oldest leaves. Leaves may drop early. Brown discoloration inside stem near soil line. See (15B) Verticillium Wilt under General Diseases.
18. *Bud Drop* (sweetpea)—Primarily an indoor problem. Young flower buds turn yellow and drop instead of opening. *Control:* Provide extra light during overcast weather. Avoid overwatering and overfeeding with fertilizer high in nitrogen. Use balanced fertilizer based on a soil test. If possible, maintain temperature below 50° F. and relative humidity below 50 per cent.
19. *Other Root-feeding Nematodes* (dagger, lance, lesion, naccobus, pin, reniform, ring, spiral, stem, sting, stubby-root, stunt)—Mostly southern states. May be associated with unthrifty, stunted plants. *Control:* Same as for Root-knot (above).
20. *Chlorosis*—Deficiency of manganese, zinc, iron, calcium, potassium, copper, or molybdenum (pages 26–28). Growth often stunted. Leaves may be yellowish, bronzed, or scorched. Lower leaves may wither and die. Have soil test made and follow instructions in the report.
21. *Bacterial Wilt*—See (15C) Bacterial Wilt under General Diseases.
22. *Black Walnut Injury*—Plants growing under black walnut trees wilt and die. *Control:* Do not grow plants within 50 feet of these trees.
23. Leaf Nematode (lupine)—See (20) Leaf Nematode under General Diseases.
24. *Seed Smut* (lupine)—See (11) Smut under General Diseases.
25. *Curly-top* (vetch)—See (19) Curly-top under General Diseases.
26. *Black Mildew* (dolichos)—See (12) Sooty Mold under General Diseases.

For additional information, read USDA Handbook 228, *Pea Diseases.*

PEACH [COMMON, CHINESE WILD or DAVID, DOUBLE (WHITE-FLOWERED, PINK-FLOWERED, RED-FLOWERED), FLOWERING], APRICOT [COMMON, DWARF FLOWERING, FLOWERING, JAPANESE, PURPLE], NECTARINE, ALMOND [DWARF FLOWERING (PINK and WHITE), FLOWERING, PINK RUSSIAN], PLUM [ALLEGHENY, AMERICAN (PURPLELEAF), BEACH or BLACK, CANADA, CHICKASAW, DAMSON TYPE or BULLACE, DOUBLE FLOWERING, GARDEN or PRUNE, HORTULAN, JAPANESE, MYROBALAN or CHERRY, PACIFIC, PURPLE or ORIENTAL, SLOE or BLACKTHORN, WILD GOOSE], PURPLELEAF BUSH, CHERRY-LAUREL or LAUROCERASUS [CAROLINA, COMMON, ENGLISH, PORTUGUESE], CHERRY [BESSEY (HANSENS-BUSH or WESTERN SAND), BLACK, BITTER, BUSH, CATALINA ISLAND, CHINESE BUSH, DOUBLE-FLOWERED SOUR, DOUBLE-FLOWERED MAZZARD, FUJI, HIGAN or ROSEBUD, HOLLYLEAF, HYBRID BUSH, EUROPEAN BIRD, KOREAN BUSH, MOUNTAIN, NANKING or MANCHU, NADEN, NANKING BUSH, ORIENTAL or JAPANESE FLOWERING (AMANOGAWA, FUGENZO, KIKU-SHIDARE, KWANZAN, KOFUGEN, NADEN, SHIROFUGEN, SHIROTAE, SHOGETSU, TEMARI, etc.), MAHALEB or ST. LUCIE, MORELLO, PURPLELEAF SAND, SAND, SARGENT'S JAPANESE, SOUR (DWARF or PIE), SWEET or MAZZARD, WEEPING HIGAN, WILD RED or PIN, YOSHINO], ROSE TREE OF CHINA, MAYDAY-TREE or HARBINGER EUROPEAN BIRD CHERRY, AMUR CHOKECHERRY, PURPLELEAF CHOKECHERRY *(Prunus)*

1. *Brown Rots, Twig Blights, Monilinia Blossom Blights*—General and serious in

Fig. 4.160. Brown rot. Note dark, shriveled, infected blossoms. (Department of Botany & Plant Pathology and Cooperative Extension Service, Michigan State University photo)

Fig. 4.161. Peach twig cankers.

warm (60° to 80° F.), humid weather. Blossoms wilt, turn brown, and rot (Figure 4.160). Leaves on twig tips suddenly wither and turn brown. Twigs may die back from sunken, brown, girdling cankers. Soft, brown, rotted areas in fruit (Figure 2.48D). Affected areas may later be covered by powdery tufts of gray to tan mold. Fruits shrivel and become dry, hard, wrinkled mummies. *Control:* Destroy wild or neglected stone fruits and other *Prunus* species nearby. Remove and destroy blighted twigs and fruit spurs when first found. Avoid dense growth, overfertilizing with nitrogen, and crowding of trees. Collect and destroy rotting fruit promptly. Handle fruit carefully to avoid skin punctures and abrasions. Dip picked fruit in captan solution (2 tablespoons per gallon). Refrigerate promptly. Follow regular spray program (Table 16 in Appendix) using captan, thiram (Thylate), sulfur,[1] or maneb. Commercial growers apply dichlone during the bloom period. Start when 5 per cent of the blossoms are open. Red Gold *peach;* Raribank and Lantz *plums;* Lexington, Nectared 2 and 3, and Redbud *nectarines;* Hemskirke, Moorpark, Tilton, and Wenatchee *apricots;* and Krassa Severa and Merton Heart *cherries* have resistance. Control insects (e.g., plum curculio, plant bugs, oriental fruit moth) that provide wounds through which brown rot fungi enter.

2. *Twig, Branch and Trunk Cankers, Dieback, Gummosis*—Widespread, may be serious. Discolored, slightly sunken areas on twigs, branches, and trunk that enlarge and may girdle affected parts. Gum may exude from cankers. Buds often killed. Twigs, branches, or entire trees, decline, wilt, and die back. See Figure 4.161. *Control:* Remove and burn all dead or dying twigs and limbs during late winter. Train young trees to avoid narrow-angled, weak crotches. Make flush pruning cuts (page 33), then paint with tree wound dressing (page 36). Avoid excessively high fertilization, especially with nitrogen, in fall. Protect trees against Winter Injury (see below), borers, scales, and other insects, sunscald, harvesting, cultivator, mowing, or other wounds. Destroy severely infected trees. Cut away affected bark and wood on large cankers (page 35) during dormant period. Paint wounds promptly with water-asphalt emulsion containing 5 tablespoons per gallon of Elgetol, or use a 1:500 solution of mercuric chloride (2 tablets in ½ pint of water

[1] Do *not* use sulfur on Apricots.

and ½ pint of glycerine). Then apply good tree wound dressing (page 36). Or apply Cerano, a mercury-containing wound dressing developed at the University of California. Follow recommended spray programs as for Brown Rots (above) and Scab (below). Varieties differ in resistance. Apply dichlone, thiram, zineb, or maneb before leaf fall, after harvest, and just after pruning in the spring.

3. *Wood Rots, Silver Leaf*—Cosmopolitan. Stone fruits are very susceptible. Often follows freezing, sunscald, drought, insect damage, untreated pruning cuts, trunk and limb bruises, or other injury. See under Birch, and (23) Wood Rot under General Diseases. See Figure 4.162. Leaves may appear silvery on one or more branches. *Control:* Follow recommended spray program in Appendix (Table 16). Control borers by spraying or painting trunk, crotches, and scaffold limbs with DDT or Thiodan. Avoid spraying fruit or foliage. Check with your county agent or extension entomologist regarding rate to use and dates of application for your area. Keep trees growing vigorously. *Plum* varieties differ in susceptibility to the several cankers and wood decays. Protect trees against Winter Injury (below). Burn all prunings.
4. *Leaf Curl, Witches'-broom, Plum Pockets*—General following cool, rainy weather in spring. Unfolding leaves may become severely curled, "blistered," swollen, pale green, yellow, reddish or purplish, and leathery. Affected leaves commonly wither and fall with second crop of leaves forming later. Severe attacks weaken trees and greatly reduce fruiting. *Peach* fruits may be irregular and knobby with dark-colored, warty spots. Such fruits usually drop early. On *cherries, flowering cherries, plums,* and *cherry-laurel*—dense clusters of long, slender, irregular twigs arise near same point on stem (Witches'-broom). Branches may be stunted. Twigs die back. Also on *plums,* extremely large, narrow, pale, hollow, wrinkled fruits that drop early (Plum Pockets). Infected shoots often distorted and swollen, later die. See Figure 2.26 under General Diseases. *Control:* Follow spray program (Table 16 in Appendix). Prune out all witches'-brooms and distorted twigs. Destroy infected fruit. Maintain trees in healthy, vigorous condition. Varieties differ in resistance. Apply *single dormant* spray before buds swell in late winter or early spring. Use lime-sulfur (1:10), bordeaux (2-2-50), phenyl mercury, dichlone, thiram, ferbam, Elgetol, captan, Daconil 2787, ziram, or dodine. Follow manufacturer's directions. Add spreader-sticker to spray to ensure wetting buds. Parkhill *peach* is somewhat resistant to Leaf Curl.
5. *Cherry Leaf Spot, Shot-hole, Yellow Leaf*—General and serious in damp 60° to 70° F. weather. Small, round to irregular, purplish to brown spots on leaves that often cause severe early leaf yellowing and dropping. Spots may drop out leaving shot-holes. Fruit may fall prematurely. Repeated leaf loss weakens or kills trees. See Figure 2.20 under General Diseases and Figure 4.163. *Control:* Collect and burn fallen leaves in autumn. Follow spray program outlined in Appendix. Use captan, thiram, ferbam, difolatan, Polyram, Daconil 2787, folpet, maneb, ziram, or sulfur. Dodine is used by commercial sour cherry growers and nurserymen. One or more sprays are needed after harvest to keep leaves from dropping prematurely. Northstar *cherry* is resistant to Leaf Spot. More resistant varieties should be available soon.
6. *Black Knot* (primarily cherries, plums, prune, Mayday-tree, apricot)—Widespread, especially in eastern half of United States. Elongated, rough, girdling, black swellings on twigs, small branches, and even trunk. Knots velvety olive-green in spring. Gradually become hard, brittle, and coal-black by fall. Affected parts may die back. Trees gradually weaken and die. See Figure 2.43 under General Diseases. *Control:* Prune and burn all infected wood in late winter. Make cuts at least 4 to 6 inches below any obvious infection, then paint wounds with tree paint. Paint large limbs with Acti-dione (200 parts per million). Follow spray program given in

Fig. 4.162. Wood rot of cherry.

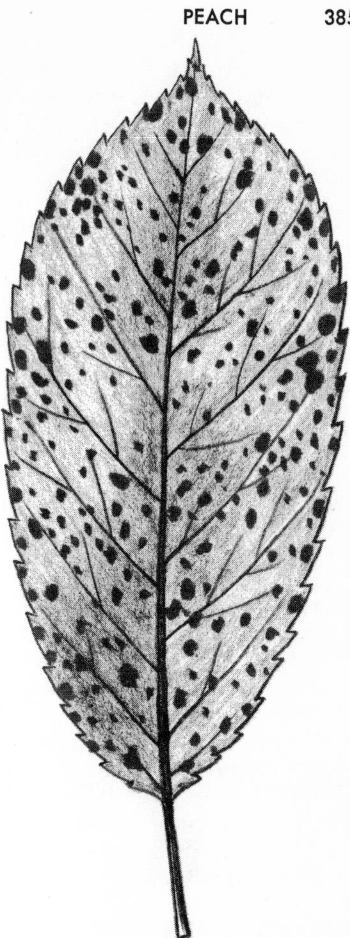

Fig. 4.163. Cherry leaf spot.

Appendix. Apply delayed dormant spray of zineb, maneb, bordeaux (6-6-50), dodine, or liquid lime-sulfur (1½ cups per gallon of water). Or apply *dormant* spray of phenyl mercuric acetate (PMA). Varieties differ in resistance. Stanley and Damson *plums* are very susceptible.

7. *Scab, Fruit Freckle*—General. Small, round, dark olive-green or black spots on fruit. If numerous, fruit may crack. Yellowish-brown blotches with gray or bluish borders develop on twigs. Twigs may die back. Leaves spotted dark green to brown and drop early. Spots may first fall out leaving shot-holes. *Almond* shoots turn brown. Leaves blacken and drop early. See Figures 2.20 and 2.30 under General Diseases. *Control:* Destroy nearby, neglected trees. A *dormant* spray (just before budbreak) on *peach* with phenyl mercury has given excellent control. Prune and burn blighted twigs before growth starts. Apply sulfur, captan, Dithane M-22 Special, or maneb at 10- to 14-day intervals. Start 10 days after bloom. Follow spray program in Table 16 of Appendix.
8. *Powdery Mildews*—General on sour cherries. Young leaves tend to crinkle, twist, and fold up. They are covered with white to light gray powdery patches that may later become rusty-brown. Mold patches also develop on young twigs and fruit. Fruit may be misshapen, stunted, cracked, and "scabby." Most serious on young,

rapidly growing trees. *Control:* If severe, apply sulfur or Karathane when first seen. Repeat in next 2 regular sprays. Also apply postharvest spray on *cherries, peaches,* and *apricots* if mildew starts to increase. Varieties differ in resistance. Manorette *cherry-plum* is moderately resistant.

9. *Bacterial Cankers, Gummosis, Bacterial Spot, Shot-hole*—General and destructive. Symptoms variable. Round to elongated, water-soaked, rough, raised or somewhat thick-edged, sunken cankers on trunk, branches, and twigs. Diseased bark is brown, gummy, and often sour-smelling. Twigs and branches are girdled and die back. Dormant buds, fruit spurs, and blossoms are blighted. May be small, sunken, brown to black spots on fruit. Fruit may become roughened with cracked, sunken spots. Small, angular, purple, reddish, or brownish to black spots on leaves; later drop out leaving shot-holes (Figure 2.20 under General Diseases and Figure 4.164). Leaves often turn yellow and drop early, weakening trees. *Control:* Grow healthy, vigorous trees in fertile soil where Bacterial Canker or Gummosis has not been present for number of years. Keep trees vigorous by proper fertilization, cultivation, and pruning of girdled or severely diseased branches. Apply 2 or 3 sprays of bordeaux (5-5-50) or phenyl mercury *after* harvest and *before* leaves fall. Follow spray program as for Brown Rots and Scab (both above). Or use zineb or zinc bordeaux ($1\frac{1}{2}$-$2\frac{1}{2}$-50). Varieties differ in resistance to Bacterial Leaf Spot *(Xanthomonas pruni)* and Bacterial Canker *(Pseudomonas syringae). Apricots*—Afred, Georgeff, Haggith, Kohl, Minnesota 36, Scout, and Sunshine; and *Peaches*—Belle of Georgia, Commanche, Dixigold, Earlired, Early-Free-Red, Garnet Beauty, Hiley Ranger, Jubilee, Olinda, Prairie Dawn, Raritan Rose, and Redhaven are resistant to Bacterial Spot. *Plums* highly resistant to Bacterial Canker—Beauty and Kelsey. *Cherry* varieties differ greatly in resistance to Bacterial Canker. (Merton Reward, Merton Heart and Mahaleb rootstock are resistant.) Check with your local nurseryman, extension horticulturist, or plant pathologist. Surgical removal of small cankers is often practical. Disinfect pruning tools between cuts by dipping in 70 per cent alcohol, household bleach, or mercuric chloride (1:1,000) solution. Best to prune during midwinter. Bacterial canker is transmitted by bud propagation from infected trees.

Fig. 4.164. Bacterial spot and shot-hole on peach leaves.

10. *X-Disease, Yellow-red Disease or Virosis* (peach, nectarine, cherries, plums, apricot, almond) —Widespread. Symptoms variable due to presence of many virus strains.

Peach—Leaves appear normal in spring for first 6 to 8 weeks. Then show diffuse, irregular, pale green to dark brown, or yellow-red to purplish areas that may drop out leaving ragged shot-holes. Leaves remain pale green to yellowish-red. Tips of leaves on most varieties bend down and margins roll up. Older leaves drop early, except for tufts of young leaves at tips. Fruits shrivel and usually drop or ripen early. Bitter to taste. Trees tend to die back gradually. X-disease may be confused with severe nitrogen deficiency and arsenic toxicity.

Sweet and Sour Cherry (little cherry, buckskin, wilt and decline) —Fruit distinctly small, poorly colored, usually pointed, and lack flavor. Foliage symptoms variable, depending on virus strain, variety, rootstock, soil conditions, and other factors. Trees may wilt, decline, and die. Seedlings of certain varieties appear resistant, e.g., *sweet cherry:* Dicke Braune Blakenburger.

On *Mazzard rootstocks*—Trees lack vigor, are dwarfed. Foliage is somewhat sparse. Leaves light green, later bronzed, may be in rosettes. Fruits on affected branches are usually small, poorly flavored, dull reddish-pink, yellow, or nearly white. Symptoms become more pronounced in later years.

On *Mahaleb rootstocks*—Bearing trees often rapidly wilt and decline in vigor and fruitfulness. Often die in less than a year. Foliage may be sparse and retarded in growth with leaves smaller, lighter green, or yellowish and narrower than normal. Leaves may drop early. Roots killed progressively from tips to trunk. Young trees may wilt and die in 1 to 2 years.

Plum, Apricot, Flowering Cherries—Most commercial varieties may be symptomless carriers.

Control: Where possible, destroy all wild cherries (especially chokecherries), plums, peaches, and other wild *Prunus* spp. within 300 yards and preferably ½ mile or further. Destroy infected trees when *first* found. Spray trunk base with Ammate or "brush-killer." Plant only certified, virus-free trees (or use virus-free budwood) from a reputable nursery. Grow only trees that are standard, well-tested varieties in your area. Follow a complete spray program (Table 16 in Appendix). Control insects, especially leafhoppers, that transmit the viruses. Use combination of methoxychlor or DDT plus malathion.

There are well over 60 stone fruit diseases caused by viruses that have been reported in the United States and adjoining areas of Canada. New ones are reported each year while others are discovered as being strains of previously known viruses. The true status of all these viruses is not known at present. For more information read a book such as USDA Handbook 10, Virus Diseases and Other Disorders With Viruslike Symptoms of Stone Fruits in North America. *If you suspect a virus disease, contact your state or extension plant pathologist.*

11. *Little-Peach, Little Plum* (peach, plum, apricot) —Roughly eastern half of United States.

Peach—Fruit stunted, ripen later than normal if at all. Fruit usually insipid. Young leaves at tips of affected branches are crinkled. Trees often stunted and bushy with pale, yellowish leaves. Leaves first darker green and more leathery than normal. Bend inward toward shoot. Short, bushy shoots with clusters of leaves arise from base of affected limbs.

Plums and Apricot—Symptoms on susceptible varieties similar to peach but usually much milder. Many plums are symptomless carriers. Japanese or Oriental plums are very susceptible.

Control: Same as for X-Disease (above).

12. *Rosette* (peach, plum, apricot, almond, cherry) —Eastern half of United States.

Peach—Symptoms variable. Young trees suddenly wilt and die. Compact rosettes,

2 to 3 inches long, composed of 200 to 400 dwarfed leaves, are conspicuous with abnormally long, straight leaves at base of shoots. Leaves turn yellow, drop in early summer. Trees usually bear no fruit and die within a year.

Plums—Growth stunted. Yellowish rosettes of mottled leaves often formed. Marianna plum appears immune; Japanese and Damson plums are susceptible.

Apricot—Symptoms variable. Trees markedly stunted. Leaves may have yellowish tinge and show mosaic mottling. Typical rosettes are formed.

Almond, Cherries—Growth stunted. Rosettes of yellowish-green leaves are formed.

Control: Same as for X-Disease (above).

13. *Peach Yellows* (peach, nectarine, plums, apricot, almond)—General over much of United States, especially in eastern states. Symptoms variable.

Peach, Apricot and Almond—Leaves pale green, yellow, or yellowish-bronze. May be mottled dark and light green. Leaves curl, roll upward and inward. Drop early, Fruit ripen prematurely. May be somewhat "bumpy," and soft. Often with reddish or purple blotches. Flavor is insipid or bitter. Reddish streaks may be seen in fruit flesh. Clusters of thin, wiry, erect shoots with narrow, yellowish, red-spotted leaves are common. Trees appear bushy. Often die in 2 to 6 years. See Figure 2.36C under General Diseases.

Plums—Certain Japanese varieties are symptomless carriers. Other plums show typical foliage symptoms, but are often milder than on peach.

Control: Same as for X-Disease (above).

14. *Sour Cherry Yellows* (Montmorency, Early Richmond, and English Morello)—Widespread. Infects all *Prunus* species. Symptoms variable, depending on virus strain, spring temperatures, variety, etc. Leaves on *sour cherry* are irregularly mottled green and yellow. Later entire leaf usually turns yellow and drops early. Failure of fruit spur development and willowy-type growth with long bare spaces on twigs are common symptoms. Yields may be reduced 20 to 70 per cent. Many ornamental, wild, and sweet cherries and other *Prunus* species—especially older trees—are symptomless carriers. Virus is probably a strain of Peach Ringspot (or Prune Dwarf); initial yellows symptoms generally appear after the first ringspot infection. *Control:* Same as for X-Disease (above). Destroy infected trees in young orchards. Infected *cherry seed* and *pollen* may carry the virus.

15. *Ringspot Complex, Tatterleaf*—Widespread on practically all *Prunus* spp. Symptoms variable. Many plants are symptomless carriers. Caused by many virus strains; some of which are seed- and pollen-transmitted. Trees do not become infected until bearing age.

Sweet Cherry—Pale green to yellow mottling, spots, ring, oakleaf, and brown line patterns may form on leaves. Dead spots (that may drop out) often develop. Leaves may appear shot-holed or lacelike (Tatterleaf). Terminal shoot growth may die back. Symptoms tend to fade as season progresses.

Sour Cherry—Finely etched rings or lines may form network on leaves. Dead spots in leaves often drop out giving "tatterleaf" condition. Young leaves may be puckered; sprouts are killed during acute phase of disease. Blossoms may be distorted. Often do not set fruit. Trees may have thin appearance. Ringspot is often accompanied by Yellows (above). Terminal shoot growth and leaf size is stunted during first 2 years of infection.

Plums—Symptoms often absent or mild. Common in Japanese and garden plums. Rings and yellowish patterns or dead flecks and spots produced on some varieties, especially during acute stage of disease. Spots may turn brown and drop out leaving shot-holes and tattered leaves. See Plum or Prune Decline (below).

Almond, Apricot—Symptoms highly variable on apricot. Most virus strains producing symptoms cause small to large, sharply defined ringspots, yellowish oakleaf patterns, yellowish mosaic or mottle patterns, twig dieback, or bark gumming and killing. Most severe during first year of infection.

Peach—Symptoms variable. Growth may be stunted. Twigs die back. Yellowish-green, yellow, or brown rings and spots or oakleaf patterns form on leaves. Centers may drop out leaving shot-holes. Buds often killed. Trees may appear to recover, show no further symptoms.

Control: Same as for X-Disease (above). Grow only standard, well-tested varieties adapted to your area. Destroy older plum trees showing evidence of decline.

16. *Plum or Prune Decline* (Line Pattern, Ringspot, Tatterleaf, Prune Dwarf)—Widespread. A virus complex that probably infects most *Prunus* spp. Symptoms variable. Foliage discolored with spotting and localized killing (reddening, yellowing, line, arc, or oakleaf patterns, ringspots, tattered and strap leaves), plus sparse foliage, witches'-brooms, and dieback of twigs and branches. Growth reduced. Individual trees or some plums decline in vigor and productivity. Leaf spots may turn brown and drop out leaving ragged and torn leaves. Common in Japanese and garden plums. Prune Dwarf is strain of Peach Ringspot virus. Striking, yellow to white lines and oakleaf patterns develop on leaves of *Oriental flowering cherries. Control:* Grow only virus-free trees of standard, well-tested varieties from a reputable nursery. Destroy older trees showing evidence of decline. Otherwise, same as for X-Disease (above).

17. *Phony Peach, Phony Disease* (peach, plums, apricot, almond)—Declining in importance. Southern half of United States, especially eastern half. Flattened, abnormally dark green leaves. Terminal leaves are close together on stunted twigs. Fruit small and ripen early. Poor in flavor. Growth is checked and trees are dwarfed. Fruit production stops after 3 to 4 years. Twigs and branches die back. The virus is in practically all wild plums. *Control:* Same as for X-Disease (above). Check with your county agent or extension plant pathologist. For additional information, read USDA Leaflet 515, *Controlling Phony Disease of Peaches.*

18. *Mosaics* (peach, nectarine, almond, apricot, plums, prune)—Declining in importance. Southwestern half of United States. Sweet and sour cherry and most cherry-like species are immune. Symptoms very variable due to many virus strains. Rugose mosaic of sweet cherry is a strain of Prunus Ringspot virus.

Peach—Disease usually proceeds through acute and chronic phases. Severely infected trees may later partially recover. Leaves may be crinkled and distorted with mottled, yellowish or yellowish-green spots, blotches, and irregular patterns. Clusters of twigs at tips of branches. Leaves often small and bunched in rosettes. Growth retarded. Flowers often streaked and spotted (broken). Fruit may be misshapen, bumpy, and dwarfed with uneven color. Freestone varieties (e.g., Elberta, J. H. Hale) are most severely attacked. A few freestones (e.g., Erly-Red-Fre, Fisher, and Valiant) are highly tolerant. Most clingstones (e.g., Paloro, Peak, Phillips, Sims) tolerate virus with little damage. *Peaches* that tend to recover readily from shock stage of infection: Blazing Gold, Dixigem, Keystone, and July Elberta. Check with your state or extension plant pathologist.

Apricot, Almond and Other Hosts—Symptoms variable depending on virus source, strain, and variety or species. Wild plums are usually symptomless carriers. Leaves may be faintly mottled or sprinkled with indistinct to brilliant, lighter green spots or splotches, rings, or oakleaf patterns. Yellowish spots, stripes, and streaks may develop in the lighter green blotches. Infected fruit may be bumpy, misshapen, and worthless. Stones of *almond* and *apricot* fruit show white rings and blotches. Some *plums, peaches,* and *sweet cherries* are symptomless carriers. Others appear immune.

Control: Same as for X-Disease (above). Control bud mites and dagger nematodes *(Xiphinema)* that help transmit certain viruses, e.g., Peach Yellow Bud Mosaic and Peach Mosaic. Check with your extension entomologist.

19. *Apricot Ring Pox*—Western states. Yellowish rings, mottling, or irregular blotches develop on leaves. Later the leaves may be ragged with shot-holes. Irregular, reddish-brown blotches or rings appear on fruit. May pit surface causing distorted, bumpy fruit. Ripe fruit often crack in discolored areas. Many affected fruit drop early. *Plums* show symptoms only on leaves. When Ring Pox and Peach Mosaic viruses occur together in *apricot,* leaf symptoms of Mosaic and fruit symptoms of Ring Pox are both intensified. *Control:* Same as for X-Disease (above).
20. *Cherry Rasp-Leaf, Leaf Enation*—Western states. Toothlike or rasplike outgrowths (enations) protrude from underleaf surface. Leaves may be smaller, narrower, markedly distorted, and longer than normal. Tree growth is stunted. Fruit crop is reduced. *Control:* Same as for X-Disease (above). Virus is transmitted by pollen.
21. *Cherry Mottle Complex* (Mild or Severe Mottle Leaf; Mild, Severe, and Necrotic Rusty Mottle [Lambert Mottle]) —Primarily in Pacific Northwest. A complex of virus strains. Symptoms differ with variety, virus strain, and other factors. *Prunus serotina* is immune. Cherry Mottle Leaf virus is spread by mites.

    Sweet Cherry—Leaves on one or more branches of certain varieties show variable, irregular, yellowish mottling. Leaves may be stunted, twisted, wrinkled, and puckered (Mottle Leaf). Later, certain leaves may become mottled bright yellow to red with islands of green (Rusty Mottle). These leaves drop early. Mottling of remaining foliage becomes more pronounced. Yellowish spots and areas may become yellowish-brown or brown and dead. Foliage from a distance has a "rusty" or bronzed look. Where heavy leaf drop occurs, foliage is sparse except for branch tips. In fall, conspicuous dark green ring and line patterns develop on background of yellow, brown, or brilliant red (Necrotic Rusty Mottle). If severe, fruit may be small, hard, ripen late, and taste insipid. Trees may be stunted and appear rosette-like. May die back and slowly decline. *Sweet cherries* that are symptomless carriers of Mottle Leaf virus: Bigarreau, Cardofer, Chinook, Fruhe, Fruhe Werdershe, MarMac, and Ranier.

    Sour and Oriental Flowering Cherries—May be symptomless carriers or show mild symptoms as for sweet cherry.

    Peach—Most varieties are symptomless. Certain varieties have leaves with striking greenish-yellow or yellow ringspots and patterns. Such leaves soon drop and tree may appear normal.

    Control: Same as for X-Disease (above).
22. *Little Cherry*—Fruit size may be reduced 50 per cent or more. Color, shape, and quality are also lowered depending on virus strain, rootstock, and variety. Symptoms much more severe on Mazzard roots than on Mahaleb rootstocks. Virus is latent in certain flowering cherries, e.g., Kwanzan and Shirofugen. *Control:* Same as for X-Disease (above).
23. *Asteroid Spot* (peach, nectarine, almond, apricot, plums) —Widespread. Small, yellow, star-shaped areas develop along veins of nearly full-grown leaves. Large, angular blotches form along veins on certain leaves. Leaves may then turn yellow with flecks and splotches becoming light green. Such leaves drop early. Symptoms may only be evident during acute stage. Apparently causes little damage. *Control:* Destroy affected trees in the nursery. Do not use for propagation. Control insects and mites. Follow spray program in Appendix (Table 16). Destroy desert peach *(Prunus andersonii),* a common symptomless carrier.
24. *Coryneum Blight, Twig Canker, Shot-hole*—Widespread. Most serious on peach, nectarine, Catalina cherry, apricot, and almond. Small, round, reddish to purple spots on twigs in early spring that may become brown, elongated, and slightly sunken. Twigs show gumming and dieback. Buds darker in color and killed. Small,

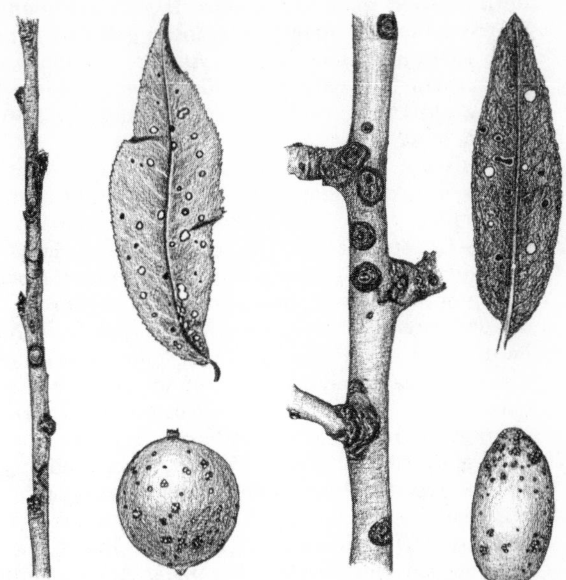

Fig. 4.165. Coryneum blight of peach (left) and almond (right).

round, tan to dark brown, reddish brown, or purplish leaf spots with dark red borders. Leaf spots may drop out leaving shot-holes. Where severe, heavy leaf drop may occur. Flowers may wither. Fruit spots small, round, and red to deep purplish-red. Later, spots become raised and roughened with fruit cracking and gumming. See Figure 4.165. *Control:* Follow spray program in Appendix. Apply one or more postharvest sprays of ziram, ferbam, fixed copper, dodine, captan, folpet, or bordeaux (5-5-50) before first heavy fall rains begin. Check with your extension plant pathologist. Lovell and Muir *peaches* are reported as blight-tolerant.

25. *Verticillium Wilt, Blackheart Wilt*—Mostly in western states. Symptoms variable. On *peach,* leaves on certain limbs are blanched, dull, wilt, and fall in early summer. Leaves on lower branches are attacked first as disease progresses upwards. Twigs die back. Gray or light to dark brown streaks are evident in sapwood of infected trunk or limbs. Trees stunted, may die gradually over period of years or very suddenly (especially *cherries*). *Apricot* trees apparently often recover naturally. On *cherry,* older leaves turn yellow and wilt in summer. Leaves later turn reddish-violet with dry, curled margins and brown patches between veins. Leaves yellow, quickly wilt, and wither. Infected branches appear scorched with leaves remaining on tree. Affected trees may live on for several years in weak condition with dead branches. Brown to black streaks in sapwood. Winter injury is common on wilt-affected trees. *Control:* Prune and burn affected limbs at trunk. Water and fertilize, where practical, to maintain good vigor. Avoid wounding trunk or roots when planting or cultivating. Destroy severely infected trees. Do not replant with susceptible crops for several years. See (15B) Verticillium Wilt under General Diseases. Control such susceptible weeds as lambsquarters, pigweed, nightshades, horsenettle, and groundcherries.

26. *Crown Gall, Hairy Root*—General. Small to large, gall-like overgrowths on roots and crown. See Figure 2.46D. Infected trees lack vigor and may die. Japanese apricot *(P. mume)* is resistant. *Control:* Plant only inspected, disease-free nursery stock in soil that has not grown crown gall-infected plants for at least 5 to 6 years. Avoid wounding trees near soil line. Destroy seriously declining trees. Remove and

burn diseased roots and stumps. If galls are found on trees and disease has not progressed too far, paint entire, intact gall and not more than ½ inch beyond gall, with following mixture: To crystalline streptomycin (2,000 parts per million) or Terramycin (200 parts per million) dissolved in a small amount of water (19 per cent), add 20 per cent iso-amyl gradually, followed by kerosene (80 per cent), melted lanolin (½ per cent), and Vaseline (½ per cent). Properly prepared, this mixture will be clear or slightly opalescent. Shake mixture before using. Nurserymen soak plants for 10 to 30 minutes in Terramycin or Agrimycin 100 before planting. S-37 *peach* rootstock is resistant to Crown Gall.

27. *Rusts*—General, especially in southern states. Small, angular, yellow spots on upper leaf surface that become purplish or bronzed. Small, reddish, reddish-brown, dark brown or black dusty pustules on underleaf surface. Leaves may drop early. Round, sunken, green spots may develop on *peach* fruit. Affected *apricot* fruit show small, hard points in the skin. *Control:* Follow spray program in Appendix. Add zineb, maneb, ferbam, thiram, or sulfur to several consecutive sprays. Start with second cover. Some varieties may be susceptible to zineb or maneb injury. Destroy alternate hosts: anemones, buttercup, hepatica, and meadowrue.
28. *Root Rots, Decline, Crown Canker*—Cosmopolitan. Trees lack vigor. Gradually die. Foliage remains stunted and sparse. Leaves often yellow and drop early. *Cherry* leaves may be red in the fall. Toadstools sometimes are evident at base of dying or dead tree. Often in clusters (Figure 2.49B). Usually associated with nematodes or water-logged soils. *Control:* See under Apple. Do not replant in same soil without first fumigating with D-D or EDB. Grow trees in fertile, well-drained soil. Avoid injuries to crown and roots. *Marianna 2624 plum* is an Armillaria root rot-resistant rootstock for *plums, apricots,* and certain *almond* varieties. *Mazzard cherry* is moderately resistant after becoming established. *Sour Cherry rootstock (Stockton Morello)* is resistant to Phytophthora Crown Rot or Canker. The Myrobalan and Marianna 2624 *plums* are also resistant and can be used as rootstocks for apricots, plums, and prune.
29. *Root-knot*—Common in southern states, especially peaches growing on sandy soils. Heavily infested trees are stunted, with light-colored foliage. Trees gradually weaken due to small, spindle-shaped or irregular "knots" on roots. Fruiting is reduced. *Control:* Grow nematode-free, resistant rootstocks, e.g., Okinawa, Nemaguard, or S-37 *peach*. Redran *peach* and Shalil and Yunnan rootstocks are resistant to certain root-knot species and susceptible to others. Grow in soil fumigated before planting with D-D, EDB, Dorlone, DBCP, or Telone (pages 538–548). Check with your nurseryman or extension plant pathologist. DBCP (Nemagon, Fumazone) is used around living trees. Follow manufacturer's directions.
30. *Other Root-feeding Nematodes* (dagger, lance, lesion, needle, pin, ring, sheath, spiral, stem, sting, stubby-root, stunt)—General in United States. Trees often stunted. Are off-color and make poor growth. Do not respond normally to water and fertilizer. Fruit production decreases. Twigs and branches may die back due to killing of small feeder roots. May result in small witches'-brooms, "brushes" of stubby roots, and brown or black rootlets. Severely infested plants may be more susceptible to Winter Injury (below) and Root Rots (above).
31. *Winter Injury* (especially peach)—Twigs, limbs, or even whole trees may die due to low winter temperatures and either low or excessively high soil moisture. Affected wood is often dark. Foliage is off-color. Fruit buds are killed. Trees lack vigor. *Control:* Grow peaches where adapted. For a list of hardy varieties for your area check with local growers, your county agent, or your extension horticulturist. Varieties differ greatly in winter hardiness.
32. *Minor Leaf Spots and Blotches, Leaf Blight, Leaf Scorch, Shot-hole*—General. Spots of various sizes, shapes, and colors. Spots may later drop out. Spots often occur on

twigs; gum may ooze. Leaves may appear scorched and withered. *Control:* Same as for Brown Rots (above). Fertilize and prune to maintain vigor.

33. *Fruit Rots, Fly Speck*—Cosmopolitan. See under Apple, and Brown Rots (above). Tan, black, green, pink, cottony, bluish-green, or brown mold may grow on rotting fruit. *Control:* Same as for Brown Rots (above). Refrigerate promptly to 32° F. Handle carefully. *Peach* Rhizopus fruit rot losses in transit, storage, and market are reduced by fruit wraps impregnated with DCNA or Botran (2,6-dichloro-4-nitroaniline).
34. *Chlorosis, Little Leaf, Mottle Leaf*—Leaves often pale and small. Yellowing varies from complete lack of green to light, yellowish-green areas or mottling between veins. Veins usually remain green. Growth may be stunted, crinkled, and rosette-like (zinc deficiency). Symptoms may appear first at tips of growing shoots where leaves are pale yellowish-green to ivory (iron deficiency). Or symptoms start on *older,* basal leaves (magnesium deficiency). Leaf margins may be scorched and rolled (potassium deficiency). Leaves may drop early beginning at bases of shoots and continuing outward. If severe, twigs and then branches die back. May be due to lack of or unavailability of iron, zinc, magnesium, potassium, or manganese (pages 26–28). *Control:* See under Maple for iron deficiency and under Walnut to control zinc deficiency. Foliar sprays of potassium sulfate check potassium deficiency. Check with a local grower, your county agent, or extension horticulturist. To correct manganese and magnesium deficiency check with your extension horticulturist. Foliar sprays may be called for. Mahaleb *sweet cherry* rootstock is more resistant to zinc deficiency than is Mazzard.
35. *Wet Feet*—Trees growing in excessively wet soils with a high water table are stunted with sparse foliage composed of dwarfed leaves, dead twigs, and branches. Leaves may yellow and drop early. Roots die back. Symptoms vary with height and persistence of high water level in soil. *Control:* Drain soil (page 36) or replant in drier spot. Marianna 2624 *plum* is tolerant to heavy, wet soil. However, Davey, Drake, and Nonpariel *almonds* are incompatible with Marianna 2624 plum.
36. *Sooty Mold*—Cosmopolitan. See under Apple. *Control:* Follow recommended spray program in Appendix (Table 16).
37. *Fire Blight* (almond, apricot, cherry, cherry-laurel, flowering cherry, plums)—See under Apple, and (24) Fire Blight under General Diseases.
38. *Thread Blight*—Southeastern states. See under Walnut.
39. *Felt Fungus*—Southern states on neglected trees. See under Hackberry. *Control:* Follow recommended spray program (Table 16 in Appendix).
40. *Mistletoe* (cherry, cherry-laurel, plums)—See (39) Mistletoe under General Diseases.
41. *2,4-D Injury*—Cherry and other *Prunus* spp. are very susceptible. Severely injured leaves are twisted or rolled, boat-shaped, or curled into "ram's-horns." *Control:* See under Grape.

PEACH BELLS — See Bellflower

PEANUT [SPANISH, VALENCIA, VIRGINIA] *(Arachis)*

1. *Southern Blight, Collar Rot, Stem and Peg Rots, White Mold, Root Rots or Soil Rot*—Widespread in warm, moist weather. Entire plant or certain branches may wilt, wither, and die from a tan to almost black rot of lower stem, crown, and roots. Fruit stems (pegs) often decay. White mold often occurs on infected parts during wet periods (Figure 4.166). *Control:* Carefully dig up and burn infected plants. Cultivate shallowly and avoid throwing soil on crowns and lateral branches. Apply PCNB-captan mixture as dust or drench in bands to soil around stems. Follow manufacturer's directions. Turn *all* plant debris under deeply (at least 4 inches) and cleanly several months before planting. Control Leaf Spots (below) and

Fig. 4.166 (above). Southern blight of peanut. (USDA photo)

Fig. 4.167. Cercospora leaf spot on Spanish peanut leaf. (Courtesy Mrs. Mary McB. Miller)

insects. Keep down weeds. Three-year rotation. If using PCNB, do *not* feed vines as hay to livestock. Peanut varieties differ in resistance to Southern Blight *(Sclerotium)* and Diplodia Collar Rot.

2. *Leaf Spots, Leaf Mold*—General in warm, humid weather. Tan to reddish-brown or black flecks, spots, or blotches on leaves. Often with light-colored margins (Figure 4.167). Elongated spots may occur on petioles, stems, and pegs. Infected leaves usually turn yellow and drop early, reducing yield of nuts. *Control:* Apply sulfur, fixed copper-sulfur mixture (10-90 dust), zineb, difolatan, dodine, liquid copper, Dithane M-45, Daconil 2787, or maneb three to six times, 10 to 14 days apart. Start when spots first appear on older leaves. Rotate. Turn plant debris under cleanly after harvest.
3. *Seed Rot, Seedling Blights*—Seeds rot. Stand is thin. Seedlings are killed or weakened. *Control:* Sow plump, large, certified, well-matured, crack-free seed treated with captan, thiram, Panogen, difolatan, or Ceresan. Plant shallow in well-prepared soil that is 70° F. or higher. Three-year rotation.
4. *Pod and Kernel Decays, Pod Rots, Seed Molds, Concealed Damage*—Cosmopolitan. Entire pod or kernel may shrivel and darken with seed blackened and oily. Black, blue-green, green, white, or tan mold may grow on pods and nuts. Pods and nuts may appear normal; but nuts have internal decay (Concealed Damage). Certain molds (e.g., *Aspergillus flavus*) produce toxic substances, called aflatoxins, that are extremely poisonous to animals and man. *Control:* Follow best local cultural practices or your state peanut-growing manual. Apply gypsum or landplaster (400 to 2,000 pounds per acre; about 10 to 50 pounds per 1,000 square feet), to plants after blooms appear. Apply in wide band over row for *bunch* peanuts and broadcast evenly for *runner* varieties. Keep plants growing vigorously until harvest. Control Leaf Spots and Southern Blight or Stem Rot (above), and Nematodes (below). Control leaf-feeding insects with sevin or DDT. Harvest when crop is mature (darkened vines, brown spots on inside of pods and light pink to red color of nut skin). Allow vines to wilt thoroughly before curing artificially for several weeks at 90° F. or above on slats or stack poles. Hold vines at least 12 inches above ground, or artificially dry peanuts in a mechanical drier or oven. (Avoid eating or feeding visibly damaged kernels from cracked or broken pods.) Cover while curing

to reduce weather and bird damage. Bunch and Spanish varieties are less subject to pod and nut decays than runner varieties.

5. *Nematode Diseases*—Widespread. Yield and quality of nuts is reduced. Plants may be stunted and yellowish-green. Tend to wilt during midday; may die prematurely. Roots, fruit stems (pegs), and peanut pods may have small to large, knotty galls (Root-knot) or enlarged, stunted, tap root with few short, stubby lateral roots (Sting Nematode). Plants affected with Lesion Nematode have small, brownish to black areas on peanut pods, pegs, and lateral roots. See Figures 4.168A and 4.168B. Other nematodes include dagger, lance, pin, reniform, ring, spiral, stubby-root, and stunt. Pods may be stunted and kernels may decay. *Control:* Three-year rotation. Fumigate with D-D, Vidden-D, or Telone about 3 weeks before planting or Zinophos at planting time. Follow manufacturer's directions. Do *not* feed vines grown on fumigated soil to dairy animals or to animals to be slaughtered later for meat.

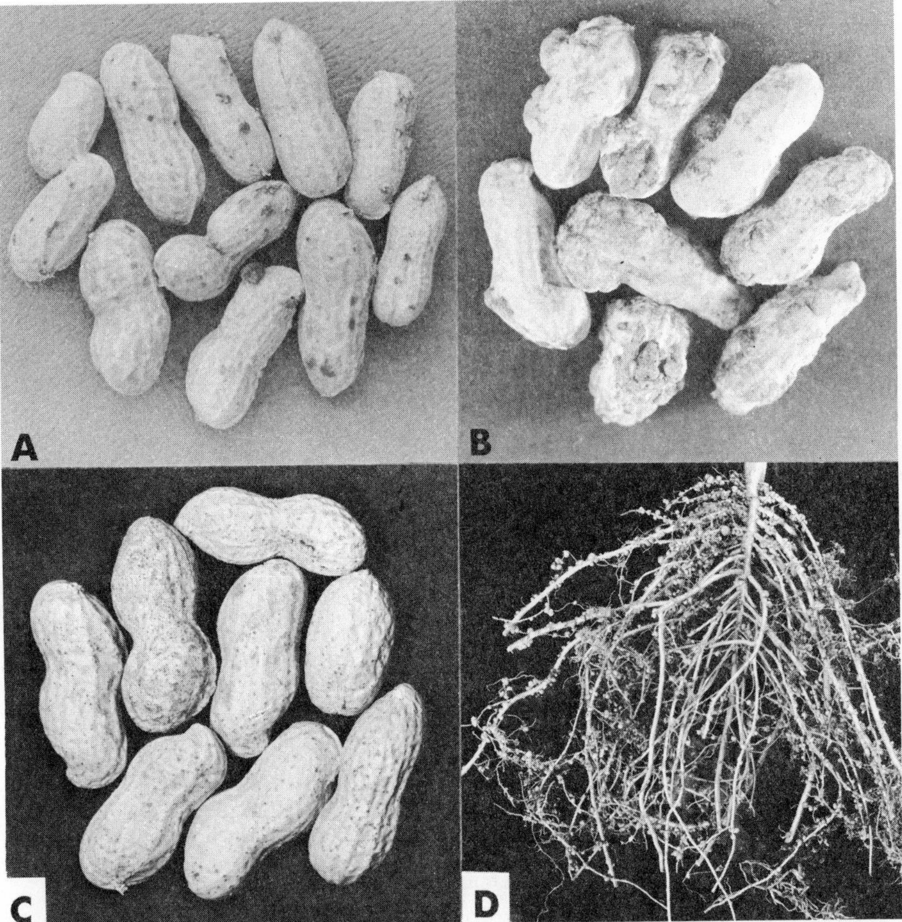

Fig. 4.168A. Nematode damage to peanuts. A. Galls and lesions on pods caused by the northern root-knot nematode. B. Misshapen fruit infected by peanut root-knot nematode. C. Very small brown lesions on pods caused by external feeding of sting nematode. D. Root of peanut plant infected with the northern root-knot nematode. (Courtesy Dr. L. I. Miller)

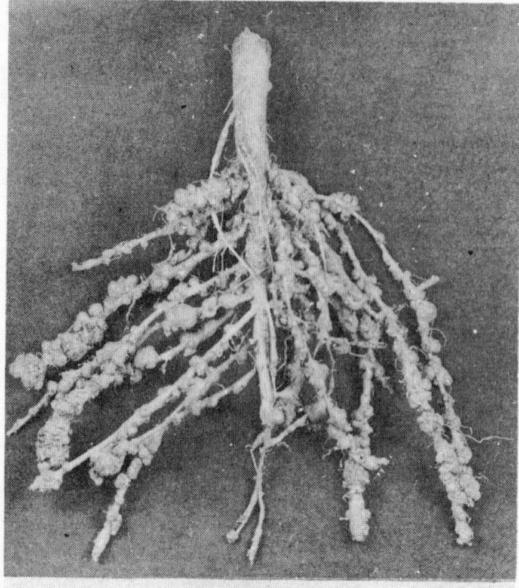

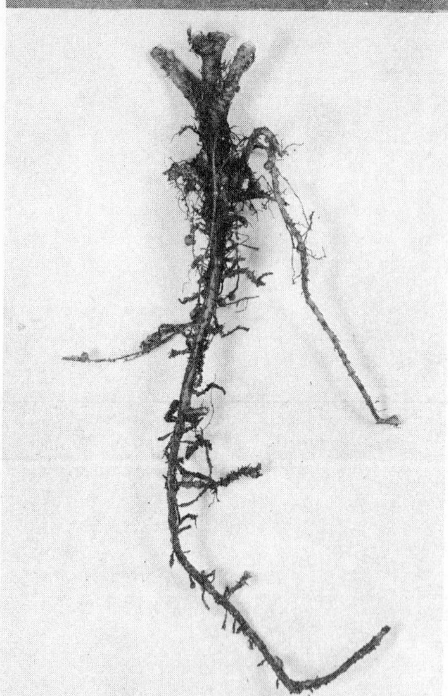

Fig. 4.168B (left). Abnormal roots of peanut plants caused by the (top) peanut root-knot nematode and (bottom) sting nematode. (Courtesy Dr. L. I. Miller)

Fig. 4.169. Ringspot on peperomia leaves. (Illinois Natural History Survey photo)

6. *Bacterial Wilt, "Yellows"*—Uncommon. Southeastern states, mostly on light, sandy soils. Plants first turn yellow, followed by death of branches or entire plants. May lead to Stem or Collar Rot (above). Peanut is quite resistant. *Control:* Rotate. Spanish varieties are more severely damaged than Virginia bunch types.
7. *Verticillium Wilt*—Leaflets turn yellow, wither, and drop early. Internal brown discoloration of stem at or below soil line. *Control:* See (15B) Verticillium Wilt under General Diseases. Bunch types are more resistant than Spanish and Valencia types.
8. *Rust*—Uncommon. Southern states. Small, orange to cinnamon-brown, dusty

pustules on leaves. If severe, lower leaves may wither and drop prematurely. Shriveled kernels may result. *Control:* If serious, same as for Leaf Spots (above).

9. *Mosaic, Mottle, Stunt*—Uncommon. Plants may be stunted. Leaves mottled, light and dark green; may be stunted, crinkled, and curled. *Stunt* causes all or part of the foliage to be severely dwarfed. Leaves are commonly curled upward and malformed. Affected parts are usually light green or yellowish. Fruit production is markedly reduced with peanuts small, malformed, and often with split shells. The Stunt virus is spread mechanically and by aphids; the Mottle virus is seed-transmitted. Some varieties may show no symptoms. See (16) Mosaic under General Diseases.
10. *Chlorosis, Manganese or Iron Deficiency*—In neutral and alkaline soils. Plants stunted. Leaves pale green between veins; may later be bronzed. *Control:* Maintain soil reaction (pH) between 5.7 and 6.2. Apply manganese sulfate to soil with acid-forming fertilizers. Or apply 2 per cent spray of manganese sulfate (or 1 per cent ferrous sulfate). See pages 26–27. On very acid soils (below pH 5.7) apply gypsum. See under Pod and Kernel Decays (above).
11. *Hopperburn*—Tips and margins of leaves are scorched. Leaves yellow. Plants dwarfed. Yield of nuts and foliage is reduced. *Control:* Apply sevin or DDT with Leaf Spot dusts or sprays to control leafhoppers.
12. *Thrips Injury*—Seedling buds and leaves may be blackened and dwarfed. Leaflets puckered and irregularly shaped with round to irregular, white spots on surface. Seedling growth delayed 2 to 3 weeks. Injury most severe in dry weather. *Control:* Apply sevin or DDT when injury is first noticed. Repeat 7 to 10 days later.

For more information, read USDA Farmers' Bulletin 2063, *Growing Peanuts.*

PEAR — See Apple
PEARLEVERLASTING — See Chrysanthemum
PEASHRUB, PEATREE — See Honeylocust
PECAN — See Walnut
PEDIOCACTUS — See Cacti
PELARGONIUM — See Geranium
PELECYPHORA — See Cacti
PELICAN-FLOWER — See Aristolochia
PELLAEA — See Ferns
PENNYROYAL — See Salvia
PENSTEMON — See Snapdragon
PEONY — See Delphinium
PEPEROMIA [EMERALD GREEN, OVAL LEAF, OVAL LEAF VARIEGATED], WATERMELON-BEGONIA *(Peperomia)*

1. *Ringspot*—Zoned, yellowish or brown rings form in leaves which later drop. Rings may fuse to form irregular patterns (Figure 4.169). Small brown spots occur on some leaves. Young leaves may be markedly cupped, curled, or twisted. Plants often stunted. *Control:* Destroy infected plants. Take cuttings only from vigorous, virus-free plants. Root in sterile medium. Control insects with malathion sprays.
2. *Corky Scab*—Indoor problem. Raised, copper-colored to dark, scablike growth on underleaf surface. *Control:* Maintain uniform soil moisture supply during moist, overcast weather. Increase air circulation.
3. *Stem and Root Rots, Cutting Rots*—Plants stunted, gradually wilt, wither, and die from dark rotting of stem at soil line or below. Roots often decay. May be associated with nematodes (e.g., burrowing, lesion, pin, root-knot, spiral). *Control:* Take cuttings from known healthy plants and grow in well-drained soil that is clean

or sterilized (pages 538–548). Avoid overwatering. If disease starts, try applying soil drenches of Dexon or PCNB-Dexon mixture. Follow manufacturer's directions.

4. *Leaf Spots, Anthracnose*—Spots of various sizes, shapes, and colors (Figure 4.170). Leaves may wither and drop early. *Control:* Pick and burn spotted leaves. Avoid syringing plants. Try spraying outdoor plants with zineb, maneb, or captan plus spreader-sticker. Make several applications, 10 days apart.

PEPPER — See Tomato

PEPPERBUSH — See Sweet-pepperbush

PEPPERGRASS — See Cabbage

PEPPERMINT — See Salvia

PEPPER-ROOT — See Cabbage

PEPPERTREE — See Sumac

PEPPERVINE — See Grape

PERENNIAL PEA — See Pea

PERISTERIA — See Orchids

PERIWINKLE — See Vinca

PERSEA — See Avocado

PERUNKILA — See Oleander

PERUVIAN DAFFODIL — See Daffodil

PERSIMMON [AMERICAN or COMMON, JAPANESE, TEXAS or BLACK], DATE PLUM *(Diospyros)*

Fig. 4.170. Leaf spot of peperomia.

1. *Cephalosporium Wilt*—Southern states on American persimmon. Top leaves suddenly curl, yellow, and wilt. Leaves drop later in summer. Trees may be completely defoliated. Brownish-black streaks appear under bark when infected wood is cut. Diseased trees soon die. *Control:* Japanese persimmons appear resistant.
2. *Leaf Spots, Leaf Blotch, Tar Spot, Scab, Anthracnose*—Spots and blotches of various colors, sizes, and shapes. Affected leaves may wither and drop early. Certain organisms also cause spotting of fruit. *Control:* If serious, collect and burn fallen leaves. Spray as for Anthracnose of Maple or use ziram.
3. *Root Rots*—Trees stunted and unthrifty with yellowish leaves. Tops die back. Roots discolored. See under Apple, and (34) Root Rot under General Diseases. May be associated with root-feeding nematodes (e.g., burrowing, citrus, stubby-root,

trophotylenchulus). Persimmon is resistant to Armillaria Root Rot. Where possible, grow trees in well-drained, sandy soils.

4. *Fruit Spots or Rots, Fly Speck*—Rot spots of various sizes and colors. May be covered with black, blue, or gray mold in damp weather. Rots may develop in storage. *Control:* Check with your extension plant pathologist or a local grower.
5. *Wood Rots*—See under Birch, and (23) Wood Rot under General Diseases.
6. *Twig Blights, Branch Canker, Dieback*—See under Apple and Elm.
7. *Crown Gall*—See under Apple, and (30) Crown Gall under General Diseases.
8. *Powdery Mildew*—See under Birch, and (7) Powdery Mildew under General Diseases.
9. *Root-knot*—See under Peach, and (37) Root-knot under General Diseases.
10. *Verticillium Wilt*—See (15B) Verticillium Wilt under General Diseases.
11. *Thread Blight*—Southeastern states. See under Walnut.
12. *Sooty Blotch*—See under Apple, and (12) Sooty Mold under General Diseases.
13. *Mistletoe*—See (39) Mistletoe under General Diseases.

PETROSELINUM — See Celery

PETUNIA — See Tomato

PHACELIA, SCORPIONWEED, CALIFORNIA-BLUEBELL, FIDDLENECK *(Phacelia)*; BABY-BLUE-EYES, FIESTA-FLOWER, FIVE-SPOT *(Nemophila)*; ROMANZOFFIA

1. *Leaf Spots*—Leaves variously spotted following rainy periods. *Control:* Pick and burn spotted leaves. If needed, spray several times, 10 days apart. Start when first spots are evident. Use zineb, captan, or maneb.
2. *Mosaic* (phacelia)—Widespread. Leaves mottled light and dark green. Often malformed and distorted. Broad green bands develop along leaf veins. *Control:* Keep down weeds. Destroy infected plants when first found. Control aphids that transmit the virus. Spray weekly with malathion or lindane.
3. *Rusts* (phacelia, romanzoffia)—Yellow-orange, reddish-brown, or black powdery pustules on leaves. Alternate hosts: wild grasses, or none. *Control:* Same as for Leaf Spots (above). Start spraying about 2 weeks before rust normally appears.
4. *Powdery Mildew* (nemophila, phacelia)—Grayish-white, powdery mold on foliage. *Control:* Apply sulfur or Karathane twice, 10 days apart.
5. *Curly-top* (phacelia)—Western states. See (19) Curly-top under General Diseases.

PHAIUS, PHALAENOPSIS — See Orchids

PHALARIS — See Ribbon Grass

PHASEOLUS — See Bean

PHILADELPHUS — See Hydrangea

PHILIBERTIA — See Butterflyweed

PHILODENDRON — See Calla Lily

PHLOX [AMOENA, ANNUAL, BLUE, CAMLA or TRAILING, CLEFT, CREEPING, DOWNY, DWARF, GARDEN, GROUND- or MOSS-PINK, MEADOW, PRAIRIE, SAND, SMOOTH, SUMMER PERENNIAL, THICK-LEAF, WILD SWEET-WILLIAM] *(Phlox)*; COLLOMIA; BLUE GILIA, SKYROCKET, BIRD'S-EYES, PRICKLY PHLOX, TREE or STANDING CYPRESS *(Gilia)*; JACOBS-LADDER or GREEK VALERIAN *(Polemonium)*; POLYGALA, MILKWORT, SENECA *(Polygala)*

1. *Powdery Mildews*—General, especially where plants are shaded or crowded. Whitish, powdery mold patches on upperside of leaves and stems from midsummer

on. Leaves may shrivel and drop early. See Figure 2.23D under General Diseases. *Control:* Space plants. Cut and burn all tops in fall. Apply sulfur, Karathane, or Acti-dione when first noticed. Repeat at 7- to 10-day intervals. Follow manufacturer's directions. *Phlox* varieties differ in resistance.

2. *Leaf Spots, Anthracnose, Blights*—General. Spots on leaves of various sizes, shapes, and colors. May merge to form irregular blotches. Usually occur first on lower leaves that wither and dry. Plants stunted. May die prematurely. Bloom is reduced. See Figure 2.19D under General Diseases. *Control:* Destroy tops and fallen leaves in autumn. Apply complete fertilizer in spring. Sow disease-free seed treated with thiram, captan, or chloranil. Water during dry periods. Divide old clumps. Keep down weeds. Apply sulfur, zineb, thiram, ferbam, or fixed copper at 7- to 10-day intervals. Start a week before leaf spot is expected. Control insects and mites, using malathion. *Phlox* varieties differ in susceptibility.
3. *Phlox Leaf Drop, Blight*—Older leaves turn brown and die from base of stem upwards. Shoots may die. Most severe on older clumps where soil nutrients or water are exhausted. Varieties differ in susceptibility. *Control:* Same as for Leaf Spots (above).
4. *Stem and Crown Rots, Southern Blight, Stem Blight or Canker, Root Rots*—General. Seedlings or older plants often stunted, wilt, wither, and may collapse. Base of stem and roots are rotted. May be covered with cottony mold. *Control:* Carefully dig up and destroy infected plants. Space plants. Three- or 4-year rotation. Avoid overwatering, overcrowding, and planting in heavy, poorly drained soil. When rot strikes, soak base of plants with 1:1,000 solution of mercuric chloride. Repeat in 2 weeks. Hill up fresh soil around infected plants.
5. *Mosaic*—Widespread on phlox. Leaves mottled light and dark green with some yellowing. *Control:* Destroy infected plants. Keep down weeds. Spray weekly with malathion to control aphids.
6. *Leaf and Stem Nematodes* (collomia, ground- or moss-pink, phlox)—Plants may be stunted. Leaves spindling, thickened, crinkled, or curled. May be split. Stems may be swollen or cracked and bent sideways near tips. Buds swollen. Shoots stunted, distorted, may die before producing flowers. Blotches on leaves are caused by Leaf Nematode. See under Chrysanthemum, and (20) Leaf Nematode under General Diseases. *Control:* Dig and burn infested clumps when *first* found. Rotate 3 to 4 years. Plant healthy seed or plants in soil fumigated with EDB, D-D, chloropicrin, or other fumigant. See "Soil Treatment Methods and Materials" in the Appendix. Varieties differ greatly in susceptibility. Nurserymen disinfest *phlox* stools by soaking 20 to 30 minutes in hot water (110° F.).
7. *Aster Yellows*—Flower petals often show white streaks and vein-banding. Clusters of flowers or individual petals may turn into green, leafy structures. Successive flowers may develop from ovaries. *Control:* Destroy infected plants. Keep down weeds. Spray weekly with mixture of DDT or sevin plus malathion. This controls leafhoppers that transmit virus.
8. *Fusarium and Verticillium Wilts*—Stems on one side of plant show wilting of lower leaves. Wilt later progresses gradually up stem. *Control:* Dig up and destroy infected plants. Rotate. Grow in well-drained soil.
9. *Streak* (phlox)—Streaks develop in leaves and stems. Leaf veins turn yellow. Later, petioles and leaves wither. *Control:* Dig up and burn infected plants.
10. *Downy Mildew*—See (6) Downy Mildew under General Diseases. *Control:* Spray as for Leaf Spots (above).
11. *Rusts*—Yellow, orange, reddish-brown, or black pustules on leaves. *Control:* Pick and burn rusted leaves. Spray as for Leaf Spots (above). Destroy tops in fall.
12. *Fasciation, Leafy Gall* (phlox)—Uncommon. See under Pea, and (20) Leafy Gall under General Diseases.

13. *Root-knot*—See (37) Root-knot under General Diseases.
14. *Gray-mold Blight*—See under Chrysanthemum, and (5) Botrytis Blight under General Diseases.
15. *Crown Gall* (phlox) —Uncommon. See (30) Crown Gall under General Diseases.
16. *Sooty Mold* (phlox) —Black mold growth on leaves following aphids or other insects. *Control:* Apply malathion to control insects.
17. *Smut* (gilia) —See (11) Smut under General Diseases.
18. *Chlorosis*—See under Rose. Resistant *phlox:* Mrs. Jenkins.

PHOENIX — See Palms

PHOENIX-TREE or CHINESE PARASOLTREE *(Firmiana)*, CALIFORNIA FREMONTIA, FLANNEL-BUSH *(Fremontia)*

1. *Twig Canker, Coral Spot*—Twigs and branches die back. Bark surface is cankered. May be covered with tiny, bright, coral-red "cushions." *Control:* Prune dead and blighted twigs. Keep trees vigorous. Fertilize and water during droughts.
2. *Web Blight*—Southeastern states. See under Bean.
3. *Root Rot*—See (34) Root Rot under General Diseases.
4. *Verticillium Wilt* (fremontia) —See (15B) Verticillium Wilt under General Diseases.
5. *Collar Rot, Stem Girdle* (fremontia) —See under Dogwood.
6. *Leaf Spot*—Unimportant. Leaves spotted. *Control:* Not necessary.

PHOTINIA — See Apple

PHYLA — See Lantana

PHYLLOSTACHYS — See Bamboos

PHYSALIS — See Tomato

PHYSOCARPUS — See Ninebark

PHYSOSTEGIA — See Salvia

PICEA — See Pine

PIERIS — See Andromeda

PIGGYBACK PLANT *(Tolmiea)*

1. *Powdery Mildew*—Grayish-white mold patches on foliage. *Control:* Spray twice, 10 days apart. Use sulfur or Karathane.

PILEA, ALUMINUM PLANT, ARTILLERY-PLANT *(Pilea)*

1. *Leaf Spots*—Round to angular, gray or tan spots. Spots may enlarge and merge to form large blotches. Older infected leaves may yellow and drop early. *Control:* Pick and burn infected leaves. Space plants. Indoors, keep water off foliage. Avoid high humidity. If needed, spray several times, 10 days apart. Use fixed copper or maneb.
2. *Powdery Mildew*—Grayish-white mold blotches on leaves. *Control:* If serious, apply sulfur or Karathane several times, 10 days apart.
3. *Root Rot*—See under Geranium, and (34) Root Rot under General Diseases. *Control:* Indoors, grow in well-drained, sterilized soil (pages 538–548). Avoid overwatering and overfertilizing.
4. *Root-knot*—See (37) Root-knot under General Diseases.

PIMPERNEL — See Primrose

PIMPINELLA — See Celery

PINCUSHION FLOWER — See Scabiosa

PINE [ALEPPO, AUSTRIAN, BLACK, BRISTLECONE or HICKORY,

CANARY, CLUSTER, COMPACT MOUNTAIN, COLORADO, COULTER, CUBAN, DIGGER, HIMALAYAN, ITALIAN STONE, JACK or NORWAY, JAPANESE (BLACK, RED, UMBRELLA, WHITE), JEFFREY, KOREAN, LACEBARK, LIMBER, LOBLOLLY, LONGLEAF, MACEDONIAN, MONTEREY, MUGHO, MOUNTAIN, PINYON, PITCH, RED, SCOTS or SCOTCH (PYRAMIDAL SCOTS), SHORE or BEACH, SHORTLEAF, SUGAR, SWISS MOUNTAIN, SWISS STONE, VIRGINIA or SCRUB, WESTERN YELLOW or PONDEROSA, WHITE (DWARF PYRAMIDAL, UMBRELLA)] *(Pinus);* FIR [ALPINE, ARIZONA, BALSAM or BALM OF GILEAD, CALIFORNIA RED, CILICIAN, COLORADO or WHITE, CONICAL, CORKBARK, GRAND, GREEK, LOWLAND WHITE or GIANT, MOMI, NIKKO, NOBLE, NORDMANN, NORMAN or CAUCASIAN, PACIFIC SILVER or CASCADE, ROCKY MOUNTAIN, SHASTA RED, SILVER, SOUTHERN BALSAM or FRASER, SPANISH, SUBALPINE, VEITCH] *(Abies);* CEDAR [ATLAS or ALGERIAN, BLUE ATLAS, DEODAR, CEDAR OF LEBANON, WEEPING BLUE ATLAS, WEEPING LEBANON] *(Cedrus);* SPRUCE [ALCOCK, BLACK, BLACK HILLS, BLUE or COLORADO (BLUE, KOSTER'S WEEPING BLUE), DRAGON, DWARF ALBERTA, DWARF ORIENTAL, DWARF SERBIAN, ENGELMANN, HEDGE-HOG, NORWAY (DWARF YELLOW, WEEPING), ORIENTAL, RED, SERBIAN, SITKA, TIGERTAIL, WESTERN WHITE or ALBERTA, WHITE or CANADIAN, WILSON, YEDDO] *(Picea);* DOUGLAS-FIR [ROCKY MOUNTAIN, WEEPING] *(Pseudotsuga);* UMBRELLA-PINE *(Sciadopitys);* REDWOOD, GIANT SEQUOIA *(Sequoia);* BALDCYPRESS *(Taxodium);* HEMLOCK [CANADA or COMMON (DARK GREEN, WEEPING), CAROLINIA, JAPANESE, MOUNTAIN, SIEBOLD, WESTERN] *(Tsuga)*

1. *Needle Casts, Leaf Blights, Tar Spots, Twig Blights*—General. May be serious, especially on young trees. Irregular, tan, yellow, red-orange, reddish-brown, brown or black specks, spots, or bands on needles (Figure 4.171). Needles later turn olive-green, yellow, red, or brown; often from tip downward. Commonly drop early. Twigs stunted. May die back. Foliage sparse and often "tufted." Growth slows down. Lower branches usually attacked first. Trees may eventually die. Often follows drought, frost, cold winter winds, or fluctuating soil water table. *Control:* Collect and burn fallen needles. Prune and burn dead twigs. Water during droughts. Fertilize trees in fall or early spring to maintain vigor. Where practical, spray new growth when "candles" ¼ emerged. Repeat 2, 4, and 6 weeks later. Use phenyl mercury, fixed copper, bordeaux (4-4-50), maneb, thiram, or folpet plus wetting agent. Some strains of *Douglas-fir* are resistant to Rhabdocline Needle Cast.
2. *Shoot or Tip Blights*—New growth stunted. Wilts. Turns pale green, droops, then yellow, finally brown, and dies back (Figure 4.172). Twigs may die. Resembles spring frost injury. If severe, trees may be browned, stunted, and disfigured. Young trees may die after repeated attacks. *Control:* Same as for Needle Casts (above).

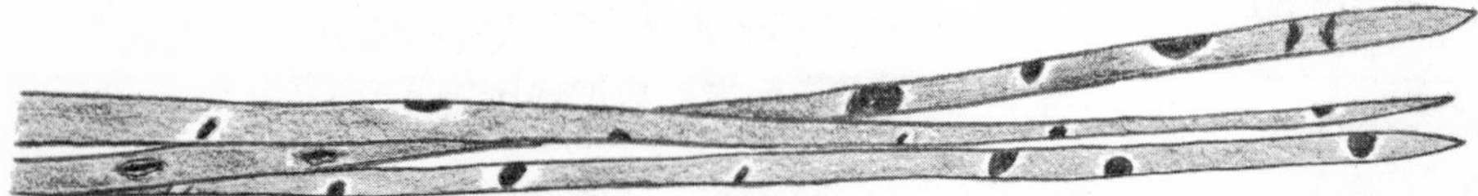

Fig. 4.171. Pine needle blight.

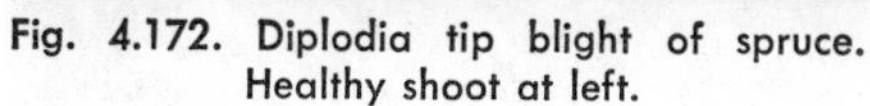

Fig. 4.172. Diplodia tip blight of spruce. Healthy shoot at left.

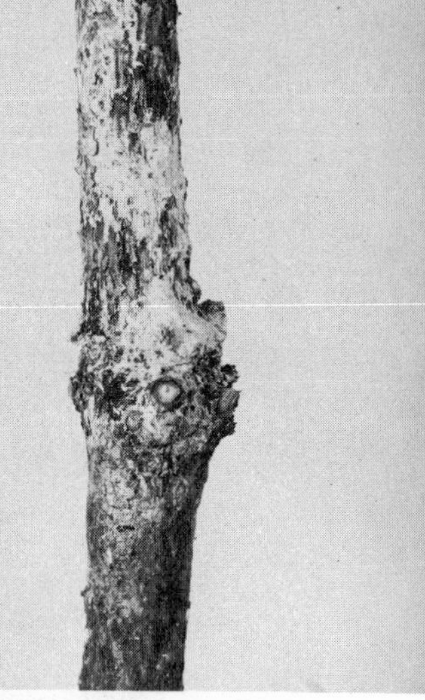

Fig. 4.173. Cytospora canker on spruce. This canker is usually indicated by conspicuous deposits of resin on the bark. The resin frequently forms a white incrustation. (Illinois Natural History Survey photo)

3. *Wood and Heart Rots, Trunk Rot, Butt Rot*—Cosmopolitan. See under Birch, and (23) Wood Rot under General Diseases. "Pecky cypress" is actually wood affected with a wood rot known as Brown Pocket Heart Rot *(Stereum)*. White Pocket or White Speck Heart Rot (caused by the fungus *Fomes pini*) attacks all conifers. "Driftwood" Douglas-fir owes its appeal to infection by *F. pini*. (These are practically the only examples of plant diseases making money for their owners.) Wood-rotting fungi frequently enter through branch stubs, other diseases (e.g., Fusiform Gall Rust), root contacts with diseased trees, or wounds made by insects (bark beetles, borers), sapsuckers, and rodents.
4. *Twig, Branch, and Trunk Cankers*—General. Trees tend to be off-color with dead or dying branches. Discolored, oval to elongate, sunken or swollen, often flattened cankers; with or without definite margins. Often found near branch stubs. Twigs and branches may be girdled and killed. Cankers later become rough and crack open. Resin streaks may flow from wounds. Weakened trees, growing on a poor site, are most susceptible. Lower branches often die first. *Control:* Remove and burn dead or dying branches and severely injured trees. Avoid bark injuries, leaving branch stubs, rodent damage, and growing in heavy, poorly drained soils. Paint wounds promptly with tree wound dressing (page 36). Control insects. Fertilize in fall or early spring and water during droughts. Spray as for Needle Casts (above). Grow adapted varieties and species in protected locations. Avoid overcrowding.
5. *Cytospora Canker, Twig Blight* (primarily spruce, Douglas-fir, and hemlock)—Widespread and serious, especially on Norway and Colorado or blue spruces. Koster's blue spruce and Douglas-fir are more resistant. Lower limbs turn brown and die back to the trunk from inconspicuous, girdling cankers. Disease later progresses upward. Tufts of needles at branch tips turn light gray or brown and die first. Large amounts of bluish-white resin are common on bark of dying branches (Figure 4.173). More prevalent on saplings and trees older than 15 years. Entire tree may die if trunk is girdled. *Control:* None very satisfactory. Promptly cut and burn dying branches and two apparently healthy ones just above, back to the trunk or nearest healthy, lateral branch. Do *not* prune when foliage is wet. Swab tools with 70 per cent alcohol between cuts. Avoid wounds. Otherwise, same as for Twig,

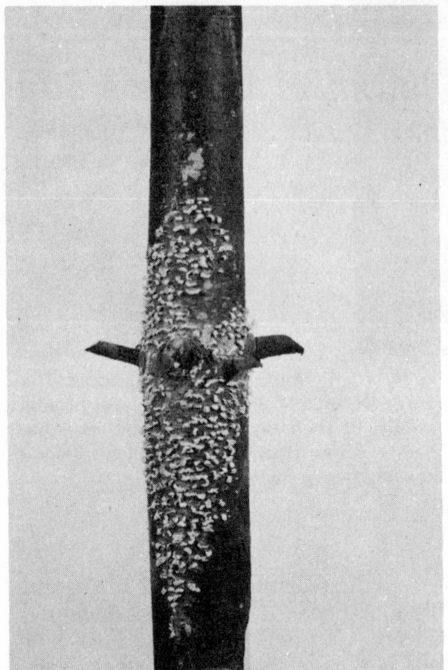

Fig. 4.174 (left). White pine blister rust. (Iowa State University photo)

Fig. 4.175. Needle rust of spruce.

Fig. 4.176. Fusiform gall rust of pine. (Courtesy Dr. V. H. Young)

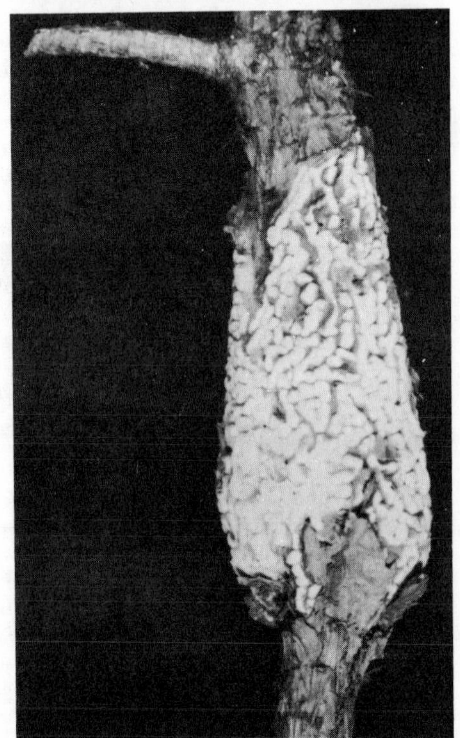

Branch, and Trunk Cankers (above). Control root-feeding insects and nematodes. Check below and with your extension entomologist or county agent.

6. *White Pine Blister Rust* (5-needle pines)—Widespread and serious, especially in forest plantings. Small, dark, oozing blisters on twigs, branches, and trunk. Blister-cankers later form larger, bright orange to pale yellow, dusty pustules over elongated, sunken or swollen cankers. See Figure 4.174. Disease spreads rapidly up and down tree killing branches over a period of several years. Entire tree may die if trunk is attacked and girdled. Alternate hosts: rust-spreading currants and gooseberries (*Ribes* spp.) and *Grossularia*. *Control:* Promptly cut and burn affected parts. Make cuts at least 4 inches below discolored bark at edges of canker. Increase to 6 inches in spring and early summer. Grow only currants and gooseberries from a reputable nursery. These will *not* become infected by the blister rust fungus. Destroy currants or gooseberries that are rust-infected.
7. *Needle and Cone Rusts*—Widespread but usually not serious. Whitish, yellow, orange-yellow, reddish or bright orange to brown, dusty pustules on needles (Figure 4.175) and cones. Needles often turn yellow; may drop early. Most serious on nursery seedlings and young trees. May cause some distortion, dwarfing; even death. Alternate hosts: amsonia, apple, arborvitae, aspens, asters, azalea, blueberry, Chilean tarweed, campanula, cottonwood, creeping snowberry, crowberry, erigeron, ferns, fireweed, flowering currant, gayfeather, goldenrod, gooseberry, groundsel, gumweed, heliopsis, huckleberry, hydrangea, iron-weed, Jerusalem-artichoke, Labrador-tea, leatherleaf or cassandra, loosestrife, *Madia* spp., marigold, Michaelmas daisy, morning-glory, mountain-ash, oaks, poplars, pyrola, ragwort, raspberry, rhododendron, silphium, sowthistle, sunflower, sweetpotato, tickseed, venus-lookingglass, willows, and wood-nymph. *Control:* If serious, spray as for Needle Casts (above). Use ferbam, zineb, maneb, or sulfur plus spreader-sticker. If practical, destroy worthless, alternate hosts (e.g., goldenrod, sowthistle, fireweed, ragwort, etc.) within 900 feet.
8. *Gall Rusts*—Perennial, more or less spherical, pear-shaped, or spindle-shaped galls (or large burls) on branches or trunk up to a foot or more in diameter. Witches'-brooms or deep cankers may form instead. Rate of growth is often slowed. Foliage beyond canker or galls is discolored. Later wilts and dies. Seedlings and saplings are frequently killed. Wood-rotting fungi may enter through rust canker wounds. Surface of galls appear whitish, yellow-orange or bright orange and blister-like in spring. See Figure 4.176. Oaks, chestnut, sweetfern, sweetgale, Indian paintbrush, peregrina, lousewort, bastard toadflax, buckleya, birds-beak, owlclover, Pacific waxmyrtle, cow-wheat, and myriagale are alternate hosts. *Control:* Where practical, remove galls by annual pruning. Destroy nearby worthless oaks and other alternate hosts. Grow rust-free nursery stock. Spray as for Needle Casts (above), but use ferbam, zineb, maneb, ziram, fixed copper, or bordeaux mixture (4-4-50). *Pines* differ in resistance to Fusiform Gall Rust and to Eastern Gall Rust.
9. *Rust Witches'-brooms* (fir, spruce)—Widespread. Young twigs stunted from clusters of compact, broomlike growths with close clusters of dwarfed leaves. Needles may yellow and drop early leaving bare witches'-brooms. Bark may show gall-like swellings. Orange blisters appear on yellowish leaves later in summer. *Control:* Where feasible, prune and burn the perennial witches'-brooms. Destroy alternate hosts of fir rust: chickweed and stickwort (*Stellaria* and *Cerastium* spp.).
10. *Seedling Blights, Damping-off, Root Rots*—Cosmopolitan. Seedlings wilt, wither, turn brown, yellow, or brick-red, and die. Roots decay. May be covered with mold. Plants easily pulled up. Often associated with damp, cold weather and root-feeding nematodes (see below). *Control:* Treat seed with thiram, dichlone, Semesan, or captan. Sow in light, well-drained soil treated with heat or chemicals (pages 538–548). Use treatment (e.g., DMTT (Mylone), SMDC (Vapam or VPM Soil

Fumigant), methyl bromide, chloropicrin, Trizone, or Vorlex) that kills weed seeds, insects, plus fungi and nematodes. Apply soil drenches of thiram, Dexon, zineb, captan, folpet, maneb, oxyquinoline sulfate (1:4,000), or ferbam plus Dyrene Monthly from May to September. Follow manufacturer's directions. Keep down weeds. Avoid overwatering and overcrowding.

11. *Root Rots*—Widespread. Trees decline in vigor. Foliage thin and weak. Needles turn yellow, wither, and drop early. Twigs and later branches die back. Probably associated with wounding by root-feeding nematodes and insects. See under Apple, and (34) Root Rot under General Diseases. Paint freshly cut stumps with tree wound dressing (page 36), creosote, borax, Borateem, or Diquat. Control root-feeding nematodes and insects. Check with your extension entomologist. *Douglas-fir* is usually resistant to Armillaria Root Rot.
12. *Root-feeding Nematodes* (awl, bud and leaf, cyst, cystoid, dagger, lance, lesion, meloidodera, needle, pin, ring, root-knot, sheath, spear, spiral, sting, stubby-root, stunt)—Plants grow more slowly. Tend to have poor color. Twigs and later branches die back. Roots stunted; may die back. See under Root Rots (above). Affected seedlings stunted with smaller needles and poor color. Many seedlings may be killed. Severely affected trees tend to winter-kill. *Control:* See under Peach, and (37) Root-knot under General Diseases. Inject DBCP into soil around nursery-size seedlings and saplings.
13. *Sooty Molds, Black Mildew*—Widespread. Heavy black mold or crust on needles and twigs (Figure 4.177). Often follows insect attacks. *Control:* Keep aphids, scales, and other insects in check by *dormant* oil or lime-sulfur spray. Or apply malathion or Dibrom (naled) when insects in "crawler" stage.

Fig. 4.177. Sooty mold, which occurs on many kinds of plants, is conspicuous because of the black, threadlike growth or crust produced on the surface of leaves and twigs, as shown (left) on pine. This disease seldom causes serious damage. (Illinois Natural History Survey photo)

14. *Littleaf Disease* (pine)—Common in southeastern states; primarily on shortleaf and loblolly pines growing in poorly drained soil. Fortunately, disease is rare in ornamental pines on home plantings. Needles, twigs, and cones are stunted. Foliage becomes thin and yellow. Feeding roots decay. Commonly associated with root-feeding nematodes (see above). Trees under 20 years of age are seldom attacked. Older trees become infected and usually die within 2 to 7 years. *Control:* Grow in fertile, light, well-drained soil. Fertilize to keep trees growing vigorously. Water during droughts. Resistant strains of *shortleaf pine* are available.
15. *Witches'-broom, Dwarf and American Mistletoes* (Douglas-fir, fir, hemlock, pine, spruce)—Primarily in western states. See Figure 2.54B, and (39) Mistletoe under General Diseases.
16. *Crown Dieback*—Tops of large trees yellow and die back, presumably because of

poor growing conditions. Foliage thins. Leader growth and needles are stunted. *Control:* Water and fertilize to maintain good vigor. Mulch and water in late fall. Grow in well-drained, fertile soil. Plant adapted species and varieties. Control insects with malathion sprays.

17. *Winter Injury, Needle Scorch*—Last year's needles turn yellow to reddish-brown and die from tip down. Usually more on exposed branches. Needles may fall during late spring or summer. Do *not* confuse with normal fall browning and dropping of oldest or innermost needles (below). *Control:* Water trees late in fall and during droughts. Mulch to prevent deep freezing and thawing (page 24). Grow adapted and recommended species in location protected from warm, dry, winter winds.
18. *Chlorotic Dwarf* (white pine) —Eastern and midwestern states. Common in young plantations. Growth tufted and dwarfed. All but current needles fall early. Trees sparsely foliated. Needles mottled and yellowish. May become tip-burned in final stages. Roots usually small and spindly. Affected trees usually die by time plantation is 15 years old. Caused by an air pollutant, possibly ozone. *Control:* Destroy affected trees. Grow resistant "varieties" and species.
19. *Sunscorch and Wind Damage* (primarily fir, hemlock, pine, spruce) —Foliage appears scorched. Needles dry and turn brown from tip down. Tips of *hemlock* branches die back. Injury follows dry, windy weather with temperatures of 95° F. or higher. Severe winter weather, mite and nematode injury may cause similar symptoms. *Control:* Keep trees well watered during hot, dry periods. Grow in protected location in rich soil at the proper time (page 19). Control mites with malathion.
20. *Douglas-fir Bacterial Gall*—California to Washington State. Round to irregular, rough galls, up to several inches in diameter on upper branches, twigs, or leader. Top may die ("spike top"). *Control:* Prune out galls. Sterilize tools between cuts and surface sterilize pruning wounds (page 35). Control aphids, which spread the causal bacteria, with malathion.
21. *Brown Felt Blights, Snow Blights*—Foliage killed under deep snow, especially at high elevations. Needles brown, matted together, and covered with white mold as snow melts (Snow Blight) or dense, brown to almost black, feltlike growth (Brown Felt Blight). See Figure 4.178. Black specks later appear on needles. Seedlings and young trees may die. *Control:* Spray nursery beds and lowest branches on other trees with lime-sulfur (1:8), PCNB, or fixed copper in late fall. Remove excess snow accumulation. Plant in fertile soil.

Fig. 4.178. Snow blight injury to spruce.

22. *Natural Leaf-browning and Shedding*—Older needles turn yellow and die on *inside* of trees and drop to the ground, usually in late fall. This is a normal shedding of leaves.
23. *Chlorosis*—Deficiency of iron, manganese, magnesium, potassium, or calcium. Usually develops as an overall yellowing or scorching of needles. If severe, needles may turn brown and die. Shoots die back. *Control:* Have soil tested and follow recommendations of report. Soil should be slightly acid.
24. *Boron Deficiency*—Where chronic, growth is stunted and off-color. Growing tips of needles, shoots, and roots may die back. *Control:* Apply borax to soil. See page 27. Follow local recommendations.
25. *Wetwood, Slime Flux* (fir, hemlock, pine) —See under Elm.
26. *Gray-mold Tip Blight* (Douglas-fir, fir, hemlock, pine, sequoia, spruce) —Common in cool, wet weather. New growth withers, curls, and dies. Coarse gray mold forms on affected parts in moist weather. *Control:* Space plants. Prune to improve ventilation. Spraying as for Needle Casts (above) should be beneficial.
27. *Crown Gall* (redwood) —See (30) Crown Gall under General Diseases.

PINEAPPLE GUAVA — See Myrtle
PINK ALMOND — See Peach
PINKS — See Carnation
PINXTERBLOOM — See Rhododendron
PIQUERIA — See Chrysanthemum
PISTACHE, PISTACHIO *(Pistacia)* — See Sumac
PISUM — See Pea
PITANGA — See Eucalyptus
PITCHERPLANT, HUNTSMANCUP, TRUMPETS *(Sarracenia)*; CALIFORNIA-PITCHERPLANT *(Darlingtonia)*

1. *Leaf Spots*—Leaves variously spotted. See (1) Fungus Leaf Spot under General Diseases. Indoors, keep water off foliage.
2. *Root Rots*—See (34) Root Rot under General Diseases. *Control:* Discard diseased plants. Grow in sterilized sandy peat or sphagnum moss that is acid (pH 4 to 5). Keep moist.
3. *Southern Blight*—See (21) Crown Rot under General Diseases. *Control:* Same as for Root Rots.

PITTOSPORUM [CAPE, JAPANESE or TOBIRA], KARO *(Pittosporum)*

1. *Leaf Spots*—Small, angular, yellow to dark brown spots. Leaves may yellow and drop early. *Control:* If practical, pick and burn spotted leaves. Apply zineb, maneb, or fixed copper, about 2 weeks apart, during rainy periods.
2. *Stem Rot, Southern Blight, Foot Rot*—Plants wilt, wither, and die from rot of stems near soil line. Affected areas may be covered with white mold in damp weather. Roots may decay. *Control:* Grow in well-drained soil. Avoid wounding bark and overwatering. Keep trees growing vigorously. Commercial nurserymen treat planting beds with steam or chemicals (pages 538–548). Post-planting treatments of PCNB-Dexon mixture may be beneficial. Follow manufacturer's directions.
3. *Chlorosis*—Most common in alkaline soils. Leaves turn yellow except for veins. Leaves may be dwarfed. *Control:* Acidify soil (page 24). Spray plants with iron sulfate or chelate (pages 26–27). Follow manufacturer's directions.
4. *Mosaic*—See (16) Mosaic under General Diseases.
5. *Verticillium Wilt*—See under Maple, and (15B) Verticillium Wilt under General Diseases.
6. *Root-knot*—See under Peach, and (37) Root-knot under General Diseases. May use

DBCP or VC-13 as a post-planting treatment. Follow manufacturer's directions.

7. *Thread Blight*—Southeastern states. Plants may be defoliated early. See under Walnut. *Control:* Spray as for Leaf Spots (above).

PITYROGRAMMA — See Ferns
PLANETREE *(Platanus)* — See Sycamore
PLANTAINLILY — See Hosta
PLATYCERIUM — See Ferns
PLATYCODON — See Bellflower
PLUM — See Peach
PLUMED THISTLE — See Chrysanthemum
PLUME HYACINTH — See Tulip
PLUMERIA — See Oleander
PLUM-YEW — See Japanese Plum-yew
POCKETBOOK PLANT — See Snapdragon
PODOCARPUS — See Yew
PODOPHYLLUM — See Mayapple
PODRANEA — See Catalpa
POINCIANA — See Honeylocust

POINSETTIA, SPURGE [CUSHION, CYPRESS, FLOWERING, PAINTED or MEXICAN FIRE-PLANT], SNOW-ON-THE-MOUNTAIN, CROWN-OF-THORNS, MILK-BUSH, SCARLET PLUME *(Euphorbia)*

1. *Stem and Cutting Rots, Foot Rots, Root Rots, Wilt*—Serious indoor problem. Plants stunted, frequently decline and die. Cuttings (or stems) rotted near soil line. Rot may be rapid, soft, and watery (Bacterial Soft Rot). Cuttings may fail to root. Stem, branches, and roots darken, eventually die. The lower leaves may gradually or suddenly wilt, wither, yellow, and drop. First noticeable at stem base, then progresses upward. Flowers stunted, may be one-sided. Longitudinal black cracks may develop in lower stem. May be associated with root-feeding nematodes (e.g., stunt). See Figure 4.179. *Control:* Avoid overcrowding, and splashing water on foliage when sprinkling. Take tip cuttings from vigorous, disease-free plants using knife dipped in 70 per cent alcohol, flamed, and cooled. Change knives between stock plants. Dip or dust in ferbam and root in light, well-drained soil that is sterilized (pages 538–548) and acid (pH 4.5). Thoroughly mix PCNB-Dexon into potting soil following manufacturer's directions. Practice strict sanitation (e.g.,

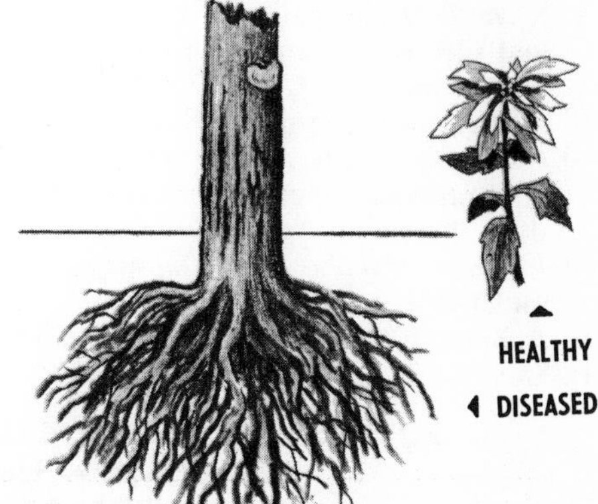

Fig. 4.179. Poinsettia root rot (*Thielaviopsis*).

use new pots or disinfect old pots, scrub hands thoroughly before taking cuttings). Indoors, maintain recommended temperature at night and regulation of day length. Keep soil on dry side and at temperature of 65° F. If misting, reduce mist period to 30 seconds and increase no-mist period to 1½ to 2 minutes. Check with your local florist or extension horticulturist. Florists drench cutting bed after cuttings are "stuck," using thiram, Semesan, or Morton Soil Drench. Follow manufacturer's directions. Drench soil monthly (1 quart per square foot) for 3 applications. Use mixture of PCNB plus Dexon, ferbam, captan, thiram, or folpet. Destroy severely infected plants or plant parts.

2. *Gray-mold Blight, Botrytis Tip Blight, Stem Canker*—Flower clusters, especially on double varieties, and colored bracts are blasted and turn brown. Brown cankers form on stem. Coarse, grayish mold grows on infected areas in cool, damp weather. *Control:* Indoors, space plants, increase air circulation, and raise temperature. Otherwise the same as for Stem and Cutting Rots (above). Maneb, zineb, Daconil 2787, or fixed copper sprays often beneficial. Carefully pick and burn affected parts.
3. *Bacterial Canker and Leaf Spot, Bacterial Blight* (poinsettia)—Long, narrow, dark, water-soaked to brownish streaks on green stems and petioles. Stems may later crack open and ooze. Growing point curves down and in. Light spots or yellowish blotches occur on leaves that later turn brown. Or spots may first be water-soaked; later chocolate-brown to rust-colored on underleaf surface. Infected leaves may drop early. Several branches may be completely blighted or entire plant may die. Plants may show no symptoms for several months after becoming infected. *Control:* Same as for Stem and Cutting Rots (above). Avoid overfertilizing plants. Fixed copper and streptomycin (Agrimycin 500) sprays, applied weekly, are beneficial. *Poinsettia* varieties differ in resistance to Bacterial Leaf Spot.
4. *Leaf and Stem Spots, Scab or Spot Anthracnose*—Small, round to elongate or irregular spots on leaves; sometimes on stems. Young stems or lower leaves may die back. Leaves may drop prematurely. *Control:* Same as for Gray-mold Blight (above).
5. *Rusts*—Powdery, yellow to cinnamon-brown pustules on both leaf surfaces. Alternate host: switchgrass *(Panicum)* or none. *Contol:* Pick and burn infected leaves. Zineb or maneb sprays should be beneficial.
6. *Chlorosis*—Plants light green to yellow. Lack vigor. Leaves may yellow and drop early. *Control:* Grow in acid soil. Apply a complete fertilizer based on a soil test. Avoid chilling and overwatering. Provide adequate lighting.
7. *Crown Gall*—Irregular galls at base of plants. Control: Destroy infected plants. Root cuttings in sterilized soil (pages 538–548). Avoid wounding stems.
8. *Stunt* (poinsettia)—Indoor problem on forced, container-grown plants. New leaves dwarfed, pale green; borne on slender stems. Leaves gradually turn yellow and wilt. Plants stunted. Roots soft, watery, and colorless. Entire plant eventually withers and dies. *Control:* Avoid overwatering; heavy, poorly drained soils; lack of soil aeration; and pots without drainage holes.
9. *Curly Leaf* (poinsettia)—Leaves and flower bracts distorted and curled. Found in landscape plantings only. *Control:* Do not propagate from affected plants. Destroy those showing symptoms.
10. *Powdery Mildew* (snow-on-the-mountain, spurges)—See (7) Powdery Mildew under General Diseases.
11. *Root-knot*—See (37) Root-knot under General Diseases. Plant growth is reduced.
12. *Stem Smut* (painted spurge)—Louisiana. See (11) Smut under General Diseases.

POINT REYES LILAC or CREEPER — See New Jersey-tea

POKERPLANT — See Redhot-pokerplant

POLEMONIUM, POLYGALA — See Phlox

POLIANTHES — See Daffodil
POLYANTHUS — See Primrose
POLYGONUM — See Silver-lacevine
POLYPODIUM, POLYSTICHUM — See Ferns
POMEGRANATE *(Punica)*

1. *Fruit Spots and Rots*—Cosmopolitan. Rot spots develop in fruit. Affected areas may be covered with gray, blue, or black mold. See under Apple.
2. *Anthacnose, Spot Anthracnose, Leaf Blotch*—See under Maple.
3. *Root-knot*—See (37) Root-knot under General Diseases.
4. *Root Rot*—See under Apple, and (34) Root Rot under General Diseases.
5. *Thread Blight*—Southeastern states. See under Walnut.

PONCIRUS — See Citrus
PONDLILY — See Waterlily
POND-SPICE — See Avocado
POPCORN — See Corn
POORMANS-ORCHID — See Tomato
POPLAR [BALM-OF-GILEAD, BALSAM, BERLIN, BLACK, BOLLEANA or TURKESTAN, CAROLINA, CHINESE, COARSE-TOOTHED, GRAY, HYBRID, JAPANESE, KOREAN, LOMBARDY, MONGOLIAN, NORWAY, RIO GRANDE, SIMON, SILVER, WHITE], ASPEN [CHINESE, EUROPEAN, GOLDEN, JAPANESE, LARGE-TOOTHED, QUAKING or QUIVERLEAF], COTTONWOOD [BLACK or WESTERN BALSAM POPLAR, COMMON or NORTHERN, COTTONLESS, EASTERN, GREAT PLAINS, SOUTHERN, SWAMP] *(Populus)*

1. *Twig, Branch, and Trunk Cankers, Dieback*—General and serious. Enlarging, round to irregular, rough, discolored, often sunken and cracked cankers on twigs, branches, and trunk (Figure 4.180). Girdled parts cause foliage beyond to turn yellow, bronze, and die. Cankers may cause swellings on trunk and forming of suckers that become infected and die. Sapwood under cankers is discolored. Twigs and small branches die back first. Trees often die from top down. *Control:* Grow poplars, cottonwoods, and aspens best adapted to your area. Check with your extension horticulturist or nurseryman. Avoid wounds (e.g., frost cracks, sunscald,

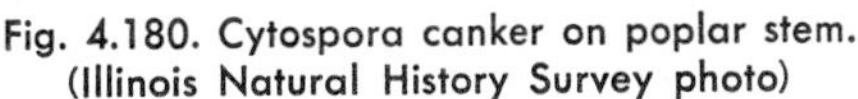
Fig. 4.180. Cytospora canker on poplar stem. (Illinois Natural History Survey photo)

fire, rodents, insects, cultivator, when planting). Paint wounds promptly with tree wound dressing. If practical, prune out all cankers on small branches. Disinfect cut surfaces with 70 per cent alcohol or a 1:1,000 solution of mercury bichloride followed by tree paint. Water during droughts. Fertilize to maintain vigorous growth. Space trees in nursery. Apply sevin plus malathion to control insects. Poplars and cottonwoods vary greatly in resistance to the several cankers. Lombardy, Norway, black, Bolleana, and Simon poplars, and eastern and western cottonwoods are very susceptible to one or more cankers. The Japanese *poplar* shows resistance to Dothichiza *(Cryptodiaporthe)* Canker; Rio Grande is resistant to Cytospora Canker. Certain cankers are so contagious that infected trees should be removed and burned. Sprays of fixed copper, zineb, or captan, when applied at 2-week intervals to nursery-size trees, has proved beneficial. For additional information check with your extension horticulturist or plant pathologist.

2. *Leaf Spots, Leaf Blotch or Blight, Ink Spot*—Common in wet seasons, but not serious. Small to large, round to irregular spots or blotches of various colors, often with dark margins. Spots may later enlarge. Certain spots may drop out leaving ragged holes. Leaves wither and fall early. Twigs may die back. *Control:* Collect and burn fallen leaves. Avoid crowding. If serious, apply two or three sprays, 10 days apart. Start when buds break open. Use phenyl mercury, zineb, fixed copper, or bordeaux (4-4-50). *Poplars* differ in resistance to Marssonina Leaf Spot (Blight).
3. *Powdery Mildews*—Widespread. White, powdery mold patches on leaves. *Control:* If serious, spray several times, 10 days apart. Use sulfur, Acti-dione, or Karathane.
4. *Leaf Rusts*—General. Bright, yellow or orange, dusty pustules, mainly on underleaf surface. Pustules may turn reddish-brown to black. Leaves may wither and fall early. Seldom causes serious damage. *Control:* Where practical, avoid planting near alternate hosts: garlic, fir, pine, Douglas-fir, hemlock, bigcone spruce, and larch. If practical, apply zineb, maneb, dichlone, ferbam, Acti-dione, bordeaux (1-1-50), fixed copper, or sulfur two or three times, 10 days apart. Start a week before rust normally appears. *Poplar* and *cottonwood* varieties and species differ in resistance to the several rusts. Siouxland *cottonwood* is resistant to common rust. Collect and burn fallen leaves.
5. *Spring Leaf Fall, Scab, Shoot Blight*—In late spring tips of young leaves turn black. Affected areas enlarge; leaves blacken, wrinkle, wilt, wither, and drop by late spring. Shoots may blight and die back causing loss of tree vigor. Olive-green mold may form on dead plant parts. *Control:* Spray as for Leaf Spots (above). Prune and burn blighted shoots. Rake and burn fallen leaves. Fertilize and water to maintain vigor.
6. *Wood Rots, Butt Rot*—Cosmopolitan. See under Birch, and (23) Wood Rot under General Diseases. Control borers by removing infested branches (dead or dying), plus monthly sprays, May to September. Use DDT, dieldrin, or Thiodan. Keep trees growing vigorously. Fertilize, and water during droughts.
7. *Yellow Leaf Blisters, Catkin Deformity*—Widespread following cold, wet, spring weather. Irregular, small to large, bright yellow, or orange-yellow to brown blisters on leaves and catkins. Underleaf surface may show whitish to bright golden "bloom." Catkins deformed. *Control:* If practical, apply fixed copper, dichlone, zineb, or ferbam *before* buds swell in early spring. Repeat as buds break open and at 10-day intervals during rainy weather. Some *poplar* species and varieties are apparently resistant.
8. *Branch Gall*—Small, rounded galls up to 1½ inches in diameter, develop at base of twigs. Some twigs and small branches may die back. *Control:* Prune infected branches and dead twigs back several inches into healthy wood. Burn prunings.
9. *Crown Gall*—Also called Bacterial Limb Gall. Rough, irregular, swollen galls

usually found at base of trunk or on roots. Trees lack vigor. Make poor growth. *Control:* See under Apple, and (30) Crown Gall under General Diseases.

10. *Root Rots*—Trees gradually decline in vigor. Foliage becomes thin and weak. Leaves may yellow and drop early. May be associated with nematodes (e.g., ring, sheath). See under Apple. *Poplars* have good resistance to Armillaria Root Rot.
11. *Verticillium Wilt*—See under Maple, and (15B) Verticillium Wilt under General Diseases.
12. *Mistletoe* (cottonwood, poplar)—See (39) Mistletoe under General Diseases.
13. *Chlorosis, Iron or Magnesium Deficiency*—Yellowing between leaf veins followed by scorching of tissue. See under Maple.
14. *Leaf Scorch*—See under Maple.
15. *Wetwood, Slime Flux*—See under Elm.
16. *Sooty Mold*—See (12) Sooty Mold under General Diseases.
17. *Seed Rot, Damping-off*—See under Pine.

POPPY [ALPINE, CORN, ICELAND, ORIENTAL, SHIRLEY, TULIP] *(Papaver)*; PRICKLY-POPPY [CRESTED, MEXICAN] *(Argemone)*; CELANDINE *(Chelidonium)*; CALIFORNIA-POPPY *(Eschscholtzia)*; TREEPOPPY or BUSHPOPPY *(Dendromecon)*; MECONOPSIS, WELSH POPPY, WIND POPPY, SATIN POPPY *(Meconopsis)*; BLOODROOT *(Sanguinaria)*

1. *Bacterial Blight, Leaf Spot* (California-poppy, poppy)—Small to large, water-soaked spots or blotches on buds, leaves, stems, flowers, seedpods, and roots (Figure 4.181). Spots soon blacken, often zoned, with slimy to gummy exudate. Leaves may wither and drop early. Stems may be girdled, causing plants to die. *Control:* Remove and burn severely infected parts. Pick and burn spotted leaves on mildly infected plants. Sow seed from disease-free plants in clean or sterilized soil (pages 538–548). Spray with copper fungicide or Agrimycin (streptomycin) during wet weather. Rotate.

Fig. 4.181. Bacterial blight on corn or shirley poppy, showing infections on a leaf, pod, and flower stalk.

2. *Stem (Foot) and Root Rots, Damping-off*—Seedlings weak, wilt, and collapse. Older plants turn pale, wilt, and wither from rot or canker of lower stem and/or roots. *Control:* Grow disease-free plants or seed in sunny spot in clean or sterilized soil that is light and well-drained. Avoid overwatering, excessive use of fertilizers high in nitrogen, wounding plants, and overcrowding. If suspicious, drench crown with mixture of PCNB plus captan, ferbam, or Dexon (page 103). Rotate. Spray as for Downy Mildew (below). Soak seed of *California-poppy* in hot water (125° F.) for 30 minutes. Cool, dry, and plant.
3. *Leaf, Stem, and Seedpod Spots*—Spots of various sizes, shapes, and colors on leaves, petioles, stems, and seedpods (capsules). Leaves may shrivel and die around blossom time. Seedlings may wilt and collapse. *Control:* Sow seed from disease-free plants. Avoid overcrowding. Maintain balanced fertility based on a soil test. Treat seed of *California-poppy* as for Stem (Foot) and Root Rots (above). Spray as for Downy Mildew (below).
4. *Downy Mildew* (meconopsis, poppy, prickly-poppy)—Seedlings blighted. May collapse and die. Yellowish to light brown spots and blotches on upper leaf surface of older plants. Blotches enlarge and become very dark. In damp weather, spots on underleaf surface are covered with white, grayish, or pale purplish mold. Leaves may yellow, wither, and be deformed. Stems distorted. Buds, flower parts, seeds, and capsules may also be infected. Plants often fail to bloom. *Control:* Destroy first infected plants. Pick and burn infected plant parts when first seen. Burn tops in fall. Sow seed from disease-free plants in well-drained soil. In cool, rainy weather apply zineb, maneb, or copper several times, 7 to 10 days apart.
5. *Verticillium Wilt*—Plants yellow, wilt, and die. Often fail to bloom. Stems may turn brown. *Control:* See (15B) Verticillium Wilt under General Diseases.
6. *Curly-top, Aster Yellows*—Plants severely stunted, bunchy, and yellowed. *Control:* Destroy infected plants when first found. Keep down weeds. Spray at least weekly with sevin or DDT plus malathion to control leafhoppers that transmit the viruses.
7. *Spotted Wilt*—Leaves stunted, twisted, and gradually turn yellow. Plants stunted, pale, and bunchy. May die suddenly. Flower stems are stunted. Often twisted and bent over. *Control:* Same as for Curly-top and Aster Yellows (above). Virus is transmitted by thrips.
8. *Leaf Smut*—Small, round, pale leaf spots that turn dark brown to black, with a reddish border. *Control:* Same as for Bacterial Blight (above). Spray or dust as for Downy Mildew (above).
9. *Powdery Mildew*—Pacific Coast. Grayish-white, powdery mold patches on leaves. Spots underneath may be purple. *Control:* If serious, apply Karathane or sulfur several times, at 10-day intervals.
10. *Black Mold*—Sootlike mold patches on foliage. Follows insect attacks. *Control:* Apply malathion as needed to keep insects in check.
11. *Rust* (pricky-poppy)—See (8) Rust under General Diseases. Spray as for Downy Mildew (above).
12. *Root-knot and Other Nematodes* (e.g., lesion)—See under Peach, and (37) Root-knot under General Diseases.
13. *Leaf Nematode* (poppy)—See (20) Leaf Nematode under General Diseases.
14. *Gray-mold Blight*—See (5) Botrytis Blight under General Diseases.
15. *Black Ringspot*—See under Cabbage.

POPPY-MALLOW — See Hollyhock
POPULUS — See Poplar
PORCELAIN BERRY — See Grape
PORT ORFORD CEDAR —See Juniper

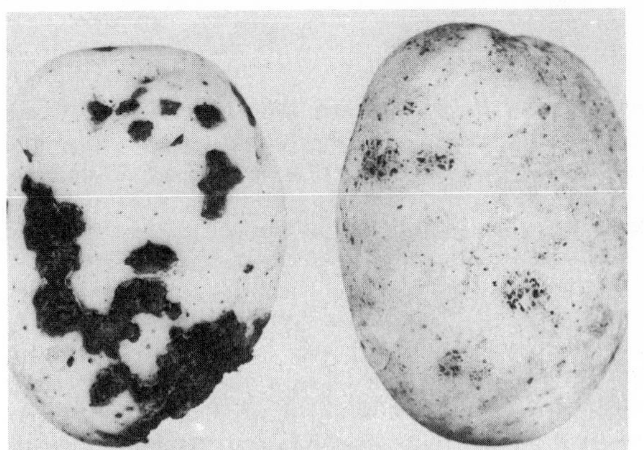

Fig. 4.182. Common scab. Left, deep pitted scab on the susceptible Chippewa variety; right, the shallow russet-type scab on a typically scab-resistant variety. (Department of Botany & Plant Pathology and Cooperative Extension Service, Michigan State University photo)

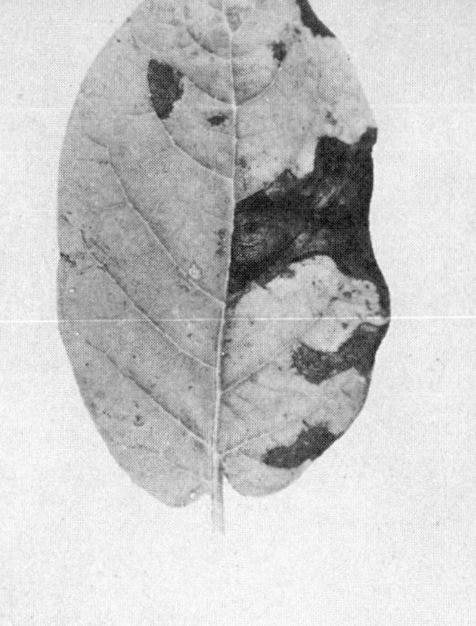

Fig. 4.183. Early blight of potato. Note concentric lines in the larger lesion. (Illinois Agricultural Experiment Station photo)

PORTULACA — See Rose-moss
POSSUMHAW — See Holly and Viburnum
POTATO *(Solanum)*

1. *Common Scab, Russet Scab*—Most common on light, sandy soils low in organic matter and soil reaction (pH) of 5.5 or above. General. Shallow, rough, russeted, raised or pitted, corky areas on tuber surface. Tubers often do not keep well in storage (Figures 2.21A and 4.182). *Control:* Plant scab-free tubers in scab-free soil. Treat seed-pieces as for Seed-piece Decays (below) or use Polyram. Four-year rotation. Include green manure crops (page 23). Somewhat resistant varieties: Anoka, Antigo, Arenac, Avon, Blanca, Catoosa, Cayuga, Cherokee, Early Gem, Emmet, Haig, Huron, Knik, LaRouge, Menominee, Navajo, Netted Gem, Norgold Russet, Norland, Ona, Onaway, Ontario, Osage, Pennchip, Plymouth, Progress, Pungo, Redkote, Redskin, Reliance, Russet Burbank, Russet Rural, Seneca, Shoshoni, Superior, Tawa, Viking, and Yampa. More highly resistant varieties should be available soon. Contact a local grower, your extension horticulturist, or potato specialist. Mixing sulfur or urea-formaldehyde (UFC-85, Uracide, N-Dure) into top 6 inches of mineral soil in row may help. Check with authorities mentioned above or your extension plant pathologist. Avoid application of alkaline materials, e.g., lime, wood ashes, and fresh manure to potato soil. Use acid-forming fertilizer (ammonium sulfate or ammonium nitrate) high in nitrogen. If practical, acidify mineral soil to pH of 4.8 to 5.2 and irrigate liberally in growing season.
2. *Early Blight, Target Spot*—General in warm, humid weather. Small, dark brown, often zoned spots on leaves. Spots may merge to kill portion of leaf (Figure 4.183). If severe, plants may be defoliated. Dark, roundish, slightly sunken spots develop on stems and tubers in storage. *Control:* Apply maneb, zineb, Dithane M-45, Polyram (metiram), Daconil 2787, or difolatan at 5- to 10-day intervals, or just before rainy periods. Start when disease is *first* seen or just after flowering. Destroy (burn or turn under deeply) volunteer potatoes, cull piles, and plant debris after harvest. Store blemish-free tubers in a dry, well-ventilated location at 40° F. Keep down weeds and insects. Dig tubers in dry weather, 2 to 3 weeks after vines die.

Grow disease-free, certified seed potatoes. Potato varieties differ in resistance. Avon, Merrimack, and Yampa have moderate resistance. More highly resistant varieties should be available soon. Three-year rotation with crops outside potato-tomato family.

3. *Late Blight*—General and serious when nights are cool (40° to 60° F.) and days are moderately warm (70° to 80° F.) and humid. Rapidly enlarging, round to irregular, dark green to grayish-purple, water-soaked areas on leaves, petioles, and stems. In damp weather, a sparse whitish mold appears—mostly on underleaf surface. Infected areas later turn dark brown or black, dry out, and die. May resemble frost or Botrytis blight (below) damage. Entire tops may die in a few days in cool, wet weather. Slightly sunken, dark brown to purplish-black blotches form on tubers. Reddish-brown granular stain of outer ¼ inch or so of potato tuber (Figures 4.184A and 4.184B). Tubers decay in storage. *Control:* Same as for Early Blight (above). If weather is cool and moist, spray every 5 to 7 days; if warm and dry, every 7 to 10 days. Maintain high, balanced fertility especially of phosphorus and potassium. Varieties resistant to several races of late blight: Ashworth, Avon, Blanca, Boone, Calrose, Canso, Catoosa, Cayuga, Chenango, Cherokee, Cortland, Delus, Empire, Essex, Fillmore, Fundy, Glenmeer, Harford, Hunter, Kennebec, Keswick, Madison, Menominee, Merrimack, Ona, Onaway, Ontario, Placid, Pennchip, Plymouth, Potomac, Pungo, Reliance, Russet Sebago, Saco, Saranac, Sebago, Seneca, Sequoia, Snowdrift, Tawa, and Virgil. More highly resistant varieties will be available soon. Check with your county agent, extension horticulturist, potato specialist, or plant pathologist.

Fig. 4.184A. Late blight of potato. Left, lesions infecting a leaflet; right, causal fungus (*Phytophthora*) fruiting on the leaf surface. (Department of Botany & Plant Pathology and Cooperative Extension Service, Michigan State University photo)

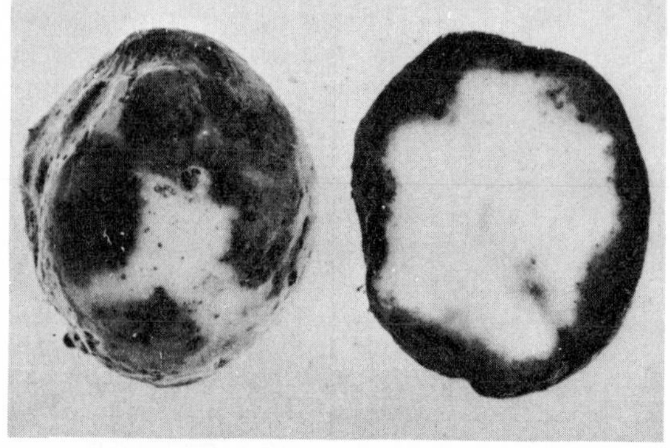

Fig. 4.184B. Potato late blight lesions on tubers. Left, external symptoms; right, internal symptoms. (Department of Botany & Plant Pathology and Cooperative Extension Service, Michigan State University photo)

4. *Tuber Rots*—Cosmopolitan. Tubers rot slowly or rapidly in field and storage. Rot may be dry and firm, or moist to watery (Fungus Rots), or slimy and foul-smelling (Bacterial Soft Rot). Rots develop rapidly under warm, moist conditions. Mold may grow on affected areas. See Figures 2.45A and 4.185. *Control:* Handle potatoes *carefully* at all times to prevent bruising, cuts, etc. Harvest in dry weather. Store only mature, sound, *dry,* clean, blemish-free tubers in clean, dark, cool, well-ventilated storage at 40° F. and 85 to 90 per cent relative humidity. First store tubers at 50° to 60° F. for 2 to 3 weeks (or 60° to 85° F. for 1 week) to heal cuts and bruises. Check with your extension horticulturist. Also control measures as for Seed-piece Decays (below). Varieties differ in resistance to various rots. Fertilize based on a soil test.
5. *Seed-piece Decays*—General. Seed-pieces rot in soil. Poor stand, especially in cold, wet soil. Emerging plants may be stunted with rolled yellowish leaves. *Control:*

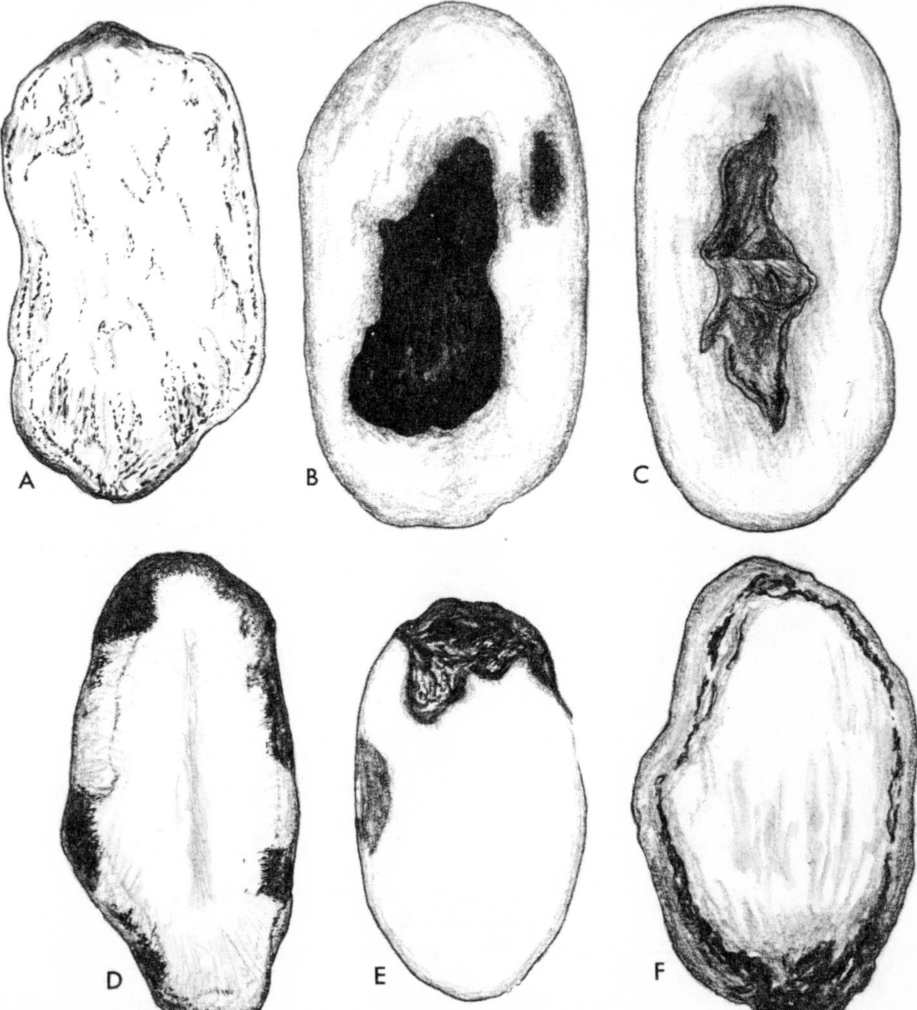

Fig. 4.185. Potato tubers cut through showing A. Net necrosis. B. Blackheart. C. Hollow heart. D. Late blight. E. Fusarium dry rot. F. Bacterial ring rot.

Plant blue tag, certified, disease-free whole tubers (if possible), well corked over and treated with dichlone, chloranil, captan, maneb, zineb, thiram, Morsodren, difolatan, Daconil 2787, Polyram, or Semesan Bel S. Follow manufacturer's directions. Addition of streptomycin (100 parts per million) often helps control Blackleg, but may reduce sprout emergence. Avoid heavy, poorly drained soils. Four- to 6-year rotation. Before planting apply Diazinon or chlordane and work into top 4 to 6 inches of soil. Follow manufacturer's directions. This treatment controls wireworms, white grubs, seed-corn maggots, tuber flea beetles, white-fringed beetle grubs, and other insects that attack seed-piece, roots, and tubers.

6. *Blackleg*—General in wet seasons or in poorly drained soils. Yellowish-green, stunted, upright, wilting plants with curled, upper leaves. Stem base usually dark green, but soon becomes dark brown to black, slimy, and rotted. Infected plants easily pulled up. Tubers often show slimy, brown to black, stem-end rot. Tuber rot develops rapidly in warm, moist storage. Often follows seed-corn maggots, borers, or other injuries. See Figure 2.45A under General Diseases. *Control:* Same as for Seed-piece Decays and Tuber Rots (both above). Control insects; see Virus Complex (below) and Seed-piece Decays (above). Hunter is resistant.
7. *Virus Complex* (mosaics, crinkle, calico (alfalfa mosaic), corky ringspot (or tobacco rattle), leafrolls, mottle, streak, curly-top or green dwarf, aster yellows or purple-top, witches'-broom, vein-banding, ringspot, yellow spot, yellow dwarf)—General. May be serious. Symptoms variable depending on viruses and strains involved, variety, age, weather conditions, and other factors. Lower leaflets roll or cup upward; are stiff and rigid (Leafroll). Leaves may show light green or yellow and dark green mottling (Mottle), streaking, crinkling, or twisting (Crinkle and Mosaic) of leaves. Or irregular, bright yellow to cream-colored areas (Calico). Leaves may be stunted, stiff, leathery, and puckered. Plants often stunted or dwarfed, and bushy. Aerial tubers often form along stem. Brown corky areas—arcs, rings, or lines—may form on and in tubers (Corky Ringspot). Shoot tips may be purplish or yellowish (Purple-top). Yield is reduced. Cultivating and hilling equipment spread certain viruses (Virus X and Spindle Tuber) from infected to healthy plants. *Control:* Apply sprays at 7- to 10-day intervals to control insects, especially aphids and leafhoppers, that transmit viruses. Mites may transmit Virus Y. Use sevin or Thiodan (endosulfan) plus malathion. Keep down weeds. Plant *only* blue tag, certified, virus-free seed tubers. Dip tuber cutting knife in trisodium phosphate or calcium hydroxide between cuts. Destroy *first* infected plants when found. Varieties resistant or immune to one or more viruses: Cherokee, Chippewa, DeSoto, Earlaine, Houma, Hunter, Katahdin, Kennebec, LaSoda, Mohawk, Menominee, Merrimack, Mohawk, Monona, Ona, Onaway, Ontario, Pennchip, Penobscot, Plymouth, Pungo, Red Beauty, Redskin, Red Warba, Reliance, Russet Sebago, Saco, Sebago, Tawa, Warba, and Yampa. More resistant varieties will be available in the future. Check with your potato specialist or extension plant pathologist regarding potato viruses prevalent in your area. Corky Ringspot virus is transmitted by the stubby-root *(Trichodorus)* nematode. Leafroll (Net Necrosis) can be cured in potato tubers by holding them at 98.6° F. for 10 to 20 days.
8. *Wilts*—Potatoes are infected by all three common wilts: Fusarium, Verticillium, and Bacterial. See (15) Wilts under General Diseases.

A. *Fusarium Wilts, Dry Rot, Rusty Dieback*—Most serious in hot, dry weather. Yellowish or bronzed leaves. Plants stunted, gradually or suddenly wilt and die. Yellowish-brown to blackish streaks inside stems and tubers. Firm, leathery, cheesy, dry rot of tubers.

B. *Verticillium Wilt*—A cool season disease. Plants often somewhat stunted, wilt gradually or suddenly around flowering time. Lower leaves become mottled yellow and droop first. Leaflets may roll upward. Tips are yellowed. Plants may die prematurely. Stem-end of tuber is discolored around the eyes. Reddish-brown streaks occur inside stems and tubers. Roots decayed. Nematodes may encourage penetration of the wilt fungus.

C. *Bacterial Wilt, Brown Rot*—Mostly in southeastern states. Plants gradually wilt and die. Stems turn brown, at first only on inside. Brown rot of tubers with ooze sticking to surface.

Control for Wilts: Plant blue tag, certified, disease-free seed grown in northern states (in soil at pH 5.5 or below for Verticillium Wilt control). Treat seed as for Seed-piece Decays (above). Four- to 6-year rotation excluding crops in potato-tomato family. Collect and burn vines before harvest, where feasible. Avoid injuring tubers. Disinfect storage locations. See under Bacterial Ring Rot (below). Increase ventilation in storage. Discard tubers with rot, discolored "eyes," or sticky ooze on surface. Varieties partly resistant to one or more types of *Verticillium Wilt:* Avon, Green Mountain, Houma, Hunter, Katahdin, Menominee, Mohawk, Mo-nona, Ona, Ontario, Pontiac, Red Beauty, Russet Burbank, Saranac, Sequoia, Shoshoni, and Tawa. More resistant varieties should be available shortly. Sebago and Katahdin have weak resistance to *Bacterial Wilt.* Check with your extension horticulturist, potato specialist, or county agent regarding adaptability of these varieties to your area. Maintain balanced soil fertility.

9. *Rhizoctonia or Black Scurf, Stem Rot and Canker*—General. Serious in cold, wet springs. Dark brown or black hard bodies (sclerotia) on tuber surface that do not rub off. Dark brown or reddish-brown, sunken cankers form at base of stem and on stolons (Figure 4.186). Plants may be stunted or killed; leaves rolled, and sickly

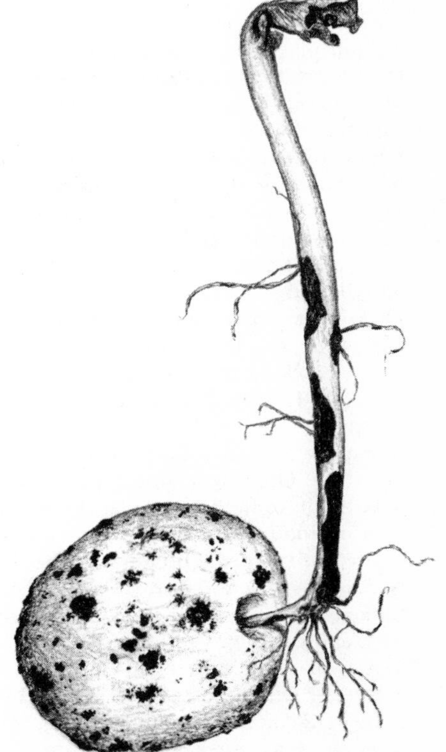

Fig. 4.186. Rhizoctonia disease of potato. Note black sclerotia on tuber surface and dark cankers on stem and stolon.

yellow or dark green. Green or reddish aerial tubers may form in leaf axils near stem base. Roots may be killed back. Stand and yield often reduced. Light, grayish-brown powdery mold (sexual stage of *Rhizoctonia*) grows on lower stem during damp summer weather. *Control:* Plant blue tag, certified, disease-free seed shallow, in warm, well-drained, fertile soil. Treat seed as for Seed-piece Decays (above). If feasible, fumigate before planting as given under Nematode Diseases (below). Four- to 6-year rotation. Harvest early. Potato varieties differ in susceptibility.

10. *Blackheart*—Internal, central, dark gray to coal-black rot of tubers. Rotted tissue is firm to flabby (Figure 4.185B). May be caused by bruising and bumping when digging, grading, sacking, etc. *Control:* Same as for Tuber Rots (above). Avoid overheating and poor ventilation. Do not store in large, solid piles. Dig promptly when soil is warm.

11. *Hollow Heart*—Large, overgrown tubers develop irregular, hollow, center heart or cavity with dark corky border (Figure 4.185C). Common in seasons favorable for late, rapid growth (excessive soil moisture and fertility). *Control:* Space plants closer together. Plant varieties with high dry-matter content. Check with your county agent, extension horticulturist, or potato specialist. Manota is resistant.

12. *Bacterial Ring Rot*—General. Foliage symptoms variable. Mottled rolled leaves on erect stems. Plants may wilt during the day and recover at night. May finally be stunted, turn yellow, and die. Tubers often have light yellowish to brownish-gray pockets or "cheesy" ring about 1/4 inch under the skin, especially after a storage period (Figure 4.185F). Rot develops rapidly in storage; often followed by Bacterial Soft Rot. *Control:* Plant *only* blue tag, certified, disease-free tubers. Disinfect cutting knives with mercuric chloride (1:500 solution) or Semesan Bel S (4 ounces per gallon). Or plant small, uncut tubers free of cuts or bruises. Stringent sanitation of all equipment, tools, and storage areas using formaldehyde (1 pint in 15 gallons of water) or copper sulfate (1 pound to 5 gallons). Use new or clean sacks, crates, and baskets for seed tubers or dip in formaldehyde solution. Merrimack, Saranac, and Teton are resistant varieties. More highly resistant varieties should be available in the future. Control insects, especially Colorado potato beetles, plant bugs, aphids, and leafhoppers that may transmit the causal bacteria. See under Virus Complex (above).

13. *Spindle Tuber*—General. Symptoms differ greatly with variety. Plants often dwarfed, spindly, and more erect with dwarfed, dark green leaves. Tubers usually elongated. Often smaller than normal, pointed at stem end and with shallow eyes. Spindle tuber is caused by a strain of Potato Virus X. Petunia, tobacco, and physalis may be symptomless hosts. *Control:* Same as for Virus Complex (above). The virus may be transmitted by cutting knives or equipment passing through a crop, also by many types of insects. Tomato seed and pollen may also transmit the virus (= Tomato Bunchy-top virus). Resistant potato varieties may be available shortly.

14. *Nematode Diseases*—Plants may become stunted and weak. Commonly pale green or yellow. Often wilt during day. Yield is often reduced. (1) Diseased plants may have small, round to irregular, knotty, warty, or pimple-like bumps, galls, or golden cysts on roots and tubers (Root-knot and Golden Nematode). (2) Tuber may be spotted with small, raised pustules that tend to merge (Potato Rot and Stem Nematodes). Infested areas may later turn gray to brown, shrink, and become sunken. (3) Roots discolored, stubby or "bushy" (dagger, lance, lesion, reniform, ring, spiral, sting, stubby-root, stunt, and sugar beet). *Control:* Prevent contamination of clean soil. Plant certified, disease-free seed. Three- to 6-year rotation excluding potato, tomato, and related plants. Keep down weeds. Fumigate soil with D-D, EDB, methyl bromide, Vorlex, or Telone (pages 538–548). Soil fumigation, however, may increase losses from Common Scab (above). Check with your

extension plant pathologist or potato specialist. For additional information, read USDA Leaflet 361, *The Golden Nematode of Potatoes and Tomatoes—How To Prevent Its Spread.* Varieties resistant to Golden Nematode may be available shortly.

15. *Tipburn, Hopperburn*—General, especially in hot, dry weather. Tips and outer margins of leaves turn yellow, then brown, and curl in hot, dry weather. Plants turn pale green or yellow. Ripen permaturely. Yield is often reduced. Commonly confused with blights. *Control:* Control leafhoppers. See under Virus Complex (above). Follow best cultural practices. Check with your county agent, extension entomologist, or potato specialist. Potato varieties differ in resistance.
16. *Psyllid Yellows*—Western half of United States. Leaves turn yellow to reddish or purplish at margins. Basal portion rolls upward (Figure 4.187). Brown dead areas later develop in leaves that wither and die early. Plants stunted and bushy. Aerial tubers sometimes form along stem. Yield may be greatly reduced. *Control:* Spray with DDT or sevin to control psyllid insects.

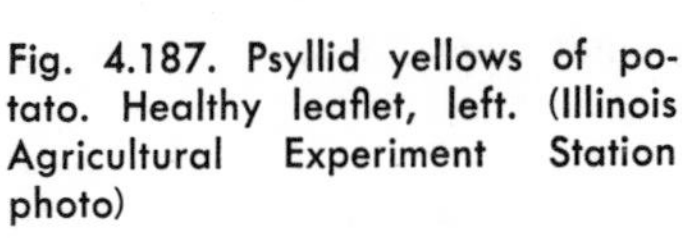

Fig. 4.187. Psyllid yellows of potato. Healthy leaflet, left. (Illinois Agricultural Experiment Station photo)

17. *Black Dot Disease, Anthracnoses*—General but minor problem. Stems cankered and girdled just below soil line by dark brown and dead areas with black dots. Foliage yellowed with rolled leaves. Tops may die early. Tubers show silvery-gray patches, covered with black dots. May develop stem-end rot. *Control:* Plant certified, disease-free seed in well-drained soil. Four- to 6-year rotation. Harvest when tubers first mature. Burn or compost plant debris after harvest.
18. *Gray Mold, Botrytis Blight*—Light brown spots, mostly on lower and inner leaves. Closely resembles Late Blight (above). Botrytis spots begin at leaf tip and enlarge to a wedge shape. Dense gray mold covers affected areas in damp weather. *Control:* Same as for Early and Late Blights (both above).
19. *Powdery Scab, Canker*—Occasional in northern states in high-moisture soils. Small, dark, slightly raised, roundish pustules on tubers that later break open and release powdery, brown masses. Large, sunken cankers may form in tubers. *Control:* Same as for Black Dot Disease (above).
20. *Silver Scurf*—Irregular, light brown spots on young tubers that may appear as watery blisters. Skin later becomes slightly loosened giving a silvery appearance. Red varieties lose color. Dry tubers become wrinkled. *Control:* Plant certified, disease-free tubers in clean soil. Treat seed pieces by soaking in hot formaldehyde (1 pint in 30 gallons) at 125° F. Do not replant potatoes where disease is severe. Harvest red varieties before full maturity. Do not leave undug potatoes in dry soil. Market or eat infected tubers as soon as possible after harvesting. Resistant

varieties: Nordak, Norgleam, and Reliance. More highly resistant varieties should be available soon. Four- to 6-year rotation.

21. *Knobbiness, "Second Growth," Malformed Tubers*—Tubers branch, produce "knobs," and grow indeterminately. May develop cracks and cavities or skin may be corky. *Control:* Plant certified, virus- and disease-free seed of good size (1½ to 2 ounce seed-piece). Maintain as uniform a soil moisture supply as possible. Control diseases and insects. Spray regularly with mixture containing maneb or zineb and sevin or DDT plus malathion.
22. *Sunburn*—Tubers exposed to light develop a green or red color. Such potatoes are bitter, and may be poisonous to some people. *Control:* Keep soil over tubers in the field. Store tubers in the dark.
23. *Stem Rots, Stalk Disease, Sclerotiniose, Southern Blight, Tuber Rots*—Plants may be stunted, wilt, and yellow, even collapse from rotting of stem. Usually at or near soil line. See under Bean, White Mold.
24. *Minor Leaf Spots and Blotches*—Spots of various colors and shapes on lower leaves, following damp weather. May be covered with mold. *Control:* Same as for Early Blight (above).
25. *Powdery Mildews*—Usually minor. Short, brown to dark brown streaks or stipled patches first appear on the stem. Later, a powdery white mold forms on both leaves and stems. Leaves yellow, wither, and die. May drop early. Vines die back; later collapse. *Control:* Where needed, apply sulfur or Karathane when symptoms *first* appear. Repeat twice more at 10-day intervals. Do not apply if temperature is 90° F. or above. Varieties differ in resistance.
26. *Leaf Scorch*—Plants often stunted or dwarfed with tips and margins of leaves scorched. May be somewhat yellowish and curled. Leaves often wither and drop early. Tubers may fail to develop. *Control:* Have soil tested. Apply fertilizer as recommended. May be due to deficiency of potassium, magnesium, or calcium, etc.
27. *Web Blight*—Southeastern states. See under Bean.
28. *Crown Gall*—See (30) Crown Gall under General Diseases.
29. *Root Rots*—See (34) Root Rot under General Diseases.
30. *Black Walnut Injury*—Plants under black walnut trees wilt and die. Roots decayed. *Control:* Avoid growing within 50 feet or more of these nut trees.

For more information read USDA Farmers' Bulletin 1881, *Potato Diseases and Their Control.*

POTATO VINE — See Tomato

POTENTILLA — See Rose

POTHOS — See Calla Lily

POT MARIGOLD, PRAIRIE-CONEFLOWER — See Chrysanthemum

POWDER-PUFF TREE — See Calliandra

PRAIRIEGENTIAN — See Gentian

PRAIRIE LILY — See Mentzelia

PRAIRIE ROCKET — See Cabbage

PRAIRIE SMOKE — See Rose

PRAYER PLANT — See Maranta

PRETTY-FACE — See Brodiaea

PRICKLY-ASH — See Hoptree

PRICKLYPEAR — See Cacti

PRICKLY-POPPY — See Poppy

PRIDE-OF-CALIFORNIA — See Pea

PRIMROSE — See Evening-primrose and below

PRIMROSE [BABY, BIRD'S-EYE, CHINESE, ENGLISH, FAIRY, POISON], AURICULA, COWSLIP, OXLIP, POLYANTHUS *(Primula)*; PIMPERNEL, SCARLET PIMPERNEL, ANAGALLIS *(Anagallis)*; ROCKJASMINE *(Androsace)*; SHOOTINGSTAR or AMERICAN COWSLIP, SIERRA SHOOTINGSTAR, MOSQUITO BILLS *(Dodecatheon)*; LOOSESTRIFE [FRINGED, GARDEN, GOLDEN, SWAMP, WATER or TUFTED, WHORLED], MONEYWORT or CREEPING CHARLIE *(Lysimachia)*

1. *Gray-mold Blight, Botrytis Blight, Flower Blight*—Common indoors, occasional outdoors. Crowns and roots rot. Large decayed spots on flower heads and leaves. Affected plant parts covered with dense gray mold in damp weather. *Control:* Avoid overwatering and overcrowding. Remove fading flowers and spotted leaves. Indoors, keep water off foliage and increase air circulation. Grow in well-drained soil, sterilized if possible (pages 538–548). Apply maneb, zineb, captan, or copper fungicide before rainy periods. Clean up and burn old tops in early spring.
2. *Fungus Leaf Spots or Blight, Anthracnose, Black Spot, Downy Mildew*—Spots or blotches of various colors, shapes, and sizes. *Control:* Same as for Gray-mold Blight (above). In addition, apply zineb or fixed copper at 10-day intervals. Start when first noticed.
3. *Bacterial Leaf Spot*—Small, irregular, water-soaked spots, sometimes with yellowish centers on older leaves. Spots later turn brown with pale yellowish halos (Figure 4.188). Spots may merge to form large, irregular, dead blotches. *Control:* Same as for Gray-mold Blight (above). Grow resistant varieties. Fixed copper or streptomycin sprays may be beneficial in damp weather.

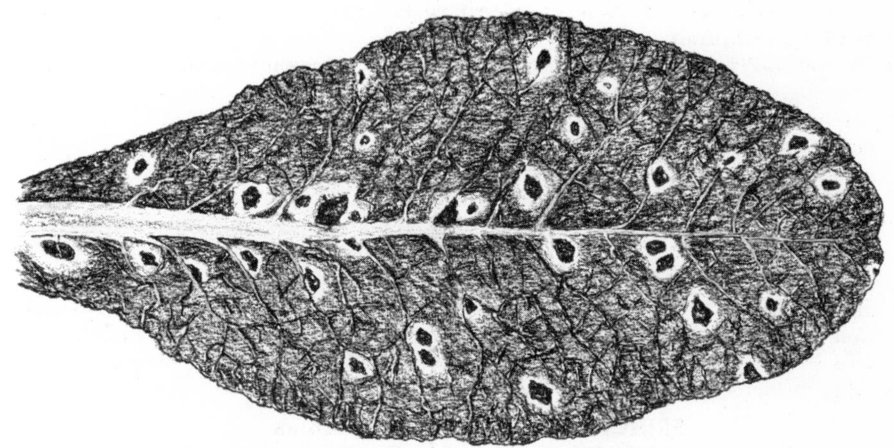

Fig. 4.188. Bacterial leaf spot of primrose.

4. *Stem (Foot) Rots, Root Rots*—Leaves mottled and pale. Later turn yellow and wither. Plants may wilt and collapse. Brown or black rotting of lower stem or roots. May be associated with root-feeding nematodes (e.g., spiral). *Control:* Avoid overwatering. Where possible grow in clean or sterilized soil (pages 538–548). A soil drench of PCNB-Dexon, applied when disease is first evident, may be beneficial.
5. *Root-knot*—Plants weak. Lack vigor. Swellings form on roots. *Control:* See (37) Root-knot under General Diseases.
6. *Damping-off*—Seedlings wilt, collapse, and die. *Control:* Same as for Gray-mold Blight (above).

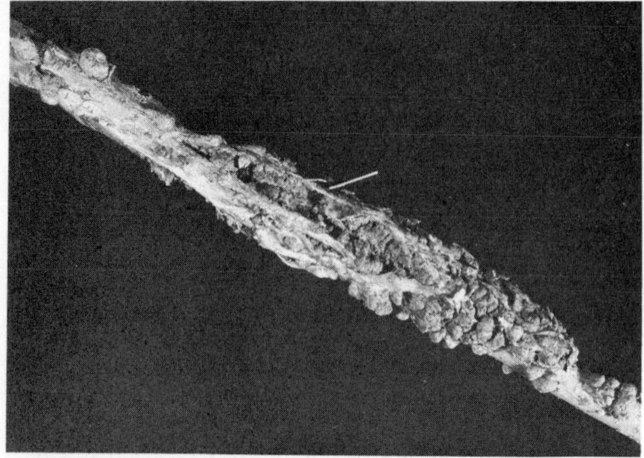

Fig. 4.189. Phomopsis gall of privet. (Pennsylvania State University Extension Service photo)

7. *Mosaics*—Leaves mottled yellow and dark green, often cupped. Young leaves and plants are stunted. Flowers flecked and streaked. *Control:* Destroy infected plants. Control insects, especially aphids, that transmit at least one virus. Use malathion or lindane.
8. *Spotted Wilt*—Leaves and plants stunted and yellowish. Few flowers produced. Irregular dead spots or blotches may develop in leaves. Entire leaf may die. *Control:* Same as for Mosaics (above). Virus is spread by thrips.
9. *Aster Yellows*—See (18) Yellows under General Diseases. Flowers may be deformed or with green petals.
10. *Chlorosis*—Primarily an indoor problem. Leaves develop yellowish or whitish mottling. Varieties differ in susceptibility. *Control:* Avoid overwatering, overfertilizing, and too acid or alkaline a soil. Follow best cultural practices. Adding iron or magnesium sulfate to soil or applying as spray is often beneficial (pages 26–28).
11. *Rusts*—See (8) Rust under General Diseases.
12. *Powdery Mildew*—See under Chrysanthemum. Foliage may be spotted, wither, and dry up.
13. *Leaf and Stem Nematode*—Leaves and stem stunted and distorted. Leaves abnormally thickened, twisted, curled, and unattractive. Plants may fail to bloom. *Control:* Destroy infested plants. Rotate plantings. Grow healthy transplants or seed in soil fumigated before planting with EDB, D-D, Vorlex, chloropicrin, or other nematocide (pages 538–548).

PRINCESFEATHER — See Cockscomb

PRINCESSTREE — See Paulownia

PRIVET [AMUR, BIGBERRY, CALIFORNIA, CHINESE, COMMON or EUROPEAN (many horticultural forms), GLOSSY or WAX, IBOLIUM, IBOTA, INDIAN, JAPANESE, LODENSE, NEPAL, QUIHOU, REGEL] *(Ligustrum)*

1. *Anthracnose, Canker, Twig Blight, Dieback*—General and serious. Leaves wilt, turn pale green then brown, shrivel, and cling to stem. Twigs blighted and killed by girdling, sunken, brown or reddish cankers at base of main stem. Common (European) privet is very susceptible. *Control:* Cut and burn infected parts. Apply zineb, captan, or ferbam at weekly intervals during wet weather. Fairly resistant privets: Amur, California, Ibota, and Regel.

2. *Wood Rots, Collar Rot*—Cosmopolitan. See under Birch, and (23) Wood Rot under General Diseases.
3. *Root Rots*—Plants in hedge gradually die in one or several places. Areas tend to increase in size each year. Plants unthrifty. Make poor growth. Foliage is stunted and thin; yellowish or brown. Roots rotted. Often covered with white mold or black, shoestring-like strands *(Armillaria).* May be associated with root-feeding nematodes (e.g., lance, lesion, needle, pin, ring, root-knot, spiral, stem, stubby-root, stunt). *Control:* Dig up and destroy all of affected plants and 2 apparently healthy ones on each side. Include all roots. Drench soil where these plants were growing. Use SMDC (Vapam or VPM Soil Fumigant), or replace with new soil.
4. *Crown Gall*—Occasional. Large, rounded, rough galls on stems near soil line. Infected plants may be more susceptible to winter-killing, drought, or high summer temperatures. *Control:* Plant only gall-free plants. Remove and destroy affected parts and seriously damaged plants.
5. *Phomopsis Stem Gall*—More or less circular galls up to 2 inches in diameter and 6 inches long at crown or base of stem (Figure 4.189). Plants may die. *Control:* Cut out and burn stems infected with gall. Do not replant in same area without sterilizing soil (pages 538–548). See under Root Rots (above). Avoid wounding stem bases.
6. *Powdery Mildew*—Powdery, whitish patches or blotches on upper leaf surface. *Control:* If serious, dust or spray with sulfur or Karathane.
7. *Minor Leaf Spots, Leaf Blights*—Prevalent during rainy seasons on overcrowded plants. Round to irregular spots of various colors and sizes. Some spots may drop out leaving shot-holes. *Control:* If serious, apply zineb, maneb, ferbam, or fixed copper before rainy periods. Prune to thin overcrowded plants.
8. *Sooty Mold*—Blackish growth on leaves. Follows insect attacks. *Control:* Apply malathion to control insects.
9. *Root-knot*—Wax privet is very susceptible. See under Peach, and (37) Root-knot under General Diseases. DBCP (Nemagon or Fumazone) may be applied around living plants. Nurserymen use *Ligustrum quihoui* (Chinese privet) as a root-knot resistant rootstock.
10. *Chlorosis*—Mineral deficiency (iron, zinc, manganese) in alkaline soils. See under Maple and pages 26–27. *Control:* Have soil tested. Treat as recommended.
11. *Mosaic, Variegation, Chlorotic Spot, Ringspot*—Leaves mildly mottled, puckered, and distorted. May be stunted, spotted, or ringed with yellow. *Control:* Dig up and burn infected plants. Set out virus-free plants from reputable nursery.
12. *Winter Injury*—Single stems or entire sections of hedge killed to ground level. New shoots arise from crown and roots. *Control:* Grow adapted varieties and species. Check with your extension horticulturist and local nurseryman.
13. *Witches'-broom*—See under Lilac.
14. *Leaf Nematode*—See (20) Leaf Nematode under General Diseases.
15. *Thread Blight*—See under Walnut.
16. *Verticillium Wilt*—See under Maple, and (15B) Verticillium Wilt under General Diseases.
17. *Algal Leaf Spot*—Far south where damp. See under Magnolia.

PROBOSCISFLOWER, DEVILSCLAW, UNICORNPLANT *(Proboscidea)*

1. *Fungus Leaf Spots*—Brown to gray spots with reddish or purple margins. *Control:* Pick and burn spotted leaves. If serious, spray several times during rainy weather, 10 days apart. Use zineb, maneb, or fixed copper.
2. *Bacterial Leaf Spot*—Minute, angular, sunken, water-soaked spots on leaves, petioles, and stems. Spots may merge to form irregular, light brown patches. Plants may die. Fruit may become brown and shriveled. *Control:* Sow disease-free seed

from healthy pods. Destroy infected plant parts and severely infected plants.
3. *Stem or Crown Rot*—Stem rots at soil line. Plants may wilt and collapse. Rotted area may be covered with cottony mold. *Control:* See under Delphinium.
4. *Mosaics*—See (16) Mosaic under General Diseases.
5. *Root Rot*—See (34) Root Rot under General Diseases.

PRUNE, PRUNUS — See Peach

PRUNELLA — See Salvia

PSEUDOLARIX — See Larch

PSEUDOTSUGA — See Pine

PSIDIUM — See Myrtle

PTELEA — See Hoptree

PTERETIS, PTERIDIUM, PTERIS — See Ferns

PUCCOON — See Mertensia

PUMPKIN — See Cucumber

PUNICA — See Pomegranate

PURPLE-CONEFLOWER, PURPLE DAISY — See Chrysanthemum

PURPLE-FLOWERED GROUNDCHERRY — See Tomato

PURPLELEAF BUSH, PURPLELEAF PLUM — See Peach

PURPLELEAF SPIDERWORT — See Rhoea

PURPLE RAGWORT — See Chrysanthemum

PURPLE ROCKCRESS — See Cabbage

PURPLE SMOKEBUSH — See Sumac

PUSCHKINIA — See Tulip

PUSSYTOES — See Chrysanthemum

PYCNANTHEMUM — See Salvia

PYRACANTHA — See Apple

PYRETHRUM — See Chrysanthemum

PYROSTEGIA — See Bignonia

PYRUS — See Apple

QUAKER BONNETS — See Pea

QUAMOCLIT — See Morning-glory

QUEEN-OF-THE-MEADOW, QUEEN-OF-THE-PRAIRIE — See Rose

QUEENS-DELIGHT — See Castorbean

QUEENS-FLOWER — See Crapemyrtle

QUERCUS — See Oak

QUINCE — See Apple

QUINCULA — See Tomato

QUININE BUSH — See Rose

QUIVERLEAF — See Poplar

RABBIT TRACKS — See Maranta

RADISH — See Cabbage

RAGGED-ROBIN — See Carnation

RAINBOW-BUSH — See Rose-moss

RAINLILY — See Daffodil

RANUNCULUS — See Delphinium

RAPE — See Cabbage

RASPBERRY [AMERICAN RED or COMMON RED, BLACK or BLACKCAP, DWARF RED, EUROPEAN RED, GOLDEN EVERGREEN, PURPLE FLOWERING, PURPLECANE, ROSELEAF, WESTERN RED,

WHITEBARK, WHITE FLOWERING or BOULDER], BLACKBERRY [ALLEGHANY, TAYLOR, or AMERICAN, CUT-LEAVED or EVERGREEN, EUROPEAN, EVERGREEN THORNLESS, HIGHBUSH, HIMALAYA, TRAILING, YANKEE], BOYSENBERRY, DEWBERRY [CALIFORNIA, CULTIVATED AMERICAN, GRAPELEAF, NORTHERN, SOUTHERN, SWAMP], CLOUDBERRY, LOGANBERRY, MAMMOTH BLACKBERRY, SALMONBERRY, THIMBLEBERRY, YOUNGBERRY or WINEBERRY *(Rubus)*

1. *Anthracnose, Spot Anthracnose, Gray Bark*—General and serious especially in raspberries, blackberry, dewberry, and loganberry. Small, reddish-brown to purple spots on young canes, shoots, and fruit spurs that enlarge, become more or less circular with sunken, light gray centers and purple borders. Heavily infected canes are stunted, gray-crusted, dry, crack, and often winter-kill. Fruits often small, deformed, dry, and seedy. Small yellowish-white spots with reddish-purple margins form on leaves. Spots may later drop out leaving shot-holes. See Figures 2.40A and 4.190. Black raspberries are very susceptible. *Control:* Remove and burn fruiting, surplus, and weak canes right after harvest. Plant certified, disease-free plants where air drainage is good. If you propagate your own stock, plant only the most vigorous tip plants or suckers. Remove old cane "handles" when setting plants. Keep down weeds. Space plants. Destroy badly infected canes when found. Destroy nearby wild brambles. Remove all weak canes in early spring. Grow in well-drained soil and avoid high rates of nitrogen fertilizer. Apply lime sulfur (⅘ pint per gallon of water) or Elgetol thoroughly as buds *begin* to swell in early spring. Repeat when new leaves are ¼ inch out. Then apply captan, ferbam, or folpet at about 10-day intervals until berries start to ripen and again right after harvest. See spray program in Appendix (Table 16). Somewhat resistant *raspberry* varieties: *Blacks*—Black Beauty, Blackhawk, Dundee, Evans, Quillen; *Purples*—Marion, Potomac, Sodus; *Reds*—Cuthbert, Indian Summer, Latham, Newburgh, Ohta (Flaming Giant), Ranere (St. Regis), Sunrise, and Turner. Check with your nurseryman or extension horticulturist regarding varieties adapted to your area.

Fig. 4.190. Raspberry leaflet infected with anthracnose. Note the dark margins to the spots and the shot-holes. (Department of Botany & Plant Pathology and Cooperative Extension Service, Michigan State University photo)

2. *Spur Blight*—General. Mainly on red raspberry. Chocolate-brown, dark blue, or purplish-brown spots and encircling bands on petioles and new canes. Affected areas are gray- to black-speckled by fall or early spring. Canes girdled. Laterals often wither and die early in season. Canes dry out and may crack. Spreading, brown, angular, or V-shaped blotches occur on leaves along a leaf vein (Figure 4.191). Such leaves wither and drop early leaving canes bare. Buds turn brown, shrivel, and die or produce yellowish leaves. *Control:* Same as for Anthracnose (above). Resistant *raspberries:* Chief, Columbian, Marcy, Ontario, Plum Farmer, and Viking. *Blackberries* are highly resistant or immune.

Fig. 4.191 (above). Spur blight of red raspberry.

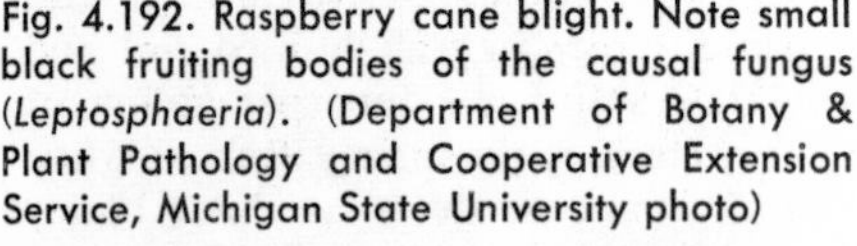

Fig. 4.192. Raspberry cane blight. Note small black fruiting bodies of the causal fungus (*Leptosphaeria*). (Department of Botany & Plant Pathology and Cooperative Extension Service, Michigan State University photo)

3. *Cane Blights, Dieback*—General in wet seasons. Fruiting canes usually wilt, wither, and die between blossoming and fruit ripening, especially in dry summers. Gray to dark brown, black-dotted, flattened cankers on canes. Canes may become cracked, brittle, easily snap off. See Figure 4.192. *Control:* Same as for Anthracnose (above). Fertilize and prune to keep plants vigorous. Prune at least 3 days before rain is predicted. Avoid wounding canes. Control insects (e.g., cane borers). Use malathion plus methoxychlor or DDT sprays. Columbian is a resistant *purple raspberry.*
4. *Crown and Root Gall, Cane Gall, Hairy Root*—General and serious. Rough, warty, white to black overgrowths or elongated galls on roots, crown, and lower parts of fruiting canes. Canes may split open and dry out. Hairy root appears as large numbers of fine roots instead of normal growth. Plants often stunted, lack vigor, gradually die. Berries small and seedy. See Figure 2.46C under General Diseases. *Control:* Plant certified, disease-free plants in clean soil. Carefully dig up completely and burn all infected plants when first noticed. Keep plants well fertilized and in good growing condition. Avoid replanting in same area for 5 or 6 years without first drenching soil with SMDC (Vapam or VPM Soil Fumigant) or formaldehyde (1 pint of 40 per cent formalin in 6 gallons of water). Apply $\frac{1}{2}$ gallon per square foot. Use 6 gallons to treat an area $3\frac{1}{2}$ by $3\frac{1}{2}$ feet. Replant the following season. Avoid injuring plants. If suspicious of these very contagious diseases, dip pruning shears in formaldehyde solution (1 part in 5 parts of water) or 70 per cent alcohol between cuts.
5. *Virus Decline* (mosaics, mottle, leaf curls, rosette or severe streak, necrosis, yellow net, ringspot, dwarf or stunt)—General and serious. Yields may be reduced 50 per cent or more. Symptoms variable; often latent (mosaic virus strains in red raspberries). Plants decline in vigor. Never recover. Leaves may be yellowish, light green to dark green, mottled, blistered or puckered (Mosaic complex), dwarfed, curled, wrinkled, and cupped downward. Plants may be slightly to greatly stunted

and "bushy" with small, dark green, bunchy, stiff, tightly curled leaves (Leaf Curl). Or foliage may be sparse with spindling canes. *Ringspot*—found in Pacific Northwest—produces yellowish blotches, ringspot, or oakleaf markings, or netlike yellowing of veins on leaves of young canes. Ringspot is transmitted by the needle (*Longidorus*) nematode. Virus infection causes a gradual decrease in fruit production over several years. Fruits often dry, seedy, crumbly, and small. Canes may be brittle. Greenish-brown to bluish-violet spots or vertical stripes may develop on canes near base (Streaks). See Figures 4.193A, 4.193B, and 4.193C. *Control:* Plant certified, virus-free plants of black, purple, and red raspberries. (Some 19 raspberry varieties and 21 blackberries are now free of all detectable viruses.) Dig up and burn infected plants when first found. Destroy all nearby (within 500 feet; or more if possible) wild brambles before setting out new plants. Control aphids

Fig. 4.193A. Mosaic on Cuthbert red raspberry. Note blistering or puckering of interveinal leaf areas. (Department of Botany & Plant Pathology and Cooperative Extension Service, Michigan State University photo)

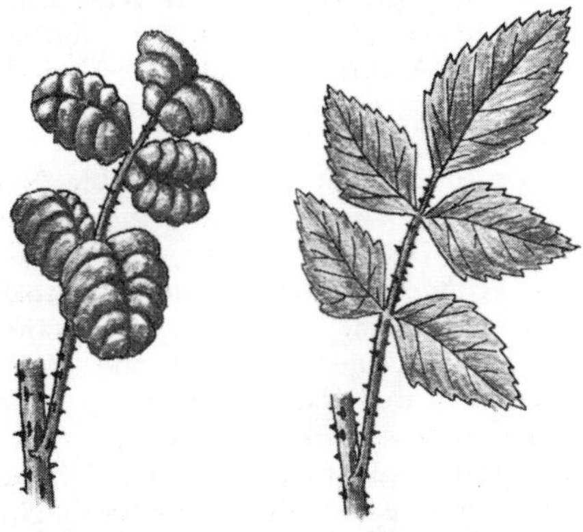

Fig. 4.193B. Raspberry leaf curl.

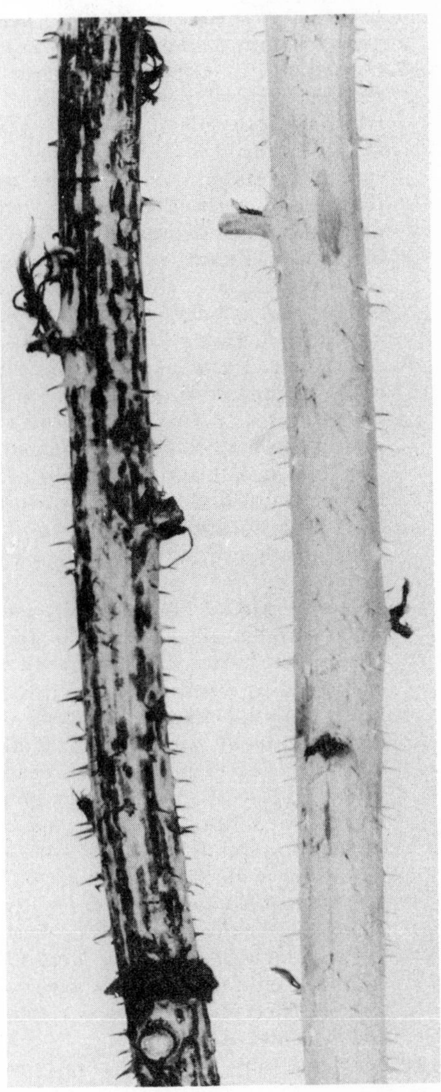

Fig. 4.193C (right). Raspberry streak. Virus infection on new cane growth, left; healthy cane, right. (Department of Botany & Plant Pathology and Cooperative Extension Service, Michigan State University photo)

that transmit certain viruses. Use malathion. Resistant *raspberries* to one or more viruses: Canby, Dike, Dundee, Indian Summer, Latham, Marcy, Milton, Newburgh, Potomac, Ranere (St. Regis), September, Tahoma, Taylor, Viking, and Washington. New Logan *raspberry* is resistant to Leaf Curl in some areas, while Plum Farmer *black raspberry* is immune to one or more forms of Leaf Curl. Advance and Himalaya *blackberries* are resistant or immune to Rosette. Mosaic virus in *red raspberry* is inactivated in 96 hours at 100° F.

6. *Fruit Rots, Molds*—Cosmopolitan when weather is warm and moist at harvest. Berries rot. Often soft and watery or may shrivel and become hard. Rotted areas may be covered with gray, tan, blue-green, white, or black mold. *Control:* Handle fruit carefully to avoid bruising. Pick at least twice a week, early in the day. Discard decaying, damaged, overripe, sunburned, or imperfect berries. Refrigerate (32° to 40° F.) promptly. Apply captan alone when blossoms open, 10 days later, and as berries start to color, especially if period is wet. Repeat at 7-day intervals if harvest period is rainy. Follow regular pruning program. Avoid overcrowding.
7. *Orange Rusts*—General on black and purple raspberries, blackberries, and dewberries. Red raspberries are immune and boysenberries are moderately resistant. Underside of leaves is covered with bright, reddish-orange, dusty pustules in late spring. Leaves may wither and drop early. See Figure 4.194. New shoots spindly or stunted with dwarfed or misshapen, yellowish-green leaves. Infected plants never recover. Produce no fruit. *Control:* Plant only certified, disease-free plants. Carefully dig up and burn rusted plants when first seen, and *before* pustules break open. Be sure to remove or kill roots. Thin out healthy canes to allow for good air drainage. Keep down weeds. Destroy nearby wild brambles, especially blackberries and dewberries. Spray as for Anthracnose (above). Resistant *blackberries:* Boysen, Ebony King, Eldorado, Evergreen, Lawton, Lowden, Russell, Snyder, and Young. Lucretia is a resistant *dewberry*. Fungicide sprays and pruning are relatively ineffective.
8. *Yellow Rusts, Leaf and Cane Rusts*—General. Small, lemon-yellow to orange, dusty pustules on both leaf surfaces, stems (canes), and petioles. In fall, pustules may be dark brown or black. Leaves curled, later wither and drop early. Infected canes become brittle and break easily. Fruit may die on canes before maturing. Alternate host: white spruce or none. *Control:* Same as for Anthracnose (above). Varieties differ greatly in resistance to these rusts. Check with your state or extension plant pathologist.
9. *Verticillium Wilt, Blue Stem*—General. Black raspberries are very susceptible. New shoots stunted, wilt, may turn bluish-black and die. Leaves dull green then yellow to chocolate-brown. Cup downward and fall early. Disease progresses upward from base. Fruiting canes may collapse or die back. Fruits dry up before ripening. Broad, blue to bluish-black streaks extend upward from soil line. Appear on older canes. Plants gradually decline. See Figure 2.32D under General Diseases and Figure 4.195. *Control:* Plant certified, disease-free plants in well-drained, fertile soil where wilted eggplant, pepper, tomato, potato, stone fruits, strawberry, or brambles have not occurred before. Dig up and destroy infected plants when found. Do not replant susceptible plants in same soil for several years without first thoroughly fumigating with SMDC (Vapam or VPM Soil Fumigant), or chloropicrin (pages 538–548). Resistant *blackberries:* Burbank Thornless, Cascade, Chehalem, Cory's Thornless, Cutleaf Evergreen, Evergreen, Himalaya, Ideal (Santiam), Jenner, Logan, Mammoth, Marion, Merton Thornless, Ollalie, Pacific, Phenomenal, and Zielinski. Wild trailing blackberries are also resistant, while Boysen and Young are severely affected. *Dewberry* is rarely affected. Resistant *red raspberries:* Cuthbert and Syracuse.
10. *Leaf and Cane Spots*—General. Small, round to irregular spots or blotches of

Fig. 4.195. Verticillium wilt of black raspberry, advanced stage. (Department of Botany & Plant Pathology and Cooperative Extension Service, Michigan State University photo)

Fig. 4.194. Orange rust of blackberry.

various colors on leaves and canes. Infected leaves may wither and drop early. Usually starts at base of cane. Canes may be stunted and weakened. Yield is often reduced. *Control:* Same as for Anthracnose (above). After harvest apply three or four sprays, 2 to 3 weeks apart. Use captan, ferbam, zineb, Dyrene, fixed copper, or bordeaux (4-4-50). Carolina, Evergreen, and Lucretia *blackberries* have resistance to Septoria *(Mycosphaerella, Sphaerulina)* Leaf and Cane Spot, as do Dixie Mandarin and Van Fleet *raspberries.* More resistant raspberries should be available soon.

11. *Root Rots, Collar Rot*—Plantings decline in vigor. Leaves small and yellowish to bronzed. Fruits small; may wither before ripening. Plants may die. Roots decay. Most serious in heavy, wet soils. See under Apple and Currant. Honey-colored or other mushrooms or fungus growth may form at crown of plants. Often associated with nematodes (e.g., dagger, lance, lesion, pin, ring, root-knot, sheath, spiral, stem, stubby-root, stunt). *Control:* Treat soil with chlordane before planting to control soil insects (e.g., root borers and weevils). Work into upper 6 inches of soil. Remove and destroy *all* of each diseased plant. Do not replant in affected spots without first sterilizing the soil (see under Verticillium Wilt above). Varieties differ in resistance. Canby is very susceptible to a wet soil, raspberry root rot *(Phytophthora).*
12. *Powdery Mildews*—Common in warm (65° to 80° F.) weather, but usually a minor problem, except on certain red raspberries. White to gray, powdery growth on underleaf surface, tips of new canes, buds, fruit spurs, and even fruit. Leaves dwarfed, mottled, and distorted with curled margins. Cane growth is stunted. Yield may be reduced. Most commercial varieties of red raspberry (especially Latham and Newburgh) and dewberry are highly susceptible as are Blackhawk and Munger black raspberries. *Control:* Space plants and prune for good ventilation and sunlight. Keep down weeds. Canby, Cumberland, Logan, Washington, and

Willamette *raspberries* have some resistance. Most *blackberries* appear to be resistant. Apply dormant or delayed dormant spray of lime-sulfur (Table 16 in Appendix). Spray with Karathane when mildew is first seen. Repeat twice more, 10 to 14 days apart. Follow manufacturer's directions. Set out disease-free plants in fertile, well-drained soil. Avoid overcrowding and more than light shade.

13. *Winter Injury*—Symptoms variable. Entire plants may die during winter or up to harvest. Tips of canes die back or buds are killed. Berries may be irregular and aborted. Presence of other diseases frequently increases severity of winter injury. *Control:* Follow recommended cultural practices for your area, e.g., mulching, fertilizing, irrigation, cultivation, pruning, selection of planting site, planting of late summer cover crop, etc. Provide for sturdy, mature wood in autumn. Prune out weak, dead, and injured canes. Control diseases and insects. Follow spray program given in Appendix (Table 16). Grow winter-hardy varieties and species recommended by your extension horticulturist.
14. *Sunscorch*—Ripening fruits are gray and dull. *Control:* Follow recommended cultural practices. Same as for Anthracnose (above).
15. *Male Berry* (blackberry sterility)—General. New canes developing from affected plants are more vigorous, but fail to set fruit or produce misshapen berries. Eldorado and older varieties are quite susceptible. *Control:* Destroy affected plants, including all roots. Do not start new plantings from "diseased" plants. Grow certified, "immune" varieties of blackberries (e.g., Bailey, Ebony King, and Hedrick).
16. *Sooty Blotches, Black Mildew*—See (12) Sooty Mold under General Diseases. *Control:* Apply malathion plus methoxychlor to control insects.
17. *Chlorosis*—Plants often weak and pale green to golden yellow (iron or manganese deficiency). Leaf margins may be scorched (potassium deficiency). Mostly in western states in alkaline soils. Due to a deficiency of one or more essential nutrients. *Control:* Maintain *balanced* fertility based on a soil test. See page 25.
18. *Calico* (raspberry)—Bright yellow areas along leaf veins that form a "network." Severely affected leaves are almost completely bright yellow. *Control:* Do not propagate from affected "mother" plants.
19. *Double Blossom, Rosette*—Primarily southeastern states. Flower buds are coarser, redder, and larger than normal. Petals increased in size and number; pink, twisted, and wrinkled. Infected buds produce short, broomlike growths without berries. Diseased canes produce no (or poor) fruits. *Control:* Remove and destroy infected blossoms, where practical. Same as for Anthracnose (above).
20. *Fire Blight, Flower and Twig Blight* (blackberry, raspberry)—See (24) Fire Blight under General Diseases.
21. *Downy Mildew*—See (6) Downy Mildew under General Diseases.
22. *Threat Blight*—Southeastern states. See under Walnut.

For additional information read USDA Farmers' Bulletins 2160, *Growing Blackberries;* 2165, *Growing Raspberries;* and 2208, *Controlling Diseases of Raspberries and Blackberries.*

RATIBIDA — See Chrysanthemum

RATTLESNAKE-MASTER — See Celery

REDBAY — See Avocado

REDBUD, RED-CARDINAL — See Honeylocust

REDCEDAR — See Juniper

RED HAW — See Apple

RED-HEART — See New Jersey-tea

REDHOT-POKERPLANT or POKER-PLANT, TORCHLILY (*Kniphofia, Tritoma*)

1. *Root-knot*—See (37) Root-knot under General Diseases. *Control:* Indoors, grow plants in sterilized soil (pages 538–548) in sterilized containers.
2. *Leaf Spots*—Small, dark spots. *Control:* Pick and burn spotted leaves. Indoors, keep water off foliage.

RED JASMINE — See Oleander
RED-ROBIN — See Cranesbill
RED-VALERIAN — See Valerian
REDWOOD — See Pine
RENANTHERA — See Orchids
RESEDA — See Mignonette
RETINOSPORA — See Juniper
REX-BEGONIA VINE — See Grape
RHAMNUS — See Buckthorn
RHAPIDOPHYLLUM, RHAPIS — See Palms
RHEUM — See Rhubarb
RHEXIA — See Deergrass
RHIPSALIDOPSIS — See Cacti
RHODODENDRON [CAROLINA, CATAWBA, COAST, DAHURIAN, EARLY, FORTUNE, KOREAN, MAYFLOWER, PIEDMONT, ROSEBAY or GREAT LAUREL, SMIRNOW], AZALEA [AMOENA, EXBURY, FLAME, GHENT, GLENN DALE, INDIAN, JAPANESE, KAEMPFERI or TORCH, KNAPHILL, KOREAN, KURUME, MACRANTHA or INDICA, MOLLIS or CHINESE, MOUNTAIN, OCONEE, PERICAT, PINKSHELL, ROSESHELL or PIEDMONT, ROYAL, SATSUKI, SNOW, SWAMP, SWEET or SMOOTH YELLOW, VUYK], PINXTERBLOOM, DOWNY PINXTERBLOOM, RHODORA *(Rhododendron)*

1. *Leaf Spots, Leaf Scorch or Blotch, Anthracnose, Spot Anthracnose, Tar Spot, Blight*—General. Small to large, round to angular or irregular, spots and blotches on leaves. Often found on leaves damaged by frost, winter injury, sunscald, drought, or insects. Spots silvery-white, yellow, gray, tan, red, reddish-brown, or dark brown. May be zonate with conspicuous margin. Centers often sprinkled with black dots or mold (Figure 4.196). Leaves may drop early. *Control:* Grow varieties and species adapted and recommended for your area. Varieties and species differ in resistance. Check with your local nurseryman, county agent, or extension horticulturist. Grow plants in partial shade, sheltered from strong, dry, winter winds. Keep soil well mulched (3 to 8 inches deep) with peatmoss, well-rotted oak leaves, pine needles, or acid leafmold. Rotted sawdust or shavings, ground corncobs, or wood chips may also be used. The soil should be light, well-drained, acid (pH 4.5 to 5.5), and high in organic matter. Avoid overapplication of fertilizer. Collect and burn fallen leaves. Water thoroughly during summer droughts and late in fall. Be sure plants get the equivalent of an inch of rainfall every 10 days. Control insects with a mixture of DDT, sevin, or lindane plus malathion. Spray with captan, zineb, maneb, thiram, ziram, ferbam, or fixed copper at 10-day intervals as leaves are expanding, and again just after flowering. Sprays may also be needed during summer or fall where humid and moist. Add about ½ teaspoonful of household detergent or commercial spreader-sticker to each gallon of spray. May combine with sprays to control insects.
2. *Winter Injury, Leaf Burn*—General where plants grown near limit of hardiness. Tips of shoots die back. Margins and tips of leaves turn brown or silvery in March or April. Flower buds may die. Usually associated with adverse growing conditions

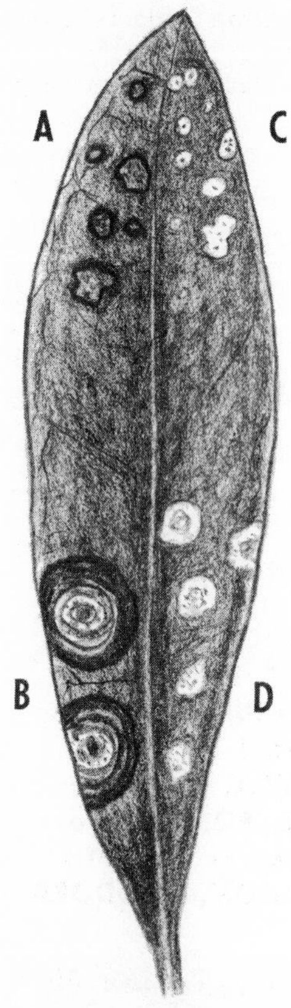

Fig. 4.196. Rhododendron leaf spots. A. Cercospora. B. Phomopsis. C. Phyllosticta. D. Exobasidium. All 4 types of spots would never be found on the same leaf.

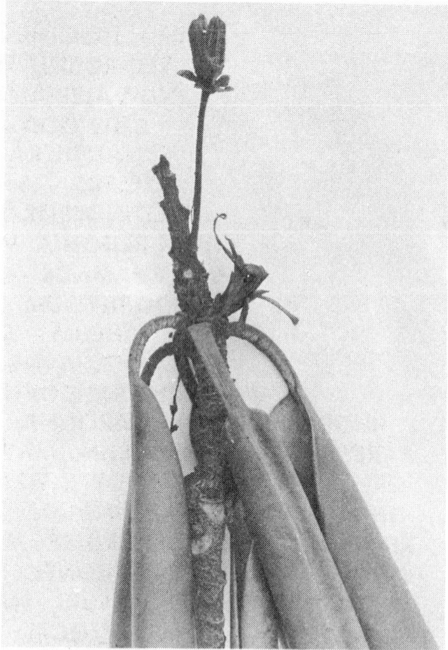

Fig. 4.197. Dieback (*Phytophthora cactorum*) of rhododendron. (Pennsylvania State University Extension Service photo)

such as exposure to cold, wind, excessive drought, overapplication of fertilizer, insect or rodent injury, etc. *Control:* Same cultural practices as for Leaf Spots (above). Fertilize in the spring.

3. *Chlorosis, Yellow Leaf*—General in neutral and alkaline soils and sometimes next to brick, stucco, or concrete foundations. New leaves first mottled and then pale yellow except for main veins (Figure 4.37). If uncorrected, next year's flower buds may not form. If severe, branches may die back. Cold injury is often mistaken for nutrient deficiency. *Control:* Make soil acid as given under Root and Stem Rots (below). For immediate relief apply sprays containing ferrous (iron) sulfate, e.g., Copperas, 2 tablespoons per gallon of water, ferbam (3 tablespoons per gallon), or iron chelate (page 27) following manufacturer's directions. Repeat sprays as needed. Grow in well-drained soil, high in organic matter, away from house walls. Yellowing may also be caused by other deficiencies (e.g., magnesium, zinc), excess fertilizer, or poor drainage.
4. *Root and Stem Rots, Wilt, Stem Canker and Dieback, Crown Rot, Twig Blights*—Widespread. Plants low in vigor. Leaves may be dull yellowish-green or water-soaked, then wilt and wither from a brown to black rot of lower stem and roots. Or terminal buds and leaves turn brown, roll up, and droop. Brown to black, sunken, girdling cankers may form on stem (Figure 4.197). Roots may decay. All parts above canker or rot later wilt and die. Small plants may die quickly. May be associated with nematodes (e.g., dagger, lance, lesion, pin, pine, ring, root-knot, sheath, sheathoid, spiral, stem, sting, stubby-root, stunt, tylenchus), borers, weevils,

or grubs. *Control:* Same as for Leaf Spots (above). Avoid overwatering, poor soil drainage, too much sun or shade, deep moist mulch, etc. Maintain an acid soil (pH 4.5 to 5.5). Add sulfur, acid fertilizer, ferrous sulfate, acid peatmoss, or oak leaves (page 24). Check with your local nurseryman or extension horticulturist. Avoid wounding roots or stems. Prune out and burn infected parts making cuts several inches below canker. Disinfect pruning tools between cuts. Control boring insects within trunk and larger branches by spraying or brushing these areas at monthly intervals starting when peonies in bloom. Use DDT, dieldrin, or Thiodan. Remove and burn severely infected plants with surrounding soil. Work chlordane into top 6 inches of soil before planting to control weevils, grubs, etc. Avoid planting close to lilacs. If feasible, sterilize soil (pages 538–548) before planting. Soak disease-free cuttings for 30 minutes before "sticking." Use Acti-dione-Thiram, folpet, or thiram (Thylate). Follow manufacturer's directions. When crown or root rot first starts, drench soil with PCNB-Dexon mixture. Follow manufacturer's directions.

5. *Ovulinia Flower Spot, Petal or Limp Blight* (primarily azalea)—Serious in southern states, especially near coast, during wet, humid weather. Indian and Kurume azaleas are very susceptible. Small, pale, round flecks or spots form on underside of flower petals. Spots are white on colored flowers and tan to brown on white flowers. When moist, spots enlarge *rapidly* and merge to form large, irregular blotches. Flowers quickly go brown, limp, and mushy (Figure 4.198). Rotted

Fig. 4.198. Azalea flower spot or limp blight (*Ovulinia*). A. Early stage of the disease. B. Later stage.

flowers covered with whitish mold; cling to leaves or stems. Hard, black objects (sclerotia) form in blighted flowers. *Control:* Space plants. Grow where air circulation is good. Where practical, pick and burn spotted flowers when *first* seen. Apply light, misty spray 2 to 4 times weekly during wet weather. Start a week before flowering and continue until blooms fall to ground. Use zineb, thiram (Thylate), Acti-dione, Daconil 2787, or maneb plus spreader-sticker. Thylate (1 tablespoon per gallon) leaves less residue on flowers than other materials. Apply Acti-dione RZ (1 ounce per square yard), or PCNB 75 (1 pound in 3 gallons of water to cover 150 square feet) or PCNB 20 per cent dust (3 ounces per square yard). Apply directly to soil, under and around azalea plants, about 4 weeks *before* bloom is expected. Follow manufacturer's directions. Then replace mulch with fresh material. Do not buy azaleas from South unless plants have bare roots. Before planting, remove and burn buds showing color.

6. *Botrytis Flower Blight*—Destructive in cool, moist weather. Progresses slightly slower than Ovulinia blight. Usually has a readily visible grayish mold growing on decaying flowers. *Control:* Apply thiram (Thylate) sprays as for Ovulinia Flower Spot (above).
7. *Bud Blast and Twig Blight*—Widespread, especially on evergreen types. Scales of terminal flower buds have brown to silvery-gray patches; later sprinkled with tiny black "bristles." Flower and leaf buds later rot, shrivel, and turn light to dark brown. Such buds remain on stem for 2 or 3 years; form rosettes. Twigs may die preventing flowering next year. *Control:* Prune and burn infected buds and twig tips when first seen. Remove and burn faded flower clusters. Destroy seedpods after blooming. Spray as for Leaf Spots and Ovulinia Flower Spot (above) or use fixed copper.

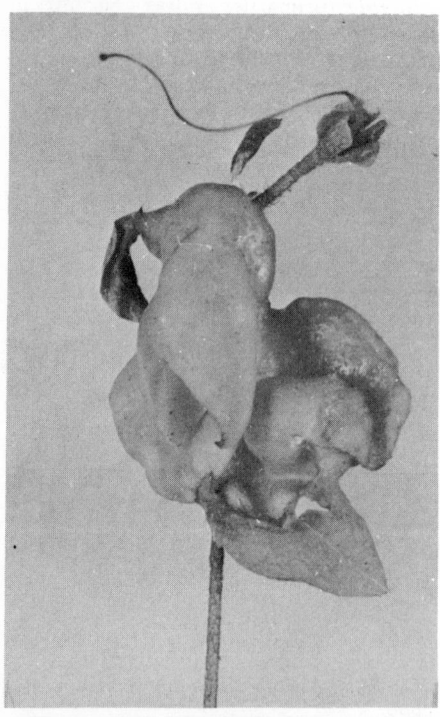

Fig. 4.199. Azalea leaf and flower gall. (Pennsylvania State University Extension Service photo)

8. *Leaf, Shoot and Flower Galls, "Rose Bloom," Yellow Spot, Witches'-broom*—General, but seldom serious. Leaves may be spotted with yellow or red, or turn light green or whitish and thickened, wholly or in part. See Figures 4.196D and 4.199. Leaf surface may be deformed, blistered (bladder-like), and curled. Whole flowers, individual petals, or seedpods may turn into thick, hard, waxy, irregular galls. Surface of affected parts becomes covered in moist weather with white to pink, powdery "bloom." Galls later turn brown and hard. Fleshy rosettes of leaves may form at a branch tip. *Control:* Handpick and burn galls when first evident, before they turn white. Otherwise spray once *before* leaves unfurl. Use captan, ferbam, or zineb (2 tablespoons per gallon) plus spreader-sticker. Repeat 2 to 3 weeks later. Spray after bloom as for Leaf Spots (above). Propagate from disease-free plants. Varieties and species differ in susceptibility.

9. *Witches'-broom*—Distinct witches'-brooms are produced on coast rhododendron and rhododendron hybrids. Leaves turn yellowish-white. Become covered with dense, mealy growth on undersurface. Infected branches die over period of years. *Control:* Destroy infected plants since causal fungus is systemic within plant.
10. *Damping-off, Cutting Rots, Crown Canker, Graft Decay*—General. Serious nursery disease in cutting beds. Stems soften and rot at soil line. Leaves may turn yellow, then darken, and drop early. Seeds rot; seedlings decay. *Control:* Increase air circulation and light. Avoid overcrowding and overwatering. Grow seeds or cuttings in sterile medium (e.g., sifted sphagnum moss). Sterilize soil (pages 538–548); or drench before planting using mixture of PCNB 75 and Dexon, folpet, thiram, or captan (1 tablespoon each per gallon). Use 1 pint of solution per square foot.
11. *Powdery Mildews*—White, powdery patches on leaves in late summer. *Control:* Apply sulfur or Karathane when mildew is first evident. Repeat as necessary.
12. *Rusts*—Eastern states and Pacific Northwest near alternate hosts, common hemlock, Sitka and Norway spruces, or unknown. Small, bright yellow or orange to dark brownish, powdery pustules on underside of leaves. Where severe, plants may be defoliated. *Control:* Avoid planting near alternate hosts. Spray as for Leaf Spots (above). Use ferbam, maneb, or zineb. Or use Acti-dione-Thiram following manufacturer's directions.
13. *Crown Gall*—Uncommon. Stem base of young plants is swollen and later corky. Found mostly on lower stems of older plants. See (30) Crown Gall under General Diseases.
14. *Verticillium Wilt*—Uncommon. See under Maple, and (15B) Verticillium Wilt under General Diseases.
15. *Sooty Mold*—Black, moldy growth on foliage. Common on shrubs under trees attacked by aphids, scales, mealybugs, and other insects. *Control:* Apply malathion plus lindane or DDT to control insects.
16. *Felt Fungus*—Deep south. Smooth, shiny, chocolate-brown to nearly black growth on foliage. See under Hackberry. Azaleas may be injured.
17. *Thread Blights*—Southeastern states. See under Walnut.

For additional information, read USDA H & G Bulletin 71, *Growing Azaleas and Rhododendrons.*

RHODORA — See Rhododendron

RHODOTYPOS — See Jetbead

RHOEA, PURPLELEAF SPIDERWORT, OYSTER-PLANT, MOSES-IN-A-BOAT *(Rhoea)*

1. *Root Rot, Crown Rot*—See under African-violet. Avoid overwatering.
2. *Root-knot*—See under African-violet, and (37) Root-knot under General Diseases.
3. *Mosaic*—See (16) Mosaic under General Diseases.

RHUBARB *(Rheum)*

1. *Root and Crown (Foot) Rots, Southern Blight, Damping-off*—Widespread. Leaves may yellow, wilt, and collapse from rot of stalk bases, crown, and roots. Brownish-black streaks may occur in lower ends of stalks. Mold may cover affected tissues in damp weather. Often associated with nematodes (e.g., cyst, dagger, root-knot, spear, spiral, stem-rot, stunt). *Control:* Grow healthy roots from disease-free fields or beds in deep, fertile, well-drained, clean soil high in organic matter. Or soil fumigated with formaldehyde, SMDC, methyl bromide, or chloropicrin (pages 538–548). Spray crowns very early in spring and again after harvest. Use fixed copper or bordeaux mixture (8-8-50). Dig up and destroy all infected plants together with 6 inches of surrounding soil. Five- to 6-year rotation. Avoid a wet mulch.
2. *Leaf and Stalk Spots and Blights, Anthracnose, Gray-mold Blight*—Widespread.

Round to irregular, variously colored spots on leaves and stalks. Spots may enlarge and blight leaf or fall out leaving ragged shot-holes. Leaves may wilt, wither, and die. Powdery gray mold may form on affected parts in damp weather. *Control:* Collect and burn tops in late fall. During harvest pick stems with spotted leaves first. If serious, spray new growth in spring with zineb, captan, thiram, maneb, folpet, or fixed copper. Apply at 7- to 10-day intervals during damp weather. Avoid applications from 10 days before harvest until cutting is complete. Apply fertilizer in spring and again after harvest. Varieties differ in resistance. Set out disease-free roots in an area where rhubarb has not grown for at least 3 years. Avoid overcrowding. Harvest carefully to avoid bruises and other injuries.

3. *Stalk Rots, Bacterial Soft Rots*—Soft and brown or slimy, foul-smelling rot of crown and stalks in field or after harvest. Often associated with nematodes. *Control:* Same as for Leaf Spots (above). Avoid wounding stalk bases. Control rhubarb curculio by applying chlordane to soil in and around rhubarb planting before growth starts in spring.
4. *Ringspots, Mosaic*—Pale yellowish to dead spots or rings on leaves, usually with pale green mottle. Young leaves show well-defined to severe light and dark green mosaic mottle. Leaves often crinkled and distorted (Figure 4.200). Symptoms may disappear in hot weather. *Control:* Dig up and destroy infected plants. Grow virus-free stock. Control aphids that transmit viruses. Use lindane, malathion, or Dibrom (naled).

Fig. 4.200. Rhubarb mosaic. (Illinois Agricultural Experiment Station photo)

5. *Curly-top*—Western states. No very characteristic symptoms are formed. See (19) Curly-top under General Diseases. *Control:* Same as for Ringspots (above). Virus is spread by leafhoppers.
6. *Downy Mildew*—Small to large, brown spots on upper leaf surface. Whitish to violet-colored mold appears on corresponding underleaf surface in cool, damp weather. *Control:* Grow healthy roots in soil that has not grown rhubarb for at least 3 years. Spraying may be needed in cool, wet weather. Apply zineb, fixed copper, or bordeaux mixture (4-4-50) at weekly intervals. Start when leaves begin to expand.

7. *Rust*—Uncommon. Large, carmine-red spots on upper leaf surface and tiny, whitish, cluster cups on underleaf surface. Alternate host: common reed grass *(Phragmites)* that grows in wet areas. *Control:* Same as for Leaf Spots (above).
8. *Verticillium Wilt*—Plants gradually wilt, wither, and die. See (15B) Verticillium Wilt under General Diseases.
9. *Bacterial Wilt, Southern Wilt*—See (15C) Bacterial Wilt under General Diseases.
10. *Root-knot and Cyst Nematode*—See (37) Root-knot under General Diseases.
11. *Crown Gall*—See (30) Crown Gall under General Diseases.
12. *Cracked Stem, Boron Deficiency*—See under Celery and page 27.

RHUS — See Sumac
RIBES — See Currant
RIBBON GRASS, CANARYGRASS *(Phalaris)*

1. *Leaf Spots, Leaf Molds*—Spots of various sizes, shapes, and colors. May later be covered with mold. Leaves may wither and die early. *Control:* Where feasible, cut off and burn severely infected leaves. Space plants. Spray weekly during rainy periods. Use zineb or maneb plus spreader-sticker.
2. *Rusts*—Yellowish-orange or rusty-brown, powdery pustules on leaves. May turn dark brown to black as leaves mature. *Control:* Same as for Leaf Spots (above). Start a week before rust normally appears.
3. *Ergot*—Horny, purple-black ergot bodies replace normal kernels. *Control:* Snip off and burn heads containing ergots.

RICINIS — See Castorbean
RIVINA — See Rougeplant
ROBINIA — See Honeylocust
ROBIN'S-PLANTAIN — See Chrysanthemum
ROCHEA — See Sedum
ROCKCRESS, ROCKET — See Cabbage
ROCKJASMINE — See Primrose
ROCK ROSE — See Sunrose
ROCKSPIREA — See Holodiscus
ROCKY MOUNTAIN GARLAND — See Fuchsia
ROLLINIA — See Pawpaw
ROMANZOFFIA — See Phacelia
ROSARYPEA — See Pea
ROSE [ALBA, APPLE, AUSTRIAN BRIER, BANKS, BOURBON, BRIER, BRISTLY or GLOSSY-LEAVED, BURNETT, BURR or ROSBURGH, CABBAGE, CAROLINA, CHEROKEE, CHINA or BENGAL, CINNAMON, DAMASK, DOG, EVERGREEN, FAIRY, FLORIBUNDA, GRANDIFLORA, HYBRID PERPETUAL or REMONTANT, HUGO ROSE or GOLDEN ROSE OF CHINA, HIMALAYAN MUSK, HYBRID TEA, JAPANESE, KOREAN, MACARTNEY, MEADOW, MINIATURE, MULTIFLORA, MUSK, NOISETTE, PASTURE, PERPETUAL BRIER or RUGOSA, POLYANTHA, PRAIRIE, PRIMROSE, PROVENCE or FRENCH, RAMBLER, RED-LEAFED, SCOTCH, SETIGERA HYBRIDS, SHRUB, SWAMP, SWEETBRIER or EGLANTINE, TEA, TREE or STANDARD, TURKESTAN, VIRGINIA, WICHURAIANA or MEMORIAL] *(Rosa)*; AGRIMONY *(Agrimonia)*; GOATSBEARD *(Aruncus)*; COWANIA, CLIFFROSE or QUININE BUSH *(Cowania)*; MOCK-STRAWBERRY or INDIAN STRAWBERRY *(Duchesnea)*; MEADOWSWEET, QUEEN-OF-THE-MEADOW, QUEEN-OF-THE-PRAIRIE, DROPWORT

(*Filipendula*); AVENS [PURPLE or WATER, WHITE], PRAIRIE SMOKE (*Geum*); OSOBERRY (*Osmaronia*); CINQUEFOIL (*Potentilla*); BURNET [AMERICAN, JAPANESE, SALAD, SITKA] (*Sanguisorba*)

1. *Rose Blackspot*—General and serious on susceptible varieties and species. Roundish black spots with irregular or frayed margins on leaves. See Figure 2.17C under General Diseases and jacket cover. Small black or purplish-red spots form on young shoots and petioles. Infected leaves often yellow and fall early, weakening plants. Defoliated plants more susceptible to winter injury, drought, Dieback, and Stem Cankers (see below). Blooming is reduced. *Control:* Buy best-quality, disease-free plants from reputable nursery. Prune and burn old canes in fall and *before* growth starts in spring. Indoors keep water off foliage and humidity down. Do not handle plants when foliage is wet. Avoid wetting foliage late in day. *Rose* varieties and species differ greatly in resistance. Check with your local nurseryman, rose grower, or extension horticulturist. Apply dormant spray of lime-sulfur (1 part in 9 parts of water) before growth starts in spring. Apply folpet, captan, zineb, maneb, Daconil 2787, ferbam, Polyram, or Dithane M-45, plus wetting agent or spreader-sticker, weekly throughout season. Spraying is more effective than dusting. Cover underside of leaves thoroughly. Collect and burn fallen leaves where practical. Space plants. Mulch plants throughout growing season. Control insects and mites. Use mixture of DDT, sevin, or methoxychlor, plus Kelthane and malathion. These materials may be safely mixed with fungicides listed. Fertilize plants based on a soil test. Keep plants growing vigorously.
2. *Powdery Mildews*—General and serious in warm (64° to 75° F.), humid weather with cool nights. Whitish-gray, powdery or mealy coating on leaves, flower buds, and young stems. Causes stunting, reduced vigor, and poor blooming. Cane tips and flower buds may be dwarfed, distorted, and killed. Leaves curl, turn reddish or purplish, wither, and drop early. Most *rose* climbers, small-flowered ramblers, some floribundas and hybrid teas are very susceptible. See Figure 2.23B under General Diseases. *Control:* Grow resistant *rose* varieties and species (with leathery, glossy leaves, e.g., wichuraiana hybrids), where possible. Check with a local nurseryman, top rose grower, or your extension horticulturist. Space and prune plants properly. Apply same dormant spray as for Blackspot (above). Apply folpet, sulfur, or Karathane, plus a wetting agent, with Blackspot sprays or dusts. Or use Acti-dione or Acti-dione-Captan mixture following manufacturer's directions. *Thorough coverage* is essential. Do *not* apply sulfur, Karathane, or Acti-dione if temperature is above 85° F. Avoid overfertilizing, especially with nitrogen, and crowding plants. Indoors avoid cold drafts, sudden temperature changes, and increase ventilation to keep moist air moving up and out. Paint steam pipes with sulfur (or vaporize sulfur).
3. *Rose Stem Cankers, Dieback, Cane Blight, Spot Anthracnose*—General and serious where plants weakened by Black Spot, winter injury, poor nutrition, or neglect. Stems die back from pruning cuts, graft unions, flower stalk stubs, and broken thorns due to tan, brown, reddish-brown, black, girdling cankers. Cankers may start as small white, red, dark reddish, pale yellow, or purple-gray spots. Several cankers develop cracks and are sprinkled with black dots, fruiting bodies of the causal fungi. Cankers often girdle stems causing foliage beyond to yellow, wilt, and die. Entire plant may die. See Figure 2.40B under General Diseases. Hybrid teas are more commonly infected than other roses. *Control:* Grow only highest quality, disease-free plants from reputable nursery. "Cut-rate" plants often infected. Prune and burn cankered canes when found. Make clean cuts 3 to 4 inches behind canker but close to a bud. See Figure 2.4. Dip or swab pruning shears or knife in 70 per cent alcohol or formaldehyde (1 ounce in 2 gallons) between cuts. Spray as for Blackspot (above). Keep plants in healthy, actively-growing condition. Cut off old

flowers. Do not fertilize plants late in season. Indoors grow in sterilized soil (pages 538–548). Protect plants for winter following local recommendations. This may mean mounding the soil 8 to 12 inches around the canes plus covering with hay or straw or covering entire plant with several inches of soil. *Manetti* rootstock is highly resistant to Stem and Graft Canker (*Leptosphaeria*).

4. *Rose Crown Gall, Stem Gall, and Hairy Root*—General. Small to large, hard, rough galls or overgrowths on roots, crown, and canes. Usually near soil line or graft union. Galls may rot and slough off. Plants often stunted, lack vigor, may die. Flowering is reduced. See Figure 2.46B under General Diseases. Hairy Root causes production of a dense mass of small, fibrous roots (1 to 10 inches long) giving a witches'-broom effect. May arise from swellings at wounds. Often associated with lesion nematodes. *Control:* Carefully remove and burn infected plant parts. Grow certified, disease-free nursery stock. Do not replant in same area within 3 years without treating soil with steam, D-D, chloropicrin, or methyl bromide (pages 538–548). Do not wound canes or roots. Dip fresh cutting wood in 0.5 per cent calcium hypochlorite (household bleach) for 15 to 20 minutes plus sanitary precautions in handling cuttings. Wrap all grafts with antiseptic tape. Vancomycin gives excellent control as a dip for bare-root plants. Follow manufacturer's directions. Disinfest benches, boxes, sacks, and tools with commercial formaldehyde, 1 part in 50 of water. Practice strict sanitation.
5. *Rusts*—General. Small, yellow, bright orange, reddish, or orange-brown, powdery pustules on petioles, young canes, underleaf surface, buds, and fruits. Pustules later turn dark brown and finally black (Figure 4.201). Leaves may wilt, wither, and drop early reducing plant vigor. Most hybrid teas and climbing roses are susceptible. *Control:* Same as for Blackspot (above). Spray with maneb, zineb, ferbam, Dithane M-45, Daconil 2787, Polyram, sulfur, or Acti-dione. Grow resistant *rose* varieties and species. Prune and burn infected canes that are elongated, gall-like, or show black and crusty areas. Where feasible, destroy nearby wild roses.

Fig. 4.201. Rose rust.

6. *Winter Injury* (primarily rose)—Canes die back from tips. Plants may be entirely killed. Varieties and species differ greatly in susceptibility. *Control:* Grow hardy varieties and species recommended for your area. Protect for winter following recommended local practices. This may mean piling a cone of soil 8 to 12 inches deep around base of plants plus adding a loose covering of pine twigs, clean straw or hay, dry seaweed, or coarse sacking (Figure 4.202). If in doubt, check with your nurseryman, a successful rose grower, your county agent, or extension horticulturist. Keep plants healthy and vigorous during growing season. Water thoroughly during prolonged summer droughts and in late fall.

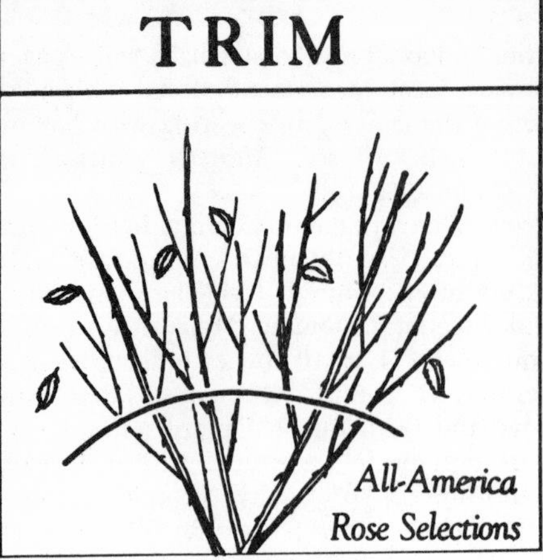

Fig. 4.202. Winter protection for roses (in climates where temperatures go below 15°F.). 1. *Trim:* After a hard frost or two, cut back canes on hybrid teas, grandifloras, and floribundas to 18 inches. Remove and burn all dead wood. Long canes left on the bush will wind-whip, become laden with snow, and thus loosen the protective mound around the plant. 2. *Clean and Spray:* Fall sanitation around bushes minimizes carryover infections into spring. Remove all fallen leaves. Give one last spray with a good fungicide-insecticide mixture to entire plant and surrounding soil. 3. *Mound:* Form cone of loose soil, 8 to 12 inches high, around canes of each plant. Use soil from another location rather than from between roses. In very cold climates (below 0°F.), add a loose covering of evergreen boughs, hay, or straw to top of mound after the soil freezes. Carefully remove mounds in spring when new growth starts.

Fig. 4.203. Rose mosaic.

7. *Fungus Leaf Spots, Leaf Blotch, Leaf Scorch, Spot Anthracnose*—Widespread. Spots of various sizes, shapes, and colors. Often with distinct border. Spots may drop out leaving shot-holes. Leaves may be distorted and ragged. Plants may be defoliated and weakened. *Control:* Same as for Blackspot (above).
8. *Gray-mold Blight, Dieback, Botrytis Blight, Blossom Blight, Bud Rot, Storage Decay*—Cosmopolitan. Buds turn brown and are blasted. May fail to open. Round, tan to light brown spots develop mostly on outer flower petals. Sunken, grayish-black cankers may grow down stems from infected buds or pruning cuts. Shoots die back. Canes turn brown and soften in cold storage. In damp weather grayish-brown mold may grow on infected tissues. *Control:* Same as for Blackspot (above). Carefully collect and burn infected buds, blossoms, and stems. Nurserymen commonly dust, dip, or spray stored roses with PCNB, captan, or Botran (dichloran). Follow manufacturer's directions. Nurserymen pack roses in boxes lined with polyethylene-coated Kraft paper. *Rose* varieties and species differ in susceptibility. Mist blooms lightly with Botran, Daconil 2787, zineb, or captan.
9. *Bacterial Blight or Blast*—Blackish-brown, sunken spots and streaks appear on petioles, calyx lobes, flower stalks, and flower receptacles. Flower buds die without opening. Follows cold, wet spring weather. *Control:* Prune and burn infected parts.
10. *Verticillium Wilt* (primarily rose)—Individual canes or entire plants gradually or suddenly wilt near blooming time. brownish to purplish streaks occur inside canes at base. Leaves on infected canes may yellow, wither, and drop early. Diseased plants may die gradually over several years. *Control:* Dig up and burn infected plants. Use disease-free budwood or plants. Grow in clean, light, well-drained soil. Avoid wounding roots, crowns, or canes and replanting in same area for 5 or 6 years without fumigating soil with chloropicrin, SMDC (Vapam or VPM Soil Fumigant), Vorlex, etc. See pages 538–548. *Manetti* rootstock is highly resistant.
11. *Crown and Root Rots*—Plants gradually decline in vigor and die. May wilt at midday or drop lower leaves. White fans of fungus growth often form between bark and wood. *Control:* Grow disease-free stock in sterilized soil.
12. *Rose Mosaics, Streak, Rosette, Infectious Chlorosis*—Widespread. Mostly on greenhouse roses. Symptoms differ greatly between varieties and species depending on virus or virus strain and seasonal conditions. Many varieties show no symptoms, especially outdoors. Ring, oakleaf, or wavy watermark patterns may develop in some leaves. Irregular, yellow or brown to reddish blotches and patterns may follow leaf veins. Leaves often distorted. Plants may be stunted, less vigorous, and produce few flowers (Figure 4.203). Common Rose Mosaic virus may be same as Prunus Ringspot virus (page 388). *Control:* Do *not* use diseased plants for propagating (budding, grafting). If practical, destroy infected plants. Replant with virus-free stock.
13. *Chlorosis*—Leaves turn pale green or yellow to ivory-colored with leaf veins remaining green until last (see Figure 4.37). Common in alkaline or very acid soils. *Control:* If soil pH is 7 or above, add sulfur, ammonium sulfate, or acid peatmoss (page 24). If soil is very acid, apply agricultural lime following a soil test. The pH should be slightly acid (5.5 to 6.5). Work iron sulfate (Copperas) into the soil using 1 to 4 ounces per square foot or use iron chelate (page 27). Follow manufacturer's directions. Water the iron sulfate or iron chelate in well. Or spray foliage with iron sulfate or chelate. Repeat as necessary.
14. *Root-knot*—Plants stunted; may die prematurely. Produce inferior blooms. Plants weak with small pale leaves. Small galls or knots found on roots. Somewhat similar root galls are produced by dagger nematodes (*Xiphinema* spp.). *Control:* Grow disease-free roses in clean or sterilized soil (pages 538–548). Keep plants vigorous. Fertilize, water, and protect plants properly for winter. *Rose* rootstocks vary greatly in resistance. Remove and destroy heavily infested plants. Nurserymen wash *rose* roots and soak in hot water (121° to 123° F.) for 10 minutes. Cool and dry.

Treatment also kills dagger, lesion, spiral, and probably other nematodes. DBCP (Nemagon or Fumazone) may be used as a soil drench around growing plants. Follow manufacturer's directions.

15. *Lesion and Other Root-Feeding Nematodes* (e.g., dagger, lance, pin, ring, sheath, spiral, stem, sting, stubby-root, stunt, tylenchus)—General. Plants stunted, slowly decline, may die back. Leaves pale or yellowish. Root system stunted with roots showing brown to black areas. May be stubby and "brushlike," with few feeding roots. May be associated with Hairy Root and Verticillium Wilt (both above). *Control:* Same as for Crown Gall and Root-knot (both above). Set plants in fumigated soil. Resistant *roses* may be available in the future.
16. *Rose Black Mold*—Primarily greenhouse and nursery disease. Serious on certain varieties (e.g., Manetti rootstock, Dr. Huey, and *Rosa odorata*) in grafting cases and in nurseries where plants are budded. Newly infected grafts (stock and scion), bruised areas, or bud unions first covered with frosty-white or grayish mold that gradually turns into a black crust. Wood at grafts is discolored. Grafts and bud unions do not take. *Control:* Use only disease-free stock for grafting and budding, or soak 2 hours in a formaldehyde solution (1 part in 320 parts of water). Use resistant or immune rootstocks where possible (e.g., Ragged Robin and *Rosa multiflora*). Grow in clean soil. Three-year rotation. Practice strict sanitation. Disinfect pots, soil, and cases with steam or formaldehyde before using.
17. *Downy Mildew* (agrimony, avens, mock-strawberry, rose)—Primarily indoor problem of rose cuttings. Irregular, yellow-gray, red to dark brown, or purplish spots on upper leaf surface with whitish-gray, downy mold on corresponding underside in damp weather. Leaves rapidly yellow, wither, and drop. Flowers may be slow or unmarketable. *Control:* Same as for Blackspot (above). Indoors, keep humidity below 85 per cent and water off foliage. Increase air circulation. Zineb or maneb sprays at weekly intervals should be used on outdoor plants.
18. *Rose Pedicel Necrosis or Topple*—Eastern states, mainly on bush roses. Bud and cane tips hang downward due to black or dark-purple shriveling of the stem. *Control:* Avoid extremes in soil moisture. Maintain balanced soil fertility based on a soil test. Keep potash level up. Mulch plants in dry weather.
19. *2,4-D Injury*—Roses are very susceptible. Leaves fernlike and twisted. See under Grape.
20. *Thread Blight* (rose)—Southeastern states. Plants may be defoliated. See under Walnut. *Control:* Same as for Blackspot (above).
21. *Crown Rot, Southern Blight*—See (21) Crown Rot under General Diseases.
22. *Aster Yellows* (avens)—See (18) Yellows under General Diseases.
23. *Fire Blight* (avens, cinquefoil, goatsbeard, rose)—See under Apple.
24. *Leaf Smut* (avens)—See (11) Smut under General Diseases.
25. *Boron deficiency* (primarily greenhouse roses)—Resembles hormone-type (2,4-D) weed-killer injury. Leaves variously distorted and elongated with irregular leaf margins. Flowering stems are multiple-branched. Also curved and distorted. Flowers aborted, darker in color than normal with irregular margins and irregularly pigmented. Tips of stems and flower shoots may die. *Control:* Apply spray containing boron (e.g., Solubor). See page 27. Or apply to soil and work in well. Repeat 2 months later and mulch with organic matter. Or use fertilizer containing boron.

For additional information on rose care, read USDA H & G Bulletin 25, *Roses for the Home.*

ROSE-ACACIA — See Honeylocust

ROSE-APPLE — See Eucalyptus

ROSEBAY — See Rhododendron

ROSELLE, ROSE-OF-SHARON, ROSEMALLOW — See Hollyhock

ROSEMARY — See Salvia and Labrador-tea

ROSE-MOSS, SUN-PLANT, RAINBOW-BUSH *(Portulaca)*

1. *White-rust*—Branches and leaves swollen and distorted with white, powdery pustules. Shoots may be spindly and erect. *Control:* Destroy infected plant parts. Keep down cruciferous weeds. See (9) White-rust under General Diseases.
2. *Damping-off, Seed Rot*—See under Beet, and (21) Crown Rot under General Diseases. Sow seed in warm, well-drained soil in full sun. Avoid overwatering.
3. *Root-knot*—See (37) Root-knot under General Diseases.
4. *Curly-top*—Western states. See (19) Curly-top under General Diseases.

ROSEOCACTUS — See Cacti

ROSE-OF-CHINA — See Hollyhock

ROSE-OF-HEAVEN — See Carnation

ROSE TREE OF CHINA — See Peach

ROSINWEED — See Chrysanthemum

ROSMARINUS — See Salvia

ROUGEPLANT *(Rivina)*

1. *Leaf Spots*—Spots of various sizes, shapes, and colors. *Control:* Pick and burn spotted leaves. If practical, spray at 10-day intervals during rainy periods. Use ferbam, zineb, or maneb. Indoors keep water off foliage. Space plants.
2. *Rust*—Small, orange-yellow spots on foliage. Pustules later may become powdery and reddish-brown to dark chocolate-brown. If severe, leaves may wither and die early. *Control:* Same as for Leaf Spots.
3. *Root Rots*—See under Geranium, and (34) Root Rot under General Diseases. May be associated with nematodes (e.g., burrowing).

ROWANTREE — See Apple

ROYSTONEA — See Palms

RUBBER-PLANT — See Fig

RUBUS — See Raspberry

RUBY GLOW — See Spirea

RUDBECKIA — See Chrysanthemum

RUE-ANEMONE — See Anemone

RUELLIA — See Clockvine

RUSSIAN-OLIVE or OLEASTER, SILVERBERRY, ELAEAGNUS [AUTUMN, CHERRY, THORNY], *(Elaeagnus)*; BUFFALOBERRY [RUSSET, SILVER] *(Shepherdia)*

1. *Leaf Spots*—Spots of various colors, sizes, and shapes. *Control:* If serious, apply zineb or maneb at 10- to 14-day intervals during wet, spring weather.
2. *Verticillium Wilt*—Leaves on one or more branches yellow, wither, and drop early. Brown discoloration may occur in wood just under bark. *Control:* See under Maple.
3. *Trunk Canker*—Oval to elongated, sunken cankers that girdle stems. May kill trees. Affected wood under bark turns brown or black. Gummy brown sap may appear at margins of some cankers. *Control:* Keep trees vigorous and free of wounds. Carefully cut out cankers and all discolored wood. Disinfect wound with household bleach and paint with tree wound dressing (page 36).
4. *Twig and Branch Cankers, Diebacks*—See under Maple.
5. *Rusts*—Yellowish spots on leaves. Alternate hosts: *Carex, Calamagrostis,* or none. *Control:* Same as for Leaf Spots (above).
6. *Powdery Mildews* (buffaloberry, Russian-olive, silverberry)—Powdery, white

patches on leaves. *Control:* If serious, apply Karathane or sulfur twice, 10 days apart.

7. *Crown Gall, Hairy Root*—See under Apple, and (30) Crown Gall under General Diseases.
8. *Thread Blights*—Southeastern states. See under Walnut.
9. *Seedling Blights, Damping-off*—See under Pine.
10. *Root Rot*—See under Apple, and (34) Root Rot under General Diseases.
11. *Wood Rot*—See under Birch, and (23) Wood Rot under General Diseases.
12. *Mistletoe* (elaeagnus) —See (39) Mistletoe under General Diseases.
13. *2,4-D Injury*—Russian-olive is quite susceptible. Leaves become thickened, may twist or roll, become boat-shaped or curled into "ram's-horns." *Control:* See under Grape.
14. *Chlorosis*—Mineral deficiency in alkaline soils. See under Maple and pages 24–28. *Control:* Have soil tested and treat as directed.

RUSTYLEAF — See Blueberry

RUTABAGA — See Cabbage

RYEGRASS — See Lawngrass

SABAL — Palms

SAFFLOWER — See Chrysanthemum

SAGE — See Salvia

SAGUARO — See Cacti

ST.-ANDREWS-CROSS — See St.-Johns-wort

ST. AUGUSTINEGRASS — See Lawngrass

ST.-JOHNS-FIRE — See Salvia

ST.-JOHNS-WORT [BUCKLEY, BUSHY, KALM, MARSH, SHRUBBY], AARONSBEARD, GOLDFLOWER, SUNSHINE SHRUB *(Hypericum);* ST.-ANDREWS-CROSS, ST.-PETERS-WORT *(Ascyrum)*

1. *Leaf and Stem Spots*—Spots of various colors, shapes, and sizes on leaves. Sometimes on stems and flower bracts. Causes little injury, but may be a nuisance. *Control:* Apply zineb or maneb during wet periods. Collect and burn fallen leaves.
2. *Rusts*—General over much of United States. Dusty pustules on leaves. Yellow to orange in spring; reddish-brown and finally black by fall. *Control:* If serious, same as for Leaf Spots.
3. *Powdery Mildew* (hypericum) —See (7) Powdery Mildew under General Diseases.
4. *Root Rot*—Plants wilt, wither, and die from rotting of roots and other underground parts. Diseased areas may be covered with dense white mold. *Control:* Grow in full sun in well-drained soil. Avoid overwatering and overfertilizing.
5. *Root-knot*—See (37) Root-knot under General Diseases.

SAINTPAULIA — See African-violet

ST.-PETERS-WORT — See St.-Johns-wort

SALAL — See Heath

SALIX — See Willow

SALMONBERRY — See Raspberry

SALPIGLOSSIS — See Tomato

SALSIFY — See Lettuce

SALVIA, SAGE [AUTUMN, BLUE, GARDEN, LYRE-LEAVED, PINEAPPLE, SCARLET, SHRUBBY BLUE, SILVER, TEXAS, VIOLET], ST.-JOHNS-FIRE, CLARY *(Salvia);* BASIL, BASILWEED *(Clinopodium);* COLEUS; DITTANY, STONEMINT *(Cunila);* DRAGONHEAD *(Dracocephalum);* HYSSOP *(Hyssopus);* LAVENDER [COMMON, ENGLISH] *(Lavandula);*

LIONS-EAR or LIONS-TAIL *(Leonotis)*; SWEET MARJORAM *(Majorana)*; HOREHOUND *(Marrubium)*; COMMON or LEMON BALM *(Melissa)*; MINT [BERGAMOT, CORN, CORSICAN, CREEPING, FIELD, HORSE, ORANGE], PEPPERMINT, PENNYROYAL, SPEARMINT *(Mentha)*; YERBA-BUENA *(Micromeria)*; BELLS OF IRELAND *(Molucella)*; BEEBALM or OSWEGO-TEA, HORSEMINT or SPOTTED MONARDA, LEMON MINT, WILDBERGAMOT *(Monarda)*; MONARDELLA; NEPETA, CATNIP, GROUND-IVY *(Nepeta)*; BASIL [AMERICAN, BUSH, SWEET] *(Ocimum)*; FALSE-DRAGONHEAD *(Physostegia)*; SELFHEAL, HEAL-ALL *(Prunella)*; MOUNTAIN-MINT *(Pycnanthemum)*; ROSEMARY *(Rosmarinus)*; SKULLCAP [HYSSOP, MAD-DOG, SHOWY] *(Scutellaria)*; BETONY, CHINESE ARTICHOKE, HEDGENETTLE, LAMBS-EARS, WOUNDWORT *(Stachys)*; GERMANDER [AMERICAN or WOOD-SAGE, SHRUBBY, TREE] *(Teucrium)*; THYME [COMMON, LEMON, WOOLLY], MOTHER-OF-THYME or CREEPING THYME *(Thymus)*

1. *Leaf Spots and Blights, Anthracnose, Spot Anthracnose, Tar Spot*—General in wet seasons. Spots of various sizes, shapes, and colors. Often with distinct margin. Some leaves may wither and drop early. Spots may occur on stem and rootstocks. *Control:* Pick and burn infected leaves. If practical, spray or dust during wet periods using zineb, maneb, Daconil 2787, or captan. Burn or cleanly turn under all tops in fall. *Mints* vary in resistance to Septoria Leaf Spot.
2. *Root-knot*—Coleus is highly susceptible. Plants may be stunted and weak with numerous, gall-like nodules on roots. Roots may be matted and "hairy." See (37) Root-knot under General Diseases.
3. *Rusts* (basilweed, dittany, germander, leonotis, mints, monarda or beebalm, monardella, mountain-mint, peppermint, physostegia, sage, salvia, spearmint, stachys, yerba-buena) —Lemon-yellow, yellow-orange, golden-brown, reddish-brown —later turn dark brown or black—dusty pustules on leaves, stem, and petioles. Leaves may wither and die early. Young *mint* shoots swollen, distorted, and twisted in spring and covered with lemon-yellow pustules. Many of these shoots do not survive to maturity. *Control:* Same as for Leaf Spots (above). Use zineb, maneb, dichlone, ferbam, Daconil 2787, or sulfur. Start *before* rust normally appears. Avoid overcrowding. *Dormant mint* roots (rhizomes or "runners") for forcing can be soaked in hot water (112° F.) for 10 minutes to destroy the rust fungus. Avoid overfertilizing with nitrogen early in spring. Rust-resistant *mint* varieties are available. Keep down weeds. Burn or cleanly turn under all tops in the fall.
4. *Crown Rots, Stem Rots, Southern Blight, Damping-off, Cutting Rots*—Seedlings, cuttings, or older plants wilt and collapse from rot at or near soil line. Stem base may be covered with cottony or gray mold. Cuttings discolored and rot at base. *Control:* Start seeds and cuttings in sterilized soil (pages 538–548). Avoid overwatering and overcrowding. Grow in well-drained soil. Where feasible, dig and burn infected plants and several inches of surrounding soil. Treat soil as for Delphinium, Stem Canker (page 254). Keep plants growing vigorously. Rotate 3 or 4 years with nonrelated plants.
5. *Root, Rhizome, and Stolon Rots*—Plants stunted, weak, may die. Roots, runners (stolons), and rhizomes discolored and decayed. Often associated with wounds, insects, and nematodes (e.g., dagger, lance, lesion, mint, needle, pin, potato-rot, ring, root-knot, sheath, spiral, stubby-root, stunt). *Control:* Same as for Crown Rots (above). If nematodes are a problem, apply DBCP before planting. Use disease-free planting stock.
6. *Mosaic* (primarily coleus and nepeta) —Symptoms differ with variety and species. Leaves mottled light and dark green, puckered, and crinkled. May show small dead

or yellow spots, splotches, ringspots, oakleaf, or other irregular markings. Leaves or entire plants may be stunted and distorted. *Control:* Destroy infected plants. Use disease-free stock or select cuttings from healthy plants.

7. *Verticillium Wilt* (coleus, mints, monarda, peppermint, spearmint) —A limiting factor in *peppermint* production. Plants stunted, yellowish to reddish or bronze; may be killed. Leaves turn yellow and drop prematurely, starting at base. Severity of Verticillium Wilt is increased by lesion nematodes. Nematode injury also speeds up symptom development. *Control:* Use wilt-free planting stock and grow in sterilized soil or where wilt has not occurred. For *mint* use Trizone, Vorlex, Telone, or chloropicrin (pages 538–548). Some *peppermint* hybrids are resistant.
8. *Blossom Blight, Gray-mold Blight, Leaf Blight*—Soft, brown spots on flowers, stems, and leaves. Stems may rot and flower clusters often collapse. Gray mold may grow on affected parts in cool, moist weather. *Control:* Cut and burn infected plant parts. Spray several times during cool, moist weather using captan or zineb.
9. *Powdery Mildews* (betony, mints, prunella, Oswego-tea, salvia, skullcap, spearmint, stachys) —Whitish to gray, powdery mold on leaves. If severe, leaves may yellow, wither, and drop early. *Control:* If serious, apply sulfur or Karathane at 7- to 10-day intervals. Sulfur may affect the flavor of mint.
10. *Leaf Nematode* (coleus, salvia) —Brown or blackish blotches on leaves, bordered by larger veins. Heavily infested leaves may die from base upward. *Control:* Collect and burn tops after harvest. Mulch plants. Grow in sterilized soil.
11. *Black Stem* (peppermint) —Dark brown to black, girdling cankers form on stems, usually at junction of lateral branches. Plant parts beyond cankers wilt and die. *Control:* Unknown. Spraying as for Leaf Spots (above) may be beneficial.
12. *Ringspot* (primarily spearmint) —Plants stunted and bushy with deformed leaves. Symptoms produced only on plants that develop from infected rhizomes. Dagger nematodes *(Xiphinema* spp.) apparently transmit the virus. Some plants show no symptoms. *Control:* Grow virus-free stolons in soil free of dagger nematodes (pages 538–548).
13. *Downy Mildew* (dragonhead, false-dragonhead, salvia) —See (6) Downy Mildew under General Diseases.
14. *Bacterial Leaf Spot* (catnip) —See (2) Bacterial Leaf Spot under General Diseases.
15. *Fusarium Wilt* (catnip) —See (15A) Fusarium Wilt under General Diseases.
16. *Aster Yellows, Spotted Wilt* (sage, salvia) —See (17) Spotted Wilt and (18) Yellows under General Diseases.

SAMBUCUS — See Snowberry
SANCHEZIA — See Clockvine
SANDMYRTLE — See Labrador-tea
SAND-VERBENA — See Four-o'clock
SANDWORT — See Carnation
SANGUINARIA — See Poppy
SANGUISORBA — See Rose

SANSEVIERIA, BOWSTRING HEMP, SNAKE PLANT [CEYLON, HAHN'S DWARF, VARIEGATED] *(Sansevieria)*

1. *Bacterial Soft Rots*—Softy, mushy, foul-smelling rot of leaves and rootstocks near soil line. Leaves become yellowish-green to pale yellow and collapse. *Control:* Avoid overwatering and overcrowding. Grow in sterilized soil (pages 538–548). Do not propagate from diseased plants.
2. *Leaf Spots*—More or less circular spots. Often with distinctive border. Spots may dry up and drop out (Figure 4.204). Leaves sometimes girdled and killed by fusing of several spots. *Control:* Cut out and destroy infected leaves. Indoors, keep water off foliage and humidity as low as practical. If needed, spray when spots first seen.

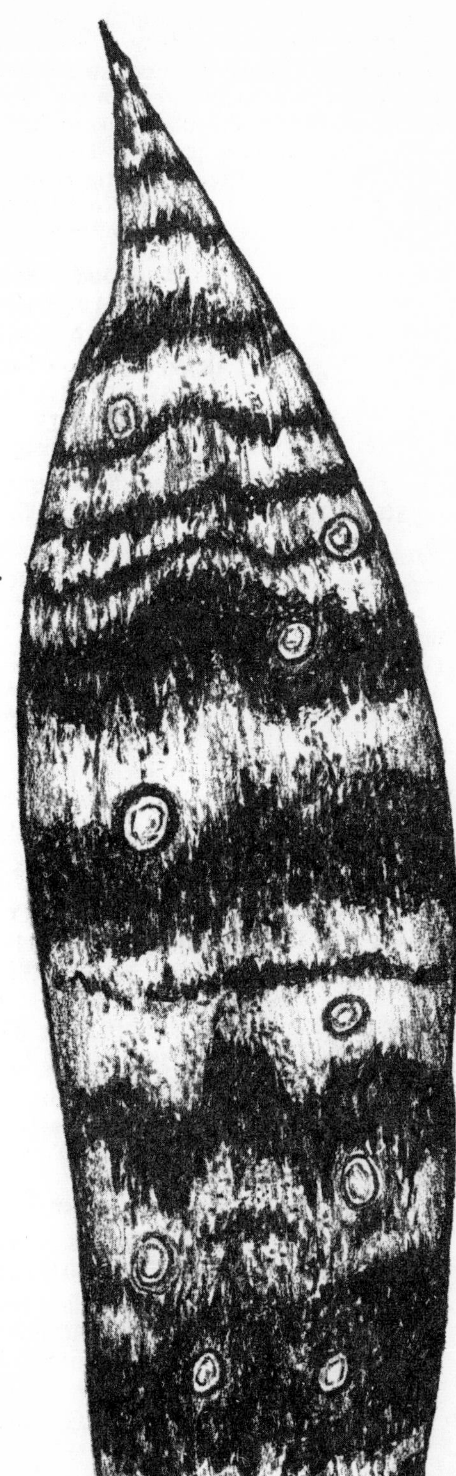

Fig. 4.204. Fusarium leaf spot of sansevieria.

For best results use zineb, captan, maneb, or fixed copper plus wetting agent.

3. *Root-knot, Lesion, and Other Root-feeding Nematodes* (e.g., lance, ring, spiral, stubby-root) —Commonly associated with weak, declining plants. See (37) Root-knot under General Diseases. *Control:* Soak bare-root plants in hot water (122° F.) for 10 minutes, or Zinophos (600 to 800 parts per million) for 30 minutes. Cool, then plant in clean or pasteurized soil.
4. *Wilt*—Leaves wilt from crown rot caused by high fertilizer level or toxic soluble salts (page 31). *Control:* Avoid overfertilizing. Flush out soluble salts with plenty of rain or tap water. Or repot in fresh soil.

SAPINDUS — See Soapberry

SAPIUM — See Castorbean

SARRACENIA — See Pitcherplant

SARSAPARILLA — See Acanthopanax

SASSAFRAS — See Avocado

SATIN-FLOWER — See Fuchsia

SAWARA-CYPRESS — See Juniper

SAXIFRAGE *(Saxifraga)* — See Hydrangea

SCABIOSA, PINCHUSION FLOWER, SWEET SCABIOUS *(Scabiosa)*

1. *Powdery Mildew*—Powdery, white coating on leaves. *Control:* Space plants. Dust or spray several times, 10 days apart. Use sulfur or Karathane.
2. *Leaf Spots, Shot-hole*—Spots of various sizes, shapes, and colors. Some may be zoned or drop out leaving shot-holes. Heavily spotted leaves wither and drop early. *Control:* Pick and burn affected parts. If serious, apply zineb or maneb at 10- to 14-day intervals. Start when first spots appear.
3. *Stem Rot, Crown Rot, Southern Blight*—See (21) Crown Rot under General Diseases.
4. *Curly-top*—Western states. See (19) Curly-top under General Diseases.
5. *Aster Yellows*—See (18) Yellows under General Diseases.
6. *Root Rots*—See under Geranium, and (34) Root Rot under General Diseases.
7. *Black Ringspot*—See under Cabbage.
8. *Mosaic*—See (16) Mosaic under General Diseases.

SCARBOROUGH-LILY — See Daffodil

SCARLET EGGPLANT — See Tomato

SCARLET PIMPERNEL — See Primrose

SCARLET PLUME — See Poinsettia

SCARLET RUNNER BEAN — See Bean

SCHEFFLERA, AUSTRALIAN UMBRELLA-TREE *(Schefflera)*

1. *Leaf Spots*—Small to large, round to irregular, brown spots and blotches. Infected leaves may wither and drop early. *Control:* Indoors keep water off foliage. Outside, spray during wet periods, using zineb, maneb, or captan to which wetting agent is added.
2. *Root-knot*—See (37) Root-knot under General Diseases.

SCHINUS — See Sumac

SCHIZANTHUS — See Tomato

SCHOLARTREE — See Honeylocust

SCHLUMBERGERA — See Cacti

SCIADOPITYS — See Pine

SCILLA — See Tulip

SCINDAPSUS — See Calla Lily

SCLEROCACTUS — See Cacti
SCORPIONWEED — See Phacelia
SCORZONERA — See Lettuce
SCOTCH BROOM — See Broom
SCREWPINE *(Pandanus)*

1. *Leaf Spots*—Small to large spots develop, working inward from leaf margin. *Control:* Prune off infected leaves and spray with fixed copper, zineb, or both. Indoors keep water off foliage. Destroy badly infected plants.
2. *Burrowing Nematode*—Associated with weak, declining plants. *Control:* See Root-knot under Peach.

SCURVYWEED, SEAKALE —See Cabbage
SCUTELLARIA — See Salvia
SEA HOLLY — See Celery
SEAKALE — See Cabbage
SEA-LAVENDER, STATICE *(Limonium)*; SEA-PINK, THRIFT *(Armeria)*

1. *Leaf Spots*—Spots of various sizes, shapes, and colors. Heavily spotted leaves drop early. *Control:* Pick and burn affected parts. If serious, apply zineb or maneb at 10- to 14-day intervals. Start when first spots appear.
2. *Gray-mold Blight, Botrytis Flower Blight*—Brown, rotting spots on shoots and flowers. Gray mold may cover affected parts in damp weather. *Control:* Same as for Leaf Spots (above).
3. *Rusts*—Not very common. Yellow, yellow-orange, reddish-brown, or black powdery pustules on both sides of leaves. *Control:* Destroy badly affected plants. Spray remainder as for Leaf Spots (above).
4. *Crown Rots* (statice) —Stem rots at or near soil line. Plants easily pulled up. Water-soaked spots, that later darken, occur on leaves and petioles where they touch soil. *Control:* Suggest soil treatment as for Cabbage Wirestem (page 188).
5. *Root Rot*—See under Geranium, and (34) Root Rot under General Diseases.
6. *Root-knot, Cyst Nematode*—See (37) Root-knot under General Diseases.
7. *Spotted Wilt, Ringspot*—See (17) Spotted Wilt under General Diseases.
8. *Aster Yellows*—See (18) Yellows under General Diseases.
9. *Powdery Mildew* (limonium) —Dense, white powdery mold on leaves; mostly underleaf surface. *Control:* Dust or spray with sulfur or Karathane. Repeat 7 to 10 days later.

SEA ONION — See Tulip
SEA-PINK — See Sea-lavender
SECHIUM — See Cucumber
SEDUM, BURRO'S-TAIL, CORAL BEADS, LIVEFOREVER, STONECROP, WALLPEPPER, WORMGRASS *(Sedum)*; CRASSULA, JADE PLANT, NECKLACE PLANT or VINE *(Crassula)*; ECHEVERIA; AIRPLANT, VELVET LEAF, PANDA PLANT *(Kalanchoë, Bryophyllum)*; ROCHEA; HOUSELEEKS, HEN-AND-CHICKENS *(Sempervivum)*

1. *Stem and Leaf Rot, Crown Rots, Southern Blight, Root Rot, Wilt*—Brown to black, water-soaked areas on stem; often at soil line. May extend upward into flower stalks and down into roots. Affected areas may be covered with dense cottony mold. Tops of plants soon wilt. Plants may die out in patches in warm, humid weather. Roots decay. *Control:* Set disease-free plants in clean, well-drained soil. Avoid overwatering, deep planting, overfertilizing, and packing soil too closely around crowns. Destroy infected plants or plant parts. Drench remainder with zineb (2 tablespoons

per gallon). Do not replant in same area without first sterilizing soil with heat or chemicals (pages 538–548). Spray weekly in cool, wet weather. Use zineb or captan.

2. *Rusts* (echeveria, hen-and-chickens, houseleek, liveforever, stonecrop) —Dark, powdery pustules on leaves. Center leaves of *houseleek* are unusually long, narrow, erect, and pale yellow. Small, dark, reddish-brown pustules break out on both leaf surfaces. Affected plants do not bloom. *Control:* Pull up and burn infected *houseleek* plants for they will not recover. Apply zineb, maneb, sulfur, or dichlone two or three times, 10 days apart, on neighboring plants or other hosts.
3. *Leaf Spots, Blotch, Anthracnose* (crassula, echeveria, kalanchoë, liveforever, rochea, stonecrop) —Round to irregular dark spots, blotches, or "scabs" on leaves. May have distinct margins. Similar spots may occur on flower petals. Leaves may drop quickly. *Control:* Collect and burn fallen leaves. Spray as for Rusts (above) when spots first appear. Repeat before rainy periods. Indoors, keep foliage dry.
4. *Powdery Mildew* (kalanchoë, sedum) —Grayish-white, powdery growth on leaves. Leaves may wither and drop early. *Control:* Indoors, keep humidity down and increase both light and air circulation. Avoid overfertilizing. Apply sulfur or Karathane twice, 10 days apart.
5. *Fusarium Wilt* (sedum) —California. Leaves turn yellow and drop, starting at base of plant. Plant may wither and die. *Control:* Same as for Stem and Leaf Rot (above).
6. *Mosaic* (kalanchoë) —Yellowish-green streaks, blotches, rings, and mosaic pattern on leaves. *Control:* Destroy affected plants. Do not propagate from diseased plants.
7. *Ringspot* (crassula) —Leaves stunted and distorted. Brown, ringlike dead areas may appear on underside of leaves. *Control:* Discard all plants showing stunting or ringspot symptoms. Propagate only from disease-free plants.
8. *Root-knot*—See (37) Root-knot under General Diseases.
9. *Crown Gall* (kalanchoë) —See (30) Crown Gall under General Diseases.
10. *Leaf Nematode* (crassula) —See (20) Leaf Nematode under General Diseases.

SELENICEREUS — See Cacti

SELF-HEAL — See Salvia

SEMPERVIVUM — See Sedum

SENECA — See Phlox

SENECIO — See Chrysanthemum

SENNA — See Honeylocust

SENSITIVE PLANT — See Pea

SEQUOIA — See Pine

SERENOA — See Palms

SERVICEBERRY, SERVICETREE, SHADBLOW, SHADBUSH — See Apple

SHALLON — See Heath

SHALLOT — See Onion

SHAMROCK — See Oxalis

SHASTA DAISY — See Chrysanthemum

SHEEP-LAUREL — See Blueberry

SHELL-FLOWER — See Gladiolus

SHEPHERDIA — See Russian-olive

SHOOTINGSTAR — See Primrose

SHORE COWBERRY — See Blueberry

SHORTIA — See Galax

SHRIMP-PLANT — See Clockvine

SHRUB-ALTHEA — See Hollyhock

SHRUB-YELLOWROOT — See Clematis

SICANA — See Cucumber
SICKLE THORN — See Asparagus
SIDA, SIDALCEA — See Hollyhock
SILENE — See Carnation
SILKGRASS — See Yucca
SILK-OAK *(Grevillea)*

1. *Dieback, Gum Disease*—See (22) Stem Blight under General Diseases. *Control:* Prune and burn dead or dying parts. Indoors keep water off foliage.
2. *Root Rot*—See (34) Root Rot under General Diseases. *Control:* Destroy badly infected plants. Sterilize potting soil and container before planting (pages 538–548). Avoid overwatering and overfertilizing. Grow in well-drained soil.
3. *Root-knot*—See (37) Root-knot under General Diseases. *Control:* Same as for Root Rot.
4. *Leaf Spot*—See (1) Fungus Leaf Spot under General Diseases.

SILKTASSEL-BUSH — See Dogwood
SILKTREE — See Honeylocust
SILPHIUM — See Chrysanthemum
SILVERBELL [CAROLINA, MOUNTAIN or TISSWOOD], SNOWDROP-TREE *(Halesia)*; SNOWBELL [AMERICAN, FRAGRANT, JAPANESE] *(Styrax)*

1. *Leaf Spot*—Large, irregular, rusty-brown blotches. *Control:* See under Maple.
2. *Wood Rot*—See under Birch, and (23) Wood Rot under General Diseases.
3. *Root-knot*—See under Peach, and (37) Root-knot under General Diseases.

SILVERBERRY — See Russian-olive
SILVER KING — See Chrysanthemum
SILVER-LACEVINE, DWARF LACE PLANT, MOUNTAIN FLEECE, FLEECEFLOWER [HIMALAYAN, REYNOUTRIA] *(Polygonum)*

1. *Leaf Spots, Tar Spot*—More or less round, gray or black spots. Leaves may wither and drop early. *Control:* Pick and burn infected plant parts. If practical, spray during wet summer weather using zineb or maneb.
2. *Smuts*—Roundish pustules or galls on leaves and fruit that break open and release black powdery masses. *Control:* Pick and burn infected parts before pustules break open. Destroy plants that show smut each year.
3. *Rusts*—General. Small, brown, powdery pustules on underside of leaves and stems. Pustules later turn black. Alternate hosts include Geranium. *Control:* If serious, apply ferbam, zineb, or maneb several times. Start 2 weeks before rust normally appears.

SILVERROD — See Chrysanthemum
SILVER THREADS *(Fittonia)*

1. *Stem and Root Rots*—Stems and roots decay. Plants stunted and weak. Often wilt and collapse. *Control:* Start plants in sterilized soil (pages 538–548). Infected plants may sometimes be cured by immersing in hot water (120° to 124° F.) for 30 minutes. Use lower temperature for unhardened plants. After treating, divide plants or take cuttings and root in steamed mixture of Perlite and peat. Keep water off foliage.
2. *Leaf and Stem Blight*—Soft, water-soaked or scalded areas form on petioles and stems. Plants may wilt and die. *Control:* Same as for Stem and Root Rots.
3. *Leaf Spot*—See (1) Fungus Leaf Spot under General Diseases.
4. *Root-knot*—See (37) Root-knot under General Diseases.

SINNINGIA — See African-violet
SISYRINCHIUM — See Iris
SKULLCAP — See Salvia
SKY-FLOWER — See Clockvine
SKYROCKET — See Phlox
SLIPPERWORT — See Snapdragon
SMELOWSKIA — See Cabbage
"SMILAX" of FLORISTS — See Asparagus
SMOKETREE — See Sumac
SMOKETREE DALEA — See Broom
SNAKE LILY — See Brodiaea
SNAKE PLANT — See Sansevieria
SNAKEROOT — See Anemone

SNAPDRAGON *(Antirrhinum)*; SLIPPERWORT, POCKETBOOK PLANT *(Calceolaria)*; PAINTED-CUP, INDIAN PAINTBRUSH *(Castilleja)*; TURTLEHEAD [PINK, RED, WHITE] *(Chelone)*; COLLINSIA, BLUELIPS, BLUE-EYED-MARY, CHINESE HOUSES *(Collinsia)*; FOXGLOVE [COMMON, YELLOW] *(Digitalis)*; LINARIA, TOADFLAX or BUTTER-AND-EGGS, KENILWORTH IVY *(Linaria)*; MAURANDYA; MONKEYFLOWER *(Mimulus)*; PENSTEMON [COBAEA, EASTERN, FOXGLOVE, LARGE-FLOWERED, SMOOTH WHITE], BEARD-TONGUE, LAVENDER QUEEN *(Penstemon)*; SYNTHYRIS; TORENIA, WISHBONE FLOWER *(Torenia)*; MULLEIN [MOTH, OLYMPIC, PURPLE] *(Verbascum)*

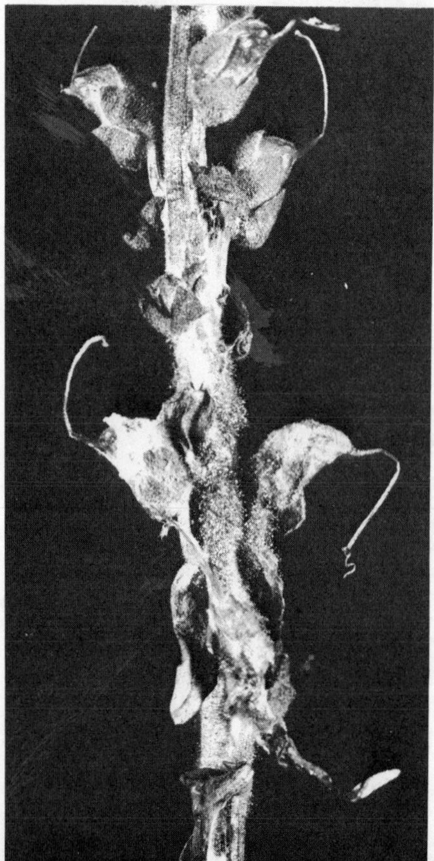

Fig. 4.205. Gray-mold Blight (*Botrytis*) on snapdragon spike. (Illinois Natural History Survey photo)

1. *Rusts* (collinsia, Indian paintbrush, monkeyflower, painted-cup, penstemon, snapdragon, synthyris, toadflax, turtlehead) —General and serious on snapdragon. Small, reddish-brown or chocolate-brown, powdery pustules on leaves (mostly underleaf surface), stems, and seedpods. *Penstemon* pustules are yellow. Infected parts may wilt, shrivel, and die. Small weak flowers are produced. See Figure 2.24A under General Diseases. Alternate hosts: Wild grasses, pines, none, or unknown. *Control:* Space plants. Keep down weeds. Collect and burn tops in fall. Plant disease-free seed, cuttings, or transplants. Use rust-resistant varieties of *snapdragon*. Indoors avoid sprinkling foliage. Keep humidity as low as practical by forced air circulation. Control insects with malathion. Apply zineb, ferbam, thiram, or maneb at 7- to 10-day intervals. Start when plants set in garden.
2. *Gray-mold Blight, Botrytis Blight*—Cosmopolitan in damp areas. Soft, tan to brown rot areas on leaves, stems, flowers, cuttings, and seedlings. Flower spike may wilt suddenly, collapse, and die. Grayish-brown mold may grow on affected areas (Figure 4.205). *Control:* Collect and burn tops in fall. Space plants. Keep down weeds. Cut flower spikes early. Indoors, control heat, increase ventilation, keep humidity below 85 per cent, and water off foliage. Apply captan, Botran, or zineb once or twice weekly in wet weather. Mix PCNB 75 (12 ounces per 100 square feet) into top 4 to 6 inches of soil a week before planting.
3. *Powdery Mildews*—General. Favored by damp, shaded conditions. White, powdery growth on leaves, stems, and flower petals. Leaves may die progressively upward starting at base. *Control:* Spray several times, 7 to 10 days apart. Use Karathane, Acti-dione, sulfur, or folpet plus wetting agent. Indoors avoid cold drafts and crowding plants.
4. *Damping-off, Southern Blight, Stem and Root Rots, Crown and Stem Canker, Wilt* —General. Seedlings yellow, wilt, and damp-off. Older plants may gradually or suddenly turn yellow, wilt, collapse, and die due to rotting of lower stem and/or roots. Water-soaked, white, yellow, or brown to black cankers or rotted areas may form at or near soil line. See Figure 4.206B. Other plants may be stunted and weak with decayed roots. Cottony mold may cover infected areas at soil line. Flower spikes pale and wilt. Flowers collapse. *Control:* Use disease-free seed or treat with Semesan and grow in clean or sterilized, well-drained soil (pages 538–548). Treat *foxglove* seed by soaking in hot water (131° F.) for 15 minutes. Avoid overwatering, deep planting, and overcrowding. Propagate from disease-free plants. Destroy infected plants and several inches of surrounding soil when first seen. Drench soil with Dexon following manufacturer's directions. Or PCNB 75 (12 ounces per 100 square feet) plus captan or ferbam (1 pound). Indoors, increase ventilation, keep water off foliage, and temperature below 60° F. Spray as for Rusts (above). Rotate.
5. *Anthracnose* (primarily foxglove, snapdragon, and toadflax) —Round to angular, enlarging, somewhat sunken, pale yellowish-green to grayish-white or purplish spots on leaves and stems. Numerous black specks may dot centers of older spots. Affected parts or entire plant may wilt, wither, and die. Seedlings wilt and collapse. *Control:* Same as for Rusts and Damping-off (both above). Pull and burn severely infected plants.
6. *Other Leaf Spots, Blights*—General. Small to large spots of various colors, sizes, and shapes. Phyllosticta spots develop concentric ridges (zoned). Centers may drop out leaving shot-holes. Leaves may appear scorched, wither, and cling to stem or drop early. Similar spots or cankers on stems may cause rapid wilting and dying (Figure 4.206B). *Control:* Same as for Anthracnose (above). Indoors, grow in sterilized soil and keep temperature as low as permits good growth.
7. *Verticillium Wilt* (foxglove, slipperwort, snapdragon) —Plants wilt slowly or suddenly, starting with certain branches. Wilting may be one-sided. Lower leaves turn yellow at margin. Wilt later progresses up stem. Greenish-brown to purplish-

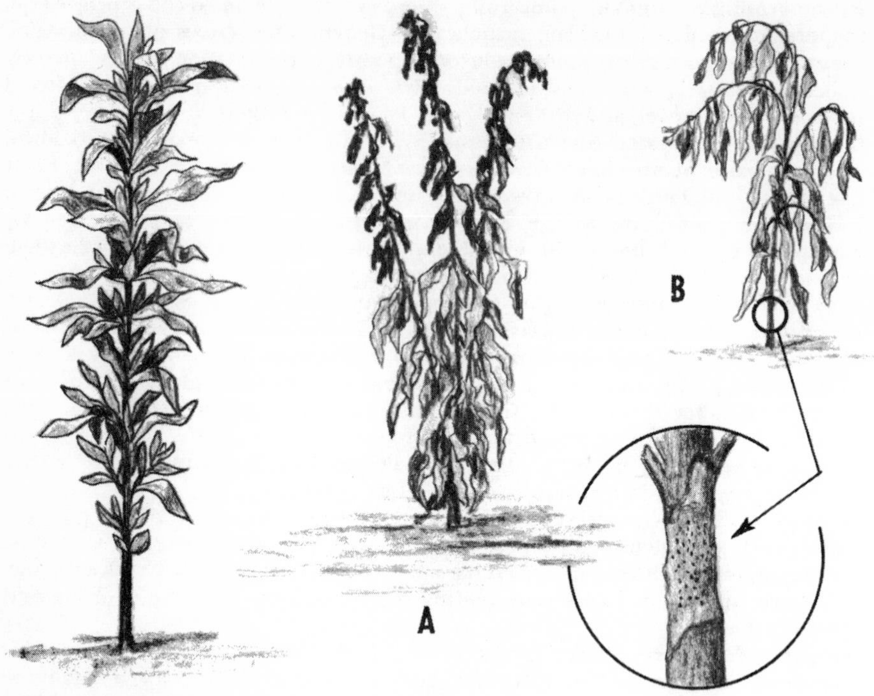

Fig. 4.206. A. Snapdragon Verticillium wilt. B. Phyllosticta blight or stem rot of snapdragon.

brown streaks inside stem near soil line (Figure 4.206A). *Control:* Rotate. Remove infected plants together with surrounding soil. Water as little as possible to obtain good growth. Indoors, grow in well-drained, clean or sterilized soil (pages 538–548). *Warning:* Do *not* use methyl bromide-chloropicrin combinations for soil fumigation.

8. *Downy Mildew* (primarily foxglove, mullein, snapdragon, and toadflax) —Seedlings and young plants stunted, bunchy, dull green, wilt, and die from top down. Older plants stunted and rosetted. Flowering is reduced. Purplish-gray to grayish-white, mealy, mold patches on underleaf surface and stems in overcast, cool, damp weather. Leaves may wither and die. *Snapdragons* may be systemically infected. *Control:* Same as for Rusts (above). No resistant varieties are available. Rotate. Indoors, keep night temperature above 52° F. Buy seed and transplants grown in disease-free localities.
9. *Mosaics* (calceolaria, foxglove, penstemon, snapdragon) —New leaves may be puckered, curled, and mottled with yellowish or light and dark green areas. Leaves and plants are stunted. *Control:* Destroy infected plants when first found. Malathion or lindane control aphids that transmit the viruses.
10. *Ringspot* (snapdragon) —Numerous zoned rings of alternating living and dead tissue. Whole spots die, then spread and merge with other spots. Whole leaf, then entire plant dies. *Control:* Same as for Mosaics (above).
11. *Spotted Wilt, Ringspot* (foxglove, penstemon, slipperwort, snapdragon) —Plants may be stunted with rosette-like growth. Leaves distorted. May show yellow, mosaic-like patterns. Flowers may be marked with pale red or yellow rings. *Control:* Same

as for Mosaics (above). Use DDT or sevin plus malathion to control virus-transmitting insects.

12. *Aster Yellows, Curly-top* (foxglove, monkeyflower, slipperwort, toadflax)—See (18) Yellows and (19) Curly-top under General Diseases. Leaves curled and dwarfed. Plants become stunted and bunchy. Leafhoppers transmit viruses. *Control:* Same as for Spotted Wilt (above).
13. *Root-knot*—Snapdragon is highly susceptible. See (37) Root-knot under General Diseases.
14. *Other Nematodes* (lesion, pin, potato-rot, stunt, tetylenchus)—Plants lack vigor. May be stunted with some dead lower leaves. Feeding roots stunted and pitted with minute, shallow brown wounds. Laboratory examination needed for positive identification. *Control:* Same as for Root-knot (above).
15. *Leaf and Stem Nematode* (butter-and-eggs, foxglove, monkeyflower, penstemon, slipperwort, toadflax)—Angular brown blotches on leaves limited by veins. *Penstemon* plants may be rosetted with stunted, unthrifty, deformed flower stalks. Few or no flowers produced. Infested plants may die early. Leaves curled, twisted, and brittle. Stems often twisted and reddish-brown where attacked. *Control:* Keep water off foliage on indoor plants. Outdoors, apply dry mulch to keep water from splashing on leaves. Pick and burn infested leaves. Burn tops in fall. Rotate.
16. *Crown Gall* (foxglove, snapdragon)—See (30) Crown Gall under General Diseases.
17. *Flower Blight*—See under Chrysanthemum, and (31) Flower Blight under General Diseases. *Control:* Same as for Gray-mold Blight (above).
18. *Snapdragon Tip Blight*—Indoor problem in cloudy, winter weather. Leaf tips wilt and die. Leaves in center of plant gradually wilt. Sunken, girdling, water-soaked spots on stem cause tip to wilt and die. *Control:* Varieties differ in susceptibility. Add lights where possible.
19. *Sooty Mold, Black Mildew* (penstemon)—Dark mold patches on foliage. See (12) Sooty Mold under General Diseases.
20. *White Smut* (butter-and-eggs, collinsia)—See (13) White Smut under General Diseases.
21. *Boron Deficiency* (calceolaria)—See page 27.
22. *Chlorosis*—Mineral deficiency in alkaline or very acid soils. See under Rose.

SNEEZEWEED, SNEEZEWORT — See Chrysanthemum
SNOWBALL — See Viburnum
SNOWBELL — See Silverbell
SNOWBERRY [COMMON or WHITE, GARDEN, MOUNTAIN, ROUND-LEAF, SPREADING], CORALBERRY or INDIANCURRANT, CHENAULT, WAXBERRY, WOLFBERRY *(Symphoricarpos)*; ABELIA [CHINESE, GLOSSY] *(Abelia)*; BUSH-HONEYSUCKLE *(Diervilla)*; HONEYSUCKLE [AMUR, BELLE, BLUELEAF, BOX, CLAVEY'S DWARF, COMMON or ITALIAN, CORALINE, EUROPEAN FLY-, EVERBLOOMING, FLY-, FRAGRANT or WINTER, GIANT BURMESE, HALL'S, HENRY, JAPANESE, JAPANESE PINK, LILAC, MANCHURIAN, MORROW, PRIVET, SAKHALIN, STANDISH, SWAMP FLY-, SWEETBERRY, TATARIAN, TRUMPET, TWINBERRY, YELLOW, ZABEL], WOODBINE *(Lonicera)*; ELDER [AMERICAN or SWEET, BLUEBERRY, CUTLEAF, EUROPEAN, EUROPEAN RED, GOLDEN, GOLDEN EUROPEAN, MEXICAN, RED-BERRIED, SCARLET] *(Sambucus)*; WEIGELA [CRIMSON, ROSE] *(Weigela)*

1. *Leaf Spots, Anthracnose, Scab or Spot Anthracnose*—Widespread. Spots of various sizes, shapes, and colors. Leaves may become distorted, wither, and drop early. Spots may also occur on stems, berries, and flowers. Berries decay, shrivel, and dry up. Twigs may die back several inches. *Control:* Destroy infected plant parts. Cut diseased *snowberry* stems to ground and burn in fall. Apply zineb, captan, ferbam, thiram, dichlone, or fixed copper at 7- to 10-day intervals from when buds swell to just before bloom. Phenyl mercury is effective against Anthracnose and Spot Anthracnose. Follow manufacturer's directions.
2. *Berry Rots* (snowberry, coralberry) —Widespread. Pink, purple, yellowish, tan, brown, or black spots on berries. Fruit may become soft, watery, rotted, and covered with gray or black mold. Or berries may shrivel and "mummify." *Control:* Spray as for Leaf Spots (above).
3. *Cankers, Twig Blights, Dieback*—Twigs die back. Often from rough, girdling cankers on twigs and branches sprinkled with coral-pink to black "pimples" (fungus fruiting bodies). *Control:* Prune and burn affected twigs and branches. Spray as for Leaf Spots (above).
4. *Honeysuckle Leaf Blight*—Widespread. Yellowish-green blotches in leaves that turn tan and then brownish-black. Whitish "bloom" often appears on underleaf surface. Leaves roll, twist, wither, and fall early (Figure 4.207). Only young leaves are infected. *Control:* Same as for Leaf Spots (above). Dwarf type may be resistant.

Fig. 4.207. Honeysuckle leaf blight (*Herpobasidium*).

5. *Rusts*—Yellowish or brownish pustules on upper leaf surface with orange "cluster cups" on underleaf surface. Alternate hosts: various wild and lawn grasses, or sedges (*Carex* spp.). *Control:* If serious, apply zineb or ferbam as for Leaf Spots (above).
6. *Collar or Crown Rot, Trunk Canker, Wood or Heart Rots*—Plants lack vigor. Die back from rot at soil line. A canker may form on trunk. *Control:* See under Dogwood, and (23) Wood Rot under General Diseases.
7. *Powdery Mildews*—General. White, powdery mold on leaves and stem tips. Tips of shoots may die back. *Snowberry* leaves may be distorted. *Control:* Spray with sulfur, Karathane, or Acti-dione several times, 10 days apart.
8. *Crown Gall and Hairy Root*—Rough, irregular, swollen, cauliflower-like galls on stem at or near soil line. Plants may appear unthrifty. *Control:* Dig up and burn affected plants. Do not replant in same soil for 4 years. Avoid injuring plants when cultivating or mowing. Grow disease-free stock.
9. *Stem Gall* (coralberry, snowberry) —Numerous, small galls may girdle stems. Parts above galls may die. *Control:* Prune and burn affected parts. Spray as for Leaf Spots (above).
10. *Root Rots*—See (34) Root Rot under General Diseases. May be associated with root-feeding nematodes (e.g., dagger, lance, pin, ring, root-knot, spiral, stem or rot, sting, stunt). Grow in full sun in well-drained, moist soil high in organic matter.

11. *Root-knot*—Weigela is very susceptible. See under Peach, and (37) Root-knot under General Diseases. Nurserymen dip *bare-root weigela* plants in DBCP (Nemagon or Fumazone) or Zinophos (cynem) for 20 minutes to rid them of nematodes. Follow manufacturer's directions. Or soak in hot water (120° F.) for 30 minutes. Cool and plant.
12. *Verticillium Wilt* (elder, honeysuckle) —See (15B) Verticillium Wilt under General Diseases.
13. *Leaf Scorch* (primarily golden elder) —Leaves scorched where hot and windy. *Control:* Grow in more protected location. Water thoroughly during hot, dry periods.
14. *Chlorosis*—In alkaline or very acid soils. See under Maple and pages 24–29.
15. *Web Blight, Thread Blight* (elder, honeysuckle) —Southeastern states. See under Bean and Walnut.
16. *Gray-mold Blight*—See (5) Botrytis Blight under General Diseases.
17. *Infectious Variegation* (honeysuckle) —Leaves mottled green and yellow with veins yellow. See (16) Mosaic under General Diseases.
18. *Ringspot* (elder) —See (17) Ringspot under General Diseases.

SNOWDROP — See Daffodil
SNOWDROP-TREE — See Silverbell
SNOWFLAKE — See Daffodil
SNOW-ON-THE-MOUNTAIN — See Poinsettia
SOAPBERRY [CHINESE, FLORIDA, SOUTHERN, WESTERN] *(Sapindus)*

1. *Leaf Spots, Leaf Blight*—Small to large, round to irregular, spots and blotches of various colors on leaves. *Control:* Collect and burn fallen leaves. Keep shrubs and trees pruned. Spray with captan or zineb during rainy periods.
2. *Canker, Dieback*—Twigs die back from discolored, girdling cankers. *Control:* Prune and burn affected parts. Otherwise same as for Leaf Spots (above).
3. *Powdery Mildew*—Grayish-white powdery mold on leaves. *Control:* If serious, apply two sprays, 10 days apart. Use sulfur or Karathane.
4. *Root Rot*—See under Apple, and (34) Root Rot under General Diseases.
5. *Mosaic*—See (16) Mosaic under General Diseases.
6. *Mistletoe*—See (39) Mistletoe under General Diseases.
7. *Thread Blight*—Southeastern states. See under Walnut.

SOAPWEED — See Yucca
SOLANUM — See Potato and Tomato
SOLIDAGO — See Chrysanthemum
SOPHORA — See Honeylocust
SOPHRONITIS — See Orchids
SORBUS — See Apple
SORRELTREE or SOURWOOD *(Oxydendrum)*

1. *Leaf Spots, Purple Blotch*—More or less round to irregular, dull red, brown, or purple blotches. Centers of spots may later turn brown and dry. *Control:* Collect and burn fallen leaves in autumn. If practical, apply fixed copper when leaves fully expanded. Repeat 2 weeks later.
2. *Twig Blight, Dieback*—Rough cankers form on dead and dying twigs and branches. Leaves at end of affected parts wither and die. *Control:* Avoid injuring trees. Remove and burn dead and dying wood. Fertilize and water to keep trees vigorous.
3. *Root Rot*—See (34) Root Rot under General Diseases. Grow in loose, moist, well-drained, acid soil in a sunny spot.
4. *Wood Rot*—See (23) Wood Rot under General Diseases.

SOUR-GUM — See Dogwood
SOURWOOD — See Sorreltree
SOUTHERN LEATHERWOOD — See Buckwheat-tree
SOUTHERN PEA — See Bean
SOUTHERN WHITE-CEDAR — See Juniper
SOUTHERNWOOD — See Chrysanthemum
SPANISH-BAYONET — See Yucca
SPANISH BLUEBELL — See Tulip
SPANISH DAGGER — See Yucca
SPANISH FLAG — See Morning-glory
SPANISH GORSE — See Broom
SPARAXIS — See Iris
SPEARMINT — See Salvia
SPECULARIA — See Bellflower

SPEEDWELL [BASTARD, CATS-TAIL, GERMANDER, ROCK] *(Veronica, Hebe)*; CULVERSROOT *(Veronicastrum)*

1. *Leaf Spots*—Common. Small, round spots on upper leaf surface. May be grayish-violet to brown with purplish margins. Spots may merge to form a leaf scorch; or drop out leaving ragged shot-holes. Leaves wither and drop early. *Control:* Collect and burn fallen leaves. Spray as for Downy Mildew (below).
2. *Downy Mildew* (speedwell)—Widespread in damp weather. Conspicuous, pale yellowish blotches on upper side of leaves with grayish or purplish mildew on corresponding underleaf surface. *Control:* Spray or dust several times, at weekly intervals or just before wet periods. Use zineb, maneb, or fixed copper.
3. *Fusarium Root and Stem Rot, Wilt*—Plants stunted and weak. Leaves progressively turn yellow then brown, starting at base of plant. *Control:* Plant in well-drained soil. Dig up and burn infected plants. Avoid replanting in same area without first sterilizing the soil (pages 538–548).
4. *Other Stem and Root Rots*—See under Geranium, and (34) Root Rot under General Diseases. May be associated with nematodes (e.g., lesion, spiral, stunt). *Control:* Same as for Fusarium Root and Stem Rot (above).
5. *Powdery Mildews*—Powdery, white mold on leaves and shoots. Young leaves and shoots may pale and wither. *Control:* Spray or dust with sulfur or Karathane.
6. *Aster Yellows*—Plants yellowish. Produce numerous secondary shoots. *Control:* Destroy affected plants. Keep down weeds. Control leafhoppers that transmit the virus. Use DDT or sevin plus malathion.
7. *Curly-top*—See (19) Curly-top under General Diseases. *Control:* Same as for Aster Yellows (above).
8. *Verticillium Wilt* (hebe)—See (15B) Verticillium Wilt under General Diseases.
9. *Leaf Smut*—See (11) Smut under General Diseases.
10. *Root-knot*—See (37) Root-knot under General Diseases.
11. *Rusts*—See under Chrysanthemum, and (8) Rust under General Diseases.

SPHAERALCEA — See Hollyhock
SPICEBUSH — See Avocado and Calycanthus
SPICELILY — See Centuryplant
SPICEWOOD — See Calycanthus

SPIDERFLOWER *(Cleome)*

1. *Leaf Spots*—Spots of various sizes, shapes, and colors. Leaves may wither and drop early. *Control:* See under Chrysanthemum.

2. *Rust*—See under Chrysanthemum, and (8) Rust under General Diseases. Alternate hosts include wild grasses.
3. *Curly-top*—Plants stunted. Do not flower normally. *Control:* Destroy infected plants. Control leafhoppers that transmit the virus. Use DDT or sevin plus malathion.
4. *Root-knot*—See (37) Root-knot under General Diseases.
5. *Downy Mildew*—See (6) Downy Mildew under General Diseases.
6. *Powdery Mildew*—See (7) Powdery Mildew under General Diseases.

SPIDERLILY — See Daffodil
SPIDERWORT — See Tradescantia
SPIKE-PRIMROSE — See Fuchsia
SPINACH, SPINACH-BEET *(Spinacea)* — See Beet
SPINDLETREE — See Bittersweet

SPIREA [ANTHONY WATERER, AZURE BLUE, BILLIARD, BLUE MIST, BRIDALWREATH or PLUMLEAF, BUMALDA, DOUGLAS, FROEBEL, GARLAND, JAPANESE WHITE, KOREAN, LILAC, NIPPON, PINK, REEVES, RUBY GLOW, SNOW-BANK, SUBALPINE, THREELOBE, THUNBERG, VANHOUTTE, VEITCHI], HARDHACK, MEADOWSWEET *(Spiraea)*

1. *Powdery Mildews*—Widespread, especially where humid and wet. White powdery mold on young leaves and shoots. Flowers may be distorted and blasted. Leaves may wither and die. *Control:* When mildew is first seen, apply sulfur or Karathane. Repeat 10 days later. Do not use when temperature is 85° F. or above.
2. *Leaf Spots*—Spots of various sizes, shapes, and colors. *Control:* Collect and burn leaves in fall. Keep plants open by annual pruning. Apply zineb, captan, or maneb at 10- to 14-day intervals during rainy periods.
3. *Fire Blight*—Blossoms blackened. Shoots die back from tip. Foliage appears wilted and dark brown or black. Dead leaves tend to cling to shoots. *Control:* See under Apple.
4. *Crown Gall and Hairy Root*—See under Apple, and (30) Crown Gall under General Diseases.
5. *Root-knot*—See (37) Root-knot under General Diseases.
6. *Root Rot*—See (34) Root Rot under General Diseases. May be associated with nematodes (e.g., lesion, root-knot, stunt)—Grow in well-drained soil in a sunny location, or in light shade.
7. *Chlorosis*—Mineral deficiency. Common in alkaline soils. See under Maple and pages 24–29.

SPLEENWORT — See Ferns
SPRENGER ASPARAGUS — See Asparagus
SPRING GLORY — See Forsythia
SPRING BELL — See Iris
SPRING STARFLOWER — See Brodiaea
SPRUCE — See Pine
SPURGE — See Poinsettia
SPURGE LAUREL — See Daphne
SQUASH — See Cucumber
SQUILL — See Tulip
SQUIRRELCORN — See Bleedingheart
STACHYS — See Salvia

STAGGERBUSH — See Blueberry
STANHOPEA — See Orchids
STANLEYA — See Cabbage
STAPHYLEA — See American Bladdernut
STARCH-HYACINTH — See Tulip
STARFIRE — See African-violet
STARGLORY — See Morning-glory
STAR-OF-BETHLEHEM — See Tulip
STARRY CAMPION — See Carnation
STARS OF PERSIA — See Onion
STAR VIOLET — See Buttonbush
STATICE — See Sea-lavender
STENANTHIUM — See Tulip
STENOLOBIUM — See Trumpet-tree
STEPHANOMERIA — See Chrysanthemum
STERNBERGIA — See Daffodil
STEVIA — See Chrysanthemum
STILLINGIA — See Castorbean
STOCK — See Cabbage
STOKES-ASTER *(Stokesia)* — See Chrysanthemum
STONECRESS — See Cabbage
STONECROP — See Sedum
STONEMINT — See Salvia
STORKSBILL — See Geranium
STRANVAESIA — See Apple

STRAWBERRY *(Fragaria)*

1. *Black Root Rot Complex*—General and serious especially in heavy, wet soils. Plants stunted, weak, and subject to drought and winterkill. Often wilt, wither, and die about fruiting time. Leaves may become purplish or bronzed with red petioles. Roots black and dead, stunted, or largely absent. Dark, reddish-brown to black (inside and out) areas scattered over the root surface (Figure 4.208). May resemble

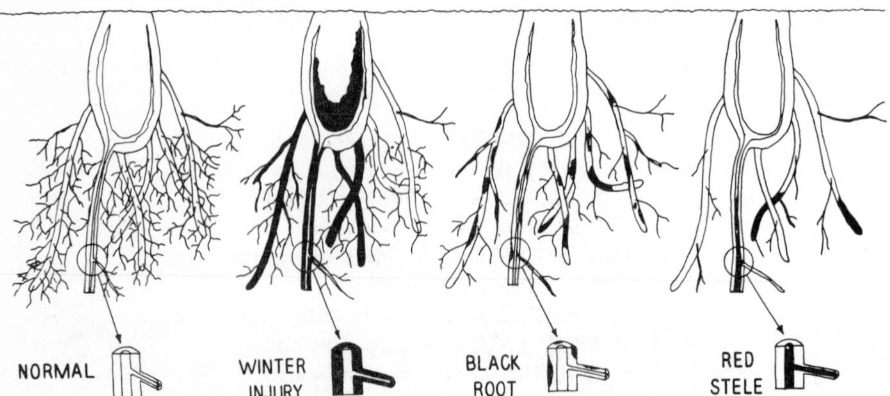

Fig. 4.208. Common symptoms of strawberry root disorders. Note close-up views for internal symptoms. (Department of Botany & Plant Pathology and Cooperative Extension Service, Michigan State University photo)

"rat-tails." Feeding roots tend to die back from tips; are often lacking. Fruit often dries up without ripening. Commonly follows winter injury (primarily alternate freezing and thawing). May be due to soil fungi, bacteria, nematodes (e.g., lesion), fertilizer burn, drought, toxic salts in soil, winter injury, or combination of two or more of these. See Figure 2.49A under General Diseases. *Control:* Purchase certified, virus-free, heat-treated plants of adapted varieties with white healthy roots. Grow in light, fertile, well-drained soil rich in organic matter. Or fumigate soil before planting as given under Verticillium Wilt (below). Control leaf and fruit diseases and insects. Follow strawberry spray program in Appendix (Table 16). Apply chlordane as preplant treatment to control rootworms, root aphids, grubs, crown borers, wireworms, root weevils, ants, armyworms, mole crickets, earwigs, crown borers, and other soil insects. Thoroughly mix with top 4 to 6 inches of soil. Control nematodes as given under Root-knot (below). Set plants deep enough to cover tops of all roots and base of crown; do *not* cover growing point. Keep plants vigorous by fertilization. Rotate beds at least every 2 years. Do not trim roots prior to setting plants. Follow local recommendations to prevent winter injury.

2. *Winter Injury*—Plants stunted. Often gradually die before fruit is all picked. Roots black and dead; but *centers* usually white (Figure 4.208). Wood under bark at crown is not discolored as with Verticillium Wilt (below). See also under Black Root Rot (above). *Control:* Apply mulch in fall, according to local recommendations. Follow other suggested cultural practices. Check with your county agent or extension horticulturist. Grow recommended, adapted varieties.
3. *Red Stele or Core*—Fairly common throughout northern ⅔ of United States in low, wet, poorly-drained areas. In spring of first bearing year, plants often severely stunted with cupped, dull, bluish-green leaves. Later, older leaves turn yellow or red, roll, wilt, and die about fruiting time. Central core (stele) of lower parts of roots is brick-red in spring; later brownish-black to red (Figure 4.208). About harvest time the roots turn brown to black, rot, and resemble rat-tails with few lateral roots. Few fruit or runners produced. *Potentilla* species are also susceptible to the red stele fungus *(Phytophthora). Control:* Plant certified, disease-free plants in light, well-drained soil. Avoid low wet spots. Varieties resistant to one or more races: Aberdeen, British Sovereign, Climax, Columbia, Early Cambridge, Fairland, Guardsman, Late Giant, Maine-55, Midland, Midway, Molalla, Monmouth, Orland, Pathfinder, Perle de Prague, Plentiful, Redcrop, Red Gauntlet, Redglow, Sparkle, Stelemaster, Sunrise, Surecrop, Talisman, Temple, and Vermilion. Races of the red stele fungus occur and a variety resistant in one area may be susceptible in another. Soil fumigation with chloropicrin, SMDC (Vapam or VPM Soil Fumigant), Trizone, and methyl bromide (pages 538–548) has given good results. Follow manufacturer's directions.
4. *Verticillium Wilt*—General and serious. Usually occurs in patches. Often confused with other root and crown rots and winter injury. Older leaves wilt, collapse, turn reddish-yellow or dark brown and often die about fruiting time. Young leaves wilt and tend to curl up along the midvein. Plants often stunted, dry, and flattened with small, yellowish leaves. Brownish streaks may occur inside the decaying crown and roots. Nematodes (e.g., lesion) increase severity of wilt. Highly *susceptible* varieties: Armore, Dixieland, Earlidawn, Jerseybelle, Lassen, Midland, Pocahontas, Redstar, Shasta, Solano, and Sparkle. *Control:* Same as for Red Stele (above). Avoid high rates of a nitrogen fertilizer. Five- or 6-year rotation excluding tomato, pepper, potato, okra, eggplant, bramble fruits, strawberry, and chrysanthemum. Somewhat *resistant* varieties: Aberdeen, Blakemore, Catskill, Cavalier, Gem, Grenadier, Howard 17 (Premier), Marshall, Robinson, Sierra, Siletz, Surecrop, Tennessee Beauty, Vermilion, and Wiltguard. Soil fumigation before planting

using chloropicrin (Table 22 in Appendix) has given excellent control as has mixture of chloropicrin and methyl bromide.

5. *Leaf Spots, Leaf Scorch, Black-Seed, Leaf Blotch*—Cosmopolitan. Favored by cool, wet weather. Small, round to angular or irregular, dark purple or reddish-purple spots on leaves (Leaf Scorch or Purple Leaf Spot); some may develop whitish centers (Leaf Spot). Leaves may later have dried, scorched look (Figures 2.8D and 4.209). Leaf-spot fungus produces black-seed on fruit. Similar spots or streaks may develop on petioles, stolons, and fruit stalks. Berry caps may turn brown. Berry yield and quality is reduced. Runner plants are weakened. *Control:* Avoid shade, spring applications of fertilizer high in nitrogen to bearing beds, and matted, crowded beds. Keep down weeds. Rotate beds every 2 years. Grow in well-drained soil. Before planting, remove older leaves and soak plants for 15 minutes in zineb or thiram solution (1 ounce in 3 gallons of water). Apply captan or folpet at 7- to 10-day intervals. Start when first leaves are unfolding. Continue until berries half size. See spray program in Appendix (Table 16). Use captan when berries start to color and during harvest season. After harvest, apply folpet, dodine, or captan as needed to keep leaf diseases in check. Varieties normally resistant to *Leaf Spot:* Albritton, Belmar,* Blakemore, Catskill, Columbia, Cyclone, Dabreak, Dorsett,* Earlibelle,* Earlidawn,* Elgin, Empire,* Fairfax,* Fletcher, Headliner, Howard 17 (Premier),* Klonmore,* Konvoy,* Majestic, Marion Bell, Massey,* Mastodon, Midland,* Midway, Missionary, Molalla, Montana Progressive, Northwest, Pocahontas, Red Rich, Redstar, Robinson, Rockhill (Wazata),* Shasta,* Siletz, Sparta Everbearing, Surecrop,* Tennessee Beauty,* Thomas,* and Vermilion. Varieties normally resistant to *Leaf Scorch:* Albritton, Blakemore,* Catskill, Dorsett, Earlibelle,* Earlidawn, Fairfax, Fairpeake, Fletcher, Gem, Howard 17 (Premier), Klonmore, Konvoy, Marion Bell, Mastodon, Midland, Midway, Red Rich, Redstar, Rockhill (Wazata), Sparkle (Paymaster), Sunrise,* Surecrop, and Vermilion. Check with your county agent or extension horticulturist to see if these varieties are adapted to your area.

Fig. 4.209. Strawberry leaf diseases. Left, Leaf blight; center, Leaf scorch; right, Leaf Spot. (Courtesy Dr. Dwight Powell)

6. *Leaf Blight*—Large, round to elongate, reddish-purple to dark brown spots or V-shaped blotches on leaves with purplish margins (Figure 4.209). Similar spots occur on sepals or cap of fruit. Older leaves are blighted and later die in large numbers. Runner plants are weak. Yield is reduced. The same fungus *(Dendrophoma)* causes a stem-end rot of the fruit. *Control:* Same as for Leaf Spots (above). Apply phenyl mercury in the fall *before* mulch is applied or in early spring after mulch is

*—Variety is normally highly resistant.

removed and *before* first leaves emerge. Somewhat resistant varieties: Earlidawn, Empire, Jerseybelle, Sparkle (Paymaster), Surecrop, and Vermilion. Howard 17 (Premier) and Tennessee Beauty are normally highly resistant.

7. *Fruit Rots* (Gray Mold, Leak, Leather Rot, Hard Rot, Tan Brown Rot, Stem-end Rot, etc.)—Cosmopolitan. Severe in wet seasons. Small, tan, or water-soaked to dark brown or black and spongy or leathery (Leather Rot) or hard (Hard Rot or Stem-end Rot), enlarging spots on ripening fruit. Rotted areas may be covered with white, blue-green, tan, gray (Gray Mold), or black mold (Figure 4.210). Fruit may collapse rapidly and juice runs (Leak or Rhizopus Rot). Fruit resting on soil or in dense plants most commonly attacked. Flowers may blast and set no fruit. Berry caps may turn brown. *Control:* Same as for Leaf Spots (above). Pick fruit frequently and early in the day, when possible. Handle carefully. Cull out all diseased berries. Dip baskets of fruit in captan 50 solution (2 tablespoons per gallon), dry rapidly, and refrigerate promptly (40° F. or below). Do *not* apply thiram or captan to fruit for canning or quick-freezing. If practical, mulch or apply polyethylene sheeting to keep fruit off soil. Apply captan at 3- to 5-day intervals through picking season if humid or rainy. See spray program in Appendix (Table 16). Red Gauntlet is resistant to Gray Mold Fruit Rot as is NC 2411 and 2355.

Fig. 4.210. Strawberry affected by the gray-mold fungus (*Botrytis*). (Department of Botany & Plant Pathology and Cooperative Extension Service, Michigan State University photo)

8. *Powdery Mildew*—General in cool, humid weather. Leaflets curl, roll upward; may be scorched at margins. Leaflets often dull gray or purplish-red. Later wither and die. Buds, flowers, fruit, fruit stems, and underleaf surface covered with thin, whitish-gray mold. Affected fruit do not develop or ripen normally. Highly *susceptible* varieties: Armore, Jerseybelle, Midland, Redglow, Robinson, Shasta, Stelemaster, and Tennessee Beauty. *Control:* Avoid growing in shaded areas or where air circulation is poor. Somewhat *resistant* strawberries: Aberdeen, Catskill, Dorsett, Dunlap, Empire, Erie, Essex, Fletcher, Florida Ninety, Marshall, Merton Princess, Missionary, Orland, Puget Beauty, Red Coat, Sierra, Siletz, Sioux, Sparkle (Paymaster), Sparta Everbearing, Sunrise, Tahoe, and Tioga. If serious, apply Karathane or sulfur thoroughly, plus wetting agent, two or three times, about 10 days apart. Start when mildew is first seen. Do not apply if weather is hot (above 85° F.). Do *not* apply sulfur to berries to be canned or put in metal containers.
9. *Crown Rots, Southern Blight*—Leaf petioles and fruit stems wilt and rot at crown. Rot may move upward causing stems to collapse. Runners may rot. Gray or white mold may grow on affected parts in damp weather. *Control:* Same as for Leaf Spots and Black Root Rot (both above). Drench affected areas with PCNB 75 per cent. Follow manufacturer's directions.
10. *Virus Complex* (Crinkles, Mottles, Curly-dwarf, Chlorotic Fleck, Mosaics, Multi-

plier, Leaf Curl, Leaf Roll, Stunt, Veinbanding, Yellows (Xanthosis or Yellow-edge), Green Petal or Aster Yellows, Witches'-broom)—General. Symptoms variable depending on viruses (and virus strains), variety, stage of plant growth, temperature, nutrition, daylight, soil moisture, and other factors. Many varieties may be virus-infected, yet show *no* symptoms. Leaves often spotted, flecked, streaked, or edged with yellow, crinkled, puckered, twisted, and cupped (or rolled) upward or downward. Leaves often dwarfed or mottled light green or yellow. Plants may be stunted, bushy, flattened, or dwarfed (Stunt, Multiplier, Green Petal, Yellows, Crinkle, Witches'-broom), spindly, low in vigor, malformed, and yellow. Stolons may be flattened, thickened, and distorted. Plants infected with Green Petal develop large, green flower petals with apparent developing berries producing elongate to leaflike seeds. Fruit size, yield, and runner production may be greatly reduced. See Figure 4.211. Dagger *(Xiphinema)* and needle *(Longidorus)* nematodes transmit several of the viruses. *Control:* Plant only certified, heat-treated, and aphid-free plants from a reputable nursery. Malathion sprays control aphids and leafhoppers that transmit viruses. Follow spray program in Table 16 in Appendix. Destroy wild and virus-infected strawberries before setting out new plants. Destroy abnormal plants when seen during first season. Locate new beds as far from old bearing beds as possible. Varieties *tolerant* to several viruses: Aroma, Blakemore, Campbell's Early, Catskill, Columbia, Dunlap, Early Ozark, Fairland, Goldsmith, Howard 17 (Premier), Institute Z4, Klonmore, Lassen, Missionary, Molalla, Montana Progressive, Northwest, Puget Beauty, Redstar, Robinson, Siletz, Solano, Surecrop, Temple, Tennessee Beauty, and Torrey. If practical, grow in fumigated soil (pages 538–548).

Fig. 4.211. Strawberry viruses. The variety Catskill severely affected. The row to the left is Catskill grown from virus-free nursery stock; row to the right was grown from regular nursery stock. (Department of Botany & Plant Pathology and Cooperative Extension Service, Michigan State University photo)

11. *Leaf or Noninfectious Variegation, Spring or June Yellows*—General. An hereditary character (or virus?) found in certain varieties. Symptoms variable. In cool springs (temperature below 50° F.), irregular green and yellow mottling, spotting, or streaking occurs in leaves of Bellmar, Blakemore, Chesapeake, Climax, Dixieland, Empire, Headliner, Howard 17 (Premier), Klonmore, Mastodon, Moneymaker, Tennessee Beauty, Van Dyke, Vermilion and most everbearing varieties. Leaves gradually show less green, become golden yellow. May be puckered and distorted. Fruit production decreases. Plants never recover. *Control:* Grow "yellows-resistant" strains of susceptible varieties that are certified virus-free. Destroy infected plants when found. Do not use runners from variegated mother plants.
12. *Rhizoctonia Bud Rot*—Widespread in low areas during cool (below 75° F.), wet weather. Flower and leaf buds turn brown and die. Outer leaves lie flat on ground;

become darker green than normal. Plants die or produce weak, spindly side growth and no fruit. Stolons or runners from affected plants may be girdled and brown at base. *Control:* Same as for Black Root Rot and Leaf Spots (both above). Do not set plants deep or cover crowns during cultivation. Cultivate frequently and shallowly to keep soil surface dry. Dig planting stock when soil moisture is moderate.

13. *Root-knot*—Widespread. Plants may be stunted, weak, pale green or yellowed, produce fewer runners. Wilt during midday, are less productive, even die. Affected plants are more susceptible to drought and winter injury. Small to large, round to irregular, knotlike galls form on roots. *Control:* Plant only certified, heat-treated plants in clean soil. Nurserymen soak *dormant* plants in hot water (124° F.) for 7½ minutes, (127° F.) for 3 minutes, or (130° F.) for 1 minute. Keep down weeds. Keep plants vigorous. Maintain high soil fertility and water during dry periods. Rotate. Keep down weeds. Cultivate very shallowly. If severe, fumigate soil in early fall before planting. Use D-D, EDB, methyl bromide, chloropicrin, Dorlone, Telone, Vidden-D, DBCP, or Mylone (pages 538–548). DBCP (Nemagon and Fumazone) has given excellent control when applied around plants. Follow manufacturer's directions. For additional information read USDA Leaflet 414, *Reducing Virus and Nematode Damage to Strawberry Plants.*
14. *Other Root-feeding Nematodes* (dagger, lance, lesion, needle, pin, ring, sheath, spear, spiral, stem, sting, stubby-root, stunt, tylenchus)—Cosmopolitan. Symptoms and control same as for Root-knot (above). Small roots may turn dark brown to black and die back. Root surface often shows dark spots or is entirely brown to black. Often associated with Black Root Rot Complex and Verticillium Wilt (both above).
15. *Spring and Summer Dwarf or "Crimp," Leaf and Bud Nematodes*—Young center leaves twisted, cupped, and distorted. May be crinkled or crimped, greatly dwarfed, narrow, glossy, stiff, and dark green. Growth stunted. Buds and plants weakened or killed. Fruit yield reduced. The nematodes live in leaf buds (up to 12,000 per bud) or crown of plants. See Figure 2.38B under General Diseases. *Control:* Dig up and burn infested plants when *first* noticed. Grow only certified, heat-treated plants. See Root-knot (above). Set in fumigated soil. Three- to 4-year rotation. DBCP may be used around plants in field. Follow manufacturer's directions.
16. *Leaf and Stem Nematode*—Mostly in Pacific Northwest following clover. Gall-like swellings on petioles, fruit stems, flower clusters, and leaves. Plants stunted, twisted, and puckered. Often appear ragged or unthrifty. Leaves stunted and creased, with leaf margins curled and veins distorted. Runners, stems, petioles, and fruit stalks shortened and swollen. Fruit production is reduced. *Control:* Same as for Spring and Summer Dwarf (above). Avoid planting where red clover has been affected.
17. *Cauliflower Disease*—A severe form of Dwarf (above). Numerous, dwarfed buds and shoots at crown; resembles a cauliflower. Due to combined attack by nematodes (*Aphelenchoides*) and a bacterium (*Corynebacterium*). *Control:* Dig up and destroy infested plants. Otherwise same as for Root-knot (above).
18. *Chlorosis, Iron Deficiency*—Common in alkaline soils. Normal leaf color gradually fades, except for veins. When severe, plants become creamy yellow. Leaves stunted. *Control:* Adjust soil reaction to pH 5.5 to 6.5. Apply ferbam to control leaf diseases or apply iron sulfate or iron chelate (pages 26–27) sprays to maintain normal green color. Or apply iron sulfate or iron chelate to the soil following manufacturer's directions. Follow local recommendations.
19. *Potassium Deficiency, Leaf Scorch*—Margins of leaflets are scorched; gradually roll upward and inward. Runners few in number; are thin and short. *Control:* Have soil test made and follow suggestions in the report.
20. *Anthracnose, Crown Rot*—Gulf Coast states during hot (80° F.), humid weather in low areas of fields. Runners and rhizomes or leaf petioles spotted, girdled, and

killed by enlarging, light to dark brown, reddish-brown, or black, sunken cankers. Runners later turn brown to black. Crowns may rot and turn reddish-brown internally causing plants to suddenly wilt and die. *Control:* Follow strawberry spray program (Table 16) in Appendix. In addition, apply bordeaux or fixed copper at 7- to 10-day intervals in warm, moist weather starting in July.

21. *Catface*—Fruit deeply furrowed, gnarled, or otherwise twisted. Due to incomplete pollination or feeding of lygus plant bugs. *Control:* Spray to control plant bugs. See Table 16 in Appendix.
22. *Tipburn*—Tips of young unfolding leaves and flowers are suddenly killed. When full-grown, such leaves are irregularly shaped, puckered, and "burned back" to about ½ normal size. Commonly follows marked increases in temperature and light intensity in spring or early summer. *Control:* Not necessary.
23. *Bacterial Angular Leaf Spot*—Minnesota and Wisconsin. Small, light green to dark green, water-soaked, angular spots on undersurface of leaves. When wet, a "milky slime" is evident. Older spots are irregular and reddish-brown above. May closely resemble Leaf Spot and Scorch (above). Petioles, runners, and flowers are blighted. *Control:* Rotate. Varieties differ in susceptibility. Gem, Robinson, Sparkle, and Trumpeter are very susceptible. If practical, avoid sprinkler irrigation.
24. *Slime Molds*—White, bluish-gray, or yellow slimy masses up to an inch or more in diameter. Grow on plants during and following damp weather. Masses usually become powdery. More unsightly than harmful. *Control:* None necessary. Quickly disappear in dry weather.
25. *Downy Mildew*—See (6) Downy Mildew under General Diseases.

For additional information read USDA Leaflet 414, *Reducing Virus and Nematode Damage to Strawberry Plants;* USDA Farmers' Bulletin 2140, *Strawberry Diseases;* USDA Farmers' Bulletin 1043, *Strawberry Varieties in the United States;* and the book, *Strawberry Diseases,* by Dr. A. G. Plakidas.

STRAWBERRY-BEGONIA — See Hydrangea
STRAWBERRY-BUSH — See Bittersweet
STRAWBERRY-SHRUB — See Calycanthus
STRAWBERRY-TOMATO — See Tomato
STRAWBERRY-TREE — See Blueberry
STRAWFLOWER — See Chrysanthemum
STRELITZIA — See Bird-of-Paradise Flower
STREPTANTHERA — See Iris
STRIPED SQUILL — See Tulip
STROMBOCACTUS — See Cacti
STYRAX — See Silverbell
SUGAR BEET — See Beet
SUGARBERRY — See Hackberry
SULTAN — See Balsam

SUMAC [CHINESE, CUTLEAF, EVERGREEN, FLAMELEAF or SHINING, FRAGRANT, ILLINOIS, JAVA, LAUREL, LITTLELEAF, LEMONADE or SKUNKBUSH, SMOOTH, STAGHORN, SUGAR], PURPLE SMOKEBUSH *(Rhus);* SMOKETREE [AMERICAN, COMMON, PURPLE] *(Cotinus);* PISTACHE, PISTACHIO [CHINESE, COMMON] *(Pistacia);* PEPPERTREE [BRAZIL, CALIFORNIA] *(Schinus)*

1. *Wilts (Fusarium and Verticillium)*—Widespread. Leaves wither, wilt, and hang down. May cling to stem. Brown, reddish-brown, or olive-brown streaks in wood just under bark. Branches may wilt suddenly or gradually with dwarfing and

yellowing of leaves. Then take on premature fall color. Wilted plants may soon die. *Control:* Destroy severely infected plants. Do *not* replant the area to the same species. Prune out wilting branches. Fertilize and water to increase vigor. Avoid wounding roots or trunk near soil line. When disease is first evident, try working zineb (1 pound to 20 square feet) into soil around plant. Then drench with water.

2. *Leaf Spots*—General. Spots of various colors, shapes, and sizes. *Control:* If serious, spray at 10- to 14-day intervals, during wet periods. Use zineb or maneb.
3. *Powdery Mildew*—Widespread on sumac. White, powdery mold on leaves from midsummer on. Leaves may wither. *Control:* When disease appears, spray several times at weekly intervals. Use sulfur, Karathane, or Acti-dione.
4. *Wood and Heart Rots*—General. See under Birch, and (23) Wood Rot under General Diseases.
5. *Root Rots*—See under Apple, and (34) Root Rot under General Diseases. Twigs may die back. May be associated with nematodes (e.g., burrowing, dagger, stem, stunt). Grow in well-drained soil.
6. *Stem Cankers, Dieback, "Umbrella Disease"*—See under Apple and Elm.
7. *Leaf Curl or Blister* (sumac)—See under Birch, Peach, and (10) Leaf Curl under General Diseases.
8. *Rusts* (smoketree, sumac)—Minor problem. See (8) Rust under General Diseases. Wild grasses are the alternate hosts.
9. *Root-knot*—See (37) Root-knot under General Diseases.
10. *Thread Blight*—Southern states. See under Walnut.
11. *Mistletoe* (smoketree)—See (39) Mistletoe under General Diseases.
12. *Crown Gall* (sumac)—See (30) Crown Gall under General Diseases.
13. *Inflorescence Blight* (sumac)—See (31) Flower Blight under General Diseases.

SUMMER-CYPRESS — See Beet

SUMMER-HYACINTH — See Tulip

SUMMER-LILAC — See Butterflybush

SUMMERSWEET — See Sweet-pepperbush

SUNDROPS — See Evening-primrose

SUNFLOWER — See Chrysanthemum

SUN-PLANT — See Rose-moss

SUNROSE, FROSTWEED or ROCK ROSE *(Helianthemum)*; FROSTWORT *(Crocanthemum)*

1. *Leaf Spots*—Small, more or less round, white or gray, reddish to purple spots with dark margins. Centers of spots may be sprinkled with black specks. *Control:* Spray several times, 10 to 14 days apart. Use zineb, maneb, or fixed copper.
2. *Root Rot*—See (34) Root Rot under General Diseases. Grow in full sun in neutral, well-drained soil.
3. *Powdery Mildew*—Dense white mold on leaves, mostly the undersurface. See (7) Powdery Mildew under General Diseases.

SURINAM-CHERRY — See Eucalyptus

SWAMPBAY — See Avocado

SWAMP-PRIVET — See Ash

SWAN RIVER DAISY — See Chrysanthemum

SWEET ALYSSUM — See Cabbage

SWEET BASIL — See Salvia

SWEETBAY — See Magnolia and Laurel

SWEETBELLS — See Labrador-tea

SWEET CORN — See Corn

SWEETFERN *(Comptonia)*

1. *Rust, Blister Rust*—Long, brown, threadlike "spore horns" extend from underleaf surface. Or curled, "ram's-horn" leaves with numerous, dusty, yellow-orange "cluster cups." Alternate hosts include 2- and 3-needle pines (where stem galls are formed) or southern white-cedar. *Control:* Do not plant near alternate hosts.

SWEETGALE — See Waxmyrtle
SWEETGUM — See Witch-hazel
SWEET-JARVIL — See Celery
SWEET MARJORAM — See Salvia
SWEETOLIVE — See Osmanthus
SWEETPEA — See Pea
SWEET-PEPPERBUSH, SUMMERSWEET or PEPPERBUSH [JAPANESE, PINK], WHITE-ALDER *(Clethra)*

1. *Root Rot*—See under Apple, and (34) Root Rot under General Diseases. Grow in acid (pH 4 to 5), moist soil high in organic matter.
2. *Leaf Spots*—See under Maple.

SWEETPOTATO *(Ipomoea)*

1. *Black Rot, Black Root, Black Shank*—General and serious. Pale, unthrifty, yellowish foliage. Roundish, small to large, black, enlarging, sunken spots on underground stems and sweetpotato roots. Potato has bitter quinine taste. Often dry and corky. Sprouts may die before emergence. See Figure 2.48C under General Diseases and Figure 4.212. *Control:* Use only certified, disease-free seed potatoes or certified healthy slips (aerial stem cuttings). A prebedding dry heat treatment of seed roots (110° F. for 3 days; or 122° F. for 6 hours) effectively controls Black Rot, Root-knot, and possibly other diseases. Before bedding, dip uncut roots in warm (60° to 80° F.) 1:1,000 solution of mercuric chloride for 8 to 10 minutes, or dip 1 minute in Semesan Bel S, Puratized Agricultural Spray, or Coromerc. Follow manufacturer's directions. Many growers dip disease-free roots and base of sprouts just before planting (bedding) in a zineb, thiram, chloranil, dichlone, or ziram solution (2 ounces per gallon for roots; 1 ounce for sprouts). Bed the roots and sprouts immediately after treating and water thoroughly. Root in well-drained, sandy loam bedding soil that is sterilized (pages 538–548) and kept moist and warm (80° to 90° F.). Grow in partial shade. Set cuttings into field or garden 10 to

Fig. 4.212. Cross section of a sweetpotato infected with black rot, showing depth of penetration by the black rot fungus. (USDA photo)

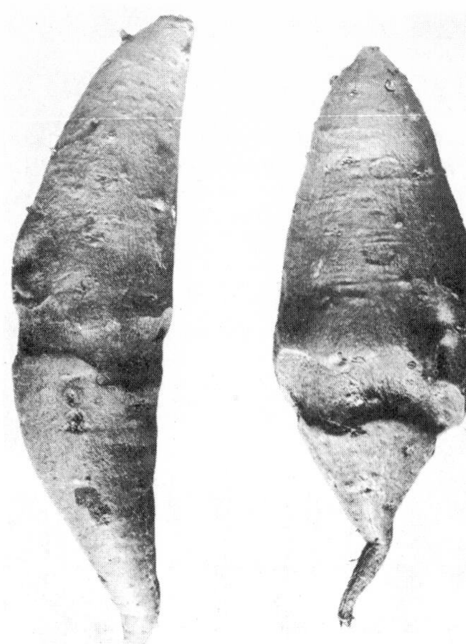

Fig. 4.213A. Ring rot of sweetpotato. (USDA photo)

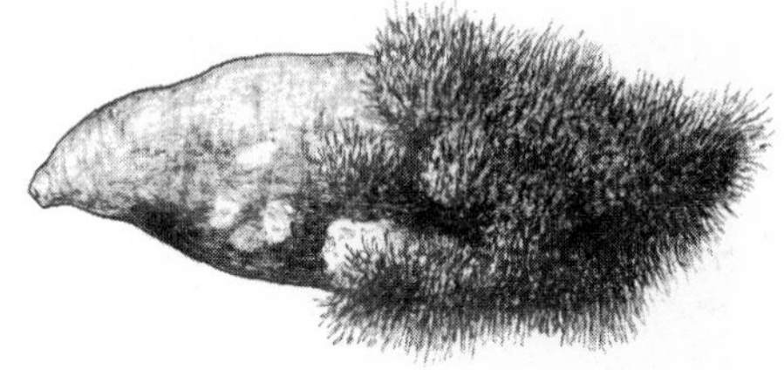

Fig. 4.213B. Soft rot of sweetpotato showing the moldy growth or "whiskers" of Rhizopus.

14 days later. Three- or four-year rotation. Keep down weeds. Control nematodes and soil insects, e.g., sweetpotato weevils, and other pests. Check with your extension entomologist or county agent regarding recommended chemicals. Discard sprouts that appear weak and unthrifty or show black spots. Store only sound, blemish-free sweetpotato roots after curing for 7 to 14 days at 80° to 95° F. and 90 per cent relative humidity. After curing, store roots in clean baskets or crates at 55° to 60° F. and a humidity of 85 to 90 per cent. *Before storage,* clean up all debris and spray storage area with copper sulfate (1 pound in 5 to 10 gallons of water) or formaldehyde (1:50 solution). Or fumigate air-tight storage houses and containers with chloropicrin (½ pound per 1,000 cubic feet) for 24 hours. Use minimum temperature of 70° F. Somewhat resistant varieties: Norin No. 1, Oklahoma 24 or Allgold, and Sunnyside. More highly resistant varieties should be available soon.

2. *Storage Rots*—General and serious. Mushy and watery, spongy, or firm and hard, yellowish, brown, pink, dark red, grayish-brown, purplish, or black rot. Potatoes shrink, may become hard and dry (mummified). Rotted areas may be covered with white, gray, blue, green, or black mold. Potatoes may rot completely in just a few days. See Figures 4.213A and 4.213B. *Control:* Same as for Black Rot (above). Harvest during warm, dry weather before frost kills vines. Handle roots carefully while digging, curing, and storing. Do *not* let potatoes lie on soil surface over an hour before curing. Control diseases in the field. A 5- to 10-second dip (or spray) in SOPP (sodium orthophenylphenate tetrahydrate; sold as Stopmold B), followed by a water or wax solution rinse, helps reduce transit and market losses due to Soft Rot (Rhizopus Rot) and Black Rot. This treatment is *not* for roots to be cured and stored. Botran (dichloran) does a better job against Soft Rot, is simpler and cheaper to use, and is much less likely to injure roots. Follow manufacturer's directions.

3. *Fusarium Wilt or Stem Rot, Yellow Blight*—General, often serious, especially in light sandy soils. Leaves turn yellow or brownish, wilt, and drop. Plants stunted, upright, and bushy. Vines later wilt and collapse. Brown to black streaks occur inside stems and a brown to blackened ring in the sweetpotato flesh (Figure 4.214). Very *susceptible* varieties: Big-Stem Jersey, Georgia, Gold Skin, Kansas 40, Little-

Fig. 4.214. Fusarium wilt or stem rot of sweetpotato. Cross section through a potato, showing the dark ring caused by Fusarium.

Stem Jersey, Maryland Golden, Nancy Gold, Nancy Hall, Porto Rico, Red Jersey, and Yellow Jersey. *Control:* Same as for Black Rot (above). Grow *resistant* varieties where adapted: Allgold, Coppergold, Dahomey, Dooley, Gem, Goldrush, Key West, Nugget, Pumpkin, Red Brazil, Southern Green, Timian, Triumph, White Yam, and Yellow Strasburg. Tanhoma is tolerant. More highly resistant varieties should be available soon.

4. *Foot Rot or Die Off*—Widespread. Enlarging, brown to black areas on stem at or near soil line. Stem is girdled causing parts beyond to turn yellow, wilt, and die starting in midsummer. A firm brown rot of sweetpotato, usually at stem end. Yield is reduced. *Control:* Same as for Black Rot (above). Control sweetpotato weevils by applying insecticide at base of plants as roots begin to enlarge. Repeat 2 weeks later. Check with your county agent or extension entomologist regarding materials and rates to apply.
5. *Leaf Spots, Leaf Blights*—Widespread. Round to angular, white, gray, or brown leaf spots. Often with definite margins. Centers of spots may later be sprinkled with black specks or mold. *Control:* Same as for Black Rot (above). If serious, apply zineb or maneb several times in rainy weather, 10 days apart.
6. *Root-knot*—Widespread. Plants stunted, unthrifty, and yellowish. Small galls form on feeder roots. Potatoes may be pitted or cracked and yellowish. Nancy Hall types are very susceptible. *Control:* Normally resistant varieties to several species and races of nematodes, where adapted: Heartogold, most Jersey varieties, Nemagold, and South Carolina E7. Porto Rico varieties are intermediate in resistance. Where serious, fumigate soil with D-D, EDB, SMDC (Vapam or VPM Soil Fumigant), chloropicrin, Vidden-D, Vorlex, Telone, or methyl bromide (pages 538–548). Grow nematode-free planting stock. Seed roots may be disinfested by dry heat (see under Black Rot above). Or soak in hot water (116° F.) for 65 minutes. Do not overheat.
7. *Scurf or Soil Stain*—General. Small, round, rusty-brown to almost black spots and blotches on potato surface. Spots may sometimes merge to form a patch over potato (Figure 4.215A). Skin may crack. Potatoes in storage may lose some water, shrink, become spongy, then dry and hard. Or secondary rot-producing organisms may cause severe losses in storage. *Control:* Same as for Black Rot (above). Or dip roots prior to planting in thiram, ferbam, ziram, or captan solution (2 pounds in 5 gallons of water). Avoid planting in heavy, wet soils high in organic matter. Practice a 4- or 5-year rotation.
8. *Soil Rot or Streptomyces Pox*—General and serious in localized areas which

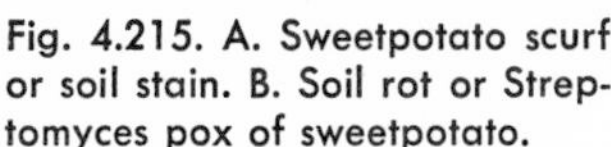
Fig. 4.215. A. Sweetpotato scurf or soil stain. B. Soil rot or Streptomyces pox of sweetpotato.

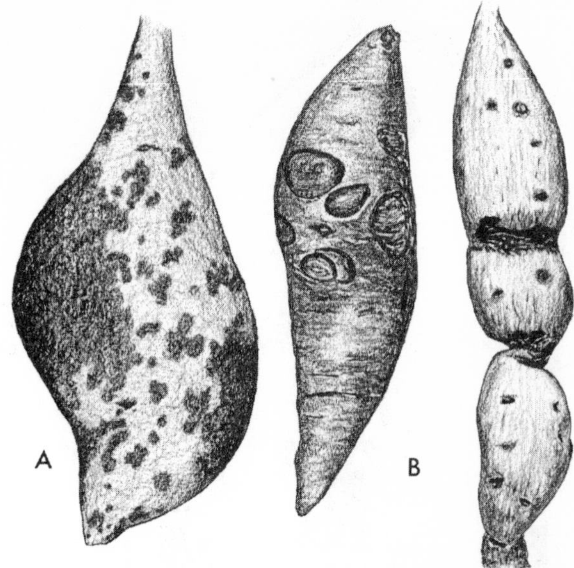

increase in size each year. Worst in dry soils in dry seasons. Plants fail to grow or may be dwarfed and lack vigor with small, thin, pale green to yellow leaves. May die before maturing. Potatoes often badly deformed and dumbbell-shaped with rough, scabby pits or corky surface areas that dry and fall out (Figure 4.215B). Nearly black girdling spots form on roots and underground stem. Yield and quality of potatoes may be greatly reduced. Roots decay. Disease does not normally develop in soils below pH 5.2. Rot organisms later attack affected potatoes. *Control:* Grow disease-free stock in uninfested soil. Or fumigate soil with chloropicrin or chloropicrin plus methyl bromide (pages 538–548). Four- to 6-year rotation. Plow under green manure crop before planting. Irrigate thoroughly as edible potatoes begin to form. Where practical, acidify infested soil (to pH 5.0) by adding sulfur, based on a soil test. The varieties Acadian, Calred, Centennial, Goldrush, Heartogold, and Nugget have some resistance. More resistant varieties should be available in the future.

9. *White-rust*—General but not serious. Irregular yellow areas form on leaves that later turn brown. Spots on underside of leaves turn pale green then brown and are covered with white, powdery pustules. Leaves and shoots may be somewhat malformed. Swellings occur on stems and petioles. *Control:* Keep down weeds, especially wild morning-glories and their relatives. Grow where air circulation is good. Spraying as for Leaf Spots (above) may be beneficial. Varieties differ in resistance.
10. *Internal Cork, Leaf Spot*—Widespread, especially in southern states. Irregular, dark brown to black, hard, corky "islands" in sweetpotato flesh. Roots usually appear normal outside except for dimples that occasionally appear above corky areas. Increases in storage. Foliage symptoms variable. See Figure 4.216. Internal cork is part of Feathery Mottle virus complex. Whitish to yellow spots that may become masked or reddish to purplish and finally dead, develop on leaves. Leaves may be slightly mottled and yellowed along the veins. Porto Rico, Goldrush, and Red Velvet are generally susceptible. Some varieties are symptomless carriers. *Control:* Plant certified, virus-free seed roots or slips as far (100 yards or more) from infected plantings as possible. Grow less-susceptible varieties, where adapted:

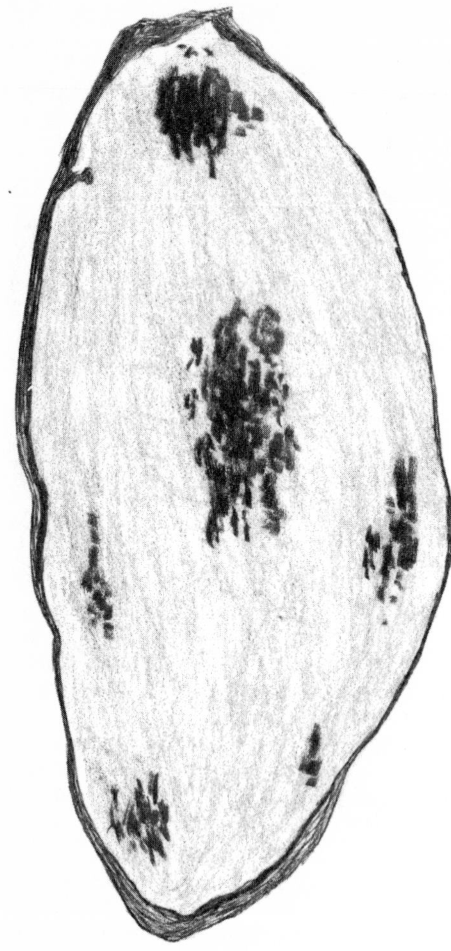

Fig. 4.216. Internal cork of sweetpotato. Lengthwise section through a potato showing the dark corky "islands" in the flesh.

Allgold, Earlyport, Gem, Golden Skin, Jersey Orange, Maryland Golden Sweet, Nancy Gold, Nancy Hall, Nemagold, Nugget, Oklahoma 2, Pelican Processor, Ranger, Redgold, Red Nancy Hall, and Yellow Jersey. Tanhoma is tolerant. More resistant varieties should be available soon. Control aphids that transmit viruses. Use malathion. Cure and store as for Black Rot (above), except for *seed stocks* that should be kept at 70° F. for the first 6 weeks of storage to better detect the virus (es).

11. *Mottle-leaf or Mosaic, Yellow Dwarf, Feathery Mottle*—Southern states. A virus complex causing variable symptoms. Foliage may or may not be mottled. Leaf shape is abnormal. Younger leaves may show feathery yellowing along smallest veins or small, diffuse, yellow spots or streaks. Small to large, pale green areas later develop. Remainder of leaf is very dark green. Plants later dwarfed and yellowed. May appear rosette-like. Yield may be greatly reduced. Nancy Hall and Porto Rico types are very susceptible. *Control:* Same as for Internal Cork (above). Destroy first infected plants. Control aphids and whiteflies that transmit viruses. Use malathion.
12. *Mottle Necrosis, Leak*—Widespread. Most severe in fairly light, sandy soils. Irregular, brownish, somewhat sunken spots on sweetpotato. When roots are cut

across, scattered, irregular, chocolate-brown, dead areas are seen, giving a marbled appearance. *Control:* Four-year rotation. Avoid planting very susceptible varieties, e.g., Big-Stem Jersey, Georgia, Triumph, and Yellow Jersey.

13. *Gray-mold Blight, Bud Rot*—Cosmopolitan on sprouts. Bedded potatoes may rot. Newly emerged sprouts are killed. *Control:* Use disease-free seed potatoes and slips. Practice good bed management (see under Black Rot above). Avoid overheating and overwatering.
14. *Root Rots*—Feeder roots decay. Sweetpotato may be completely rotted. Plants stunted and unthrifty. Often associated with soil insects and nematodes (e.g., awl, burrowing, dagger, lance, lesion, pin, reniform, ring, rot, spiral, sting, stubby-root, stunt). *Control:* Same as for Black Rot and Root-knot (both above). Grow on tops of ridges in sandy loam soil that is well-drained. Varieties differ somewhat in resistance. Control wireworms and white grubs by applying chlordane or Diazinon to soil before planting. Work into top 4 inches of soil. Check with your extension entomologist as regards latest insect control recommendations.
15. *Russet Crack* (Maryland, New Jersey, Louisiana and probably other states)—A new virus disease. Fleshy potatoes banded or spotted by transverse, dark brown to gray areas composed of shallow, fine cracks. Bands vary in width and potatoes may be constricted in these areas. Inner flesh is unaffected. Dark gray to black spots and streaks occur on the sprouts, vines, and petioles. Leaf veins may be blackened. Leaf spots may be light green or dark gray to yellowish; often with a distinct margin. *Control:* Use only certified, disease-free seed potatoes or certified healthy slips. Control aphids that transmit the virus. Use malathion.
16. *Rust*—Mostly in southern states. Colorless to deep orange-red, dusty pustules on underside of leaves. If severe, leaves may wither and die. Pines are the alternate host. *Control:* Not usually necessary. Try spraying as for Leaf Spots (above).
17. *2,4-D Injury*—Sweetpotato is very susceptible. See under Grape.
18. *Boron Deficiency, Internal Brown Spot*—Vines stunted and gnarled with curled petioles. Older leaves turn yellow and drop early. Roots rough and malformed (dumbbell-shaped, lopsided, spindle-shaped), may show dark external or internal discoloration. *Control:* See under Beet and page 27.
19. *Magnesium Deficiency*—Common on sandy, acid soils. Yellow areas develop between leaf veins on older leaves (reddish-purple instead of yellow in Porto Rico type of sweetpotato), while veins remain green. If severe, older leaves turn brown and dry up while yellowing appears in new leaves at tips of vines. Yield may be cut in half. *Control:* Have a soil test made and follow suggestions in the report. See page 28.
20. *Verticillium Wilt*—See (15B) Verticillium Wilt under General Diseases.
21. *Curly-top*—Western states. See (19) Curly-top under General Diseases.
22. *Thread Blight*—Southeastern states. See under Walnut.
23. *Crown Rot, Southern Blight*—See under Hollyhock.
24. *Slime Molds*—Cottony-white, yellow to purplish, slimy coating on any aboveground plant parts. Later these become powdery and whitish to grayish in color. Common in hotbed where foliage is dense. *Control:* None necessary. Avoid overwatering and increase ventilation.
25. *Sooty Mold*—See (12) Sooty Mold under General Diseases.

For additional information read USDA Farmers' Bulletin 1059, *Sweetpotato Diseases.*

SWEET SCABIOUS — See Scabiosa

SWEETSHRUB — See Calycanthus

SWEET SULTAN — See Chrysanthemum

SWEET-WILLIAM — See Carnation

SWISS CHARD — See Beet

SWORD BEAN — See Bean
SYAGRUS — See Palms
SYCAMORE [AMERICAN or BUTTONWOOD, ARIZONA, CALIFORNIA], PLANETREE [LONDON, ORIENTAL or EUROPEAN] *(Platanus)*

1. *Anthracnose, Leaf Blight, Twig Canker*—General and damaging on American and Western sycamores and Oriental planetree in cold (below 60° F.), wet, spring weather. London plane is rather resistant. Buds and young expanding leaves turn brown and die in early spring. Often resembles frost injury. Small to large, brown dead areas develop along and between leaf veins of older leaves. Affected leaves often drop early, followed by a new crop of leaves within several weeks (Figure 4.217). Weakened trees are more susceptible to borers, drought, winter injury, and other diseases. Sunken, discolored cankers may girdle and kill twigs and smaller branches. Often results in witches'-brooms. Young trees may be killed by trunk cankers. *Control:* Collect and burn fallen leaves and twigs. Remove dead and severely cankered twigs and limbs. Thin out crown for better air circulation. Fertilize severely infected trees to increase vigor. Apply phenyl mercury (e.g., Coromerc, Puratized, Phix, Tag 331), dodine, fixed copper, zineb, thiram, dichlone, maneb, or captan one to three times, 7 to 10 days apart. Start as buds *begin* to swell.

Fig. 4.217. Sycamore anthracnose.

2. *London Plane Blight, Cankerstain*—Serious in eastern and southern states and localized areas of Midwest (e.g., St. Louis). Much less common on American sycamore. Foliage becomes thin. Leaves dwarfed, later wither, turn yellow, and drop in large numbers in early summer. Trees die within several years. Elongated, rough, sunken, dark brown or black cankers develop on trunk and larger branches. Girdling cankers kill parts beyond. Wood underneath cankers shows reddish-brown to bluish-black streaks or wedge-shaped sections when cut through. The causal fungus (*Endoconidiophora* or *Ceratostomella*) enters through bark injuries of all kinds. *Control:* Remove and burn infected trees as soon as possible. Avoid injuring bark. Pruning tools and ladder parts that touch wood should be thoroughly washed in 70 per cent alcohol or household bleach between trees. Disinfect climbing ropes by exposure to formaldehyde fumes for 3 hours in a *closed* container (¼ pound of 40 per cent commercial formaldehyde in a 10-gallon can). Apply tree wound dressing containing a fungicide (e.g., gilsonite-varnish type paint containing 0.25 per cent phenyl mercury nitrate, 6 per cent phenols, or 0.5 per cent dichlone, ferbam, folpet, or thiram. *Examples:* Gilbert's Tree Wound Dressing, Flozon Tree

Wound Paint, and Cerano). Whenever possible, prune trees in winter. Control wood-boring insects as given under Wood Rot (below).

3. *Leaf Scorch*—Browning or scorching of leaves between veins or along margins. Follows hot, dry, windy weather in July and August. *Control:* Water during summer droughts. Prune and fertilize trees to increase vigor.
4. *Wood Rot, Trunk Rot, Heart Rot*—See under Birch, and (23) Wood Rot under General Diseases. Keep trees growing vigorously. Control borers by pruning and burning dead and dying branches. Spray trunk monthly, May to September, using DDT, dieldrin, or Thiodan.
5. *Powdery Mildews*—Widespread. Powdery, grayish-white mold on leaves and young shoots (Figure 4.218). Leaves yellow, wither, and fall early. If severe, leaves may be stunted and malformed. *Control:* Where practical, spray two or more times, 10 to 14 days apart. Use sulfur or Karathane.

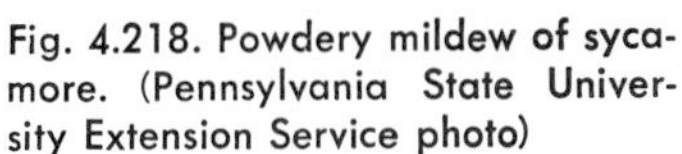

Fig. 4.218. Powdery mildew of sycamore. (Pennsylvania State University Extension Service photo)

6. *Leaf Spots and Blights*—Small, round to irregular, brown, gray, or black spots. If severe, leaves may blight and drop early. *Control:* Same as for Anthracnose (above).
7. *Crown Gall*—Rough, irregular, swollen galls form at base of trunk or on roots. Trees lack vigor. Make poor growth. Young trees may die. *Control:* See under Apple.
8. *Winter Injury, Frost Crack*—See under Elm.
9. *Root Rots*—Trees gradually decline in vigor. Foliage thin and weak. Leaves may turn yellow, wither, and drop early. *Control:* See under Apple and Oak.
10. *Cankers, Dieback, Twig Blight*—Twigs and branches discolored, wilt, and die back from cankers on twigs and limbs. Coral-pink to black pimples may form in affected bark. If severe, trees may die prematurely. *Control:* Avoid bark injuries. Maintain tree vigor. Fertilize and water during dry periods. Remove and burn cankered parts. Make cuts well below any sign of infection. Destroy severely infected trees. Control insects with malathion.
11. *2,4-D Injury*—Sycamore is quite susceptible. Leaves become thickened, twist or roll. May appear boat-shaped or curled into "ram's-horns." *Control:* See under Grape.
12. *Verticillium Wilt*—See under Maple, and (15B) Verticillium Wilt under General Diseases.
13. *Wetwood, Slime Flux*—See under Elm.
14. *Chlorosis*—Common in alkaline soils. See under Maple and Walnut.
15. *Rosy Canker of London Planetree*—Long, narrow cankers (inner bark water-soaked and pink to red) develop on trunks when roots are injured by manufactured illuminating gas. If severe, foliage may suddenly wilt and turn brown. Branches die

back. *Control:* Call gas company and have leak repaired. Follow suggestions outlined by gas company officials. See page 44.

16. *Sooty Blotch*—See under Apple, and (12) Sooty Mold under General Diseases.
17. *Mistletoe*—See (39) Mistletoe under General Diseases.

SYMPHORICARPOS — See Snowberry
SYNGONIUM — See Calla Lily
SYNTHYRIS — See Snapdragon
SYRINGA — See Lilac
TABEBUIA — See Trumpettree
TABERNAEMONTANA — See Oleander
TAENIDIA — See Angelica
TAGETES — See Chrysanthemum
TALLOWTREE — See Castorbean
TAMARACK — See Larch
TAMARISK [COMMON, FIVESTAMEN, FRENCH, KASHGAR, ODESSA, SMALL-FLOWERED], ATHEL TREE *(Tamarix)*

1. *Twig Blight, Canker*—Twigs and branches die back from discolored cankers. Foliage beyond withers and dies. *Control:* Prune out and burn dead and cankered wood. Keep trees vigorous. Fertilize and water during droughts.
2. *Wood Rot*—See under Birch.
3. *Root Rot*—See under Apple. Grow in sandy soil in full sun.
4. *Chlorosis*—Mineral deficiency in alkaline soils. See under Maple. Have soil tested and treat as recommended.
5. *Powdery Mildew*—See under Birch.

TAMPALA — See Cockscomb
TANBARK-OAK — See Oak
TANSY *(Tanacetum)*, TARRAGON, TARAXACUM, TASSELFLOWER — See Chrysanthemum
TASMANIA STRINGY-BARK — See Eucalyptus
TASSELTREE — See Dogwood
TAXODIUM — See Pine
TAXUS — See Yew
TEABERRY — See Heath
TEA OLIVE — See Osmanthus
TEASEL [COMMON, FULLERS] *(Dipsacus)*

1. *Leaf Spot*—Indefinite, often moldy spots. *Control:* Pick and burn spotted leaves. Burn tops in fall. If serious, spray several times, 10 days apart. Use zineb or maneb.
2. *Stem Rot, Crown Rot, Southern Blight*—Stems rot at soil line. May be covered with cottony mold. See under Delphinium.
3. *Mosaics*—Distinct, light and dark green mottling of leaves. Leaves may be somewhat malformed. Plants often partly stunted. May die early. *Control:* Keep down weeds. Destroy infected plants when first seen. Control aphids that transmit the viruses. Use malathion or lindane.
4. *Powdery Mildew*—Grayish-white blotches on foliage. *Control:* Pick and burn infected leaves. If practical, spray twice, 10 days apart. Use sulfur or Karathane.
5. *Leaf and Stem Nematode*—Flower heads dwarfed and distorted. See (20) Leaf Nematode under General Diseases. Sow disease-free seed.

6. *Root Rot*—See (34) Root Rot under General Diseases. May be associated with nematodes (e.g., lesion, stem and bulb).
7. *Downy Mildew*—See (6) Downy Mildew under General Diseases.

TECOMARIA — See Trumpettree

TETRAGONIA — See Beet

TEUCRIUM — See Salvia

TEXAS-BLUEBELL — See Gentian

TEXAS SILVERLEAF, TEXAS RANGER *(Leucophyllum)*

1. *Twig Canker*—Twigs die back from girdling cankers. See (22) Twig Blight under General Diseases.
2. *Root Rot*—See under Apple, and (34) Root Rot under General Diseases.

THALICTRUM — See Delphinium

THELOCACTUS — See Cacti

THERMOPSIS — See Pea

THIMBLEBERRY — See Raspberry

THIMBLEFLOWER — See Chrysanthemum

THISTLE — See Chrysanthemum and Lettuce

THOROUGHWORT — See Chrysanthemum

THRIFT — See Sea-lavender

THUJA, THUJOPSIS — See Juniper

THUNBERGIA — See Clockvine

THYME *(Thymus)* — See Salvia

TIARELLA — See Hydrangea

TICKSEED, TIDYTIPS — See Chrysanthemum

TIGERFLOWER *(Tigridia)* — See Gladiolus

TILIA — See Linden

TI PLANT — See Dracaena

TISSWOOD — See Silverbell

TITHONIA — See Chrysanthemum

TOADFLAX — See Snapdragon

TOBACCO — See Tomato

TOBIRA — See Pittosporum

TOLMIEA — See Piggyback Plant

TOMATO [CHERRY, COMMON, CLIMBING, CURRANT, GRAPE, MAYAN HUSK, PEAR] *(Lycopersicon)*; BROWALLIA; PEPPER [CAYENNE, CHERRY, CHILE or BIRD, CONE, PAPRIKA, PIMIENTO, RED, SWEET or BELL, TABASCO] *(Capsicum)*; TREE-TOMATO *(Cyphomandra)*; DATURA [COMMON, HINDU], ANGELS-TRUMPET *(Datura)*; APPLE-OF-PERU *(Nicandra)*; TOBACCO [FLOWERING, JASMINE, TREE] *(Nicotiana)*; CUPFLOWER or WHITECUP *(Nierembergia)*; PETUNIA [DWARF, GARDEN, SEASIDE, VIOLET, WHITE] *(Petunia)*; GROUNDCHERRY, HUSK-TOMATO, CHINESE LANTERNPLANT or WINTERCHERRY, TOMATILLO, STRAWBERRY-TOMATO, CAPE-GOOSEBERRY *(Physalis)*; PURPLE-FLOWERED GROUNDCHERRY *(Quincula)*; PAINTED-TONGUE *(Salpiglossis)*; BUTTERFLY-FLOWER, FRINGEFLOWER, POORMANS-ORCHID *(Schizanthus)*; EGGPLANT, SCARLET or TOMATO EGGPLANT, NIGHTSHADE, JERUSALEM-CHERRY or CHRISTMAS CHERRY, POTATO VINE *(Solanum)*

1. *Tomato Septoria Leaf Spot or Blight*—General over most of United States in warm (60° to 80° F.), moist weather. Numerous, small, roundish, whitish spots with dark margins. Spots later sprinkled with one or more black dots. Leaves may yellow and drop in large numbers starting at base of plant. Spots also occur on petioles, stems, blossoms, and fruit stems. Fruit size and quality are reduced. Exposed fruit may sunscald. See Figure 2.17B under General Diseases. *Control:* Turn under or burn plant debris after harvest. Three- to 5-year rotation excluding potato, pepper, eggplant, and tomato. Grow in well-drained soil where air circulation is good. Seedbed soil should be clean or pasteurized (pages 538–548). Treat seed as for Bacterial Spot (below) or plant disease-free, certified transplants. Spray in seedbed as for Seed Rots (below). Space plants widely or stake. Keep down weeds in and around (within 50 feet) garden area (e.g., Jimsonweed, horsenettle, groundcherries, and nightshades). Indoors, increase air circulation; keep night temperatures up (65° F. or higher than outdoors). Keep plants vigorous. Fertilize and water during dry periods. Grow only *tomato* varieties recommended for your area. Check with your county agent or extension horticulturist. Apply maneb, Dithane M-45, zineb, Miller 658, Daconil 2787, difolatan, Polyram, or fixed copper every 5 to 10 days. Start when *first* fruits are pea- to walnut-size. Thorough coverage is essential. *Do not use copper on eggplant.* Cultivate and handle plants only when they are dry. Somewhat resistant *tomatoes:* Earliana, Globe, John Baer, Perfection, Ponderosa, Rockingham, Sparks, Stone, and Trophy.
2. *Early Blight, Collar Rot, Alternaria Fruit Rot* (datura, eggplant, Jerusalem-cherry, nightshade, pepper, petunia, tobacco, tomato)—General. Brown to almost black spots on leaves and lower stem (Collar Rot). Plants may be stunted or even collapse. Most leaf spots are roundish to angular with concentric rings (target spots). Leaves turn yellow to brown and fall early, starting at base of plant. See Figure 2.19A under General Diseases. Fruit often sunscald. Brown to black, sunken, leathery spots may form near stem end of *tomato* and *pepper* fruit that may be covered with dark brown mold. Such fruit may drop early. Size and quality of fruit are poor. Spots on flower stems may cause blossoms to drop. *Control:* Same as for Septoria Leaf Spot (above) and Seed Rots (below). Avoid wounding fruit. Fixed copper is *not* effective against Early Blight. *Tomato* varieties having some resistance to Early Blight: Floralou, Indian River, Manalucie, Manapal, Marglobe, New Hampshire Surecrop, Norton, PA-103, Southland, Stone, Texto 2, and W-R Jubilee. *Tomatoes* resistant to Collar Rot: Southland and Urbana.
3. *Late Blight* (eggplant, Jerusalem-cherry, nightshade, petunia, tomato)—General and serious in humid regions and in cool, wet seasons. Irregular, greenish-black, water-soaked, rapidly enlarging, greasy spots on leaves, petioles, and stems. In moist weather, a whitish-gray downy growth appears, mostly on underside of leaves (Figure 4.219). Infected foliage soon dries, turns brown, and withers. *Tomato* fruit spots are greasy and dark green to brown or nearly black. Fruit are firm with a corrugated appearance. Bacterial Soft Rot often follows. Rotting of vines is very rapid in moist weather with cool nights (40° to 60° F.) and moderately warm, humid days (70° to 85° F.). Severely affected plants appear as if killed by frost. *Control:* Same as for Septoria Leaf Spot (above). If weather is cool and moist, spray every 5 to 7 days; if warm and dry, every 7 to 10 days. Sprays are much more effective than dusts. *Tomatoes* that may appear resistant: Garden State, New Hampshire Surecrop, Rockingham, Southland, and West Virginia 63.
4. *Other Leaf Spots and Blights, Leaf Mold, Anthracnoses*—Widespread in rainy seasons. Small to large, round to irregular spots of various colors; often with borders of different color. Spots also occur on petioles and stems. Leaves may turn yellow to brown and fall early. Injury is most evident on lower leaves in moist weather. *Control:* Same as for Septoria Leaf Spot (above). Or spray with Polyram.

Fig. 4.219. Tomato late blight on leaves. (Illinois Agricultural Experiment Station photo)

Tomato varieties resistant to Gray Leaf Spot *(Stemphylium)*, where adapted: Chico, Chico 66, El Monte, Floralou, Florida-Hawaii Hybrid, Indian River, La Pinta, Manalee, Manalucie, Manapal, Marion, N-64, Ohio W-R Red Jubilee, PA-103, Rio Grande, Supermarket, Tecumseh, Weshaven, and W-R Jubilee. *Tomato* varieties frequently resistant to one or more races of Leaf Mold *(Cladosporium)*, a common indoor problem: Bay State Forcing, Florida-Hawaii Hybrid, Improved Bay State, Improved Vetomold, Indian River, Leaf-Mold Resistant Marglobe, Manalucie, Manapal, Mold-Resistant Waltham Forcing, Ohio W-R 25, P-115, Pioneer, Stirling Castle, Tuckcross (Hybrid, K, O, V, W), V-121, Vantage, V.F. N8, Vinequeen, Waltham Improved, Waltham Mold-Proof Forcing, and Weibull's Immuna. Indoors, keep humidity below 85 per cent and increase ventilation. Maintain temperature of 65° F. Keep night temperature higher than that outdoors so no dew forms on foliage. Keep water off foliage. Water in morning. Space plants to avoid unnecessary shading. Appleblossom *petunia* is resistant to Botrytis Blight.

5. *Fruit Spots and Rots, Anthracnoses*—Cosmopolitan and serious on eggplant, pepper, and tomato. Small to large, circular to irregular fruit spots. Usually sunken, water-soaked, dark yellow or green (pepper), tan or brown to black, often with concentric zones. Fruit may shrivel and darken or become watery and collapse. Most common on ripening fruit. Fruit may later be covered with white, pink to red, black, tan, gray, green or yellow to orange mold. Bacterial Soft Rot may follow. Internal rot may be moldy or slimy and pink, gray, or brown to black. Some fruit-rotting fungi produce yellow, brown, or gray spots on leaves and stems. See Other Leaf Spots (above), Phytophthora Blight (below), and Figure 2.48B under General Diseases. *Control:* Destroy all rotting fruit. Keep *tomato* fruit off soil by staking or mulching. Handle carefully to avoid cuts and bruises. Keep fruit dry and cool. Grow crack-free *tomato* varieties. Otherwise same as for Septoria Leaf Spot

(above). Spray with maneb, zineb, thiram, captan, difolatan, ziram, Dyrene, folpet, or Daconil 2787. Follow manufacturer's directions. *Botrytis* on indoor *tomatoes* can often be checked by liming acid soils to a pH of about 7. Grow in well-drained, disease-free soil, if possible. Control insects using sevin plus malathion. Treat seed as for Bacterial Spot (below). Indoors, same as for Other Leaf Spots (above). Resistant *eggplant* varieties to Phomopsis Blight: Florida Beauty, Florida Market, and Florida Market 10. *Pepper* and *tomato* varieties also differ in resistance to certain rots (e.g., *tomato* to Nailhead Spot: Break O'Day, Glovel, Marglobe, and Pritchard; and to Buckeye Rot).

6. *Sunscald* (primarily pepper and tomato)—General in hot, dry weather where blights, wilts, and leaf spots are uncontrolled. Most common on staked tomatoes. Large, irregular, slightly sunken, whitish areas on fruit with papery texture. Black, gray, or dark green mold may develop on affected surfaces (Figures 1.1A and 4.220). Entire fruit may later rot. White or reddish spots may develop in middle of leaves on young plants. *Control:* Same as for Septoria Leaf Spot (above). Water and fertilize to keep plants vigorous. Avoid varieties with sparse foliage. Handle fruit carefully. Grow wilt-resistant varieties, where practical.

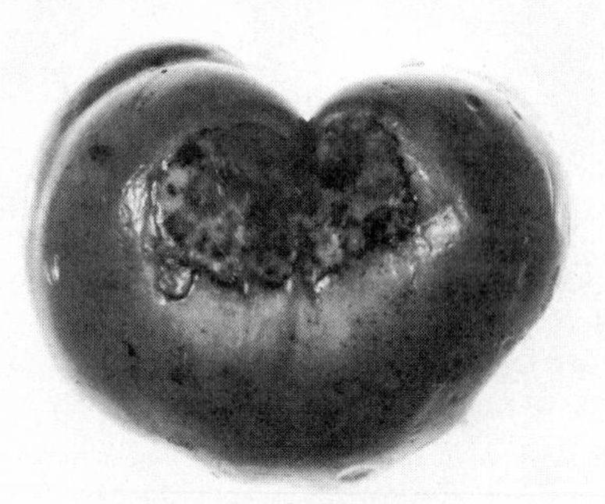

Fig. 4.220. Tomato sunscald. Note growth of secondary fungi on scalded area. (Illinois Agricultural Experiment Station photo)

7. *Blossom-end Rot, Leather-end* (primarily pepper and tomato)—General and may be serious. Slight water-soaked area forms at or near bottom of fruit (blossom-end) that later enlarges and darkens. Becomes dry, sunken, flat, leathery, and finally dark brown to black (tomato); or light-colored and papery (pepper). Spots vary in size. May affect ½ of fruit. Various molds and fruit rots may follow. See Figure 1.1A. *Control:* Grow in well-drained soil. Water during dry periods to maintain *even* soil moisture supply and promote *steady* growth. Mulch or cultivate shallowly during dry periods. Fertilize adequately, based on a soil test. Keep phosphorus level up. Avoid overfertilizing (especially with nitrogen and potassium), and too deep or close cultivation. Use sufficient calcium-containing fertilizer, gypsum, or lime. Or spray when rot is anticipated (first fruits size of golf balls). Some states recommend 4 weekly sprays of calcium nitrate or anhydrous calcium chloride (½ to ¾ ounce per gallon of water). Before using, check with a local grower, your county agent, or extension horticulturist. Avoid severe pruning and too close planting. Control leaf spots and blights as given under Septoria Leaf Spot (above). Varieties differ in susceptibility. Bradley and Summertime *tomatoes* are tolerant.
8. *Tomato Fruit Cracking*—Radial cracks grow out from stem of fruit like a "star" or concentric cracks occur as rings around center or shoulders of fruit (Figure 4.221).

Fig. 4.221. Tomato fruit cracking.

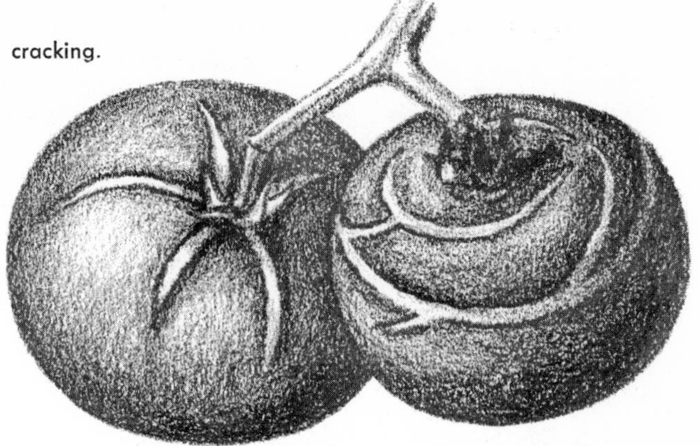

Cracking greatly increases the possibility of fruit decays. *Control:* Same as for Blossom-end Rot (above). Numerous resistant varieties are available. Where practical, do not stake plants.

9. *Bacterial Soft Rots* (eggplant, Jerusalem-cherry, nightshade, pepper, tomato) — Cosmopolitan. Water-soaked, greasy areas on fruit or stem that rapidly enlarge. Usually accompanied by a foul odor. In 3 to 10 days entire fruit is soft, watery, and hangs limply. Follows other diseases, growth cracks, and insect or mechanical injuries. See Figure 4.222. *Control:* Same as for Septoria Leaf Spot and Fruit Rots (both above). Handle fruit carefully. Refrigerate promptly after harvest. Control insects with malathion, rotenone, or sevin sprays applied at least weekly.
10. *Bacterial Spot, Canker, Speck, Wildfire* (primarily eggplant, groundcherry, pepper, and tomato) —General over most of United States in wet seasons. Small, "scabby," whitish, brown spots or black specks and spots develop on green fruit. See Figures 4.222, 4.223 and 4.225C. Fruits may be spotted, roughened, cracked, and distorted.

Fig. 4.222 (below). Bacterial soft rot of tomato (left). Bacterial canker of tomato (right).

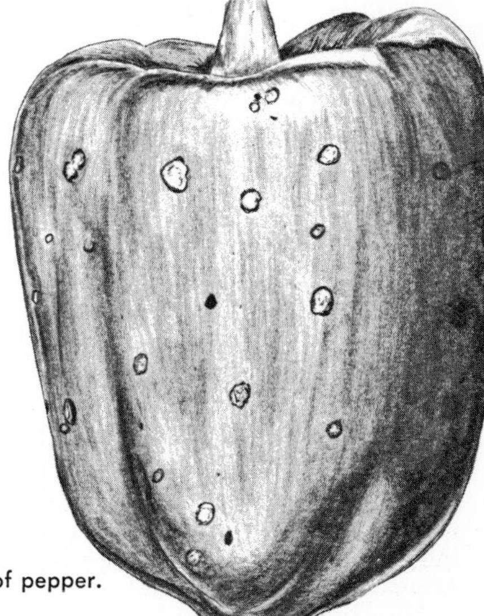

Fig. 4.223. Bacterial spot of pepper.

Such fruits later rot. Flowers may be blasted and drop off. Plants may be stunted. Small, yellowish-green, brown or black, greasy-appearing leaf spots, often with yellowish margins. Or small, round, pale green "warts" may form. Spots later may enlarge. Leaves may roll, turn yellow or brown, wilt, and drop off, usually starting at base of plant. This often results in sunscald. *Canker* produces elongated, yellowish to yellow-brown stripes or cracks on some *tomato* stems. Upper parts of older plants suddenly wilt. Later, lower leaves wilt, curl, wither, and die. Calyx end of certain fruit may be brown with calyx brown, withered, and cracked. Small, brown, cracked areas—surrounded by a white halo—may form in fruit (Figure 4.222). Seedlings may be destroyed or remain stunted. The *bacterial spot* organism may be present externally and internally without producing symptoms. *Control:* Plant certified, disease-free seed or transplants in disease-free soil. Soak uncertified *tomato* and *eggplant* seed in hot water (exactly 122° F. for 25 minutes), cool, dry, and dust with thiram, captan, or Semesan before planting. See Table 20 in Appendix. Soak *pepper* seed for 5 minutes in a warm (70° F.), 1:2,000 solution of mercuric chloride, rinse for 15 minutes in running water, dry, and dust as above. Or soak *pepper* seed in hot water (125° F. for 30 minutes), cool, dry, and dust as above. Destroy plant debris after harvest. Avoid overhead irrigation whenever possible. Cultivate and handle plants only when they are dry. Four-year rotation. Exclude pepper, tomato, and related weeds in tomato-potato family. Several seedbed or field sprays of Agrimycin 500 or fixed copper (1½ tablespoons per gallon) plus maneb or zineb (1 tablespoon) or streptomycin (200 parts per million) should help. Do *not* use streptomycin *after* fruits start to form. Check with your county agent or extension plant pathologist. Maintain *high balanced* fertility based on a soil test. *Pepper* varieties somewhat resistant to Bacterial Spot: Anaheim, Blight-resistant World Beater, Calcom, Mexican Woolley, Long Red Cayenne, Red Chili, Santanka, Squash, Sunnybrook 833, Sunnybrook Sweet Cheese, Sweet Yellow, Waltham Beauty, Wonder, and Yellow Oshkosh.

11. *2,4-D or 2,4,5-T Injury*—General. Tomato and related plants are very sensitive. Fumes and spray drift may injure plants ½ mile or more away. Leaves and stems often considerably curled, distorted, and twisted. Leaf margins are wavy or frilled. Some blossoms drop without setting fruit. Tomato fruit often cone-shaped, crack open, or are cat-faced (see below). Contaminated sprayers and pesticide containers often responsible for injury. *Control:* See under Grape.
12. *Mosaics, Calico, Streaks, Yellow Net, Tobacco Etch*—General. May be serious. Symptoms variable depending on virus, virus strain, plant, variety, time of year, and environmental conditions. Leaves may be mildly to severely mottled with yellowish-green to light and dark green mosaic patterns (Mosaics). Or develop irregular, oakleaf or ring patterns. Affected leaves usually curled, crinkled, puckered, and deformed (Figures 4.224A and 4.224B). May resemble 2,4-D injury with narrow (fernleaf), spindly, shoestring-like leaves (Cucumber Mosaic virus). Brown or purple streaks may appear on stems, petioles, and along leaf veins (Streaks, Potato Virus Y). Plants commonly stunted, yellowed, and bushy. Fruit often reduced in size and number. May be roughened or malformed and whitish, yellow and green. Dead blotches, streaks, rings, mottled patterns or corky brown areas (Internal Browning) may develop under the skin. Blossoming is reduced. *Petunia* flower petals may be mottled, flecked, streaked, or show ring patterns. May not open completely. Long brown streaks may develop on *tomato* and *pepper* stems and fruit, accompanied by dropping of leaves, blossoms, and fruit—especially on pepper. Stems may be very brittle. *Eggplant* is often killed. *Control:* Destroy infected plants when first found. Keep down *all* weeds around plant-growing area (especially wild groundcherries, plantains, horsenettle, Jimsonweed, nightshades, catnip, Jerusalem-cherry, milkweeds, motherwort, white cockle, burdock, wild

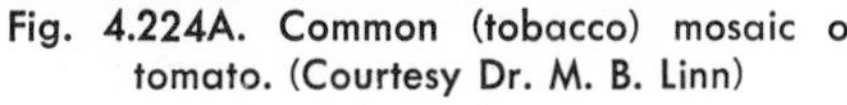

Fig. 4.224A. Common (tobacco) mosaic of tomato. (Courtesy Dr. M. B. Linn)

Fig. 4.224B. Pepper mosaic.

cucumber, flowering spurge, pokeweed, bittersweet, and matrimony vine). See also under Cucumber Mosaic, page 241. Set out disease-free, certified transplants. Or start with virus-free seed (seed 1-year-old is free of virus). Scrub hands thoroughly with soap and hot, running water before touching healthy plants. Avoid use of tobacco while working with plants. Losses from certain viruses may be reduced by spraying tomato and pepper plants several hours before transplanting. Use 1 gallon of whole or skim milk or 1 pound of dried skim milk mixed with a gallon of water. Apply to 20 square yards of garden or plant bed. Dip hands in whole or skim milk, or 4 ounces of dried skim milk per quart of water, every few minutes to keep them moist, whenever handling plants (e.g., plant pulling, transplanting, pruning, tying). Control insects, especially aphids, cucumber beetles, flea beetles, and grasshoppers that transmit certain viruses. Use malathion or Diazinon plus sevin. *Pepper* varieties usually resistant to Tobacco Mosaic: Allbig, Burlington, Calcom, Caldel, California Wonder-300, Delaware Belle, Delbell, Early Wonder, Emerald Giant, Florida Giant Resistant, Fresno Chili, Illinois Bell, Improved Yolo, Keystone Resistant Giant, Liberty Bell, Long Thin and Thick Cayenne, Merced, Mexican Chili M4, Mosaic-Resistant Wonder Giant, Pacific Bell Sweet, Paul's Jersey Giant, Pepperoncini, Rutgers Worldbeater No. 13, Santa Fe Gem, Thickwall World Beater, Titan, and Yolo Wonder (A, B, L, Improved B M.R.). N-64 *tomato* is resistant to Tobacco Mosaic. Italian El and Neopolitan *peppers* are resistant to Potato Virus Y. Certain strains of Tobacco Mosaic virus may be carried on *tomato* and *pepper* seed. Seed may be freed of virus by extracting *tomato* seed with hydrochloric acid (HCl). Use 25 milliliters of commercial HCl to 5 pounds of fruit pulp.

13. *Double-Virus Streaks* (tomato)—Symptoms variable depending on age of plant, time of year, virus combination, amount and duration of light, etc. Leaves often mottled green and yellow. Numerous, irregular, grayish-brown to black, papery

spots on young leaves. Stems and petioles develop numerous, dark brown or black streaks. Top of plant may die rapidly or plants become dwarfed, spindly, yellowish, and nonproductive. Fruit may be deformed with irregular, dark brown, greasy to "glassy" blotches. *Control:* Same as for Mosaics (above).

14. *Tomato Graywall, "Blotchy Ripening," Internal Browning*—Symptoms variable. Ripening fruit have uneven coloring. Gray to grayish-brown blotches and streaks develop on surface of green fruits when there is internal browning of tissues in outer wall. Common on fruits from interior of plants with very heavy foliage. *Control:* Maintain adequate potash levels in soil, based on a soil test. Indoors, maintain even, moderate temperature. Avoid succulent plants and Tobacco Mosaic infection (see above). Resistant varieties: Floralou, Globe A strain, Indian River, Manalucie, Manapal, Ohio W-R 7, 25, and 29.
15. *Spotted Wilt, Tip Blight*—Widespread. Symptoms differ greatly with plant, variety, age, and virus strain. Numerous tiny, brown, yellowish-green, ringlike spots or bronzing of young leaves and petioles commonly occur. Growing *pepper, tomato,* and *Schizanthus* tips may show dark streaks, wilt, and wither. Leaves may be mottled yellow and distorted. Plants often stunted and bushy with drooping leaves. Occasionally die. Small to large pale red, yellowish, white or dark brown to almost black, irregularly mottled blotches (often composed of concentric rings) may develop on *tomato* and *pepper* fruit. Fruit may be roughened and distorted. See Figure 2.35A under General Diseases. *Petunia* flowers may show irregular light streaks and blotches. *Control:* Same as for Mosaics (above). Control thrips that transmit the virus. Use malathion plus sevin. *Tomato* varieties differ in resistance. Florida-Hawaii Hybrid and Pearl Harbor are resistant to certain strains of the virus.
16. *Ringspot Complex*—Tip leaves are curled and brown. Yellowish-green, dark green, gray or brownish ring patterns (also line and oakleaf markings) form on leaves, petioles, and fruits. Plants may be yellowed and stunted, with puckered, mottled, and malformed leaves. *Pepper* and *tomato* fruit are often misshapen. Grasshoppers, thrips, and flea beetles transmit the viruses. *Control:* Same as for Mosaics (above). Control dagger nematodes *(Xiphinema)* that transmit at least 2 of the viruses (Tomato and Tobacco Ringspots), and stubby-root nematodes *(Trichodorus)* that spread Tobacco Rattle virus of *pepper.* See under Root-knot (below).
17. *Curly-top, Western Yellow Blight*—Normally western half of United States. Plants stunted and spindling or dwarfed and bushy. Seedlings may turn yellow and die. A pronounced twisting, upward curling, or cupping of young leaves is common. Foliage becomes stiff, brittle, and droops. Flowers often malformed. Numerous secondary shoots are often formed. *Pepper* leaf stems curve sharply downward. Plants eventually become yellowed and dwarfed; stiffly erect. Few deformed pepper and tomato fruit are produced after infection occurs. *Tomato* fruits ripen regardless of size. *Control:* Same as for Mosaics (above), and (19) Curly-top under General Diseases. Control leafhoppers that transmit the virus, using malathion plus sevin. Partially resistant *tomatoes:* Line 193, Owyhee, and Payette. More resistant varieties should be available shortly. Shading young plants under "hot caps," muslin, or cheesecloth may be practical in the home garden. Plant early and heavy. Avoid planting near beets.
18. *Eggplant Yellows*—Caused by Tobacco Ringspot virus. Plants turn yellow starting at top and progressing downward. Oakleaf patterns may form and leaf drop may occur. Plants stunted. Yield is greatly reduced. *Control:* Same as for Ringspots (above).
19. *Fusarium Wilts* (browallia, painted-tongue, pepper, petunia, tomato)—Serious in hot weather (80° to 95° F.), especially in central, southern, and western states. Seedlings wilt and die. Older plants are stunted. Leaves wilt, turn yellow, wither,

and drop off starting at base of plant. Wilting and yellowing or browning may first be only on one side of plant. Brown to almost black streaks occur inside lower stem and roots (Figures 2.31A and 4.225A). A brown canker may girdle stem near soil

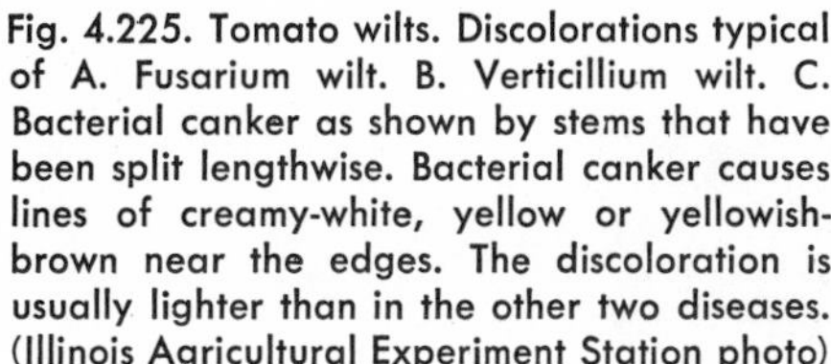

Fig. 4.225. Tomato wilts. Discolorations typical of A. Fusarium wilt. B. Verticillium wilt. C. Bacterial canker as shown by stems that have been split lengthwise. Bacterial canker causes lines of creamy-white, yellow or yellowish-brown near the edges. The discoloration is usually lighter than in the other two diseases. (Illinois Agricultural Experiment Station photo)

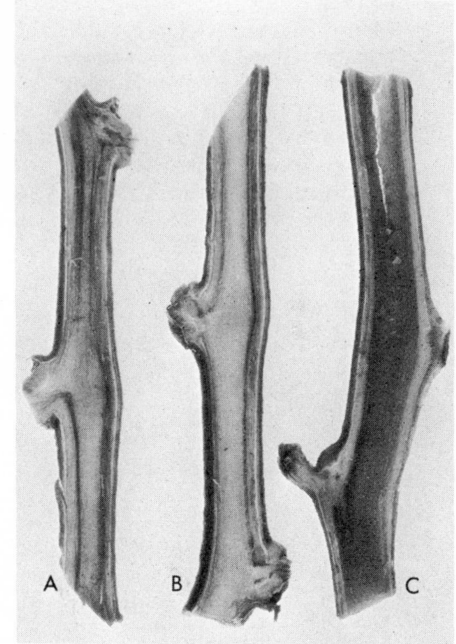

line. Roots discolored and rotted. Fruit size and yield are reduced. Injury may be increased by root-feeding nematodes (e.g., lesion, root-knot). *Control:* Plant certified, disease-free seed or transplants in fertile, well-drained, wilt-free soil. If practical, sterilize or fumigate soil in seedbeds, cold frames, and greenhouse beds with heat, chloropicrin, methyl bromide, or Vorlex (pages 538–548). Practice 3- or 4-year rotation. Use normally disease-resistant, adapted *tomato* varieties in infested soil: Ace 55 (VF), Alamo, Avalanche, Blackhawk, Blair Forcing, Boone, Bradley, Brookston, Buckeye State, Campbell 1327, Chesapeake, Chico 66, Coronet (OW 6), CPC-2, Delsher, Enterpriser, Epoch, Floralou, Florida-Hawaii Hybrid, Fortune, Garden State, Golden Marglobe, Grothen's Globe WR, Heinz 1350 and 1370, Homestead, Illinois Pride, Imperial, Indark, Indian River, Jefferson, Kanora, KC-146, Kokomo, Kopiah, Manalee, Manalucie, Manapal, Manasota, Marion, Marvana, Merbein Early and Midseason, Michigan-Ohio Hybrid, Michigan State Forcing, Mocross Supreme and Surprise, Nemared, N-64, Ohio W-R Globe, Ohio W-R Red Jubilee, Ohio W-R (7, 25, and 29), PA-103, Pearson VF-6 and VF-11, Pink Prize, Pinkshipper, Porte, Red Global, Red Top V9, Rio Grande, Riverside, Rockingham, Roma, Roma VF, Rutgers Hybrid, Scarlet Knight, September Dawn, Simi, Solid Red Strain B, Southland, Spartan Pink No. 10, Stair, Sunray, Sun-up, Superman, Supermarket, Surprise, Tecumseh, Texto 2, Tippecanoe, Tipton, Tom Boy, Tom Tom, Tucker's Forcing, Tuckcross (Hybrid, K, O, W), VF (8, 13L, 14, 36, 145), West Virginia 63, Wiltmaster, W-R Jubilee, and Young. Resistant *peppers:* College No. 6, and Mexican Chili No. 9. Resistance may break down at high temperatures (over 90° F.). Work soil around plants as season progresses.

20. *Verticillium Wilt*—Widespread during cool, moist seasons, especially in northern

states. Favored by temperatures of 70° to 75° F., retarded by higher temperatures. Symptoms very similar to Fusarium Wilt (above). *Eggplant* is commonly attacked. Midday wilting and evening recovery is common for a time. Root-feeding nematodes (e.g., lesion) often encourage penetration of the wilt fungus. See Figures 2.32A and B under General Diseases and Figures 4.225B and 4.226. *Control:* Long rotation excluding tomato, pepper, eggplant, potato, melons, okra, bramble fruits, and strawberry. Use disease-free soil (pages 538–548), seed, and transplants. Normally resistant *tomato* varieties where adapted: Ace 55 (VF), Campbell 1327, CPC-2, Earlypak, Enterpriser, Essar, FM 428 VF, G 790, Galaxy, Geneva 790, Grand Pak, Heinz 1350, Highlander, Loran Blood, Manhattan, NR-13, Pearson strains, Porte, Red Top VR-9, Riverside, Roma VF, Rutgers Hybrid, Simi, Superman, Tom Tom, Utah-13, VF (6, 11, 13L, 14, 36, 145, 365, N8), VR Moscow, and West Virginia 63. Resistant *eggplant* and *pepper* varieties should be available soon.

Fig. 4.226. Verticillium wilt of tomato. (Illinois Agricultural Experiment Station photo)

21. *Bacterial Wilt, Brown Rot*—Mostly southern states or where southern-grown transplants are used. Whole plant is stunted, usually wilts rapidly. Plants die, usually without yellowing or spotting of leaves. Inside of stem dark brown, water-soaked, and decayed near ground line. Root-feeding nematodes often provide entrance wounds. *Control:* Plant disease-free seed or certified transplants early in season in well-drained, disease-free soil. Destroy first infected plants. Collect and burn crop debris after harvest. Four- or 5-year rotation in southern states. Exclude tomato, pepper, eggplant, potato, beans, peanut, groundcherry, and related plants. Grow resistant varieties where available and adapted. Somewhat resistant *eggplant:* Florida Market. Resistant *tomato* varieties may be available soon.
22. *Seed Rots, Damping-off, Stem and Collar Rots, Foot Rots*—Cosmopolitan. Seeds rot. Stand is thin. Seedlings wilt, wither, may collapse and die from rot at soil line

Fig. 4.227. Damping-off of pepper seedlings.

(Figure 4.227). Canker at base of stem girdles and kills foliage beyond. Mold may develop on affected parts. *Control:* Treat seed as for Bacterial Spot (above) then plant in warm (70° F.), fertile, well-drained, clean or sterilized soil (pages 538–548). Spray seedlings in seedbed or flats at 4- to 7-day intervals. Use captan or zineb (1½ tablespoons per gallon), ziram, or ferbam (2½ tablespoons per gallon). Spray both soil and plants using *low* pressure. Use 1 gallon of spray to cover 20 square feet of surface. For *pepper* and *tomato* use 50:50 mixture of captan 50 and PCNB 75. See page 104. Use ½ pint per plant at transplanting time. Keep seedbed soil on dry side. Avoid overcrowding and overwatering. After harvest burn old plant debris. Otherwise, same as for Septoria Leaf Spot (above).

23. *Root-knot and Other Root-feeding Nematodes* (e.g., awl, burrowing, cyst, dagger, lance, lesion, naccobus, pin, reniform, ring, sheath, spiral, stem-rot, sting, stubby-root, stunt) —Widespread and serious, especially in southern states. Plants may lack vigor, yellowed, and stunted with variously sized galls and swellings on roots (Root-knot, Cyst nematodes). Roots may be stunted, stubby, "bushy," and discolored. Plants may wilt in dry weather but recover at night for a time, then die. See Figure 2.52D under General Diseases. Tomato, eggplant, most sweet peppers, and petunia are highly susceptible. *Control:* Plant certified, disease-free transplants grown in sterilized or fumigated soil, or sow seed in similarly treated soil. Use heat, EDB, D-D, Vidden-D, chloropicrin, Telone, Vorlex, or methyl bromide. Follow manufacturer's directions. Many home gardens in the south are fumigated each fall to protect against Root-knot and other nematodes (page 548). Check with a local grower, your county agent, extension horticulturist, or plant pathologist. *Pepper* varieties resistant to certain species of Root-knot: Anaheim Chili, Bush Red, Cayenne, Italian Pickling, Nemaheart, and Santanka. Resistant *tomatoes* to one or more Root-knot species and biotypes: Anahu, Florida-Hawaii Hybrid, Gilestar, Hawaii 55, Kolea, Kolohi, Merbein (Canner, Early, Midseason, Monarch), N-64, Nemared, Tuckcross K, and VF-N8.

24. *Aster Yellows*—Plants yellowish, bushy, and dwarfed with many rosette-like, secondary shoots. Flowers often malformed and greenish. *Tomato* leaves may curl; turn purple or yellow. *Control:* Same as for Curly-top and Mosaics (both above).
25. *Southern Blight, Watery Soft Rot, Stem Rots or Cankers*—General and destructive in warm, moist weather. Plants may gradually wilt and die. Stem rots at or below soil line. Cottony mold may be evident on lower branches and stem in which irregular, brown to black fungus bodies (sclerotia) are embedded. Fruits on or near soil may become soft and watery. *Control:* Carefully dig up and burn wilting plants together with several inches of surrounding soil, when first seen. Plant seedbed on clean or fumigated soil using SMDC (Vapam or VPM Soil Fumigant), chloropicrin, Mylone, or Vorlex (pages 538–548). Four-year rotation or longer. Use disease-free transplants. Resistant *peppers:* Santanka and Tabasco. Keep calcium level high in the soil.
26. *Phytophthora Blight, Fruit Rots* (eggplant, pepper, petunia, tomato)—Slight yellowing, sudden wilting and drying of leaves. Irregular, small to large, water-soaked areas on fruit, that are wrinkled, sunken, and light gray or brown. Fine, cottony mold may grow on diseased areas. Pale to dark brown or dark green, water-soaked cankers may girdle stems causing withering and dying of parts above. Dark green spots on leaves enlarge and become bleached. Seedlings damp-off. *Petunias* develop black, dry rot of crown and lateral branches near soil line. *Control:* Same as for Septoria Leaf Spot and Fruit Rots (both above). Grow in clean or sterilized soil. Resistant *eggplant:* Colossal Florida High Bush. Resistant *pepper:* Oakview Wonder. *Tomatoes* also differ in resistance. Avoid sprinkling foliage. A weekly soil drench of Dexon is effective on *petunia.*
27. *Root Rots*—Plants stunted. Tend to wilt in hot weather or in bright sun. Roots and stem base discolored; later rot away. Yield reduced. Often associated with root-feeding nematodes. *Control:* Same as for Seed Rots and Root-knot (both above). Rotate. Avoid deep and close cultivations. Hilling up soil around base of plant may help. Keep plants vigorous through proper fertilization. Water during droughts. Spray plants in field as for Septoria Leaf Spot (above). An in-furrow application or spray of captan-PCNB or copper-PCNB is effective on *pepper.* Zineb or Dithane A-40 is effective for certain *tomato* root rots. Follow manufacturer's directions.
28. *Chlorosis*—Common in very acid or in alkaline soils. Foliage turns yellow, yellow-green or pale green, especially between veins. Affected leaves may wither and die. Plants often stunted. *Control:* Usually due to deficiency of iron, zinc, manganese, or magnesium. Add a tablespoon of the sulfate form of these elements in a gallon of water to one or more pest sprays. Or use corresponding chelate (pages 26–27). Follow manufacturer's directions. Maintain balanced soil fertility based on a soil test. The soil reaction (pH) should be near neutral.
29. *Potassium Deficiency*—Plants woody and slow-growing. Some leaves dark bluish-green. Margins and between veins of *lower* leaves turn yellow, then scorched, and curled. Drop early. Fruit often fail to ripen evenly. May show blotchy ripening and drop early. *Control:* Have soil test made. Follow suggestions in the report.
30. *Downy Mildew, Blue Mold* (eggplant, pepper, tobacco, tomato)—Southern states, chiefly in seedbed during cool, wet weather. Pale green to yellow or brown spots on upper leaf surface of seedlings. A delicate, white to pale blue mold covers corresponding spots on lower leaf surface. Seedlings killed. Seedbed may appear "scalded." Leaves on older plants shrivel and drop early. Plants usually recover when weather turns warm and dry. *Control:* Same as for Seed Rots (above). Grow eggplant, pepper, and tomato seedlings as far from tobacco seedbeds as possible.
31. *Psyllid Yellows, Purple Top*—Primarily western half of United States. Plants stunted. New growth yellowed and spindly after feeding of psyllid insects. Prior growth remains normal. Leaves formed later are purplish, stunted, twisted and

curled upward, and leathery. Terminal leaves turn reddish-purple. Fruit yield and quality is reduced. See also under Potato. *Control:* Spray malathion plus sevin or methoxychlor to control psyllids.

32. *Cloudy Spot* (tomato)—Irregular, whitish to yellow spots, up to ½ inch in diameter, develop just under skin of green and ripe fruit. These are caused by feeding punctures of stink bugs. *Control:* Follow regular spray program using malathion plus sevin or methoxychlor.
33. *Puffing, "Pops"* (tomato)—Gulf states and California, especially on winter and spring crops. Also an indoor problem. Irregularly shaped fruits with large, air-filled cavities inside. Angular "corners" occur on green fruit. *Control:* Avoid planting such susceptible varieties as Marglobe, Pritchard, and Urbana. Fertilize based on a soil test. Maintain uniform soil moisture supply. Resistant varieties: Hotset, Porter, Summer Prolific, and Summerset.
34. *Blossom Drop* (primarily tomato and pepper)—Fruits fail to set, especially on early flower clusters. Yield reduced. Usually caused by high or low temperatures, low humidity, hot dry winds, lack of rain, beating rains, 2,4-D injury, diseases, or an overabundance or lack of nitrogenous fertilizer. *Control:* Set plants early. Plant adapted varieties least sensitive to temperature changes. Check with a local grower, your county agent, or extension horticulturist. Maintain an adequate supply of soil nutrients and moisture. If feasible, apply a blossom-set hormone spray. Follow manufacturer's directions.
35. *Catface* (tomato)—Fruit extremely malformed and scarred at blossom end, with irregular, swollen protuberances and bands of scar tissue. Starts with deformation of the stigma in developing flower several days before pollination. *Control:* Grow resistant varieties, e.g., Manapal, Young. May be caused by extreme heat or cold, drought, and contact with hormone-type herbicide.
36. *Physiological Leafroll* (tomato)—Older and lower leaves roll upward and inward following extended periods of wet or dry weather. Leaves become stiff and leathery (Figure 4.228). The "disease" later progresses up plant. *Control:* Same as for Blossom-end Rot (above). Avoid severe pruning. Grow in well-drained soil. Varieties differ greatly in resistance. Fruiting is unaffected.

Fig. 4.228. Physiological leafroll of tomato.

37. *Virus Leafroll* (tomato)—Young leaflets curl downward while older ones roll upward. Margins yellow. Terminal leaflets stunted. Fruits reduced in size and number. *Control:* Same as for Mosaics and Curly-top (both above).
38. *Pseudo Curly-top* (tomato)—Florida. Plants stop growing; become stiff and brittle.

Leaves roll upward, become leathery, yellowed, and purple on undersurface. Axillary shoots sometimes produced. Infection occurs mainly in August and September. *Control:* Keep down weeds, e.g., nightshades and ragweed. Spray margins of garden or field with malathion and sevin to kill treehoppers that spread virus.

39. *Blight, Anthracnose* (butterfly-flower)—Small, water-soaked spots on stems and petioles. Rapidly growing shoots later wilt, turn brown, and die back from tips. Girdling cankers often develop on older stems causing yellowing and death of foliage beyond. Small brown spots occur on leaves. *Control:* Prune and burn affected parts. Spray as for Septoria Leaf Spot (above). Start when first spots appear.
40. *Powdery Mildew* (primarily butterfly-flower, eggplant, petunia, salpiglossis, tobacco, tomato, tree-tomato)—Powdery, white blotches on leaves. *Control:* If necessary, apply sulfur or Karathane. Avoid overfertilizing with nitrogen.
41. *Leafy Gall, Fasciation* (butterfly-flower, petunia, tobacco)—Uncommon. Abnormal, swollen, fleshy shoots with numerous, deformed, leaflike structures. See (20) Leafy Gall under General Diseases. *Control:* Plant seed or seedlings in soil that has been heat- or chemical-treated (pages 538–548).
42. *Crown Gall, Hairy Root* (Jerusalem-cherry, nightshade, tomato)—Rough, swollen galls on stem near soil line. *Control:* Carefully dig up and burn infected plants. Do not replant in same location for 5 years without first drenching soil with SMDC (Vapam or VPM Soil Fumigant).
43. *Rust* (datura, eggplant, petunia, purple-flowered groundcherry)—Small, yellowish pustules on leaves. *Control:* Spray as for Septoria Leaf Spot (above). Keep down wild grasses that are alternate hosts.
44. *White Smut* (browallia, Chinese lanternplant, groundcherry)—See (13) White Smut under General Diseases.
45. *Web Blight*—Southeastern states. See under Bean.
46. *Scab* (eggplant)—See under Potato.
47. *Black Walnut Injury*—Plants growing under black walnut trees wilt and die. May closely resemble Fusarium Wilt (above). *Control:* Avoid growing within 50 feet of these trees.
48. *Tomato Crease Stem*—Southern states. Plants dwarfed, rigidly upright, and bunchy. Upper parts of certain main stems develop flattened areas with deep longitudinal creases. Brown discoloration commonly occurs near creases. *Control:* Grow in well-drained soil. Avoid use of excessive amounts of nitrogenous fertilizer before plants have many fruits larger than a golfball. Resistant varieties: Firesteel, Indian River, Manapal, and Stokesdale.
49. *Stem Nematode* (Chinese lanternplant, groundcherry, schizanthus, tomato)—See under Phlox.
50. *Oedema*—Indoor problem. Small, grayish, raised eruptions usually extend along the veins on underside of leaves. *Control:* Increase ventilation and heat.
51. *Blossom Blight*—See (31) Flower Blight under General Diseases.

For additional information read USDA Handbook 203, *Tomato Diseases and Their Control,* and Info. Bulletin 276, *Pepper Production.*

TORCH FLOWER — See Chrysanthemum

TORCHLILY — See Redhot-pokerplant

TORENIA — See Snapdragon

TOTAI — See Palms

TOYON — See Apple

TRACHELOSPERMUM — See Oleander

TRACHYCARPUS — See Palms

TRACHYMENE — See Celery
TRADESCANTIA, WANDERING-JEW, SPIDERWORT *(Tradescantia);* DAYFLOWER [CREEPING, VIRGINIA] *(Commelina);* WANDERING-JEW *(Zebrina)*

1. *Leaf Spots*—Spots of various colors, sizes, and shapes. *Control:* Pick and burn spotted leaves. If practical, spray several times, 10 days apart, during rainy periods. Use zineb or maneb.
2. *Gray-mold Leaf Blight*—See (5) Botrytis Blight under General Diseases.
3. *Root-knot*—Southern states and indoor problem in northern states. See (37) Root-knot under General Diseases. Galls may form in leaves of Tradescantia, causing puckering and distortion. *Control:* Dip infested plants in hot water (122° F.) for 15 minutes. Replant in sterilized soil.
4. *Rust*—Reddish-brown to black, powdery pustules on leaves. *Control:* Same as for Leaf Spots (above).
5. *Other Root-feeding Nematodes* (e.g., burrowing, lesion)—Often associated with unthrifty, declining plants. *Control:* Same as for Root-knot (above).
6. *Mosaic* (wandering-Jew)—See (16) Mosaic under General Diseases.

TRAGOPOGON — See Lettuce
TRAILING-ARBUTUS — See Heath
TRAILING FOUR-O'CLOCK — See Four-o'clock
TRANSVAAL DAISY — See Chrysanthemum
TREE CYPRESS — See Phlox
TREE-OF-HEAVEN *(Ailanthus)*

1. *Verticillium Wilt*—Serious. Leaves wilt, yellow, and drop prematurely. Often on only one side of tree. Branches slowly die. Trees may die suddenly or die back gradually over period of years. Winter injury may follow wilt. Yellowish-brown streaks occur in wood just under bark. *Control:* Cut down and burn severely infected trees including as many roots as possible. Where only 1 to 6 limbs are dead or show wilt, promptly prune off affected parts. Fertilize and water trees heavily to stimulate vigor.
2. *Leaf Spots*—Widespread. Leaves variously spotted. *Control:* See under Maple.
3. *Wood Rots*—General. See under Birch, and (23) Wood Rot under General Diseases.
4. *Twig Blights, Cankers, Dieback*—Occasional. See under Maple.
5. *Root Rots*—See under Apple.
6. *Black Mildew*—See (12) Sooty Mold under General Diseases.
7. *Seedling Blight, Damping-off*—See under Pine.

TREEMALLOW — See Hollyhock
TREE PEONY — See Delphinium
TREEPOPPY — See Poppy
TREE-TOMATO — See Tomato
TRICHOCEREUS — See Cacti
TRICHOSANTHES — See Cucumber
TRILLIUM [CATESBY, COAST, DWARF WHITE, GREAT WHITE, LARGE-FLOWERED, NODDING, PAINTED, PRAIRIE, PURPLE, RED, SNOW, TOAD, YELLOW], WAKEROBIN *(Trillium)*

1. *Leaf Spots*—Spots of various sizes, shapes, and colors. If severe, leaves may wither and drop early. *Control:* If practical, apply zineb or maneb at about 10-day intervals during rainy periods.

2. *Stem Rot*—Plants wilt and wither. Stems discolor and rot at base. *Control:* See under Delphinium, Stem Canker.
3. *Rust*—Small yellowish pustules on leaves. Cutgrass *(Leersia)* is the alternate host. *Control:* Same as for Leaf Spots (above).
4. *Leaf Smut*—Black, sooty pustules on leaves. See (11) Smut under General Diseases.

TRIPLET LILY — See Brodiaea
TRITOMA — See Redhot-pokerplant
TRITONIA — See Gladiolus
TROLLIUS — See Anemone
TROPAEOLUM — See Nasturtium
TROPICAL LILAC — See Lantana
TROUT-LILY — See Erythronium
TRUMPETCREEPER — See Trumpetvine
TRUMPETFLOWER — See Bignonia
TRUMPETTREE [SILVER, WHITEWOOD] *(Tabebuia)*; FLORIDA YELLOWTRUMPET or YELLOW-ELDER *(Stenolobium)*; CAPE-HONEYSUCKLE *(Tecomaria)*

1. *Rust* (Florida yellowtrumpet, trumpettree)—Yellow, yellowish-orange, reddish-brown or black, powdery pustules on leaves. *Control:* Collect and burn fallen leaves. If needed, spray several times, 10 to 14 days apart. Start a week before rust normally appears. Use ferbam, zineb, or maneb.
2. *Root Rots*—See under Apple, and (34) Root Rot under General Diseases.
3. *Anthracnose* (cape-honeysuckle)—Spots develop on leaves. *Control:* Collect and burn spotted leaves. If needed, spray as for Rust (above).

TRUMPETVINE or TRUMPETCREEPER *(Campsis)*; CAT'S-CLAW *(Doxantha)*

1. *Leaf Spots, Leaf Blight*—General. Spots of various sizes, shapes, and colors. Leaves may wither and drop early. *Control:* Pick and burn spotted leaves. If practical, spray during rainy periods, about 10 days apart. Use zineb, ferbam, thiram, maneb, captan, or fixed copper.
2. *Powdery Mildew*—White, powdery mold may cover foliage in late summer and fall. *Control:* When disease appears, apply sulfur, Karathane, or Acti-dione several times, 7 to 10 days apart.
3. *Root Rot*—See (34) Root Rot under General Diseases.
4. *Verticillium Wilt*—See (15B) Verticillium Wilt under General Diseases.
5. *Mistletoe*—See (39) Mistletoe under General Diseases.

TSUGA — See Pine
TUBEROSE — See Daffodil
TULIP [BREEDER, COTTAGE, DARWIN, DOUBLE, DUC VAN TOL, FOSTERIANA, FRINGED, GREIGII, HORNED, LADY, LILY-FLOWERED, MENDEL, MULTI-FLOWERED, PARROT, REMBRANDT, SINGLE and DOUBLE EARLY, TRIUMPH, TURKISH, VIRIDIFLORA, WATERLILY or KAUFMANNIANA, WOOD] *(Tulipa)*; AFRICAN-LILY or LILY-OF-THE-NILE *(Agapanthus)*; GLORY-OF-THE-SNOW *(Chionodoxa)*; SUMMER-HYACINTH *(Galtonia)*; FRITILLARY [PINK, SCARLET, YELLOW], GUINEA-HEN FLOWER, CHECKERED-LILY, CROWN IMPERIAL *(Fritillaria)*; HYACINTH [COMMON, DUTCH, MINIATURE, ROMAN] *(Hyacinthus)*; CAPE-COWSLIP *(Lachenalia)*; GRAPE-HYACINTH, STARCH-HYACINTH,

TASSEL-HYACINTH, FEATHER-HYACINTH, BLUEBELLS *(Muscari)*; STAR-OF-BETHLEHEM, CHINCHERINCHEE *(Ornithogalum)*; STRIPED SQUILL, LEBANON SQUILL *(Puschkinia)*; SQUILL [BLUE-FLOWERED, BYZANTINE, SIBERIAN], STAR OR WILD-HYACINTH, SPANISH BLUEBELL, SEA ONION, CUBAN LILY, BLUEBELLS OF ENGLAND *(Scilla, Endymion)*; STENANTHIUM, FEATHERBELLS *(Stenanthium)*

1. *Fire, Botrytis Blights* (fritillary, hyacinth, tulip)—General on tulip in cool, wet spring weather. Emerging leaves and shoot (or "firehead") may be twisted and rotted. *Hyacinth* leaf tips are dark brown to almost black and shriveled. Petioles rot and may collapse. Small, soft, yellowish-brown, sunken spots develop on *tulip* and *hyacinth* leaves and flower petals (Measles), with water-soaked margins. Spots enlarge and merge to form whitish-gray to brown blotches. *Tulip* leaves and flower stalk may collapse. Flower buds may not open. Grayish-brown mold grows on affected parts in damp weather. Often follows frost or hail injury. *Tulip* bulbs develop yellow or brown spots, usually round and scabby, on outer white scales. Small, shiny, black bodies (sclerotia) may develop on outer brown husks of affected bulbs. See Figure 2.47A under General Diseases and Figure 4.229. *Control:* Plant only sound, blemish-free, healthy bulbs (without husks) in deep, light, very well-drained, fertile soil in sunny spot. Three- or 4-year rotation. Avoid bruising, freezing, sunburning, or otherwise injuring bulbs. Avoid a wet mulch, overwatering, and overfertilizing, especially with nitrogen or manure. Carefully remove and burn all infected parts as they occur plus 3 inches of surrounding soil. Dig bulbs in early summer during dry weather, after tops turn yellow. Dry, clean, and store bulbs at 40° F. and low humidity. Handle bulbs carefully to avoid injuries. Do not work among wet plants. Indoors, keep water off foliage. Avoid forcing at too high a temperature. Increase air circulation. Remove fading flower heads and burn all tops as leaves turn yellow. Spray once or twice weekly from when leaves emerge to just before bloom. Use captan, dichlone, thiram, zineb, maneb, Botran, ferbam, or PCNB plus detergent. Dust or spray bed and planting hole with PCNB just before planting. See Bulb Rots (below). Yellow *tulips* have been reported as being less susceptible than reds.

Fig. 4.229. Tulip fire.

2. *Bulb, Crown, and Root Rots*—Cosmopolitan. May be serious with plants dying in clumps. Shoots fail to emerge or produce stunted, yellowish, purplish, or reddish leaves that later wilt, rot at base, wither, and die. Roots often decay; may fail to produce flowers. Bulbs spotted; rot in field, storage, or indoors. Affected areas may be covered with black, blue-green, gray, pink, brown, or white mold. Rot may be

chalky, powdery and dry, and firm (fungus rots); or rapid, mushy, slimy, and foul-smelling (Bacterial Soft Rot). Bacterial Soft Rot is common in indoor bulb bowls. Often associated with mites and nematodes. See Figure 2.51D under General Diseases. *Control:* Same as for Fire (above). Cure bulbs after digging and store in thin layers in a dry, well-ventilated location at 40° F. Follow recommendations of a commercial grower or your extension horticulturist. Discard spotted, rotted, or damaged bulbs. Varieties may differ in resistance. Avoid forcing bulbs too early and at high temperatures. Before planting or storage, dust bulbs with thiram, zineb, chloranil, or captan plus lindane. Control bulb mites by dipping *dormant tulip* bulbs in hot water (122° F.) for several minutes. Or soak as for Leaf, Bulb and Stem Nematode (below). Before planting, dust hole and edges with PCNB. Use 1 pound of 20 per cent dust to 40 square feet. Follow manufacturer's directions. Avoid overfertilizing with nitrogen and excessively wet, humid conditions. Four- to 5-year rotation.

3. *Flower Breaking, Mosaics* (African-lily, cape-cowslip, fritillaria, hyacinth, squill, star-of-Bethlehem, summer-hyacinth, tulip) —General. Irregular stripes, bars, streaks, spots, and blotches in tulip and hyacinth flower petals. *Tulip* petals may be feather-edged. Leaves and flower stem often mottled or streaked with light and dark green, gray, or yellow areas. Plants lack vigor; often stunted. Flowering may be prevented. Rembrandt, Bizarre, Bijbloemen and Peppermint Stick *tulips* are naturally infected. Double-flowered *tulips* are more susceptible than single ones. Most pure white *tulip* flowers appear normal; some turn pink or red. Viruses easily spread by cutting infected then healthy flower stems. See Figure 2.34B under General Diseases and Figure 4.230A. *Control:* Destroy infected plants promptly, where feasible. Control aphids that transmit viruses. Use lindane or malathion. Avoid planting close to gladiolus, cucurbits, lilies, and solid-color tulips near Rembrandts, Bizarre, Bijbloemen, and Peppermint Sticks. Keep down weeds.

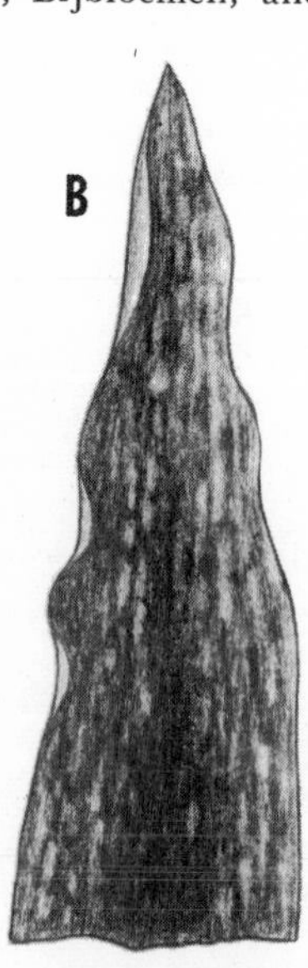

Fig. 4.230. A. Mottle-streaking of tulip. B. Tulip chlorosis.

4. *Leaf, Bulb and Stem Nematode, Ring Disease* (glory-of-the-snow, grape-hyacinth, hyacinth, squill, summer-hyacinth, tulip) —Leaves stunted and deformed with yellowish to brown flecks and spots. Leaves often later shrivel, split, and die. Flowering reduced. Bulb scales are swollen, spongy, may turn brown, gummy, and

rot. *Hyacinth* bulbs, when cut across, show dark rings. Badly infested bulbs do not sprout. See Figures 2.53B and 4.231. *Control:* Plant only large, firm, nematode-free bulbs in uninfested soil. If suspicious, soak *dormant* bulbs for 3 hours in hot water (110° F.) and formalin (1 part formaldehyde to 200 parts of water). Plant immediately. Avoid heavy, poorly drained soil. Dig up and burn infested bulbs and adjacent ones which appear healthy. Dig up and destroy 6 inches of soil surrounding these bulbs. Three-year rotation with nonbulb crops. See (20) Leaf Nematode and (38) Bulb Nematode under General Diseases.

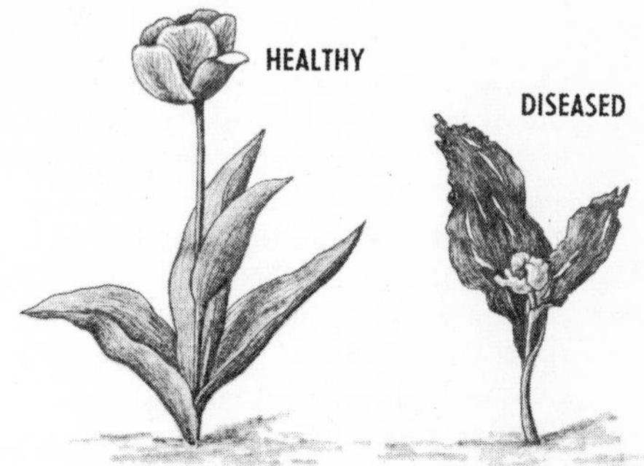

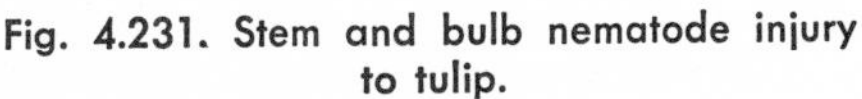
Fig. 4.231. Stem and bulb nematode injury to tulip.

5. *Hyacinth Yellows, Yellow Rot*—Widespread, often serious. Most common on forced plants. Elongated, yellow, water-soaked stripes on leaves and flower stalks that turn brown and shrivel. Infected bulbs show sunken, slimy, yellow pockets when cut across. Such bulbs are soft when squeezed. Leaves and flower stalks easily pulled up. Flower heads distorted and may not open. Bulbs often rot in bulb bowl or soil; never send up leaves. Often followed by Bacterial Soft Rot. See Bulb Rots (above). *Control:* Same as for Fire and Bulb Rots (both above). Cure bulbs properly. Keep away from wet plants. Varieties differ in resistance. Commercial growers *promote* disease by storing hyacinth bulbs for several months at 70° to 86° F. and one month at 98° to 110° F. The soft, rotted bulbs are then discarded.
6. *Stem Rot, Southern Blight, Flower Spot or Blossom Blight* (squill, star-of-Bethlehem, tulip)—Flower stalk shrivels and rots below the flower. Flower or entire plant may collapse, wither, and die. May be covered with whitish mold in cool, moist weather. Serious on double varieties growing in shady, wet areas. *Control:* Same as for Fire (above).
7. *Flower Stalk Collapse, "Topple," Loose Bud* (hyacinth, tulip)—Widespread. Primarily an indoor problem. Glassy, water-soaked spots form on flower stalk or neck. Flower stalks may crack, shrivel, and collapse. Pull away easily before flower bud opens. Varieties differ greatly in resistance. *Control:* When forcing, avoid overwatering, sudden shifts in temperature, high temperatures and humidity, and high level of potash in the soil. Ripen bulbs thoroughly before planting, especially after cold, wet seasons. Do not force bulbs too early. Grow shallow in well-drained soil. Foliage sprays of a 2 per cent calcium nitrate solution, plus spreader-sticker, have given control.
8. *Flower and Leaf Smut* (grape-hyacinth, plume hyacinth, squill, tulip)—Dark, purplish-black, powdery pustules replace flower anthers or form black streaks in leaves. Flowers spoiled. *Control:* Dig and burn infected plants.
9. *Frost Injury*—Late spring frosts may cause numerous, small, pale brown spots to form on leaves and flower stalk. Spots may merge to form broad bands. Leaves may show slits and be torn and ragged. *Control:* Protect against severe frosts.
10. *Winter Injury* (primarily tulip)—Shoots twisted and abnormal. Bulbs rot.

Control: Avoid planting bulbs in late fall in heavy soil where drainage is poor. See recommended cultural practices under Fire and Bulb Rots (both above).

11. *Leaf Spots* (cape-cowslip, fritillary, squill, star-of-Bethlehem)—Pale brown to sootlike spots. If severe, leaves may turn yellow, then wither and die. *Control:* Same as for Fire (above).
12. *Rust* (fritillaria, hyacinth, scilla, stenanthium)—Pale yellow, then brown, powdery pustules on leaves. *Control:* Same as for Fire (above).
13. *Tobacco Necrosis or Streak* (tulip)—Only few varieties affected. Small to large, whitish to brown spots and streaks, surrounded by purple lines. Form on poorly developed leaves, stems, and flowers. Leaves may be twisted, stunted, and shriveled. Plants usually die in patches. Seldom produce bulbs. *Control:* Grow in clean soil that has not grown tobacco or potato. Carefully sterilize soil for indoor plants, since causal virus is soil-borne (pages 538–548). Do *not* replant in outdoor beds without first sterilizing soil with heat or chemicals.
14. *Sunscald* (tulip)—Flower petals dry and shrivel, especially at upper margins. Shoots twisted and abnormal. Bulbs rot. *Control:* If practical, shade plants in hot, dry weather. Avoid planting in late fall in heavy, poorly drained soil.
15. *Anthracnose* (tulip)—California. Small to large, oval to elongated, water-soaked spots on leaves and flower stems. Spots later dry out and develop black margins. Centers of spots sprinkled with black specks. *Control:* Same as for Fire (above).
16. *Blindness*—Failure to flower. May be due to disease (see above), root failure in dry soil, too early forcing, or heating of bulbs in storage or transit.
17. *Ringspot* (tulip)—Yellowish-green, mottled areas in leaves. Plants may be somewhat stunted. *Control:* Destroy infected plants.

TULIPTREE — See Magnolia

TUNA — See Cacti

TUPELO — See Dogwood

TURFING DAISY — See Chrysanthemum

TURKEYCORN — See Bleedingheart

TURNIP — See Cabbage

TURQUOISE BERRY — See Grape

TURTLEHEAD — See Snapdragon

TUSSILAGO — See Chrysanthemum

TWINFLOWER — See Pea and below

TWINFLOWER *(Linnaea)*

1. *Leaf Spots, Tar Spot*—Spots of various sizes, shapes, and colors on leaves during rainy seasons. *Control:* Collect and burn tops in fall. Apply zineb, maneb, or captan at about 10-day intervals during rainy periods.
2. *Black Mildew*—See (12) Sooty Mold under General Diseases.

UDO — See Acanthopanax

ULMUS — See Elm

UMBELLULARIA — See Avocado

UMBRELLA-PINE — See Pine

UMBRELLAPLANT, EGYPTIAN PAPER PLANT, CHUFA *(Cyperus)*

1. *Leaf Spot*—See (1) Fungus Leaf Spot under General Diseases.
2. *Root-knot*—See (37) Root-knot under General Diseases. *Control:* Pot plants in sterilized soil (pages 538–548).

UMBRELLA-TREE — See Schefflera

UMBRELLAWORT — See Four-o'clock

UNICORNPLANT — See Proboscisflower

UVULARIA — See Lily

VACCINIUM — See Blueberry

VALERIAN [COMMON or GARDEN-HELIOTROPE, EDIBLE] *(Valeriana);* RED-VALERIAN or JUPITERS-BEARD *(Centranthus);* CORNSALAD [BEAKED, COMMON, ITALIAN], LAMBSLETTUCE *(Valerianella)*

1. *Leaf Spots*—Round to irregular spots of various sizes and colors. *Control:* Pick and burn infected leaves. Collect and burn tops in fall. If practical, spray several times during rainy weather at 10-day intervals. Use zineb or maneb.
2. *Rusts* (valerian) —Yellow, yellowish-orange or dark, powdery pustules on foliage. Alternate hosts: none or *Carex* spp. *Control:* Same as for Leaf Spots (above). Destroy sedges *(Carex).*
3. *Powdery Mildew*—Grayish-white, powdery patches on upper leaf surface. *Control:* Apply sulfur or Karathane several times, 10 days apart.
4. *Stem Rot, Root Rots*—Plants lack vigor, wilt. Stems rot at soil line or below. Roots may decay. See under Delphinium, Stem Canker.
5. *Curly-top*—See (19) Curly-top under General Diseases.

VALERIANELLA — See Valerian

VALLOTA — See Daffodil

VANDA — See Orchids

VANILLALEAF — See Barberry

VEGETABLE-MARROW — See Cucumber

VEGETABLE OYSTER — See Lettuce

VEGETABLE SPONGE — See Cucumber

VELVET LEAF — See Sedum

VENUS-LOOKINGGLASS — See Bellflower

VERBASCUM — See Snapdragon

VERBENA — See Lantana

VERBESINA — See Chrysanthemum

VERONICA, VERONICASTRUM — See Speedwell

VERVAIN — See Lantana

VETCH — See Pea

VETCHLEAF SOPHORA — See Honeylocust

VIBURNUM [BIRCHLEAF, BURKWOOD, DAVIDI, DOUBLEFILE, FRAGRANT (CARLESI or KOREAN SPICE), ICHANG, JAPANESE, LAURESTINUS, LEATHERY, MAPLELEAF or DOCKMACKIE, PINK LEATHERY, RHYTIDOPHYLLUM, SANDANKWA, SIEBOLD, SWEET, TEA], SNOWBALL [CHENAULT, CHINESE, COMMON, DWARF, JAPANESE], AMERICAN HIGHBUSH CRANBERRY, BLACKHAW, DWARF WITHE-ROD, EUROPEAN CRANBERRY-BUSH or HIGHBUSH CRANBERRY, ARROWWOOD, WAYFARING-TREE, HOBBLEBUSH, WITHE-ROD, NANNYBERRY, POSSUMHAW *(Viburnum);* BEAUTYBUSH or CHINESE BEAUTYBUSH *(Kolkwitzia)*

1. *Powdery Mildew* (viburnum) —General. White, powdery mold may cover foliage in late summer and fall. Leaves may wither. *Control:* Spray two or three times, 10 days apart. Use Karathane. *Warning:* Do *not* use sulfur on viburnums, which may cause black spots on leaves followed shortly by heavy leaf fall. Resistant *viburnums:* Cayuga and Mohawk.

2. *Bacterial Leaf Spot* (viburnum) —Widespread on various viburnums especially the species *carlesi.* Round to irregular, water-soaked spots that turn dark and sunken. Irregular, sunken, brownish-black cankers form on young stems. Affected leaves may wither and drop early. *Control:* Carefully remove and burn infected plant parts as they occur. Spray 1 or 2 times weekly in cool, moist weather. Use bordeaux (4-4-50), fixed copper and spray lime (2 tablespoons each per gallon of water), or streptomycin (50 to 100 parts per million). Start when leaves begin to unfurl. Continue for 3 weeks. Or use Agrimycin following manufacturer's directions. Resistant *viburnums:* Cayuga and Mohawk.
3. *Fungus Leaf Spots, Leaf Blight, Spot Anthracnose, Downy Mildew*—General. Spots of various colors, shapes, and sizes. If severe, leaves may wither and drop early. *Control:* If necessary, spray weekly with fixed copper and spray lime, zineb, or maneb.
4. *Gray-mold Blight, Shoot Blight* (viburnum) —Grayish-brown rotted spots on leaves. Spots start at margin. May later blight whole leaf. Flower parts often blighted. Twigs killed. *Control:* Spray weekly in cool, damp weather as for Fungus Leaf Spots (above).
5. *Verticillium Wilt, Dieback* (viburnum) —See under Maple, and (15B) Verticillium Wilt under General Diseases.
6. *Rusts*—Yellow, yellow-orange, reddish-brown, or black pustules on leaves. *Control:* Not usually needed. Spray as for Fungus Leaf Spots (above).
7. *Root Rots*—Plants decline, lack vigor, may die back. Often associated with nematodes (e.g., dagger, lesion, pin, ring, root-knot, spiral, stem or rot, stunt). See under Apple, and (34) Root Rot under General Diseases. Grow in full sun or very light shade in well-drained yet moist soil high in organic matter.
8. *Wood Rot*—See under Birch, and (23) Wood Rot under General Diseases.
9. *Root-knot*—Viburnums are quite susceptible. See under Peach, and (37) Root-knot under General Diseases.
10. *Crown Gall*—See under Apple, and (30) Crown Gall under General Diseases.
11. *Collar Rot, Stem Canker or Girdle*—See under Dogwood and Currant.
12. *Stem or Branch Gall, Twig Canker, Dieback*—Unimportant. Knoblike galls on stems. Stems may die back. *Control:* Prune out and burn galls. Swab or dip pruning shears in 70 per cent alcohol between cuts.
13. *Chlorosis*—Occurs in very acid or in alkaline soils. See under Maple.
14. *Thread Blight*—Southeastern states. See under Walnut.
15. *Sooty Mold*—Black moldy patches on foliage. *Control:* Apply lindane or malathion to control aphids.

VICIA — See Pea

VIGNA — See Bean

VINCA, PERIWINKLE [BIGLEAF, COMMON or SMALL, DWARF, MADAGASCAR], GROUND-MYRTLE *(Vinca);* AMSONIA

1. *Dieback, Wilt, Stem Canker, Leaf Spots, Leaf Mold*—Prevalent in wet seasons. Leaves spotted. Drop early. Dark brown to black stem cankers may kill out plants in patches. Shoot tips beyond cankers wilt, turn dark brown to black, and die back. *Control:* Remove and burn seriously infected plants or plant parts when first seen. Use only disease-free stock from a reputable nursery. Spray at 10- to 14-day intervals. Start when buds open, using captan, zineb, ferbam, thiram, maneb, dichlone, or fixed copper. Otherwise, same as for Gray-mold Blight (below). Drenching soil in affected spots with a 1:1,000 solution of corrosive sublimate, Semesan (1 tablespoonful per gallon), or PCNB 75 (1 tablespoonful per gallon) plus Dexon, following manufacturer's directions, should be beneficial. Or fumigate soil before planting (pages 538–548).

2. *Gray-mold Blight, Botrytis Blight*—Brown or black spots on leaves. Extend inward from margin. May later cover whole leaf. Affected areas covered by coarse gray mold in damp weather. *Control:* Keep water off foliage when sprinkling. Grow in well-drained soil where air circulation is good. Avoid wet mulch and overcrowding. Spray or dust as for Dieback (above).
3. *Root and Stem Rot*—Shoot tips usually wilt and die back. Roots and stems decay. May closely resemble Dieback and Stem Canker (above). Sometimes associated with nematodes (e.g., burrowing, root-knot). *Control:* Same as for Dieback (above).
4. *Aster Yellows, Curly-top, Mosaic*—Leaves streaked or mottled light and dark green and yellow. May curve downward. Plants and flowers are dwarfed. Shoots may wilt and die in hot weather. Blue flowers may show white streaks (Flower Breaking). Do not confuse with normal variegated periwinkle. *Control:* Pull and burn infected plants when first found. Keep down weeds. Spray with DDT or sevin plus malathion to control aphids and leafhoppers that spread these viruses. Potted *periwinkle* plants can be cured of Aster Yellows by holding at a temperature of 100° to 107° F. for 2 to 3 weeks.
5. *Root-knot*—See (37) Root-knot under General Diseases.
6. *Rusts*—Rather rare. Once in a bed, may reappear each year. Underside of leaves sprinkled with yellow-orange, reddish-brown, or nearly black powdery pustules. Alternate hosts: Pines, marsh and cord grasses *(Spartina)*, or none. *Control:* Pull and burn infected plants. Spray as for Dieback (above). Use ferbam, zineb, thiram, or maneb.
7. *Black Ringspot*—See under Cabbage.

VINE — See Grape

VINE BOWER — See Clematis

VIOLA, VIOLET — See Pansy

VIPER'S-BUGLOSS, VIRGINIA COWSLIP — See Mertensia

VIRGINIA-CREEPER — See Grape

VIRGINIA SNAKEROOT — See Aristolochia

VIRGINSBOWER — See Clematis

VITEX — See Lantana

VITUS — See Grape

WAFER ASH — See Hoptree

WAHOO — See Bittersweet

WAKEROBIN — See Trillium

WALKINGLEAF — See Ferns

WALLCRESS, WALLFLOWER — See Cabbage

WALLPEPPER — See Sedum

WALNUT [BLACK, CALIFORNIA, CALIFORNIA BLACK or HINDS, EASTERN BLACK, PERSIAN or ENGLISH, HYBRID, JAPANESE], BUTTERNUT or WHITE WALNUT *(Juglans)*; PECAN, HICKORY [BITTERNUT, KINGNUT or BIG SHELLBARK, MOCKERNUT or WHITE, PIGNUT, SHAGBARK, SHELLBARK, WATER] *(Carya)*

1. *Anthracnose, Leaf Blotch, Leaf Spots, Nursery Blight*—General in wet seasons. Anthracnose is important in the Southeast. Spots and blotches on leaves of various sizes, shapes, and colors. Some spots may drop out leaving ragged holes. If serious, leaves and fruit may wither and drop early. *Control:* Spray *pecan* when buds burst open; leaves 1/4 to 1/2 grown; as *tips* of small nuts turn brown; and 2, 4, and 6 weeks later. Use dodine, Polyram. Dithane M-45, or zineb. Do *not* apply dodine to such

susceptible *pecan* varieties as Moore and Van Deman. Follow manufacturer's directions. Add DDT, sevin, or methoxychlor plus malathion to all sprays to control insects (e.g., leaf and nut casebearers, aphids, plant bugs, weevils, scales, etc.) and mites. Where economical, spray *butternut, hickories,* and *walnuts* two to four times, 10 to 14 days apart. Start when buds begin to open. Use zineb, Polyram, ziram, Dithane M-45, maneb, phenyl mercury, dodine, or fixed copper. Collect and burn fallen leaves. Keep trees vigorous. Prune, water, and fertilize where practical. Avoid crowding trees. Varieties differ in resistance. Schley *pecan* is resistant to Liver Spot *(Gnomonia).*

2. *Scab* (primarily pecan and hickories)—General and serious during cool, rainy periods, especially in southeastern and Gulf Coast states. Dark brown to black circular patches on nuts that become slightly sunken. Enlarging, round to irregular, olive-brown to black spots occur on leaves. Often associated with the veins. If severe, infected leaves and nuts may drop early. See Figure 4.232. *Control:* Knock off and burn old shucks, leaves, and leaf stems after harvest. Apply spray program for *pecan* as given under Anthracnose (above). Spray at about 2-week intervals or just before rainy periods. Prune off low limbs. Space trees. Burkett, Delmas, Halbert, Moore, and Schley pecans are highly susceptible. Resistant *pecan* varieties: Brake, Curtis, Desirable, Elliot, Farley, Kernodle, Mahan, Moneymaker, Success, and Stuart.
3. *Downy Spot, White Mold, Witches'-broom*—Widespread. Leaflets stunted, turn yellow, then blacken and fall prematurely. White, glistening, frosty "mildew" appears on underleaf surface in early summer. Compact, swollen, bunchy clusters of upright shoots (witches'-brooms), 2 to 3 feet across, occur on branches and trunk of shagbark hickory and certain walnuts. Most evident during winter. *Control:* Cut out and burn witches'-brooms. Collect and burn infected leaves. If practical, spray as for Anthracnose and Scab (both above). Schley and a few other *pecan* varieties are normally highly resistant. Burkett, Moneymaker, and Stuart are very susceptible.
4. *Bunch Disease* (butternut, Japanese, Persian and eastern black walnuts, pecan, hickories)—One or more branches, or entire tree, show bushy clusters of slender, willowy twigs (witches'-brooms). "Bunch" growths most conspicuous in spring and early summer since diseased branches often leaf out 2 weeks earlier than healthy ones. Shoots and branches die back. Leaflets narrow, curled, and turn yellow. Few, if any, nuts are produced. Stuart *pecan* has some resistance while Desirable, Mahan, and Schley are very susceptible. *Control:* Destroy all infected trees. Plant and propagate only from disease-free trees.
5. *Twig, Branch and Trunk Cankers, Dieback, Branch-wilt*—Widespread. Twigs and later main branches or even the entire tree may gradually die back from small to large, sunken cankers. Leaves on certain branches wilt, wither, and remain attached. New shoots or suckers may develop on trunk (Figure 4.233). *Control:* Promptly cut out and burn dead and discolored, blighted wood well below any sign of infection. Keep trees vigorous. See under Anthracnose (above). Destroy badly infected trees. Both northern and southern *California black walnuts* are highly resistant to Branch-wilt *(Hendersonula).* Whitewash on exposed limbs reduces sunscald and Branch-wilt that follows. Prepare by taking 25 pounds of processed lime (ground hot lime or quick lime) plus 2½ pounds of salt and 1 to 1¼ pounds of dusting sulfur. Add water to make 25 gallons. Avoid bark injuries. Treat promptly with tree wound dressing (pages 34–36).
6. *Crown Gall*—Widespread. Rough, irregularly swollen galls at base of trunk, bud or graft union on lower stem, and roots. Trees lack vigor. Make poor growth. Leaves may turn yellow. Trees may die. Young galls are easily confused with callus overgrowths formed at wounds, and graft or bud unions. *Control:* Destroy severely

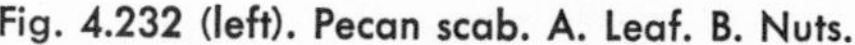

Fig. 4.232 (left). Pecan scab. A. Leaf. B. Nuts.

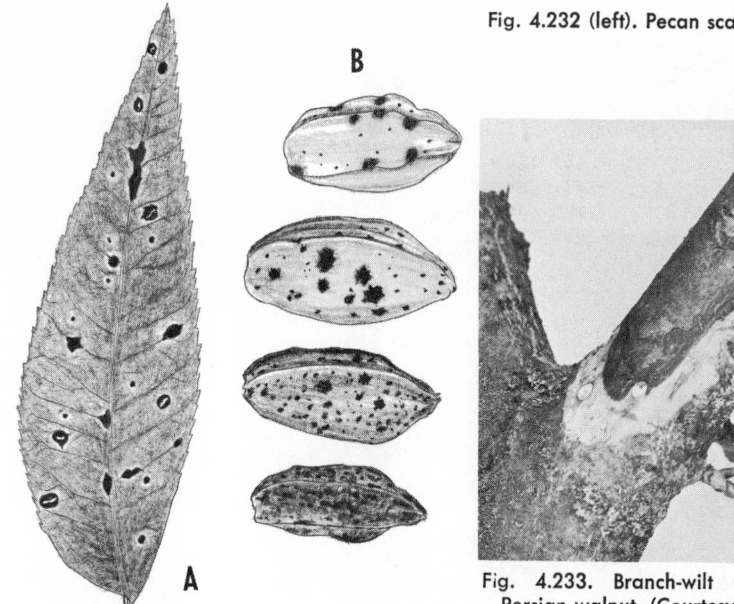

Fig. 4.233. Branch-wilt (*Hendersonula*) of Persian walnut. (Courtesy Dr. J. D. Paxton)

infected trees; preferably by burning. Buy only disease-free nursery stock. If galls are found on young nursery stock, soak 1 hour in Terramycin solution (400 parts per million). Avoid cultivating near trunks of infected trees. See (30) Crown Gall under General Diseases. In *walnuts,* remove all gall tissue well back into healthy wood underneath. Then paint entire area with bordeaux or solution of Elgetol (1 part) and methanol (4 or 5 parts) diluted 9 times with water. Keep trees vigorous. Water during droughts. Fertilize based on a soil test.

7. *Rosette, Zinc Deficiency, Little Leaf, Yellows*—Widespread. Young leaves on tips of upper branches are crinkled and yellow-mottled. Later, leaflets are dwarfed and twisted, become narrowed, harden, turn reddish-brown and drop. Healthy growth usually occurs below these affected tips. Symptoms may later appear on lower branches. Sometimes only certain limbs are affected. Shoots may die back from tips and form bunchy "rosettes." Nut production is decreased. Affected trees are so weakened they become susceptible to killing by borers or winter injury. See Figure 4.234. *Control:* Spray *pecans* three times with 36 per cent zinc sulfate (3 tablespoons per gallon) 1, 2, and 3 months after buds open. Or apply zinc chelate sprays according to manufacturer's directions (page 27). May apply with regular pest sprays. Check with a local grower, nurseryman, or your extension horticulturist. Zinc sulfate or zinc chelate may also be applied to *acid* soil under trees in February or March. Use 1 to 15 pounds per tree (1 to 5 pounds for young trees; 10 to 15 pounds for large, mature trees. Use at rate of about ½ pound per inch in trunk diameter) of zinc sulfate or 1 to 2 pounds of zinc chelate. Alkaline and neutral soils require much heavier rates than acid soils for a comparable response. Moneymaker *pecan* is quite resistant.

To correct zinc deficiency in *walnuts* use strips of galvanized 22- or 24-gauge iron sheet metal, 2 inches long and about ¾ inch wide. Drive these at 2-inch intervals into the wood of either limbs *or* main trunk. Insert in horizontal bands around the tree parallel to the wood grain. Six bands, about 2 inches apart, should be inserted per tree trunk or limb. On trees larger than 10 inches in diameter, drive the strips

Fig. 4.234. Zinc deficiency on pecan. (Courtesy Dr. David W. Rosberg)

into limbs, not trunks. Headless, galvanized nails or glaziers' points should be used in place of strips on young trees. Use 5 or 6 nails to each strip of metal. Or inject zinc sulfate into soil under trees (1 pound of 36 per cent per gallon). Apply 10 to 25 pounds of zinc sulfate per tree, depending on size. Varieties may differ in susceptibility. Check with your extension horticulturist or plant pathologist.

8. *Root Rots, Blackline*—Cosmopolitan. Leaves on certain branches or the whole crown may yellow, then brown, and drop prematurely. Trees decline in vigor—gradually or may suddenly die. Trees stunted; branches die back. Several rots are more frequent on Persian walnuts grafted on black walnut than on their own rootstocks. Often associated with nematodes (e.g., awl, cacopaurus, dagger, lance, lesion, pin, ring, spiral, stubby-root, stunt). See under Apple, and (34) Root Rot under General Diseases. The Northern California *black walnut* is resistant to Armillaria Root Rot. Certain more vigorous *Persian walnut* seedlings, e.g., Mangregian, are recommended for rootstocks in place of black walnut, at least in Oregon. Avoid growing *pecans* in soil known to be infested with the Cotton Root Rot fungus *(Phymatotrichum)*. If nematodes are the primary problem, apply DBCP (Nemagon, Fumazone) to the root-zone area. Follow manufacturer's directions.
9. *Wood and Heart Rots, Trunk Decay, Crown Rot*—Cosmopolitan. See under Birch, Apple, and (23) Wood Rot under General Diseases. Decay of wood occurs around and through wounds. Trees may gradually lose vigor, become stunted, show sparse yellow foliage. May set a heavy crop of small-sized nuts. *Control:* Cut broken branches flush with a main limb or trunk (page 33). Paint over wounds promptly with tree wound dressing. Grow in well-drained soil.
10. *Nut Molds, Storage Molds*—Pink, gray, black, bluish-green or brown molds appear

first on shell (shucks) but soon penetrate and destroy the kernel inside. *Control:* Harvest early, dry promptly, and store only sound nuts at about 38° F. and 90 per cent relative humidity. Spray as for Scab (above). Control plant bugs and other insects with DDT or sevin plus malathion. Check with your county agent or extension entomologist.

11. *Powdery Mildews*—General. Powdery, grayish-white mold on leaves, young shoots, and nuts. If severe, leaves may drop early. *Pecan* nuts infected early may split prematurely, causing shriveled kernels. *Control:* Where serious, apply sulfur, Karathane, Polyram, or bordeaux (3-1-50) twice, 10 days apart. Most *pecan* varieties are resistant; but Curtis, Moore, and Pabst are susceptible.

12. *Bacterial Blight, Blast of Walnuts*—Most serious on Persian or English walnut along foggy coasts, or during rainy seasons. Small, angular, water-soaked spots form on young leaves and petioles. Spots turn reddish-brown to black, may merge, distorting and curling leaves. Black, elongated, slightly sunken spots or streaks occur on young shoots. If girdled, shoots die back. Infected buds turn dark brown to black and die. Blackish-brown, sunken blotches form on green nuts (Figure 2.18D). Kernels within such nuts may darken and shrivel. Young infected nuts usually drop early. Male catkins may be killed or produce contaminated pollen. See also Figure 4.30. *Control:* If practical, apply fixed copper or bordeaux (2-1-50), plus spreader-sticker, when about half the buds are open (pistillate flowers *not* yet visible). Then follow with 3 or more weekly applications; or apply just before rainy periods. Thorough coverage is essential. If you live in a commercial nut-growing area, follow spray program published by your cooperative extension service. Cut and burn badly infected shoots. Plant disease-free trees. Varieties differ in resistance. *Resistant Persian walnuts:* Earhardt, Eureka, Howe, and San Jose. *Susceptible* varieties include Chase, Franquette, Hartley, Payne, Placentia, and Santa Rosa. Early-blooming varieties are more apt to be severely infected than late-blooming ones.

13. *Thread Blight*—Southeastern states in damp areas where plants crowded and neglected. Small, dark masses (sclerotia) may form on twigs and leaf stems. Whitish fungus threads (hyphae) spread from these to lower surface of leaflets. Infected leaflets discolor and wither. Dead leaves become matted and hang by spiderweb-like threads until frost. *Control:* Same as for Anthracnose and Scab (both above).

14. *Collar or Crown Rots*—Irregular, brown to black, girdling cankers form at crown. Trees stunted with thin, yellowish-green foliage. Nut crop may be heavy year before death occurs. *Control:* See under Dogwood. Grow walnuts grafted on Persian or Paradox rootstocks. Graft unions should be well above ground.

15. *"Mouse Ear" (Little Leaf), Manganese Deficiency*—Mainly southeastern states, in neutral or alkaline soils. *Pecan* leaves are blunt at tips instead of pointed. Where severe, leaflets crinkled, rounded and dwarfed (Mouse Ear). *Walnut* leaves pale green except for veins that are dark green. Only certain limbs may be affected. Nut production and tree growth may be greatly decreased. *Control:* Spray trees with 1 or 2 per cent manganese sulfate or apply 2 to 4 pounds of manganese sulfate to root zone of mature trees. Manganese chelate materials (page 27) may also be used. Check with your extension horticulturist.

16. *Boron Deficiency, Dieback, "Snake Head"*—Western states. Large, irregular, dark brown spots develop along margins and between leaf veins. Elongate, leafless shoots often evident in tops of affected trees. These shoots (Snake Heads) may be curved; die back over winter. Nuts may drop when size of large peas. *Control:* Apply borax to surface of soil under the "drip line." Use 2 to 3 pounds for each tree 12 to 14 years old and 4 to 5 pounds for trees 18 to 25 years old. If only partial control is obtained, use ½ the original amount of borax the following year. Check page 27 and with your extension horticulturist.

17. *Leaf Scorch, Sunscald*—Brown dead spots in leaves between the veins. Severely affected leaves may wither and drop early. Most common on trees exposed to hot, dry winds or growing in soil underlaid with sand or gravel. *Control:* See under Maple.
18. *Winter Injury, Sunscald*—Branches, large limbs, or even entire tree may fail to start growth in spring. Young trees are most commonly affected. Trees often die back from the top. See under Apple and Elm. *Control:* Grow where trees are adapted. Avoid heavy, late applications of nitrogen.
19. *2,4-D Injury*—Hickories and walnuts are very susceptible. Leaves become thickened, twist or roll. May be boat-shaped or curl into "ram's-horns." *Control:* See under Grape.
20. *Verticillium Wilt* (pecan, walnuts)—See under Maple, and (15B) Verticillium Wilt under General Diseases.
21. *Root-knot*—See under Peach, and (37) Root-knot under General Diseases.
22. *Sooty Mold*—See (12) Sooty Mold under General Diseases.
23. *Mistletoe*—See (39) Mistletoe under General Diseases.
24. *Felt Fungus*—Southeastern states on neglected trees. Purple-black, feltlike growth on bark. Associated with scale insects. *Control:* See under Hackberry.
25. *Spanish-Moss, Gray Moss*—Pest in southern coastal areas, especially near live oak trees. Most troublesome in neglected orchards in poorly drained locations. *Control:* Keep trees vigorous through fertilization and thorough pest control program. Apply *dormant* spray of copper sulfate and calcium arsenate (1 pound of each per 10 gallons of water) for 3 consecutive years.
26. *Wetwood, Slime Flux* (hickory)—See under Elm.

For additional information read USDA Handbook 240, *Controlling Insects and Diseases of the Pecan,* and USDA Leaflet 525, *Growing Black Walnuts for Home Use.*

WANDERING-JEW — See Tradescantia
WANDFLOWER — See Iris
WART PLANT — See Aloe
WASHINGTONIA — See Palms
WATERCRESS — See Cabbage
WATERLILY [AMERICAN or FRAGRANT, CAPE BLUE, EUROPEAN WHITE, MAGNOLIA or TUBEROUS, MEXICAN, PYGMY, YELLOW], YELLOW or COW PONDLILY *(Nuphar, Nymphaea)*; LOTUS [AMERICAN, EGYPTIAN or HINDU] *(Nelumbo)*

1. *Leaf Spots*—Spots of various sizes, shapes, and colors. Leaves may curl up and wither. Plants may be defoliated. *Control:* Pick and burn spotted leaves. If practical, spray several times during wet periods. Use zineb, maneb, folpet, captan, or fixed copper.
2. *White Smut* (pondlily, waterlily)—Pale yellowish spots on leaves filled with dark brown to black, powdery masses. *Control:* Same as for Leaf Spots (above).
3. *Leaf and Stem Rot* (waterlily)—Foliage wilts and withers from rotting of leaf bases and stem. *Control:* Where feasible, grow in sterilized soil (pages 538–548).

WATERMELON — See Cucumber
WATERMELON-BEGONIA — See Peperomia
WATSONIA — See Gladiolus
WATTLE — See Honeylocust
WAXBERRY — See Snowberry
WAXGOURD — See Cucumber

WAXMYRTLE [COMMON or CANDLEBERRY, PACIFIC or CALIFORNIA BAYBERRY], BAYBERRY, EVERGREEN BAYBERRY, SWEETGALE *(Myrica)*

1. *Leaf Spots*—Spots of various sizes, shapes, and colors. If severe, leaves may wither and drop early. *Control:* Pick and burn spotted leaves. If practical, spray several times during rainy periods, at 10-day intervals. Use ferbam, zineb, maneb, or fixed copper.
2. *Rusts*—Bright orange, reddish-brown or black, powdery pustules on leaves. Alternate hosts: white-cedar *(Chamaecyparis)* or pines. *Control:* Spray as for Leaf Spots.
3. *Root Rots*—See (34) Root Rot under General Diseases. May be associated with nematodes (e.g., lesion, ring).
4. *Sooty Mold, Black Mildews* (waxmyrtle)—Gulf states. Black, moldy patches on leaves. *Control:* Spray with malathion plus a fungicide as for Leaf Spots (above).
5. *Twig Blight* (sweetgale)—Twigs blighted and die back. *Control:* Cut and burn affected parts. Spray as for Leaf Spots (above).
6. *Bayberry Virus Yellows*—Plants stunted and bushy with few or no fruits. Older leaves pale, yellowish, stunted, and leathery. Young leaves have wavy margins and tips. *Control:* Destroy infected plants.
7. *Seedling Blight*—See under Pine.

WAYFARING-TREE — See Viburnum

WEEPING MYALL — See Honeylocust

WEIGELA — See Snowberry

WELSH POPPY — See Poppy

WEST INDIA GHERKIN — See Cucumber

WHITE-ALDER — See Sweet-pepperbush

WHITE BEAMTREE — See Apple

WHITEBRUSH — See Lantana

WHITE-CEDAR — See Juniper

WHITECUP — See Tomato

WHITE KERRIA — See Jetbead

WHITE SNAKEROOT — See Chrysanthemum

WHITLOWGRASS — See Cabbage

WHORTLEBERRY — See Blueberry

WICOPY — See Leatherwood

WILDBERGAMOT — See Salvia

WILD CUCUMBER — See Cucumber

WILD-HYACINTH — See Brodiaea, Colchicum, and Tulip

WILD INDIGO — See False-indigo

WILDOLIVE — See Osmanthus

WILD ROSEMARY — See Labrador-tea

WILD SWEET-WILLIAM — See Phlox

WILD TUBEROSE — See Centuryplant

WILLOW [BABYLON WEEPING, BASKET or COMMON OSIER, BAY (LEAF) or LAUREL-LEAVED, BEARBERRY, BEBB, BLACK, BLUESTEM, COLORADO, CONTORTED HANKOW, CRACK, DIAMOND, DWARF or SAGE, GOAT, GOLDEN or YELLOWSTEM, GRAY, HEARTLEAF, KOREAN, NIOBE, PACIFIC, PEACHLEAF, PRAIRIE, PUSSY, RED, REDSTEM WHITE, RING-LEAF, ROSE-GOLD PUSSY,

ROSEMARY, SHARPLEAF, SHINING, SILKY, THURLOW WEEPING, WEEPING, WEEPING WHITE, WHITE or HUNTINGDON, WISCONSIN WEEPING, YELLOW, YELLOWSTEM WHITE], GOLDEN OSIER, PURPLE OSIER or URAL, DWARF PURPLE OSIER *(Salix)*

1. *Powdery Mildews*—General, but not serious. White, powdery mold on leaves. Some leaves may wither and drop early. In autumn, the mildew may be sprinkled with tiny black dots (fruiting bodies of mildew fungus). Leaves curled, may shrivel and drop early. *Control:* Spray two or three times, 10 to 14 days apart. Start when mildew first appears. Use sulfur or Karathane.
2. *Leaf Spots, Spot Anthracnose or Gray Scab, Tar Spot, Scorch*—Widespread, not serious. Small to large, round to irregular spots of various colors on leaves. Some also on shoots. Leaves may wither and drop early. *Control:* Collect and burn fallen leaves. Spray several times, 10 to 14 days apart. Start when buds begin to swell. Use zineb, captan, thiram, maneb, dichlone, phenyl mercury, fixed copper, or bordeaux (3-3-50). Later sprays may be needed if prolonged wet weather continues.
3. *Leaf Blight, Physalospora Black Canker, Scab* (*Venturia* or *Pollacia*)—Serious in northeastern states. Irregular, reddish-brown or black spots on leaves. Dense brown or olive-green mold may form on underside of leaves. Leaves quickly wilt, blacken, and fall. Tips of young twigs may shrivel, bend over, and die back from brown to black cankers. Young trees may die if defoliated several years in a row. *Control:* Prune and burn dead or blighted twigs. Keep plants vigorous. Fertilize and water during dry periods. Spray as for Leaf Spots (above). Weeping, pussy, bay, basket (common osier), and purple osier are resistant to Scab or Leaf Blight (*Fusicladium*). Protect trees against Winter Injury (pages 41–43).
4. *Leaf Rusts*—General. Lemon- to orange-yellow or reddish-brown, powdery pustules on underside of leaves. Later, dark brown to black, crustlike areas appear on both leaf surfaces. Leaves may be distorted and drop early. Rusts may produce orange-yellow or reddish-brown pustules and black cankers on twigs. Alternate hosts: larch, firs (alpine, balsam, and white), garlic, gooseberries and currants, wild orchids, saxifrage, spindletree, or none. *Control:* Same as for Leaf Spots (above). Spray using zineb, ferbam, maneb, fixed copper, or dichlone. Repeat 10 and 20 days later.
5. *Twig and Branch Cankers, Dieback, Twig Blights*—Widespread. Discolored, sunken, often sharply defined areas (cankers) develop on twigs, branches, or trunk. Cankers enlarge and gradually girdle infected part causing death to portions beyond. Trees may die after repeated attacks over several years. *Control:* Cut and burn dead and cankered parts. Spray as for Leaf Spots (above), using fixed copper or bordeaux (4-4-50). Willows vary greatly in susceptibility. Check with your nurseryman or extension plant pathologist. Black and peachleaf willows are reported resistant to *Cystospora canker;* bay, basket, and weeping willows are resistant to *Black canker (Physalospora).* Keep trees vigorous. Fertilize and water during drought periods.
6. *Crown Gall*—Widespread. Rough, irregular, "cauliflower-like" overgrowths on roots, trunk, or branches. Usually found near soil line. Trees lack vigor. Make poor growth. Leaves may yellow or branches and roots may die. Young trees may be killed. *Control:* Destroy young infected trees and nursery stock. Do not replant area with susceptible plants (see (30) Crown Gall under General Diseases) for at least 4 to 5 years. Avoid injuring stem and roots of healthy trees. Infections occur through wounds.
7. *Wood and Heart Rots*—Cosmopolitan. See under Birch, and (23) Wood Rot under General Diseases. Spraying as for Leaf Spots (above) may be beneficial. Control borers by pruning and burning dead and dying branches. Spray trunk and scaffold

branches at monthly intervals, May to September. Use DDT, dieldrin, or Thiodan. Keep trees growing vigorously.

8. *Root Rots, Dieback, Cutting Rot*—Plants make poor growth. Foliage becomes pale. Tops die back. Roots or base of cuttings often decay. Internal root tissues may be discolored. Often associated with root-feeding nematodes (e.g., dagger, lance, lesion, root-knot, stunt). *Control:* Grow in clean or pasteurized soil (pages 538–548) that is well-drained.
9. *Root-knot*—See under Peach, and (37) Root-knot under General Diseases. Weeping willow is very susceptible.
10. *Chlorosis*—Common in alkaline soils. See under Maple.
11. *Wetwood, Slime Flux*—See under Elm.
12. *Sunscald, Winter Injury*—See under Apple and Elm.
13. *Sooty Mold*—See under Linden, and (12) Sooty Mold under General Diseases.
14. *Bleeding Canker*—Northeastern states. See under Beech and Maple.
15. *Felt Fungus*—Southern states on neglected trees. Smooth, shiny, chocolate-brown to almost black growth over bark. See under Hackberry.
16. *Leaf Blister*—See under Birch, and (10) Leaf Curl under General Diseases.
17. *Mistletoe*—See (39) Mistletoe under General Diseases.

WINDFLOWER — See Anemone
WINEBERRY — See Raspberry
WINTERBERRY — See Bittersweet, Currant, and Holly
WINTERCHERRY — See Tomato
WINTERCREEPER — See Bittersweet
WINTER DAFFODIL — See Daffodil
WINTERGREEN — See Heath
WINTER MELON — See Cucumber
WIRELETTUCE — See Chrysanthemum
WISHBONE FLOWER — See Snapdragon
WISTERIA — See Honeylocust
WITCH-HAZEL [CHINESE, COMMON or VIRGINIAN, JAPANESE, SPRING, VERNAL] *(Hamamelis)*; SWEETGUM *(Liquidambar)*

1. *Leaf Spots*—Spots of various sizes, shapes, and colors. Spots may enlarge and merge to form irregular blotches. Centers may drop out giving leaves a ragged appearance. Premature leaf fall may be heavy. *Control:* Where serious, spray three times, 2 weeks apart. Start when leaves are 1/4 inch long. Use zineb, maneb, ferbam, or fixed copper and spray lime (2 level tablespoons each per gallon).
2. *Powdery Mildews* (witch-hazel)—Eastern half of United States. Grayish-white, powdery mold on foliage in late summer and fall. *Control:* If serious, spray with sulfur or Karathane several times a week apart. Start when mildew first appears.
3. *Wood Rots*—Cosmopolitan. See under Birch, and (23) Wood Rot under General Diseases.
4. *Root Rots*—See under Apple, and (34) Root Rot under General Diseases. May be associated with root-feeding nematodes (e.g., lesion, ring, root-knot, sheath, spiral, stubby-root, stunt). Grow in well-drained soil that is light, moist, and acid.
5. *Root-knot*—See under Peach, and (37) Root-knot under General Diseases.
6. *Chlorosis* (sweetgum)—Yellowing of leaf tissue between veins. Most severe on youngest tissue. Where severe, shoot and leaf growth may be dwarfed. *Control:* See under Maple.
7. *Crown Gall*—See under Apple, and (30) Crown Gall under General Diseases.
8. *Bleeding Necrosis of Sweetgum*—Eastern states. Trees "bleed" profusely. Resembles oil on bark of trunk and larger branches. Inner bark and sapwood underneath

oozing areas is brown and dead. Foliage is thin. Terminal branches die back. Trees usually die in a short time. See also under Maple. *Control:* Destroy affected trees.

9. *Sweetgum Blight, Leader Dieback*—Southeastern states. Leaves dwarfed and fall early. Foliage appears thin. Leader branches die back. Trees usually die progressively. *Control:* Water thoroughly during droughts. Remove and burn dead branches. Check with your extension plant pathologist.
10. *Twig Cankers, Dieback* (sweetgum) —See under Maple.
11. *Thread Blight* (sweetgum) —Southeastern states. See under Walnut.
12. *Felt Fungi* (sweetgum) —Southern states. See under Hackberry.
13. *Mistletoe* (sweetgum) —See (39) Mistletoe under General Diseases.

WITHE-ROD — See Viburnum

WITLOOF CHICORY — See Lettuce

WOLFBERRY — See Snowberry

WOLFSBANE — See Delphinium

WOOD ANEMONE — See Anemone

WOODBINE — See Grape and Snowberry

WOOD-SAGE — See Salvia

WOODSIA — See Ferns

WOODSORREL — See Oxalis

WOODWARDIA — See Ferns

WOODWAXEN — See Broom

WORMGRASS — See Sedum

WORMWOOD — See Chrysanthemum

WOUNDWORT — See Salvia

WYETHIA — See Chrysanthemum

XANTHORHIZA — See Clematis

XANTHOSOMA — See Calla Lily

YAM [CHINESE or CINNAMONVINE, WILD], AIR POTATO, YAMPEE *(Dioscorea)*

(This is the *true* yam. For sweetpotato see page 470.)

1. *Leaf Spots or Blotch, Anthracnose*—Small to large spots or blotches. Often with conspicuous margin. *Control:* Not usually needed. Pick and burn spotted leaves. Spray during rainy periods, at about 10-day intervals. Use zineb, maneb, or captan.
2. *Root-knot and Other Nematodes* (aphelenchus, burrowing, dagger, lesion, pin, ring, spiral, stem, stunt, tylenchus) —If severe, vines may be stunted and yellowish. Tuber surface is discolored, cracked, and roughened. Tubers may rot. Roots and tops make poor growth. *Control:* Disinfest tubers of Root-knot and possibly other nematodes by soaking in hot water (113° F.) for 30 minutes. See also (37) Root-knot under General Diseases.
3. *Storage Rots*—See under Sweetpotato. Affected areas may be covered with blue-green or black fuzzy mold.
4. *Southern Blight, Crown Rot*—See (21) Crown Rot under General Diseases.

YAMPEE — See Yam

YARDLONGBEAN — See Bean

YARROW — See Chrysanthemum

YATE-TREE — See Eucalyptus

YAUPON — See Holly

YAUTIA — See Calla Lily

YELLOW-CEDAR — See Juniper

YELLOW-ELDER — See Trumpettree

YELLOW IRONWEED — See Chrysanthemum

YELLOW-JESSAMINE — See Butterflybush

YELLOW-POPLAR — See Magnolia

YELLOWROOT — See Clematis

YELLOW STAR — See Chrysanthemum

YELLOWTRUMPET — See Trumpettree

YELLOWTUFT — See Cabbage

YELLOWWOOD — See Honeylocust

YERBA-BUENA — See Salvia

YEW [CANADA or GROUND-HEMLOCK, CUSHION, ENGLISH (many horticultural forms), HATFIELD, HICKS, IRISH, JAPANESE (many horticultural forms), PACIFIC or WESTERN] *(Taxus)*; PODOCARPUS, AFRICAN YELLOW-WOOD *(Podocarpus)*

1. *Winter Injury*—Small twigs at ends of branches turn yellow to reddish-brown or brown in late winter or early spring, especially on south and southwest sides where plants are exposed. Tips of needles appear scorched (Figure 4.235). *Control:* Shake off heavy snow or ice loads promptly. Protect from drying, winter winds and sun. Grow in protected location; erect canvas, burlap, or slat screens; or spray with Wilt Pruf, Plantcote, Good-rite latex VL-600, D-Wax, Stop-Wilt, or Foli-gard. Check with your extension horticulturist. Apply mulch in late fall to prevent deep freezing plus alternating of freezing and thawing. Water plants thoroughly before applying mulch. Shear off browned twigs in late spring. Fertilize only in early spring.

Fig. 4.235. Leaf scorch of yew. (Shade Tree Laboratories, University of Massachusetts photo)

2. *Dieback, Root Rot*—Growing tips turn yellow to light brown, later wilt, wither, and die. Plants usually slowly decline in vigor and die. Foliage becomes thin. A whitish mold or black, rather brittle "shoestrings" *(Armillaria)* may be evident beneath bark under soil surface. Bark on deeper roots decays and falls away easily. May be associated with root-feeding nematodes (e.g., bud and leaf, burrowing, dagger, lance, lesion, pin, ring, root-knot, sheath, spear, spiral, sting, stubby-root,

stunt, tylenchus). *Control:* Grow healthy, injury-free nursery stock in light, well-drained soil that is near neutral (pH 6.5). Make soil lighter if possible. Add lime to acid soil to raise pH to 6.5. Completely remove and destroy severely affected plants. Before replanting in same area, drench soil thoroughly with SMDC (Vapam or VPM Soil Fumigant). Follow manufacturer's directions. Avoid injuries to roots or trunk. Maintain vigorous growth. Fertilize, water, and protect adequately for winter. Apply 2 sprays of dieldrin or chlordane to lower parts of plants, as well as soil surface underneath. Check with your extension entomologist as regards timing of sprays for your area. These sprays control root weevils (Black Vine and Strawberry) that feed on roots and foliage. To control weevils for 5 years, grub-proof soil before planting. Use chlordane, heptachlor, or dieldrin.

3. *Twig Blights*—See under Juniper. Yew twigs are easily damaged—splitting of crotch or bark removed—by snow or ice, sloppy pruning, mowing, cultivating, etc. Results in dead tips and branches. *Control:* Prune out and burn infected twigs. During rainy springs apply copper sprays at 2-week intervals. Prevent damage to bark and twigs.
4. *Needle Blights*—Needles turn brown and die back. May be sprinkled with black dots (fungus fruiting bodies). Drop prematurely. Shoots may die back. *Control:* Rarely necessary. Spray as for Twig Blights (above).
5. *Wet Feet*—Plants gradually decline; may die. Foliage is yellow. All roots in lower part of root ball are dead. Common on yews in poorly drained soils, plants growing near downspouts, or in overwatered areas. *Control:* Correct the condition. Connect downspout to an underground pipe. Soil pH should be about 6.5.
6. *Crown Gall*—Rough, irregular, swollen galls at base of trunk or on roots. Plant lacks vigor. Makes poor growth. *Control:* See under Apple, and (30) Crown Gall under General Diseases.
7. *Wood and Heart Rots*—See under Birch, and (23) Wood Rot under General Diseases.
8. *Brown Felt Blight* (yew)—Western states at high altitudes where snow is deep. *Control:* See under Pine.
9. *Seedling Blight, Damping-off*—See under Pine.
10. *Natural Leaf-browning and Shedding*—Older and inner leaves suddenly turn yellow in late summer or early fall. Remain on plant for several weeks before dropping. This is natural.

YOUNGBERRY — See Raspberry

YUCCA: SPANISH-BAYONET, JOSHUA-TREE, ADAMS-NEEDLE or SILKGRASS, MOUNDLILY or SPANISH DAGGER, SOAPWEED or BEARGRASS, CANDLES OF THE LORD *(Yucca)*

1. *Leaf Spots, Leaf Blight, Leaf Mold, Flower Blight*—General. Small to large, more or less round spots. Often zonate with purple margin. If severe, large portions of leaves may die. Spots may occur on flowers. *Control:* Prune and burn affected parts. Indoors, keep water off foliage. If necessary, spray with zineb, maneb, or copper fungicide, plus spreader-sticker, before rainy periods.
2. *Stem Rot*—Base of stem rots. May be covered with white mold. See (21) Crown Rot under General Diseases.
3. *Root-knot*—See (37) Root-knot under General Diseases.
4. *Rust*—Small, yellowish pustules on leaves. *Control:* Same as for Leaf Spots.
5. *Crown Gall*—See (30) Crown Gall under General Diseases.

ZANTEDESCHIA — See Calla Lily

ZANTHOXYLUM — See Hoptree

ZAUSCHNERIA — See Evening-primrose

ZEBRA-PLANT — See Maranta
ZEBRINA — See Tradescantia
ZELKOVA — See Elm
ZEPHYRANTHES, ZEPHYRLILY — See Daffodil
ZINNIA — See Chrysanthemum
ZOYSIA, ZOYSIAGRASS — See Lawngrass
ZYGOCACTUS — See Cacti
ZYGOPETALUM — See Orchids

Appendix

Useful units of measure 516
Approximate rates of application equivalents 517
Converting temperature from Fahrenheit to Centigrade 517
Conversion factors 517
Modern fungicides 518
Equivalent volumes (liquid) for common measures 520
Amount (volume) of liquids required to prepare different amounts of spray mixtures at different dilutions . . 520
Amount (weight) of wettable powder required for preparing different amounts of spray mixture at different dosage levels 520
Amount (grams) of chemical required to prepare different amounts of spray mixture 521
Conversion table for use of materials on small areas 521
Level tablespoons of fungicide for use in gallon lots of spray 522
Small amounts of liquid fungicide . . 522
Streptomycin formulations 522
Percentage solution 522
Spray or dust schedules for home-grown fruit 523
Approximate amount of spray material required for fruit trees of different sizes 526
Gallons per acre required to spray orchards of different planting distances: square planting 526
Plants needed for an acre of land when set the indicated number of inches or feet apart 527
Seed treatment 527
Precautions 527
Types of treatments 528
Hot water treatment times (in minutes) for seed and other plant parts 529
Seed treatment methods and materials for vegetables, flowers, trees, and shrubs 532
Soil treatment methods and materials . 538
General precautions and suggestions 538
Treatments using heat 538
Effects of soil disinfestation by steam 542
Temperatures necessary to kill various pests 542
Treatments using chemicals . . . 543
Soil-disinfesting chemicals—materials, brands, controls, application, and remarks . . 544
Methods of applying fumigants
liquid 543
gaseous 547
granular 548
Home garden nematode control . 548
Trees relatively free of troublesome diseases 549
How much fertilizer for 100 feet of row for vegetables and flowers? . . . 549
Rates of application of sprays to row crops 550
Operating chart for tractor boom sprayers 551
Tractor speed conversions 551
How to figure tractor speed and spray rates 551

Approximate number of feet of row per acre at given distances 552
Compatibility chart for fungicides, insecticides, and miticides 552
Glossary 553
Bibliography 571

USEFUL UNITS OF MEASURE

A teaspoonful (tsp.) or tablespoonful (tbsp.) throughout this book refers to a level, standard measuring teaspoon or tablespoon.

80 drops = 1 teaspoonful (tsp.) = 0.17 fluid ounce (fl. oz.) = 5 milliliters (ml.)

1 tablespoon (tbsp.) = 3 teaspoons (tsp.) = 15 milliliters (ml.) or cubic centimeters (cc.) = ½ fluid ounce (fl. oz.) = 0.902 cubic inch (cu. in.)

1 cup = 16 tbsp. = 8 fl. ozs. = 236.6 cc. = ½ pint = 2 gills

1 pint (pt.) = 16 fl. ozs. = 473.2 cc. = 1.04 pounds (lbs.) = 28.87 cu. in. = 0.473 liters (l.) = 32 tbsp. = 2 cups = 0.125 gallon (gal.) (*Note:* 1 pint or quart dry measure is about 16 per cent larger than 1 pint or quart liquid measure.)

1 quart (liquid) = 2 pints = 946.3 ml. = 32 fl. ozs. = 57.75 cu. in. (liquid) = 0.25 gal. = 0.946 liter

1 quart (dry) = 67.2 cu. in.

1 U.S. gallon = 4 quarts (qts.) = 8 pints (pts.) = 128 fl. ozs. = 0.134 cu. ft. = 0.83 British gallon = 3,785.3 ml. or cc. = 231 cu. in. (liquid) = 8.3358 lbs. water capacity = 269 cu. in. (dry)

1 liter = 1,000 ml. or cc. = approximately 1 qt., 1 fl. oz. (or 1.08 qts.) = 0.035 cu. ft. = 61.02 cu. in. = 2.1 pints (liquid)

1 gram (gm.) = 1,000 milligrams (mg.) = 0.035 oz. avoirdupois = 15.432 grains

1 pound (lb.) = 16 ozs. = 453.59 gms. = 7,000 grains

1 kilogram = 1,000 gms. = approximately 2.2 lbs.

1 ounce (oz.) = 28.35 gms. = 437.5 grains

1 fluid ounce (fl. oz.) = 2 tbsp. = 29.57 ml. or cc. = 1.805 cu. in. = 8 fluid drams

1 peck = 0.25 bushel (bu.) = 16 pints = 8 qts. = 8.8 liters = 32 cups = 537.6 cu. in.

1 bushel = 4 pecks = 32 qts. = 64 pints = 128 cups = 40 seed flats (16 × 23 × 3 in.) = 2150.4 cu. in. = 1.24 cu. ft. = approximately 1/20 cubic yard (cu. yd.) = 35.24 liters

1 cubic foot (cu. ft.) = 0.804 bu. = 1,728 cu. in. = 0.031 cu. yd. = 7.48 gal. = 59.84 pints (liquid) = 28.32 liters = 29.92 qts. (liquid) = 25.71 qts. (dry)

1 cubic foot of water = 62.2 lbs. (One pound of water = 27.68 cu. in. = 0.016 cu. ft.)

1 cubic yard (cu. yd.) = 27 cu. ft. = 46,656 cu. in. = 202 gal. = 1,616 pints (liquid) = 808 qts. (liquid) = 21.71 bu.

1 square foot (sq. ft.) = 144 sq. in. = 0.111 sq. yd.

1 square yard (sq. yd.) = 1,296 sq. in. = 9 sq. ft.

1 square rod = 30¼ sq. yds. = 272¼ sq. ft.

1 acre = 43,560 sq. ft. = 4,840 sq. yd. = 160 sq. rods (rds.) = 0.4047 hectare

1 mile = 5,280 ft. = 1,760 yds. = 1,609.35 meters = 320 rds.

1 mile, square = 640 acres = 1 section = 2.59 sq. kilometers

10 millimeters (mm.) = 1 centimeter (cm.) = 0.3937 inch

100 centimeters = 1 meter (m.) = 39.37 inches

1 inch = 25.4 mm. = 2.54 cm. = 0.0254 meter

1 yard = 3 feet = 0.9144 meter

1 rod = 5½ yards = 16½ feet

1 cubic foot (cu. ft.) *dry* soil (approximate) = sandy (90 lbs.), loamy (80 lbs.), clayey (75 lbs.)

1 bushel (bu.) *dry* soil (approximate) = sandy (112 lbs.), loamy (100 lbs.), clayey (94 lbs.)

1 part per million (ppm.) = 1 milligram per liter or kilogram = 0.0001 per cent = 0.013 ounce by weight in 100 gallons = 0.379 gm. in 100 gal. = 1 inch in nearly 16 miles = 1 oz. of salt in 62,500 lbs. of sugar = 1 minute in about 2 years = a 1-gm. needle in a 1-ton haystack = 1 oz. of sand in 31¼ tons of cement = 1 oz. of dye in 7,530 gal. of water = 1 lb. in 500 tons = 1 penny of $10,000.

1 part per billion (ppb.) = 1 inch in nearly 1,600 miles = 1 drop in 20,000 gal.

1 per cent = 10,000 parts per million (ppm.) = 10 g. per liter = 1.28 ozs. by weight per gallon = 8.336 lbs. per 100 gallons

APPROXIMATE RATES OF APPLICATION EQUIVALENTS

(U.S. Measures)

1 ounce per square foot = 2,722.5 pounds per acre
1 ounce per square yard = 302.5 pounds per acre
1 ounce per 100 square feet = 27.2 pounds per acre
1 pound per 100 square feet = 435.6 pounds per acre
1 pound per 1,000 square feet = 43.6 pounds per acre
1 pound per acre = 1 ounce per 2,733 square feet (0.37 oz./1,000 sq. ft.) = 4.5 grams per gallon = 0.0104 gm. per sq. ft.
100 pounds per acre = 2½ pounds per 1,000 square feet = 4 ounces per 1,000 square feet = 1.04 gms. per sq. ft.
5 gallons per acre = 1 pint per 1,000 square feet = 0.43 ml. per square foot
100 gallons per acre = 2.5 gallons per 1,000 square feet = 1 quart per 100 square feet

CONVERTING TEMPERATURE FROM FAHRENHEIT TO CENTIGRADE AND VICE VERSA

To convert from *Fahrenheit* to *Centigrade:* Subtract 32 from the Fahrenheit reading, multiply by 5, and divide the product by 9. *Example:* 131° F. − 32 = 99 × 5 = 495; 495 ÷ 9 = 55° C.

To convert from *Centigrade* to *Fahrenheit:* Multiply the Centigrade reading by 9, divide the product by 5 and add 32. *Example:* 25° C. × 9 = 225 ÷ 5 = 45; 45 + 32 = 77° F.

TABLE 5
CONVERSION FACTORS

To Change	To	Multiply By
Inches	Centimeters	2.54
Centimeters	Inches	0.394
Feet	Meters	0.305
Meters	Inches	39.37
Miles	Kilometers	1.609
Kilometers	Miles	0.621
Square Inches	Square Centimeters	6.452
Square Centimeters	Square Inches	0.155
Square Yards	Square Meters	0.836
Square Meters	Square Yards	1.196
Cubic Yards	Cubic Meters	0.765
Cubic Meters	Cubic Yards	1.308
Cubic Inches	Cubic Centimeters	16.387
Cubic Centimeters	Cubic Inches	0.061
Cubic Centimeters	Fluid Ounces	0.034
Fluid Ounces	Cubic Centimeters	29.57
Quarts	Liters	0.946
Liters	Quarts	1.057
Grams	Ounces (avoirdupois)	0.0352
Ounces (avoirdupois)	Grams	28.35
Grains	Milligrams	64.799
Pounds (avoirdupois)	Kilograms	0.454
Pounds (apothecary)	Kilograms	0.373
Kilograms	Pounds (apothecary)	2.205
Ounces (apothecary)	Grams	31.103
Grams	Grains	15.432

TABLE 6

Modern Fungicides

Common Name and Active Ingredient	Trade Names and Distributors	Principal Uses and Remarks
CAPTAN N-(trichloromethylthio)-4-cyclohexene-1,2-dicarboximide *or* N-(trichloromethylthio)-tetrahydrophthalimide	Captan 50-W, Captan 75 Seed Protectant, Captan-Dieldrin 60–15 Seed Protectant, Captan Garden Spray, Captan 80 Spray-Dip (Stauffer), Orthocide 50 or 80 Wettable, Orthocide Fruit and Vegetable Wash, Orthocide 75 Seed Protectant, Orthocide Garden Fungicide (Chevron), Chipman Captan Dust (Chipman), Agway Captan 5D and 7.5D (Agway), F&B Captan 7.5 Dust (Faesy & Besthoff), Miller Captan Dust (Miller), Captan 7.5 Dust (Chevron), etc.	Excellent, safe fungicide to control leafspots, blights, fruit rots, etc. on fruits, ornamentals, vegetables, and turf. Seed protectant (often mixed with insecticide) for vegetables, flowers, and grasses. Postharvest dip for fruits and vegetables. Does *not* control powdery mildews and rusts. Soil treatment on plant beds to control crown rot and seedling blights. Widely used in multipurpose sprays and dusts. Both a protectant and an eradicant.
CHLORANIL Tetrachloro-*p*-benzoquinone	Spergon, Spergon Wettable, Spergon Seed Protectant (U.S. Rubber), Spergon Spray Powder (Niagara, General Chem.), Geigy SP 50 (Geigy), Spergon Dust (General), etc.	Seed and bulb treatment for flowers, vegetables, and grasses. Soil drench for crown rot of flowers. Corm and bulb dip for flowers. Sprays and dusts for certain foliage diseases.
DICHLONE 2,3-Dichloro-1,4-naphthoquinone	Phygon, Phygon-XL, Phygon Seed Protectant (U.S. Rubber), Corona Phygon-XL (Pittsburgh Plate Glass), Niagara Phygon (Niagara), Green Cross Phygon-XL Fungicide (Green Cross), Stauffer Phygon-XL (Stauffer), etc.	Seed treatment for certain vegetables and flowers. Spray for certain blights and fruit rots of vegetables and fruits. Soil drench to control damping-off. Treat as directed. Injurious at 85° F. or above. Mostly eradicative.
DODINE N-dodecylguanidine acetate	Cyprex Dodine 65-W, Cyprex Dodine Dust (Cyanamid), Miller's Cyprex Dusts (Miller), Pennsalt Cyprex 4% Dust (Pennsalt), etc.	Controls certain foliage diseases of apple, cherry, strawberry, pecan, and roses. Gives long-lasting protection. Good eradicant.
FERBAM Ferric dimethyldithiocarbamate	Fermate Ferbam Fungicide (DuPont), Karbam Black (Sherwin-Williams), Carbamate (Niagara), Ortho Ferbam 76 (Chevron), Orchard Brand Ferbam (General), Coromate Ferbam Fungicide (Pittsburgh Plate Glass), Chipman Ferbam W-76 (Chipman), etc.	General, safe fungicide to control many foliage diseases of flowers, trees, shrubs, and fruits. Soil drench to control damping-off and seedling blights. Used in some multipurpose sprays. May leave objectionable black spray deposit on flowers, woodwork, etc. Mostly protective.

TABLE 6 (*continued*)

MODERN FUNGICIDES

Common Name and Active Ingredient	Trade Names and Distributors	Principal Uses and Remarks
FOLPET (Phaltan) N-trichloromethylthiophthalimide	Corona Folpet 50 Wettable (Pittsburgh Plate Glass), Ortho Phaltan Rose and Garden Fungicide, Ortho Phaltan 50 Wettable (Chevron), Niagara Phaltan 50 Wettable (Niagara), Stauffer Folpet (Stauffer), etc.	A close relative of captan and used for many of the same purposes on fruits, flowers, turf, vegetables, trees, and shrubs. Controls many powdery mildews. Excellent for roses. Both a protectant and eradicant. In multipurpose mixes.
MANEB Manganese ethylenebis (dithiocarbamate) also special formulation containing zinc	Manzate Maneb Fungicide, Manzate D (DuPont), Dithane M-22 and M-22 Special (Rohm & Haas), Kilgore's Maneb 80 Wettable (Kilgore), Dithane M-45 (Rohm & Haas), Twin Light Maneb Dust (Seacoast), 2% Maneb Dust (Carolina), Maneb 4.5D (Agway), etc.	Excellent general fungicide to control foliage and fruit diseases of vegetables, trees, turf, flowers, and some fruits. Very useful for tomato, potato, and vine crops. In multipurpose mixes. Controls rusts but not powdery mildews. Mostly protective.
THIRAM (TMTD) Bis(dimethylthiocarbamoyl) disulfide *or* Tetramethylthiuram disulfide	Tersan 75, Thylate Thiram Fungicide, Delsan A-D, Arasan 50-Red and 75 (DuPont), Panoram 75 and D-31 (Morton), Thiram 50 Dust (U.S. Rubber, DuPont), Thiram 4.8D (Agway), Thiram-65 and 75, Thiram S-42 (Pennsalt)	Seed and bulb treatment for vegetables, flowers, and grasses. Controls certain lawn, fruit, and vegetable diseases. Controls rusts. Soil drench for crown rot and damping-off. Only protective.
ZINEB Zinc ethylenebis (dithiocarbamate)	Dithane Z-78 (Rohm & Haas), Parzate Zineb Fungicide, Parzate C (DuPont), Ortho Zineb 75 Wettable, Ortho Dust (Chevron), Niagara Zineb (Niagara), Stauffer Zineb (Stauffer), Chipman Zineb (Chipman), Corona Zineb (Pittsburgh Plate Glass), Flight Brand 10% Dithane Dust (Carolina), Zineb 19.5D TF (Agway), 6.5% Zineb Dust (Flag Sulphur)	Excellent, safe fungicide for vegetables, fruits, flowers, trees, and shrubs to control leaf spots, blights, fruit rots, etc. Also useful on lawns as a soil drench to control crown and root rots. Controls rusts but not powdery mildews. In many multipurpose mixes for vegetables and flowers. Only protective.
ZIRAM Zinc dimethyldithiocarbamate	Zerlate Ziram Fungicide (DuPont), Karbam White (Sherwin-Williams), Z-C Spray or Dust (Niagara), Orchard Brand Ziram (Gen.), Ziram (Chevron, Chipman, Stauffer), etc.	General, safe fungicide. Useful for vegetables and ornamentals, especially tender seedlings. In many vegetable and flower multipurpose mixtures. Only protective.

TABLE 7

EQUIVALENT VOLUMES (LIQUID) FOR COMMON MEASURES

Measuring Unit Used	Number of Units to Fill Measure in Column 1					
	Tsp.	Tbsp.	Cup	Pint	cc.	Liter
1 Teaspoonful	1.00	0.33	0.021	0.010	4.9	0.0049
1 Tablespoonful	3.00	1.00	0.663	0.031	14.8	0.0148
1 Fluid Ounce	6.00	2.00	0.125	0.062	29.6	0.0296
1 Cup	48.00	16.00	1.000	0.500	236.6	0.2366
1 Pint	96.00	32.00	2.000	1.000	473.2	0.4732
1 Quart	192.00	64.00	4.000	2.000	946.3	0.9463
1 Gallon	768.00	256.00	16.000	8.000	3,785.3	3.7853
1 Liter	202.88	67.63	4.328	2.164	1,000.0	1.0000
1 Milliliter (cc.)	0.20	0.068	0.0042	0.0021	1.0	0.0010

TABLE 8

AMOUNT (VOLUME) OF LIQUIDS REQUIRED TO PREPARE DIFFERENT AMOUNTS OF SPRAY MIXTURES AT DIFFERENT DILUTIONS

Dilution of Spray Required	Recommended Dosage of Chemical in 100 Gallons of Water			Amount of Material Required To Prepare Spray for					
				20 gallons		5 gallons		1 gallon	
	cups	pints	qts.	pints	cc.	cc.	tsp.	cc.	tsp.
1–3200	0.5	0.25	0.12	0.050	23.7	5.9	1.2	1.18	0.2
1–1600	1.0	0.50	0.25	0.100	47.7	11.8	2.4	2.37	0.5
1–800	2.0	1.00	0.50	0.200	94.6	23.7	4.8	4.73	1.0
1–400	4.0	2.00	1.00	0.400	189.3	47.3	9.6	9.46	1.9
1–200	8.0	4.00	2.00	0.800	378.6	94.6	19.2	18.93	3.8
1–100	16.0	8.00	4.00	1.600	757.1	189.3	38.3	37.86	7.7
1–50	32.0	16.00	8.00	3.200	1,514.2	378.6	76.6	75.71	15.3
1–25	64.0	32.00	16.00	6.400	3,028.5	757.1	153.2	151.42	30.6
1–10	160.0	80.00	40.00	16.000	7,571.0	1,893.0	383.0	378.60	76.5

TABLE 9

AMOUNT (WEIGHT) OF WETTABLE POWDER REQUIRED FOR PREPARING DIFFERENT AMOUNTS OF SPRAY MIXTURE AT DIFFERENT DOSAGE LEVELS

Recommended Dosages per 100 Gallons			Amount of Material Required To Prepare Spray Mixture							
			50 gallons		20 gallons		5 gallons		1 gallon	
lbs.	ozs.	gms.	ozs.	gms.	ozs.	gms.	ozs.	gms.	ozs.	gms.
0.25	4	113	2	56	0.8	23	0.20	6	0.04	1
0.50	8	227	4	113	1.6	45	0.40	11	0.08	2
1.00	16	454	8	227	3.2	91	0.80	23	0.16	5
1.50	24	681	12	340	4.8	136	1.20	34	0.24	7
2.00	32	908	16	454	6.4	182	1.60	45	0.32	9
3.00	48	1,362	24	681	9.6	272	2.40	68	0.48	14
4.00	64	1,816	32	908	12.8	363	3.20	91	0.64	18
5.00	80	2,270	40	1,135	16.0	454	4.00	113	0.80	23

TABLE 10

AMOUNT (GRAMS) OF CHEMICAL REQUIRED TO PREPARE DIFFERENT AMOUNTS OF SPRAY MIXTURE

Recommended Pound Dose per 100 Gallons	Grams of Material Required To Prepare*				
	100 gal.	50 gal.	20 gal.	5 gal.	1 gal.
0.25	113.4	56.7	22.7	5.7	1.1
0.50	226.8	113.4	45.4	11.3	2.3
0.75	340.2	170.1	68.0	17.0	3.4
1.00	453.6	226.8	90.7	22.7	4.5
1.25	566.9	283.5	113.4	28.4	5.7
1.50	680.3	340.1	136.1	34.0	6.8
1.75	793.7	396.8	158.7	39.7	7.9
2.00	907.1	453.6	181.4	43.4	9.1
2.50	1,133.8	566.9	226.8	56.7	11.3
3.00	1,360.6	680.3	272.2	68.0	13.6
4.00	1,814.1	907.1	362.8	90.7	18.1
5.00	2,267.7	1,133.8	453.6	113.4	22.7

* To convert to ounces or pounds divide by 28.35 or 453.59 respectively.

TABLE 11

CONVERSION TABLE FOR USE OF MATERIALS ON SMALL AREAS

Rate Per Acre	Rate Per 1,000 Square Feet	Rate Per 100 Square Feet
	Liquid Materials	
1 pt.	¾ tbsp.	¼ tsp.
1 qt.	1½ tbsp.	½ tsp.
1 gal.	6 tbsp.	2 tsp.
25 gal.	4½ pts.	1 cup
50 gal.	4½ qts.	1 pt.
75 gal.	6½ qts.	1½ pts.
100 gal.	9 qts.	1 qt.
	Dry Materials	
1 lb.	2½ tsp.	¼ tsp.
3 lbs.	2¼ tbsp.	¾ tsp.
4 lbs.	3 tbsp.	1 tsp.
5 lbs.	4 tbsp.	1¼ tsp.
6 lbs.	4½ tbsp.	1½ tsp.
8 lbs.	⅖ cup	1¾ tsp.
10 lbs.	½ cup	2 tsp.
100 lbs.	2¼ lbs.	¼ lb.

Dry: 1 lb./100 gal. = 1 oz./6¼ gal. = ½ oz./3 gal. = 4.5 gm./gal.
Liquid: 1 pt./100 gal. = 1 fl. oz./6¼ gal. = ½ fl. oz./3 gal. = 1 tsp./gal.
1 qt./100 gal. = 10 tsp./5 gal. = 2 tbsp./3 gal. = 2 tsp./gal.
1 gal./100 gal. = ¾ cup & 4 tsp./5 gal. = ½ cup/3 gal. = 8 tsp./gal.

TABLE 12

Level Tablespoons of Fungicides for Use in Gallon Lots of Spray

Fungicide	Pounds per 100 Gallons						
	½	1	1½	2	4	6	8
	Level tablespoons to use in 1 gallon of spray*						
Captan, 50% wettable powder	⅜	¾	1⅛	1½	3	4½	6
Chloranil, 96% wettable powder (Spergon)	½	1	1½	2	4	6	8
Copper Sulfate (snow)	⅙	⅓	½	⅝	1¼	1⅞	2½
Dichlone (Phygon)	⅓	⅔	1	1⅓	2⅔	4	5⅓
Dinocap (Karathane-WD)	⅓	⅔	1	1⅓	2⅔	4	5⅓
Dodine (Cyprex 65-W)	⅙	⅓	½	⅔	1½	2	3
Dyrene, 50% wettable powder	⅜	¾	1⅛	1½	3	4½	6
Ferbam, 76% wettable powder	⅝	1¼	1¾	2½	5	7½	10
Fixed Copper, 50% metallic copper	¼	½	¾	1	2	3	4
Folpet (Phaltan, 50% wp)	⅜	¾	1⅛	1½	3	4½	6
Maneb, 80% wettable powder	¼	½	¾	1	2	3	4
PCNB (Terraclor, 75% wp)	½	1	1½	2	4	6	8
Semesan	¼	½	¾	1	2	3	4
Spray Lime	½	1	1½	2	4	6	8
Thiram, 65% (Thylate)	⅓	⅔	1	1⅓	2⅔	4	5⅓
Thiram, 75% wettable powder	⅜	¾	1⅛	1½	3	4½	6
Wettable Sulfur (dry), 95%	¼	½	¾	1	2	3	4
Zineb, 65–75% wettable powder	⅓	⅔	1	1⅓	2⅔	4	5⅓
Ziram, 76% wettable powder	⅝	1¼	1¾	2½	5	7½	10

* Tablespoon amounts are based on a *level* tablespoon containing ½ ounce.

TABLE 13

Small Amounts of Liquid Fungicide

Amount of Liquid Fungicide Recommended for 100 Gallons of Spray	Amount of Fungicide To Use in 1 Gallon of Spray
12 gallons	32 tablespoons or 1 pint
10 gallons	26¾ tablespoons or ⅘ pint
1 gallon	2½ tablespoons
1 quart	⅝ tablespoon

TABLE 14

Streptomycin Formulations: Number of Teaspoons Per Gallon of Different Formulations To Make a 100-Parts-per-Million Solution

Formulation	Teaspoons
Agrimycin 17, 17% wettable powder	1½
Agrimycin 100, 15% wettable powder	1½
Agri-strep, 21.2% wettable powder	¾
Miller Streptomycin Antibiotic Spray Powder, 8.5% wettable powder	2½
Ortho Streptomycin Spray, 21.2% wettable powder	1½
Phytomycin, 20% liquid	½

TABLE 15

Percentage Solution

Per Cent	Dilution or Rate	Parts Per Million (ppm.)	Grams Per Liter
10.0	1:10	100,000	100.0
1.0	1:100	10,000	10.0
0.1	1:1,000	1,000	1.0
0.01	1:10,000	100	0.1
0.001	1:100,000	10	0.01
0.0001	1:1,000,000	1	0.001

SPRAY OR DUST SCHEDULES FOR HOME-GROWN FRUIT

The schedules outlined below for tree, bush, and bramble fruits are *average* home fruit programs. Weather, disease, and pest conditions vary greatly from region to region in the United States. This makes absolute, fixed schedules impossible. Most commercial fruit growers use 20 to 40 per cent more sprays than are listed below. *There are no short cuts!* Supplement these fruit programs with those published by your state and available at your county extension office.

The multipurpose fruit spray (page 111) containing captan, methoxychlor, and malathion is recommended for all sprays or dusts except as noted below. This mixture, sold under a variety of trade names (e.g., Acme Fruit Tree Spray, Breck's Many Purpose Spray, De-Pester Fruit Tree Spray, Eastern States Garden and Orchard Dust, E-Z-Flo Fruit Guard, F & B Multi-Purpose Spray or Dust, Green Gro Home Garden Fruit Spray, Hopkins Strawberry Dust, Lebanon Fruit & Berry Spray, Miller's New Fruit Spray or Dust, and Ortho Home Orchard Spray), can be used safely on most ornamentals, shrubs, flowers, vines, small trees, and lawns. If it does not give satisfactory results, ask local authorities (e.g., your county agent, extension plant pathologist, horticulturist, or entomologist) for more information.

The combination of methoxychlor (or sevin), malathion, and captan may occasionally cause leaf injury and leaf drop on peaches, plums, and cherries; plus spotting or scorching of leaves of certain apple and pear varieties when spray is applied too heavily or when the spray film dries very slowly (wet or foggy weather). Sevin may cause early apple fruit drop. Injury is more severe in cool, humid, wet or foggy weather when spray dries slowly or when sprays are applied too heavily. Where the separate ingredients are purchased and mixed in the spray tank, thiram or ferbam can often be substituted for captan, and sevin for methoxychlor.

Follow a complete spray or dust program. Two or 3 applications in the spring will not give satisfactory disease and insect control of most fruits.

Follow local, recommended, cultural practices regarding pruning, cultivation, fertilization, mulching, and rodent control. Plant only enough certified, disease-free, top quality nursery stock that you can properly take care of. Unsprayed and uncared for plants make good breeding places for disease-producing organisms and insect pests. Thin out trees, brambles, and bush fruits to make spraying and dusting easy. Cut back tops of taller trees if they cannot be reached with spray.

Remember to:

1. *Spray thoroughly*—Cover all above-ground plant parts, including both leaf surfaces, with each spray or dust. Spray *tops* of trees with special care. Continue spraying until a noticeable amount of dripping occurs. See Table 17. Use adequate pressure. Keep spray mixture stirred to prevent settling. Wash out sprayer and hose thoroughly after using. Prepare a *fresh* spray mixture each time you spray. Be sure to mix materials in a clean container.

2. *Spray often*—Additional sprays or dusts will be needed during long wet periods. An inch of rainfall removes or reduces effectiveness of the spray film. Temperatures above 90° F. shorten the period when the spray film is protective.

3. *Measure carefully. Use the amounts suggested*—A more concentrated spray may cause severe injury or leave residues at harvest that exceed safe tolerances.

Apply dusts or sprays when air is calm and foliage is dry. Never make applications at temperatures below 40° F. or when freezing temperatures are expected. Follow all safety precautions outlined in Section 3 (pages 107–109). *Always read, understand, and follow all precautions listed on the package label.*

Dispose of dropped and wormy or infected fruits promptly. Remove and destroy all dead or diseased leaves, twigs, prunings, trash, and other debris from fruit plantings. Leave flush pruning cuts (page 33).

TABLE 16

SPRAY OR DUST SCHEDULES FOR HOME-GROWN FRUIT

When To Spray	Apple Crabapple	Pear Quince	Peach Plum Prune Cherries Nectarine Apricot Almond	Grape	Raspberry Blackberry Dewberry Boysenberry	Strawberry	Blueberry Huckleberry	Currant Gooseberry
DORMANT SPRAY Apply *before* buds swell; *not* later			Spray[1]				Before growth starts in spring	
GREEN TIP SPRAY When buds "broken" to show green tips	Spray[3]	Spray			Buds[8] starting to swell	Buds swelling	Buds swelling	As first leaves unfold[3]
PREBLOOM SPRAY When most blossom buds show color and *before* blossoms open	Spray[2]	Spray[2,3]	Spray[3]	Shoots ½ to 1 inch long	Buds breaking (shoots ¼ inch out)	Just before 1st blooms open[3,4]	Just before bloom[2]	Just before bloom[2]
BLOOM SPRAY When flowers are open	Do *not* spray during bloom, or use *just* captan plus another fungicide[4]			Shoots 4 to 8 inches long[3]	Shoots 4 to 8 inches long[2,3]	When 10% blooms are open	Petal-fall	Captan alone during bloom
PETAL-FALL SPRAY When 90 per cent of petals have fallen	Spray[5,7]	Spray	Spray[2,7]		Just before bloom	Captan alone during bloom		Spray
1ST COVER SPRAY 5 to 10 days after petal-fall	Spray	Spray	Spray	Just before bloom	Just after bloom	Early berries half grown	Berries ¼ inch[3]	
2ND COVER SPRAY 7 to 10 days after 1st cover	Spray	Spray	Spray	Just after bloom	Berries half grown	Fruit growing		Spray[5]
3RD COVER SPRAY 7 to 10 days after 2nd cover	Spray	Spray	Spray	Fruit pea-size[3]	Fruit beginning to ripen[5]	Berries start to ripen[5]	Early berries blue[5]	
4TH COVER SPRAY 7 to 10 days after 3rd cover	Spray		Spray[5]	Fruit touching in clusters			Spray[5]	Just after harvest[6]

TABLE 16 (*continued*)

SPRAY OR DUST SCHEDULES FOR HOME-GROWN FRUIT

When To Spray	Apple Crabapple	Pear Quince	Peach Plum Prune Cherries Nectarine Apricot Almond	Grape	Raspberry Blackberry Dewberry Boysenberry	Strawberry	Blueberry Huckleberry	Currant Gooseberry
5TH COVER SPRAY 7 to 10 days after 4th cover	Spray	Spray[5]	Spray[5,6]	Fruit growing	Just after harvest[6]	Just after harvest and renovation[6]	Just after harvest[6]	
6TH COVER SPRAY 7 to 10 days after 5th cover	Spray[5]		Spray[5,6]	Fruit growing[5]				
7TH COVER SPRAY 7 to 10 days after 6th cover	Spray[5]	Spray[5]	Spray[5,6]					
ADDITIONAL COVER SPRAYS Late maturing varieties	Spray[5]		Spray[5,6]					

[1] Use ferbam, 2½ tablespoons per gallon (¾ cup in 5 gallons) on stone fruits as a dormant spray to control *Leaf Curl*, *Black Knot*, and *Plum Pockets*. Outdoor temperature must be 40° F. or above when spray is applied.

[2] If *Rust* is a problem, add ferbam, zineb, or thiram (1 tablespoon per gallon; ⅓ cup in 5 gallons) to multipurpose spray. See Section 4 regarding timing of applications.

[3] If *Powdery Mildew* is a problem, add Karathane-WD (⅔ teaspoon per gallon; 1 tablespoon in 5 gallons) or wettable sulfur (2 tablespoons per gallon; ½ cup in 5 gallons). Some fruits like *blueberries*, *grapes*, *raspberries*, and certain *apple* varieties are sulfur-sensitive. Use sulfur with caution on these crops. Fixed copper may be used on *grapes*.

[4] Do *not* apply multipurpose spray during bloom or pollinating insects will be killed. To control *Rust*, *Brown Rot*, *Scab*, *Blossom Blight*, *Graymold Blight*, and other diseases that attack during bloom, apply mixture of captan plus ferbam, zineb, or thiram (1 tablespoon of each per gallon; ⅓ cup in 5 gallons). Streptomycin (100 ppm.) can be added to control *Fire Blight* of *Apples*, *Pears*, *Quinces*, *and Crabapples*. Spray *just before* wet periods, where possible.

[5] *Do not apply multipurpose spray within a week or two of harvest*. *Captan* alone may be used up to harvest and even as dip after picking to protect against *Fruit Rots*. A mixture of captan and zineb (1½ tablespoons of each per gallon) is recommended for cover sprays on *apples* to control summer diseases.

[6] *Cherries*, *raspberries*, other brambles, *strawberries*, *currants*, *gooseberries*, and *blueberries* should be sprayed one or more times *after harvest* to protect growth for next year's crop. See under plant involved in Section 3.

[7] To control *peach* and *apple* borers, thoroughly spray or paint the trunk and scaffold branches 4 times, at 14-day intervals. Use DDT (4 heaping tablespoons of 50 per cent powder in a gallon of water) or use Thiodan following manufacturer's directions. *Keep spray off fruit and foliage*. Check with your county agent or extension entomologist regarding timing of these applications. Keep trees vigorous.

[8] Use *liquid lime-sulfur* alone in first two sprays for *raspberries*, *dewberries*, and *blackberries* to control *Anthracnose*, *Cane Blight*, and *Spur Blight;* also aids in control of scale insects, mites, etc. Use ⅘ pint per gallon; 2 quarts in 5 gallons of spray. Use multipurpose spray for *all* later applications.

TABLE 17

APPROXIMATE AMOUNT OF SPRAY MATERIAL REQUIRED FOR FRUIT TREES OF DIFFERENT SIZES

Height in Feet	Spread in Feet	Gallons Per Application
4	3	up to ½
5	6	3
6	8	4
7	10	5
8	12	6
9	14	7
10	16	8
11	18	10
13	20	12
16	24	13
19	26	14
25	30	15
30	35	20

A Simple Way To Figure Amount of Spray for Apple Trees (*Normal* not *Dwarf*)
For *green tip* and earlier sprays, divide *age* of tree by 4 to find gallons needed per tree;
For *prebloom* and *bloom* sprays, divide by 3;
For *petal-fall* spray, divide by 2;
For *all cover* sprays, divide by 1.5.

Example: A 10-year-old tree should be given the following amounts: green tip and earlier stages — 2.5 gallons; prebloom and bloom stages — 3.3 gallons; petal-fall stage — 5 gallons; each cover spray — 6.7 gallons.

TABLE 18

GALLONS PER ACRE REQUIRED TO SPRAY ORCHARDS OF DIFFERENT PLANTING DISTANCES: SQUARE PLANTING

Distance Between Trees in Feet	Desired Gallons Per Tree							
	2	5	7	9	10	12	15	20
	Gallons Per Acre Required							
18	268	670	938	1,206	1,340	1,608	2,010	2,680
20	218	545	763	981	1,090	1,308	1,635	2,180
22	180	450	630	810	900	1,080	1,350	1,800
24	150	375	525	675	750	900	1,125	1,500
25	140	350	490	630	700	840	1,050	1,400
30	96	240	336	432	480	576	720	960
35	72	180	252	324	360	432	540	720
40	54	135	189	243	270	324	405	540
45	44	110	154	198	220	264	330	440
50	34	85	119	153	170	204	255	340

TABLE 19

PLANTS NEEDED FOR AN ACRE OF LAND WHEN SET THE INDICATED NUMBER OF INCHES OR FEET APART (NOTE: TO ESTIMATE FOR SMALLER AREAS, FIGURE THAT A PLOT 33x66 FEET = 1/20 OF AN ACRE)

Inches	Plants	Inches	Plants	Feet	Plants	Feet	Plants
1x1	6,272,640	8x12	65,340	1x1	43,560	6x6	1,210
1x3	2,090,880	9x9	77,440	1x2	21,780	6x12	605
1x6	1,045,440	9x12	58,080	1x3	14,520	7x7	888
1x12	522,720	10x10	62,726	1x4	10,890	7x12	518
2x2	1,568,160	10x12	52,272	1x5	8,712	8x8	680
2x3	1,045,440	10x18	34,848	1x6	7,260	8x12	453
2x6	522,720	10x24	26,132	1x8	5,445	9x9	537
2x12	261,360	10x30	20,908	1x12	3,630	9x12	403
3x3	696,960	10x48	13,068	2x2	10,890	9x18	268
3x6	348,480	10x60	10,454	2x4	5,445	10x10	435
3x12	174,240	15x15	27,878	2x6	3,630	10x20	217
4x4	392,040	15x20	20,908	2x8	2,722	12x12	302
4x6	261,360	15x36	11,616	2x12	1,815	12x18	201
4x12	130,680	15x48	8,712	3x3	4,840	12x24	151
5x5	250,905	15x60	6,969	3x6	2,420	12x30	121
5x9	139,392	18x18	19,360	3x12	1,210	15x15	103
5x12	104,544	18x24	14,520	4x4	2,722	15x20	145
6x6	174,240	18x36	9,680	4x6	1,185	15x30	96
6x12	87,120	18x48	7,260	4x8	1,361	18x24	100
7x7	128,013	20x20	15,681	4x12	907	18x30	80
7x12	74,674	20x36	8,712	5x5	1,742	20x30	72
8x8	98,010	20x48	6,534	5x12	726	24x30	60

SEED TREATMENT*

The value of chemical treatment of garden seed has been proved repeatedly. Disease-producing organisms on and in seed are killed. Protection is also provided against certain seed-rotting and seedling-blight fungi in the soil (Figure A.1). Seed treatments give maximum insurance benefits when cold, wet weather follows planting. Pretreated vegetable and lawngrass seed can be purchased from seedsmen or seed supply houses.

Commercial seed treating is done by special machines using a slurry, mist-type liquid, or ready-mix liquid disinfestant.

Garden seed can be treated inside the seed packet or in larger quantities by using Mason jars or a rotating drum.

To treat in a seed packet, place a small mound of the treatment chemical on the tip (¼ inch) of a penknife or broad end of a toothpick; dump into the seed packet, fold the top tightly shut, and shake thoroughly for a minute or two. Excess seed protectant may be sifted out before planting.

For larger amounts, fill Mason jar ½ full (or less) of seed, add ½ to 1 teaspoon of seed treatment chemical, screw lid on tightly, and roll jar on floor for 5 minutes until seed is evenly coated.

* Also included are treatments for bulbs, corms, tubers, rhizomes, roots, and other propagative plant parts, to control disease-producing organisms

Precautions

Remember that 1 to 2 ounces of seed protectant (treatments 3–8) are enough to treat a whole *bushel* of seed. Do not overdose!

All seed treatment chemicals are toxic or poisonous. Carefully mark treated seed. Do *not* use for feed or food. Be sure containers used for treated seed are thoroughly cleaned before reuse.

Avoid inhaling dusts or fumes when treating. Treat outdoors or in a well-ventilated room.

Follow manufacturer's directions when handling seed-treatment materials.

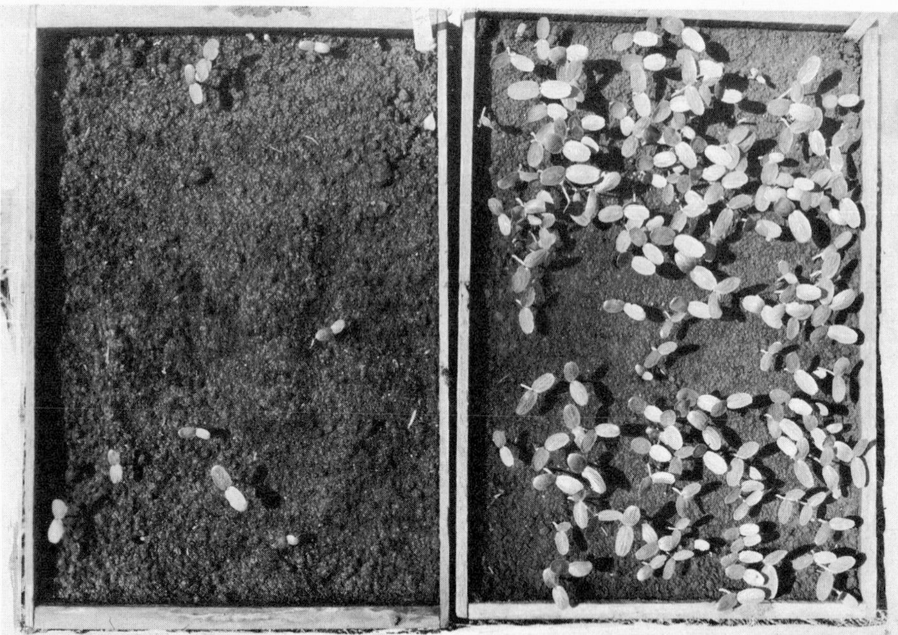

Fig. A.1. Cucumber seed treatment showing damping-off control. Left, seed untreated; right, treated with a fungicide. (Illinois Agricultural Experiment Station photo)

Types of Treatments

Seed treatments are of two general types: (1) *eradicative* seed treatment which destroys disease-causing fungi and bacteria carried on and within the seed, and (2) *protective* seed treatment which applies a coating to the seed surface, protecting against seed rot and damping-off caused by soil organisms. Both types are important in producing disease-free vegetable and flower plants. Since seed-treatment materials do *not* serve both as eradicants and as protectants, it is usually advisable to follow the eradicative treatment with a protective treatment.

A. Eradicative Treatments

Treatment 1—Mercuric chloride (also called corrosive sublimate or bichloride of mercury) soak. Very poisonous. Your pharmacist may require that a "poison register" be signed before he will sell the chemical. An effective treatment for certain diseases of cucumber, muskmelon (cantaloup), pepper, pumpkin, squash, sweetpotato, watermelon, amaryllis, calla lily, canna, China-aster, gladiolus, iris, snowflake, and sweetpea. May also be used as a general disinfectant and soil drench. Sold as a white powder or as blue tablets. Commonly prepared to make a 1 in 1:1,000 solution by dissolving 1 ounce of mercuric chloride powder in 7½ gallons of water (one 7.3 or 8 grain tablet in a pint of water). Use a wood, glass, enamel, or earthenware container as mercury "eats away" at metals. Dissolve the mercury in a small amount of hot water and add to cold water to bring to the specified volume. Use at least a quart of solution for each 4 ounces of seed.

For *cucumber, melons, pumpkin,* and *squash* place the seed in a loose-mesh cotton bag—not over ½ full—and suspend it in a warm (60° to 80° F.) 1:1,000 solution of mercuric chloride for 5 to 10 minutes. Remove and rinse in cold, running water for 15 minutes. Then dry and dust with treatment 3, 4, 5, 6, 7, or 8.

Presoak *sweetpea* seed 1 minute in alcohol. Follow by a 20-minute soak in a 1:1,000 solution. Then dry and dust with treatment 3, 4, 5, or 7.

For *pepper* seed soak 5 minutes in a 1:2,000 solution (1 tablet in 2 pints of water). Wash 15 minutes in running water, dry, and apply treatment 3 or 4.

Soak dormant *canna* tubers (rootstocks) or *amaryllis* bulbs for 2 hours in a 1:1,000 solution; soak *dormant calla lily* corms or rhizomes for 30 to 60 minutes, then wash with running water, and plant. Soak *snowflake* bulbs or *China-aster* seed in a 1:1,000 solution for 30 minutes. Wash thoroughly for 10 minutes in running water, dry, and dust with treatment 3, 4, or 5. For *iris* bulbs and rhizomes soak only 10 minutes.

For *sweetpotato* dip 8 to 10 minutes in a 1:1,000 solution.

Soak *gladiolus* corms 2 hours in a 1:1,000 solution just before planting.

Do *not* use the mercury solution more than 3 times. Make sure all washed seed is dried thoroughly before storing or planting.

To protect against damping-off and seed rot, treat seed with a protective treatment (see below and Table 20) before planting.

Treatment 2—Hot water soak for many types of fresh, vigorous, vegetable and flower seed, bulbs, corms, rhizomes, tubers, and other propagative parts. Properly applied just before planting, this treatment kills most internal and external disease-causing organisms. Also useful for disinfesting cuttings, bare-root nursery stock, and certain potted plants of nematodes.

An accurate thermometer is essential. Temperatures 1 to 2 degrees higher than recommended may seriously injure or kill the seed, and 1–2 degrees lower may not kill the disease-causing organisms.

Place seed loosely in a cloth bag and soak for 10 minutes at 110° F. to warm the seed. Then remove to water bath held *constantly* at exactly the recommended temperature. Stir the water slowly but constantly during treatment. Hot or cold water may be added to adjust the temperature.

After treating, cool in cold water, and dry seed in a thin layer. Then apply protective fungicide treatment (see below and Table 20) to control seed rot, damping-off, and other diseases.

Hot Water Treatment Times (in Minutes) for Seed and Other Plant Parts

VEGETABLE Plant Part	Temperature of Water Bath in Degrees Fahrenheit									
	110–111	112–113	115–116	118	120	122	125	127	130–131	135
Broccoli (seed)						20				
Brussel Sprouts (seed)						25				
Cabbage (seed)						25				
Carrot (seed)						15–20				
Cauliflower (seed)						20				
Celery, Celeriac (seed)				30						
Chinese Cabbage (seed)						20				
Collard (seed)						20				
Coriander (seed)								30		
Cress (seed)						15				
Cucumber (seed)						20				
Eggplant (seed)						25				
Garlic (cloves)	180				20					
Kale, Kohlrabi (seed)						20				
Lettuce (seed)				30						
Mint (roots)		10								
Mustard, Radish (seed)						15				
New Zealand Spinach (seed)					60–120					
Onion (sets)	120									
Pepper (seed)							30			
Rape, Rutabaga (seed)						20				
Shallot (cloves)			60							
Spinach (seed)						25				
Sweetpotato (roots)			65							
Tomato (seed)						25				
Turnip (seed)						20				
Yam (tubers)		30								

Hot Water Treatment Times (in Minutes) for Seed and Other Plant Parts (*continued*)

FLOWER Plant Part	Temperature of Water Bath in Degrees Fahrenheit									
	110–111	112–113	115–116	118	120	122	125	127	130–131	135
African-violet (potted plants)	30									
Aloe (plants)			20–40							
Begonia (potted plants)			5	3	1					
(tubers)					30					
Bird-of-paradise-flower (seed)										30
Caladium (tubers)						30				
California-poppy (seed)							30			
Calla lily (rhizomes or corms)						60				
Chinese Evergreen (canes)					30	10				
Chrysanthemum (dormant plants)		30		15						
Delphinium, Larkspur (seed)						10	10	10		
Dieffenbachia (hardened canes)					40–60					
Ferns (potted plants)	10–15									
Foxglove (seed)									15	
Gladiolus (corms)	240									
(dormant cormels)										30
Glory-of-the-snow (dormant bulbs)	180									
Haworthia (plants)			20–40							
Hyacinth (dormant bulbs)	180									
Iris (dormant bulbs)	180									
Lily (dormant bulblets)	60									
Croft Easter lily (bulbs)			120							
Narcissus (dormant bulbs)	180–240									
Nasturtium, Garden (seed)							30			
Nephthytis (bare-root)					30					
Orchid, *Vanda* (cuttings)			10							
Pansy, Violet, Viola (potted plants)	30									
Peony (dormant roots)					30					
Philodendron (canes)					30					
Phlox (stools)	20–30									
Rose (bare-root)						10				
Sansevieria (bare-root)						10				
Silver Threads (plants)						30				
Snowdrop (dormant bulbs)	180									
Scilla, Squill (dormant bulbs)	180									
Stock (seed)									10	
Tradescantia (potted plants)						15				
Transvaal daisy (bare-root)				30						
Tuberose (tubers, offsets, or "seed")					60					
Tulip (dormant bulbs)	180									
Zinnia (seed)							30			
FRUIT, SHRUB, TREE										
Avocado (seed)						30				
Boxwood (bare-root)				30						

Hot Water Treatment Times (in Minutes) for Seed and Other Plant Parts (*continued*)

FRUIT, SHRUB, TREE (*cont.*) Plant Part	Temperature of Water Bath in Degrees Fahrenheit									
	110–111	112–113	115–116	118	120	122	125	127	130–131	135
Citrus (bare-root)	.....	25	.....	.....	.....	10				
Gooseberry (cuttings)	30									
Grape (rootings)	.....	.....	.....	.....	.....	10	5	3		
Locust, Black (dormant trees)	.....	.....	.....	30						
Redcedar, Eastern (transplants)	.....	.....	.....	.....	.....	.....	2			
Strawberry (dormant plants)	.....	.....	.....	.....	.....	.....	.....	2–3	1	
Weigela (bare-root)	.....	.....	.....	.....	30					

Dormant, properly cured bulbs of *daffodil, glory-of-the-snow, garlic, hyacinth, iris, lily, narcissus, snowdrop, squill,* and *tulip* should be soaked in hot water and formaldehyde (1 part of 40 per cent commercial formalin in 200 parts of water) to control root- and bulb-rotting fungi, nematodes, bulbs mites, etc.

B. Protective Treatments

Treatment 3—Thiram 50 or 75 per cent. Used for protecting most vegetable, flower, grass, tree, and shrub seed against seed rot, seedling blights (damping-off), and surface-borne smuts. Also a flower bulb dust and drench in flats or seedbeds for controlling seed rot and damping-off. Kills surface-borne organisms. Available in 1-ounce packets.

Thiram is sold as Arasan Thiram Seed Protectant (75, 50-Red, SF-X, SF-M); Panoram 75; Thiram 75W, and SF-75. Mostly used at ½ to ⅔ teaspoonful per pound of seed. But check package labels for directions. Often combined with an insecticide, e.g., dieldrin, and sold as Delsan A-D Seed Protectant and Panoram D-31. These fungicide-insecticide combinations have proved definitely superior on beans, carrot, corn, cucurbit, and pea seed.

Treatment 4—Captan 75 per cent. Same uses and rates as for treatment 3, but check package labels. Sold as Stauffer Captan 75-W, Orthocide Seed Protectant, Orthocide 75, and Captan 75. Often sold combined with an insecticide, e.g., dieldrin: Captan-Dieldrin 60-15 Seed Protectant and Orthocide-Dieldrin 60-15 Seed Protectant. These combinations, like treatment 3, have proved definitely superior on beans, carrot, corn, cucurbit, and pea seed.

Treatment 5—Chloranil. Has same uses as treatments 3 and 4. Sold as Spergon, Spergon-SL Seed Protectant, and Niagara Seed Protectant. Check trade labels for precise directions. Available in 1-ounce packets.

Treatment 6—Dichlone. Used for treating various vegetables and flowers to control seed rot and seedling blights. Sold as Phygon Seed Protectant, Phygon-XL, and Phygon Naugets.

Treatment 7—Semesan. A mercury-containing dust or dip treatment for many vegetable and flower seed, also for flower bulbs, corms, roots, tubers, and cacti. Semesan Bel S is used as a dip for sweetpotato roots and white (Irish) potato seed tubers.

Treatment 8—Dexon. Sold as a 70 per cent wettable powder to control seed decay and damping-off of seedlings caused by *Pythium, Phytophthora,* and *Aphanomyces* (water molds). Compatible with other fungicides and insecticides recommended for seed treatment (treatments 3–7 above). May be applied in combination with them. Apply Dexon solutions immediately after mixing as they decompose upon exposure to light.

TABLE 20

SEED TREATMENT METHODS AND MATERIALS FOR VEGETABLES, FLOWERS, TREES, AND SHRUBS*

Crop	Diseases	Treatment Number	Method of Treatment†	Time and Remarks
VEGETABLES				
Anise, Caraway, Chicory, Dill, Endive, Escarole, Fennel, Parsley, Salsify	Seed rot, damping-off	3, 7	D or S	Any time
Asparagus	Damping-off	3, Calogreen 4 ozs. or Ceresan M ⅓ oz. per lb. of seed	D	Just before planting
Beans	Seed rot, damping-off, stem blights, root rots	3,4,5,8	D or S	Any time
Beet, Swiss Chard, Mangel, Mangold	Seed rot, damping-off, leaf spot	3, 4, 6, 8		Any time. See Beet
Cabbage, Cauliflower, Broccoli, Brussels Sprouts, Chinese Cabbage, Collard, Cress, Kale, Kohlrabi, Mustard, Radish, Rape, Rutabaga, Turnip	Seed rot, damping-off, blackleg, black rot, downy mildew, Alternaria, scab, yellows, anthracnose, Cercosporella and Mycosphaerella leaf spots	2 then 3, 4, 5, 7	Dip then D	Just before planting. See Cabbage
Carrot	Seed rot, damping-off	3, 4, 5, 6	D or S	Any time
	Bacterial blight	2 then 3, 4, 5, 6	Dip then D or S	Just before planting
	Storage rots	3, 4	D	Before storage
Celery, Celeriac	Leaf blights, seed rot, damping-off, root rots	2 then 3, 4, 5, 7	Dip 30 min. then D or S	Just before planting
Coriander	Bacterial blight	2 then 3, 4, 5, 7	Dip 30 min. then D or S	Just before planting
Corn (Broom, Ornamental, Pop, Sweet)	Seed rot, seedling blights, root rots, leaf spots, and blights	3, 4, 5, 6, 8	D or S	Any time
Cucumber, Melons, Pumpkin, Squash, Chayote, Gherkin, Gourds	Seed rots, damping-off, black rot and gummy stem blight, angular leaf spot, anthracnoses, leaf blight, scab, Fusarium wilt	1 or 2, then 3, 4, 5, 7	Dip 5 min. then D or S	Just before planting. See Cucumber
Lettuce	Seed rot, damping-off, leaf spots and blights	3, 4, 5, 7	D or S	Any time
	Septoria and Marssonina leaf spots	2 then 3, 4, 5, 7	Dip 30 min. then D or S	See Lettuce

* Follow manufacturer's directions regarding rate, precautions, and other factors.
† Treatment method: D — dust; S — slurry; Dip — soak or dip; P — pelleting.

TABLE 20 (*continued*)

SEED TREATMENT METHODS AND MATERIALS FOR VEGETABLES, FLOWERS, TREES, AND SHRUBS*

Crop	Diseases	Treatment Number	Method of Treatment†	Time and Remarks
Okra	Seed rot, damping-off	3, 4, 5, 6	D or S	Any time
Onion, Chives, Garlic, Leek, Shallot	Seed rot, damping-off, smut, purple blotch	3, 4	D or P	See Onion
	Nematodes	2	Dip 20 min. to 3 hours	See Onion
Parsnip	Seed rot, damping-off	3, 7	D or S	Any time
Pea	Seed rot, root rots, damping-off, Ascochyta and Mycosphaerella blights, Fusarium wilts, bacterial blights	3, 4, 5, 6, 7, 8	D or S	Any time
Peanut	Seed rot, seedling blights	3, 4, Ceresan, Panogen	D or S	Any time
Potato, Irish	Seed-piece decays	3, 4, 5, 6, 7, maneb, zineb	D or Dip	Treat cut seed
	Blackleg	Streptomycin 100 parts per million	Dip	Treat cut seed
Shallot	Bulb nematode	2	Dip 1 hour	Just before planting
	Bulb rots	Dowicide B	Dip 15 min.	See Onion
Spinach	Anthracnose, seed rot, damping-off	3, 4, 5, 6, 7	D or S	Any time
	Downy mildew	2	Dip 25 min.	See Beet
Sweetpotato	Black rot, scurf, seed decay	1, 6, 7	Dip	See Sweetpotato
	Root-knot	2	Dip 65 min.	See Sweetpotato
Tomato, Eggplant, Pepper	Bacterial spot, canker and speck, seed rot, damping-off, anthracnose, ripe rot, Cercospora and Septoria leaf spots, Phytophthora blight, early blight, Phomopsis blight, Rhizoctonia, Verticillium wilt, Didymella stem rot	1 or 2 then 3, 4, 7	Dip then D or S	Just before planting. See Tomato
Yam	Root-knot	2	Dip 30 min.	See Yam

* Follow manufacturer's directions regarding rate, precautions, and other factors.
† Treatment method: D — dust; S — slurry; Dip — soak or dip; P — pelleting.

TABLE 20 (*continued*)

SEED TREATMENT METHODS AND MATERIALS FOR VEGETABLES, FLOWERS, TREES, AND SHRUBS*

Crop	Diseases	Treatment Number	Method of Treatment†	Time and Remarks
FLOWERS				
Most flowers both annuals and perennials	Seed rot, certain leaf spots	3, 4, 5, 7	D or S	Any time
	Damping-off, cutting rots	3, 4, 7, PCNB and 4 or 8	Soil drench. Follow manufacturer's directions	Apply 1 pint per square foot
African-violet	Leaf nematode	2	Dip 30 min.	Potted plants
Aloe, Haworthia	Root rot (*Pythium*)	2	Dip 20 to 40 min.	Just before planting. See Aloe
Amaryllis	Bulb rots, leaf scorch	1	Dip 2 hrs.	Just before planting
Aster, China-	Leaf spots, Fusarium wilt, seed rot, damping-off, stem rot	1, 7 then 3, 4	Dip 30 min. then D	Just before planting
Begonia	Leaf nematode	2	Dip	Potted plants
	Root-knot	2	Dip 30 min.	Tubers. Just before planting
Bird-of-paradise-flower	Root and seed rots	2	Dip 30 min.	Just before planting
Cacti	Slimy collar rot	7	Dip 5 min.	Just before planting
Caladium	Sclerotium tuber rot, root-knot	2	Dip 30 min.	Just before planting
California-poppy	Stem and root rots, Heterosporium spot	2	Dip 30 min.	Just before planting
Calla Lily	Bacterial soft rot, root rot, corm or rhizome rot	1 or 2	Dip 30 to 60 min.	See Calla Lily
Canna	Bacterial bud rot	1	Dip 2 hrs.	See Canna
Castorbean	Seed rot, seedling blight	3, 4, 5	D or S	Any time
Chinese Evergreen	Root-knot, Rhizoctonia, water molds	2	Dip 10 or 30 min.	See Calla Lily
Chrysanthemum	Leaf nematode	2	Dip 15 to 30 min.	See Chrysanthemum

* Follow manufacturer's directions regarding rate, precautions, and other factors.
† Treatment method: D — dust; S — slurry; Dip — soak or dip; P — pelleting.

TABLE 20 (*continued*)

SEED TREATMENT METHODS AND MATERIALS FOR VEGETABLES, FLOWERS, TREES, AND SHRUBS*

Crop	Diseases	Treatment Number	Method of Treatment†	Time and Remarks
Delphinium, Larkspur	Bacterial stem rot, Diaporthe blight	2	Dip 10 min.	See Delphinium
Dieffenbachia (hardened canes)	Root and stem rots, bacterial leaf spot	2	Dip 40 to 60 min.	See Calla Lily
Ferns	Leaf nematode	2	Dip 10 to 15 min.	Potted plants
Foxglove	Anthracnose	2 then 3, 4, 5	Dip 15 min. then D	Just before planting
Gladiolus	Corm rots, yellows, neck rot, bacterial leaf blight, scab, leaf and flower spots	1, 2, 3, mercury dip	D or Dip	See Gladiolus
	Nematodes, smut	2 plus formalin 1:200	Dip 4 hrs.	See Gladiolus
Hollyhock, Hibiscus, Lavatera, Mallow	Seed rot, damping-off	3, 4, 5, 6	D or S	See Hollyhock
Iris	Crown rots, bacterial soft rot, rhizome and bulb rots	1, 7, phenyl mercury	Dip 10 min.	See Iris
	Storage rots or molds	3, 5	D	Before storage
	Bulb, root, and stem nematodes, ink disease, bulb and rhizome rots	2 plus formalin 1·200	Dip 3 hrs.	See Iris
Lily	Nematodes (leaf and bud; lesion), Rhizoctonia	2 plus formalin 1:200	Dip 1 to 2 hrs. then 5	See Lily
	Bulb and crown rots	PCNB plus 4, 8, or ferbam	Dip 30 min.	See Lily
Mint	Rust	2	Dip 10 min.	See Salvia
Narcissus, Daffodil, Jonquil, Snowdrop	Bulb rots, root rots, neck rot, Botrytis blight	Puratized, Mersolite 8, Dowicide B	Dip following manufacturer's directions	Just before planting. See Daffodil
	Stem, bulb, and root nematodes, browning disease, bulb rots, root rots	2 plus formalin 1:200	Dip 3 to 4 hours	See Daffodil
Nasturtium, Garden	Heterosporium leaf spot	2	Dip 30 min.	Just before planting

* Follow manufacturer's directions regarding rate, precautions, and other factors.
† Treatment method: D — dust; S — slurry; Dip — soak or dip; P — pelleting.

TABLE 20 (*continued*)

SEED TREATMENT METHODS AND MATERIALS FOR VEGETABLES, FLOWERS, TREES, AND SHRUBS*

Crop	Diseases	Treatment Number	Method of Treatment†	Time and Remarks
Nephthytis, Syngonium	Root rot, black cane rot	2	Dip 30 min.	See Calla Lily
Orchid, *Vanda*	Leaf and bud nematode	2	Dip 15 min.	See Orchids
Orchids	Seed rot, seedling blight, mold	chlorine	Dip	See Orchids
	Leaf blight, black rot, stem and root rots, bacterial soft rot	Bioquin 700 or Natriphene	Dip	Potted plants. See Orchids
Pansy, Violet, Viola	Anthracnose, scab, wilts, damping-off, smuts	3, 4, 5, 6, 7	D or S	Any time
	Leaf and bud nematodes	2	Dip 30 min.	See Pansy
Peony	Root-knot	2	Dip 30 min.	See Delphinium
Philodendron	Bacterial stem rot, cane rots, root-knot	2	Dip 10 to 30 min.	See Calla Lily
Phlox	Stem and bulb nematode	2	Dip 20 to 30 min.	See Phlox
Rose	Botrytis storage decay	4, PCNB	Dip or D	See Rose
	Nematodes	2	Dip 10 min.	See Rose
Safflower	Rust, Alternaria leaf spot, seed rot, damping-off	Panogen, Ceresan, Acti-dione	Dip or D	See Chrysanthemum
Sansevieria	Root-knot, other nematodes	2	Dip 10 min.	See Sansevieria
Silver Threads	Stem and root rots	2	Dip 30 min.	See Silver Threads
Snapdragon	Seed rot, damping-off	7	D or S	Any time
Snowflake	Gray-mold blight	1	Dip 30 min.	Just before planting
Stock	Bacterial blight	2 then 7	Dip 10 min.	See Cabbage
Sweetpea	Anthracnose, wilt, seed rot, damping-off, streak, root rots, fasciation	1 then 3, 4, 6, 7	Dip 20 min. then D	Pre-dip 1 min. in alcohol. See Pea
Transvaal Daisy	Root-knot, other nematodes	2	Dip 30 min.	See Chrysanthemum

* Follow manufacturer's directions regarding rate, precautions, and other factors.
† Treatment method: D — dust; S — slurry; Dip — soak or dip; P — pelleting.

TABLE 20 (*continued*)

SEED TREATMENT METHODS AND MATERIALS FOR VEGETABLES, FLOWERS, TREES, AND SHRUBS*

Crop	Diseases	Treatment Number	Method of Treatment†	Time and Remarks
Tradescantia	Root-knot, other nematodes	2	Dip 15 min.	See Tradescantia
Tuberose	Root-knot, other nematodes	2	Dip 60 min.	See Daffodil
Tulip, Glory-of-the snow, Grape-hyacinth,	Bulb and root rots	3, 4, 5, zineb	D	See Tulip
Hyacinth, Scilla, Snowdrop	Nematodes	2 plus formalin 1:200	Dip 3 hrs.	See Tulip
Zinnia	Alternaria blight, damping-off, Rhizoctonia	2 then 3, 4, 5	Dip 30 min.	Just before planting
TREES, FRUITS, SHRUBS, AND VINES				
Practically all	Seed rot	3, 4, 6, 7	D or S	Any time
	Damping-off	3, 4, 7, 8, ferbam, PCNB-Dexon	Soil drench	1 pint per square foot
Avocado	Seed and root rots	2	Dip 30 min.	Just before planting
Boxwood	Root-knot	2	Dip 30 min.	See Boxwood
Citrus	Root rot, nematodes	2	Dip 10 min.	See Citrus
Gooseberry	Bud nematode	2	Dip 30 min.	See Currant
Grape	Root-knot, other nematodes	2	Dip 3 to 10 min.	See Grape
Locust, Black	Root-knot	2	Dip 30 min.	See Honeylocust
Redcedar, Eastern	Nematodes	2	Dip 2 min.	See Juniper
Strawberry	Nematodes	2	Dip 1 to 7½ min.	See Strawberry
	Leaf spots	zineb, 3	Dip	See Strawberry
Weigela	Root-knot	2	Dip 30 min.	See Snowberry
LAWNGRASSES	Seed rot, seedling blights	3, 4	D or S	See Lawngrasses

* Follow manufacturer's directions regarding rate, precautions, and other factors.
† Treatment method: D — dust; S — slurry; Dip — soak or dip; P — pelleting.

SOIL TREATMENT METHODS AND MATERIALS

The purpose of soil treatment (disinfestation) is to kill disease-inducing organisms (bacteria, fungi, and nematodes), viruses, insects, and weed seeds in the soil. This eliminates the need to change soil in greenhouses, cold frames, hot beds, and other plant beds.

Soil can be sterilized (better called disinfested or pasteurized) or fumigated easily using either heat or chemicals. Heat is generally the most effective since it kills all types of pests. Many chemicals are quite selective and will kill only nematodes or fungi at normal rates of application (Table 22).

In recent years a number of chemicals have been formulated as liquids to be applied in the soil. Most of these chemicals become gases and diffuse in the soil to effect the kill.

Soil disinfestation should be an important part of your sanitation program. It will help produce healthy, vigorous plants.

General Precautions and Suggestions

Soil condition. Soil must be loose and easily crumbled to a depth of *at least* 8 to 10 inches so it can be thoroughly penetrated by heat or chemicals. All lumps, trash, and clods should be broken up, and old plant debris—especially large diseased roots—should be removed. Soil should be well mixed and in good planting condition when treated. It should be moist enough to permit good seed germination or so it will just hold its shape when squeezed in the hand. Do *not* treat when soil is excessively wet or dry.

Soil amendments. All soil amendments, e.g., manure, peat, compost, other humus material, and sand, must be added before treating. It is particularly important that organic matter is well decomposed.

When using soil fumigants do *not* add fertilizers containing ammonia or ammonium salts to soil at or near time of treating. Use only fertilizers high in nitrate nitrogen until soil temperature is above 70° F. and crop is well established. Maintain adequate amounts of calcium, magnesium, and phosphorus in the soil based on a soil test.

Treating tools. When using steam or methyl bromide, treat tools (hoes, rakes, trowels, markers, shovels, spading forks), clay pots, flats, and rubber footwear by laying them on top of the soil and under the gas-tight cover. Otherwise, dip in a 1:20 formaldehyde solution after each use in contaminated soil and before using in treated soil. Boards or concrete at bed edges should also be treated.

Avoid reinfestation of treated soil. Do not transplant seedlings, cuttings, or other plants from untreated or contaminated soil into disinfested soil. Soil is easily recontaminated by nonsterilized flats or pots, tools containing small bits of untreated soil, and contaminated water spattered by careless watering. Also, guard against disease-causing organisms in and on seed and other plant material, unsterilized compost or manure, or gardeners' hands and feet.

Wait before planting. After steaming, wait a day or two before seeding or planting. When chemicals are used, it may take 2 to 4 weeks to aerate soil before it is safe to seed or plant. (See "Application and Remarks" under specific chemicals in Table 22.) Soils high in organic matter or clay, excessively wet, or treated at low temperatures may retain the chemical at toxic levels for even longer periods. *Follow the manufacturer's directions on the label!* A week after treating with a chemical, work soil at least once to a depth of several inches to allow gas to escape.

Other precautions. After deciding whether to use heat or chemicals to disinfest soil, read the special precautions outlined under "Treatments Using Heat," below, and "Treatments Using Chemicals," page 543.

Treatments Using Heat

Precautions

Temperature and time. When using steam, do *not* let the soil temperature rise much above 180° F. or leave steam on too long. Use an accurate thermometer. Do not guess. Hold the coolest spot in the soil at a temperature of 180° F. for 30 minutes. Or turn steam off when soil temperature reaches 200° F. Soil will not heat up uniformly, so it is essential that the soil temperature be measured in many locations to make sure it is up to 180°. Check corners and along edges of benches and ground beds, especially near where steam is injected.

Pressure. When steaming large quantities of soil, use a pressure between 15 and 100 pounds per square inch. Allow at least a $1\frac{1}{2}$-inch opening for steam.
Boiler compounds. Chromate materials are especially toxic to plants and should not be added to a boiler used for soil disinfestation.

Methods

There are numerous ways in which heat can be used to disinfest or pasteurize soil. Some methods (1–3 below) are suitable for treating only a few flower pots or flats. Others are probably practical only for the commercial grower (4–10).

1. *Oven Sterilization*—Suitable for disinfesting small amounts of soil. Place soil in a small greenhouse flat, deep baking pan, kettle, or roaster (aluminum, glass, or iron). Soil should be level, moist (but not wet), and not over 4 inches deep. Bury a raw potato, about $1\frac{1}{2}$ inches in diameter, in the center. Then cover container with heavy aluminum foil and seal down edges. Punch a small hole through the center of the foil and insert the bulb end of a meat or candy thermometer into the soil center. Place in a low-heat oven held at 180° to 200° F. Keep in the oven 30 minutes after the temperature reaches 180° F. Then remove and let cool for 24 hours before using. The potato should be well cooked. *Avoid oven temperatures above 200° F.* that might burn organic matter and destroy soil structure.
2. *Pressure Cooker With Pressure*—Use a home-canning-type pressure cooker. Put several cups of water in the bottom of the cooker. Place soil in shallow pans (no more than 3 or 4 inches deep). Level soil but do not tamp or firm. Stack pans on rack inside cooker. Separate each pan with clothespins or lath strips for free circulation of steam. Close the lid, but do not tighten steam valve completely until all air is forced out and live steam is escaping. When pressure has reached 10 pounds, run at this level for 15 minutes. Then turn off heat. Remove pans of soil when cool.
3. *Pressure Cooker Without Pressure*—Pour about a gallon of water into a pressure cooker, laundry boiler, or large kettle. Use a rack to hold soil pans out of water. Put in shallow pans of soil (see 2 above) and clamp on lid. Leave steam valve slightly open. Apply sufficient heat to keep water boiling. Open valve or lid just enough to hold in steam but prevent much pressure from building up. When live steam begins to be forced out, continue to apply heat for 30 minutes. Keep cooker closed and do not remove soil until it is cold.
4. *Tank or Vault Steaming*—Flats full of soil can be placed in a large tank, vault, or steam chest and sealed in. (The tank or vault is constructed of tile, concrete, building tile, or sheet metal.) The tank is then filled with steam at a pressure of several pounds and held for an hour or longer. Flats should be stacked in racks or at least be separated by blocks of wood. This allows free circulation of steam.
5. *Underground Tile Method*—A steam boiler is needed to make steam under pressure. A 50-horsepower boiler will steam about 500 square feet of bed satisfactorily to a depth of 16 to 22 inches in 4 to 8 hours. In 1 or 2 hours, sterilization can be effected to 9 to 15 inches. The quantity of steam needed to raise the soil temperature to 180° F. varies with many factors. A generally accepted working average is 6.5 pounds *per cubic foot* of soil, or 42 B.T.U. per cubic foot per degree of rise. Heating pipes in the greenhouse may be used for carrying steam from the boiler to a group of buried tile lines.

Use 3- or 4-inch cement or agricultural drain tile laid in parallel rows 12 to 20 inches apart at a depth of 10 to 12 inches in the soil. Distance and depth depend principally on the nature of the soil and depth of cultivation. The heavier the soil, the closer the drain tile should be placed. Pipes from the steam boiler are connected to the tile lines by a short branch pipe extension. The pipe should reach several feet into the drain. Valves should be placed in the conducting pipe so a combination of drains can be treated at once, depending on the boiler capacity. Crushed stone placed on each joint aids the steam in penetrating the soil. The opposite ends of each string of tiles should be left open until steam flows freely and then they should be closed. For best results, cross-headers should be placed approximately every 50 feet.

The soil temperature farthest removed from the steam inlet (or the coldest spot) should be kept at 180° F. for at least 30 minutes (160° for 1 to 2 hours). A longer time is better for heavier soils. Determine temperature by inserting thermometers into soil at different places and at various depths.

In ground beds, it is usually sufficient to raise the temperature at the soil surface to 180° F. if soil is well loosened before steaming. Heat is retained longer by covering soil with polyethylene, synthetic rubber sheets, especially made sterilizing cloths, canvas, or sisalkraft paper (Figure A.2).

6. *Underground Pipe Method*—This method is especially adapted to raised benches, although it may also be used in ground beds. It is basically the same as the underground tile method. Use 1½- to 2-inch pipes bored with 3/16- to ¼-inch holes from 6 to 12 inches apart. Or use agricultural tile.

In benches, lay pipes or tile about 1 inch from the bottom (6 to 8 inches deep), and 12 to 14 inches apart, with the perforations on the underside. If placed in ground beds, the pipes or tile should be about 15 inches deep and about 18 inches apart. Install pipes as a permanent fixture or remove after each treatment to avoid deterioration. For easy removal, fasten wires at 10- to 15-foot intervals when lines are being laid. Extend wires far enough above the soil surface to make it possible to grasp them easily. For steaming, see 5 above.

7. *Aboveground Tubing Method*—Lay 4- to 10-foot sections of aluminum alloy tubing on top of the soil. Leave 1-inch spacings between sections. Lead steam directly from boiler line into tubing. Drill small holes alternately on opposite sides of the tubing. The only way to determine the exact size and spacing of holes is by trial. As a rule, the nearer holes are to the steam inlet, the larger or closer together they must be. In some installations the holes are ¼ inch in diameter, from 18 to 24 inches apart near the steam inlet, and up to 36 inches apart at the opposite end.

Usually it is best to treat no more than 100 linear feet of bench or bed space at a time. In a long bed, lead steam into a T-coupling at the center of the bed, just above the soil surface.

When tubes are laid in position—on top of the soil, 12 to 15 inches apart—cover the soil and tubes with a plastic (vinyl or polyethylene) cover, plastic-impregnated fabric, sisalkraft paper, or

Fig. A.2. Steaming a greenhouse bench. (Illinois Natural History Survey photo)

rubberized sterilizing cloth, and tack or weight along margins to keep steam in. Use the same temperature and time as for other steam-sterilizing methods (see 5 above).

8. *Thomas Method*—This labor-saving and effective method utilizes strips of porous canvas hose laid about 18 inches apart on top of the soil. The hose may run the full length of the bench and is connected to a steam outlet. The bench is covered with specially treated, steam-proof canvas that hangs about 1 foot over each side of the bench and battened to the sides of the bench where necessary. Steam is turned on under pressure to fill the full length of the hose and is allowed to penetrate into the soil. Use the same temperature and timing as for 5 above.

9. *Inverted-pan Method*—Especially suitable for treating soil in shallow benches and flats. Also useful for small ground beds if soil is well loosened before treatment. The reinforced steel, 20- to 24-gauge galvanized iron or wooden pan may be any desired shape to fit over a section of bench or bed area. It should not be more than 70 to 75 square feet in area, but at least 8 inches deep. Before steaming, press pan firmly into position. You may have to weight the pan down with stones, concrete blocks, or sandbags to hold it in place.

A flexible 1-inch hose connects the steam line or portable boiler to a short pipe, fitted with hose connections. The pipe is inserted through the wall of the pan (Figure A.3). Use the same temperature and timing as for other steam-sterilizing methods (see 5 above). Usually ½ to 1½ hours, at steam pressure of 80 to 100 pounds per inch, is sufficient. The method is *not* dependable below 8 or 9 inches in depth.

10. *Flash-flame Pasteurizers*—Popular with commercial growers, pasteurizing about 2 cubic yards of soil an hour. The equipment—a heated cylinder 8 feet long and 20 inches in diameter—is adapted from a tar-melting machine used in road construction. A kerosene-burning blowtorch throws a continuous flame into the lower end of the slightly sloping cylinder. The cylinder is turned about 40 times per minute by means of a 1½ horsepower motor (gasoline or electric). Soil is shoveled into the upper cylinder end just fast enough to bring its temperature between

Fig. A.3. An inverted-pan sterilizer in use on a greenhouse bench. (Courtesy Dr. J. L. Forsberg)

175° and 190° F. when it drops out pasteurized at the lower end. For details, read Cornell University Bulletin 875, *New Flash-Flame Soil Pasteurizer.*

Effects of Soil Disinfestation by Steam

When soil is "sterilized" by steaming at 180° F. for 30 minutes, all organisms living in the soil are *not* killed (Table 21), and therefore, a better term to use is soil disinfestation. Fortunately, most pathogenic (disease-causing) organisms are destroyed at relatively low temperatures so there really is no need for complete sterilization. In fact, "over-sterilization" or "overcooking" soil may cause harmful results.

Physical effects. After steaming, heavy soils in particular may become more granular, with improvement in drainage and aeration. These soils, however, may be more difficult to wet thoroughly since water flows rapidly through noncapillary pores and may not wet the capillary pores. Therefore, several applications of water may be needed to thoroughly wet all soil particles.

Biological effects. Steaming at 180° F. for 30 minutes kills most soil insects, weed seeds, and pathogenic organisms (Table 21). Unfortunately, many beneficial bacteria are also destroyed. Bacteria that convert ammonium forms of nitrogen to nitrate nitrogen are killed, but spore-forming, ammonifying organisms that convert organic matter to ammonium forms of nitrogen are difficult to kill. As a result, these organisms increase rapidly after steaming. Even 180° F. for 6 hours will not kill all ammonifying organisms.

Soil disinfestation greatly reduces the number of soil microorganisms for several days after treatment. Populations of surviving organisms gradually increase until they *exceed* that in untreated soil. Also, because the first organisms that return after treatment meet no severe competition, *it is important that the soil does not become recontaminated* with plant pathogens. Through careless introduction of a disease-causing organism into "sterilized" soil, losses can become *more* severe than they were originally.

Chemical effects. Rapid accumulation of ammonium nitrogen (NH_4+) occurs in soils after steaming because all nitrifying bacteria, but not ammonifying bacteria, are killed. Although ammonium nitrogen can be utilized by plants, it may build up in quantities high enough to "burn" roots and foliage. Large amounts of ammonia may accumulate in soils high in organic matter or soils to which manure has recently been added. The fresher the manure the greater the problem.

Soluble salts increase in soils after steaming due to a release of adsorbed salts into the soil solution. This release is greater for soils high in humus or manure or where soil was heavily fertilized prior to steaming.

In some soils steam treatment greatly increases the exchangeable (soluble)-manganese content of soil. Oxidizing bacteria that fix manganese in soil are destroyed by sterilization. High soluble-manganese levels following steaming may cause manganese toxicity.

TABLE 21

TEMPERATURES NECESSARY TO KILL PLANT PATHOGENS AND OTHER HARMFUL ORGANISMS

Temperature, Fahrenheit*	Organisms Killed
115°	Water molds
120	Nematodes
125	*Rhizoctonia solani* (causes seedling damping-off)
130	*Botrytis* gray mold
140	Most plant-pathogenic fungi and bacteria, worms, slugs, centipedes
140–160	Soil insects
160	All plant-pathogenic bacteria; most plant viruses
160–180	Most weed seeds
200–212	Resistant weed seeds and resistant plant viruses

* Most temperatures indicated are for 30-minute exposures with moist conditions. Adapted from: Baker, K. F., and Roistacher, C. N., Heat Treatment of Soil, Sec. 8, *The U. C. System for Producing Healthy Container-Grown Plants.* California Exp. Sta. Ext. Serv. Manual 23, pp. 123–137, 1957.

In poorly aerated or wet soils, nitrites (NO_2-) which are toxic to plant growth can accumulate. Such injury is rare.

Water-soluble organic compounds, toxic to plants, may be formed by partial breakdown of organic matter. Little information is available on the nature of such organic-matter residues.

Avoid toxicity problems. Do not steam longer or at a higher temperature than given under Precautions, Treatments Using Heat, page 538. Oversterilization increases the problems due to soluble salts, manganese toxicity, and toxic organic compounds.

Leaching heavily with water is the only way to reduce hazards due to soluble-manganese levels, soluble salts, and soluble organic compounds. However, leaching also removes soil nutrients, increases danger of recontamination, may delay planting (especially in winter months), and increases costs.

Broadcasting 5 pounds of superphosphate or gypsum (calcium sulfate) per 100 square feet, either before or after steaming, followed by a light watering, helps reduce release of ammonia.

Transplanting immediately after soil is cooled (below 85° F. at 6 inches) will reinoculate soil with nitrifying bacteria before toxic levels of ammonia can accumulate. New plants should also be watered well and soil kept moist so that the soluble salt concentration will be diluted.

Aging soil for several weeks enables a "natural biological balance" to develop. This practice, however, increases storage costs of disinfested potting soil. Also, a delay may be brought about when seedlings and cuttings are ready for planting.

Treatments Using Chemicals

A number of volatile chemicals (Table 22) are available for treating soil to control nematodes, fungi, bacteria, and weed seeds. Some fumigants control all these pests; others are specifically for nematodes or fungi.

Excellent literature on calibration of equipment is available from fumigant suppliers. Fumigants are marketed as liquids, granules, and gases.

Precautions

Temperature. For chemical treatment, the soil temperature 4 to 6 inches deep should be 50° to 75° or 80° F. to permit effective gas dispersion. Certain soil fumigants containing dichloropropenes can be used successfully at soil temperatures as low as 40° F.

Time for treatment. Late summer or fall is usually the ideal time for chemical treatment of soil; crops have been harvested and soil temperatures are best for fumigation.

Safety. When using soil fumigants observe the following safety precautions:

1. *Avoid inhaling fumes.* The chemicals are often irritating to the membranes of mouth, nose, and throat.
2. *Wear safety goggles* to protect the eyes.
3. *Avoid spilling chemicals on skin, clothing, or shoes.* If this should happen, wash skin promptly with generous amounts of soap and water. Remove affected clothing or shoes immediately. Clothes should be washed and shoes aired until all odor of fumigant has disappeared.
4. *Corrosion.* These materials are corrosive to certain metals, e.g., aluminum, magnesium, and their alloys. Applicators should be rinsed with a 50:50 mixture of kerosene and fuel oil immediately after use. Do *not* use water.
5. *Proper storage.* Store fumigants in original, well-labeled, tightly closed containers in a cool, dry place where they cannot contaminate food and feed, and where children and pets cannot reach them. Keep out of inhabited dwellings and away from heat and open flame. Avoid freezing.
6. *Always follow manufacturer's directions carefully!* Note the recommended dosage and method of application for a given crop.
7. *Avoid plant damage.* Most chemical fumigants, especially gaseous ones (e.g., SMDC, DMTT, Vorlex, chloropicrin, and methyl bromide) *cannot be used* in a greenhouse with living plants. Fumigants used outdoors should not be applied close to valuable plant materials. Growers should post warning signs when using dangerous chemicals in a greenhouse or confined area. Use of soil disinfestants on alkaline soils may cause a phosphorus deficiency in certain plants.

Methods of Applying Liquid Fumigants

1. *A hand applicator* is suitable for treating small areas. It consists of a container for the fumigant, a long, hollow, pointed base for penetrating the soil, and a plunger or trip device. When the plunger is pressed, an exact amount of fumigant

TABLE 22

SOIL-DISINFESTING CHEMICALS—MATERIALS, BRANDS, CONTROLS, APPLICATION, AND REMARKS

Treatments		
Materials, Brands	Controls	Application and Remarks
Formaldehyde (40 per cent commercial formaldehyde solution in water and methanol)	Damping-off, seedling blights, other soil-inhabiting disease organisms, soil insects, soft or germinating weed seed. Disinfectant for tools, equipment, and storage areas. Also a seed disinfestant. Does *not* control nematodes.	Sprinkle 3 tablespoons of formaldehyde diluted with 4 to 6 times that much water on bushel of soil (32 quarts); 1 tablespoon in ½ cup of water treats florist's flat of soil. Mix in thoroughly with shovel or hoe on flat surface. Put treated soil in flats, pots, boxes, or cans, or leave in compact pile and cover with plastic, wet burlap, or canvas for 48 hours. Drench soil in plant beds or seed flats. Use 1 gallon in 49 gallons of water (1 cup in 3 gal.). Apply slowly, ½ to 1½ gal. per square foot, using sprinkling can. Cover soil. After 2 to 4 days remove cover, work soil, and plant when all odor is gone. *Never use in greenhouse where plants are growing.*
Captan, 50 per cent WP; Thiram, 65 or 75 per cent WP; Zineb, 75 per cent WP	Seed rot, damping-off, seedling blights in greenhouse benches, flats, pots, hotbeds, and cold frames	Soil should be loose, fine, and fairly dry. Apply ½ ounce dust per square foot. Mix thoroughly with top 2½ to 3 inches of soil. Seed can be planted immediately. May also use as soil drench, 1 tablespoon per gallon. Use 1 pint per square foot. Repeat at 5-day intervals if disease persists.
Semesan, Pano-drench, Morton Soil Drench C, Panogen Soil Drench (all contain mercury)	Seed rot, damping-off, seedling blights in greenhouse benches, flats, pots, hotbeds, and cold frames	Mix with water and apply as drench to loose, level, fairly dry soil. Follow manufacturer's directions. Apply with sprinkling can. Plant when treated soil has dried sufficiently if package label does not state otherwise.
PCNB, Terraclor, Terracap 10-10 Dust, Ortho PCNB 20 Dust, Miller's Terraclor Dusts, Monsanto PCNB 80% Dust Concentrate, etc. (pentachloronitrobenzene)	Certain disease-causing fungi, e.g., *Rhizoctonia*, *Botrytis*, *Sclerotinia*, *Sclerotium*, etc.	Various application methods including suspension in transplant water, soil surface sprays or dusts, and dry mixing in upper 4 to 6 inches of soil. Sometimes mixed with ferbam, captan, thiram, Dexon, folpet, maneb, ziram, etc. *Thorough* mixing with soil is essential. Follow manufacturer's directions.
EDB, Ethylene Dibromide, Dowfume W-85, Soil-Fume 80-20 and 60-40; Trona Bromofume 40 and 85; Ortho Ethylene Dibromide 83 Soil Fumigant (1,2-dibromoethane)	Nematodes, certain soil insects, garden centipedes	Apply 6 to 8 inches deep, at 10- to 12-inch intervals, with special tractor-mounted equipment. Do *not* use where *onions* will be grown within 3 or 4 years. Wait 2 to 3 weeks before planting. Follow manufacturer's directions. EDB is recommended for *fall* treatment only. See also note on Home Garden Nematode Control (below).

TABLE 22 (*continued*)

Soil-Disinfesting Chemicals — Materials, Brands, Controls, Application, and Remarks

Treatments Materials, Brands	Controls	Application and Remarks
D-D Soil Fumigant, Telone, Stauffer D-D Soil Fumigant, Vidden-D, Ortho D-D Soil Fumigant, Olin OMA-D, Nemafume (mixed dichloropropenes)	Nematodes, certain soil insects	Apply under soil surface like EDB at 10- to 12-inch intervals. Do not plant until 2 to 4 weeks after treatment. Use at heavier rate on muck (peat) soil. Follow manufacturer's directions. See also note on Home Garden Nematode Control (below).
Chloropicrin, Larvacide 100, Picfume, Tri-clor, etc. (tear gas or trichloronitromethane). Larvacide Soil Fumigant contains 93.5 per cent chloropicrin and 6.5 per cent EDB; Nemex is a 50–50 mixture of chloropicrin and dichloropropenes	Damping-off, seedling blights, Verticillium wilt, other soil-inhabiting disease organisms, weed seeds, nematodes, soil insects	Apply with special applicator equipment in holes 6 to 8 inches deep in rows 8 to 10 inches apart. Space holes 10 to 12 inches apart in rows. Inject chemical into each hole and close by stepping on hole. After treating, apply gas-proof cover or sufficient water to soak upper inch of soil to seal in gas. Maintain water seal for 3 days. Do not plant in treated soil until *all* traces of chloropicrin have disappeared (12 days to 4 weeks). Use chloropicrin mask and canister while working. Not for treating flats. Follow manufacturer's directions.
Dorlone, Olin OMAfume D-EDB (mixture of 75.2 per cent Telone and 18.7 per cent EDB)	Nematodes	Same as for D-D above. Use at rate of 12 to 20 gallons per acre. Do *not* use where *onions* will be grown within 3 or 4 years.
Methyl Bromide, Dowfume MC-2, Dowfume MC-33, Brozone, Pano-Brome, Pano-Brome CL, Pestmaster Methyl Bromide, Picride, Profume, Trizone, Great Lakes Bromo-O-Gas, Weedfume, Nemaster, MBC Fumigant, etc. (methyl bromide, usually with 2 per cent chloropicrin added)	Nematodes, grubs, cutworms, and other soil insects, weed seeds, damping-off, seedling blights, wilts, other soil-inhabiting disease organisms	Gas in pressure cans or cylinders. Must apply with special applicator under *gas-proof* plastic cover. To kill soil fungi use 3 to 4 pounds per 100 sq. ft. For other pests use 1 to 2 lbs. For field application, use about 400 lbs. per acre when applied by continuous-surface, multiple-row machine. A 7- to 10-day wait is needed between treating and planting. Good in cold frames, greenhouses, and outdoor beds. *Extremely poisonous!* Do *not* use before planting *onions*, *garlic*, *celery*, *cauliflower*, *carnation*, *salvia*, *delphinium*, *and snapdragon*. Germination of many types of flower seed may be reduced when sowed in treated soil high in clay or organic matter.
SMDC, Stauffer Vapam Soil Fumigant, Chem-Vape, DuPont VPM Soil Fumigant (32.7 per cent sodium N-methyldithiocarbamate)	Soil-inhabiting disease organisms, nematodes, many germinating weed seeds, certain soil insects	For clay soils use 1½ to 2 quarts per 100 square feet; for light- and medium-textured soils 1 quart. Sprinkle uniformly over soil with sprinkling can, hose proportioner, sprayer, or irrigation system. Cover treated area with gas-proof plastic film for 4 days after treating; or apply water seal to upper inch of treated soil (15 to 20 gallons per 100 square feet). Do not treat more than 100 sq. ft. before applying water seal. When top-treated soil has dried sufficiently, cultivate 1 to 2 inches deep. Do not plant until 3 weeks after treating. Follow manufacturer's directions. Fall treatment is best.

TABLE 22 (*continued*)

SOIL-DISINFESTING CHEMICALS — MATERIALS, BRANDS, CONTROLS, APPLICATION, AND REMARKS

Treatments: Materials, Brands	Controls	Application and Remarks
DMTT; Mylone 25 per cent WP, 50 per cent WP, 85 per cent WP, Dust 50; Soil Fumigant M; Miller Mico-Fume 25-D (Mylone) (3,5-dimethyltetrahydro-1, 3, 5, 2H-thiadiazine-2-thione)	Soil fungi, weed seeds, nematodes, soil insects	Apply at rate of ¾ to 2⅔ pounds per 100 square feet depending on formulation. Wait 3 to 4 weeks before planting. Apply as drench, powder, or granules. Disc, rake, or rototill into soil. Use for seed and plant beds. Cover treated area with plastic cover or apply water seal as for SMDC (above). Follow manufacturer's directions. Fall treatment is best.
MIT; Vorlex Soil Fumigant (20 per cent methyl isothiocyanate and 80 per cent chlorinated C_3 hydrocarbons including dichloropropenes)	Nematodes, soil insects, weed seeds, soil-borne fungi	Apply like chloropicrin (above) but use chisel spacings of 8 inches. Pack treated soil and apply water seal or plastic cover. Leave soil undisturbed for at least 4 days. Then cultivate to prevent soil crusting. Do not plant for at least 2 weeks or until *all* odor is gone. Follow manufacturer's directions.
DBCP; Nemagon Soil Fumigant; Ortho Nemagon 45 Soil Fumigant; Miller Nemagon 10 per cent and 25 per cent Granular or No. 2 EC; Stauffer Nemagon 8.6E; Nemafume, Nemadrench, Fumazone, Nema-X, Nema-Kill (1,2-dibromo-3-chloropropane)	Nematodes, damping-off (*Pythium, Rhizoctonia*)	Apply like D-D and EDB, or with sprinkling can, hose proportioner, granules mixed with fertilizer, or apply alone. Very slow acting. May be safely applied to soil around certain living plants. Seedlings more easily injured than older plants. For some crops safe to apply in row at planting time or later as a side dressing. *Plants susceptible to DBCP injury:* beans, beets, potato, tomato, eggplant, peppers, tobacco, sweetpotato, onions, garlic, spinach, lettuce, carnation, chrysanthemum, dwarf palms, etc. Follow manufacturer's directions.
V-C 13 Nemacide, V-C 13 (75 per cent 0-2,4-dichlorophenyl 0, 0-diethyl phosphorothioate)	Nematodes, certain soil insects	May be used as a drench or soak on turf and around certain ornamentals, for treating potting soil (1 teaspoon per quart of water treats 1 cubic foot of soil), or preplanting treatment. Apply as emulsifiable concentrate (liquid drench) or granules. Work into top 6 inches of soil. Follow manufacturer's directions.
Bedrench (81 per cent allyl alcohol and 11.5 per cent EDB)	Nematodes, weed seeds, some damping-off fungi	Apply like SMDC (above). Considerable water and 2-week wait period needed. A seedbed drench. Follow manufacturer's directions.
Cynem; Zinophos, Nemaphos (0, 0-diethyl 0-2-pyrazinyl phosphorothioate)	Nematodes, certain soil insects	Available as liquid or as granules. May apply to soil with hose-end sprayer. Then water in thoroughly. Controls nematodes within living roots of certain plants. Useful as dip to bare-root plants. *Highly toxic to mammals*. Follow manufacturer's directions.
TCTP; Penphene (tetrachlorothiophene)	Nematodes	Available as emulsifiable concentrate, liquid, or granules. Follow manufacturer's directions.

is placed in each location. Several models are available which can be accurately calibrated to deliver the exact dosage recommended by the manufacturer.

Before treatment, the soil surface is usually marked off into 8- to 12-inch squares. Application is in a diamond pattern by marking injections at junctions of the cross in the first row, halfway between the second row, junctions in the third row, halfway between in the fourth row, and so on across the garden or plant bed.

2. *Plow-furrow equipment* may be purchased from several companies. The machinery is mounted on a plow. A control block can be mounted on the tractor steering post to regulate delivery rate that varies with tractor speed. The fumigant outlet (spray nozzle) is in front of each plowshare. Chemical is placed in the bottom of the furrow and should be covered immediately by soil from the moldboard, to a depth of 5 to 10 inches. A chain or spike-tooth harrow, cultipacker, drag, float, bedshaper, or roller should be dragged behind to close soil openings and level ridges. After treatment, soil is compacted and leveled with a roller or spike-tooth harrow.

3. *Tractor-mounted shank or chisel applicators* are widely used by commercial growers. The liquid fumigant enters soil through tubes attached to the rear of a staggered row of cultivator shanks, spaced 8 to 12 inches apart. Chemical is placed in the bottom of the furrow (usually 5 to 10 inches deep) made by the shank or chisel as it is pulled through the soil. These machines must be carefully calibrated to deliver the recommended dosage uniformly. A chain harrow, cultipacker, roller, or heavy drag pulled behind the applicator seals the furrows. Soil should be plowed and worked thoroughly before fumigant is applied. Small garden tractors equipped with chisels, e.g., the Morton Soil Fumigator, are useful for treating soil in greenhouses, seedbeds, and limited field areas.

4. *In-row, shank-injector method.* An economical treatment suggested for certain row crops growing on light, sandy soils. One-third the normal broadcast dosage of fumigant is injected into the row. Soil is then usually thrown by sweeps into a ridge over the fumigant. Spaces between ridges are usually 36 or 42 inches for sweetpotatoes, and 4–6 feet for tomatoes. Provides for a treated strip 1 foot or more wide in which young plants start. Gives good protection at a relatively low cost per acre for fumigant. The treated strip must be carefully marked for planting which should be done in 2 or more weeks.

5. *Blade applicator* (developed at Oregon State University). A continuous sheet of chemical is sprayed into the soil through an injection boom mounted in a protected recess under a blade 4 to 6 feet wide that runs 6 to 8 inches beneath the soil behind a tractor. Soil should be firmed with a roller or float after treatment.

6. *Application in 8-inch ridges using SMDC* (Vapam, VPM Soil Fumigant, Chem-Vape). A convenient method in certain areas where soil is loose and sandy. Fumigant is dropped from an applicator mounted just in front of the ridging disc wheels. Ridges are then rolled flat except for a shallow, 2-inch furrow in which crops are planted 2 or more weeks later, when chemical fumes have escaped. Fifty pounds or more of chemical is applied per acre in rows 30 to 42 inches apart. Check with the manufacturer or your chemical dealer regarding recommended rates.

Methods of Applying Gaseous Fumigants

Methyl bromide is an extremely poisonous, odorless gas. A gas mask fitted with a black canister must be worn during application and when cover is removed. Treatment is suggested *only* for commercial growers who are properly equipped. All necessary precautions must be followed to the letter!

The chemical is available in 1-pound cans or in steel cylinders of various sizes. Most formulations now contain 98 per cent methyl bromide and 2 per cent chloropicrin. The latter is added as a warning agent. Treatment is made under gas-proof covers (e.g., polyethylene or vinyl plastic). For small soil lots, metal-lined vaults or steel drums may be used (Figure A.4). Canvas must *not* be used.

1. *Disinfestation of tools, containers, and machinery.* All materials to be disinfested should be carefully covered with a gas-proof cover leaving *no* place for possible leaks. Cover edges are weighted down. Leave enough room under the top so that gas can circulate freely. Place end of plastic or copper tube from methyl bromide container under the cover, with

Fig. A.4. Releasing methyl bromide under a polyethylene cover to kill old vegetation. Area is then seeded to establish a new lawn. (Courtesy Dow Chemical Co.)

end passing into open trays where the liquid methyl bromide is vaporized. Dowfume MC-2 can be vaporized without using evaporating pans. Before treating, the cubic footage of the space must be calculated accurately. Use fumigant at rate of 1 to 4 pounds for every 100 cubic feet. The cover is left in place for 48 hours if temperature is cool (55° to 65° F.). Twenty-four hours is sufficient in hot weather (66° or above). Do *not* disturb soil for another 4 days.

2. *Treatment of seedbeds, coldframes, transplant beds, and other small areas.* Prepare seedbed as you would for planting. Place cover supports (e.g., boxes or baskets, clay pots, bags of mulching material, etc.) at regular intervals to allow free circulation of gas. Then put evaporating pans (trays) on the prepared soil area and attach the applicator tubing securely to the pans. The other end of the tubing, connected to the applicator, should be outside the area to be treated. When using 1-pound cans, place one evaporator pan in the center of each 100 square feet. Place the gas-proof cover carefully over the soil area and bury the cover edges to a depth of 6 inches. When cover is securely in place, release gas with the specially designed applicator. Follow manufacturer's directions. Leave cover in place for 24 to 48 hours.

3. *Field-scale application.* Tractor-mounted machines are available to apply highly volatile soil fumigants while simultaneously laying a gas-proof cover. Follow manufacturer's directions implicitly.

Methods of Applying Granular Fumigants

DBCP (primarily Fumazone and Nemagon), DMTT (Mylone), or other fumigant may be applied with any grain drill or fertilizer or granule distributor that delivers granular material 5 to 6 inches below the soil surface. A spacing of 6 to 12 inches is desirable. In-row treatment of widely spaced crops is most economical and gives satisfactory control. Carefully follow manufacturer's directions.

Home Garden Nematode Control

Buy EDB, D-D, Telone, or DBCP and make application from a quart Mason jar with 2 tenpenny nail holes punched in opposite sides of the lid. One hole for pouring the fumigant; the other to provide air to the jar. With a hoe, make furrows 6 to 8 inches deep and 9 to 10 inches apart across the garden. Walk along the furrow dribbling undiluted fumigant in the furrow (Figure A.5). Use about 1 cup of Telone or D-D Soil Fumigant or ½ cup of EDB-40 (¼ cup of EDB-85) for each 75 feet of row. DBCP (Fumazone or Nemagon) granules or liquid can be sprinkled into the furrow by hand. Follow manufacturer's directions. After application of *no more* than 100 feet, stop, rake over, and cover furrow. Tamp soil to seal in the gas. Continue the 9- to 10-inch spacing across the garden. Wait 2–3 weeks before planting. If soil temperature during waiting period is below 60° F., or if rainfall has been heavy, wait 4 weeks. The ideal time

Fig. A.5. Applying a soil fumigant by pouring it from a fruit jar into a furrow. After application the furrow must be filled with soil immediately. (USDA photo)

for fumigation is in late summer or fall after crops have been harvested and soil temperature is still moderate (60° F. or more).

For additional information, read USDA Farmers' Bulletin 2048, *Controlling Nematodes in the Home Garden.*

TABLE 23

TREES RELATIVELY FREE OF TROUBLESOME DISEASES

Alder, European	Franklin-tree	Kentucky Coffeetree	Sassafras
Arborvitae	Ginkgo	Laburnum	Silverbell
Beech	Goldenlarch	Larch	Snowball, Japanese
Buckthorn, Sea	Goldenrain-tree	Locust, Black	Snowbell
Cedar, Atlas	Hackberry	Magnolia	Sophora
Cedar of Lebanon	Hazelnut, Turkish	Mulberry	Sorreltree (Sourwood)
Cedrela	Hemlock	Osage-orange	Stewartia
China-fir	Hickory	Parrotia	Sweetgum
Corktree, Amur	Honeylocust	Paulownia	Torreya, Japanese
Cucumber-tree	Hophornbeam	Pawpaw	Tuliptree
Cypress, Bald	Hornbeam	Persimmon	Tupelo
Douglas-fir	Japanese plum-yew	Phoenix-tree	Umbrella-pine
Dove-tree	Japanese Raisin-tree	Privet, Glossy	Yellowwood
Eucommia	Kalopanax	Quince	Yew
Fir	Katsura-tree	Russian-olive	Zelkova

TABLE 24

HOW MUCH FERTILIZER FOR 100 FEET OF ROW FOR VEGETABLES AND FLOWERS?

If Rate Given Per 100 Square Feet Is*	Then Use This Much Per Row If Distance Between Rows Is:			
Pounds	*1 Foot*	*2 Feet*	*3 Feet*	*4 Feet*
½	8 ozs.†	1 lb.	1½ lbs.	2 lbs.
1	1 lb.	2 lbs.	3 lbs.	4 lbs.
2	2 lbs.	4 lbs.	6 lbs.	8 lbs.
3	3 lbs.	6 lbs.	9 lbs.	12 lbs.
4	4 lbs.	8 lbs.	12 lbs.	16 lbs.
5	5 lbs.	10 lbs.	15 lbs.	20 lbs.

* 1 pound (lb.) per 100 square feet = 435.6 pounds per acre
† 1 ounce (oz.) = 2 tablespoons (tbsp.)

TABLE 25

RATES OF APPLICATION OF SPRAYS TO ROW CROPS

Distance Between Rows	Gallons Per Acre	Quarts Per 100 Ft. of Row	Feet of Row Covered by One Gallon
1 Foot	75	2/3	600
	100	1	400
	125	1–1/6	341
	150	1 1/3	300
	175	1 2/3	240
	200	2	218
3 Feet	75	2	194
	100	3	145
	125	3 1/2	116
	150	4	97
	175	5	83
	200	6	73

Row Applications

Rows 12 inches apart—43,560 feet of row per acre.

Rows 24 inches apart—21,780 feet of row per acre.

Rows 36 inches apart—14,520 feet of row per acre.

Rows 48 inches apart—10,890 feet of row per acre.

Example: How much spray does one put on 100 feet of row if the nozzles on the spray boom are 24 inches apart and the recommended application rate is 150 gallons per acre? Calculation: 150 gallons for 21,780 feet = X gallons on 100 feet.

$$X = \frac{(150)(100)}{21,780} = 0.688 \text{ gallons per nozzle}$$

per 100 feet (slightly less than 2½ quarts)

TABLE 26

Operating Chart for Tractor Boom Sprayers

Teejet Tip No.	Pressure (pounds)	Gallons Per Acre			
		3 MPH	4 MPH	5 MPH	7½ MPH
¼ T 80015	20	10.5	7.8	6.3	4.3
or	25	11.7	8.8	7.1	4.7
¼ TT 80015	30	12.9	9.7	7.7	5.2
100 mesh screen	35	13.9	10.4	8.3	5.6
	40	14.9	11.1	8.9	6.0
¼ T 8002	20	14.0	10.5	8.4	5.1
or	25	15.7	11.8	9.4	6.3
¼ TT 8002	30	17.2	12.9	10.3	6.9
50 mesh screen	35	18.5	13.8	11.1	7.4
	40	19.8	14.8	11.8	7.9
½ T 8003	20	21.0	15.7	12.6	8.4
or	25	23.4	17.6	14.1	9.4
¼ TT 8003	30	25.8	19.3	15.4	10.3
50 mesh screen	35	27.7	20.7	16.6	11.1
	40	29.6	22.2	17.8	11.8

TABLE 27

Tractor Speed Conversions

Miles Per Hour	Feet Per Minute	Time Required in Seconds To Travel		
		100 feet	200 feet	300 feet
1	88	68	136	205
2	176	34	68	102
2.5	220	27	54	81
3	264	23	46	68
3.5	308	20	40	60
4	352	17	34	51
5	440	13.6	27	41
6	528	11.3	23	34
7	618	9.7	20	29
8	704	8.5	17	26

How to Figure Tractor Speed and Spray Rates

$$\text{Tractor Speed (MPH)} = \frac{\text{Gallons/Minute} \times 495}{\text{Gallons/Acre} \times \text{Width of Spray}}$$

$$\text{Gallons Per Acre} = \frac{\text{Gallons/Minute} \times 495}{\text{MPH} \times \text{Width of Spray}}$$

$$\text{Tractor Speed (MPH)} = \frac{0.682 \times \text{Distance}}{\text{Seconds}}$$

TABLE 28

Approximate Number of Feet of Row Per Acre at Given Distances

Distances Between Rows (Inches)	(Feet)	Feet of Row Per Acre
12	(1 foot)	43,560
18		29,010
24	(2 feet)	21,758
30		17,427
36	(3 feet)	14,526
42		12,439
48	(4 feet)	10,853
54		9,684
60	(5 feet)	8,714
72	(6 feet)	7,263

COMPATIBILITY CHART FOR FUNGICIDES, INSECTICIDES, AND MITICIDES

How To Use This Chart

Use as you would a road mileage chart. For example, if you wish to know whether captan may be safely combined with DDT, read down the vertical column headed "captan" until you come to the horizontal column marked "DDT, etc." The □ sign where the two columns meet tells you that captan and DDT may be used together safely. **Warning:** Do not mix fungicides in **wettable powder form** with **liquid concentrates of insecticides.**

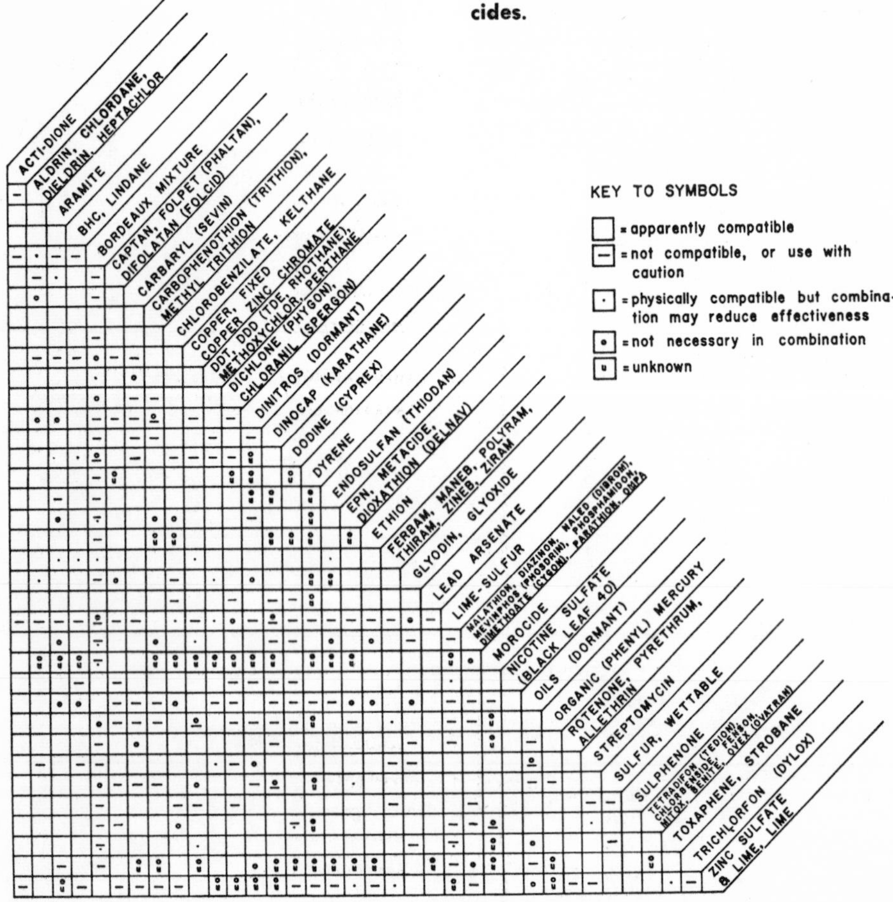

Glossary

Acervulus (pl. **acervuli**) —A saucer-shaped, spore-producing body of a fungus.

Acid soil—A soil with an acid reaction or pH. Measurable only by a delicate test. See pH.

Acme Garden Fungicide—A general fungicide for fruit, ornamentals, and vegetables. Contains 30 per cent captan and 3 per cent Karathane.

Actinomycetes—A group of microorganisms apparently intermediate between bacteria and fungi.

Active ingredient—The actual toxic agent present in a pesticide.

Adventitious roots—Roots that appear in an unusual place or position.

Aerate—Referring to soil means the loosening of hard compact soil by incorporating organic matter, sand, or other material into it to allow passage of air around soil particles.

Agar—A gelatin-like material extracted from seaweed. Used for preparing culture media on which microorganisms are grown and studied.

Air drainage—Air outlets and convection currents that prevent "dead" air and frost "pockets."

Albinism—The white appearance of plant parts resulting from the failure of chlorophyll to develop.

Alkaline soil—A soil with a basic or "sweet" reaction. See pH.

Alternate host—One of two kinds of plants on which a parasitic fungus (e.g., rust) must develop to complete its life cycle.

Aluminum sulfate—Used to acidify soils: up to 5 pounds per 100 square feet. Plants may develop aluminum toxicity with indiscriminate use. Sulfur or an acid fertilizer is safer to use. See page 24.

Amobam—A liquid fungicide for use on vegetables. Contains 42 per cent ethylenebis (dithiocarbamate).

Annual—A plant that completes its life cycle from seed in one year and then dies. See also Winter Annual.

Anther—Pollen-bearing portion of the flower.

Anthracnose—A disease caused by fungi that produces spores in a special type of fruiting body (sunken and saucer-shaped). Usually characterized by limited ulcer-like areas on stem, leaf, or fruit.

Antibiotic—Damaging to life. Especially a complex chemical substance produced by one microorganism which inhibits or kills other microorganisms (e.g., streptomycin, Acti-dione). Most antibiotics on the market are effective against bacteria (antibacterial), although a few are antifungal in action.

Asexual reproduction—Vegetative reproduction from plant parts other than seeds (or from spores), produced by simple budding as the imperfect stage of certain fungi (e.g., yeasts). Also by grafting, cuttings ("slips"), and division of the root system.

Autoecious—A term used with rusts. Completing the life cycle on only one host plant.

Axil—Angle between the upper side of a leaf or stem and the supporting stem or branch.

Axillary—In the angle formed by the stem and leaf.

Bactericide—Any chemical or physical agent that kills or protects from bacteria.

Bacterium (pl. **bacteria**)—Microscopic, generally one-celled plants that lack chlorophyll. Bacteria, like fungi, cannot manufacture their own food. Some feed on dead, organic matter and keep the earth from becoming a "junk yard" of plant and animal remains. Many live in the bodies of plants or animals and cause disease. Bacteria that cause plant disease usually enter through natural openings (water pores or stomates) or through wounds. Bacteria may kill plant cells or cause "cancerous" development of them. See Figure 1.2. Bacteria reproduce by simple fission or dividing in half. Some have whiplike flagella that may aid them to swim.

Band application—An application of spray, dust, or granules to a continuous restricted area such as to or along a crop row rather than over the entire field area (broadcast).

Biennial—A plant having a life span of more than 1 year but not more than 2 years.

Biological control—Control of pests by means of predators, parasites, and disease-producing organisms.

Blade—The flat part of a leaf or petal.

Blast, Blasting—Sudden death of young buds, flowers, or young fruits.

Blight—A general term that may include spotting, discoloration, sudden wilting, or death of leaves, flowers, fruit, stems, or the entire plant. Usually young growing tissues are attacked. May be coupled with name of host part affected, leading to such common names as blossom blight, twig blight, cane blight, and tip blight.

Blotch—A blot or spot usually superficial and irregular in shape and size on leaves, shoots, and fruit. There is no sharp distinction between leaf blight, leaf blotch, and leaf spot.

Bluestone—See Copper Sulfate.

Bordeaux (4-4-50) mixture—The figures following the name of this ancient fungicide discovered near Bordeaux, France, indicate the amounts of copper sulfate, hydrated lime, and water to be mixed. In this case, 4 pounds of copper sulfate and 4 pounds of lime in 50 gallons of water. Homemade mixtures are desirable, but prepared mixtures are available. Leaves a conspicuous residue on plant surfaces. May be injurious to copper-sensitive plants in cold, wet weather. Examples: Acme Bordeaux Mixture, Bor-dox, Copper Hydro Bordo, and Ortho Bordo Mixture.

Bordeaux paint—A fungicidal tree dressing. Stir raw linseed oil into dry, wettable bordeaux powder until you get a paste that can be painted on a tree wound.

Botrytis—Genus name for a widespread fungus that causes blights of peony, tulip, and lilies. See (5) Botrytis Blight under General Diseases. Occurs on many fading flower heads. The gray mold is composed largely of grapelike clusters of spores. Small, hard, black, resting bodies called sclerotia, remain alive in old plant parts or in the soil.

Bract—A modified leaf in a flower cluster (e.g., poinsettia); a small leaf at the base of a flower or flower stalk.

Bramble—A cane bush (e.g., raspberry, blackberry) with spines. Fruit is a berry.

Breaking—General term for discoloration of a flower; usually caused by a virus. The color may be either darker or lighter and in variegated streaks or blotches.

Broadcast application—An application of a spray, dust, or granules over an entire area. See also Band Application.

Broadleaf—Any plant with a flat leaf. Usually applied to evergreens (e.g., azalea, boxwood, holly, rhododendron) with that sort of leaf rather than the needles found on pines, spruces, and firs.

Broad-spectrum fungicide—One that controls a wide range of diseases when applied correctly.

Bud-break—Resting buds resume growth.

Budding—A special type of grafting using a single bud as a scion.

Bulb—A short, flattened, or disc-shaped, underground stem composed of concentric layers of fleshy scale leaves attached to a stem plate at the base.

Bulblet (bulbil, bulbel)—Small, immature bulbs produced on the stem at or below the soil line or at the base of an older, "mother" bulb.

Caddy—A liquid turf fungicide containing 20 per cent cadmium chloride (12.3% cadmium).

Cadminate—A wettable powder turf fungicide containing 60 per cent cadmium succinate (29% cadmium).

Cad-Trete—A turf fungicide containing 75 per cent thiram and 8.3 per cent cadmium chloride hydrate.

Calcareous—Containing lime or lime compounds.

Calcium cyanamide—A high nitrogen fertilizer used in soil sterilization.

Callus—Tissue overgrowth around a wound or canker. Develops from cambium or other exposed meristem.

Calo-clor, Calocure—Turf fungicides containing a mixture of calomel (2 parts) and corrosive sublimate (1 part).

Calogreen—A turf fungicide containing mercury; also used to treat asparagus seed.

Calomel—Mercurous chloride. Used in seed and corm treatments, as a soil drench, and as a turf fungicide.

Calyx—Outermost flower whorl; sepals collectively.

Cambium—A thin layer or cylinder of living cells (meristematic tissue) that divide to form new plant tissues. Normally extends over the plant body except at growing tips. If the cambium is destroyed, as may occur in banding trees, the plant dies.

Cane—The externally woody, internally pithy stem of brambles and vines.

Canker—A definite, localized, dead, often sunken or cracked area on a stem, twig, limb, or trunk surrounded by living tissues. Cankers may girdle affected parts resulting in a dieback starting from the tip.

Capsule—A dry fruit of the pod type developed from a compound pistil that consists of 2 or more united carpels.

Carbon disulfide—An inflammable volatile liquid. Used as a fumigant for borers and a soil disinfectant for Armillaria root rot.

Carpel—A simple pistil, or a member of a compound pistil.

Carrier—Plant or animal carrying internally an infectious disease agent (e.g., virus), but not showing marked symptoms. A carrier plant can be a source of infection to others. An insect contaminated externally with an infectious agent (e.g., bacterium, virus, fungus, nematode) is sometimes called a carrier. Also the liquid or solid material added to a chemical or formulation to facilitate its field use.

Catkin—A type of flower cluster, usually bearing only female (pistillate) flowers or only male (staminate) flowers.

Causal organism—The organism that produces a given disease.

Cedar apple—A popular term given to the hard, brown gall produced on junipers. See (8) Rust under General Diseases.

Cell—The structural and functional unit of all plant and animal life. The living organism may have from one cell (bacteria to billions (a large tree).

Cellu-quin—A wood preservative for picking boxes, hampers, etc. Use as a brush or dip. Contains 0.72 per cent copper-8-quinolinolate.

Ceresan 2%—Contains 2 per cent ethyl mercury chloride; a bulb treatment for narcissus.

Ceresan L—A red liquid containing 2.89 per cent methyl mercury, 2,2-dihydroxy propyl mercaptide and 0.62 per cent methyl mercury acetate. Used as a seed and corm treatment.

Ceresan M—Contains 7.7 per cent ethylmercuri-p-toluene sulfonanilide; a powder used as a seed, bulb, and corm treatment.

Certification of seed, transplants, cuttings, or other plant parts—Seed or plants produced and sold under inspection control to maintain varietal purity, freedom from harmful diseases, insect and mite pests.

Chelates—Metal-containing compounds useful in supplying deficient minerals to plants. See pages 27–28.

Chemotherapy—Treatment of disease by chemicals (chemotherapeutants) working internally. The chemical agent has a toxic effect directly or indirectly on the pathogen without injury to the host plant.

Chipcote 25—Contains 5.41 per cent methylmercury nitrile, a liquid for treating seed and corms.

Chlamydospore—A thick-walled asexual spore formed by the modification of a fungus hypha. The term is also applied to the spores produced by smuts.

Chlorophyll—The green materials found in leaves and other green plant parts by means of which the plant converts water and carbon dioxide from the air into food utilizing the energy of sunlight in a process called photosynthesis.

Chlorosis (adjective, **chlorotic**)—Yellowing or whitening of normally green tissues because of partial to complete failure of chlorophyll to develop. Chlorosis is a common disease symptom. May be due to a virus, the lack of or unavailability of some nutrient (e.g., iron, manganese, zinc, nitrogen, boron, magnesium), lack of oxygen in a water-logged soil, alkali injury, or some other factor.

Clone—A group of plants or horticultural variety derived from one original plant by means of vegetative propagation (e.g., rooting of cuttings or slips, budding, grafting, bulblets). All plants have the same heredity and are quite uniform when grown under the same conditions. Also the vegetative progeny from a single seedling.

Clove—Segment of a garlic or shallot bulb used for propagation.

CM-19—A fungicide for controlling *Botrytis* of plants. Contains 17 per cent phenylphenols and related aryl phenols and 2 per cent octyl- and related alkylphenols.

Coalesce—The growing or fusing together into one body or spot; sometimes to form a blight or blotch.

Cold frame—A plant bed on the ground enclosed by side walls which are usually 12 to 24 inches high and covered with transparent material. The heat comes from sunlight.

Community pot—A pot to which seedlings or "plantlets" are first transferred before being potted or set out as individuals.

Compatible—Refers to chemicals that can be mixed together without changing their effects adversely on pests or plants. A compatibility chart for fungicides, insecticides, and miticides is given on page 552, also different kinds or varieties of plants that set fruit when cross-pollinated, or make a successful graft union when intergrafted.

Complete fertilizer—Any fertilizer containing the three basic elements usually lacking in soil: nitrogen (N), phosphorus (P), and potassium (K).

Compost—Decomposed vegetable matter, important to both outdoor and indoor plants.

Concentric—One circle within another with a common center. A common symptom of numerous diseases caused by fungi, viruses, and bacteria.

Conidium (pl. **conidia**)—An asexual type of fungus spore formed from the end of a special spore-bearing hypha.

Conk—A forestry term for fruiting bodies (sporophores) of wood-rotting fungi formed on tree stumps, branches, or trunks. See (23) Wood Rot under General Diseases.

Contagious—Spreading from one to another.

Control of plant diseases—Prevention or alleviation from plant disease. There are four principal methods of control: a. *Exclusion*—Keeping the pathogen away from a disease-free area, through quarantines or disinfection of plants, seeds, or other plant parts. b. *Eradication*—Destruction (roguing) of infected plants or plant parts or killing of the pathogen or agent on or in the host by use of chemicals. c. *Protection*—Application of sprays or dusts to plants to prevent entrance of the pathogen into the plant. d. *Immunization*—Production of resistant or immune plant varieties, chemotherapy, or other treatment to inactivate or nullify the effect of the pathogen inside the plant. See also Section 3.

Copper or Cupric oleate—A green liquid used to control certain foliage diseases, e.g., rose powdery mildew and pansy leaf spot. Sold as Bord Oil, Destruxol Rose Spray, Koppersol, KXL, K-3, Extrax Copper Spray, J & P Mildew King, etc.

Copper naphthenate—A liquid wood preservative. Sold as Cop-R-Nap, Cuprinol Green, Ferro Copper Naphthenate 8%, Naptox, Rot-Not, Tuscopper, Copper-cure, and Copper-Curo Coppernate. Apply by brush or dip. Injurious to seedlings 1 to 2 inches from treated surface. Excellent residual qualities.

Copper sulfate (bluestone)—A blue crystalline material sold as a powder, "snow," small and large crystals. Contains about 25 per cent metallic copper. Used to make bordeaux mixture, as a disinfectant for potato sacks and storage areas, and in ponds to kill algae.

Corm—A short, enlarged, solid, underground stem (e.g., crocus, gladiolus). True bulbs (e.g., onion, tulip, hyacinth) are composed of fleshy scales. Corms are solid.

Cormel—A tiny, secondary corm produced around the base of the mother corm.

Corolla—The flower petals; collectively.

Cotyledon—The first leaf (or leaves) of a growing embryo.

Cover crop—Plants grown to improve and maintain soil structure, add organic matter, and prevent soil erosion.

Crown gall—A tumor-like enlargement of roots or stem caused by bacteria. See (30) Crown Gall under General Diseases.

Crown rot (Southern blight)—A disease caused by a fungus that attacks hundreds of different ornamentals and vegetables in warm, moist weather. White wefts of mycelium spread fanwise up the stem from the crown and also out into soil. See (21) Crown Rot under General Diseases.

Crucifers—Members of the cabbage family including cabbage, broccoli, Brussels sprouts, cauliflower, horseradish, rape, kohlrabi, turnip, alyssum, honesty, and stock.

Cucurbits—Members of the cucumber family including cucumber, melons, squash, pumpkins, gourds, and watermelon.

Cultivar—Term used to describe a plant maintained only in cultivation; usually propagated asexually.

Culture—To artificially grow organisms (e.g., fungi, bacteria, and nematodes) on a prepared food material such as agar or broth (culture medium), or on living plant tissue. The entire process of obtaining an organism (such as a tree wilt-producing fungus) on prepared media is often called culturing.

Cuprinol—See Copper Naphthenate and Zinc Naphthenate.

Curl—The distortion, puffing, and crinkling of a leaf resulting from the unequal growth of its two sides. See (10) Leaf Curl under General Diseases.

Curly-top—A common virus disease in western states. Plants are stunted with curled and mottled leaves. See (19) Curly-top under General Diseases.

Cutting—A vegetative plant part removed for purpose of growing another plant, identical to the parent.

Cyanamid—See Calcium Cyanamide.

Cynem—A nematocide-insecticide containing 0,0-diethyl 0-2-pyrazinyl phosphorothioate. Effective in controlling nematodes within roots of many plants. Available as granules and liquids. Sold as Zinophos and Nemaphos.

Damping-off—Decay of seeds in the soil or young seedlings before or after emergence. Most evident in young seedlings that suddenly wilt, topple over, and die from rot at the stem base. Woody seedlings often wilt and remain upright. Generally caused by seed- and soil-borne fungi. See (21) Crown Rot under General Diseases.

Deciduous—Plants that drop their leaves in the fall, or once a year, as compared to evergreens that retain their leaves (needles) for two years or longer.

Defoliate—To lose or become stripped of leaves.

Delayed dormant spray—One applied to fruit trees, raspberries, other fruits, and shade trees when the new green tips are 1/8 to 1/4 inch out.

Desiccation—Drying out.

Diagnosis—Identification of the nature and cause of a plant trouble.

Dieback—Progressive death of shoots, branches, and roots generally starting at the tips. May be due to cankers, stem or root rots, borers, nematodes, winter injury, deficiency or excess of moisture or nutrients, or some other factor.

Dinitro materials—Derivatives of cresol and phenol. Semi-liquid eradicant fungicides. Example: Elgetol (contains 19 per cent sodium dinitro-o-cresol). Useful as a dormant and delayed dormant spray for control of certain insects and diseases of apple, peach, pear, apricot, ornamental shrubs, and trees; and pre-emergence ground spray to eradicate certain soil-borne fungi.

Dioecious—Male and female flowers on different plants.

Disbudding—Pinching off of unwanted flower buds.

Disease, plant—A continuously affected condition in which any part of a living plant is abnormal (e.g., structure, function, or economic value) or which interferes with the normal activity of the plant's cells or organs. Injury, in contrast, results from a momentary damage. This is, perhaps, not the best definition, but it is one that is generally accepted with minor changes. Diseases may be caused by living pathogens, viruses, improper environmental conditions, and few higher parasitic plants such as dodder, broom rape, mistletoes, and witchweed. See also Section 2 which covers symptoms and controls.

Disinfectant—A material that kills microorganisms (fungi, bacteria, nematodes) once a plant, or any of its parts, has become infected or infested. See also Chemotherapy.

Disinfection—Freeing a diseased plant or plant parts from infection. Or the destruction of a disease agent or disease-inducing organism in the immediate environment of the host plant.

Disinfestant—A material that removes, kills, or inactivates disease-causing organisms *before* they can cause infection. It may be applied on the surface of a seed, other plant part, or in the soil.

Dissemination—The spread of infectious material (inoculum) from a diseased to a healthy plant by wind, water, man, insects, animals, machinery, or other means.

Dithane A-40—A soluble form of nabam used with zinc sulfate or other metallic salts as a spray for control of diseases of vegetables, fruits, and ornamentals. Contains 93 per cent nabam. Will replace Dithane D-14.

Dodder—Also called goldthread, strangleweed, hell-bind, and love vine. See (40) Dodder under General Diseases.

Dormant period—Time during which no growth occurs.

Dormant spray—A spray applied when plants are in a dormant condition. The temperature should be 40° to 45° F. or above.

Dosage—See Rate.

Double working—Grafted twice. A variety is grafted to an intermediate stock.

Du-ter W-20—A new fungicide containing 20 per cent triphenyl tin hydroxide. Distributed by Thompson-Hayward.

Dwarfing—The underdevelopment of a plant or plant organs. May be caused by any type of disease agent under certain conditions, or by faulty nutrition.

Dyrene—A fungicide useful for vegetables, fruits, lawns, and ornamentals. Available as a wettable powder, dust, and granules. Contains 2,4-dichloro-6-o-chloroanilo-s-triazene. Sold as Dyrene. Asgrow Turf Fungicide, Turftox, etc. May cause skin irritation unless gloves and protective clothing are used. Sold in combination with Dexon.

Eelworms—See Nematodes.

Elcide 73—A liquid fungicide used as a preplanting treatment for gladiolus corms and spring-flowering bulbs. Contains 12 per cent sodium ethylmercurithiosalicylate (Thimerosal).

Elgetol—See Dinitro Materials.

Embryo—A beginning young plant, usually contained in a seed or surrounded by protective tissue.

Emergence—Appearance of the shoot above the soil surface.

Emulsifiable liquid—One that forms an emulsion when mixed with water. May cause plant injury when combined with a wettable powder in a single spray. See compatibility chart of pesticides, page 552.

Emulsion—A mixture in which one liquid is dispersed as minute globules in another liquid.

Enphytotic—A plant disease that causes about the same amount of damage each year.

Epidermis—The outermost layer of cells of a leaf, stem, or other young plant organs.

Epiphyte—A plant that finds lodging (not for parasitic purposes) in nature on another, usually larger, plant. Examples: Orchids and bromeliads.

Epiphytotic—The sudden and destructive development of a plant disease, usually over large areas. Corresponds to an epidemic of a human disease.

Eradicant fungicide—A chemical or physical agent that destroys a fungus at its source, e.g., after its establishment within a plant host.

Eradication—Control of disease by eliminating the pathogen after it is already established. See also Control of Plant Diseases.

Escape—Plants in a given population (e.g., field or garden) of a species or variety that remain free of disease where it is prevalent, although they possess no natural inherent resistance to the disease. Plants may escape attack because of the way they grow (e.g., early-maturing plants escape late-season diseases).

Etiolation—Excessive yellowing (often spindliness) in plants due to a lack of light.

Etiology—The study or description of the cause of disease; together with the relations of the causal factor to the host.

Evergreen—Plants that retain their functional leaves throughout the year.

Exanthema—A name for copper deficiency in fruits.

Exclusion—Control of disease by preventing its introduction (e.g., by quarantines) into disease-free areas. See Control of Plant Diseases.

Exudate—A substance (usually liquid) formed inside a plant and discharged from diseased or injured tissues. The presence of an exudate often aids in diagnosis (e.g., fire blight).

F_1—The first generation progeny of a cross.

Facultative parasite—See Parasite.

Family—Botanical grouping of plant genera according to resemblance and similarity. Plants in the same family commonly share many of the same diseases.

Fasciation—A distortion of a plant caused by an injury or infection that results in thin, flattened, and sometimes curved shoots. The plant may look as if several of its stems were fused.

Feeder roots—Fine roots with a large absorbing area (root hairs).

Fertilizer—Any material containing nutrients available to plants. To be labeled and sold as such it must be state-licensed with the analysis printed on the package label.

Fire blight—A common bacterial disease that attacks pomes and many ornamental shrubs in the rose or apple family. Affected portions turn black or brown and appear to be scorched by fire. See (24) Fire Blight under General Diseases.

Fire Blight Canker Paint No. 1 (G.L.F.)—A liquid applied with a brush for eradicating fire blight from the branches that have become recently infected. Contains 14 per cent cadmium sulfate.

Flag (flagging)—A branch with drooping or dead leaves on an otherwise healthy-appearing tree.

Flagellum (pl. **flagella**)—A tiny, whiplike filament produced by a cell (certain bacteria and spores of the lower fungi) that enable the cell to swim through a liquid. See Figure 1.2.

Flat—Shallow box used for starting plants. Commonly 16 x 22 inches and 2 to 4 inches deep.

Fleck—A small, white to translucent lesion (spot) visible through a leaf.

Flozon Tree Wound Paint—See Gilsonite-varnish Wound Dressing.

Foliage—The leaves of a plant.

Foliar feeding—Applying liquid nutrients to the leaves. Often applied in pesticide sprays.

Forcing crops—Those induced to reach maturity out of season as indoors.

Fore—A coordination product of zinc ion and manganese ethylenebis (dithiocarbamate). Controls many turf diseases, black spot of roses, and Botrytis petal spot of chrysanthemum.

Formaldehyde (formalin)—A 40 per cent solution in water and methanol; used for soil and seed treatments and for disinfesting potato sacks, bins, equipment, and storage areas. Examples: DuPont Formaldehyde Solution, Parsons U.S.P. Formaldehyde, and B. and A. Formaldehyde.

Foundation planting stock—A stock of high quality, grown separately, and very carefully rogued to produce daughter plants for sale.

Frass—The wet or dry sawdust-like material excreted by borers. Usually evident at their exit holes.

Frond—The divided leaf of a fern. Also used for palm.

Fruiting body—A complex fungus structure that contains or bears spores. There are numerous types.

Fulex A-D-O (anti-damping formula)—A soil drench for control of damping-off. Contains 25 per cent hydroxyquinoline sulfate.

Fumigant—A volatile disinfectant that destroys organisms by a gas or vapor. See Table 22 in the Appendix.

Fungchex—A turf fungicide containing 65.4 per cent calomel and 32.6 per cent corrosive sublimate.

Fungicide—A chemical or physical agent that kills or inhibits fungi (often used in a broad sense to include bacteria). Captan, zineb, bordeaux mixture, fixed copper, maneb, ferbam, dodine, sulfur, and lime-sulfur are fungicides. May be used as disinfectants or eradicants to kill fungi in soil or seed or rarely in plants. Usually applied as protectants covering susceptible plant parts before the pathogen can infect.

Fungistat—A chemical or physical agent that prevents development of fungi without killing them.

Fungus (pl. **fungi**)—A low form of plant life that, lacking chlorophyll and being incapable of manufacturing its own food, feeds on dead or living plant or animal matter. The body of a fungus consists of delicate, microscopic threads known as *hyphae,* many of which form branched systems called *mycelia* often evident to the naked eye. The mycelia which may form inside or on the surface of the plant host have different branching habits and structures that help to identify the fungus. Many fungi multiply by forming *spores* at the ends of, within, or on specialized hyphae. The spores are microscopic bodies that function like the seeds of higher plants and are carried by water, wind, man, insects, animals, and machinery. A spore landing on a plant under the proper conditions (usually moderate temperature and a film of moisture) can produce a new fungus body (Figure 1.3). Many fungi produce both sexual and nonsexual (asexual) spores. The manner of production of the sexually formed spores is the basis of classification of fungi into three of their main groups: Phycomycetes, Ascomycetes, and Basidiomycetes. Sexually produced spores have not been found in the fourth main group, the Fungi Imperfecti. Spores are not known for some fungi, which have been classified in a fifth group, the Mycelia Sterilia. See Figure 1.3.

Gall—An abnormal outgrowth or swelling of plant tissue (often of unorganized plant cells) due to irritation by insects, mites, bacteria, fungi, viruses, or nematodes. Often more or less spherical in shape. Examples: black knot, crown gall, cedar-apple gall, root-knot.

Gene—The unit of inheritance transmitted from parent to offspring that controls the development in the offspring of one or more characteristics in the parent.

Genus (pl. **genera**)—A group of related species. See Species and Family.

Germ tube—The hyphal thread produced by a germinating fungus spore. It may grow into a plant through a natural opening or wound or penetrate directly through the unbroken epidermis. The hyphal thread grows, branches, and becomes the new fungus body. See Figure 1.3.

Germicide—A substance that kills microorganisms.

Germination—The beginning of growth. See Germ Tube.

Gerox—A soil drench for control of certain ornamental diseases. Contains 25.3 per cent streptomycin sulfate and 31.8 per cent 8-hydroxyquinoline citrate.

Gilbert's Tree Wound Dressing—Contains 6 per cent phenols. For use on plane trees where canker stain is a problem.

Gilsonite-Varnish Wound Dressing—Used as a wound dressing on plane trees. Federal specification TT-V-51 Gilsonite Varnish plus 0.2 per cent phenylmercury nitrate. Example: Flozon Tree Wound Paint.

Girdle—A canker that surrounds a stem, completely cutting off water or the nutrient supply and thus causing death. Girdling roots of trees may also cause death.

Glyodin—A liquid fungicide used as a spray to control certain fruit diseases. Contains 2-heptadecyl glyoxalidine acetate. Examples: Crag Liquid Glyodin and Glyoxide Dry.

Glyoxide Dry—A 70 per cent, micronized, dry, wettable powder containing glyodin. See Glyodin.

Graft indexing—A plant is grafted to another plant to determine the presence or absence of a virus. The method detects the presence of viruses not readily transmitted mechanically.

Graftage—Method of inserting buds, twigs, or shoots in other stems or roots for fusion of tissues.

Ground cover—Any plant used to cover the ground, hold soil, and give foliage texture (e.g., ivy, pachysandra, vinca), usually as a substitute for grass.

Growing season—The period between commencing of growth in the spring to cessation of growth in the fall.

Growth regulator (plant regulator)—A hormonal substance capable of changing the growth characteristics of plants.

Grub—The larva of a beetle.

Gummosis—Exuding of sap, gum, or latex from inside a plant. Often due to a parasite working within the plant. May also be due to unfavorable growing conditions or other environmental factors.

Guttation—The normal forcing out or exuding of moisture (cell sap) from an uncut plant surface. Microorganisms thrive in this moisture and many enter plants through the opening as the guttation drop is "reabsorbed" by the plant after sunup.

Hairy root—The development of large numbers of small roots on a limited area of a root. See (30) Crown Gall under General Diseases.

Hardening, Hardening-off—Subjecting plants to unfavorable conditions to hasten maturing of tissues for increasing hardiness.

Hardiness—The quality that causes plants to resist injury from unfavorable temperatures.

Hardpan—An impervious layer of soil or rock that prevents downward drainage of water.

Haustorium (pl. **haustoria**)—Special rootlike, food-absorbing, sucking organs produced by dodders, mistletoes, and certain fungi that grow into a plant or host cell by means of which food is obtained from the host plant.

Hazard—The probability that injury will result from the use of a substance in the quantity and manner proposed.

Heading back—Pruning off the terminal part of a twig or branch.

Healing over—The process whereby a wound is closed or protected by a new growth (callus) without replacing the lost parts.

Heartwood—The central cylinder of xylem tissue in a wood stem or trunk.

Heel in—Temporary planting for nursery stock. Roots are placed in a shallow trench, covered with soil and watered.

Herb—Botanically, a plant with fleshy, not woody, stems. Commonly a plant used for flavoring, fragrance, or medicinal purposes.

Herbaceous—Plants with soft, nonwoody stems (e.g., annuals, biennials, and perennials) that normally die back to the ground in the winter.

Herbicide—Any chemical or agent used for killing or inhibiting the growth of weeds.

Heteroecious—Pertains to the rust fungi that require two or more unrelated hosts for completion of the life cycle. See (8) Rust under General Diseases.

Heterozygous—Having mixed hereditary factors. Not a pure line.

Homozygous—Purity of type. A pure line.

Honeydew—A sweet sticky secretion given off by aphids, whiteflies, scales, and other insects. An attractant for ants, and a favorable medium for black, sooty mold fungi. See (12) Sooty Mold under General Diseases.

Hopperburn—Marginal yellowing, scorching, and curling of potato, dahlia, and other foliage due to the feeding of leafhopper insects.

Hormone—A naturally occurring or synthetic compound that stimulates plants in a specific manner.

Horticulture—The art and science dealing with fruits, vegetables, flowers, ornamentals, shrubs, and trees.

Host—Any plant attacked by (or harboring) a living parasite and from which the invader is obtaining its nourishment. See Suscept.

Host indexing—A procedure to determine whether a given plant is a carrier of a virus disease. Material is taken from one plant and transferred to another plant that will develop characteristic symptoms if affected by the virus disease in question.

Host range—The various kinds of plants attacked by a given parasite.

Hotbed—Similar to cold frames but provided with a source of heat (electricity, fermenting organic matter, steam, hot air flues) to supplement sunlight.

Humidity, relative—The weight of water vapor in the air as compared to the total weight of water vapor which the air is capable of holding at a given temperature.

Humus—Decomposing organic matter from any source that may become fine, rich, black soil. May come from vegetable refuse, leaf mold, manure, peat, or animal matter.

Hyaline—Clear, translucent.

Hybrid—First-generation progeny from a cross of different varieties, strains, inbred lines, or species.

Hydathode—Special structures through which water of guttation can easily escape. Microorganisms can also enter through these natural openings.

Hydrogen-ion concentration—A measure of the acidity of a chemical in solution. It is expressed in terms of the pH of the solution. See pH.

Hygiene—See Sanitation.

Hyperplasia (adjective, **hyperplastic**) —A term applied to a disease producing an abnormal increase in the *number of cells* (without their enlargement) resulting in the forming of galls or tumors.

Hypertrophy (adjective, **hypertrophic**) —An abnormal increase in the *size* of an organ or tissue brought about by enlargement of the component cells or by an increase in cell division or both.

Hypha (pl. **hyphae**) —A single thread or filament that constitutes the body (mycelium) of a fungus. It may be divided into cells by cross walls or be one long cell. Some hyphae are specialized for producing spores, penetrating host tissues, overwintering, or trapping nematodes.

Hypoplasia (adjective, **hypoplastic**) —A term applied to a disease resulting in the underdevelopment of plant cells, tissues, or organs due to subnormal cell production.

Immune (immunity)—The ability of a plant to remain exempt from disease due to inherent properties of the plant (e.g., tough outer wall, hairiness, nature of natural openings, waxy coating) .

Immunization—The process of increasing the resistance of a living organism.

Incompatible—Pertains to different kinds or varieties of plants that do not successfully cross-pollinate or intergraft.

Incubation period—The time between inoculation of a plant by a disease-producing agent and the appearance of visible symptoms. This period may vary from a few hours to as long as a year or more. Another meaning—the maintaining of inoculated plants or pathogens in an environment favorable for disease development.

Indexing—Determining the presence of a disease. See Graft Indexing and Host Indexing.

Infect (infection)—The process of becoming established in a parasitic relationship with a host plant.

Infection court—Any place where an infection may take place (e.g., leaf, fruit, petal, stem, root) .

Infest—To be present in numbers (e.g., insects, mites, nematodes) . Do not confuse with infect (infection) that applies only to living, diseased plants or animals.

Inflorescence—The flowering structure of a plant.

Injury—Momentary damage to a plant by an adverse factor (e.g., insect or rodent bite, action of a chemical, physical, or electrical agent) .

Inoculate (inoculation)—Bringing infectious material (inoculum) in contact with a host plant (infection court) .

Inoculum—The infectious agent, pathogen, or its part (e.g., spores, mycelium, virus particles, nematode cysts) that is capable of infecting plants.

Insect vector—An insect that transmits a disease-inducing organism or agent.

Insecticide—A chemical or physical agent that kills, inhibits, or protects against insects. Examples: DDT, lindane, malathion, methoxychlor, sevin, Dibrom, and rotenone.

Internode—The part of a stem between the nodes or two leaf axils.

Intumescence—A knoblike blister or pustule formed by outgrowths of elongated cells on leaves, stems, or other plant parts that have burst from sudden water excess following dry periods.

Invasion—Growth or movement of an infectious agent into a plant and its establishment in it.

Knot—Knoblike overgrowth on roots or

stems with an imperfect vascular system.

Knudson's Formula "C"—The most frequently used formula for sowing orchid seeds in culture flasks or tubes:

Calcium nitrate	1.00 gram
Monobasic potassium phosphate	0.25 gram
Magnesium sulfate	0.25 gram
Ammonium sulfate	0.50 gram
Sucrose	20.00 grams
Ferrous sulfate	0.025 gram
Manganese sulfate	0.0075 gram

Add above ingredients, one at a time, to one liter (1000 cc.) of distilled water, and dissolve completely. Fifteen grams of agar is then added and the mixture warmed in a double boiler until all agar has dissolved. The pH must then be checked very accurately, and adjusted to that required by the particular seed being sown. (Most orchid seeds germinate and grow best at a pH of 5.0 to 5.2.) Prepared flasks are placed in an autoclave (or pressure cooker) for a sterilization period of about 20 minutes, at 15 to 20 pounds pressure. The hot flasks are then placed on their sides to cool. The seed may be sown as soon as the flasks have thoroughly cooled and the agar medium has solidified.

Kromad—A broad-spectrum turf fungicide containing 5 per cent cadmium sebacate, 5 per cent potassium chromate, 1 per cent malachite green, 0.5 per cent auramine, and 16 per cent thiram.

Larva (pl. **larvae**) —The immature form of certain insects. Larvae hatch from eggs, are wingless, often wormlike or grublike, and develop into a pupal or chrysalis stage.

Leaching—The washing of soluble nutrients down through the soil.

Leaf mold—Partially decomposed leaves used in potting mixtures or to add organic matter to garden soil.

Leaf spot—A definitely delimited lesion on a leaf. There are thousands of different kinds of leaf spots caused by hundreds of fungi plus some bacteria and viruses. Usually the damage is not severe enough to warrant special control measures. See (1) Fungus Leaf Spot and (2) Bacterial Leaf Spot under General Diseases.

Leafstalk—A petiole.

Lesion—A localized area of diseased tissue. Spots, cankers, blisters, pustules, and scabs are lesions.

Lichen—A fungus living together with a green or blue-green alga in a symbiotic relationship. The fungus receives food from the alga which in turn gets protection and food from the fungus. Lichens grow on living trees and shrubs. Most abundant in the south where they flourish in shady, damp locations in neglected garden plantings. Lichens may be disfiguring and when abundant interfere with normal light and gas exchange. Remove, where necessary, by rubbing when moist or by spraying with a copper-containing fungicide.

Life cycle—The complete succession of developmental stages in the life of an organism.

Lime—Hydrated spray lime is sometimes used in combination with a pesticide (especially bordeaux mixture and fixed copper). Ground limestone is used to check soil acidity. Lime-induced chlorosis may occur when acid-loving plants (e.g., azaleas, rhododendrons, blueberries) grow near a concrete foundation or in alkaline soil.

Lime-sulfur (liquid)—Formed by boiling sulfur and milk of lime together. An old fungicide now largely replaced by safer materials, e.g., captan, zineb, maneb, ferbam, folpet, dodine, and ziram. Liquid lime-sulfur is still used as a dormant or delayed dormant spray on certain stone fruits, apples, and bramble fruits. See Section 3. Contains 26 to 30 per cent solution of calcium polysulfides. Examples: F. & B. Lime Sulphur Solution, Miller Lime Sulfur Solution, Orthorix Spray.

Lister—A plow used for ridging.

Loam—A mellow soil composed of about equal parts silt and sand and less than 20 per cent clay.

Local infection—Infection involving only a limited or localized part of a plant (e.g., leaf spot, fleck, scab).

Lodging—Falling over, lying down, or broken over, as in plants beaten down by wind, rain, disease or insects, or combinations of these.

Lysol—A liquid mixture of crude cresols used for treating gladiolus corms or as a disinfectant.

Macroscopic—Visible to the naked eye without the aid of a microscope. See also Microscopic.

Maturity—The state of ripeness. Usually refers to that stage of development which results in maximum quality.

Mersolite 8—Contains phenylmercury acetate. Used as a soak for narcissus bulbs.

Methocel—Methyl cellulose 15 C.P.S. Available as a powder or fiber. Used as a sticker in seed treatment and for pelleting seed (e.g., onion).

Mico-Ban 531—A foliage and turf fungicide containing cadmium calcium copper zinc sulfate-chromate, 95 per cent wettable powder.

Micron—A unit of length equal to 1/1,000 of a millimeter or 0.00003937 of an inch long. Used for measuring fungus parts, bacteria, nematodes, and other microscopic objects.

Microscopic—Visible only under the microscope. See also Macroscopic.

Mildew—A plant disease characterized by a thin, whitish coating of mycelial growth and spores on the surface of infected plant parts (downy and powdery mildews).

Mildew King—A liquid fungicide for control

of rose powdery mildew. Contains 30 per cent copper oleate.

Miller 658 Fungicide—A wettable spray powder fungicide containing 95 per cent copper zinc chromate complex. Used on turf, ornamentals, and vegetables. Exempt from tolerance.

Mistletoe—See (39) Mistletoe under General Diseases in Section 2.

Mold—Any fungus with conspicuous, profuse, or woolly growth (mycelium or spore masses). Occurs most commonly on damp or decaying matter and on surface of plant tissue.

Monoecious—Plants having separate staminate (male) and pistillate (female) flowers or reproductive organs on the same individual. In rusts, all stages of the life cycle occur on a single species of plant. See also Heteroecious, and (8) Rust under General Diseases.

Morsodren (Morton Soil Drench)—A liquid fungicide containing 2.2 per cent methylmercury dicyandiamide. Applied as a soil drench or spray to control damping-off, root, stem, and crown rots of ornamentals and vegetables; bulb and corm dip; disinfectant for tools, pots, benches, and walkways.

Mosaic—A virus disease characterized by a mottling of the foliage or by variegated patterns of dark green to yellow which form a mosaic.

Mottle—An irregular pattern of light and dark areas.

Mulch—A protective layer of some substance such as straw, dry leaves, compost, peatmoss, gravel or small stones, ground corncobs, buckwheat hulls, or sawdust on top of the soil. Often used to catch rainfall, prevent splashing, retain moisture, control weeds, keep the soil temperature down in summer, keep produce clean, or some other reason.

Multipurpose spray or dust—One that controls a wide range of pests. See in Section 3.

Mummy—A dried, shriveled fruit, the result of some fungus disease such as brown rot or black rot. The mummy may hang on the tree or fall to the ground where it survives the winter and is the source of infection for next year's crop.

Mushroom (toadstool)—A conspicuous fleshy fungus, especially one with gills. The aboveground part is the reproductive part of the fungus. The word properly applies to all fruiting bodies whether edible, poisonous, tough and unpalatable, or leathery.

Mutation (bud sport)—A genetic change within an organism or its parts which changes its characteristics.

Mycelium (pl. **mycelia**)—The mass of interwoven threads (hyphae) making up the vegetative body of a fungus. The mycelia of fungi show great variation in appearance and structure.

Mycology—The science dealing with fungi.

Mycorrhiza (pl. **mycorrhizae**) —A generally mutually beneficial relationship (symbiosis) between roots and fungi. Many plants cannot grow normally without the presence of mycorrhizal fungi.

Mycostatin—An antifungal antibiotic. Used as a dip for control of postharvest rots of fruits and flowers.

Nabam—A liquid fungicide containing 22 per cent disodium ethylenebis (dithiocarbamate). Used with zinc sulfate as a spray or soil drench for control of diseases of vegetables, turf, and ornamentals. Sold as Chem-Bam, Dithane D-14, F & B Nabam 22, Nabam Liquid Fungicide, Ortho Nabam Liquid Spray, Niagara Nabam Solution, Parzate Liquid Nabam Fungicide, etc.

Natriphene—A fungicide containing 100 per cent sodium salt of 2-hydroxy diphenyl. Used as a drench to control damping-off and other diseases of ornamentals—especially orchids.

Necrosis (adjective, **necrotic**) Localized or general death and disintegration of plant cells or tissues. When several cells die together a spot or lesion is formed. If the lesion is sunken and ugly, it is often called a canker; if small, white, and translucent it is called a fleck.

Needle cast—Disease of evergreens that results in a large drop of needles. Often called needle blight.

Nematocide—A chemical or physical agent that kills, inhibits, or protects against nematodes. Some give off gases after application to soil and are called soil fumigants.

Nematodes (nemas, eelworms, or roundworms)—Generally microscopic tubular animals usually living free in moist soil, water, and decaying matter or as parasites of plants and animals. Responsible for many plant diseases. Nematodes that cause plant disease pierce the cells of a plant with a stylet and suck up juices. Nematodes also play a role in providing wounds by which other pathogens may enter and also transmit disease-producing organisms and viruses into plants. See (20) Leaf Nematode, (37) Root-knot, and (38) Bulb Nematode under General Diseases. Controlled by soil fumigation. See Table 22 in the Appendix. Rotation and hot water treatment of bulbs or other plant parts are other control measures.

Neutral soil—One whose reaction is neither acid or alkaline. See pH.

Niacide A—A fungicide mixture containing 35 per cent ferbam, 24 per cent manganous dimethyldithiocarbamate, 6.2 per cent thiram, 1.2 per cent manganous benzothiazylmercaptide, and 1.1 per cent 2,2′-dithiobenzothiazole.

Niacide M—Mixture similar to Niacide A, but contains no ferbam and the percentages of other ingredients are doubled. Used by commercial apple growers in cover sprays.

Node—A slightly enlarged portion of the stem (joint) where leaves and buds arise, and where branches originate.

Nodule—A lump, knot, or tubercle.

Noninfectious disease—A disease that cannot be transmitted from one plant to another. A disease caused by a physiogenic agent (physical or environmental factors).

Nursery—An establishment for growing, handling, or retailing plants.

Nutrients—Elements available to plants through soil, air, and water that are utilized in growth.

Obligate parasite—See Parasite.

Oedema—A swelling or a disease in plants sometimes caused by overwatering in cloudy weather when there is reduced evaporation (transpiration). See page 39.

Omazene—A wettable powder fungicide used as a spray to control powdery mildew of roses. Contains 50 per cent copper dihydrazinium sulfate.

On the dry side—Keeping soil or planting medium barely moist as compared to rather wet. A recommended procedure to keep losses from seed rot, damping-off, and seedling blight at a minimum.

Organic matter—Any plant or animal material that is decomposed, partially decomposed, or undecomposed.

Ornamental plants—Those grown for accent, attraction, beautification, color, screening, specimen, and other aesthetic reasons.

Ortho Lawn Disease Control—A multipurpose turf fungicide containing a mixture of captan, PCNB, and cadmium carbonate. Mostly used in southern states.

Ortho Lawn and Turf Fungicide—A multipurpose turf fungicide containing 60 per cent folpet, 5 per cent cadmium carbonate, and 10 per cent thiram. Used in northern states.

Ortho LM Seed Protectant—Contains 2.25 per cent methylmercury 8-hydroxyquinolinolate; a liquid seed and corm treatment.

Orthocide Soil Treater "X"—A fungicide for the control of soil-borne diseases. Contains 10 per cent captan and 10 per cent PCNB.

Orthorix Spray—See Lime-sulfur.

Oxyquinoline sulfate (also **benzoate and citrate**)—A soil fungicide. Used as a drench to control damping-off and other soil-borne diseases. Used at rate of about 1:4,000 (1 level teaspoonful in 3 gallons of water). Follow manufacturer's directions. Examples: Sunox and Fulex A-D-O.

Panicle—A loose, open-flower cluster.

Parasite—An organism that lives on or in some living plant or animal (called a host) and obtains all or part of its nutrients from it. Many fungi have both parasitic and saprophytic stages. True rusts, white-rusts, and powdery mildews are *obligate* parasites, having no saprophytic stage. Rusts and powdery mildews can live only in living tissues. *Facultative parasites* are organisms that can grow either on living or dead organic matter. *Obligate parasites,* on the other hand, can live only as parasites.

Parasitic plant—One that derives all or part of its nutrients from another on which it lives. See Parasite.

Pathogen—Any organism or agent capable of causing disease. Most pathogens are parasites but there are a few exceptions.

Pathogenic—Capable of causing disease.

Pathology—The science of disease.

Peanut peg—The fruit-bearing stalk.

Peatmoss—An excellent mulch. Widely used to help "lighten" a heavy soil. Difficult to wet. Care must be taken to prevent drying out entirely for then the mulch absorbs water away from plant roots. A valuable ingredient in potting mixtures.

Pedicel—The support for a single flower.

Peduncle—A primary flower stalk supporting either a solitary flower or cluster.

Pentachlorophenol—A liquid wood preservative for preventing wood rots. Sold as a 4 to 5 per cent solution under such trade names as Bonide Pentide, Miller PCP-10, Pentox, and Wood Tox. Also sold as a concentrate under various trade names.

Perennial—Any plant that has a life span of more than two years.

Perianth—The external envelope of a flower consisting of sepals and petals.

Perlite—An almost weightless, gritty, white substance used as a substitute for sand or vermiculite.

Pest—Any organism injuring plants or plant products.

Pesticide—Any chemical or physical agent that destroy pests (e.g., fungicide, insecticide, miticide, nematocide, herbicide, etc.)

Pesticide tolerance—The established quantity of a pesticde that can legally remain on harvested, edible products in interstate commerce. Tolerances are set by the Pure Food and Drug Administration (FDA) of the Department of Health, Education, and Welfare, Washington, D.C., to protect the health of humans or animals that will eat the treated plants.

Petiole—The stalk of a leaf.

pH—A symbol of a scale used to designate the relative acidity of a solution. The scale ranges from 1 to 14. pH 7.0, the midpoint, represents a neutral solution. Numbers less than 7 indicate increasing acidity; those more than 7, increasing alkalinity. You can test soil yourself with a simple test kit, available at many garden supply stores. Or you can have it done at your county extension office or state agricultural experiment station.

Phenmad—A turf fungicide containing 10 per cent phenyl mercury acetate (PMA).

Phenyl mercury compounds—Liquids and wettable powders containing various salts of phenyl or organic mercury. Used as spray for control of diseases of apple, ornamentals, and turf and as a seed treatment. See also in Section 3 and Table 22 in the Appendix.

Phloem—Tissue in plants through which foods are transported from leaves to roots. See Xylem. Phloem is the inner bark of woody plants.

Photoperiodism—Response of plants to the daily length of light.

Photosynthesis—The complicated processes by which green plants make sugar from water and carbon dioxide in the air through the energy of sunlight. See Chlorophyll.

Phyllody—Change of a normal flower to leafy structures. Characteristic of certain virus infections that cause witches'-broom disease.

Physiogenic disease—A disease produced by some unfavorable physical or environmental factor (e.g., excess or deficiency of light, temperature, water, soil nutrients, chemical, physical, or mechanical injury, etc.)

Physiologic race—Subdivision within a species of fungus that differs in virulence, symptom expression, or to some extent in host range, from other races (or strains) and the rest of the species. Frequently after a new variety of a plant (e.g., potato) has been bred for resistance to a species of fungus (e.g., late blight), the organism in turn develops (principally through hybridization or mutation) a new race that attacks the variety at will.

Phytopathology—Plant pathology or the science of plant disease.

Phytotoxic—Injurious to plants.

Pinch, Pinching back—Removing the tip of a stem, an extra flower bud, or tip bud using fingernails, knife, or shears. This stimulates lateral growth.

Pistillate flower—One that contains pistils (female parts) but no stamens (male parts).

Polytrap—A spray for control of powdery mildew of roses. Contains 75 per cent polyisobutylene.

Pome—Fruit with an embedded core like apple, pear, and quince.

Powdery mildew—Fungi that form a white coating on the surface of leaves, stems, buds, and flowers. See (7) Powdery Mildew under General Diseases.

PPM—Parts per million.

Primary infection—The first infection by a pathogen after it has gone through a resting or dormant period. See Secondary Infection.

Procumbent—Nearly prostrate. Spreading.

Progeny—The young or seedlings of a plant.

Protectant—A chemical applied to a plant surface in advance of the pathogen to prevent infection.

Prothallus—The small, flat development of a spore on which the two sexes appear in the development of ferns and other lower plant groups.

Pruning—The judicious removal of leaves, shoots, twigs, branches, or roots of a plant to increase its usefulness, vigor, or productivity.

Pseudobulb—The thickened or bulblike stems of certain orchids borne above the ground or substratum.

Pupa—An intermediate resting stage during which an insect changes from a larva to an adult.

Pustule—A small, raised, blister-like or pimple-like structure that may rupture the epidermis and expose the causal agent (e.g., rust, smut, white-rust).

Pycnidium (pl. **pycnidia**)—A closed, asexual type of spore-producing body of a fungus. A pycnidium appears as a minute speck to the unaided eye. Commonly found in diseased tissue.

Quarantine—Regulation forbidding sale or shipment of plants or plant parts, usually to prevent disease, insect, nematode, or weed invasion of an area.

Race or Strain—A subgroup within a species of fungi, bacteria, nematodes, or viruses that differs in virulence, symptom expression, or to some extent in host range from other races (or strains) and the rest of the species. See Physiologic Race.

Rate or Dosage—Synonymous terms. Usually refers to the amount of active ingredient applied to a unit area (e.g., acre or 1,000 square feet) regardless of percentage of chemical in the carrier.

Receptacle—The more or less expanded portion of the stem on which flower parts are borne.

Renewal (replacement) spurs—Grape canes near the trunk cut back to two buds to provide new fruiting wood in a desired location.

Renovation—To invigorate or rejuvenate, thin plants, remove weeds, and form new plants (e.g., in a strawberry bed).

Resistant (resistance)—The sum of the inherent qualities of a host plant that retard the activities of the causal agent. A plant may be slightly, moderately, or highly resistant. The ability of a host plant to suppress or retard the activity of a pathogen.

Respiration—Oxidation or utilization of foods by plants and animals resulting in energy release. Carbon dioxide, water, and other materials are liberated.

Rhizoid—A rootlike structure.

Rhizome—An elongate, fleshy, usually underground, horizontal stem that forms both roots (below) and shoots (above) at its nodes.

Rhizomorph—A cordlike or rootlike strand composed of a bundle of fungus hyphae by which a fungus makes its way for considerable distances through soil, along under the bark of woody plants, or elsewhere. Example: Armillaria root rot fungus.

Ringspot—Symptom of disease characterized by yellowish or dead (necrotic) rings with green tissue inside ring as in certain virus diseases. See (17) Spotted Wilt under General Diseases.

Rogue (roguing)—To remove and destroy undesired or diseased individual plants from a planting on the basis of disease infection, not true-to-type, insect infestation, or other reason.

Root-knot—A nematode-caused disease characterized by galls on the roots. Usually most commonly found in sandy soils attacking hundreds of kinds of plants. See (37) Root-knot under General Diseases. Also the name of several species of nematodes.

Rootone F—Contains growth-promoting substances and 4 per cent thiram. Used on cuttings, seeds, and bulbs.

Rootstock—The fleshy root of a herbaceous perennial plant with buds and eyes.

Rosette—Symptom of disease with stems shortened to produce a bunchy growth habit.

Rot—State of decomposition and putrefaction. May be dry and firm to mushy and slimy. Caused by an organism disintegrating large numbers of living cells. Usually caused by fungi and bacteria.

Rot-Not—See Copper Naphthenate.

Rugose—Rough. Used as part of the names of certain virus diseases characterized by warty, roughened, or severely crinkled leaves or other plant parts.

Runner—A slender, horizontal stem that grows close to the soil surface. Rooting occurs at the joints (e.g., strawberry).

Russet—Brownish, roughened, or corky areas on the skin of leaves, fruit, or tubers, as a result of disease, insects, or spray injury.

Rust—A disease caused by a rust fungus, or the fungus itself. The life cycle of a rust fungus may involve up to five different types of spores. Rusts may parasitize one species of plant during their lives (*autoecious* or *monoecious*) or two types of species (*heteroecious*). A rust that is heteroecious and has five spore forms (numbered 0 to IV) is the stem rust of grasses. The five spore types are the reddish-brown *urediospores* (II) that spread the rust from grass plant to grass plant, the dark *teliospores* (III) that infect nothing but remain on straw or stubble resisting winter temperatures, and germinate in the spring to produce the *basidiospores* (IV) which carry rust to barberry, infect it, germinate, and produce *pycniospores* (0) that fuse sexually to produce *aeciospores* (I) which blow to and infect the grass plant to complete the life cycle. See (8) Rust under General Diseases.

Safety—The practical certainty that injury will not result from the use of a substance in a proposed quantity and manner.

Sanitation—Keeping the garden clean. Destroying all infested and infected plant parts during the season. Removing and composting or burning all plant tops in the fall, together with surrounding weeds. Often the most important part of disease control, especially in the home garden. See Section 3.

Saprophyte—An organism that feeds on dead organic matter, as opposed to a parasite that feeds on living tissue. See Parasite.

Sapwood—The outer part of xylem tissue in a woody stem.

Scab—A roughened, crustlike, diseased area (lesion) on the surface of a plant part. Also the disease in which scab is a symptom. See (14) Scab under General Diseases.

Scaffold branches—The primary branches of a tree which arise from the trunk.

Scald—A blanched appearance of plant parts.

Scape—A leafless flower stalk.

Scion—A piece of twig or shoots inserted on another in grafting.

Sclerotium (pl. **sclerotia**)—A small, compact, fungus resting body composed of an interwoven mass of mycelial threads with a hard outer rind. Sclerotia are generally dark colored, more or less round or flat, and vary greatly in size. Sclerotia may remain viable in the soil, in plant refuse, or in seeds for many years and are capable of germinating or bearing fruiting bodies that infect new plants under favorable conditions of temperature and moisture.

Scorch—"Burning" of plant tissue from infection, lack or excess of some nutrient, or weather conditions. Often appears as dead areas along or between the veins, margins, and tips of leaves.

Secondary infection—Infection resulting from the spread of infectious material that has been produced following a primary infection (the first infection by a disease-producing organism after a resting period) or from other secondary infections without an intervening inactive period.

Seed disinfectant or disinfestant—A chemical that destroys certain disease-causing organisms carried *in* (disinfectant) or *on* (disinfectant or disinfestant) the seed. It is not necessarily a seed protectant.

Seed protectant—A chemical applied to seed before planting to prevent seed decay and damping-off. See pages 527–537.

Semeson—Contains hydroxymercurichlorophenol. A dust seed disinfectant for vegetable and flower seeds, bulbs, corms, roots, and tubers. Or a pink powder (Semesan Bel S, Semesan Seed Disinfectant) used for potato and sweetpotato seed treatment; also as a soil drench and a turf fungicide (Semesan Turf Fungicide). See Tables 20 and 22 in the Appendix.

Sepal—A division of the outer floral envelope or calyx. The sepals may be united into one piece, or separate. May be green and inconspicuous or be the brightly colored part of the flower.

Sexual propagation—To increase plant numbers by seed.

Shade tolerant—Plants that can grow on reduced sunlight.

Shell SD 4741—An experimental soil fungicide containing 0,0,0-trimethylphosphorothioate. Useful in controlling root and crown rots, and damping-off, of numerous ornamentals.

Shoot—A leafy stem of current season's growth.

Shot-hole—A symptom of disease in which small diseased fragments of leaves drop out leaving small holes making them look as if riddled by shot. See (4) Shot-hole under General Diseases.

Shrub—A woody plant, usually less than 12 feet tall and under 2 inches stem diameter, which branches from the crown into several stems.

Side-dressing—Fertilizer applied to the side of a row crop during growth. Often below the soil surface.

Sign—Evidence of disease indicated by the presence of disease-causing organisms or of any of their parts and products (e.g., spores, mycelium, exudate, fruiting bodies of the pathogen).

Slime flux or Wetwood—Found on certain trees that bleed freely when wounded and sometimes forced out of previously unwounded bark by pressure in the vascular system. Slime flux of elm is known to be caused by bacteria.

Slime mold—Primitive fungi whose plasmodia "flow" over low-lying vegetation like an amoeba. Found commonly on lawns, strawberry beds, seed beds, rotting logs, and tree trunks. The fruiting stage is powdery. See under Lawngrass.

Slip—A herbaceous or softwood cutting.

Slurry—A thick suspension of a finely divided material in a liquid. A common method of commercial seed treatment.

Smut—A disease caused by a smut fungus, or the fungus itself. Characterized by resting spores (chlamydospores or teliospores) that generally accumulate in black, powdery masses (sori). The black spore masses may break up into a fine dustlike powder readily scattered by the wind or remain firm and more or less covered. See (11) Smut and (13) White Smut under General Diseases.

Soil conditioners—Chemicals that aggregate soil particles for improved soil structure.

Soil sterilization—Treating soil by heat or chemicals to kill living organisms in it. See pages 538–548 in Appendix.

Soilless culture—Also called hydroponics. The growing of plants in nutrient solution without soil.

Solanaceous—Plants in a family that include potato, tomato, eggplant, pepper, tobacco, and Chinese lanternplant.

Sol-Kop 10—A completely water-soluble copper fungicide containing 10 per cent metallic copper. Controls angular leaf spot and Alternaria leaf spot of cucurbits, leaf blights of carrot, potato, and tomato, and leaf spot of beets.

Sooty molds—Fungi with dark hyphae that live in the honeydew secreted by aphids, mealybugs, scales, and whiteflies, that forms a sooty coating on foliage and fruit. See (12) Sooty Mold under General Diseases.

Sorus (pl. **sori**)—A compact mass of spores produced in, or on, the host plant by fungi such as rusts and smuts.

Southern blight—See Crown Rot.

Spathe—A large bract or pair of bracts sheathing a flower cluster (e.g., Jack-in-the-pulpit).

Species—Any one kind of life subordinate to a genus but above a race, strain, or variety. See Genus and also Race.

Sphagnum moss—Gray or tan bog mosses dried and used as a planting medium. Milled sphagnum is sterile and fine-textured.

Spike—A simple, slender flower cluster of stalkless flowers (e.g., rye).

Spore—A part of a fungus corresponding to the seed of higher plants. A microscopic, one- to many-celled body serving to reproduce and disseminate a fungus. Spores may be either nonsexual (asexual), formed directly from vegetative hyphae, but often in special fruiting structures (e.g., pycnidia, acervuli, sporodochia); or sexual, formed from a union of two cells representing a difference in sex. See Figure 1.3. Some, called resting spores, have thick walls that enable them to survive unfavorable growing conditions. Some spores are very light and can be blown hundreds of miles by the wind. Others are transported easily by water, insects, animals, man, and machinery. When conditions are favorable, the spore germinates to produce a hyphal tube that later develops into a new fungus body.

Sporodochium (pl. **sporodochia**)—A cushion-shaped spore-producing body of a fungus.

Sporophore—A stalklike structure on which spores are borne.

Sporulate—To form spores.

Spot—A definite, localized, diseased area. See (1) Fungus Leaf Spot and (2) Bacterial Leaf Spot under General Diseases.

Spotrete—A lawn fungicide containing 75 per cent thiram.

Spray lime—Spray grade, freshly made, finely divided calcium hydroxide. Used in making bordeaux mixture, a diluent for dusts, and as a safener in certain sprays. Must be fresh and uncarbonated to be effective.

Spreaders and stickers—Materials used to reduce surface tension and retain a uniform deposit of pesticide on plant surfaces and to make them adhere for a longer period of time. Some common *spreaders* are Santomerse, Tween-20, soap, and household detergents. Common materials used primarily for their sticking qualities are the following: Goodrite PEPS, casein, powdered skim milk, wheat flour, soybean flour, and fish oils. Some commercial *spreader-stickers* include: Triton B-1956, Ortho Spreader-Sticker, DuPont Spreader-Sticker, Plyac Spreader-Sticker, Miller Nu-Film, X-77 Spreader-Activator, Filmfast, and Spred-Rite. See also page 118.

Spur—A short, woody stem (branch). The principal fruiting area of many fruit trees.

Staghead—Dying of a tree from the top downward. Characteristic of certain diseases (e.g., oak wilt on white and bur oaks).

Staminate flower—One that has stamens (male parts) but no pistils (female parts).

Starter solution—Fertilizer (possibly also containing pesticides) dissolved in water and applied immediately following transplanting. See also page 29.

Stauffer Captan-Folpet—A fungicide combination of captan and folpet. Formulated as a 30:30 mixture for spraying, or a 10:10 mixture for dusting. A soil fungicide for use as an in-furrow treatment.

Stele—The central cylinder in the stems and roots of vascular plants.

Sterilant—Any agent or chemical that destroys all living organisms in a substance (e.g., soil). See Table 22 in the Appendix.

Sterilization (soil)—Use of steam or chemicals to prevent disease, insect, nematode, or weed problems. Often better called pasteurization.

Stickers—See Spreaders and page 118.

Stock—The root or stem on which a graft is made.

Stolon (runner)—A creeping, trailing, horizontal stem or runner, usually underground, that may produce roots, leaves, or new stems from nodes and become an independent plant.

Stoma (pl. **stomata**)—A minute pore opening in the leaf or stem of plants, utilized in the exchange of gases for respiration, photosynthesis, and transpiration. Many disease-producing microorganisms enter plants through pores. See Figure 1.3.

Strain—An organism (plant) or group of organisms (or virus) that differs in origin or minor aspects from other organisms of the same species or variety. See Race. Also a special type of plant selected from a variety.

Streak—An elongated lesion with irregular sides.

Streptomycin—An antibacterial antibiotic. Available as powder or liquid to control various bacterial diseases of plants. See page 105. The material should be 'freshly manufactured" during the current season.

Stroma (pl. **stromata**)—A cushion-like, fungus mass of tissue on or in which spore production usually occurs—primarily in reproductive bodies.

Stunted—An unthrifty plant reduced in size and vigor due to unfavorable environmental conditions. May be due to a wide range of parasitic and nonparasitic agents.

Substrate—The material or substance on which a saprophytic organism feeds and develops.

Succulent—Watery (e.g., stems and leaves high in water content). Also refers to a group of drought-resistant ornamental plants with thick, fleshy leaves or stems and high in water content.

Sucker—A shoot arising from a root.

Sulphuric acid—Sometimes used to control damping-off in plant beds of forest-tree seedlings, especially conifers. Prepare by mixing 4 fluid ounces of commercial sulphuric acid in $2\frac{1}{2}$ gallons of water. Apply directly to plant beds at time of seeding. *Be sure to add the acid to the water* and not the other way around.

Sunox—See Oxyquinoline Sulfate.

Sunscald—Plant tissues burned or scorched by too much sun and other unfavorable conditions.

Suscept—Usually a more precise term for host. Any living organism liable to infection by a given disease-producing agent. See Host.

Susceptibility—Lacking inherent ability to resist disease or attack by a pathogen.

Symbiosis—A mutually beneficial association of two or more organisms (e.g., powdery mildew fungus and mite = witches'-broom of hackberry).

Symptoms—External or internal expressions of plant disease produced by the plant. Symptoms may be:

(a) **necrotic**—disease resulting in death of tissues. Results in formation of blights, damping-off, wilt, or rots.
(b) **hypoplastic**—underdevelopment in size or number of plant organs. May result in chlorosis, mosaic, mottling, yellows, or dwarfing.
(c) **hyperplastic**—overgrowth in size or number of plant organs. Results in formation of galls, witches'-brooms, scab, callus growth, or curl.

Synergist—Any substance that increases the toxic effects of a pesticide.

Systemic—Applies (1) to a disease in which the pathogen (or a single infection) spreads generally throughout the plant body or (2) to chemicals that spread through a plant internally.

TC-90—A liquid (emulsifiable concentrate) fungicide containing 48 per cent copper salts of fatty and rosin acids—metallic copper equals 4 per cent. Useful against leaf spots of peanut and carrot; downy mildew, powdery mildew, and angular leaf spot of cucurbits; bacterial leaf spot of tomato and pepper; leaf blights of beets, tomato, potato, and celery.

Tenacity—The tendency of a deposit to resist removal by weathering.

Tendril (rust)—Slender, leafless organs that serve a climbing plant (e.g., pea, grape) as a means of attachment. Also the spore "horns" of certain cedar-rust fungi.

Terminal—The end of a shoot, twig, or branch.

Terramycin—An antibiotic containing oxytetracycline. Often combined with streptomycin (as in Agrimycin) to control some bacterial diseases, e.g., Crown Gall.

Tersan OM—A multipurpose turf fungicide containing 45 per cent thiram and 10 per cent hydroxymercurichlorophenol.

Thimer—A multipurpose turf fungicide containing 75 per cent thiram and 3 per cent phenyl mercury acetate.

Thiramad—A turf fungicide containing 75 per cent thiram.

Thylate Thiram Fungicide—Contains 65 per cent thiram. Used to control diseases of ornamentals, apples, peaches, strawberries, celery, tomatoes, and other plants.

Tissue—A group of cells of similar structure that perform a special function.

Tissue test—Determination of plant food needs by a chemical analysis of leaves or stems.

Toadstool—See Mushroom.

Tolerance (tolerant)—The degree of endurance of a plant to the effects of adverse conditions, chemicals, or parasites. A tolerant plant is capable of sustaining a disease without serious injury or crop loss. Also refers to the amount of toxic residue allowable in or on edible plant parts under the law. See Pesticide Tolerance.

Topworking—Changing the variety (or varieties) of a tree by inserting buds or grafts on its branches.

Toxicity—The capacity of a substance to produce injury.

Toxin—A "poison" produced by an organism.

Trade names—Names given to company products to distinguish them from similar competitive products. See Tables 6 and 22.

Translocation—The movement of water, minerals, and food within a plant.

Translucent—So clear that light rays may pass through.

Transpiration—The loss of water by evaporation from a living plant. Occurs largely from internal leaf surfaces.

Trapex—A soil fumigant containing 20 per cent methyl isothiocyanate. Controls nematodes, soil insects, weed seeds, and certain disease-producing fungi.

Tree surgery—The art of removing large limbs, cleaning and treating of wounds, cabling and bracing weak trunks, crotches, and branches.

Tuber—A short, fleshy, much enlarged, mostly underground stem having numerous buds or "eyes" (e.g., Irish potato, cyclamen, winter-aconite, anemone).

Tuberous-rooted—A swollen root (e.g., dahlia, sweetpotato) distinguished from a tuber that is a swollen underground stem.

Turf—Grass used for a lawn.

Twig—One-year-old stem or branch of a woody plant.

Tylosis (pl. **tyloses**)—A cell outgrowth into the cavity of a xylem vessel, plugging it. Important in certain wilt diseases of woody plants (e.g., Dutch elm disease and oak wilt).

Variegation—A general term for discoloration (presence of two or more colors) of foliage or flowers from genetic causes—*not* from virus infection.

Variety—A group of closely related plants of common origin that differ from each other in certain minor details such as form, color, flower, and fruit. Special types are selected from the variety and are called strains.

Vascular—Refers to plant tissues that conduct fluids. See Xylem and Phloem.

Vector—An agent that transmits a pathogen (e.g., man, insects, mites, birds, nematodes).

Vegetative—Nonsexual. Also refers to a significant increase in size.

Vein banding—A symptom of a virus disease in which areas along leaf veins are darker green than tissue between the veins.

Vein clearing—Disappearance of green color in or along the leaf veins.

Vermiculite (horticultural grade)—A sterile planting medium, like mica, used for rooting cuttings. Frequently combined with loam and peatmoss as a substitute for sand.

Verticillium wilt—A widespread, systemic, vascular disease that attacks hundreds of different kinds of plants. See (15B) Verticillium Wilt under General Diseases.

Viability—State of being alive (e.g., ability of seeds to germinate).

Virulent—Strong ability to produce disease. Highly pathogenic.

Viruliferous—Containing or carrying a virus. Usually pertains to an insect that carries a virus and can infect a plant with it.

Virus—Submicroscopic, filterable, infectious agents (bodies) too small to be seen with a compound microscope. Viruses have characteristics of both living and nonliving matter. They are large, high molecular weight nucleoproteins capable of multiplying (replicating) and acting like living organisms when in specific living plant and/or animal cells. Usually recognizable by the symptoms they produce in infected hosts. See page 13.

Water-logged—Without soil aeration due to a lack of or poor soil drainage.

Watersprouts—Rapidly growing shoots arising from adventitious or latent buds on the trunk or branches.

Weed—Any unwanted plant.

Well-drained soil—Draining out of all excess moisture from a soil. May occur naturally or through the aid of agricultural drain tile, lightening with peatmoss and sand or gravel. See page 36.

Well-prepared soil—A soil that has ample organic matter and plant nutrients, which has been well mixed to remove large rocks and objectionable plant debris to give maximum facilities for good plant growth.

Wettable powder—One that is easily wetted by water and will go into suspension.

Wetwood—See Slime Flux.

Whorl—Three or more leaves, flowers, twigs, or other plant parts arranged in a circle and radiating from one point.

Wilson's Anti-Damp—Contains 2.5 per cent 8-hydroquinoline benzoate. Used as a fungicide drench to control damping-off, wilts, crown and root rots, etc.; primarily of ornamentals.

Wilt—Lack of freshness and drooping of leaves from lack of water (inadequate water supply or excessive transpiration) ; a vascular disease that interrupts the normal uptake and distribution of water by a plant; or a toxin produced by an organism. See (15) Wilts under General Diseases.

Winter annual—Plants that customarily germinate in the fall, overwinter as small plants, and complete growth, flowering, and seed production in the following season before dying. Examples: pansies, common chickweed, mustards, numerous "annual" grasses.

Witches'-broom—A symptom of disease where an abnormal brushlike development of many weak shoots arises at or close to the same point. Hackberry trees are commonly affected with this disease. Witches'-brooms may be caused by fungi, mites, insects, viruses, bacteria, and mistletoes.

Wood Tox—See Pentachlorophenol.

Woodridge Mixture "21"—A turf fungicide containing 66.7 per cent calomel and 33.3 per cent corrosive sublimate.

Woody—Hard, tough, and fibrous. Nonherbaceous.

Xylem—The complex conducting tissue in plants by which water and minerals move up the stem from roots to leaves. Furnishes mechanical support for the plant.

Yellows—A disease (caused by a fungus, virus, bacterium, or a deficiency of one or more essential elements) characterized by yellowing and stunting of affected parts.

Zinc napththenate—A liquid wood preservative applied by brush or as a dip. Sold as Cuprinol Clear No. 10.

Zinc sulfate—Used to control zinc deficiencies in fruits, trees, and shrubs. Used with nabam to form zineb. Flake form containing 25.5 per cent is more easily dissolved but tends to harden in storage; 36 per cent monohydrated form is more available.

Zinophos—Nematocide-insecticide containing 0,0-diethyl 0-2-pyrazinyl phosphorothioate. Highly toxic to mammals. See page 546.

Zonate—Marked with zones.

Bibliography

In preparing this second edition of *How To Control Plant Diseases . . . in home and garden* thousands of references have been reviewed that have been accumulating in my files for the past 20 years. The bibliography presented here is but a selected, small sampling of the rapidly expanding fields of plant pathology (including nematology, virology, mycology, and bacteriology), horticulture, botany, forestry, and entomology. Emphasis has been placed on listing the most readily available sources of information for the layman. Articles and abstracts from such scientific journals as *The Plant Disease Reporter, Phytopathology, Review of Applied Mycology, Journal of Economic Entomology, Biological Abstracts, NAC News, Arborist's News, Proceedings of the National Shade Tree Conference, Canadian Plant Disease Survey,* etc. have been omitted.

The bibliography is divided into two sections: A, Books and state experiment station or cooperative extension service bulletins, circulars, and spray programs, and B, USDA publications. Some of the older references are out of print but many public libraries maintain bound copies of state experiment station and cooperative extension service publications as well as those published by the U.S. Department of Agriculture.

A. BOOKS AND PUBLICATIONS OF LAND-GRANT INSTITUTIONS*

AINSWORTH, G. C., AND G. R. BISBY. 1963. A dictionary of the fungi. 5th ed. Commonwealth Mycological Institute, Kew, Surrey, England.

ALEXOPOULOS, C. J. 1962. Introductory mycology. 2nd ed. John Wiley & Sons, Inc., New York, N.Y.

AMERICAN SOCIETY FOR HORTICULTURAL SCIENCE. 1954. The care and feeding of garden plants. National Fertilizer Association, Washington, D.C.

ANDERSON, H. W. 1956. Diseases of fruit crops. McGraw-Hill Book Co., Inc., New York, N.Y.

ANONYMOUS. 1956. Plant pest handbook. Conn. Agr. Exp. Sta. Bul. 600.

———. 1959. Compendium of plant diseases. Rohm & Haas Co., Philadelphia, Pa.

———. Effect of natural gas on house plants and vegetation. American Gas Assn. 420 Lexington Ave., New York, N.Y.

———. Outdoor housekeeping. National Sprayer & Duster Assn. 850 Wrigley Building North, 410 N. Michigan, Chicago, Ill. 60611.

* To obtain a copy of an experiment station or cooperative extension service bulletin, circular, or spray program contact your local county extension office or write to the Bulletin Room, College of Agriculture at the land-grant institution listed (see pages 6–7).

———. Sprayer and duster manual. National Sprayer & Duster Assn. 850 Wrigley Building North, 410 N. Michigan, Chicago, Ill. 60611.

ARTHUR, J. C. 1934. Manual of the rusts in United States and Canada. Purdue Research Foundation, Lafayette, Ind.

BAILEY, L. H. 1949. Manual of cultivated plants. Revised ed. The Macmillan Co., New York, N.Y.

———, and ETHEL ZOE BAILEY, compilers. 1941. Hortus second. The Macmillan Co., New York, N.Y.

BAKER, K. F. (ed.). 1957. The U.C. system for producing healthy container-grown plants. Univ. of California Manual 23.

———, and W. C. SNYDER (eds.). 1965. Ecology of soil-borne plant pathogens, prelude to biological control. Univ. of California Press, Berkeley, Calif.

BARNETT, H. L. 1960. Illustrated genera of imperfect fungi. 2nd ed. Burgess Publishing Co., Minneapolis, Minn.

BAWDEN, F. C. 1964. Plant viruses and virus diseases. 4th ed. The Ronald Press Co., New York, N.Y.

BEAUMONT, A. 1956. Diseases of garden plants. Transatlantic Arts, Inc., Forest Hills, N.Y.

BESSEY, E. A. 1950. Morphology and taxonomy of fungi. The Blakiston Div., McGraw-Hill Book Co., Inc., New York, N.Y.

BISBY, G. R. 1945. An introduction to the taxonomy and nomenclature of fungi. Commonwealth Mycological Institute, Kew, Surrey, England.

BLODGETT, E. C., AND A. E. RICH. 1949. Potato tuber diseases, defects, and insect injuries in the Pacific Northwest. The State College of Washington Pop. Bul. 195.

BOYCE, J. S. 1961. Forest pathology. 3rd ed. McGraw-Hill Book Co., Inc., New York, N.Y.

BRAY, D. F. 1958. Gas injury to shade trees. Scientific Tree Topics 2:19–22. Bartlett Tree Research Labs., Stamford, Conn.

BROGDON, J. E., M. E. MARVEL, AND R. S. MULLIN. Commercial vegetable insect and disease control guide. Univ. of Florida Ext. Circ. 193E (revised annually).

BURNETT, F. M., AND W. M. STANLEY (eds.) 1959. The viruses. 2 volumes. Academic Press, New York, N.Y.

BURNETT, H. C. 1965. Orchid diseases. State of Florida, Dept. of Agriculture, Div. of Plant Industry Vol. I, No. 3.

BUTLER, E. J., AND S. G. JONES. 1949. Plant pathology. Macmillan & Co. Ltd., London, England.

CARBONNEAU, M. C. 1965. Geraniums for the home and garden. Univ. of Illinois Ext. Circ. 904.

CAROSELLI, N. E. 1957. Verticillium wilt of maples. R.I. Agr. Exp. Sta. Bul. 335.

CARTER, J. C. 1964. The wetwood disease of elm. Ill. Nat. Hist. Survey Circ. 50.

———. 1964. Illinois trees: their diseases. 3rd printing. Ill. Nat. Hist. Survey Circ. 46.

CARTER, W. 1962. Insects in relation to plant diseases. Interscience Publishers, New York, N.Y.

CHITWOOD, B. G., AND W. BIRCHFIELD. 1956. Nematodes, their kinds and characteristics. State Plant Board of Fla. Vol. II. Bul. 9.

CHRISTENSEN, C. M. 1965. Common fleshy fungi. 2nd ed. Burgess Publishing Co., Minneapolis, Minn.

———. 1965. 3rd. ed. The molds and man—an introduction to the fungi. Univ. of Minnesota Press, Minneapolis, Minn.

CHRISTIE, J. R. 1959. Plant nematodes, their bionomics and control. Univ. of Florida, Gainesville, Fla.

CHUPP, C. 1953. A monograph of the fungus genus *Cercospora*. Published by the author, Ithaca, N.Y.

———, AND A. F. SHERF. 1960. Vegetable diseases and their control. The Ronald Press Co., New York, N.Y.

CLEMENTS, F. E., AND C. L. SHEAR. 1931. The genera of fungi. The Hafner Publishing Co., Inc., New York, N.Y.

COOK, R. L., AND C. E. MILLAR. 1953. Plant nutrient deficiencies diagnosed by plant symptoms, tissue tests, soil tests. Michigan State College Spec. Bul. 353.

CORBETT, M. K., AND H. D. SISLER (eds.). 1964. Plant virology. Univ. of Florida Press, Gainesville, Fla.

COUCH, H. B. 1962. Diseases of turfgrasses. Reinhold Publ. Corp., New York, N.Y.

COX, R. E. 1954. Some common diseases of flowering dogwood in Delaware. Del. Agr. Exp. Sta. Circ. 27.

CUMMINS, G. B. 1959. Illustrated genera of rust fungi. Burgess Publishing Co., Minneapolis, Minn.

———. 1962. Supplement to Arthur's Manual of the rusts in United States and Canada. The Hafner Publishing Co., Inc., New York, N.Y.

DAVIS, S. H., JR., AND F. C. SWIFT. 1962. Diseases and insect pests of rhododendron and azalea. Rutgers—The State Univ. Ext. Bul. 345.

DIMOCK, A. W. 1953. The gardener's ABC of pest and disease. M. Barrows and Co., Inc., New York, N.Y.

———. 1961. Diseases of bearded irises. Cornell Univ. Ext. Bul. 1050.

DOWSON, W. J. 1957. Plant diseases due to bacteria. 2nd ed. Cambridge University Press, New York, N.Y.

EDDINS, A. H. 1952. Diseases, deficiencies and injuries of cabbage and other crucifers in Florida. Fla. Agr. Exp. Stas. Bul. 492.

ELLIOTT, CHARLOTTE. 1951. Manual of bacterial plant pathogens. 2nd ed. Chronica Botanica Co., Waltham, Mass.

ESAU, KATHERINE. 1961. Plants, viruses, and insects. Harvard University Press, Cambridge, Mass.

FAWCETT, H. S. 1936. Citrus diseases and their control. 2nd ed. McGraw-Hill Book Co., Inc., New York, N.Y.

FENSKA, R. R. 1954. Tree experts manual. 2nd ed. A. T. De La Mare Co., Inc., New York, N.Y.

FISCHER, G. W. 1953. Manual of the North American smut fungi. The Ronald Press Co., New York, N.Y.

———, AND C. S. HOLTON. 1957. Biology and control of the smut fungi. The Ronald Press Co., New York, N.Y.

FITZPATRICK, H. M. 1930. The lower fungi—Phycomycetes. McGraw-Hill Book Co., Inc., New York, N.Y.

FORSBERG, J. L. 1963. Diseases of ornamental plants. Univ. of Illinois College of Agriculture Special Publ. No. 3, Urbana, Ill.

FOSLER, G. M. 1962. Soil sterilization methods for the indoor gardener. Univ. of Illinois Ext. Circ. 793.

FREAR, D. E. H. 1955. Chemistry of the pesticides. 3rd ed. D. Van Nostrand Co., Inc., Princeton, N.J.

——— (ed.). Pesticide handbook—Entoma. College Science Publishers, State College, Pa. (Revised each year; lists over 9,500 pesticides by trade names and formulae.)

FREEMAN, T. E., AND R. S. MULLIN. 1964. Turfgrass diseases and their control. Univ. of Florida Ext. Circ. 221A.

GAMBRELL, F. L., AND R. M. GILMER. 1956. Insects and diseases of fruit nursery stocks and their control. N.Y. (Geneva) Agr. Exp. Sta. Bul. 776.

GARRETT, S. D. 1956. Biology of root-infecting fungi. Cambridge University Press, New York, N.Y.

GIANFAGNA, A. U., AND J. K. RATHMELL, JR. 1962. Growing better house plants. The Pennsylvania State University Ext. Circ. 491.

GILMAN, J. C. 1957. A manual of soil fungi. 2nd ed. Iowa State College Press, Ames, Iowa.

GOODEY, T. 1963. Soil and freshwater nematodes. Revised edition, J. B. Goodey (ed.). Methuen & Co. Ltd., London, England.

GOULD, C. J. 1946. Narcissus diseases in Washington. Washington Agr. Exp. Sta. Bul. 480.

———. 1950. Diseases of bulbous iris. Washington Agr. Ext. Bul. 424.

——— (ed.). 1957. Handbook on bulb growing and forcing. Northwest Bulb Growers Assn., Mt. Vernon, Wash.

———, M. R. HARRIS, AND G. W. EADE. Diseases of ornamentals. State Coll. Wash. Agr. Ext. Mimeo. 683.

GRAM, E., AND ANNA WEBER. 1953. Plant diseases in orchard, nursery and garden crops. Ed. and adapted by R. W. G. Dennis. Tr. from the Danish by Evelyn Ramsden. Philosophical Library, New York, N.Y.

GROGAN, R. G., W. C. SNYDER, AND R. BARDIN. 1955. Diseases of lettuce. Univ. of California Circ. 448.

GUBA, E. F. 1945. Carnation wilt diseases and their control. Mass. Agr. Exp. Sta. Bul. 427.

———, AND J. A. STEVENSON. 1963. Fungus and nematode inhabitants and diseases of holly *(Ilex)*. Univ. of Mass. Exp. Sta. Bul. 530.

GUSTAFSON, C. D. 1956. How to identify avocado diseases. Univ. of Calif. Ext. Service Leaflet 61.

GUTTERMAN, C. E. F. 1935. Diseases of iris. Cornell Univ. Ext. Bul. 324.

———. 1935. Peony diseases. Cornell Univ. Ext. Bul. 321.

HANNA, L. W. 1958. Hanna's handbook of agricultural chemicals. 2nd ed. Published by the author, Forest Grove, Oreg.

HARRIS, M. R., AND D. H. BRANNON. 1962. Diseases and insect pests of potatoes. Washington State Univ. Ext. Bul. 553.

HEALD, F. D. 1933. Manual of plant diseases. 2nd ed. McGraw-Hill Book Co., Inc., New York, N.Y.

HOPPER, B. E., AND E. J. CAIRNS. 1959. Taxonomic keys to plant, soil and aquatic nematodes. Alabama Polytechnic Institute, Auburn, Ala.

HORSFALL, J. G. 1956. Principles of fungicidal action. Chronica Botanica Co., Waltham, Mass.

———, AND A. E. DIMOND (eds.). 1959–60. Plant pathology, an advanced treatise. 3 volumes. Academic Press, New York, N.Y.

HOUGH, W. S., AND A. F. MASON. 1951. Spraying, dusting and fumigating of plants. Rev. ed. The Macmillan Co., New York, N.Y.

HOWARD, F. L., J. B. ROWELL, AND H. L. KEIL. 1951. Fungus diseases of turf grasses. R.I. Agr. Exp. Sta. Bul. 308.

HUTCHINSON, M. T. *et al.* 1961. Plant parasitic nematodes of New Jersey. N.J. (Rutgers) Agr. Exp. Sta. Bul. 796.

JANES, R. L., H. S. POTTER, AND G. E. GUYER. Chemical control of insects and diseases of commercial vegetables. Mich. State Univ. Ext. Bul. 312 (revised yearly).

JENKINS, W. R., D. P. TAYLOR, R. A. RHODE, AND B. W. COURSEN. 1957. Nematodes associated with crop plants in Maryland. Univ. Md. Bul. A-89.

KLOS, E. J. 1959. Tree fruit diseases in Michigan. Mich. State Univ. Ext. Bul. 361.

KLOTZ, L. J. 1961. Color handbook of citrus diseases. 3rd ed. Univ. of California Press, Berkeley, Cal.

LARGE, E. C. 1940. The advance of the fungi. Henry Holt and Co., New York, N.Y. (republished 1962, Dover Publishing Inc., New York, N.Y.)

LAWRENCE, F. P., AND J. E. BROGDON. 1955. Control of insects and diseases of dooryard citrus trees. Univ. Fla. Agr. Ext. Serv. Circ. 139.

LEACH, J. G. 1940. Insect transmission of plant diseases. McGraw-Hill Book Co., Inc., New York, N.Y.

———. 1965. Diseases of the iris in West Virginia and their control. West Virginia Univ. Exp. Sta. Bul. 509.

LINN, M. B. 1964. Vegetable diseases. Revised ed. Univ. of Illinois Ext. Circ. 802.

———, AND W. H. LUCKMANN. 1965. Tomato diseases and insect pests. Univ. of Illinois Ext. Circ. 912.

LYLE, E. W. 1944. Rose diseases. Texas Agr. Exp. Sta. Circ. 87.

McCRUM, R. C., J. G. BARRAT, M. T. HILBORN, AND A. E. RICH. 1960. Apple virus diseases, an illustrated review. Maine Agr. Exp. Sta. Bul. 595 and N.H. Agr. Exp. Sta. Tech. Bul. 101.

McELWEE, E. W. 1962. Growing camellias in Florida. Univ. of Florida Ext. Bul. 161A.

MAGIE, R. O., AND W. G. COWPERTHWAITE. 1954. Commercial gladiolus production in Florida. Fla. Agr. Exp. Stas. Bul. 535.

———, A. J. OVERMAN, AND W. E. WATERS. 1964. Gladiolus corm production in Florida. Florida Agr. Exp. Stas. Bul. 664.

METCALF, C. L., AND W. P. FLINT (revised by R. L. Metcalf). 1962. Destructive and useful insects, their habits and control. 4th ed. McGraw-Hill Book Co., Inc., New York, N.Y.

MOORE, W. C. 1939. Diseases of bulbs. Ministry of Agriculture and Fisheries (Great Britain) Bul. 117.

NELSON, R. 1948. Diseases of gladiolus. Mich. Agr. Exp. Sta. Spec. Bul. 350.

NEWHALL, A. G., AND W. T. SCHROEDER. 1951. New flash-flame soil pasteurizer. Cornell Univ. Agr. Exp. Sta. Bul. 875.

NICHOLS, L. P. 1962. Diseases of ornamental shrubs and vines. The Pennsylvania State Univ. Ext. Circ. 502.

———. 1962. Tree diseases, description and control. The Pennsylvania State Univ. Spec. Circ. 46.

———, AND O. D. BURKE. 1963. Diseases of commercial florist crops. The Pennsylvania State Univ. Ext. Circ. 517.

———, AND J. O. PEPPER. 1963. Diseases and insects of the flower garden and their control. The Pennsylvania State Univ. Ext. Circ. 347.

NOBLE, MARY, J. DE TEMPE, AND P. NEERGAARD. 1958. An annotated list of seed-borne diseases. Commonwealth Mycological Institute, Kew, Surrey, England.

NORTHEN, H. T., AND REBECCA T. NORTHEN. 1956. The complete book of greenhouse gardening. The Ronald Press Co., New York, N.Y.

O'REILLY, H. J. 1963. Armillaria root rot of deciduous fruits, nuts, and grapevines. Univ. of California Circ. 525.

PARKINSON, D., AND T. S. WAID (eds.). 1960. The ecology of soil fungi—an international symposium. Univ. of Liverpool Press, Liverpool, England.

PARRIS, G. K. 1952. Diseases of watermelons. Fla. Agr. Exp. Stas. Bul. 491.

PARTYKA, R. E., AND L. J. ALEXANDER. 1965. Greenhouse tomatoes—disease control. The Ohio State Univ. Bul. SB-16.

PAULUS, A. O., D. E. MUNNECKE, AND P. A. CHANDLER. 1960. Diseases of geraniums in California. Univ. of Calif. Leaflet 130.

PEACE, T. R. 1962. Pathology of trees and shrubs, with special reference to Britain. Oxford University Press, New York, N.Y.

PHILLIPS, A. M., J. R. LARGE, AND J. R. COLE. 1964. Insects and diseases of the pecan in Florida. Fla. Agr. Exp. Stas. Bul. 619A.

PIRONE, P. P. 1959. Tree maintenance. 3rd ed. Oxford University Press, New York, N.Y.

———. 1940. Diseases of the gardenia. N.J. Agr. Exp. Sta. Bul. 679.

———, B. O. DODGE, AND H. W. RICKETT. 1960. Diseases and pests of ornamental plants. 3rd ed. The Ronald Press Co., New York, N.Y.

PLAKIDAS, A. G. 1964. Strawberry diseases. Louisiana State University Press, Baton Rouge, La.

POWELL, D., R. H. MEYER, AND F. W. OWEN. 1966. Pest control in commercial fruit plantings. Univ. of Illinois Ext. Circ. 936.

———, B. JANSON, AND E. G. SHARVELLE. 1965. Diseases of apples and pears in the Midwest. North Central Regional Ext. Publ. 16, and Univ. of Illinois Ext. Circ. 909.

RAABE, R. D., A. O. PAULUS, AND A. H. MCCAIN. 1963. Diseases of camellia in California. Univ. of Calif. Leaflet 151.

RADER, W. E. 1952. Diseases of stored carrots in New York State. Cornell Univ. Bul. 889.

REHDER, A. 1940. Manual of cultivated trees and shrubs. The Macmillan Co., New York, N.Y.

ROHDE, R. A., AND W. R. JENKINS. 1958. Basis for resistance of *Asparagus officinalis* var. *altilis* L. to the stubby-root nematode *Trichodorus christiei* Allen. Univ. Md. Bul. A-97.

ROSSBERG, D. W., AND D. R. KING. 1961. Pecan diseases and insects and their control. Texas A & M College Bul. MP-313.

RUELE, G. D., AND R. B. LEDIN. 1960. Mango growing in Florida. Fla. Ext. Bul. 174.

RUSSELL, E. W. 1961. Soil conditions and plant growth. 9th ed. John Wiley & Sons, Inc., New York, N.Y.

SASSER, J. N. 1954. Identification and host-parasite relationships of certain root-knot nematodes. Univ. Md. Agr. Exp. Sta. Tech. Bul. A-77.

———, AND W. R. JENKINS (eds.). 1960. Nematology, fundamentals and recent advances with emphasis on plant parasitic and soil forms. Univ. of North Carolina Press, Chapel Hill, N.C.

———, AND E. J. CAIRNS (eds.). 1958. Plant nematology notes, from Workshops, 1954 and 1955. Southern Regional Nematode Project (S-19). 2nd ed.

SEYMOUR, A. B. 1929. Host index of the fungi of North America. Harvard University Press, Cambridge, Mass.

SHARVELLE, E. G. 1961. The nature and uses of modern fungicides. Burgess Publishing Co., Minneapolis, Minn.

SHERF, A. F. 1959. Vegetable diseases. Cornell Univ. Ext. Bul. 1034.

———, AND A. A. MUKA. 1964. Insects and diseases in the home vegetable garden. Cornell Univ. Misc. Bul. 59.

SHURTLEFF, M. C., D. P. TAYLOR, J. W. COURTER, AND H. B. PETTY, JR. 1964. Soil disinfestation, methods and materials. Univ. of Illinois Ext. Circ. 893.

SILVERBORG, S. B., AND R. L. GILBERTSON. 1962. Tree diseases in New York State plantations, a field manual. State Univ. College of Forestry, Syracuse Univ. Bul. 44.

SMITH, K. M. 1957. A textbook of plant virus diseases. 2nd ed. Little, Brown and Co., Boston, Mass.

———. 1960. Plant viruses. 3rd ed. John Wiley & Sons Inc., New York, N.Y.

SMITH, R. E. 1940. Diseases of flowers and other ornamentals. Calif. Agr. Ext. Circ. 118.

———. 1940. Diseases of truck crops. Calif. Agr. Ext. Circ. 119.

———. 1941. Diseases of fruits and nuts. Calif. Agr. Ext. Circ. 120.

SPRAGUE, H. B., editor-in-chief. 1964. Hunger signs in crops. 3rd ed. David McKay Co., Inc., New York, N.Y.

STAKMAN, E. C., AND J. G. HARRAR. 1957. Principles of plant pathology. The Ronald Press Co., New York, N.Y.

STAPP, C. 1961. Bacterial plant pathogens (translated by A. Schoenfeld). Oxford University Press, New York, N.Y.

STEINER, G. 1956. Plant nematodes the grower should know. Fla. Dept. Agr. Bul. 131.

STERN, A. C. (ed.). 1962. Air pollution. Vol. I. Academic Press, New York, N.Y.

STEVENS, F. L., AND J. G. HALL. 1933. Diseases of economic plants. The Macmillan Co., New York, N.Y.

STEVENS, N. E., AND R. B. STEVENS. 1952. Disease in plants. Chronica Botanica Co., Waltham, Mass.

STREETS, R. B. 1937. Phymatotrichum (cotton or Texas) root rot in Arizona. Univ. of Arizona Tech. Bul. 71.

SUIT, R. F. 1949. Parasitic diseases of citrus in Florida. Fla. Agr. Exp. Stas. Bul. 463.

TAYLOR, N. (ed.). 1961. Taylor's encyclopedia of gardening. Houghton Mifflin Co., Boston, Mass.

THORNE, G. 1961. Principles of nematology. McGraw-Hill Book Co., Inc., New York, N.Y.

TORGENSEON, D. C., R. A. YOUNG, AND J. A. MILBRATH. 1954. Phytophthora root rot diseases of Lawson cypress and other ornamentals. Oreg. Agr. Exp. Sta. Bul. 537.

VANDEMARK, J. S., M. C. SHURTLEFF, AND W. H. LUCKMANN. 1966. Illinois vegetable garden guide. Univ. of Illinois Ext. Circ. 882

WALKER, J. C. 1952. Diseases of vegetable crops. McGraw-Hill Book Co., Inc., New York, N.Y.

———. 1957. Plant pathology. 2nd ed. McGraw-Hill Book Co., Inc., New York, N.Y.

WALLACE, H. R. 1963. The biology of plant parasitic nematodes. Edward Arnold Publishers Ltd., London, England.

WESTCOTT, CYNTHIA. 1960. Plant disease handbook. 2nd ed. D. Van Nostrand Co., Inc., Princeton, N. J.

———. 1960. Anyone can grow roses. 3rd ed. D. Van Nostrand Co., Inc., Princeton, N.J.

———. 1961. Are you your garden's worst pest? Doubleday and Co., Inc., New York, N.Y.

———. 1964. The gardener's bug book. 3rd ed. Doubleday and Co., Inc., New York, N.Y.

———, AND P. K. NELSON (eds.). 1955. Handbook of plant pests and diseases. Brooklyn Botanic Garden, Brooklyn, N.Y.

WILHELM, S. 1961. Diseases of strawberry, a guide for the commercial grower. Univ. of Calif. Circ. 494.

WILSON, C. L., AND W. E. LOOMIS. 1967. Botany. 4th ed. Holt, Rinehart and Winston Inc., New York, N.Y.

WORMALD, H. 1955. Diseases of fruit and hops. Crosby Lockwood and Son, London, England.

ZYCH, C. C., AND D. POWELL. 1960. Strawberry growing in Illinois. Univ. of Illinois Ext. Circ. 819.

B. USDA PUBLICATIONS*

1. Flowers, House Plants, Lawns, and Shrubs

	*Order No.***
Controlling insects on flowers	AB 237
Roses for the home	G 25
Better lawns: establishment, maintenance, renovation, lawn problems, grasses	G 51
Lawn insects: how to control them	G 53
Lawn diseases: how to control them	G 61
Growing chrysanthemums in the home garden	G 65
Growing iris in the home garden	G 66
Insects and related pests of house plants: how to control them	G 67
Growing azaleas and rhododendrons	G 71
Selecting and growing house plants	G 82
Growing camellias	G 86
Growing flowering annuals	G 91
Growing flowering perennials	G 114
Gardenia culture—indoors and out	L 199
Barberry eradication in stem rust control	L 416
Spring-flowering bulbs	L 439
Narcissus bulb fly: how to prevent its damage in home gardens	L 444
Indoor garden for decorative plants	Title

2. Fruits

Virus diseases and other disorders with virus-like symptoms of stone fruits in North America	AH 10
Fig growing in the south	AH 196
Strawberry culture: South Atlantic and Gulf Coast regions	F 1026
Strawberry culture: eastern United States	F 1028
Strawberry varieties in the United States	F 1043
Control of grape diseases and insects in the eastern United States	F 1893
Blueberry growing	F 1951
Growing American bunch grapes	F 2123
Strawberry diseases	F 2140
Muscadine grapes—a fruit for the south	F 2157
Growing blackberries	F 2160
Growing raspberries	F 2165

---

* All USDA publications mentioned in this book are available by writing to:
Office of Information
U.S. Department of Agriculture
Washington, D.C. 20250.

Schools, clubs, and other organizations may purchase bulk quantities of publications from the Superintendent of Documents, Government Printing Office, Washington, D.C., 20402. When ordering publications be sure to print your name, address, and zip code clearly. Please order using *only* the letter and number following the title of each publication you desire. For example, the leaflet, *Apple Bitter Rot* (see page 138), should be ordered by L 406. There is a small handling charge (usually 5 to 30 cents) for a number of USDA publications. For a complete listing of U.S. government bulletins and other publications, call at your county extension office.

** Letter symbols: AB = Agriculture Information Bulletin; AH = Agriculture Handbook; F = Farmers' Bulletin; FPL = Forest Pest Leaflet; G = Home and Garden Bulletin; L = Leaflet; M = Miscellaneous Publication; and TB = Technical Bulletin.

	*Order No.***
Growing cherries east of the Rocky Mountains	F 2185
Growing peaches east of the Rocky Mountains	F 2205
Controlling diseases of raspberries and blackberries	F 2208
Growing figs in the south for home use	G 87
Quince growing	L 158
Blight of pears, apples, and quinces	L 187
Home fruit garden in the Southeastern and Central Southern states	L 219
Home fruit garden in the Central Southwestern states	L 221
Home fruit garden in the Pacific Coast states and Arizona	L 224
Home fruit garden in the Northeastern and North Central states	L 227
Apple bitter rot	L 406
Dwarf fruit trees: selection and care	L 407
Reducing virus and nematode damage to strawberry plants	L 414
Cherry leaf-spot and its control	L 489
Bridge grafting and inarching damaged fruit trees	L 508
Controlling phony disease of peaches	L 515
Burrowing nematode—a pest of citrus	PA 566

3. Trees

The Dutch elm disease and its control	AB 193
Protecting trees against damage from construction work	AB 285
Diseases of shade and ornamental maples	AH 211
Controlling insects and diseases of the pecan	AH 240
Culture, diseases, and pests of the box tree	F 1855
Reducing damage to trees from construction work	F 1957
Common diseases of important shade trees	F 1987
Chestnut blight and resistant chestnuts	F 2068
Annosus root rot in eastern pines	FPL 76
Armillaria root rot	FPL 78
Eastern gall rust	FPL 80
Nectria canker of hardwoods	FPL 84
Walnut anthracnose	FPL 85
White trunk rot of hardwoods	FPL 88
Chestnut blight	FPL 94
Maple diseases and their control—a guide for homeowners	G 81
Pruning shade trees and repairing their injuries	G 83
Growing the flowering dogwood	G 88
Protecting shade trees during home construction	G 104
Elm bark beetles	L 185
Growing black walnuts for home use	L 525

4. Vegetables

Commercial watermelon growing	AB 259
Pepper production	AB 276
Diseases of cabbage and related plants	AH 144
Corn diseases in the United States	AH 199
Tomato diseases and their control	AH 203
Onion diseases and their control	AH 208
Muskmelon culture	AH 216
Lettuce and its production	AH 221
Bean diseases—how to control them	AH 225
Pea diseases	AH 228

	Order No.**
Sweetpotato diseases	F 1059
Storage of sweetpotatoes	F 1442
Asparagus culture	F 1646
Potato diseases and their control	F 1881
Home storage of vegetables and fruits	F 1939
Pepper production, disease and insect control	F 2051
Growing peanuts	F 2063
Growing pumpkins and squashes	F 2086
Bacterial wilt and Stewart's leaf blight of corn	F 2092
Potato growing in the south	F 2098
Controlling potato insects	F 2168
Controlling tomato diseases	F 2200 (Rev.)
Suburban and farm vegetable gardens	G 9
Insects and diseases of vegetables in the home garden	G 46
Storing perishable foods in the home	G 78
Production of turnips and rutabagas	L 142
Watch out for witchweed	L 331
Growing eggplant	L 351
Rhubarb production—outdoors and in	L 354
Growing table beets	L 360
Golden nematode of potatoes and tomatoes: how to prevent its spread	L 361
The sugar beet nematode and its control	L 486
Zinc deficiency of field and vegetable crops in the west	L 495
Muskmelons for the garden	L 509
Watermelons for the garden	L 528
A monographic study of bean diseases and methods for their control	TB 868

5. Miscellaneous

Boron injury to plants	AB 211
Commercial storage of fruits, vegetables, florist, and nursery stocks	AH 66
Index of plant diseases in the United States	AH 165
Chemical control of plant-parasitic nematodes	AH 286
Losses in agriculture	AH 291
Hotbeds and coldframes	F 1743
Controlling nematodes in the home garden	F 2048
Using phenoxy herbicides effectively	F 2183
Hand sprayers and dusters	G 63
Selecting fertilizers for lawns and gardens	G 89
Iron deficiency in plants: how to control it in yards and gardens	G 102
How much fertilizer shall I use?	L 307
Electric heating of hotbeds	L 445
Plant hardiness zone map	M 814
Hotbed and propagating frame, plan no. 5971	M 986
Protecting farms and gardens through plant quarantines	PA 541
Safe use of pesticides . . . in the home, in the garden	PA 589
Farmers' checklist for pesticide safety	PA 622
Trees, the yearbook of agriculture 1949	
Plant Diseases, the yearbook of agriculture 1953	

* * *

The few selected references are just a minute sampling of the vast amount of material published on diseases of flowers, fruits, house plants, lawns and turf, nut trees, shrubs, trees, vegetables, and vines.

A great deal of new information on plant diseases and their control, including some experimental or controversial work, may be found in the bulletins, magazines, and yearbooks of plant societies, e.g., African Violet Society of America, Inc., American Amaryllis Society, American Begonia Society, American Camellia Society, American Daffodil Society, American Dahlia Society, American Delphinium Society, American Fern Society, American Gloxinia Society, American Hemerocallis Society, American Iris Society, American Orchid Society, American Peony Society, American Poinsettia Society, American Rhododendron Society, American Rock Garden Society, American Rose Society, American Association of Botanical Gardens and Arboretums, and many others. A list of horticultural organizations—national and international; botanical gardens, arboretums, public and private parks; federal and state agricultural agencies; horticultural periodicals, reference books, and libraries; specialized nurseries and suppliers; and flower shows and year-round horticultural events—is given in *The Gardener's Directory,* compiled by J. W. Stephenson, and published by Hanover House, Garden City, New York.

Index

Illustrations are indicated by page numbers in brackets, thus: [357].

Aarons-rod, 377; *see* Thermopsis
Aaronsbeard, 446; *see* St.-Johns-wort
Abelia, 457
  chlorosis, 24–29, 344, 459
  leaf spot, 48, 458
  powdery mildew, 56, 458
  root-knot, 90, 459
  root rot, 87, 458
Abelmosk mallow, 297; *see* Hibiscus (arborescent forms)
Abies, 402; *see* Fir
Abronia, 274; *see* Sand-verbena
Abrus, 377; *see also* Pea
  root-knot, 90, 381
Abutilon, 297
  infectious chlorosis (variegation), 70, 299
  leaf blight, 51, 298
  leaf spot, 48, 298
  light and temperature requirements, 40
  mosaic, 70, 299
  root-knot, 90, 299
  root nematode, 299
  root rot, 87, 298
  rust, 58, 298
  stem rot, 76, 298
  Verticillium wilt, 68, 299
  virus-infected, 299
Abyssinian wildflower, 281; *see also* Gladiolus
  dry rot (*Stromatinia*), 90, 282
Acacia, 300
  algal leaf spot, 302, 341
  canker, dieback, 77, 300
  chlorosis, 24–29, 302
  iron deficiency, 24, 26, 27, 302
  leaf spot, 48, 301
  mistletoe, 93, 302
  powdery mildew, 56, 301
  root-knot, 90, 301, 392
  root nematode, 301
  root rot, 87, 137, 301
  rust, 58, 301
  witches'-broom, rust, 58, 301
  wood rot, 79, 170, 301
  zinc deficiency, 26, 27, 302
Acalypha, 126
  downy mildew, 55, 126
  leaf spot, 48, 126
  oedema, 39, 126
  powdery mildew, 56, 126
  red leaf gall, 126
  root-knot, 90, 126
  root nematode, 126
  root rot, 87, 126, 278
Acanthopanax, 126
  leaf spot, 48, 127
  root rot, 87, 127
  rust, 58, 127
  Verticillium wilt, 68, 127
Acer, 343; *see* Maple, Boxelder
Acervulus, 553
Achillea, 219
  crown gall, 83, 226
  petal blight, 86, 221, 224
  powdery mildew, 56, 221
  root-knot, 90, 225
  root rot, 87, 221
  rust, 58, 223
  stem rot, 76, 221
Achimenes, 127; *see* Cupid's-bower
Achlys, 153
  leaf spot, 48, 153
Acid soil, 24, 553
Acidanthera, 281; *see also* Gladiolus
  dry rot (*Stromatinia*), 90, 282
Acidifying soil, 24, 415, 435, 443
Acme Quality Paints, Incorporated, 120
  Bordeaux Mixture, 105, 554
  Fruit Tree Spray, 523
  Garden Fungicide, 553
Aconite, 254
  bacterial leaf spot, black blotch, 50, 255
  chlorosis, 24–29, 257, 443
  downy mildew, 55, 257
  leaf and stem smut, 62, 257
  mosaic, 70, 255
  powdery mildew, 56, 256
  root-knot, 90, 256
  root nematode, 256
  root rot, 87, 254
  rust, 58, 256
  soil drench, 254
  stem or crown rot, 76, 254, 255
  Verticillium wilt, 68, 256
Aconitum, 254; *see also* Aconite
  mosaic, 70, 255
  rust, 58, 256
  smut, 62, 257
  Verticillium wilt, 68, 256
Acrocomia, 371; *see* Palms
Actaea, 131; *see* Baneberry
Acti-dione
  eradicant and protective fungicide, 140, 146, 313, 314, 435, 441
  formulations, 106
  injury, 45
  powdery mildew control, 58, 106, 168, 171, 173, 221, 236, 256, 304, 321, 347, 359, 400, 440, 458
  properties, 100, 106
  uses, 106
  wound treatment, 384
Acti-dione-Captan, 106, 440
  -Ferrated, 106, 325
  -PM, 106
  -RZ, 106, 321, 435
  -S, 106
  -Thiram, 106, 199, 320, 321, 437
Actinomeris, 219; *see* Yellow ironweed
Actinomycetes, 553
Actispray, 106
Active ingredient, 553
Adam-and-Eve, 267; *see* Erythronium
Adams-needle, 512; *see* Yucca
Adderstongue, 267; *see* Erythronium
Adiantum, 269; *see* Ferns
Adoxa, rust, 152
Aerate soil, 553
Aerides, 366, *see* Orchids
Aesculus, 302; *see* Horsechestnut, Buckeye
Aethionema, 186; *see* Candytuft
Aflatoxin, 394
African daisy, 220; *see* Arctotis, Gazania
African forget-me-not, 348; *see* Anchusa
African-lily, 494; *see also* Tulip
  mosaic, 70, 496
African-violet, 127
  bacterial leaf blight, 50, 129
  Botrytis blight, gray-mold blight, 52, 86, 127
  bud drop, 39, 129
  chlorosis, leaf scorch, 24–29, 128
  crown and stem rot, 76, [77], 127
  flower blight, 86, 127
  leaf (foliar) nematode, 75, 129
  light and temperature requirements, 39, 40, 41
  low air humidity injury, 39
  mosaic, 70, 129
  petiole rot, 128
  plant soak, 129, 530, 534
  powdery mildew, 46, [128]
  ringspot, [128]
  root-knot, 90, 127, 128
  root nematode, 127
  root rot, 87, 127
  soil drench, 127
  soil mix, 22
  soluble salt injury, 31, 128
  stunt disease, 73, 129
  watering, excess, injury from, 38
African yellow-wood, 511; *see also* Yew
  root nematode, 511–12
  root rot, 87, 511–12
Agapanthus, 496; *see also* Tulip
  mosaic, 70, 496
Agarita, 153; *see* Mahonia, Oregon-grape
Agathea, light and temperature requirements, 40
Agave, 216; *see* Centuryplant
Ageratum, 219
  crown gall, 83, 226
  powdery mildew, 56, 221
  root rot, 87, 221
  rust, 58, 223
  southern blight, 76, 221
  stem rot, 76, 221
Aglaonema, 195; *see* Chinese evergreen
Agribor, 27
Agricultural drain tile, 36
Agricultural experiment stations
  help by, 97, 101
  listing off, 6–7
Agrimony (*Agrimonia*), 439
  downy mildew, 55, 444

Agrimony (*continued*)
leaf spot, 48, 443
mosaic, 70, 443
powdery mildew, 56, 440
root rot, 87, 443
rust, 58, 441
Agrimycin, 105, 413, 468
-17, 105, 522
-100, 83, 105, 197, 282, 392, 522
-500, 105, 410, 484
Agri-strep, 83, 105, 522
Agropyron, 319; *see also* Wheatgrass
rust, 183, 321
Agrostemma, 203; *see* Corncockle
Agrostis, 142; *see also* Bentgrass, Redtop
rust, 152, 321
Agway, Incorporated, 102, 103, 120, 518, 519
Captan 5D and 7.5D, 102, 518
Maneb 4.5D, 103, 519
Ailanthus, 493; *see* Tree-of-Heaven
Air drainage, 553
Air humidity, 39–40
control, 40
Air layering, in transmitting viruses, 13
Air-plant, light and temperature requirements, 40
Air pollution, 43–44
control, 44
damage to plants, 43
Atlantic seaboard, 43
California, 43
U.S. losses, 43
injury, extent of, 43
confused with, 43
governed by, 43
pollutants, 43–44
ethylene, 43, 44
fluorides, 43
oxidant, 45
ozone, 43, 44
peroxyacetyl nitrate (PAN), 44
"smog," 43, 44
sulfur dioxide, [43]
Air potato, 510; *see* Yam
Airplant, 451; *see* Kalanchoë
Ajuga, 129
crown rot, 76, 130
root-knot, 90, 130
soil drench, 130
southern blight, 76, 130
Albizzia, 300; *see* "Mimosa" tree
Alcohol
crown gall treatment, 86
as disinfectant, 35, 78, 81, 83, 98–99, 134, 171, 271, 276, 278, 289, 301, 332, 343, 360, 366, 368, 370, 386, 403, 409, 428, 440, 500
seed treatment, 81, 379, 528
Alder, 170
canker, dieback, 77, 171, 172
catkin gall, 171
gray-mold (leaves), 52
leaf blister or curl, 62, 171
leaf spot, 48, 171
mistletoe, 93, 172
powdery mildew, 56, 171
root-knot, 90
root rot, 87, 137, 171–72
rust, 58, 171, 318
sooty mold, 64, 172
witches'-broom, 62, 171
wood rot, heart rot, butt rot, 79, 171
Aldrin, as soil insecticide, 23, 82, 98, 178, 282
Alexanders, 132; *see* Angelica
Alfilaria, 236; *see* Heronsbill
Algal leaf spot, algae, green scum, 201–2, 297, 341, 342
Alkali injury, 24
Alkaline soil, 24, 553
Alkanet, 348; *see* Anchusa
Allied Chemical Corporation, General Chemical Division, 120
Allionia, 274; *see* Trailing four-o'clock
Allium, 361; *see also* Onion, Chives, Garlic, Leek, Shallot
bulb rot, 90, 361, 362
leaf blight, 51, 361, 362, 363
leaf spot, 48, 365
mosaic, 70, 365
rust, 58, 365
yellows, 73, 365
Allspice, 198; *see* Calycanthus
Allyl alcohol and EDB, 546
Almond, 382; *see also* Flowering almond
anthracnose, 51, 392
asteroid spot, 390
bacterial leaf spot, 50, 386
bacterial shoot blight, canker, 50, 79, 386
blossom blight, 86, 382
brown rot, 86, 382, 525
chlorosis, 24–29, 393
Coryneum blight, 77, 390, [391]
crown gall, 83, 391
crown rot or canker, 76, 383, 392
dieback, 77, 383, 385
fire blight, 79, 393
fruit spot or rot, 86, 382, 384, 385, 392
gray-mold, 52, 86, 393
leaf blight, 51, 392
leaf curl, 62, 384
leaf spot, 48, 392
little leaf, 26, 27, 393
mosaic, 70, 389
peach yellows, 73, 388
phony peach, 389
powdery mildew, 56, 392
ringspot, 72, 388
root-knot, 90, 392
root nematode, 392
root rot, 87, 392
rosette, 73, 387
rust, 58, 392
scab, 65, 385
shot-hole, 52, 386, 390, 392
spray schedule, 524–25
thread blight, 393, 505
Verticillium wilt, 68, 391
wood rot, 79, 170, 384
X-disease, 387
Alnus, 170; *see* Alder
Aloe, 130
light and temperature requirements, 40
plant soak, 130, 530, 534
root rot, 87, 130, 278
Alpine currant, 245; *see* Flowering currant
Alternanthera, 230
Fusarium wilt, 66, 230
leaf blight, spot, 51, 230
root and crown rot, 87, 230
root-knot, 90, 230
root nematode, 230
soil drench, 230
white-rust, 62, 230
Alternaria blight, 204, [217], [239], [363], 480
Alternate host, 58, 553
Althaea, 297; *see* Hollyhock
Aluminum foil, as tree wrap, [42]
Aluminum plant, 401; *see* Pilea
light and temperature requirements, 40
Aluminum sulfate, 24, 553
Aluminum toxicity, 24
Alumroot, 304; *see* Heuchera
Alyssum, 186; *see also* Sweet alyssum
aster yellows, 73, 193
chlorosis, 24–29, 194
clubroot, 89, 189
curly-top, 75, 192
damping-off, 76, 188
downy mildew, 55, 189
iron deficiency, 24, 26, 27, 194
stem rot, 76, 188, 191
white-rust, 62, 191
Amaranth, 229
aster yellows, 73, 230
blossom blight, 86, 230
curly-top, 75, 162, 230
damping-off (seed rot, blight), 76, 162, 230
leaf blight, spot, 48, 51, 230
leaf roll, 230, 418
mosaic, 70
ringspot, 72, 230
root and crown rot, 76, 87, 230
root-knot, 90, 230
root nematode, 230
soil drench, 230
white-rust, 62, 230
Amaranthus, 230; *see* Amaranth, Love-lies-bleeding
Amaryllis, 249; *see* Lycoris for Hardy amaryllis
bulb rot, 90, 250
bulb soak, 250, 528, 534
chlorosis, 24–29
gray-mold blight, 52, 86, 251
iron deficiency, 24, 26, 27
leaf spot, 48, 253
light and temperature requirements, 40
mosaic, 70, [71], 250–51
root nematode, 253
root rot, 87, 250
southern blight, 76, 250
spotted wilt, ringspot, 72, 253
Stagonospora leaf scorch, red blotch, red fire disease, 51, 251
Amazon-lily, 249
gray-mold blight, 52, 251
mosaic, 70, 250–51
ringspot, 72, 253
root rot, 87, 250
Stagonospora leaf scorch or red spot, 48, 250
Amchem Products, Incorporated, 120
Amelanchier, 133; *see also* Apple
black mildew, witches'-broom, 64, 141
blossom blight, 79, 86, 133–34
canker, dieback, 77, 138
fire blight, 79, 133–34
fruit rot, 86, 138
leaf blight, 51, 140
leaf blister, witches'-broom, 62, 143
leaf spot, 48, 140
powdery mildew, 56, 135–36
root rot, 87, 137–38
rust, 58, 135
Verticillium wilt, 68, 143
wood rot, 79, 139
American bladdernut, 130
leaf spot, 48, 130
sooty blotch, 64, 130
twig blight, dieback, 77, 130
American cowslip, 423; *see* Shooting-star
American Cyanamid Company, 102, 120, 518
American Gas Association, 45
American highbush cranberry, 499; *see* Viburnum
American mistletoe, [93], 94
American Potash & Chemical Corporation, 120
American Potash Institute, 120
American spikenard, 126
leaf spot, 48, 127
rust, 58, 127
Verticillium wilt, 68, 127
American yellowwood, 300; *see* Yellow-wood
Amitrol injury, 45
Ammate, 356, 387
injury, 45
Ammonium molybdate, 28, 193
nitrate, 415
sulfate, 29, 415, 443
Amobam, 106, 553
Amorpha, 269; *see also* Indigobush
powdery mildew, 56, 269
root nematode, 269
root rot, 87, 269
rust, 58, 269
twig canker, 77, 269
Amorphophallis, light and temperature requirements, 40
Ampelopsis, 286; *see also* Grape
canker, dieback, 77, 290
downy mildew, 55, 288
leaf spot, 48, 287, 291
powdery mildew, 56, 289

root nematode, 291
root rot, 87, 137, 289
rust, 58, 291
thread blight, 291, 505
Amphyl solution, as disinfectant, 81, 134, 360, 370
Amsonia, 500
leaf spot, 48, 500
mosaic, 70, 501
rust, 58, 405, 501
Anacahuita, 231; *see* Cordia
Anagallis, 423
aster yellows, 73, 424
leaf spot, 48, 423
root-knot, 90, 423
Anaphalis, 219; *see* Pearleverlasting
Anchusa, 348
aster yellows, 73, 349
curly-top, 75, 349
damping-off, 76, 349
mosaic, 70, 349
powdery mildew, 56, 349
rust, 58, 349
stem rot, 76, 254, 349
Ancistrocactus, 194; *see* Cacti, Echinocactus
Androc Chemical Company, 120
Andromeda, 131
chlorosis, 24–29
dieback, 77, 131
iron deficiency, 24, 26, 27
leaf spot, 48, 131
root nematode, 131
root rot, 87, 131, 434
tar spot, 48, 131
twig blight, 77, 131
Andropogon (wild), rust, 302, 375, 381
Androsace, 423; *see* Rockjasmine
Anemone, 131
aster yellows, 73, 132
blossom blight, 86, 131
Botrytis blight, collar rot, 52, 76, 131
crown rot, 76, 132
downy mildew, 55, 131
flower breaking, 70, 132
leaf gall, spot disease, 48, 132
leaf spot, 48, 131
leaf and stem nematode, 75, 132
leaf and stem smut, 62, [131]
mosaic, 70, 132
powdery mildew, 56, 132
rhizome rot, 76, 132
root-knot, 90, 132
root rot, 87, 132
rust, 58, 131, 392
southern blight, 76, 132
spotted wilt, 72, 132
white smut, 65, 132
Anemonella, 131; *see* Rue-anemone
Anethum, 213; *see* Dill
Angelica, 132
leaf spot, 48, 132
root rot, 87, 132
rust, 58, 132
Angels-trumpet, 479: *see* Datura
Angraecum, 366; *see* Orchids
Angular leaf spot, [240]
Animal
injuries, 42–43, 45
spread of mistletoes, 93
Anise, 213
aster yellows, 73, 214
leaf spot, 48, 213
root rot, 87, 215
rust, 58, 215
seed rot, damping-off, 76, 214
seed treatment, 213, 532
stem rot, 76, 214
Anise-root, 213; *see also* Sweet-jarvil
rust, 58, 215
seed rot, damping-off, 76, 214
seed treatment, 213
Anisetree, 341, *see also* Magnolia
algal leaf spot, 341
black mildew, 64, 341
sooty mold, 64, 341
Annual aster, 220; *see* China-aster
Annual blanket-flower, 220; *see* Gaillardia
Anoda, 297
powdery mildew, 56, 298
rust, 58, 298
Antennaria, 219; *see also* Everlasting
white-rust, 62, 226
Anthemis, 219; *see* Camomile
Anther, 553
Anther smut, 62, 206
Anthony waterer, 461; *see* Spirea
Anthracnose, 48, [51], 52, [78], [157], [238], [246], 264, 343, [357], [427], [476], 553
Anthriscus, 213; *see* Celery, Salad chervil
Anthurium, 195
anthracnose, 51, 197
leaf spot, 48, 197
light and temperature requirements, 40
root nematode, 197
Antibiotic, 553
Antibiotics, to control plant diseases, 105–6
Anticarie, 80, 362
Antidesiccants, 21, 38, 41, 42, 182, 296, 511
Antirrhinum, 454; *see* Snapdragon
Antrol Garden Products, Boyle-Midway, 120
Ants, insect control, 321, 463
Aphelandra, light and temperature requirements, 40
Aphids, 13, [242]
as bacteria carriers, 407
control, 72, 73, 74, 99, 161, 242, 277, 329, 360, 365, 373, 406, 430, 437, 438, 463, 474, 485, 496, 500, 501
secretion of "honeydew," 64
as virus carriers, 13, 70, 73, 74, 98, 192, 329, 430, 474, 485, 496, 500, 501
Apium, 213; *see* Celeriac, Celery
Appendix, 516–52
Apple, 133
Alternaria cork rot, 86, 138
Apple Bitter Rot, 138
anthracnose, 51, 86, 140
bacterial blast, 50, 79, 86, 133–34
baldwin spot, 140
bark canker, 76, 138–39, 140
bitter pit or stippen, 140
bitter rot, [87], 138
black end, 140
black pox, 138
black rot, 48, 77, 86, 136
Blight of Pears, Apples, and Quinces, 579
blossom blight, 86, 133–34
blotch, 51, 77, 86, [137]
borer control, 139, 525
boron deficiency, "drought spot" or dieback, 27, 141
brown core or heart, 140
burr knot, 137
butt rot, 79, 139
calcium deficiency, 28, 141–42
canker (twig, branch, trunk), 77, 138–39
chlorosis, 24–29, 143
collar rot, 76, 139
copper deficiency, 28, 141–42
crown gall, 83, [84], 136–37
dapple apple, 143
dieback, 77, 136, 138–39, 141, 142
dry heat virus treatment, 142
felt fungus, 77, 138–39, 293
fire blight, 77, 79, [80], 86, 133–34, 525
flat limb, 142
fly speck, 137
freezing injury, 42, 140
frost crack, 41
fruit breakdown, 140
fruit rot, 86, 138
fruit spot, 86, 138
gray-mold rot (*Botrytis*), 52, 86, 138
green crinkle or false sting, 143
hairy root, 83, 136–37
heart rot, 79, 139
Helminthosporium leaf spot and bark canker, 48, 140
internal breakdown, 140
internal cork, 27, 141
iron deficiency, 26, 27, 143
jonathan spot, 140
leaf blight, 51, 140
leaf scorch, 41, 141–42
leaf spot, 48, 140
limb blight, 77, 138–39
little leaf, 26, 27, 141
magnesium deficiency, 28, 141–42
manganese deficiency, 26, 27, 143
mistletoe, [93], 94, 141
mosaic, 70, 142
plant-food utilization, 26
potassium deficiency, 26, 141–42
powdery mildew, 56, 135, [136]
pruning, 32
resistant varieties, 134, 135, 136, 137, 138, 139, 140
rodent protection, 42
root nematode, 141
root rot, 87, 137–38
rubbery wood, 142
russet ring, 142
rust, 58, [59], 135, 405
scab, 48, 65, [66], 86, 106, 134–35
scald, 140
scar skin, 143
soggy breakdown or soft scald, 140
sooty mold, blotch, [64], 137
spray schedule, 524–25
amount of spray required, 526
stayman spot, 140
stem- or wood-pitting, 142
storage rot, 86, 138
sulfur dioxide injury, 43
sulfur-sensitive, 525
sunscald, 41, 139–40
thread blight, 143, 505
twig canker, 77, 136, [138], 139
2, 4-D injury, 45, 143, 288
water core, 140
wetwood, 143, 264
winter injury, 41, [139]
protection, 42
wood rot, silver leaf, 79, 139, 170
woolly knot, 137
zinc deficiency, little leaf, rosette, 26, 27, 141
Apple-of-Peru, 479
leaf spot, 48, 480
mosaic, 70, 484
ringspot, 72, 486
root rot, 87, 490
spotted wilt, 72, 486
streak, 484
Application
equivalents, 517
soil fumigants, 538–47, [548], [549]
sprays to row crops, 550
Apricot, 382
asteroid spot, 390
bacterial canker, gummosis, 79, 386
bacterial leaf spot, 50, 386
black knot, 81, 384
blossom blight, 86, 382
brown rot, 86, 382
canker, 77, 383, 386, 390, 392
chlorosis, 24–29, 393
Coryneum blight, 51, 77, 390
crown gall, 38, 391
dieback, 77, 383
felt fungus, 293, 393
fire blight, 79, 393
fluorides injury, 43–44
fruit spot, rot, 86, 382, 384, 385, 392
leaf curl, 62, 384
leaf spot, 48, 392
little leaf, 26, 27, 393
little-peach, 387
mosaic, 70, 389
oxidant or PAN injury, 44
peach yellows, 73, 388
phony peach, 389
powdery mildew, 56, 385
ring pox, 72, 390
ringspot, 72, 388

Apricot (*continued*)
root-knot, 90, 392
root nematode, 392
root rot, 87, 392
rosette, 73, 387
rust, 58, 392
scab, 65, 385
shot-hole, 52, 386, 390, 392
silver leaf, 79, 170, 384
spray schedule, 524–25
twig blight, 77, 382
Verticillium wilt, 68, 391
wood rot, 79, 170, 384
X-disease, 387
Aquilegia, 254; *see* Columbine
Arabian-tea, 173
leaf tip blight, 51, 173
Arabis, 186; *see also* Rockcress
mosaic, 70, 192
Arachis, 393; *see* Peanut
Aralia, 126; *see also* Acanthopanax, American spikenard, Hercules-club, Sarsaparilla, Udo
canker, dieback, 77, 127
leaf spot, 48, 127
powdery mildew, 56, 127
root-knot, 90
root rot, 87, 127
rust, 58, 127
spot anthracnose or scab, 51, 127
stem rot, 76, 127
Verticillium wilt, 68, 127
wood rot, 79, 127, 170
Arasan
50-Red, 103, 519, 531
-75, 103, 519, 531
SF-M, 531
SF-X, 531
Thiram Seed Protectant, 103, 519, 531
Araucaria, 143
branch blight, 77, 143
crown gall, 83, 143
dieback, 77, 143
leaf spot, 48, 143
root rot, 87, 137, 143
seedling blight, damping-off, 76, 143 405
Arboretems, help by, 5
Arborist, help by, 9, 33, 34, 35, 45, 46, 83–86, 138, 263, 266, 345
fertilizing trees, 30
Arborvitae, 312; *see also* Hiba arborvitae
brown felt blight, 315, 407
canker, dieback, 77, 313, 314
chlorosis, 24–29, 315
Coryneum or Berkman's blight, 51, 76, 314
damping-off, seedling blight, 76, 314, 405
gray-mold blight, 52, 314
leaf blight, needle cast, 51, 314
leaf-browning, shedding, 314
needle blight, spot, 51, 313
nursery blight, 77, 314
pruning, 33
root nematode, 315
root rot, 87, 314
rust, 58, 313, 405
snow blight, 315, 407
sooty mold, 64, 315
twig blight, dieback, 77, 313
winter injury, browning, 41, 314
wood, trunk, or butt rot, 79, 170, 314
Arbutus, 175
chlorosis, 24–29, 178
crown gall, 83, 176
iron deficiency, 24, 26, 27, 178
leaf blight, blotch, 51, 178
leaf spot, 48, 178
red leaf gall, spot, 62, 176
root rot, 87, 178
rust, 58, 177
spot anthracnose, 51, 178
tar spot, 48, 178
trunk canker, 77, 178, 259
wood rot, 79, 170, 178
Archontophoenix, 370; *see* Palms
Arctostaphylos, 175; *see* Bearberry, Manzanita
Arctotis, 220
leaf blotch, 51, 221
leaf spot, 48, 221
root-knot, 90, 225
root rot, 87, 221
Ardisia, 144
algal leaf spot, 144, 341
black mildew, 64, 341
light and temperature requirements, 40
Arecastrum, 371; *see* Palms
Arenaria, 203
anther smut, 62, 206
leaf spot, 48, 205
powdery mildew, 56, 128, 206
root rot, 87, 204
rust, 58, 204
Arenga, 371; *see* Palms
Argemone, 413; *see* Prickly-poppy
Argyreia, 351
root-knot, 90, 351
root rot, 87, 351
Ariocarpus, 194; *see* Cacti
Arisaema, 195; *see* Jack-in-the-pulpit
Aristida, rust, 166, 186, 354
Aristolochia, 144
gray-mold blight, 52, 144
leaf spot, 48, 144
root rot, 87, 144, 278
Armeria, 451
leaf spot, 48, 451
rust, 58, 451
Armillaria (or shoestring) root rot, 87, [88], 137–38, [358], 425, 511
resistant plants, 89, 138, 268, 271, 392, 398–99, 406, 413, 504
Armoracia, 186; *see* Horseradish
Armyworms, insect control, 463
Arnica, 220
leaf spot, 48, 221
powdery mildew, 56, 221
rust, 58, 223
white smut, 65, 226
Aronia, 133; *see* Chokeberry
Arrowhead-plant, light and temperature requirements, 40
Arrowroot, 347
leaf spot, 48, 347
rust, 58, 347
Arrowwood, 499; *see also* Viburnum
bacterial leaf spot, 50, 79, 500
canker, dieback, 77, 500
downy mildew, 55, 500
fungus leaf spot, 48, 500
powdery mildew, 56, 499
root-knot, 90, 391, 500
root rot, 87, 137, 500
rust, 58, 500
Arsenical injury, 45, 387
Artemisia, 220
crown gall, 83, 226
downy mildew, 55, 225
leaf blight, 51, 221
leaf spot, 48, 221
powdery mildew, 56, 221
root-knot, 90, 225
root rot, 87, 221
rust, 58, 223
white-rust, 62, 226
Arctic daisy, 219; *see* Chrysanthemum
Artichoke; *see* Globe artichoke, Jerusalem-artichoke
Artillery-plant, 401
leaf spot, 48, 401
light and temperature requirements, 40
powdery mildew, 56, 401
root-knot, 90, 401
root rot, 87, 278, 401
Arum-lily, 195; *see* Calla lily
Aruncus, 439; *see* Goatsbeard
Arwell, Incorporated, 120
Asclepias, 185; *see* Butterflyweed, Milkweed
Ascochyta blight, 379, [380], [381]
Ascorbic acid, for air pollution control, 44
Ascyrum, 446
leaf spot, 48, 446
rust, 58, 446
Asexual reproduction, 553
Ash, 144; *see* Hoptree for Wafer ash
anthracnose, 51, 145
black mildew, sooty mold, 64, 145
borer control, 145
canker, (branch, trunk), 77, 145
crown gall, hairy root, 83, 136, 145
dieback, 77, 145, 146
felt fungus, 146, 293
flower gall of white ash, 145
leaf scorch, 41, 145
leaf spot, 48, 145
mistletoe, 93, 145
powdery mildew, 56, 145
ringspot of white ash, 72, 146
root-knot, 90, 145, 392
root rot, 87, 137, 145
rust, 58, 144, [145]
seedling blight, 77, 146, 405
twig blight, 77, 145
Verticillium wilt, 68, 145
wood rot, 79, 145, [147], 170
Asimina, 377; *see* Pawpaw
Asparagus, garden, 146
anthracnose, 51, 146
bacterial soft rot, 81, 146, [147]
branchlet (stem) blight, spot, 48, 76, 146
chlorosis, 24–29, 148
crown gall, 83, 146
damping-off, 76, 148
Fusarium wilt, yellows, 66, 146
rot, 146, [147]
gray-mold blight, 52, 146
leaf spot, 48, 146
Phytophthora rot, [147]
root and foot rot, 76, 87, 146, 148
root-knot, 90, 148
root nematode, 148
rust, 58, 146
seed treatment, 148, 532
spear rot, 146, [147]
stem canker, dieback, 77, 146
stem or crown rot, 76, 146
Verticillium wilt, 68, 148
Asparagus-bean, 155; *see also* Black-eyed Pea
bacterial spot, 50, 155
Botrytis blight, gray-mold blight, 52, 159–60
chlorosis, 24–29, 160
leaf spot, 48, 159–60
mosaic, 70, 155
pod spot, 86, 159–60
powdery mildew, 56, 158
root-knot, 90, 159
southern blight, 76, 156–57
stem rot, 76, 156
Verticillium wilt, 68, 160
Asparagus-fern, 146; *see also* Asparagus, garden
anthracnose, 51, 146
canker, dieback, 77, 146
crown gall, 83, 146
Fusarium wilt, 66, 146
leaf drop, 39, 148
leaf mold, 48, 146
leaf spot, blight, 48, 146
root or foot rot, 76, 146
root-knot, 90, 148
root nematode, 148
Aspasia, 366; *see* Orchids
Aspen, 411; *see also* Poplar
canker, dieback, 77, 411
catkin deformity, 412
ink spot, 48, 412
leaf blight, 51, 412
leaf scorch, 41, 344, 413
leaf spot, 48, 412
powdery mildew, 56, 412
rust, 58, 318, 405, 412
shoot blight, 51, 65, 412
twig blight, 77, 411
Verticillium wilt, 68, 343, 413
wood rot, 79, 170, 412
Aspergillus flavus, 394

Asphalt-varnish tree paint, 36
Asphyxiation, 37
Aspidistra, 148
  anthracnose, 51, 148
  chlorosis, 24–29, 148
  leaf blight, 51, 148
  leaf spot, 48, 148
  root rot, 87, 148
Asplenium, 269; *see* Ferns
Aster, perennial, 220; *see also* China-aster, Golden-aster, Stokes-aster
  bacterial wilt, 69, 223
  chlorosis, 24–29, 227
  crown gall, 83, 226
  damping-off, seed rot, 76, 221
  dodder, [94]
  downy mildew, 55, 225
  foliar or leaf nematode, 75, 226
  foot rot, 76, 221
  gray-mold blight, 52, [54], 86, 224
  leaf blight or spot, 48, 51, 221
  light, effect on flowering, 38
  mosaic, 70, 222
  petal blight, 86, 221, 224
  powdery mildew, 56, 221
  ringspot, 72, 222
  root-knot, 90, 225
  root nematode, 226–27
  root rot, 87, 221
  rust, 58, 223, 405
  seed treatment, 221
  soil drench, 222
  sooty mold, 64, 226
  spotted wilt, 72, 222
  stem canker or rot, 76, 77, 221, 225
  tar spot, 48, 221
  Verticillium wilt, 68, 223
  white smut, 65, 226
Aster daisy, 219; *see* Chrysanthemum
Aster yellows, 73, [74], 75, 98, [222], 501
  control by dry heat, 501
  weeds as virus reservoir, 98
Astilbe, 148
  Fusarium wilt, 66, 148
  powdery mildew, 56, 148
  root nematode, 148
Astrophytum, 194; *see* Cacti
Atamasco-lily, 250; *see also* Zephyranthes
  rust, 58, 253
Athel tree, 478; *see* Tamarisk
Athyrium, 269; *see* Ferns
Aubretia, 186; *see also* Cabbage, Rock-cress
  white-rust, 62, 191
Aucuba, 148
  anthracnose, 51, 148
  blossom blight, 86, 148
  frost or winter injury, 149
  gray-mold blight, 52, 148
  leaf spot, 48, 148
  light and temperature requirements, 40
  powdery mildew, 56
  Verticillium wilt, 68, 149
  wither tip, 76, 148
Auricula, 423; *see* Primrose
Australian brush-cherry, 268; *see* Eugenia, Eucalyptus
Australian pea, 377; *see* Hyacinth-bean
Australian-pine, 212; *see* Casuarina
Australian umbrella-tree, 450; *see* Schefflera
Autoecious rust, 58, 553
Automobile injury, 45
Autumn-crocus, 230; *see* Colchicum
Avens, 440
  aster yellows, 73, 444
  downy mildew, 55, 444
  fire blight, 79, 444
  leaf blotch, spot, 48, 51, 443
  leaf smut, 65, 444
  powdery mildew, 56, 440
  root-knot, 90, 443
  root rot, 87, 443
  rust, 58, 441
Avocado, 149
  anthracnose, black spot, 51, 149
  Avocado Root Rot, 149
  bacterial blast, 50, 79, 86, 151
  black mildew, 64, 151
  branch canker, 77, 149–50
  Cercospora spot, 48, 149
  chlorosis, 24–29, 151, 344
  collar rot, 76, 149, 170, 259
  crown gall, 83, 151, 391
  dieback, 77, 149–50
  downy mildew, 55, 151
  flower spot, 86, 149
  fruit spot or rot, 86, 149, 150–51
  fruit storage, 151
  leaf blight, blotch, 51, 149
  leaf spot, 48, 149
  mottle leaf, 151
  oedema, 39, 150
  powdery mildew, 56, 150
  resistant varieties, 149
  root-knot, 90, 151
  root nematode, 151
  root rot, 87, 149
  scab, 65, 149, [150]
  seed rot, 86, 151
  seed treatment, 149, 530, 537
  seedling blight, 76, 151, 405
  sunblotch, [150]
  trunk canker, 77, 149, 259
  twig canker, 77, 149–50
  Verticillium wilt, 68, 150
  wood rot, 79, 149, 170
  zinc deficiency, little leaf, rosette, 26, 27, 150
Axe injury, 45
Axonopus, 319; *see* Carpetgrass
Azalea, 433; *see also* Rhododendron
  acidify soil, 435
  anthracnose, 51, 433
  bud blast, 436
  chlorosis, yellow leaf, 24–29, 434
  crown gall, 83, 437
  crown rot or wilt, 76, 434, 437
  cutting rot, 76, 437
  cutting soak, 435
  damping-off, 76, 437
  felt fungus, 293, 437
  flower and leaf gall, 62, [436]
  flower spray, 435, 436
  gray-mold (*Botrytis*) blight, 52, 86, 436
  Growing Azaleas and Rhododendrons, 437
  insect control, 433, 435, 437
  iron deficiency, 24, 26, 27, 434
  leaf scorch, angular leaf spot, 51, 433
  leaf spot, 48, 433
  leaf and stem gall, "rose bloom," 62, 436
  light and temperature requirements, 38, 40
  Ovulinia flower spot, petal or limp blight, 86, [434]
  powdery mildew, 56, 437
  root-knot, 90, 434
  root rot, wilt, 87, 434
  rust, 58, 405, 437
  shoot blight, 51, 433
  soil mixture for, 22
  soil treatment, 435, 437
  sooty mold, 64, 437
  stem rot, wilt, 76, 434
  tar spot, 48, 433
  thread blight, 437, 505
  twig blight, 77, 434, 436
  Verticillium wilt, 68, 343, 437
  winter injury, leaf burn, 41, 433–34
Azara, 151
  stem rot, 76, 151
Aztec lily, 282; *see* Tigerflower

B

B & A Formaldehyde, 558
Babiana, 307; *see also* Iris
  mosaic, 70, 308
  root nematode, 310
Baby-blue-eyes, 399; *see* Nemophila
Babysbreath, 203; *see also* Carnation
  aster yellows, 73, 205
  damping-off, 76, 204
  fasciation, 62, 206
  gray-mold blight, 52, 86, 205
  ringspot, 72, 205
  root-knot, 90, 206
  root nematode, 206
  root rot, 87, 204
  root and stem gall, 83, 206
  rust, 58, 204
  stem rot, 76, 204
Babytears vine, 151
  leaf spot, 48, 151
  powdery mildew, 56, 151
  rust, 58, 151
Bachelors-button, 220; *see also* Centaurea
  aster yellows, 73, 222
  downy mildew, 55, 225
  Fusarium wilt, 66, 223
  powdery mildew, 56, 221
  rust, 58, 223
  Verticillium wilt, 68, 223
  white-rust, 62, 226
Bacteria, 9, 10, [11], 553–54
  diseases caused by, 10
    bacterial canker, 77, 386, [483], 484, [487]
    black rot, [188]
    blackleg, [82]
    blight, [50], 51, [157], [172], [188], [225], [381], [413], 505
    brown rot, 50, 69, 488
    bud rot, 50, [202]
    cane gall, 83, [84]; 98, 136, 146, [173], [176], 391, 428, 488
    canker, 77, 79, 410, [483], 484, [487]
    collar rot, 76, 81
    crown gall, 83, [84], 98, 136, 146, 168, [173], [176], 391, 428, 478, 508,556
    fire blight, 79, [80], 81, 86, 133–34, 525
    fruit spot, [240], [483]
    gall or knot, 360, 370, 407
    gummosis, 79, 386
    hairy root, 83, 136–37
    leaf blotch, [50], 51
    leaf spot, [50], 51, [153], [196], [211], [240], [281], [309], [311], [386], 410, [423], 468, 483
    ring rot, [417], 420
    root gall, 83
    scab, 65, [283]
    shoot or twig blight, 79, [153], [333]
    soft rot, 10, 81, [82], 83, [147], [190], [248], [483]
    spot, [483]
    wetwood, slime flux, 264, [265]
    wilt, 10, 50, 66, 69, [70], 204, 211, 231, [232], 240, 354, 488
  enter plants, 10, 70, 81, 83, 137, 362, 488
  flagella, 10
  lethal temperature, 11, 542
  movement, 10
  numbers, 9–10, 11
  overwinter (oversummer), 10–11
  physiologic races, 97
  reproduction, 10
  size, 9
  spread, 10, 79–80, 155, 231, 240, 280, 386, 407
Bag-flower, 317; *see* Clerodendron
Bahiagrass, 319; *see also* Bluegrass
  brown patch, Rhizoctonia disease, 321
  fairy ring, 322
  leaf spot, blight, 48, 319
  Sclerotinia dollar spot, blight, 321
  seed treatment, 326, 537
  seedling blight, 76, 326
  slime mold, 322
Baldcypress, 402
  felt fungus, 293, 407

Baldcypress (*continued*)
root rot, 87, 405, 406
twig blight, 77, 402, 403
wood rot, 79, 170, 403
Balloonflower, 168
blight, 51, 76, 169
leaf spot, 48, 169
root rot, 87, 167
Balm, 447
leaf spot, 48, 447
Balm-of-Gilead, 411; *see* Poplar
Balsam, garden, 152
anthracnose, 51, 152
bacterial wilt, 69, 152
damping-off, 76, 152
downy mildew, 55, 152
leaf spot, 48, 152
mosaic, 70, 152
root-knot, 90, 152
root nematode, 152
root rot, 87, 152
rust, 58, 152
stem rot, 76, 152
Verticillium wilt, 68, 152
Balsam-apple, 238; *see also* Cucurbit
anthracnose, 51, 238
downy mildew, 55, 243
leaf blight, 51, 239
mosaic, 70, 241
powdery mildew, 56, 242
root-knot, 90, 244–45
Balsam-pear, 238; *see* Balsam-apple
Balsamroot (*Balsamorhiza*), 220
leaf gall nematode, 75, 226
leaf spot, 48, 221
powdery mildew, 56, 221
rust, 58, 223
Baltic ivy, 310; *see* Ivy
Bamboos (*Bambusa*), 152
bacterial leaf spot, 50, 152
black mildew, 64, 152
culm spot, 76, 152
fungus leaf spot, 48, 152
rust, 58, 152
stem smut, 62, 152
Band application, 554
Baneberry, 131
leaf spot, 48, 131
rust, 58, 131
smut, 62, 131
Bangalay, 268; *see* Eucalyptus
Baptisia, 269; *see* False-indigo
Barbados cherry, 152
algal leaf spot, green scurf, 152, 341
chlorosis, 24–29, 152
leaf spot, 48, 152
root-knot, 90, 152
Barber "Pre-Plant" 50-D, 106
Barberry, 153
anthracnose 51, 153
bacterial leaf spot, twig blight, 50, [153]
Barberry Eradication in Stem Rust Control, 154
canker, 77, 154, 345
damping-off, 76, 155, 405
dieback, 77, 154
gray-mold blight, blossom blight, fruit rot, 52, 86, 154
identifying Japanese and rust-spreading barberries, [154]
leaf spot, blotch, 48, 153
mosaic, 70, 154
powdery mildew, 56, 153
quarantine regulations, 154
root-knot, 90, 153
root nematode, 154
root rot, 87, 137, 154
rust, 56, [153]
twig blight, 77, [153], 154
Verticillium wilt, 68, 153
wood or heart rot, 79, 154, 170
Barco Manufacturing Company, Incorporated, 120
Bark beetles, insect control, 263
wounds, entrance for fungi, 403
Barley, rust, 256
Barrel cactus, 194; *see* Cacti, Echinocactus
Barrel sprayer, 113–14
Bartlett, N. M., Manufacturing Company, Incorporated, 120
Basi-Cop, 104
Basic Copper Fungicide, 104
Basil, 446; *see* Basilweed, Ocimum
Basilweed, 446
leaf spot, 48, 447
root-knot, 90, 447
rust, 58, 447
Basketflower, 220; *see also* Centaurea, Spiderlily
aster yellows, 73, 222
powdery mildew, 56, 221
Basswood, 338; *see* Linden
Bastard toadflax, rust, 405
Bauhinia, 300
leaf spot, 48, 301
powdery mildew, 56, 301
Bayberry, 507
chlorosis, 24–29
iron deficiency, 24, 26, 27
leaf spot, 48, 507
rust, 58, 313, 507
virus yellows, 73, 507
Bayer 25141, 373
Beach aster, 220; *see* Erigeron
Beach-pea, 377; *see* Sweetpea
Beaked cornsalad, 499; *see* Cornsalad
Beamtree, 133; *see* Mountain-ash
Bean, garden types, 155; *see also* Lima bean, Jackbean, Scarlet runner bean, Tepary bean
anthracnose, 51, 97, [157]
ashy stem blight, 76, 158
aster yellows, 73
bacterial blight, [50], 97, 155, [157]
bacterial soft rot, 81, 86, 160
bacterial wilt, 69, 159
baldhead or snakehead, 160
Bean Diseases — How to Control Them, 160
blossom and pod drop, 41, 160
boron toxicity, 28
brown spot, 50, 155
charcoal rot, 76, 158
chlorosis, 24–29, 160
chlorotic mottle, 155
copper deficiency, 28, 160
crown gall, 83, 160
crown rot, 76, 156–57
curly-top, 75, [159]
damping-off, 76, 158, 188
downy mildew, 55, 158
fertilizing, 29
Fusarium wilt or yellows, 66, 160
root rot, 87, 156
gray-mold blight, 52, 86, 159–60
iron deficiency, 24, 26, 27, 160
leaf scorch, 41, 160
leaf spot, 48, 159–60
magnesium deficiency, 28, 160
manganese deficiency, 28, 160
mosaic, 70, [71], 155
mottle, 70, 155
oxidant or PAN injury, 44
ozone injury, 44
plant-food utilization, 29
pod blight or spot, 86, [157], 159–60
pod drop, 41, 160
pod mottle, 155
potassium deficiency, 26, 160
powdery mildew, 56, 158
pseudo curly-top, 160
resistant varieties, 155, 156, 157
Rhizoctonia root rot, 87, 156
ringspot, 72, 160
root-knot, cyst nematode, 80, 159
root nematode, 160
root rot, 87, [156]
rust, 58, [59], 158
scab, 65, 159–60
seed rot, 86, 158
seed treatment, 158, 531, 532
soil deficiency, 160
southern blight, 76, 156
spotted wilt, 72, 160
stem anthracnose, 159–60
stem canker, 77, [156]
stem rot, 76, 156
stipple, 155
streak, 70, 155
sunscald, 41, 160
temperature, effect on, 41
2,4-D injury, 45, 160, 288
Verticillium wilt, 68, 160
web blight, 160
white mold, watery soft rot, sclerotiniose, Sclerotinia wilt, 76, 156–57
yellow dot, 155
zinc deficiency, 26, 27, 160
Bean, John, Division, FMC Corporation, 120
Beantree, 285; *see* Goldenchain
Bearberry, 175; *see also* Manzanita
black mildew, 64, 179
powdery mildew, 56, 176
red leaf gall, 62, 176
rust, 58, 176
shoot gall or hypertrophy, 62, 176
Beard-tongue, 454; *see* Penstemon
Beargrass; *see* Camass, Yucca
Beautyberry, 317; *see* Callicarpa
Beautybush, 499
leaf spot, 48, 500
pruning, 32
Bed Fume, 106
Bedrench, 106, 188, 546
active ingredient, 546
application, 546
uses, 546
Bedstraw, 185
chlorosis, 24–29, 186, 443
downy mildew, 55, 186
leaf spot, 48, 186
powdery mildew, 56, 186
root rot, 87, 137, 186
rust, 58, 169, 186
Beebalm, 447; *see* Monarda
Beech, 161
anthracnose, 51, 162, 343
bleeding or Phytophthora canker, 77, 161
canker (twig, branch), 77, 161
chlorosis, 24–29, 162, 359
dieback, 77, 161
felt fungus, 162, 293
flooding injury, 37
leaf scorch, 41, 161–62
leaf spot, 48, 162
mistletoe, 93, 162
mottle leaf, 162
powdery mildew, 56, 161
pruning, 32
root-knot, 90, 161
root nematode, 161
root rot, 87, 137, 161
seedling blight, damping-off, 76, 162 405
soil grade change injury, 46
sooty mold, 64, 162, 266
Verticillium wilt, 68, 162, 343
wood or heart rot, butt rot, 79, [161], 170
Beefwood, 212; *see* Casuarina
Beet, garden and sugar, 162
anthracnose, 51, 166
bacterial pocket or beet gall, 83, 166
bacterial soft rot, 81, 166
bacterial (black) streak, leaf spot, 50, 166
bacterial wilt, 69, 166
beet or virus yellows, 73, 164–65
black root rot, 87, 162
blackleg, 76, 164
boron deficiency, [10], 27, 164
Cercospora leaf spot, 48, 162, [163]
chlorosis, 24–29, 166
cracked stem, 27, 164
crown gall, 83, 166
crown rot, 76, 164, 166
curly-top, 75, 162
beet leafhopper, spread by, 162, [163]
damping-off, 76, 162
downy mildew, blue mold, 55, 165
dry rot, 76, 176

fertilizing, 29
Fusarium wilt, yellows, 66, 164
gray-mold blight, 52, 166
heart or crown rot, [10], 76, 164
iron deficiency, 24, 26, 27, 168
leaf spot, rot, 48, 166
leafhopper, 159, [163]
manganese deficiency, 26, 27, 166
mosaic, 70, 164
weed hosts, 165
Phoma rot, 87, 164
phosphorus deficiency, blackheart, 25, 164
plant-food utilization, 26
powdery mildew, 56, 167
resistant varieties, 162, 164, 165, 167
ringspot, 72, 164, 166
root gall, 90, 165
root-knot, cyst nematode, 90, 165
root nematode, 165, 167
root rot, 87, 156, 166
rust, 58, 166
savoy (virus), 70, 164
scab, 65, 165, 415
seed rot, 162
seed treatment, 162, 532
southern blight, 76, 166
storage rot, 86, 164, 207
The Sugar Beet Nematode and Its Control, 165
Verticillium wilt, 68, 167
watery soft rot, drop, stem rot, wilt, 76, 166, 207
web blight, 160, 166
white-rust, 62, 166
yellow net, 70, 164
yellow vein, 70, 164
zinc deficiency, 26, 27, 166
Beet leafhopper, [163]
Beetles, cucumber, [240]
Beetles, as virus carriers, 156, 192, 240
Begonia, 167
anthracnose, 51, 168
aster yellows, 73, 168
bacterial leaf spot or bacteriosis, [50], 168
bud drop, 39
crown gall, 83, 168
damping-off, cutting rot, 76, 167
flower blight, 52, 86, 167
gray-mold (*Botrytis*) blight, blotch, 52, 167
leaf nematode blight, 75, 168
leaf spot, 48, 168
light, effect on flowering, 38
light and temperature requirements, 39, 40
low humidity, leaf scorch, 39
mosaic, 70, 168
oedema, corky scab, 39, 168
plant dip, 168, 530, 534
powdery mildew, 56, [167]
root-knot, 90, 168
root nematode, 167
root rot, 87, 167
soil drench, 167
soluble salt injury, 31
spotted wilt, ringspot, 72, 168
stem, crown rot, 76, 167
tuber soak, 168, 530, 534
Verticillium wilt, 68, 168
watering, excess, injury, 38
Begonia-leaf, 195; *see* Anthurium
Belamcanda, 307; *see also* Iris
bacterial leaf spot, 50, 308
leaf spot, scorch, 48, 308
mosaic, 70, 308
Belladonna-lily, 249; *see* Amaryllis
Bellflower, 168; *see* Balloonflower for Chinese bellflower
aster yellows, 73, 169
gray-mold blight, 52, 76, 167, 169
leaf spot, 48, 169
leaf and stem nematode, 75, 169
mosaic, 70, 169
powdery mildew, 56, 169
root-knot, 90, 169
root rot, 87, 160
rust, 58, 169
soil drench, 169
southern blight, 76, 169
stem or crown rot, 76, 169
Verticillium wilt, 68, 169
Bellis, 220; *see* English (or true) daisy
Bellows duster, 116
Bells of Ireland, 447
crown rot, 76, 447
Bellwort, 335
leaf spot, 48, 338
rust, 58, 337
Beloperone, 229; *see* Chuperosa, Shrimp-plant
Benincasa, 238; *see* Chinese waxgourd
Bent, Bentgrass, 319; *see also* Bluegrass
anthracnose, 51, 319
brown patch, Rhizoctonia disease, 321
chlorosis, yellowing, 24–29, 325
compaction, 326
copper spot, 326
cottony blight, Pythium disease, 324
damping-off, 76, 326
fairy ring, 322
Fusarium blight, 326
Fusarium patch, pink snow mold, 322
grease spot, spot blight (*Pythium*), 324
insect injury, 327
leaf blight, 51, 319
leaf smut, 62, 325
leaf spot, 48, 319
melting-out, 319
Ophiobolus patch, 327
red thread or pink patch, 326
resistant varieties, 324
root-knot, 90, 324
root nematode, 324
root rot, 87, 319
rust, 58, 321
Sclerotinia dollar spot or blight, 321, [322]
seed treatment, 326, 537
slime mold, 322
smut, 62, 325
snow scald, Typhula blight, 322
soil drench, 320–21
yellow tuft (nematode), 324
Benzoin, 149; *see* Spicebush
Berberis, 153; *see* Barberry
Bermuda buttercup, 370; *see* Oxalis
Bermudagrass, 319
brown patch, Rhizoctonia disease 321
chlorosis, 24–29, 325
downy mildew, 55, 327
fairy ring, 322
Fusarium patch, 322
insect injury, 327
leaf blight, 51, 319
leaf spot, 48, 319
powdery mildew, 56, 321
resistant varieties, 320, 324
root-knot, 90, 324
root nematode, 324
root rot, 87, 319, 326
rust, 58, 321
Sclerotinia dollar spot or blight, 321
seed treatment, 326, 537
seedling blight, 76, 326
slime mold, 322
smut, 62, 325
southern blight, 76, 319
spot blight, cottony blight (*Pythium*), 324
spring dead spot, 326
Berry moth, insect control, 289
Berry spot or rot, 86, [260], [465]
Beta, 162; *see* Beet, Mangel, Mangold, Swiss Chard
Betony, 447; *see* Stachys
Betula, 170; *see* Birch
Bibliography, 571–81
Bichloride of mercury, 101, 528; *see also* Mercuric chloride
Bidens, 220; *see* Bur-marigold
Big vein, [330]
Bignonia, 169
black mildew, 64, 170
canker, dieback, 77, 138–39, 170
gray-mold blight, 52
leaf spot, 48, 170, 221
root-knot, 90, 170
sooty mold 64, 170
spot anthracnose, 51, 170
Bioquin 700, 366, 367
Birch, 170
anthracnose, 51, 171
bark canker, 77, 172
bleeding canker, 77, 161, 345
borer control, 171
canker (twig, branch), 77, 171, 172
dieback, 77, 171, 172
leaf blister, 62, 171
leaf spot, 48, 171
mistletoe, 93, 172
powdery mildew, 56, 171
pruning, 32
root rot, 87, 137, 171–72
rust, 58, 171, 318
soil grade change injury, 46
twig blight, 77, 172
2,4-D injury, 45, 288
wetwood, slime flux, 172, 264
witches'-broom, 62, 171
wood rot, heart rot, butt rot, 79, [170], 171
Bird
injury, 45
spread of fungi, 217
spread of mistletoes, 93, 94
as virus carriers, 13
Bird-of-paradise-flower, 172
bacterial wilt, 69, 173
root nematode, 172
root rot, 87, 172–73
seed rot, 172–73
seed treatment, 173, 531, 534
Birds-beak, rust, 405
Bird's-eyes, 399; *see* Gilia
Birdsnest-hemp, light and temperature requirements, 40
Birthwort, 144; *see* Aristolochia
Bis (dimethylthiocarbamoyl) disulfide, 103, 519
Bishopscap, 304; *see* Mitella
Bistort, alpine (*Polygonum viviparum*), rust, 256
Bittersweet, 173
canker, dieback, 77, 173, 345
crown gall, 83, 173
leaf spot, 48, 173
powdery mildew, 56, 173
root-knot, 90, 174
root nematode, 174
root rot, 87, 174
Black-alder, 295; *see* Holly
Black bundle, 234
Black cohosh, 131; *see* Cimicifuga
Black gum, 258; *see also* Tupelo
canker, branch and trunk, 77, 259
mistletoe, 93, 261
rust, 58, 261
Verticillium wilt, 68, 261, 343
wood rot, 79, 170, 259
Black knot, [81], 384–85, 525
Black Leaf Company, 120
Black locust, 300; *see* Locust
Black mildew or mold [64], 65, 200
Black root rot, [88], [462]
Black rot, [188], [207], [287], [470]
Black-salsify, 328
aster yellows, 73, 328–29
root-knot, 90, 332
white-rust, 62, 331
Black sampson, 220; *see* Echinacea
Black-snakeroot, 131; *see* Cimicifuga
Black spot, 48, [49], [199], [256]
Black walnut, 501; *see* Walnut
Blackberry, 427
anthracnose, 51, 427, 525
cane blight, 76, 428, 525
cane and crown gall, hairy root, 83, 428
cane spot, 430–31
canker, dieback, 77, 428
collar rot, 76, 431
Controlling Diseases of Raspberries and Blackberries, 432

Blackberry (*continued*)
double blossom, 432
downy mildew, 55, 432
dwarf or stunt, 73, 428
fire blight, flower blight, 79, 86, 432
fruit rot, spot, or mold, 52, 86, 430
gray-mold blight, 52, 86, 430
Growing Blackberries, 432
leaf curl, 70, 428
leaf spot, 48, 427, 430–31
male berry, blackberry sterility, 432
mosaic, 70, 428
orange rust, 58, 430, [431]
powdery mildew, 56, 431
resistant varieties, 430, 431, 432
rodent protection, 42
root rot, 87, 137, 247, 431
rosette, 73, 428, 432
sooty blotch, 64, 432
spot anthracnose, 51, 427
spray schedule, 524–25
spur blight, 77, 427, 525
streak, 70, 428
sulfur dioxide injury, 43
sunscorch, 41, 432
thread blight, 432, 505
Verticillium wilt, 68, 431
winter injury, 41, 432
yellow or cane rust, 58, 430
Blackberry-lily, 307; *see* Belamcanda
Black-eyed pea, 155
anthracnose, 51, 157
Ascochyta leaf and pod spot, 48, 86, 159–60
bacterial blight or spot, 50, 155
chlorosis, 24–29, 160
Cladosporium spot, 159–60
curly-top, 75, 159
damping-off, 76, 158, 188
downy mildew, 55, 158
Fusarium wilt, 66, 160
gray-mold blight, 52, 86, 159–60
leaf bronzing or burn, 41, 160
leaf spot, 48, 159–60
mosaic, 70, 155
powdery mildew, 56, 158
resistant varieties, 160
root nematode, 160
rust, 58, 158
southern blight, 76, 156
stem canker, rot, 77, 156
Verticillium wilt, 68, 160
web blight, 160
Black-eyed-Susan, 220; *see* Rudbeckia, Clockvine
Blackhaw, 499; *see* Viburnum
Blackheart, [417], 420
Blackleg, 81, [82], 278, 418
Blackthorn, 382; *see* Plum
Bladdernut, 130; *see* American bladdernut
Bladder-senna, 300
powdery mildew, 56, 301
root rot, 87, 137, 301
rust, 58, 301
seedling blight, 76, 302
twig blight, 77, 300
Blanket-flower, 220; *see* Gaillardia
Blast, blasting, 553
Blazing-star; *see* Liatris, Mentzelia, Tritonia
Bleach, household, as disinfectant, 23, 35, 79, 81, 99, 134, 138–39, 195, 196, 204, 206, 259, 360, 366, 368, 370, 386, 441, 445
Blechnum, 269; *see* Ferns
Bleeding canker, 161, 259, 345
Bleedingheart, 174
chlorosis, 24–29
Fusarium wilt, 66, 174
gray-mold blight, 52, 174
iron deficiency, 24, 26, 27
leaf spot, 48, 174
root-knot, 90, 174
soil drench, 174
stem or crown rot, wilt, 76, 174
Blessedthistle, 220
southern blight, 76, 221
Blight, 553
bacterial shoot, 79, [153], [333]
blossom, 52, [85], 86, [133], [199], [224], [260], [383], [435]
Botrytis, 52, [54], [85], [207], 224, 227, [454], [465], 495
brown felt, 77, 293, 407
cane, 77, [247], [428]
fire, 79, [80], 81, [133], 525
flower, [85], 86, [133], [199], [224], [260], [383], [435]
gray-mold, 52, [54], [85], 167, 224, 408, [454], 525
inflorescence, 52, [85], 86, [133], [199], [224], [260], [383], [435]
leaf, 48, [50], [51], 52, [140], [179], [188], [217], [224], [233], [239], [252], [256], [262], [324], [329], [415], [458], [464]
limb, 77
needle, 51, 313, 314, [402]
ray, [85], 86, [224]
snow [407]
southern, 76, 393, [394], 490, 556
stem, 77
twig, 52, 77, [153], [313], [317], 345
western yellow, 75, 486
Bloat, onion, [92], 93, 364
Bloodflower, 185; *see* Butterflyweed, Milkweed
Bloodleaf, 230
inflorescence smut, 62, 230
leaf spot, 48, 230
light and temperature requirements, 40
root and crown rot, 76, 87, 230
root-knot, 90, 230
Bloodroot, 413; *see also* Poppy
gray-mold blight, 52, 414
leaf spot, 48, 414
root rot, 87, 414
Bloom spray, 524
precautions, 524, 525
Blossom blight, 52, [85], 86, 133, [199], [224], [260], [383], 435, 525
Blossom drop, 41, 491
Blossom-end rot, [10], [244]
Blotch, 554
fruit, 86
leaf, 48, [51], 52, [256], [303], [363]
sooty, [64], 65, 100, 137, 200, 266, 277, 369, [406], 566
twig, [137]
Blue cohosh, 153
leaf blight, gray-mold (*Botrytis*) blight, 51, 52, 154
leaf spot, 48, 153
Blue daisy, 220
powdery mildew, 56, 221
Blue dawn-flower, 351; *see* Morning-glory
Blue dicks, 183; *see* Brodiaea
Blue flag, 307; *see* Iris
Blue gilia, 399; *see* Gilia
Blue laceflower, 213
aster yellows, 73, 214
root-knot, 90, 214
root rot, 87, 215
seed rot, damping-off, 76, 214
seed treatment, 213
stem, crown rot, 76, 214
Blue mold, [309]; *see also* Downy mildew
Blue spirea, 317; *see* Verbena
Bluebeard, 317; *see* Verbena
Bluebells; *see* Mertensia, Grape-hyacinth
Bluebells of England, 495; *see* Squill
Bluebells-of-Scotland, 168; *see* Bellflower, Campanula
Blueberry, 175
anthracnose, 51, 178
bacterial stem canker, 50, 77, 179
black mildew, 64, 179
blossom blight, 52, 86, 175
Botryosphaeria stem canker, 77, 178
Botrytis or gray-mold blight, 52, 86, 175–76
bud gall, 180
cane canker, dieback, 77, 178
chlorosis, 24–29, 178
crown or cane call, 83, [176]
fruit or berry rot, 86, [175], 179–80
grub control, 178
iron deficiency, chlorosis, 24, 26, 27, 178
leaf blight, 51, 178
leaf rust, 58, 177
leaf spot, 48, 178
magnesium deficiency, 28, 180
mosaic, 70, 177
mummy berry, brown rot, 86, [175]
powdery mildew, 56, [176]
red leaf disease, 62, 180
red leaf gall, "swamp cheese," rose bloom, 62, 176
resistant varieties, 175, 176, 177, 178, 179, 180
ringspot, 72, 177–78
rodent protection, 42
root gall, 90, 180
root nematode, 180
root rot, 87, 178
rust, 58, 177, 178, 405
shoestring, 177, [179]
spot anthracnose, 51, 178
spray schedule, 524–25
stunt, 73, [177]
sulfur-sensitive, 525
tar spot, 48, 178
twig or stem blight, 77, 178
twig canker, 77, 178, 259
winter injury, 41, 179
protection, 42, 179
witches'-broom, 177
wood rot, 79, 170, 178–79
Blue-blossom, 355; *see* New Jersey-tea
Bluebonnet, 377; *see* Lupine
Blue-eyed grass, 307; *see also* Iris
leaf blight, 51, 308
root nematode, 310
rust, 58, 308
Blue-eyed-Mary, 454; *see* Collinsia
Bluegrass, 319
algae, green scum, 326
anthracnose, 51, 319
brown patch, Rhizoctonia disease, 321
buried debris, 327
chlorosis, yellowing, 24–29, 325
compaction, 326
damping-off or seedling blight, 76, 326
dog injury, 327
fairy ring, 322, [323]
Fusarium blight, 326
Fusarium patch, pink snow mold, 322
Helminthosporium leaf spot, blight, 48, 319, [320]
insect injury, control 321, 327
iron deficiency, 24, 26, 27, 325
leaf blight, blotch, 51, 319
leaf spot, 48, 319, [320], 321
magnesium deficiency, 28, 324
melting-out, 319
mosaic, 70, 327
moss, 326
mushrooms or toadstools, 322, [323]
Ophiobolus patch, 327
powdery mildew, 56, [320], 321
puffballs, 322
Pythium disease, 324
red thread, pink patch, 326
resistant varieties (clones), 320, 321, 325, 327
root-knot, 90, 324
root nematode, 324
root rot, 87, 319
rust, 58, [320], 321
Sclerotinia dollar spot or blight, 321, [322]
seed rot, 326
seed treatment, 326, 537
Septoria leaf spot, 48, 319–20
slime mold, [322]
smut, 62, 325
soil drench, 320–21
stem and culm rot, 76, 319

Typhula blight, snow scald, gray snow mold, 322, [323]
Bluelips, 454; *see* Collinsia
Bluestone; *see* Copper sulfate
Bluets, 185; *see* Houstonia
Bog-laurel, 175; *see* Mountain-laurel
Bog-rosemary, rust, 58
Boisduvalia, 275
rust, 58, 275
Boltonia, 220
leaf spot, 48, 221
powdery mildew, 56, 221
rust, 58, 223
white-rust, 62, 226
white or leaf smut, 65, 226
Boneset, 220; *see* Eupatorium
Bonide Pentide, 563
Borage (*Borago*), 348–49
leaf spot, 48, 221, 349
Borateem, cut stump treatment, 406
Borax, control boron deficiency, 27, 141, 164, 191, 206, 209, 285, 290, 297, 332, 408, 505
cut stump treatment, 406
Bord oil, 556
Bordeaux mixture, 104–5, 554
as disinfectant, 105
dry products, 105
formula, 104, 554
injury from, 104
preparation of, 105
as protective fungicide, 144, 153, 179, 182, 183, 209, 212, 272, 310, 313, 318, 332, 335, 342, 384, 385, 386, 405, 437, 438, 468, 500, 505, 508
tree paint, 36, 105, 259, 370, 503, 554
uses, 105
Bor-dox, 105, 554
Borers
cane, control, 428
crown, control, 463
iris, control, 307
tree, control, 79, 134, 139, 145, 171, 259, 266, 300–301, 332, 346, 358, 384, 435, 477, 525
wounds, entrance for fungi, 402
Boric acid, control boron deficiency, 27, 129, 141
Boron
Boron Injury to Plants, 28
deficiency, [10], 27, 141, 164, 209, 214, 285, 290, 297, 408, 444, 475, 505, 508–9
control of, 27, 141, 214
promotes, 27
toxicity, 28, 290
Boro-Spray, 27, 141
Boston ivy, 287
canker, dieback, or wilt, 77, 290
downy mildew, 55, 288
leaf blight, 51, 291
leaf spot, 48, 287, 291
powdery mildew, 56, 289
root nematode, 291
root rot, 87, 137, 289
Botanical gardens, help by, 5
Botran (DCNA, dichloran)
active ingredient, 104, 393
formulations, 104
post-harvest fruit wrap or dip, 104, 393, 443, 471
as protective fungicide, 55, 86, 157, 167, 205, 214, 224, 277, 280, 305, 308, 330, 335, 393, 455, 495
soil application, 104, 167, 282, 284, 362
uses, 104
Botrytis (fungus), 554
blight, 52–53, [54], [85], 224, 277, 495
fruit rot, [465]
leaf spot, 52, [283]
lethal temperature (fungus), 542
neck rot, [361]
Bougainvillea, 180
Cercospora leaf spot, 48, 180
chlorosis, 24–29, 180, 443
leaf spot, 48, 180
mosaic, 70, 180
Boussingaultia, 340; *see also* Lythrum
root-knot, 90, 340
Bouteloua, rust, 185
Bouvardia, 181
leaf nematode, 75, 181
root-knot, 90, 181
rust, 58, 181
Bowstring hemp, 448; *see* Sansevieria
Box sandmyrtle, 316
leaf gall, 62, 316
Boxelder, 343; *see also* Maple
anthracnose, 51, 343
bacterial leaf spot, 50, 347
chlorosis, 24–29, 344–45
dieback, 77, 345
felt fungus, 293, 347
ice injury, 43
leaf blight, 51, 343
leaf spot, 48, 346
powdery mildew, 56, 346
root rot, 87, 137, 346
seedling blight, 76, 347, 405
sooty mold, 64, 346
tar spot, 48, 346
twig blight, canker, 77, 345
2,4-D injury, 288, 347
Verticillium wilt, 68, 343
wood rot, 79, 170, 346
Boxwood, Box, 181
canker, dieback, 77, [181], 182
crown gall, 83, 183
Culture, Diseases and Pests of the Box Tree, 183
decline, 87, [181]
heart or trunk rot, 79, 170, 182
iron deficiency, 24, 26, 27
leaf blight, cast, 51, 182
leaf spot, 48, [182]
Phytophthora root rot, 87, 181–82
plant dip, 182, 530, 537
root-knot, 90, 182
root nematode, 182
root rot, 87, 181–82
soil drench, 182
spray program, 181–82
sunscald, 41, 182
thread blight, 182, 272
tip blight, 51, 182
twig blight, 77, 181–82
Verticillium wilt, 68
windburn, 182
winter injury, 41–42, 182
Boyle-Midway, Incorporated, 120
Boysenberry, 427; *see also* Blackberry, Raspberry
anthracnose, 51, 427
cane canker, dieback, 77, 428
cane and crown gall, 83, 428
downy mildew, 55, 432
fruit rot, 86, 430
gray mold of fruit, 52, 86, 430
leaf spot, 48, 427, 430
mosaic, 70, 428
powdery mildew, 56, 431
rust, 58, 430
spray schedule, 524–25
Verticillium wilt, 68, 430
winter injury, 41, 432
Brachycome, 220
aster yellows, 73, 222
stem rot, 76, 221
Bracket, [79], [147], [161], [170]
Bradson Company, 120
Brake, 270; *see* Ferns
Bramble, 554
Branch canker, 77, [217], [265], [344], [358], [403], [411], [503]
dieback, 77
stub, entrance for fungi, 79, 403
-wilt (*Hendersonula*), 502, [503]
Brassavola, 366; *see* Orchids
Brassia, 366; *see* Orchids
Brassica, 185; *see* Broccoli, Brussels Sprouts, Cabbage, Cauliflower, Chinese Cabbage, Collard, Kale, Kohlrabi, Mustard, Rape, Rutabaga, Turnip
Brassicol, 101
Brayton Chemicals, Incorporated, 120
Breaking (flower), 70, [71], 72, [192], [250], [284], 368, [375], [378], 469, 554
Breck's Many Purpose Spray, 523
Bridalwreath, 461; *see* Spirea
Broad-spectrum fungicide, 554
Broadcast application, 554
Broadleaf, 554
Broccoli, 186; *see also* Cabbage
aster yellows, 73, 193
bacterial leaf spot, 50, 190
bacterial soft rot, 76, 81, 190
black ringspot, 72, 192
black rot, 50, 187
blackleg, 76, 187
boron deficiency, 27, 191
chlorosis, magnesium deficiency, 24, 28, 194
clubroot, 89, 189
curly-top, 75, 192
damping-off, 76, 188, 190, 191, 193
downy mildew, 55, [189]
drop, cottony rot, 76, 191
fertilizing, 29
Fusarium yellows, 66, 186–87
gray-mold blight, 52, 86, 190, 191
leaf mold, 48, 190, 191
leaf spot, 48, 190, 191
mosaic, 70, 192
oedema, 39, 194
plant-food utilization, 29
powdery mildew, 56, 193
premature flowering, 41
resistant varieties, 187, 190
root-knot, 90, 191
root nematode, 194
root rot, 87, 187, 190, 193
seed rot, 188
seed treatment, 187, 529, 532
southern blight, 76, 191
spotted wilt, 72, 192
tipburn, 191
Verticillium wilt, 68, 193
whiptail, molybdenum deficiency, 28, 193
white-rust, 62, 191
Brodiaea, 183
rust, 58, 183
Bromeliads, light and temperature requirements, 40
Bromex, 106
Bromo-O-Gas, 545
Broom, 183
chlorosis, 24–29
dieback, 77, 183
leaf blight, 51, 183
leaf spot, 48, 183
powdery mildew, 56, 183
root nematode, 183
root rot, 87, 137, 183
rust, 58, 183
Brother Juniper, 19, [20]
Broussonetia, 271; *see* Paper-mulberry
Browallia, 479; *see also* Tomato
aster yellows, 73, 490
Fusarium wilt, 66, 486
light and temperature requirements, 40
root-knot, 90, 489
spotted wilt, 72, 486
white smut, 65, 492
Brown felt blight or canker, 77, 293, 407
Brown patch, 321
Brown rot or blight (bacterial), 50, 69 488
Brown rot (fungal), 86, [87], 382, [383], 528
Brown-eyed-Susan, 220; *see* Goldenglow
Brozone, 106, 545
Brush-cherry, 268; *see* Eugenia
Brush-killer, 356, 387
Brussels sprouts, 185; *see also* Cabbage
bacterial leaf spot, 50, 190
bacterial soft rot, 76, 81, 190
black ringspot, 72, 192
black rot, 50, 187
blackleg, 77, 187
chlorosis, 24–29, 194
clubroot, 89, 189

Brussels sprouts (*continued*)
damping-off, 76, 188, 190, 191, 193
downy mildew, 55, 189
drop, cottony rot, 76, 191
Fusarium yellows, 66, 186–87
gray-mold blight, 52, 86, 190, 191
leaf spot, 48, 190, 191
mosaic, 70, 192
oedema, 39, 194
root-knot, cyst nematode, 90, 191
root nematode, 194
root rot, 87, 187, 190, 193
seed treatment, 187, 529, 532
spotted wilt, 72, 192
Verticillium wilt, 68, 193
whiptail, molybdenum deficiency, 28, 193
white-rust, 62, 191
Bryonopsis, 238; *see also* Cucurbit
bacterial spot, 50, 239
downy mildew, 55, 243
Bryophyllum, 451; *see* Kalanchoë
Buchloë, 319; *see* Buffalograss
Buckeye, 302; *see also* Horsechestnut
leaf blister, yellow, 62, 171, 303–4
leaf blotch, 51, 302, [303]
leaf scorch, 41, 303
leaf spot, 48, 303
mistletoe, 93, 304
powdery mildew, 56, 303
rust, 58, 304
witches'-broom, 62, 171, 303–4
wood rot, 79, 170, 303
Buckleya, rust, 58, 405
Buckthorn, 183
leaf spot, 48, 184
as noxious weed, 184
powdery mildew, 56, 184
root rot, 87, 184
rust, 58, [184]
sooty mold, 64, 184
wood rot, 79, 170, 184
Buckwheat-tree, 184
black mildew, 64, 184
leaf spot, 48, 184
Bud
blast, 50, 52, 255, 256, 436
blight, 50, 52
-break, 554
drop, 25, 39, 200, 277, 299
nematode, 16, 75, [76], 467
rot, 50, 52, [54], 86, [202], 253, 466–67
Budding, in transmitting viruses, 13, 70
Buddleia, 185; *see* Butterflybush
Buffalo Turbine Agricultural Equipment Company, Incorporated, 120
Buffaloberry, 445
damping-off, 76, 405, 446
leaf spot, 48, 445
powdery mildew, 56, 445
root rot, 87, 137, 445
rust, 58, 445
wood rot, 79, 170, 446
Buffalograss, 319; *see also* Bluegrass
anthracnose, 51, 319
leaf spot, 48, 319
root nematode, 324
root rot, 87, 319
rust, 58, 321
seed treatment, 326, 537
seedling blight, 76, 326
smut, 62, 325
tar spot, 48, 319
Bugbane, 131; *see* Cimicifuga
Bugleweed, 129; *see* Ajuga
Bugloss, 348; *see* Anchusa
Bulb, 554
diseases, 87–93
control, 528–37
flies, control, 250, 252
mites, control, 252, 282, 496, 497
nematodes, 16, [92], 93, [309], [497]
ring disease, [92], 93, 252, 364, 496–97
rot, [77], [90], 250, [309], 497
treatment, 250, 361, 362, 495–96, 528–37
Bulldozer injury, 46
Bulls-eye leaf spot, 48, [346]
Bulrush, rust, 210
Bunchberry, 258; *see* Dogwood
Bunchy top, 337
Bundleflower, 183
leaf spot, 48, 183
powdery mildew, 56, 183
rust, 58, 183
Bunya-bunya, 143; *see* Araucaria
Burgundy mixture, 105
composition, 105
precautions, 105
preparation of, 105
uses, 105
Burlap strips, for wrapping trees, [42], 140, 162, 264
Bur-marigold, 220; *see also* Chrysanthemum
aster yellows, 73, 222
leaf spot, 48, 221
powdery mildew, 56, 221
root-knot, 90, 225
rust, 58, 223
Burnet, 440
leaf spot, 48, 443
powdery mildew, 56, 440
rust, 58, 441
Burning-bush; *see* Euonymus, Kochia
Burro's-tail, 451; *see* Sedum
Bush-honeysuckle, 457
leaf spot, 48, 458
powdery mildew, 56, 458
root-knot, 90, 459
root rot, 87, 458
Bush-mallow, 297; *see* Red false-mallow
Bush morning-glory, 351; *see* California-rose
Bush-pea, 377; *see* Thermopsis
Bushcherry, 382; *see* Cherry
Bushpoppy, 413
leaf smut, 62, 414
Butia, 371; *see* Palms
Butt rot, 79, [116], 145, 147, [161], [170], 171, [385]
Butter-and-eggs, 454; *see* Toadflax
Buttercup, 254
aster yellows, 73, 256
chlorosis, 24–29, 257, 443
curly-top, 75, 257
downy mildew, 55, 257
gray-mold blight, 52, 255
leaf and stem nematode, 75, 257
leaf rot, 51, 256
leaf smut, white smut, 62, 65, 257
leaf spot, 48, 256
mosaic, 70, 255
powdery mildew, 56, 256
ringspot, spotted wilt, 72, 257
root rot, 87, 254
rust, 58, 256, 392
stem rot, 76, 254
Butterflybush, 185
chlorosis, 24–29, 185, 443
leaf spot, 48, 185
mosaic, 70, 185
pruning, 32
root-knot, 90, 185
root rot, 87, 137, 185
sooty mold, 64, 185
stem or twig canker, 77, 185
Butterfly-flower, 479
anthracnose, 51, 492
aster yellows, 73, 490
blight, 51, 492
damping-off, 76, 488
fasciation or leafy gall, 62, 492
leaf and stem nematode, 75, 400, 492
light and temperature requirements, 40
powdery mildew, 56, 492
root-knot, 90, 489
root nematode, 489
root rot, 87, 490
spotted wilt, 72, 486
stem or foot rot, 76, 488
Verticillium wilt, 68, 487–88
Butterfly-pea, 377
leaf spot, 48, 380
root rot, 87, 378
Butterflyweed, 220
leaf spot, 48, 185, 221
mosaic, 70, 185
powdery mildew, 56, 185
root rot, 87, 185, 278
rust, 58, 185
Butternut, 501; *see also* Walnut
anthracnose, 51, 501
bacterial blight, 50, 505
bunch disease, 502
canker (twig, branch), dieback, 77, 502
downy spot, 48, 502
leaf blight, blotch, 51, 501
leaf spot, 48, 501
nut mold, 86, 504
root-knot, 90, 392, 506
root rot, 87, 137, 504
sooty mold, 64, 506
trunk canker, 77, 502
wood rot, 79, 170, 504
Button snakeroot, 220; *see* Liatris
Buttonbush, 185
chlorosis, 24–29, 186, 443
leaf blight, 51, 186
leaf spot, 48, 186
powdery mildew, 56, 186
rust, 58, 169, 186
thread blight, 186, 505
Buttonwood, 476; *see* Sycamore
Buxus, 181; *see* Boxwood

C

Cabbage, 186
anthracnose, 51, 190
aster yellows, 73, 193
bacterial leaf spot, 50, 190
bacterial soft rot, wet rot, stump rot, 76, 81, [190]
black (*Alternaria*) leaf spot, 48, 190
black ringspot, ring necrosis, 72, 192
black rot or blight, 50, 97, 187, [188]
blackleg, canker, dry rot, 77, 97, [187]
boron deficiency, brown heart or rot, 27, 191
chlorosis, 24–29, 194
clubroot, [89], 90, 98, 189
crown gall, 83, 146, 193
curly-top, 75, 192
damping-off, 76, 188, 190, 191, 193
Diseases of Cabbage and Related Plants, 194
downy mildew, 55, 189
drop, cottony rot, 76, 191
fertilizing, 29
foot or collar rot, 76, 188
Fusarium wilt or yellows, 66, [67], 186–87
gray-mold (*Botrytis*) blight, 52, 86, 190, 191
head rot, 86, 190, 191, 207
insect control, 187, 192
leaf spot, 48, 190, 191
mosaic, 70, 192
Mycosphaerella leaf spot, 48, 190
oedema, intumescence, 39, 194
plant-food utilization, 26
powdery mildew, 56, 193
resistant varieties, 186–87, 188, 189, 191, 192, 193
Rhizoctonia bottom rot, 188
root-knot, cyst nematode, 90, 191
root nematode, 194
root rot, 87, 187, 190, 193
scab, 65, 165, 193, 415
seed rot, 188
seed treatment, 187, 529, 532
seedbed treatment, 188
soil drench, 188, 189
southern blight, 76, 191
storage rot, 81, 86, 188, 190, 191
sulfur dioxide injury, 43
temperature, effect on, 41
tipburn, 191
transplant water, 189, 191, 193

Verticillium wilt, 68, 193
whiptail, molybdenum deficiency, 28, 193
white mold or blight, 51, 190
white-rust, white blister, [61], 62, 191
white spot, 48, 190
wirestem, Rhizoctonia disease, 77, [188]
Cabbage worms, as virus carriers, 192
Cacti, 194
anthracnose, 51, 194
bacterial blight, 50, 195
bacterial necrosis, 195
bacterial soft rot, 76, 81, 194
black mildew, 64, 195
bud drop, 25, 195
cladode rot, 76, 194
collar rot, 76, 194
corky scab, 39, 194
crown gall, 83, 195
cutting rot, 76, 194
Fusarium wilt, 66, 194
glassiness, 39, 195
gray-mold blight, rot, 52, 195
Helminthosporium stem rot, 76, 194–95
kinds
barrel (*Echinocactus*), 194
biznaga (*Astrophytum*), 194
cholla (*Opuntia*), 194
Christmas (*Schlumbergera*, *Zygocactus*), 194
crab (*Epiphyllum*, *Zygocactus*), 194
Easter (*Schlumbergera*), 194
fishhook (*Ancistrocactus*), 194
hairbrush (*Pachycereus*), 194
hatchet (*Pelecyphora*), 194
hook (*Hamatocactus*), 194
Indian fig (*Opuntia*), 194
living rock (*Ariocarpus*), 194
manco caballo (*Homalocephala*), 194
mitra (*Astrophytum*), 194
night-blooming (*Lylocereus*, *Nyctocereus*, *Selenicereus*, *Trichocereus*), 194
nopal (*Opuntia*), 194
old-man (*Cephalocereus*), 194
orchid (*Epiphyllum*), 194
organ-pipe (*Pachycereus*), 194
pincushion (*Mammillaria*) ,194
saguaro or giant (*Carnegiea*), 194
sea-urchin (*Echinopsis*), 194
seven-sisters (*Ariocarpus*), 194
snowball (*Pediocactus*), 194
star (*Astrophytum*, *Echinocactus*), 194
tasajillo (*Opuntia*), 194
tuna (*Opuntia*), 194
Turk's-head (*Melocactus*), 194
leaf spot, scorch, 48, 195
light and temperature requirements, 40
root-knot, cyst nematode, 90, 194, 195
root nematode, 194
root rot, 87, 194
scab, 65, 195
scorch, "sunscald," 41, 195
seed and cutting treatment, 194
seedling blight, 76, 194
soil mix for, 22
stem and root rot, 76, 87, 194
zonate spot, 48, 195
Caddy, 106, 321, 554
Cadminate, 106, 321, 322, 554
Cadtrete, 106, 321, 554
Caesalpinia, 300
anthracnose, 51, 301
canker, dieback, 77, 300
crown gall, 83, 136, 301
root rot, 87, 137, 301
rust, 58, 301
wood rot, 79, 170, 301
Calabash, 238; *see* Gourds
Caladium, 195
bacterial soft rot, 81, 86, 196
downy mildew, 55, 199
gray-mold blight, 52, 198
leaf spot, 48, 197
light and temperature requirements, 39, 40
root-knot, 90, 197
root nematode, 197
root rot, 87, 196–7
southern blight, stem rot, 76, 196–7
tuber rot, 86, 196–7
tuber soak, 197, 530, 534
Calamagrostis, rust, 439, 445
Calamondin, 227; *see* Citrus, Orange
Calanthe, 366; *see* Orchids
Calathea, 347
leaf spot, 48, 347
Calcareous, 554
Calceolaria, 454
aster yellows, 73, 457
boron deficiency, 27, 457
flower blight, 52, 86, 455
gray-mold (*Botrytis*) blight, 52, 86, 455
leaf nematode, 75, 457
mosaic, 70, 456
root rot, 87, 455
seed treatment, 455
soil drench, 455
spotted wilt, ringspot, 72, 456
stem or collar rot, 76, 455
Verticillium wilt, 68, 455
Calcium
carbonate (ground limestone), 28
chloride, 331, 482
cyanamide, 175, 554
deficiency, 28, 214, 497
control of, 28
hypochlorite; *see* Household bleach
nitrate, 28, 215, 482, 497
oxide, to prevent fluorides injury, 44
promotes, 28
sulfate (gypsum), 28, 29
toxicity, 28
Calendula, 220
aster yellows, 73, 222
black mold, 64, 226
chlorosis, 24–29, 227
crown gall, 83, 226
foliar or leaf nematode, 75, 226
gray-mold blight, 52, 224
leaf blight, 51, 221
leaf spot, 48, 221
mosaic, 70, 222
powdery mildew, 56, 221
ringspot, 72, 222
root-knot, 90, 225
root nematode, 226–27
root rot, 87, 221
rust, 58, 223
soil drench, 222
southern blight, 76, 221
spotted wilt, ringspot, 72, 223
stem rot, 76, 221
white smut, [65], 226
Calico, 70, 214
Calico-flower, 144; *see* Aristolochia
California barberry, 153; *see* Mahonia, Oregon-grape
California-bluebell, 398; *see* Phacelia
California buckeye, 302; *see* Buckeye
California fremontia, 401; *see* Fremontia
California fuchsia; *see* Fuchsia, Zauschneria
California-hyacinth, 183; *see* Brodiaea
California-laurel, 149
bacterial leaf spot, 50, 151
black mildew, sooty mold, 64, 151
canker, dieback, 77, 149–50
leaf blight, 51, 149
root nematode, 151
wood rot, 79, 149, 170, 259
California-pitcherplant, 408
leaf spot, 48, 408
California-poppy, 413
aster yellows, 73, 414
bacterial blight, 50, 413
gray-mold blight, 52, 414
leaf mold or spot, 48, 414
leaf smut, 62, 414
powdery mildew, 56, 414
root-knot, 90, 414
seed treatment, 414, 530, 534
soil drench, 414
spotted wilt, 72, 414
stem or foot rot, 76, 414
Verticillium wilt, 68, 414
California-rose, 351
leaf spot, 48, 351
root rot, 87, 351
rust, 58, 351
white-rust, 62, 191, 351
California sweetshrub, 198; *see* Calycanthus
Calla (wild), 195
gray-mold blight, 52, 198
leaf spot, 48, 197
mosaic, 70, 197
root rot, 87, 197
Calla lily, 195
bacterial soft rot, leafstalk rot, 76, 81, 83, 90, 196
corm (rhizome) treatment, 83, 196, 197, 528, 529, 530, 534
flower blight, 86, 197
gray-mold blight, 52, 198
leaf spot, blight, 48, 51, 197
light and temperature requirements, 40
mosaic, 70, 197
rhizome or corm rot, 83, 90, 196, 197
root-knot, 90, 197
root nematode, 197
root rot, 87, [196], 197
soil drench, 197
southern blight, 76, 196–97
spotted wilt, ringspot, 72, 197
Calliandra, 198
root rot, 87, 198
Callicarpa, 317
black mildew, 64, 318
dieback, canker, 77, 138, 318
leaf spot, 48, 317
root nematode, 318
Callirhoë, 297; *see* Poppy-mallow
Callistephus, 220; *see* China-aster
Calluna, 294; *see* Heather
Callus, 554
Callus growth, [35]
Calochortus, 347
rust, 58, 347
Calo-clor, 101, 106, 322, 554
Calocure, 101, 106, 321, 322, 324, 554
Calogreen, 148, 554
Calonyction, 351; *see* Moonflower
Calycanthus, 198
canker (twig, branch), 77, 198, 218, 345
crown gall, 83, 136, 198
powdery mildew, 56, 198
Calyx, 554
Camass (*Camassia*), 230
Botrytis blight, 52, 230, 495
leaf spot, 48, 230
root rot, 87, 231
smut, 62, 230
Cambium, 554
Camellia, 198
algal leaf spot, 201–2
angular leaf spot, 48, [199], 200
black mold or mildew, 64, 200
black spot, 48, [199], 200
bud drop 39, 200
bud and flower blight (*Botrytis*), 52, 86, 200
bud rot, 52, 200
chlorosis, 24–29, 200–201
concentric spot, 48, [199], 200
crown gall, 83, 201
damping-off, 76, 199
dieback, canker, 77, 199
flower blight, 52, 86, 198, [199]
graft blight, 199
Growing Camellias, 201
iron deficiency, 24, 26, 27, 200–201
leaf blight, blotch, 51, 199
leaf curl, 62, 200
leaf and flower variegation, 70, 200
leaf scorch, 41, 201
leaf spot, 48, [199], 200
leaf, bud, and stem gall, 62, 200 [201]

Camellia (*continued*)
light and temperature requirements, 40
magnesium deficiency, 28, 200–201
manganese deficiency, 26, 27, 200–201
nitrogen deficiency, 25, 200–201
oedema, corky scab, scurf, 39, 201
Phytophthora root rot, 87, 201
root-knot, 90, 201
root nematode, 201
root rot, 87, 201
soil drench, 201
sooty mold, 64, 200
spot anthracnose or scab, 51, 199
sunscald, 41, 200
white spot, 48, [199], 200
yellow mottle leaf, 70, 200
zinc deficiency, 26, 27, 200–201
Camomile, 219
aster yellows, 73, 223
damping-off, seed rot, 76, 221
petal blight, 86, 221, 224
root-knot, 90, 221
root rot, 87, 221
seed treatment, 221
Campanula, 168; *see also* Bellflower, Canterbury-bells
aster yellows, 73, 169
crown rot, 76, 169
curly-top, 75, 169
damping-off, 76, 169
leaf spot, 48, 169
leaf and stem nematode, 75, 169
powdery mildew, 56, 169
root-knot, 90, 169
root rot, 87, 169
rust, 58, 169, 405
soil drench, 169
southern blight, 76, 169
spotted wilt, ringspot, 72, 169
stem rot or blight, 76, 169
Verticillium wilt, 68, 169
Camphor-tree, 149
anthracnose, 51, 149
black mildew, 64, 151
canker, dieback, 77, 149–50
chlorosis, 24–29, 151, 344
leaf spot, 48, 149
mistletoe, 93, 151
powdery mildew, 56, 150
root nematode, 151
root rot, 87, 149
spot anthracnose or scab, 51, 149
Verticilium wilt, 68, 150
Campion (*Silene*), 203; *see also* Evening campion, Red campion
Alternaria blight, collar and branch rot, 51, 76, 204
anther or flower smut, 62, 206
rust, 58, 204
Campsis, 494; *see* Trumpetvine
Camptosorus, 269; *see* Ferns
Canada, help by Science Service of the Department of Agriculture, 5
Canada Rex Spray Company, Limited, 120
Canary ivy, 310; *see* Ivy
Canarybirdflower, 354; *see* Nasturtium
Canarygrass, 439; *see* Ribbon grass
Canavalia, 155; *see* Jackbean and Bean, garden types
Canby, 173
leaf spot 48, 173
Candleberry, 507; *see* Waxmyrtle
Candles of the Lord, 512; *see* Yucca
Candytuft, 186; *see also* Cabbage
clubroot, 89, 189
damping-off, 76, 188
downy mildew, 55, 189
gray-mold blight, 52, 86, 190–1
mosaic, 70, 192
powdery mildew, 56, 193
ringspot, 72, 192
root-knot, 90, 191
root rot, 87, 193
white-rust, 62, 191
Cane, 554
blight, 77, [247], [428]
canker, [78], [288]
gall, 83, [84], [176]
Canker, [78], [80], 554
bacterial, 77, 79, 386, [483], 484, [487]
bleeding, 161, 259, 345
branch, 77, [217], [265], [344], [358], [403], [411], [503]
cane, [78], [288]
stem, 77, [78], 79, [156], [187], [188], [276], [419], 428
trunk, 77, [258], 259, 445
twig, 77, [138], 218, 345, [383], [391]
Canna, 202
aster yellows, 73, 203
bacterial bud rot and stalk rot, 50, [202], 534
bacterial wilt, 69, 203, 488
leaf spot, 48, 203
mosaic, 70, 202–3
petal blight, 86, 203
rhizome or tuber rot, 90, 196, 203
root nematode, 203
rootstock (tuber) soak, 202, 528, 529, 534
rust, 58, 203
southern blight, crown rot, 76, 196, 203
Cantaloup, 237; *see also* Cucurbit, Muskmelon
anthracnose, 51, 86, [238]
bacterial leaf spot, 50, 239, 245
bacterial soft rot, 83, 86, 238, 244
bacterial wilt, 69, 240
curly-top, 75, 243
downy mildew, 55, 243
fruit rot, 51, 52, 65, 81, 86, [238], 244
Fusarium wilt, 66, 241
leaf blight, 51, [238]
mosaic, 70, 241
powdery mildew, 56, 242
ringspot, 72, 243
root-knot, 90, 244–45
root nematode, 244
root rot, 87, 156, 244
scab, 65, 238–39
seed treatment, 238, 528, 532
seedbed spray, 244
Verticillium wilt, 68, 245
Canterbury-bells, 168
aster yellows, 73, 169
leaf spot, 48, 169
powdery mildew, 56, 169
root-knot, 90, 169
root rot, 87, 169
rust, 58, 169
soil drench, 169
southern blight, 76, 169
spotted wilt, 72, 169
stem rot, 76, 169
Cape coast lily, 249; *see* Crinum
Cape-cowslip, 494; *see also* Tulip
leaf spot, 48, 498
mosaic, 70, 496
Cape-gooseberry, 479; *see* Ground-cherry
Cape-honeysuckle, 494
anthracnose, 51, 494
root rot, 87, 137, 494
Cape-jasmine, 275; *see* Gardenia
Cape-marigold, 220
aster yellows, 73, 222
damping-off, 76, 221
downy mildew, 55, 225
Fusarium wilt, 66, 223
gray-mold blight, 52, 186, 224
mosaic, 70, 222
root-knot, 90, 225
root rot, 87, 221
rust, 58, 223
seed treatment, 221
soil drench, 222
stem rot, 76, 221
Verticillium wilt, 68, 223
Capsicum, 479; *see* Pepper
Capsule, 554
Captan
Acti-dione mixtures, 106
active ingredient, 102, 518
bulb dip, 250, 336, 337
disinfectant, 199, 204, 221, 278
-Folpet mixture, 567
gallon lots, 522
in multipurpose mixes, 102, 111, 523
-PCNB mixtures, 77, 104, 188, 204, 328, 330, 337, 410, 432, 475, 489, 490
post-harvest dip or spray, 102, 208, 383, 465, 518
properties, 100, 102, 518
as protective fungicide, 48, 55, 65, 86, 100, 205, 224, 238, 248, 277, 278, 280, 287, 301, 305, 307, 314, 318, 320, 326, 343, 349, 362, 367, 383, 423, 430, 433, 436, 440, 464, 476, 482, 495, 500, 508, 518, 523, 525
seed treatment, 77, 81, 97, 102, 278, 282, 298, 318, 326, 330, 362, 375, 378, 394, 405, 418, 472, 484, 518, 531
soil application, 77, 97, 102, 106, 162, 167, 188, 248, 255, 261, 320, 326, 330, 406, 489, 490, 518, 544
trade names and distributors, 102, 518
uses, 102, 188, 518
Captan 50-W, 102, 518
-Dieldrin 60–15 Seed Protectant, 102, 518, 531
80-WP Spray-Dip, 102, 518
Garden Spray, 102, 518
7.5 Dust, 102, 188, 518
75 Seed Protectant, 102, 518, 531
Caragana, 300; *see* Peashrub
Caranda, 360; *see* Carissa
Caraway, 213
aster yellows, 73, 214
mosaic, 70, 214
root-knot, 90, 214
seed rot, damping-off, 76, 214
seed treatment, 213, 532
stem rot, 76, 214
Carbamate, 102
Carbon disulfide, 137–38, 555
Cardinal climber, 351; *see* Cypressvine
Cardinalflower, 339; *see* Lobelia
Cardoon, 328; *see also* Globe artichoke
leaf spot, 48, 331
powdery mildew, 56, 331
yellows, 73, 328–29
Carex, rust, 58, 246, 327, 331, 445, 458, 499
Carissa, 360
canker, dieback, 77, 360
leaf spot, 48, 360
root-knot, 90, 360, 392
root rot, 87, 137, 361
Carnation, 203
Alternaria blight, 51, 76, 204
anther smut, 62, 206
anthracnose, 51, 205
aster yellows, 73, 204
bacterial leaf spot, 50, 206
bacterial wilt, 69, 204
boron deficiency, 27, 206
branch rot, 76, 204
chlorosis, 24–29, 206, 443
"cultured" disease-free cuttings, 97, 203
crown gall, root and stem gall, 83, 206
curly-top, [75], 162, 205
cutting dip, 204
cutting rot, 76, 204
damping-off, 76, 204
downy mildew, 55, 206
dry heat treatment, 205
etched ring, 72, 204
ethylene injury, 44
fasciation or leafy gall, witches'-broom, 81, 206
flower blight, 52, 86, 205
foot rot, 76, 204
Fusarium bud rot, 205
Fusarium wilt or yellows, 66, [202], 203–4
gray-mold (*Botrytis*) blight, flower blight, 52, 86, 205

greasy blotch, 51, 205
leaf blight, rot, 51, 205
leaf spot, 48, 205
leaf and stem nematode, 75, 206
mosaic, 70, [71], 205
mottle, 70, 205
Phialophora wilt, 68, 204
powdery mildew, 56, 128, 206
ringspot, 72, 204
root-knot, cyst nematode, 90, 203, 204, 206
root nematode, 203, 204, 206
root rot, 87, 204
rust, 58, 204, [205]
sleepiness, 44
soil drench, 204
southern blight, 76, 204
stem, collar rot, 76, 204
streak, 70, 204
Verticillium wilt, 68, 204
web blight, 160, 206
Carnegiea, 194; *see* Saguaro, Cacti
Carolina allspice, 198; *see* Calycanthus
Carolina Chemicals, Incorporated, 103, 120, 519
Carolina-jessamine, 185
black mildew, 64, 185
black spot, 48, 185
chlorosis, 24–29, 185, 443
iron deficiency, 24, 26, 27
leaf spot, 48, 185
root rot, 87, 137, 185
silky thread blight, 185, 505
sooty mold, 64, 185
Carolina moonseed, 351
burrowing nematode, 351
leaf spot, 48, 351
root rot, 87, 351
Carosel, 161
Carpetgrass, 319; *see also* Bluegrass
brown patch, Rhizoctonia disease, 321
chlorosis, 24–29, 325
fairy ring, 322
leaf blight, 51, 319
leaf spot, 48, 319
Pythium disease, 324
root nematode, 324
rust, 58, 321
Sclerotinia dollar spot or blight, 321
seed treatment, 326, 537
seedling blight, 76, 326
Carpinus, 170; *see* Hornbeam
Carrier, 555
Carrot, 206
Alternaria leaf blight, 51, 206–7
aster yellows, 73, [74], 207
bacterial blight, 50, 206–7
bacterial soft rot, 81, 87, 207, 209
bacterial or southern wilt, 69, 209, 488
black rot, 86, [207]
boron deficiency, root cracking, 27, 209
copper deficiency, 28
crown gall, 83, 209
crown and root canker, 77, 207
curly-top, 75, 162, 209
damping-off, 76, 209
downy mildew, 55, 210
dry rot (Fusarium), 86, 207
fertilizing, 29
gray-mold rot, 52, 86, 206, 207
insect control, 207
irregular roots, 209
leaf blight, 51, 206–7
leaf spot, 48, 206–7
mosaic, 70, 209
motley dwarf, 209
Phymatotrichum root rot, 87, [208]
plant-food utilization, 26
potassium deficiency, 26, 210
resistant varieties, 207, 209
ringspot, 72, 209
root-knot, cyst nematode, 90, [91], 209
root nematode, 208
root rot, 87, [208]
rust, 58, 210
scab, 65, 165, 209, 415
seed rot, 209
seed treatment, 207, 529, 531, 532
sour rot, 87, 207
southern blight, 76, 209
stem nematode, 208
storage rot, 86, [207], 208
cleanliness, 208
violet root rot, 87, 208
watery soft rot, cottony rot, Sclerotinia rot, 76, 86, 156, [207]
web blight, 160, 210
Carrot rust fly, insect control, 208, 214
weevil, insect control, 208
Cart sprayer, 113–14
Carthamus, 220; *see* Safflower
Carum, 213; *see* Caraway
Carya, 501; *see* Hickory, Pecan
Caryopteris, 317; *see* Verbena
Caryota, 371; *see* Palms
Casaba, 237; *see also* Cucurbit, Muskmelon
anthracnose, 51, 238
bacterial leaf spot, 50, 239
bacterial soft rot, 81, 86, 238, 244
bacterial wilt, 69, 240
curly-top, 75, 243
downy mildew, 55, 243
fruit rot, 86, 238, 240, 241, 244
Fusarium wilt, 66, 241
leaf blight, 51, 239
mosaic, 70, 241
ringspot, 72, 243
scab, 65, 238
seed treatment, 238
Verticillium wilt, 68, 245
Casein, as sticker, 119
Cashmere bouquet, 317; *see* Clerodendron
Cassabanana, 238; *see* Sicana
Cassandra, 175; *see also* Chamaedaphne
rust, 58, 176, 405
Cassia, 300
dieback, 77, 300
leaf spot, 48, 301
powdery mildew, 56, 301
root-knot, 90, 301, 392
root rot, 87, 137, 301
Cassiope, 175
leaf gall, 62, 176
Castanea, 216; *see* Chestnut, Chinquapin
Castanopsis, 216; *see* Golden chinquapin
Castilleja, 454
powdery mildew, 56, 455
rust, 58, 455
Cast-iron plant, 148; *see* Aspidistra
light and temperature requirements, 40
Castle Chemical Company, 120
Castor aralia, 126; *see* Acanthopanax
Castorbean, 210
bacterial leaf spot, 50, 210, [211]
bacterial wilt, brown rot, 69, 211
Botrytis capsule mold, 86, 210
capsule drop, 210
capsule mold, 86, 210
cotton (*Phymatotrichum*) root rot, 87, [211]
crown gall, 83, 211–12
gray-mold blight, flower blight, Botrytis blight, 52, 86, 210
leaf spot, 48, 210
red gall, seedling, 212
resistant varieties, 210, 211
root-knot, 90, 211
root nematode, 211
root rot, 87, 137, 156, [211]
seed rot, 211
seed treatment, 211
seedling blight, 76, 211
southern blight, 76, 211
stem, crown rot, 76, 211
Thielaviopsis root rot, 87, 210–11
Verticillium wilt, 68, 211
Casuarina, 212
root-knot, 90, 212, 392
root rot, 87, 137, 212
Catalpa, 212
anthracnose, 51, 212
Botrytis seedling blight, 52, 213, 405
chlorosis, 24–29, 213, 344
crown gall, 83, 213
damping-off, southern blight, 76, 213, 405
dieback, canker, 77, 138, 213, 345
leaf scorch, 41, 213, 344
leaf spot, 48, 212
powdery mildew, 56, 212
root-knot, 90, 213, 392
root nematode, 212
root rot, 87, 137, 212
sooty mold, 64, 213, 266
spot anthracnose, 51, 212
Verticillium wilt, 68, [212], 343
wood rot, 79, 170, 212
Catasetum, 366; *see* Orchids
Catchfly, 203; *see* Silene
Catclaw, 300; *see* Acacia
Catface, 468, 491
Catha, 173
leaf tip blight, 51, 173
Catkin, 555
Catnip, 447
bacterial leaf spot, 50, 448
Fusarium wilt, 66, 448
leaf spot, 48, 447
mosaic, 70, 447
root rot, 87, 447
southern blight, 76, 447
Cat's-claw, 494
root rot, 87, 494
Cattleya, 366; *see* Orchids
Cauliflower, 185; *see also* Cabbage
aster yellows, 73, 193
bacterial leaf spot, 50, 190
bacterial soft rot, 76, 81, 190
black ringspot, 72, 192
black rot, 50, 187
blackleg, 77, 187
boron deficiency, 27, 191
chlorosis, magnesium deficiency, 24–28, 194
clubroot, 89, 189
curly-top, 75, 192
damping-off, 76, 188, 190, 191, 193
downy mildew, 55, 189
drop, cottony rot, 76, 191
Fusarium yellows, 66, 186–87
gray-mold blight, 52, 86, 190, 191
head browning (*Alternaria*), 48, 190
leaf blight, 51, 190
leaf mold, 48, 190, 191
leaf spot, 48, 190, 191
mosaic, 70, 192
oedema, 39, 194
powdery mildew, 56, 193
resistant varieties, 187, 190, 193
root-knot, 90, 191
root nematode, 194
root rot, 87, 187, 190, 193
seed treatment, 187, 529, 532
southern blight, 76, 191
spotted wilt, 72, 192
storage rot, 86, 188, 190, 191
temperature requirements, 41
tipburn, 191
Verticillium wilt, 68, 193
whiptail, molybdenum deficiency, 28, 193
white-rust, 62, 191
Cauliflower disease, 467
Caulophyllum, 153; *see* Blue cohosh
Causal organism, 555
Ceanothus, 355; *see* New Jersey-tea
Cedar (*Cedrus*), 402; *see also* Chamaecyparis for Incense-cedar, Port Orford or Lawson, White-cedar, and Yellow-cedar; *see* Juniper for Redcedar
canker, dieback, 77, 403
root rot, 87, 405, 406
tip blight, 402, 408
wood rot, 79, 170, 403

Cedar apple, [59], 313, 555
Cedar of Lebanon, 402; *see* Cedar
Cedrela, 218; *see also* Chinaberry
  wood rot, 87, 219
Cedrus, 402; *see* Cedar
Celandine, 413
  leaf spot, 48, 414
  root rot, 87, 414
  soil drench, 414
Celastrus, 173; *see* Bittersweet
Celeriac, 213; *see also* Celery
  aster yellows, 73, 214
  bacterial soft rot, 81, 214
  boron deficiency, 27, 215
  curly-top, 75, 215
  damping-off, 76, 214
  leaf blight, 51, 213
  mosaic, 70, 214
  ringspot, 72, 215
  seed treatment, 213, 529, 532
  spotted wilt, 72, 215
  stem nematode, 75, 215
  Verticillium wilt, 68, 215
Celery, 213
  anthracnose, 51, 213
  aster yellows, 73, 214
  bacterial petiole spot, 50, 215
  bacterial soft rot, 81, 214
  bacterial spot, blight, 50, 213
  blackheart or heart rot, calcium deficiency, 28, 214–15
  boron deficiency, stem-cracking, brown checking, 27, 215
  calico, 70, 214
  crown or basal rot, 76, 215
  curly-top, 75, 215
  damping-off, 76, 214
  downy mildew, 55, 210, 215
  early blight, 51, 213
  foot rot, 76, 215
  Fusarium wilt or yellows, 66, 214
  gray-mold rot, 52, 214
  insect control 214, 215
  late blight, [51], 213
  leaf blight, 51, 213
  leaf spot, 48, 213
  magnesium deficiency, 28, 215
  mosaic, 70, 214
  motley dwarf, 73, 214
  pink rot, 76, 214
  plant-food utilization, 26
  resistant varieties, 213, 214, 215
  ringspot, 72, 215
  root-knot, 90, 214
  root nematode, 215
  root rot, 87, 215
  seed rot, 214
  seed treatment, 213, 529, 532
  spotted wilt, 72, 215
  stem nematode, 75, 215
  stem or stalk rot, 76, 214
  storage rot, 86, 214, 215
  streak, 214
  temperature, effect on, 41
  transplant dip, 214
  Verticillium wilt, 68, 215
  yellow spot, 214
Cellu-quin, 555
Celosia, 229; *see* Cockscomb
Celtis, 292; *see* Hackberry
Celtuce, 328; *see* Lettuce
Centaurea, 220
  aster yellows, 73, 222
  curly dwarf, 73, 222
  downy mildew, 55, 225
  Fusarium wilt, 66, 223
  gray-mold blight, 52, 86, 224
  powdery mildew, 56, 221
  ringspot, 72, 223
  root-knot, 90, 225
  root nematode, 226–27
  root rot, 87, 221
  rust, 58, 223
  soil drench, 222
  southern blight, 76, 221
  stem or crown rot, 76, 221
  Verticillium wilt, 68, 223
  white-rust, 62, 226
Centerchem Mercuric Chloride, 101
Centigrade, conversion to Fahrenheit, 517
Centipedegrass, 319; *see also* Bluegrass
  anthracnose, 51, 319
  brown patch, Rhizoctonia disease, 321
  chlorosis, 24–29, 325
  fairy ring, 322
  leaf spot, blotch, 48, 319
  root nematode, 324
  Sclerotinia dollar spot or blight, 321
  seed treatment, 326, 537
  seedling blight, 76, 326
  slime mold, 322
Centipedes, lethal temperature, 542
Centranthus, 499
  leaf spot, 48, 499
Centrosema, 377
  leaf spot, 48, 380
  root rot, 87, 378
Centuryplant, 216
  anthracnose, black rot, 51, 216
  gray-mold blight, 52, 167, 216
  leaf scorch, blight, 51, 216
  leaf spot, 48, 216
  light and temperature requirements, 40
  root-knot, 90, 216
Cephalanthus, 185; *see* Buttonbush
Cephalocereus, 194; *see* Cacti, Cereus
Cephalosporium wilt, 398
Cephalotaxus, 311
  twig blight, 51, 77, 311, 313
Cerano, as tree wound dressing, 36, 384, 477
Cercis, 300; *see* Redbud
Cercospora leaf spot, 149, 162, [163], 180, [394], [434]
Ceresan, 106, 394, 555
  L, 106, 282, 555
  M, 148, 555
  -200, 106
  2%, 555
Cereus, 194; *see also* Cacti
  anthracnose, 51, 194
  bacterial soft rot, 76, 81, 194
  bud drop, 25, 195
  collar rot, 76, 194
  corky scab, 39, 194
  Fusarium wilt, 66, 194
  glassiness, 39, 195
  gray-mold blight, 52, 195
  leaf scorch, 48, 195
  leaf spot, 48, 195
  root-knot, cyst nematode, 90, 194 195
  root nematode, 194
  root and stem rot, wilt, 76, 87, 194–95
  stem and branch rot, 76, 194
Ceriman, 196
  leaf spot, blight, 48, 51, 197
Certification (seed, transplants, cuttings, etc.), 555
Chaenomeles, 133; *see* Flowering quince
Chaerophyllum, 213; *see* Celery, Salad chervil
Chamaecyparis, 312
  canker, twig and branch, 77, 314
  chlorosis, 24–29, 315
  crown gall, 83, 136, 315
  dieback, 77, 313
  gray-mold blight, 52, 314
  iron chlorosis, 24, 26, 27, 315
  leaf blight, 51, 314
  nursery or juniper blight, 51, 77, 313
  root rot, 87, 314, [315]
  rust
    gall, 56, 313
    leaf, 56, 313
    witches'-broom, 56, 313
  twig blight, 77, 314
  winter injury, sunscorch, 41, 314
  wood rot, 79, 170, 314
Chamaedaphne, 175
  chlorosis, 24–29, 178
  iron deficiency, 24, 26, 27, 178
  leaf gall, 62, 178
  leaf spot, 48, 178
  rust, 58, 176, 470
Chamaerops, 371; *see* Palms
Chamberlain Corporation, Dobbins Division, 120
Champion Sprayer Company, 120
Chapin, R. E. Manufacturing Works, Incorporated, 120
Chapman Chemical Company, 120
Chaste-tree, 317
  leaf spot, 48, 317
  root rot, 87, 318
Chayote, 238; *see also* Cucurbit
  anthracnose, 51, 238
  fruit rot, 86, 244
  leaf spot, 48, 239
  mosaic, 70, 241
  root-knot, 90, 244–45
  root nematode, 244
  seed treatment, 244, 532
  southern blight, 76, 243
  Verticillium wilt, 68, 245
Checkerberry, 294
  fruit spot, 86, 294
  leaf spot, 48, 294
  powdery mildew, 56, 294
  red leaf gall, 294
  sooty mold or blotch, 64, 294
Checkered-lily, 494; *see* Fritillary
Checkermallow, 297; *see* Sidalcea
Cheeseweed, rust, 298
Cheiranthus, 186; *see* Wallflower
Chelated Nutramin, 27
Chelates, 27, 555
  copper, 28
  iron, 27
  manganese, 27
  zinc, 27
Chelidonium, 413; *see* Celandine
Chelone, 454; *see* Turtlehead
Chemagro Corporation, 120
Chem-Bam, 562
Chemical Formulators, Incorporated, 120
Chemical injury, 43–45
Chemical soil treatments, 98, 106, 538–49
Chemley Products Company, 120
Chemotherapeutant, 100
Chemotherapy, 100, 555
Chem-vape, 106, 545, 547
Chenille plant, 126; *see* Acalypha
Cherry, 382; *see also* Flowering cherry
  bacterial canker, 77, 386
  bacterial leaf spot, 50, 386
  black knot, [81], 384
  blossom blight, 86, 382
  brown rot, 51, 86, 382
  chlorosis, 24–29, 393
  Coryneum blight, 77, 390
  crown gall, 83, 391
  dieback, 77, 383
  felt fungus, 293, 393
  fire blight, 79, 133, 393
  fruit spot or rot, 86, 382, 384, 385, 392
  leaf blister, 62, 384
  leaf spot, yellow leaf, 48, 384, [385]
  little cherry, 390
  mistletoe, 93, 393
  mottle complex, 70, 390
  powdery mildew, 56, 385
  prune dwarf, 73, 389
  pruning, 32, 383, 384
  rasp leaf, leaf enation, 390
  resistant varieties, 383, 384, 386, 387
  ringspot, tatterleaf, 72, 388
  root-knot, 90, 392
  root nematode, 392
  root rot, 87, 392
  rosette, 73, 387
  rust, 58, 392
  scab, 65, 385
  shot-hole, 52, [53], 384, 386, 390, 392
  silver leaf, 79, 170, 384
  sour cherry yellows, 73, 388
  spray schedule, 524–25
  trunk canker, 77, 383
  twig blight, 77, 382
  2, 4-D injury, 288, 393
  Verticillium wilt, 68, 391
  winter injury, 42, 392

witches'-broom, 62, [63], 384
wood rot, 79, 170, 384, [385]
X-disease, 387
Cherry-laurel, 382; *see also* Cherry
bacterial spot, 50, 386
blossom blight, 86, 382
brown rot, 51, 77, 382
fire blight, 79, 393
leaf spot, 48, 393
mistletoe, 93, 393
powdery mildew, 56, 385
root rot, 87, 392
shot-hole, 52, 386, 390, 392
thread blight, 393, 505
twig blight, 77, 382
Verticillium wilt, 68, 391
witches'-broom, 62, 384
Chervil, 213; *see* Salad chervil
Chestnut, 216
anthracnose, 51, 218, 356
blossom-end rot of nuts, 86, 218
Botryosphaeria twig canker, 77, 218
Chestnut Blight and Resistant Chestnuts, 217
Cryptodiaporthe twig canker, 77, 218
dieback, 77, 218
Endothia canker, blight, 77, 216, [217]
leaf spot, 48, 218
mistletoe, 93, 218
nut rot, 86, 218
oak wilt, 218, 355
Phytophthora root rot, 87, 218
powdery mildew, 56, 218
resistant varieties, 217, 218
root rot, 87, 218, 358
rust, 58, 218, 405
twig blight, 77, 218
Verticillium wilt, 68, 218, 343
wood rot, 79, 218
Chevron Chemical Company, Ortho Division, 102, 103, 105, 120, 518, 519
Chicken gizzard plant, 230; *see* Bloodleaf
Chickweed
rust, 58, 405
as virus source, 73
Chicory, 328
anthracnose, 51, 331
aster yellows, 73, 328–29
bacterial rot, 50, 330
bacterial soft rot, 81, 86, 328
boron deficiency, 27, 331
bottom rot, 76, 328
damping-off, 76, 330
downy mildew, 55, 329
drop, sclerotiniose, watery soft rot, 76, 328
gray-mold blight, rot, 52, 328, 329
leaf spot, 48, 331
mosaic, 70, 329
powdery mildew, 56, 331
ringspot, 72, 331
root-knot, 90, 332
root nematode, 332
root rot, 87, 331
rust, 58, 330–31
seed rot, 330
seed treatment, 330, 532
slime mold, 322, 332
southern blight, 76, 328
spotted wilt, 72, 331
tipburn, 328
Verticillium wilt, 68, 332
white smut, 65
Chiggers, mite control, 321
Chilean tarweed, rust, 405
Chilopus, 212; *see* Desert-willow
China-aster or annual aster, 221
anthracnose, 51, 221
aster yellows, 73, [74], 222
blossom blight, 86, 221, 224
Botrytis blight or gray-mold, 52, 86, 224
chlorosis, 24–29, 227
curly dwarf, 73, 222
curly-top, 75, 223
downy mildew, 55, 225
Fusarium wilt, stem rot, 66, [67], 76, 223
leaf blight, 51, 221
leaf spot, 48, 221
mosaic, 70, 222
powdery mildew, 56, 221
resistant varieties, 223, 224
ringspot, 72, 222
root-knot, 90, 225
root nematode, 226–27
root rot, 87, 221
rust, 58, 223
seed treatment, 221, 223, 528, 529, 534
soil drench, 222
southern blight, 76, 221
spotted wilt, 72, 223
stem canker, 77, 221, 225
stem or foot rot, 76, 221
Verticillium wilt, 68, 223
China-tree, 218; *see* Chinaberry
Chinaberry, 218
black mildew, 64, 218
canker, limb blight, 77, 218
downy mildew, 55, 218
leaf spot, 48, 218
mistletoe, 93, 219
powdery mildew, 56, 218
root-knot, 90, 218, 392
root rot, 87, 137, 219
sooty mold, 64, 218
thread blight, 219, 505
twig blight, 77, 218
Verticillium wilt, 68, 219
wood rot, 79, 219
Chincherinchee, 495; *see* Star-of-Bethlehem
Chinese angelica, 126; *see* Aralia
Chinese artichoke, 447; *see* Stachys
Chinese beautybush, 499; *see* Beautybush
Chinese bellflower, 168; *see* Balloonflower
Chinese bittersweet, 173; *see* Bittersweet
Chinese cabbage, 186; *see also* Cabbage
anthracnose, 51, 190
aster yellows, 73, 193
bacterial leaf spot, 50, 190
bacterial soft rot, 76, 81, 190
black rot, 50, 187
blackleg, 77, 187
chlorosis, 24–29, 194
clubroot, 89, 189
curly-top, 75, 192
damping-off, 76, 188, 190, 191, 193
downy mildew, 55, 189
leaf spot, 48, 190, 191
mosaic, 70, 192
powdery mildew, 56, 193
resistant varieties, 190
root-knot, 90, 191
root nematode, 194
seed treatment, 187, 529, 532
southern blight, 76, 191
white-rust, 82, 191
Chinese cedrela, 218; *see* Cedrela
Chinese chives, 261; *see* Chives, Onion
Chinese evergreen, 195
bacterial leaf spot, 50, 198
bacterial soft rot, 81, 196, 198
leaf rot, 51, 197
light and temperature requirements, 40
plant (cane) soak, 197, 530, 534
root-knot, 90, 197
root nematode, 197
root rot, 87, 196
soil drench, 197
stem (cane) rot, 76, 81, 196–97
Chinese fan, 370–71; *see* Palms
Chinese forget-me-not, 349; *see* Houndstongue
Chinese hibiscus, 297; *see* Hibiscus (arborescent forms)
Chinese hollygrape, 153; *see* Mahonia, Oregon-grape
Chinese houses; 454; *see* Collinsia
Chinese lanternplant, 479
angular leaf spot, 50, 483
bacterial wilt, 69, 488
chlorosis, 24–29, 490
curly-top, 75, 486
leaf spot, 48, 480
leaf and stem nematode, 76, 400, 492
mosaic, 70, 484
ringspot, 72, 486
root-knot, 90, 489
root nematode, 489
root rot, 87, 490
rust, 58, 492
southern blight, 76, 490
spotted wilt, 72, 486
Verticillium wilt, 68, 487
white smut, 65, 492
wildfire, 50, 483
Chinese parasoltree, 401; *see* Phoenixtree
Chinese primrose, 423; *see* Primrose
Chinese redbud, 300; *see* Redbud
Chinese sacred lily, 249; *see* Daffodil
Chinese scholartree, 300; *see* Sophora
Chinese stranvaesia, 133; *see* Stranvaesia
Chinese tallowtree, 210
leaf spot, 48, 210
root rot, 87, 137, 156, 210–11
Chinese trumpetcreeper, 169; *see* Bignonia
Chinese waxgourd, 238; *see also* Cucurbit
anthracnose, 51, 238
downy mildew, 55, 243
root-knot, 90, 244–45
seed treatment, 238
Chinese wolfberry, 347; *see* Matrimonyvine
Chinosol, 367
Chinquapin, 216; *see also* Chestnut, Golden chinquapin, Oak
blight, canker, 77, 216–17
brown felt canker, 218, 293
leaf blister, 62, 218, 357
leaf spot, 48, 218
oak wilt, 218, 355
powdery mildew, 56, 218
root rot, 87, 218, 358
wood rot, 79, 170, 218
Chionanthus, 144; *see* Fringetree
Chionodoxa, 494; *see also* Tulip
bulb nematode, 92, 496–97
bulb soak, 497, 530, 531, 537
Chipcote, 106, 282, 555
Chipman Chemical Company, Incorporated, 102, 103, 105, 120, 518, 519
Captan Dust, 102, 518
Chipcote, 106, 282, 555
Ferbam W-76, 102, 518
Spreader-Activator, 119
Zineb, 103, 519
Ziram, 103, 519
Chives, 361; *see also* Onion
bulb nematode, 92, 364
bulb rot, 90, 361, 362
downy mildew, 55, 363
gray-mold blight, 52, 361, 362
leaf spot, blotch, 48, 51, 361, 362, 363, 365
rust, 58, 365
seed treatment, 362, 533
smut, 62, 362
soil treatment, 362
Verticillium wilt, 68, 365
Chloranil
active ingredient, 102, 518
seed, corm and bulb treatment, 77, 97, 102, 165, 207, 210, 213, 221, 222, 233, 238, 273, 282, 299, 330, 375, 378, 418, 470, 518, 531
soil drench, 102, 106, 518
spray or dust, 102, 188, 518
trade names and distributors, 102, 518
uses, 102, 518
Chlordane, as soil insecticide, 23, 82, 86, 98, 166, 178, 187, 208, 215, 233, 250, 321, 418, 431, 435, 438, 463, 475, 512

Chlorine, as disinfectant, 190, 366
as seed treatment, 233
Chlorobenzilate, injury, 45
Chloropicrin (tear gas)
active ingredient, 545
disinfesting soil, 83, 91, 106, 188, 214, 253, 308, 322, 336, 400, 406, 424, 430, 437, 441, 448, 463, 467, 472, 473, 487, 489, 490, 545
mixture with methyl bromide, 473, 545
precautions, 543, 545
storage fumigation, 471
trade names, 106, 545
Chlorosis, 24–29, 44, [179], 276, 336, 344, [359], 382, 393, 408, 432, 434, 467, [496]
controlling, 24, 26–28, 276, 344, 359
defined, 555
iron, 24, 26, 27, [179], [359]
Chokeberry, 133; *see also* Apple
blossom blight, 86, 133–34
canker, dieback, 77, 138
chlorosis, 24–29, 143
fire blight, 79, 133–34
fruit rot, 86, 138
iron deficiency, 24, 26, 27, 143
leaf spot, 48, 140
rust, 58, 135
Chokecherry, 382; *see* Cherry, Peach
Cholla, 194; *see* Cacti, Opuntia
Christmas cactus, 194; *see* Cacti, Epiphyllum
Christmas cherry, 479; *see* Jerusalem-cherry, Eggplant
Christmas-rose, 254
black spot, blight, 48, 51, 86, [256]
downy mildew, 55, 257
flower spot, 86, 255, 256
gray-mold blight, 52, 86, 255, 257
leaf spot, blotch, 48, 51, 256
soil drench, 254, 255
stem or crown rot, 76, 254
Christmasberry, 133; *see* Photinia
Chrysalidocarpus, 371; *see* Palms
Chrysanthemum, 219
aster yellows, 73, 222
bacterial blight or stem rot, 50, [225]
bacterial soft rot, 81, 221
bacterial wilt or stem rot, 69, 223
blossom blight, 52, 86, 221, 224
chlorosis, 24–29, 227
crown gall, 83, 226
crown rot, 76, [77], 221
"cultured" (disease-free) cuttings, 97, 222, 223, 225
curly-top, 75, 223
cutting dip, 221
cutting rot, 76, 221
damping-off, seed rot, 76, 221
fasciation or leafy gall, 81, 226
flower distortion, 223
Fusarium wilt, 66, 223
gray-mold blight, 52, 86, 224
Growing Chrysanthemums in the Home Garden, 227
head blight, [85], 86, 224
Itersonilia petal blight, 86, 224–25
leaf blight, 51, [217], 221, [224]
leaf or foliar nematode, 75, [76], 226
leaf spot, 48, [217], 221
light, effect on flowering, 39
and temperature requirements, 40
mosaic, leaf curl, mottle, 70, 222
plant soak, 226, 530, 534
powdery mildew, 56, [221]
ray blight, speck, 86, 221, [224]
resistant varieties, 221, 223, 225
ringspot, 72, 222, 223
root-knot, 90, 225
root nematode, 226–27
root rot, [77], 87, 221
rosette, 73, 222
rust, 58, [222], 223
seed treatment, 221
Septoria leaf spot and blight, [217], 221
soil drench, 222
southern blight, 76, 221
spotted wilt, ringspot, 72, 223
stem (foot) rot, canker, 77, 221, 225
Stemphylium petal specking and leaf blight, [224]
stunt, 73, 222
tomato aspermy, 223
2, 4-D injury, 226, 288
Verticillium (*Phialophora*) wilt, 68, 223
white-rust, 62, 226
white or leaf smut, 65, 226
yellow dwarf, 73, 222
yellow strapleaf, 227
Chrysobalanus, 227
algal spot, 227, 341
leaf spot, 48, 227
Chrysopsis, 220; *see* Golden-aster
Chufa, 498; *see* Umbrellaplant
Chuperosa, 229
rust, 58, 229
Cibotium, 269; *see* Ferns
Cicada killer, insect control, 321
Cichorium, 328; *see* Endive, Escarole, Chicory
Cigarflower, 227
gray-mold blight, 52, 167, 227
leaf spot, 48, 227
powdery mildew, 56, 227
root-knot, 90, 227
root rot, 87, 227, 278
Cimicifuga, 131
black mildew, 64
chlorosis, 24–29
downy mildew, 55, 131
iron deficiency, 24, 26, 27
leaf spot, 48, 131
root-knot, 90, 132
rust, 58, 131
smut, 62, 131
Cinchona, 185
leaf spot, 48, 186
root-knot, 90, 186
root rot, 87, 137, 186
Cineraria, Florists', 220; *see also* Senecio
aster yellows, 73, 222
damping-off, seed rot, 76, 221
downy mildew, 55, 225
Fusarium wilt, 66, 223
gray-mold blight, 52, 86, 224
leaf spot, 48, 221
light and temperature requirements, 40
mosaic, 70, 222
powdery mildew, 56, 221
root-knot, 90, 225
root rot, 87, 221
seed treatment, 221
soil drench, 222
spotted wilt, 72, 223
stem rot, 76, 221
streak, 72, 223
Verticillium wilt, 68, 223
Cinnamomum, 149; *see* Camphor-tree, Cinnamon-tree
Cinnamon-tree, 149; *see also* Camphor-tree
anthracnose, 51, 149
leaf spot, 48, 149
Cinnamonvine, 510; *see* Yam
Cinquefoil, 440; *see* Potentilla
Cirsium, 220; *see* Thistle, Plumed thistle
Cissus, 287; *see also* Grape
leaf spot, 48, 287, 291
light and temperature requirements, 40
root rot, 87, 137, 289
rust, 58, 291
smut, 62, 290
Citraldin, 227; *see* Hardy orange, Citrus
Citron, 238; *see also* Cucurbit, Watermelon
anthracnose, 51, 238
curly-top, 75, 243
downy mildew, 55, 243
fruit rot, 81, 86, 238, 244
Fusarium wilt, 66, 241
leaf spot, 48, 238, 239, 244
mosaic, 70, 241
powdery mildew, 56, 242
scab, 48, 65, 86, 238–39
seed rot, 244
seed treatment, 238, 528, 532
Verticillium wilt, 68, 245
Citrullus, 238; *see* Citron, Watermelon, Cucurbit
Citrus, 227
anthracnose, withertip, 51, 228
bacterial blast, 50
chlorosis, 24–29, 228
collar or foot rot, 76, 227–28
Color Handbook of Citrus Diseases, 228
crown gall, 83, 136, 228
damping-off, 76
Florida Guide to Citrus Insects, Diseases, and Nutritional Disorders, 228
fluorides injury, 43–44
fruit rot, 86
Handbook of Citrus Diseases in Florida, 228
iron deficiency, 24, 26, 27, 228
leaf spot, 48, 228
leaf yellowing, mottle-leaf, 24–29, 228
light and temperature requirements, 40
mistletoe, 93
other diseases, 228
Phytophthora root rot, 87, 227–28
plant soak, 228, 531, 537
root-knot, 90, 228
root nematode, 228
root rot, 87, 227–28
scab or spot anthracnose, 65, 228
sooty blotch or mold, 64, 228
Thielaviopsis root rot, 87, 227–28
tree decline, 87, 227–28
twig blight, 77, 228
withertip, 51, 228
wood rot, 79, 170
zinc deficiency, 26, 27, 228
Cladosporium leaf blight or blotch, [256]
Cladrastis, 300; *see* Yellowwood
Clarkia, 275
anthracnose, 51, 275
aster yellows, 73, 275
chlorosis, 24–29
damping-off, 76, 275
downy mildew, 55, 225, 275
Fusarium wilt, 66, 275
gray-mold blight, 52, 275
iron deficiency, 24, 26, 27
leaf spot, 48, 221, 275
rust, 58, 275
soil drench, 275
stem canker, 77, 275
stem (foot) rot, 76, 275
Verticillium wilt, 68, 275
Clary, 446; *see* Salvia
Cleary, W. A. Corporation, 120
Cleistothecia, [57]
Clematis, 228
black mildew, 64, 229
crown gall, 83, 229
dieback, 77, 228
leaf blight, 51, 228
leaf spot, 48, 228
mosaic, 70, 229
powdery mildew, 56, 229
root-knot, cyst nematode, 90, 229
root nematode, 229
rust, 58, 229
smut, 62, 229
stem rot, wilt, 76, 228
Cleome, 460; *see* Spiderflower
Clerodendron, 317
leaf gall, 62, 317
leaf spot, 48, 317
root-knot, 90, 317
Clethra, 470
chlorosis, 24–29
iron deficiency, 24, 26, 27
leaf spot, 48, 346, 470
root rot, 87, 137, 470
Cliffgreen, 173
leaf spot, 48, 173
Cliffrose, 439
rust, 58, 441
Cliftonia, 184; *see* Buckwheat-tree

Climbing hempweed, 220
leaf spot, 48, 221
rust, 58, 223
Climbing mignonette, 340; *see also* Lythrum
root-knot, 90, 340
Clinopodium, 446
leaf spot, 48, 447
root-knot, 90, 447
rust, 58, 447
Clitoria, 377
leaf spot, 48, 380
root rot, 87, 378
Clivia, light and temperature requirements, 40
Clockvine, 229
aster yellows, 73, 229
crown gall, 83, 229
root-knot, 90, 229
root nematode, 229
Clone, 555
Clorox, as disinfectant, 35, 99, 134
Cloth-of-gold, 281; *see* Crocus
Cloudberry, 427: *see* Blackberry, Raspberry
Clove, 555
Clove currant, 245; *see* Flowering currant
Clovetree, 268; *see* Eugenia, Eucalyptus
Clubroot, [89], 90, [189]
susceptible weeds, 89–90
CM-19, 106, 555
Cnicus, 220
southern blight, 76, 221
Coalesce, 555
Cocculus, 351; *see* Carolina moonseed
Cochlearia, 186; *see also* Cabbage
white-rust, 62, 191
Cockscomb, 229
black ringspot, 72, 192, 230
curly-top, 75, 162, 230
damping-off, seed rot, 76, 162, 230
leaf blight, spot, 48, 51, 230
leaf roll, 230, 418
mosaic, 70, 230
root-knot, 90, 230
root nematode, 230
root rot, 87, 230
soil drench, 230
stem (crown) rot, 76, 230
Verticillium wilt, 68, 230
Coconut (*Cocos*), 371; *see* Palms
Cocoplum, 227
algal spot, 227, 341
leaf spot, 48, 227
C-O-C-S, 104
Codiaeum, 237; *see* Croton
Coelogyne, 366; *see* Orchids
Coffeeberry, 182; *see also* Buckthorn
leaf spot, 48, 184
rust, 58, 184
sooty mold, 64, 184
Coffeetree, 300; *see* Kentucky coffeetree
Cohosh; *see* Baneberry, Blue cohosh, Cimicifuga
Colchicum, 230
corm or bulb rot, 90, 230, 495
leaf spot, 48, 230
smut, 62, 230
tip blight (*Botrytis*), 52, 230, 495
Cold frame, 555
Coleus, 446
blossom blight, 86, 448
cutting rot, 76, 447
damping-off, 76, 447
gray-mold rot, 52, 86, 447
leaf nematode, 75, 448
leaf spot, blight, 48, 51, 447
light and temperature requirements, 40
mosaic, 70, 447
root-knot, 90, 447
root nematode, 447
Verticillium wilt, 68, 448
Collar rot, 76, 81, [258], 259
Collard, 185; *see also* Cabbage
anthracnose, 51, 190
bacterial leaf spot, 50, 190
bacterial soft rot, 76, 81, 190
chlorosis, 24–29, 194
clubroot, 89, 189
curly-top, 75, 192
downy mildew, 55, 189
Fusarium yellows, 66, 186–87
leaf spot, 48, 190, 191
powdery mildew, 56, 193
seed treatment, 187, 528, 532
Verticillium wilt, 68, 193
white-rust, 62, 191
Colleges, Land-grant,
help by, 5, 6
listing of, 6–7
Collinsia, 454
leaf spot, 48, 455
root rot, 87, 455
rust, 58, 455
soil drench, 455
white smut, 65, 457
Collomia, 399
powdery mildew, 56, 399–400
rust, 58, 400
stem nematode, 75, 400
Colocasia, 195; *see* Elephants-ear, Dasheen
Coltsfoot, 221
leaf spot, 48, 221
Columbine, 254
chlorosis, 24–29, 257, 443
damping-off, 76, 256
gray-mold blight, 52, 86, 255
leaf blotch, blight, 51, 256
leaf spot, 48, 256
leaf and stem smut, 62, 257
mosaic, 70, 255
powdery mildew, 56, 256
ringspot, 72, 257
root-knot, 90, 256
root nematode, 256
root rot, 87, 254
rust, 58, 256
soil drench, 254, 255, 256
stem or crown rot, 76, 254
Columbo, 278
black mildew, 64, 278
leaf spot, 48, 278
rust, 58, 278
Colutea, 300; *see* Bladder-senna
Commelina, 493; *see* Dayflower
Commercial Storage of Fruits, Vegetables, Florist, and Nursery Stocks, 100
Community pot, 555
Compassplant, 220; *see* Silphium
Compatible, 555
Compatibility chart, for pesticides, 552
Complete fertilizer, 555
Compost, 23, 555
Compressed air sprayer, 112, [113]
Comptonia, 470; *see* Sweetfern
Concentric, 555
Coneflower, 220; *see* Rudbeckia, Prairie-coneflower, Purple-coneflower
Confederate-jasmine, 360
black mildew, 64, 360
leaf spot, 48, 360
root rot, 87, 137, 361
sooty mold, 64, 360
Confederate-rose, 297; *see* Hibiscus (arborescent forms)
Conifers, injury by
changing soil grade, 46
fluorides, 43–44
oxidant or PAN injury, 44
ozone, 44
soil drainage, poor, 36
sulfur dioxide, 43
Conk, [79], [147], [161], [170], 555
Construction damage, 46
protection against, 46
Protecting Shade Trees During Home Construction, 46
Protecting Trees Against Damage From Construction Work, 46
Control methods, plant diseases, 39–94, 96–111, 556; *see also* Cultural practices
air circulation, increase, 39, 50, 54, 58, 62
air humidity
decrease, 39, 48, 55, 56, 62, 204
increase, 39
alternate host, destroy or remove, 62, 99
basic methods, 97
cover crop, 92
crop rotation, 16, 48, 56, 65, 68, 70, 77, 81, 83, 86, 89, 90, 92, 93, 98
cropping practice, change, 65, 98, 222
cultivating, 41, 89, 98, 99, 175, 233, 523
curing bulbs, corms, tubers, 52–53
insect and mite control, 48, 65, 68, 70, 72, 73, 74, 79, 81, 82, 83, 86, 87, 98, 187, 192, 208, 231, 232, 233, 242, 433, 435, 437
mulch, 52, 76, 86–87, 98, 99, 523
nematocides, 16, 76, 89, 91–92, 538–49
organic, matter, add, 92, 165, 193, 415, 473
plant disease-free seed, corms, cuttings, bulbs, tubers, nursery stock, 48, 52, 55, 56, 62, 64, 65, 68, 70, 72, 73, 76, 77, 81, 83, 89, 90, 92, 93, 94, 97, 164, 238, 418, 420, [466], 523
in disease-free soil, 55, 68, 70, 77, 81, 83, 89, 91, 93, 94
hot water soaked seed, tubers, plants, etc., 76, 93, 97, 168, 173, 187, 196, 197, 198, 207, 213, 221, 225, 226, 228, 238, 248, 252, 254, 256, 270, 282, 283, 285, 290, 301, 307, 310, 315, 331, 338, 354, 363–64, 369, 375, 400, 443, 450, 453, 455, 458, 467, 472, 484, 493, 497, 529–31
quarantines, 99, 154, 246
at recommended time (early or late), 75, 92, 98, 99, 164, 486
resistant varieties or plants, 48, 56, 58, 62, 64, 65, 67–68, 70, 72, 75, 76, 79, 87, 89, 92, 94, 97, 186–87, 238–43
shallow, 41, 99
in sterile rooting medium, 55, 68, 77, 78, 81, 83, 89, 90, 91, 93, 214
virus-free stock, 52, 72, 73, 177, 222, 250, 387, 419, 429, [466], 473, 475, 485
in well-prepared, well-drained soil, 68, 70, 77, 78, 81, 82, 86, 89, 93, 98, 210, 276, 433, 463
propagate from healthy stock, 55, 68, 70, 72, 76, 77, 470
proper handling and storage of fruit and vegetables, 55, 83, 86, 99, 207–8
pruning, [32], [33], [34], 52, 62, 65, 68, 78, 79, 80, 81, 89, 99, 523
rodent control, 42, 523
sanitation, 48, 52, 54, 56, 58, 62, 64, 65, 68, 70, 72, 73, 76, 77, 78, 81, 82, 83, 87, 89, 90, 92, 93, 94, 98, 99, 428, [442], 528
seed, corm, bulb treatment, 64, 65, 68, 70, 77, 78, 81, 83, 86, 87, 89, 90, 93, 97, 100, 238, 418
soil drench, 77, 82, 89, 97, 188, 189
fertility, balanced, 54, 56, 68, 80, 89, 191, 245, 416, 432, 484
fertilizing, 48, 65, 68, 77, 79, 92, 96, 99, 523
grade change, [46], [47]
pH, adjust, 65, 83, 89, 178, 193, 200, 299, 415, 419, 433, 435, 473, 482
treatment, 77, 82, 83, 91–92, 93, 94, 98, 538–49
space plants, 50, 54, 56, 58, 77, 86, 89
spray or dust program, 48, 52, 55, 56, 58, 62, 65, 68, 70, 72, 73, 74, 76, 78, 81, 82, 86, 87, 98, 100, [442], 523

Control methods (*continued*)
  stake trees and shrubs, 36, [37]
  temperature, change, 58, 62, 70, 83, 190, 214, 410
  water control, 39, 48, 54, 56, 58, 62, 68, 76, 77, 79, 86, 89–90, 98, 99, 155, 157, 209, 239
  weed control, 52, 54, 62, 65, 68, 70, 72, 73, 89, 92, 93, 98, 162, 165, 223, 241, 328, 368, 429, 480, 484
  winter protection, [42], 441, [442], 463
  wound dressing, apply, 65, 68, 79, 81, 86, 476–77
    treatment, [34], [35], 36, 78, 79
  wounding, avoid, 65, 78, 79, 83, 86, 89, 98, 411
Convallaria, 335; *see* Lily-of-the-valley
Conversion
  factors, 517
  Fahrenheit to Centigrade and vice versa, 517
  gallon lots of spray, 522
  pesticide materials on small areas, 521
Convolvulus, 351; *see* California-rose
Cooperative Extension Service, help by 97, 101; *see also* County agent, and Extension entomologist, horticulturist, plant pathologist
Cooperia, 249
  leaf spot, 48, 253
  rust, 58, 253
Cootamundra, 300; *see* Acacia
Copoloid, 239
Coposil, 104
Copper
  chelates, 28
  -Cure, 556
  -Curo Coppernate, 556
  deficiency, 28
  dihydrazinium sulfate, 563
  disease control, 28, 48, 51, 55, 56, 62, 86, 104, 153, 155, 162, 168, 173, 176, 201, 215, 228, 238, 372, 413, 512
  Hydro Bordo, 105, 554
  injury, 28, 45
  naphthenate, 556
  (or cupric) oleate, 556
  promotes, 28
  sulfate, 105, 556
    algae, control, 326
    copper deficiency, control, 28
    disinfectant, 83, 208, 420, 471
    dormant spray, 134
    gallon lots, 522
Copper, Fixed, fungicides, 48, 51, 55, 56, 62, 65, 86, 104, 153, 155, 162, 168, 172, 176, 179, 182, 183, 186, 228, 239, 242, 243, 246, 264, 272, 280, 287, 289, 300, 303, 310, 314, 318, 332, 341, 349, 350, 351, 353, 354, 370, 376, 381, 400, 401, 405, 407, 410, 423, 425, 433, 436, 437, 438, 451, 458, 460, 468, 469, 484, 500, 502, 505, 506, 508, 509, 522
  dormant spray, 176
  gallon lots, 522
  injury, 104, 158
  soil application, 104, 255, 490
  spray or dust, 104
  -sulfur mixture, 394
  trade names, 104
  uses, 104
  water-soluble products, 104, 162, 207 213, 239, 242, 394
Copperas, 292, 325, 345, 434, 443
Copperleaf, 126; *see* Acalypha
Copper-tip, 282; *see* Gladiolus
Cop-R-Nap, 556
Coptis, 254
  chlorosis, 24–29, 257, 443
  leaf spot, 48, 256
Coral beads, 451; *see* Sedum
Coralbean, 300; *see* Erythrina
Coralbells, 304
  leaf and stem nematode, 75, 306
  leaf spot, 48, 305
  leafy gall, 81, 306
  powdery mildew, 56, 304
  root-knot, 90, 305
  rust, 58, 305
  smut (leaf and stem), 62, 306
  stem or crown rot, 76, 305–6
Coralberry, 457; *see also* Snowberry
  berry rot, 86, 458
  canker, 77, 458
  gray-mold blight, 52, 86, 459
  leaf spot, 48, 458
  powdery mildew, 56, 458
  root rot, 87, 458
  rust, 58, 458
  spot anthracnose or scab, 51, 86, 458
  stem gall, 458
Coral-tree, 300; *see* Erythrina
Cord grasses (*Spartina*), rust, 144, 185, 186, 501
Cordia, 231
  powdery mildew, 56, 231
  root rot, 87, 137, 231
  rust, 58, 231
Cordyline, 261; *see also* Dracaena
  leaf spot, 48, 261
  root-knot, 90, 261
  root rot, 87, 261
Coreopsis, 220
  aster yellows, 73, 222
  curly-top, 75, 223
  leaf spot, 48, 221
  powdery mildew, 56, 221
  root-knot, 90, 225
  root rot, 87, 221
  rust, 58, 223
  scab, 65, 226
  soil drench, 222
  southern blight, 76, 221
  stem rot, 76, 221
  Verticillium wilt, 68, 223
Coriander (*Coriandrum*), 213; *see also* Celery
  anthracnose, leaf blight, 61, 213
  crown, stem, or foot rot, 76, 214
  curly-top, 75, 215
  Fusarium wilt, 66, 214
  mosaic, 70, 214
  motley dwarf, 214
  root-knot, 90, 214
  root rot, 87, 215
  seed rot, damping-off, 76, 214
  seed treatment, 213, 529, 532
Corky scab, 194, 397; *see also* Oedema
Corm, Cormel, 556
  diseases, 87–93
  rot, 90, 282, [283]
  scab, 65, 282, [283]
  treatment, 282, 285, 528–37
Corn, 231
  bacterial blight, leaf spot, stalk rot, 50, 76, 234
  bacterial wilt or Stewart's disease, 69, [70], 231, [232]
  black bundle, 76, 234
  boron toxicity, 28
  celery stripe, 70, 234
  chlorosis, 24–29, 235
  copper deficiency, 28, 235
  Corn Diseases in the United States, 235
  crazy top or downy mildew, 55, [235]
  ear or kernel rot, 86, 233
  fluorides injury, 43–44
  Helminthosporium leaf blight, 51, [233]
  insect control, 231, 232, 233
  iron deficiency, 26, 27, 235
  leaf blight, 51, 233, 234
  leaf fleck, 234
  leaf spot, 48, 233, 234
  light, effect on flowering, 38
  magnesium deficiency, 28, 235
  maize dwarf mosaic, 70, 234
  manganese deficiency, 26, 27, 235
  mosaic, 70, 234
  nitrogen deficiency, 25, 235
  phosphorus deficiency, 25, 235
  plant-food utilization, 26
  potassium deficiency, 26, 235
  purple sheath spot, 48, 234
  resistant varieties (hybrids), 231, 232, 233, 234
  root nematode, 234–35
  root rot, 87, 232
  rust, 58, 234, 370
  seed rot, seedling blight, 76, 233
  seed treatment, 233, 531, 532
  smut, 62, [63], 232
  stalk rot, 76, [232], 234
  stunt, 73, 234
  sulfur deficiency, 28, 235
  Watch out for Witchweed, 234
  witchweed (*Striga*), 234
  zinc deficiency, 26, 27, 235
Corn borer, European, control, 232
Corn earworm, insect control, 232, 233
Corn-marigold, 219; *see also* Chrysanthemum
  aster yellows, 73, 222
  damping-off, seed rot, 76, 221
  leaf spot, 48, 221
  seed treatment, 221
Corncockle, 203
  leaf spot, 48, 205
  root rot, 87, 204
  stem rot, 76, 204
Cornel, 258; *see also* Dogwood
  gray-mold blight, 52, 86, 259
  rust, 58, 261
  twig blight, 77, 259
Cornelian-cherry, 258; *see* Dogwood
Cornell University Peat-lite Mix A, 22
Cornflag, 281; *see* Gladiolus
Cornflower, 220; *see also* Centaurea
  aster yellows, 73, 222
  downy mildew, 55, 225
  Fusarium wilt 66, 223
  powdery mildew, 56, 221
  root-knot, 90, 225
  root rot, 87, 221
  rust, 58, 223
  southern blight, 76, 221
  stem rot, 76, 221
  Verticillium wilt, 68, 223
  white-rust, 62, 226
Cornflower aster, 220; *see also* Stokesaster
  downy mildew, 55, 225
  head blight (*Botrytis*), 52, 86, 224
  mosaic, 70, 222
Cornsalad, 499
  curly-top, 73, 499
  leaf spot, 48, 499
  powdery mildew, 56, 499
  root rot, 87, 499
  stem rot, 76, 254, 499
Cornus, 258; *see* Dogwood, Cornel
Coromate Ferbam Fungicide, 102, 518
Coromerc, 106, 470, 476
Coromerc Liquid, 106
Corona
  Folpet 50 Wettable, 103, 519
  Micronized Tri-Basic Copper Sulfate, 104
  Phygon-XL, 102, 518
  "26" Copper Fungicide, 104
  Zineb, 103, 519
Coronilla, 377; *see also* Pea
  root-knot, 90, 381
Corrosive sublimate, 101, 528; *see also* Mercuric chloride
Corydalis, 174
  downy mildew, 55, 174
  leaf spot, 48, 174
  root-knot, 90, 174
  rust, 58, 174
Corylus, 170; *see* Filbert, Hazelnut
Coryneum blight, 390, [391]
Coryphantha, 194; *see* Cacti
Cosmos, 220
  aster yellows, 73, 222
  bacterial wilt, 69, 223
  curly-top, 75, 223
  Fusarium wilt, 66, 223
  leaf spot, 48, 221
  light, effect on flowering, 38
  mosaic, 70, 222
  powdery mildew, 56, 221
  root-knot, 90, 225

root rot, 87, 221
rust, 58, 223
soil drench, 222
southern blight, 76, 221
stem canker, blight, 51, 77, 225
stem rot, 76, 221
Costmary, 219; *see* Chrysanthemum
Cotinus, 468; *see* Smoketree
Cotoneaster, 133
black rot, 86, 136
canker, 77, 138
chlorosis, 24–29, 143
collar rot, 76, 139
fire blight, 79, 133–34
hairy root, 83, 137
leaf spot, blight, 48, 51, 136, 140
powdery mildew, 56, 135–36
pruning, 32
root rot, 87, 137–38
scab, 65, 134–35
twig blight, 77, 138
Verticillium wilt, 68, 143
Cotton (*Phymatotrichum*) root rot, 88–89 [208], [211], 504
list of resistant plants, 89
Cotton-rose, 297; *see* Hibiscus (arborescent forms)
Cottonwood, 411; *see also* Poplar
canker (twig, branch, trunk), dieback, 77, 411–12
catkin deformity, 412
chlorosis, 24–29, 413
crown gall, 83, 136, 412
ink spot, 48, 412
leaf blister, yellow, 62, 412
leaf spot, 48, 412
mistletoe, 93, 413
powdery mildew, 56, 412
resistant cottonwoods, 412
root nematode, 413
rust, 58, 405, 412
sooty mold, 64, 413
wood rot, 79, 170, 412
Cottony blight (*Pythium*), [324]
Cotyledon, 556
County agent, 3
help by, 5, 24, 25, 36, 65, 79, 94, 98, 99, 101, 107, 135, 137, 139, 141, 150, 151, 159, 165, 187, 193, 209, 228, 231, 232, 233, 234, 259, 263, 264, 271, 287, 288, 291, 293, 332, 341, 356, 384, 389, 392, 393, 404, 416, 419, 420, 421, 433, 441, 463, 464, 471, 472, 480, 482, 484, 489, 491, 505
County agricultural agent, 5; *see* County agent
County extension director, 5; *see* County agent
County extension office; *see also* County agent
information, 17, 19, 27, 29, 33, 42, 97, 112
soil tests, 24–25
USDA bulletins, 578
Coventry-bells, 168; *see* Bellflower, Campanula
Cover crop, 23, 556
Cover sprays, 524–25
Cowania, 439
rust, 58, 441
Cowberry, 175; *see* Blueberry
Cowslip, 423; *see* Primrose
Cow-wheat, rust, 405
Crab cactus, 194; *see* Cacti, Epiphyllum
Crabapple, flowering, 133; *see also* Apple
anthracnose, 51, 86, 140
black rot, 48, 77, 86, 136
blossom blight, 86, 133–34
blotch, 48, 77, 86, [137]
canker (twig, branch, trunk), 77, 138–39
chlorosis, 24–29, 143
collar rot, 76, 134
crown gall, hairy root, 83, 136–37
dieback, 77, 136, 138–39
fire blight, 77, 79, 86, 133–34, 525
fruit rot or spot, 86, 138
leaf scorch, 41, 140, 141–42
leaf spot, 48, 140
mosaic, 70, 142
powdery mildew, 56, 135–36
pruning, 32
resistant varieties, 134, 135, 136
rodent protection, 42
root-knot, 90, 141
root nematode, 141
root rot, 87, 137–38
rust, 56, 135
scab, 48, 65, 86, 134–35
sooty mold, 64, 137
spray schedule, 524–25
stem- or wood-pitting, 142
sulfur dioxide injury, 43
winter injury, 42, 139
wood rot, silver leaf, 79, 139, 170
Crag Glyodin, 106
Crambe, 186; *see* Seakale
Cranberry-bush, 499; *see* Viburnum
Cranesbill (*Geranium*), 236
bacterial leaf spot, stem rot, 50, 76, 236
Botrytis leaf spot, 52, 236
downy mildew, 55, 236
fungus leaf spot, 48, 236
mosaic, 70, 236
powdery mildew, 56, 236
rhizome rot, 90, 236
root-knot, 90, 236
root nematode, 236
root rot, 87, 236, 238
rust, 58, 223, 236, 453
stem or crown rot, 76, 221, 236
Crank duster, 116, [121]
Crape-jasmine, 360
leaf spot, mold, 48, 360
root rot, 87, 137, 361
Crapemyrtle, 236
black spot, 48, 237
chlorosis, 24–29, 237
leaf spot, blotch, 48, 51, 237
powdery mildew, 56, 236–37
root-knot, 90, 237
root rot, 87, 237
sooty mold, 64, 237
thread blight, 237, 505
tip blight, 51, 237
Crassula, 451
anthracnose, 51, 452
flower blight, 86, 452
leaf nematode, 75, 452
leaf spot, 48, 452
light and temperature requirements, 40
ringspot, 72, 452
root rot, 87, 451
soil drench, 451–52
Crataegus, 133; *see* Hawthorn
Crazy top, [235]
Creeping cedar, 312; *see* Juniper
Creeping Charlie, 423; *see* Loosestrife
Creeping laredo, 153; *see* Mahonia, Oregon-grape
Creeping mint, 447; *see* Mint
Creeping snowberry, 294; *see also* Gaultheria
rust, 58
Creeping thyme, 447
root rot, 87, 278, 447
Creosote, cut stump treatment, 406
Crepis, 220; *see* Hawksbeard
Cress, 186; *see* Garden cress
Crickets, mole, insect control, 463
Crinkle, 70, 418, 465–66
Crinkleroot, 186; *see* Toothwort
Crinum, 249
leaf spot, 48, 253
mosaic, 70, 250–51
Stagonospora leaf scorch, red spot, 51, 251
Crocanthemum, 469
leaf spot, 48, 469
Crocosmia, 281; *see* Gladiolus
Crocus, 281
bacterial scab, 51, 65, 92, 283
blind buds, 285
corm or bulb rot, 92, 282
mosaic, 70, 284
rodent injury, 98
root rot, 87, 284
Crop rotation, in disease control, 16, 48, 56, 65, 68, 70, 77, 81, 82, 83, 86, 89, 90, 92, 93, 98
Crossvine, 169; *see* Bignonia
Crotalaria, 377
anthracnose, 51, 380
Fusarium wilt, 66, 377
gray-mold blight, 52, 86, 381
leaf spot, 48, 379, 380
mosaic, 70, 378
powdery mildew, 56, 378
root-knot, 90, 381
root rot, 87, 377, 378
seed treatment, 379
soil drench, 377, 379
stem canker, 77, 378
stem rot, 76, 378, 379
Croton, 237
anthracnose, 51, 237
bacterial wilt, 69, 237
leaf and stem spot, 48, 237
light and temperature requirements, 40
root nematode, 237
root rot, 87, 237
rust, 58
Crowberry, rust, 405
Crowfoot, 254; *see* Buttercup
Crown gall, 83, [84], 98, 136, 146, [173], [176], 391, 428, 508, 556
confused with, 83
immune plants, 83
paint, 86, 392, 503
plant soak, 392, 441, 503
severity increased by nematodes, 91
Crown imperial, 494; *see* Fritillary
Crown rot, 76, [77], 98, [309], 556
Crownbeard, 221
downy mildew, 55, 225
leaf spot, 48, 221
powdery mildew, 56, 221
root-knot, 90, 225
root rot, 87, 221
rust, 58, 223
stem spot, 76, 221
Crown-of-thorns, 409; *see* Spurge
light and temperature requirements, 40
Crownvetch, 377; *see also* Pea
root-knot, 90, 381
Crucifers, 185–86, 556
Cryophytum, 306; *see* Figmarigold, Ice plant
Cryptogramma, 269; *see* Ferns
Cryptomeria, 312
Botrytis blight, 52, 314
leaf blight, 51, 314
leaf spot, 48, 314
root rot, 87, 314
twig blight, 77, 313
winter injury, 41, 314
wood rot, 79, 170, 314
Cuban laurel, 271; *see* Fig, Rubber-plant
Cuban lily, 495; *see* Squill
Cuckoo-flower, 203; *see* Lychnis
Cucumber, 237; *see also* Cucurbit
angular leaf spot, 50, 239, [240]
anthracnose, 51, 238
aster yellows, 73, 243
bacterial soft rot, 81, 238, 241
bacterial spot, blight, 50, 86, 239, 245
bacterial wilt, 69, [70], 240
beetles, [240]
blossom blight, 86, 245
chlorosis, 24–29, 245
crown gall, 83, 245
curly-top, 75, 243
damping-off, 76, 244, [528]
downy mildew, [55], 243
fruit spot, rot, 86, 238, 240, 241, 244
Fusarium wilt, 66, 241
gray-mold rot, 52, 244, 245
gummy stem blight, 76, 244
leaf blight, 51, 239
leaf spot, 48, 238, 239, 244

Cucumber (*continued*)
mosaic, 70, 164, [241], 242
weed hosts, 165
powdery mildew, 56, 242
resistant varieties, 238, 239, 240, 242, 243, 245
ringspot, 72, 243
root-knot, cyst nematode, 90, 244–45
root nematode, 244
root rot, 87, 156, 244
scab, spot rot, pox, 65, [66], 238–39
seed rot, 244
seed treatment, 238, [528], 529, 532
seedbed spray, 244
southern blight, 76, 243
stem or foot rot, 76, 243
2, 4-D injury, 245, 288
Verticillium wilt, 68, 245
watery soft rot, white wilt, cottony rot, 76, 243
web blight, 160, 245
Cucumber beetles, [240]
control of, 240, 485
as virus carriers, 242, 485
Cucumber-tree, 340; *see* Magnolia
Cucumis, 238; *see* Cucumber, Muskmelon, Gourds, Cantaloup, Honeydew melon, Casaba, West India gherkin, Cucurbit
Cucurbit, 237–38, 556
Alternaria blight, 51, [239]
angular leaf spot, 50, 239, [240]
anthracnose, 51, 86, [238]
aphids, [242]
aster yellows, 73, 243
bacterial soft rot, 81, 86, 238, 244
bacterial spot or blight, 50, 86, 239, [240], 245
bacterial wilt, 69, [70], 240
black rot (stem-end rot), 86, 244, 245
blossom blight, 86, 245
blossom-end rot, [244], 245
boron deficiency, 27, 245
chlorosis, 24–29, 245
Commercial Watermelon Growing, 245
crown gall, 83, 245
cucumber beetles, [240]
curly-top, 75, 243
damping-off, 76, 244, [528]
downy mildew, [55], 243
fruit spot, rot, 86, 238, [240], 241, 244
Fusarium wilt, 66, [67], 241
gray-mold rot, 52, 86, 244, 245
Growing Pumpkins and Squashes, 245
gummy stem blight, 76, 86, 244
iron deficiency, 26, 27, 245
leaf blight, 51, [239]
leaf spot, 48, 238, 239, 244
magnesium deficiency, 28, 245
manganese deficiency, 26, 27, 245
mosaic, 70, [241], 242
Muskmelon Culture, 245
Muskmelons for the Garden, 245
plant-food utilization chart, 26
potassium deficiency, 26, 245
powdery mildew, 56, [242], 243
Pumpkins and Squashes, 245
ringspot, 72, 243
root-knot, cyst nematode, 90, [91], 244–45
root nematode, 244
root rot, 87, 156, 244
scab, spot rot, pox, 48, 65, [66], 86, 238–39
seed rot, 244
seed treatment, 238, [528], 529, 531, 532
seedbed spray, 244
sooty mold, 64, 245
southern blight, 76, 243
stem rot, 76, 243
stem streak, dieback, 245
storage rot, 86, 244
2, 4-D injury, 45, 245, 288
Verticillium wilt, 68, 245
Watermelons for the Garden, 245
watery soft rot, white wilt, cottony rot, 76, 243
web blight, 160, 245
zinc deficiency, 26, 27, 245
Cucurbita, 238; *see* Squash, Pumpkin, Gourds, Vegetable-marrow
Cultivar, 556
Cultivator wounds, entrance for bacteria and fungi, 81, 98
Cultural practices, 9, 19–48, 96–100; *see also* Control methods, plant diseases
air humidity, control, 39–40, 48, 50, 54, 55, 56, 58, 62, 204
The Care and Feeding of Garden Plants, 19, 25
cultivating, shallow, 41
fertilizing, 29, 30, [31], 48, 54, 56, 65, 68, 77, 79, 80, 89, 92, 96
help on, 5, 9, 97, 98
light, 38
loosening soil, 23–24
mulching, 21, 24, 42, 52, 76, 86–87
nutrient deficiencies, 25–29
organic matter, 23, 92
pesticides, precautions in using, 45, 101, 105, 106, 107, 108, 109, 112, 523, 525, 527, 538, 543
planting, 19, 20, [21], 98
pruning, [32], [33], [34], 52, 62, 65, 68, 78, 79, 80, 81, 83, 89
resistant varieties, 48, 56, 58, 62, 64, 65, 67, 68, 70, 72, 75, 76, 79, 87, 89, 92, 94, 97
rotation, 16, 48, 56, 65, 68, 70, 77, 81, 82, 83, 86, 89, 90, 92, 93, 98
sanitation, 48, 52, 54, 56, 58, 62, 64, 65, 68, 70, 72, 73, 76, 77, 78, 81, 86, 92, 93, 94
soil, 21–22
drainage, 36–37
staking trees and shrubs, 36, [37]
sunscorch, 41
temperature, 40–41, 58, 62, 70, 83
tree wound treatment, [34], [35], 36, 78, 79
watering, 37–38, 39, 48, 54, 56, 58, 63, 68, 76, 77, 79, 86, 89–90, 98
winter protection, [42], 43
Culture, 556
Culversroot, 460
leaf spot, 48, 460
powdery mildew, 56, 460
root nematode, 460
root rot, 87, 278, 460
rust, 58, 223, 460
stem rot, 76, 278, 460
Cunila, 446; *see* Dittany
Cupflower, 479; *see* Tomato
Cuphea, 227; *see* Cigarflower
Cupid's-bower, 127
light and temperature requirements, 40
root rot, 87, 127
tuber or rhizome rot, 86, 127
Cup-plant, 220; *see* Silphium
Cupressus, 312; *see* Cypress
Cuprinol Green, 556
Cuprocide, 104
Curl, 556
Curly-top, [75], 98, [159], 162, [192], 556
weeds as virus reservoir, 98
Currant, 245; *see also* Flowering currant
anthracnose, 51, [246]
cane blight, 76, [247]
canker, black pustule, cane-knot canker, 77, 247
chlorosis, 24–29, 247
collar rot, 76, 247
dieback, 77, 247
downy mildew, dryberry, 55, 247
fruit spot, rot, 86, 246
gray-mold blight, 52, 86, 246, 247
iron deficiency, 26, 27, 247
leaf scorch, potassium deficiency, 26, 248
leaf spot, 48, 246
manganese deficiency, 26, 27, 247
mosaic, 70, 247
powdery mildew, 56, 246
quarantine regulations, 246
resistant varieties, 246
ringspot, 72
root-knot, 90, 247
root nematode, 247
root rot, 87, 247
rust, 58, 246, 405, 508
spray schedule, 524–25
thread blight, 247, 505
twig canker, 77, 247
Verticillium wilt, 68, 153, 247
Curuba, 238; *see* Sicana
Curvularia leaf spot, [283]
Cushaw, 238; *see* Pumpkin
Cushion aloe, 130; *see* Aloe
Cushion-pink, 203; *see also* Silene; Pinks, garden; Mullein-pink
damping-off, 76, 204
downy mildew, 55, 206
flower or anther smut, 62, 206
Cutting, 556
Cutting rot, 86, [279]
dip, 529–37
Cuttings, in transmitting viruses, 13
Cutworms, insect control, 187, 215, 321
Cyathea, 269; *see* Ferns
Cyclamen, 248
bacterial soft rot, 81, [248]
Botrytis bud and leaf rot, 52, 248
crown rot, 76, 248
Fusarium wilt, 66, 249
gray-mold blight, 52, 248
leaf and bud blight, 51, 249
leaf nematode, 75, 249
leaf spot, 48, 249
light and temperature requirements, 39, 40
petal spot or rot, 52, 86, 248
root clump dip, 249
root-knot, 90, 249
root nematode, 249
root rot, 87, 249
seedling blight, damping-off, 76, 249
soil treatment, 248, 249
stunt, wilt, Ramularia leaf disease, 76, 248
tuber rot, 86, 248
white mold, 48, 249
Cycloheximide, 106; *see also* Acti-dione
Cyclophorus, 269; *see* Ferns
Cycnoches, 366; *see* Orchids
Cydonia, 133; *see* Quince
Cymbidium, 366; *see* Orchids
Cynara, 328; *see* Globe artichoke
Cynem, 89, 546, 556; *see also* Zinophos
Cynodon, 319; *see* Bermudagrass
Cynoglossum, 349; *see* Houndstongue
Cyperus, 498
leaf spot, 48, 498
root-knot, 90, 498
Cyphomandra, 479
bacterial canker, 50, 483–84
powdery mildew, 56, 492
Cypress, 312; *see* Chamaecyparis for Hinoki cypress and Sawara-cypress
canker, dieback, 77, 314
chlorosis, 24–29, 315
crown gall, 83, 136, 315
gray-mold blight, 52, 314
iron deficiency, 24, 26, 27, 315
mistletoe, 93, 315
needle cast, 51, 314
nursery or juniper blight, 51, 313, 314
pruning, 32
root nematode, 315
root rot, 87, 314
rust, 58, 313
seedling blight, 76, 314
twig blight, 77, 314
wood rot, 79, 170, 314
Cypressvine, 351
root-knot, 90, 351
root rot, 87, 351
rust, 58, 351
white-rust, 62, 191, 351
Cyprex, 102; *see also* Dodine
Dodine Dust, 102, 518
Dodine 65-W, 102, 518
gallon lots, 522

Cypripedium, 366; *see* Orchids
Cyrilla, 184; *see* Southern leatherwood
Cyrtomium, 269; *see* Ferns
Cyst nematode, 90
Cystopteris, 269; *see* Ferns
Cytisus, 183; *see* Broom, Genista
Cytospora canker, 77, [403], [411]

D

Daconil 2787
  active ingredient, 104
  post-harvest dip or dust, 443
  as protective fungicide, 55, 58, 86, 158, 162, 189, 190, 207, 213, 221, 224, 238, 239, 243, 280, 307, 320, 321, 326, 329, 362, 363, 384, 394, 410, 415, 435, 440, 441, 447, 480, 482
  seed-piece treatment, 418
  in transplant water, 89, 189
  uses, 104
Daffodil, 249
  bacterial blight, stem rot, 50, 76, 253
  basal rot, [90], 250
  bud blast, 253
  bulb and bud nematode, 92, 252
  bulb rot, [90], 250
  bulb soak, 250, 252, 530, 531, 535
  decline, 251
  ethylene injury, 44
  fire, neck rot, 76, 251
  flower spot, 52, 86, 251
  flower streak, 70, 250
  gray-mold blight, Botrytis blight, 52, 86, 251
  leaf blight, 51, 251
  leaf spot, 48, 253
  light and temperature requirements, 40
  mosaic, 70, [250], 251
  root and crown rot, 76, 87, 250
  root nematode, 250, 253
  "smoulder" or neck rot, 76, 252
  Stagonospora leaf scorch, 51, [251]
  stem, leaf, and bulb nematode, browning or "ring disease," 75, 92, [252]
  white mold or Ramularia blight, 51, [251], 252
  white streak or paper tip, 251
  yellow dwarf, 73, 253, 365
  yellow stripe, gray disease, 250
Dahlia, 220
  aster yellows, 73, 222
  bacterial soft rot, 81, 221, 226
  bacterial wilt, 69, 223
  blossom blight, 86, 221, 224
  bud rot, 224
  bulb (potato-rot) nematode, 92, 226–27
  crown gall, 83, 226
  fasciation, 81, 226
  Fusarium wilt, 66, 223
  gray-mold blight, 52, 86, 224
  hopperburn, 227
  leaf blight, 51, 221
  leaf nematode, 75, 226
  leaf smut, 62, 226
  leaf spot, 48, 221
  light, effect on flowering, 38
  mosaic, 70, [74], 222
  powdery mildew, 56, 221
  resistant varieties, 226
  ringspot, oakleaf disease, 72, [73], 222
  root-knot, 90, [91], 225
  root nematode, 226–27
  root rot, 87, 221
  scab, 65, 165, 226
  soil drench, 222
  southern blight, 76, 221
  spotted wilt, 72, 223
  stem or cutting rot, 76, 221
  storage or tuber rot, 81, 86, 221, 226
  stunt, dwarf, 73, [74], 222
  tip cuttings, 223
  Verticillium wilt, 68, 223
Dahoon, 295; *see also* Holly
  black mildew, 64, 296
  sooty mold, 64, 296
Daisy, 219–20
  African; *see* Arctotis, Gazania
  blue; *see* Blue daisy
  English; *see* English daisy
  giant; *see* Chrysanthemum
  Michaelmas; *see* Aster, perennial
  oxeye; *see* Chrysanthemum
  painted, *see* Pyrethrum, Chrysanthemum
  Paris; *see* Marguerite
  Shasta; *see* Chrysanthemum, Shasta daisy
  Swan River; *see* Swan River daisy
  Transvaal; *see* Transvaal daisy
  turfing; *see* Matricaria
Dalea, 183
  leaf spot, 48, 183
  mistletoe, 93, 183
  root rot, 87, 137, 183
  rust, 58, 183
Damesrocket, 186; *see also* Cabbage
  clubroot, 89, 189
  downy mildew, 55, 189
  mosaic, flower breaking, 70, 192
  Verticillium wilt, 68, 193
  white-rust, 62, 191
Damping-off, 52, 76, [163], 167, [489], [528], 556
  control, 77, 97, 405–6, [528], 531, 544–45
Dandelion, 221
  bacterial leaf spot, 50, 225
  bacterial soft rot, 81, 221
  crown rot, 76, 221
  gray-mold blight, 52, 224
  leaf and stem gall, 48, 221
  leaf spot, 48, 221
  powdery mildew, 56, 221
  ringspot, 72, 223
  root-knot, 90, 225
  root nematode, 226–27
  root rot, 87, 221
  rust, 58, 223
  stem and leaf nematode, 75, 226
  yellows, 73, 222
Dangleberry, 175; *see* Huckleberry
Danitra, 27, 28, 345
Daphne, 253
  anthracnose, 51, 253
  canker, 77, 253
  dieback, 77, 253
  Fusarium wilt, 66, 254
  leaf spot, 48, 253
  mosaic, 70, 253–54
  root rot, 87, 253
  stem, crown, or collar rot, wilt, 76, 253
  twig blight (*Botrytis*), 52, 253
  Verticillium wilt, 68, 254
  winter injury, 41, 254
Darlingtonia, 408; *see* California-pitcherplant
Darworth, Incorporated, Chemical Products Division, 120
Dasheen, 195; *see also* Elephants-ear
  bacterial soft rot, 76, 81, 86, 196
  tuber rot, 86, 196–97
Date plum, 398; *see* Persimmon
Datura, 479, *see also* Tomato
  aster yellows, 73, 490
  bacterial wilt, 69, 488
  early blight, 51, 480
  fruit rot, pod blight, 86, 481
  leaf roll (virus), 491
  leaf spot, 48, 480
  mosaic, 70, 484
  ringspot, 72, 486
  root rot, 87, 490
  rust, 58, 492
  southern blight, 76, 490
  spotted wilt, 72, 486
Daucus, 206; *see* Carrot
Davallia, 269; *see* Ferns
Dayflower, 493; *see also* Tradescantia
  leaf spot, 48, 493
  root-knot, 90, 493
  rust, 58, 493
Daylily, 295
  blight, 295
  gray-mold blight, 52, 295
  leaf spot, leaf blight, 48, 51, 295
  root-knot, 90, 295
  root nematode, 295
  root rot, 87, 278, 295
  russet spot, 48, 295
  winter or frost injury, 41, 295
DBCP, 546
  active ingredient, 546
  application, 545, 548
  disinfesting soil, 89, 91, 92, 128, 138, 141, 148, 159, 182, 191, 197, 201, 209, 244, 249, 266, 271, 284, 285, 290, 317, 324, 341, 346, 360, 364, 365, 392, 395, 400, 406, 409, 420, 425, 444, 447, 467, 504
    home garden, 548, [549]
  plant soak, 459
  trade names, 546
  uses, 546
DCNA, 104; *see also* Botran
D-D
  active ingredient, 545
  application, 545
  disinfesting soil, 83, 159, 165, 252, 253, 271, 290, 308, 392, 421, 441, 467, 472, 489
    home garden, 548–49
  precautions, 545
  trade names, 545
D-D Soil Fumigant, 545, 548
DDT, 48, 65, 82, 98, 129, 132, 161, 168, 169, 182, 197, 205, 209, 222, 227, 231, 253, 256, 263, 264, 266, 284, 301, 346, 353, 363, 387, 394, 397, 400, 421, 428, 433, 437, 440, 457, 460, 461, 501, 502, 505
  borer control, 79, 139, 145, 171, 259, 266, 307, 332, 346, 358, 384, 435, 477, 508–9
  corm, bulb treatment, 282
  injury, 45, 200
  in multipurpose mixes, 111
Dead-arm, [288], 289
Dealer, garden supply, help by, 112
"Decline," 87, [181]
Decumaria, 304
  leaf spot, 48, 305
Deergrass, 254
  leaf spot, 48, 254
Deer's-tongue, 278; *see* Columbo
Deficiency, nutrient, 25–29
  oxygen, 45
Delayed dormant spray, 556
Delisle ceanothus, 355; *see* New Jersey-tea
Delonix, 300; *see* Caesalpinia
Delphinium, 254
  bacterial collar rot, soft crown rot, 76, 81, 254
  black leaf spot, blotch, [50], 255
  "blacks," 257
  chlorosis, 24–29, 257, 443
  collar rot, 76, 254
  crown gall, 83, 168, 257
  curly-top, 75, 257
  damping-off, 76, 256
  flower blight, 86, 255, 256, 257
  Fusarium wilt, 66, 256
  gray-mold (*Botrytis*) blight, 52, 86, 255
  "greens," 75, 256
  leaf blight or blotch, 51, 256
  leaf spot, 48, 256
  leaf and stem nematode, 75, 257
  leaf and stem smut, 62, 257
  mosaic, 70, 255
  powdery mildew, 56, 256
  resistant varieties, 256
  ringspot, 72, 255, 257
  root-knot, 90, 256
  root nematode, 256
  root rot, 87, 254
  rust, 58, 256
  seed rot, 256

Delphinium (*continued*)
seed treatment, 254, 530, 535
soil drench, 254, 255, 256
southern blight, 76, 254
spotted wilt, 72, 254, 257
stem canker, 77, 254
stem (crown) rot, wilt, 76, 254
thread blight, 257, 272
Verticillium wilt, 68, 256
white smut, 65, 257
yellows, stunt, 73, 256
Delsan A-D Seed Protectant, 103, 282, 519, 531
Dendrobium, 366; *see* Orchids
Dendromecon, 413
leaf smut, 62, 413
Dennstaedtia, 269; *see* Ferns
Dentaria, 186; *see* Toothwort
DePester Fruit Tree Spray, 523
Spreader-Activator, 119
Sticker, 119
Desertplume, 186; *see also* Cabbage
leaf spot, 48, 190
rust, 58, 193
Desert-willow, 212
damping-off, 76, 213, 405
leaf spot, 48, 212
root rot, 87, 137, 212
Desmanthus, 183; *see* Bundleflower
Destruxol Rose Spray, 556
Deutzia, 304
leaf spot, 48, 305
pruning, 32
root-knot, 90, 305
root rot, 87, 306
Verticillium wilt, 68, 306
Devilsclaw, 425; *see* Proboscisflower
Devil's ivy, 196; *see* Pothos
Devilwood, 370; *see* Osmanthus
Dewberry, 427; *see also* Boysenberry, Raspberry
anthracnose, 51, 427, 525
black mildew, 64, 432
cane blight, 77, 428, 525
cane and crown gall, hairy root, 83, 428
canker, dieback, 77, 428
chlorosis, 24–29, 432
double blossom, 432
downy mildew, 55, 432
dwarf, 73, 428
fruit rot, spot, or mold, 86, 430
gray-mold blight, 52, 86, 430
iron deficiency, 24, 26, 27, 432
leaf curl, 70, 428
leaf spot, 48, 427, 430–31
mosaic, 70, 428
orange rust, 58, 430
powdery mildew, 56, 431
root rot, 87, 137, 247, 431
rosette, 73, 428, 432
sooty blotch, 64, 432
spot anthracnose, 51, 427
spray schedule, 524–25
spur blight, 76, 427, 525
Verticillium wilt, 68, 430
winter injury, 41, 432
yellow rust, 58, 430
Dexon
active ingredient, 104
bulb soak, 336
-dieldrin mixture, 158
-Dyrene mixtures, 104, 324
formulations, 104, 531
-PCNB mixtures, 89, 104, 138, 162, 197, 222, 249, 254, 297, 398, 408, 409, 410, 423, 435, 437, 500
seed treatment, 158, 162, 379, 531
soil application, 77, 89, 97, 104, 106, 127, 138, 149, 162, 167, 182, 197, 201, 222, 249, 254, 256, 275, 278, 297, 324, 326, 336, 337, 379, 398, 406, 408, 409, 410, 423, 435, 437, 455, 490, 500
uses, 104
Diagnosis of disease; see Disease
Diamond Alkali Company, 122
Diamond Poison Antiseptic, 101
Dianthus, 203; *see also* Carnation; Pinks, garden; Sweet-william
anther smut, 62, 206
aster yellows, 73, 205
Diazinon, 72, 164, 165, 192, 214, 362, 363, 365, 485
injury, 270
soil application, 23, 82, 86, 98, 166, 187, 209, 233, 324, 362, 371, 418, 475
Dibrom (naled), 70, 72, 73, 74, 82, 86, 98, 164, 165, 187, 192, 329, 406, 438
Dicentra, 174; *see* Bleedingheart, Dutchmans-breeches, Squirrelcorn
Dichlone
active ingredient, 102, 518
as "eradicant" fungicide, 55, 100, 102, 134, 144, 157, 158, 171, 175, 204, 224, 264, 275, 294, 296, 298, 301, 303, 305, 308, 318, 321, 337, 343, 357, 383, 384, 447, 458, 476, 495, 500, 508, 518
injury, 45
oxidant or PAN prevention, 44
-PCNB mixtures, 104
seed, bulb treatment, 77, 102, 162, 165, 207, 233, 299, 318, 375, 378, 405, 418, 470, 518, 531
soil drench, 102, 106
trade names, 102, 518
tree wound dressing, 476
uses, 102, 518
Dichloran (DCNA), 104; *see also* Botran
Dichloropropenes, mixed, 545; *see also* D-D, Telone, Vidden-D
Dicksonia, 269; *see* Ferns
Dictyosperma, 371; *see* Palms
Didiscus, 213; *see* Blue laceflower
Dieback, 77, [181], 345, [434], 556
Dieffenbachia, 196
anthracnose, 51, 197
bacterial leaf spot or rot, 50, [196], 197
bacterial soft rot, 76, 81, 196
bacterial stem rot, 76, 196
cane soak, 196, 197, 530, 535
leaf spot, 48, 197
light and temperature requirements, 40
root rot, 87, 196–97
soil drench, 197
southern blight, 76, 196–97
stem (cane) rot, 76, 196–97
Dieldrin
seed, bulb, corm treatment, 58, 233, 238, 250, 282, 362, 379, 531
soil insecticide, 23, 82, 98, 178, 282, 512
tree borer control, 79, 171, 301, 332, 341, 358, 435, 477, 508–9
Diervilla, 457; *see* Bush-honeysuckle
Difolatan (Folcid)
active ingredient, 104
formulations, 104
lawn fungicide, 320, 321
as protective fungicide, 134, 136, 140, 190, 207, 239, 243, 363, 384, 394, 415, 480, 482
seed treatment, 104, 379, 394, 418
soil application, 104, 156
uses, 104
Digitalis, 454; *see* Foxglove
Dill, 213; *see also* Celery
aster yellows, 73, 214
curly-top, 75, 215
damping-off, seed rot, 76, 214
Fusarium wilt, yellows, 76, 214
leaf blight, 51, 214
leaf spot, 48, 213
mosaic, 70, 214
motley dwarf, 214
root-knot, 90, 214
root nematode, 215
root rot, 87, 215
seed treatment, 213, 532
stem spot, rot, 76, 214
Dimethyltetrahydro-1,3,5,2H-thiadiazine-2-thione, 546
Dimorphotheca, 220; *see* Cape-marigold
Dinitro materials, 556
injury, 45
Dinocap, 101; *see also* Karathane
Dioscorea, 510; *see* Yam
Diospyros, 398; *see* Persimmon
Diphenylamine (DPA), 140
Diplodia tip blight, 402, [403]
Dipsacus, 478; *see* Teasel
Diquat, cut stump treatment, 406
Dirca, 327; *see* Leatherwood
Directory of Poison Control Centers, 108
Disease; *see also* Control methods
causes, 9
unfavorable growing conditions, 9
parasites, 9–17
classification, 8
control, 96–111
help on, 5, 6, 106
defined, 8, 557
diagnosis, 7, 9, 19
help on, 7, 9
"equation," 96
extent of, 8
factors affecting development, 96
fossils, 8
loss in the United States, 8
to forests, 8
by nematodes, 17
numbers of in United States, 8
types of, 48–94
Disinfectants, 98–99
defined, 557
Disinfestant, 557
Disinfesting soil, 538–39, [540], [541], 542–47, [548]
precautions, 538–39, 543
tools, containers, machinery, 547–48
Disodium ethylenebis (dithiocarbamate), 562
Distichlis, rusts, 166, 186, 354
Dithane A-40, 106, 490, 557
Dithane D-14, 562
Dithane M-22 Special, 103, 385, 519
Dithane M-45 (Fore), 103, 146, 233, 238, 239, 243, 320, 321, 326, 329, 362, 394, 415, 440, 441, 501, 502, 519
as protective fungicide, 62, 145, 162, 189
Dithane S-31, 106, 146
Dithane Z-78, 103, 519; *see also* Zineb
Dittany, 446
leaf spot, 48, 447
rust, 58, 447
DMTT, 545; *see also* Mylone
active ingredient, 545
application, 545, 548
disinfesting soil, 91, 106, 188, 193, 214, 405, 467, 490, 543, 545
precautions, 543
trade names, 106, 545
uses, 545
Dockmackie, 499; *see* Viburnum
Dodder, [94]
Dodecatheon, 423; *see* Shootingstar
Dodine
active ingredient, 102, 518
eradicant and protective fungicide, 100, 102, 134, 302, 384, 385, 394, 464, 476, 501, 502, 518
gallon lots, 522
injury, 45
trade names and distributors, 102, 518
uses, 102, 518
Dog injury, 315, 327
Doggett-Pfeil Company, 122
Dogstooth-violet, 267; *see also* Erythronium
Botrytis blight, 52, 267
leaf smut, 62, 267
rust, 58, 267
Dogwood, 258
canker, dieback, 77, 259
chlorosis, 24–29
collar rot, bleeding canker, 76, 77, [258], 259

crown gall, 83, 136, 261
felt fungus, 261, 293
gray-mold blight, flower and shoot blight, bud blight, 52, 86, 259
Growing the Flowering Dogwood, 261
heat or drought injury, 260
herbicide injury, 45, 261, 288
iron deficiency, 24, 26, 27
leaf scorch, 41, 260
leaf spot, 48, 259
mistletoe, 93, 261
powdery mildew, 56, 259
pruning, 32
root-knot, 90, 259
root nematode, 259
root rot, 87, 137, 259
rust, 58, 261
soil drench, 259
sooty mold, black mildew, 64, 261
spot anthracnose, 51, 86, 259, [260]
sunscald, 41, 260
thread blight, 261, 505
tree borer control, 259
twig blight, 77, 259
Verticillium wilt, 68, 261, 343
wood rot, heart rot, 79, 170, 259
Dolichos, 377; *see* Hyacinth-bean
Dollar spot (*Sclerotinia*), 321, [322]
Dolomitic limestone, 28
Dorlone
active ingredient, 545
disinfesting soil, 209, 392, 467
precautions, 545
uses, 545
Dormant spray, 161, 176, 384, 385, 524, 525, 545, 557
oil injury, 45
Doronicum, 220; *see* Leopardsbane
Dothiorella canker, dieback, wilt, 264, [265], [358]
Double-flowered Dyer's greenweed, 183; *see* Woodwaxen, Broom
Double orange daisy, 220; *see* Erigeron
Douglas Chemical Company, 122
Douglas-fir, 402
bacterial gall, 83, 407
brown felt blight, 407
canker (twig, branch, trunk), dieback, 77, 403
chlorosis, 24–29, 408
crown dieback, 77, 406–7
Cytospora canker, twig blight, 77, 403
damping-off, 76, 405
gray-mold blight, 52, 408
leaf blight, 51, 402
mistletoe, 93, 406
needle cast, 51, 402
root nematode, 406
root rot, 87, 405, 406
rust, 58, 405
seed rot, 405
seed treatment, 405
seedling blight, 76, 405
snow blight, 407
soil treatment, 405, 406
stump treatment, 406
tip blight, 76, 402
twig blight, 77, 402, 403
wood rot, 79, 170, 403
Dow Chemical Company, Agricultural Chemical Division, 122
Dowfume MC-2, 106, 545, 548
Dowfume MC-33, 545
Dowfume W-85, 544
Dowicide A, 208
Dowicide B, 250, 282, 362
Downy mildew, Blue mold, [55], 56, [189], [235], [363], 379, [381], 490
Downy pinxterbloom, 433; *see* Azalea
Doxantha, 494
root rot, 87, 494
DPA, 140
Draba, 186; *see* Whitlowgrass
Dracaena, 261
anthracnose, 51, 261
chlorosis, 24–29, 261
cutting dip, 261
gray-mold blight, 52, 261
iron deficiency, 24, 26, 27
leaf spot, 48, 261
light and temperature requirements, 40
root-knot, 90, 261
root nematode, 261
root rot, 87, 261
stem rot, 76, 261
tip blight, 51, 261, [262]
Dragonhead (*Dracocephalum*), 447
downy mildew, 55, 448
leaf spot, 48, 447
southern blight, 76, 447
Dragonroot, 195; *see* Jack-in-the-pulpit
Dragontree, 261; *see* Dracaena
Drainage, soil, 36–37
Dreft, as spreader, 119
Dropwort, 439; *see* Meadowsweet
Drought injury, 38, 78, 260
crack, 41
entrance for fungi, 79, 402
Dryopteris, 269; *see* Ferns
Duchesnea, 439; *see* Cowania and Mock-strawberry
Dumbcane, 196; *see* Dieffenbachia
Duo Copper, 104
E. I. du Pont de Nemours & Company, 102, 103, 122, 518, 519
Formaldehyde Solution, 558
Spreader-Sticker, 119
VPM Soil Fumigant, 106, 545
Duranta, 317
black leaf spot, 48, 317
black mold or mildew, 64, 317
seedling blight, 76, 318
stem or crown rot, 76, 278, 318
Dusters, 115–23
accessories, 115, 116, 117
bellows, 116
crank, 116, [121]
Hand Sprayers and Dusters, 118
knapsack, 116, [122]
bellows, 117
rotary-fan, 117
maintenance of, 117
manufacturers, 120–24
Outdoor Housekeeping, 118
plunger type, 115, [120]
power, 117, [123]
precautions, 115
rotary-fan, 116
Sprayer and Duster Manual, 118
uses, 118
Dusting
advantages and disadvantages, 111
coverage, 109–10
equipment, 115–23
manufacturers, 120–24
multipurpose mixes, 111
precautions, 45, 105, 106, 107, 108
publications, 118
tips, 109
vs. spraying, 111
Dusts, multipurpose mixes, 110–11
Dusty-miller, 220; *see* Artemisia, Centaurea, Senecio
Dutch elm disease, [262], 263
Dutchmans-breeches, 174
downy mildew, 55, 174
leaf spot, 48, 174
rust, 58, 174
Dutchmans-pipe, 144; *see* Aristolochia
Du-Ter, 557
Dwarf cape-jasmine, 275; *see* Gardenia
Dwarf cornel, 258; *see* Dogwood
Dwarf indigo, 269; *see* Indigobush
Dwarf lace plant, 453; *see* Silver-lacevine
Dwarf mistletoe, [93], 94
cankers, entrance for fungi, 79
Dwarf morning-glory, 357; *see* California-rose
Dwarf (virus), 73, [74], 75
Dwarf withe-rod, 499; *see* Viburnum
Dwarf Yaupon, 295; *see* Yaupon
Dwarfing, 557
D-Wax, 38, 511
Dyer's greenweed, 183; *see* Woodwaxen, Broom
Dylox (trichlorfon), 187
Dyrene, 55, 106, 557
active ingredient, 557
formulations, 557
-Dexon mixtures, 104, 324, 557
gallon lots, 522
as protective fungicide, 158, 162, 178, 207, 213, 238, 243, 320, 321, 322, 362, 363, 482
soil application, 406
trade names, 557
uses, 557
Dyschoriste, 229
rust, 58, 229

E

Earth-star, light and temperature requirements, 40
Earthworms, control, 321
Earwigs, insect control, 463
Eastern States Garden and Orchard Dust, 523
Echeveria, 451
leaf spot, 48, 452
light and temperature requirements, 40
root-knot, 90, 452
rust, 58, 452
Echinacea, 220
leaf spot, 48, 221
mosaic, 70, 222
root rot, 87, 221
Echinocactus, 194, *see also* Cacti
anthracnose, 51, 194
black leaf spot, 48, 195
corky scab, 39, 194
leaf scorch, 51, 195
root rot, 87, 194–95
scald, 41, 195
stem rot, 76, 194–95
Echinocereus, 194; *see* Cacti, Cereus
Echinocystis, 238; *see* Mock-cucumber
Echinomastus, 194; *see* Cacti, Echinocactus
Echinops, 220; *see* Globethistle
Echinopsis, 194; *see* Cacti, Cereus
Echium, 349
leaf spot, 48, 221, 349
root rot, 87, 278, 349
EDB, 544
active ingredient, 544
disinfesting soil, 83, 91, 159, 209, 244, 252, 253, 271, 308, 392, 400, 420, 424, 467, 472, 489
home garden, 548–49
precautions, 364, 544
trade names, 544
EDB-40, 548
EDB-85, 548
Edco Corporation, 122
MBX, 106
Nemadrench, 546
Eelworms; *see* Nematodes
Eggplant, 479; *see also* Tomato
anthracnose, 51, 480, 481
aster yellows, 73, 490
bacterial soft rot, 81, 86, 481, 483
bacterial spot, 50, 483
bacterial wilt, 69, 488
curly-top, 75, 486
damping-off, 76, 488
downy mildew, 55, 490
early blight, 51, 480
fruit spot, rot, 86, 480, 481, 490
Fusarium wilt, 66, 486
gray leaf spot, 48, 480–81
gray-mold rot, 52, 86, 481
insect control, 483, 485, 486
late blight, 51, 480
leaf spot, 48, 480
mosaic, 70, 484
Phytophthora blight, 51, 490
powdery mildew, 56, 492
resistant varieties, 482, 488
root-knot, 90, 489
root nematode, 489

Eggplant (*continued*)
root rot, 87, 490
rust, 58, 492
scab, 65, 415, 492
seed rot, 488
seed treatment, 484, 528, 533
seedbed treatment, 489
Septoria leaf spot, 48, 480
southern blight, 76, 490
spotted wilt, 72, 486
stem or collar rot, 76, 488
streak, 484
Verticillium wilt, 68, [69], 487
web blight, 160, 492
yellows, 73, 486
Egyptian paper plant, 498; *see* Umbrellaplant
8-hydroxyquinoline sulfate (or benzoate), 366, 367
trade names, 367
Elaeagnus, 445; *see* Russian-olive, Silverberry
Elanco Products Company, Division of Eli Lilly & Company, 122
Elcide-73, 106, 282, 557
Elder, 457
canker, twig blight, dieback, 77, 458
crown rot, 77, 458
flooding injury, 37
leaf scorch, 41, 459
leaf spot, 48, 458
powdery mildew, 56, 458
ringspot, 72, 459
root rot, 87, 458
rust, 58, 458
spot anthracnose, 51, 458
thread blight, 459, 505
trunk canker, 77, 259, 458
Verticillium wilt, 68, 459
web blight, 160, 459
wood rot, 79, 170, 458
Elecampane, 220; *see also* Inula
powdery mildew, 56, 221
rust, 58, 223
Electrical injuries, 46, 48
Elephants-ear, 195
bacterial soft rot, 81, 196
leaf spot, 48, 197
root-knot, 90, 197
root rot, 87, 196–97
southern blight, 76, 196–97
tuber rot, 86, 196–97
Elgetol, 556
bramble dormant spray, 427
crown gall treatment, 83, 502
rust control, 146, 313
tree wound treatment, 81, 383, 502
uses, 556–57
Eli Lilly & Company, 122
Elm, 263
anthracnose, 51, 264
bark beetle control, 263
bleeding canker, 77, 161, 345
canker, dieback, 77, 266
chlorosis, 24–29, 266, 344
Dothiorella wilt, dieback, canker, 77, 264, [265]
Dutch elm disease (Ceratostomella, Ceratocystis wilt), [262], 263
soil treatment, 263
The Dutch Elm Disease and Its Control, 263
Elm Bark Beetles, 263
flooding injury, 37
freezing injury, 41, 264
frost crack, canker, 41, 264
girdling roots, 45
ice injury, 43
leaf blister, 62, 266
leaf spot, 48, 264
lightning injury, 46
mistletoe, 93, 266
mosaic, mottle-leaf, witches'-broom, 70, 265
Nectria canker, 77, 266
phloem necrosis, 264
physiological leaf scorch, 41, 266
powdery mildew, 56, 266
pruning, 32, 33
resistant varieties, 263, 264, 266
root-knot, 90, 266, 392
root nematode, 266, 392
root rot, 87, 137, 266, 358
seedling blight, damping-off, seed rot, 76, 266, 405
soil grade change injury, 46
sooty mold, 64, 266
sunscald, 264
thread blight, 266, 505
tree borer control, 266
twig blight, 77, 264, 266
2, 4-D injury, 45, 266, [267], 288
Verticillium wilt, 68, 264
virus leaf scorch, 266
wetwood or slime flux, 264, [265]
The Wetwood Disease of Elm, 265
winter injury, drying, 41, 264
wood, heart rot, 79, 170, 266
Elymus, rust, 152, 183
Emilia, 220
mosaic, 70, 222
root-knot, 90, 225
root nematode, 226–27
rust, 58, 223
spotted wilt, yellow spot, 72, 223
stem rot, 76, 221
Emjeo, 28
Empresstree, 376; *see* Paulownia
Emulsifiable liquid, 557
Encelia, 220
root-knot, 90, 225
rust, 58, 223
Endive, 328; *see also* Chicory
anthracnose, 51, 331
bacterial rot, 50, 330
bacterial soft rot, 81, 86, 328
bottom rot, foot rot, 76, 328
brown heart, 331
downy mildew, 55, 329
gray-mold blight, rot, 52, 328, 329
leaf spot, 48, 331
mosaic, 70, 329
oxidant or PAN injury, 44
powdery mildew, 56, 331
root-knot, 90, 332
rust, 58, 330–31
seed treatment, 330, 532
spotted wilt, 72, 331
Verticillium wilt, 68, 332
watery soft rot, 76, 328
Endosulfan; *see* Thiodan
Endothia canker, 216, [217]
Endymion, 495; *see* Squill
Engelmann ivy, 287; *see* Boston ivy, Grape
English (or true) daisy, 220
aster yellows, 73, 222
crown rot, 76, 221
gray-mold blight, 52, 86, 224
leaf spot, 48, 221
root-knot, 90, 225
root rot, 87, 221
English ivy, 310; *see* Ivy
English marigold, 220; *see* Calendula
Entomologist, extension, help by, 5, 6, 23, 65, 79, 98, 106, 107, 128, 137, 139, 159, 161, 162, 182, 187, 231, 233, 257, 259, 264, 293, 332, 341, 384, 390, 404, 406, 421, 471, 472, 475, 505, 512, 523, 525
Environment, effect on disease development, 96
Epidendrum, 366; *see* Orchids
Epigaea, 294; *see* Trailing-arbutus
Epilobium, rust, 275
Epiphyllum, 194; *see also* Cacti
corky scab, glassiness, 39, 194, 195
Epiphytotic, 557
Episcia, 127; *see* African-violet
Epithelantha, 194; *see* Cacti
Epsom salts, 28, 325
Equipment
dusting, 115–23
fumigating soil, 543, 547, 548
spraying, 112–19
spread of viruses, 421
Equivalent volumes, liquid, 520
Eradicant fungicide, 100, 557
Eradication, 557
Eradicative seed treatment, 528–31
Eranthemum, 229
leaf spot, 48, 229
oedema, 39, 229
Eremochloa, 319; *see* Centipedegrass
Erica, 294; *see* Heath
Erigeron, 220
aster yellows, 73, 222
downy mildew, 55, 225–26
gray-mold blight, 52, 224
leaf gall, 62, 221
leaf spot, 48, 221
mosaic, 70, 222
powdery mildew, 56, 221
rust, 58, 223, 405
spotted wilt, 72, 223
stem rot, 76, 221
Verticillium wilt, 68, 223
white smut, 65, 226
Eriobotrya, 133; *see* Loquat
Erodium, 236; *see* Heronsbill
Ervatamia, 360; *see* Crape-jasmine
Eryngium, 213
leaf spot, 48, 213
root rot, 87, 215
seed rot, damping-off, 76, 214
seed treatment, 213
stem rot, 76, 214
white or leaf smut, 65, 215
Eryngo, 213; *see* Eryngium
Erysimum, 186; *see also* Wallflower, western
clubroot, 89, 189
downy mildew, 55, 189
mosaic, 70, 192
powdery mildew, 56, 193
rust, 58, 193
spotted wilt, 72, 192
white-rust, 62, 191
Erythrina, 300
leaf spot, 48, 301
root-knot, 90, 301, 392
root rot, 87, 137, 301
thread blight, 302, 505
Verticillium wilt, 68, 301, 343
Erythronium, 264
black spot, 48, 267
Botrytis blight, 52, 267
leaf blight, 51, 267
leaf smut, 62, 267
leaf spot, 48, 267
rust, 58, 267
Escape (plant), 557
Escarole, 328; *see also* Chicory
bacterial rot, leaf spot, 50, 330
bacterial soft rot, 81, 86, 328
bottom rot, 76, 328
damping-off, 76, 330
downy mildew, 55, 329
gray mold blight, rot, 52, 328, 329
oxidant or PAN injury, 44
seed treatment, 330, 532
spotted wilt, 72, 331
Eschscholtzia, 413; *see* California-poppy
Escobaria, 194; *see* Cacti
Es-Min-El, 27
Ethanol, as disinfectant, 98
Ethoxyquin, 140
Ethylene, as air pollutant, 43, 44
control, 44
resistant plants, 44
susceptible plants, 44
symptoms, 44
Ethylene Dibromide, 544; *see also* EDB
Etiolation, 557
Etiology, 557
Eucalyptus, 267–68
crown gall, 83, 268
gray-mold blight, 52, 268
heart rot, 79, 268
leaf spot, 48, 268
mosaic, 70, 268
oedema, 39, 268
root rot, 87, 137, 268
seedling blight, 76, 268, 405
shoot dieback, 77, 268
winter and frost injury, 41, 268
wood rot, 79, 268

Eucharis, 249; *see* Amazon-lily
Eugenia, 268; *see also* Eucalyptus
  black mildew, 64, 268
  leaf spot, 48, 268
  root rot, 87, 137, 268
Euonymus, 173
  anthracnose, 51, 173
  canker, dieback, 77, 173, 345
  crown gall, 83, [173]
  leaf scab, 65, 173
  leaf spot, 48, 173
  mosaic, infectious variegation, 70, 174
  powdery mildew, 56, 173
  root-knot, 90, 174
  root nematode, 174
  root rot, 87, 174
  thread blight, 174, 505
Eupatorium, 220
  aster yellows, 73, 222
  downy mildew, 55, 225
  Fusarium wilt, 56, 223
  gray-mold blight, 52, 86, 224
  leaf spot, 48, 221
  powdery mildew, 56, 221
  root-knot, 90, 225
  root nematode, 226–27
  root rot, 87, 221
  rust, 58, 223
  stem rot, 76, 221
  white-rust, 62, 226
  white smut, 65, 226
Euphorbia, 409; *see* Poinsettia, Spurge
European corn borer, insect control, 232
European cranberry-bush, 499; *see also* Viburnum
  bacterial leaf spot, 50, 79, 500
  dieback, 77, 500
  downy mildew, 55, 500
  gray-mold blight, 52, 77, 86, 500
  powdery mildew, 56, 500
Eustoma, 278; *see* Prairiegentian
Evening campion (*Lychnis*), 203; *see also* Carnation, Silene
  leaf spot, 48, 205
  root rot, 87, 204
  rust, 58, 204
  southern blight, 76, 204
Evening-primrose, 268
  anthracnose, 51, 268
  downy mildew, 55, 268
  leaf gall, 268
  leaf spot, 48, 268
  mosaic, 70, 268
  powdery mildew, 56, 268
  root rot, 87, 268
  rust, 58, 268
  stem nematode, 75, 268, 400
Evening star, 249; *see* Cooperia
Evergreen
  drought injury, 38
  fertilizing, 30
  flooding injury, 37
  planting, 20
  pruning, 33
  soil drainage, 36
  watering, 37–38
  winter injury, 41, 42
    protection, 42
Everlasting, 219
  downy mildew, 55, 225
  leaf spot, 48, 221
  white-rust, 62, 226
Everlasting pea; 377; *see* Sweetpea
Exacum, 278
  Botrytis blight, stem canker, 52, 77, 278
  damping-off, 76, 278
Exanthema, 557
Excess care (injury), 45
Exclusion, 557–58
Exobasidium leaf spot, 48, [434]
Experiment stations, listing of, 6–7
Extension agent, 5; *see* County agent
Extrax Copper Spray, 556
Exudate, 558
E-Z-Flo Chemical Company, 122
  Fruit Guard, 523

F

Factors affecting disease development, 96
Faesy & Besthoff, Incorporated, 102, 122, 518
  Captan 7.5 Dust, 102, 518
  Lime Sulphur Solution, 561
  Multi-Purpose Spray or Dust, 523
  Nabam-22, 562
Fagus, 161; *see* Beech
Fahrenheit, conversion to Centigrade, 517
Fairfield-Niagara, FMC Corporation, 122
Fairy lily, 249; *see* Cooperia
Fairy ring, 322, [323]
Fall-daffodil, 250; *see also* Daffodil
  Stagonospora leaf scorch or red spot, 48, 251
Falling stars, 282; *see* Gladiolus
False-acacia, 300; *see* Locust
False-aloe, 216; *see* Centuryplant
False-boneset, 220
  leaf spot, 48, 221
  root rot, 87, 221
  rust, 58, 223
False-camomile, 220; *see* Boltonia, Camomile, Matricaria
False-dragonhead, 447
  crown or stem rot, 76, 448
  downy mildew, 55, 558
  leaf spot, 48, 447
  rust, 58, 447
  southern blight, 76, 447
False-garlic, 361; *see also* Onion
  anthracnose, 51, 365
  mosaic, 70, 365
  rust, 58, 365
False holly, 370; *see* Osmanthus
False-indigo, 269
  leaf spot, 48, 269
  powdery mildew, 56, 269
  root nematode, 269
  root rot, 87, 269
  rust, 58, 269
False-mallow, 297
  leaf spot, 48, 298
  root rot, 87, 299
  rust, 58, 298
False-mesquite, 198
  rust, 58, 198
False sunflower, 220; *see* Heliopsis
Farewell-to-spring, 275; *see* Godetia
Farm advisor (er), 5; *see* County agent
Farmrite M-53 Fixed Copper, 104
Fasciation, 81, [82], 226, 381, 558
Fawn-lily, 267; *see* Erythronium
Feather-hyacinth, 495; *see* Grape-hyacinth
Featherbells, 495
  rust, 58, 498
February daphne, 253; *see* Daphne
Federal Insecticide, Fungicide, and Rodenticide Act, 107
Feet of row per acre at given distances, 552
Feijoa, 353
  fruit rot, 51, 86, 353
  root nematode, 353
  root rot, 87, 137, 353
  spot anthracnose, 51, 353
  thread blight, 353
Felicia, 220
  powdery mildew, 56, 221
Fendlera, 304
  rust, 58, 305
Fennel, 213
  aster yellows, 73, 214
  bacterial soft rot, 81, 214
  blackheart, calcium deficiency, 28, 214–15
  chlorosis, 24–29
  curly-top, 75, 215
  damping-off, 76, 214
  downy mildew, 55, 210, 215
  gray-mold rot, 52, 214
  iron deficiency, 24, 26, 27
  leaf spot, 48, 213
  mosaic, 70, 214
  root-knot, 90, 214
  root rot, 87, 215
  seed treatment, 213, 532
  stem rot, canker, 76, 77, 214
Ferbam
  active ingredient, 102, 518
  bulb treatment, 336, 337, 338
  control of iron deficiency, 336, 434, 467
  as "disinfectant," 199, 204, 221, 261, 276, 278, 409, 472
  dormant spray, 176, 525
  gallon lots, 522
  in multipurpose mixes, 102, 111, 518, 523
  oxidant or PAN prevention, 44
  as protective fungicide, 55, 62, 86, 100, 102, 145, 153, 157, 167, 169, 174, 175, 183, 186, 197, 205, 212, 213, 214, 221, 228, 264, 287, 296, 298, 301, 308, 313, 318, 328, 337, 343, 351, 365, 367, 384, 405, 424, 425, 433, 436, 440, 458, 495, 500, 508, 509, 518, 525
  soil drench, 77, 97, 102, 106, 127, 167, 197, 255, 261, 406, 489, 518
  trade names and distributors, 102, 518
  in tree wound dressing, 476
  uses, 102, 518
Fermate Ferbam Fungicide, 102, 518
Ferns, 269–70; *see also* Asparagus-fern
  anthracnose, tip blight, of *Nephrolepis Pteris*, 51, 270
  bacterial leaf spot or blight of *Asplenium*, 50, 270
  Botrytis blight, 52, 270
  chlorosis, 24–29
  damping-off, 76, 270
  inflorescence smut, of *Osmunda*, 62, 270
  leaf blister or gall, of *Cystopteris*, *Dryopteris*, *Onoclea*, *Osmunda*, *Polystichum*, *Pteretis*, 62, 270
  leaf nematode, of *Adiantum*, *Asplenium*, *Blechnum*, *Dryopteris*, *Polypodium*, *Polystichum*, *Pteris*, 75, [270]
  leaf spot or blight, of *Adiantum*, *Asplenium*, *Athyrium*, *Camptosorus*, *Dryopteris*, *Nephrolepis*, *Ophioglossum*, *Osmunda*, *Polypodium*, *Polystichum*, *Pteridium*, 48, 51, 270
  light and temperature requirements, 39, 40
  low humidity, leaf scorch, 39, 41, 270
  plant soak, 270, 530, 535
  rust, of *Athyrium*, *Camptosorus*, *Cryptogramma*, *Cystopteris*, *Dennstaedtia*, *Dryopteris*, *Nephrolepis*, *Onoclea*, *Osmunda*, *Pellaea*, *Polypodium*, *Polystichum*, *Pteretis*, *Pteridium*, *Woodsia*, *Woodwardia*, 58, 270, 405
  soluble salts injury, 31
  sooty mold, black mildew, 64, 270
  tar spot, of *Dryopteris*, *Polystichum*, *Pteretis*, *Pteridium*, 48, 270
Ferocactus, 194; *see* Echinocactus
Ferric dimethyldithiocarbamate, 102, 518
Ferro Copper Naphthenate 8%, 556
Ferrous ammonium sulfate, 327
Ferrous sulfate; *see* Iron sulfate
Fertilizer, 29–31, 558
  analysis, 29
Fertilizer Borate 46 and 65, 27
Fertilizing plants, 29, 30, [31]
  The Care and Feeding of Garden Plants, 19, 25
  foliar application, 30
  how much fertilizer for 100 feet of row for vegetables and flowers?, 549
  injury, 29, 32, 45
  starter solution, 29–30
Fertminal, 27

Fescue, Fescuegrass (*Festuca*), 319; *see also* Bluegrass
anthracnose, 51, 319
brown patch, Rhizoctonia disease, 321
chlorosis, 24–29, 325
damping-off, 76, 326
fairy ring, 322
foot rot, 76, 319
Fusarium blight, 326
Fusarium patch, pink snow mold, 322
insect injury, control, 321, 327
leaf blight, 51, 319
leaf spot, 48, 319
melting-out, 319
powdery mildew, 56, 321
red thread, pink patch, 326
resistant grasses, 320, 327
root nematode, 324
root rot, 87, 319
rust, 58, 321
Sclerotinia dollar spot, 321
seed treatment, 326, 537
slime mold, 322
smut, 62, 325
soil drench, 320–21
stem or culm rot, 76, 319
tar spot, 48, 319
Typhula blight, snow scald, 322
Fetterbush, 175; *see* Lyonia
Feverfew, 219; *see also* Chrysanthemum
damping-off, seed rot, 76, 221
powdery mildew, 56, 221
root rot, 87, 221
seed treatment, 221
stem rot, 76, 221
white-rust, 62, 226
Ficus, 270; *see* Fig
Fiddleneck, 398; *see* Phacelia
Fiesta-flower, 399; *see* Nemophila
Fig, 271
anthracnose, 51, 271
canker, dieback, 77, 138, 266, 271
chlorosis, 24–29, 272, 503–4, 505
crown gall, 83, 136, 168, 272
felt fungus, 272, 293
Fig Growing in the South, 272
fruit spot or rot, 86, 271
Fusarium wilt, 66, 272
Growing Figs in the South for Home Use, 272
leaf blotch, blight, 51, 272
leaf scorch, fall, [41], 271
leaf spot, 48, 271, 272
light and temperature requirements, 40
limb blight, 77, 271
manganese deficiency, 26, 27, 272, 505
mosaic, 70, 272
resistant varieties, 271, 272
root-knot, cyst nematode, 90, 271, 392
root nematode, 271
root rot, 87, 137, 271
rust, 58, 272
rusty leaf, 272
sooty mold, 64, 137, 272
souring, 271
southern blight, 76, 272
sunscald, 139, 272
thread blight, 272, 505
twig blight, 77, 271
web blight, leaf blight, 51, 160, 272
winter injury, sunscald, 41, 139, 272
wood rot, 79, 170, 272
zinc deficiency, 26, 27, 272, 503–4
Figmarigold, 306
root-knot, 90, 306
sooty mold, 64, 306
Filaree, 236; *see* Heronsbill
Filbert, 170; *see also* Hazelnut
bacterial blight, 50, 79, 171, [172]
canker (twig, branch, trunk), 77, 171, 172
crown gall, 83, 136, 172
kernel bitter rot, brown spot or stain, 86, 172
leaf blister, 62, 171
leaf spot, 48, 171
powdery mildew, 56, 171
resistant varieties, 171
root rot, 87, 137, 171–72
sunscald, 41, 172
twig blight, 77, 172
wood rot, heart rot, 79, 170
Filipendula, 440; *see* Meadowsweet
Finocchio, 213; *see also* Fennel
bacterial soft rot, 81, 214
Fir, 402
black mildew, 64, 406
brown felt blight, 407
canker (twig, branch, trunk), 77, 403
chlorosis, 24–29, 408
crown dieback, 77, 406–7
gray-mold blight, 52, 408
insect and mite control, 406, 407
iron deficiency, 24, 26, 27, 408
mistletoe, 93, 406
needle blight, cast, 51, 402
root rot, 87, 405, 406
rust, needle, 58, 177, 405, 508
witches'-broom, 58, 405
seed treatment, 405
seedling blight, 76, 405
snow blight, 407
soil treatment, 405, 406
stump treatment, 406
sunscorch, wind injury, 41, 407
tar spot, 48, 402
tip blight, 76, 402
twig blight, 77, 402, 403
wetwood, 264, 408
wood, trunk, butt rot, 79, 170, 403
Fire (*Botrytis*), 251, [495]
Fire blight, 79, [80], 81, [133], 134, 525
Blight of Pears, Apples, and Quinces, 579
Fire Blight Canker Paint No. 1 (G.L.F.), 558
Fire-chalice, 268
rust, 58, 268
Fire injury, 45
entrance for fungi, 79
Fire-pink, 203; *see* Silene
Firecracker plant, 227; *see* Cigarflower
Firethorn, 133; *see* Pyracantha, Apple
Fireweed (*Epilobium*), rust, 275, 405
Firewheel, 220; *see also* Gaillardia
aster yellows, 73, 222
white smut, 65, 226
Firmiana, 401; *see* Phoenix-tree
Fish oil, as sticker, 119
Fishhook cactus, 194; *see* Cacti, Mammillaria
Fittonia, 451; *see* Silver threads
Five-leaf or Five-fingered aralia, 126; *see* Acanthopanax
Five-spot, 399; *see* Phacelia
Fixed copper fungicides; *see* Copper, fixed fungicides
Flag (flagging), 558
Flag Sulphur & Chemical Company, 103, 122, 519
Flagella, 10, [11], 558
Flag's Citrus Nutritional Spray, 27
Flame-vine, 169; *see* Bignonia
Flame violets, 127; *see* African-violet light and temperature requirements, 40
Flamingo-flower, 195; *see* Anthurium
Flannel-bush, 401; *see* Fremontia
Flash-flame pasteurizers, 541–42
Flax, flowering, 273
aster yellows, 73, 273
curly-top, 75, 273
damping-off, 76, 273
gray-mold, 52
root-knot, 90, 168, 273
root rot, 87, 273
rust, 58
seed treatment, 273
stem rot, 76, 273
Flea beetles,
as bacteria carriers, 231
insect control, 231, 289, 485
as virus carriers, 72, 485, 486
Fleabane, 220; *see also* Erigeron
aster yellows, 73, 222
downy mildew, 55, 225
mosaic, 70, 222
powdery mildew, 56, 221
rust, 58, 223
spotted wilt, 72, 223
Verticillium wilt, 68, 223
white smut, 65, 226
Fleas, insect control, 321
Fleck, 558
Fleeceflower, 453; *see* Silver-lacevine
Flight Brand 10% Dithane Dust, 103, 519
Floras-paintbrush, 228; *see* Emilia
Florida yellowtrumpet, 494
root rot, 87, 137, 494
rust, 58, 494
Floridin Company, 122
Florist, help by, 5, 9, 23, 32, 40, 65, 129, 206, 369, 410, 496
Commercial Storage of Fruits, Vegetables, Florist, and Nursery Stocks, 100
Florist Products, 122
Flower
blight, spot, 53, [85], 86, [133], [199], [224], [260], [435]
breaking, 70, [71], 72, [192], [250], [284], 368, [375], [378], 496
diseases, 86–87
distortion, 223
drop, 41, 484
fertilizing, 29–30
gall, [436]
multipurpose mixes, 110–11
spray, 192, 199, 205, 224, 248, 280, 308, 335, 367, 435, 436, 443
starter solution, 29–30
Flowering almond, 382; *see also* Almond
bacterial leaf spot, 50, 386
blossom blight, 86, 382
brown rot, 86, 382
fire blight, 79, 133, 393
gray-mold blight, rot, 52, 86, 393
powdery mildew, 56, 385
root rot, 87, 392
shot-hole, 52, 386, 390, 392
thread blight, 393, 505
twig blight, 77, 382
Flowering apricot, 382; *see* Apricot
Flowering cherry, 382; *see also* Cherry
bacterial leaf spot, 50, 386
blossom blight, 86, 382
dieback, 77, 383
fire blight, 79, 133, 393
leaf blister, 62, 384
leaf spot, shot-hole, 48, 384
mottle complex, 70, 390
plum decline (virus), 389
powdery mildew, 56, 385
pruning, 32, 383, 384
root nematode, 392
scab, 65, 385
sooty mold, 64, 137, 393
twig blight, 77, 382
wet feet, 393
winter injury, 41, 392
witches'-broom, 62, 384
Flowering crabapple, 133; *see* Apple, Crabapple, flowering
Flowering currant, 245; *see also* Currant
anthracnose, 51, [246]
bacterial spot, 50, 248
dieback, 77, 247
fruit rot, 86, 246
gray-mold blight, dieback, 52, 247
leaf spot, 48, [246]
rust, 58, 246, 405
Flowering flax, 273; *see* Flax, flowering
Flowering kale, 185; *see* Kale, Cabbage
Flowering maple, 297; *see* Abutilon
Flowering peach, 382; *see* Peach
Flowering quince, 133; *see also* Apple, Quince
black rot, 51, 77, 86, 136
blossom blight, 79, 86, [133], 134
canker, dieback, 77, 138
crown gall, 83, 136–37
felt fungus, 138, 293
fire blight, 79, 86, [133], 134

fruit spot, rot, 86, 138
hairy root, 83, 137
leaf blight, 51, 140
leaf spot, 48, 140
mosaic, 70, 142
root-knot, 90, 141
root rot, 87, 137–38
rust, 58, 135
twig blight, 77, 138
Flowering raspberry, 427; *see also* Raspberry
cane blight, 77, 428
leaf curl, 70, 428
leaf spot, 48, 427, 430–31
mosaic, 70, 428
powdery mildew, 56, 431
rust, 58, 430
Flowering tobacco, 479; *see also* Tomato
aster yellows, 73, 490
bacterial wilt, 69, 488
crown gall, 83, 492
curly-top, 75, 486
downy mildew, blue mold, 55, 490
early blight, 51, 480
fasciation, 81, 492
leaf spot, 48, 480
mosaic, 70, 484
powdery mildew, 56, 492
ringspot, 72, 486
root-knot, 90, 489
root nematode, 489
root rot, 87, 490
spotted wilt, 72, 486
tobacco mosaic virus, [14]
Flower-of-an-hour, 297; *see also* Roselle
leaf spot, 48, 298
root-knot, 90, 299
root rot, 87, 299
Flower-of-Jove, 203; *see* Maltese cross, Lychnis
Flozon Tree Wound Paint, 476–77
Fluorides, as air pollutants, 43–44
control, 44
resistant plants, 43
sensitive plants, 43
symptoms, 43
Fluxit, 119
Fly-honeysuckle, 457; *see* Honeysuckle
FMC Corporation, 102, 103, 105, 106, 120, 123, 518, 519
Foamflower, 304
powdery mildew, 56, 304
rust, 58, 305
Foeniculum, 213; *see* Fennel, Finocchio
Fogfruit, 317; *see* Lemon-verbena
Folcid, 104; *see also* Difolatan
Foliage diseases, 48–76
Foliar feeding, 30, 558
Foliar nematodes, 75, [76], [270], 337, 369, 375, 400, 456, 467, 497
Foli-gard, 21, 38, 41, 42, 182, 296, 511
Folpet (Phaltan)
active ingredient, 103, 519
-Captan, 567
eradicant and protective qualities, 103, 519
gallon lots, 522
in multipurpose mixes, 103, 111, 519
-PCNB mixtures, 77, 104, 410, 437
as protective fungicide, 55, 58, 62, 103, 134, 137, 138, 167, 199, 204, 207, 221, 239, 242, 246, 259, 289, 308, 320, 384, 440, 464, 482, 506, 519
soil drench, 77, 97, 106, 162, 326, 406, 410, 437
trade names and distributors, 103, 519
tree wound dressing, 476
uses, 103, 519
Foot rot, 76, 188
Forcing crops, 558
Fore, 106, 320, 321, 326, 558
Forester, help by, 263, 264
Forestiera, 144
leaf spot, 48, 145
mistletoe, 93, 145
powdery mildew, 56, 145
root rot, 87, 137, 145
rust, 58, 144
Forget-me-not, 349
aster yellows, 73, 349
black ringspot, 72, 192, 349
crown rot, wilt, 76, 254, 349
downy mildew, 55, 349
gray-mold blight, 52, 349
powdery mildew, 56, 349
rust, 58, 349
Formaldehyde, Formalin
bulb, corm, or rhizome soak, 93, 196, 250, 252, 282, 307, 310, 338, 364, 421, 497, 531
disinfectant, 83, 99, 158, 196, 208, 214, 226, 314, 366, 367, 370, 420, 428, 440, 441, 444, 471, 476
precautions, 544
soil application, 83, 193, 259, 322, 362, 437, 544
trade names, 558
uses, 558
Forsythia, 273
anthracnose, 51, 273
bacterial blight, shoot blight, 50, 81, 273
blossom blight, 86, 273
cane blight, dieback, 77, 273
chlorosis, 24–29, 274, 344
crown gall, 83, 136, 273
leaf spot, 48, 273
Phomopsis stem gall, 77, 273
pruning, 32
root-knot, 90, 274
root nematode, 274
root rot, 87, 274
southern blight, 76, 273
twig blight, dieback, 77, 273
Fortunella, 228; *see* Kumquat, Citrus
Foundation planting stock, 558
Four-o'clock, 274
curly-top, 75, 162, 274
downy mildew, 55, 274
leaf spot, 48, 274
root-knot, 90, 159, 274
root rot, 87, 274
rust, 58, 274
white-rust, 62, 274
Foxglove, 454
anthracnose, 51, 455
chlorosis, 24–29, 443, 457
crown gall, 83, 457
curly-top, 75, 457
downy mildew, 55, 456
flower blight, 52, 86, 455
Fusarium wilt, 66, 455
gray-mold (*Botrytis*) blight, 52, 86, 455
insect control, 456, 457
leaf blight, 51, 455
leaf spot, 48, 455
leaf and stem nematode, 75, 455
mosaic, 70, 456
ringspot, 72, 456
root-knot, 90, 457
root rot, 87, 455
seed treatment, 455, 530, 535
soil drench, 455
stem rot, wilt, 76, 455
Verticillium wilt, 68, 455
Fragaria, 462; *see* Strawberry
Fragrant glad, 281; *see also* Gladiolus
dry rot (*Stromatinia*), 90, 282
Frangipani, 360
chlorosis, 24–29, 344
leaf spot, 48, 360
mistletoe, 93, 361
root rot, 87, 137, 361
rust, 58, 361
Franklin-tree, 274
black mildew, 64, 274
chlorosis, 24–29
iron deficiency, 24, 26, 27
leaf spot, 48, 274
root rot, 87, 274
Franklinia, 274; *see* Franklin-tree, Loblolly-bay
Frasera, 278; *see* Columbo
Frass, 558
Fraxinus, 144; *see* Ash
Freesia, 281; *see also* Gladiolus
bacterial scab, 50, 65, 90, 283
corm rot, 90, 282, [283]
gray-mold blight, 52, 282
leaf spot, 48, 282
mosaic, 70, 284
root-knot, 90, 285
yellows (*Fusarium*), wilt, 66, 282
Freezing injury, 42
entrance for bacteria, 81, 362, 365
Fremontia, 401
collar rot, stem girdle, 77, 259, 401
leaf spot, 48, 401
Verticillium wilt, 68, 401
French endive, 328; *see* Chicory
French-mulberry, 317; *see* Callicarpa
Frijolito, 300; *see* Sophora
Fringed hibiscus, 297; *see* Hibiscus (arborescent forms)
Fringeflower, 479; *see* Butterfly-flower
Fringetree, 144
chlorosis, 24–29
iron deficiency, 24, 26, 27
leaf spot, 48, 145
powdery mildew, 56, 145
root rot, 87, 137, 145
wood rot, 79, 145, 170
Fritillary (*Fritillaria*), 494; *see also* Tulip
Botrytis blight, 52, 495
bulb rot, 90, 495
leaf spot, 48, 498
mosaic, 70, 496
rust, 58, 498
Froelichia, 230
damping-off, seed rot, 76, 230
leaf spot, 48, 230
root-knot, 90, 230
root and crown rot, 76, 87, 230
soil drench, 230
white-rust, 62, 230
Frogeye leaf spot, 48
Frost injury, 318–19, 497
crack, 41
entrance for fungi, 68, 79, 402
Frostweed, 469; *see* Sunrose
Frostwort, 469
leaf spot, 48, 469
Fruit
cracking, 422, [483]
dip or spray, 102, 208, 383, 518
diseases, 86–87
fertilizing, small, 29, 30
grower, help by, 5, 140, 141, 291, 392
multipurpose mixes, 111, 523
spot, speck, rot, or blotch, 86, [87], [238], [287], [391], 525
spray amounts, needed, 526
spray schedules, 523–25
storage, 138
Commercial Storage of Fruits, Vegetables, Florist, and Nursery Stocks, 100
Home Storage of Vegetables and Fruits, 87, 99
rots, 86, 138
wraps, 140, 393
Fruiting body (fungus), 558
FTE, 27
Fuchsia, 275
gray-mold blight, 52, 275
leaf spot, 48, 221, 275
light and temperature requirements, 40
powdery mildew, 56, 275
root-knot, 90, 128, 275
root rot, 87, 275
rust, 58, 275
soil drench, 275
spotted wilt, 72, 168, 275
Verticillium wilt, 68, 275
Fulex A-D-O, 106, 367, 558, 563
Fumazone, 545, 548; *see also* DBCP
Fumigant, Fumigation, soil
application of, 538–48, [549]
gaseous, 547
granular, 548
hand, 543, 547, 548, [549]

Fumigant (*continued*)
equipment, 543, 547, 548
fungi control, 538–48
nematode control, 16, 76, 538–48, [549]
precautions in using, 106, 538, 543–48
trade names, 106, 543–48
uses, 98, 106, 538, 544–46
waiting period, 543–48
Fungchex, 101, 106, 558
Fungi, 11, [12], 13, 558
diseases caused by, 13
fruiting bodies, 12, 45, 56, [57], [79], [187], [247], [265], [287], [358], [428], 440, 458, 490, 508, 558, 562
lethal temperatures, 542
life cycles, 12, 13
numbers, 11
overwinter, 13
parasites, 11
penetration, [12]
physiologic races (strains), 13, 97
saprophytes, 12, 565
sclerotia, 13, 76, 89, 97, 157, 174, 198, [207], 243, 250, 252, 361, [419], 490, 505
spores, 11, [12], 79, 558
germination, 11, [12]
movement of, 12
numbers, 79
size, 12
stages, 12, 565
variety, [12]
as virus carriers, 13, 330
Fungicide, 558; *see also* Pesticide
accidents, 108
Acti-dione, 58, 106
active ingredients, 101, 102–3, 518–19
application equivalents, 517
Bordeaux mixture, 36, 104
Botran (dichloran), 55, 86
Burgundy mixture, 105
captan, 48, 55, 65, 77, 81, 86, 97, 102, 518, 522
chemotherapeutant, 100
chloranil, 77, 97, 102, 518, 522
common names, 101, 102–3, 518–19
compatibility chart, 552
conversion table, 422
copper, fixed, 48, 51, 55, 56, 62, 65, 86, 104, 522
Daconil 2787, 55, 58, 86, 89, 104
definition, 100, 558
Dexon, 77, 89, 97, 104
dichlone, 44, 55, 102, 518, 522
difolatan (Folcid), 104
disinfesting soil, 77, 82, 89, 97, 101, 106, 538–49
distributors and manufacturers, 120–24
dodine (Cyprex), 102, 518, 522
Dyrene, 55, 104, 106, 522
equipment, 109, [110], 112–23
eradicant, 100, 557
ferbam, 44, 55, 62, 77, 86, 97, 102, 518, 522
folpet (Phaltan), 55, 58, 62, 77, 103, 519, 522
formulations, 102–3, 112
fruit spray schedules, 523–25
gallon lots, 522
Karathane (dinocap), 58, 101, 522
label, 107
caution or warning statement, 107
as legal document, 108
must show, 107
Lanstan, 104
lawn, 106, 319–27
liquid, preparation of small amounts, 522
maneb, 44, 48, 55, 56, 62, 65, 86, 103, 519, 523
measuring apparatus, 109
mercuric chloride, 35, 77, 79, 81, 82, 83, 89, 97, 101, 528–29
miscellaneous, 106
modern, 100–103, 518–19
multipurpose mixes, 110–11, 523
injury from, 523
notes, 523
trade names, 523
PCNB (Terraclor), 77, 89, 97, 101, 104, 111, 522
percentage solution, 522
phenyl (organic) mercury, 36, 106, 564
Polyram (metiram), 55, 62, 86, 104
precautions, 45, 101, 105, 106, 107, 108, 109, 112, 523, 525, 527, 538, 543–49
preparation of spray mixtures, 108–9, 520–21
preventive schedule, 100
protective, 100, 102–3
purchasing, 523
safety, 45, 107
schedule, 100
seed treatment, 93, 97, 527–37
Semesan, 97, 106, 522
"shot-gun" soil drench, 111
soil application, 77, 82, 89, 97, 101, 106, 544
spraying and dusting tips, 108, 109, 523
streptomycin, 51, 81, 83, 105, 522
sulfur, 58, 62, 65, 105
thiram, 44, 55, 62, 77, 81, 86, 97, 103, 519, 522, 531
trade names, 101, 102–3, 105, 106, 518–19
uses, 100, 102–3, 105, 106
zineb, 44, 48, 55, 56, 62, 65, 77, 86, 97, 103, 519, 522
ziram, 48, 65, 77, 103, 519, 522
Fungiclor, 101
Fungistat, 558
Fungus leaf spot, 48, [49]
Furcraea, 216
leaf scorch, 51, 216
leaf spot, 48, 216
root-knot, 90, 216
Fusarium dry rot, [417]
leaf spot, [449]
wilt or yellows, 66, [67], 68, 91, 98, 146, 186, [202], 203, 214, 241, [472], [487]
following nematode wounds, 66, 91, 203, 301, 377, 487
Fusiform gall rust, [404], 405
wounds, entrance for wood rot fungi, 402

G

Gaillardia, 220
aster yellows, 73, 222
downy mildew, 55, 225
leaf spot, 48, 221
mosaic, 70, 222
powdery mildew, 56, 221
root nematode, 226–27
root rot, 87, 221
rust, 58, 223
spotted wilt, 72, 223
white smut, 65, 226
Galanthus, 249; *see* Snowdrop
Galax, 275
chlorosis, 24–29
leaf spot, 48, 275
Galinsoga, as virus source, 73
Galium, 185; *see* Bedstraw
Gall, 558
bacterial root, 83
black knot, [81], 384–85, 525
cane, 83, [84], 176
cedar gall, [58], 313, 555
crown, 83, [84], 98, 136, 146, [173], [176], 391, 428, 508, 556
entrance for bacteria, 83
nematodes, 91
leaf, 13, 62, [201], [436]
leaf nematode, 75, [76], 226, [270]
nematode root, 16, 90, [91], [395], [396]
rust, [404], 405
Gallowhur Chemicals, Canada Limited, 122
Galtonia, 494; *see* Summer-hyacinth
Gama (*Tripsacum*), as virus source, 234
Garden alternanthera, 230; *see* Alternanthera
Garden balsam, 152; *see* Balsam, garden
Garden clubs, help by, 9
Garden cress, 186; *see also* Peppergrass
chlorosis, 24–29, 194
clubroot, 89, 189
curly-top, 75, 192
damping-off, 76, 188
downy mildew, 55, 189
leaf blight, 51, 190
leaf and stem nematode, 75, 194
mosaic, 70, 192
resistant varieties, 189
ringspot, 72, 192
rust, 58, 193
seed treatment, 187, 188, 529–32
white-rust, 62, 191
Garden-heliotrope, 499; *see* Valerian
Garden hose sprayers, 114, [117]
Garden huckleberry, 175
Garden pinks, 203; *see* Pinks, garden
Garden supply dealer, help by, 5, 9, 369
Garden verbena, 317; *see* Verbena
Gardenia, 275
bacterial leaf spot, 50, 276
bud rot, drop, 277
chlorosis, 24–29, 276
crown gall, 83, 277
cutting dip, 276
dieback, 77, 277
fungus leaf spot, 48, [276]
Gardenia Culture, 277
gray-mold (*Botrytis*) blight, petal blight, 52, 86, 277
insect control, 277
iron deficiency, 24, 26, 27, 276
light and temperature requirements, 40, 41
Phomopsis (*Diaporthe*) canker, 77, 275–76
plant dip, 277
powdery mildew, 56, 277
resistant varieties, 276
root-knot, 90, 277
root nematode, 277
root rot, 87, 277
soil mixture for, 22
sooty mold, 64, 277
stem canker, gall, 77, 275, [276]
tipburn, 276–77
Garland-flower, 253; *see* Daphne
Garlic, 361; *see also* Onion
aster yellows, 73, 365
bacterial soft rot, 81, 90, 362
blast, 52, 362
bulb or clove rot, 90, 361, 362
canker, 77
clove, seed treatment, 362, 364, 529, 531, 533
downy mildew, 55, 363
gray-mold neck rot, 52, 90, 361–62
insect control, 362, 363, 365
leaf blight, blotch, 51, 361, 362, 363, 365
mosaic, 70, 365
pink root, 87, 363
root-knot, 90, 365
rust, 58, 365, 508
smut, 62, 362
soil treatment, 362
southern blight, 76, 366
stem and bulb nematode, 75, 364
Verticillium wilt, 68, 365
white rot, 90, 362
Garrya, 258; *see* Silktassel-bush, Tassel-tree
Gas injury, 44–45, 477–78
control, 45
Effect of Natural Gas on House Plants and Vegetation, 45
symptoms, 44, 477–78
Gaultheria, 294; *see also* Checkerberry, Salal
black mildew, 64, 294

fruit spot, 86, 294
leaf spot, 48, 294
powdery mildew, 56, 294
red leaf gall, 294
sooty mold or blotch, 64, 294
spot anthracnose, 51, 294
Gaura, 275
aster yellows, 73, 222, 275
downy mildew, 55, 225, 275
leaf spot, 48, 221, 275
powdery mildew, 56, 275
root rot, 87, 275
rust, 58, 275
soil drench, 275
Gayfeather, 220; *see also* Liatris
rust, 58, 223, 405
Gaylussacia, 175; *see* Huckleberry
Gazania, 220
crown rot, 76, 221
Geiger-tree, 231; *see* Cordia
Geigy Chemical Corporation, Agricultural Chemical Division, 102, 122
SP-50, 102, 518
Gelsemium, 185; *see* Carolina-jessamine, Yellow-jessamine
General Aniline & Film Corporation, 122
General Chemical Division, Allied Chemical Corporation, 102, 103, 120, 518, 519
General diseases, 48–94
Genista, 183; *see also* Broom, Woodwaxen
leaf spot, 48, 183
powdery mildew, 56, 183
root rot, 87, 137, 183
rust, 58, 183
Gentian (*Gentiana*), 277
Botrytis blight, 52, 278
damping-off, 76, 278
leaf spot, blotch, 48, 51, 278
root rot, 87, 278
rust, 58, 278
stem (crown) canker, rot, 52, 76, 77, 278
Geranium of florists (*Pelargonium*), 278; *see also* Cranesbill
bacterial leaf spot and stem rot (wilt), 50, 69, 280, [281]
blackleg (*Pythium, Fusarium*), 76, [77], 278, [279], 280
blossom blight, 52, [85], 86, 280
chlorotic spot, 278
crown gall, 83, 280
curly-top (leaf cupping), 75, 278
cutting dip, 278, 280
cutting rot, 76, [279]
disease-free, "cultured" cuttings, 97, 278, 279
fungus leaf spot, 48, 280
gray-mold (*Botrytis*) blight, 52, [54], [85], 86, 280
leaf breaking, 278–79
leaf curl (crinkle), 70, 278
leaf nematode, 75, 280
leafy gall or fasciation, 81, [82], 280
light, effect on flowering, 38
light and temperature requirements, 40
measles, 70, 278, [279]
mosaic, 70, 278
mottle, 70, 278
oedema or dropsy, 39, 280, [281]
potassium deficiency, 26, 280–81
ringspot, 72, 278
root-knot, 90, 278, 280
root nematode, 278
root rot, 87, 278
soil drench, 278
soil mix for, 22
spotted wilt, 72, 278
stem rot, 76, 278–79, 280
Verticillium wilt, 68, 280
virus complex, 278–79
watering, excess, injury by, 38
yellow net vein, 278
Gerbera, 220; *see* Transvaal daisy
Germ tube, 559
German camomile, 220; *see* Matricaria
German catchfly, 203; *see* Lychnis
German ivy, 220; *see* Senecio
Germander, 447
downy mildew, 55, 448
leaf spot, 48, 447
powdery mildew, 56, 448
root-knot, 90, 447
rust, 58, 447
Gerox, 106, 559
Gesneria, 127; *see* African-violet
Geum, 440; *see* Avens
Gherkin, 237; *see* West India gherkin
Giant chinquapin, 216; *see* Golden chinquapin
Giant daisy, 219; *see* Chrysanthemum
Giant night white bloomer, 351; *see* Morning-glory
Giant sequoia, 402; *see* Sequoia
Gilbert's Tree Wound Dressing, 476, 559
Gilia, 399
aster yellows, 73, 400
downy mildew, 55, 400
leaf spot, 48, 400
mosaic, 70, 400
powdery mildew, 56, 399–400
root-knot, 90, 401
root rot, 87, 400
rust, 58, 400
smut, 62, 401
soil drench, 400
Gillette Inhibitor Company, Incorporated, 122
Gilsonite-varnish wound dressing, 36, 476, 559
Ginkgo, 281
anthracnose, 51, 281
leaf scorch, 41, 281, 344
leaf spot, 48, 281
root-knot, 90, 281, 392
root nematode, 281
root rot, 87, 281
wood rot, 79, 170, 281
Girdle, 559
Girdling tree roots, [45]
control of, 45
Glacier-lily, 267; *see* Erythronium
Gladiolus, 281
aster yellows, grassy-top, 73, 284
bacterial leaf spot and blight, 50, 282, [283]
bacterial scab, 51, 65, 90, [283]
boron deficiency, 27, 285
Botrytis leaf spot, 52, [283]
bulb and stem nematode, 75, 92, 285
chlorosis, 24–29, 284
collar rot, 65, 76, 282
corm rot, 52, 81, 86, [88], 282
corm treatment, 282, 283, 285, 528, 529, 530, 535
cormel soak, 282, 528, 530, 535
crown gall, 83, 285
Curvularia leaf spot, 48, [283]
fluorides injury, 44
flower spike dip, 284
flower spot, blight, 52, [85], 86, 283–84
fungus leaf spot, 48, [283]
Fusarium yellows, 66, 282
insect and mite control, 282
iron deficiency, 24, 26, 27, 284
leaf blight, 51, 283
mosaic, flower breaking, 70, [284]
neck or collar rot, 76, 282, [283], 284
ringspot, 72
root-knot, 90, 285
root nematode, 285
root rot, 87, [88], 284
rust, 58, 285
Septoria leaf spot, 48, [283]
smut, 62, 285
soil treatment, 282, 284, 285
southern blight, 76, 282
Stemphylium leaf spot, 48, [283]
"topple," calcium deficiency, 28, 284, 497
white break, stunt, 70, [284]
Gleditsia, 300; *see* Honeylocust
Globe-amaranth, 229; *see also* Amaranth
curly-top, 75, 162, 230
gray-mold blight, 52, 230
leaf spot, 48, 230
ringspot, 72, 230
root-knot, 90, 230
root nematode, 230
white-rust, 62, 230
yellows, 73, 230
Globe artichoke, 328
curly dwarf, 73, 328
gray-mold blight, 52, 86, 329
leaf spot, 48, 331
powdery mildew, 56, 331
root-knot, 90, 332
southern blight, 76, 323
stem rot, 76, 328
yellows (virus), 73, 328–29
Globe lily, 347
rust, 58, 347
Globe-tulip, 347
rust, 58, 347
Globeflower, 131
leaf spot, 48, 131
smut, 62, 131
Globemallow, 297
powdery mildew, 56, 298
root rot, 87, 299
rust, 58, 298
Globethistle, 220
crown and root rot, 76, 87, 221
Gloriosa daisy, 220; *see* Rudbeckia
Glory-of-the-snow, 496; *see also* Tulip
bulb nematode, 92, 496–97
bulb soak, 497, 530, 531, 537
Glorybower, 317; see Clerodendron
Gloryvine, 286; *see* Grape
Glossary, 553–69
Gloxinia, 127
aster yellows, 73, 129
bacterial leaf blight, 50, 129
boron deficiency, 27, 129
bud drop, 39, 129
bud rot (*Botrytis*), 52, 127
crown, stem rot, 76, 127
flower blight, 52, 86, 127
leaf nematode, 75, 129
leaf rot, 127
light, effect on flowering, 38
light and temperature requirements, 40
plant soak, 129
powdery mildew, 56, 128
root nematode, 127
root rot, 87, 127
Sclerotinia blight, 76, 129
soil drench, 127, 129
spotted wilt, ringspot, 72, 129
watering excess, injury by, 38
Glyodin, 106, 559
as protective fungicide, 100, 134
Glyoxide Dry, 106, 559
Goatsbeard, 439
fire blight, 79, 444
leaf spot, 48, 443
stem canker, 77, 440
Godetia, 275
aster yellows, 73, 222, 275
curly-top, 75, 223, 275
damping-off, 76, 275
downy mildew, 55, 225, 275
gray-mold blight, 52, 275
root-knot, 90, 128, 275
root rot, 87, 275
rust, 58, 275
soil drench, 275
spotted wilt, 72, 168, 275
stem rot, 76, 275
Golddust, 186; *see* Alyssum
Golddust-tree, 148; *see* Aucuba
Golden-aster, 220
leaf spot, 48, 221
powdery mildew, 56, 221
rust, 58, 223
Golden chinquapin, 216
blight, canker, 77, 216–17
brown felt canker, 218, 293
leaf blister, 62, 218, 357
leaf spot, 48, 218

Golden chinquapin (*continued*)
oak wilt, 218, 355
powdery mildew, 56, 218
root rot, 87, 218, 358
twig blight, 77, 218
wood rot, 79, 218
Golden currant, 245; *see* Flowering currant, Currant
Golden-dewdrop, 317; *see* Duranta
Golden eardrops, 174; *see* Bleedingheart
Golden-eyed grass, 307; *see* Blue-eyed grass
Golden marguerite, 219; *see* Camomile
Golden-pea, 377; *see* Thermopsis
Golden queen, 131; *see* Globeflower
Golden rose of China, 439; *see* Rose
Golden-shower, 300; *see* Cassia
Golden spiderlily, 250; *see* Lycoris
Golden-star, 219; *see* Chrysanthemum, Golden-aster
Golden-wave, *220; see* Coreopsis
Golden wreath, 300; *see* Acacia
Goldenbells, 273; *see* Forsythia
Goldenchain, 285
Botrytis blight, gray-mold blight, 52, 286
dieback, 77, 285–86
leaf spot, 48, 285
mosaic, infectious variegation, 70, 286
root-knot, 90, 286
root nematode, 286
root rot, 87, 137, 286
twig blight, dieback, 77, 285–86
Goldeneggs, 268; *see* Evening-primrose
Goldenglow (*Rudbeckia*), 220
aster yellows, 73, 222
downy mildew, 55, 225
leaf spot, 48, 221
mosaic, 70, 222
powdery mildew, 56, 221
root rot, 87, 221
rust, 58, 223
soil drench, 222
southern blight, 76, 221
stem or crown rot, 76, 221
white smut, 65, 226
yellow dwarf, 73, 222
Goldenlarch, 318; *see also* Larch
canker, 77, 319
Goldenrain-tree, 286
coral spot, twig canker, 77, 286
leaf spot, 48, 286
pruning, 32
Verticillium wilt, 68, 286, 343
Goldenrod, 220
downy mildew, 55, 225
leaf spot, 48, 221
mosaic, 70, 222
powdery mildew, 56, 221
root nematode, 226–27
root rot, 87, 221
rust, 58, 223, 405
spot anthracnose, 51, 221
stem canker, 77, 221, 225
Goldentuft, 186; *see* Alyssum
Goldflower, 446; *see* St.-Johns-wort
Goldthread (*Coptis*), 254
chlorosis, 24–29, 257, 443
iron deficiency, 24, 26, 27, 257
leaf spot, 48, 256
Goldthread (dodder), 94
Gomphrena, 229; *see* Globe-amaranth
Goodrite PEPS, as sticker, 119
latex VL-600, 38, 511
Gooseberry, 245; *see also* Currant
anthracnose, 51, 246
bud nematode, 75, 248
cane blight, 77, 247
chlorosis, 24–29, 247
crown gall, 83, 248
cutting dip, 248, 531, 537
dieback, 77, 247
downy mildew, 55, 247
fruit rot, 86, 246
leaf scorch, potassium deficiency, 26, 248
leaf spot, 48, 246
mosaic, 70, 247
powdery mildew, 56, 246
quarantine regulations, 246
resistant varieties, 246–47
root-knot, 90, 247
root nematode, 247
root rot, 87, 247
rust, 58, 246, 405, 508
scab or spot anthracnose, 51, 65, 246
spray schedule, 524–25
sunscald, 247
thread blight, 247, 505
Verticillium wilt, 68, 153, 247
Gordonia, 274; *see* Franklin-tree, Loblolly-bay
Gorse, 183; *see* Genista
Gourds, 238; *see also* Cucurbit
angular leaf spot, 50, 239
anthracnose, 51, 86, 238
bacterial soft rot, 81, 86, 238, 244
bacterial spot, 50, 245
bacterial wilt, 69, 240
blossom blight, 86, 245
damping-off, seed rot, 76, 244
downy mildew, 55, 243
fruit spot or rot, 86, 238, 244
Fusarium wilt, 66, 241
gummy stem blight, black rot, 76, 86, 244
leaf spot, 48, 238, 239, 244
mosaic, 70, 241
powdery mildew, 56, 242
resistant varieties, 239
ringspot, 72, 243
root-knot, 90, 244–45
root rot, 87, 156, 244
scab, 48, 65, 86, 238–39
seed treatment, 238, 532
Grace, W. R. and Company, Davison Chemical Division, 120
Grade changes, [46], [47]
Graft incompatibility blight, 334, [335]
Graft indexing, 559
Grafting, in transmitting viruses, 13, 70
Grama (*Bouteloua*), rust, 185
Grammatophyllum, 366; *see* Orchids
Granadilla, 376; *see* Passionflower
Grand dutchess, 370; *see* Oxalis
Grape, 286
anthracnose, leaf scab, 51, 289
black rot, 48, 86, [287], 288
boron deficiency, 27, 290
chlorosis, 24–29, 292
Control of Grape Diseases and Insects in the Eastern United States, 292
crown gall, 83, 289
dead arm, 77, [288], 289
dieback, canker, 77, 289
downy mildew, [55], 288
fanleaf, infectious degeneration, 291
fluorides injury, 43–44
fly speck, 289
fruit rot, spot, 86, 287, 288, 289
sulfur dioxide fumigation, control by, 289
gray-mold blight, 52, 86, 289
insect control, 289
iron chlorosis, deficiency, 24, 26, 27, 292
leaf blotch, 51, 291
leaf spot, 48, 287, 289, 291
leafroll or red-leaf, white emperor disease, 291
magnesium deficiency, 28, 290
manganese deficiency, 26, 27, 290
mosaic, yellow or infectious yellows, 70, 291
Mycosphaerella leaf spot, 48, 291
ozone injury, 44
Pierce's disease, grape decline, 290–91
plant-food utilization, 26
plant, heat treatment, 291
potassium deficiency, leaf scorch, 26, 290
powdery mildew, 56, 289
pruning, 32
resistant varieties, 287, 288, 289, 291
root dip, 290, 531, 537
root-knot, 90, 290
root nematode, 291
root rot, 87, 137, 289
rust, 58, 291
shoot or cane blight, wilt, 77, 290
sooty mold, 64, 291
spot anthracnose, bird's eye rot, 86, 291
spray schedule, 524–25
sulfur-sensitive, 525
2, 4-D injury, 45, [288]
Verticillium wilt, 68, 291
wood rot, 79, 289–90
zinc deficiency, little leaf, 26, 27, 290
Grape berry moth, insect control, 289
Grape-hyacinth, 494–95; *see also* Hyacinth
bacterial soft rot, 81, 90, 495–96
bulb rot, 81, 90, 495–96
bulb and stem nematode, 92, 496–97
bulb treatment, 497, 537
flower smut, 62, 497
light and temperature requirements, 40
Grape-ivy, 287; *see* Cissus
Grapefruit, 227; *see also* Citrus
anthracnose, 51, 228
bacterial blast, 50
chlorosis, 24–29, 228
crown gall, 83, 136, 228
fruit rot, 86
leaf spot, 48, 228
root nematode, 228
root rot, 87, 227–28
scab or spot anthracnose, 65, 228
sooty blotch, 64, 228
twig blight, 77, 228
Grass clippings, as mulch, 21
for loosening soil, 23
Grass diseases, 319–27
Grasses, air pollutant damage, 43–44
fluorides, 43
oxidant or PAN, 44
ozone, 44
sulfur dioxide, 43
Grasses, wild, rusts, 152, 166, 174, 183, 186, 193, 212, 223, 256, 298, 302, 337, 349, 354, 370, 375, 381, 399, 410, 455, 458, 461, 492, 494
Grasshoppers, as virus carriers, 72, 192, 242, 485, 486
Gray leaf spot, [325]
Gray-mold (*Botrytis*) blight, 52–53, [54], [85], 167, 224, 408, [454], 525
neck rot, [361]
rot, [207], [465]
Great Lakes Bromo-O-Gas, 545
Great laurel, 433; *see* Rhododendron
Greek valerian, 399; *see* Polemonium
Green Cross Phygon-XL Fungicide, 102, 518
Green Cross Products Division, Sherwin-Williams Company, of Canada, Limited, 102, 122, 518
Green Gro Home Garden Fruit Spray, 523
Green manure crop, 23, 473
Green tip spray, 524
Greenol Iron Chelate, 345
Grevillea, 453; *see* Silk-oak
Gromwell, 349; *see* Lithospermum
Grossularia, rust, 405
Ground cover, 559
Ground goldflower, 220; *see* Golden-aster
Ground-hemlock, 511; *see* Yew
Ground-ivy, 447; *see also* Catnip
leaf spot, 48, 447
mosaic, 70, 447
Ground-myrtle, 500; *see* Vinca
Ground-pink, 399; *see* Phlox
Ground-rattan, 371; *see* Palms
Groundcherry, 479
angular leaf spot, 50, 483
bacterial wilt, 69, 488
chlorosis, 24–29, 490
curly-top, 75, 486
leaf spot, 48, 480
leaf and stem nematode, 75, 400, 492
mosaic, 70, 484
ringspot, 72, 486
root-knot, 90, 489

root rot, 87, 490
rust, 58, 492
southern blight, 76, 490
white smut, 65, 492
wildfire, 50, 483
Groundsel, 220; *see also* Senecio
aster yellows, 73, 222
damping-off, seed rot, 76, 221
downy mildew, 55, 225
Fusarium wilt, 66, 223
leaf nematode, 75, 226
powdery mildew, 56, 221
rust, 58, 223, 405
seed treatment, 221
Verticillium wilt, 68, 223
white-rust, 62, 226
white smut, 65, 226
Grubs, insect control, 178, 234, 321, 418, 435, 463, 475
Gru-gru, 371; *see* Palms
Guava, 353
algal leaf spot, 341, 353
anthracnose, 51, 353
damping-off, 76, 353
fruit spot or rot, 86, 353
leaf spot, 48, 353
root-knot, 90, 353
root nematode, 353
root rot, 87, 137, 353
seed treatment, 353
spot anthracnose or scab, 51, 353
thread blight, 353, 505
wood rot, 79, 353
zinc deficiency, little leaf, 26, 27, 353
Guernsey-lily, 250
root nematode, 250, 253
Stagonospora leaf scorch, red spot, 48, 251
Guinea bean, 238; *see* Cucurbit
Guinea gold, 221; *see* Marigold
Guinea-hen flower, 494; *see* Fritillary
Gum tree, 267–68; *see* Eucalyptus
Gumweed, rust, 405
Gumbo, 297; *see* Okra
Gummosis (bacterial), 79
Gummosis (defined), 559
Gymnocladus, 300; *see* Kentucky coffeetree
Gypsophila, 203; *see* Babysbreath
Gypsum (land plaster), 28, 394, 482

H

Hackberry, 292
chlorosis, 24–29, 293, 344
downy mildew, 55, 293
felt fungus, canker, 77, 293
leaf blight, 51, 264, 292
leaf spot, 48, 264, 292
mistletoe, 93, 293
mosaic, 70, 265, 292
powdery mildew, 56, 292
root rot, 87, 137, 292
seedling blight, 76, 293, 405
thread blight, 293, 505
2, 4-D injury, 45, 266, 288, 293
winter injury, 41, 264, 292
witches'-broom, 292, [293]
wood rot, 79, 170, 292
Hail injury, 45
entrance for bacteria, 81, 362
entrance for fungi, 78
Hairy root, 83, 136–37, 559
Halesia, 453
chlorosis, 24–29
leaf spot, 48, 453
root-knot, 90, 453
wood rot, 79, 170, 453
Halo blight, [157]
Hamamelis, 509; *see* Witch-hazel
Hamatocactus, 194; *see* Cacti, Echinocactus
Harbinger European bird cherry, 382; *see* Cherry, Peach
Hardhack, 461; *see also* Spirea
chlorosis, 24–29, 461
iron chlorosis, 24, 26, 27, 461
leaf spot, 48, 461
powdery mildew, 56, 461
stem girdle, canker, 77
Hardie Manufacturing Company, 122
Hardiness, 559
Hardpan, 559
Hardy amaryllis, 250; *see* Lycoris, Amaryllis
Hardy aster, 220; *see* Aster, perennial
Hardy orange, 227; *see also* Citrus
anthracnose, 51, 228
canker, dieback, 77, 228
chlorosis, leaf yellowing, 24–29, 228
fruit rot, 86
root nematode, 228
root rot, 87, 136, 228
spot anthracnose or scab, 50, 228
twig blight, 76, 228
Harebell, 168; *see also* Bellflower, Campanula
rust, 58, 169
Harvest wounds, entrance for bacteria, 81
Hawaiian ti plant, 261; *see* Cordyline
Hawksbeard, 220
leaf spot, 48, 221
powdery mildew, 56, 221
rust, 58, 223
Haworthia, 130; *see* Aloe
plant soak, 130, 530, 534
Hawthorn, 133
black rot, 51, 77, 86, 136
blossom blight, 79, 86, 133–34
felt fungus canker, 77, 138, 293
fire blight, 77, 79, 86, 133–34
fruit spot or rot, 79, 86, 138
gray-mold rot, 52, 138
leaf blight, 51, [140]
leaf scorch, 41, 140, 141
leaf spot, 48, 140
mistletoe, 93, 141
powdery mildew, 56, 135–36
root rot, 87, 137–38
rust, 58, [134], 135
scab, 65, 134–35
seedling blight, 76, 143, 405
sooty mold, 64, 137
sulfur dioxide injury, 43
wood rot, 79, 139, 170
Hayes Spray Gun Company, 122
Hazelnut, 170
bacterial spot, blight, bacteriosis, 70, 79, 171
canker, dieback, 77, 171, 172
crown gall, 83, 136, 172
dieback, 77, 171, 172
kernel bitter rot, brown stain, 86, 172
leaf blister, 62, 171
leaf spot, 48, 171
powdery mildew, 56, 171
root rot, 87, 137, 171–72
sooty mold, 64, 172
twig blight, 77, 172
wood rot, heart rot, 79, 171
Head blight, [85], 224
Heal-all, 447; *see* Prunella
Healing over, [35], 559
Heart rot, 79
Heartwood, 559
Hearts and honey vine, 351; *see* Morning-glory
Heat treatments, soil, 98, 538–39, [540], [541], 542–43
Heath, 294
chlorosis, "yellows," 24–29, 294
damping-off, cutting rot, 76, 294
gray-mold blight, 52, 294
iron deficiency, 24, 26, 27, 294
powdery mildew, twist, 56, 294
rust, 58, 294
soil drench, 294
stem or collar rot, 76, 294
Verticillium wilt, 68, 294
Heather, 294
chlorosis, 24–29, 294
collar rot, 76, 294
iron deficiency, 24, 26, 27, 294
root nematode, 294
root rot, 87, 294
soil drench, 294
Heavenly bamboo, 353; *see* Nandina
Hebe, 460; *see also* Speedwell
Fusarium wilt, 66, 460
leaf spot, 48, 460
root rot, 87, 278, 460
Verticillium wilt, 68, 460
Hedera, 310; *see* Ivy
Hedge, pruning, 32
Hedgenettle, 447; *see* Stachys
Hedgethorn, 360; *see* Carissa
Hedyscepe, 371; *see* Palms
Helenium, 220; *see* Sneezeweed
Helianthemum, 469; *see* Sunrose
Helianthus, 220; *see* Jerusalem-artichoke, Sunflower
Helichrysum, 220; *see* Strawflower
Heliopsis, 220
black patch, 51, 221
leaf spot, 48, 221
mosaic, 70, 220
powdery mildew, 56, 221
root rot, 87, 221
rust, 58, 223, 405
Heliotrope (*Heliotropium*), 349
curly-top, 75, 349
damping-off, 76, 349
gray-mold blight, shoot blight, 52, 349
leaf scorch, 41, 349
leaf spot, blight, 48, 51, 349
mosaic, 70, 349
root-knot, 90, 349
rust, 58, 349
southern blight, 76, 254, 349
Verticillium wilt, 68, 349
Helleborus, 254; *see* Christmas-rose
Helminthosporium leaf blight, [233], [320]
Helxine, 151; *see* Babytears vine
Hemerocallis, 295; *see* Daylily
Hemlock, 402
canker (twig, branch, trunk), 77, 403
chlorosis, 24–29, 408
Cytospora canker, twig blight, 77, 403
damping-off, 76, 405
gray-mold blight, 52, 408
iron deficiency, 24, 26, 27, 408
mistletoe, dwarf, 93, 406
needle or leaf blight, 51, 402
root rot, 87, 405, 406
rust, needle and cone, 58, 177, 405
seed treatment, 405
seedling blight, 76, 405
snow blight, 407
soil treatment, 405, 406
sunscorch, 41, 407
twig blight, 77, 402, 403
wetwood, 264, 408
wind damage, sunscorch, 41, 407
wood rot, 79, 170, 403
Hemptree, 317
leaf spot, 48, 317
root rot, 87, 318
Hen-and-chickens, 451
leaf and stem rot, 51, 76, 451
root rot, 87, 451
rust, 58, 452
soil drench, 541–42
Hepatica, 131; *see also* Liverleaf
rust, 58, 392
smut, 62, 131
Heptachlor, soil application, 250, 282, 512
Herb, 559
Herb-Robert, 236; *see* Cranesbill
Herbicide, 559
Herbicide injury, [10], 45, [267], [288], *Using Phenoxy Herbicides Effectively*, 45, 112
Hercules-club (*Aralia*), 126
canker, dieback, 77, 127
leaf spot, 48, 127
root-knot, 90, 126
root rot, 87, 127
spot anthracnose or scab, 51, 65, 127
wood rot, 79, 127, 170

Hercules-club (*Zanthoxylum*), 302; *see also* Prickly-ash
canker (twig, branch), dieback, 77, 302
leaf spot, 48, 302
mistletoe, 93, 302
root-knot, 90, 302, 392
rust, 58, 302
tar spot, 48, 302
wood rot, 79, 170, 302
Hercules Powder Company, 122
Heronsbill, 236
aster yellows, 73, 236
bacterial leaf spot, 50, 236
curly-top, 75, 236
downy mildew, 55, 236
root-knot, 90, 236
root nematode, 236
root rot, 87, 221, 236
stem (crown) rot, southern blight, 76, 221, 236
Hesperis, 186; *see* Damesrocket
Hess and Clark, 122
Heteroecious rust, 58, 559
Heuchera, 304
leaf and stem nematode, 75, 306
leaf spot, 48, 305
leafy gall, 81, 306
powdery mildew, 56, 304
root-knot, 90, 305
root nematode, 306
root rot, 87, 306
rust, 58, 305
smut (leaf and stem), 62, 306
stem or crown rot, 76, 305–6
Hexachlorobenzene, 362
Hiba arborvitae, 312; *see also* Arborvitae
twig blight, 76, 313
Hibiscus, 297; *see* Hibiscus (arborescent forms), Okra, Roselle, Rose-mallows, Rose-of-Sharon, Flower-of-an-hour
Hibiscus (arborescent forms), 297
bacterial leaf spot, 50, 299
bacterial wilt, 69, 299, 488
blossom blight, 52, 86, 298
chlorosis, 24–29, 299
crown gall, 83, 298
damping-off, seed rot, 76, 299
dieback, 77, 298
flower bud drop, 299
gray-mold blight, 52, 86, 299
iron deficiency, 24, 26, 27, 299
leaf blight, 51, 298
leaf spot, 48, 298
magnesium deficiency, 28, 299
manganese deficiency, 26, 27, 299
mosaic, 70, 299
pruning, 32
root-knot, 90, 299
root nematode, 299
root rot, 87, 299
rust, 58, 298
seed treatment, 299, 535
stem or crown rot, 76, 298
strapleaf, molybdenum deficiency, 28, 193, 299
thread blight, 299, 505
twig blight, 76, 299, 345
web blight, 160, 299
zinc deficiency, 26, 27, 299
Hickory, 501; *see also* Pecan
anthracnose, 51, 501
bunch disease, 502
canker (twig, branch), dieback, 77, 502
crown gall, 83, 136, 502
felt fungus, 293, 506
leaf blotch, 51, 501
leaf scorch, sunscald, 41, 344, 502
leaf spot, 48, 501
mistletoe, 93, 506
nut mold, 86, 504
powdery mildew, 56, 505
root-knot, 90, 392, 506
root rot, 87, 137, 504
rosette, little leaf, zinc deficiency, 26, 27, 503–4
scab, 65, 502
soil grade change injury, 46
thread blight, 505
2, 4-D injury, 288, 506
wetwood, slime flux, 264, 506
witches'-broom, 502
wood rot, heart rot, trunk decay, 79, 170, 504
Highbush cranberry, 499; *see also* European cranberry-bush, Viburnum
canker, dieback, 77, 500
powdery mildew, 56, 499
Hinoki cypress, 312; *see* Chamaecyparis
Hippeastrum, 249; *see* Amaryllis
Hobblebush, 499; *see* Viburnum
Hollow heart, [417], 420
Holly, 295
anthracnose, 51, 295
bacterial leaf and twig blight, 50, 79 297
black mildew, 64, 296
boron deficiency, 27, 297
Botrytis flower blight, 52, 86, 297
canker, dieback, 77, 296
chlorosis, 24, 297, 344
crown gall, 83, 297
felt fungus, 293, 297
green algae, 297
leaf blight, 51, 295
leaf rot, drop of cuttings, 76, 297
leaf scorch, 41, [296]
leaf spot, 48, 295
powdery mildew, 56, 296
root-knot, 90, 296
root nematode, 296
root rot, 87, 137, 296
rust, 58, 297
scald, 41, 296
soil drench, 297
sooty mold, 64, 296
spine spot, purple spot, 296
spot anthracnose, 51, 295
tar spot, 48, 295
thread blight, 297, 505
twig blight, 77, 295, 296
winter injury, 41, 297
wood rot, 79, 170, 297
Holly-osmanthus, 370; *see* Osmanthus
Hollygrape, 153; *see* Oregon-grape
Hollyhock, 297
anthracnose, 51, 298
bacterial wilt, 69, 299, 488
canker, 77, 298
chlorosis, 24–29, 299
fasciation, leafy gall, 81
hairy root, crown gall, 83, 298
leaf spot, 48, 298
mosaic, 70, 299
powdery mildew, 56, 298
root-knot, 90, 299
root nematode, 299
root rot, 87, 299
rust, 58, [298]
seed treatment, 299, 535
seedling blight, seed rot, 76, 298
southern blight, 76, 298
stem or crown rot, 76, 298
web blight, 160, 299
Holodiscus, 299
fire blight, 79, 133–34, 300
leaf blight, 51, 300
leaf spot, 48, 300
powdery mildew, 56, 300
twig canker, dieback, coral spot, 77, 300
witches'-broom, 300
Homalocephala, 194; *see* Cacti, Echinocactus
Homalomena, 196
leaf spot, 48, 197
Home garden
Controlling Nematodes in the Home Garden, 549
fruit disease control, 523–25
Home Storage of Vegetables and Fruits, 87, 99
nematode control, 548–49
Honesty, 186; *see also* Cabbage
clubroot, 89, 189
leaf spot, 48, 190
mosaic, 70, 192
ringspot, 72, 192
white-rust, 62, 191
Honeydew
growing in flowing sap, 64
secretions by insects, 64, 559
Honeydew melon, 237; *see also* Cucurbit
bacterial leaf spot, 50, 239, 245
bacterial soft rot, 81, 244
curly-top, 75, 243
fruit spot, rot, 86, 244
mosaic, 70, 241
powdery mildew, 56, 242
scab, 65, 238–39
seed treatment, 238
stem blight, 77, 243
Honeylocust, 300
borer control, 300–301
canker (twig, branch, trunk), dieback, 77, 300
chlorosis, 24–29, 302
crown or collar rot, 76, 301
felt fungus, 293, 302
hairy root, 83, 136, 301
iron deficiency, 24, 26, 27, 302, 344
leaf spot, 48, 301
mistletoe, 93, 302
pod blight, 86, 301
powdery mildew, 56, 301
pruning, 33
root-knot, 90, 301, 392
root nematode, 301
root rot, 87, 137, 301
rust, 56, 301
tar spot, 48, 301
2, 4-D injury, 45, 288, 301
Verticillium wilt, 68, 301, 343
winter injury, sunscald, 41, 139, 301
witches'-broom, 301
wood or heart rot, 79, 170, 301
zinc deficiency, 26, 27, 302, 503–4
Honeysuckle, 457
anthracnose, 51, 458
canker, dieback, 77, 458
crown gall, hairy root, 83, 458
gray-mold blight, 52, 458
infectious variegation, 70, 459
leaf blight, 51, [458]
leaf spot, 48, 458
powdery mildew, 56, 458
pruning, 32
root-knot, 90, 459
root nematode, 458
root rot, 87, 458
rust, 58, 458
thread blight, 459, 505
twig blight, 77, 458
Verticillium wilt, 68, 459
wood or collar rot, 79, 259, 458
Hoop pine, 143; *see* Araucaria
Hophornbeam, 170
canker (twig, branch, trunk), 77, 171, 172
leaf blister, 62, 171
leaf spot, 48, 171
powdery mildew, 56, 171
root rot, 87, 137, 171–72
rust, 58, 171
wood rot, heart rot, 79, 171
Hopkins Strawberry Dust, 523
Hopperburn, 227, 397, 421, 560
Hoptree, 302
leaf spot, 48, 302
powdery mildew, 56, 302, 303
root rot, 87, 302
rust, 58, 302
Horehound, 447
leaf spot, 48, 447
root-knot, 90, 447
Hormone, 560
Hornbeam, 170
canker (twig, branch, trunk), 77, 171, 172
felt fungus, 172, 293
leaf blister, 62, 171
leaf spot, 48, 171
powdery mildew, 56, 171
root rot, 87, 137, 171–72

twig blight, 77, 172
wood rot, heart rot, 79, 171
Horsechestnut, 302
anthracnose, leaf blight, 51, 303
bleeding canker, 77, 161, 304, 345
Botrytis twig blight, 52, 303
canker (branch, trunk), dieback, 77, 266, 303, 345
frost crack, 41
leaf blister, yellow, 62, 171, 303–4
leaf blotch, 51, 302, [303]
leaf scorch, 41, 303
leaf spot, 48, 303
mistletoe, 93, 304
powdery mildew, 56, 303
root rot, 87, 137, 303
rust, 58, 304
twig blight, 77, 266, 303, 345
Verticillium wilt, 68, 303, 343
wetwood, slime flux, 264, 303
witches'-broom, 62, 171, 303–4
wood rot, 79, 170, 303
Horsemint, 447; *see* Monarda
Horseradish, 186; *see also* Cabbage
bacterial blight, 50, 187
bacterial leaf spot, 50, 190
bacterial soft rot, 81, 90, 190
black ringspot, 72, 192
brittle root, curly-top, 75, 192
clubroot, 89, 189
crown gall, 83, 146, 193
downy mildew, 55, 189
leaf blight, 51, 190
leaf spot, 48, 190
mosaic, 70, 192
powdery mildew, 56, 193
resistant strains, 191
root-knot, 90, 191
root rot, 87, 188, 190
Verticillium wilt, 68, 193
white-rust, white blister, [61], 62, 191
Horsetail-tree, 212; *see* Casuarina
Hortensia, 304; *see* Hydrangea
Horticulture, 560
Horticulturist, extension, help by, 5, 6, 23, 24, 27, 29, 30, 32, 33, 42, 44, 87, 94, 99, 134, 135, 138, 140, 141, 149, 150, 151, 165, 182, 193, 199, 206, 208, 209, 214, 232, 244, 268, 288, 292, 296, 300, 301, 311, 313, 369, 382, 386, 392, 393, 410, 411, 415, 416, 417, 419, 420, 425, 427, 432, 433, 435, 440, 441, 463, 464, 480, 482, 489, 491, 496, 503, 504, 505, 511, 523
Host, 560
Host indexing, 560
Host range, 560
Host resistance, effect on disease development, 96
Hosta, 304
anthracnose, 51, 304
crown rot, 76, 304
leaf spot, 48, 304
root rot, 87, 304
Hot water soak, for seed, bulbs, tubers, potted plants, rhizomes, roots, 76, 93, 97, 168, 173, 187, 196, 197, 207, 213, 221, 225, 226, 228, 238, 248, 252, 254, 256, 270, 282, 283, 285, 290, 301, 307, 310, 315, 331, 338, 354, 363–64, 369, 375, 400, 443, 450, 453, 455, 458, 467, 472, 484, 493, 497, 529–31
precautions, 529
treatment times and temperatures, 529–31
Houndstongue, 349
downy mildew, 55, 349
leaf spot, 48, 221, 349
mosaic, 70, 349
powdery mildew, 56, 349
root-knot, 90, 349
root rot, 87, 278, 349
southern blight, 76, 254, 349
stem rot, 76, 349
House paint as tree wound dressing, 36
House plants
air humidity control, 39
drought injury, 38
fertilizing, 31
flooding injury, 38
illumination in foot-candles (Table), 39
light and temperature requirements, 39, 40
low humidity injury, 39
moving outdoors, 39
Selecting and Growing House Plants, 32
soil mixtures, 22
soluble salts injury, 31–32
sterilizing soil, 537–49
temperature, 40
watering, 38
Household bleach, as disinfectant, 23, 35, 79, 81, 99, 134, 138–39, 195, 196, 204, 206, 259, 360, 366, 370, 380, 441, 445
Household detergent or soap, as spreader, 109, 118
Household sprayers, [112]
Houseleek, 451
leaf and stem rot, 51, 76, 451
root rot, 87, 451
rust, 58, 452
soil drench, 451–52
Houstonia, 185
chlorosis, 24–29, 186, 443
downy mildew, 55, 186
leaf spot, 48, 186
root rot, 87, 137, 186
rust, 58, 169, 186
Howard, A. H. Chemical Company, Limited, 122
Howea, 371; *see* Palms
Hubbard-Hall Chemical Company, 122
Huckleberry, 175
black mildew, 64, 179
chlorosis, 24–29, 178
fruit or berry rot, 86, 179–80
leaf blight, spot, 48, 51, 178
leaf gall, 62, 176
powdery mildew, 56, 176
red leaf gall, spot, 62, 176
root-knot, 90, 180
rust, 58, 177, 405
spray schedule, 524–25
tar spot, 48, 178
H. D. Hudson Manufacturing Company, 122
Huisache, 300; *see* Acacia
Humidity, air, 39
Humidity (air), relative, 560
Hunger Signs in Crops, 25
Huntsmanscup, 408; *see* Pitcherplant
Husk-tomato, 479; *see* Groundcherry
Hyacinth, 494
bacterial soft rot, 81, 90, 495–96
Botrytis blight, fire, 52, 86, 90, 495
bulb rot, 81, 90, 495–96
bulb soak, 496–97, 530, 531, 537
bulb storage, 497
light and temperature requirements, 40
mosaic, 70, 496
ring disease or bulb nematode, 75, [92], 93, 496–97
root rot, 87, 495–96
rust, 58, 498
"topple" or loose bud, 28, 497
yellows, yellow rot, 50, 90, 497
Hyacinth-bean, 377; *see also* Pea
bacterial spot, 50, 380
black mildew, 64, 382
leaf spot, 48, 379, 380
mosaic, 70, 378
powdery mildew, 56, 378
root-knot, 90, 381
root rot, 87, 378
Hyacinthus, 494; *see* Hyacinth
Hyaline, 560
Hybrid, 560
Hydathode, 560
Hydrangea, 304
bacterial wilt, 69, 305
bud blight, 52, 304
chlorosis, 24–29, 305
damping-off, cutting rot, 76, 306
flower and shoot blight, 52, 86, 304
flower spot, 86, 305
gray-mold (*Botrytis*), blight, 52, 86, 304
iron deficiency, 24, 26, 27, 305
leaf nematode, 75, 306
leaf spot, 48, 305
light and temperature requirements, 40
magnesium deficiency, 28, 305
plant dip or spray, 305
powdery mildew, 56, 304, [305]
pruning, 32
ringspot, 72, 306
root-knot, 90, 305
root nematode, 306
root rot, 87, 306
rust, 58, 305, 405
southern blight, 76, 305–6
stem, crown rot, 76, 305–6
stem nematode, 75, 306
sunscald, 41, 139, 264, 306
twig canker, dieback, 77, 305
wood rot, 79, 170, 306
Hydrocarbons, oxidized, injury, 44
Hydrochloric acid (HCl), extracting tomato seed, 485
Hydrogen ion concentration (pH), 24, 560
Hydrogen peroxide, as disinfectant, 366
Hydrosome, light and temperature requirements, 40
Hydroxymercurichlorophenol, 565
Hygro, 30
Hymenocallis, 252; *see* Spiderlily
Hypericum, 446; *see* St.-Johns-wort
Hyperplasia (hyperplastic), 560, 567
Hypertrophy, 560
Hypha, [12], 560
Hypoplasia (hypoplastic), 560, 567
Hyssop (*Hyssopus*), 447
root-knot, 90, 447

I

Iberis, 186; *see* Candytuft
Ice injury, [43], 45
entrance for fungi, 79
Iceplant, 306
root-knot, 90, 306
sooty mold, 64, 306
I-F-N Mixture, 27
Ilex, 295; *see* Holly, Inkberry, Yaupon
Illicium, 341; *see* Anisetree
Immune, immunity, 560
Impatiens, 152; *see* Balsam, garden
Imperial Chemical Company, 122
Importing plants, 99
Incense-cedar, 312
brown felt blight, 315, 407
canker, 77, 314
crown gall, 83, 136, 315
mistletoe, 93, 315
needle cast, 51, 315
root rot, 87, 314
rust, gall, 58, 313
witches'-broom, 58, 313
wood rot, 79, 170, 314
Incompatible, 560
Incubation period, 560
Index of Plant Diseases in the United States, 3
Indexing (plants), 560
India rubber tree, 271; *see also* Rubber-plant
anthracnose, 51, 271
canker, dieback, 77, 138, 266, 271
crown gall, 83, 168, 272
Indian cherry, 183; *see* Buckthorn
Indian corn, 231; *see* Corn
Indian-cup, 220; *see* Silphium
Indian fig, 194; *see* Opuntia
Indian laurel, 271; *see* Fig, Rubber-plant
Indian paintbrush, 454; *see also* Painted-cup
rust, 58, 405, 455
Indian shot, 202; *see* Canna

Indian strawberry, 439; *see* Mock-strawberry
Indian-tobacco, 339; *see* Lobelia
Indiancurrant, 457; *see* Coralberry
Indigo, 269
powdery mildew, 56, 269
root rot, 87, 269
rust, 58, 269
Indigobush, 269
canker (twig), 77, 269
leaf spot, 48, 269
powdery mildew, 56, 269
root nematode, 269
root rot, 87, 269
rust, 58, 269
Indigofera, 269; *see* Indigo
Indoor clover, 230; *see* Alternanthera
Industrial Fumigant Company, 122
Infect (infection), 560
Infection court, [12], 560
Infectious variegation, 70, 83, 174, 200, 299
Infest, 560
Inflorescence, 560
Inflorescence blight, 52, [85], 86, [133], [199], [224], [260], [435]
Injuries
entrance for, bacteria, 34, 70, 81, 83, 98, 134, 204, 208, 301, 362, 482, 489, 508
disease organisms, 87, 208
fungi, 34, 46, 66, 68, 79, 98, 203, 208, 232, 353, 356, 377, 383, 402, 463, 482, 504
Ink spot, 309
Inkberry, 295; *see also* Holly
black mildew, 64, 296
canker, dieback, 77, 296
chlorosis, 24–29, 297, 344
felt fungus, 293, 297
iron deficiency, 24, 26, 27, 297, 344
sooty mold, 64, 296
twig blight, 77, 296
Inoculate (inoculation), 560
Inoculum, 560
Insecticide
compatibility chart, 552
The Federal Insecticide, Fungicide, and Rodenticide Act, 107
injury, 45
soil, 23, 82, 86
Insects, [163], [240], [242]
buildup of, 96
carriers of
bacteria, 70
fungi, 48
viruses, 13, 70, 73, 74, 98
control of, 23, 48, 65, 68, 70, 73, 74, 98, 99, 161, 180, 187, 192, 209, 231, 232, 233, 240, 394, 397, 483, 485, 502
injuries
entrance for, bacteria, 70, 81, 362, 483
disease organisms, 87
fungi, 77, 79, 263, 383, 403
lethal temperature, 542
as plant-weakening agents, 78, 264
spread of diseases, 48, 79, 87, 98, 133–34, 162, 164, 217, 239, 356
vector, 560
Internal cork, 473, [474]
Intumescence, 560
Inula, 220
leaf spot, 48, 221
petal blight, 86, 224
powdery mildew, 56, 221
rust, 58, 223
Invasion (plant), 560
Inverted-pan sterilizer, [541]
Ipomoea; *see* Morning-glory, Sweet-potato
Iresine, 230; *see* Bloodleaf
Iris, 307
bacterial leaf spot or blight, 50, 308, [309]
bacterial scab, 65, 283
bacterial soft rot, 81, [82],83, 307
blindness, blasting, 310
blossom blight, 52, 86, 308
blue mold, 90, 307, [309]
bulb and stem nematode, 75, 92, 308, [309]
bulb rot, 83, [90], 307
chlorosis, 24–29, 310, 443
crown rot, 76, [77], 307, [309]
Didymellina (*Heterosporium*) leaf spot, 48, [49], 308
flower spot, 86, 308
fluorides injury, 43–44
gray-mold blight, 52, 86, 308
Growing Iris in the Home Garden, 310
ink disease, 90, 307, [309], 310
insect control, 307
iron deficiency, 24, 26, 27, 310, 443
leaf blight, blotch, 51, 308, 310
leaf spot, 48, [49], 308
leaf streak, stunt, 72, 310
mosaic, stripe, 70, 308, [309]
rhizome or bulb soak, 83, 307, 310, 528, 530, 531, 535
rhizome rot, 76, [77], 81, 83, 307
ringspot, 72, 310
root-knot, 90, 310
root nematode, 310
root rot, 87, 307
rust, 58, 308, [309]
sclerotium (*Sclerotinia*) rot, 76, 307
scorch, red fire, 51, 308
soil treatment, 307
southern blight, 76, 307
"topple," calcium deficiency, 28, 310, 497
Iris borers, control, 307
Irn-Gro, 26
Iron
chelates, 27, 143, 178, 200, 261, 276, 284, 299, 305, 336, 345, 408, 443, 467
chlorosis, 24, 26, 27, 143, 170, [179], 276, 284, 292, 294, 344, [359], 467
citrate, 345
deficiency, 26, 27, 141, 325
injury, 26
phosphate, 345
sulfate, 26, 141, 166, 178, 186, 200, 201, 247, 261, 276, 284, 292, 299, 305, 325, 336, 345, 397, 408, 424, 434, 435, 443, 467
tartrate, 345
Iron Deficiency in Plants: How to Control It in Yards and Gardens, 24
Ironbark, 268; *see* Eucalyptus
Ironweed, 219; *see also* Yellow ironweed
rust, 58, 223, 405
Ironwood, 170; *see* Hophornbeam
Irregular roots, 209, 210
Ivy (Albany, Algerian or Canary, Baltic, Clustered, English, Heart-leaved, Irish (*Hedera*), 310; *see also* Boston ivy; Cissus for Grape- and Marine-ivy; Senecio for German-ivy; Toadflax for Kenilworth ivy
anthracnose, 51, 310
bacterial leaf spot, 50, 310, [311]
fungus leaf spot, 48, 310, [311]
light and temperature requirements, 40
powdery mildew, 56, 311
root nematode, 311
root rot, 87, 311
sooty mold, 64, 311
spot anthracnose or scab, 65, 310
stem canker, 77, 310
stem spot, dieback, twig blight, 77, 310
winter injury, sunscald, 41, 311
Ivy-arum, 196; *see* Pothos
Ixia, 307
corm rot, yellows (*Fusarium*), 66, 307
gray-mold blight, 52, 308
mosaic, 70, 308
Ixora, 185
chlorosis, 24–29, 186, 443
root-knot, 90, 186
root nematode, 186
root rot, 87, 137, 186
sooty mold, 64, 186

J

J & P Mildew King, 556
Jacaranda, 212
root rot, 87, 137, 212
Jack-in-the-pulpit, 195
chlorosis, 24–29
downy mildew, 55, 198
leaf spot, mold, 48, 197
leaf and stalk blight, 51, 197
rust, 58, 198
Jackbean, 155; *see also* Bean, garden types
leaf spot, 48, 159–60
pod spot, 86, 159–60
root-knot, 90, 159
Jacobs-ladder, 399; *see* Polemonium
Jacquemontia, 351
leaf spot, 48, 351
root-knot, 90, 351
root nematode, 351
rust, 58, 351
thread blight, 351, 505
white-rust, 62, 191, 351
Jade plant, 451; *see* Crassula
Japanese beetle, insect control, 233, 289
Japanese cornel, 258; *see* Dogwood
Japanese cornelian-cherry, 258; *see* Dogwood
Japanese honeylocust, 300; *see* Honeylocust
Japanese Juneberry, 133; *see* Serviceberry, Amelanchier
Japanese lawngrass, 319; *see* Zoysia
Japanese pagodatree, 300; *see* Sophora
Japanese plum-yew, 311
twig blight, 51, 77, 311, 313
Japanese quince, 133; *see* Flowering quince
Japanese spurge, 371; *see* Pachysandra
Japanese winterberry, 295; *see* Holly
Japanese zelkova, 263; *see* Elm
Jasmine (*Jasminum*), 311
algal leaf spot, 312, 341
blossom blight, 86, 311
Botrytis shoot blight, 52, 312
crown gall, 83, 311
crown rot, 76, 311
leaf spot, 48, 311
root-knot, 90, 311, 392
root nematode, 311
root rot, 87, 137, 311
rust, 58, 312
southern blight, 76, 311
spot anthracnose, scab, 51, 65, 311
stem gall, 76, 312
variegation, infectious chlorosis, 70, 299, 312
Jersey-tea, 355; *see* New Jersey-tea
Jerusalem-artichoke, 220
bacterial spot, 50, 225
crown gall, 83, 226
downy mildew, 55, 225
leaf spot, 48, 221
powdery mildew, 56, 221
root-knot, 90, 225
root and tuber rot, 86, 87, 221, 226
rust, 58, 223, 405
southern blight, 76, 221
stem rot, 76, 221
storage rot, 86, 221, 226
Jerusalem-cherry, 479; *see also* Eggplant
bacterial soft rot, 81, 483
crown gall, 83, 492
early blight, 51, 480
gray leaf spot, 48, 480–81
late blight, 51, 480
light and temperature requirements, 40
mosaic, 70, 484
spotted wilt, 72, 486
Verticillium wilt, 68, 487–88

Jerusalem-cross, 203; *see* Maltese cross, Carnation
Jerusalem-thorn, 300; *see* Parkinsonia
Jessamine, 185; *see* Carolina jessamine
Jetbead, 312
anthracnose, 51, 312
fire blight, 79, 133–34
leaf spot, 48, 312
twig blight, coral spot, 77, 312
Jewelberry, 317; *see* Callicarpa
Joe-pye-weed, 220; *see also* Eupatorium
downy mildew, 55, 225
powdery mildew, 56, 221
rust, 58, 223
John Innes Composts, 22–23
potting, 23
seed, 22–23
John Taylor Fertilizer Company, 36
Johnsongrass, as virus source, 234
Jonquil, 249; *see* Daffodil
Josephscoat, 230; *see* Amaranth
Joshua-tree, 512; *see* Yucca
Jubaea, 371; *see* Palms
Judas-tree, 300; *see* Redbud
Juglans, 501, *see* Walnut, Butternut
Juneberry, 133; *see* Serviceberry, Amelanchier
Jungleflame, 185; *see* Ixora
Juniper, Redcedar, 312
black mildew, 64, 315
Botrytis blight, 42, 314
brown felt blight, 315, 407
canker (twig, branch), 77, 314
chlorosis, 24–29, 315
crown gall, 83, 136, 315
damping-off, seedling blight, 76, 314, 405
dog injury, 315
iron deficiency, 24, 26, 27, 315
leaf browning and shedding, 314
mistletoe, 93, 315
needle or leaf blight, cast, 51, 313, 314
nursery or juniper blight, 51, 77, [313]
pruning, 33
root nematode, 315
root rot, 87, 314
rust, 58, 313
gall, [59], 313
needle, 58, 313
witches'-broom, 58, 313
snow blight, 315, 407
sooty mold, 64, 315
transplant dip, 315, 531, 537
twig blight (*Phomopsis*), 77, [313]
winter injury, 41, [42], 314
wood rot, 79, 170, 314
Juniperus, 312; *see* Juniper
Jupiters-beard, 499
leaf spot, 48, 499

K

Kafir-lily, light and temperature requirements, 40
Kalanchoë, 451
crown gall, 83, 452
leaf spot, 48, 452
light, effect on flowering, 38
light and temperature requirements, 40
mosaic, 70, 452
powdery mildew, 56, 452
soil drench, 451–52
stem (crown) rot, wilt, 76, 451
Kale, 185; *see also* Flowering kale, Cabbage
bacterial leaf spot, 50, 190
bacterial soft rot, 76, 81, 190
black ringspot, 72, 192
black rot, 50, 187
blackleg, 77, 187
chlorosis, 24–29, 194
clubroot, 89, 189
damping-off, 76, 188, 190, 191, 193
downy mildew, 55, 189
drop, cottony rot, 76, 191
Fusarium yellows, 66, 186–87
gray-mold blight, 52, 86, 190, 191
leaf spot, 48, 190, 191
mosaic, 70, 192
oedema, 39, 194
powdery mildew, 56, 193
resistant varieties, 187, 189, 190, 193
root-knot, 90, 191
root rot, 87, 187, 190, 193
seed treatment, 187, 529, 532
southern blight, 76, 191
temperature requirements, 41
Verticillium wilt, 68, 193
white-rust, 62, 191
Kalmia, 175; *see* Mountain-laurel
Kalopanax, 126; *see* Acanthopanax
Kangaroo thorn, 300; *see* Acacia
Kangaroo vine, 289, *see* Cissus
Karanda, 360; *see* Carissa
Karathane (dinocap)
active ingredient, 101
compatibility, 101
eradicant action, 101
formulations (-WD, Liquid Concentrate, Dust), 101
gallon lots, 522
injury, 45
in multipurpose mixes, 101, 111
powdery mildew, control, 58, 101, 128, 144, 153, 158, 161, 167, 169, 171, 173, 176, 185, 212, 218, 221, 228, 236, 242, 246, 256, 259, 266, 289, 292, 294, 296, 298, 300, 301, 304, 318, 321, 342, 347, 349, 359, 370, 400, 401, 425, 432, 437, 440, 448, 451, 452, 458, 459, 477, 505, 508, 525
precautions, 101
Karbam Black, 102, 518
Karbam White, 103, 519
Karo, 408; *see* Pittosporum
Kelthane, 440
Kenilworth ivy; 454; *see* Toadflax
Kentia, 371; *see* Palms
Kentucky coffeetree, 300
leaf spot, 48, 301
root rot, 87, 137, 301
sooty mold, 64, 302
Verticillium wilt, 68, 301, 343
wood rot, 79, 170, 301
Kentucky yellowwood, 300; *see* Yellowwood
Kerria, 315; *see* Jetbead for White kerria
canker, 77, 316
fire blight, 79, 133, 316
leaf blight, 51, 316
leaf spot, 48, 316
root rot, 87, 316
twig blight, 77, 316, [317]
Kilgore's Citrus Nutritional Sprays, 27
Maneb 80 Wettable, 103, 519
Kingfisher daisy, 220
powdery mildew, 56, 221
K-Mag, 27
Knapsack duster, 116, [122]
sprayer, 113, [114]
Knife injury, 45
Kniphofia, 432; *see* Redhot-pokerplant
Knot, 560–61; *see also* Root-knot
Knudson's Formula "C," for orchids, 561
Kochia, 162
curly-top, 75, 162
damping-off, 76, 162
root rot, 87, 156, 166
rust, 58, 166
virus yellows, 73, 165
Koeleria, rust, 154
Koelreuteria, 286; *see* Goldenrain-tree
Kohlrabi, 185; *see also* Cabbage
bacterial soft rot, 76, 81, 190
black ringspot, 72, 192
black rot, 50, 187
chlorosis, 24–29, 194
clubroot, 89, 189
damping-off, 76, 188, 190, 191, 193
downy mildew, 55, 189
drop, cottony rot, 76, 191
Fusarium yellows, 66, 186–87
gray-mold blight, 52, 86, 190, 191
leaf spot, 48, 190, 191
mosaic, 70, 192
powdery mildew, 56, 193
root-knot, 90, 191
root nematode, 194
root rot, 87, 187, 190, 193
seed treatment, 187, 529, 532
southern blight, 76, 191
Verticillium wilt, 68, 193
Kolker Methyl Bromide, 106
Kolkwitzia, 499; *see* Beautybush
Koppersol, 556
Krenite, 146
Kromad, 106, 314, 321, 325, 326, 561
K-3, 556
Kuhnia, 220; *see* False-boneset
Kumquat, 227; *see also* Citrus
fruit rot, 86
leaf spot, 48, 228
root nematode, 228
KXL, 556

L

Labrador-tea, 316
chlorosis, 24–29
iron deficiency, 24, 26, 27
leaf gall, 62, 316
leaf spot, 48, 316
powdery mildew, 56, 316
rust, 58, 316, 405
spot anthracnose, 51, 316
tar spot, 48, 316
Laburnum, 285; *see* Goldenchain
Lacebugs, as virus carriers, 164
Lace-fern, 146; *see* Asparagus-fern
Laceflower, 213; *see* Blue laceflower
Lachenalia, 494; *see* Cape-cowslip
Ladies'-tobacco, 219; *see* Everlasting
Ladys-sorrel, 370; *see* Oxalis
Laelia, 366; *see* Orchids
Lagenaria, 238; *see* Gourds, Cucurbit
Lagerstroemia, 236; *see* Crapemyrtle
Lambkill, 175; *see* Mountain-laurel
Lambs-ears, 447; *see* Stachys
Lambslettuce, 499; *see* Cornsalad
Lambsquarters, white-rust, 62
Land-grant institutions, listing of, 6–7
help by, 5, 6
Land plaster, 394
Landscape architect, help by, 46
Lanstan
active ingredient, 104
soil application, 104, 106, 156, 157, 162, 379
uses, 104
Lantana, 317
black mildew, 64, 318
Fusarium wilt, 66, 318
leaf nematode, 75, 226, 317
leaf spot, 48, 317
light and temperature requirements, 40
mosaic, 70, 318
root-knot, 90, 317
root rot, 87, 318
rust, 58, 318
Larch, 318
canker, 77, 319
chlorosis, 24–29, 319
damping-off, 76, 319
frost injury, 41, 318–19
gray-mold, 52, 318, 319
leaf or needle cast, 51, 318
mistletoe, dwarf, 93, 319
needle blight, 51, 318
root nematode, 318
root rot, 87, 137, 318
rust, 58, 171, 318, 508
seed treatment, 319
seedling blight, 77, 319
shoot or twig blight, 79, 318
sulfur dioxide injury, 43
wood rot, 79, 170, 318
Laredo, 153; *see* Mahonia, Oregon-grape

Larix, 318; *see* Larch
Larkspur, 254; *see also* Delphinium
bacterial bud rot, stem rot, 50, 76, 81, 254
bacterial leaf spot, 50, 255
bacterial wilt, 69
curly-top, 75, 257
damping-off, seed rot, 76, 256
leaf blotch, 51, 256
mosaic, 70, 255
seed treatment, 254, 530, 535
soil drench, 254, 255, 256
Verticillium wilt, 68, 256
Larvacide-100, 106, 545
Larvacide Soil Fumigant, 545
Late blight (*Phytophthora*), [416], [417], 480, [481]
Lathyrus, 377; *see* Sweetpea
Latuca, 328; *see* Lettuce
Laurel, true, 319; Sweetbay (*Laurus*); *see also* Mountain-laurel for Bog, Mountain-, Pale, and Sheep-laurel
anthracnose, 51, 319
leaf spot, 48, 319
light requirements, 39
thread blight, 319, 505
Laurocerasus, 382; *see* Cherry-laurel
Laurus, 319; *see* Laurel, true
Lavandula, 446; *see* Lavender
Lavatera, 297
anthracnose, 51, 298
damping-off, seed rot, 76, 299
infectious variegation, 70, 299
leaf spot, 48, 298
root rot, 87, 299
rust, 58, 298
seed treatment, 299, 535
Lavender, 446
leaf spot, 48, 447
root-knot, 90, 447
root rot, 87, 447
Lavender queen, 454; *see* Penstemon
Lawn
air pollution injury, 43–44
fluorides, 43
ozone, 44
PAN or oxidant damage, 44
sulfur dioxide, 43
diseases, 319–27; *see under* name of grass, e.g., Bluegrass, Bermuda-grass, St. Augustine grass, etc.
fertilizing, 29
fungicides, broad-spectrum, 106
trade names, 106
insects, control, 321, 327
mower injury, 45
plant-food utilization, 26
seed treatment, 326, 537
Lawn Insects; How to Control Them, 327
Lawrence, B. M. & Company, 122
Laws, governing pesticides, 107
Lawson cypress or cedar, 312; *see* Chamaecyparis
Layia, 220; *see* Tidytips
Leaching, 561
Lead arsenate, tree borer control, 139
Leadplant, 269; *see* Indigobush
Leadtree, 300
root rot, 87, 137, 301
rust, 58, 301
Leaf
blight, 48, [51], 52, [140], [179], [188], [217], [224], [233], [239], [252], [256], [262], [324], [329], [415], [458], [464]
blister, 62, [357]
blotch, 48, [51], 52, [137], [256], [303], [363]
browning, shedding (evergreens), 314, 408, 512
curl, 62, [63], 384
curl (virus), 70, [429]
diseases, 48–76
folder, insect control, 289
gall, 13, 62, [201], [436]
gall nematode, 75, 226, [270]
mold, 23, 480
nematode, 75, [76], [270], 337, 369, 375, 400, 456, 467, 497
roll (physiological), [491]
roll (virus), 418, 491
rust, 58, [59], [134], [145], [205], [222], [298], [309], [320], [404], [441]
scald, [41]
scorch, 41, [251], 260, 266, [296], 344, [359], 373, [464], 506, [511]
smut, 62, [63], 65, [131]
spot, 13, 561
bacterial, 50, [51], [153], [196], [211], [240], [281], [283], [309], [311], [386], [413], [423]
fungus, 48, [49], [134], [163], [179], [182], [187], [199], [217], [246], [256], [260], [265], [276], [283], [287], [311], [317], [320], [325], [346], [372], [385], [398], [434], [449], [464]
Leaf beet, 162; *see* Beet, Swiss Chard
Leaf mold, 561
Leafhoppers, 13, [163]
control, 74, 99, 289, 321, 387, 397, 400, 460, 501
as virus carriers, 13, 73, 74, 98, 159, 162, 177, 284, 387, 400, 460, 501
Leafy gall, 81, [82], 492
Leather flower, 228; *see* Clematis
Leatherleaf; *see* Chamaedaphne, Mahonia
Leatherwood, 327; *see also* Southern leatherwood
rust, 58, 327
sooty mold, 64, 327
Lebanon Fruit & Berry Spray, 523
Lebanon squill, 495; *see* Tulip
Lebbek, 300; *see also* "Mimosa" tree
rust, 58
Ledum, 316; *see* Labrador-tea
Leek, 361; *see also* Onion
aster yellows, 73, 365
bacterial soft rot, 81, 90, 362
bulb rot, 81, 90, 361, 362
downy mildew, 55, 363
gray-mold neck rot, 52, 90, 361
insect control, 362, 363, 365
leaf blight, blotch, 51, 361, 362, 363
mosaic, 70, 365
pink root, 87, 363
purple blotch, 51, 363
root-knot, 90, 365
rust, 58, 365
seed treatment, 362, 532
smudge, 90, 364
smut, 62, 362
soil treatment, 362
southern blight, 76, 366
tip blight, 52, 362
Verticillium wilt, 68, 365
white rot, 90, 362
Leersia, rust, 494
Leffingwell Chemical Company, 122
Leiophyllum, 316
leaf gall, 62, 316
Lemaireocereus, 194; *see* Cacti, Cereus
Lemon, 227; *see also* Citrus
anthracnose, wither tip, 51, 228
chlorosis, 24–29, 228
crown gall, 83, 136, 228
fruit spot, rot, 86
leaf spot, 48, 228
root nematode, 228
root rot, 87, 227–28
scab, 65
sooty blotch, 64, 228
twig blight, 77, 228
Lemon balm, 447
leaf spot, 48, 447
Lemon mint, 447; *see* Monarda
Lemon-verbena, 317
black mildew, 64, 318
crown gall, 83, 318
leaf spot, 48, 317
root-knot, 90, 317
root rot, 87, 318
southern blight, 76, 318
spot anthracnose, 51, 317
Lens, 377; *see* Lentil
Lenten rose, 254; *see* Christmas-rose
Lentil, 377; *see also* Pea
Ascochyta blight, 51, 76, 379
damping-off, 76, 379
Fusarium wilt, root rot, 66, 377
gray-mold blight, pod rot, 52, 86, 381
leaf blight, 51, 379
mosaic, 70, 378
root-knot, cyst nematode, 90, 381
root rot, 87, 377, 378
seed treatment, 379
soil drench, 377, 379
stem and crown rot, 76, 378
Verticillium wilt, 68, 382
white mold, 76, 378
wilt (virus), 378
Leonotis, 447
leaf spot, 48, 447
rust, 58, 447
Leopardsbane, 220
leaf nematode, 75, 226
powdery mildew, 56, 221
root-knot, 90, 225
Lepidium, 186; *see* Garden cress, Peppergrass
Lesion, 561
Lettuce, 328
anthracnose, 51, 331
aster yellows, white heart, 73, 328–29
bacterial leaf blight, rot, 50, 330
bacterial soft rot, 81, 86, 328
bacterial wilt, 69, 330
big vein, [330]
boron deficiency, 27, 331
bottom rot, 76, 86, 328
brown blight, 331
crown gall, 83, 332
curly-top, 75, 332
damping-off, 76, 330
downy mildew, [55], 329
drop (sclerotiniose), watery soft rot, 76, 86, 328
fertilizing, 29
Fusarium wilt, yellows, 66, 329
gray-mold blight, 52, 86, 328, 329
head rot, 328
insect control, 328, 329
leaf spot, 48, 331
Lettuce and Its Production, 332
light, effect on flowering, 38
marginal blight, 50, 330
mosaic, 70, 329
oxidant or PAN injury, 44
plant-food utilization, 26
powdery mildew, 56, 331
premature flowering, 41
resistant varieties, 328, 329, 330, 331
rib blight, brown or red rib, 331
root-knot, 90, 332
root nematode, 332
root rot, stunt, 87, 331
rust, 58, 330–31
seed rot, 330
seed treatment, 330, 331, 529, 532
seedbed treatment, 330
slime mold, 322, 332
southern blight, 76, 328
spotted wilt, ringspot, 72, 331
temperature, effect on, 41
tipburn, 328, [329]
Verticillium wilt, 68, 332
white-rust, 62, 331
Leucaena, 300
root rot, 87, 137, 301
rust, 58, 301
Leucojum, 250; *see* Snowflake
Leucophyllum, 479; *see* Texas silverleaf
Leucothoë, 316
black mildew, 64, 316
black spot, 48, 316
chlorosis, 24–29
felt fungus, 293, 316
iron deficiency, 24, 26, 27
leaf gall, 62, 316
leaf spot, 48, 316, [317]
spot anthracnose, 51, 316
tar spot, 48, 316

Liatris, 220
chlorosis, 24–29, 227
iron deficiency, 24, 26, 27, 227
leaf spot, 48, 221
powdery mildew, 56, 221
root-knot, 90, 225
root rot, 87, 221
rust, 58, 223
stem rot, 76, 221
Verticillium wilt, 68, 223
Libocedrus, 312; *see* Incense-cedar
Lichen, 561
Life cycle, 561
Light, 38–39
injury to plants, 38, 39
and temperature requirements for house plants, 39, 40
Lightning injury to trees, 46
entrance for fungi, 79
tree protection equipment, 46–48
Lightwood, 300; *see* Acacia
Ligustrum, 424; *see* Privet
Lilac, 332
anthracnose, 51, 334
bacterial blight, 50, 332, [333]
blossom blight, 52, 86, 332, 334
borer control, 332
canker (stem, twig), 77, 332
crown gall, 83, 334
dieback, 77, 332
frost injury, 334
graft blight, 334, [335]
gray-mold blight, flower blight, 52, 86, 334
leaf blight or blotch, 51, 334
leaf spot, 48, 334
light requirements of, 39
mosaic, 70, 334
Phytophthora blight, 332
powdery mildew, 56, [57], 332
pruning, 32
ringspot, 72, 334
root-knot, 90, 334, 392
root nematode, 334
root rot, 87, 137, 334
shoot blight, 76, 79, 332, [333]
Verticillium wilt, 68, 334
witches'-broom, 334
wood rot, 79, 170, 334
Lily (*Lilium*), 335
bacterial soft rot, 81, 335
Botrytis or gray-mold blight, 52, [54], 335
brown scale rot, [90], 335
bulb rot, 81, [90], 335
bulb (bublet) treatment, 336, 337, 338, 530, 531, 535
bunchy top, dieback, 75, 337
chlorosis (noninfectious), 24–29, 336
damping-off, 76, 162, 335
fleck, 70, 336
flower breaking, 70, 336
frost injury, 41, 338
insect control, 336
iron deficiency, 26, 27, 336
leaf and bud nematode, 75, 337
leaf scorch, tipburn, 336
leaf spot, 48, 338
mosaic, mottle, 70, 336, [337]
resistant lilies, 335, 336, 337
ringspot, 336
rodent injury, to bulbs, 98
root-knot, 90, 338
root nematode, 338
root rot, 87, 335, 337
rosette, yellow flat, 73, 336
rust, 58, 337
scale tip rot, [90], 335
soft mealy rot, [90], 335
soil treatment, 336, 337
southern blight, 76, 337
stem canker, 77, 337
stem or foot rot, stump rot, 76, 335
Lily leek, 361; *see* Onion
Lily-of-the-Nile, 494; *see also* Tulip
mosaic, 70, 496
Lily-of-the-valley, 335
anthracnose, 51, 338
gray-mold blight, fire, 52, 86, 335
leaf blotch, 51, 338
leaf spot, 48, 338
rhizome or crown rot, 90, 335
root-knot, 90, 338
root nematode, 338
southern blight, 76, 337
Lima bean, 155; *see also* Bean, garden types
bacterial spot, 50, 155
chlorosis, 24–29, 160
curly-top, 75, 159
damping-off, 76, 158
downy mildew, 55, 158
leaf spot, 48, 159–60
mosaic, 70, 155
pod blight or spot, 86, 158–59
pod drop, 160
resistant varieties, 158, 159
root-knot, 90, 159
root nematode, 160
scab, 65, [157], 159–60
seed rot, 86, 158
seed treatment, 158, 531, 532
stem anthracnose, 51, 159–60
white mold, watery soft rot, Sclerotinia wilt, 76, 156–57
Limb blight, 77
Lime, 561
Lime (Citrus), 187; *see also* Linden
anthracnose, withertip, 51, 228
bacterial blast, 50
chlorosis, 24–29, 228
fruit rot, 86
root nematode, 228
root rot, 87, 227–28
scab, 65
twig blight, 76, 228
Lime-sulfur (liquid), 105, 561
active ingredients, 105, 561
compatibility, 105, 552
as fungicide, 105, 144, 161, 182, 237, 292, 318, 347, 359, 384, 385, 406, 407, 427, 432, 525
injury, 105
precautions, 105
trade names, 561
uses, 105, 525
Liming soil, 24, 89
Limonium, 451; *see* Sea-lavender, Statice
Limp blight (*Ovulinia*), [435]
Linaria, 454; *see* Toadflax
Lindane, 65, 72, 74, 98, 132, 152, 168, 182, 185, 200, 205, 222, 242, 251, 266, 308, 336, 339, 349, 353, 360, 373, 424, 433, 437, 438, 456, 496, 500
in multipurpose mixes, 111
seed treatment, 162, 233, 238, 379
Linden, 338
anthracnose, 51, 338
bleeding canker, 77, 161, 339, 345
canker, (twig, branch, trunk), 77, 266, 338, 345
damping-off, 76, 339, 405
felt fungus, 293, 339
frost crack, 41
ice injury, 43
leaf blight, blotch, 51, 338
leaf scorch, 41, 338
leaf spot, 48, 338
mistletoe, 93, 339
powdery mildew, 56, 338
root rot, 87, 137, 339
seed rot, 339
sooty mold, [64], 65, 339
spot anthracnose, 51, 338
Verticillium wilt, 68, 339, 343
wetwood, slime flux, 264, 339
winter injury, sunscald, 41–43, 139, 264, 339
wood rot, 79, 170, 339
Lindera, 149; *see* Spicebush
Lingonberry, 175; *see* Blueberry
Linnaea, 498
black mildew, 64, 498
leaf spot, 48, 498
tar spot, 48, 498
Linum, 273; *see* Flax, flowering
Lions' ear, 447
leaf spot, 48, 447
rust, 58, 447
Lions-tail, 447
leaf spot, 48, 447
rust, 58, 447
Lippia, 317; *see* Lemon-verbena
Liquid Lux, as spreader, 119
Liquidambar, 509; *see* Sweetgum
Liriodendron, 341; *see* Tuliptree
Lithocarpus, 355; *see* Tanbark-oak
Lithospermum, 349
leaf spot, 48, 349
mosaic, 70, 349
powdery mildew, 56, 349
root rot, 87, 278, 349
rust, 58, 349
Litsea, 149; *see* Pond-spice
Liveforever, 451; *see* Sedum
Liverleaf, 131
downy mildew, 55, 131
leaf spot, 48, 131
rust, 58, 131
smut, 62, 131
Livistona, 371–72; *see* Palms
Lobelia, 339
curly-top, 75, 340
damping-off, 76, 339
gray-mold blight, 52, 224, 339
leaf smut, 62, 340
leaf spot, 48, 339
mosaic, 70, 339
powdery mildew, 56, 169, 340
root-knot, 90, 340
root rot, 87, 339
rust, 58, 223, 339
spotted wilt, 72, 340
stem and crown rot, 76, 339
Loblolly-bay, 274
black mildew, 64, 274
chlorosis, 24–29
iron deficiency, 24, 26, 27
leaf spot, 48, 274
root rot, 87, 274
Lobularia, 186; *see* Sweet alyssum
Local infection, 561
Lockhartia, 366; *see* Orchids
Locust, 300
borer control, 300–301
canker (twig, branch, trunk), dieback, 77, 300
chlorosis, 24–29, 302
collar rot, 76, 301
damping-off, 76, 302, 405
iron deficiency, 24, 26, 27, 302
leaf blight, 51, 301
leaf spot, 48, 301
mistletoe, 93, 301
powdery mildew, 56, 301
root-knot, 90, 301, 392
root nematode, 301
root rot, 87, 137, 301
rust, 58, 301
seedling blight, 76, 302, 405
tree dip (black locust), 301, 531, 537
2,4-D injury, 288, 301
Verticillium wilt, 68, 301, 343
wetwood, slime flux, 264, 302
winter injury, sunscald, 41, 139, 301
witches'-broom, 301
wood or heart rot, 79, 170, 301
zinc deficiency, 26, 27, 302, 503–4
Lodging, 561
Loganberry, 427; *see also* Blackberry, Raspberry
crown gall, 83, 428
mosaic, 70, 428
rust, 58, 430
Lolium, 319; *see* Ryegrass
London plane, 476; *see also* Sycamore
anthracnose, 51, 476
blight, cankerstain, 77, 476
canker (twig, branch), dieback, 77, 476
frost crack, 41, 264, 477
leaf spot, 48, 477
powdery mildew, 56, 477
rosy canker, 44–45, 477–78

London plane (*continued*)
tree wound dressing, 476–77
twig blight, 77, 476
wood rot, 79, 170, 477
Long-day plants, 38
Lonicera, 457; *see* Honeysuckle
Loosestrife (*Lysimachia*), 423; *see* Lythrum for Winged and Purple Loosestrife
leaf blight and stem necrosis, 51, 423
leaf spot, 48, 423
root-knot, 90, 423
root rot, 87, 423
rust, 58, 405, 424
soil drench, 423
stem or crown rot, 76, 423
stem nematode, 75, 424
Lophophora, 194; *see* Cacti
Loquat, 133; *see also* Apple
anthracnose, 51, 140
crown, collar rot, 76, 77, 139
crown gall, 83, 136–37
fire blight, 79, 133–34
flower blight, 79, 86, 133–34
fruit spot or rot, 86, 138
leaf blotch, 51, 140
leaf spot, 48, 140
root-knot, 90, 141
root nematode, 141
root rot, 87, 137–38
scab, 65, 134–35
Lorenz Chemical Company, 122
Los Angeles Chemical Company, 122
Lotus. 506
leaf spot, 48, 506
Lousewort, rust, 405
Love-lies-bleeding, 230; *see also* Amaranth
aster yellows, 73, 230
Love vine (dodder), 94
Luffa, 238; *see* Gourds, Cucurbit
Lunaria, 186; *see* Honesty
Lungwort, 348; *see* Mertensia
Lupine (*Lupinus*), 377; *see also* Pea
Ascochyta blight, 51, 76, 379
chlorosis, 24–29, 382
crown gall, 83, 381
damping-off, 76, 379
downy mildew, 55, 379
Fusarium wilt, 66, 377
gray-mold blight, 52, 86, 381
leaf blight, 51, 379
leaf nematode, 75, 382
leaf spot, 48, 379, 380
mosaic, 70, 378
powdery mildew, 56, 378
ringspot, 72, 379
root-knot, 90, 381
root nematode, 90, 382
root rot, 87, 377, 378
rust, 58, 381
seed smut, 62, 382
seed treatment, 379
seedling blight, 76, 379
soil drench, 377, 378, 379
southern blight, 76, 378
spotted wilt, 72, 378
stem or crown rot, 76, 378
Verticillium wilt, 68, 382
Lycaste, 366; *see* Orchids
Lychnis, 203; *see also* Maltese cross, Evening campion, Mullein-pink, Red campion
Alternaria blight, collar and branch rot, 51, 76, 204
anther smut, 62, 206
aster yellows, 73, 204
gray-mold blight, Botrytis flower blight, 52, 86, 205
leaf spot, 48, 205
ringspot, 72, 205
root rot, 87, 204
rust, 58, 204
southern blight, 76, 204
stem rot, 76, 204
Lycium, 347; *see* Matrimony-vine
Lycopersicon, 479; *see* Tomato
Lycoris, 250
bulb rot, 90, 250
root nematode, 250, 253
Stagonospora leaf scorch, red spot, 48, 251
stem and bulb nematode, 75, 92, 252
Lylocereus, 194; *see* Cacti, Cereus
Lyonia, 175
black mildew, 64, 179
leaf blight, blotch, 51, 178
leaf gall, shoot hypertrophy, 62, 176
leaf spot, 48, 178
powdery mildew, 56, 176
rust, 58, 177
tar spot, 48, 178
wood rot, 79, 170, 178
Lysimachia, 423; *see* Loosestrife
Lysol, as disinfectant, 81, 134, 360, 370, 561
Lythrum, 340; *see also* Loosestrife
leaf spot, 48, 340
root-knot, 90, 340
root rot, 87, 221, 340

M

Maackia, 300
root rot, 87, 137, 301
Maclura, 369; *see* Osage-orange
Madeira-vine, 340; *see also* Lythrum
root-knot, 90, 340
Madia, rust, 405
Madrone (*Madrona*), 175; *see also* Arbutus
dieback, canker, 77, 178
leaf blight, blotch, 51, 178
leaf spot, 48, 178
red leaf gall, spot, 62, 176
rust, 58, 177
tar spot, 48, 178
trunk canker, 77, 178, 259
Maggots, insect control, 187, 209
Magic lily, 249; *see* Amaryllis
Magnesium deficiency, 28, 180, 215, 325, 475
control, 27, 28, 180, 215, 299, 305, 325
chelates, 27, 215, 299
sulfate (Epsom salts), 26, 28, 180, 201, 215, 299, 305, 325
promotes, 28
Magnolia, 340–41
algal leaf spot, 341
bacterial leaf spot, 50, 341
black mildew, 64, 341
bud rot, 52, 341
canker, 77, 341
chlorosis, 24–29, 341, 344
cutting rot, 341
dieback, 77, 341
felt fungus, 293, 341
iron deficiency, 24, 26, 27, 341
leaf spot, 48, 341
petal rot, 52, 86, 341
powdery mildew, 56, 341
pruning, 32
root-knot, 90, 341
root nematode, 341
root rot, 87, 137, 341
seedling blight, 76, 341, 405
sooty mold, [64], 65, 341
spot anthracnose, 51, 341
sunscald, leaf scorch, 41, 341
thread blight, 341, 505
twig blight, 77, 341
Verticillium wilt, 68, 341, 343
wood rot, heart rot, 79, 170, 341
Mahonia, 153; *see also* Oregon-grape
canker, 77, 154, 345
leaf blotch, 51, 153
leaf scorch, scald, 41, 154
leaf spot, 48, 153
powdery mildew, 56, 153
quarantine regulations, 154
root-knot, 90, 153
root nematode, 154
root rot, 87, 137, 154
rust, 58, 153–54
Maidenhair-tree, 281; *see* Ginkgo
Maintenance
dusters, 117
sprayers, 114–15
Majorana, 447; *see* Salvia
Malabar-plum, 268; *see* Eugenia, Eucalyptus
Malacothrix, 220
rust, 58, 223
Malanga, 196; *see* Xanthosoma
Malathion, 48, 65, 70, 72, 73, 74, 82, 83, 87, 98
injury, 45
in multipurpose mixes, 110–11, 523
Maleberry, 175; *see* Lyonia
Mallinckrodt Chemical Works, 101, 122
Corrosive Sublimate, 101
Mallow, 297
anthracnose, 51, 298
aster yellows, 73, 299
chlorosis, 24–29, 299
crown gall, 83, 298
curly-top, 75, 299
leaf spot, 48, 298
mosaic, 70, 299
powdery mildew, 56, 298
root-knot, 90, 299
root nematode, 299
root rot, 87, 299
rust, 58, 298
seed treatment, 299, 535
spotted wilt, 72, 299
stem canker, 77, 298
Malpighia, 152; *see* Barbados cherry
Maltese cross, 203; *see also* Carnation
Alternaria blight, collar and branch rot, 51, 76, 204
flower smut, 62, 206
gray-mold (*Botrytis*) blight, flower blight, 52, 86, 205
leaf spot, 48, 205
ringspot, 72, 205
root rot, 87, 204
rust, 58, 204
Malus, 133; *see* Apple, Crabapple, flowering
Malva, 297; *see* Mallow
Malvastrum, 297; *see* Red false-mallow
Mammillaria, 194; *see also* Cacti
anthracnose, 51, 194
root-knot, 90, 194, 195
root rot, 87, 194
zonate spot, 48, 195
Mammoth blackberry, 427; *see* Blackberry, Raspberry
Mandrake, 348; *see* Mayapple
Maneb
active ingredient, 103, 519
algae control, 297
4.5D, 103, 519
gallon lots, 522
in multipurpose mixes, 103, 111, 519
oxidant or PAN prevention, 44
-PCNB mixtures, 104
properties, 100, 103, 519
as protective fungicide, 48, 55, 56, 62, 65, 86, 103, 167, 224, 248, 275, 280, 305, 308, 310, 314, 318, 320, 321, 329, 337, 349, 353, 362, 370, 381, 383, 384, 385, 394, 401, 405, 410, 415, 423, 424, 425, 433, 435, 440, 460, 476, 480, 482, 484, 495, 500, 502, 508, 509, 512, 519
seed-piece treatment, 418
soil application, 162, 167, 248, 255, 406
trade names and distributors, 103, 519
2% Dust, 103, 519
uses, 103, 519
Manfreda, 216
leaf spot, 48, 216
rust, 58, 216
Manganese
chelates, 27, 143, 299, 505
deficiency, 26, 27, 143, 290, 374, 397, 505
ethylenebis (dithiocarbamate), 103, 519
oxide, 26

sulfate, 26, 143, 166, 186, 201, 247, 290, 299, 397, 424, 505
toxicity, 24, 27
Mangel, 162; *see also* Beet
bacterial soft rot, 81, 86, 166
boron deficiency, 27, 164
crown gall, 83, 166
crown rot, 76, 164, 166
curly-top, 75, 162
damping-off, 76, 162
heart rot, 27, 164
leaf spot, 48, 162, 166
mosaic, 70, 164
root-knot, 90, 165
root rot, 87, 156, 166
rust, 58, 166
scab, 65, 165, 415
seed rot, 162
seed treatment, 162, 532
southern blight, 76, 166
virus yellows, 73, 164
yellow net, 70, 164
Mango (*Mangifera*), 342
anthracnose, 51, 342
blossom blight, 86, 342
brown felt fungus, 293, 343
fruit rot, 86, 342, 343
fruit soak, 342
leaf spot, 48, 342
Macrophoma blight, 342
powdery mildew, 56, 342
"red rust," green scurf, algal leaf spot, 342–43
root nematode, 343
root rot, 87, 137, 343
sooty mold, 64, 342
spot anthracnose, scab, 51, 65, 342
tipburn, 343
twig blight, dieback, 77, 343, 345
wood rot, 79, 170, 343
zinc deficiency, little leaf, 26, 27, 343
Mango melon, 237; *see* Cucurbit, Muskmelon
Mangold, 162; *see also* Beet, Mangel
blackheart, phosphorus deficiency, 25, 164
blackleg, 76, 164
Cercospora leaf spot, 48, 162
crown gall, 83, 166
damping-off, 76, 162
downy mildew, 55, 165
leaf spot, 48, 166
mosaic, 70, 164
root-knot, cyst nematode, 90, 165
rust, 58, 166
scab, 65, 165, 415
seed rot, 162
seed treatment, 162, 532
virus yellows, 73, 164
Manilagrass, 319; *see* Zoysia
Manzanita, 175
black mildew, 64, 179
leaf spot, 64, 178
mistletoe, 93, 179
red leaf spot, gall, 62, 176
root rot, 87, 178
rust, 58, 176
shoot gall, hypertrophy, 62, 176
wood rot, 79, 170, 178
Manzate D Maneb, 103, 519
Manzate Maneb Fungicide, 103, 519
Maple, 343
anthracnose, [51], 343
bacterial leaf spot, 50, 347
black mildew, 64, 346
bleeding or Phytophthora canker, 77, 161, 345
bulls-eye spot, 48, [346]
canker (twig, branch, trunk), 77, [344], 345
chlorosis, 24–9, 344–5
crown gall, 83, 136, 346
dieback, 77, 345
Diseases of Shade and Ornamental Maples, 347
felt fungus, 293, 347
frost crack, 41
gall mites, 347
girdling roots, [45]
ice injury, 43
inflorescence or flower blight, 86, 347
insect and mite control, 346, 347
leaf blight, 51, 343
leaf blister, 62, 346, 384
leaf scab, 65, 346
leaf scorch, physiological, [10], 41, 344
leaf spot, 48, [346]
lightning injury, 46
Maple Diseases and Their Control—A Guide for Homeowners, 347
mistletoe, 93, 347
Nectria canker, [344], 345
Phyllosticta leaf spot, 48, [346]
powdery mildew, 56, 346
pruning, 32, 33
root nematode, 346
root rot, 87, 137, 346
salt injury, 44
seedling blight, 76, 347, 405
Septoria leaf spot, 48, [346]
slime flux, wetwood, 264, 346
soil grade change injury, 46
sooty mold, 64, 346
tar spot, 48, [346]
thread blight, 347, 505
twig blight, 77, 345
2,4-D injury, 288, 347
Verticillium wilt, 68, [69], 343, [344]
winter injury, sunscald, 41, 139, 264, 346
wood rot, 79, 170, 346
Maranta, 347
leaf spot, 48, 347
light and temperature requirements, 40
root-knot, 90, 347
root nematode, 347
root rot, 87, 347
rust, 58, 347
Marble queen, 196; *see* Pothos
Marbleseed, 349
root rot, 87, 278, 349
rust, 58, 349
Marguerite, 219; *see also* Chrysanthemum, Camomile
aster yellows, 73, 223
crown gall, 83, 226
curly-top, 75, 223
damping-off, seed rot, 76, 221
light and temperature requirements, 40
powdery mildew, 56, 221
root-knot, 90, 225
root nematode, 226–27
seed treatment, 221
Verticillium wilt, 68, 223
Marigold, 220
aster yellows, 73, [222]
bacterial wilt, 69, 223
damping-off, 76, 221
flower spot, 86, 221, 224
Fusarium wilt, 66, 223
head blight, 52, 86, 224
leaf spot, 48, 221
mosaic, 70, 222
resistant varieties, 223
root-knot, 90, 225
root nematode, 226–27
root rot, 87, 221
rust, 58, 223, 405
seed treatment, 221
soil drench, 222
southern blight, 76, 227
stem (crown) rot, wilt, 76, 221, 223, 224
Verticillium wilt, 68, 223
Marine-ivy, 287; *see* Cissus
Mariposa lily, 347
rust, 58, 347
Marl, 24
Marrubium, 447
leaf spot, 48, 447
root-knot, 90, 447
Marsh grasses (*Spartina*), rust, 144, 185, 186, 501
Mason jar, for treating seed, 97, 527
disinfesting soil, 548, [549]
Matricaria, 220
aster yellows, 73, 222
petal blight, 86, 224
powdery mildew, 56, 221
root-knot, 90, 225
rust, 58, 223
white-rust, 62, 226
Matrimony-vine, 347
leaf spot, 48, 347
mosaic, 70, 348
powdery mildew, 56, 347
rust, 58, 348
Matthiola, 186; *see* Stock
Maurandya, 454
leaf spot, 48, 455
Mayapple, 348
gray-mold blight, 52, 348
leaf blight, 51, 348
leaf spot, 48, 348
ringspot, 72, 348
rust, 58, 348
stem rot, 76, 348
Mayday-tree, 382; *see* Peach, Plum
Mayflower, 294; *see* Trailing-arbutus
MBC Fumigant, 545
McLaughlin Gormley King Company, 122
M.C.P.A. injury, 45
Meadow saffron, 230; *see* Colchicum
Meadowbeauty, 254
leaf spot, 48, 254
Meadowrue, 254
downy mildew, 55, 257
leaf spot, 48, 256
leaf and stem smut, 62, 257
powdery mildew, 56, 256
rust, 58, 256, 392
white smut, 65, 257
Meadowsweet (*Filipendula*), 439–40; *see also* Spirea
leaf spot, 48, 443
powdery mildew, 56, 440
rust, 58, 441
seedling blight, 76, 405
Mealybugs, insect control, 277, 360, 373, 437
secretion of "honeydew," 64
as virus carriers, 13, 98
Measles (virus), [279]
Measurement
sprays and dusts, 520–22
units of, 516
Measuring apparatus, 109
Mechanical injuries, [43], 45–46
treatment [35], 45
Meconopsis, 413; *see also* Poppy
black mold, 64, 414
downy mildew, 55, 414
powdery mildew, 56, 414
root rot, 87, 414
soil drench, 414
stem canker or rot, 76, 77, 414
Medlar, 133; *see also* Apple
fire blight, 79, 133–34
leaf blight, spot, 48, 51, 140
rust, 58, 135
Melia, 218; *see* Chinaberry
Melissa, 447
leaf spot, 48, 447
Melocactus, 194; *see* Cacti
Melon, 237; *see* Honeydew melon, Muskmelon, Watermelon
Melon aphid, [242]
Melothria, 238; *see also* Cucurbit
downy mildew, 55, 243
powdery mildew, 56, 242
root-knot, 90, 244–45
Memmi, 106, 250
Menispermum, 351; *see* Moonseed
Mentha, 447; *see* Mint
Mentzelia, 348
leaf spot, 48, 348
root and stem rot, 76, 87, 348
rust, 58, 348
Menziesia, 175
leaf gall, 62, 176
leaf spot, 48, 178

Menziesia (*continued*)
  powdery mildew, 56, 176
  rust, 58, 177
  tar spot, 48, 178
Merck & Company, Incorporated, Merck Chemical Division, 105, 122
Mercuram, 321
Mercuric chloride
  combined with mercurous chloride, 101
  disinfectant, 35, 79, 81, 83, 101, 134, 139, 171, 176, 259, 271, 278, 301, 367, 386, 420, 528
  plant dip, 83
  precautions, 101, 528
  preparation of solution, 101, 528
  seed, bulb, corm, tuber, rhizome treatment, 81, 83, 97, 101, 196, 202, 238, 250, 282, 307, 379, 470, 484, 528–29
  soil drench, 77, 82, 89, 101, 130, 169, 174, 221, 254, 256, 307, 339, 377, 378, 400, 500, 528
  trade names, 101
  tree wound treatment, 101, 383
  uses, 101
Mercurous chloride, combined with mercuric chloride, 101
Mercury; *see* Mercuric chloride, Phenyl (organic) mercury
Merrybells, 335
  leaf spot, 48, 338
  rust, 58, 337
Mersolite-8, 561
Mersolite-W, 250
Mertensia, 348
  downy mildew, 55, 349
  gray-mold blight, 52, 349
  leaf smut, 62, 349
  leaf spot, 48, 221, 349
  mosaic, 70, 349
  powdery mildew, 56, 349
  rust, 58, 349
  stem rot, 76, 349
Mescalbean, 300; *see* Sophora
Mesembryanthemum, 306; *see* Figmarigold, Iceplant
Mespilus, 133; *see* Medlar
Metalsalts Corporation, 122
Methanol, as crown gall treatment, 83
  as disinfectant, 98
Methocel, 561
Methoxychlor, 48, 65, 73, 83, 87, 98, 263, 264, 289, 387, 428, 440, 491
  in multipurpose mixes, 111, 523
Methyl bromide, 545
  applying, 545, 547, [548]
  disinfesting soil, 83, 91, 106, 214, 322, 336, 406, 420, 437, 441, 467, 472, 487, 489, 502
  formulations, 545, 547
  mixtures with chloropicrin, 543, 547
  precautions, 543, 545, 547
  trade names, 106, 545
  uses, 545
Methyl isothiocyanate, 568; and 80 per cent chlorinated C_3 hydrocarbons including dichloropropenes, 546
Metiram, 104; *see also* Polyram
Mexican fire-plant, 409; *see* Spurge
Mexican flame-vine, 220; *see* Senecio
Mexican sunflower, 220; *see* Torch flower
Mezereon, 253; *see* Daphne
Mice, protection against, 42
Michaelmas daisy, 220; *see also* Aster, perennial
  rust, 58, 223, 405
Mico-Ban 531, 106, 561
Micro Nu-Cop, 104
Microgel, 104
Micromeria, 447
  rust, 58, 447
Micron, 561
Midvale Chemical Company, 122
Mignonette, 350; *see* Boussingaultia for Climbing mignonette
  aster yellows, 73, 350
  black ringspot, 192, 350
  curly-top, 75, 350
  damping-off, 76, 350
  downy mildew, 55, 350
  leaf blight, 51, 350
  leaf spot, 48, 350
  root-knot, 90, 350
  root rot, 87, 350
  Verticillium wilt, 68, 350
Mikania, 220
  leaf spot, 48, 221
  rust, 58, 223
Mildew, 561
  black, [64], 65, 200
  downy, [55], 56, [189], [235], [363], 379, [381], 490
  powdery, 56, [57], 58, 100, 118, [128], [136], [167], 173, [176], [221], [242], [305], [320], [378], [477], 525
Mildew King, 106, 556, 561–62
Milk, for virus control, 485
  powdered skim, as sticker, 119
Milk-bush, 409; *see* Poinsettia, Spurge
Milkweed, 185
  anthracnose, 51, 185
  leaf spot, 48, 185, 221
  mosaic, 70, 185
  powdery mildew, 56, 185
  root rot, 87, 185, 278
  rust, 58, 185
  stem blight, 77
Milkwort, 399; *see* Polygala
Miller Chemical Company, 122
Miller Chemical & Fertilizer Corporation, 102, 105, 122, 518
  Captan Dust, 102, 518
  Cyprex Dusts, 102, 518
  Lime Sulfur Solution, 561
  Mico-Fume 25-D (Mylone), 106, 546
  Nemagon 10 per cent; 25 per cent Granular; No. 2 EC, 545
  New Fruit Spray or Dust, 523
  Nu-Film, 119
  PCP-10, 563
  658 Fungicide, 106, 213, 480, 562
  Streptomycin Antibiotic Spray Powder, 105, 522
  Terraclor Dusts, 544
Miller Products Company, 122
Miltonia, 366; *see* Orchids
Mimosa, 377; *see* Sensitive plant and "Mimosa" tree
"Mimosa" tree (*Albizzia*), 300
  canker (twig, branch, trunk), dieback, 77, 300
  Fusarium wilt, 66, 301
  leaf spot, 48, 301
  root-knot, 90, 301
  root nematode, 301
  root rot, 87, 137, 301
  2,4-D injury, 45, 288, 301
  winter injury, sunscald, 41, 139, 301
Mimulus, 454; *see* Monkeyflower
Miniature Josephscoat, 230; *see* Alternanthera
Miniature orange, 227; *see* Citrus, Orange
Minnie-bush, 175; *see* Menziesia
Minor elements, 27
Mint, 447
  black stem, 77, 448
  leaf spot, 48, 447
  powdery mildew, 56, 447
  rhizome or root soak, 447, 529, 535
  ringspot, 72, 448
  root-knot, 90, 447
  root nematode, 447
  rust, 58, 447
  spot anthracnose, 51, 447
  stem canker, rot, 76, 77, 447
  Verticillium wilt, 68, 448
Mirabilis, 274; *see* Four-o'clock
Miracle-Gro, 30
Missouri primrose, 268; *see* Evening-primrose
Mistflower, 220; *see* Ageratum, Eupatorium
Mistletoes, [93], 94
  American, [93], 94
  Dwarf, [93], 94
MIT, 546; *see also* Vorlex Soil Fumigant
  active ingredient, 546
  application, 546
  disinfesting soil, 83, 91, 106, 188, 214, 244, 336, 364, 365, 406, 420, 424, 443, 448, 472, 487, 489, 490
  precautions, 543, 546
  trade names, 546
  uses, 546
Mitchella, 185; *see* Partridgeberry
Mitella, 304
  leaf rot, 51, 305
  leaf spot, 48, 305
  powdery mildew, 56, 304
  rust, 58, 305
Mites
  as bacteria carriers, 280, 282
  as cause of witches'-broom, 171
  control, 98, 145, 321, 407
  as fungus carriers, 48
  in spread of diseases, 13, 98
  as virus carriers, 13, 98, 272, 389–90, 390
Miticides, compatibility chart, 552
Mitrewort, 304; *see* Mitella
Mock-cucumber, 238; *see also* Cucurbit
  anthracnose, 51, 86, 238
  curly-top, 75, 243
  downy mildew, 55, 243
  fruit spot, 86, 238, 241, 244
  Fusarium wilt, 66, 241
  leaf spot, 48, 238, 239, 244
  mosaic, 70, 241
  powdery mildew, 56, 242
  seed treatment, 238
Mock-strawberry, 439
  downy mildew, 55, 444
  leaf spot, 48, 443
  rust, 58, 441
Mockorange, 304
  chlorosis, 24–29, 305
  flower and shoot blight (*Botrytis*), 52, 86, 304
  flower spot, 52, 86, 305
  iron deficiency, 24, 26, 27, 305
  leafy gall, 81, 306
  leaf spot, 48, 305
  magnesium deficiency, 28, 305
  powdery mildew, 56, 304
  pruning, 32
  root-knot, 90, 305
  root nematode, 306
  root rot, 87, 306
  rust, 58, 305
  sooty mold or blotch, 64, 306
  twig canker, dieback, 77, 305
  Verticillium wilt, 68, 306
  wood rot, 79, 170, 306
Mold, 562
Molucella, 447
  crown rot, 76, 447
Molybdenum deficiency, 28, 193, 299
Momordica, 238; *see* Balsam-apple
Monarda, 447
  leaf spot, 48, 447
  mosaic, 70, 447
  powdery mildew, 56, 448
  rust, 58, 447
  southern blight, 76, 447
  Verticillium wilt, 68, 448
Monardella, 447
  leaf spot, 48, 447
  rust, 58, 447
Moneywort, 423; *see* Loosestrife
Monkey-coconut, 371; *see* Palms
Monkeyflower, 454
  aster yellows, 73, 457
  leaf nematode, 75, 457
  leaf spot, 48, 455
  powdery mildew, 56, 455
  rust, 58, 455
Monkeypuzzle tree, 143; *see* Araucaria
Monkshood, 254; *see also* Aconite
  bacterial leaf spot, 50, 255
  leaf and stem smut, 62, 257
  mosaic, 70, 255

rust, 58, 256
soil drench, 254
stem or crown rot, 76, 254
Verticillium wilt, 68, 256
Monkshood-vine, 286; *see also* Ampelopsis
canker, dieback, 77, 290
downy mildew, 55, 288
powdery mildew, 56, 289
Monoecious (rust), 58, 562
Monsanto Chemical Company, Agricultural Chemical Division, 122
PCNB 80% Dust Concentrate, 544
Monstera, 196
leaf blight, spot, 48, 51, 197
Montbretia, 282; *see* Tritonia
Monuron, weed control, 368
Mooney Chemicals, Incorporated, 122–23
Moonflower, 351
leaf nematode, 75, 226, 351
leaf spot, 48, 351
ringspot, 72, 351
root-knot, 90, 351
root nematode, 351
root rot, 87, 351
rust, 58, 351
white-rust, 62, 191, 351
Moonseed, 351
leaf smut, 65, 351
leaf spot, 48, 351
powdery mildew, 56, 351
Morestan, powdery mildew control, 135, 242, 289
Moreton bay pine, 143; *see* Araucaria
Morning-glory, 351
blossom blight, 86, 351
curly-top, 75, 351
Fusarium wilt, 66, 351
leaf spot, 48, 351
mosaic, 70, 351
ringspot, 72, 351
root-knot, 90, 351
root nematode, 351
root rot, 87, 351
rust, 58, 351, 405
southern blight, 76, 351
stem canker, 77, 351
thread blight, 351, 505
white-rust, 62, 191, 351
Morocide, powdery mildew control, 58, 135, 242, 289
Morsodren, 106, 418, 562; *see also* Morton Soil Drench
Morton Chemical Company, Division of Morton International, Incorporated, 103, 123, 519
Morton Soil Drench (Morsodren), 77, 106, 169, 204, 222, 378, 410, 544, 562
Morton Soil Fumigator, 547
Morus, 352; *see* Mulberry
Mosaic, 70, [71], 98, [165], [192], [241], [250], [284], [309], [337], [375], [378], [429], [438], [442], [485], [496]
nematodes, as virus carriers, 192
weeds as virus reservoir, 72, 98, 165, 241, 329, 368, 484
Moses-in-a-boat, 437; *see* Rhoea
Mosquito bills, 423; *see* Shootingstar
Moss, control, 326–27
Moss campion, 203; *see* Silene
Moss-pink, 399; *see* Phlox
Mother-of-thyme, 447
root rot, 87, 278, 447
Mottle, 70, [496], 562
Moundlily, 512; *see* Yucca
Mountain-ash, 133; *see also* Apple
black rot, canker, 48, 77, 136
blossom blight, 86, 133–34
canker, dieback, 77, 138–39
crown gall, 83, 136–37
fire blight, 77, 79, 86, 133–34
fruit spot or rot, 86, 138
leaf blight, 51, 140
leaf spot, 48, 140
powdery mildew, 56, 135–36
root rot, 87, 137–38
rust, 58, 135, 405
scab, 65, 134–35
sooty blotch, 64, 137
sunscald, 41, 139–40
twig blight, 77, 138–39
winter injury, 41, 139
wood rot, 79, 139, 170
Mountain bluet, 220; *see* Centaurea
Mountain cranberry, 175; *see* Blueberry
Mountain creeper, 229; *see also* Clockvine
root-knot, 90, 229
Mountain-ebony, 300; *see* Bauhinia
Mountain fleece, 453; *see* Silver-lacevine
Mountain-holly, 295; *see also* Holly
leaf spot, 48, 295
powdery mildew, 56, 296
tar spot, 48, 295
wood rot, 79, 170, 297
Mountain-laurel, 175
canker, 77, 178
chlorosis, 24–29, 178, [179]
drought injury, 179
flower blight, 86, 179
iron deficiency, 24, 26, 27, 178, [179]
leaf blight, blotch, 51, 178, [179]
leaf spot, 48, 178, [179]
magnesium deficiency, 28, 180
mummy berry, 175
powdery mildew, 56, 176
red leaf gall, spot, 62, 176
root rot, 87, 178
tar spot, 48, 178
winter injury, 41, 178
witches'-broom, 177
wood rot, 79, 170, 178
Mountain-mint, 447
leaf spot, 48, 447
powdery mildew, 56, 448
rust, 58, 447
Mountain spicewood, 198; *see* Calycanthus
Mountain spurge, 371; *see* Pachysandra
Mouse injury to trees, 42
protection, 42
Mowing wounds, 45
entrance for fungi, 46, 79
Mugwort, 220; *see* Artemisia
Muhlenbergia, rust, 298
Mulberry, 352
bacterial spot, blight, 50, 352
canker, 77, 352
dieback, 77, 352
false mildew, 352
hairy root, 83, 136, 353
leaf spot, 48, 352
Mycosphaerella leaf spot, 48, 352
"popcorn" disease, 352
powdery mildew, 56, 352
root-knot, 90, 352, 292
root rot, 87, 137, 393
rust, 58, 353
twig blight, 77, 352
wetwood, slime flux, 264, 353
wood, heart rot, 79, 352
Mulch, 21, 433, 562
advantages of, 24, 42, 99, 407, 465, 562
loose soil surface, 24
organic matter, 21, 42, 43
types
corn cobs (ground), 42, 99, 433
leaves, 42, 99, 296, 433
paper, 24
peatmoss, 42, 433
pine needles, 296, 433
plastic, 24, 328, 465
salt hay, 42
sawdust, 42, 296, 433
straw, 99
wood shavings (or chips), 42, 433
winter protection, 42, 99, 407, 463
Mullein, 454
downy mildew, 55, 456
leaf spot, 48, 455
powdery mildew, 56, 455
root-knot, 90, 457
root rot, 87, 455
soil drench, 455
Mullein-pink, 203; *see also* Lychnis
aster yellows, 73, 204
leaf spot, 48, 205
Multifilm X-77, 119
Multipurpose sprays and dusts, flowers, fruit, shrubs, trees, vegetables, 110–11, 523
injury from, 523
notes on, 111, 523
trade names, 523
Mumfume, 106
Mummy (fruit), 562
Mummy berry, [175]
Muscari, 495; *see* Grape-hyacinth
Mushrooms (toadstools), 322, [323], 562
Musk mallow, 297; *see* Hibiscus (arborescent forms)
Muskmelon, 237; *see also* Cucurbit
Alternaria blight, 51, [239]
angular leaf spot, 50, 239
anthracnose, 51, [238]
aster yellows, 73, 243
bacterial leaf spot, 50, 239, 245
bacterial soft rot, 81, 238, 244
bacterial wilt, 69, 240
blossom blight, 86, 245
chlorosis, 24–29, 245
crown gall, 83, 245
curly-top, 75, 243
damping-off, 76, 244
downy mildew, 55, 243
fruit rot, 86, [238], 244
Fusarium wilt, 66, 241
gummy stem blight, 76, 86, 244
leaf blight, 51, [238]
leaf spot, 48, 238, 239, 244
mosaic, 70, 241
Muskmelon Culture, 245
Muskmelons for the Garden, 245
powdery mildew, 56, 242
resistant varieties, 239, 240, 241, 242, 243
ringspot, 72, 243
root-knot, 90, 244–45
root nematode, 244
root rot, 87, 156, 244
scab, 65, 238–39
seed treatment, 238, 528, 532
seedbed spray, 244
southern blight, 76, 243
stem rot, 76, 243
2,4-D injury, 45, 245, 288
Verticillium wilt, 68, 245
Mustard, 186; *see also* Cabbage
anthracnose, 51, 190
aster yellows, 73, 193
bacterial soft rot, 76, 81, 190
bacterial spot, 50, 190
black rot, 50, 187
chlorosis, 24–29, 194
clubroot, [89], 90, 189
crown rot, drop, 76, 191
curly-top, 75, 192
damping-off, 76, 188, 190, 191, 193
downy mildew, 55, 189
Fusarium wilt, 66, 186–87
leaf spot, 48, 190, 191
mosaic, 70, 192
powdery mildew, 56, 193
resistant varieties, 187, 190
ringspot, 72, 192
root-knot, root gall, 90, 191
root nematode, 194
root rot, 87, 187, 190, 193
rust, 58, 193
seed treatment, 187, 188, 529, 532
stem (crown) rot, wilt, 76, 191
temperature requirements, 41
Verticillium wilt, 68, 193
web blight, 160, 194
white-rust, 62, 191
Mutation (bud sport), 562
Mycelium, 562
Mycoban, 362
Mycorrhiza (mycorrhizae), 562
Mycosphaerella blight, 379, [380], [381]
Mycostatin, 562
Myers, F. E. & Bro. Company, Subsidiary of McNeil Corporation, 123

Mylone (DMTT), disinfesting soil, 91, 106, 188, 193, 214, 405, 467, 490, 545, 548
Dust-50, 543
85% WP, 545
25% WP, 545
Myosotis, 349; *see* Forget-me-not
Myriagale, rust, 405
Myrica, 507; *see* Bayberry, Sweetgale, Waxmyrtle
Myrtle, 353
leaf spot, 48, 353
powdery mildew, 56, 353
stem or crown rot, 76, 254, 353
Myrtle boxleaf, 173
leaf spot, 48, 173
Myrtus, 353; *see* Myrtle

N

Nabam, 326, 562
active ingredient, 562
plus urea-formaldehyde, 362
properties, 100
trade names, 562
uses, 562
Nabam Liquid Fungicide, 562
Naled; *see* Dibrom
Nandina, 353
anthracnose, 51, 353
chlorosis, 24–29, 344
leaf spot, 48, 353
root-knot, 90, 354
root nematode, 354
root rot, 87, 354
Nannyberry, 499; *see* Viburnum
Naptox, 556
Narcissus, 249; *see also* Daffodil
bacterial blight, stem rot, 50, 76, 253
bud blast, 253
bulb and bud nematode, 92, 252
bulb rot, [90], 250
bulb soak, 250, 252, 530, 531, 535
decline, 251
ethylene injury, 44
flower spot, 52, 86, 251
gray-mold blight, 52, 86, 251
leaf blight, 51, 251
leaf scorch, 51, [251]
leaf spot, 48, 253
leaf, stem, and bulb nematode, 75, 92, [252]
light and temperature requirements, 40
mosaic, 70, [250], 251
paper tip, white streak, 251
resistant varieties, 250, 251
root nematode, 253
root rot, 87, 250
"smoulder," neck rot, 76, [252]
Stagonospora leaf scorch, 51, [251]
stripe, 250
white mold, Ramularia blight, 51, [251], 252
yellow dwarf, 73, 253, 365
Narcissus bulb flies, control, 250
Nasturtium, garden, 354; *see also* Watercress
aster yellows, 73, 354
bacterial leaf spot, 50, 354
bacterial wilt, 69, 354
curly-top, 73, 354
fasciation, 81, 354
gray-mold blight, 52, 354
Heterosporium leaf spot, 48, 354, 535
leaf spot, 48, 354
mosaic, 70, 354
ringspot, 72, 354
root-knot, root gall, 90, 354
rust, 58, 354
seed treatment, 354, 530, 535
spotted wilt, 72, 354
stem rot, 76, 354
Natal-plum, 360; *see* Carissa
National Agricultural Chemicals Association, 107
National Sprayer and Duster Association, 118
Nationwide Chemical Company, 123
Natriphene, 106, 366–67, 562
Natriphene Company, 122
N-dodecylguanidine acetate, 102, 518
N-dure, 362, 415
Neck rot, [252], [283], [361]
Necklace plant or vine, 451; *see* Crassula
Necrosis (necrotic), 562, 567
Nectarine, 382; *see also* Peach
asteroid spot, 390
bacterial leaf spot, 50, 386
bacterial shoot blight, 70, 79, 386
blossom blight, 86, 382
brown rot, 76, 86, 382, 525
canker (twig, branch, trunk), 77, 383, 386, 390, 392
crown gall, hairy root, 83, 391
dieback, 77, 383
fruit rot, 86, 382, 384, 385, 392
leaf blight, 51, 392
leaf curl, 62, 384
mosaic, 70, 389
peach yellows, 73, 388
powdery mildew, 56, 385
pruning, 32, 383, 384
ringspot, 72, 388
rust, 58, 392
scab, 65, 385
shot-hole, 52, 386, 390, 392
silver leaf, 79, 384
spray schedule, 524–25
twig blight, 77, 382
wood rot, 79, 170, 384
X-disease, 387
Nectria canker, [344], 345
Needle blight (evergreens), 51, 313, 314, [402]
cast (evergreens), 51, 314, 402, 562
rust, 58, [404], 405
Nelumbo, 506
leaf spot, 48, 506
Nema-Kill, 546
Nemadrench, 546
Nemafume, 546
Nemagon, 546, 548; *see also* DBCP 8.6E, 546
Soil Fumigant, 546
Nemaphos, 546
Nemas; *see* Nematodes
Nemaster, 545
Nematocides, 16, 76, 324, 538, 543–47, [548], [549], 562
Nematodes, 14, [15], 16–17, 562
body parts, 14, [15]
bud, leaf, leaf gall, stem, 17, 75, [76], 226, [252], [270], 337, 360, 375, 400, 456, 467, [497]
bulb, ring disease, onion bloat, [92], 93, 252, 364
control, 16, 76, 538–49
chemical, 16, 76, 538, 543–47, [548], [549]
cultural, 16, 76
by hot water soak, 76, 168, 270, 290, 301, 315, 364, 376, 400, 443, 450, 458, 467, 472. 493, 497, 510
in home garden, 548, [549]
Controlling Nematodes in the Home Garden, 549
cyst, 90, 92
dwarf or crimp, 75, [76], 467
feed on, 14, [15]
found, 14
identifying, 7, 17
collection of samples, 7, 17
life cycle, 16, 75
losses due to, 8, 17
numbers, 14, 16, 252, 467
over winter, 75
penetration, 16
physiologic races, 93, 97
as plant-weakening agents, 78
predator [15]
reproduction, 16
root-knot, 16, 90, [91], 92, 392, [395], [396], 467, 472, 565
size, 14
specimens, diagnosing, 7
how and what to send, 7
mailing, 7
spiral, [15]
spread by, 16, 75, 91, 93
sting, [395], [396]
symptoms of, 16–17, 324
temperatures
effect on growth, reproduction, 16
lethal, 16, 529–31, 542
types, 14
as virus carriers, 13, 72, 166, 178, 192, 243, 291, 331, 389–90, 418, 429, 448, 466, 486
wounds, entrance for bacteria and fungi, 14, 66, 68, 70, 76, 81, 88, 91, 167, 203, 208, 301, 366, 377, 448, 463, 487, 488
Nematologist, extension, help by, 106, 107, 324
Nema-X, 546
Nemex, 545
Nemopanthus, 295; *see* Mountain-holly
Nemophila, 399
leaf spot, 48, 399
powdery mildew, 56, 399
Nepeta, 447; *see* Catnip, Ground-ivy
Nephrolepis, 269; *see* Ferns
Nephthytis, 196
bacterial leaf spot, 50, 197
cane rot, 76, 196–97
leaf spot, 48, 197
plant (cane) soak, 197, 530, 536
root-knot, 90, 197
root nematode, 197
root rot, 87, 196–97
soil drench, 197
Nerine, 250; *see* Guernsey-lily
Nerium, 360; *see* Oleander
Net necrosis, [417]
Nettle, as virus source, 73
Neutral copper; *see* Fixed copper fungicides
Neutral soil, 24, 562
New Guinea bean, 238; *see* Cucurbit
New Jersey-tea, 355
canker, dieback, 77, 138, 345, 355
chlorosis, 24–29
crown gall, 83, 136, 355
iron deficiency, 24, 26, 27
leaf spot, 48, 355
powdery mildew, 56, 355
root rot, 87, 137, 355
rust, 58, 355
wood rot, 79, 170, 355
New Zealand spinach, 162
bacterial leaf spot, 50, 166
Cercospora leaf spot, 48, 162
chlorosis, 24–29, 166
curly-top, 75, 162
leaf spot, 48, 166
mosaic, 70, 164
ringspot, 72, 164, 166
root-knot, cyst nematode, 90, 165
seed treatment, 162, 529
Verticillium wilt, 68, 167
virus yellows, 73, 164
Niacide A, 562
M, 100, 134, 313, 562
Niagara Chemical Division, Food Machinery & Chemical Corporation, 102, 103, 105, 106, 123, 518, 519
Actispray, 106
Agri-strep, 83, 105
Carbamate, 102, 518
C-O-C-S, 104
Nabam Solution, 562
Phaltan 50 Wettable, 103, 519
Phygon, 102, 518
Seed Protectant, 102, 518, 531
Z-C Spray or Dust, 103, 519
Zineb, 103, 519
Nicandra, 479; *see* Apple-of-Peru
Nicotiana, 479; *see* Flowering tobacco
Nicotine sulfate, injury, 45
Nierembergia, 479; *see* Tomato
Nightshade, 379; *see also* Jerusalem-cherry, Eggplant

bacterial soft rot, 81, 480, 481
crown gall, 83, 492
late blight, 51, 480
mosaic, 70, 484
ringspot, 72, 486
Ninebark, 355
fire blight, 79, 133–34, 355
leaf spot, 48, 346, 355
powdery mildew, 56, 171, 355
root rot, 87, 355
wood rot, 79, 170, 355
Nitrogen deficiency, 25, 235
control, 24, 235
injury, 25
promotes, 25
N-(1,1,2,2,-tetrachloroethylsulfenyl)-cis-△-4-cyclohexene-1,2-dicarboximide, 104
Noninfectious (nonparasitic) diseases, ailments, 9, [10], 24–29, 31–32, 39, [41], [42], [43], 44, [45], [128], 140, 145, 160, 164, [179], 182, 194, 214, 260, [267], [281], [288], 303, 318, 326, 327, 341, [417], 420, 433–34, 447–48, 563
Norfolk island pine, 143; *see* Araucaria
light and temperature requirements, 40
Northern white-cedar, 312; *see* Arborvitae
Norwich Pharmaceutical Company, 123
No Scald DPA, 140
Nothoscordum, 361; *see* False-garlic
Nott Manufacturing Company, 124
Nozzles, spray, 109, [110]
pattern, 109, [110]
types, 109, [110]
N-(trichloromethylthio)-4-cyclohexene-1-2, dicarboximide, 102, 518
N-trichloromethylthiophthalimide, 103, 519
N-(trichloromethylthio) tetrahydrophthalimide, 102, 518
Nu-Film, 119
Nu-Iron, 27, 345
Nu-Manese, 26
Nuphar, 506; *see* Waterlily
Nursery, 563
Nursery stock
Commercial Storage of Fruits, Vegetables, Florist and Nursery Stocks, 100
dip, 529–37
inspection of, 99
Nurseryman, help by, 5, 9, 24, 32, 44, 65, 79, 97, 134, 135, 138, 149, 154, 162, 178, 182, 199, 268, 290, 296, 300, 301, 311, 313, 356, 386, 411, 425, 427, 433, 435, 440, 441, 503, 508
Nutramin 6, 27
Nutra-Phos 3–25, 27
Nutra-Spray, 27
Nutri-Sperse, 27
Nutrient deficiencies, in plants, 25–9
The Care and Feeding of Garden Plants, 19, 25
confused with, 29, 70
The Diagnosis of Mineral Deficiencies in Plants by Visual Symptoms, 25
Hunger Signs in Crops, 25
as plant-weakening agents, 78
Nutrients (plant), 563
Nu-Z, 27
Nyctocereus, 194; *see* Cacti, Cereus
Nymphaea, 506; *see* Waterlily
Nyssa, 258; *see* Tupelo, Black gum, Sour-gum

O

Oak, 355
anthracnose, 51, 356, [357]
bark patch, 359
black mildew, 64, 359
bleeding canker, 77, 161, 345, 360
borer control, 358
canker (twig, branch), dieback, 77, 266, [358]
chlorosis, iron deficiency, 24, 344, [359]
crown gall, 83, 136, 359
Dothiorella canker, 77, [358]
felt fungus, 293, 360
flooding injury, 37
frosty mildew, 358
girdling roots, 45
iron deficiency, 24, 26, 27, [359]
leaf blight, 51, 356
leaf blister or curl, 62, [357]
leaf scorch, physiological, 41, [359]
leaf spot, 48, 358
lightning injury, 46
mistletoe, 93, 360
powdery mildew, 56, 359
root nematode, 360, 392
root rot, 87, 137, [358]
rust, 58, 359, 405
seed (acorn) rot, 360, 405
soil grade change injury, 46
soil treatment (oak wilt), 263, 356
sooty mold, 64, 359
spot anthracnose, 51, 358
twig blight, 77, 356, 358
2,4-D injury, 45, 288, 360
Verticillium wilt, 68, 343, 360
wetwood, slime flux, 264, 360
wilt, 355, [356], [357]
wood, heart, or butt rot, 79, 170, 358
Oakes Manufacturing Company, 124
Oceanspray, 299; *see* Holodiscus
Ocimum, 447
mosaic, 70, 447
root-knot, 90, 447
Oconee-bells, 275
chlorosis, 24–29
leaf spot, 48, 275
Octopus-plant, light and temperature requirements, 40
Odontoglossum, 366; *see* Orchids
Oedema, 39, 168, 194, [228], 375, 563
Oenothera, 268; *see* Evening-primrose
Oil, dormant, injury, 45
spray, 342, 347, 406
Okra, 297
anthracnose, 51, 298
bacterial soft rot, 81, 298
bacterial wilt, 69, 299, 488
blossom blight, 86, 298
curly-top, 75, 299
damping-off, seed rot, 76, 299
dieback, 77, 298
Fusarium wilt, 66, 299
leaf spot, 48, 298
pod spot and rot, 52, 86, 298
powdery mildew, 56, 298
ringspot, spotted wilt, 72, 299
root-knot, 90, 299
root nematode, 299
root rot, 87, 299
rust, 58, 298
seed treatment, 299, 533
southern blight, 76, 298
stem rot, 76, 298
Verticillium wilt, 68, 299
web blight, 160, 299
Olea, 370; *see* Olive
Oleander, 360
anthracnose, 51, 360
bacterial knot or gall, 360
cutting dip, 360
canker, witches'-broom, 77, 360
chlorosis, 24–9, 344, 361
leaf spot, 48, 360
root nematode, 361
root rot, 87, 137, 361
sooty mold, 64, 360
spot anthracnose, scab, 65, 360
Oleaster, 445; *see* Russian-olive
Olin-Mathieson Chemical Corporation, Agricultural Division, 105, 124
OMA-D, 545
OMAfume D-EDB, 545
Olive, 370
anthracnose, 51, 370
bacterial knot, 360, 370
bitter pit, dry rot, 370
black mildew, 64, 370
fruit spot, rot, 86, 370
leaf spot, 48, 370
root-knot, 90, 370, 392
root nematode, 370
root rot, 87, 137, 370
soft or blue nose, 370
Verticillium wilt, 68, 370
OMA-D, 545
OMAfume D-EDB, 545
Omazene, 106, 563
Oncidium, 366, *see* Orchids
1-chloro-2-nitropropane, 104
1,2,dibromo-3-chloropropane, 546
Onion, 361
aphid control, 365
aster yellows, 73, 365
bacterial leaf streak, rot, 50, 366
bacterial soft rot, 81, 90, 362, 365
blast, tip blight, 51, 52, 362
bloat (stem and bulb nematode), [92], 93, 364
bulb curing and storage, 361
bulb rot, 81, [90], 361, 362
chlorosis, 24–29, 366
damping-off, 76, 365
downy mildew, 55, [363]
freezing injury, 365
gray-mold rot, 52, 90, 361
leaf blight, 51, 362, 363
leaf fleck, 58, 362
leaf spot, 48, 365
maggot control, 362
mosaic, 70, 365
neck rot, [90], [361], 362
Onion Diseases and Their Control, 366
pink root, 87, 363, [364]
potassium deficiency, 26, 366
purple blotch, 51, [363]
resistant varieties, 361, 362, 363, 364, 365
root-knot, 90, 365
root nematode, 366
root rot, 87, 363
rust, 58, 365
scab, 65, 365
seed and set treatment, 362, 365, 529, 533
smudge, 90, 364, [365]
smut, 62, [63], 362
soil treatment, 362
southern blight, 76, 366
sunscald, 365
thrips control, 363
Verticillium wilt, 68, 365
white rot, 87, [90], 362
yellow dwarf, 73, 365
Onoclea, 269; *see* Ferns
Onosmodium, 349
root rot, 87, 278, 349
rust, 58, 349
Onychium, 269; *see* Ferns
Ophioglossum, 269; *see* Ferns
Opuntia, 194; *see also* Cacti
anthracnose, 51, 194
bacterial soft rot, 76, 81, 194
black mildew, 64, 195
cladode rot, spot, 76, 194
glassiness, 39, 195
leaf scorch, 41, 195
leaf spot, black, 48, 195
oedema, 39, 194
root-knot, 90, 194, 195
scab, 65, 195
stem rot, 76, 194
"sunscald," 41, 195
Orange, 227; *see also* Citrus, Hardy orange
anthracnose, wither tip, 51, 228
bacterial blast, 50
chlorosis, 24–29, 228
crown gall, 83, 136, 228
fruit rot, 86
plant-food utilization, 26
powdery mildew, 56
root nematode, 228

Orange (*continued*)
root rot, 87, 227–28
scab, 65
sooty blotch, 64, 228
twig blight, 77, 228
Orange-eye butterflybush, 185; *see* Butterflybush
Orange rust, 430, [431]
Orange sunflower, 220; *see* Heliopsis
Orchard Brand Ferbam, 102, 518
Filmfast Spreader-Sticker, 119
-530, 104
Ziram, 103, 519
Orchards, spray requirements for, 526
Orchid supply dealer, help by, 369
Orchid-tree, 300; *see* Bauhinia
Orchids, 366
anthracnose, 51, 367
bacterial brown spot or rot, 50, 367
bacterial soft rot, 81, 367
black rot (*Pythium* and *Phytophthora*), 48, 51, 366
black spot, 48, 367
Cattleya blossom brown necrotic streak, 368
chlorosis, 24–29
cutting (*Vanda*) dip, 369, 530, 536
damping-off, 76, 366
dry sepal, ethylene injury, 44
flower breaking, 70, 368
fly speck, 369
Fusarium wilt, 66, 367–68
gray-mold blight, 52, 367
insect control, 368
Knudson formula, 366, 561
leaf blight or blotch, 51, 366, 367
leaf necrosis, 367
leaf nematode, yellow bud blight, 75, 369
leaf spot, 48, 367
light and temperature requirements, 40
mold, 366, 367, 369
mosaic, 70, 368
mottle, 70, 368
Orchid Diseases, 369
petal spot or blight, 86, 367
plant soak, 366–67, 367, 369, 530, 536
ringspot, 72, 368
root nematode, 369
root rot, 87, 367
rust, 58, 367, 508
seed rot, 366
seed treatment, 366, 536
seedling blight, 76, 366
snow mold, potting fiber mold, 369
sooty blotch or mold, 64, 369
stem, collar, or crown rot, 76, 367
sunburn, 369
tipburn, 369
Oregon-grape, 153; *see also* Mahonia
canker, 77, 154
leaf blotch, 51, 153
leaf scorch, scald, 41, 154
leaf spot, 48, 153
root-knot, 90, 153
root nematode, 154
root rot, 87, 137, 154
rust, 58, 153, 370
Organ-pipe cactus, 194; *see* Cacti, Cereus
Organic matter, 23, 92, 563
compost pile, 23
types of, 23
Organic mercury; *see* Phenyl mercury
Oriental orange, 227; *see* Hardy orange, Citrus
Ornamental alliums, 361; *see* Onion, Allium
Ornithogalum, 495; *see* Star-of-Bethlehem
Ortho
Bordo Mixture, 105, 554
Copper Fungicide "53," 104
D-D Soil Fumigant, 545, 548
Dust, 103
Ethylene Dibromide 83 Soil Fumigant, 544
Ferbam-76, 102, 518
Greenol Iron Chelate, 345
Home Orchard Spray, 523
-K, 104
Lawn Disease Control, 321, 563
Lawn and Turf Fungicide, 106, 320, 321, 322, 325, 563
LM Apple Spray, 106
LM Seed Protectant, 282, 563
Nabam Liquid Spray, 562
Nemagon 45 Soil Fumigant, 546
PCNB 20 Dust, 544
Phaltan 50 Wettable, 103, 519
Phaltan Rose Garden Fungicide, 102, 518
Soil Treater "X," 563
Spray-Sticker, 119
Spreader Sticker, 119
Streptomycin Spray, 105, 522
X-77 Spreader, 119
Zineb 75 Wettable, 103, 519
Ziram, 103, 519
Orthocide
-Dieldrin 60–15 Seed Protectant, 102, 518, 531
80 Wettable, 102, 518
50 Wettable, 102
Fruit and Vegetable Wash, 102, 518
Garden Fungicide, 102, 518
75, 531
75 Seed Protectant, 102, 518, 531
Soil Treater "X," 104, 188, 328
Orthorix Spray, 561
0,0-diethyl 0-2-pyrazinyl phosphorothioate, 546
0-2,4-dichlorophenyl 0,0-diethyl phosphorothioate, 546
Osage-orange, 369
damping-off, 76, 369, 405
leaf blight, 51, 369
leaf spot, 48, 369
mistletoe, 93, 369
root rot, 87, 369
rust, 58, 369
Verticillium wilt, 68, 369
Osier; *see* Dogwood, Willow
Osmanthus, 370
bacterial knot, 360, 370
black leaf spot, 48, 370
black mildew, 64, 370
leaf spot, 48, 370
mistletoe, 93, 370
root-knot, 90, 370, 392
root rot, 87, 137, 370
sooty mold, 64, 370
Verticillium wilt, 68, 370
Osmaronia, 440
leaf spot, 48, 443
powdery mildew, 56, 440
Osmorhiza, 213; *see* Sweet-jarvil, Aniseroot
Osmunda, 269; *see* Ferns
Osoberry, 440
leaf spot, 48, 443
powdery mildew, 56, 440
Ostrya, 170; *see* Hophornbeam
Oswego-tea, 447; *see* Monarda
Oven, for treating soil, 539
Owlclover, rust, 405
Oxalis, 370
curly-top, 75, 162, 371
leaf spot, 48, 370
light and temperature requirements, 40
powdery mildew, 56, 370
root rot, 87, 278, 371
rust, 58, 154, 370
seed smut, 62, 371
stem nematode, 75, 371
tar spot, 48, 370
Oxeye, 220; *see* Heliopsis
Oxeye daisy, 219; *see also* Chrysanthemum
aster yellows, 73, 222
leaf nematode, 75, 226
mosaic, 70, 222
seed rot, damping-off, 76, 221
seed treatment, 221
Oxidant, as air pollutant, 44
control, 44
resistant plants, 44
sensitive plants, 44
symptoms, 44
Oxlip, 423; *see* Primrose
Oxybaphus, 274; *see* Umbrellawort
Oxydendrum, 459; *see* Sorreltree
Oxygen deficiency, in soil, 37
Oxyquinoline materials (sulfate), 100, 280, 563; *see also* 8-hydroxyquinoline sulfate
soil drench, 97, 127, 167, 204, 371, 406
Oxytetracycline, 104; *see also* Terramycin
Oyster-plant, 437; *see* Rhoea
Ozoban, 44
Ozone, as air pollutant, 43–44
control, 44
resistant plants, 44
sensitive plants, 44
symptoms, 44

P

Pachistima, 173
leaf spot, 48, 173
Pachycereus, 194; *see* Cacti, Cereus
Pachysandra, 371
insect control, 371
leaf spot, 48, 371
root-knot, 90, 371
root nematode, 371
root rot, 87, 371
soil drench, 371
Volutella leaf blight, stem canker, dieback, 51, 77, 371
Pacific Supply Cooperative, Chemicals Division, 124
Paeonia, 254; *see* Peony
Pagoda-tree, 300; *see* Sophora
Painted-cup, 454
powdery mildew, 56, 455
rust, 58, 455
Painted daisy, 219; *see* Pyrethrum, Chrysanthemum
Painted-tongue, 479
aster yellows, 73, 490
Fusarium wilt, 66, 486
leaf blight, 51, 486
powdery mildew, 56, 492
root-knot, 90, 489
root nematode, 489
spotted wilt, 72, 486
stem, foot rot, 76, 488
Verticillium wilt, 68, 487
Pak-choi, 185; *see* Chinese cabbage
Pale laurel, 175; *see* Mountain-laurel
Palmetto, 372; *see* Palms
Palms, 371–72
anthracnose, 51, 372
bacterial leaf spot, 50, 372
bacterial wilt, 69, 373
black mildew, 64, 373
black scorch, 51, 373
bud rot, 373
canker, gummosis, 77, 373
chlorosis, manganese deficiency, 24, 26, 27, 373, 374, 505
damping-off, seed decay, 76, 374
false smut, 372
felt fungus, 293, 373
fruit spot, rot, 86, 373
gummosis, 373
insect control, 373
leaf blight, dieback, 51, 372, 373
leaf drop, 373
leaf scab, 65, 372
leaf scorch, withered leaf tip, 41, 373
leaf spot, 48, [372]
light and temperature requirements, 40
plant dip, 373
potting mix, 374
root-knot, 90, 373
root nematode, 373
root rot, decline, 87, 372
sunscald, 373
thread blight, 373, 505

  trunk (stem) or butt rot, 76, 372, 373
  wilt, 372, 373
  wood rot, 79, 372, 373
PAN, as air pollutant, 44
  control, 44
  resistant plants, 44
  sensitive plants, 44
  symptoms, 44
Panda plant, 451; *see* Kalanchoë
Pandanus, 451
  burrowing nematode, 392, 451
  leaf spot, 48, 451
  light and temperature requirements, 40
Panicum, rust, 212, 410
Pano-Brome, 106, 545
  CL, 545
Pano-drench, 77, 97, 106, 204, 544
Pano-Fume, 106
Panogen, 106, 282, 373
  seed, corm, bulb treatment, 106, 282, 394
  Soil Drench, 544
  Turf Spray, 106, 320, 321, 322
Panoram 75, 103, 519, 531
  D-31, 103, 282, 519, 531
Pansy, 374; *see also* Violet
  anthracnose, 51, 374
  aster yellows, 73, 376
  bacterial soft rot, 81, 375
  calico, 70, 375
  crown gall, 83
  crown, stem rot, 76, 374, 375
  curly-top, 75, 376
  damping-off, 76, 375
  downy mildew, 55, 375
  flower blight, 86, 374
  flower breaking, 70, [375]
  Fusarium wilt, 66, 375
  gray-mold (*Botrytis*) blight, rot, 52, 86, 374
  insect control, 374, 375, 376
  leaf nematode, 75, 375–76
  leaf spot, 48, 374
  light, effect on flowering, 38
  mercury injury, 98
  mosaic, 70, [375]
  oedema, corky scab, 39, 375
  plant soak, 376, 530, 536
  powdery mildew, 56, 374
  ringspot, 72, 376
  root-knot, 90, 375
  root nematode, 375
  root rot, wilt, 87, 375
  rust, 58, 375
  seed smut, 62, 375
  seed treatment, 375, 536
  sooty mold, 64, 376
  southern blight, 76, 375
  spot anthracnose, scab, 65, [374]
Papaver, 413; *see* Poppy
Paper-mulberry, 271
  canker, dieback, 77, 138, 266, 271
  leaf spot, 48, 271, 272
  mistletoe, 93, 272
  root-knot, 90, 271, 392
  root nematode, 271
  root rot, 87, 137, 271
Parasites, 9–17, 563
Parasitic flowering plants, 93–94
  dodder, [94]
  mistletoes, [93], 94
  witchweed (*Striga*), 234
Parasoltree, 401; *see* Phoenix-tree
Parathion, injury, 45
Paris daisy, 219; *see* Marguerite
Park department, help by, 263, 266
Parkinsonia, 300
  leaf spot, 48, 301
  mistletoe, 93, 302
  root rot, 87, 137, 301
  sooty mold, 64, 302
Parlor-ivy, 220; *see* Senecio
Parsley, 213
  aster yellows, 73, 214
  bacterial soft rot, 81, 214
  blackheart, heart rot, calcium deficiency, 28, 214–15
  boron deficiency, 27, 215
  crown rot, 76, 215
  curly-top, 75, 215
  damping-off, 76, 214
  downy mildew, 55, 210, 215
  Fusarium wilt, yellows, 66, 214
  leaf blight, 51, 213
  leaf spot, 48, 213
  mosaic, 70, 214
  root-knot, 90, 214
  root nematode, 215
  root rot, 87, 215
  seed treatment, 213, 532
  stem nematode, 75, 215
  stem rot, 76, 214
  temperature requirements, 41
  Verticillium wilt, 68, 215
Parsnip, 206
  aster yellows, 73, 207
  bacterial soft rot, 81, 87, 207, 209
  boron deficiency, 27, 209
  canker, 77, 207
  crown gall, 83, 209
  crown rot, 76, 209
  curly-top, 75, 162, 209
  damping-off, 76, 209
  downy mildew, 55, 210
  gray-mold rot, 52, 86, 206, [207]
  irregular roots, 210
  Itersonilia leaf spot, 48, 206–7, 209
  leaf spot, 48, 206–7
  mosaic, 70, 209
  petiole rot, 76, 207, 209
  potassium deficiency, 26, 210
  powdery mildew, 56, 210
  resistant varieties, 209
  ringspot, 72, 209
  root canker, "sore head," 77, 209
  root-knot, 90, 209
  root nematode, 208
  root rot, 87, 208
  scab, 65, 165, 209, 415
  seed rot, 209
  seed treatment, 207, 532
  stem nematode, 75, 208
  storage rot, 86, [207], 208
    cleanliness, 208
  watery soft rot, cottony rot, Sclerotinia rot, 76, 86, 156, 207–8
  white-rust, 62, 210
Parsons U.S.P. Formaldehyde, 558
Parthenocissus, 287; *see* Boston ivy, Virginia-creeper
Partridge-breast, 130; *see* Aloe
Partridgeberry, 185
  black mildew, 64, 186
  chlorosis, 24–29, 186, 443
  stem rot, 76, 186
Parzate
  C, 103, 519
  Liquid Nabam Fungicide, 562
  Zineb Fungicide, 103, 519
Paspalum, 319; *see* Bahiagrass
Pasqueflower, 131; *see* Anemone
Passionflower (*Passiflora*), 376
  anthracnose, 51, 376
  gray-mold blight, 52, 167, 376
  leaf and stem spot, 48, 376
  mosaic, woodiness disease, 70, 376
  root-knot, 90, 376
  root rot, 87, 376
  seedling wilt, 76, 376
  southern blight, stem (collar) rot, 76, 376
Pasteurization of soil, 538–49; *see* Steam disinfestation of soil
Pastinaca, 206; *see* Parsnip
Pathogen, 563
Pathogenic, 563
Pathology, 563
Patience plant, 152; *see* Balsam, garden
Paulownia, 376
  leaf spot, 48, 376
  root rot, 87, 170, 377
  wood rot, 79, 377
Pawpaw, 377
  canker (twig, branch), dieback, 77, 345
  fruit rot, 86, 377
  leaf blotch, 51, 377
  leaf spot, 48, 377
  sooty mold, 64, 377
  wood rot, 79, 170, 377
PCNB (pentachloronitrobenzene), 101
  active ingredient, 101, 544
  bulb, corm treatment, 336, 362
  formulations, 101, 188
  gallon lots, 522
  mixtures with other fungicides, 77, 89, 104, 138, 162, 188, 197, 204, 222, 240, 254, 297, 328, 330, 393, 398, 408, 409, 410, 423, 435, 455, 489, 490, 500
  precautions, 189
  "shot-gun" soil drench, 111
  soil application, 77, 89, 97, 101, 106, 129, 130, 138, 146, 156, 157, 162, 188, 197, 198–99, 204, 214, 222, 224, 249, 250, 254, 282, 284, 297, 298, 304, 307, 318, 321, 328, 330, 336, 339, 348, 362, 367, 393, 398, 407, 408, 409, 410, 423, 435, 437, 455, 465, 489, 490, 495, 496, 500, 544
  storage treatment, 443
  trade names, 101, 188, 222, 544
  in transplant water, 89, 189
  uses, 101, 104
PCNB-Captan, 188, 328
  10-10 Dust, 104
  25-25 Wettable Powder, 104
PCNB-Dexon 35-35, 222
P-(dimethylamino) benzenediazo sodium sulfonate, 104
Pea, garden, 377; *see also* Sweetpea
  anthracnose, 51, 380
  Ascochyta blight, 51, 76, 379, [380]
  bacterial leaf spot, blight, 50, 86, 380, [381]
  bacterial soft rot, 81, 86, 381
  bacterial wilt, 69, 382
  black walnut injury, 382
  blossom blight, 86, 381
  boron toxicity, 28
  chlorosis, 24–29, 382
  damping-off, 76, 377, 379, 381
  downy mildew, 55, 379, [381]
  fasciation, 81, 381
  foot rot, 76, 378, 379
  Fusarium wilt, 66, 377
  gray-mold blight, 52, 86, 381
  insect control, 379
  leaf spot, blotch, 48, 379, 380
  light, effect on flowering, 38
  mosaic, 70, 378
  mottle, 70, 378
  Mycosphaerella blight, 51, 76, 379, [380]
  near wilt, 66, 377
  Pea Diseases, 382
  pod spot, rot, 86, 379, 380, [381]
  powdery mildew, 56, 378
  resistant varieties, 377, 378, 379, 380
  ringspot, 72, 379
  root-knot, cyst nematode, 90, 381
  root nematode, 382
  root rot, 87, [88], 377, 378
  rust, 58, 381
  scab, 65, 380, [381]
  Sclerotinia white mold, 76, 378
  seed rot, 379
  seed treatment, 379, 531, 532
  Septoria blight, 51, 379
  southern blight, 76, 378
  spotted wilt, 72, 379
  stem and bulb nematode, 75, 92, 382
  stem canker, 77, 378
  stem (crown) rot, 76, [88], 378, 379
  streak, 378
  stunt, 73, 378
  temperature, effect on, 41
  Verticillium wilt, 68, 382
  wilt (virus), 378
Peach, 382
  asteroid spot, 390
  bacterial canker, 79, 386
  bacterial leaf spot, 50, [386]
  bacterial shoot blight, 50, 79, 386
  black knot, 81, 384

Peach (*continued*)
blossom blight, 86, 382, [383]
borer control, 384, 525
brown rot, 77, 86, 382, [383], 525
canker (twig, branch, trunk), 77, [383], 386, 390, 392
chlorosis, mottle leaf, 24–27, 393
Controlling Phony Disease of Peaches, 389
Coryneum blight, 51, 77, 390, [391]
crown gall, hairy root, 83, [84], 391
decline, 389, 392
dieback, 77, 383
fly speck, 86, 393
fruit spot, rot, 86, 382, 384, 385, 392
gummosis, 383, 386, 390
insect control, 383, 384, 387
leaf blight, 51, 392
leaf curl, 62, [63], 384, 525
leaf spot, 48, 392
little leaf, 26, 27, 393
little-peach, 387
mosaic, 70, 389
mottle, 70, 390
phony peach, phony disease, 389
plant-food utilization, 26
plant soak, 392
powdery mildew, 56, 385
pruning, 32, 383, 384
resistant varieties, 383, 384
ringspot complex, 72, 388
root-knot, 90, 392
root nematode, 392
root rot, 87, 392
rosette, 73, 387
rust, 58, 392
scab, fruit freckle, 65, [66], 385
shot-hole, 52, [53], [386], 390, 392
silver leaf, 79, 384
sooty mold, 64, 137, 393
spray schedule, 524–25
twig blight, 77, 382
Verticillium or blackheart wilt, 68, 391
wet feet, 393
winter injury, 42, 392
wood rot, 79, 170, 384
wound treatment, 383, 384, 386, 392
X-disease, yellow-red disease, 387
yellows, 73, [74], 388
Peanut, 393
bacterial wilt, "yellows," 69, 396
chlorosis, manganese deficiency, 24, 26, 27, 397
concealed damage, 86, 394
Diplodia collar rot, 76, 393–94
Fusarium wilt, 66
gray-mold leaf rot, 52, 394
Growing Peanuts, 397
hopperburn, 397
insect control, 394, 397
iron deficiency, 24, 26, 27, 397
leaf mold, 394
leaf spot, 48, [394]
mosaic, 70, 397
mottle, 370, 397
nematode injury, 90, [395], [396]
plant-food utilization, 26
pod, kernel decay, 86, 394
pod rot, 393
resistant varieties, 394, 395, 396
root-knot, 90, [395], [396]
root nematode, 395–96
root rot, 87, 393
rust, 58, 396–97
seed mold, rot, 394
seed treatment, 394, 532
seedling blight, 76, 394
soil rot, 393
soil treatment, 393
southern blight, collar rot, 76, 393, [394]
stem blight or rot, 76, 393
sting nematode, [395], [396]
stunt, 73, 397
thrips injury, 397
Verticillium wilt, 68, 396
white mold, 76, 393
Peanut peg, 563
Pear, 133; *see also* Apple
anthracnose, 51, 140
bacterial blast, 50, 79, 86, 134–35
bitter pit, 140
black end, 140
black rot, canker, 48, 77, 86, 136
Blight of Pears, Apples, and Quinces, 579
blossom blight or blast, 52, 79, 86, 133–34, 141
boron deficiency, cork, drought spot, 27, 141
brown core, heart, 140
calcium deficiency, 38, 141–42
canker (twig, branch, trunk), 77, 138–39
chlorosis, 24–29, 143
copper deficiency, 28, 141–42
crown gall, 83, 136–37
decline, 142
dieback, 77, 136, 138–39, 141
felt fungus, 138, 293
fire blight, 77, 79, [80], 133–34, 525
fly speck, 86, 137
fruit spot, rot, 86, 138
gray-mold (*Botrytis*) rot, 52, 86, 138
leaf blight, 51, 140
leaf blister, 62, 143
leaf scorch, physiological, 41, 141–42
leaf spot or fleck, 48, 140
magnesium deficiency, 28, 141–42
mistletoe, 93, 141
mosaic, 70, 142
Mycosphaerella leaf spot, 48, 140
potassium deficiency, 26, 141–42
powdery mildew, 56, 135–36
pruning, 32
root-knot, 90, 141
root nematode, 141
root rot, 87, 137–38
rubbery wood, 142
rust, 58, 135
scab, 48, 65, 86, 134–35
scald, 140
silver leaf, 79, 139
sooty blotch, 64, 137
spot anthracnose, 51, 140
spray schedule, 524–25
stem- or wood-pitting, 142
stony pit, 141
storage rot, 86, 138
sunscald, 41, 139–40
thread blight, 143, 505
twig blight, 77, 138–39
wood rot, 79, 139, 170
zinc deficiency, 26, 27, 141
Pearleverlasting, 219
leaf spot, 48, 221
rust, 58, 223
Pearson-Ferguson Chemical Company, Incorporated, 124
Peashrub, 300
damping-off, seedling blight, 76, 302, 405
hairy root, 83, 136, 301
leaf blight, 51, 301
leaf spot, 48, 301
pod blight, 86, 301
root nematode, 301
root rot, 87, 137, 301
Peatmoss, 21, 24, 563
acid, 24, 434
as mulch, 21, 42, 433
Peatree, 300; *see* Peashrub
Pecan, 501
anthracnose, 51, 501
boron deficiency, dieback, "snakehead," 27, 505
bunch disease, 502
Controlling Insects and Diseases of the Pecan, 506
crown gall, 83, 502
dieback, canker (twig, branch), 77, 502
downy spot, white mold, 51, 502
felt fungus, 293, 506
insect control, 502, 505
leaf scorch, 41, 506
leaf spot, 48, 501
lightning injury, 46
liver spot, 48, 501–2
mistletoe, 93, 506
"mouse ear," manganese deficiency, 26, 27, 505
nursery blight, 51, 501
nut mold or rot, 86, 504
powdery mildew, 56, 505
resistant varieties, 502, 503, 505
root-knot, 90, 392, 506
root nematode, 504
root rot, 87, 137, 504
scab, 65, 86, 502, [503]
Spanish-moss, gray moss, 506
spray program, 501, 502, 503
storage mold, 86, 504
sunscald, 41, 139, 502
thread blight, 505
twig and trunk canker, 77, 502
2, 4-D injury, 288, 506
Verticillium wilt, 68, 343, 506
winter injury, 41, 139, 264, 506
wood rot, 79, 170, 504
zinc deficiency, rosette, little leaf, yellows, 26, 27, 503, [504]
Pediocactus, 194; *see* Cacti
Pelargonium, 278; *see* Geranium of florists
Pelecyphora, 194; *see* Cacti
Pelican-flower, 144; *see* Aristolochia
Pellaea, 269; *see* Ferns
Penick, S. B. & Company, 124
Pennsalt Chemicals Corporation, 102, 103, 124, 518, 519
Cyprex 4% Dust, 102, 518
Pennyroyal, 447; *see* Mint
Penphene, 546
Penstemon, 454
black mildew, crust, 64, 457
insect control, 456, 457
leaf and stem nematode, 75, 457
leaf spot, 48, 455
mosaic, 70, 456
powdery mildew, 56, 455
root-knot, 90, 457
root nematode, 457
root rot, 87, 455
rust, 58, 455
seed treatment, 455
soil drench, 455
sooty mold, 64, 457
spotted wilt, 72, 456
stem (crown) rot, spot, 76, 455
Pentachloronitrobenzene, 101, 544; *see also* PCNB
Pentachlorophenol, 563
trade names, 563
Pentox, 563
Peony, 254
anthracnose, 51, 256
Botrytis blight, gray-mold, bud blast, 52, [54], 76, 86, 255, 256, 257
bud blast, 52, 255, 256
chlorosis, 24–29, 257
crown elongation, 257
crown gall, 83, 168, 257
crown rot, 76, 254
flower blight, 52, 86, 255, 256, 257
leaf blotch, blight, 51, [256]
leaf curl, 257
leaf spot, 48, 256
leaf and stem nematode, 75, 257
LeMoine disease, 257
mosaic, 70, 255
powdery mildew, 56, 256
ringspot, 72, [255], 257
root-knot, 90, 256
root nematode, 256
root rot, 87, 254
root soak, 256, 530, 536
soil drench, 254, 255, 256
southern blight, 76, 254
stem canker, 77, 254
tip blight, 51, 254
Verticillium wilt, 68, 256
witches'-broom, 257
Peperomia, 397
anthracnose, 51, 398
corky scab, 39, 397
cutting rot, 76, 397

foliar nematode, 75
leaf spot, 48, [398]
light and temperature requirements, 40
ringspot, 72, [396], 397
root-knot, 90, 397
root nematode, 397
root rot, 87, 397
soil drench, 398
stem rot, 76, 397
Pepper, 479; *see also* Tomato
anthracnose, 51, 86, 480, 481
aster yellows, 73, 490
bacterial soft rot, 81, 86, 480, 481, 483
bacterial spot, 50, 86, [483], 484
bacterial wilt, 69, 488
blossom blight or rot, 86, 492
blossom drop, 41, 484
blossom-end rot, 482
chlorosis, 24–29, 490
curly-top, 75, 486
damping-off, 76, 488, [489]
downy mildew, blue mold, 55, 490
early blight, 51, 480
fruit spot, rot, 86, 480, 481, 483
Fusarium wilt, 66, 486
gray-mold rot, 52, 86, 480
insect control, 483, 485, 486
leaf mold, 480–81
leaf spot, 48, 480
milk, virus control, 485
mosaic, 70, 484, [485]
oxidant or PAN injury, 44
Pepper Production, 492
Phytophthora blight, 51, 490
powdery mildew, 56, 492
resistant varieties, 484, 485, 487, 489, 490
ringspot, 72, 486
root-knot, 90, 489
root nematode, 489
root rot, 87, 490
seed rot, 488
seed treatment, 484, 528, 529, 533
seedbed treatment, 489, 490
southern blight, 76, 490
spotted wilt, 72, 486
stem or collar rot, 76, 480, 488, 490
streak, 70, 484
sunscald, 482
temperature, effect on, 41
tobacco etch, 484
Verticillium wilt, 68, 487
web blight, 160, 492
Pepper-root, 186; *see* Toothwort
Pepperbush, 470; *see* Clethra
Peppergrass, 186; *see also* Garden cress, Cabbage
chlorosis, 24–29, 194
clubroot, 89, 189
crown rot, 76, 191
curly-top, 75, 192
damping-off, 76, 188
downy mildew, 55, 189
leaf spot, 48, 190
mosaic, 70, 192
ringspot, 72, 192
root rot, 87, 193
rust, 58, 193
stem and leaf nematode, 75, 194
white-rust, 62, 191
Peppermint, 447; *see* Mint
Peppermint stick, 220; *see* Senecio
Peppertree, 468
root-knot, 90, 469
root nematode, 469
root rot, 87, 137, 469
Verticillium wilt, 68, 468–69
wood rot, 79, 170, 469
Peppervine, 286; *see also* Ampelopsis
leaf spot, 48, 287, 291
thread blight, 291, 505
PEPS, 119
Percentage solution, 522
Peregrina, rust, 405
Perennial honesty, 186; *see* Honesty
Perennial pea, 377; *see* Sweetpea
Peristeria, 366; *see* Orchids
Periwinkle, 500; *see* Vinca
"PERK" Nutritional, 27
Perlite, 563
Peroxyacetyl nitrate (PAN), as air pollutant, 44
control, 44
resistant plants, 44
susceptible plants, 44
symptoms, 44
Persea, 149; *see* Avocado, Redbay, Swampbay
Persimmon, 398
anthracnose, 51, 398
canker, dieback, 77, 138, 266, 399
Cephalosporium wilt, 398
crown gall, 83, 136, 399
fly speck, 86, 399
fruit spot, rot, 86, 399
gray-mold rot, 52, 86, 399
leaf blotch, 51, 398
leaf spot, 48, 398
mistletoe, 93, 399
powdery mildew, 56, 171, 399
root-knot, 90, 392, 399
root nematode, 398–99
root rot, 87, 137, 398
scab, 65, 398
sooty blotch, 64, 137, 399
tar spot, 48, 398
thread blight, 399, 505
twig blight, 77, 138, 266, 399
Verticillium wilt, 68, 399
wood rot, 79, 170, 399
Perunkila, 360; *see* Carissa
Peruvian daffodil, 250; *see* Spiderlily
Pest (plant), 563
Pesticide; *see also* Fungicide
accidents, 108
application equivalents, 517
chemicals, 101–6
compatibility chart, 552
defined, 563
distributors and manufacturers, 120–24
equipment, 109, [110], 112–23, 543–49
formulations, 102–3, 112
fruit spray schedules, 523–25
injury, [10], 45, 97–98, [267], [288], 523
from herbicides, 45, 109
label, 107
as legal document, 107
must show, 107
warning or caution statement, 107
laws, 107
measuring apparatus, 109
multipurpose mixes, 110–11, 523
notes, 523
injury from, 523
trade names, 523
percentage solution, 522
Poison Control Centers, Directory of, 108
precautions, 45, 101, 105, 106, 107, 108, 109, 112, 523, 525, 527, 538, 543
preparation of spray mixtures, 108–9, 520–21
purchasing, 523
safety, 45, 106, 108, 543
Open Door to Plenty, 107
Pesticides and Public Policy, 107
The Search for Abundance, 107
seed treatment, 97, 527–37
spraying and dusting tips, 108, 109, 523
spreaders, 118–19
stickers, 118–19
tolerance, 563
Using Phenoxy Herbicides Effectively, 45, 112
wetting agents, 118–19
Pestmaster Methyl Bromide, 106, 545
Petal-fall spray, 524
Petroselinum, 213; *see* Parsley
Petunia, 479
aster yellows, 73, 490
bacterial wilt, 69, 488
chlorosis, 24–29, 490
curly-top, 75, 486
damping-off, 76, 488
early blight, 51, 480
fasciation, leafy gall, 81, 492
flower breaking, 70, 484, 486
Fusarium wilt, 66, 486
gray-mold blight, 52, 86, 480
late blight, 51, 480
leaf blotch, 51, 480
leaf spot, 48, 480
mercury injury, 98
mosaic, 70, 484
oxidant or PAN injury, 44
powdery mildew, 56, 492
ringspot, 72, 486
root-knot, 90, 489
root rot, 87, 490
rust, 58, 492
soil drench, 490
spotted wilt, 72, 486
stem rot, crown rot, 76, 488, 490
Verticillium wilt, 68, 487
Pfizer, Charles A. & Company, Incorporated, 44, 105, 124
pH, 563
scale, 24, 563
soil, 24, 193, 415
test, 25, 325
Phacelia, 399
curly-top, 75, 399
leaf spot, 48, 399
mosaic, 70, 399
powdery mildew, 56, 399
rust, 58, 399
Phaius, 366; *see* Orchids
Phalaenopsis, 366; *see* Orchids
Phalaris, 439; *see* Ribbon grass
Phaltan, 103, 519; *see also* Folpet
50 Wettable, 103, 519
gallon lots, 522
Rose and Garden Fungicide, 103, 519
75-WP, 103, 519
Phaseolus, 155; *see* Bean, garden types; Lima bean; Scarlet runner bean; Tepary bean
Phelps Dodge Refining Corporation, 124
Phenmad, 563
Phenol, in tree wound dressings, 36, 376
Phenyl mercuric acetate (PMA), 250, 321, 385, 563
10% Phenyl Mercury Acetate, 106
Phenyl mercury lactate, 106
Phenyl (organic) mercury materials, 106, 564
algae control, 297
as disinfectants, 36, 106, 271, 276, 297
formulations, 106
as fungicides, 100, 106, 134, 140, 144, 172, 182, 206, 212, 264, 301, 308, 314, 320, 321, 322, 324, 327, 341, 343, 357, 384, 385, 386, 458, 476, 502, 508
rhizome, bulb, corm soak, 307
soil drench, 77, 106, 169, 204, 222, 307, 378, 410
trade names, 106, 378, 476
in tree wound dressings, 36, 476–77
uses, 106
Philadelphus, 304; *see* Mockorange
Philibertia, 185
powdery mildew, 56, 185
rust, 58, 185
Philodendron, 196
bacterial leaf and stem rot, 50, 81, 197–98
cane, plant, or root soak, 197, 198, 530, 536
fungus leaf spot, 48, 197
leaf spot (physiological), 198
leaf yellowing, dieback, 198
light and temperature requirements, 40
root nematode, 197
root rot, 87, 196–97
soil drench, 197
soluble salts injury, 31, 198
sooty mold, 64, 198
southern blight, 76, 196–97
stem rot, 76, 196–97
Phix, 106, 476

Phloem, 564
Phloem necrosis, 264
Phlox, 399
aster yellows, 73, 400
chlorosis, 24–29, 401, 443
crown, stem rot, 76, 400
crown gall, 83, 401
downy mildew, 55, 400
fasciation, leafy gall, 81, 381, 400
gray-mold blight, 52, 224, 401
leaf blight, [51], 400
leaf drop, blight, 400
leaf nematode, 75, 400
leaf spot, 48, 400
mosaic, 70, 400
plant soak, 400, 530, 536
powdery mildew, 56, [57], 399–400
root-knot, 90, 401
root rot, 87, 400
rust, 58, 400
soil drench, 400
sooty mold, 64, 401
southern blight, 76, 400
stem blight, canker, 77, 400
stem nematode, 75, 400
streak, 400
Verticillium wilt, 68, 400
Phoenix, 372; *see* Palms
Phoenix-tree, 401
leaf spot, 48, 401
root rot, 87, 401
twig canker, coral spot, 77, 401
web blight, 160, 401
Phomopsis, leaf spot, [434]
stem gall, [424], 425
twig blight, 77, [313], [317]
Phosphorus
deficiency, 25, 164, 235
control, 25
injury, 25
promotes, 25
Photinia, 133; *see also* Apple
anthracnose, 51, 140
fire blight, 79, 133–34
leaf blight, 51, 140
leaf spot, 48, [134], 140
powdery mildew, 56, 135–36
root rot, 87, 137–38
rust, 58, 135
scab, 65, [134], 135
twig blight, canker, 77, 138
Phragmites, rust, 439
Phygon, 102, 518, 531; *see also* Dichlone
gallon lots, 522
Naugets, 531
Seed Protectant, 102, 518, 531
-XL, 102, 518, 531
Phyla, 317
gray patch, 318
Phyllody, 564
Phyllostachys, 152; *see* Bamboos
Phyllosticta blight, 76, [456]
leaf spot, 48, [311], [346], [434]
Phymatotrichum root rot, 88–89, [208], [211], 504
list of resistant plants, 89
Physalis, 479; *see* Groundcherry, Chinese lanternplant
Physiogenic disease, 564; *see also* Noninfectious diseases
Physiologic races (strains), 13, 97, 380, 564
Physocarpus, 355; *see* Ninebark
Physostegia, 447; *see* False-dragonhead
Phytomycin, 105, 522
Phytopathology, 564
Phytophthora root rot, [315]
Phytotoxic, 564
Picea, 402; *see* Spruce
Picfume, 106, 545
Picride, 545
Pieris, 131; *see* Andromeda
Piggy-back plant, 401
light and temperature requirements, 40
powdery mildew, 56, 401
Pigweed, white-rust, 62
Pilea, 401
leaf spot, 48, 401
powdery mildew, 56, 401
root-knot, 90, 401
root rot, 87, 278, 401
Pimpernel, 423
aster yellows, 73, 424
leaf spot, 48, 423
mosaic, 70, 424
root-knot, 90, 423
Pimpinella, 213; *see* Anise
Pincushion cactus, 194; *see* Cacti, Mammillaria
Pincushion flower, 450; *see* Scabiosa
Pine, 401–2
black mildew, 64, 406
boron deficiency, 27, 408
brown felt blight, 407
canker (twig, branch, trunk), 77, 403
chlorosis, 24–29, 408
chlorotic dwarf, 44, 407
crown dieback, 77, 406–7
Cytospora canker, 77, 403
damping-off, 76, 405
dieback, 77, 406
fluorides injury, 43–44
girdling roots, 45
gray-mold blight, 52, 408
iron deficiency, 24, 26, 27, 408
insect and mite control, 406, 407
leaf blight, 51, 402
leaf browning and shedding, 408
lightning injury, 46,
littleaf disease, 406
mistletoe, dwarf, [93], 94, 406
needle blight, cast, 51, [402]
needle scorch, 41, 407
oxidant or PAN injury, 44
pruning, 32, 33
rodent protection, 42
root-knot, 90, 406
root nematode, 406
root rot, 87, 405, 406
rust
blister, 58, [404], 405
gall, fusiform, 58, [404], 405, 455, 470
needle, cone, 58, 169, 351, 359, 405, 475, 501
seed treatment, 405
seedling blight, 76, 405
shoot or tip blight, 402, 408
snow blight, 407
soil treatment, 405, 406
sooty mold, 64, [406]
stump treatment, 406
sunscorch, 41, 407
tar spot, 48, 402
twig blight, 77, 402, 403
wetwood, 264, 408
wind damage, 407
winter injury, needle scorch, 41, 407
wood rot, trunk or butt rot, 79, 170, 403
Pineapple guava, 353; *see* Feijoa
Pink root, 363, [364]
Pinks, garden, 203; *see also* Carnation, Cushion-pink, Silene, Mullein-pink
Alternaria blight, 51, 76, 204
anthracnose, 51, 205
boron deficiency, 27, 206
chlorosis, 24–29, 206, 443
curly-top, 75, 162, 205
cutting dip, 204
damping-off, 76, 204
Fusarium wilt, 66, 203–4
gray-mold blight, 52, 86, 205
leaf spot, 48, 205
leaf and stem nematode, 75, 206
mosaic, 70, 204
ringspot, 72, 205
root-knot, 90, 203, 204, 206
root nematode, 206
root rot, 87, 204
rust, 58, 204
southern blight, 76, 203, 204
stem rot, blight, 76, 204
Pinus, 402; *see* Pine
Pinxterbloom, 433; *see* Azalea
Piqueria, 220
aster yellows, 73, 222
damping-off, 76, 221
fasciation, 81, 226
powdery mildew, 56, 221
root-knot, 90, 225
root nematode, 226–27
seed treatment, 221
stem rot, 76, 221
Pistache, 468; *see* Pistachio
Pistachio (*Pistacia*), 468
leaf spot, 48, 469
root-knot, 90, 469
root nematode, 469
root rot, 87, 137, 469
thread blight, 469, 505
Verticillium wilt, 68, 468–69
wood rot, 79, 170, 469
Pisum, 377; *see* Pea
Pitanga cherry, 268; *see* Eugenia, Eucalyptus
Pitcherplant, 408; *see also* California-pitcherplant
leaf spot, 48, 408
root rot, 87, 408
southern blight, 76, 408
Pittosporum, 408
chlorosis, iron deficiency, 24–29, 408
foot rot, 76, 408
leaf spot, 48, 408
light and temperature requirements, 40
mosaic, 70, 408
root-knot, 90, 392, 408–9
soil treatment, 408
southern blight, stem rot, 76, 408
thread blight, 409, 505
Verticillium wilt, 68, 343, 408
Pittsburgh Plate Glass Company, Corona Chemical Division, 102, 103, 124, 518, 519
Pityrogramma, 269; *see* Ferns
Planetree, 476; *see also* London plane, Sycamore
anthracnose, 51, 476
blight, cankerstain, 77, 476
canker, (twig, branch), dieback, 77, 476
mistletoe, 93, 478
powdery mildew, 56, 477
twig blight, 77, 476
Verticillium wilt, 68, 343, 477
wood rot, 79, 170, 477
Plant bugs, as virus carriers, 164
control, 505
Plant disease control, 96–124; *see also* Control methods
Plant Diseases, Index of in the United States, 3
Plant-food utilization chart, 26
Plant Hardiness Zone Map, 43
Plant hoppers, as virus carriers, 13
Plant pathologist, extension or state, help by, 5, 6, 25, 44, 89, 106, 107, 126, 135, 138, 142, 150, 151, 154, 178, 182, 199, 228, 234, 271, 287, 290, 291, 292, 314, 319, 326, 356, 370, 373, 382, 386, 387, 389, 391, 415, 416, 418, 421, 430, 484, 489, 504, 508, 523
Plant Quarantine Division, 99
Plant Quarantines, 99
Plant societies, help by, 9, 581
Plant tissue test, 25, 290
Plantainlily, 304; *see* Hosta
Plantcote, 21, 38, 41, 42, 182, 296, 511
Planting, 19–21, 98
adapted varieties and types, 97
distances, number of plants required, 527
rose, [21]
shock, 21
stock, in transmitting viruses, 13
Platanus, 476; *see* Sycamore, Planetree, London plane
Platycerium, 269; *see* Ferns
Platycodon, 168; *see* Balloonflower
Plum, 382
asteroid spot, 390
bacterial leaf spot, black spot, 50, 386

bacterial shoot blight, gummosis, 50, 79, 386
black knot, [81], 384
blossom blight, 86, 382
brown rot, 77, 86, [87], 382, 525
chlorosis, 24–29, 393
crown gall, 83, 391
decline (virus complex), 70, 389
fire blight, 79, 393
fluorides injury, 43–44
fruit spot, rot, 86, 382, 384, 385, 392
leaf blotch, 51, 392
leaf curl, 62, 384
leaf spot, shot-hole, 48, 392
line pattern, 389
little-plum, 387
mistletoe, 93, 393
mosaic, 70, 389
phony peach, 389
pockets, 62, [63], 384, 525
powdery mildew, 56, 385
prune dwarf, 73, 389
resistant varieties, 383, 384, 386
ringspot complex, 72, 388, 389
root nematode, 392
root rot, 87, 392
rosette, 73, 387
rust, 58, 392
scab, 65, 385
shot-hole, 52, 386, 390, 392
sooty mold, 64, 137, 393
spray schedule, 524–25
tatterleaf, 388, 389
thread blight, 393, 505
twig blight, 77, 382
Verticillium wilt, 68, 391
witches'-broom, 62, 384
wood rot, 79, 170, 384
X-disease, 387
yellows, 73, 388
zinc deficiency, 26, 27, 393

Plum pockets, 62, [63], 384, 525
Plum-yew, 311; *see* Japanese plum-yew
Plume hyacinth; *see* Virginia-creeper, Grape-hyacinth
Plumed thistle, 220; *see also* Thistle
inflorescence smut, 62, 206, 227
powdery mildew, 56, 221
rust, 58, 223
white-rust, 62, 226
Plumeria, 360; *see* Frangipani
Plumy coconut, 371; *see* Palms
Plunger-type duster, 115, [120]
Plyac Spreader-Sticker, 119
PMA; *see* Phenyl mercuric acetate
PMAS, 106, 378
Poa, 319; *see* Bluegrass
Pocketbook plant, 454; *see* Calceolaria
Pod drop, 41, 160
Pod spot or rot, [157], [381]
Podocarpus, 511; *see also* Yew
root nematode, 511–12
root rot, 87, 511–12
Podophyllum, 348; *see* Mayapple
Podranea, 212
root-knot, 90, 213, 392
Poinciana, 300; *see also* Caesalpinia
anthracnose, 51, 301
canker, dieback, 77, 301
crown gall, 83, 136, 301
root rot, 87, 137, 301
rust, 58, 301
Poinsettia, 409
bacterial canker, leaf spot, 50, 79, 410
bacterial soft rot, 81, 409
blossom blight, 52, 86, 410
chlorosis, 24–29, 410
crown gall, 83, 410
curly leaf, 410
cutting rot, 76, 409
cutting treatment, 409
gray-mold (*Botrytis*) tip blight, stem canker, 52, 77, 86, 410
leaf spot, 48, 410
light, effect on flowering, 38
light and temperature requirements, 40
root-knot, 90, 410
root nematode, 409
root rot, 87, [409]
rust, 58, 410
soil treatment, 409, 410
spot anthracnose or scab, 51, 65, 410
stem or foot rot, 76, 409
stunt, 410
wilt, 409
Point Reyes lilac or creeper, 355; *see* New Jersey–tea
Poison Control Centers, Directory of, 108
Poker-plant, 432; *see* Redhot-poker-plant
Polemonium, 399
Fusarium wilt, 66, 400
leaf spot, 48, 400
powdery mildew, 56, 399–400
rust, 58, 400
Verticillium wilt, 68, 400
Polianthes, 250; *see* Tuberose
Pollen, as virus carrier, 13, 72, 155, 265, 388, 390, 420
Polyanthus, 423; *see* Primrose
Polybor, 27, 141, 297
Polycide, 322
Polygala, 399
anthracnose, 51, 400
chlorosis, 24–29, 401, 443
leaf spot, 48, 400
rust, 58, 400
Polygonum, 453; *see also* Silver-lacevine
viviparum, rust, 256
Polypodium, 269; *see* Ferns
Polyram (metiram)
active ingredient, 104
as protective fungicide, 55, 62, 86, 100, 104, 134, 146, 158, 162, 190, 243, 314, 329, 384, 415, 440, 441, 480, 501, 502, 505
seed treatment, 104, 418
soil application, 104
uses, 104
Polystichum, 270; *see* Ferns
Polytrap, 106, 564
Pomegranate, 411
anthracnose, 51, 411
fruit spot, rot, 52, 86, 138, 411
gray-mold rot, 52, 86, 411
leaf blotch, 51, 411
root-knot, 90, 411
root rot, 87, 137, 411
spot anthracnose, 51, 411
thread blight, 411, 505
Pomegranate melon, 237; *see* Cucurbit, Muskmelon
Poncirus, 227; *see* Hardy orange, Citrus
Pond-spice, 149
leaf spot, 48, 149
Pondlily, 506
leaf spot, 48, 506
white smut, 65, 506
Poormans-orchid, 479; *see* Butterfly-flower
Popcorn, 231; *see* Corn
Poplar, 411
bacterial limb gall, 83, 136, 412
branch gall, 412
canker (twig, branch, trunk), dieback, 77, [411]
catkin deformity, 412
chlorosis, iron or magnesium deficiency, 24–29, 344, 413
crown gall, 83, 136, 412
Cytospora canker, 77, [411], 412
damping-off, 76, 405, 413
Dothichiza canker, 77, 411, 412
frost crack, 41
ink spot, 48, 412
leaf blister, yellow, 62, 412
leaf blotch, blight, 51, 412
leaf scorch, 41, 344, 413
leaf spot, 48, 412
lightning injury, 46
mistletoe, 93, 413
powdery mildew, 56, 412
resistant poplars, 412
root nematode, 413
root rot, 87, 137, 413
rust, 58, 318, 405, 412
seed rot, 405, 413
shoot blight, scab, 65, 412
sooty mold, 64, 413
spring leaf fall, 412
Verticillium wilt, 68, 343, 413
wetwood, slime flux, 264, 413
wood or butt rot, 79, 170, 412
Poppy, 413
bacterial blight, leaf spot, 50, [413]
black ringspot, 192, 414
curly-top, 75, 414
damping-off, 76, 414
downy mildew, 55, 414
gray-mold blight, 52, 414
leaf nematode, 75, 414
leaf smut, 62, 414
leaf spot, 48, 414
powdery mildew, 56, 414
root-knot, 90, 414
root nematode, 414
root and stem rot, 76, 87, 414
seedpod spot, 414
soil drench, 414
spotted wilt, 72, 414
Verticillium wilt, 68, 414
Poppy-mallow, 297
leaf spot, 48, 298
root rot, 87, 299
rust, 58, 298
Verticillium wilt, 68, 298
Populus, 411; *see* Poplar, Cottonwood
Porcelain berry, 286; *see* Ampelopsis
Port Orford cedar, 312; *see* Chamaecyparis
Portulaca, 445; *see* Rose-moss
Possumhaw; *see* Holly, Viburnum
Pot marigold, 220; *see* Calendula
Potash, 210, 248, 277
Potassium
deficiency, 26, 235, 248, 290, 467, 490
control, 26, 248, 393
permanganate, as disinfectant, 276
promotes, 26
Potato (white or Irish), 415
aster yellows, purple-top, 73, 418
bacterial soft rot, 81, [82], 417
bacterial wilt, brown rot, 69, 419
black dot disease, anthracnose, 421
black scurf (*Rhizoctonia*), 77, 86, [419]
black walnut injury, 422
blackheart, [417], 420
blackleg, 81, [82], 418
calico, 70, 418
canker, 77, [419]
corky ringspot, 418
crinkle, 70, 418
crown gall, 83, 422
curly-top, green dwarf, 75, 418
early blight, target spot, 51, [415]
fertilizing, 29
Fusarium wilt, dry rot, rusty dieback 61, [417], 418
golden nematode, 420
Golden Nematode of Potatoes and Tomatoes — How To Prevent Its Spread, 421
gray-mold (*Botrytis*) blight, 52, 421
hollow heart, [417], 420
hopperburn, 421
insect and mite control, 418, 421
knobbiness, malformed tubers, 422
late blight, 51, [416], [417]
leaf blotch, 51, 422
leaf scorch, 24–29, 422
leaf spot, 48, 422
leafroll, net necrosis, [417], 418
mosaic, 70, 418
mottle, 70, 418
ozone injury, 44
plant-food utilization, 26
Potato Diseases and Their Control, 422
powdery mildew, 56, 422
psyllid yellows, [421]
resistant varieties, 415, 416, 418, 419, 420, 422
Rhizoctonia disease, 77, 86, [419]
ring rot, bacterial, 50, 86, [417], 420
ringspot, 72, 418
root-knot, 90, 420
root rot, 87, 422

Potato (*continued*)
root and rot nematodes, 92, 420
scab, common, 65 [66], [415]
powdery, or canker, 86, 421
sclerotiniose, 76, 156, 422
scurf, 86, [419]
seed-piece decay, 417
seed-piece treatment, 418, 421, 531, 532
silver scurf, 421
soil treatment, 415, 420
southern blight, 76, 422
spindle tuber, 418, 420
stem or stalk rot, canker, 76, 77, 419, 422
storage, 415, 417, 420
streak, 70, 418
sunburn, 422
tuber rot, 86, [416], [417], 422
vein-banding, 418
Verticillium wilt, "pink eye," 68, 418–19
web blight, 160, 422
witches'-broom, 418
yellow dwarf, 73, 418
yellow spot, 418
Potato specialist, help by, 416, 418, 419, 420, 421
Potato vine, 479; *see* Eggplant, Jerusalem-cherry
Potentilla, 440
crown gall, 83, 441
downy mildew, 55, 444
fire blight, 79, 444
leaf blight, 51, 443
leaf spot, 48, 443
powdery mildew, 56, 440
rust, 58, 441
Pothos, 196
leaf spot, 48, 197
light and temperature requirements, 40
root nematode, 197
root rot, 87, 196–97
soluble salts injury, 31
Potting fiber mold, 369
Powder-puff, 198; *see* Calliandra
Powdery mildew, 56, [57], 58, 100, 118, [128], [136], [167], 173, [176], [221], [242], [305], [320], [378], [477], 525, 564
Power duster, 117, [123]
sprayer, 114, [118], [119]
Practices, 9, 19–48, 96–100; *see also* Cultural practices
Prairie-coneflower, 220
downy mildew, 55, 225
leaf spot, 48, 221
powdery mildew, 56, 221
root rot, 87, 221
rust, 58, 223
white smut, 65, 226
Prairie lily, 348; *see* Mentzelia
Prairie rocket, 186; *see* Erysimum and Wallflower, western
Prairie smoke, 440; *see* Avens
Prairiegentian, 278
leaf spot, 48, 278
root rot, 87, 278
stem blight, 76, 278
Prayer plant, 347; *see* Maranta
Prebloom spray, 524
Precautions, when handling pesticides, 45, 101, 105, 106, 107, 108, 109, 112, 527, 538, 543, 549
spraying fruit, 523, 525
treating seed, 527, 528
treating soil, 538, 543–49
Pressure cooker, for treating soil, 539
Pretty-face, 183; *see* Brodiaea
Prickly-ash, 302
canker (twig, stem), dieback, 77, 302
leaf spot, 48, 302
mistletoe, 93, 302
powdery mildew, 56, 302, 303
rust, 58, 302
sooty blotch, 64, 137, 302
tar spot, 48, 302
wood rot, 79, 170, 302
Prickly phlox, 399; *see* Gilia
Prickly-poppy, 413
downy mildew, 55, 414
leaf spot, 48, 414
root rot, 87, 414
rust, 58, 414
soil drench, 414
Pricklypear, 194; *see* Cacti, Opuntia
Pride-of-California, 377; *see* Sweetpea
Primary infection, 564
Primrose (*Primula*), 423
anthracnose, 51, 423
aster yellows, 73, 424
bacterial leaf spot, 50, [423]
blackspot, 48, 423
chlorosis, 24–29, 424
damping-off, 76, 423
downy mildew, 55, 423
flower blight, 86, 423
gray-mold (*Botrytis*) blight, 52, 86, 423
leaf blight, 51, 423
leaf spot, 48, 423
leaf and stem nematode, 75, 424
light and temperature requirements, 40
mosaic, 70, 424
powdery mildew, 56, 221, 424
root-knot, 90, 423
root nematode, 423
root rot, 87, 423
rust, 58, 424
soil drench, 423
spotted wilt, 72, 424
stem rot, 76, 423
Princesfeather, 230; *see* Amaranth
Princesstree, 376; *see* Paulownia
Privet, 424
algal leaf spot, 341, 425
anthracnose, 51, 424
canker, dieback, 77, 424
chlorosis, 24–29, 344, 425
chlorotic spot, 425
crown gall, 83, 425
leaf blight, 51, 425
leaf nematode, 75, 425
leaf scorch, 41
leaf spot, 48, 425
mosaic, 70, 425
powdery mildew, 56, 425
ringspot, 72, 425
root-knot, 90, 425
root nematode, 425
root rot, 87, 425
soil drench, 425
sooty mold, 64, 425
stem gall (*Phomopsis*), [424], 425
thread blight, 425, 505
twig blight, 77, 424
variegation, 425
Verticillium wilt, 68, 343, 425
winter injury, 41, 425
witches'-broom, 334, 425
wood or collar rot, 79, 425
Proboscisflower (*Proboscidea*), 425
bacterial leaf spot, 50, 425
fungus leaf spot, 48, 425
mosaic, 70, 426
root rot, 87, 426
southern blight, 76, 426
stem (crown) rot, 76, 254, 426
Profume, 545
Propagation, in transmitting viruses, 13, 70, 73
Protectant, 564
Protective seed treatment, 527–28, 531–37
Prune, 382; *see* Plum
Prunella, 447
leaf spot, 48, 447
powdery mildew, 56, 448
root rot, 87, 447
southern blight, 76, 447
tar spot, 48, 447
Pruning, [32], [33], [34], 564
cuts, entrance for bacteria and fungi, 79, 80
evergreens, 33
flowers, 33
fruit trees, 32, 383, 523
hedges, 32
potted plants, 33
Pruning Shade Trees and Repairing Their Injuries, 36
in relation to buds, [32]
shrubs, 32
at transplanting, 32
trees, 32–33, [34]
correct procedure, 33
Prunus, 382; *see* Almond, Apricot, Cherry, Cherry-laurel, Flowering Almond, Flowering Cherry, Nectarine, Peach, Plum
rust, 58, 256, 392
Pseudobulb, 564
Pseudolarix, 318
canker, 77, 319
Pseudotsuga, 402; *see* Douglas-fir
Psidium, 353; *see* Feijoa, Guava
Psyllid yellows, [421], 490–91
Ptelea, 302; *see* Hoptree
Pteretis, 270; *see* Ferns
Pteridium, 270; *see* Ferns
Pteris, 270; *see* Ferns
Puccoon, 349; *see* Lithospermum
Puffballs, 322
Pummelo, 227; *see* Citrus, Grapefruit
Pumpkin, 238; *see also* Cucurbit
angular leaf spot, 50, 239
anthracnose, 51, 238
aster yellows, 73, 243
bacterial soft rot, 81, 86, 238, 244
bacterial spot, 50, 86, [240], 245
bacterial wilt, 69, 240
black rot, 86, 244, 245
blossom blight, 86, 245
chlorosis, 24–29, 245
curly-top, 75, 243
damping-off, seed rot, 76, 244
downy mildew, 55, 243
fruit spot, rot, 81, 86, 241, 244
Fusarium wilt, 66, 241
gray-mold rot, 52, 86, 244, 245
Growing Pumpkins and Squashes, 245
gummy stem blight, 76, 86, 244
leaf blight, 51, 239
leaf spot, 48, 238, 239, 245
mosaic, 70, 241
potassium deficiency, 26, 245
powdery mildew, 56, 242
resistant varieties, 243
ringspot, 72, 243
root-knot, cyst nematode, 90, 244–45
root nematode, 244
root rot, 87, 156, 244
scab, 65, 238–39
seed treatment, 238, 528, 532
seedbed spray, 244
southern blight, 76, 243
stem rot, 76, 243
storage rot, 86, 244
Verticillium wilt, 68, 245
Punica, 411; *see* Pomegranate
Puratized
Agricultural Spray, 106, 338, 470, 476
Apple Spray, 106
root or bulb dip, 250
Pure Food and Drug Administration (FDA), 101, 563
regulations, 107
Purex, as disinfectant, 35, 99
Purple-coneflower, 220; *see also* Echinacea
mosaic, 70, 222
Purple daisy, 220; *see* Echinacea
Purple-flowered groundcherry, 479
leaf spot, 48, 480
powdery mildew, 56, 492
rust, 58, 492
Purple loosestrife, 340; *see* Lythrum
Purple osier, 507–8; *see* Willow
Purple rockcress, 186; *see* Cabbage, Rockcress
Purple sheath spot, 234
Purple smokebush, 468; *see* Sumac, and Smoketree for Purple smoketree
Purpleleaf bush, 382; *see* Peach, Plum

Purpleleaf plum, 382; *see* Plum
Purpleleaf spiderwort, 437; *see* Rhoea
Puschkinia, 495; *see* Tulip
Pussytoes, 219; *see* Everlasting
Pustule, 564
Pycnanthemum, 447; *see* Mountain-mint
Pycnidium, 564
Pyracantha, 133; *see also* Apple
  canker, dieback, 77, 138–39
  chlorosis, 24–29, 143
  felt fungus, 138, 293
  fire blight, 79, 133–34
  flat limb, 142
  fruit spot, rot, 86, 138
  leaf blight, 51, 140
  pruning, 32
  root rot, 87, 137–38
  scab, 65, 134–35
  thread blight, silky, 143, 505
  twig blight, 77, 138–39
Pyrethrum, 219; *see also* Chrysanthemum
  aster yellows, 73, 222
  damping-off, 76, 221
  fasciation, 81, 226
  gray-mold blight, 52, 86, 224
  leaf blight, 51, 221
  leaf nematode, 75, 226
  petal blight, 52, 86, 221, 224
  powdery mildew, 56, 221
  root-knot, 90, 225
  root nematode, 226–27
  root rot, 87, 221
  rosette, 73, 222
  seed treatment, 221
  stem rot, 76, 221
Pyrola, rust, 58, 405
Pyrostegia, 169; *see* Bignonia
Pyrus, 133; *see* Pear
Pythium, 91, [324]

Q

Quaker bonnets, 377; *see* Lupine
Quamoclit, 351; *see also* Cypressvine
  rust, 58, 351
Quarantine laws and regulations, 99, 154, 564
  for importing plants, permit, 99
  inspectors, 99
  Plant Quarantine Division, address, 99
  prohibited plant materials, 99
  savings, 99
  special provisions, 99
Queen-of-the-meadow, 439; *see* Meadowsweet
Queen-of-the-prairie, 439; *see* Meadowsweet
Queens-delight, 210
  leaf spot, 48, 210
  root rot, 87, 137, 156, 210–11
  rust, 58, 212
Queen's-flower, 236; *see* Crapemyrtle
Quercus, 355; *see* Oak
Quince, 133; *see also* Apple, Flowering quince
  anthracnose, 51, 140
  black rot, 51, 77, 86, 136
  Blight of Pears, Apples, and Quinces, 579
  blossom blight, 79, 86, 133–34
  boron deficiency, 27, 141
  Botrytis blight, rot, 52, 86, 138
  brown rot, 77, 86, 138
  canker, dieback, 77, 138
  crown gall, hairy root, 83, 136–37
  fire blight, 79, 133–34, 525
  fruit spot, rot, 86, 138
  gray-mold rot, 52, 86, 138
  leaf blight, 51, 140
  leaf scorch, 41, 140
  leaf spot, 48, 140
  mosaic, 70, 142
  powdery mildew, 56, 135–36
  root-knot, 90, 141
  root nematode, 141
  root rot, 87, 137–38
  rust, 58, 135
  scab, 65, 134–35
  silver leaf, 79, 139
  spot anthracnose, 51, 86, 140
  spray schedule, 524–25
  stem- or wood-pitting, 142
  thread blight, 143, 505
  twig blight, 77, 138–39
  Verticillium wilt, 68, 143
  wood rot, 79, 139, 170
Quincula, 479; *see* Purple-flowered groundcherry
Quinine bush, 539
  rust, 58, 441
Quiverleaf, 411; *see* Poplar

R

Rabbit injury, control, 42
  use of Arasan 42-S, 42
  fence, 42
  repellent, 42
Rabbit tracks, 347; *see* Maranta
Race or Strain, 564; *see also* Physiologic race
Radish, 186; *see also* Cabbage
  aster yellows, 73, 193
  bacterial soft rot, 81, 190
  bacterial spot, black rot, 50, 190
  black root, 87, 193
  boron deficiency, 27, 191
  clubroot, 89, 189
  crown gall, 83, 146, 193
  crown (foot), stem rot, 76, 191
  curly-top, 75, 192
  damping-off, seedling blight, 76, 188, 191, 193
  downy mildew, 55, 189
  Fusarium wilt, yellows, 66, 186–87
  leaf blotch, 51, 190
  leaf spot, 48, 190
  leaf and stem nematode, 75, 194
  mosaic, 70, 192
  powdery mildew, 56, 193
  premature flowering, 41
  resistant varieties, 187, 189, 190, 191, 193
  Rhizoctonia disease, [188]
  ringspot, 72, 192
  root-knot, 90, 191
  root rot, 87, 188, 190–91, 193
  rust, 58, 193
  scab, 65, 165, [193], 415
  seed rot, 188
  seed treatment, 187, 188, 529, 532
  Verticillium wilt, 68, 193
  web blight, 160, 194
  white-rust, [61], 62, 191
Ragged-robin, 203; *see* Lychnis
Ragwort, 220; *see also* Senecio
  rust, 58, 223, 405
Rainbow-bush, 445; *see* Rose-moss
Rainlily, 249
  leaf spot, 48, 253
  rust, 58, 253
Ranunculus, 254; *see* Buttercup
Rape, 185; *see also* Cabbage
  anthracnose, 51, 190
  aster yellows, 73, 193
  black rot, 50, 187
  blackleg, 77, 187
  canker, 77, 187
  clubroot, 89, 189
  damping-off, 76, 188, 190, 191, 193
  downy mildew, 55, 189
  gray-mold rot, 52, 86, 190, 191
  leaf blight, 51, 190
  leaf spot, 48, 190, 191
  mosaic, 70, 192
  powdery mildew, 56, 193
  resistant varieties, 189
  ringspot, 72, 192
  root-knot, 90, 191
  root rot, 87, 187, 190, 193
  scab, 65, 165, 193, 415
  seed treatment, 187, 529, 532
  slimy soft rot, 81, 86, 190
  watery soft rot, 76, 86, 191
  white mold, 51, 190
  white-rust, 62, 191
Raphanus, 186; *see* Radish
Ra-Pid-Gro, 30
  Corporation, 124
Raplex, 27
  -Fe, 27
  -Mn, 27
  -Zn, 27
Raspberry, 426–27; *see also* Flowering raspberry
  anthracnose, 51, [78], [427], 525
  black mildew, 64, 432
  calico, 70, 432
  cane blight, dieback, 77, [428], 525
  cane and crown gall, hairy root, 83, [84], 428
  cane spot, 430–31
  canker, dieback, 77, 428
  chlorosis, 24–29, 432
  Controlling Diseases of Raspberries and Blackberries, 432
  double blossom, 432
  downy mildew, 55, 432
  dwarf, stunt, 73, 428
  fire blight, flower blight, 79, 86, 432
  fruit spot, rot, or mold, 86, 430
  gray-mold blight, 52, 86, 430
  Growing Raspberries, 432
  insect control, 428, 430, 431, 432
  leaf curl, 70, 428, [429]
  leaf rust, 58, 405, 430
  leaf spot, 48, 427, 430–31
  mosaic, 70, 428, [429]
  mottle, 70, 428
  necrosis, 428
  orange rust, 58, 430
  powdery mildew, 56, 431
  pruning, 32–33
  resistant varieties, 427, 428, 430, 431
  ringspot, 72, 428–29
  rodent protection, 42
  root-knot, 90, 431
  root nematode, 431
  root rot, 87, 137, 247, 431
  rosette, 73, 428, 432
  soil drench, 428, 430
  spot anthracnose, 51, 427
  spray schedule, 524–25
  spur blight, 77, 427, [428]
  streak, 70, 428, [429]
  sulfur dioxide injury, 43
  sulfur-sensitive, 525
  sunscorch, 41, 432
  Verticillium wilt, blue stem, 68, [69], 430, [431]
  virus decline, 428
  winter injury, 41, 432
  yellow net, 428
  yellow rust, 58, 430
Rate or Dosage (pesticide), 564
Ratibida, 220; *see* Prairie-coneflower
Rattlesnake-master, 213; *see* Eryngium
Ray blight, [85], 86, [224]
Red campion (*Lychnis*), 203; *see also* Lychnis
  flower or anther smut, 62, 206
  leaf spot, 48, 205
  rust, 58, 204
Red-cardinal, 300; *see* Erythrina
Red dracaena, 261; *see* Cordyline
Red false-mallow, 297
Red haw, 133; *see* Hawthorn
Red-heart, 355; *see* New Jersey-tea
Red jasmine, 360
  leaf spot, 48, 360
  root rot, 87, 137, 361
Red-leaf hibiscus, 297; *see* Hibiscus (arborescent forms)
Red stele or core, [462], 463
Redbay, 149; *see also* Avocado
  black mildew, 64, 151
  dieback, 77, 149–50
  leaf spot, black leaf spot, 48, 149
  root-knot, 90, 151
  wood rot, 79, 149
Redbud, 300
  borer control, 300–301

Redbud (*continued*)
canker (twig, branch, trunk), dieback, 77, 300
crown gall, 83, 136, 301
downy mildew, 55, 302
leaf spot, 48, 301
root rot, 87, 137, 301
2,4-D injury, [10], 288, 301
Verticillium wilt, 68, 301, 343
wetwood, slime flux, 264, 302
wood rot, 79, 170, 301
Redcedar, 312; *see* Juniper
Redhot-pokerplant, 432
leaf spot, 48, 433
root-knot, 90, 433
Redrobin, 236; *see* Cranesbill
Redtop, 319; *see also* Bent, Bentgrass
anthracnose, 51, 319
brown patch, Rhizoctonia disease, 321
chlorosis, 24–29, 325
copper spot, 326
damping-off, 76, 326
fairy ring, 322
foot rot, 76, 319
Fusarium patch, pink snow mold, 322
leaf rot, 50, 319
leaf scald, 50, 319
leaf smut, 62, 325
leaf spot, 48, 319
powdery mildew, 56, 321
Pythium disease, 324
root rot, 87, 319
rust, 58, 321
Sclerotinia dollar spot or blight, 321
seed treatment, 326, 537
smut, 62, 325
soil drench, 320–21
tar spot, 48, 319
Typhula blight, snow scald, 322
Redwood, 402
bark canker, 77, 403
crown gall, 83, 408
needle blight, 51, 402
root rot, 87, 405, 406
seed treatment, 405
seedling blight, 76, 405
soil treatment, 405, 406
twig blight, 77, 402, 403
wood rot, 79, 170, 403
Reedgrass (*Calamagrostis*) rust, 439, 445
Ree-Green, 27
References, 571–81
Renanthera, 366; *see* Orchids
Renovation, 564
Re-Nu, 27, 345
Reseda, 350; *see* Mignonette
Resistant (resistance), 564
Retinospora, 312; *see* Chamaecyparis
Rex-begonia vine, 287; *see* Cissus
Rhamnus, 183; *see* Buckthorn
Rhapidophyllum, 372; *see* Palms
Rhapis, 372; *see* Palms
Rheum, 437; *see* Rhubarb
Rhexia, 254
leaf spot, 48, 254
Rhizoctonia (fungus)
disease, [188], 321, [419], 466
lethal temperature, 542
Rhizome, 564
Rhizome divisions, in transmitting viruses, 13
Rhizome rot, 76, [77], 90
treatment, 528–37
Rhizomorph, 564–65
Rhododendron, 433; *see also* Azalea
acidify soil, 435
bud blast, 436
canker, dieback, 77, [434]
Cercospora leaf spot, 48, 433, [434]
chlorosis, 24–29, 434
crown gall, 83, 437
crown or collar rot, wilt, 76, 434, 437
cutting soak, 435
damping-off, 76, 437
Exobasidium leaf spot, 48, 433, [434]
felt fungus, 293, 437
flower gall, 62, 436
flower spot, blight, 52, 86, 435, 436
Growing Azaleas and Rhododendrons, 437
insect control, 433, 435, 437
leaf scorch, angular leaf spot, 51, 433
leaf and shoot gall, 62, 436
leaf spot, 48, 433, 434
light requirements of, 39
mulch, 433
Ovulinia flower spot, petal or limp blight, 86, 435
Phomopsis leaf spot, 48, 433, [434]
Phyllosticta leaf spot, 48, 433, [434]
powdery mildew, 56, 437
root-knot, 90, 434
root nematode, 434
root rot, 87, 434
rust, 58, 405, 437
seedling blight, 76, 437
shoot blight, 51, 433
soil treatment, 435, 437
sooty mold, 64, 437
spot anthracnose, 51, 433
sunscald, leaf burn, 41, 433
thread blight, 437, 505
twig blight, 77, 434, 436
winter injury, leaf burn, 41, 433–34
witches'-broom, 62, 436
yellow spot, 62, 436
Rhodora, 433; *see* Azalea
Rhodotypos, 312; *see* Jetbead
Rhoea, 437
crown rot, 76, 127, 437
light and temperature requirements, 40
mosaic, 70, 437
root-knot, 90, 128, 437
root rot, 87, 127, 437
Rhubarb, 437
anthracnose, 51, 437
bacterial soft rot, crown rot, 81, 438
bacterial wilt, southern wilt, 69, 439
cracked stem, boron deficiency, 27, 215, 439
crown gall, 83, 439
crown (foot) rot, 76, 437
curly-top, 75, 438
damping-off, 76, 437
downy mildew, 55, 438
gray-mold blight, rot, 52, 437
insect control, 438
leaf spot, blight, 48, 51, 437
mosaic, 70, [438]
ringspot, 72, 438
root-knot, cyst nematode, 90, 437, 439
root nematode, 437
root rot, 87, 437
rust, 58, 439
southern blight, 76, 437
stalk spot or rot, 76, 437, 438
stem nematode, 75, 437
sulfur dioxide injury, [43]
Verticillium wilt, 68, 439
Rhus, 468; *see* Sumac
Ribbon Grass, 439
ergot, 439
leaf mold, 439
leaf spot, 48, 439
rust, 58, 439
Ribes, 245; *see* Currant, Flowering currant, Gooseberry
rust, 58, 246, 405
Ricinis, 210; *see* Castorbean
Ring disease, [92], 93, [252]
Ring rot, [417], 420
Ringspot, 72, [73], [255], 388, [396], 428–29, 565
nematodes, as virus carriers, 166, 178, 243, 429
seed, as virus carrier, 388
weeds, as virus reservoir, 73
Rivina, 445; *see* Rougeplant
Roberts Chemical Company, 124
Robinia, 300; *see* Locust
Robin's-plantain, 220; *see* Erigeron
Rochea, 451; *see* Crassula
Rock rose, 469; *see* Sunrose
Rockcress, 186; *see also* Cabbage
black ringspot, 72, 192
chlorosis, 24–29, 194
clubroot, 89, 189
damping-off, 76, 188
downy mildew, 55, 189
gray-mold blight, 52, 86, 190–91
leaf spot, 48, 190
mosaic, 70, 192
transmitted by nematodes, 192
root-knot, 90, 191
root nematode, 194
root rot, 87, 193
rust, 58, 193
white-rust, 62, 191
Rocket, 186; *see* Damesrocket
Rockjasmine, 423
downy mildew, 55, 423
leaf spot, 48, 423
rust, 58, 424
Rockspirea, 299; *see* Holodiscus
Rocky Mountain garland, 275; *see* Clarkia, Fuchsia
Rodent
control, 42
injury, 42
as spreaders of disease, 98
wounds, entrance for fungi, 402
Rogue (roguing), 565
Rohm & Haas Company, Agricultural & Sanitary Chemicals Department, 103, 124, 519
Rollinia, 377
dieback, 77, 345
fruit rot, 86, 377
leaf blotch, 51, 377
Romanzoffia, 399
leaf spot, 48, 399
rust, 58, 399
Root diseases, 87–93
confused with, 88
control by hot water soak, 182, 252
gall (nematode), 90, [91]
-knot nematode, 16, 90, [91], 392, [395], [396], 467, 472, 565
and crown gall, cankers, root rots, wilts, 66, 78, 91
distributed and spread by, 91
resistant plants, 92, 159, 489
rot, [77], 87, [88], 98, 137, [156], [196], [207], [208], [211], 278, [315], [358], [364], 392, [409], 434, [462], [470], [471], [473]
Root injuries, 45–46
Root-Lowell Corporation, Division of Root-Lowell Manufacturing Company, 124
Root maggots, insect control, 187
Root weevils, insect control, 463, 512
Rootone F, 565
Rootstock, 565
Rootworms, insect control, 233, 463
Rorippa, 186; *see* Watercress
Rosa, 439; *see* Rose
Rosarypea, 377; *see also* Pea
root-knot, 90, 381
Rose, 439
bacterial blight, blast, 50, 443
black mold, 444
blackspot, 48, [49], 440
blossom blight, 52, 86, 443
boron deficiency, 27, 445
bud blast, ethylene injury, 44
bud drop, 39
bud rot, 52, 443
cane blight, 77, 440
canker, dieback, 77, [78], 440
chlorosis, 24–29, 443
crown (collar) rot, 76, 444
crown or stem gall, hairy root, 83, [84], 441
cutting dip, 441, 444
dieback, 77, 440, 443
downy mildew, 55, 444
fertilizing, [31]
fire blight, 79, 133–34, 444
flower spray, 443
gray-mold (*Botrytis*) blight, 52, 86, 443
infectious chlorosis, 443
insect and mite control, 440
iron deficiency, 24, 26, 27, 443
leaf blotch, scorch, 51, 443
leaf spot, 48, 443
mercury injury, 98
mosaic, 70, [442], 443

pedicel necrosis, topple, 444
plant soak, 441, 443, 530, 536
planting, [21]
powdery mildew, 56, [57], 440
pruning, 440
resistant varieties and types, 440, 441, 443, 444
root-knot, 90, 443
root nematode, 441, 444
root rot, 87, 443
Roses for the Home, 444
rosette, 73, 443
rust, 58, [441]
soil drainage, 36
southern blight, 76, 444
spot anthracnose, 51, 440, 443
spray or dust schedule, 440, 441, 443
storage decay, 76, 443
storage treatment, 443
streak, 443
thread blight, 444, 505
2,4-D injury, 45, 288, 444
Verticillium wilt, 68, 443
winter injury, 41, 441
protection, 441, [442]
Rose-acacia, 300; *see* Locust
Rose-apple, 268; *see* Eugenia, Eucalyptus
Rose campion, 203; *see* Lychnis
Rose daphne, 253; *see* Daphne
Rose-moss, 445
curly-top, 75, 445
damping-off, 76, 162, 445
root-knot, 90, 445
seed rot, 445
white-rust, 62, 445
Rose-of-China, 297; *see* Hibiscus (arborescent forms)
Rose-of-Heaven, 203; *see* Lychnis
Rose-of-Sharon, 297, *see also* Hibiscus (arborescent forms)
bacterial spot, 50, 299
blossom blight, 52, 86, 298
gray-mold blight, 52, 86, 299
pruning, 32
rust, 58, 298
Rose tree of China, 382; *see* Peach
Rosebay, 433; *see* Rhododendron
Roselle, 297
anthracnose, 51, 298
blossom blight, 52, 86, 298
chlorosis, 24–29, 299
damping-off, seed rot, 76, 299
dieback, 77, 298
gray-mold blight, 52, 86, 298, 299
leaf spot, 48, 298
pod spot, rot, 86, 298
powdery mildew, 56, 298
root nematode, 299
root-knot, 90, 299
root rot, 87, 299
seed treatment, 299
southern blight, 76, 298
stem rot, 76, 298
Rosemallow, 297
chlorosis, 24–29, 299
crown gall, 83, 298
damping-off, seed rot, 76, 299
dieback, 77, 298
leaf spot, 48, 298
pod spot, 86, 298
powdery mildew, 56, 298
root-knot, 90, 299
root rot, 87, 299
rust, 58, 298
seed treatment, 299
Rosemary, 447
root rot, 87, 447
Roseocactus, 194; *see* Cacti
Rosette, 73, 141, 336, 387–88, 565
Rosinweed, 220; *see* Golden-aster
Rosmarinus, 447
root rot, 87, 447
Rot, 565
bacterial soft, 10, 81, [82], 83, [147], [190], [248], [483]
bacterial stem, 81, [225], 280
berry, 86, [175], [260], [465]
blossom, 52, [85], 86, [133], [199], [383], [435]
blossom-end, [10], [244]
brown rot (bacterial), 50, 69
bud, 50, 52, [54], 86, 200, [202], 253, 466–67
bulb, [77], [90], 250, [309], 495
butt, [79], [116], [161], [170]
collar, 76, 81, [258], 259
corm, 90, [283]
crown, 76, [77], 98, [309], [394], 556
cutting, 86, [279]
foot, 76, 188
fruit, 86, [87], [238], [287], [391], 525
heart, 79
inflorescence or ray, 52, 53, 85, [86], [133], [199], [224], [260], [435]
rhizome, 76, [77], 90
ring, [417], 420
root, [77], 87, [88], 98, [156], [196], [207], [208], [211], [315], [358], [364], [409], [462], [470], [471], [473]
sapwood, 79, [147], [161], [170], [385]
seed, 86, 223, 438, 527, [528]
stalk, 76, [232]
stem, 76, 81, [225]
storage, 86, [87], [147], [207], [416], [417], [470], [473]
tuber, 86, [416], [417]
wood, [79], [147], [161], [170], [385]
wound, 79
Rot-Not, 556
Rotary-fan duster, 116
Rotation; *see* Crop rotation
Rotenone, 98, 157, 483
in multipurpose mixes, 111
Rougeplant, 445
leaf spot, 48, 445
root nematode, 445
root rot, 87, 278, 445
rust, 58, 445
Round-leaf mallow, rust, 298
Row crops, rates of application of sprays, 550
Rowantree, 133; *see* Mountain-ash
Royal poinciana, 300; *see* Caesalpinia
Rubber-plant, 272; *see also* Fig
anthracnose, 51, 271
canker, dieback, 77, 138, 261, 271
crown gall, 83, 168, 272
leaf scorch, fall, [41], 271
leaf spot, 48, 271
light and temperature requirements, 40
low humidity injury, 39, [41], 271
oedema, 39, 272
root-knot, cyst nematode, 90, 271, 392
Rubus, 427; *see* Blackberry, Boysenberry, Dewberry, Flowering Raspberry, Loganberry, Raspberry, Thimbleberry
Ruby glow, 461; *see* Spirea
Rudbeckia, 220
aster yellows, 73, 222
downy mildew, 55, 225
leaf spot, 48, 221
mosaic, 70, 222
powdery mildew, 56, 221
root rot, 87, 221
rust, 58, 223
soil drench, 222
southern blight, 76, 221
stem (crown) rot, 76, 221
Verticillium wilt, 68, 223
white smut, 65, 226
yellow dwarf, 73, 222
Rue-anemone, 131
leaf spot, 48, 131
powdery mildew, 56, 132
rust, 58, 131
smut, 62, 131
Ruellia, 229
leaf spot, 48, 229
root nematode, 229
root rot, 87, 229
rust, 58, 229
Rugose, 565
Runner (plant), 565
Rushes, rust, 223
Russet, 565
Russian-olive, 445
canker (twig, branch), dieback, 77, 345, 445
chlorosis, 24–29, 344, 446
crown gall, hairy root, 83, 136, 446
leaf spot, 48, 445
mistletoe, 93, 446
powdery mildew, 56, 445
root rot, 87, 137, 446
rust, 58, 445
seedling blight, 76, 405, 446
thread blight, 445, 505
trunk canker, 77, 445
2,4-D injury, 288, 446
Verticillium wilt, 68, 343, 445
wood rot, 79, 170, 446
Rust, 13, 58, [59], 60, 62, [134], [145], [153], 169, [184], [205], [222], [298], [309], [320], [404], [431], [441], 508, 525, 565
autoecious, 58
gall, [404]
leaf, stem, needle, 58, [59], [320], [404], [441]
white-, [61], 62
Rustyleaf, 175; *see* Menziesia
Rutabaga, 185; *see also* Cabbage
anthracnose, 51, 190
aster yellows, 73, 193
bacterial leaf spot, 50, 190
bacterial soft rot, 76, 81, 190
black rot, 50, 187
blackleg, 77, 187
boron deficiency, 27, 191
chlorosis, 24–29, 194
clubroot, 89, 189
crown gall, 83, 146, 193
curly-top, 75, 192
downy mildew, 55, 189
Fusarium wilt, 66, 186–87
gray-mold rot, 52, 86, 190, 191
leaf spot, 48, 190, 191
mosaic, 70, 192
powdery mildew, 56, 193
resistant varieties, 189, 190
root-knot, 90, 191
root nematode, 194
root rot, 87, 187, 190, 191, 193
scab, 65, 165, 193, 415
seed treatment, 187, 529, 532
Verticillium wilt, 68, 193
watery soft rot, 76, 191
white-rust, 62, 191
Rye, rust, 349
Ryegrass, 319; *see also* Bluegrass
anthracnose, 51, 319
bacterial spot, 50
brown patch, Rhizoctonia disease, 321
cottony blight, Pythium disease, [324]
fairy ring, 322
leaf blight, blotch, or scald, 51, 319
leaf spot, 48, 319
red thread, pink patch, 326
root nematode, 324
root rot, 87, 319
rust, 58, 321
seed treatment, 326, 537
seedling blight, 76, 326
slime mold, 322
smut, 62, 325
Typhula blight, snow scald, 322

S

Sabal, 372; *see* Palms
Safety, when handling pesticides, 107, 108
Open Door to Plenty, 107
Pesticides and Public Policy, 107
Poison Control Centers, Directory of, 108
The Search for Abundance, 107
Safflower, 220
anthracnose, blight, 51, 221
aster yellows, 73, 223

Safflower (*continued*)
bacterial blight, 50, 225
crown (stem) rot, 76, 221
damping-off, 76, 221
Fusarium wilt, 66, 223
gray-mold blight, 52, 86, 224
leaf spot, 48, 221
mosaic, 70, 222
powdery mildew, 56, 221
resistant varieties, 223, 224
root-knot, 90, 225
root nematode, 226–27
root rot, 87, 221
rust, 58, 223
seed treatment, 221, 224, 536
stem rot, 76, 221
Verticillium wilt, 68, 223
Sage, 446; *see* Salvia
Saguaro, 194; *see also* Cacti, Cereus
bacterial necrosis, 195
crown gall, 83, 195
dry rot, 76, 194
St.-Andrews-cross, 446; *see* St.-Peters-wort
St. Augustinegrass, 319; *see also* Bluegrass
brown patch, Rhizoctonia disease, 321
chlorosis, 24–29, 325
foot rot, 76, 319
gray leaf spot, 48, [325]
leaf spot, blotch, 48, 319
root nematode, 324
rust, 58, 321
Sclerotinia dollar spot, blight, 321
seed treatment, 326, 537
seedling blight, 76, 326
slime mold, 322
St.-Johns-fire, 446; *see* Salvia
St.-Johns-wort, 446
leaf spot, 48, 446
powdery mildew, 56, 446
root-knot, 90, 446
root rot, 87, 446
stem spot, 446
St. Paul ivy, 287; *see* Boston ivy, Grape
St.-Peters-wort, 446
leaf spot, 48, 446
rust, 58, 446
Saintpaulia, 127; *see* African-violet
Sal soda, 105
Salad chervil, 213
crown rot, 76, 214
curly-top, 75, 214
downy mildew, 55, 210, 215
mosaic, 70, 214
root-knot, 90, 214
root rot, 87, 215
seed rot, damping-off, 76, 214
seed treatment, 213
streak, 214
Verticillium wilt, 68, 215
Salal, 294
black mildew, 64, 294
leaf spot, 48, 294
powdery mildew, 56, 294
spot anthracnose, 51, 294
Salinity, 31
control, 32
Salix, 508; *see* Willow
Salmonberry, 427; *see* Blackberry, Raspberry
Salpiglossis, 479; *see* Painted-tongue
Salsify, 328; *see also* Black-salsify
aster yellows, 73, 328–29
bacterial soft rot, 81, 328
curly-top, 75, 332
leaf blight, 51, 331
leaf spot, 48, 331
leaf and stem nematode, 75, 332
powdery mildew, 56, 331
root-knot, 90, 332
root nematode, 332
root rot, 87, 331
rust, 58, 330–31
scab, 65, 332
seed rot, 86, 330
seed treatment, 330, 531
southern blight, 76, 328
stem rot, 76, 328
Verticillium wilt, 68, 332
white-rust, 62, 331
Salt injury, 31–32, 44
confused with, 44
control, 32
excess soluble salts, 31
Salvia, 446
aster yellows, 73, 448
damping-off, 76, 447
downy mildew, 55, 448
leaf nematode, 75, 448
leaf spot, 48, 447
mosaic, 70, 447
powdery mildew, 56, 448
root-knot, 90, 447
root nematode, 447
root rot, 87, 278, 447
rust, 58, 447
southern blight, 76, 447
spotted wilt, 72, 448
stem rot, 76, 447
Sambucus, 457; *see* Elder
Sanchezia, 229
root rot, 87, 229
Sand-verbena, 274
downy mildew, 55, 274
leaf spot, 48, 274
rust, 58, 274
Sandmyrtle, 316
leaf gall, 62, 316
Sandwort, 203; *see also* Arenaria
powdery mildew, 56, 206
Sanguinaria, 413; *see* Bloodroot
Sanguisorba, 440; *see* Burnet
Saniclor, as disinfectant, 35, 99
Sanitation, as disease control practice; *see also* Control methods
48, 52, 54, 56, 58, 62, 64, 65, 68, 70, 72, 73, 76, 77, 78, 81, 82, 83, 87, 89, 90, 92, 93, 94, 98, 99, [442], 565
Sansevieria, 449
bacterial soft rot, 81, 448
cutting rot, 87
Fusarium leaf spot, 48, 448, [449]
light and temperature requirements, 40
plant soak, 450, 530, 536
root-knot, 90, 450
root nematode, 450
wilt, 31, 450
Santomerse-80, 118
Santoquin, 140
Sap beetle, insect control, 233
Sapindus, 459; *see* Soapberry
Sapium, 210; *see* Chinese tallowtree
Saprophyte, 565
Sapsucker punctures, entrance for fungi, 79, 403
Sapwood, 565
Sapwood rot, 79
Sarolex, 324
Sarracenia, 408; *see* Pitcherplant
Sarsaparilla, 126
leaf spot, 48, 127
powdery mildew, 56, 127
rust, 58, 127
Sassafras, 149
canker, dieback, 77, 149–50
leaf spot, 48, 149
mistletoe, 93, 151
mosaic, 70
powdery mildew, 56, 150
root rot, 87, 149
sooty blotch, 64, 151
Verticillium wilt, 68, 150
wood rot, 79, 149, 170, 259
yellows, 73, 151
Satin-flower, 275; *see also* Godetia
aster yellows, 73, 222, 275
rust, 58, 275
spotted wilt, 72, 168, 275
Satin poppy, 413; *see* Meconopsis
Savoy (virus), 164
Saw palmetto, 372; *see* Palms
Sawara-cypress, 312; *see* Chamaecyparis
Saxifrage (*Saxifraga*), 304
leaf spot, 48, 305
powdery mildew, 56, 304
rust, 58, 305, 508
Scab, 13, 65, [66], [134], [150], [157], 165, [193], [374], [381], 385, [415], [503], 565
control, 65, 165, 193, 290, 385, 415, 502
Scabiosa, 450
aster yellows, 73, 450
black ringspot, 72, 192, 450
curly-top, 75, 450
leaf spot, shot-hole, 48, 450
mosaic, 70, 450
powdery mildew, 56, 450
root rot, 87, 278, 450
southern blight, 76, 450
stem (crown) rot, 76, 450
Scaffold branches, 565
Scale insects, control, 161, 277, 293, 342, 360, 373, 406, 437
secretion of "honeydew," 64
Scarborough-lily, 250
Stagonospora leaf scorch, red spot, 48, 251
Scarlet eggplant, 479; *see* Jerusalem-cherry, Eggplant
Scarlet lychnis, 203; *see* Maltese cross
Scarlet pimpernel, 423; *see* Pimpernel
Scarlet plume, 409; *see* Poinsettia, Spurge
Scarlet runner bean, 155; *see also* Bean, garden types
anthracnose, 51, 157
bacterial blight, 50, 155
leaf spot, 48, 159–60
powdery mildew, 56, 158
root rot, 87, 156
rust, 58, 158
Scentless camomile, 220; *see* Matricaria
Schedules, for spraying home-grown fruit, 523–25
Schefflera, 450
leaf scorch, 41
leaf spot, 48, 450
light and temperature requirements, 40
root-knot, 90, 450
soluble salts injury, 31
Schinus, 468; *see* Peppertree
Schizanthus, 479; *see also* Butterfly-flower
leaf and stem nematode, 75, 400, 492
Schlumbergera, 194; *see* Cacti
Scholartree, 300; *see* Sophora
Sciadopitys, 402; *see* Umbrella-pine
Scilla, 495; *see* Squill
Scindapsus, 196; *see* Pothos
light and temperature requirements, 40
Scion, 565
Scirpus, rust, 210
Sclerocactus, 194; *see* Cacti, Echinocactus
Sclerotia, 13, 76, 89, 97, 157, 174, 198, [207], 243, 250, 252, 361, [419], 490, 505, 565
Scorch, 41, 565
Scorpionweed, 399; *see* Phacelia
Scorzonera, 328; *see* Black-salsify
Scotch broom, 183; *see* Broom
Scotch laburnum, 285; *see* Goldenchain
Screwpine, 451
burrowing nematode, 392, 451
leaf spot, 48, 451
light and temperature requirements, 40
Scurf or soil stain, 472, [473]
Scurvyweed, 186; *see also* Cabbage
white-rust, 62, 191
Scutellaria, 447; *see* Skullcap
Sea holly, 213; *see* Eryngium
Sea-lavender, 451
aster yellows, 73, 451
flower blight, 86, 451
gray-mold (*Botrytis*) blight, 52, 451
leaf spot, 48, 451
powdery mildew, 56, 451
root-knot, cyst nematode, 90, 451

root rot, 87, 278, 451
rust, 58, 451
spotted wilt, ringspot, 72, 451
Sea onion, 495; *see* Squill
Sea-pink, 451; *see* Armeria
Sea-urchin cactus, 194; *see* Cacti, Echinocactus
Seacoast Laboratories, Incorporated, 103, 124, 519
Seakale, 186; *see also* Cabbage
bacterial blight, black rot, 50, 187
clubroot, 89, 189
Fusarium wilt, yellows, 66, 186–87
mosaic, 70, 192
Sechium, 238; *see* Chayote
Secondary infection, 565
Sedges, rusts, 58, 223, 246, 327, 331, 445, 458, 499
Sedum, 451
Fusarium wilt, 66, 452
leaf blotch, 51, 452
leaf spot, 48, 452
light and temperature requirements, 40
powdery mildew, 56, 452
root-knot, 90, 452
rust, 58, 452
soil drench, 451–52
southern blight, 76, 451
stem (crown) rot, 76, 451
Seed corn
beetle, insect control, 233
maggot, insect control, 233, 418
Seed disinfectant (or disinfestant), 527–37, 565
Seed protectant, 531–37, 565
Seed rot (decay), 86, 233, 488, 527, [528], 531
smut, 62
treatment, 97, 527–37
advantages, 527
eradicative, 528–31
hot water, 529–31
mercuric chloride, 528–29
methods, 97, 527–31
precautions, 527
protective, 531–37
types of, 528
as virus carriers, 13, 72, 155, 242, 243, 265, 388, 397, 420, 485
Seedling blight, 52, 55, 76, [232], 233, 405, [489], 527, [528]
Selenicereus, 194; *see* Cacti, Cereus
Selfheal, 447; *see* Prunella
Semesan, 565
active ingredient, 565
disinfectant, 204, 276, 367
gallon lots, 522
seed, bulb, corm, rhizome, tuber treatment, 106, 165, 194, 213, 221, 238, 330, 375, 378, 405, 484, 531, 565
Seed Disinfectant, 565
soil application, 97, 106, 254, 256, 410, 500, 544
Semesan Bel S, 565
disinfectant, 420
soil drench, 307
tuber or root dip, 307, 418, 470, 531
Semesan Turf Fungicide, 565
Sempervivum, 451; *see* Houseleek, Hen-and-chickens
Seneca, 399; *see* Polygala
Senecio, 220
aster yellows, 73, 222
damping-off, seed rot, 76, 221
downy mildew, 55, 225
Fusarium wilt, 66, 223
gray-mold blight, 52, 224
leaf nematode, 75, 226
leaf spot, 48, 221
light and temperature requirements, 40
mosaic, 70, 222
powdery mildew, 56, 221
root-knot, 90, 225
root rot, 87, 221
rust, 58, 223
seed treatment, 221
soil drench, 222
spotted wilt, 72, 223
stem rot, 76, 221
Verticillium wilt, 68, 223
white-rust, 62, 226
white smut, 65, 226
Senna, 300; *see* Cassia
Sensitive plant, 377
leaf spot, 48, 379, 380
root rot, 87
rust, 58, 381
Septoria leaf spot and/or blight, [217], [283], [346], 379–80, 480
Sequestrene, 27, 28, 345
NaFe Iron Chelates, 27, 345
Na_2Cu Copper Chelate, 28
Na_2Mn Manganese Chelate, 27
Na_2Zn Zinc Chelate, 27
138 Fe, 27, 345
330 Fe, 27, 345
Sequoia, 402
bark canker, 77, 403
gray-mold blight, 52, 408
needle blight, 51, 402
root rot, 87, 405, 406
seed treatment, 405
seedling blight, 76, 405
soil treatment, 405, 406
twig blight, 77, 402, 403
winter injury, 41, 407
wood rot, 79, 170, 403
Serenoa, 372; *see* Palms
Serviceberry, 133; *see also* Amelanchier, Apple
black mildew, 64, 141
blossom blight, 79, 86, 133–34
canker, 77, 138
dieback, 77, 138
fire blight, 79, 133–34
fruit rot, 86, 138
leaf blight, 51, 140
leaf blister, 62, 143
leaf spot, 48, 140
powdery mildew, 56, 135–36
root rot, 87, 137–38
rust, 58, 135
Verticillium wilt, 68, 143
Servicetree, 133; *see* Mountain-ash
Sevin, 48, 65, 70, 73, 74, 82, 83, 87, 98, 187, 222, 232, 233, 242, 289, 307, 346, 394, 397, 400, 418, 421, 433, 440, 457, 460, 461, 483, 485, 486, 491, 492, 500, 502, 505
injury, 45, 523
in multipurpose mixes, 111, 523
Shadblow, 133; *see* Serviceberry, Amelanchier
Shadbush, 133; *see* Serviceberry, Amelanchier
Shallon, 294; *see* Salal
Shallot, 361; *see also* Onion
aphid control, 365
aster yellows, 73, 365
bacterial soft rot, 81, 90, 362
bulb nematode, 92, 364
bulb rot, 81, 90, 361, 362
chlorosis, 24–29, 366
clove, bulb, or bulblet treatment, 362, 364, 529, 533
downy mildew, 55, 363
Fusarium root rot, 87, 363
gray-mold, 52, 361
leaf spot, blotch, 48, 51, 361, 362, 363, 365
maggot control, 362
mosaic, 70, 365
neck rot, 76, 361–62
pink root, 87, 363
purple blotch, 51, 363
resistant varieties, 364, 365
root-knot, 90, 365
root rot, 87, 363
rust, 58, 365
smudge, 90, 364
smut, 62, 362
soil treatment, 362
southern blight, 76, 366
thrips control, 363
white rot, 90, 362
Shamrock, 370; *see* Oxalis
Shasta daisy (*Chrysanthemum maximum*), 219; *see also* Chrysanthemum
crown gall, 83, 226
curly-top, 75, 223
fasciation, leafy gall, 81, 226
leaf blotch, 51, 221
leaf spot, 48, 221
root-knot, 90, 225
root nematode, 226–27
root rot, 87, 221
stem rot, 76, 221
Verticillium wilt, 68, 223
Sheep-laurel, 175; *see* Mountain-laurel
Shell Chemical Company, Agricultural Chemicals Division, 124
D-D Soil Fumigant, 545, 548
SD-4741, 566
Shell-flower, 282; *see* Tigerflower
Shellac, for treating wounds, 35, 36, 161, 181, 259
Shepherdia, 445; *see* Buffaloberry
Sherwin-Williams Company, 102, 103, 124, 518, 519
Spred-Rite, 119
Shield, 369
Shoestring (virus), 178
Shooflyplant, 479; *see* Apple-of-Peru
Shoot, 566
Shoot blight (bacterial), 332, [333]
Shootingstar, 423
leaf spot, 48, 423
rust, 58, 424
Shore cowberry, 175; *see* Blueberry, Whortleberry
Short-day plants, 38
Shortia, 275
leaf spot, 48, 275
"Shot-gun" soil drench, 111, 379
Shot-hole, 52, [53], [386], [391], [427], 566
Shrimp-plant, 229; *see* Chuperosa
light and temperature requirements, 40
Shrub, 566
Shrub-althea, 297; *see* Rose-of-Sharon
Shrubs
drought injury, 38
fertilizing, 30, [31]
multipurpose mixes, 111
pruning, [32], [33]
staking, 36
watering, 37–38
winter injury, [42]
Sicana, 238; *see also* Cucurbit
anthracnose, 51, 238
seed treatment, 238
Sickle thorn, 146; *see* Asparagus, garden and Asparagus-fern
Sida, 297
leaf spot, 48, 298
mosaic, 70, 299
root-knot, 90, 299
root rot, 87, 299
rust, 56, 298
southern blight, 76, 298
Sidalcea, 297
leaf spot, 48, 298
mosaic, 70, 299
root-knot, 90, 299
rust, 58, 298
southern blight, 76, 298
Side dressing, 29, 566
Sign, of disease, 566
Silene, 203; *see also* Carnation
chlorosis, 24–29, 206, 443
damping-off, 76, 204
downy mildew, 55, 206
flower or anther smut, 62, 206
leaf spot, 48, 205
root-knot, 90, 206
root rot, 87, 204
rust, 58, 204
Silk-oak, 453
dieback, gum disease, 77, 453
leaf spot, 48, 453
light and temperature requirements 40

Silk-oak (*continued*)
root-knot, 90, 453
root rot, 87, 453
Silkgrass, 512; *see* Yucca
Silktassel-bush, 258
black mildew, 64, 261
leaf spot, 48, 259
root rot, 87, 137, 259
Silktree, 300; *see* "Mimosa" tree
Silky sophora, 300; *see* Sophora
Silphium, 220
downy mildew, 55, 225
leaf spot, 48, 221
powdery mildew, 56, 221
root rot, 87, 221
rust, 58, 223, 405
white smut, 65, 226
Silver king, 220; *see* Artemisia
Silver-lacevine, 453
leaf spot, 48, 453
rust, 58, 453
smut, 62, 453
tar spot, 48, 453
Silver threads, 453
bacterial soft rot, 81, 453
leaf spot, 48, 453
leaf and stem blight, 51, 77, 453
plant soak, 453, 530, 536
root-knot, 90, 453
root and stem rot, 76, 87, 453
Silverbell, 453; *see* Halesia
Silverberry, 445; *see also* Russian-olive
canker (twig, branch), dieback, 77, 345, 445
chlorosis, 24–29, 344, 446
leaf spot, 48, 445
powdery mildew, 56, 445
rust, 58, 445
Silverrod, 220; *see* Goldenrod
Silvex, injury, 45, 261
Sinningia, 127; *see* Gloxinia
Sisalkraft paper, for wrapping trees, [42], 140, 162, 264
Sisyrinchium, 307; *see also* Blue-eyed grass
rust, 186
Skullcap, 447
leaf spot, 48, 447
powdery mildew, 56, 448
root rot, 87, 447
stem rot (*Botrytis*), 52, 76, 448
Skunkbush, 468; *see* Sumac
Sky-flower, 229; *see also* Clockvine
crown gall, 83, 229
root-knot, 90, 229
Skyrocket, 399; *see* Gilia
Slide pump sprayers, 113, [115]
Slime flux (wetwood), 264, [265], 566
Slime mold, [322], 468, 475, 566
Slip, 566
Slipperwort, 454; *see* Calceolaria
Slugs
as fungus carriers, 158
lethal temperature, 542
as virus carriers, 13
Slurry seed treatment, 527, 566
SMDC
active ingredient, 545
applying, 545, 547
disinfesting soil, 106, 188, 189, 193, 214, 261, 271, 290, 356, 405, 425, 428, 430, 437, 443, 463, 472, 490, 492, 512, 543, 547
precautions, 543, 545
soil drench, 189
trade names, 106, 545
uses, 545
"Smilax" of florists, 146; *see also* Asparagus
Fusarium wilt, 66, 146
Smith, D. B., & Company, Incorporated, 124
"Smog," as air pollutant, 44
control, 44
resistant plants, 44
sensitive plants, 44
symptoms, 44
Smokebush, 468; *see* Sumac
Smoketree, 468
canker, dieback, 77, 138, 266, 469
leaf spot, 48, 469
mistletoe, 93, 469
root nematode, 469
root rot, 87, 137, 469
rust, 58, 469
Verticillium wilt, 68, 468
"Smoulder," [252]
Smudge, [364], 365
Smut, 13, 62, [63], 64, [131], 206, 232, 566
anther, 62, 206
leaf, 62, [63], 362
physiologic races, 97
seed, 62
stem, 62
white, [65]
Snails, as virus carriers, 13
Snake lily, 183; *see* Brodiaea
Snake melon, 237; *see* Cucurbit, Muskmelon
Snake plant, 448; *see* Sansevieria
Snakeroot; *see* Aristolochia for Virginia snakeroot, Eupatorium for White snakeroot, Liatris for Button snakeroot, Cimicifuga for Black-snakeroot
Snapdragon, 454
anthracnose, 51, 455
canker, 77, 455
chlorosis, 24–29, 443, 457
crown gall, 83, 457
crown rot, 76, 455
crown and stem canker, 77, 455
damping-off, 76, 455
downy mildew, 55, 456
ethylene injury, 44
flower blight, 86, 221, 455
Fusarium wilt, 66, 455
gray-mold (*Botrytis*) blight, 52, 86, [454], 455
insect control, 456, 457
leaf blight, 51, 455
leaf spot, 48, 455
mercury injury, 98
mosaic, 70, 456
Phyllosticta leaf blight, stem rot, 51, 76, 455, [456]
powdery mildew, 56, 455
ringspot, 72, 456
root-knot, 90, 457
root nematode, 457
root rot, 87, 455
rust, 58, [59], 455
seed treatment, 455, 536
soil drench, 455
southern blight, 76, 455
spotted wilt, 72, 456
stem (collar) rot, wilt, 76, 455, [456]
tip blight, 51, 457
Verticillium wilt, 68, 455, [456]
Snapweed, light and temperature requirements, 40
Sneezeweed, 220
aster yellows, 73, 222
leaf smut, 62, 226
leaf spot, 48, 221
powdery mildew, 56, 221
root rot, 87, 221
rust, 58, 223
Verticillium wilt, 68, 223
Sneezewort, 219; *see* Yarrow, Achillea
Snow blight, [407]
Snow injury, 45
Snow mold, 322, [323]
potting fiber mold, 369
Snow-on-the-mountain, 409; see Spurge
Snowball, 499; *see also* Viburnum
bacterial leaf spot, 50, 500
canker, dieback, 77, 500
crown gall, 83, 136, 500
downy mildew, 55, 500
fungus leaf spot, 48, 500
gray-mold blight, 52, 77, 86, 500
powdery mildew, 56, 499
root-knot, 90, 391, 500
root rot, 87, 137, 500
spot anthracnose, 51, 500
thread blight, 500, 505
Verticillium wilt, 68, 343, 500
Snowbell, 453
chlorosis, 24–29
leaf spot, 48, 453
root-knot, 90, 392, 453
Snowberry, 457
anthracnose, 51, 458
berry rot, 86, 458
chlorosis, 24–29, 344, 459
collar rot, 76, 458
crown gall, hairy root, 83, 458
flower spot, 52, 86
gray-mold blight, 52, 86, 458
leaf spot, 48, 458
powdery mildew, 56, 458
pruning, 32, 458
root nematode, 458
root rot, 87, 458
rust, 58, 405, 458
spot anthracnose, scab, 51, 86, 458
stem gall, 458
twig canker, 77, 458
Snowdrop, 249; *see also* Daffodil
Botrytis blight, 52, 251
bulb rot, 90, 250
bulb soak, 250, 530, 531, 535, 537
smoulder, neck rot, 90, 252
stem and bulb nematode, 75, 92, 252
Snowdrop-tree, 453; *see* Halesia
Snowdrop windflower, 131; *see* Anemone
Snowflake, 250
bulb rot, 90, 250
bulb soak, 528, 529, 536
root nematode, 250, 253
Stagonospora leaf scorch, red blotch, 51, 251
Soapberry, 459
canker, dieback, 77, 459
leaf blight, 51, 459
leaf spot, 48, 459
mistletoe, 93, 459
mosaic, 70, 459
powdery mildew, 56, 459
root rot, 87, 137, 459
thread blight, 459, 505
Soapweed, 512; *see* Yucca
Sod webworm, insect control, 321
Sodium
carbonate, 105
dinitro-o-cresol, 556
molybdate, 28
N-methyldithiocarbamate, 545
o-hydroxyphenyl (2-hydroxydiphenyl), 366–67
orthophenylphenate tetrahydrate, 471
penta borate, 141
Soft rot (bacterial), 10, 81, [82], 83, [147], [190], [248], [483]
Soft rot (*Rhizopus*), [471]
Soil, 21
acid, 24, 553
acidifying, 24, 325, 415, 435, 443
alkaline, 24, 553
deficiencies, 25–29
drainage, 36–37
drench, 77, 97, 102, 111, 360
fertility, as affects disease development, 96
fill, preventing injury, 46, [47]
fumigants, 106, 538–49
fungicides, 106
grade change, [46], [47]
insecticide, 23, 82, 86, 90, 98, 166, 178
liming, 24
loosening, 23
mixes, 22–23
for cacti, 22
Cornell University Peat-lite Mix A, 22
John Innes Composts, 22–23
UC Soil Mix C, 22
pasteurization, 538–49
pH (reaction), 24, 193, 415, 563
rot, [157], [473]
sterilization, 538–49
surface, loose, 24

test, 25
treatments, 538–49
as virus carrier, 13
waterlogged, 36–37
Soil Fumigant M, 106, 545
Soilfume 60–40, 544
Soilfume 80–20, 544
Solanaceous, 566
Solanum; *see* Potato, Eggplant, Jerusalem-cherry
Solidago, 220; *see* Goldenrod
Sol-Kop-10, 104, 207, 213, 239, 566
Soluble salts, excess, 31–32
Solubor, 141, 285, 444
Solution, percentage, 522
Sooty mold or blotch, [64], 65, 100, 137, 200, 266, 277, 369, [406], 566
Sophora, 300
brooming disease, 301
canker (twig, branch), dieback, 77, 300
damping-off, 76, 302
leaf spot, 48, 301
mistletoe, 93, 302
powdery mildew, 56, 301
root-knot, 90, 301, 392
root rot, 87, 137, 301
rust, 58, 301
twig blight, 77, 301
Verticillium wilt, 68, 301, 343
Sophronitis, 366; *see* Orchids
SOPP (sodium orthophenylphenate tetrahydrate), 471
Sorbistat K, 362
Sorbus, 133; *see* Mountain-ash
Sorreltree, 459
chlorosis, 24–29
dieback, 77, 459
iron deficiency, 24, 26, 27
leaf spot, 48, 459
purple blotch, 51, 459
root rot, 87, 459
twig blight, 77, 459
wood rot, 79, 459
Sorus, 566
Sour-gum, 258; *see also* Tupelo
chlorosis, 24–29
iron deficiency, 24, 26, 27
rust, 58, 261
wood rot, 79, 170, 259
Sourwood, 459; *see* Sorreltree
Southern blight, 76, 393, [394], 490, 556
Southern leatherwood, 184
brown felt canker, 77, 184, 293
leaf spot, 48, 184
rust, 58, 184
Southern pea, 155; *see* Black-eyed pea
Southern white-cedar, 312; *see* Chamaecyparis
Southernwood, 220; *see* Artemisia
Sowthistle, rust, 405
Soybean flour, as sticker, 119
Spanish-bayonet, 512; *see* Yucca
Spanish bluebell, 495; *see* Squill
Spanish broom, 183; *see* Genista
Spanish dagger, 512; *see* Yucca
Spanish flag, 351; *see* Cypressvine
Spanish gorse, 183; *see* Broom
Spanish-moss, control, 506
Sparaxis, 307
mosaic, 70, 308
Spartina, rust, 144, 185, 186, 501
Spathe, 566
Spearmint, 447; *see* Mint
Species, 566
Specimens, diagnosing, 7
how and what to send, 7
mailing, 7
for nematode determination, 7
Speck blight, [224]
Specularia, 168; *see* Venus-lookingglass
Speedwell, 460
aster yellows, 73, 460
curly-top, 75, 460
downy mildew, 55, 460
Fusarium root and stem rot, wilt, 76, 87, 460
leaf smut, 62, 460
leaf spot, 48, 460
powdery mildew, 56, 460
root-knot, 90, 460
root nematode, 460
root rot, 87, 278, 460
rust, 58, 223, 460
stem (crown) rot, 76, 278, 460
Spencer Chemical Company, 124
Spergon, 102, 518, 531; *see also* Chloranil
Dust, 102, 518
gallon lots, 522
Seed Protectant, 102, 518, 531
—SL Seed Protectant, 531
Spray Powder, 102, 518
Wettable, 102, 518
Sphaeralcea, 297; *see* Globemallow
Sphag-lite, 22
Sphagnum moss, 566
Spicebush, 149; *see also* Calycanthus
canker, 77, 149–50
leaf spot, 48, 149
mistletoe, 93, 151
root rot, 87, 149
sooty blotch, 64, 151
Spicelily, 216; *see* Manfreda
Spicewood, 198; *see* Calycanthus
Spiderflower, 460
curly-top, 75, 461
downy mildew, 55, 461
leaf spot, 48, 221, 460
powdery mildew, 56, 461
root-knot, 90, 461
rust, 58, 223, 461
Spiderlily, 250
leaf spot, 48, 253
mosaic, 70, 250–51
root nematode, 250, 253
Stagonospora leaf blotch, red spot, 48, 251
Spiderwort; *see* Tradescantia, *and* Rhoea for Purpleleaf spiderwort
Spike-primrose, 275
rust, 58, 275
Spinach (*Spinacea*), 162; *see also* New Zealand spinach
anthracnose, 51, 166
bacterial soft rot, 81, 166
black root rot, 87, 162
boron deficiency, 27, 164
Cercospora leaf spot, 48, 162
chlorosis, 24–29, 166
crown rot, wilt, 76, 164, 166
curly-top, 75, 162
damping-off, 76, 162, [163]
downy mildew, blue mold, [55', 165
fertilizing, 29
Fusarium wilt, yellows, 66, 164
heart rot, 76, 164
leaf or white smut, 65, 166
leaf spot, 48, 166
mosaic, blight, 70, 164
oxidant or PAN injury, 44
ozone injury, 44
phosphorus deficiency, 25, 164
premature flowering, 41
resistant varieties, 164, 165
ringspot, 72, 164, 166
root-knot, cyst nematode, 90, 165
root nematode, 165, 167
root rot, 87, 156, 166
rust, 58, 166
scab, 65, 165, 415
seed rot, 162
seed treatment, 162, 165, 529, 533
spotted wilt, 72, 166
sulfur dioxide injury, 43
temperature requirements, 41
Verticillium wilt, 68, 167
watery soft rot, 76, 166
white-rust, [61], 62, 166
yellow dwarf, 73, 166
yellows, "blight," 70, 164, [165]
weed hosts, 165
Spinach beet, 162; *see* Beet, Swiss chard
Spindle-tree, 173; *see also* Euonymus
rust, 58, 508
Spindle tuber, 420
Spirea (*Spiraea*), 461
chlorosis, 24–29, 344, 461
crown gall, hairy root, 83, 136, 461
fire blight, 79, 133–34, 461
leaf spot, 48, 461
light requirements of, 39
powdery mildew, 56, 461
pruning, 32
root-knot, 90, 461
root nematode, 461
root rot, 87, 461
seedling blight, 76, 405
Spleenwort, 269; *see* Ferns
S-P-M (Sul-Po-Mag), 27
Spores, fungus, 11, [12], 566
Sporobolus, rust, 298, 337
Sporodochium, 566
Spot anthracnose, 48, 51, [260], [374]
Spotrete, 566
Spotted monarda, 447; *see* Monarda
Spotted wilt, 13, 72, [73], 168, 379, 486
weed hosts, 73
Spray injury, 45, 104, 105, 523
Spray lime, 566
gallon lots, 522
Spraycop, 104
Sprayer contamination, 109, 112
Sprayers, 112–19
accessories, 112, 113, 114
barrel, 113–14
cart, 113–14
clogging of nozzles, 112
compressed air, 112, [113]
decontamination of, 112
garden hose, 114, [117]
Hand Sprayers and Dusters, 118
household, [112]
knapsack, 113, [114]
maintenance of, 114–15
manufacturers, 120–24
nozzles, 109, [110]
Outdoor Housekeeping, 118
power, 114, [118], [119]
slide pump, 113, [115]
Sprayer and Duster Manual, 118
tractor boom, operating chart, 551
trombone, 113, [115]
uses, 118
Using Phenoxy Herbicides Effectively, 45, 112
wheelbarrow, 113–14, [116]
Spraying
advantages and disadvantages, 111
application to row crops, 550
compatibility chart, 552
contamination of sprayers, 109, 112
coverage, 109, [110], 112
equipment, 112–19
manufacturers, 120–24
fruits, 523–26
materials needed for fruit trees, 523
measuring apparatus, 109
multipurpose mixes, 110–11, 523
injury from, 523
notes on, 111, 523
trade names, 523
nozzles, 109, [110]
clogging, 112
orchards, 526
pattern, 109, [110]
percentage solution, 522
precautions, 45, 105, 106, 107, 108, 109, 112, 523
preparation of mixes, 108, 109, 520–21
publications, 118
tips, 109, 523
vs. dusting, 111
Spreader-Activator, 119
Spreader-stickers, 109, 566–67
rates, 119
trade names, 118–19, 567
uses, 118–19, 566–67
Spred-Rite, 119, 567
Sprenger asparagus, 141; *see also* Asparagus, garden and Asparagus-fern
crown gall, 83, 146
Spring-Bak, 326
Spring bell, 307; *see* Blue-eyed grass
Spring glory, 273; *see* Forsythia

Spring starflower, 183; *see* Brodiaea
Spruce, 402
brown felt blight, 407
canker (twig, branch, trunk), dieback, 77, 403
chlorosis, 24–29, 408
Cytospora canker, 77, [403]
damping-off, 76, 405
Diplodia tip blight, 76, 402, [403]
gray-mold blight, 52, 408
iron deficiency, 24, 26, 27, 408
mistletoe, dwarf, 93, 406
needle cast, 51, 402
root nematode, 406
root rot, 87, 405, 406
rust, needle, cone, 58, 177, [404], 405
witches'-broom, 58, 405
seed treatment, 405
seedling blight, 76, 405
snow blight, [407]
stump treatment, 406
sunscorch, wind damage, 41, 407
tar spot, 48, 402
twig blight, 77, 402, 403
wood rot, trunk, butt rot, 79, 170, 403
Spur, 567
Spur blight, 427, [428]
Spurge (*Euphorbia*), 409; *see also* Poinsettia, and *see* Pachysandra for Japanese and Mountain spurge
gray-mold (*Botrytis*) blight, 52, 86, 410
leaf spot, 48, 410
light and temperature requirements, 40
powdery mildew, 56, 410
root nematode, 409
root rot, 87, 409
rust, 58, 381, 410
soil treatment, 409, 410
stem smut, 62, 410
Verticillium wilt, 68, 409
Spurge laurel, 253; *see* Daphne
Squash, 238; *see also* Cucurbit
angular leaf spot, 50, 239
anthracnose, 51, 238
aster yellows, 73, 243
bacterial soft rot, 81, 86, 238, 244
bacterial spot, 50, 245
bacterial wilt, 69, 240
black rot, 86, 244
blossom blight, 52, 86, 245
blossom-end rot, 245
boron deficiency, 27, 245
chlorosis, 24–29, 245
curly-top, 75, 243
damping-off, seed rot, 76, 244
downy mildew, 55, 243
fruit spot, rot, 86, 244
Fusarium wilt, 66, 241
gray-mold rot, 52, 86, 244, 245
Growing Pumpkins and Squashes, 245
gummy stem blight, 76, 86, 244
leaf blight, 51, 239
leaf spot, 48, 238, 239, 244
mosaic, 70, 241
potassium deficiency, 26, 245
powdery mildew, 56, [242], 243
resistant varieties, 240, 242, 243, 245
ringspot, 72, 243
root-knot, cyst nematode, 90, 244–45
root nematode, 244
root rot, 87, 156, 244
scab, 65, 86, 238–39
seed rot, 244
seed treatment, 238, 528, 532
seedbed treatment, 244
southern blight, 76, 243
stem (crown) rot, 76, 244
storage rot, 86, 244
Verticillium wilt, 68, 245
Squaw-apple, rust, 58
Squibb Institute for Medical Research, 124
Squill, 495; *see also* Hyacinth
bulb nematode, 92, 496
bulb rot, 90, 495–96
bulb treatment, 496, 497, 530, 531, 537
crown and stem rot, 76, 497
leaf spot, 48, 498
mosaic, 70, 494
rust, 58, 498
smut, 62, 497
Squirrelcorn, 174
downy mildew, 55, 174
rust, 58, 174
Stachys, 447
leaf spot, 48, 447
powdery mildew, 56, 448
root-knot, 90, 447
rust, 58, 447
Staggerbush, 175; *see* Lyonia
Staghead, [357], 567
Staking trees and shrubs, 36, [37]
Stalk rot, 76, [232]
Standing cypress, 399; *see* Gilia
Stanhopea, 366; *see* Orchids
Stanleya, 186; *see also* Desertplume
leaf spot, 48, 190
rust, 58, 193
Staphylea, 130; *see* American bladdernut
Star cactus, 194; *see* Cacti, Echinocactus
Star hyacinth, 495; *see* Squill
Star-of-Bethlehem, 495; *see also* Tulip
leaf spot, 48, 498
light and temperature requirements, 40
mosaic, 70, 496
southern blight, 76, 497
Star violet, 185; *see* Houstonia
Starch-hyacinth, 494; *see* Grape-hyacinth
Starfire, 127; *see* African-violet
Starglory, 351; *see* Cypressvine
Starry campion, 203; *see* Silene
Stars of Persia, 361; *see* Allium, Onion
Starter solution, 29, 30, 567
Statice, 451; *see also* Sea-lavender
aster yellows, 73, 451
crown rot, 76, 451
flower blight, 86, 451
gray-mold (*Botrytis*) blight, 52, 86, 451
powdery mildew, 56, 451
root rot, 87, 278, 451
rust, 58, 451
Stauffer Chemical Company, Agricultural Chemical Division, 27, 102, 103, 105, 124, 518, 519
Agri-strep, 83, 105, 522
Captan-Folpet, 567
Captan 75-W, 531
D-D Soil Fumigant, 545, 548
Folpet 50-WP and 75-WP, 103, 519
Phygon-XL, 102, 518; *see also* Dichlone
Vapam Soil Fumigant, 106, 188, 189, 193, 214, 263, 271, 290, 356, 405, 425, 428, 430, 443, 463, 472, 490, 492, 512, 545, 547; *see also* SMDC
Zineb 75-W, 103, 519
Ziram, 103, 519
Steam disinfestation of soil, 538–43
effects, 542–43
methods, 539–42
aboveground tubing method, 540–41
flash-flame pasteurizers, 541–42
inverted-pan, [541]
oven, 539
pressure cooker, 539
tank or vault, 539
Thomas method, [540], 541
underground pipe, 540
underground tile, 539–40
precautions, 538–39, 542–43
pressure, 538–41
temperature, 538–42
time, 538–41
toxicity problems, 542–43
waiting period, 538
Stele, 567
Stem
blight, 77
canker, 77, [78], 79, [156], [187], [188], [276], [419]
diseases, 76–86
gall, 62, [201], [424], 458
nematode, 16, 75, [76], 467, [497]
rot, 13, 76, [77], 81, [147], [456]
rust, 58
smut, 62, [63]
spot, [260]
Stemphylium, petal specking and leaf blight, [224]
leaf spot, [283]
Stenanthium, 495; *see also* Hyacinth
rust, 58, 498
Stenolobium, 494
root rot, 87, 137, 494
rust, 58, 494
Stenotaphrum, 319; *see* St. Augustinegrass
Stephanomeria, 220
leaf spot, 48, 221
rust, 58, 223
Sterilizing soil, 106, 538–49, 567
methods, 539–41, 543–49
chemicals, 543–49
heat, 539–41
effects, 542–43
precautions, 538, 543, 544–48
Sterilizing wounds, 34–36
Sternbergia, 250
Stagonospora leaf scorch, red spot, 48, 251
Stevia, 220; *see* Piqueria
Stickers, trade names, 119, 567
uses, 118–19, 566–67
Stickwort, rust, 405
Stillingia, 210; *see* Queens-delight
Sting nematode, [395], [396]
Stink bugs, insect control, 491
Stipa, rust, 298
Stock, 567
Stock, 186; *see also* Cabbage
anthracnose, 51, 190
black ringspot, 72, 192
black rot, bacterial blight, 50, 187
clubroot, 89, 189
curly-top, 75, [192]
damping-off, 76, 188
downy mildew, 55, 189
foot rot, 76, 188
Fusarium wilt, 66, 186–87
gray-mold blight, 52, 86, 190–91
leaf spot, 48, 190
mosaic, flower breaking, 70, [192]
resistant varieties, 189
root-knot, 90, 191
root nematode, 194
root rot, 87, 193
seed treatment, 187, 530, 536
spotted wilt, 72, 192
stem (crown) rot, 76, 191
Verticillium wilt, 68, 193
white-rust, 62, 191
wirestem, 77, 188
Stokes-aster (*Stokesia*), 220; *see also* Cornflower aster
head blight (*Botrytis*), 52, 86, 224
leaf spot, 48, 221
mosaic, 70, 222
powdery mildew, 56, 221
Stolon (runner), 567
Stoma (stomata), [12], 567
Stonecress, 186; *see* Candytuft
Stonecrop, 451; *see* Sedum
Stonemint, 446
leaf spot, 48, 447
rust, 58, 447
Stop-Scald, 140
Stop-Wilt, 21, 38, 41, 42, 182, 296, 511
Stopmold B, 471
Storage (fruits and vegetables)
cleanliness, 83, 99, 208, 471
Home Storage of Vegetables and Fruits, 87, 99
rot, 86, [87], [147], [207], 244, [416], [417], [470], [473]
Storksbill, 278; *see* Geranium of florists
Strain, 567; *see also* Physiologic race
Strangleweed (dodder), 94

Stranvaesia, 133; *see also* Apple
fire blight, 79, 133–34
root rot, 87, 137–38
Strawberry, 462
angular leaf spot (bacterial), 50, 468
anthracnose, 51, 467
aster yellows, green petal, 73, 466
bacterial soft rot, 81, 86, 465
black root rot, 87, [88], [462], 463
black-seed, 464
bud rot (*Rhizoctonia*), 466–67
bulb and stem nematode, 75, 92, 467
catface, 468
cauliflower disease, 467
chlorosis, 24–29, 467
chlorotic fleck, 70, 465
crinkle, 70, 465–66
crown rot, 76, 465, 467
curly-dwarf, 73, 465–66
downy mildew, 55, 468
dwarf or "crimp" (spring and summer), 75, [76], 467
fruit dip, 465
fruit rot, 52, 86, 465
gray-mold rot, blossom blight, 52, 86, [465], 525
hard rot, 86, 465
insect control, 463, 466
iron deficiency, 24, 26, 27, 467
leaf blotch, blight, 51, [464]
leaf and bud nematode, 75, [76], 467
leaf curl, 70, 466
leaf roll, 466
leaf scorch, 51, [464]
leaf spot, 48, [49], [464]
leaf variegation (genetic), 466
leak, Rhizopus rot, 86, 465
leather rot, 86, 465
mosaic, 70, 465–66
mottle, 70, 465–66
multiplier, 465–66
PCNB, soil drench, 465
plant dip or soak, 464, 467, 531, 537
potassium deficiency, leaf scorch, 26, 467
powdery mildew, 56, 465
red stele, core, 87, [462], 463
Reducing Virus and Nematode Damage to Strawberry Plants, 467, 468
resistant varieties, 463, 464, 465, 466
root-knot, 90, 467
root nematode, 463, 467
root rot, 87, 462
slime mold, 468
soil fumigation, 463, 467, 543–49
southern blight, 76, 465
spray schedule, 468, 524–25
spring or June yellows, 466
Strawberry Diseases, 468
Strawberry Varieties in the United States, 468
stunt, 73, 466
tipburn, 468
veinbanding, 466
Verticillium wilt, 68, 463–64
virus-free stock, [466]
winter injury, 41, [462], 463
witches'-broom, 466
yellow-edge, xanthosis, 466
yellows (virus), 73, 466
Strawberry-begonia, 304; *see* Saxifrage
Strawberry-bush, 173; *see* Euonymus
Strawberry-shrub, 198; *see* Calycanthus
Strawberry-tomato, 479; *see* Ground-cherry
Strawberry-tree, 175; *see also* Arbutus
crown gall, 83, 176
Strawflower, 220
aster yellows, 73, 222
curly dwarf, 73, 222
curly-top, 73, 223
downy mildew, 55, 225
ringspot, 72, 222
root-knot, 90, 225
stem rot, 76, 221
Verticillium wilt, 68, 223
Streak, 13, 70, [429], 567
Strelitzia, 172; *see* Bird-of-paradise-flower
Streptanthera, 307
mosaic, 70, 308
Streptomyces pox, 472, [473]
Streptomycin, 105, 567
disease control, 51, 81, 83, 134, 153, 168, 191, 196, 197, 202, 280, 283, 310, 360, 392, 410, 413, 423, 484, 500
formulations, 105, 522
injury, 105
preparing solutions, 522
sulfate, 105
trade names and distributors, 105, 522
uses, 105
Streptomycin Antibiotic Spray Powder, 105, 522
Streptomycin Spray, 105, 522
Striga, 234
Striped squill, 465; *see* Tulip
Stroma, 567
Strombocactus, 194; *see* Cacti, Echinocactus
Stunt (virus), 13, 73, [74], 75, [177]
Styrax, 453
chlorosis, 24–29
leaf spot, 48, 453
root-knot, 90, 453
Substrate, 567
Succulent, 567
Sugar beet, 162; *see* Beet, garden
The Sugar Beet Nematode and Its Control, 165
weed hosts, 165
Sugarberry, 292; *see also* Hackberry
chlorosis, 24–29
downy mildew, 55, 293
felt fungus, 77, 293
leaf blight, 51, 264, 293
leaf spot, 48, 264, 293
mistletoe, 93, 293
powdery mildew, 56, 292
root rot, 87, 137, 293
seedling blight, 76, 293, 405
thread blight, 293, 505
witches'-broom, 292, [293]
wood rot, 79, 170, 292
Sulfasoil, 325
Sulfur
for acidifying soil, 24, 415, 435, 443, 473
deficiency, 28
control, 29
resistant plants, 43
susceptible plants, 43
symptoms, 43
dioxide, as air pollutant, 43
control, 44
-fixed copper mixture, 394
as fumigant, 289
in *Iliad* and *Odyssey*, 105
particle size, 105
promotes, 29
as protective fungicide, 58, 62, 65, 100, 128, 440, 525
formulations, 105
gallon lots, 522
injury, 45, 105, 499
in multipurpose mixes, 111
uses, 105
Sulphuric acid, damping-off control, 567
Sul-Po-Mag, 27
Sultan, 152; *see* Balsam, garden
Sumac, 468
canker, dieback, 77, 469
crown gall, 83, 469
Fusarium wilt, 66, 468–69
gray-mold, inflorescence blight, 52, 86, 469
inflorescence blight, 86, 469
leaf curl, blister, 62, 171, 384, 469
leaf spot, mold, 48, 469
powdery mildew, 56, 469
root nematode, 469
root rot, 87, 137, 469
rust, 58, 469
sulfur dioxide injury, 43
"umbrella disease," 77, 138, 469
Verticillium wilt, 68, 468–69
wood rot, 79, 170, 469
Summer-cypress, 162, *see also* Kochia
virus yellows, 73, 164–65
Summer-hyacinth, 496; *see also* Hyacinth
bulb nematode, 92, 496–97
leaf spot, 48, 498
mosaic, 70, 496
Summer-lilac, 185; *see* Butterflybush
Summersweet, 470; *see* Clethra
Sun-plant, 445; *see* Rose-moss
Sunblotch, [150]
Sundrops, 268; *see* Evening-primrose
Sunflower, 220; *see* Heliopsis for Orange sunflower
aster yellows, 73, 222
bacterial leaf spot, blight, 50, 225
bacterial wilt, 69, 223
crown gall, 83, 226
curly dwarf, 73, 220
damping-off, seed rot, 76, 221
downy mildew, 55, 225
Fusarium wilt, 66, 223
gray-mold blight, bud rot, 52, 86, 224
leaf gall nematode, 75, 226
leaf smut, 62, 226
leaf spot, 48, 221
mosaic, 70, 222
petal blight, 86, 221, 224
powdery mildew, 56, 221
resistant varieties, 224
ringspot, 72, 222
root-knot, 90, 225
root rot, 87, 221
rust, 58, 223, 405
seed treatment, 221
southern blight, stem rot, 76, 221
stem (crown) rot, 76, 221
Verticillium wilt, leaf mottle, 68, 223
white-rust, 62, 226
white smut, 65, 226
Sunox, 106, 563
Sunrose, 469
gray-mold blight, 52
leaf spot, 48, 369
powdery mildew, 56, 469
root rot, 87, 469
Sunscald or burn, [41], 42, 182, 200, 260, 369, 498, 567
control, 42, 498, 502
fruits and vegetables, [10], 365, [482]
trees and shrubs, 41, 42, 172, 182, 260, 264, 502
entrance for fungi, 79, 383, [503]
Sunscorch, 41, 42
control, 41, [42]
evergreens, 42
Sunshine shrub; 446; *see* St.-Johns-wort
Superintendent of Documents, 578
Superphosphate, 29
Surinam-cherry, 268; *see* Eugenia, Eucalyptus
Sur-Ten Wetting Agent, 119
Suscept, 567
Susceptibility, 567
Swamp-privet, 144; *see* Forestiera
Swampbay, 149; *see also* Avocado
black mildew, 64, 151
Swan River daisy, 220
aster yellows, 73, 222
Swede, 185; *see* Rutabaga
Sweet alyssum, 186, *see also* Alyssum, Cabbage
aster yellows, 73, 193
black ringspot, 72, 192
blackleg, 77, 187
chlorosis, 24–29, 194
clubroot, 89, 189
crown rot, 76, 191
damping-off, 76, 188
downy mildew, 55, 189
mosaic, flower breaking, 70, 192
powdery mildew, 56, 193
root-knot, 90, 191
root nematode, 194
root rot, 87, 193
white-rust, 62, 191
Sweet basil, 447; *see* Ocimum

Sweet corn, 231; *see* Corn
Sweet-jarvil, 213
  damping-off, seed rot, 76, 86, 214
  leaf spot, 48, 213
  rust, 58, 215
  seed treatment, 213
Sweet marjoram, 347; *see* Salvia
Sweet-pepperbush, 470; *see* Clethra
Sweet scabious, 450; *see* Scabiosa
Sweet sultan, 220; *see* Centaurea
Sweet-william, 203; *see also* Carnation
  Alternaria blight, 51, 76, 204
  anther smut, 62, 206
  anthracnose, 51, 205
  aster yellows, 73, 204
  chlorosis, 24–29, 206, 443
  curly-top, 75, 162, 205
  damping-off, 76, 204
  Fusarium wilt, 66, 203–4
  gray-mold blight, 52, 86, 205
  leaf spot, 48, 205
  leaf and stem nematode, 75, 206
  mosaic, 70, 204
  ringspot, 72, 205
  root-knot, 90, 206
  root nematode, 206
  root rot, 87, 204
  rust, 58, 204
  southern blight, 76, 204
  stem rot, 76, 204
Sweetbay; *see* Magnolia and Laurel, true
Sweetbells, 316; *see* Leucothoë
Sweetfern, 470
  rust, 58, 313, 405, 470
Sweetgale, 507
  leaf spot, 48, 507
  rust, 58, 405, 507
  twig blight, 77, 507
Sweetgum, 509
  bleeding necrosis, 509–10
  canker (twig, branch), dieback, 77, 345, 510
  chlorosis, 24–29, 344, 509
  felt fungi, 293, 510
  iron deficiency, 24, 26, 27, 344, 509
  leader dieback, blight, 77, 510
  leaf spot, 48, 509
  mistletoe, 93, 510
  root-knot, 90, 392, 509
  root nematode, 509
  root rot, 87, 137, 509
  thread blight, 505, 510
  twig canker, 77, 345, 510
  wood rot, 79, 170, 509
Sweetolive, 370; *see* Osmanthus
Sweetpea, 377
  anthracnose, 51, 380
  Ascochyta blight, 51, 379
  bacterial leaf spot, 50, 380
  blossom blight, 52, 86, 381
  bud drop, 39, 382
  chlorosis, 24–29, 382
  crown gall, 83, 381
  damping-off, 76, 379
  downy mildew, 55, 379
  ethylene injury, 44
  fasciation, leafy gall, 81, [82], 381
  Fusarium wilt, root rot, 66, 87, 377
  gray-mold blight, 52, 86, 381
  leaf spot, 48, 379, 380
  mosaics, flower breaking, 70, [378]
  Mycosphaerella blight, 51, 379
  powdery mildew, 56, [378]
  Ramularia white blight, 51, 380
  root-knot, 90, 381
  root nematode, 382
  root rot, 87, 377, 378
  rust, 58, 381
  seed treatment, 379, 528, 536
  soil drench, 377, 378, 379
  southern blight, 76, 378
  spotted wilt, ringspot, 72, 379
  stem (crown) rot, 76, 378, 379
  streak, 70, 378, 379
  Verticillium wilt, 68, 382
Sweetpotato, 470
  bacterial soft rot, 81, 86, 471
  black rot, black root, black shank, 76 86, [87], [470], 471
  boron deficiency, internal brown spot, 27, 475
  bud rot, 52, 475
  crown rot, 76, 298, 475
  curly-top, 75, 475
  damping-off, 76, 470, 472
  feathery mottle, 474
  fertilizing, 29
  foot (crown) rot, die off, 76, 472
  Fusarium wilt, stem rot, yellow blight, 66, 471, [472]
  gray-mold blight, bud rot, 52, 87, 475
  insect control, 471, 474, 475
  internal cork, leaf spot, 473, [474]
  leaf blight, 51, 472
  leaf spot, 48, 472
  magnesium deficiency, 28, 475
  mottle-leaf, mosaic, 70, 474
  mottle necrosis, leak, 474
  plant-food utilization, 26
  resistant varieties, 471, 472, 473, 474
  Rhizopus rot, soft rot, 86, [470], 471
  ring rot, [471]
  ringspot, 72
  root-knot, 90, 472
  root nematode, 475
  root rot, 87, [470], 475
  root or seed treatment, 470, 471, 472, 528, 529, 531, 533
  russet crack, 475
  rust, 58, 405, 475
  scurf or soil stain, 86, [473]
  seedbed management, 470, 475
  slime mold, 475
  soil rot, Streptomyces pox, 86, 472, [473]
  sooty mold, 64, 475
  southern blight, 76, 298, 475
  sprout dip, 470
  stem nematode, 75
  storage, 471, 474
    rot, 86, [471]
  Sweetpotato Diseases, 475
  thread blight, 475, 505
  2,4-D injury, 45, 288, 475
  Verticillium wilt, 68, 475
  white-rust, 62, 473
  yellow dwarf, 73, 474
Sweetshrub, 198; *see* Calycanthus
Swerl, as spreader, 119
Swift & Company, Agricultural Chemical Division, 124
Swiss chard, 162; *see also* Beet
  boron deficiency, 27, 164
  chlorosis, 24–29, 166
  curly-top, 75, 162
  damping-off, 76, 162
  downy mildew, 55, 165
  heart rot, 166
  leaf spot, 48, 166
  mosaic, 70, 184
  oxidant or PAN injury, 44
  ringspot, 72, 164, 166
  root-knot, cyst nematode, 90, 165
  root nematode, 165, 167
  root rot, 87, 156, 166
  rust, 58, 166
  scab, 65, 165, 415
  seed rot, 86, 162
  seed treatment, 162, 532
  southern blight, 76, 164, 207
  virus yellows, 73, 164
  yellow net, 70, 164
Swiss-cheese-plant, light and temperature requirements, 40
Swordbean, 155; *see* Jackbean
Sycamore, 476
  anthracnose, 51, [476]
  borer control, 477
  canker (twig, branch), dieback, 77, 476
  chlorosis, 24–29, 343, 477, 503–4
  crown gall, 83, 136, 477
  leaf blight, 51, 476, 477
  leaf scorch, 41, 477
  leaf spot, 48, 477
  mistletoe, 93, 478
  powdery mildew, 56, [477]
  root rot, 87, 137, 358, 477
  sooty blotch, 64, 137, 478
  twig blight, 77, 476
  2,4-D injury, 45, 288, 477
  Verticillium wilt, 68, 343, 477
  wetwood, slime flux, 264, 477
  winter injury, 41, 264, 477
  wood, trunk, heart rot, 79, 170, 477
Symbiosis, 567
Symphoricarpos, 457; *see* Snowberry, Coralberry
Symptoms of disease, 48–94, 567
Synergist, 567
Syngonium, 196; *see* Nephthytis
Synthyris, 454
  leaf spot, 48, 455
  rust, 58, 455
Syringa, 332; *see* Lilac
Systemic, 567

T

Tabebuia, 494
  rust, 58, 494
Tabernaemontana, 360
  leaf spot, mold, 48, 360
  root rot, 87, 137, 361
Taenidia, 132
  leaf spot, 48, 132
  rust, 58, 132
Tag Fungicide, 106, 476
Tagetes, 221; *see* Marigold
Tallowtree, 210; *see* Chinese tallowtree
Tamarack, 318; *see* Larch
Tamarisk (*Tamarix*), 478
  canker, 77, 478
  chlorosis, 24–29, 343, 478
  powdery mildew, 56, 171, 478
  root rot, 87, 137, 478
  twig blight, 77, 478
  wood rot, 79, 170, 478
Tampala, 230; *see* Amaranth
Tanacetum, 221; *see* Tansy
Tanbark-oak, 355
  leaf blight, 51, 356
  leaf spot, 48, 358
  rust, 58, 359
  wood rot, 79, 170, 358
Tangelo, 227; *see* Citrus, Orange
Tangerine, 227, *see* Citrus, Orange
Tansy, 221
  leaf spot, 48, 221
  powdery mildew, 56, 221
  root-knot, 90, 225
  rust, 58, 223
Tar spot, 48, [346]
Taraxacum, 221; *see* Dandelion
Tarragon, 220; *see* Artemisia
Tasmania string-bark, 268; *see* Eucalyptus
Tassel-hyacinth, 495; *see* Grape-hyacinth
Tasselflower, 220; *see* Emilia
Tasseltree, 258
  leaf spot, 48, 259
  root rot, 87, 137, 259
  sooty mold, 64, 137, 261
Taxodium, 402; *see* Baldcypress
Taxus, 511; *see* Yew
TC-90, 104, 207, 213, 239, 567–68
TCTP, 546
  active ingredient, 546
  formulations, 546
  trade names, 546
  uses, 546
Tea olive, 370; *see* Osmanthus
Teaberry, 294; *see* Checkerberry
Tear gas, 106, 545; *see* Chloropicrin
Teasel, 478
  downy mildew, 55, 479
  leaf spot, 48, 478
  leaf and stem nematode, 75, 478
  mosaic, 70, 478
  powdery mildew, 56, 478
  root nematode, 479
  root rot, 87, 479

southern blight, 76, 254, 478
stem (crown) rot, 76, 254, 478
Tecomaria, 494
anthracnose, 51, 494
root rot, 87, 137, 494
Telone, disinfesting soil, 91, 159, 165, 209, 214, 244, 271, 290, 364, 365, 392, 395, 420, 448, 467, 472, 489, 545
home garden, 548, [549]
Telvar, weed control, 368
Temperature requirements of house plants, 40
Tenacity (pesticide), 568
Tennessee Corporation, 124
"26" Copper Fungicide, 104
Tepary bean, 155; *see also* Bean, garden types
curly-top, 75, 159
leaf spot, 48, 159–60
powdery mildew, 56, 158
root rot, 87, 156
rust, 58, 158
southern blight, 76, 156–7
Terminal (plant), 568
Terra Tox-10, 101
Terracap, 104, 188, 328, 544
Terraclor, 101, 544; *see also* PCNB
gallon lots, 522
Terrafume-2, 101
Terramycin
active ingredient, 105, 568
-basic copper sulfate, 105
in combination with streptomycin, 105
plant dip, 83, 392, 503
Tersan
OM, 106, 320, 321, 322, 324, 325–26, 326, 327, 568
-75, 103, 519
Tetrachloroisophthalonitrile, 104
Tetrachloro-p-benzoquinone, 102, 518
Tetrachlorothiophene, 546
Tetragonia, 162; *see* New Zealand spinach
Tetramethylthiuram disulfide, 103, 519
Teucrium, 447; *see* Germander
Texas bean, 155; *see* Bean, garden types, Tepary bean
Texas-bluebell, 278; *see* Prairiegentian
Texas ranger, 479; *see* Texas silverleaf
Texas (*Phymatotrichum*) root rot, 88–89, [208], [211], 504
list of resistant plants, 89
Texas silverleaf, 479
root rot, 87, 137, 479
twig canker, 77, 479
Thalictrum, 254; *see* Meadowrue
Thanksgiving cactus, 194; *see* Cacti, Epiphyllum
Thelocactus, 194; *see* Cacti, Echinocactus
Thermopsis, 377; *see also* Pea
leaf spot, 48, 379, 380
powdery mildew, 56, 378
Thielaviopsis root rot, [409]
Thimbleberry, 427; *see also* Blackberry, Raspberry
canker, blight, 77, 428
fruit rot, 86, 430
gray-mold blight, 52, 86, 430
mosaic, 70, 428
rust, 58, 430
Verticillium wilt, 68, 430
Thimbleflower, 220; *see* Rudbeckia
Thimer, 106, 320, 321, 322, 325, 568
Thiodan (endosulfan)
spray, 418
tree borer control, 72, 79, 139, 144, 171, 259, 266, 301, 332, 346, 358, 384, 435, 477, 508–9, 525
Thiram (TMTD)
-Acti-dione mixtures, 106
active ingredient, 103, 519
as "disinfectant," 221, 464, 470, 472
lawn diseases, 103, 519
multipurpose mixes, 111, 519, 523
oxidant or PAN prevention, 44
-PCNB mixtures, 77, 104, 410, 437
post-harvest dip or spray, 208
as protective fungicide, 55, 62, 86, 100, 167, 169, 175, 182, 204, 210, 228, 294, 301, 302, 310, 313, 321, 322, 326, 328, 329, 343, 356, 367, 383, 384, 433, 458, 476, 482, 495, 500, 508, 519, 525
seed and bulb treatment, 77, 81, 97, 103, 238, 273, 282, 299, 319, 326, 330, 362, 375, 379, 394, 405, 418, 484, 519, 531
soil application, 77, 97, 103, 167, 261, 294, 326, 406, 410, 519, 544
trade names and distributors, 103, 519
tree wound dressing, 476
uses, 103, 519
Thiram
50 Dust, 103, 519
4.8D, 103, 519
-75, 103, 519, 531
-65, 103, 519
S-42, 103, 519
SF-75, 531
Thiramad, 568
Thistle, 220; *see also* Plumed thistle, Globethistle
inflorescence smut, 62, 206, 227
leaf spot, 48, 221
powdery mildew, 56, 221
root rot, 87, 221
rust, 58, 223
stem (crown) rot, 76, 221
white-rust, 62, 226
Thompson-Hayward Chemical Company, 124
Thorn, 133; *see* Hawthorn
Thoroughwort, 220; *see* Eupatorium
Thread blight, 272, 505
3,5-dimethyltetrahydro-1,3,5 2H-thiadiazine-2-thione, 546
Thrift, 451; *see* Armeria
Thrips
control, 72, 73, 168, 363, 397, 486
injury, 397
as virus carriers, 13, 72, 73, 98, 168, 486
Thuja, 312; *see* Arborvitae
Thujopsis, 312; *see* Hiba arborvitae
Thunbergia, 229; *see* Clockvine
Thylate Thiram Fungicide, 103, 134, 199, 435, 436, 519, 568
gallon lots, 522
Thyme (*Thymus*), 447
root rot, 87, 278, 447
Ti plant, 261; *see* Cordyline
Tiarella, 304
powdery mildew, 56, 304
rust, 58, 305
Tickseed, 220; *see also* Coreopsis
aster yellows, 73, 222
curly-top, 75, 223
powdery mildew, 56, 221
rust, 58, 223, 405
scab, 65, 165, 226
Verticillium wilt, 68, 223
Tide, as spreader, 119
Tidytips, 220
powdery mildew, 56, 221
spotted wilt, 72, 223
Tigerflower (*Tigridia*), 282; *see also* Gladiolus
bacterial scab, 50, 65, 90, 282
bulb nematode, 92, 285
bulb (storage) rot, 90, 282
mosaic, 70, 284
Tile, agricultural drain, 36
Tilia, 338; *see* Linden
Tip blight, 261, [262], [403]
Tip burn, 191, 328, [329], 343
Tissue, 568
Tissue test, 568
Tisswood, 453; *see* Halesia
Tithonia, 220; *see* Torch flower
TMTD; *see* Thiram
Toadflax, 454
anthracnose, 51, 455
aster yellows, 73, 457
downy mildew, 55, 456
insect control, 456, 457
leaf spot, 48, 455
leaf and stem nematode, 75, 457
powdery mildew, 56, 455
root-knot, 90, 457
root rot, 87, 455
rust, 58, 455
seed treatment, 455
soil drench, 455
southern blight, 76, 455
stem rot, 76, 455
white smut, 65, 457
Toadstools (mushrooms), 79, [88], 359, 562
Tobacco, 479; *see also* Flowering tobacco
mosaic virus, [14]
oxidant or PAN injury, 44
ozone injury, 44
Tobira, 408; *see* Pittosporum
Tolerance (tolerant), 568
Tolmiea, 401
powdery mildew, 56, 401
Tomatillo, 479; *see* Groundcherry
Tomato, 379
anthracnose, 51, [87], 480, 481
aster yellows, 73, 490
bacterial canker, 51, 77, 81, [483], 484, [487]
bacterial soft rot, 81, 86, 480, 481, [483]
bacterial speck, 50, 86, 483
bacterial spot, 50, 86, 483
bacterial wilt, 69, 488
black walnut injury, 492
blossom blight, 86, 492
blossom drop, 41, 484
blossom-end rot, [10], 482
blotchy ripening, 486
bud drop, 39, 484
bunchy-top, 420
calico, 70, 484
catface, 491
chlorosis, 24–29, 490
cloudy spot, 491
crease stem, 492
crown gall, hairy root, 83, 492
curly-top, 75, 486
damping-off, 76, 488
double-virus streak, 485–86
downy mildew, 55, 490
early blight (*Alternaria*), [51], 480
ethylene injury, 44
fertilizing, 29
fruit cracking, 482, [483]
fruit spot, rot, 86, [87], 480, 481, 482, 483, 490
Fusarium wilt, 66, [67], 486, [487]
gray leaf spot, 48, 480–81
gray wall, 486
insect control, 483, 485, 486, 491, 492
internal browning, 484, 486
late blight, 51, 86, 480, [481]
leaf mold, 480–81
leaf roll (physiological), [491]
leaf roll (virus), 491
leaf spot, 48, 480
leaf and stem nematode, 75, 400, 492
milk, virus control, 485
mosaic, 70, 484, [485]
oedema, 39, 492
oxidant or PAN injury, 44
ozone injury, 44
Phytophthora blight, 51, 490
plant-food utilization, 26
potassium deficiency, 26, 490
powdery mildew, 56, 492
pseudo curly-top, 491
psyllid yellows, 490
puffing, "pops," 491
purple top, 73, 490
resistant varieties, 480, 481, 482, 483, 485, 486, 487, 488, 489, 491, 492
ringspot complex, 72, 486
root-knot, 90, [91], 489
root nematode, 489
root rot, 87, 490
seed rot, 488
seed treatment, 484, 485, 529, 533
seedbed treatment, 489, 490

Tomato (*continued*)
Septoria leaf spot, blight, [49], 480
soil fumigation, 487, 489, 490
southern blight, 76, 490
spotted wilt, tip blight, 72, [73], 486
staking vines, 99
stem (collar) rot, 76, 480, 488
streak, 484
sunscald, [10], [482]
temperature, effect on, 41
tobacco etch, 484
tobacco mosaic virus, [14]
Tomato Diseases and Their Control, 492
2,4-D or 2,4,5-T injury, 45, 288, 484
Verticillium wilt, 68, [69], [487], [488]
web blight, 160, 492
western yellow blight, 75, 486
wildfire, 50, 483
yellow net, 484
Tomato aspermy, 223
Tomato eggplant, 479; *see* Jerusalem-cherry, Eggplant
Toothwort, 186; *see also* Cabbage
downy mildew, 55, 189
leaf spot, 48, 190
rust, 58, 193
white-rust, 62, 191
Topworking, 568
Torch flower, 220; *see also* Chrysanthemum
Fusarium wilt, 66, 223
root-knot, 90, 225
Torch plant, 130; *see* Aloe
Torchlily, 432; *see* Redhot-pokerplant
Torenia, 454
root-knot, 90, 457
Totai, 371; *see* Palms
Toxicity, 568
Toxin, 568
Toyon, 133; *see* Photinia
Trace elements, 27
Trachelospermum, 360; *see* Confederate-jasmine
Trachycarpus, 372; *see* Palms
Trachymene, 213; *see* Blue laceflower
Tractor, boom sprayer operating chart, 551
injury, 46
speed conversions, 551
Trade names, 568
Tradescantia, 493; *see also* Dayflower
gray-mold leaf blight, 52, 493
leaf spot, 48, 493
light and temperature requirements, 40
mosaic, 70, 493
plant soak, 493, 530, 537
root-knot, 90, 493
root nematode, 493
rust, 58, 493
Tragopogon, 328; *see* Salsify
Trailing-arbutus, 294
leaf spot, 48, 294
powdery mildew, 56, 294
soil drench, 294
wilt, crown (collar) rot, 76, 294
Trailing four-o'clock, 274
downy mildew, 55, 274
leaf spot, 48, 274
root-knot, 90, 159, 274
root rot, 87, 274
rust, 58, 274
white-rust, 62, 274
Translucent, 568
Transpiration, 568
Transvaal daisy, 220
gray-mold blight, 52, 86, 224
leaf spot, 48, 221
plant soak, 225, 530, 536
powdery mildew, 56, 221
ray speck, 86, 221
ringspot, 72, 223
root-knot, 90, 225
root rot, 87, 221
soil drench, 222
stem (crown) rot, 76, 221
Verticillium wilt, 68, 223
Trapex, 568
Treating, potted plants, 529, 530–31, 534–37
seed, bulbs, corms, tubers, rhizomes, 527–37
soil, 538–49
Tree(s)
borer control, 79, 139, 145, 171, 259, 266, 300–301, 332, 346, 358, 384, 435, 477, 508–9
"butchery," [34]
construction damage, 46
drought injury, 38
fertilizing, 30, [31], 32–33
girdling roots, [45]
guying, 36, [37]
ice injury, [43]
injury from soil grade change, [46], [47]
lightning injury, 46
multipurpose spray mixtures, 111
paint, [35], 36, 476–77
Protecting Shade Trees During Home Construction, 46
Protecting Trees Against Damage From Construction Work, 46
pruning, [32], [34]
Pruning Shade Trees and Repairing Their Injuries, 36
relatively free of troublesome diseases, 549
removal, 33–34
roots, strangling injury, [45]
staking, 36, [37]
Tree Bracing, 36
watering, 37–38
winter injury, 41, 42
protection, [42]
wound, dressing, 35, 36, 139, 161, 171, 476–7
treatment, 34, [35], 383–4
Tree cypress (*Gilia*), 399; *see also* Cypress and Chamaecyparis
Tree-of-Heaven, 493
black mildew, 64, 493
canker, dieback, 77, 345, 493
gray-mold blight, 52
ice injury, 43
leaf spot, 48, 246, 493
root rot, 87, 137, 493
seedling blight, damping-off, 76, 405, 493
twig blight, 77, 493
Verticillium wilt, 68, 493
wood rot, 79, 170, 493
Tree peony, 254; *see* Peony
Tree surgeon, help by, 218, 265
Tree surgery, 568
Tree-tomato, 479; *see also* Tomato
bacterial canker, 50, 483–4
powdery mildew, 56, 492
Treehoppers, as virus carriers, 160, 492
control, 492
Treemallow, 297; *see* Lavatera
Treepoppy, 413
leaf smut, 62, 414
Triangle Iron Chelate Granules, 27
Tri-Basic Copper Sulphate, 104
Tri-clor, 545
Trichlorfon (Dylox), 187
Trichloronitromethane, 545
Trichocereus, 194; *see* Cacti, Cereus
Trichosanthes, 238; *see* Gourds, Cucurbit
Tricop, 104
Tridens, rust, 302
Trifoliate orange, 227; *see* Hardy orange, Citrus
Trifume, 106
Trillium, 493
leaf smut, 62, 494
leaf spot, 48, 493
rust, 58, 494
stem rot, 76, 254, 494
Triplet lily, 183
rust, 58, 183
Tritoma, 432; *see* Redhot-pokerplant
Triton B-1956, 119
Tritonia, 282
corm rot, 90, 282
leaf spot, blight, 48, 51, 283
mosaic, 70, 284
southern blight, 76, 282
yellows (*Fusarium*), 66, 90, 282
Trizone, 106, 406, 448, 545
Trollius, 131; *see* Globeflower
Trombone sprayer, 113, [115]
Trona Bromofume 40 and 85, 544
Tronabor, 27
Tropaeolum, 354; *see* Nasturtium
Tropical lilac, 317; *see* Duranta
Troutlily, 267; *see* Erythronium
True daisy, 220; *see* English daisy
True mistletoe, [93], 94
Trumpetcreeper; *see* Trumpetvine, Bignonia
Trumpetflower, 169; *see* Bignonia
Trumpets, 408; *see* Pitcherplant
Trumpettree, 494
rust, 58, 494
Trumpetvine, 494
leaf blight, 51, 494
leaf spot, 48, 494
mistletoe, 93, 494
powdery mildew, 56, 494
root rot, 87, 494
Verticillium wilt, 68, 494
Trunk canker, 77, [258], 259, 445
Tsuga, 402; *see* Hemlock
Tuber, 568
Tuber rot, 86, 196, [416], [417]
Tuberose, 250
bacterial soft rot, 81, 86, 250
Botrytis blight, flower spot, 52, 86, 251
leaf and stem spot, 48, 253
root-knot, 90, 253
root rot, 87, 250
tuber, offset, or "seed" soak, 253, 530, 537
Tuberous-rooted, 568
Tuco Products Company, Division of The Upjohn Company, 106, 124
Tulip (*Tulipa*), 494
anthracnose, 51, 498
bacterial soft rot, 81, 90, 495–96
blindness, 498
bulb dust, 496
bulb soak, 496, 497, 530, 531, 537
bulb spot, rot, 81, [90], 495–96
bulb and stem nematode, 92, 496, [497]
bulb storage, 496
chlorosis, 24–29, [496]
fire, Botrytis blight, 52, [85], 86, 90, [495]
flower spot, 86, 497
flower stalk collapse, loose bud, 28, 497
frost injury, 497
insect and bulb mite control, 496, 497
light and temperature requirements, 40
mosaic, mottle-streaking, flower breaking, 70, [71], [496]
Rembrandt, virus-infected, 496
ringspot, 72, 498
rodent injury, to bulbs, 98
root rot, 87, 495
smut, 62, 497
soil treatment, 495, 496
southern blight, 76, 497
stem rot, 76, 497
sulfur dioxide injury, 43
sunscald, 41, 498
tobacco necrosis, streak, 70, 498
"topple," calcium deficiency, 28, 497
winter injury, 41, 497–98
Tuliptree, 341
anthracnose, 51, 341
canker (twig, branch), 77, 341
leaf spot, 48, 341
leaf yellowing, scorch, 41, 341
lightning injury (yellow-poplar), 46
powdery mildew, 56, 341
root nematode, 341
root rot, 87, 137, 341
seedling blight, 76, 341

soil grade change injury, 46
sooty mold, 64, 341
tar spot, 48, 341
Verticillium wilt, 68, 341
wetwood, slime flux, 264, 341
wood rot, 79, 170, 341
Tuna, 194; *see* Cacti, Opuntia
Tupelo; 258; *see also* Dogwood
canker (branch, trunk), 76, 259
felt fungus, 261, 293
leaf spot, blotch, 48, 51, 259
mistletoe, 93, 261
root nematode, 259
root rot, 87, 137, 259
rust, 58, 261
thread blight, 261, 505
Verticillium wilt, 68, 261, 343
wood rot, heart rot, 79, 170, 259
Turf specialist, help by, 5
Turf-Tec T, 101
Turfing daisy, 220; *see* Matricaria
Turkey corn, 174; *see* Bleedingheart, Squirrelcorn
Turnip, 185; *see also* Cabbage
anthracnose, 51, 190
bacterial soft rot, 76, 81, 190
bacterial spot, black rot, 50, 187
blackleg, 77, 187
boron deficiency, brown heart, 27, 191
chlorosis, 24–29, 194
clubroot, 89, [189]
crown gall, 83, 146, 193
curly-top, 75, 192
damping-off, 76, 188, 190, 191, 193
downy mildew, 55, 189
fertilizing, 29
Fusarium wilt, yellows, 66, 186–87
gray-mold blight, 52, 86, 190, 191
leaf spot, 48, 190, 191
mosaic, 70, 192
powdery mildew, 56, 193
resistant varieties, 189, 190
ringspot, 72, 192
root-knot, 90, 191
root nematode, 194
root rot, 87, 187, 190, 191, 193
scab, 65, 165, 193, 415
seed treatment, 187, 529, 532
southern blight, 76, 191
storage rot, 81, 86, 188, 190, 191
Verticillium wilt, 68, 193
watery soft rot, 76, 191
web blight, 160, 194
white-rust, 62, 191
Turnip-rooted chervil, 213; *see* Celery, Salad chervil
Turquoise berry, 286; *see* Ampelopsis
Turtlehead, 454
leaf spot, 48, 455
powdery mildew, 56, 455
rust, 58, 455
Tuscopper, 556
Tussilago, 221
leaf spot, 48, 221
Tween-20, 118
Twig blight, 52, 77, [153], [313], [317], 345
canker, 77, [138], 218, 345, [383], [391]
Twin Light Maneb Dust, 103, 519
Twinflower (*Linnaea*), 498; *see also* Hyacinth-bean
black mildew, 64, 498
leaf spot, 48, 498
tar spot, 48, 498
2,4-D injury, [10], 45, 160, 261, [267], [288], 484, 506
control of dodder, 94
using safely, 45, 112
2,4-dichloro-6-o-chloroanilo-s-triazene, 557
2,4,5-T injury, 45, 261, 484
using safely, 112
2-(1-methylheptyl)-4-6-dinitrophenyl crotonate, 101
2,6-dichloro-4-nitroaniline, 104, 393
2,3-dichloro-1, 4-naphthoquinone, 102, 518
Tylosis (tyloses), 568

U

UC Soil Mix C, 22
The UC System for Producing Healthy Container-Grown Plants, 22, 542, 572
Udo, 126
blight, 51, 127
stem rot, 76, 127
Verticillium wilt, 68, 127
watery soft rot, 76, 127
UFC-85, 362, 415
Ulmus, 263; *see* Elm
Umbellularia, 149; *see* California-laurel
Umbrella-pine, 402
Cytospora stem canker, 77, 403
damping-off, 76, 405
leaf spot, 48, 402
root rot, 87, 405, 406
seed rot, 405
seed treatment, 405
soil treatment, 405, 406
twig blight, 77, 402, 403
Umbrella-tree, 450; *see* Schefflera
Umbrellaplant, 498
leaf spot, 48, 498
root-knot, 90, 498
Umbrellawort, 274
downy mildew, 55, 274
leaf spot, 48, 274
root rot, 87, 274
white-rust, 62, 274
Unicornplant, 425; *see* Proboscisflower
Union Carbide Chemicals Corporation, Division of Union Carbide Corporation, 124
United Cooperatives, Incorporated, 124
U.S. Borax Chemical Corporation, 27, 124
USDA printed matter, 571, 578–80
how to obtain, 578
United States Rubber Company, 102, 103, 124, 518, 519
Units of measure, 516
Universal Metal Products Division, Leigh Products, Incorporated, 124
Universities, land-grant, help by, 5, 6
listing of, 6–7
Upjohn Company, The, 104, 126
Uracide, 362, 415
Urea-formaldehyde, 362
and nabam, 362
trade names, 415
Useful units of measure, 516
Using Phenoxy Herbicides Effectively, 45
Uvularia, 335
leaf spot, 48, 338
rust, 58, 337

V

Vaccinium, 175; *see* Blueberry
Valerian (*Valeriana*), 499
leaf spot, 48, 499
powdery mildew, 56, 499
root rot, 87, 499
rust, 56, 499
stem rot, 76, 254, 499
Valerianella, 499; *see* Cornsalad
Valley Chemical Company, 124
Vallota, 250
Stagonospora leaf scorch, red spot, 48, 251
Vancomycin, 441
Vanda, 366; *see* Orchids
hot water soak, 369, 530, 536
Vanderbilt, R. T., Company, Incorporated, 124
Vandermolen Export Company, 124
Vanillaleaf, 153
leaf spot, 48, 153
Vapam, disinfesting soil, 106, 188, 189, 193, 214, 263, 271, 290, 356, 405, 425, 428, 430, 443, 463, 472, 490, 492, 512, 545, 547; *see also* SMDC
Variegation (genetic), 568
Variegation, infectious (virus), 70, 83, 200, 299
Variety, 568
Vascular tissue, or system, 66, [67], 68, [69], [262], [487]
Vatsol, 119
Vaughan's Seed Company, 124
VC-13 Nemacide, 546
active ingredient, 546
application, 546
disinfesting soil, 89, 92, 148, 182, 201, 271, 324, 409, 545
precautions, 546
trade names, 546
uses, 546
Vector, 568
Vegetable grower, help by, 5, 193, 482, 489, 491
Vegetable-marrow, 238; *see also* Cucurbit
angular leaf spot, 50, 239
anthracnose, 51, 86, 238
aster yellows, 73, 243
bacterial soft rot, 81, 86, 238, 244
bacterial wilt, 69, 240
blossom blight, 52, 86, 245
chlorosis, 24–29, 245
curly-top, 75, 243
damping-off, seed rot, 76, 244
downy mildew, 55, 243
fruit spot, rot, 86, 238, 241, 244
Fusarium wilt, 66, 241
gray-mold rot, 52, 86, 244, 245
gummy stem blight, 76, 86, 244
leaf blight, 51, 239
leaf spot, 48, 238, 239, 244
mosaic, 70, 241
powdery mildew, 56, 242
resistant varieties, 243
ringspot, 72, 243
root-knot, 90, 244–45
root nematode, 244
root rot, 87, 156, 244
seed rot, 244
seed treatment, 238
southern blight, 76, 243
stem, crown rot, 76, 243
Verticillium wilt, 68, 245
Vegetable oyster, 328; *see* Salsify
Vegetable sponge, 238; *see* Gourds
Vegetables
fertilizing, 29
starter solution, 29–30
light and temperature requirements, 38, 41
multipurpose mixes, 110–1
storage, 83, 207, 470
Commercial Storage of Fruits, Vegetables, Florist, and Nursery Stocks, 100
Home Storage of Vegetables and Fruits, 87, 99
rots, 86
Vegetative, 568
Vein banding, 568
Vel, as spreader, 119
Velsicol Chemical Corporation, 124
Velvet plant or leaf; *see* Kalanchoë
light and temperature requirements, 40
Venus-lookingglass, 168
leaf spot, 48, 169
root rot, 87, 169
rust, 58, 169, 405
seed smut, 62, 169
Verbascum, 454; *see* Mullein
Verbena, 317
bacterial wilt, 69, 318
downy mildew, 55, 318
flower blight (*Botrytis*), 52, 86, 318
leaf nematode, 75, 226, 317
mosaic, 70, 318
powdery mildew, 56, 318
root-knot, 90, 317
root rot, 87, 318

Verbena (*continued*)
rust, 58, 318
spotted wilt, 72, 318
stem blight (*Botrytis*), 52, 318
stem rot, 76, 318
Verbesina, 221; *see* Crownbeard
Vermiculite (horticultural grade), 568
Veronica, 460; *see* Speedwell
Veronicastrum, 460; *see* Culversroot
Versene-OL, 345
Versenol Iron Chelate, 345
Verticillium wilt, 66, 68, [69], 91, 98, [212], [344], 370, 391, [431], [456], 463, [487], [488], 493, 568
control, by adjusting soil pH, 419
following nematode or other wounds, 68, 91, 448, 463, 488
resistant (or immune) plants, 68–69
weed hosts, 391
Vervain, 317; *see* Verbena
Vetch, 377; *see also* Crownvetch, Pea
anthracnose, 51, 380
Ascochyta blight, 51, 379
curly-top, 75, 382
damping-off, 76, 379
downy mildew, 55, 379
Fusarium wilt, 66, 377
leaf spot, 48, 379, 380
mosaic, 70, 378
Mycosphaerella blight, 51, 76, 379
powdery mildew, 56, 378
root-knot, 90, 381
root nematode, 382
root rot, 87, 377, 378
rust, 58, 381
seed treatment, 379
soil drench, 377, 378, 379
southern blight, crown rot, 76, 378
spotted wilt, 72, 379
stem rot, 76, 378, 379
Vetchleaf sophora, 300; *see* Sophora
Viability, 568
Viburnum, 499
bacterial leaf spot, 50, 79, 500
blossom blight, 52, 86, 500
chlorosis, 24–29, 344, 500
collar rot, 76, 500
crown gall, 83, 136, 500
dieback, 77, 500
downy mildew, 55, 500
fungus leaf spot, mold, 48, 500
gray-mold or shoot blight, 52, 77, 500
light, requirements of, 39
powdery mildew, 56, 499
pruning, 32
root-knot, 90, 391, 500
root nematode, 500
root rot, 87, 137, 500
rust, 58, 500
sooty mold, 64, 500
spot anthracnose, 51, 500
stem canker, girdle, 77, 247, 259, 500
thread blight, 500, 505
twig canker, stem girdle, 77, 500
Verticillium wilt, dieback, 68, 343, 500
wood rot, 79, 170, 500
Vicia, 377; *see* Vetch
Vidden-D, disinfesting soil, 159, 209, 290, 395, 467, 489, 545
Vigna, 155; *see* Asparagus-bean, Black-eyed pea
Vinca, 500
aster yellows, 73, 501
black ringspot, 72, 192, 501
canker, dieback, wilt, 77, 500
curly-top, 75, 501
flower breaking, 70, 501
gray-mold (*Botrytis*) blight, 52, 501
leaf mold, 51, 500
leaf spot, 48, 500
mosaic, 70, 501
root-knot, 90, 501
root nematode, 501
root rot, 87, 501
rust, 58, 501
soil drench, 500
stem rot, blight, 76, 500, 501
Vine, 286; *see* Grape
Vine bower, 228; *see* Clematis
Viola, 374; *see also* Pansy, Violet
plant soak, 376, 530, 536
Violet, 374; *see also* Pansy
anthracnose, 51, 374
crown, stem rot, 76, 374, 375
curly-top, 75, 376
downy mildew, 55, 375
flower blight, 86, 374
gray-mold blight, rot, 52, 86, 374
insect control, 374, 375, 376
leaf nematode, 75, 375–76
leaf spot, 48, 374
light, effect on flowering, 38
mercury injury, 98
mosaic, 70, 375
oedema, corky scab, 39, 375
plant soak, 376, 530, 536
powdery mildew, 56, 374
ringspot, 72, 376
root-knot, 90, 375
root nematode, 375
root rot, 87, 375
rust, 58, 375
seed treatment, 375, 536
smut, 62, 375
sooty mold, 64, 376
southern blight, 76, 375
spot anthracnose, scab, 65, [374]
sulfur dioxide injury, 43
Viper's-bugloss, 349
leaf spot, 48, 221, 349
root rot, 87, 278, 349
Virginia-Carolina Chemical Corporation, 124
Virginia cowslip, 348; *see* Mertensia
Virginia-creeper, 287
canker, dieback, 77, 290
downy mildew, 55, 288
leaf spot, 48, 287, 291
powdery mildew, 56, 289
root nematode, 291
root rot, 87, 137, 289
spot anthracnose, leaf scab, 51, 289, 291
thread blight, 291, 505
Virginia Smelting Company, 124
Virginia snakeroot, 144; *see* Aristolochia
Virgins-bower, 228; *see* Clematis
Virulent, 568
Viruses, 13–14, [14], 568
composition, 13, 568
control, by dry heat, 142, 205, 418, 438, 501
definition, 13
diseases caused by, 13
aster yellows, 73, [74], 75, 98, 222, 501
big vein, [330]
crinkle, 70, 418
curly-top, 13, [75], 98, [159], 162, 164, [192], 486
dwarf, 73, [74], 75, 418
flower breaking, 70, [71], 72, [192], [250], [284], [375], [378], 496
infectious variegation, 70, 83, 200
leaf curl, 70, [429]
leafroll, 418, 491
little-peach, little plum, 387
mosaic, 13, 70, [71], 98, 155, 164, [165], [192], 205, [241], [250], [284], [309], [337], [375], [378], 389, [429], [438], [442], [485], [496]
mottle, 13, 70, 205, 496
phloem necrosis of elm, 13, 264
ring pox, 390
ringspot, 13, 72, [73], 205, [255], 331, 388, [396], 418, 486
rosette, 73, 387–88
spindle tuber, 420
spotted wilt, 13, 72, [73], 168, 379, 486
streak, 13, 70, [429]
stunt, 13, 73, [74], 75, [177]
western yellow blight, 75, 486
yellows, 13, 73, [74], 75, 222, 388
lethal (or inactivation) temperature, **142, 205, 291, 542**
multiplication, 13
over winter, 14, 72, 73, 98
size, 13
spread by, 13, 70, 72, 73, 142, 155, 156, 160, 162, 168, 177, 178, 192, 205, 234, 242, 243, 264, 272, 284, 291, 299, 300, 328, 330, 388, 389, 390, 397, 418, 420, 430–31, 466, 485, 486, 496, 501
strains or physiologic races, 13, 97, 380
Vitex, 317
leaf spot, 48, 317
root rot, 87, 318
Vitus, 286; *see* Grape
Vocational agriculture teacher, help by, 3
Vorlex, disinfesting soil, 83, 91, 106, 188, 214, 244, 336, 364, 365, 406, 420, 424, 443, 448, 472, 487, 489, 490, 546; *see also* MIT
VPM Soil Fumigant, disinfesting soil, 83, 106, 188, 189, 193, 214, 263, 271, 290, 356, 405–6, 425, 428, 430, 443, 463, 472, 490, 492, 512, 545, 547; *see also* SMDC

W

Wafer ash, 302; *see* Hoptree
Wahoo, 173; *see* Euonymus
Wakerobin, 493; *see* Trillium
Wallcress, 186; *see* Rockcress
Wallflower (*Cherianthus*), 186; *see also* Wallflower, western and Erysimum
aster yellows, 73, 193
bacterial rot, 50, 187
black ringspot, 72, 192
chlorosis, 24–29, 194
clubroot, 89, 189
crown gall, 83, 146, 193
crown or stem rot, 76, 191
damping-off, 76, 188
downy mildew, 55, 189
gray-mold blight, 52, 190–91
leaf spot, 48, 190
leafy gall, 81
mosaic, flower breaking, 70, 192
powdery mildew, 56, 193
resistant varieties, 189
rust, 58, 193
stem and bulb nematode, 75, 92, 194
white-rust, 62, 191
Wallflower, western (*Erysimum*), 186; *also includes* Alpine and Siberian wallflower; *see also* Cabbage
bacterial wilt, 69, 187–88
chlorosis, 24–29, 194
clubroot, 89, 189
downy mildew, 55, 189
leaf spot, 48, 190
powdery mildew, 56, 193
root rot, 87, 193
rust, 58, 193
spotted wilt, 72, 192
white-rust, 62, 191
Wallpepper, 451; *see* Sedum
Walnut, 501
anthracnose, 51, 501
bacterial blight, blast, [50], 81, 505
boron deficiency, dieback, "snake head," 27, 505
branch-wilt, 77, [503]
bunch disease, 502
canker (twig, branch), dieback, 77, 502
collar rot, 77, 504, 505
crown gall, 83, 502
downy spot, 51, 502
felt fungus, 293, 506
Growing Black Walnuts for Home Use, 506
insect control, 502, 505
leaf blotch, 51, 501
leaf scorch, sunscald, 41, 344, 506
leaf spot, 48, 501

mistletoe, 93, 506
"mouse ear," manganese deficiency, 26, 27, 505
nut mold, rot, 87, 504, 505
powdery mildew, 56, 505
resistant walnuts, 505
root-knot, 90, 392, 506
root nematode, 504
root rot, blackline, 87, 137, 504
scab, 65, 502
sooty mold, 64, 506
sunscald, winter injury, 41, 139, 264, 502
thread blight, 505
trunk canker, 77, 502
2,4-D injury, 288, 506
Verticillium wilt, 68, 343, 506
wood rot, heart rot, trunk decay, 79, 170, 504
zinc deficiency, rosette, little leaf, yellows, 26, 27, 503–4
Wandering-jew, 493; *see* Tradescantia
Wandflower, 307
mosaic, 70, 308
Wart plant, 130; *see* Aloe
Washing soda, 105
Washingtonia, 372; *see* Palms
Water-conducting tissue, 66, [67], 68, [69]
Water-logged, 568
Water molds, control, 531
lethal temperature, 542
Watercress, 186; *see also* Cabbage
aster yellows, 73, 193
chlorosis, 24–29, 194
clubroot, 89, 189
damping-off, 76, 188
downy mildew, 55, 189
leaf spot, 48, 190
mosaic, 70, 192
root-knot, 90, 191
root rot, 87, 193
rust, 58, 193
white-rust, 62, 191
Watering plants, 37–38
to prevent foliage diseases, 38, 98
cold injury, [128]
Waterlily, 506
leaf spot, 48, 506
leaf and stem rot, 51, 506
white smut, 65, 506
Watermelon, 238; *see also* Cucurbit
anthracnose, 51, 86, 238
bacterial soft rot, 81, 86, 238, 244
bacterial spot, 50, 245
bacterial wilt, 69, 240
blossom-end rot, [244], 245
chlorosis, 24–29, 245
Commercial Watermelon Growing, 245
cottony rot, 76, 243
curly-top, 75, 243
damping-off, 76, 244
downy mildew, 55, 243
fruit spot, rot, 86, 238, 244
Fusarium wilt, 66, [67], 241
gummy stem blight, 76, 244
leaf spot, 48, 238, 239, 244
mosaic, 70, 241
powdery mildew, 56, 242
resistant varieties, 238, 240, 241, 243
ringspot, 72, 243
root-knot, 90, [91], 244–45
root nematode, 244
root rot, 87, 156, 244
scab, 48, 65, 86, 238–39
seed rot, 244
seed treatment, 238, 528, 532
seedbed drench, 244
southern blight, 76, 243
stem (crown) rot, 76, 243
2,4-D injury, 45, 245, 288
Verticillium wilt, 68, 245
Watermelons for the Garden, 245
Watermelon-begonia, 397; *see* Peperomia
Watersprouts, 568
Watery soft rot, 156, [207]
Watsonia, 282
mosaic, 70, 284
root rot, 87, 284
Wattle, 300; *see* Acacia
Wax-plant, light and temperature requirements, 40
Waxberry, 457; *see* Snowberry
Waxgourd, 238; *see* Chinese waxgourd
Waxmyrtle, 507
black mildew, 64, 507
leaf spot, 48, 507
root nematode, 507
root rot, 87, 507
rust, 58, 313, 405, 507
seedling blight, 76, 405, 507
sooty mold, 64, 507
Wayfaring-tree, 499; *see* Viburnum
Web blight, 160, 272
Webworms, insect control, 321
Weed control, 62, 98, 368
as disease control measure, 52, 54, 56, 58, 62, 68, 76, 77, 79, 86, 89–90, 98, 162, 165, 214, 233–34, 241, 328, 368, 429, 480
Weed seeds, lethal temperature, 542
Weed specialist, extension, help by, 94, 98, 234
Weedfume, 106, 545
Weedkiller, injury, [10], 45, 109, 112, [267], [288]
contamination of sprayers, 109, 112, 484
Weeds, as virus reservoir, 98
Weeping myall, 300; *see* Acacia
Weeping tupelo, 258; *see* Tupelo
Weevils, insect control, 161, 209, 435, 463
Weigela, 457
crown gall, 83, 458
leaf spot, 48, 458
plant soak, 459, 531, 537
powdery mildew, 56, 458
pruning, 32
root-knot, 90, 392, 459
root nematode, 458
root rot, 87, 458
twig blight, 77, 458
Well-drained soil, 569
Well-prepared soil, 569
Welsh poppy, 413; *see* Meconopsis
West India gherkin, 237; *see also* Cucurbit
angular leaf spot, 50, 239
bacterial soft rot, 81, 238, 244
bacterial wilt, 69, 240
curly-top, 75, 243
downy mildew, 55, 243
leaf blight, 51, 239
mosaic, 70, 241
powdery mildew, 56, 242
scab, 65, 238–39
seed rot, 244
seed treatment, 238, 532
Westbrook Manufacturing Company, 124
Westcott, Dr. Cynthia, 3, 48
Western yellow blight, 75, 486
"Wet feet," 393, 512
Wettable powder, 569
Wetting agents, trade names, 118–19
rates, 119
uses, 118–19
Wetwood (slime flux), 264, [265], 566
Wheat flour, as sticker, 119
Wheatgrass, 319; *see also* Bluegrass
anthracnose, 51, 319
bacterial spot, blight, 50
chlorosis, yellowing, 24–29, 325
fairy ring, 322
foot rot, 76, 319
iron deficiency, 24, 26, 27, 325
leaf blotch, scald, 51, 319
leaf spot, 48, 319
magnesium deficiency, 28, 325
mosaic, 70, 327
powdery mildew, 56, 321
root nematode, 324
root rot, 87, 319
rust, 58, 321, 381
seed rot, 326
seed treatment, 326, 537
seedling blight, 76, 326
smut, 62, 325
snow mold, 322
stem (culm) rot, 76, 319
tar spot, 48, 319
Wheelbarrow sprayer, 113–14, [116]
Whiptail, 28, 193
White-alder, 470; *see* Clethra
White beamtree, 133; *see* Mountain-ash
White blister, 62
White-cedar, 312; *see* Chamaecyparis
White daisy, 220; *see* Tidytips
White kerria, 312; *see* Jetbead
White mold (Ramularia blight), [251]
White pine blister rust, [404], 405
White-rust, [61], 62, 166, 191, 473
White smut, [65]
White snakeroot, 220; *see* Eupatorium
White walnut, 501; *see* Walnut
Whitebrush, 317; *see* Lemon-verbena
Whitecup, 479; *see* Tomato
Whiteflies, as bacteria carriers, 280
insect control, 277, 474
secretion of "honeydew," 64
as virus carriers, 13, 98, 299, 474
Whitewash, control of sunscald, 502
Whitewood, 494; *see* Trumpettree
Whitlowgrass, 186; *see also* Cabbage
downy mildew, 55, 189
powdery mildew, 56, 193
rust, 58, 193
white-rust, 62, 191
Wicopy, 327; *see* Leatherwood
Wild bleedingheart, 174; *see* Bleeding-heart
Wild cucumber, 238; *see* Mock-cucumber
Wild hyacinth; *see* Camass, Brodiaea, Squill
Wild indigo, 269; *see* False indigo
Wild pink, 203; *see* Silene
Wild rosemary, 316; *see* Labrador-tea
Wild sweet-william, 399; *see* Phlox
Wild tuberose, 216; *see* Manfreda
Wildbergamot, 447; *see* Monarda
Wildolive, 370; *see* Osmanthus, Olive
Willow, 507–8
bleeding canker, 77, 161, 345, 509
borer, control, 508–9
canker (twig, branch), dieback, 77, 508
chlorosis, 24–29, 344, 509
crown gall, 83, 508
cutting rot, 76, 509
Cytospora canker, 77, 508
felt fungus, 293, 509
frost crack, 41
leaf blight, 51, 508
leaf blister, 62, 171, 509
leaf scorch, 51, 508
leaf spot, 48, 508
mistletoe, 93, 509
Physalospora black canker, 77, 508
powdery mildew, 56, 508
pruning, 33
resistant varieties or species, 508
root-knot, 90, 392, 509
root nematode, 509
root rot, dieback, 87, 509
rust, 58, 318, 405, 508
scab (*Venturia* or *Pollacia*), 65, 508
sooty mold, 64, 339, 509
spot anthracnose, gray scab, 51, 508
sunscald, winter injury, 41, 139, 264, 509
tar spot, 48, 508
twig blight, 77, 508
wetwood, slime flux, 264, 509
wood rot, heart rot, 79, 170, 508–9
Wilson's Anti-Damp, 106, 367, 569
Wilt, 569
Wilt Pruf, 21, 38, 41, 42, 182, 296, 511
Wilts, 65–70, 72
bacterial or brown rot, 10, 50, 66, 69, [70], 204, 211, 231, [232], 240, 354, 488
Fusarium, or yellows, 66, [67], 68, 91, 98, [202], 203, 241, [472], [487]

Wilts (*continued*)
severity increased by nematodes, 68, 91, 203, 377, 448, 463, 487, 488
spotted, 72, [73]
Verticillium, 66, 68, [69], 91, 98, [212], [344], 370, 391, [431], [456], 463, [487], [488], 493, 568
Wind injury, 45, 182
entrance for fungi, 79
Wind poppy, 413; *see* Meconopsis
Windflower, 131; *see* Anemone
Wineberry, 427; *see* Dewberry, Raspberry
Winged loosestrife, 340; *see* Lythrum
Winter-daffodil, 250; *see* Fall-daffodil
Winter injury, 41–43, [139], 182, 254, 295, 311, 314, 392, 407, 432, 433–34, [462], 463, 511
as plant weakening agent, 78
protection, [42], 182, 264, 407, 432, [442], 463, 511
Winter melon, 237; *see* Cucurbit, Muskmelon
Winterberry; *see* Holly, Currant, Euonymus
Wintercherry, 479; *see* Chinese lantern-plant
Wintercreeper, 173; *see* Euonymus
Wintergreen, 294; *see* Checkerberry
Wire injury, 45
Wirelettuce, 220
leaf spot, 48, 221
rust, 58, 223
Wirestem, [188]
Wireworms, insect control, 209, 215, 233, 282, 463, 475
Wishbone flower, 454
root-knot, 90, 457
Wisteria, 300
canker, dieback, 77, 300
crown gall, 83, 136, 301
leaf spot, 48, 301
mosaic, 70, 302
powdery mildew, 56, 301
root-knot, 90, 301, 392
root rot, 87, 137, 301
wood, heart rot, 87, 170, 301
Witch-hazel, 509
crown gall, 83, 136, 509
leaf spot, 48, 509
powdery mildew, 56, 509
root rot, 87, 137, 509
wood rot, 79, 170, 509
Witches'-broom, 62, [63], 81, [93], 94, 141, 171, 177, 265, 292, [293], 300, 301, 359, 405, 436, 437, 476, 569
Witchweed (*Striga*), 234
Withe-rod, 499; *see* Viburnum
Witloof chicory, 328; *see* Chicory
Wolfberry, 457; *see also* Snowberry, *and see* Matrimony-vine for Chinese wolfberry
collar rot, 76, 259, 458
leaf spot, 48, 458
powdery mildew, 56, 458
rust, 58, 458
twig blight, canker, 77, 458
Wolfsbane, 254; *see* Aconitum, Monkshood
Wood anemone, 131; *see also* Anemone
white smut, 65, 132
Wood-nettle, rust, 174
Wood Ridge Chemical Corporation, 124
Wood rot, [79], [88], [147], [161], [170], 259, [385]
Wood shavings, as mulch, 42
control of Verticillium wilt, 370
Wood Tox, 563
Woodbine; *see* Virginia-creeper, Honeysuckle
Woodbury Chemical Company, 124
Woodnymph, rust, 58, 405
Woodridge Mixture "21," 101, 106, 569
Woodsia, 270; *see* Ferns
Woodsorrel, 370; *see* Oxalis
Woodwardia, 270; *see* Ferns
Woodwaxen, 183; *see also* Broom
dieback, 77, 183
powdery mildew, 56, 183
rust, 58, 183
Woolfolk Chemical Works, Incorporated, 124
Woolly manzanita, 175; *see* Manzanita
Wormgrass, 451; *see* Sedum
Wormwood, 220; *see* Artemisia
Worms, lethal temperature, 542
Wortleberry, 175; *see also* Blueberry
leaf and shoot (stem) gall, 62, 176
leaf rust, 58, 177
leaf spot, 48, 178
powdery mildew, 56, 176
witches'-broom, 62, 177
Wound, 34–36, 45–46
dressing, 35, 36, 161, 171, 476–77
entrance for bacteria, 34, 70, 81, 83, 98, 134, 204, 208, 301, 362, 482, 489, 508
entrance for disease organisms, 87, 208
entrance for fungi, 34, 46, 66, 68, 79, 98, 203, 208, 353, 356, 377, 383, 402, 463, 482, 504
rot, 79
shaping, [35]
sterilizing, 35
treatment, trees, 34, 35, 161, 383, 384, 392
Woundwort, 447; *see* Stachys
Wyethia, 221
leaf gall nematode, 75, 226
leaf spot, 48, 221
rust, 58, 223

X

Xanthorhiza, 228
leaf spot, 48, 228
Xanthosoma, 196
bacterial soft rot, 76, 81, 196
leaf spot, 48, 197
powdery gray rot, 87, 196–97
root rot, 87, 162, 196
X-disease, 387
Xterminator Products Corporation, 124
XXX Mineral Mix, 27
Xylem, 569

Y

Yam (true), 510; *see also* Sweetpotato
anthracnose, 51, 510
crown rot, 76, 510
leaf blotch, 51, 510
leaf spot, 48, 510
root-knot, 90, 510
root nematode, 510
southern blight, 76, 510
storage rot, 86, 471, 510
tuber treatment, 510, 529, 533
Yampee, 510; *see* Yam
Yardlongbean, 155; *see* Asparagus-bean
Yarrow, 219
crown gall, 83, 226
powdery mildew, 56, 221
root-knot, 90, 225
root rot, 87, 221
rust, 58, 223
stem rot, 76, 221
Yate-tree, 268; *see* Eucalyptus
Yaupon, 295; *see also* Holly
black mildew, 64, 296
leaf spot, tar spot, 48, 295
root rot, 87, 137, 296
sooty mold, 64, 296
Yautia, 196; *see also* Xanthosoma
powdery gray rot, 87, 196–97
Yellow adderstongue, 267; *see* Erythronium
Yellow blight, western, 75, 486
Yellow-cedar, 312; *see* Chamaecyparis
Yellow cherry, 183; *see* Buckthorn
Yellow cypress, 312; *see* Chamaecyparis
Yellow dwarf, 365
Yellow-elder, 494; *see* Stenolobium
Yellow ironweed, 219
leaf spot, 48, 221
powdery mildew, 56, 221
ringspot, 72, 223
rust, 58, 223
Yellow-jasmine, 185
black spot, 48, 185
leaf spot, 48, 185
root rot, 87, 137, 185
sooty mold, 64, 185
Yellow-poplar, 341; *see* Tuliptree
Yellow star, 220; *see* Sneezeweed
Yellow vein (virus), 164, 278–79
Yellow wild indigo, 269; *see* False-indigo
Yellowroot, 228
leaf spot, 48, 228
Yellows, (*Fusarium*), 66, [67], 68
virus, 73, [74], 75, 164, 388
Yellowwood, 300
canker, dieback, 77, 300
powdery mildew, 56, 301
pruning, 32
Verticillium wilt, 68, 301, 343
wood rot, 79, 170, 301
Yerba-buena, 447
rust, 58, 447
Yew, 511
brown felt blight, 407, 512
crown gall, 83, 136, 512
damping-off, seedling blight, 76, 405, 512
dieback, 77, 511
natural leaf browning, shedding, 512
needle, leaf blight, 51, 512
pruning, 32, 33
root nematode, 511–12
root rot, 87, 511
root weevil, control, 512
twig blight, 77, 313, 512
wet feet, 512
winter injury, leaf scorch, 41, 42, [511]
protection, 511
wood and heart rot, 79, 170, 512
Youngberry, 427; *see* Dewberry, Raspberry
Yucca, 512
crown gall, 83, 512
flower blight, 86, 512
leaf blight, 51, 512
leaf spot or mold, 48, 512
root-knot, 90, 512
rust, 58, 512
stem rot, 76, 512

Z

Zantedeschia, 195; *see* Calla lily
Zanthoxylum, 302; *see* Prickly-ash, Hercules-club
Zauschneria, 268
rust, 58, 268
Z-C Spray or Dust, 103, 519
Zea, 231; *see* Corn
Zebra-plant, 347
leaf spot, 48, 347
Zebrina, 493; *see* Tradescantia
Zelkova, 263; *see* Elm
Zephyrlily (*Zephyranthes*), 250
bulb rot, 90, 250
leaf spot, 48, 253
rust, 58, 253
Stagonospora leaf scorch, red spot, 48, 251
Zerlate Ziram Fungicide, 103, 503, 519
Zinc
bordeaux, 386
chelates, 27, 150, 298
deficiency, 26, 27, 150, 290, 342, 503, [504]
control, 27, 141, 150, 201, 298, 299, 343, 353, 503–4
dimethyl dithiocarbamate, 103, 519
ethylene bis(dithiocarbamate), 103, 519
napththenate, 569
oxysulfate, 27

polyethylene thiuram disulfide complex, 104
sulfate, 27, 141, 150, 166, 186, 201, 290, 299, 343, 353, 503–4
add to nabam (= Zineb), 562, 569
Zineb
active ingredient, 103, 519
algae control, 297
as "disinfectant," 204, 221, 278, 464, 470
Dust, 103
gallon lots, 522
as lawn fungicide, 103, 106, 519
in multipurpose mixes, 103, 111, 519
19.5D TF, 103, 519
PAN or oxidant prevention, 44
as protective fungicide, 48, 55, 56, 62, 65, 86, 100, 103, 167, 199, 205, 224, 248, 275, 277, 280, 287, 294, 296, 305, 308, 310, 313, 314, 318, 320, 329, 335, 337, 343, 349, 351, 353, 357, 362, 363, 365, 367, 370, 381, 384, 385, 386, 394, 405, 410, 415, 423, 424, 425, 433, 436, 437, 440, 443, 451, 458, 476, 480, 482, 484, 495, 500, 501, 502, 508, 509, 512, 519, 525
seed-piece treatment, 418
6.5% Dust, 103, 519
soil application, 77, 97, 103, 106, 167, 248, 255, 259, 261, 278, 294, 320, 326, 406, 451–52, 489, 490, 519, 544
trade names and distributors, 103, 519
uses, 103, 519
Zinnia, 221
aster yellows, 73, 222
bacterial wilt, 69, 223
blossom or head blight, 52, 86, 224
curly dwarf, 73, 222
curly-top, 75, 223
damping-off, seed rot, 76, 221
gray mold (*Botrytis*), 52, 86, 224
leaf blight (*Alternaria*), 51, [217], 221
leaf nematode, 75, 226
leaf spot, 48, 221
mosaic, 70, 222
powdery mildew, 56, [57], 221
resistant varieties, 221, 223
ringspot, 72, 222
root-knot, 90, 225
root nematode, 226–27
root rot, 87, 221
seed treatment, 221, 530, 537
soil drench, 222
southern blight, 76, 221
spotted wilt, 72, 223
stem canker, 77, 225
stem rot, wilt, 76, 221
Verticillium wilt, 68, 223
Zinophos (Cynem), 89, 546, 569
active ingredient, 546, 556, 569
application, 546
disinfesting soil, 182, 244, 324, 395, 546
plant dip or soak, 277, 373, 450, 459, 546
precautions, 546
trade names, 546, 556
uses, 546, 556
Ziram
active ingredient, 103, 519
as "disinfectant," 221, 470, 472
gallon lots, 522
in multipurpose mixes, 103, 519
as protective fungicide, 48, 65, 100, 103, 144, 158, 167, 175, 182, 188, 197, 204, 207, 213, 221, 238, 302, 384, 398, 405, 433, 482, 502, 519
soil application, 77, 103, 106, 167, 489, 519
trade names and distributors, 103, 519
uses, 103, 519
Zoysia, Zoysiagrass, 319; *see also* Bluegrass
brown patch, Rhizoctonia disease, 321
chlorosis, 24–29, 325
fairy ring, 322
leaf spot, blotch, 48, 319
root nematode, 324
root rot, 87, 319
rust, 58, 321
Sclerotinia dollar spot, blight, 321
slime mold, 322
Zygocactus, 194; *see* Cacti
Zygopetalum, 366; *see* Orchids